Lecture Notes in Civil Engineering

Volume 196

Lecture Notes in Civil Engineering (LNCE) publishes the latest developments in Civil Engineering - quickly, informally and in top quality. Though original research reported in proceedings and post-proceedings represents the core of LNCE, edited volumes of exceptionally high quality and interest may also be considered for publication. Volumes published in LNCE embrace all aspects and subfields of, as well as new challenges in, Civil Engineering. Topics in the series include:

- Construction and Structural Mechanics
- Building Materials
- Concrete, Steel and Timber Structures
- Geotechnical Engineering
- Earthquake Engineering
- Coastal Engineering
- Ocean and Offshore Engineering; Ships and Floating Structures
- Hydraulics, Hydrology and Water Resources Engineering
- Environmental Engineering and Sustainability
- Structural Health and Monitoring
- Surveying and Geographical Information Systems
- Indoor Environments
- Transportation and Traffic
- Risk Analysis
- Safety and Security

To submit a proposal or request further information, please contact the appropriate Springer Editor:

- Pierpaolo Riva at pierpaolo.riva@springer.com (Europe and Americas);
- Swati Meherishi at swati.meherishi@springer.com (Asia - except China, and Australia, New Zealand);
- Wayne Hu at wayne.hu@springer.com (China).

All books in the series now indexed by Scopus and EI Compendex database!

More information about this series at https://link.springer.com/bookseries/15087

Editors
Ashok Kumar Gupta
Department of Civil Engineering
Jaypee University of Information Technology
Solan, India

Sanjay Kumar Shukla
School of Engineering
Edith Cowan University
Joondalup, WA, Australia

Hazi Azamathulla
Department of Civil and Environmental Engineering
University of the West Indies
St. Augustine, Trinidad and Tobago

ISSN 2366-2557 ISSN 2366-2565 (electronic)
Lecture Notes in Civil Engineering
ISBN 978-981-16-6556-1 ISBN 978-981-16-6557-8 (eBook)
https://doi.org/10.1007/978-981-16-6557-8

This Springer imprint is published by the registered company Springer Nature Singapore Pte Ltd.
The registered company address is: 152 Beach Road, #21-01/04 Gateway East, Singapore 189721, Singapore

Ashok Kumar Gupta · Sanjay Kumar Shukla ·
Hazi Azamathulla
Editors

Advances in Construction Materials and Sustainable Environment

Select Proceedings of ICCME 2020

Springer

Preface

With the increasing population worldwide, depletion of major resources with subsequent increase in waste generation has become inevitable. There is a need to understand that development at the stake of resources and its contribution to mass waste production require immediate attention. One such solution is the adoption of sustainability in our materials which are used for construction. Sustainability is often defined as the ability to fulfill the present socioeconomic demands of the society without compromising the naturally available resources required for future use and still maintaining the ecological balance. Therefore, in the present scenario, selection, planning, design, construction, and maintenance methods as used by civil engineers require a sustainable approach. The growing importance of attaining sustainable environment can significantly be realized by incorporating use of sustainable materials for construction. A new generation of stronger, lighter, and more sustainable building materials can help solve many problems in construction and drive the current practices to be more sustainable. These materials have the added benefit of protecting the environment by reducing the carbon footprint and promote a cleaner Earth while being aesthetically appealing.

Civil engineers play a decisive role in designing, constructing, and ensuring a sustainable future by bridging the gap between science and society. In this role, engineers are required to actively promote and participate with multidisciplinary teams within their professional obligation to extend their knowledge base and be a part of the policy decisions at all levels. This first international conference, held during June 3–4, 2021, achieved its aims by establishing a knowledge sharing platform regarding clean technologies and sustainable construction materials for achieving the transition from conventional to sustainable construction practices. Out of a total of 170 participants who attended this conference, about 100 were experts in field of construction materials from different areas, including academicians, field engineers, and researchers apart from practitioners and students. The conference has created exchange of thoughts and paved way for new ideas among participants from various sectors through sharing of concepts, theory, methodology, and application from various disciplines about developing and implementing sustainable construction materials and attaining sustainable environment.

There were deliberations on new initiatives in the fields of advance materials and methods, geotechnics and geoenvironmentally sustainable materials, bio-inspired engineering materials, and developments in transportation and urban planning. This exposed the participants to the developed and progressive strategies with optimal utilization of available resources for developing the materials and methods and consolidating the suggestions. Key recommendations were made during the conference for field applications. This further led to disseminating the knowledge on the conference themes.

We do hope that this book will create awareness and appreciation among academicians, scientists, researchers, and practitioners from various disciplines and sectors about the need and the new initiatives toward development of sustainable construction materials and environment. The comments and suggestions from the readers and users of this book are most welcome.

Solan, India
Perth, Australia
St. Augustine, Trinidad and Tobago/West Indies

Ashok Kumar Gupta
Sanjay Kumar Shukla
Hazi Azamathulla

Acknowledgements

This book entitled Advances in Construction Materials and Sustainable Environment comprises select papers presented at the 1st International Conference on Construction Materials and Environment (ICCME 2021). Notably, the book would have not been possible without the support from the contributing researchers/authors, sponsors, and all those who appreciated this conference for making it a milestone in the area of construction materials and environment sustainability. Furthermore, I would like to express my gratitude to the anonymous reviewers for their valuable comments and suggestions, which certainly enhanced the manuscript. Besides them, there are numerous other people who supported us in this endeavor. I would like to extend my deep-felt thanks to Prof. Vinod Kumar, Vice-Chancellor of Jaypee University of Information Technology, Waknaghat, Solan, for supporting us during ICCME 2021; Prof. Sanjay Kumar Shukla of Edith Cowan University and Prof. Hazi Azamathulla of University of the West Indies at St. Augustine, Trinidad, co-editors of this proceedings book who helped me with the drafting and reviewing process. I also convey my earnest gratitude to Dr. Akash Chakraborty, Ms. Sushmitha Shanmuga Sundaram, and the team of Springer for their full support and cooperation at all the stages of the publication of this book. Last but not least, I would also like to thank Dr. Saurabh Rawat, General Chair of ICCME 2021, and Assistant Editor for this proceeding book of ICCME who worked relentlessly and completed ICCME 2021 successfully. He has significantly assisted in drafting, reviewing, and editing the proceedings book. His continuous and sincere effort has made it possible for achieving this book possible.

Solan, India Ashok Kumar Gupta

Contents

Use of Biochar for Sustainable Environmental Remediation 1
Krishna R. Reddy and Jyoti K. Chetri

Macro and microscale Engineering Response of Rigid-Soft Gravel-Rubber Inclusions: Insights from Detailed Laboratory and DEM Numerical Investigations 11
Gabriele Chiaro, Ali Tasalloti, Kevin Chew, Jayan S. Vinod, and Krishna Allulakshmi

Structural Health Monitoring of Heritage Structures Using Geotechnical Instruments 29
Abhinav Maloo and Parth Thaker

Zinc-Based Anodes for Cathodic Protection of Reinforced Concrete Structures 45
Arpit Goyal

An Overview: Supplementary Cementitious Materials 53
Pooja Jha, A. K. Sachan, and R. P. Singh

Degradation and Decoloration of RB5 Dye via UV Radiation Using Fe-TiO_2 Composite Photocatalyst in Fixed-Mode 65
Lavneet Kumar, Ina Thakur, Anoop Verma, B. S. Bhatia, and Charanjit Kaur Mangat

Study on Concrete Developed with Recycled Fine Aggregate 77
Nancy Soni and Dharmendra Kumar Shukla

Examination of Platooning Variables on Two-Lane Highways Having Mixed Traffic Situation 95
Amardeep Boora, Indrajit Ghosh, Satish Chandra, and Kavita Rani

Walkability Analysis of an Urban Area: Gender-Based and Combined Model Approach 111
Kavita Rani, Amardeep Boora, and Manoranjan Parida

State of Art: Review for Sustainable Application of Waste Material in Rigid Pavement 127
Amardeep Boora and Ankit Dharma

Study on Cost Modeling and Economical Design of Superstructure 143
Peerzada Danish, Kamil Ashraf Bhat, S. Ganesh, and J. Anita Jessie

Structural Assessment of an Overhead Storage Reservoir Using NDT: A Case Study 151
Ram Prakash and Sunita Bansal

Fracture and Impact Studies on Steel Fibre and Wire Mesh Embedded Concrete 163
S. Kanchidurai, P. Jaishankar, R. Vidya, and Prakhash Neelamegam

Physico-Mechanical and Thermal Properties of Lightweight Structural Concrete with Light Expanded Clay Aggregate for Energy-Efficient Buildings 175
Rajesh Kumar, Rajni Lakhani, and Ashok Kumar

Agro-Industrial Wastes Incorporated Cement Stabilized Mud Composites for Roof and Wall Assembly in Energy Efficient Building Envelope 187
Rajesh Kumar, Rajni Lakhani, Bibhakar Kumar Singh, Mahesh Sharma, and S. K. Negi

Seismic Response of Reinforced Concrete Frames with Masonry Infills 197
Vinayak Sharma and Sushil Kumar Madan

Analysis of Seismic Behavior of Buildings With and Without Shear Walls in Various Seismic Zones and Soil Types 207
Diptanshu Lal, Biplav Regmi, Haris Farooq Bhat, and S. Anbu Kumar

Review on the Durability Parameters of Self-compacting Concrete 219
Reshul Raj, Mayur Bhat, Achal Agrawal, and Narayan Chandak

An Approach Towards Zero-Waste Building Construction 239
Anil Soharu, Naveen BP, and Arjun Sil

Improvement of Hard Water Characteristics and Scale Formation Under the Effect of Pulsating Electromagnetic Field 259
Amrit Anand Dosar and Vivek Srivastava

Critical Overview of Reinforcing Sand Using Geocell for Shallow Foundation 271
K. Anusha Raj, Pragya Sinha, Sanjeev Kumar, and Davinder Singh

Road Bridges Across Cooum and Adyar Rivers in Chennai City—Need for Structural Health Monitoring 281
A. Rose Enid Teresa, S. Stella, M. Goutham Priya, P. Gajalakshmi, and J. Revathy

Evaluating Factors Affecting Red Mud Interfacial Strength Using Binder Cement Kiln Dust and Polypropylene Fiber 295
Pankaj Sharma, Saurabh Rawat, and Ashok Kumar Gupta

Review on Thermal Energy Efficiency Using Gypsum Integrated Phase Change Materials in Buildings 305
Kavita Vaishnav, Gift Pon Lazarus, Sunita Bansal, and Yaman Hooda

Coastline Protection Using a Rubble Mound Seawall: A Case Study 321
Prakhar Joshi, Prashant, Pritesh Goyal, and Pradeep K. Goyal

Prediction of Compressive Strength of Rubberized Concrete Using Ordinary Least Squares Regression Model 331
Prabhat Kala, Shivam Upadhya, Pradhyumna Asthana, and Pradeep K. Goyal

Effect of Soil Fill on the Load Distribution Characteristics of RC Skew Box Culverts for Road Under Bridge Design 341
Shimol Philip, R. Rakendu, and Rajesh Lal

Chloride Ion Penetration of GGBS-Based Geopolymer Concrete with Different Molarities of NaOH 355
N. Sailaja, M. Naveen, S. K. Amir Basha, B. Sarath Chandra Kumar, C. Ravi Kumar Reddy, Y. Himath Kumar, and J. D. Chaitanya Kumar

Application of RSM in the Optimization of GGBS and Metakaoline Based Geopolymer Concrete 365
Ch. Pavan Kalyan, D. Anil Kumar, K. Saloman Raju, B. Sarath Chandra Kumar, C. Ravi Kumar Reddy, Y. Himath Kumar, and J. D. Chaitanya Kumar

Evaluation of Use of Plastic and Rubber in Road Construction 375
Neeraj Kumar, Nikhil, Ashutosh Kumar, and A. R. Kongan

Evaluation of Cloth Bag and Gunny Bag as Potential Reinforcing Materials for Pond Ash 385
Sujit Kumar Pradhan, Anwesha Rath, and Goutam Kumar Pothal

The Effect of Sisal Fiber on Mechanical Strength of Concrete M20 Grade 395
Celso Januário Baúque, Ankit Thakur, and Bhartesh

Waste Plastic Management via Pyrolysis as Sustainable Route 409
Sahil Chauhan, Subhankar Basu, Sk Aakash Hossain, and Arasavilli Srija

Seismic Fragility of Buildings Subjected to Pounding Effects with Soil–Structure Interaction 425
Rajan L. Wankhade, Ajinkya Sawarkar, Ayush Chandwani, Shahaji Chavan, Pratik Malkar, and Gaurav Sawarkar

Sustainable Bituminous Pavement: A Study on Low-Density Polymer Modified Bituminous Binder 435
Vishnu Vijayan, Jeevan Mathew Tharayil, R. Rakhil Krishna, Jiji Saji, Divya S. Shaji, and G. Lakshmi

Identification of Parking Sites in the Kukatpally Region Using GIS and AHP 447
Ramu Penki, T. Srinivasa Rao, G. Vinod Naik, and Rapaka Aparna

Minimization of Risks in Highway Projects Using Buffer 457
Rahul Garg and Saurabh Rawat

Trend Modeling for Air Quality—An Approach 467
M. Goutham Priya and S. Jayalakshmi

Traffic Analysis on Intersection Using PTV Vissim 481
Pranjal Sharma, Ashok Kumar Gupta, and Akash Bhardwaj

Parametric Strength of Sustainable Concrete Using Fly Ash, GGBS and Recycled Aggregates as Per Taguchi's Approach 491
Yaman Hooda, Sunita Bansal, and Anjali Gupta

Numerical Analysis on Voided Slab with Different Reinforcement on ANSYS 2020R1 505
Nikita Jain and Asif Hussain

Laboratory-Based Study of Flexible Facing in Soil Nailed Slope 517
Mohammad Farhad Ayazi, Samrity Jalota, and Amanpreet Tangri

Summer Vs Winter Air Pollutants Variation for Year 2019 and Lockdown Effect for CRRI Mathura Road Station Delhi 531
Shahbaz Ahsan, Shashi Tiwari, S. M. Huzaifa Abbadullah, Deepak Nader, and Gaurav Kumar

Use of Fly Ash—A Resourceful Byproduct in Road Embankment: A Review 539
Deepak Kumar Sahay and Sunita Bansal

Application of Industrial Wastes for Soil Strength Improvement 551
S. Muthu Lakshmi, S. Geetha, M. Selvakumar, S. Revathy, and K. M. Shri Varshini

Physical and Mechanical Characteristics of Cement Mortar with Coal Bottom Ash as Fine Aggregate Under Elevated Temperature 561
Abhishek Srivastava, S. K. Singh, and Rajesh Kumar

Prediction of Air Pollution Due to Mobile Sources Using Line Source Models 573
M. Selvakumar, S. Geetha, and S. Muthu Lakshmi

Estimation of Methane Generation from Municipal Solid Waste of Mohali Landfill Site 585
Rishi Rana, Abhinav Choudhary, Karma Yangzom, and Kaushal Kumar

Evaluation of Water Quality Index to Assess the Impact of River Pollution on Vembanad Lake—A Ramsar Site 597
Rohan Nair, K. V. Hariprasad, S. Ashwin Shenoi, M. P. Amrutesh, Kiran V. Gireesh, Gevargis Muramthookil Thomas, and S. N. Jyothi

A Scientometric Analysis on Bio-Bitumen 607
Ramu Penki, Banna Madhavi, K. Akhilesh Patnaik, and A Sri Divya

A Scientometric Analysis on Aggregate Blending 621
Kota Komal Kumar and Ramu Penki

Contextual Analysis on Antenatal and Postnatal Effects of 2018 Flood in Ernakulam, Kerala 635
Athira B. Menon, Devi priya, Krishna Rajeev, Geena Prasad, and Gevargis Muramthookil Thomas

Mechanical Properties of Concrete Containing Plastic Fiber 647
Shubham Sharma and Amardeep Boora

Structural Health Monitoring Through the Application of Piezoelectric Sensors – State of the Art Review 657
Aishwarya Thakur and Saurav

The Inhibitive Effect of Vitamin-C on the Corrosive Performance of Mild Steel in Ground Granulated Blast Furnace Slag-Based Concrete 675
Imran Qasim and Khushpreet Singh

Hydrodynamic Modeling for Identifying Flood Vulnerability Zones in Mahi Lower Sub-basin 687
R. Rathod Krina and Sudhanshu Dixit

Influence of Distinctive Parameters on Fundamental Time Period of the Building 699
Shubam Sharma and Aditya Kumar Tiwary

Experimental Study of Fiber-Reinforced Concrete Prepared with Recycled Coarse Aggregate Bagasse Ash and Polypropylene Fiber ... 711
Harish Kumar and Aditya Kumar Tiwary

Effect of Shear Walls on Tall Buildings with Different Corner Configuration Subjected to Wind Loads ... 723
Saransh Mahajan, Vikramaditya Yadav, Rahul Raj, and Ritu Raj

Analytical Investigation of Moment Resisting Frame Structure—A Case Study on Performance-Based Capacity Spectrum Method ... 735
Ajay Singh Thakur and Tanmay Gupta

Study of Slope Stability Using Flexible Facing ... 747
Amanpreet Tangri and Saurabh Rawat

Use of Waste Polymers in a Plastic Bricks as Sustainable Building and Construction Materials ... 757
Prajwal Madghe, Himanshu Berad, Abhijeet Roy, Nishant Vaidya, Nivesh Sakharwade, and Rajan L. Wankhade

Determination of the Probability of Collapse for Existing Building Using Rapid Visual Screening as Tool ... 767
Salil Jha and Shilpa Pal

Prioritizing Buildings for Seismic Retrofit on the Basis of RVS Score ... 779
Salil Jha and Shilpa Pal

Utilization of Rice Straw Ash as a Replacement of Cement and Fine Aggregate in Mortar Mixes ... 793
Mohammad Ihtesham Hadizai and Aditya Kumar Tiwary

Analytical Behavior of Concrete-Filled Single-Skin and Double-Skin Tube Columns Subjected to Axial Loading ... 809
Sakshi Bhatia and Aditya Kumar Tiwary

Corrosion Monitoring in Reinforced Concrete Structures by Impressed Current Technique ... 823
Meenakshi Dixit and Ashok Kumar Gupta

Evaluation of Model 3D Printer and Design Mix for 3D Concrete Printing ... 837
Ashutosh Dwivedi, Ankit Pal, Shiv Singh Patel, Ajay Chourasia, and A. K. Jain

Axial Compression Behavior of Single-Skin and Double-Skin Concrete-Filled Steel Tube Columns: A Review ... 849
Sakshi Bhatia and Aditya Kumar Tiwary

Comparative Seismic Analysis of Multi-storied Building with and Without Floating Columns 863
Arvind Thakur and Amreen Khatun

Axial Loading Behaviour of Concrete Filled Steel Tube (CFST) Columns: A Parametric Study 873
Aditya Kumar Tiwary and Ashok Kumar Gupta

Comparative Investigation on Mode Shapes and Natural Frequency of Low-Rise RC Frame Building 885
Pushkar Sharma and Tanmay Gupta

Study of Behavior of the Masonry Infill Structures Subjected to Lateral Loads 899
Shilpa Pal, Sahil Yadav, Omkarnath Thakur, and Rohit Kashyap

Quantification of Municipal Solid Waste and Effect of Open Dumping on Soil in Smart City Dharamshala, Himachal Pradesh 913
Anchal Sharma, Love Sharma, and Rajiv Ganguly

Strength Characteristics of Clayey Soil Stabilized with Brick Kiln Dust and Sisal Fiber 927
Mandeep Singh, Kanwarpreet Singh, and Abhishek Sharma

Numerical Investigation of Wind Load on Side Ratio of High-Rise Buildings 937
Rahul Kumar Meena, Ritu Raj, and S. Anbukumar

Effective Remediation Techniques for Twin Shallow Lakes in Panchkula, Haryana 953
Prachi Vasistha and Rajiv Ganguly

Influence of Pond Ash on Compaction and Strength of Clayey Soil Mixed with Terrazyme 973
Nitish Kumar, Abhishek Sharma, and Kanwarpreet Singh

Seismic Analysis of Vertically Regular and Irregular Buildings with Shear Walls and RCC X-Bracing System 981
Mohd Zahid, Md Miraz, Mohd Faizan Saifi Warsi, and Shilpa Pal

Crumb Rubber Concrete Formation by Partial Replacement of Fine Aggregates 995
Pradeep K. Goyal and Akhilesh Chauhan

Influence of Mineral Admixture on Acceleration Carbonation Curing of Concrete: A Review 1005
Ishfaq Ahmad Bhat, Khushpreet Singh, and Nittin Sharma

Pre-Engineered Buildings—A Cost Saving Approach 1019
Abhijeet Roy, Rohit Motwani, Aditya Jaiswal, Aishwarya Raipurkar, and Kshitija Kadam

Effect of Curing on Compressive and Shear Strength Parameters of Liming Waste Ash Stabilized Expansive Soil 1035
Niraj Singh Parihar and Ashok Kumar Gupta

Analysis of Critical Factors Affecting Labor Productivity of Construction Projects in Himachal Pradesh 1047
Kaushal Kumar and Rishi Rana

Effect of Adding Fly Ash and Metakaolin on Mechanical Properties of Concrete 1059
S. K. Singh, Maninder Singh, and Rajesh Goyal

About the Editors

Dr. Ashok Kumar Gupta is currently a Professor of Civil Engineering and Dean (Academics and Research), Jaypee University of Information Technology, Waknaghat, Solan, Himachal Pradesh, India. He obtained his B.E. (Civil) with honours and M.E. (Geotechnical Engineering) from University of Roorkee which is now Indian Institute of Technology, Roorkee. He completed his Ph.D. from Indian institute of Technology, Delhi. His major areas of research includes testing and modeling of geotechnical materials, finite element method and its applications to geotechnical engineering, continuum damage mechanics and its application to rockfill materials modeling, and environmental geotechnique. He has published 57 papers in international journals of repute and co-authored 2 books and seven book chapters. He is also a reviewer for the Journal of Geotechnical and Geoenvironmental Engineering, ASCE and International Journal of Geomechanics, ASCE for which he has been awarded the Best Reviewer Award for three consecutive years. Professor Gupta is also the Founder Chairman, Indian Geotechnical Society (IGS) Shimla Chapter. He is a lifetime member of Indian Geotechnical Society, Indian Society of Rock Mechanics and Tunneling Technology (ISRMTT) and Indian Society of Technical Education (ISTE).

Dr. Sanjay Kumar Shukla is Founding Editor-in-Chief of the International Journal of Geosynthetics and Ground Engineering, Springer Nature, Switzerland. He is Founding Research Group Leader (Geotechnical and Geoenvironmental Engineering) at the School of Engineering, Edith Cowan University, Joondalup, Perth, Australia. He holds the Distinguished Professorship in Civil Engineering at Delhi Technological University, Delhi, VIT University, Vellore, Chitkara University, Solan, Himachal Pradesh, VR Siddhartha Engineering College, Vijayawada, Amity University, Noida, and Amrita University, Coimbatore, India. He has over 25 years of experience in teaching, research and consultancy in the field of Civil (Geotechnical) Engineering. He collaborates with several world-class universities, research institutions, industries and individuals on academic and field projects. As a consulting geotechnical engineer, he has successfully provided solutions to the challenging field problems faced by many engineering organisations. He has authored more than 280 research papers and technical articles, including over 175 refereed journal publications. He is also author/editor of 23 books, including 7 textbooks and 22 book chapters. In 2020/2021, his ICE textbooks, namely *Core Principles of Soil Mechanics* and *Core Concepts of Geotechnical Engineering*, have been ranked #1 by Amazon. His research and academic works have been cited well. Shukla's generalized expressions for seismic active thrust (2015) and seismic passive resistance (2013) are routinely used by practising engineers worldwide for designing the retaining structures. Shukla's wraparound reinforcement technique, developed during 2007–2008, is a well-established ground improvement technique. He has been honoured with several awards, including 2021 ECU Aspire Award from the Business Events Perth, Australia, and the most prestigious IGS Award 2018 from the International Geosynthetics Society (IGS), USA, in recognition of his outstanding contribution to the development and use of geosynthetics. He serves on the editorial boards of several international journals. He is a fellow of American Society of Civil Engineers and Engineers Australia, a life fellow of the Institution of Engineers (India) and Indian Geotechnical Society, and a member of several other professional bodies.

Dr. Hazi Azamathulla is currently a Professor of Civil and Environmental Engineering at the University of the West Indies at St. Augustine, Trinidad. He obtained his B.Tech. (Civil Engineering) from G. Pulla Reddy Engineering College, Kurnool, Sri Krishna Devaraya University (SKDU), Anantapur, India. He did his M.E. (Water Resources Engineering) from SGSITS, Devi Ahilya Vishwa Vidyalaya (DAVV), Indore, India and Ph.D. in Hydraulic Engineering from Indian Institute of Technology Bombay. His major research areas include water resources engineering hydraulics, physical hydraulic model studies, hydro informatics and climate change. He has authored/co-authored more than 120 research papers in reputed refereed journals. He is/has been a member of the editorial board of several high-ranked journals: *Water Science and Technology*, *Water Science and Technology: Water Supply*, *Journal of Pipeline Systems Engineering—ASCE* (2009–2013), *Dam Engineering Journal.* He is also a Life Member of Indian Society for Hydraulics (ISH) and Associate Member in Institution of Engineers (India). He is a fellow member of International Flood Network (IFNet), Japan and International Association of Hydrological Sciences (IAHS), UK. He is the Associate Editor of *Journal of Hydrology.*

Use of Biochar for Sustainable Environmental Remediation

Krishna R. Reddy and Jyoti K. Chetri

Abstract Biochar is a stable carbon-rich material produced from biomass (wastes) during bioenergy production. Since it is produced as a by-product from waste biomass during production of fuel, it can be considered a sustainable material. The unique properties of biochar such as high internal porosity, high intrinsic surface area, high moisture and nutrient retention capacity, and high adsorption capacity make it appealing for wide range of applications. Biochar has found use in variety of applications ranging from soil amendment for improved crop yield in agriculture, soil stabilizer in construction, to environmental remediation media. Biochar is shown to be effective for nutrient removal from wastewater, mixed contaminant removal from stormwater, contaminated soil remediation, and fugitive landfill gas mitigation. Although biochars have amenable properties for wide range of applications, their properties may vary significantly depending upon type of feedstock, production process, processing temperature, and residence time. Hence, all the biochars are not the same and cannot be used in all applications. It is crucial to perform complete characterization (including contamination content, if present) of the biochar before planning to use in any application. Moreover, biochars can also be customized as per the project requirements which offers potential for unlimited use of biochars. One of the challenges for economic and sustainable application of biochar could be local availability in required quantities, specifically for large-scale applications. Overall, biochar is shown to offer a sustainable solution to wide range of environmental problems.

Keywords Biochar · Biomass · Pyrolysis · Soil amendment · Wastewater treatment · Stormwater treatment · Soil remediation · Landfill biocover · Sustainability

K. R. Reddy (✉) · J. K. Chetri
Deptartment of Civil, Materials, and Environmental Engineering, University of Illinois at Chicago, Chicago IL60607, USA
e-mail: kreddy@uic.edu

J. K. Chetri
e-mail: jkc4@uic.edu

A. K. Gupta et al. (eds.), *Advances in Construction Materials and Sustainable Environment*, Lecture Notes in Civil Engineering 196,
https://doi.org/10.1007/978-981-16-6557-8_1

1 Introduction

Biochar is a solid material obtained from the thermochemical conversion of biomass under oxygen-deficient or oxygen-deprived conditions. The number of publications on biochar have been increasing every year, demonstrating the growing interest of researchers on biochar. Much of research is being pursued in the fields of environmental science, energy fuels, environmental engineering, chemical engineering, soil science, applied biotechnology, agricultural engineering, agronomy, green/sustainable science and engineering, among others. There are different methods and feedstock used for producing biochar, and the physical, chemical, and engineering properties of biochar can vary depending on the production method and feedstock [1]. Over past decade, many engineering applications of biochar are being actively explored by many researchers [1, 2]. This paper provides an overview of production of biochar, properties of biochars, and applications of biochars. Examples of environmental remediation applications based on the authors' research are highlighted.

2 Biochar Production

Biochar is a solid material by-product produced when any biomass is subjected to thermochemical process to generate bioenergy (bio-oil or syngas) under oxygen-limited conditions. Pyrolysis and gasification are the two commonly used thermochemical processes. In pyrolysis, chemical degradation of biomass occurs at higher temperatures in the absence of oxygen. The pyrolysis process uses kilns and retorts and other specialized equipment to contain the biomass while excluding the oxygen. Pyrolysis can be fast pyrolysis in which the biomass is heated at very high temperatures (nearly 450–600 °C) with low residence time, producing more oils and liquids and less syngas. On the other hand, slow pyrolysis involves heating biomass under low temperature with relatively high residence time, producing more syngas and less bio-oil. More biochar is produced in slow pyrolysis conditions as compared to high pyrolysis conditions. Gasification is the process of heating the biomass directly in reaction unit with introduced air, excluding more air produces more biochar. The biochar yield and its carbon content vary based on the production process. Other thermochemical conversion processes, including their typical process temperature, residence time, biochar yield, and carbon content, are presented by Xie et al. [1].

Many researchers use laboratory-scale pyrolysis and gasification units which allow producing biochar under controlled feedstock and process temperature/residence time conditions. Often, these units are limited to produce only small quantities of biochar for use in research experiments. However, large-scale commercial pyrolysis/gasification units use large quantities of biomass under less controlled temperature/residence time conditions. Hence, the quality of biochar could significantly differ from that of laboratory-produced biochar. Even personal top-lit up draft

(TLUD) gasifiers are developed to produce biochar, and TLUD gasifier cookstoves are being promoted in developing countries to be safe and beneficial.

The feedstock could be any biodegradable products, wastes, or residues. The common feedstocks include agricultural residues such as corn cobs, corn stover, rice and peanut hulls, and animal manure; yard wastes such as tree bark and wood chips; animal wastes such as poultry litter, cat litter, and livestock manure; industrial wastes such as paper mill sludge; municipal solid wastes; and by-products from bioenergy such as defatted dried distiller grains [1]. Although biochar can be produced from almost any type of biomass, the yield and properties of biochar vary drastically based on the feedstock and also on the production process. Therefore, hundreds of different types of biochars are possible depending on the combinations of feedstock and production process conditions used.

3 Physical and Chemical Properties

The physical and chemical properties of biochar are dictated by the feedstock used and the production process conditions. The most common physical properties include moisture content, specific gravity, dry density, bulk density, particle size distribution, porosity, and surface area. The chemical properties of interest include CHN elemental composition, pH/OP/EC, organic matter, volatile matter, ash content, fixed carbon content, zeta potential, inorganic metals content, and organic chemicals (PAHs, PCBs, etc.) content. Generally, increase in pyrolysis temperature is shown to result biochar with high specific surface area, high carbon and nitrogen contents, and low hydrogen and oxygen contents [1]. The biochars made from the same feedstock but using different production conditions are shown to exhibit significant different physical and chemical properties [3, 4]. Some biochars contain significant quantities of toxic metals and polyaromatic hydrocarbons (PAHs), which are not desirable for use in engineering applications [5].

For instance, Yargicoglu et al. [3] reported physical and chemical properties of waste-wood-derived biochars produced commercially by different vendors (Fig. 1). These biochars had various particle sizes ranging from powder form to pellets. The porosity of these biochars varied between 30 and 55%, and the mean particle size ranged from 0.2 to 7 mm. Biochars are lightweight with specific gravity mostly less

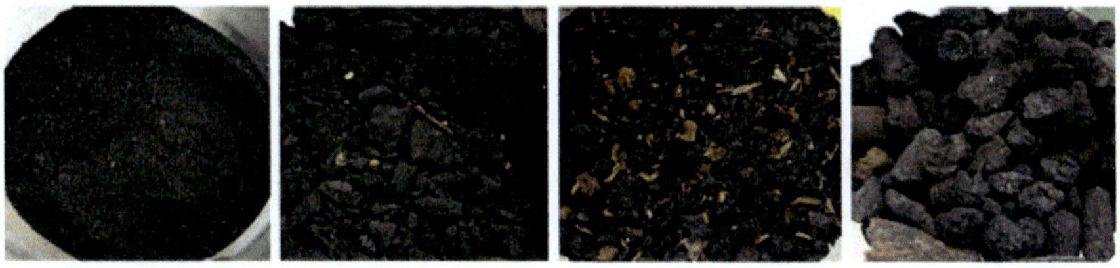

Fig. 1 Images of different waste-wood-derived commercially produced biochars

than 1. Biochars have high intrinsic porosity and high specific surface area. These biochars have low ash content and higher fixed carbon. Biochars are also characterized with high organic matter content as they are produced from the biomass. The pH of the wood-derived biochars can range from 6.2 to 8.7. Some biochars may have elevated polyaromatic hydrocarbons (PAHs) as they are produced during incomplete combustion of biomass [3]. Since PAHs are highly toxic in nature, biochars should be selected with due consideration before any field application. In addition to PAHs, biochars may also contain toxic metals whose leachability could be a concern in environmental applications. Higher PAHs and trace metals have been associated with lower H:C ratios (extent of carbonization) and high ash content. In the study by Yargicoglu et al. [3], PAHs were below detection limit for wood-derived biochars produced via gasification. Biochars are commonly characterized with high cation exchange capacity (CEC), partly due to their negative surface charge and resultant affinity for soil cations including heavy metals (such as Pb^{2+}, Cr^{3+}) [1].

4 Engineering/Geotechnical Properties

In addition to the physical and chemical properties, engineering properties of biochars are needed to be evaluated prior to using in engineering applications. These engineering properties include: (1) hydraulic properties such as water holding capacity, saturated hydraulic conductivity, and moisture-suction-hydraulic conductivity; (2) compressibility characteristics such as compression ratio; and (3) shear strength properties. Several studies have reported engineering properties of biochars. These properties are shown to be highly dependent on the type of biochar (how it is produced). For instance, waste-wood-derived biochars are shown to have high water holding capacity, high hydraulic conductivity, lower compressibility, and high shear strength, making them suitable for engineering applications [6–8].

The properties of waste-wood-derived biochar-amended soils are also shown to be favorable for engineering applications [2, 6–8]. For instance, Sadasivam and Reddy [6] showed that biochar-amended soils possess greater cohesion and friction angle than the soil under wet conditions. Biochar amendment has also shown to improve slope stability which could be attributed to the improved soil shear strength properties due to biochar amendment [6, 7]. However, the degree of improvement in soil strength is dependent on the type of biochar. Wood-based biochar with low ash content has shown better strength properties.

5 Biochar Applications

Biochars have been commonly used in agriculture for improving the soil's fertility and crop yield [2]. The unique properties of biochar such as long-term stability or recalcitrance, high internal porosity, and high moisture and nutrient retention make it amenable for agricultural applications. Especially, biochar's better moisture and nutrient retention capacity improves nutrient stock in the root zone of crops, minimizes nutrient leaching, and ultimately enhances crop production [9]. In addition, biochar itself can release important nutrients nitrogen (N), phosphorous (P), and potassium (K) in soil which can be absorbed by the plants [9].

Applications of biochar in environmental engineering applications are rapidly growing. High porosity and specific surface area of biochar make it an excellent candidate for sorption of many harmful contaminants including heavy metals, nutrients and organic pollutants from soil, sediments, stormwater, and groundwater. Since biochar is derived from waste biomass and produced as a by-product in the process of production of bioenergy, it offers the cost benefit when compared to using other sorptive materials such as activated carbon [1]. Hence, biochar is increasingly being recognized and used in the environmental engineering applications.

In the recent years, there has been an increasing trend of using biochar in various infrastructure engineering applications such as ground modification and improving aggregate stability and soil shear strength properties. Biochar is being explored as a sustainable material for treating expansive soils. High moisture retention and absorption property of biochar preclude water from coming in contact with soil thereby preventing problematic swelling of the soil. The major mechanisms postulated for reduction in swell potential of expansive clay soil with biochar amendment include adsorption, flocculation, and cation exchange between biochar and clay particles preventing the entry of water into the soil and reducing the diffuse double layer thickness. Some researchers even made attempts to use biochar in concrete mixes.

6 Environmental Remediation Applications

The use of biochar in environmental engineering applications has been investigated by the authors for over a decade. Selected projects that use biochar to address the problems of wastewater pollution, urban stormwater pollution, soil pollution, and landfill gas emissions are briefly presented here.

6.1 Nutrient Removal from Wastewater

The wastewater may contain wide range of contaminants including nutrients such as N, P, and K which when discharged to local water bodies may cause eutrophication and affect aquatic ecosystem. The high surface area and sorptive capacity of biochar make it a suitable adsorbent for nutrients including N, P, and K. Biochar has shown promising N and P recovery efficiencies from wastewater. Combining magnesium oxide (MgO) with biochar has shown to improve N and P recoveries from wastewater by biochar sorption and struvite precipitation mechanisms [9, 10]. Increasing the MgO and biochar mixing ratios has shown significant improvement in N and P removal efficiencies. The mixing of MgO with biochar (1:10 by weight) resulted in nearly 100% removal of P and 50% removal of N. The exhausted biochar after nutrient adsorption is shown to be an excellent fertilizer for plant growth. Hence, biochar and MgO mixture serves as an excellent adsorbent for recovery of nutrient (N and P) from wastewater. In addition, mixture of biochar and struvite has good nutrient qualities and can be directly used as potential fertilizer [10].

6.2 *Polluted Stormwater Treatment*

Stormwaters are the most common non-point source of water pollution to rivers, lakes, estuaries, and beaches. Urban stormwater runoffs may carry a wide range of contaminants endangering the surface water bodies. Beach closures are becoming a growing concern due to the presence of pathogens, harmful microorganisms, and other pollutants in stormwater runoffs that can cause serious public health problems. Contaminants mainly include nutrients (N and P), heavy metals, PAHs, as well as pathogens such as *E. Coli.* Use of filtration system has become a popular choice in urban setting due to its ability to remove suspended and dissolved pollutants in a smaller footprint without requirement of huge space. Reddy et al. [11] proposed an in-ground permeable reactive filter system to treat mixed contaminants in urban stormwater runoff. Biochar was tested as one of the potential filter media for the permeable reactive barrier [11]. Column tests were performed with biochar to evaluate its removal potential for mixed contaminants from stormwater. The results illustrated that the biochar could remove total suspended solids and nitrate by 86% and phosphate by 47%. Similarly, biochar was successful in removing some PAHs like phenanthrene and naphthalene. In addition, biochar was able to remove pathogens like *E. Coli.* Hence, biochar has potential to be an effective medium for the treatment of selected contaminants found in stormwater.

6.3 *Contaminated Soil Remediation*

Phytoremediation is a green and sustainable option for remediation of contaminated soils. The plant's ability to absorb, degrade, extract, or immobilize contaminants is utilized in phytoremediation by growing suitable plants in the contaminated soil [12]. However, contaminant removal efficiency is limited by plant's ability to grow on the contaminated soil site. The use of biochar for soil improvement has long been identified and has been in practice. Hence, Chirakarra and Reddy [12] studied the effect of biomass amendments using compost, biochar, and nutrient solution and their combinations on the survival and contaminant removal potential of oat plant and sunflower through laboratory pot experiments. Altogether ten soil treatments were tested which involved: Clean unamended (T1), Contaminated unamended (T2), Biochar (T4), Compost (T4), Nutrient solution (T5), EDTA (T6), Igepal CA-720

(T7), EDTA + Igepal CA-720 (T8), Compost + EDTA (T9), and Compost + Igepal CA 720. Sunflower showed improved germination, survival, and growth with biochar amendment. The biochar amendment also showed removal potential for metals like Cd and Pb and PAHs. Overall, biochar is shown great potential to create favorable conditions for the seed germination and plant growth, thereby creating favorable conditions for contaminant stabilization or degradation.

6.4 *Fugitive Landfill Gas Mitigation*

Municipal solid waste landfills are the third-largest source of anthropogenic methane emissions in the USA, contributing about 15.1% of these emissions in 2019. Methane is flammable gas and can pose explosion hazard if present between its lower explosive limit of 5% (v/v) and upper explosive limit of 15% (v/v). Methane is a highly potent greenhouse gas and an asphyxiant. In modern engineered landfills, gas collection system is installed to capture landfill gas emissions. However, the efficiency of the gas collection system is affected by various factors such as gas production rates, climatic conditions such as precipitation, and temperatures and local site conditions, and hence, the gas collection efficiency is found to be about 60%, which increases the likelihood of fugitive emissions. In addition, it is not practically and economically feasible to install gas collection systems at old and abandoned landfills, and therefore, fugitive emissions is a major issue at such landfills. In the recent years, organics rich biocovers have emerged as environment friendly and sustainable alternative to mitigate fugitive landfill emissions through microbial oxidation process. Landfill cover soils have microorganisms which can oxidize methane to carbon dioxide in the presence of oxygen (Fig. 2). Recently, biochar has garnered attention for use as an organic amendment for biocover to enhance microbial methane oxidation due to its unique properties such as high long-term stability, high internal porosity, high

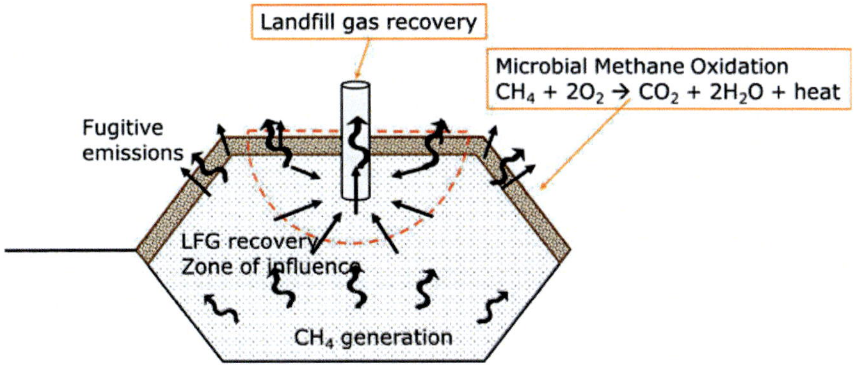

Fig. 2 Landfill methane oxidation with biochar-amended soil cover

moisture retention, high adsorption potential for methane, and ability to enhance microbial colonization [13–16].

Various laboratory batch and column incubation studies as well as field scale testing showed significantly higher methane sorption and oxidation rates and efficiencies in biochar-amended soil under landfill gas conditions [14, 15, 17, 18]. The biochar-amended soils were inhabited by methane-oxidizing microbes such as methanotrophs. Methanotrophic genera such as *Methylobacter, Methylocystis, and Methylosinus* were found in abundance in the biochar-amended soils [16]. Biochar is an inert material; hence, it does not perform microbial oxidation by itself but helps to enhance the microbial activity by creating conducive environment for microbial proliferation. Recently, attempts have been made to improve the performance of biochar-amended soil by activating the biochar with methane-oxidizing bacterial (MOB) consortium before amending the soil [19]. Biochar activation showed very high methane oxidation rates alleviating the initial lag phase which is generally encountered upon busing biochar-amended soil due to microbial acclimation phase. Biochar activation not only helped to reduce the initial lag phase but also enhanced the microbial methane oxidation rates (Fig. 3). Hence, biochar activation appears to be a promising technique to shorten the microbial lag phase and enhance the microbial methane oxidation in biochar-amended landfill cover soils.

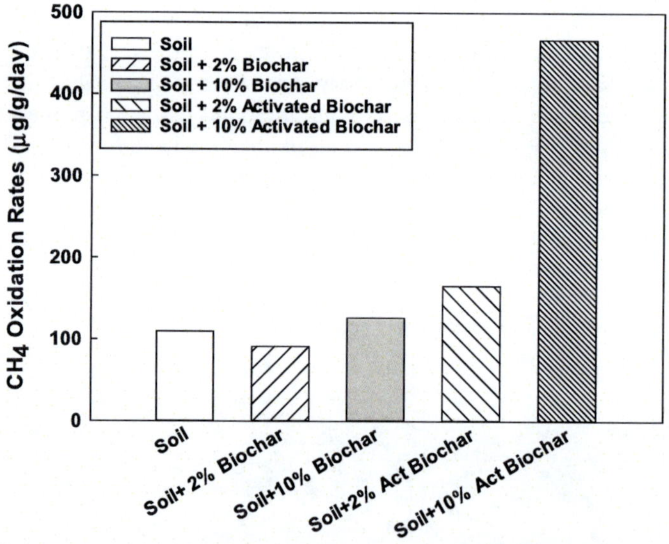

Fig. 3 Methane oxidation rates of soil, non-activated biochar-amended soils, and activated biochar-amended soils

7 Conclusion

Biochar is a stabilized, carbon-rich recalcitrant compound, produced from thermochemical conversion of biomass (waste) under oxygen limited condition. Biochar can be produced from wide variety of feedstocks such as agricultural residues, yard wastes, animal wastes, industrial wastes, and municipal solid wastes. Yield and properties of biochar vary dramatically based on the type of feedstock and production process, and therefore, it is important to characterize the biochar completely before using in any applications. Biochar has gained popularity due to its unique properties such as high specific surface area, microporosity, and sorption capacities. The use of charred materials in soil nourishment has long been in practice, even before introduction of biochar. Apart from soil amendment in agriculture, biochar is used in environmental engineering applications such as wastewater treatment, stormwater treatment groundwater remediation and infrastructure engineering application such as soil/waste stabilization and concrete admixtures. Moreover, new and innovative customization of biochar offers potential for use in wider range of applications.

References

1. Xie, T., Reddy, K. R., Wang, C., Yargicoglu, E., & Spokas, K. (2015). Characteristics and applications of biochar for environmental remediation: A review. *Critical Reviews in Environment Science and Technology, 45*(9), 939–969. https://doi.org/10.1080/10643389.2014.924180
2. Xie, T., Sadasivam, B. Y., Reddy, K. R., Wang, C., & Spokas, K. (2016). Review of the effects of biochar amendment on soil properties and carbon sequestration. *J Hazard Toxic Radioact Waste, 20*(1), 04015013. https://doi.org/10.1061/(ASCE)HZ.2153-5515.0000293
3. Yargicoglu, E., Sadasivam, B. Y., Reddy, K. R., & Spokas, K. (2015). Physical and chemical characterization of waste wood derived biochars. *Waste Management, 36*(2), 256–268. https://doi.org/10.1016/j.wasman.2014.10.029
4. Yargicoglu, E. N., & Reddy, K.R. (2015). Characterization and surface analysis of commercially available biochars for geoenvironmental applications. In: International Foundations Congress and Equipment Expo (IFCEE) 2015, ASCE, San Antonio, TX, pp. 2637–2646. https://doi.org/10.1061/9780784479087.245
5. Yargicoglu, E. N., & Reddy, K. R. (2014). Evaluation of PAH and metal contents of different biochars for use in climate change mitigation systems. In: *International Conference on Sustainable Infrastructure*, Long Beach, CA, November 6–8, ASCE, Reston, VA. https://doi.org/10.1061/9780784478745.011
6. Sadasivam, B. Y., & Reddy, K. R. (2015). Engineering properties of waste-wood derived biochars and biochar-amended soils. *Int J Geotech, 9*(5), 521–535. https://doi.org/10.1179/1939787915Y.0000000004
7. Reddy, K. R., Yaghoubi, P., & Yukselen-Aksoy, Y. (2015). Effects of biochar amendment on geotechnical properties of landfill cover soil. *Waste Manag Res, 33*(6), 524–532. https://doi.org/10.1177/0734242X15580192
8. Kumar, G., Yepes, J. E., Hoyos, L. R., & Reddy, K. R. (2017). Unsaturated hydraulic properties of biochar and biochar-amended soils for landfill covers. In: *PanAm-UNSAT 2017: The Second Pan-American Conference on Unsaturated Soils*. ASCE Press. https://doi.org/10.1061/9780784481691.041
9. Xie, T., Reddy, K. R., Wang, C., Xu, K. (2014). Effects of amendment of biochar produced from woody biomass on soil quality and crop yield. In: K. R. Reddy, S. Feng (Eds.), *Geotechnical*

Special Publication 241-Geoenvironmental Engineering, Geo-Shanghai 2014, pp.170–180. American Society of Civil Engineers, Reston, VA. https://doi.org/10.1061/9780784413432.018

10. Zheng, M., Xie, T., Li, J., Xu, K., & Wang, C. (2018). Biochar as a carrier of struvite precipitation for nitrogen and phosphorus recovery from urine. *Journal of Environmental Engineering, 144*(10), 04018101. https://doi.org/10.1061/(ASCE)EE.1943-7870.0001450
11. Reddy, K., Xie, T., & Dastgheibi, S. (2014). Evaluation of biochar as a potential filter media for the removal of mixed contaminants from urban storm water runoff. *Journal of Environmental Engineering, 140*(12), 04014043. https://doi.org/10.1061/(ASCE)EE.1943-7870.0000872
12. Chirakkara, R. A., & Reddy, K. R. (2015). Biomass and chemical amendments for enhanced phytoremediation of mixed contaminated soils. *Ecological Engineering, 85*, 265–274. https://doi.org/10.1016/j.ecoleng.2015.09.029
13. Reddy, K. R., Yargicoglu, E. N., Yue, D., & Yaghoubi, P. (2014). Enhanced microbial methane oxidation in landfill cover soil amended with biochar. *J Geotech Geoenviron, 140*(9), 04014047. https://doi.org/10.1061/(ASCE)GT.1943-5606.0001148
14. Reddy, K. R., Yargicoglu, E. N., & Chetri, J. K. (2021). Effects of biochar on methane oxidation and properties of landfill cover soil: Long-term column incubation tests. *Journal of Environmental Engineering, 147*(1), 04020144. https://doi.org/10.1061/(ASCE)EE.1943-7870.0001829
15. Reddy, K. R., Yargicoglu, E. N., & Chetri, J. K. (2021). Field-scale performance of biochar-amended soil covers for landfill methane oxidation. *Biomass Conversion and Biorefinery*. https://doi.org/10.1007/s13399-021-01487-w
16. Yargicoglu, E. Y., & Reddy, K. R. (2017). Microbial abundance and activity in biochar-amended landfill cover soils: Evidences from large-scale column and field experiments. *Journal of Environmental Engineering, 143*(9), 04017058. https://doi.org/10.1061/(ASCE)EE.1943-7870.0001254
17. Sadasivam, B. Y., & Reddy, K. R. (2015). Adsorption and transport of methane in biochars derived from waste wood. *Waste Management, 43*(9), 218–229. https://doi.org/10.1016/j.wasman.2015.04.025
18. Sadasivam, B. Y., & Reddy, K. R. (2015). Adsorption and transport of methane in landfill cover soil amended with waste-wood biochars. *Journal of Environmental Management, 158*(8), 11–23. https://doi.org/10.1016/j.jenvman.2015.04.032
19. Rai, R. K., Chetri, J. K., & Reddy, K. R. (2019). Effect of methanotrophic-activated biochar-amended soil in mitigating CH_4 emissions from landfills. In: *4th International Conference on Civil and Environmental Geology and Mining Engineering*, Trabzon, Turkey, April 20–22, 2019.

Macro and microscale Engineering Response of Rigid-Soft Gravel-Rubber Inclusions: Insights from Detailed Laboratory and DEM Numerical Investigations

Gabriele Chiaro, Ali Tasalloti, Kevin Chew, Jayan S. Vinod, and Krishna Allulakshmi

Abstract This paper summarizes the main results of a broader research activity aimed at facilitating the use of gravel-rubber mixtures (GRMs) in geotechnical engineering applications. Precisely, the results of direct shear tests carried out on three GRMs having different aspect ratios ($AR = 0.28$, 0.57, and 0.66) and prepared at different volumetric rubber contents ($VRC = 0$, 10, 25, 40, and 100%) are reported and discussed. Additionally, an original 3D hybrid DEM numerical model (named DEM4GRM model) for rigid-soft particle mixtures is presented, and its performance is evaluated. Lastly, using newly proposed micromechanical equations, factors such as fabric and force anisotropy, and strong-force chains are systematically examined throughout the shearing process to gain insight on the particle-level behavior of mixtures with varying *VRC*.

Keywords Gravel-rubber mixtures · Direct shear · DEM · Micromechanics

1 Introduction

In Australasia, geotechnical engineers are increasingly expected to meet requirements for environmentally friendly design, construction, and innovation [1]. As a result, reuse and recycling of construction and demolition materials, commercial wastes and industrial by-products in geotechnical engineering applications are progressively sought as it provides substantial benefits in terms of increased sustainability and reduced environmental impacts [2–6; among others]. Specifically, to the New Zealand context, due to the large number of waste tires produced annually

G. Chiaro (✉) · A. Tasalloti · K. Chew
Department of Civil and Natural Resources Engineering, University of Canterbury, Christchurch, New Zealand
e-mail: gabriele.chiaro@canterbury.ac.nz

J. S. Vinod · K. Allulakshmi
School Civil Mining Environmental Engineering, University of Wollongong, Wollongong, Australia

A. K. Gupta et al. (eds.), *Advances in Construction Materials and Sustainable Environment*, Lecture Notes in Civil Engineering 196,
https://doi.org/10.1007/978-981-16-6557-8_2

[7] and the need to design more seismic resilient structures [8], the recycling of tire-derived aggregates (TDAs) mixed with granular soils to produce lightweight fill materials with excellent mechanical and energy adsorption properties is of great significance [9].

A comprehensive literature review compiled by Tasalloti et al. [10] has shown that much of the previous research on granular mixtures has focused on the mechanical characterization of sand-rubber mixtures. Yet, in the selection of the soil type and recycled rubber size to form soil-rubber mixtures for use in geotechnical applications, the availability and the cost efficiency of both materials should be carefully considered [11]. In fact, to avoid inherent segregation of binary mixtures made of large and small particles [12, 13], the recycled rubber should be cut into smaller (sand size-like) pieces when mixed with sandy soils, which unavoidably increases the implementation costs. Consequently, as an alternative, the use of gravel-rubber mixtures (GRMs) has been increasingly recommended by researchers [14–20].

Nevertheless, compared to sand-rubber mixtures, the mechanical behavior of GRMs is still largely unknown. Therefore, as a part of a research activity aimed at facilitating, the use of GRMs is geotechnical applications in New Zealand [9, 15, 18, 20, 21], two subsequent steps were undertaken by the authors: (i) a number of direct shear investigations were carried out on three gravel-granulated rubber composites mixed at volumetric rubber content (*VRC*) of 0, 10, 25, 40 and 100% to evaluate the combined effects of *VRC*, aspect ratio ($AR = D_{50,R}/D_{50,G}$) and applied normal stress on the strength properties of GRMs; and (ii) a novel three-dimensional hybrid discrete element method model (named DEM4GRM) able to accurately describes the macromechanical direct shear response of rigid-soft particle mixtures was developed. Moreover, ad hoc equations defining key micromechanical features of such synthetic materials throughout the shearing processes were established for GRMs. In this paper, experimental findings are first summarized. Then, the new DEM4GRM model [22] is briefly introduced and simulation results are shown. Finally, using the newly proposed equations, microscale aspects such as fabric and force anisotropy and strong-force chains are scrutinized throughout the shearing process and compared among mixtures with varying *VRC*.

2 Experimental Study

2.1 Materials and Procedure

The strength properties of GRMs were estimated by means of a medium-size direct shear box (100 × 100 mm in cross-section and 53 mm in height) under three normal stress (σ_n) levels of 30, 60 and 100 kPa. The horizontal displacement rate was 1 mm/min.

Three types of rigid-soft granular mixtures were prepared by mixing a uniformly-graded rounded gravel (G) and three recycled rubber particle sizes: large (R1, $D_{50} =$

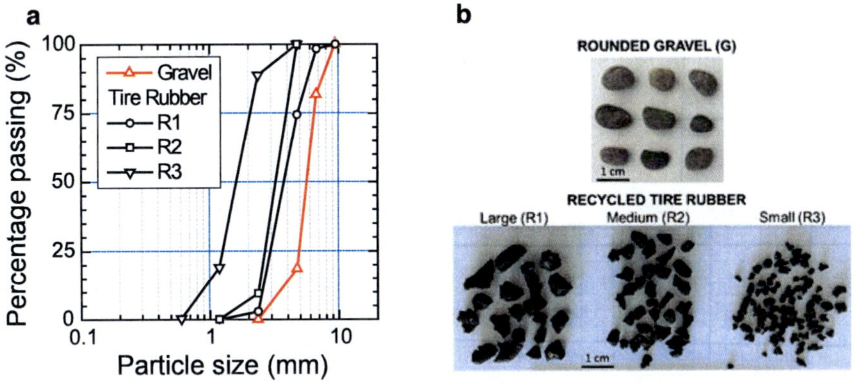

Fig. 1 **a** Particle size distribution curves and **b** photographic images of the gravel and granulated recycled tire rubber used in this study

3.74 mm), medium (R2, $D_{50} = 3.23$ mm), and small (R3, $D_{50} = 1.61$ mm). Figure 1 shows the particle size distribution curves and photographic images of each material. The material properties are listed in Table 1. The aspect ratio of the mixtures is G-R1 ($AR = 0.66$), G-R2 ($AR = 0.57$), and G-R3 ($AR = 0.28$). The GRMs were prepared at volumetric rubber contents (*VRC*) of 0, 10, 25, 40, and 100%, where *VRC* is defined as the ratio between the volume of rubber and the total volume of solids.

All specimens were tested in dry conditions and prepared at a degree of compaction of 90–95% (by dry tamping method), which was calculated based on the values of maximum dry density (ρ_{max}—obtained by standard proctor compaction tests). For completeness, the minimum dry density (ρ_{min}) was also determined by carefully pouring the materials in the compaction mold with zero depositional height. Note that vibratory table tests were found not effective to obtain ρ_{max} for GRMs. This is obviously in contrast to the response of typical hard-grained stiff granular materials for which vibratory compaction is most effective [23], but it can be attributed to the high energy absorption nature of the rubber particles in the mixtures.

Table 1 Properties of gravel and recycled tire rubber

Material	Gravel**	Recycled tire rubber		
		Large**	Medium***	Small**
Symbol	G	R1	R2	R3
Mean diameter, D_{50} (mm)	5.67	3.74	3.23	1.61
Specific gravity, G_s	2.71	1.15	1.15	1.14
Maximum dry density, ρ_{max} (kg/m^3)*	1753	649	634	602
Minimum dry density, ρ_{min} (kg/m^3)	1571	510	460	429

*by proctor compaction tests
**experimental data from Chiaro et al. [9, 21] and Tasalloti et al. [18, 20]
***experimental data from this study

Segregation is an inherent difficulty in granular mixtures that may be triggered by size, density, stiffness, and shape features [12]. Segregation in the GRM specimens was prevented by minimizing any vibration and avoiding granular flow.

2.2 *Test Results*

The materials investigated in this study are essentially binary granular mixtures consisting of rigid and soft particles of different size and shape. Their packing, density, and mechanical behavior are influenced by the inherent properties of the materials (i.e., size and shape), the proportion as well as the size ratio of large/small and rigid/soft particles in the total volume of solids.

2.2.1 Packing and Density Characteristics of GRMs

The variation of dry density (ρ_d) is shown in Fig. 2 for G-R1 and G-R3 mixtures. As the rubber particles are much lighter than the gravel grains (i.e., the G_s of gravel grains is more than twice than that of rubber particles), both the minimum and maximum dry densities of GRMs decrease almost linearly by increasing *VRC*. However, at $VRC \leq 40\%$, the dry density values of G-R3 are slightly above the linear trends. This is due to the fact that at lower *VRC*, small rubber particles can occupy the voids between large gravel grains (Fig. 2b) which results in an increase of the density state of the mixtures [12, 13]. In contrast, because the size of R1 particles is almost similar to that of the gravel, the rubber particles cannot fit in the voids between the gravel grains, even at lower *VRC*, but rather their replace the gravel grains in the mixtures (Fig. 2b).

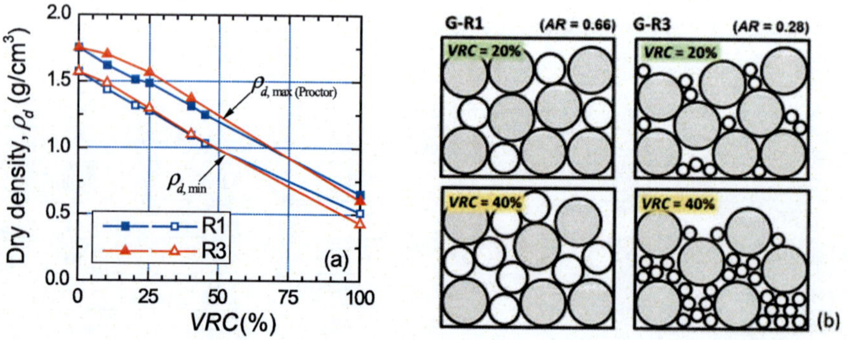

Fig. 2 Variation of **a** minimum and maximum dry density; and **b** idealized illustration of volume-biased packing of GRMs for $VRC = 20$ and 40%

2.2.2 Shear Strength Characteristics of GRMs

The stress–strain-volumetric behavior of G-R2 obtained by direct shear tests at σ_n = 30, 60, and 100 kPa is reported in Fig. 3. It can be seen that irrespective of σ_n, by increasing *VRC*, the material response gradually changes from dilative with a clear peak shear state to contractive without peak shear state. Moreover, the peak shear stress (τ_{max}) increases with increasing σ_n and decreases with decreasing *VRC*. Similar tendencies were observed for G-R1 and G-R3 (due to page limitation, not all the experimental data could be reported here but can be found in Tasalloti et al. [18, 20]). The summary plot reported in Fig. 4a shows that τ_{max} decreases with *AR*. Moreover, it indicates that the effects of *AR* on the GRM strength are less significant as compared to those due to *VRC* and σ_n.

Figure 4b reports the values of the Mohr–Coulomb friction angle (ϕ') for all mixtures. Essentially, ϕ' decreases significantly with increasing *VRC* (and only slightly with *AR*) from about 54° (gravel) to 29° (graduated rubber). Notably, excluding G-R2 and G-R3 with *VRC* > 85% and G-R1 with *VRC* > 95%, most of the GRMs have a high strength (i.e., ϕ'> 30°) irrespective of the *VRC* and rubber particle size making them suitable structural fill materials for many geotechnical applications [2, 9].

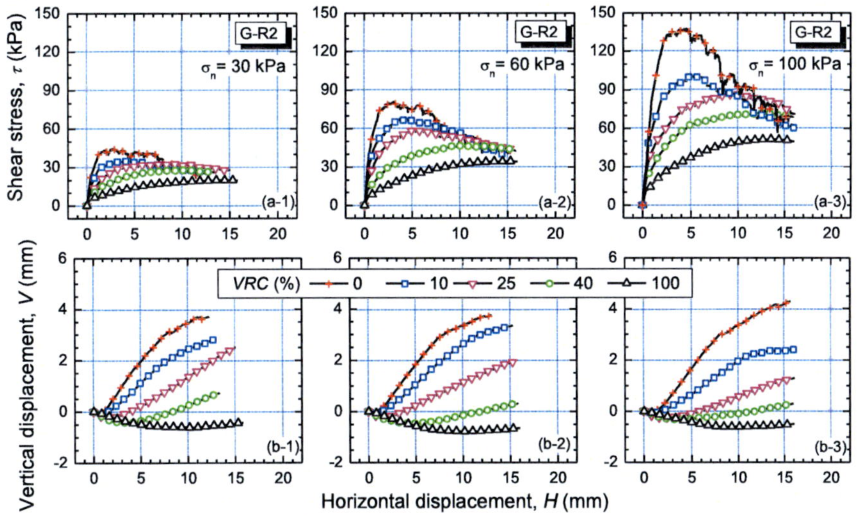

Fig. 3 G-R2 behavior in direct shear tests at 30, 60, and 100 kPa normal stress

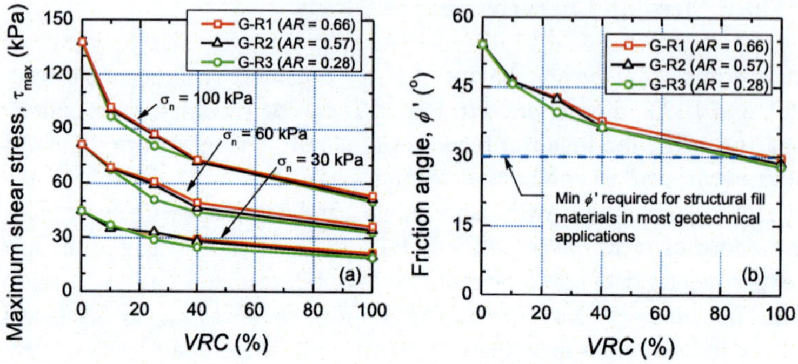

Fig. 4 **a** Maximum shear stress and **b** friction angle of GRMs in direct shear tests

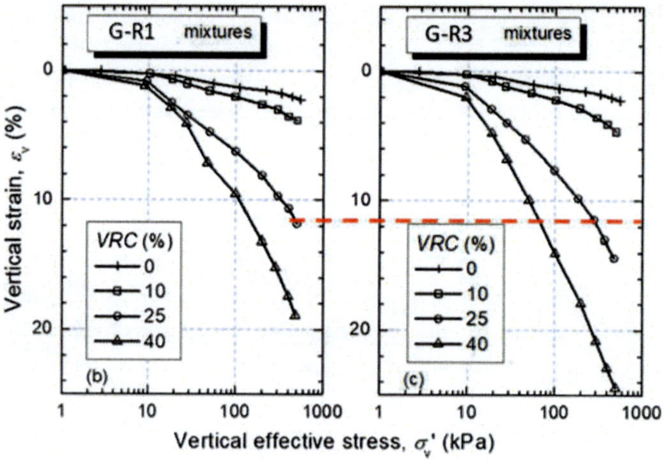

Fig. 5 Compressibility of GRMs in one-dimensional compression tests with two-hour creep loading stages

2.3 *Key Aspects to Consider in the Selection of GRMs*

2.3.1 Compressibility

Despite the adequate strength of GRMs, the ultimate adoption of GRMs as structural fills in geotechnical application would depend also on their compressibility under sustained loads. Typical results obtained for G-R1 and G-R3 by one-dimensional compression tests with two-hour creep loading stages [21]. It is obvious that the higher is the vertical stress applied on GRMs, the higher is the vertical strain developed, and the lower is the *VRC* that may be accepted in the mixtures to satisfy compressibility requirements [18, 20]. Moreover, *AR* also plays a key role; in fact, for any combination of *VRC* and vertical stress, it can be seen that the mixtures with small rubber particles (G-R3) experience large vertical strain as compared to those with large rubber particles (G-R1). Therefore, to have GRMs with enhanced compressibility, gravel and rubber particles with similar grain size should be used.

2.3.2 Environmental Implications

The introduction of new or alternative (recycled waste) materials in geotechnical applications may have benefits in terms of cost reductions and increased performance. However, it is essential to ensure that such innovations do not result in long-term negative impacts on the environment, e.g., through the leaching of toxic chemicals into the surrounding soil environment, groundwater, and surface water. The results of leaching tests conducted by Banasiak et al. [24] on GRMs have indicated that the leachate from smaller tire particles (R3) had a higher content of metals (e.g., Zn), as compared to the large rubber one (R1), implying that particle size and surface area influence the concentration of elements in tire-gravel mix leachate. To minimize the leaching of metals, therefore, the use of large rubber particle size is desirable.

3 DEM Numerical Modeling of GRMs

The discrete element method (DEM) is a very useful tool to investigate particle-level interactions and develop a better understanding of the macro and micromechanical behavior of granular matters [25]. Due to the granular and discrete nature of GRMs, DEM is indeed an ideal numerical modeling method to investigate the micromechanical behavior of such materials. Yet, the soft and low modulus nature of the rubber particles challenge one of the fundamental hypothesis of DEM by particle rigidity. Based on this hypothesis, contact forces are determined from the overlap between contacting bodies but the particles do not deform [25]. While the rubber softness may be accounted for by prescribing a low material stiffness, the volumetric behavior of pure rubber ($\nu^{\text{rubber}} = 0.5$) cannot be replicated easily in DEM using the particle rigidity method.

To address this issue, Asadi et al. [26] implemented a deformable agglomerate to model more accurate changes in the shape (deformation) of rubber particles by using weak internal bonds between tire rubber particles, allowing the spheres within a cluster to move and rearrange. In the following, Ren et al. [27] implemented a deformable single rubber particle model that is able to describe the volumetric behavior of rubber. It has been indicated that with sufficient rows of bonded spheres packed in a body-centered cubic (BCC) arrangement, this model can consistently and precisely capture the strength and volumetric response of a single rubber particle as the nonlinear mechanical behavior of rubber is described by a piecewise linear contact/bond model.

Taking advantage of both the well-established particle rigidity approach applicable to rigid gravel grains and the possibility to capture the soft rubber behavior using

BCC deformable particles, in this study, a 3D hybrid DEM model for GRMs (called DEM4GRM model) was developed using the particle flow code three-dimensional program (PFC3D, version 6.1 [28]). Such model makes it possible to explore the micromechanical direct shear behavior of GRMs under varying *VRC* and normal stress levels.

The framework in PFC3D is based on the DEM formulation by Cundall and Strack [29], where an explicit numerical scheme is used to monitor the interaction of particles individually and its corresponding contacts with neighboring particles. Moreover, based on the particle rigidity approach, contact forces are evaluated from the overlap between contacting bodies but the particles do not deform [25]. In this study, the Hertz-Mindlin contact law [30] was used to define the particle interaction at each contact. It is based on a nonlinear contact force-overlap relationship. The DEM4GRM model input parameters defining the contacts are provided in Table 2. As the gravel and rubber particles have different engineering properties, the surface property inheritance functionality in PFC3D [28] was used to determine hybrid contact properties between the gravel and rubber particles.

Table 2 DEM4GRM model: input parameters for G-R1 mixtures

Item	Parameter	Units	Value
Gravel	Particle density	kg/m^3	2710
	Coefficient of friction		0.72
	Shear modulus	MPa	90
	Poisson's ratio		0.3
Rubber	Particle density	kg/m^3	1150
	Coefficient of friction		0.27
Rubber–bonds	Shear modulus bonded spheres	MPa	12
	Bond elastic modulus	MPa	20
	Normal-to-shear stiffness ratio (k_n/k_s)		4.0
	Tensile strength	Pa	30E200
	Shear strength	Pa	30E200
	Critical damping ratio		0.25
Wall	Coefficient of friction		0.70
	Shear modulus	MPa	80,000
	Poisson's ratio		0.25
Global	Damping coefficient		0.2

3.1 Features of the DEM4GRM Model

3.1.1 Single Rubber Particles Model

Knowing the importance of accounting for the compressive volumetric behavior of rubber, an improved tire rubber particle model was proposed in this study, which builds on the studies from Asadi et al. [26] and Ren et al. [27]. Precisely, to attain an accurate simulation setup and reduced computation time, the BCC rubber model by Ren et al. [27] was enhanced and optimized. Figure 6a displays the proposed rubber model consisting of a cluster of 35 balls, BCC packed, and bonded together by linear parallel bonds [31]. The intercluster nonbond contacts, which would form when in contact with external particles, are defined using the Hertz-Mindlin model [30]. To guarantee that the compressive load is applied more evenly, an outer shell was positioned on each side of the cubic assembly. The model elastic parameters were defined using the results of a series of uniaxial compression tests on single

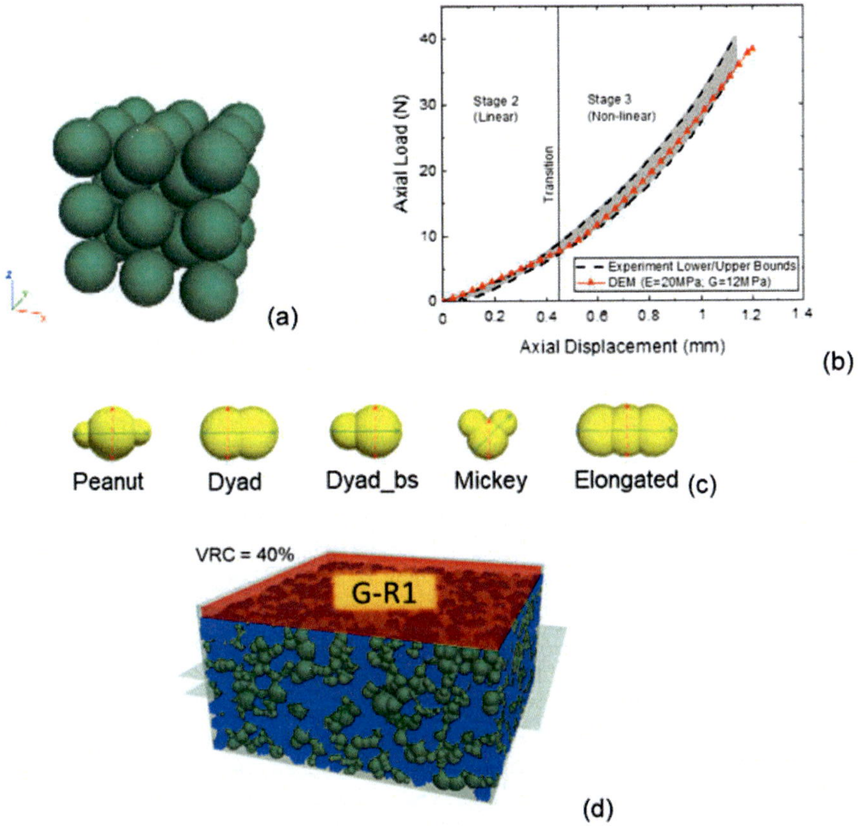

Fig. 6 Characteristics of the newly proposed DEM4GRM model

tire rubber particles as shown in Fig. 6b. Full details of the proposed single rubber particle model are available from Chew [32].

3.1.2 Gravel Particle Model

Due to the gravel particles size and shape variability, gravel grains were modeled using a simple clump shapes (Fig. 6c) as those considered by Garzia and Bray [33].

3.1.3 DEM Modeling of GRMs

The GRM specimens were generated using the overlapping method [28, 34] based on the particle size distribution of gravel and large rubber particles (R1) used in the laboratory experiments (Fig. 1). The specimens were quasi-statically compressed until reaching the required target degree of compaction. The normal stress was applied to the specimens by means of a force applied on the top plate. In the following, the specimens were sheared by moving the lower box at a rate of 0.002 m/s (to maintain quasi-static condition). Frictionless lateral walls were adopted to minimize k_0 effects. Figure 6d shows a typical DEM specimen obtained for $VRC = 40\%$.

The model parameters for the reference materials were determined by simulating the gravel ($VRC = 0\%$) and rubber ($VRC = 100\%$) behavior in direct shear tests and fitting the shear and volumetric response that best match the experimental direct shear results. No additional model parameters were defined for mixtures with $VRC = 10$, 25, and 40%, but rather the shear and volumetric trends in the simulations were merely defined by the proportion of gravel and rubber particles in the mixtures and the normal stress level applied on the specimens.

3.2 *Simulation Results*

In Fig. 7a, b, typical DEM simulation results are presented for gravel and granulated rubber specimens, respectively, sheared at $\sigma_n = 30$, 60, and 100 kPa. It can be seen that irrespective of the stress level, there is a very good agreement with the experimental tests results, both in terms of stress and volumetric behaviors.

Moreover, the results obtained for GRMs sheared at $\sigma_n = 100$ kPa are reported in Fig. 7c. Despite no explicit model parameters were defined for the mixtures, there is a good agreement with the experimental tests results, being the model capable of capturing key futures of dilative (rigid gravel-like) and compressive (soft rubber-like) mixture responses with varying *VRC*. Yet, while the stress response is perfectly described, the volumetric response appears to be slightly overpredicted, although the trends are correct (i.e., showing a more contractive behavior with increasing *VRC*).

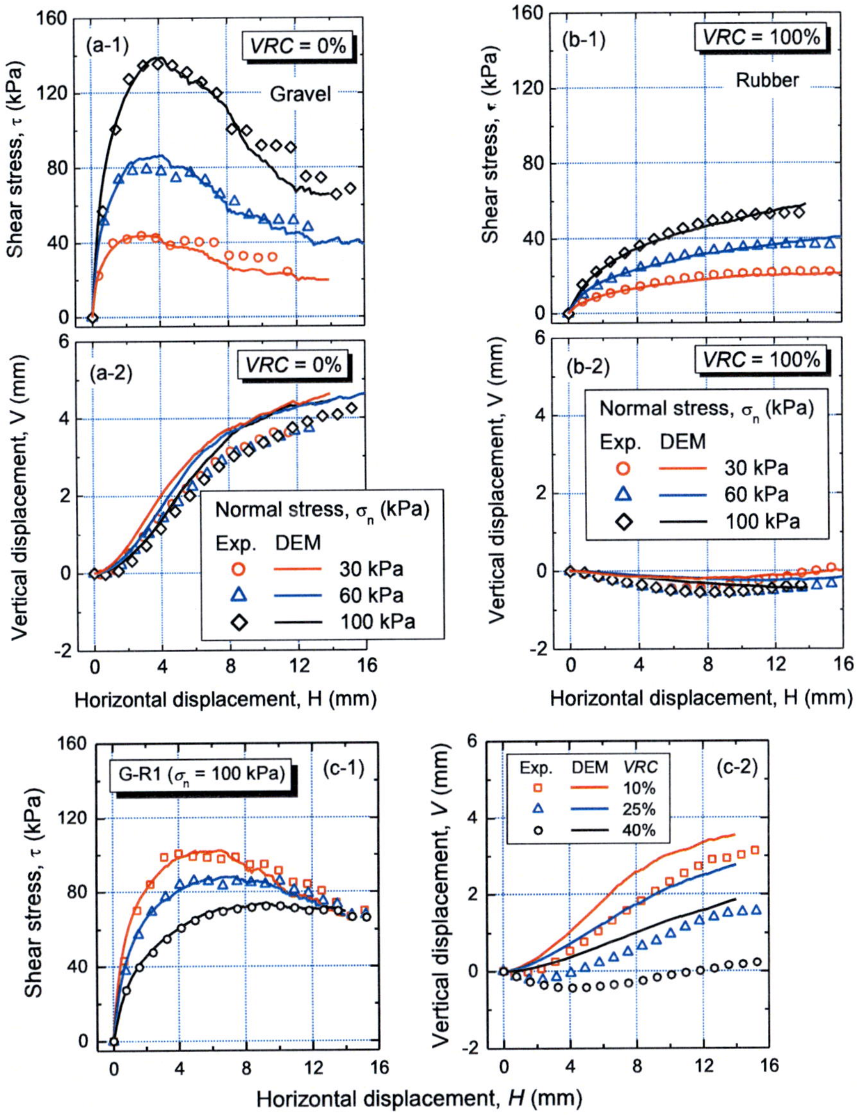

Fig. 7 DEM4GRM simulation results for: **a** gravel specimens; **b** granulated rubber specimens; and **c** G-R1 specimens

4 Micromechanics Analyzes and Discussion

4.1 *Fabric and Force Anisotropy*

The methodology proposed by Rothenburg and Bathurst [35] was used to quantify the fabric and force anisotropy in the DEM4GRM model simulations during shearing with some modifications in line with the latest research outcomes (e.g., [36]). A curve fitting approach using second-order Fourier series yielded the following expressions that best describe the (polar) distribution of different contact types considered in this study. The shape of the distribution of the contact vector—fabric can be approximated using the harmonic function:

$$E(\theta) = \frac{\{1 + a\cos 2(\theta - \theta_a)\}}{2\pi} \tag{2.1}$$

where a represents the magnitude or coefficient of anisotropy of the contact orientations, and θ_a is the principal direction of this anisotropy. Similarly, the contact normal force distribution can be represented by

$$\overline{f}_n(\theta) = \overline{f}_n^o\{1 + a_n\cos 2(\theta - \theta_n)\} \tag{2.2}$$

where $\overline{f}_n^o$ is the average contact normal force, a_n is the parameter describing the anisotropy of the contact normal force component, and θ_n is the direction of the maximum average normal forces. The expressions in Eqs. (2.1) and (2.2) are found to be similar to those used by Rothenburg and Bathurst [35].

The following expression was proposed to best approximate the tangential force distribution for the GRMs considered in this study:

$$\overline{f}_t(\theta) = \overline{f}_t^o\{1 + a_t\cos 4(\theta - \theta_t)\} \tag{2.3}$$

where $\overline{f}_t^o$ is the average contact shear force, a_t is the parameter describing the anisotropy of the contact tangential force component, and θ_t defines the direction of the peak contact shear force. Note that this expression differs to that reported by Rothenburg and Bathurst [35], by which the tangential force approaches zero between the peaks. This is due to the fact that Rothenburg and Bathurst [35] considered perfect spheres whereas in this study clumps were used. The interlocking effects introduced by the clumps give rise to "background" tangential forces that are randomly orientated, and thus, the tangential forces do not diminish along orthogonal planes. The resultant tangential force distribution is akin to a superposition of an isotropic "background" tangential forces (Eq. 3), which forms a slightly different distribution to that observed by Rothenburg and Bathurst [35].

Figure 8 shows the comparison between polar histograms of the fabric, normal force and tangential force distribution at different shearing stages obtained by the proposed approximation functions, and DEM simulations for GRMs (*VRC* = 25% at $\sigma_n = 60$ kPa). For inclusiveness, the variation during the shearing process of the force anisotropy coefficients (a, a_n, and a_t) is also reported. Although not presented here, it is important to mention that very similar trends were obtained at 30 and 100 kPa normal stress.

Generally, the fabric of the rubber (*VRC* = 100%) specimens is significantly different from that of the gravel (*VRC* = 0%) and that of the mixtures look like one or the other depending on the *VRC* content. Specifically, at *VRC* = 40%, the trends in a, a_n, and a_t begin to be similar to that of *VRC* = 100%, suggesting that the soft-like rubber-dominated behavior is expected for the mixtures with *VRC* ≥ 40%. Contrarily, the peaks in a, a_n, and a_t coefficients trends during shearing are obvious up to *VRC* = 25%, suggesting that rigid-like gravel-dominated behavior is expected for the mixtures with *VRC* ≤ 25%. This was confirmed by the analysis of the force network as described in the next section.

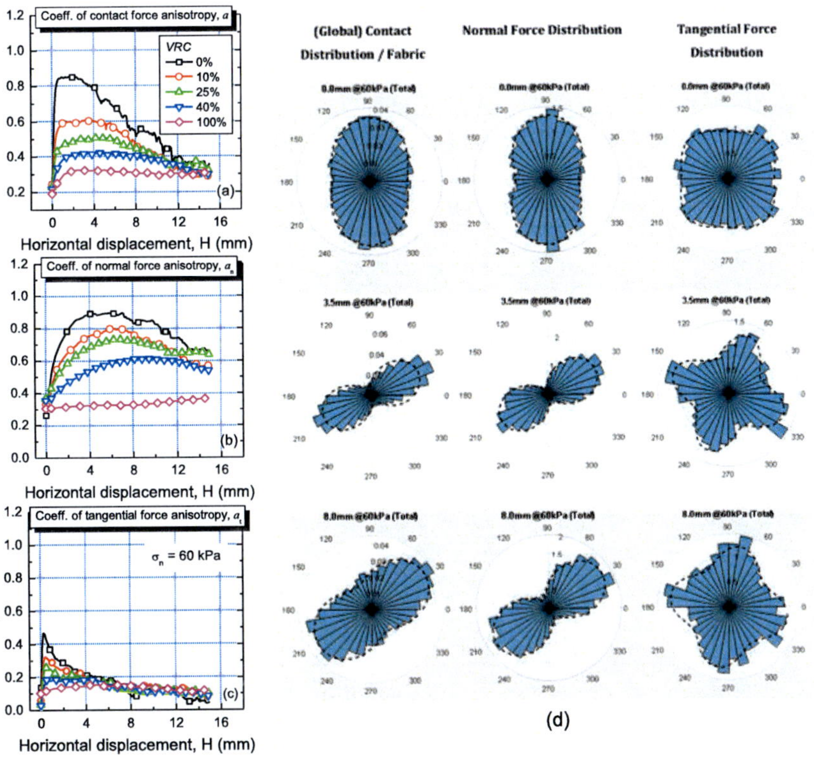

Fig. 8 Variation of force anisotropy coefficients with *VRC* at 60 kPa normal stress: **a** contact force; **b** normal force; **c** tangential force; and **d** polar histograms (*VRC* =25%)

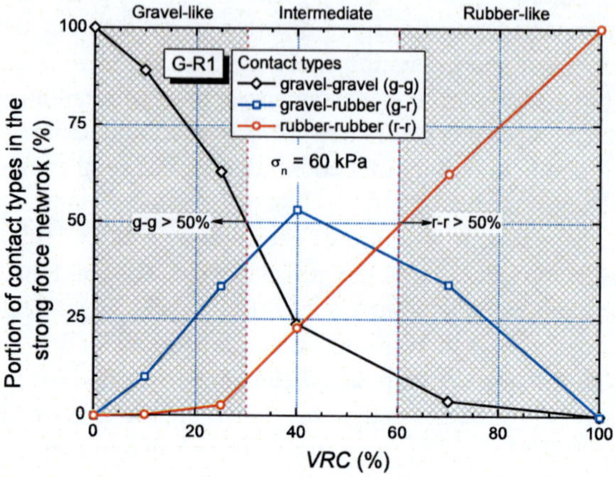

Fig. 9 Behavioral zones for GRMs in direct shear tests

4.2 *Force Network*

It has been well established that the contact network can be separated into two categories: the strong force network and the weak force network [37]. The strong force network consists of contact forces greater than the average force in the assembly and represents the load-bearing force chains that form during shearing and is responsible for the fabric and force anisotropy observed [36, 38, 39].

Figure 9 reports a typical distribution of the amount of gravel-gravel (g-g), gravel-rubber (g-r), and rubber-rubber (r-r) contacts participating in the strong force network for various GRMs. It can be observed that the proportion of r-r contacts in the strong force network is irrelevant up to *VRC* = 25–30%, and more than half of the strong contacts are g-g contacts. From *VRC* = 25% onward, the proportion of strong r-r contacts increases significantly while the proportion of strong g-g force chains decrease rapidly. At *VRC* = 40%, both the r-r and g-g contacts carry an equal proportion of the strong contacts. Yet, the strong forces chains are primarily transmitted via the g-r contacts, giving rise to an intermediate behavior where the specimen displays primarily a dilative behavior during shearing. The peak shear stress is also less pronounced. Above *VRC* = 60%, r-r contacts sustain most of the strong-force chains in the assembly.

5 Concluding Remarks

In this paper, the results of direct shear tests carried out on gravel-rubber mixtures (GRMs)—having different aspect ratios (AR = 0.28, 0.57, and 0.66), prepared at different volumetric rubber contents (*VRC* = 0, 10, 25, 40, and 100%) and sheared at 30, 60, and 100 kPa normal stress—were reported and discussed. Moreover, an original 3D hybrid DEM numerical model for rigid-soft particle mixtures was presented, and its performance was assessed. Finally, using newly proposed constitutive equations, microscale factors such as fabric, force anisotropy, and strong force chains were systematically examined throughout the shearing process to gain insight on the micromechanical behavior of mixtures with varying *VRC*.

The following main conclusions can be drawn from this study:

Macroscale

(i) Irrespective of the applied normal stress level and rubber particle size (or aspect ratio, *AR*), the response of GRMs changed rapidly from dilative-like to contractive-like with the addition of rubber in the mixtures;

(ii) Independently from *AR*, the friction angle (ϕ') was found to be between 54° (gravel) and 29° (rubber) indicating that the majority of GRMs are suitable structural fill materials for most geotechnical applications ($\phi' \geq 30°$);

(iii) The ultimate adoption of GRMs in geotechnical application depends also on their compressibility that may limit the *VRC* in the mixtures, especially when small rubber particles (or aspect ratios) are used to form GRMs%;

(iv) From an environmental viewpoint, the use of GRMs with large rubber particle (free from steel wires) is desirable since it would minimize the leaching of toxic metals (e.g., Zn) from granulated rubber.

Microscale

(v) Micro fabric and force anisotropy generally decreased with increasing *VRC*, following a similar pathway observed for the macroscale strength behavior;

(vi) Based on the strong force network analysis, three distinct behavioral zones were identified for GRMs:

- gravel-like rigid behavior zone: VRC $\leq$ 30%;
- intermediate behavior/transition zone: 30% < VRC < 60%; and
- rubber-like soft behavior zone: VRC $\geq$ 60%.

It is important to mention that the above is valid for the level of normal stress considered in this study. In fact, at higher stress levels, the rubber-like behavior could become more predominant even at lower *VRC* levels.

Acknowledgements The authors would like to acknowledge the research support provided by Ministry of Business, Innovation and Employment of New Zealand (MBIE Smart Ideas Research Grant No. 56289). The laboratory assistance provided by Dr Sean Rees is greatly appreciated. The views and opinion expressed in this paper are those of the authors and do not necessarily reflect those of MBIE.

References

1. Arulrajah, A., Narsilio, G., Kodikara, J., & Orense, R. P. (2015). Key issues in environmental geotechnics: Australia-New Zealand. *Environmental Geotechnics, 6*, 326–330.

2. Chiaro, G., Indraratna, B., Tasalloti, S. M. A., & Rujikiatkamjorn, C. (2015). Optimisation of coal wash–slag blend as a structural fill. *Ground Improvement, 168*(1), 33–44.
3. Tasalloti, S. M. A., Indraratna, B., Rujikiatkamjorn, C., Heitor, A., & Chiaro, G. (2015). A laboratory study on the shear behavior of mixtures of coal wash and steel furnace slag as potential structural fill. *Geotechnical Testing Journal, 38*(4), 361–372.
4. Mashiri, M. S., Vinod, J. S., Sheich, M. N., & Tsang, H. H. (2015). Shear strength and dilatancy behaviour of sand–tyre chip mixtures. *Soils and Foundations, 55*(3), 517–528.
5. Qi, Y., Indraratna, B., & Twak, M. (2020). Use of recycled rubber elements in track stabilisation. *Geotechnical Special Publication, GSP 319*, 49–59.
6. Ghorbani, B., Arulrajah, A., Narsilio, G., Horpibulsuk, S., & Bo, M. W. (2021). Dynamic characterization of recycled glass-recycled concrete blends using experimental analysis and artificial neural network modeling. *Soil Dynamic Earthquake Engineering 142*, 106544.
7. Ministry for the Environment. (2015). Waste tyres economic research. *Report 3*, pp. 87.
8. Hernandez, E., Palermo, A., Granello, G., Chiaro, G., & Banasiak, L. (2020). Eco-rubber seismic isolation foundation systems, a sustainable solution for New Zealand context. *Structural Engineering International, 30*(2), 192–200.
9. Chiaro, G., Tasalloti, A., Banasiak, L., Palermo, A., Granello, G., & Rees, S. (2020). Sustainable recycling of end-of-life tyres in civil (geotechnical) engineering applications: Turning issues into opportunities in the New Zealand context. *New Zealand Geomechanical News, 99*, 38–47.
10. Tasalloti, A., Chiaro, G., Murali, A., & Banasiak, L. (2021). Physical and mechanical properties of granulated rubber mixed with granular soils—a literature review. *Sustainability, 13*(8), 1–28, 4309.
11. Hazarika, H., & Abdullah, A. (2016). Improvement effects of two and three dimensional geosynthetics used in liquefaction countermeasures. *Japan Geotechnical Society Special Publication, 2*(68), 2336–2341.
12. Kim, H. K., & Santamarina, J. C. (2008). Sand-rubber mixtures (large rubber chips). *Canadian Geotechnical Journal, 45*, 1457–1466.
13. Lee, J. S., Doods, J., & Santamarina, J. C. (2007). Behavior of rigid-soft particle mixtures. *Journal of Materials in Civil Engineering, 19*(2), 179–184.
14. Pasha, S. M. K., Hazarika, K., & Yoshimoto, N. (2019). Physical and mechanical properties of gravel-tire chips mixture (GTCM). *Geosynthetics International, 26*(1), 92–110.
15. Chiaro, G., Palermo, A., Banasiak, L. J., & Granello, G. (2019). Direct shear behaviour of gravel-granulated tyre rubber mixtures. In: Proceedings of 13th ANZ Geomechanical Conference, Perth, Australia, 221–226.
16. Hazarika, H., Pasha, S. M. K., Ishibashi, I., Yoshimoto, N., Kinoshita, T., Endo, S., Karmokar, A. K., & Hitosugi, T. (2020). Tire-chip reinforced foundation as liquefaction countermeasure for residential buildings. *Soils and Foundations, 60*(2), 315–326.
17. Balaji, P. Dashti, S., & Liel, A. B. (2020). In-ground gravel–rubber panel walls to mitigate and base isolate shallow-founded structures on liquefiable ground. *Journal Geotechnical Geoenvironmental Engineering, 146*(9). https://doi.org/10.1061/(ASCE)GT.1943-5606.0002310
18. Tasalloti, A., Chiaro, G., Banasaik, L., & Palermo, A. (2021). Experimental investigation of the mechanical behaviour of gravel-granulated tyre rubber mixtures. *Construction and Building Matererials, 273*, 127149. https://doi.org/10.1016/j.conbuildmat.2020.121749
19. Pitilakis, D., Anastasiadis, A., Vratsikidis, A., Kapouniaris, A., Massimino, M. R., Abate, G., & Corsico, S. (2021). Large-scale field testing of geotechnical seismic isolation of structures using gravel-rubber mixtures. Earthquake Engineering and Structural Dynamics. https://doi.org/10.5281/zenodo.4557538
20. Tasalloti, A., Chiaro, G., Palermo, A., & Banasiak, L. J. (2020). Effect of rubber crumbs volumetric content on the shear strength of gravelly soil in direct shear apparatus. *Geotechnical Special Publications GSP, 319*, 259–266.
21. Chiaro, G., Palermo, A., Granello, G., Tasalloti, A., & Banasiak, L. J. (2021). Reuse of waste tires to develop eco-rubber seismic-isolation foundation systems: Preliminary results. *Lecture Notes in Civil Engineering, 144*, 159–169.

22. Chew, K., Chiaro, G., Tasalloti, A., Vinod, J., & Allulakshmi, K. (2021). Response of gravel-rubber mixtures under direct shear testing: experimental and DEM numerical investigation. In: Proceedings of 21st NZ Geotechnical Social Symposium, Dunedin, New Zealand, p. 10.
23. Iolli, S., Modoni, G., Chiaro, G., & Salvatore, E. (2015). Predictive correlations for the compaction of clean sands. *Transportation Geotechnics, 4*, 38–49.
24. Banasiak, L. J., Chiaro, G., Palermo, A., & Granello, G. (2021). Environmental implications of the recycling of end-of-life tires in seismic-isolation foundation systems. *Lecture Notes in Civil Engineering, 144*, 43–52.
25. O'Sullivan, C. (2011). *Particulate discrete element modelling: A geomechanics perspective.* Taylor & Francis.
26. Asadi, M., Mahboubi, A., & Thoeni, K. (2018). Discrete modeling of sand–tire mixture considering grain-scale deformability. *Granular Matter, 20*, 1–13.
27. Ren, Z. L., Cheng, Y. P., & Xu, X. (2020). A DEM method for simulating rubber tyres. *Géotechnique Letters, 10*, 73–79.
28. Itasca Consulting Group: Particle flow code in three dimensions—PFC3D (2018).
29. Cundall, P. A., & Strack, O. D. L. (1979). Discrete numerical model for granular assemblies. *Géotechnique, 29*(1), 47–65.
30. Mindlin, R. D., & Deresiewicz, H. (1953). Elastic spheres in contact under varying oblique forces. *Journal of Applied Mechanics, 20*, 327–344.
31. Potyondy, D. O., & Cundall, P. A. (2004). A bonded-particle model for rock. *International Journal of Rock Mechanics and Mining Sciences, 41*, 1329–1364.
32. Chew, K. (2021). *The mechanical behaviour of gravel-rubber mixtures: insights from DEM numerical investigations.* MEng Thesis, University of Canterbury, pp. 112.
33. Garcia, F. E., & Bray, J. D. (2019). Modeling the shear response of granular materials with discrete element assemblages of sphere-clusters. *Computers and Geotechnics, 106*, 99–107.
34. Zhang, M., Yang, Y., Zhang, H., & Yu, H. S. (2019). DEM and experimental study of bi-directional simple shear. *Granular Matter, 21*, 1–13.
35. Rothenburg, L., & Bathurst, R. J. (1989). Analytical study of induced anisotropy in idealized granular materials. *Géotechnique, 39*, 601–614.
36. Guo, N., & Zhao, J. (2013). The signature of shear-induced anisotropy in granular media. *Computers and Geotechnics, 47*, 1–15.
37. Radjai, F., Wolf, D. E., Jean, M., & Moreau, J. J. (1998). Bimodal character of stress transmission in granular packings. *Physical Review Letters, 80*, 61–64.
38. Huang, X., Hanley, K. J., O'Sullivan, C., & Kwok, C. Y. (2014). Exploring the influence of interparticle friction on critical state behaviour using DEM. *International Journal for Numerical and Analytical Methods in Geomechanics, 38*, 1276–1297.
39. Lopera Perez, J. C., Kwok, C. Y., & Senetakis, K. (2017). Investigation of the micro-mechanics of sand–rubber mixtures at very small strains. *Geosynthetics International, 24*, 30–44.

Structural Health Monitoring of Heritage Structures Using Geotechnical Instruments

Abhinav Maloo and Parth Thaker

Abstract Geotechnical engineering plays major role in imparting safety of structures. Geotechnical instruments find their application in monitoring the safety of structures during construction in their proximity. This paper aims to present geotechnical instrumentation as a tool to assist on field. Geotechnical instrumentation is a surging discipline in the construction industry when structural health monitoring is thought off. Considering the safety prospect, it discusses the importance of instrumentation during all phases of a project, be it design, construction, or operation phase. Observations obtained by field instruments provide the means by which a project can be designed safe and efficient. Planning, installation, and monitoring hold the efficient performance of instrumentation. Special attention is given to the classification of geotechnical instruments and monitoring of the existing important structures. Classifications have been made on the structural parameters which attributes the failures in structure. The effective planning for instrumentation, including selection, monitoring, and analysis has been also discussed. An approach toward the effective use of instruments in the existing important structures for their preservation during process of construction activity in its vicinity has been presented. This paper highlights the usefulness of instrumentation monitoring in achieving better control of construction, safety of the structures, and design authentication.

Keyword Geotechnical instruments · Heritage structures · Instrumentation planning · Structural health monitoring

A. Maloo (✉)
CEPT University, Ahmedabad, India

P. Thaker
Fuji Silvertech Concrete Private Limited, Ahmedabad, India

A. K. Gupta et al. (eds.), *Advances in Construction Materials and Sustainable Environment*, Lecture Notes in Civil Engineering 196,
https://doi.org/10.1007/978-981-16-6557-8_3

1 Introduction

The youngest branch of civil engineering, geotechnical engineering, has provided explanations to a range of different issues over years. The first society devoted to the discipline of geotechnical engineering, international society for soil mechanics, and geotechnical engineering (ISSMGE) did focused the structure stability [1]. Geotechnical designs are assumptive to some extent, and every construction job associating earth or rock tends to encounter surprises. The uncertainties in earth/rock behavior invites deformations, which may affect the structures above it. Observational methods are helpful in certain conditions.

Instrumentation as a tool to assist in the field is more than 60 years old. Initially, it comprised of general mechanical and hydraulic themed instruments which were less precise and sensitive. Recently, the role of geotechnical instruments has been secured in the industry, for the health monitoring of structures [2]. This was followed by upgradation of the traditional instruments with new technologies giving boost to geotechnical engineering profession.

Many cities around the world are densely populated. To support the demand and due to less availability of space, underground construction is a common practice. Underground construction is prone to failures due to inability to check the health of lower structures, limitations in soil investigation techniques or presence of former structures, utilities, etc. [3]. Failures are the third main challenge which the construction industry is facing. Unforeseen ground conditions are one of the main reasons for failures, which costs sky high [4]. An aspect of risk management arises here, which can reduce this cost. This can be achieved by the observational method of monitoring.

Heritage and the important existing buildings are an integral part of historical memory. Their safeguarding is thus a chief concern [5]. Monitoring structural behavior of the structures under any shift from their balanced state is quite essential. Geotechnical instruments complement the building by monitoring the structural behaviors. Instrumentation has been demonstrated to be a complementary tool which is beneficial in every phase of the project.

2 Previous Works

Natural and man-made structures need to be monitored for assurance of safety, design validations, and cost overruns. Monitoring the performance of the structure has been beneficial, concerned to design and safety. Instrumentation monitoring keeps the record of the health of the building and thus proving a vital component in construction [6]. Soil strata play a decisive role while designing the structure. Though it is full of uncertainties, but by monitoring the shifts, the failures can be avoided. Field instrumentations find its application in controlling the construction by showcasing the impacts due to the construction activities. Special attention is needed to protect the instruments for their life and monitoring accuracy [7]. Field instrumentation combination with numeric modeling have proved to mitigate the differential movement problems in railroad bridges [8].

Geotechnical instrumentation helps in evaluating and monitoring the performance of the structure. It facilitates in the detection of the distress areas thus aiding in corrective remedial measures. Geotechnical instrumentation helps in to keep a check on the structural design [9]. By measuring, the settlements with the settlement instruments minimize the reliance on surveying and also increase the accuracy of measuring the settlements [10].

Field instrumentation is an evolving genre in geotechnical engineering. Future advancements and upgradations attribute the success of the projects by monitoring. Observation method tends to avoid the failures by observing the health of the structures under the loads, pressures, etc. The popularity of instrumentation is still a question in the construction industry. This is supported by inadequate knowledge, minimum accessibility, less credible works, and studies.

3 Structural Health Monitoring

Structural health monitoring (SHM) is a technical term which have been associated to the health of the structure over its life. The science behind the term got emerged from structural monitoring or structural integrity monitoring [11]. SHM is interpreted as a process of enforcing a damage detection approach which furnishes a potential of time, safety, and economic benefits [12]. There has been significant growth to SHM, from the on-field issues and concerns. Inappropriate knowledge and relaxations are still the key challenges down the lane [13]. Establishing a robust SHM technology is a tough test for the engineering community. Modification in the structures, monitoring of the structures, fatigue assessments, and structures subjected to movements are some of the cases where the structural monitoring finds its application [14].

4 Monitoring Methods

Geodetic (surveying) and geotechnical techniques are the two broad classes in which monitoring is practically classified. Geodetic methods frequently measure the interest points on the structure with the application of surveying instruments such as theodolites, levels, total stations, and GPS. Geotechnical methods use the geotechnical instruments such as tilt plates, inclinometers, and piezometers to monitor the different parameters such as deformations, stress, and water levels [15]. Geotechnical instruments are exposed to more parameters over geodetic methods such as load, stress, and water pressures. Most of the geodetic measurements are subjected to operator's accessibility. Maintenance, installation, and range are some limitations of geotechnical instruments [16]. Geotechnical instruments are the integral tools to assist field monitoring with geodetic methods [17].

5 Geotechnical Instrumentation

5.1 Role of Geotechnical Instrumentation

The soil plays an important role in defining safety of structure and is full of inherent uncertainties. There is no manual to trace its actual behavior under the impact of loads and stress generated from the activities close to it. So, to track these variations and avoiding failures at the initiation, instrumentation and its monitoring are essential.

Instrumentation and monitoring play a decisive role in reducing the risks associated with project in safeguarding the existing important structures nearby. It helps in identifying the possible geological hazards to proposed structures. Instrumentation is a useful tool for identifying the trends of behaviors and quantifying the changes of crucial parameters. A good instrumentation should have one or more of the following purposes.

Structure Safety

Instruments can provide the early warning of approaching failure. This allows for employment of precautionary actions well within time.

Design Verifications

Designs are assumptive in nature to some extent. Instruments are used to validate these assumptions. This helps in verification of the initial designs.

Construction Control

Instruments are capable to monitor physical parameters such as settlements, lateral movements, and pressure. The observations help to determine the speed at which the construction may proceed without any adverse effects on structure.

In-Service Performance of Structure

Instrument monitors the in-service performance of the structure. This indicates the status of the structure, by which any future failure can be avoided.

Geotechnical instruments form an integral part of the construction process. Instrumentation and its monitoring followed by evaluation are highly beneficial in determining the performance of a structure and in keeping a check on its health.

Table 1 Geotechnical instrumentation benefits

Design phase
• Characterizing initial site conditions • Characterizing and delineating the geotechnical properties of the site • Formulating the remedial measures by identifying risks
Execution phase
• Controlling and monitoring project agility • Public safety • Eliminating the unwanted cost overruns evoked due to construction activities • Tracking the uncertainties during the construction procedures and fixing them
Post-execution phase
• Tracking the health of the structure and thus improving it • Tracking the substructure properties and tracing any deformations or settlements • Measuring the trends in the structures over time and thus employing the conclusions further in designs

5.2 *Benefits*

Geotechnical instrumentation and monitoring have been a complementary tool to assist on the field and has demonstrated its significance by updating the site conditions [2]. If planned adequately, it supports the progress in all phases of the construction (Table 1).

5.3 *Instruments Type*

In the broad sense, instrument may be used to delineate an arrangement which is capable of detecting, recording, and conveying different parameters such as deformation, stress, loads, pore water pressures, and so on. To trace the parameters as

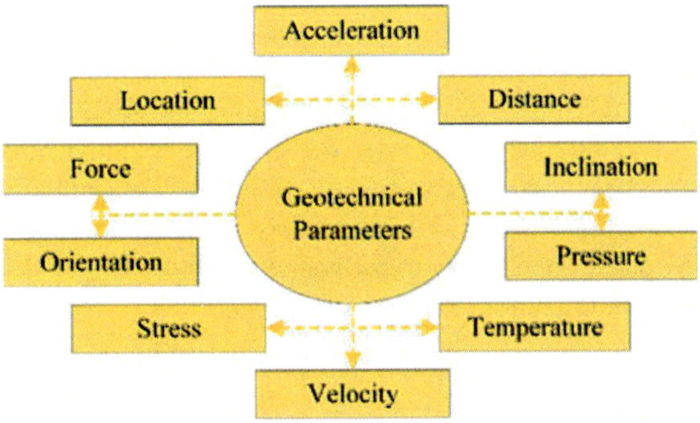

Fig. 1 Geotechnical parameters

shown in Fig. 1, instruments have been classified based on their specific purposes and working principles.

In the seam of the data acquisition system and communication system, transducers form the basis for geotechnical instruments. These are the electronic devices which convert the signals gathered while monitoring into the digital data which can be further analyzed. Transducers are further classified into (a) mechanical (dial indicators, micrometers), (b) hydraulic (mainly used to measure pressures), (c) pneumatic (measure pressure by application of gas), and (d) electrical transducers. All the classified types are being used in the industry as per the requirement [6].

Measurement of Deformations

For the measurement of deformation, tilt meters, inclinometers, and extensometers are the geotechnical instruments that are specifically developed. Tilt meters are installed on the structures for measuring the rotational deformation at a specific location. Inclinometers generally measure the lateral deformation by and measuring the angles with the plane while being pulled upward at suitable intervals using a lowered probe into the borehole. Extensometers are used to measure the relative movements or displacements between the points of pendulums [17].

Measurement of Load, Stress, and Strains

Instruments classified under this category mainly rely on the use of transducers to sense the smallest possible deformations and compressions caused by the loads acting there upon. Mainly, the instruments fall under three classes: (a) load cells, (b) pressure cells, and (c) strain gages.

Load cells are inserted in such a way that the structure forces through the cells and are used to monitor the temporary and permanent loads in the structure. Pressure cells are installed within fills or against structures to measure the stresses, direction of stresses, and their magnitudes. Strain gauges are normally installed on the surfaces or embedded as per the geometry of the structural members to sense the compressions and to determine stresses in the members.

Measurement of Pore Water Pressures and Water Level

Maintaining the water levels during construction phase is important. Piezometers, steel tapes, audio readers, and electrical dipmeters are some instruments which assist in the measurement of water pressures and water levels.

Piezometers are devices for the measurement of ground water pressures. They are used in an open standpipe, fixed in filled embankment, or driven into ground. The essential part of a piezometer is a pressure gauge, a mechanical, electrical, hydraulic / pneumatic transducer. Steel tapes and audio readers are the other available instruments for measuring water levels.

Measurement of Temperature

Temperature fluctuations have an adverse impact on the structural elements and are thus vital to measure. Temperature can be measured by various instruments such as.

Thermometers, which are used for spot measurements of temperatures such as mercury thermometers or bi-metal thermometers.

Thermistors contain semiconductors which change their resistance on temperature fluctuations. Resistance temperature device (RTD) works on the principle of change that electrical resistance of the wire is proportional to temperature change. Vibrating wire transducers are the ones which are also available for temperature measurement. These are expensive compared with the above-mentioned instruments.

Table 2 represents the commonly used geotechnical instruments with their application, operations, and remarks

6 Instrumentation Planning

The use of geotechnical instrumentation is not solely the selection, but a comprehensive step by step process. Proper planning and qualified interpretation by the personnel are essential in ensuring the effectiveness of geotechnical program [18].

Installation of the instrument without any planned program for its monitoring is quite obviously a failure. Hence, it is important for planning monitoring programs to ensure accurate handling and the performance of the instrument during its construction and operational phase. Planning must incorporate the essential dots and potential link between the dots [2].

Planning should progress through the steps listed below.

6.1 Project Definition

During project initiation, the project characteristics such as topography, soil properties, ground water conditions, environmental conditions, nearby structures status, and construction methodologies need to be studied for their magnitude and potential impact.

6.2 Identifying Mechanism

Identifying the mechanism to control behavior of soil and purpose of instrumentation based on the project conditions is an important part of the process. This is required in order to answer the various geotechnical questions that may arise and also in order to mitigate the impact of construction activities.

Table 2 Commonly used geotechnical instruments

Instrument	Application	Operation	Remarks
Building settling point	They are installed within the zone of influence to monitor the settlements of critical buildings and structures	The elevation of the settlement point are measured with digital levels and are than compared with the pilot readings	Reference points should be selected accurately
Tilt meter	It is used to measure the horizontal and vertical tilt in the buildings	Tilt meter is fixed on the pegs of the plates, and the horizontal or vertical tilts are observed in the read out unit in terms of angle. To find the tilt, the initial observation is subtracted with the current tilt, which is either represented in degrees or displacement	Manual reading with read out units Should be installed and supported well on the walls, buildings Sensitive to temperature changes
Bi-reflex targets	Bi-reflex targets are generally installed on the point of interest for displacement monitoring	With the total station, the coordinates of the targets are observed and compared with the last observations. This occurrence of the displacement is captured by subtracting the current and initial observations	Location of the target should be accurately determined before installation
Crack meters	They are utilized for measuring the movement and width of surface cracks in the concrete structures	Subsequent readings with intervals decided are taken and are compared with the initial readings. Thus, the change in the width of crack is determined	Should be installed after analyzing the condition of the buildings, walls
Settlement monitoring point	SMPs are the instruments which measures the vertical settlement of the pavements	The elevations of the settlement point are measured with digital levels and are than compared with the pilot readings	Need to be robust as is subjected to vertical loads

(continued)

Table 2 (continued)

Instrument	Application	Operation	Remarks
Inclinometer	Inclinometers are installed on the surface to measure the lateral deformations of ground evoked by construction activities	The probe is lowered in the well, and the inclination is measured at intervals across the length in two directions. The observations are then processed, and the data is reported as a chart of cumulative deflection against depth	Should have extra length above for easy operations Probe needs to be lowered orderly Casing quality is important
Piezometer	Installed to determine the ground water behavior and levels during and after the construction activities such as excavation and dewatering	Water level sounder probe is lowered in the PVC pipe casing; when the probe tip touches water, there is buzzer which sounds and the observation is noted	Due to moisture there are some fellable beeps which should be ignored Casing quality is important

6.3 *Purpose Identification and Justification*

On the basis of the project conditions and its severity, purpose for instrumentation is identified and the instrument types to be incorporated are decided. Also the role of each instrument to be installed on the site is justified. Parameters to be monitored are selected as per the site claims.

6.4 *Estimating Limits and Remedial Actions*

The limits for the parameters to be monitored are predicted, so that remedial actions may be drafted, and the required instrument range may be selected. These estimates are purely based on the project definition and calculations. Remedial actions, for solving the problem which may arise by the results of observations, are worked in advance.

6.5 *Monitoring Program*

This is an essential aspect in the context of geotechnical instrumentation. This incorporates the planning of monitoring program and establishing a framework to monitor the instruments. The frequency of monitoring is also decided based on the severity of the construction and nearby conditions. The required number of labors and material are also planned in advance to avoid any delays.

6.6 Planning Installation

After defining the basics, the reliable instruments are selected, locations are identified, installation mechanism is decided, and installation is made. Each instrument's purpose is attributed and listed so as to track during the field works.

6.7 Field Works

The most crucial part is the field works which includes calibration of the installed instruments, monitoring, recording data, and regular instrument maintenance. These are the pivotal activities, and the results are inclined to the excellency of field works.

6.8 Data Collection and Analysis

This includes planning the data collection as per the frequencies expressed in the monitoring program and processing the collected data, for interpretation as per the guidelines planned.

6.9 Reporting and Implementation

Interpreted and analyzed values are reported to the concerned authorities for the implementation of remedial measures against the results.

7 Typical Instrumentation Scheme

Geotechnical instrumentation benefits as an observational support for monitoring of engineering parameters [19]. The use of geotechnical instrumentation to assist on field is particularly important to the uncertainties related to the ground. Planning is a vital component when we talk about geotechnical instrumentation apart from installation and monitoring. Planning results all the possible dimensions, established on the characteristics of the project.

A typical scheme for instrumentation is illustrated below.

7.1 Instrumentation

Instrumentation is a long process, survived by planning, installation, and monitoring. Figure 2 represents the layout of the project, which shows the construction area and the nearby existing structures. It is fundamental to know the status of the site before initiation, as there may be hidden uncertainties which may result in time and cost overruns if not identified. Project characteristics provides the ground for the effective planning of instrumentation.

Ahead of instrumentation, the settlement contour as shown in Fig. 3 is outlined. The anticipated settlement contours are determined majorly on the basis of loads and impacts of the construction activities and, status of the structures in the vicinity. This provides the base for predicting the mechanism and how the earth or soil will behave under loads, stresses, and pressure.

Generally, contours planning is carried out by defining the zone of influence. These zones indicate that the intensity by which the neighboring area may be affected

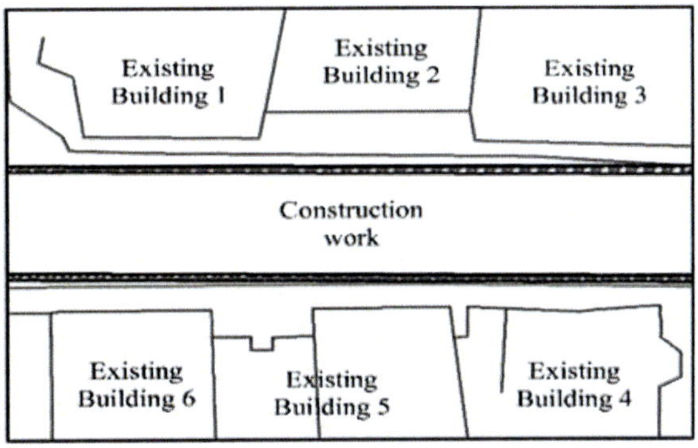

Fig. 2 Construction layout

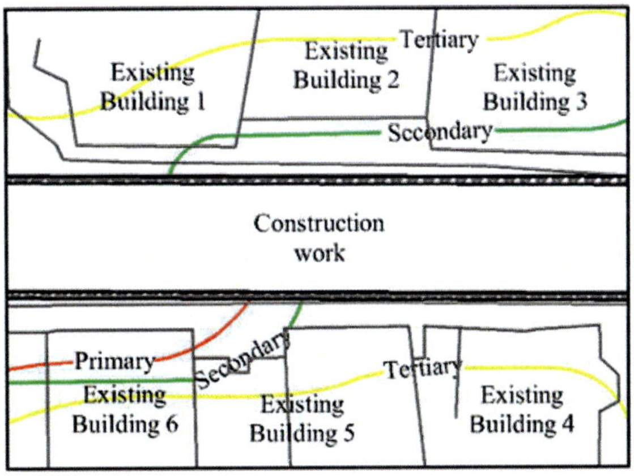

Fig. 3 Settlement contour

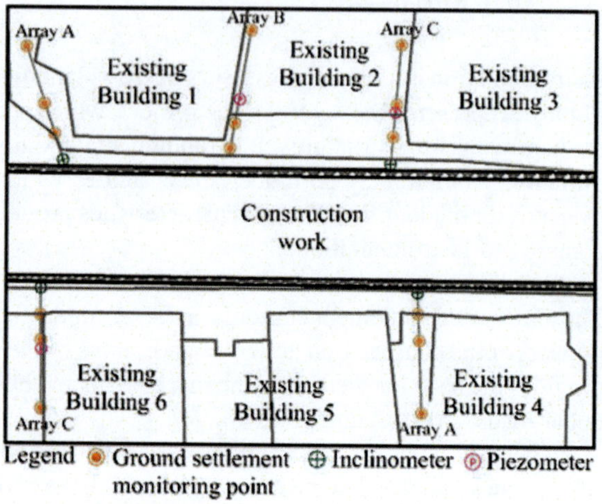

Fig. 4 Field instrumentation near existing structures

during construction. Red depicts the primary zone; green depicts secondary zone, and yellow depicts the tertiary zone of influence.

Installation of instruments is carried out in monitoring arrays, after the selection of instruments. The selection is made after the identification of the risks and the parameters which are to be monitored. Figure 4 shows the installation of the instruments in arrays.

Spacing between arrays and their exact number depends upon the ground conditions and the construction designs. Arrays should be at 90° to the planned construction wherever attainable. All arrays extend the tertiary anticipated settlement contours. However, instruments without arrays can also be installed if planned or subjected to installation constraints. The purpose of arrays is to monitor ground movements from the construction area at an approximate distance ranging from 150 to 200 m. Monitoring locations are typically identified within the range of 150–400 m approximately.

Existing buildings (special attention to the heritage buildings and other important structures) assessment is made before the construction is initiated. All the buildings falling in the primary zone of influence are precisely monitored with the instruments such as building settlement points, tilt plates, crack meters, and settlement monitoring points. Generally secondary and the tertiary zones do not require instrumentation for building. Instruments are installed at an approximate distance of 5 m along the façade of the buildings toward the construction site. Figure 5 depicts the instrument installed around the building. The different instruments (installed in the same building) are used to validate the results with each other as they all depict the deformations caused by construction activities. The installation of the instruments is done in a way that it serves the purpose and damage to the building is minimum. Further, if there is any requirement to install more equipments, than it can be installed on the decisions of the personnel at the site.

Fig. 5 Field instrumentation in existing structures

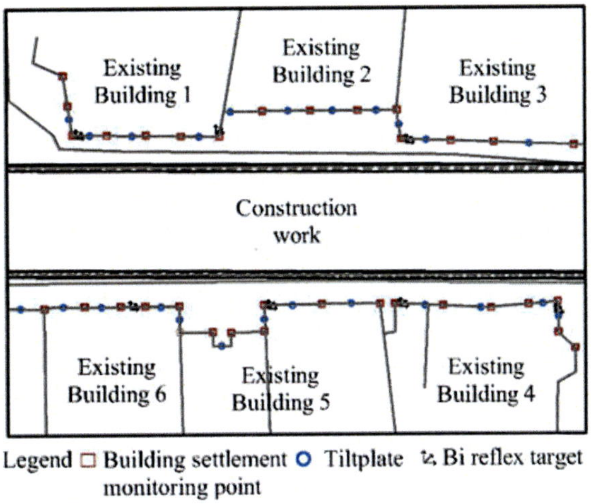

7.2 *Instrumentation Monitoring*

A monitoring plan is a prerequisite while planning for geotechnical instrumentation. It consists of the instrument monitoring frequencies and monitoring criteria. Monitoring frequencies are generally based on the equipment type, construction activities, and work schedule and instrument locations. Monitoring frequencies for any equipment may be increased or decreased as per the situations on the site.

Base Readings

After the installation, the instrument are calibrated as per their classification. Base readings for the instrument are taken in advance of all the construction activities. These base reading are used for comparisons with the current readings. Minimum four sets of readings are taken to confirm the base readings.

Monitoring Criteria

The readings collected are due to the ground movements or impacts due to construction activities. Each instrument is assigned control values, which are further bifurcated as alert, action, and alarm values. These values are defined on the level of serviceability limit. Each instrument type have their limits. The instruments readings are thus placed in defined values. If the defined values are crossed, then it results in increased monitoring frequencies, preventive measures, activity stopping, etc.

Measurements from the instruments are foremost, the personnel's who carried out the installation and familiarized with purpose of monitoring need to be deployed. In conjunction with measurements, instrument name, temperature, date, and time shall be recorded for every instrument. An immediate check for the plausibility of the data should be made, for the occurrence of any unexpected or contradicting data.

An immediate data processing is carried out for any change which expect immediate actions. It comprises of the raw data transformations via calibration factors into relevant engineering units which can be compared. Data processing includes corrections, transformations, and verifications of the measured data. All the processed data results should be presented in graphs and tables specifying the instrument details such as name, identification, location, date, and time.

An installation, calibration, and monitoring report covering all the aspects should be submitted after the process. Monitoring documents should be attached in the report for reference. Reports should point out the differences and the alarm values for the instrument results. If identified any differences, then the situation should be hinted in the report for the mitigation processes. Also the probable reasons should be concluded.

8 Conclusions

Geotechnical instruments are the tools which assist by way of monitoring and analyzing a number of parameters based on subsoil movements due to construction activities in its vicinity. This paper provides the overview of the instrumentation for assessing the health of existing important structure during construction in its vicinity. For ensuring the effectiveness of the monitoring system, proper control of construction, accuracy of instruments, and performance of instruments, proper planning of instrumentation is necessary. There is no minimum or maximum number of instruments which should be installed; it depends upon the condition of the structure, non-destructive testing, construction methodology, and the vulnerability of the nearby construction.

Typical scheme for instrumentation presented in the paper can be used as a tool to understand the process of instrumentation. Application of geotechnical instrument on projects is essential, as it acts as an early warning system to the approaching failures. Sufficient time is required for installation, initialization, and calibration of the instruments to enable base readings to be taken. Monitoring procedure come with no shortcuts or substitutes for getting reliable data from instruments. The key personnel's or man power involved in the planning, installation, monitoring, and maintenance process needs to be top-notch. Geotechnical instruments are tools, which require eyes for better performance.

Geotechnical instruments are of immense value and forms an integral part during all the phases of construction.

References

1. Caputo, V. (2003). The role of geotechnical engineering in the preservation of our architectural heritage. *WIT Transactions on The Built Environment, 66.*
2. Dunnicliff, J. (1993). *Geotechnical instrumentation for monitoring field performance.* Wiley.
3. Korff, M. (2009). Deformations and damage to buildings adjacent to deep excavations in soft soils. *Delft Cluster.*
4. Van Staveren, M. (2018). *Uncertainty and ground conditions: A risk management approach.* CRC Press.
5. Ceriotti, M., Mottola, L., Picco, G. P., Murphy, A. L., Guna, S., Corra, M., & Zanon, P. (2009, April). Monitoring heritage buildings with wireless sensor networks: The Torre Aquila deployment. In *2009 International Conference on Information Processing in Sensor Networks* (pp. 277–288). IEEE.
6. Chuaqui, M., Ford, S., & Janes, M. (2007). Field Instrumentation for an innovative design-build excavation adjacent to heritage structures. In *7th FMGM 2007: Field Measurements in Geomechanics* (pp. 1–11).
7. Arulrajah, A., Bo, M. W., Chu, J., & Nikraz, H. (2009). Instrumentation at Changi land reclamation project, Singapore. *Proceedings of the Institution of Civil Engineers-Geotechnical Engineering, 162*(1), 33–40.
8. Mishra, D., Tutumluer, E., Stark, T. D., Hyslip, J. P., Chrismer, S. M., & Tomas, M. (2012). Investigation of differential movement at railroad bridge approaches through geotechnical instrumentation. *Journal of Zhejiang University Science A, 13*(11), 814–824.
9. Mauriya, K. V. (2010). Geotechnical Instrumentation in Earth and Rock-Fill Dams. In *Indian Geotechnical Conference* (pp. 16–18).
10. Saye, S. R., & Ladd, C. C. (2004). Analysis of geotechnical instrumentation to assess foundation performance of I-15. In *Geotechnical Engineering for Transportation Projects* (pp. 2103–2114).
11. Brownjohn, J. M. (2007). Structural health monitoring of civil infrastructure. *Philosophical Transactions of the Royal Society A: Mathematical, Physical and Engineering Sciences, 365*(1851), 589–622.
12. Sohn, H., Farrar, C. R., Hemez, F. M., Shunk, D. D., Stinemates, D. W., Nadler, B. R., & Czarnecki, J. J. (2003). *A review of structural health monitoring literature: 1996–2001* (p. 1). Los Alamos National Laboratory.
13. Farrar, C. R., & Worden, K. (2010). An introduction to structural health monitoring. *New Trends in Vibration Based Structural Health Monitoring*, 1–17.
14. Moss, R. M., & Matthews, S. L. (1995). In-service structural monitoring. A state of the art review. *Structural Engineer, 73*(2).
15. Chrzanowski, A., & Kurz, B. (1983). A telemetric system for monitoring deformations in difficult terrain and climate conditions. In *Deformation measurements. International Symposium on Deformation Measurements by Geodetic Methods* (Vol. 3, pp. 245–261).
16. Ding, X., Ren, D., Montgomery, B., & Swindells, C. (2000). Automatic monitoring of slope deformations using geotechnical instruments. *Journal of Surveying Engineering, 126*(2), 57–68.
17. Ding, X., & Qin, H. (2000). Geotechnical instruments in structural monitoring. *Journal of Geospatial Engineering, 2*(1), 45–56.
18. Kong, S. K. (2003). *Application of geotechnical instruments for safety control in basement construction works.*
19. Authority, A. S. (2016). *Geotechnical instrumentation and monitoring guidelines.* NSW Government.

Zinc-Based Anodes for Cathodic Protection of Reinforced Concrete Structures

Arpit Goyal

Abstract Corrosion of steel is the biggest durability issue for the reinforced concrete structure and results in structural failure. Out of the available protection techniques, cathodic protection has proven to be the most efficient corrosion prevention technique in chloride contaminated structure. The most vital component of the technique is applied anodes. Zinc-based anodes are preferred in reinforced concrete structures. This paper reviews several types of zinc-based cathodic protection anodes available, their use, application, and service life.

Keywords Cathodic protection · Reinforced concrete · Zinc-based anodes · Corrosion

1 Introduction

Corrosion of steel reinforcement is the leading cause of failure of reinforced concrete (RC) structures [1]. It is the dominant degradation mechanism for premature failure of 70–90% of RC structures [1]. Premature failure of RC structures can lead to loss of life, time, and money. Today, the need for an effective corrosion prevention technique is much higher than ever due to its high societal relevance.

Corrosion leads to direct and indirect economic losses. As per statistics, every 90 s, approximately 1 ton of steel is converted into rust worldwide [2]. The global economic cost of corrosion losses is estimated to be 3.4% of the world's gross domestic product (GDP), i.e., USD 2.5 trillion. In India, as per the 2011–2012 study, the annual cost of corrosion to the construction sector is about \$8015 million including \$6472 million for new construction and \$1543 million for repair and maintenance [3]. Over the approaching decades, this condition is expected to worsen as most of the countries have enormous aging infrastructures. This will increase the demand to rehabilitate and repair the structures by a factor of two to five by 2050 [4].

A. Goyal (✉)
Thapar Institute of Engineering and Technology, Patiala, Punjab 147004, India

A. K. Gupta et al. (eds.), *Advances in Construction Materials and Sustainable Environment*, Lecture Notes in Civil Engineering 196,
https://doi.org/10.1007/978-981-16-6557-8_4

One of the main causes of steel corrosion is the exposure of the concrete to external aggressive agents such as chlorides present in deicing salt and seawater. Out of the available corrosion protection techniques, cathodic protection (CP) has shown to be the most successful in preventing corrosion in chloride contaminated RC structures [5]. Compared to other electrochemical techniques, CP is cost-effective in the long run [6].

This paper discusses the theoretical background of the cathodic protection technique in atmospherically exposed reinforced concrete structures and reviews zinc-based anodes used for the technique, highlighting the latest development.

2 Cathodic Protection of Reinforced Concrete Structures

CP is based on the principle of delivering sufficient cathodic polarization current to the structure under protection, in order to negatively shift the steel potentials to either reduce the corrosion rate or passivate the steel [6]. Based on the method of application, CP can be of two types, i.e., sacrificial anode cathodic protection (SACP) and impressed current cathodic protection (ICCP) [7]. The schematics of the two methods are shown in Fig. 1. In SACP, steel bars (cathode) in concrete are connected to less noble metals (anode) than steel like zinc or aluminium. As a result, a potential difference is established and the anode sacrifices itself to protect the steel bar. However, in ICCP, a permanent anode is fixed on the protected structure and a small constant current is supplied from anode to steel through the concrete electrolyte using a permanent external power source [7].

The most critical component of any CP system is the anode used. The general requirements of the effective anode are good ionic conductivity, good adhesion to the concrete surface, durable and produce adequate weight addition, and change of appearance of the structure. Some of the commonly available anodes are thermal sprayed zinc, titanium anodes, organic carbon coatings, and conductive cementitious overlay anodes. However, there is still a need for anodes with improved performance characteristics such as a stronger anode-concrete bond, low cost, and longer life and are easy to install. The most common method of monitoring of atmospherically exposed CP structures is as per criteria given in BS EN ISO 12696, i.e., (a)

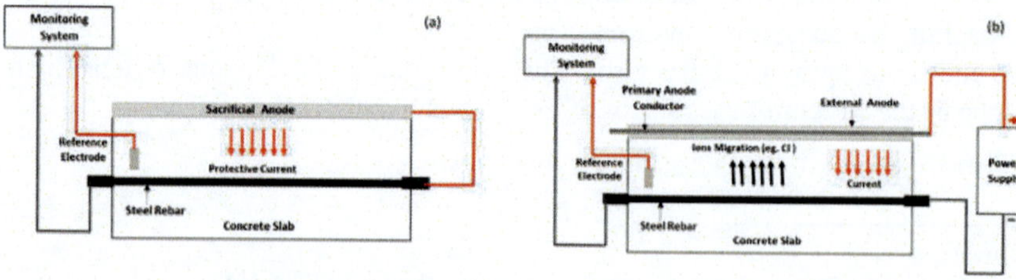

Fig. 1 Schematics of **a** SACP and **b** ICCP [7]

Instantaneous OFF potential more negative than −720 mV versus Ag/AgCl/0.5MKCl (silver-silver chloride) or (b) 100 mV decay criterion [8].

3 Zinc-Based Anodes for Cathodic Protection

Over the past few years, the use of zinc and zinc alloy anodes for CP of steel in RC structures has been substantially increased. However, they are mostly used as SACP anodes [8].

The potential gradient between zinc anode and the reinforcing steel bars generates a macrocell current in the concrete volume, corresponding to galvanic protection current [9]. There are various forms of zinc (Zn) anodes currently available in the market such as

- Thermally or arc sprayed coating of Zn, Zn–Al, or Al–Zn–In,
- Rolled zinc sheets,
- Discrete anodes, and
- Anode jackets.

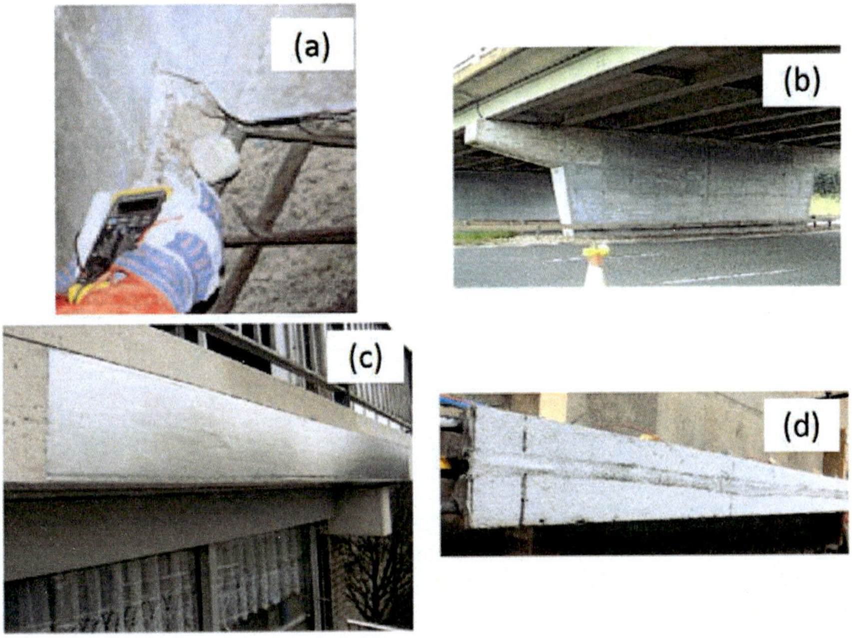

Fig. 2 Installation of **a** localized point anodes, **b** thermally sprayed zinc anode, **c** rolled zinc sheet, and **d** zinc rich paint [10, 11]

The Fig. 2 shows different zinc anodes applied to the RC structures and are explained further.

3.1 *Localized ("Point") Anodes*

These are installed in combination with a patch repair system to enhance the durability of the repairs with an estimated service life of ten years. Patch repairing the chloride or carbonated concrete leads to accelerated corrosion around the surrounding repair limiting its life to 3–5 years [12]. This is called as incipient anode or halo effect. Installation of localized point CP anodes in the patch repair, corrodes sacrificially, making the potential less noble in the patch repair and eliminating adverse cell effect [12].

Sergi and Whitmore analyzed the performance of sacrificial zinc anodes in RC and reported that anode was capable of successfully protecting the structure for ten years [13]. The anode can be installed easily and rapidly without any special equipment at a low cost by simply tying it to the reinforcing steel in the area to be patched before casting the repaired concrete.

Major drawbacks include degradation of throwing distance with aging, thus requiring multiple anodes resulting in increased cost. In addition, issues related to protection in highly carbonated, highly chloride, and highly resistive concrete [12]. Rincon et al. [14] studied the effect of environmental exposure (95% RH and 77% RH) on the performance of point anode and concluded that anode did not provide sufficient galvanic protection when samples were exposed to the tropical environment, i.e., 77% RH and 28 °C temperature and rebars continued to corrode. However, in a 95% RH environment, only rebars close to the anode were protected. Moreover, only after one year of exposure, leaching of zinc activation products was observed [14].

Hence, point anodes are only suitable for patch repair and exposed to the high humid environment due to low concrete resistivity.

3.2 *Thermally or Arc Sprayed Zinc (TSZ)*

In the United States, the application of TSZ anodes on concrete surfaces has been actively supported by the Department of Transportation (DOT) of Oregon. Between 1995–2005, more than 80,000 m^2 TSZ anodes have been installed in Oregon alone [15].

This type of Zn anodes had been used as SACP and ICCP anode and found the later appears to be more effective. The Al–Zn and Al–Zn–In alloys are used as SACP anodes and Zn coating for both ICCP and SACP. These coatings are generally about 50 μm thick and are applied by melting metal wires with an electric arc and sprayed over concrete elements using pressurized air. Thus, the application of TSZ requires specialist operators and sophisticated plant and equipment's involving "high energy input" [16]. Thus, the TSZ anode system could be viewed as a "high carbon footprint" system together with health and safety issues for the site operatives.

The technique has been applied to several structures, especially in the USA, and the results showed that the system works effectively in high RH and low resistive concrete environment. Reduction in durability may occur in application to wet structures where self-corrosion of zinc occurs [17].

For thermal sprayed zinc, ICCP systems generally fail when the anode operating voltage exceeds the specified operating limits or when the anode begins to delaminate [18]. When anode age electrochemically, chemical and physical changes were observed at Zn–concrete interface such as acidification, oxidation, and secondary mineralization leading to its failure.

Several studies have been carried out to study the effect of electrochemical aging on bond strength of TSZ and the eventual adhesive failure of TSZ anodes. When TSZ was used as SACP anode, studies showed the formation of white corrosion products at the Zn-cement paste interface. Dugarte and Sagues [19] observed the microstructure of this white deposit and concluded that the deposits were corrosion products of zinc, in the form of zinc oxides and hydroxides [19]. Furthermore, it was observed that on aging of the anode electrochemically, zinc replaces calcium in the cement paste, thus further weakens the interface bond [12].

Covino et al. [20] studied the effect of concrete surface preparation and electrochemical aging on bond behavior of TSZ anodes. The authors observed that preheating the samples before the anode application showed an 80% higher initial bond but does not affect the long-term bond strength of the coating anode [20]. The average initial bond for preheated samples was 2.06 MPa, compared to 1.15 MPa for unheated samples. The bond strength initially decreased with the amount of charge passed until 50 kC/m^2 of charge and then increases to as high as 3.0 MPa for 600 kC/m^2 of charge and again decreases with further increase in the charge [20]. Legoux and Dallaire [21] showed a high moisture level, low concrete surface temperature, surface contamination, and amount of aggregates exposed to lead to lower initial adhesion strength of TSZ anodes [21]. The main reason for increased adhesion strength is the formation of zinc oxidation products. With progress in the formation of zinc corrosion products, the Zn to concrete adhesive bond initially improves but then decreases due to an increase in the thickness of the corrosion product layer.

Moisture at the Zn-concrete interface is essential for the successful performance and long-term service of the TSZ anode. The presence of moisture lowers the driving voltage required for the effective performance of the ICCP system, increases the flow of current for the SACP system and also redistributes the anode dissolution products in concrete pore solution [22].

Furthermore, research showed that TSZ requires periodical wetting to deliver adequate and uniform current when operated as a SACP anode. Sagüés and Powers [23] reported that the anode will not be able to deliver the required current in the absence of direct wetting, irrespective of high humidity exposed conditions [23].

However, this could be reestablished by either direct wetting the anode [24] or by the application of humectants [25]. Humectant solutions maintain the moisture level at the anode-concrete interface, thus reducing the electrical resistivity of concrete, increase the current for galvanic system and redistribute coating's dissolution product in concrete [12, 26]. Several humectants have been studied overtime such as LiBr, $LiNO_3$, and $KC_2H_3O_2$. The results showed that LiBr performed best among all to improve the performance of Zn anodes [12]. Hence, TSZ anodes should only be used as SACP system when concrete is exposed to aggressive environments such as high humidity or wet/dry condition because it is only in this environment that they can deliver sufficient current to satisfy the 100 mV decay performance criterion [27].

Hence, TSZ can only provide effective CP for structures in a marine environment with tidal and splash zones provided concrete has low resistivity. In a very humid environment, the decrease in corrosion activity is observed mainly due to TSZ acting as a barrier coating to steel bars.

3.3 Rolled Zinc Sheets or Zinc Layer Anodes (ZLAs)

This anode is used as a galvanic anode and consists of rolls of high purity zinc foil, coated on one side with a nominally thick low resistance ionic conductive gel (hydrogel) [9]. They are applied to concrete surfaces with a direct or indirect connection to steel to form the galvanic system [28]. Anodes generally fail due to loss of adhesion or leaking of hydrogel which can be caused by excessive moisture intrusion [28]. Garcia et al. [9] studied the polarization response of a ZLA in the concrete electrolyte and its evolution overtime and concluded that ZLA should not be used as electrodes that are inert overtime.

ZLA is unsuitable for ICCP applications as they lead to blistering and loss of adhesion [29]. The driving voltage required to pass effective current to steel from the anode was observed to be very high, and hence is not suitable for CP of atmospherically exposed reinforced concrete and results in debonding [30].

3.4 Zinc-Rich Paint (ZRP) Conductive Coating

This is the new conductive coating anode being developed at Coventry University, United Kingdom. The anode is suitable for the ICCP system and offers the advantage of easy application, longer durability, and effective performance. They can be easily applied by roller or brush without any need for thermal treatment and can last for 15 years [5]. The anode offers effective performance and can even act as a barrier coating to prevent the ingress of aggressive chemicals such as chlorides. In addition, the coating is breathable, i.e., it allows moisture to evaporate and prevents long-term debondment and premature failure [5, 6]. Mainly, the pull-off failure of the ZRP anode (2.73 MPa) is much higher than the available coating anodes (1.0–1.5 MPa) and thus have a longer service life. The coating is still at the developing stage to improve its conductivity and thus offering better performance and longer service life.

4 Conclusion

Cathodic protection is the most successful technique to prevent chloride-induced corrosion in reinforced concrete. Various forms of CP anodes are available, however, there is still demand for new anodes with improved durability, service life, and performance with ease of application. Zinc in various forms of anodes has proved to be effective to protect RC structures. Point anodes are of limited effectiveness and are only suitable for patch repair, exposed to the high humid environment due to low concrete resistivity. Thermally, sprayed zinc is effective but requires a sophisticated plant. Equipment and specialist for its application have huge health and safety concerns. Moreover, they are most effective in tidal and splash zone considering concrete has good conductivity. A new zinc-based conductive coating anode is under development and offers the advantage of easy application, longer service life, and better durability and performance compared to available coating anodes.

References

1. Angst, U. M. (2018). Challenges and opportunities in corrosion of steel in concrete. *Materials and Structures Constructions, 51*, 1–20.
2. Goni, K. M. O. L., & Mazumder, M. A. J. (2019). Green corrosion inhibitors. In: *Corrosion Inhibitors*
3. Koch, G., Varney, J., Thompson, N., Moghissi, O., Gould, M., & Payer, J. (2016). *International Measures of Prevention, Application, and Economics of Corrosion Technologies Study, Houston.*
4. Angst, U. M. (2019). A critical review of the science and engineering of cathodic protection of steel in soil and concrete. *Corrosion, 75*, 1420–1433.
5. Goyal, A., Pouya, H. S., & Ganjian, E. (2019). Performance assessment of specialist conductive paint for cathodic protection of steel in reinforced concrete structures. *Construction and Building Materials, 223*, 1083–1094.
6. Goyal, A., Pouya, H. S., Ganjian, E., Olubanwo, A. O., & Khorami, M. (2019). Predicting the corrosion rate of steel in cathodically protected concrete using potential shift. *Construction and Building Materials, 194.*
7. Goyal, A., Sadeghi, H., Eshmaiel, P., & Peter, G. (2018). A Review of Corrosion and Protection of Steel in Concrete. *Arabian Journal for Science and Engineering, 43*, 5035–5055.
8. BSI: BS EN ISO 12696. (2016). *Cathodic protection of steel in concrete, London.*
9. Garcia, D., Laurens, S., & Panin, S. (2018). Electrochemical behavior of zinc layer anodes used for galvanic protection of steel in reinforced concrete. *RILEM Technical Letters, 3*, 59–65.
10. Broomfield, J. P. (2007). The principles and practice of galvanic cathodic protection for reinforced concrete structures-technical notes 6. *Corrosion Preventive Association.*
11. Broomfield, J. P., & Belbol, S. E. (2011). Impressed current anodes for the cathodic protection of atmospherically exposed reinforced concrete-technical notes 11. *Corrosion Preventive Association.*

12. Troconis de Rincón, O., Sagüés, A., Torrres-Acosta, A., & Martinez-Madrid, M. (2018). Galvanic anodes for reinforced concrete structures: A review. *Corrosion, 74*, 2613.
13. Sergi, G., & Whitmore, D. (2010). Performance of zinc sacrificial anodes for long-term control of reinforcement corrosion. *NACE International*, pp. 1–17.
14. de Rincon, O. T., Hernandez-Lopez, Y., de Valle-Moreno, A., Torres-acosta, A., Barrios, F., Montero, P., Oidor-salinas, P., & Montero, J. R. (2008). Environmental influence on point anodes performance in reinforced concrete. *Construction and Building Materials, 22*, 494–503.
15. Das, S. C., Pouya, H. S., & Ganjian, E. (2015). Zinc-Rich paint As anode for cathodic protection of steel in concrete. *Journal of Materials in Civil Engineering, 27*, 1–9.
16. Wilson, K., Jawed, M., & Ngala, V. (2013). The selection and use of cathodic protection systems for the repair of reinforced concrete structures. In: *Construction and Building Materials.*
17. Broomfield, J. P., & Belbol, S. E. (2006). Impressed current anodes for the cathodic protection of atmospherically exposed reinforced concrete—technical note 11. *Corrosion Preventive Association.*
18. Bullard, S. J., Cramer, S., & Covino, B. (2009). *Effectiveness of Cathodic Protection, Albany.*
19. Dugarte, M. J., & Sagues, A. A. (2014). Sacrificial point anodes for cathodic prevention of reinforcing steel in concrete repairs: Part 1—polarization behavior. *Corrosion, 70*, 303–317.
20. Covino, B. S., Bullard, S. J., Holcomb, G. R., Cramer, S. D., Mcgill, G. E., & Cryer, C. B. (1996). Bond strength of electrochemically aged arc sprayed zinc coatings on concrete. *Corrosion.*
21. Legoux, J. G., & Dallaire, S. (1995). Adhesion mechanisms of arc-sprayed zinc on concrete. *Journal of Thermal Spray Technology, 4*, 395–400.
22. Covino, B. S., Holcomb, G. R., Bullard, S. J., Russell, J. H., Cramer, S. D., Bennett, J. E., & Laylor, H. M. (1999). Electrochemical aging of humectant-treated thermal-sprayed zinc anodes for cathodic protection. *Corrosion.*
23. Sagüés, A. A., & Powers, R. G. (1996). Sprayed-Zinc sacrificial anodes for reinforced concrete in marine service. *Corrosion, 52*, 508–522.
24. Bennett, J., Bushman, J. B., Costa, J., & Noyce, P. (2000). Field application of performance enhancing chemicals to metallized zinc anodes. *Corrosion.*
25. Darowicki, K., Orlikowski, J., Cebulski, S., & Krakowiak, S. (2003). Conducting coatings as anodes in cathodic protection. *Progress in organic coatings, 46*, 191–196.
26. Broomfield, J. P. (2011). Cathodic protection using thermal sprayed metals. In: M. Granthom (Ed.) *Concrete Repair, a Practical Guide.* Routledge: London and New York.
27. Holmes, S. P., Wilcox, G. D., Robins, P. J., Glass, G. K., & Roberts, A. C. (2011). Responsive behaviour of galvanic anodes in concrete and the basis for its utilisation. *Corrosion Science, 53*, 3450–3454.
28. The Concrete Society. (2011). *Cathodic protection of steel in concrete—Appendices.* Surrey.
29. Anwar, M. S., Sujitha, B., & Vedalakshmi, R. (2014). Light-weight cementitious conductive anode for impressed current cathodic protection of steel reinforced concrete application. *Construction and Building Materials, 71*, 167–180.
30. Brousseau, R., Arnott, M., & Baldock, B. (1995). Laboratory performance of zinc anodes for impressed current cathodic protection of reinforced concrete. *Corrosion, 51*, 639–644.

An Overview: Supplementary Cementitious Materials

Pooja Jha, A. K. Sachan, and R. P. Singh

Abstract In the present scenario, there is a large production of agriculture wastes (AWs) and industrial waste (IW) have produced severe environmental problems related to their safe disposal. This review paper deals with the feasible usage of different types of debris like AW and IW in the production of mortar and concrete. These are used as supplementary cementitious material (SCM) to enhance the workability (WA), strength along with the durability properties of the concrete. It reviews on the evaluation of various physical properties of these wastes (AW and IW) includes fly ash (FA), ground granulated blast furnace slag (GGBFS), and silica fume (SF) and usefulness in the concrete production. It is used as SCM and can be advantageous in the strength and durability properties of concrete. It describes the influence of the addition of SCMs on the fresh properties (FP) and hardened properties (HP) of a concrete mortar without affecting the quality of concrete. Due to the similar properties of cement, these wastes may be used as a cement substitute and cement additive in the concrete industry. It also describes the utilization of the new emerging wastes like ground waste expanded perlite (WEP), and it is used as pozzolanic material (PM) and valuable SCMs. Due to its substantial activity, WEP can be used as a cement additive as well as a cement substitute.

Keyword Wastes · Concrete · High-performance concrete · Pozzolanic material · Waste expanded perlite

P. Jha (✉) · A. K. Sachan · R. P. Singh
Department of Civil Engineering, Motilal Nehru National Institute of Technology, Allahabad 211004, India
e-mail: rce1607@mnnit.ac.in

A. K. Sachan
e-mail: sachan@mnnit.ac.in

R. P. Singh
e-mail: rps@mnnit.ac.in

A. K. Gupta et al. (eds.), *Advances in Construction Materials and Sustainable Environment*, Lecture Notes in Civil Engineering 196,
https://doi.org/10.1007/978-981-16-6557-8_5

1 Introduction

Nowadays, the most vital building material used in the construction industry is concrete. It has reported that portland cement (PC) utilization improved drastically over a period little more than a century (1880–1990). Therefore, the process of cement manufacturing is the main reason for CO_2 emission and worldwide, it is the third-largest CO_2 producer. Cement industry alone generated seven percent of all CO_2 in the world [1]. There are drastic increases in the emission of CO_2 from cement production has been seen [2, 3]. To overcome these problems, cement can be replaced by producing a new emerging material that is the SCMs. These are the PM, which can be classified as natural pozzolana (NP) as well as artificial pozzolana (AP). The NP can be generally found in volcanic tuffs, and the AP can be obtained from FA and metallurgical slags, etc. [4, 5]. Many researchers [6–8] have observed that these SCMs are by-products materials that enhance concrete construction properties and also protect environmental resources (included sustainability of concrete). Naik and Singh [9] have studied that these SCMs may decrease the early strength (ES) of concrete, mainly if the cement replacement rate (CRR) is more. Still, at optimum replacement percentage, it is producing valuable, strong, and durable concrete. SCM is by-product of silicon (Si) and aluminum (Al). Al and Si contents have various benefits such as reduced permeability, reduced segregation, resistance against the freeze, and resistance against sulfate attack of concrete. Not only this, but it has also improved the compressive strength (CS) as well as durability (DB) of concrete.

In this paper, several AW and IW have been described and introduced. In this paper, the introduction and explanation of the AW and IW materials as SCMs for concrete. It also describes its influence of the addition of SCMs on the properties of concrete and mortar. It has also been described as expanded perlite (EP), which is used as the new effective SCM. It is used mostly in horticulture and agriculture as well as in the building materials technology (acoustic, lightweight composites, fire insulation, and thermal insulation). Various techniques like abrasives and filtration use, specially prepared expanded perlite.

2 Waste Material as SCM

The classification and specifications of different SCMs like FA and GGBFS are given below in Tables 1, 2, and 3.

The various SCMs as per the standards along with the uses are given in Table 3.

The uses of SCMs in concrete are beneficial in many ways. Their purposes enhance and accelerate the strength of concrete, improve the resistance against sulfate attack, resistance against chloride ions, and making concrete easier to pump. SCMs also play a useful role in reducing the water permeability and other fluids, deleterious expansion, and risk of delayed ettringite formation.

Table 1 Classification and specifications for FA as per [10]

Specifications	Types
Volcanic ashes or pumicites and tuffs	Class N (raw pozzolana)
Diatomaceous earth	
Opaline cherts and shales	
Calcined clays	
Pozzolanic properties	Class F
Pozzolanic and cementitious properties	Class C

Table 2 Classification and specifications for GGBFS as per [11]

Specifications	Types
Low activity index (LAI)	Grade 80
Moderate activity index (MAI)	Grade 100
High activity index (HAI)	Grade 120

Table 3 Classification, uses, and specifications of various SCMs

Standards	SCMs	Uses
Standard specification for coal FA and raw or calcined natural pozzolana for use as a mineral admixture in PC concrete [10]	Coal FA and raw or calcined natural pozzolan	Concrete
Specification for GGBFS for use in concrete and mortar [11]	GGBFS	Concrete and mortar
Specification for SF for use in concrete and mortar [12]	SF	Cementitious mixture

2.1 Different Types of SCMs

FA: FA is a siliceous or alumina siliceous material which can be used as cement replacement material (CRM) due to similar properties of cement. It improves workability, strength in long-term basis, resistance against sulfate attack, and DB in concrete. FA is formed from the coal burning in electric power generation plants, and it has high pozzolanic activity [13]. Due to its chemical properties and mineral constituents, the color of FA may change from tan to dark gray. There are various predominant areas of FA applications. They are the production of concrete [14], cement clinkers [15], waste solidification [16], more geopolymer concrete in the fresh state [17], and road basement material [18]. Cement is replaced by FA, which is used as a CRM, makes the conventional and high-performance concrete, which is used as an SCM in the construction industry. Similarly, the environmental advantages of waste disposal and CO_2 sequestration [19, 20]. In the fresh properties of concrete, it means at early ages, FA improves workability, thermal cracking reduces and heat of hydration also lowers in concrete and in the hardened state such that at the later periods, it enhances the various properties like durability as well mechanical properties of concrete [21]. There are many limitations of FA, and one of the flaws reported by Vargas and Halog [22] is that the full utilization of FA is not achieved, and it is partially replaced by the cement. Lam et al. [23] have described the various properties of FA concrete-like mechanical, fracture, and durable properties, and its effect of different type of FA on multiple other properties like freeze–thaw resistance has been reported by Uysal and Akyuncu [24]. The maximum percentage of FA used is restricted to 35% in the manufacturing of portland pozzolana cement as per the code IS-1489, 2000. From the literature review, it has seen that when FA beyond 35% replaces cement, the strength characteristics are not increases and show a decreasing rate after attaining the optimum replacement value. In most cases, like FA cement needs improvement to enhance the strength in the mix by making more products of hydration. It is possible to attain more significant than 50% replacement of FA by the proper engineering procedure.

GGBFS: GGBFS is made of the material used to make iron and produced by the blast furnace. At a temperature of approximately 1600 °C, different products like molten slag (MS) and molten iron (MI) have formed by the combination of coke, limestone, and iron ore in the furnace. Malhotra et al. [6] have reported that the leading country in the world which produced GGBFS was Germany. For public purposes, it has also been used in North America. Molten slag is made mostly of silicon dioxide ranges from 30 to 40% and calcium dioxide ranges from 40%. By using high-pressure water jets, silicates, and alumina, which are the essential components in molten slag has cooled down [25]. Therefore, granular glassy material is formed during rapid cooling, which has latent hydraulic properties at temperature ranges from 900–800 °C results in noncrystalline slag. 35–65% replacement level of GGBFS may prove to be advantageous in concrete, which also helps in reducing the carbon dioxide production. There are three strength grades of GGBFS (Grade 80, 100, and 120) as per ASTM C 98911.

SF: SF is manufactured from silicon metal, ferrosilicon alloy, which is collected from the oxidized vapor on the top of the electric arc furnaces, and it is being used as supplementary cementing material for concrete elements. SF also known as condensed silica fumes (CSF), microsilica, silica dust, volatilized silica, and micropores (trademark name). Most of the silica fume particles are ultra-fine particles and spherical. Due to its high fineness and glass content, SF shows a high pozzolanic reactivity, which is very constructive when used in concrete. The SF replacement level is 5–10% when replacing the cement with SF [26]. Mechanical properties of concrete are affected substantially as consequences of strengthening the interfacial zone. Amoudi et al. [27] have studied that SF has a vital influence on the aggregate-cement interface (ACI). Due to high pozzolanic and extreme fineness, its addition produces less permeability concrete. Nguyen et al. [28] have documented that the effects of rice husk ash (RHA) and SF in both binary systems as well as ternary system on the property of cement pastes, and the CS of concrete was studied. The various physical properties (PP) and

Table 4 Typical physical properties (PP) of various SCMs

PP	FA (Range) [29]	GGBFS (Range) [30]	SF (Range) [29]
Particles size (PS)	<1 μm to >100 μm	<45 μm	<1 μm
Diameter of the particles (D)	<20 μm in size	–	
Surface area (SA)	300–500 m^2/kg (min surface are 200 m^2/kg Max surface area 700 m^2/kg)	400–600 m^2/kg	13.000–30.000 m^2/kg
Density (p)	540–860 m^2/kg	–	481–720 kg/m^3
Max bulk density under close-packed storage (BD)	1120–1500 kg/m^3	–	131–430 kg/m^3
Specific gravity (SG)	1.9–2.9	2.61	2.22

Table 5 Typical chemical properties (CP) of various SCMs

Chemical composition	FA % by mass [31]	GGBFS % by mass [32]	SF % by mass [33]
SiO_2	27.88–59.40	35	95.3
CaO	0.37–27.68	40	0.3
Al_2O_3	5.23–33.99	13	0.6
Fe_2O_3	1.21–29.63	–	0.3
MgO	0.42–8.79	8	0.4
Na_2O	0.20–6.90	–	0.3
SO_3	0.04–4.71	–	–
K_2O	0.64–6.68	–	0.8
TiO_2	0.24–1.73	–	–
LOI	0.21–28.37	–	–

chemical properties (CP) of various SCMs such as FA, GGBFS, and SF as shown in Tables 4 and 5.

2.2 *The Influence of SCMs on the FP of Concrete*

The slump test (ST) can be performed to evaluate the FP of the concrete. Fine SCMs, mainly metakaolin and SF, are used to reduce the slump and expand the water consumption [34, 35]. However, not all SCMs increase in water consumption. For example, FA as well as GGBFS reduce the water demand and also enhancing the properties of fresh concrete at the same time [35]. Therefore, Gesoglu et al. [36] have found that that the higher replacement levels are necessary if better results are desired, which was manifest in the study carried out by. They could also add the mineral admixtures and increase the filling and passing ability of self-compacting concrete by [37]. They had also documented that there was higher in the water reduction effect when FA was used as an SCM with a 40% replacement level. Similar outcomes can also be seen by integrating FA and superplasticizers in HPC. The same approach has seen because the consequences of their study showed that the inclusion of these two materials enhances the WA as well as the concrete performance [38].

2.3 The Influence of SCMs on the HP and DP of Concrete

SCMs lower the porosity of the concrete. Filling the available voids in the cement to enhance CS and DB by introducing SCMs in it. Due to hydration, the inclusion of FA to concrete cannot only minimize the dense packing and water content (WC) but also increases the hydration and pozzolanic reactions, which consequences to decelerate the permeability of concrete. The calcium hydroxide $Ca(OH)_2$ can develop voids permeable in nature in the hardened concrete during the hydration process. By adding calcium hydroxide during the pozzolanic reaction, the leaching of calcium hydroxide can be minimized. Voids can be choked by calcium silicate hydrate gel in the chemical reaction and added to the density of the concrete, which in turn lowers the permeability. SCMs such as SF, RHA, and metakaolin play an essential role in early age as well as later-age strength improvement in concrete [35, 39]. On the other hands of FA and GGBFS, this strength improvement does not occur at an early age [35]. The proportions of slag, SF, and FA had increased to attain better CS [40]; these proportions studied as 17% for slag, 15% for SF, and 10% for FA.

During the winter season, the CS of concrete produced showed an increase of approximately 5% in emissions of CO_2 compared with concrete produced in the season concluded by [41]. Besides, they showed that the amount of CO_2 emitted for concrete containing SCM was lowered by as much as 47% compared with concrete without SCM. The reason for these consequences is due to the cement replacement and admixtures that have a remarkable amount of carbon dioxide with materials such as FA or GGBFS, which have a lower amount of CO_2. Due to the combined effects of the multi binder on high-performance concrete, the CS significantly reduced. A considerable part of SF and FA reported to be the essential factor affecting properties included the drying shrinkage. Specimens of silica fume and metakaolin have more CS. Borhan et al. [42] have studied the porosity, CS, permeability, and resistance to chemical agents of multi blended mortar (MB mortar) containing FA and SF. The outcomes show that strength was 20% lower for the MB concrete at an early age, while at the final-age, strength of both the control mortar and the MB concrete was approximately the same. Therefore, the MB concrete outperformed the control mortar in terms of low permeability. SCMs improve against the sulfate resistance due to good pozzolanic activity that enhanced the microstructure and prevent the formation of ettringite which is major factor of sulfate attack [43].

Table 6 Properties of WEP [44]

WEP	Range
PS	<100 μm
BD	50–150 kg/m^3

Table 7 Variation of CS by partially replacing the cement with WEP

% Replacement of WEP	0	10	20	30	40	References
CS (N/mm^2) at 28 days	43.3	40	39	40	–	[45]
CS (N/mm^2) at 28 days	28.8	–	17.3	–	10.9	[46]

2.4 Applications of WEP as New Constructive SCM

There was worldwide usage of the expanded perlite (EP) as a cementitious material. EP used as valuable lightweight building materials used in the agriculture industry. Fined grained WEP is being formed during both productions as well as the processing of EP. By the incorporation of ground WEP, consequences of strength tests showed that strength roughly to 50%. Ground WEP can be used as a cement additive as well as a cement substitute due to its high activity. Some properties of WEP are shown in Table 6, and variation of CS. by partially replacing the cement with WEP shown in Table 7.

The possibility of usage of raw perlite rock as a cement additive [47]. Ramezanipour et al. [48] showed to evaluate the application of WEP as SCM was conducted. He also studied the use of WEP as SCM. Many researchers worked on the various SCMs used in cement as well as the concrete industry; there are calcined clays, GGBFS [15], limestone [49], FA [22], and natural zeolites [50]. In an industry like binding materials science, SCMs are one of the most predominant topics [16]. The various content of portland clinker lowers by the addition of SCM in cement, and it also reduces the total amount of CO_2 liberates [51]. In industries like autoclaved aerated concrete manufacturing, EP can be used as quartz sand replacement. Thermal conductivity lowers by 15% without an essential reduction of strength by replacement of 10% sand [51]. The cement replaced by calcined raw perlite rock, which increases the properties of concrete in the HS includes the DB properties [51].

3 Conclusions

In the concrete mixture, the SCM will be a vital interest and a practical solution for sustainable construction and also construct greening in respect to the environment. The following conclusions are derived below:

1. The SCMs have been proven to crucial materials to enhance the strength and performance of concrete. These materials have a positive influence on the

concrete performance included FP, HP (CS and tensile strength (TS)), and drying shrinkage.

2. Various issues (environmental, technical, and economic) caused by cement production has neutralized or minimized by the introduction of SCMs in cement or concrete. Most of these SCMs are by-products, and their addition serves as a crucial means to save environmental resources, which may result in more viable constructions in future.
3. Usage of cement has minimized through the SCMs by the introduction of SCMs are added to concrete, which may lead to the environmental benefits of lower emission of CO_2.
4. There is growth in the manufacturing of FA to lower the effect on the environment and to revise the possibility in the sector of construction; there is a worldwide demand to understand the various advantage of FA usage in the concrete industry. While new procedures and technique are settling to engineer, the concrete with vast volumes of FA to create superior outcomes, due to the various types of problems like increased shrinkage, high carbonation, and slow development of strength, utilization of vast volumes of FA remain incomplete. Researchers focus on the concept of the green economy, which is significant to society as well as the environment. Therefore, the main principle binder used in concrete is portland cement (energy-intensive). It is also responsible for significant emissions of CO_2 gas called greenhouse gas, and the manufacturing of cement remarkably leads to global warming, which leads to the change in the climate. Hence, the improvement of existing knowledge and examination of further convenient IW as well as AW to be used as SCM in the mixture of concrete.
5. Cement mortar improvement along with ground WEP enables obtaining strength upgrade up to over fifty percent. The modification of the strength development rate mainly depends on the WEP content. WEP can be utilized as an additive and substitute for portland cement. The inclusion of WEP consequences is a higher strength development all over the hydration time, while replacement of cement at 28 days leads to a less reduction in early strength.

Acknowledgements The authors are grateful to Director MNNIT Allahabad for providing the necessary help and support required. One of the authors, Miss Pooja, a research scholar, is also appreciative to MHRD New Delhi, India, for providing financial assistance for research work.

Conflict of Interest: There is no conflict of interest.

References

1. Malhotra, V. M. (2002). Introduction: sustainable development and concrete technology. *Concrete International, 24*(7).

2. Taylor, M., Tam, C., & Gielen, D. (2006). Energy efficiency and CO2 emissions from the global cement industry. *Korea, 50*(2.2), 61–67.
3. Muga, H., Betz, K., Walker, J., Pranger, C., & Vidor, A. (2005). *Development of appropriate and sustainable construction materials*. Michigan Technological University, Michigan, USA.
4. Mehta, P. K. (1989) Pozzolanic and cementitious by-products in concrete—Another look. In: V. M. Malhotra (Ed.) *Proceedings 3rd International Conference on the Use of Fly Ash, Silica Fume, Slag, and Natural Pozzolans in Concrete, ACI SP-114, Trondheim* (pp. 1–43).
5. Malhotra, V. M. (1987). *Supplementary Cementing Materials for Concrete*. Centre for Mineral and Energy Technology, Ottawa, Canada, p. 428.
6. Malhotra, V. M. (1996). *Pozzolanic and Cementitious Materials*. Gordon and Breach Publishers, Amsterdam, p. 208.
7. Aitcin, P. C. (1998). *High Performance Concrete*. Taylor and Francis, USA, ISBN-13: 9780419192701, p. 624.
8. Mehta, P. K., & Monteiro, P. J. M. (2006). *Concrete-Microstructure Properties and Materials* (3rd ed., p. 659). Mc Graw-Hill.
9. Naik, T. R., & Singh, S. S. (1998). Fly ash generation and utilization-an overview. In A. K. Suri & A. B. Harapanahalli (Eds.), *Recent Trends in Fly Ash Utilization* (pp. 1–25). SOFEM Publisher.
10. ASTM C 618-94. (1994). *Standard specification for coal fly ash and raw or calcined natural pozzolana for use as a mineral admixture in Portland cement concrete*. http://www.astm.org/Standards/C618.htm
11. ASTM C 989-93. (1993). *Specification for Ground Granulated Blast-Furnace Slag for Use in Concrete and Mortars*. http://www.astm.org/Standards/C989.htm
12. ASTM C 1240-93. (1993). *Specification for Silica Fume for Use in Concrete and Mortars*. http://www.astm.org/Standards/C1240.htm
13. Haque, M. N., & Kayali, O. (1998). Properties of high-strength concrete using a fine fly ash. *Cement Concrete Research, 28*, 1445–1452.
14. Sivasundaram, V., Carette, G. G., & Malhotra, V. M. (1990). Long-term strength development of high-volume fly ash concrete. *Cement and Concrete Composites, 12*(4), 263–270.
15. Tironi, A., Trezza, M. A., Scian, N., & Irassar, E. F. (2013). Assessment of pozzolanic activity of different calcined clays. *Cement and Concrete Composites, 37*, 319–327.
16. Puertas, F., Varga, C., del Mar Alonso, M., Aranzazu Diaz-Batista, M., & Lizarraga, S. (2015). New technology for alternative pozzolanic additions for Portland cement from abandoned landfills Cement Wapno Beton, 2/2015, pp. 88–105.
17. Embong, R., Kusbiantoro, A., Shafiq, N., & Nuruddin, M. F. (2016). Strength and microstructural properties of fly ash based geopolymer concrete containing high-calcium and water-absorptive aggregate. *Journal of Cleaner Production, 112*, 816–822.
18. Sobolev, K., Vivian, I. F., Saha, R., Wasiuddin, N. M., & Saltibus, N. E. (2014). The effect of fly ash on the rheological properties of bituminous materials. *Fuel, 116*, 471–477.
19. Ukwattage, N., Ranjith, P., Yellishetty, M., Bui, H., & Xu, T. (2015). A laboratory-scale study of the aqueous mineral carbonation of coal fly ash for CO2 sequestration. *Journal of Cleaner Production, 103*, 665–674.
20. Dananjayan, R. R. T., Kandasamy, P., & Andimuthu, R. (2016). Direct mineral carbonation of coal fly ash for CO2 sequestration. *Journal of Cleaner Production, 112*, 4173–4182.
21. Aahmaran, M., & Li, V. C. (2009). Durability properties of micro-cracked ECC containing high volumes fly ash. *Cement and Concrete Research, 39*(11), 1033–1043.
22. Vargas, J., & Halog, A. (2015). Effective carbon emission reductions from using upgraded fly ash in the cement industry. *Journal of Cleaner Production, 103*, 948–995.
23. Lam, L., Wong, Y. L., & Poon, C. S. (2000). Degree of hydration and gel/space ratio of high-volume fly ash/cement systems. *Cement and Concrete Research, 30*, 747–756.
24. Uysal, M., & Akyuncu, V. (2012). Durability performance of concrete incorporating Class F and Class C fly ashes. *Construction and Building Materials, 34*, 170–178.
25. Higgins, D. (2007). Briefing: GGBS and sustainability. *Proceedings of the Institution of Civil Engineers Construction Materials, 160*, 99–101.

26. Kosmatka, S. H., Kerkhoff, B., & William, C. (2003). *Design and Control of Concrete Mixtures* (14th ed.). Portland Cement Association.
27. Al-Amoudi, O. S. B., Al-Kutti, W. A., Ahmad, S., & Maslehuddin, M. (2009). Correlation between compressive strength and certain durability indices of plain and blended cement concretes. *Cement and Concrete Composites, 31*(9), 672–676.
28. Nguyen, V. T. (2011). *Rice husk ash as a mineral admixture for ultra-high performance concrete.*
29. Terence, C. H. (2005). *Silica Fume User's Manual, Lovettsville, VA.*
30. Shumuye, E. D., & Jun, Z. (2018). A Review on Ground Granulated Blast Slag (GGBS) in Concrete. In: *Proceedings of the Eighth International Conference on Advances in Civil and Structural Engineering, CSE.*
31. Feng, X., & Clark, B. (2011). Evaluation of the physical and chemical properties of fly ash products for use in Portland cement concrete. In: *2011 World of Coal Ash Conference, Denver, CO, USA* (Vol. 9).
32. Suresh, D., & Nagaraju, K. (2015). Ground granulated blast slag (GGBS) in concrete–a review. *IOSR journal of mechanical and civil engineering, 12*(4), 76–82.
33. Nochaiya, T., Wongkeo, W., & Chaipanich, A. (2010). Utilization of fly ash with silica fume and properties of Portland cement–fly ash–silica fume concrete. *Fuel, 89*(3), 768–774.
34. Neville, A. M. (2005). *Properties of Concrete*, 4th Edn. Pearson Education Ltd., Essex, England.
35. Safiuddin, M., & Zain, M. F. M. (2006). Supplementary cementing materials for high-performance concrete. *BRAC University Journal, 3*(2), 47–57.
36. Gesoglu, M., Guneyisi, E., & Ozbay, E. (2009). Properties of self-compacting concrete made with binary, ternary, and quaternary cementitious blends of fly ash, blast furnace slag, and silica fume. *Construction and Building Materials, 23*, 1847–1854.
37. Wu, Z., & Naik, T. R. (2002). Properties of concrete produced from multicomponent blended cement. *Cement and Concrete Research, 32*, 1937–1942.
38. Yin, J., Zhou, S., Xie, Y., Chen, Y., & Yan, Q. (2002). Investigation on compounding and application of C80–C100 high-performance concrete. *Cement Concrete Research, 32*, 173–177.
39. Malhotra, V. M. (1993). Fly ash, slag, silica fume, and rice husk ash in concrete: A review. *Concrete International, 15*(4), 23–28.
40. Penga, Y., Hu, S., & Ding, Q. (2009). Dense packing properties of mineral admixtures in the cementitious material. *Particuology, 7*, 399–402.
41. Park, J., Tae, S., & Kim, T. (2012). Life cycle CO_2 assessment of concrete by compressive strength on the construction site in Korea. *Renewable Sustainable Energy Review, 16*, 2940–2946.
42. Borhan, M. N., Ismail, A., & Rahmat, R. A. (2010). Evaluation of palm oil fuel ash (POFA) on asphalt mixtures. *Australian Journal of Basic and Applied Sciences, 4*(10), 5456–5463.
43. Saha, A. K., & Sarker, P. K. (2020). Effect of sulphate exposure on mortar consisting of ferronickel slag aggregate and supplementary cementitious materials. *Journal of Building Engineering, 28*, 101012.
44. Kotwica, L., Pichor, W., Kapeluszna, E., & Rozycka, A. (2017). Utilization of waste expanded perlite as new effective supplementary cementitious material. *Journal of Cleaner Production, 140*, 1344–1352.
45. Ramezanianpour, A. A., Karein, S. M. M., Vosoughi, P., Pilvar, A., Isapour, S., & Moodi, F. (2014). Effects of calcined perlite powder as a SCM on the strength and permeability of concrete. *Construction and Building Materials, 66*, 222–228.
46. Sengul, O., Azizi, S., Karaosmanoglu, F., & Tasdemir, M. A. (2011). Effect of expanded perlite on the mechanical properties and thermal conductivity of lightweight concrete. *Energy and Buildings, 43*(2–3), 671–676.
47. Erdem, T. K., Meral, C., Tokyay, M., & Erdogan, T. Y. (2007).Use of perlite as a pozzolanic addition in producing blended cement. *Cement and Concrete Composites, 29*, 13–21.
48. Ramezanianpour, A. A., Mahmoud Motahari Karein, S., Vosoughi, P., Pilvar, A., Isapour, S., Faramarz Moodi Beddar, M., Meddah, A., Boubakria, M., & Haddad, N. (2014). A study of the effects of partial replacement of clinker by limestone in the cement manufacture. *Cement Wapno Beton*, 3/2104, 185–193.

49. Markiv, T., Sobolev, K., Franus, M., & Franus, W. (2016). Mechanical and durability properties of concretes incorporating natural zeolite. *Archives of Civil and Mechanical Engineering, 16*, 554–562.
50. Juenger, M. C. G., & Siddique, R. (2015). Recent advances in understanding the role of supplementary cementitious materials in concrete. *Cement and Concrete Research, 78*, 71–80.
51. Rozycka, A., & Pichor, W. (2016). Effect of perlite waste addition on the properties of autoclaved aerated concrete. *Construction and Building Materials, 120*, 65–71.

Degradation and Decoloration of RB5 Dye via UV Radiation Using Fe-TiO_2 Composite Photocatalyst in Fixed-Mode

Lavneet Kumar, Ina Thakur, Anoop Verma, B. S. Bhatia, and Charanjit Kaur Mangat

Abstract The present study demonstrates the utilization of innovative and economical Fe-TiO_2 composite beads for the degradation and color removal of RB5 dye. The composite beads were made up of waste foundry sand (FS), and fly ash (FA) mixed with clay from which, at acidic conditions, Fe was leaching out, leading to the photo-Fenton process. The beads were coated using the TiO_2 catalyst leading to the photocatalysis process. Thus the incorporation of in-situ hybrid processes of the photo-Fenton and the photocatalysis lead to synergistic degradation of the compound. The best-operating conditions for the degradation of RB5 in fixed-mode were attained by optimizing the different parameters such as oxidant dose (H_2O_2), number of beads, and UV intensity of the batch reactor. With all conditions optimized ($H_2O_2 = 300$ mg L^{-1}, UV intensity = 25 W m^{-2} and number of beads (% area of reactor covered) = 100%), approximately 91% degradation and 97% decoloration of RB5 was achieved after 45 min. The beads were recycled 100 times to evaluate the stability and durability of the catalyst.

Keywords RB5 · Degradation · Photo-Fenton · Photocatalysis · Synergy · Dual process

1 Introduction

The rigorous development in textile industries over the past decades has increased the use of potentially toxic chemicals such as dyes. This extensive use of dyes in the textile industries is because of its high stability to light, solubility, easy dyeing

L. Kumar (✉) · I. Thakur · A. Verma
School of Energy and Environment, Thapar Institute of Engineering and Technology, Patiala 147004, India

A. Verma
e-mail: anoop.kumar@thapar.edu

B. S. Bhatia · C. K. Mangat
RIMT University, Mandi Gobindgarh, Punjab, India

A. K. Gupta et al. (eds.), *Advances in Construction Materials and Sustainable Environment*, Lecture Notes in Civil Engineering 196,
https://doi.org/10.1007/978-981-16-6557-8_6

procedure, and color variety [1, 2]. Every year more than 70 × 10^4 tons of dye is produced throughout the world, and during manufacturing, approximately 10–15% of the dye is lost in wastewater [3, 4]. Releasing such toxic, mutagenic, and carcinogenic compounds into the water bodies affects human health as well as aquatic lives [5–7]. Therefore, the degradation of such dyes in the wastewater is the foremost concern to the environmental attenuation. In recent year literature, many different technologies such as microfiltration [8, 9], chemical treatment (coagulation/flocculation) [10, 11], and biological treatment (aerobic and anaerobic) [12, 13] have been demonstrated to treat the dyes present in water and wastewater to reduce its impact on the environment. Those as mentioned earlier, physical and chemical methods don't degrade the color but only reduce the concentration of the dye via converting one form to another [6]. Advanced oxidation processes (AOP) such as photo-Fenton [14–16], photocatalysis [17, 18], and electrodialysis [19, 20] have been extensively used and are elaborately stated in the literature for the removal of dyes. Such treatment technologies are also found to have proved their effectiveness over conventional methods. AOPs can be used solely as a single process or in combination with other conventional methods. It is clear from the literature review that, titanium oxide-based photocatalysis and photo-Fenton have attracted considerable attention in recent years [1, 7, 13, 15, 18]. The mechanism of these two studies (photo-Fenton and photocatalysis) is well established in literature where a series of reactions ultimately lead to the generation of reactive oxygen species like $HO^{\cdot}$, $O_2^{\cdot -}$, $HO_2^{\cdot}$, etc.

A lot of research has been done using AOPs regarding the degradation of dyes such as TiO_2 assisted photocatalytic and photo-Fenton processes in homogeneous and heterogeneous phases. These studies usually face problems like separation of catalyst, high recovery cost, mass transfer limitation, electron–hole recombination in photocatalysis, and problems like iron sludge formation in the photo-Fenton process. However, these drawbacks may be overcome by hybridizing the two processes (photo-Fenton and photocatalysis) together in a fixed bed reactor. This is a relatively new approach where the novel concept of the hybrid process (photocatalysis and photo-Fenton) is being incorporated in a single place, moreover in fixed form. Over and above the most important concept is the application of this novel technology in the degradation of dyes is not widely reported. In an overview, this technology has the potential to remove all the technical hitches for scaling up the treatment reactor. The purpose of the present study is to evaluate the feasibility of fixed bed dual process for dye degradation using a batch reactor. The main emphasis would be given to enhancing the kinetic rate constant alongside a decrease in degradation time. Moreover, the use of waste materials like FS, FA for making the composite structure makes the research more interesting from the social point of view.

2 Materials and Method

2.1 Chemicals

For performing the tests, RB5 dye was acquired from the local market in Patiala, Punjab, India. Sodium acetate ($C_2H_3NaO_2$) and acetic acid (CH_3COOH) were bought from the Tokyo chemical industry, India. 30% W/V H_2O_2 was obtained from Ranbaxy Ltd, Gurugram, India. All the chemicals were formed using double distilled water. FS and FA were obtained from the local industries nearby Patiala city, Punjab, India. P25 Degussa TiO_2 catalyst was procured from Evonik Industries, Mumbai, India.

2.2 Composite Beads Preparation and TiO_2 Coating

In this study foundry sand, fly ash, and clay (support material), were mixed and spherical beads were formed manually. A bead size of 1.5 cm was selected and made by measuring the diameter of the bead from at least four different directions using a screw gauge. Then the beads were dried at ambient temperature and after that calcined for 2 h at 800 °C. Afterward, the curing of the beads for 48 h was done that increases the solidity of the composite beads. The composite beads were then covered by TiO_2 was using the dip-coating process as described in the previous study [21]. At least two times the coating of TiO_2 was performed for the appropriate coverage of the bead surface. Then the beads were heated for 1.5 h at 350 °C in the muffle furnace to reach proper binding of the catalyst over the composite beads. After that, the coated composite beads were rinsed with double distilled water to separate the roughly attached catalyst and then dried in the oven.

2.3 Experimental Set-Up and Analysis

The RB5 dye degradation and decoloration experiments were conducted under a closed rectangular chamber having 7 UV tubes (36 W Philips, 365 nm) fixed under the roof of the chamber. Borosilicate glass reactors (0.06 m height and 0.21 m diameter) were used with a maximum working volume of 1000 mL. The glass reactor was supplied with 200 mL of 50 mg/L RB5 and covered with an appropriate number of Fe-TiO_2 composite beads. The pH of the reaction was set to 3.5 and H_2O_2 oxidant was added using Auto pipette into the reactor. After that, the reactor was kept on a jack to manage the height (to see the change in intensity) of the reactor under the chamber having UV lights. The UV intensity of the reactor was varied from 10 to 25 W m^{-2}. To maintain the homogeneity of the reaction, aeration (3 L min^{-1}) was provided using two spargers as shown in Fig. 1. After setting up the system the dual

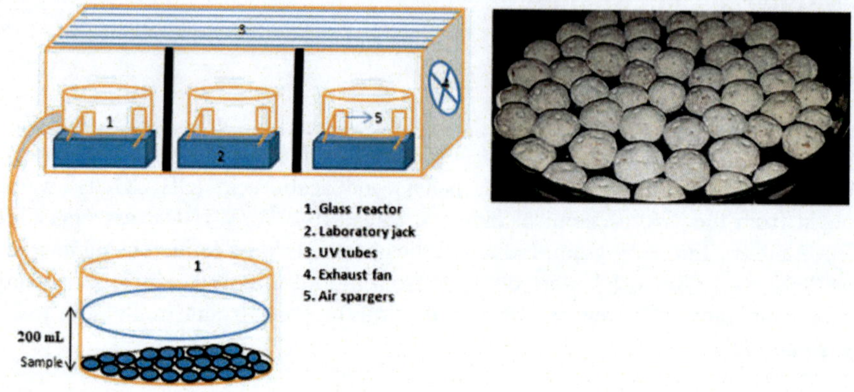

Fig. 1 Experimental set-up and coated beads

effect of the technique was studied by treating the RB5 dye. From the reactor, the samples were collected regularly. Different parameters were varied such as H_2O_2, the number of beads in terms of % area of reactor covered, and UV intensity to optimize the process. The samples obtained were analyzed using a UV–vis spectrophotometer (model no. T60U), LABINDIA at 597 nm. The tests were executed three times and the average of the results with error bars was shown in the figures.

3 Results and Discussion

3.1 Preliminary Experiments

In a batch reactor, the blank experiments were performed without TiO_2 coated beads under UV radiation (Photolytic studies) that revealed a minute extent of degradation (12%) and decoloration (9%) of RB5. Then in adsorption experiment with TiO_2 covered beads, only 8 and 4% removal of degradable compounds and color was observed, respectively. It is because of the formation of slime layer over the surface of the catalyst by the pollutant that obstructs the operating sites, on the account of which no additional degradation of RB5 occurred. The H_2O_2 only experiment showed 6% degradation and 8% decoloration due to very small production of OH radical. H_2O_2 with UV light showed 14% of degradation and 11% of decoloration of RB5 after 45 min of reaction time (Fig. 2). In photo-Fenton process, the addition of H_2O_2 (300 mg L^{-1}) under acidic conditions (pH less than 4) yielded 27% of RB5 degradation and 40% of RB5 decoloration in 45 min of reaction time. On the contrary, in photocatalysis process using TiO_2 coated beads 41% of RB5 degradation with 35% of decoloration after 45 min of reaction time. Whereas, the dual-effect of photo-Fenton and photocatalysis in fixed-mode accelerated the degradation efficiency up to 95% and 98% of decoloration of RB5 in 45 min of reaction.

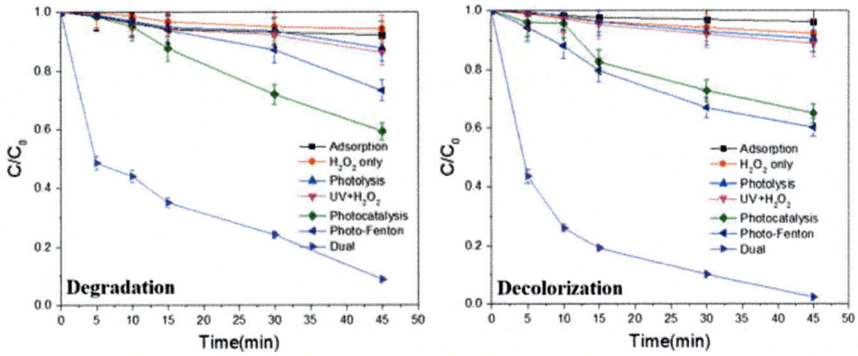

Fig. 2 Preliminary experiment under different conditions for RB5 degradation and decoloration

As expected the rate constant (first order) in case of dual technique (after optimization) is substantially high (degradation = 0.057 min^{-1} and decoloration = 0.089 min^{-1}) in comparison to photo-Fenton (degradation = 0.006 min^{-1} and decoloration = 0.012 min^{-1}) and to TiO_2 assisted photocatalysis (degradation = 0.011 $\text{min}^{-1}$and decoloration = 0.01 min^{-1}), which confirmed the viability of our technique (Fig. 2). Thus this dual technique exhibited the synergistic effect for the degradation and decoloration of RB5 in fixed bed reactor as shown in Eqs. (1–3):

The synergy of dual process on the photocatalytic process

$$\begin{aligned}
\%\,\text{synergy} &= 100 \times \left\{\left(k_{\text{dual}} - k_{\text{photocatalysis}}\right)\right\}/k_{\text{dual}}\\
\%\,\text{synergy} &= 100 \times \{(0.057 - 0.011)\}/0.057\\
\%\,\text{synergy} &= 80\%\ (\textbf{degradation of RB5})\\
\%\,\text{synergy} &= 100 \times \{(0.089 - 0.01)\}/0.089\\
\%\,\text{synergy} &= 88\%\ (\textbf{decoloration of RB5})
\end{aligned} \tag{1}$$

The synergistic effect of dual process over the photo-Fenton process

$$\begin{aligned}
\%\,\text{synergy} &= 100 \times \left\{\left(k_{\text{dual}} - k_{\text{photo-Fenton}}\right)\right\}/k_{\text{dual}}\\
\%\,\text{synergy} &= 100 \times \{(0.057 - 0.006\}/0.057\\
\%\,\text{synergy} &= 89\%\ (\textbf{degradation of RB5})\\
\%\,\text{synergy} &= 100 \times \{(0.089 - 0.012\}/0.089\\
\%\,\text{synergy} &= 86\%\ (\textbf{decoloration of RB5})
\end{aligned} \tag{2}$$

The synergistic effect of dual process on the photocatalytic and the photo-Fenton process together, i.e., Total synergy:

$$\begin{aligned}
\%\,\text{synergy} &= 100 \times \left\{(k_{\text{dual}} - \left(k_{\text{photocatalysis}} + k_{\text{photo-Fenton}}\right)\right\}/k_{\text{dual}}\\
\%\,\text{synergy} &= 100 \times \{(0.057 - (0.011 + 0.006)\}/0.057
\end{aligned}$$

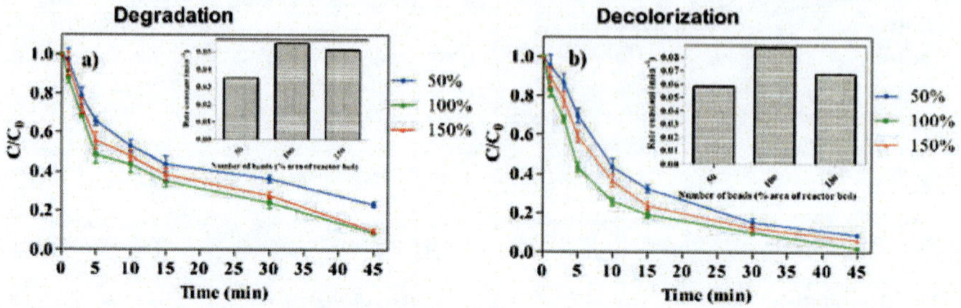

Fig. 3 Effect of variation of no. of beads on the degradation and decoloration of RB5

$$\begin{aligned}
\%\ \text{synergy} &= 59\%\ (\textbf{degradation of RB5}) \\
\%\ \text{synergy} &= 100 \times \{(0.089 - (0.01 + 0.012)\}/0.089 \\
\%\ \text{synergy} &= 75\%\ (\textbf{decoloration of RB5})
\end{aligned} \tag{3}$$

3.2 *Effect of Number of Beads*

For a particular reactor design, the coverage of reactors surface using the number of catalyst beads is one of the most essential factors for achieving the required degradation rate. In this study, the number of beads was varied and results were mentioned in the form of % area of reactor bed covered. The best degradation (91%) and decoloration (97%) of RB5 were attained with 100% area of reactor covered having all the sites active leading to efficient degradation of the pollutant. Subsequently, rate of degradation and decoloration also increased from 0.035 min^{-1} to 0.055 min^{-1} and from 0.058 min^{-1} to 0.086 min^{-1} respectively as shown in Fig. 3a, b. When % area of reactor covered reduced to 50%, degradation (76%) and decoloration (91%) efficiency also decreased as the number of available sites that are active over the catalyst gets decreased. On increasing the % area of the reactor to 150% having overlapping of beads covered there was a reduction in degradation (90%) and decoloration (93%) efficiency as shown in Fig. 3a, b. This reduction in efficiency is because of the fact that the scavenging of OH radicals predominates due to an increase in the concentration of both TiO_2 and iron that led to the increase in the production of OH radicals. Thus scavenging effect predominates in that case [22].

3.3 *UV Intensity Effect*

The utilization of solar radiation regarding technical feasibility and degradation is the main essential factor for the commercialized feature of this technology. As it is

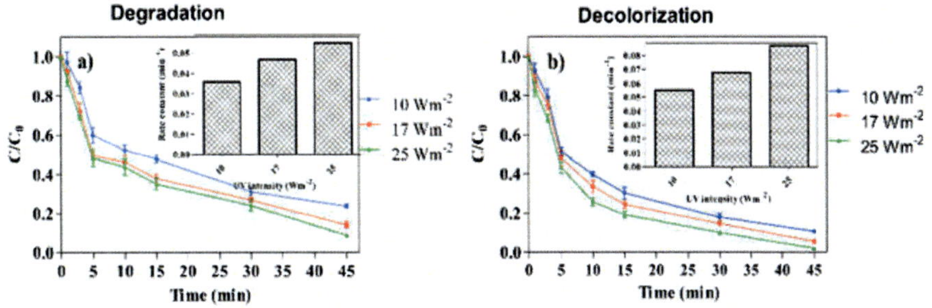

Fig. 4 Effect of variation of UV intensity on the degradation and decoloration of RB5

very difficult to get constant solar radiation reaching the earth for the entire time, the variation of UV irradiation in terms of intensity (10–25 W m^{-2}) at the lab scale was studied for degradation and decoloration of RB5.

It was observed that when UV intensity of 25 W m^{-2} was provided to the system for 45 min, 92% of degradation in RB5 and 97% of decoloration was found (Fig. 4a, b) because of the increment in OH radicals. The reason behind this might be the high UV intensity absorption by H_2O_2 which further enhances the OH radical production. The rate of degradation and decoloration increased from 0.036 min^{-1} to 0.055 min^{-1} and from 0.055 min^{1} to 0.087 min^{-1} respectively with an increase in the UV intensity. The results showed a reasonable decrease in RB5 concentration at different UV intensities.

3.4 Oxidant Concentration Effect

The current study shows that the oxidant doze is simultaneously consumed by both processes (photocatalysis and photo-Fenton), thus played a dual role in the system. The addition of H_2O_2 in the TiO_2-photocatalytic process prevents the electron–hole recombination by reacting with an electron present in the conduction band, as shown in Eq. (4). Hence, the optimization of the oxidant (H_2O_2) dose is a significant parameter for both the techniques taking place. This dual technique also increased the concentration of hydroxide radicals and oxidizing species in the system. In the current study, the optimum concentration of oxidant dose was obtained by varying its concentration from 150 mg L^{-1} to 900 mg L^{-1} with 200 mL of RB5 solution (50 ppm), and the optimum dose was found out to be 300 mg L^{-1} at pH 3.5. 95% of RB5 degradation was observed within 45 min of reaction time at an optimum dose of H_2O_2, as given in Fig. 5a. It can be observed from (inset Fig. 5a) that initially, the degradation rate constant of RB5 increased from 0.035 to 0.057 min^{-1} and then decreased afterward. The decoloration of RB5 was also studied, as shown in Fig. 5b and 98% in color reduction was obtained within 45 min of reaction time.

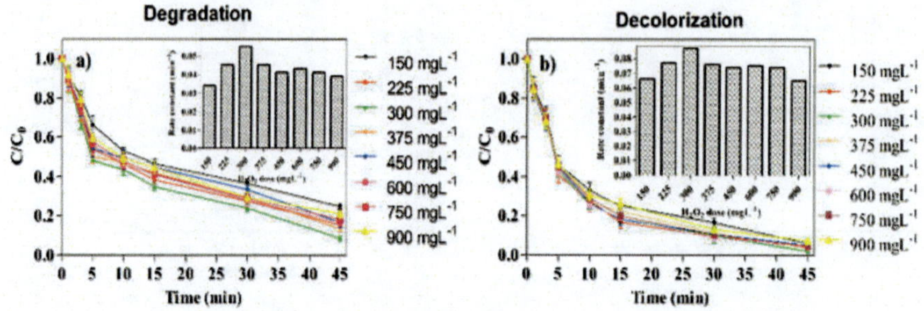

Fig. 5 Effect of variation of oxidant dose on the degradation and decoloration of RB5

The maximum decoloration rate constant was 0.089 min^{-1} and then decreased, as shown in (inset Fig. 5b). This decrease in the degradation and decoloration rate might be the H_2O_2 scavenging effect. Because when we add a higher concentration of H_2O_2, it started using OH radical converting into water, as given in Eqs. (5) and (6)

$$TiO_2(e-) + H_2O_2 \rightarrow TiO_2 + OH^* + OH^- \quad (4)$$

$$H_2O_2 + OH^* \rightarrow H_2O + HO_2^* \quad (5)$$

$$HO_2 + OH^* \rightarrow H_2O + O_2 \quad (6)$$

3.5 *Durability Study*

Another essential factor for deciding the economic viability of this technique is the durability of the support material. It can be seen in Fig. 6a that the Fe-TiO_2 composite beads were reused 100 times to analyze its durability. Only 19% and 10% reduction in the beads activity for RB5 degradation and decoloration was observed, respectively. This slight loss in degradation and decoloration of RB5 is because of the scrapping of the coated catalyst from the beads. The occurrence of TiO_2-photocatalysis in the system was confirmed by different characterization (XRD analysis, SEM–EDS) reported in [21] different studies of ours.

Further, the concentration of iron leaching from the beads in the solution was estimated before and after 100 cycles to confirm the presence of the photo-Fenton process during the treatment of RB5 as shown in Fig. 6b. These findings show the substantial chances of this technique for field-scale trials.

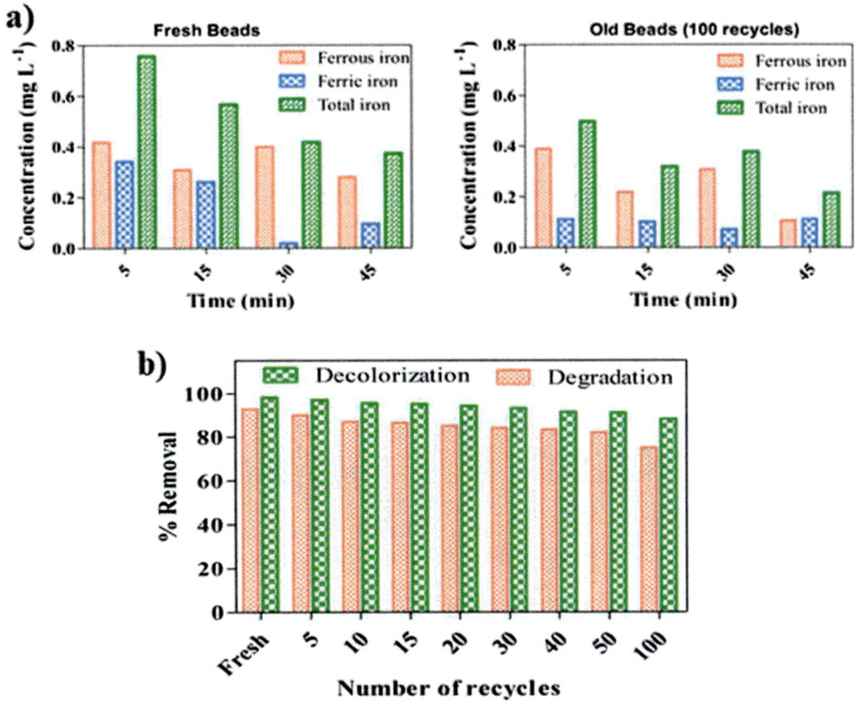

Fig. 6 **a** Concentration of iron in fresh and old beads. **b** Durability study for 100 recycles

4 Conclusion

The present work demonstrates the efficiency and durability of the beads composed of iron and TiO_2 for the removal of RB5 from the distilled water under acidic conditions under UV irradiation. RB5 dye degradation and decoloration levels of 91 and 97% were achieved with an oxidant dose of 300 mg L^1, 25 W/m^{-2} of UV irradiation, and 45 min of treatment time. From the results, it was found that the aforementioned operating parameters showed their potential in the degradation and decoloration of RB5 dye. The significant increment in the rate constant in the case of dual process confirms the viability of the dual technique. Therefore, this technique in fixed-mode has proven to be a reasonable solution for the effective treatment of contaminated water. The beads were reused 100 times and found only 19 and 10% reduction in the degradation and decoloration efficiency.

References

1. Soltani, T., & Entezari, M. H. (2013). Solar photocatalytic degradation of RB5 by ferrite bismuth nanoparticles synthesized via ultrasound. *Ultrasonics Sonochemistry, 20*(5), 1245–1253. https://doi.org/10.1016/j.ultsonch.2013.01.012
2. Mahmoodi, N. M., & Arami, M. (2006). Bulk phase degradation of Acid Red 14 by nanophotocatalysis using immobilized titanium (IV) oxide nanoparticles. *Journal of Photochemistry and Photobiology, A: Chemistry, 182*, 60–66. https://doi.org/10.1016/j.jphotochem.2006.01.014
3. Natarajan, T. S., Natarajan, K., Bajaj, H. C., & Tayade, R. J. (2013). Enhanced photocatalytic activity of bismuth-doped TiO_2 nanotubes under direct sunlight irradiation for degradation of Rhodamine B dye. *Journal of Nanoparticle Research, 15*. https://doi.org/10.1007/s11051-013-1669-3
4. Idriss, H., Ajmala, A., Majeedb, I., & Malika, R. N. (2014). Principles and mechanisms of photocatalytic dye degradation on TiO_2 based photocatalysts: A comparative overview. *Rsc Advances, 4*(70), 37003–37026. https://doi.org/10.1039/b000000x
5. Mahmood, S., Khalid, A., Arshad, M., Mahmood, T., & Crowley, D. E. (2016). Detoxification of azo dyes by bacterial oxidoreductase enzymes. *Critical Reviews in Biotechnology, 36*, 639–651. https://doi.org/10.3109/07388551.2015.1004518
6. Natarajan, S., Bajaj, H. C., & Tayade, R. J. (2018). Recent advances based on the synergetic effect of adsorption for removal of dyes from waste water using photocatalytic process. *Journal of Environmental Sciences (China), 65*, 201–222. https://doi.org/10.1016/j.jes.2017.03.011
7. Lucas, M. S., & Peres, J. A. (2006). Decoloration of the azo dye reactive black 5 by Fenton and photo-Fenton oxidation. *Dyes and Pigments, 71*, 236–244. https://doi.org/10.1016/j.dyepig.2005.07.007
8. Kertèsz, S., & Jiránková, H. (2014). Submerged hollow fiber micro filtration as a part of hybrid photocatalytic process for dye wastewater treatment. *Desalination, 343*, 106–112. https://doi.org/10.1016/j.desal.2013.11.013
9. Juang, Y., Nurhayati, E., Huang, C., Pan, J. R., & Huang, S. (2013). A hybrid electrochemical advanced oxidation/microfiltration system using BDD/Ti anode for acid yellow 36 dye wastewater treatment. *Separation and Purification Technology, 120*, 289–295. https://doi.org/10.1016/j.seppur.2013.09.042
10. Guibal, E., & Roussy, J. (2007). Coagulation and flocculation of dye-containing solutions using a biopolymer (Chitosan). *Reactive and Functional Polymers, 67*, 33–42. https://doi.org/10.1016/j.reactfunctpolym.2006.08.008
11. Lau, Y., Wong, Y., Teng, T., Morad, N., & Rafatullah, M. (2014). Coagulation-flocculation of azo dye Acid Orange 7 with green refined laterite soil. *Chemical Engineering Journal, 246*, 383–390. https://doi.org/10.1016/j.cej.2014.02.100
12. Quezada, M., & Moreno, G. (2004). Aerobic degradation of the azo dye acid red 151 in a sequencing batch biofilter. *Bioresource Technology, 92*, 143–149. https://doi.org/10.1016/j.biortech.2003.09.001
13. O'Neill, C., Lopez, A., Esteves, S., Hawkes, F. R., Hawkes, D. L., & Wilcox, S. (2000). Azo-dye degradation in an anaerobic-aerobic treatment system operating on simulated textile effluent. *Applied Microbiology and Biotechnology, 53*(2), 249–254.
14. Velmurugan, S., Ganesh, B., Babuponnusami, A., & Rajasekaran, R. (2016). Decolourisation of reactive blue 28 from dye waste water by photo fenton process and sono fenton processes. *International Journal of Chemical Sciences, 14*, 1433–1446.
15. Severo, E. D. C., Anchieta, C. G., Foletto, V. S., Kuhn, R. C., Collazzo, G. C., Mazutti, M. A., & Foletto, E. L. (2016). Degradation of Amaranth azo dye in water by heterogeneous photo-Fenton process using $FeWO_4$ catalyst prepared by microwave irradiation. *Water Science and Technology, 73*, 88–94. https://doi.org/10.2166/wst.2015.469
16. Tarkwa, J. B., Oturan, N., Acayanka, E., Laminsi, S., & Oturan, M. A. (2019). Photo-Fenton oxidation of Orange G azo dye: Process optimization and mineralization mechanism. *Environmental Chemistry Letters, 17*, 473–479. https://doi.org/10.1007/s10311-018-0773-0

17. H. Sudrajat, B. Sandhya, S. Hiroshi, T. Satoshi, Rapid enhaced photocatalytic degradation of dyes using novel N-doped ZrO_2. *Journal of Environmental Management, 165*, 224–234. https://doi.org/10.1016/j.jenvman.2015.09.036
18. Akpan, U. G., & Hameed, B. H. (2009). Parameters affecting the photocatalytic degradation of dyes using TiO_2-based photocatalysts: A review. *Journal of Hazardous Materials, 170*, 520–529. https://doi.org/10.1016/j.jhazmat.2009.05.039
19. Sindelar, F. W., Silva, L. F. O., Machado, V. R., dos Santos, L. C. M., & Stülp, S. (2015). Treatment of effluent from the agate dyeing industry using photodegradation and electrodialysis processes. *Separation Science and Technology, 50*, 142–147. https://doi.org/10.1080/01496395.2014.947519
20. Li, X., Jin, X., Zhao, N., Angelidaki, I., & Zhang, Y. (2017). Novel bio-electro-Fenton technology for azo dye wastewater treatment using microbial reverse-electrodialysis electrolysis cell. *Bioresource Technology, 228*, 322–329. https://doi.org/10.1016/j.biortech.2016.12.114
21. Thakur, I., Örmeci, B., & Verma, A. (2020). Inactivation of E. coli in water employing Fe-TiO_2 composite incorporating in-situ dual process of photocatalysis and photo-Fenton in fixed-mode. *Journal of Water Process Engineering, 33*, 101085. https://doi.org/10.1016/j.jwpe.2019.101085
22. Bansal, P., & Verma, A. (2017). Synergistic effect of dual process (photocatalysis and photo-Fenton) for the degradation of Cephalexin using TiO_2 immobilized novel clay beads with waste fly ash/foundry sand. *Journal of Photochemistry and Photobiology, A: Chemistry, 342*, 131–142. https://doi.org/10.1016/j.jphotochem.2017.04.010

Study on Concrete Developed with Recycled Fine Aggregate

Nancy Soni and Dharmendra Kumar Shukla

Abstract Limited resources and exponentially growing demand of natural fine aggregate (NFA) in construction industry is creating environmental anarchy. Concurrently, construction and demolition waste (CD-W) that has potential of providing aggregates for construction activities is abundantly consuming valuable land spaces for years. Thus, current study aims to investigate feasibility of recycled fine aggregates (RFA) obtained from CD-W as NFA saver in concrete. Conventional drawbacks of RFA that limited its utilization in concrete are resolved in the study by adopting dissimilar RFA procurement by crushing recycled coarse aggregates (RCA). Different percentages of NFA were replaced with RFA in the concrete with identical proportions of water–cement ratio, coarse aggregate, fine aggregate, super-plasticizer and silica fume to observe effects of replacement on slump value, compressive strength, water absorption and bulk density. Workability was found to be reduced at higher NFA replacements with RFA due to its higher water absorption characteristics. To counter the raised water requirement, copper slag (CS) which is known as low water absorbing industrial waste was also used as a replacement of NFA in combination with RFA. Concrete with increasing replacements have presented superior compressive strength, which is also justified by XRD and SEM analysis that showed relatively more C–S–H gel formation in concrete with RFA. Sustainability analysis of concrete with RFA exhibited lower embodied energy (EE) and CO_2 equivalent along with superior mechanical strength.

Keywords Sustainable concrete · Micro-structure · Energy conservation · Recycling and reuse of material

N. Soni (✉) · D. K. Shukla
Department of Civil Engineering, Jaypee University of Engineering and Technology, Guna, India

D. K. Shukla
e-mail: dharmendra.shukla@juet.ac.in

A. K. Gupta et al. (eds.), *Advances in Construction Materials and Sustainable Environment*, Lecture Notes in Civil Engineering 196,
https://doi.org/10.1007/978-981-16-6557-8_7

1 Introduction

Expanding infrastructural development led to higher demand of concrete and thus to its ingredients. Aggregates in total (fine and coarse) contributes to 60–70% of concrete volume, out of which 35–45% is fine aggregate [1]. Aggregates are inextricable element of concrete and, therefore, cause prominent impact on fresh and hardened properties of concrete. Natural aggregates, gravels and sand are the largely used raw material on earth after water [2]. India has known to be consuming 750 million tons of natural sand in 2018 [3]. Desert seems endless source of sand but is not appropriate to be used in construction due to its deficiency in silicon dioxide, and smooth and fine texture [4]. Hence, river sand being economical locally available best-suited source is exploited for construction activities [4]. Natural process of sand production through weathering in rivers takes millions of years, and its utilization is far rapid than its rate of replenishment. This vast difference between rate of production and consumption is rooting many environmental imbalances [5]. Aftereffects of sand mining from rivers are serious threats to physical and coastal ecosystem. Rigorous refining of sand extraction from rivers containing organic impurities, and its transportation disrupts the environment [5, 6]. Sand is not just a valuable mineral used in concrete, mortar and brick making, but also acts as barrier to strong tidal waves and storms and supports habitat to arthropods [7]. Therefore, sustainable sand substitutes are to be induced and optimized in construction industry to debar the environmental concerns created by excessive sand mining and restore better Earth to live.

Concept of sustainable development was first coined in 1992 in Earth Summit in Rio de Janeiro city, and construction industries being measurelessly vast are now focusing on the prescribed guidelines to develop sustainably [8]. Also, several countries have accepted supplementing natural aggregates with technically feasible, economical and eco-friendly materials. On the other hand, structural deterioration, city rearrangement, traffic expansion and natural calamities produce large quantity of construction and demolition waste (CD-W) [9]. Closest approximation of CD-W in India is estimated to be 100 million tons [10]. Usually, CD-W is left unattended in landfill, creating significant environmental issues and health hazard [11]. Additionally, it is evident that use of re-processed CD-W trashes in concrete production is a judiciously feasible, economic and environmental-friendly alternative of natural fine aggregates [12]. Concept of utilizing recycled CD-W in concrete production was scientifically discovered by the countries suffered in World War II [13], and the first paper stating the feasibility of using recycled fine aggregates (RFA) in concrete was published by Nixon in 1978 [14], which concluded that compressive strength of concrete containing fine recycled aggregate does not reduce as much as its workability. Studies have justified the utilization of CD-W in production of concrete [11, 15]. Hansen in 1986 [16] proposed the use of finer fraction of recycled fines (<2 mm) as binding alternative in concrete as soil stabilizer and as filtration medium in water treating plants.

Usually, fine share of recycled CD-W is considered as useless secondary product due to its higher contamination. But, it is proven that composition and properties of RFA highly depend on its origin and comminuting process [17]. RFA commonly have less particle density and loose bulk density due to presence of mortar or concrete [13]. RFA contains approximately 20% of older hydrated cement which may be responsible for raised water absorption percentage [16]. Despite all the inadequacy of RFA, hindering its utilization in concrete, if designed properly, can provide similar or even higher mechanical strengths as compared to concrete with natural fine aggregate (NFA) [18].

Innovative combinations of non-conventional waste material have to be involved in construction industry to promote sustainable development, hence to conserve environment by compensating deficiency of natural resources. Thereby, a viable solution of replacing NFA with RFA in concrete is proposed and investigated in the present paper so that the required characteristic strength can be achieved without losing the desired slump.

This study focuses on the effects of replacing full/partial NFA with RFA on fresh and hardened concrete. Concrete with OPC-53 as binder and silica fume as mineral filler is designed for targeted mean strength of 40 MPa. The effective water binder ratio of 0.4 and dosage of superplasticizer, i.e., 2% of cement, are kept same throughout the experiments. Higher dosage of superplasticizer is selected to ensure compensation of expected increase of water requirements due to RFA addition. Superplasticizer (SP) with polycarboxylate ether base is used as high-range water reducer in the concrete mixes. The free water content in the concrete may reduce with the introduction of RFA and may influence in its microstructure causing hike in water absorption and thus lower workability [19]. Therefore, a water demand compensating material, copper slag (CS), is added in combination with RFA against NFA replacement. It has higher specific gravity than sand, therefore, results in higher density concrete which have lower water absorption percentages [20]. Production of copper produces 2–3 times CS as a by-product than the copper [21] and copper slag with less than 0.8% of copper is abandoned as waste and sold at cheap prices [22]. Therefore, recycling them to produce some worthy product is a necessity of the time. CS with specific properties has been a promising alternative of fine aggregate in concrete [23, 24]. Hence, a combination of RFA and CS may be successfully utilized for 100% NFA replacement.

A case study by Hammond et al. 2008 [25] showed that concrete contributes around 35% EE and 55% CO_2eq among all major building material, respectively. Although operational energy (OE) of the building is an important integral part of building energy evaluation, it can be controlled by efficient designing. Building regulations of current time are such that they make efforts to reduce OE for sustainability; and as the OE reduces, EE and CO_2eq gain a significant importance for evaluation of lifetime carbon footprint of building [26]. OE of domestic building are 1/10th time of its EE, whereas for commercial building, the OE can be as low as 1/30th time of EE [25]. Thus, the energy used in construction is of great importance that adds up to the assortment of building material, and EE and Co_2eq of material are the important parameters for future saving decision making while construction planning [27].

Usually, M-40 grade of concrete is used in mass concreting and thus has vast scope of RFA utilization. Therefore, concrete mix of M-40 grade is altered for its NFA content with different RFA percentage. Reference concrete mix, referred to as R_0 has 100% NFA. Concrete mixes are named after their replacement percentage with RFA as R_{20}, R_{40}, R_{60}, R_{80} and R_{100} for 20, 40, 60, 80 and 100% replacement. Water-binder ratio and SP dosage are maintained constant in all the mixes to observe true deviations in results of workability, water absorption, bulk density and compressive strength. The concretes mixes with drier consistency are supported with 5, 10, 15 and 20% of copper slag (CS) and indicated as C_5, C_{10}, C_{15} and C_{20}. Concrete mixes of present study are also evaluated on the basis of their cost, EE and Co_2eq for their economic and sustainable viability.

2 Experimental Program

2.1 Production of Fine Recycled Aggregate

Recycled fine aggregates (RFA) used in research are obtained by double-stage crushing of CD-W collected from a single source constructed in 2002 with M-30 grade of concrete. Initially sorted demolished concrete from whole CD-W is grinded in jaw crusher followed by reduction in horizontal-axis impact crusher. Impact crushing being most efficient in aggregate size reduction, also is likely to produces grains with most finished surfaces [17]. Crushing process imparts angularity to the grains with rough surface texture, increasing its specific area and hence water absorption [28]. Also, it is observed that the attrition of particles in impact crusher tends to remove adhered mortar from the surface which consequently reduces the water absorption capacity of recycled sand [29]. Impact crusher has high reduction ratio and capabilities of crushing hard materials that produces fine shaped cubical particles with no cracks [30]. A schematic diagram showing all the important CD-W crushing steps is shown in Fig. 1.

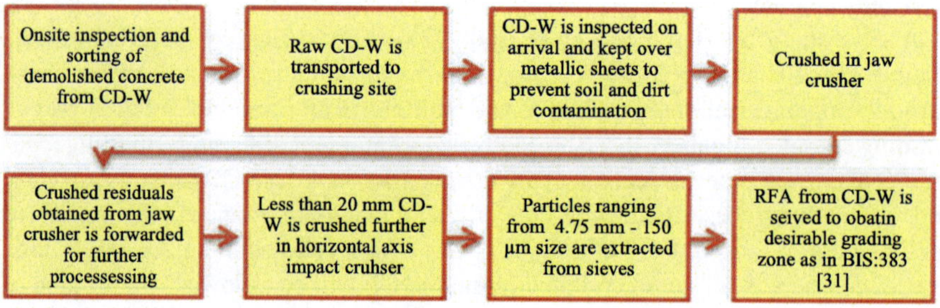

Fig. 1 Process of recycled fine aggregate procurement

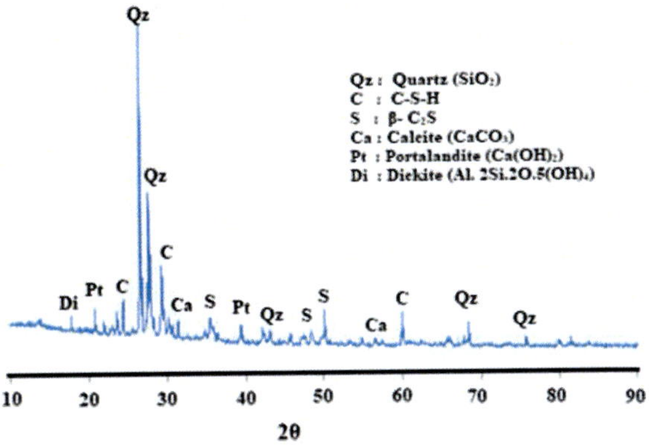

Fig. 2 P-XRD of RFA obtained from CD-W

Powder X-ray diffraction (PXRD) of RFA is performed to obtain major peaks for distinguishing mineralogical compositions, as shown in Fig. 2.

Peaks at dickite show the presence of crushed natural coarse aggregate, primarily involved from CD-W. The major peaks at C–S–H and portlandite, and minor peaks at β-C_2S indicates unhydrated cement grains that may provide extra hydration products on hydration [32].

Particle size distribution is a major controlling factor governing suitability of RFA in concrete [33]. "Sieve analysis method" as detailed in BIS: 2386-1 [34] was used for granulometric analysis of NFA and RFA. NFA is found to be in grading zone-II as specified in BIS: 383 [31]. Graphical representation of particle size distribution of NFA and RFA used in the current study is shown in Fig. 3.

After sieving RFA, all the particles of different sizes are kept separately and mixed again to create same grading curve as that of NFA, which may ensure same fineness modulus, uniformity during replacement, thus preventing effects that may arise due to this parameter.

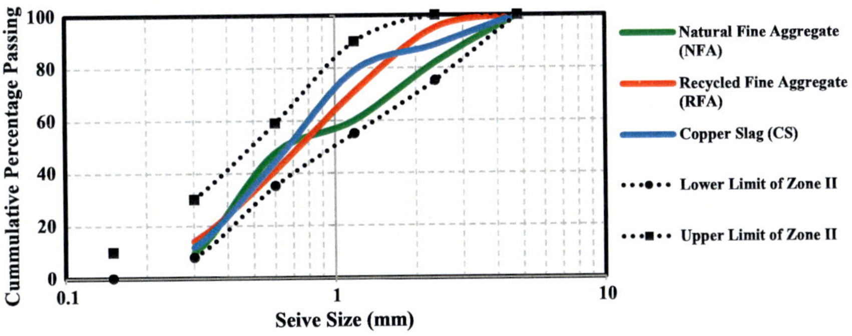

Fig. 3 Particle size distribution curve of fine aggregates

3 Material Used

In the research, following materials were used, like cement, fine aggregate, coarse aggregate, admixture, filler and water.

3.1 Cement, Water, Filler and Super-Plasticizer

The binder used in the experiment is OPC-53 conforming to BIS: 12269 [35]. The physical characteristics are tested following BIS: 4031-6 [36] and chemical testing is performed as per BIS: 4032 [37]. Tap water from laboratory at normal room temperature, free form visible contaminants and pH ranging from 6.8 to 8.0 is taken for concrete mixing. Silica fume conforming to BIS: 15388 [38] is used as the fine pozzolanic filler material. The physical and chemical properties of cement and silica fumes used for the experiments are shown in Table 1.

The concrete with higher workability is prepared using high range water reducing poly-carboxylate ether-based ultraplasticizer, Addage Plast PCE 850 manufactured by Sakshichem Sciences Pvt. Ltd. Addage Plast PCE 850 possesses properties for high water reduction in concrete with improved workability and increase in strength by almost 20–30%. It is a clear white liquid with relative density 1.06 ± 0.02 at 25 °C, chloride ion content of $<0.2\%$ and $pH \geq 6$.

3.2 Coarse Aggregate, Natural Fine Aggregate, Copper Slag and Recycled Fine Aggregate

Crushed granite aggregates of maximum nominal size 20 mm is taken as natural coarse aggregate (NCA) for the experiments. Natural river sand is taken as natural fine aggregate and specification of all the natural aggregates are confirming to BIS: 383 [31]. Copper slag is used in concrete with RFA to increase free water content and hence to increase workability that may get lowered due to higher percentage of water absorption of RFA. Specific gravity and water absorption of all the coarse and fine aggregates are determined as per BIS: 2386-3 [39]. The physical properties of natural coarse aggregate (NCA), natural fine aggregate (NFA), recycled fine aggregate (RFA) and copper slag (CS) are listed in Table 2.

Table 1 Physical and chemical characteristics of cement and silica fume

Properties	Cement	Silica fume
Fineness (m^2/kg)	288 (Blaine's)	20,000 (Nitrogen adsorption)
Specific gravity	3.20	2.26
Compressive strength		
3 days (MPa)	43.2	–
7 days (MPa)	55.3	–
28 days (MPa)	67.2	–
Initial setting time (min)	118	–
Final setting time (min)	192	–
Soundness (mm)	1	–
Silica (SiO_2) (%)	20.89	89.67
Alumina (Al_2O_3) (%)	6.10	1.14
Iron oxide (Fe_2O_3) (%)	3.75	0.81
Calcium oxide (CaO) (%)	63.91	2.34
Magnesium oxide (MgO) (%)	2.50	0.69
Loss of ignition (%)	1.70	1.71
Total sulfur (%)	1.74	0.58
Insoluble residue (%)	2.78	3.01
Chloride (%)	0.015	0.01

Table 2 Properties of aggregates

Properties	NCA	NFA	RFA	CS
Color	Dark gray	Whitish	Grayish	Glassy black
Water absorption (%)	0.5	0.90	2.6	0.3
Specific gravity	2.64	2.62	2.70	3.36
Particle shape	Angular	Sub-angular	Angular	Asymmetrical
SiO_2 (%)	–	99.90	95	26
Zone	–	II	II	II
Fineness modulus	6.46	2.76	2.75	2.77
Los angles abrasion value (%)	14.73	–	–	–
Aggregate impact value (%)	13.52	–	–	–
Aggregate crushing value (%)	28.0	–	–	–
Bulk density (kg/m^3)	1611	1541	1596	2082

4 Mix Proportioning

The concrete mix design is prepared by referring the guidelines specified in BIS: 10262 [40] for "concrete mix proportioning." The designed concrete mix proportions and quantities for the concrete with 100% natural sand are given in Table 3.

4.1 *Casting, Curing and Testing of Specimens*

All the proportioned material used for concrete production is batched manually using a weighing machine with precision of 10^{-3} kg and mixed in a semi-automatic pan mixer with hydraulically functioning actuator as prescribed in BIS: 4925 [41]. Freshly prepared concrete mix is subjected to slump cone test of workability as prescribed in BIS: 1199-3 [42].

Sánchez-Roldán et al. [43] experimentally observed that the water absorption of hardened concrete reduces if aggregates are pre-soaked with the part of water from the effective water required for mixing concrete. Therefore, concept of pre-moistening the aggregates while mixing concrete is adopted due to the fact that RFA has water absorbing properties and may absorb water required for hydration of cement, thus rendering the concrete non-uniformly hydrated leading to low characteristic strength.

Six specimens of each concrete mix design was casted in 150 mm size cubical molds as specified in BIS: 516 [44] for testing hardened concrete properties. All the specimens were demoded after 24 h of drying in humid surrounding and then kept immersed in water tank with controlled temperature of 20 ± 2 °C for curing up to the age of testing.

4.2 *Tests on Concrete*

In the current paper, the influence of varying percentage of NFA replacements with RFA and CS on slump flow of freshly mixed concrete and on compressive

Table 3 Mix proportions for 1 m^3 concrete

Material	Cement (OPC-53)	Water	Coarse aggregate (NCA)	Fine aggregate	Super-plasticizer (SP)	Silica fumes (SF)
Quantity (kg/m^3)	450	200	864	662	9	45
Percentage (%)	20	9	39	30	0.4	2

strength, water absorption and bulk density of hardened concrete were experimentally analyzed. The impacts of crushed RFA on the various characteristics of concrete are reported. The slump value determination test is performed in accordance with BIS: 1199-3 [42].

Water absorption in concrete by immersion and bulk density is important parameters that govern the durability offered by concrete [33]. The procedure adopted for determination of water absorption and bulk density on 28th day is as specified in BIS: 2386-3 [39].

5 Results and Discussion

Results of slump value, compressive strength, water absorption percentage and bulk density of concrete with various percentages of NFA replaced with RFA and CS are shown in Table 4.

Mixes with 80% and 100% replacement of NFA with RFA were discarded as dry consistency of concrete was obtained. The highest slump of 150 mm was recorded 100% NFA concrete mix. It is evident from Table 4 that the slump goes on reducing as the percentage of recycled sand in concrete is increased and the reduced slump value recovers with incremental addition of CS. The percentage of CS in concrete is limited to 20% because the primary aim of the research is to maximize use of recycled fine aggregate in concrete and generating retrenchment opportunities for natural sand.

The minimum permissible slump required for concrete is 25 mm as given in BIS-456 [45]. However, this slump is considered as low and has its successful applications in mass concreting, lightly reinforced structural concrete sections, canal lining, cement concrete pavements, decks of bridges etc. with proper controls. RFA causes slump declination, thus to prepare concrete with higher required slumps either additional water or some water saving processes has to be adopted. Here, in this study, CS is the low water requiring fine aggregate which is responsible for possible substitution of 100% NFA with desired workability.

Figure 4 shows the graphical representation of 7th day and 28th day compressive strength on primary vertical axis against various mixes with variable fine aggregate content on horizontal axis, along with the slump value plotted on secondary vertical axis. The dotted lines denote the outcomes of reference concrete mix, R_0, which has NFA as its fine aggregate content. Percentage values given over the bars in Fig. 4 are indicating the increase in compressive strength as compared to the reference concrete mix, R_0.

It is evident from Fig. 4 that CS increases the workability with increased RFA contents and also significantly contributing to the compressive strength. The increased slump of concrete with CS addition is ensuring homogeneity of mix and enabling proper compaction during casting, thereby increasing density and strength, and reducing water absorption percentages. However, strength enhancement due to

Table 4 Test results

Mix	Slump (mm)	Compressive strength (MPa)		Water absorption (%)	Bulk density (kg/m^3)
		7th day	28th day		
Variable NFA% and RFA%					
R_0	150	35.6	49.8	1.41	2307
R_{20}	135	36.1	50.4	1.36	2242
R_{40}	103	41.5	53.7	1.38	2273
R_{60}	70	42.4	54.7	1.3	2293
$^aR_{80}$	–	–	–	–	–
$^aR_{100}$	–	–	–	–	–
40% NFA, variable RFA and CS					
$R_{55}C_5$	104	38.8	59.6	0.38	2315
$R_{50}C_{10}$	117	39.9	59.4	0.375	2306
$R_{45}C_{15}$	133	40.8	58.4	0.368	2324
$R_{40}C_{20}$	140	40.8	57.5	0.35	2328
20% NFA, variable RFA and CS					
$R_{75}C_5$	62	36.9	63.8	0.33	2315
$R_{70}C_{10}$	70	37.6	62.8	0.35	2324
$R_{65}C_{15}$	88	38	62.5	0.35	2322
$R_{60}C_{20}$	103	38.9	61.2	0.39	2317
0% NFA, variable RFA and CS					
$R_{95}C_5$	41	36.4	62.3	0.38	2318
$R_{90}C_{10}$	50	37.3	61.6	0.49	2315
$R_{85}C_{15}$	58	38.3	60.6	0.53	2312
$R_{80}C_{20}$	66	38.1	60.1	0.48	2317

[a]Mix was so dry that concrete cannot be casted

CS is gradually deteriorating as properties of CS is comparable to that of NFA and increased NFA will reduce strength, unlike RFA.

Despite highest workability of controlled concrete mix with all natural ingredients (R_0), its compressive strength is found to be least among all the mixes with varying percentages of RFA, or in combination with RFA and CS. Water absorbing coarse aggregates are porous and hence considered undesirable to be used for strengthening of concrete. RFA being tiny particles are not adversely affecting strength of concrete and also its water absorption characteristic is making water available for better cement hydration and thus increasing early strength of concrete. Maximum 19.27% of increase in 7th day compressive strength is attained at $R_{60}C_{20}$ concrete mix which justifies early strength gain, suggesting early shuttering removal and thereby providing economy and speed to construction.

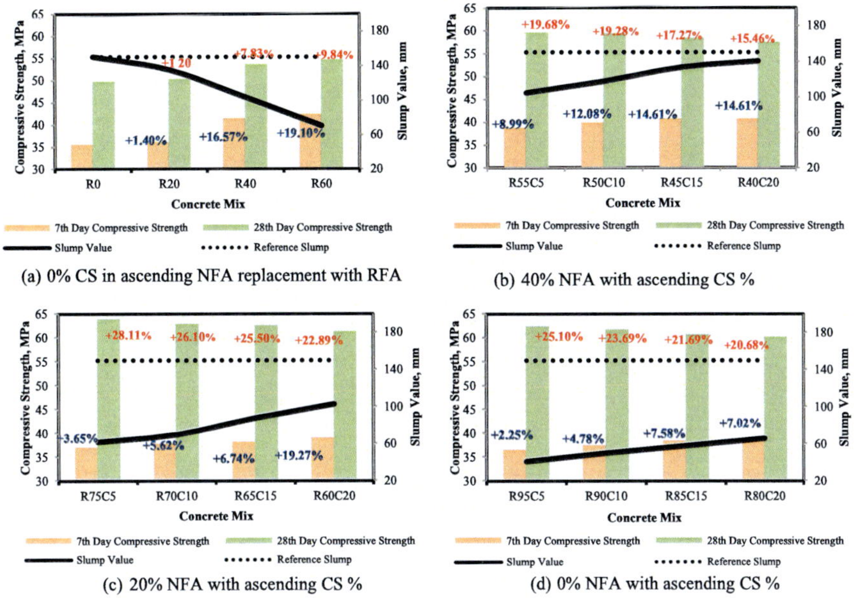

(a) 0% CS in ascending NFA replacement with RFA

(b) 40% NFA with ascending CS %

(c) 20% NFA with ascending CS %

(d) 0% NFA with ascending CS %

Fig. 4 Graphical representation of 7th day and 28th day compressive strength in MPa, and slump value in mm for various fine aggregate contents and combinations

$R_{75}C_5$ saved 80% NFA and registered highest 28th day compressive strength which is 28.11% more than that of reference mix (R_0). However, 100% NFA can also be saved with significant hike in compressive strength with proper control and vibrations for superior compaction. Cutting back 100% of natural sand with recycled sand without strength compromise is an incomparable achievement that will open many corridors toward sustainable development. The rough surface texture and angular shape of RFA formed by attrition crushing provides extra bonding surfaces for binder adheration in concrete and better aptness in resisting compressive loads which increased compressive strength of concrete.

A negative relationship is obtained between water absorption and compressive strength of hardened concretes with variable fine aggregate compositions as shown in Fig. 5. Absorption of water by RFA controls the heat of hydration of cement, thereby reducing shrinkage micro-cracks and resulted in durable and low water absorbing concrete. Regardless of higher water absorption of RFA than NFA, concrete made with RFA is having less water absorption percentage as compared to the concrete with 100% NFA. Therefore, it can be concluded that the concrete of improved quality, i.e., lesser water absorption and higher compressive strength is obtained with RFA as fine aggregate.

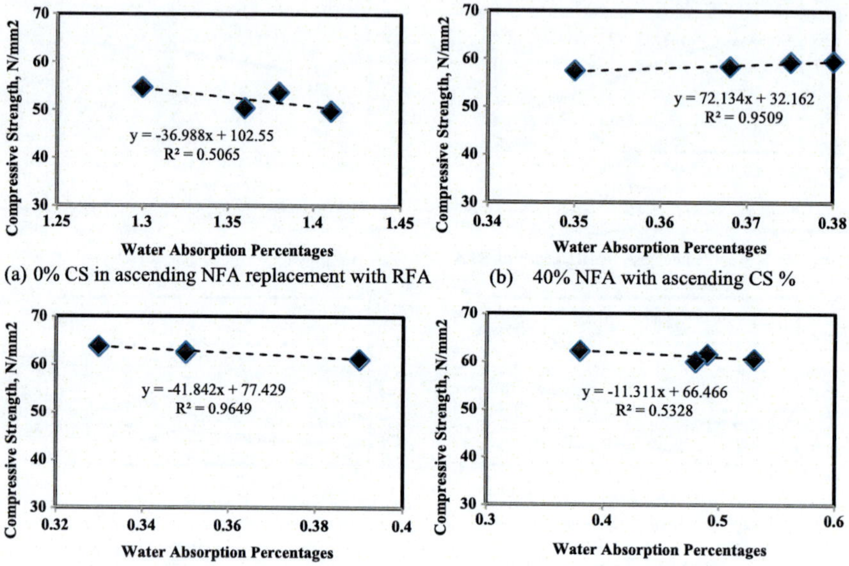

(a) 0% CS in ascending NFA replacement with RFA (b) 40% NFA with ascending CS %
(c) 20% NFA with ascending CS % (d) 0% NFA with ascending CS %

Fig. 5 Graphical representation of relation between water absorption percentage and compressive strength on 28th day for various hardened concrete mixes

5.1 Micro-structure Analysis

Micro-structural investigation of concrete made with variable fine aggregate combinations are analyzed after 28 days of curing for result validation. Scanning electron microscopy (SEM) image of a concrete core of R_0, R_{60} and $R_{60}C_{20}$ made up with NFA, NFA and RFA, and NFA, RFA and CS, respectively, as its fine aggregate are shown in Fig. 6a–c at 20 μm magnification.

Three shades of gray are observed in SEM images and are identified in Fig. 6, lighter shade or white spots as unhydrated cement, gray area as C–S–H gel and dark hollows as pores; however, in Fig. 6c, dark gray spots are copper slag particles.

Unhydrated cement particles are evidently more in Fig. 6a of R_0 specifying relatively low hydration. A denser and more homogeneous micro-structure is observed at concrete mixes with RFA as shown in Fig. 6b, c, i.e., for R_{60} and $R_{60}C_{20}$. Figure 6c shows weak interfacial transition zone (ITZ) around CS due to its spherical and smooth surface and thereby restricted its use. Also, branching micro-cracks are visible near ITZ at $R_{60}C_{20}$. Interpretations observed from Fig. 6 at microscopic level are clearly contributing for macroscopic behavior of concrete.

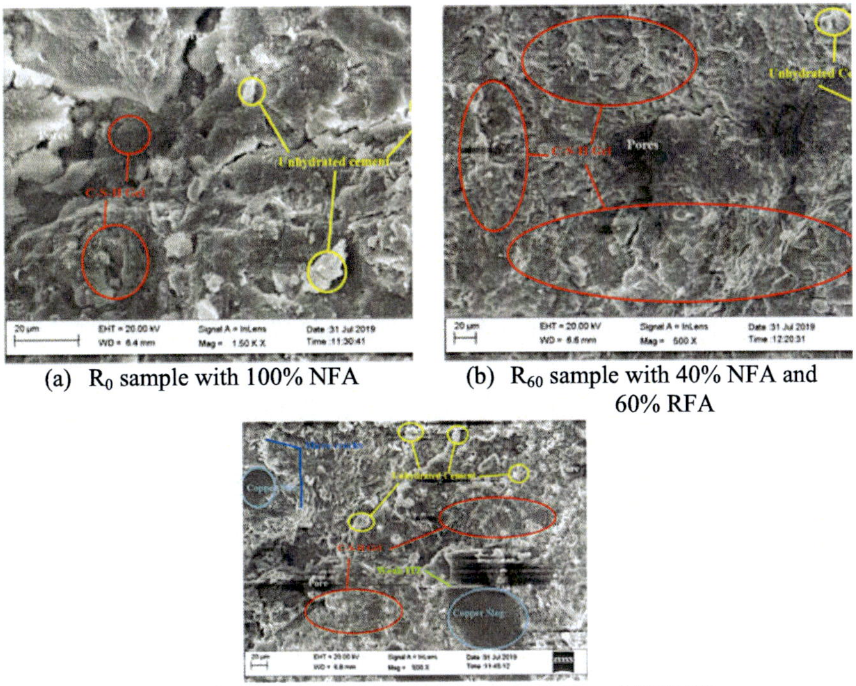

(a) R_0 sample with 100% NFA

(b) R_{60} sample with 40% NFA and 60% RFA

(c) $R_{60}C_{20}$ sample with 20% NFA, 60% RFA and 20% CS

Fig. 6 SEM images for concrete core with variable fine aggregate compositions

5.2 Sustainability Evaluation

Sustainability of concrete is governed by EE and CO_2eq of its compositional materials as every material has its own EE and CO_2eq. The total amount of EE and CO_2eq of concrete is calculated by adding together EE and CO_2eq of all the materials used in its production weight-wise in 1m^3 of volume. The data of individual EE and CO_2eq of all the materials used in the production of eco-friendly concrete proposed in this study is taken from previous literature [46–50].

Despite having 9.121 MJ/kg of EE and 1.47236 CO_2eq /kg, EE and CO_2eq of CS is not accounted in calculations. Technically, high EE and CO_2eq of CS is due to the process of extracting refined copper metal by smelting process, and as CS is an involuntary by-product of copper industry, it does not have EE and CO_2eq of its own. However, CS utilization increases the EE and CO_2eq of concrete, but being used as a waste in concrete, it does not imposes any harmful effects on environment. In fact, CS in concrete is a better way of dumping it harmlessly as a promising quality enhancing material [48]. The unit EE, ECO_2 eq/kg and cost in INR per kg is given in Table 5. The results after calculation of energy and cost as per the data given in Table 5, is given in Table 6.

Table 5 Unit values of EE, ECO_2 eq/kg and cost material wise

Material	OPC	SF	NFA	RFA	NCA	SP	CS	Water
Unit EE (MJ/kg)	4.8	0.036	0.081	0.046	0.083	11.5	–	0.2
ECO_2eq/kg	0.93	0.014	0.0023	0.0012	0.0048	0.6	–	0.0008
Unit Cost (INR per kg)	7.5	19	3	0.8	1	55	0.05	0.1

Table 6 Energy and cost analysis of concrete samples

Sample	Total EE (MJ/m^3)	Total (CO_2 eq/m^3)	Total cost (INR/m^3)
R_0	2430.45	430.33	7585
R_{20}	2425.82	430.23	7293.7
R_{40}	2421.18	430.03	7002.4
R_{60}	2416.55	429.93	6711.2
$R_{55}C_5$	2415.03	429.83	6686.4
$R_{50}C_{10}$	2413.51	429.83	6661.5
$R_{45}C_{15}$	2411.98	429.83	6636.7
$R_{40}C_{20}$	2410.46	429.73	6611.8
$R_{75}C_5$	2410.39	429.73	6395.1
$R_{70}C_{10}$	2408.87	429.73	6370.2
$R_{65}C_{15}$	2407.34	429.63	6345.4
$R_{60}C_{20}$	2405.82	429.63	6320.6
$R_{95}C_5$	2405.76	429.63	6103.8
$R_{90}C_{10}$	2404.24	429.53	6078.9
$R_{85}C_{15}$	2402.71	429.53	6054.2
$R_{80}C_{20}$	2401.19	429.43	6029.3

As evident from Table 6, EE of NFA is 43.2% and 47.8% higher than that of RFA, respectively. Therefore, by replacing NFA with RFA, a significant amount of energy can be saved. Cost analysis of concrete mixes with variable fine aggregate contents produced for the current study is performed to establish an economic viability of concrete along with the ecological benefits.

RFA and CS being waste products are cheaply available and require less or no processing cost. Thus, a maximum saving of 20.5% can be achieved with use of RFA and CS in concrete as replacement of NFA.

Figure 7 shows all the aspects of concrete with RFA and CS that may help in decision making. The graph represents the percentage increase or decrease achieved in terms of EE, CO_2eq, compressive strength, water absorption and cost of various concrete mixes when compared with R_0. The dotted lines are indicators of sustainability decider, and solid lines are indicators of strength, durability and economic viability.

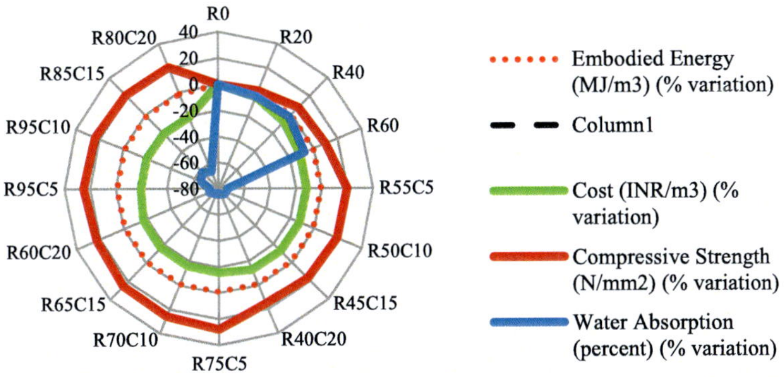

Fig. 7 Comparative analysis of EE, Co_2eq, cost, water absorption and compressive strength with reference to R_0

As evident from Fig. 7, compressive strength is the only parameter that is exceeding the limits of reference (R_0), concretes with replacement of NFA can be called sustainable concretes that does not compromised the strength and durability of structure.

6 Conclusion

The experiment of the research is designed to identify and maximize the feasibility of using RFA as a replacement of 100% NFA in concrete and its effect on the fresh and hardened properties of concrete. Concrete mixing with RFA above 60% content as fine aggregate was difficult due to excessive slump loss resulted from high water absorption of RFA. Consequently, CS was introduced as a slump-enhancing material. The comparative study on various concrete mixes having NFA, RFA and CS in varying percentages was conducted to draw following conclusions:

- RFA obtained from CD-W is perfectly suitable to replace NFA used in concrete production to establish sustainable balance between CD-W generation and natural sand extraction from rivers.
- XRD analysis of RFA showed presence of unhydrated cement grains, which provide additional strength to concrete.
- Substitution of NFA with RFA in concrete produces superior quality concrete with high load-carrying capacity. Thus, not only saving natural sand but also leads to save significant quantity of cement used for higher strength concrete production.
- When added water and dose of super-plasticizer is kept same in concrete design, replacement up to 60% NFA is possible with only RFA at a dropped slump value of 70 mm, which is significant to be proposed for almost all the structural applications in absence of copper slag.

- To extend NFA replacement up to 100%, a combination of RFA and CS is proposed, where RFA contributes to sustainability, strength and rapid hydration of cement and CS is slump and strength enhancing material.
- BIS: 456[63] states that 25 mm is the least slump value for fresh concrete to be applicable in mass concreting and in lightly reinforced sections with proper compaction. Therefore, 95% RFA and 5% CS can replace absolute NFA and with 25% more compressive strength.
- Micro-structural analysis validated the superior properties of concrete with RFA than that of NFA, and weak ITZ around CS justifying limited usage in concrete.
- The energy and cost analysis of all the concrete mixes by varying NFA percentages with RFA, and combination of RFA and CS, show that concrete with RFA are more sustainable, economical and durable as compared with concrete with NFA.

Acknowledgements Author, Nancy Soni has received grant from Government of Madhya-Pradesh, India through Environmental Planning and Co-ordination Organization (EPCO) to carry out this research. This research work is copyright to Environmental Planning and Co-ordination Organization (EPCO). The authors declare no conflict of interest.

References

1. Kosmatka, S. H., Kerkhoff, B., & Panarese, W. C. (2002). *Design and control of concrete mixtures* (Vol. 5420, pp. 60077–61083). Portland Cement Association.
2. Villioth, J. (2014). *Building an economy on quicksand*. See http://www.ejolt.org/2014/08/building-an-economy-on-quicksand/. Accessed January 2020
3. Government of India (Ministry of Housing and Urban Affairs). (2018). *Strategy for promoting processing of construction and demolition (C&D) waste and utilization of recycled products.*
4. Zhang, G., Song, J., Yang, J., & Liu, X. (2006). Performance of mortar and concrete made with a fine aggregate of desert sand. *Building and Environment, 41*(11), 1478–1481.
5. Chilamkurthy, K., Marckson, A. V., Chopperla, S. T., & Santhanam, M. (2016). A statistical overview of sand demand in Asia and Europe. In *International conference UKIERE CTMC* (Vol. 16).
6. Gavriletea, M. D. (2017). Environmental impacts of sand exploitation. Analysis of sand market. *Sustainability, 9*(7), 1118.
7. Saviour, M. N., & Stalin, P. (2012). Soil and sand mining: Causes, consequences and management. *IOSR Journal of Pharmacy (IOSRPHR), 2*(4), 01–06.
8. Limbachiya, M. C., Koulouris, A., Roberts, J. J., & Fried, A. N. (2004). Performance of recycled aggregate concrete. In *Proceeding of RILEM International Symposium on Environment-Conscious Materials and Systems for Sustainable Development* (pp. 127–136).
9. Safiuddin, M., Alengaram, U. J., Rahman, M. M., Salam, M. A., & Jumaat, M. Z. (2013). Use of recycled concrete aggregate in concrete: A review. *Journal of Civil Engineering and Management, 19*(6), 796–810.
10. Government of India (Building, Materials and Technology Promotion Council Ministry of Housing & Urban Affairs). (2018). *A ready Rockner—Utilisation of recycled produce of construction and demolition waste.*
11. Silva, R. V., De Brito, J., & Dhir, R. K. (2014). Properties and composition of recycled aggregates from construction and demolition waste suitable for concrete production. *Construction and Building Materials, 65*, 201–217.

12. Behera, M., Bhattacharyya, S. K., Minocha, A. K., Deoliya, R., & Maiti, S. (2014). Recycled aggregate from C&D waste & its use in concrete–A breakthrough towards sustainability in construction sector: A review. *Construction and building materials, 68*, 501–516.
13. Solyman, M. (2005). *Classification of recycled sands and their applications as fine aggregates for concrete and bituminous mixtures* [Doctoral dissertation].
14. Nixon, P. J. (1978). Recycled concrete as an aggregate for concrete—A review. *Matériaux et Construction, 11*(5), 371.
15. Bravo, M., De Brito, J., Pontes, J., & Evangelista, L. (2015). Durability performance of concrete with recycled aggregates from construction and demolition waste plants. *Construction and Building Materials, 77*, 357–369.
16. Hansen, T. C. (1986). Recycled aggregates and recycled aggregate concrete second state-of-the-art report developments 1945–1985. *Materials and Structures, 19*(3), 201–246.
17. Ulsen, C. (2011). *Characterization and separability of fine aggregates produced from CDW* [Ph.D. thesis in mineral engineering]. Polytechnic School of the University of São Paulo, Brazil (in Portuguese).
18. Evangelista, L., & De Brito, J. (2014). Concrete with fine recycled aggregates: A review. *European Journal of Environmental and Civil Engineering, 18*(2), 129–172.
19. Evangelista, L., & De Brito, J. (2010). Durability performance of concrete made with fine recycled concrete aggregates. *Cement and Concrete Composites, 32*(1), 9–14.
20. Ambily, P. S., Umarani, C., Ravisankar, K., Prem, P. R., Bharatkumar, B. H., & Iyer, N. R. (2015). Studies on ultra high performance concrete incorporating copper slag as fine aggregate. *Construction and Building Materials, 77*, 233–240.
21. Caliskan, S., & Behnood, A. (2004). Recycling copper slag as coarse aggregate: Hardened properties of concrete. In *Proceedings of Seventh International Conference on Concrete Technology in Developing Countries* (pp. 91–98).
22. Shi, C., Meyer, C., & Behnood, A. (2008). Utilization of copper slag in cement and concrete. *Resources, Conservation and recycling, 52*(10), 1115–1120.
23. Gorai, B., & Jana, R. K. (2003). Characteristics and utilisation of copper slag—A review. *Resources, Conservation and Recycling, 39*(4), 299–313.
24. Alnuaimi, A. S. (2012). Effects of copper slag as a replacement for fine aggregate on the behavior and ultimate strength of reinforced concrete slender columns. *The Journal of Engineering Research (TJER), 9*(2), 90–102.
25. Hammond, G. P., & Jones, C. I. (2008). Embodied energy and carbon in construction materials. *Proceedings of the Institution of Civil Engineers-Energy, 161*(2), 87–98.
26. Vukotic, L., Fenner, R. A., & Symons, K. (2010, September). Assessing embodied energy of building structural elements. *Proceedings of the Institution of Civil Engineers-Engineering Sustainability, 163*(3), 147–158.
27. Wyatt, T. (2011). Energy used in buildings. In *Whole life energy rating for buildings* (pp. 2–8). Thomas Telford Publishing.
28. Fumoto, T., & Yamada, M. (2002). Influence of quality of recycled fine aggregate on properties of concrete. *Memoirs of the Faculty of Engineering, Osaka City University, 43*, 97–103.
29. Pepe, M., Toledo Filho, R. D., Koenders, E. A., & Martinelli, E. (2014). Alternative processing procedures for recycled aggregates in structural concrete. *Construction and Building Materials, 69*, 124–132.
30. Balasubramanian, A. (2017). *Size reduction by crushing methods—Technical report.*
31. Bureau Indian Standard—383. (2016). *Coarse and fine aggregate for concrete—Specification.*
32. Poon, C. S., Qiao, X. C., & Chan, D. (2006). The cause and influence of self-cementing properties of fine recycled concrete aggregates on the properties of unbound sub-base. *Waste management, 26*(10), 1166–1172.
33. Kurda, R., de Brito, J., & Silvestre, J. D. (2018). Indirect evaluation of the compressive strength of recycled aggregate concrete with high fly ash ratios. *Magazine of Concrete Research, 70*(4), 204–216.
34. Bureau Indian Standard—2386:1. (1963). *Method of test for aggregate for concrete, particle size and shape.*

35. Bureau Indian Standard—12269. (2013). *Ordinary Portland cement, 53 grade—Specification.*
36. Bureau Indian Standard—4031:6. (1988). *Method of physical tests for hydraulic cement, determination of compressive strength of hydraulic cement other than masonry cement.*
37. Bureau Indian Standard—4032. (1985). *Method of chemical analysis of hydraulic cement.*
38. Bureau Indian Standard—15388. (2003). *Silica fume—Specification.*
39. Bureau Indian Standard—2386:3. (1963). *Methods of test for aggregates for concrete, measuring mortar making properties of fine aggregate.*
40. Bureau Indian Standard—10262. (2019). *Concrete mix proportioning—Guidelines.*
41. Bureau Indian Standard—4925. (2004). *Concrete batching & mixing plant—Specification.*
42. Bureau Indian Standard—1199-3. (2018). *Fresh concrete—Method of sampling, testing & analysis, determination of consistency of fresh concrete.*
43. Sánchez-Roldán, Z., Martín-Morales, M., Valverde-Palacios, I., Valverde-Espinosa, I., & Zamorano, M. (2016). Study of potential advantages of pre-soaking on the properties of pre-cast concrete made with recycled coarse aggregate. *Materiales de Construcción, 66*(321), 076.
44. Bureau Indian Standard—516. (1959). *Methods of tests for strength of concrete.*
45. Bureau Indian Standard—456. (2000). *Plain & reinforced concrete—Code of practice.*
46. Hossain, M. U., Poon, C. S., Lo, I. M., & Cheng, J. C. (2016). Comparative environmental evaluation of aggregate production from recycled waste materials and virgin sources by LCA. *Resources, Conservation and Recycling, 109*, 67–77.
47. Jones, R., McCarthy, M., & Newlands, M. (2011, May). Fly ash route to low embodied CO_2 and implications for concrete construction. In *World of Coal Ash Conference*. Denver, Colorado, USA.
48. Mithun, B. M., & Narasimhan, M. C. (2016). Performance of alkali activated slag concrete mixes incorporating copper slag as fine aggregate. *Journal of Cleaner Production, 112*, 837–844.
49. Prem, P. R., Verma, M., & Ambily, P. S. (2018). Sustainable cleaner production of concrete with high volume copper slag. *Journal of Cleaner Production, 193*, 43–58.
50. Siddique, S., Chaudhary, S., Shrivastava, S., & Gupta, T. (2019). Sustainable utilisation of ceramic waste in concrete: Exposure to adverse conditions. *Journal of cleaner production, 210*, 246–255.

Examination of Platooning Variables on Two-Lane Highways Having Mixed Traffic Situation

Amardeep Boora, Indrajit Ghosh, Satish Chandra, and Kavita Rani

Abstract Frequent formation of platoon on a road exhibit the poor performance of a specific roadway facility. Present study examined different platooning variables to identify the principal variable which is responsible for frequent formation of platoon and poor level of service (LOS) on two-lane roads. Consequently, a speed difference (SD) range of −4 to +10 km/h between successive vehicles was found appropriate to identify the follower and non-follower. Different gap values (i.e. between successive vehicles) ranging from 1.9 to 4.3 s were observed across all the selected locations by utilizing gap acceptance curve method ahead of which the possibility of vehicles to travel at their desired speed will increase. A number of relationships were developed among all the platooning variables and performance parameters. Results are depicting that two-way traffic volume is the principal parameter that affects the LOS of the specific roadway facility. Among the entire performance parameters, some specific performance parameters which signify the following vehicles were found to exhibit a significant correlation with principal platooning variable. Consequently, different LOS ranges were proposed by considering NFPC and FD as appropriate parameters for two-lane intercity highways under mixed traffic situations.

Keywords Following vehicle · Speed difference · Platooning · Cut-off gap value · Level of service

---

A. Boora (✉)
Department of Civil Engineering, Jaypee University of Information Technology, Waknaghat, Solan, Himachal Pradesh 173234, India

I. Ghosh · K. Rani
Department of Civil Engineering, Indian Institute of Technology (IIT) Roorkee, Roorkee, Uttarakhand 247667, India
e-mail: indrafce@iitr.ac.in

S. Chandra
Central Research Road Institute (CRRI) New Delhi, New Delhi, Delhi 110025, India

A. K. Gupta et al. (eds.), *Advances in Construction Materials and Sustainable Environment*, Lecture Notes in Civil Engineering 196,
https://doi.org/10.1007/978-981-16-6557-8_8

1 Introduction

Two-lane highways have significant place in most countries highway systems as these types of highway facilities provide a variety of transportation allied functions to different types of vehicles. In India, two-lane highways (one lane in each direction for two-directional traffic movements) comprise 53% of the National Highways (NH) [1]. These highways 'performance can be examined by proposing different LOS thresholds', which helps to determine whether or not the public funds are being used in sufficient ways [2]. Because of the accommodation of several vehicles types having variations in their static and dynamic features, the performance assessment of these types of highway facilities becomes more challenging. It is to note that several variables already have been utilized in previous studies for the same purpose. The Highway Capacity Manual of United States [3] proposed various parameters namely, average travel speed (ATS), percent time spent following (PTSF) and average travel speed as a percentage of free flow speed (PFFS) to evaluate the performance of different types of two-lane highways. However, use of PTSF as performance parameters was found problematic [4, 5]. Consequently, new parameters and guidelines were introduced to appraise the performance of these types of highways around the world, namely, follower density (FD) [6–9], percent impeded (PI) [10]. It is to be noted that aforementioned studies were carried out in homogeneous traffic situations. While in India, mixed traffic situation is observed. Therefore, directly implementation of these guidelines on the highways of developed countries (having mixed traffic) can be problematic. Nonetheless, under mixed traffic scenario, a similar study [11] was performed to suggest specific LOS ranges by using the criteria suggested by the U.S. HCM [3]. Unfortunately, no guidelines are provided by the Indian standard [12] in this context. Due to lack of such course of action, transportation planners and engineers in India are forced to adopt the previously recommended guidelines by HCM [3]. The present study is focused to identify the key platooning variable for two-lane highways and assess different LOS ranges under mixed traffic situations.

2 Data Collection

Five different study sections were selected for data collection purposes. Out of which, two sections (Section 1-NH 47) and Section 2-NH 58) were situated in the Western and Southern regions of India respectively, while Section 3 (NH 4), Section 4 (SH 31) and Section 5 (SH 59) were situated in the Northern region. All the sections were finalized after considering several factors which were following:

- Selected study sections should be facilitated with a minimum carriageway width of 7.0 m along with the shoulder.
- All the study sections should be free from intersectional and curves control.

Table 1 Variation in different traffic features experienced across all the study sections

Section	Total observed vehicles	Passenger car (%)	Two wheeler (%)	Auto rickshaw (%)	Heavy vehicles (%)	Volume (veh/h)
Section-1	8367	34	46	14	6	2789
Section-2	3547	39	41	10	10	1773
Section-3	1892	46	22	7	25	946
Section-4	1637	15	73	5	7	818
Section-5	1197	39	39	7	15	598

Data was collected with the help of high-resolution video camera (3840 × 2160 pixel) on weekdays only. The camera was mounted on a viewpoint across each section to identify each type of vehicle at the trap's exit and entry point (50–60 m). Data extraction was conducted manually for different time intervals on big television screen in Civil Engineering Lab. Different vehicle types were found travelling across all the study sections which were categorized into different categories. Table 1 is exhibiting the variation in different traffic characteristics found in all the study sections.

3 Performance Parameters Used in the Study

Before finalizing the performance parameters it should be taken care of that the chosen parameters should be recognized in the field without any difficulty [2]. In earlier studies [8, 11, 13], speed related parameters were not found to a suitable one in order to evaluate the performance of the specific roadway stretch. Therefore, different follower related performance parameters were selected for the present study which is listed below.

- NF,
- PF,
- FD and
- NFPC.

All the aforementioned parameters and two-way traffic volume were considered for each five min period. Recently [14], stream equivalency factor (SEF) was proposed to transform mixed traffic into homogeneous traffic in terms of PCU only. Identification of free moving vehicles is required to examine all the following related parameters. Different studies define free moving vehicles in several ways. U.S. HCM [3] specify that the FFS can be calculated at low traffic volume only. While, few studies [6, 8] identified vehicles in following conditions travelling with a headway value less than eight sec. After that, the same methodology was utilized (as proposed by HCM [3]) in another study [15] to calculate the FFS. It was observed that the interaction between vehicles gets diminished while travelling with a headway threshold value more than seven sec [13]. On the other hand, a cut-off headway value of five sec was identified more suitable to define a vehicle in following or non-following situation [16]. Another study [4] characterize following vehicle travelling with a headway value less than six sec. Earlier [17], free moving vehicles were identified with five sec headway and three sec tailway. On the basis of the past studies it was revealed that the FFS was defined by considering headway values only. All these guidelines were examined (i.e. headway with ATS and 85th percentile speed) for the present study also but did not find appropriate for Indian traffic which is mixed in nature. As a result, it was concluded that these previous methodologies (i.e. headway) cannot be implemented for Indian mixed traffic situations as headway value gets influenced due to the presence of different vehicle types. Thereafter, two parameters namely, SD and gap value were selected to identify the following and non-following vehicles. For the same purpose speed was calculated by using Eq. (1). After establishing several relationships between SD and gap threshold values across all the study sections a gap value of ten sec was found suitable to define FFS under mixed traffic situations. Figure 1a is showing such relationship for a study section NH-58.

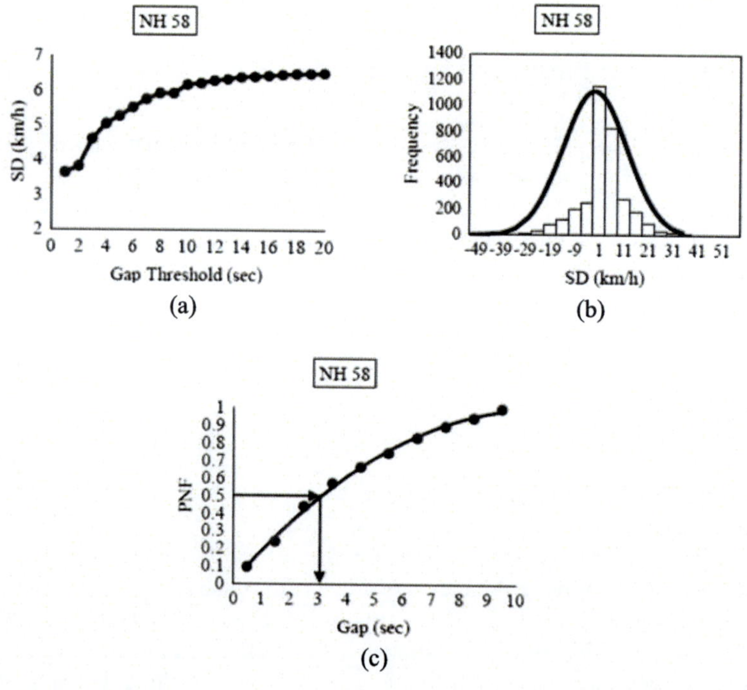

Fig. 1 **a** Relationship between SD and gap threshold. **b** Identification of different SD limits by superimposing normal curve on the histogram. **c** Probability of not following at NH-58

Table 2 Variation in observed gap values across all the sections

Sections	Observed gap values (s)	Speed difference limit (km/h)
Section-1	1.9	−2.20 to +7.42
Section-2	3	−0.43 to +4.09
Section-3	3.4	−0.69 to +9
Section-4	3.7	−2.53 to +10
Section-5	4.3	−3.41 to +6.12

$$V_i = \frac{T_L}{\frac{F_{E2-F_{E1}}}{25}} \times 3.6 \tag{1}$$

where, V_i = speed of ith vehicle (km/h); T_L = total trap length (m); F_{E1} = frame number at entry point; and F_{E2} = frame number at exit point.

In order to strengthen the results and remove the confusion regarding the definition of free moving vehicle (i.e. 10 s gap value), a normal curve of free flowing vehicles was superimposed on the histogram of SD [18] as shown in Fig. 1b. Same graphs were plotted for all the study sections. After examining all the graphs, an SD limit ranging from −4 to +10 km/h was found appropriate for identifying the followers under mixed traffic situations on two-lane highways. Different cut-off gap values were calculated with the help of acceptance curve method [19] and the definition of follower and non-followers as proposed in the present study (i.e. gap value of ten sec and SD limit range of −4 to +10 km/h) as shown in Table 2. Previously, a cut-off gap value of 2.6 s was proposed to identify the follower under mixed traffic situation [11]. But the actual shortcoming of the study was to neglect the dependency of gap value on traffic volume (i.e. difference between two consecutive vehicles from rare bumper to front bumper). Therefore, it fails to exhibit the actual field situation under mixed traffic and cannot be used to identify the following and non-following vehicles under mixed traffic situations. On the other hand, present study successfully investigated the impact of traffic volume on gap value (i.e. increased with reduction in traffic volume) as shown in Table 2. On the basis of the detailed study of traffic data obtained from all the selected sections, different platooning parameters were finalized for examining their relationships with all of the performance parameters (i.e. the proportion of 2W, AR, HV and two-way traffic volume).

Another platooning variable named directional traffic volume was also tried to establish the relationship with all performance parameters. Since no association was found between directional percentage of traffic volume and performance parameters, directional % of traffic volume was not considered for this study. It is worth noting that NFPC is the number of followers as a percentage of capacity. First of all number of followers were identified using speed difference limit −4 to +10 km/h and gap value ranging from 1.9 to 4.3 s as shown in Table 2. Later, capacity was calculated using the Greenshields model. Earlier [20] an attempt was made to calculate the capacity of two-lane roads in India. Same procedure was followed in the present study. However, models other than Greenshields were also tried in the present study but not found appropriate as speed of the section was found to be affected by volume right from the beginning of the curve. Subsequently, effects of different platooning variables on all of the parameters which are used in the present study to evaluate the performance of two-lane highways were examined as shown in Figs. 2, 3, 4 and 5.

It was experienced in the field that the lower speed and increment in the proportion of AR is responsible to increase the following situation. Due to their high speed and fast manoeuvrability, an improvement in the 2W ratio was supposed to decrease the proportion of followers. Unlike the condition seen in the field, no such trend was detected from the field data as shown in Figs. 2 and 3. Similar type of hypothesis was made in case of the HV as made with the proportion of AR as the HV appears to

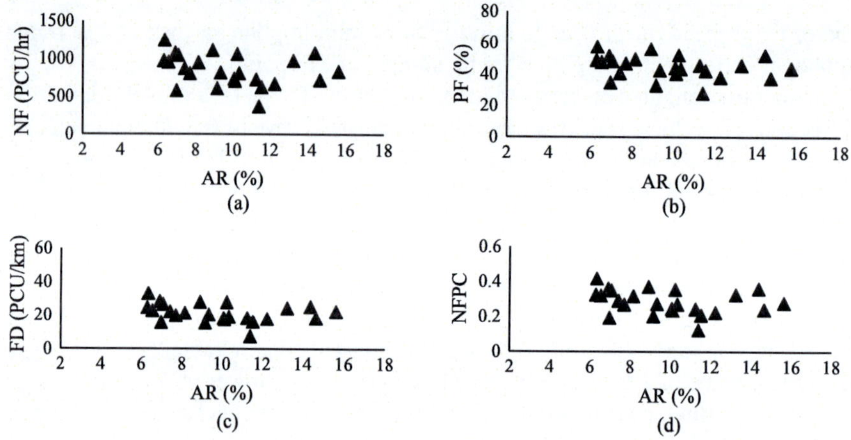

Fig. 2 Examination of the variation in the different parameters with the proportion of AR at Section-2

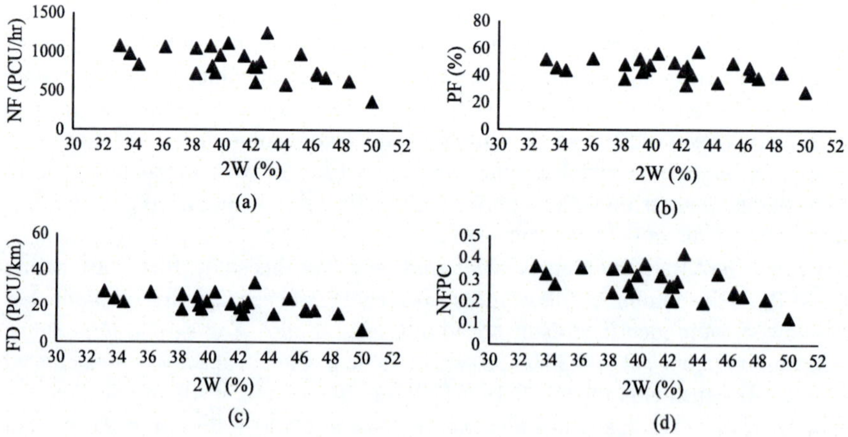

Fig. 3 Examination of the variation in different parameters with the proportion of 2W at Section-2

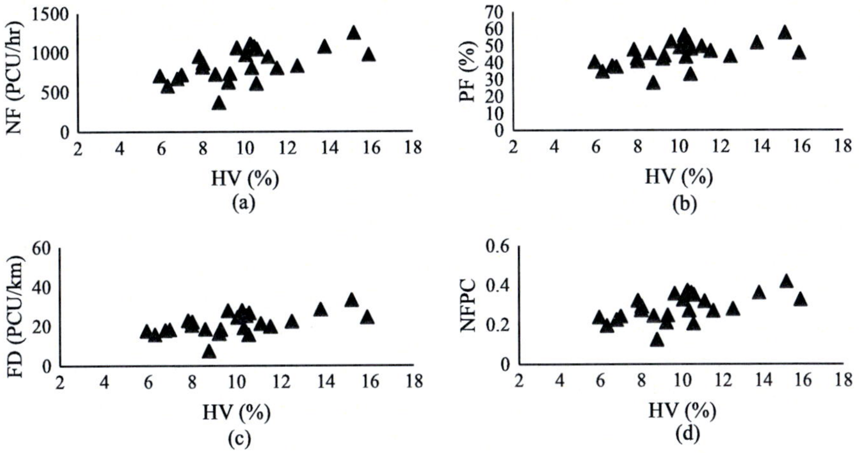

Fig. 4 Examination of the variation in different parameters with the proportion of HV at Section-2

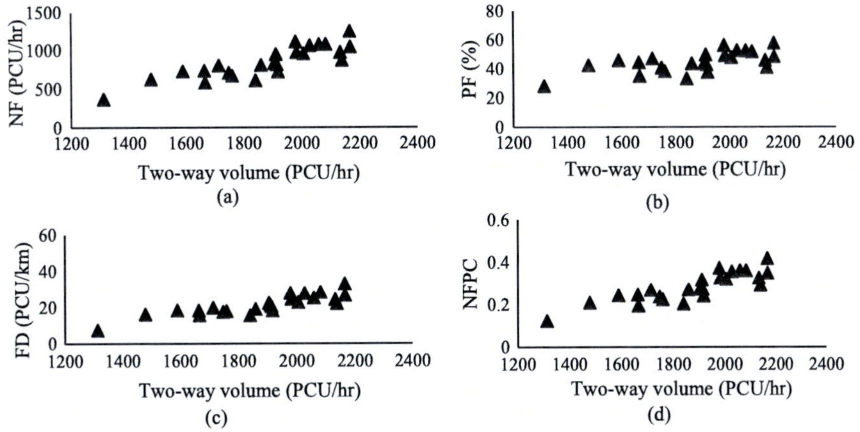

Fig. 5 Examination of the variation in different parameters with the proportion of two-way volume at Section-2

operate at lesser speed with poor manoeuvrability relative to the rest of the vehicle types. However, no trend was observed from the relationships of proportion of HV with all the parameters as shown in Fig. 4. Similar trends were found out across each and every one of the study sections. Finally, similar types of relationships were tried between all the performance parameters and traffic volume for NH-58 study section as shown in Fig. 5. From Fig. 5 it is evident that all the parameters tend to increase with increase in traffic volume. Secondly, two-way traffic volume was identified as the key platooning variable as it exhibits a good association with all the parameters as shown in Fig. 5. Same results were obtained across all the study sections.

3.1 Correlation Analysis

To reinforce the findings, correlation among all parameters and platooning variables were analyzed based on R^2 vale along with t and p-value. Correlation analysis was carried out at a confidence level of 95% as shown in Table 3. Underscore values in Table 3 are representing incoherent relationships between different parameters and platooning variables, while other correlation coefficients were found as per the expectation. All the parameters namely FD, PF, NF and NFPC were found positively associated with traffic volume. Out of each performance parameter, NFPC was identified as the key parameter which affects the performance of two-lane highways followed by FD. On the other hand, NF and traffic volume was also showing good association. Nevertheless, NF also cannot be used as a single measure as it does not reflect the actual situation of congestion [11].

4 Statistical Analysis

Regression analyzes were carried out in order to examine the significance of different relationships established between different performance parameters and platooning variables. On the basis of the findings, only two-way traffic volume (V) was identified as the suitable platooning variable as shown in Eqs. (2)–(5). Bold faces values are found to be significant at 95% confidence level, while underlined values are exhibiting contradicting outcomes.

Later, a goodness of fit test was conducted in SPSS software in order to classify the appropriate model. Resultant, a power model was found suitable for all the follower related parameters (i.e. based on R^2, F and P-value) as shown in Table 4. All the parameters were showing good correlation with traffic volume and were found to be significant based on P-value as shown in Table 4. Earlier, it was concluded that the NF can't utilize as a sole performance parameter (as it doesn't reflect the actual congestion situation, i.e. NF can be changed from one location to another) [11]. Therefore, a new parameter named NFPC (i.e. combination of NF and the capacity of a particular highway) was introduced in order to explain the congested situation logically. Figure 6 is depicting that the NFPC is the major performance parameter followed by the FD which has a good correlation with all other parameters including two-way traffic volume. On the other hand, the correlation of PF with traffic volume was not found up to the mark.

$$\mathrm{NF} = 2.182(2\mathrm{W}) + 3.97(\mathrm{AR}) - 5.468(\mathrm{HV}) + 0.626(V) - \mathbf{448.64}\ \mathbf{R^2 = 0.83} \tag{2}$$

$$t(P\text{-value}) = \underline{1.04(0.29)}, 0.69(0.49), \underline{-1.27(0.205)}, 0.035(0.00), -2.45(0.01)$$

Table 3 Examination of association among all the performance parameters and platooning factors of all section

Parameters	Percentage share of AR					Percentage share of 2W				
	Section-1	Section-2	Section-3	Section-5	Section-4	Section-1	Section-2	Section-3	Section-5	Section-4
NF *t*-value (*p*-value)	−0.22 −1.32 (0.20)	−0.29 −1.41 (0.17)	−0.06 −0.31 (0.76)	0.01 0.06 (0.95)	0.34 1.72 (0.10)	**−0.45*** **−2.96** **(0.01)**	**−0.58*** **−3.39** **(0.00)**	**−0.49*** **−2.66** **(0.01)**	−0.14 −0.71 (0.49)	−0.16 −0.77 (0.45)
PF *t*-value (*p*-value)	−0.24 −1.47 (0.15)	−0.26 −1.31 (0.20)	0.004 0.02 (0.98)	0.06 0.30 (0.77)	0.19 0.93 (0.36	**−0.36*** **−2.30** **(0.03)**	**−0.44*** **−2.32** **(0.03)**	**−0.51*** **−2.84** **(0.01)**	0.04 0.23 (0.82)	0.22 1.08 (0.29)
FD *t*-value (*p*-value)	−0.21 −1.26 (0.21)	−0.28 −1.41 (0.17)	−0.03 −0.19 (0.85)	0.05 0.24 (0.81)	0.36 1.85 (0.08)	**−0.43*** **−2.79** **(0.01)**	**−0.59*** **−3.48** **(0.00)**	**−0.49*** **−2.64** **(0.01)**	−0.16 −0.79 (0.44)	−0.17 −0.81 (0.42)
NFPC *t*-value (*p*-value)	−0.22 −1.32 (0.20)	−0.28 −1.41 (0.17)	−0.06 −0.31 (0.76)	0.01 0.06 (0.95)	0.34 1.72 (0.10)	**−0.45*** **−2.96** **(0.01)**	**−0.58*** **−3.39** **(0.00)**	**−0.49*** **−2.66** **(0.01)**	−0.14 −0.71 (0.49)	−0.16 −0.77 (0.45)
Parameters	Percentage share of HV					Two-way volume				
	Section-1	Section-2	Section-3	Section-5	Section-4	Section-1	Section-2	Section-3	Section-5	Section-4
NF *t*-value (*p*-value)	**0.61*** 4.58 (0.00)	**0.57*** 3.31 (0.00)	0.09 0.43 (0.67)	**0.47*** **2.52** **(0.02)**	0.14 0.71 (0.49)	**0.38*** **2.45** **(0.02)**	**0.85*** **7.87** **(0.00)**	**0.89*** **9.49** **(0.00)**	**0.68*** **4.38** **(0.00)**	**0.48*** **2.63** **(0.02)**
PF *t*-value (*p*-value)	**0.66*** 5.14 (0.00)	**0.53*** 2.97 (0.01)	0.04 0.19 (0.85)	0.29 1.45 (0.16)	−0.17 −0.85 (0.41)	−0.07 −0.45 (0.66)	**0.63*** **3.84** **(0.00)**	**0.64*** **3.91** **(0.00)**	0.39 2.05 (0.05)	0.007 0.03 (0.97)
FD *t*-value (*p*-value)	**0.53*** **3.71** **(0.00)**	**0.53*** **3.31** **(0.00)**	0.10 0.49 (0.63)	0.49* 2.64 (0.01)	0.16 0.80 (0.43)	**0.48*** **3.20** **(0.00)**	**0.85*** **7.87** **(0.00)**	**0.87*** **8.67** **(0.00)**	**0.67*** **4.28** **(0.00)**	**0.47*** **2.51** **(0.02)**

(continued)

Table 3 (continued)

Parameters	Percentage share of HV					Two-way volume				
	Section-1	Section-2	Section-3	Section-5	Section-4	Section-1	Section-2	Section-3	Section-5	Section-4
NFPC	**0.61***	**0.57***	0.09	0.47*	0.14	**0.38***	**0.85***	**0.89***	**0.68***	**0.48***
t-value	**4.58**	**3.31**	0.43	2.52	0.71	**2.45**	**7.87**	**9.49**	**4.38**	**2.63**
(*p*-value)	**(0.00)**	**(0.00)**	(0.67)	(0.02)	(0.49)	**(0.02)**	**(0.00)**	**(0.00)**	**(0.00)**	**(0.02)**

*Correlation is significant at the 5% confidence level

Table 4 Goodness of fit test among all the parameters and platooning variables

Parameters	Model	Model detail			Parameter estimates	
		R^2	*F*	Sig. (*P*-value)	Constant	*b*1
NF	Power	0.87	873.80	0.00	0.0002	1.965
PF	Power	0.61	210.29	0.00	0.025	0.964
FD	Power	0.82	626.16	0.00	3.77E−06	2.017
NFPC	Power	0.85	17.39	0.00	1.312E−07	1.899

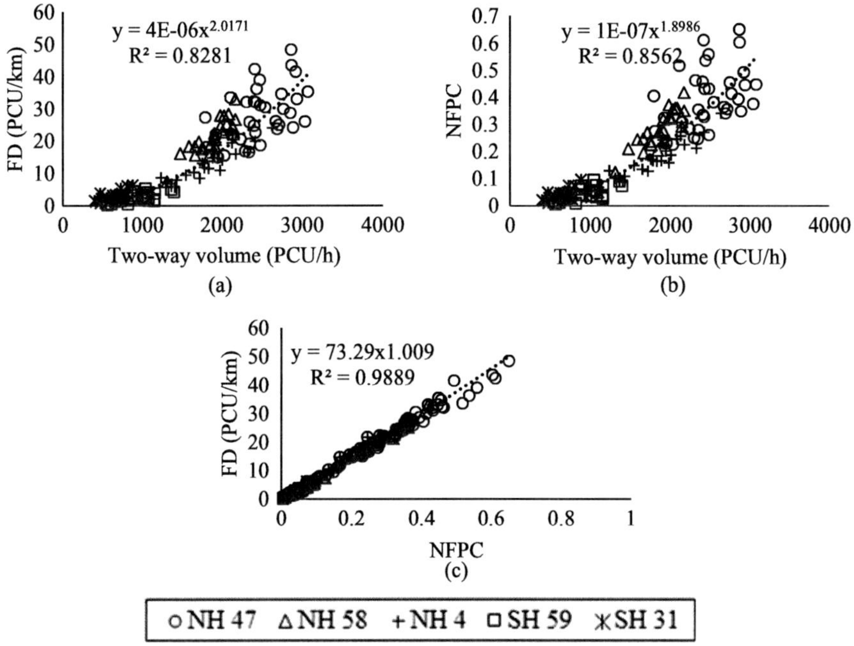

Fig. 6 Variation in NFPC and FD

$$\mathrm{PF} = 0.111(2\mathrm{W}) + 0.261(\mathrm{AR}) - 0.103(\mathrm{HV}) + \mathbf{0.018}(\boldsymbol{V}) - 3.773\,\mathbf{R^2} = \mathbf{0.63} \tag{3}$$

$$t(P\text{-value}) = \underline{0.10(0.29)}, 0.90(0.36), \underline{-0.47(0.64)}, \mathbf{10.54(0.00)}, -0.406(0.68)$$

$$\mathrm{FD} = 0.056(2\mathrm{W}) + 0.038(\mathrm{AR}) - 0.165(\mathrm{HV}) + \mathbf{0.0151}(\boldsymbol{V}) - \mathbf{0.0151}\,\mathbf{R^2} = \mathbf{0.82} \tag{4}$$

$$t(P\text{-value}) = \underline{1.14(0.25)}, 0.28(0.77), \underline{-1.60(0.11)}, \mathbf{18.00(0.00)}, 18.00(0.00)$$

$$\text{NFPC} = 0.0007(2\text{W}) + 0.0013(\text{AR}) - 0.002(\text{HV}) + \mathbf{0.0002}(V) - \mathbf{0.135}\ \mathbf{R}^2 = \mathbf{0.82} \quad (5)$$

$$t(P\text{-value}) = \underline{1.04(0.30)}, 0.70(0.48), \underline{-1.63(0.10)}, \mathbf{17.39(0.00)}, -2.24(0.02)$$

5 Calibration of LOS Thresholds

The present study was focusing to calibrate different LOS ranges in order to evaluate the performance of two-lane highways. Proposing various LOS ranges is a characterization related issue. Therefore, method of clustering analysis was used to attain diverse LOS ranges. Earlier [21], it was concluded that the clustering analysis technique can be utilized to separate various data set or objects into classes of comparable data or objects. Clustering analysis was carried out by considering FD and NFPC as the performance parameters and with assistance of various distances namely; Euclidean distance, Squared-Euclidean and Cityblock distance. Before that, a descriptive analysis examination was directed and data were found to be positively skewed. Consequently, *K*-median clustering analysis technique was utilized rather than *K*-mean clustering technique. Five numbers clusters were picked to adjust six surely understood LOS ranges [3]. Clustering analysis was directed in MATLAB. While, so as to get ideal limits among the LOS ranges, the quantity of cycle was kept up to 100. Silhouette plot [22] was utilized so as to reinforce the results of the examination and to validate the distance selection criteria for clustering analysis. For each level of service, the importance of silhouette plot can be analyzed with the help of thickness and width of the silhouette plot curve. Silhouette value interpretation can be represented using the universal thumb rule shown in Table 5. After evaluation of the silhouette plots, a typical estimation of silhouette 0.80 gained by virtue of square-Euclidean distance which seemed, by all accounts, to be reasonable as showed up in Fig. 7a. Initially, different ranges of FD were adjusted by using *K*-median clustering technique while Eq. (6) (which is the numerical depiction of the association among NFPC and FD) was used to assess different LOS ranges of NFPC.

Table 5 Cluster validation criteria

Average silhouette value range	Meaning
0.71–1.0	Exhibiting strong structure
0.51–0.70	Exhibiting reasonable structure
0.26–0.50	Exhibiting weak structure
<0.25	No substantial structure has been found

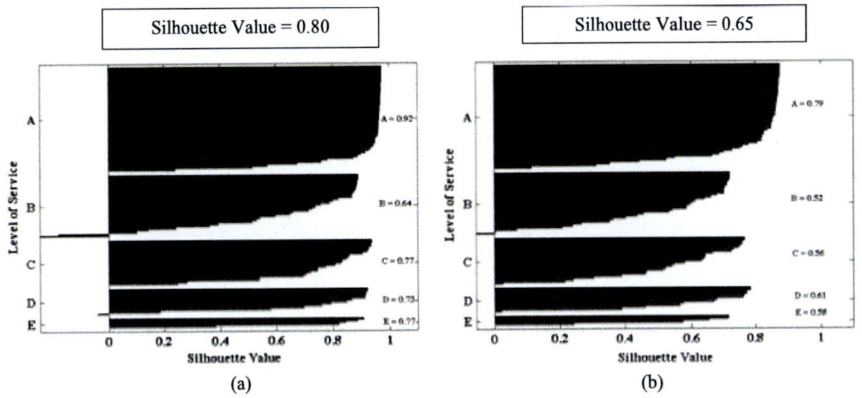

Fig. 7 Silhouette plot with the help of **a** square-euclidean distance and **b** cityblock distance

$$FD = 73.29(NFPC)^{1.009} \quad (6)$$

Relationship of NFPC and FD was found measurably critical at 95% certainty level having 0.9889 R^2 esteem. In the present examination, a similar relationship was utilized to analyze distinctive LOS ranges appeared in Table 6. Distinctive LOS ranges proposed in the present study are found up to 9% lower than the ranges proposed in previous study [11] as tabulated in Table 6. The principle purpose for such perceptions is that the proposed diverse LOS ranges for two-lane highways were calibrated by utilizing the PTSF values as recommended by U.S. HCM [3]. Issues related to the utilization of PTSF as a performance parameter were recognized in the past examinations [4, 5]. Alongside this, PTSF is extremely monotonous to compute under mixed traffic situations (i.e. because of the nearness of various vehicle types with various dynamic and static attributes which makes the activity progressively monotonous). Therefore, PTSF was not utilized as a performance parameter for two-lane highways under mixed traffic situations. Whereas, the LOS ranges prescribed in the present examination depend on the relationship of FD and NFPC (as PTSF was not seen as a critical measure under mixed traffic situations). Additionally, the present investigation considers the impacts of two-way traffic volume to distinguish

Table 6 Comparison between previous and proposed LOS ranges

	Number of follower per capacity [11]	Projected number of follower per capacity
A	≤0.15	≤0.13
B	>0.15–0.31	>0.13–0.28
C	>0.31–0.51	>0.28–0.39
D	>0.51–0.75	>0.39–0.50
E	>0.75	>0.50–0.66

a cut-off gap value for characterizing the followers. While in the past investigation's impact on traffic volume was thought to be unimportant.

6 Discussion

In the present study, different platooning variables and performance parameters were examined to classify the principal platooning variable and performance parameters to calibrate varying LOS ranges for two-lane highways under mixed traffic situations. For the same purpose, data were obtained from different selected sections of the present study. SEF was utilized to switch the mixed traffic into identical traffic. Previously used methodologies were examined to identify the FFC but found irrelevant to mixed traffic because of the existence of various types of vehicles with specific static and dynamic characteristics. Consequently, a new measure named SD and gap value rather than headway between two consecutive vehicles were utilized for the same. A ten sec gap value was acknowledged beyond which vehicles will travel under FFC at their desired speed. Later, an SD limit ranged from −4 to + 10 km/h and different cut-off gap values were estimated to classify the following and non-following vehicles. Current study suggests new guidelines to recognize the followers as it concluded that the two-way traffic volume has a significant effect on gap value (i.e. which is a unique feature of mixed traffic situation) and recognized as the key contributor which affects the performance of two-lane highways. While in an earlier study [11], the effect of traffic volume was not considered. By considering the proposed SD limit and cut-off gap values, different parameters were examined. Among all the performance parameters, speed related parameters were found inappropriate as the other parameters were depicting good association with two-way traffic volume. The main reason for this was the absence of a reference point in case of ATS and ATS_{PC} while other parameters showed poor correlation with follower related parameters as reported by several preceding examinations [4, 11, 16]. Accordingly, different LOS ranges were proposed by taking NFPC and FD as the main parameters by utilizing clustering technique. In order to use the proposed guidelines of the present study to define different LOS ranges, traffic engineers and planners have to measure the capacity of a particular highway facility as well as the number of following vehicles to calculate the NFPC values. Consequently, proposed LOS ranges of the present study can be used to define LOS provided to the road user at a particular roadway facility.

7 Conclusions and Recommendation

In the present study, traffic volume was identified as the major contributor to affect the performance of two-lane intercity highways. Consequently, different LOS ranges are proposed for two-lane highways under mixed traffic situations. The most important findings of the study are as follows:

1. A new parameter named SD between two successive vehicles is proposed in the present study to classify the followers along with FFC.
2. A gap value of ten sec and an SD range of −4 km/h to +10 km/h were found suitable for the identification of the followers.
3. Two-way traffic volume was recognized as the principal platooning variable which affects the performance of these highways. While other platooning variables did not show so much correlation with performance parameters.
4. It is also concluded that the volume of both directions affects the gap value maintained by two consecutive vehicles in the same lane of the traffic stream.
5. On the premise of the graphical and statistical analysis, it was concluded that previously used speed related parameters couldn't be used to assess performance of two-lane highways under mixed traffic situations.
6. Among all the follower related parameters NFPC was recognized as the best parameter followed by FD to assess the performance of two-lane highways under mixed traffic situations.
7. LOS ranges proposed in a previous study [11] were found higher than the LOS ranges suggested in the present research.
8. In the present research two-way traffic volume was recognized as the key factor which influences the no. of followers and thus performance of two-lane intercity highways, whereas traffic demand was not taken into account in the past studies.

Guidelines proposed in the present examination so as to recognize the followers and non-followers can be adequately used for assessing diverse LOS ranges for developing countries also. The proposed approach will be demonstrated as a guide for traffic architects and organizers (i.e. to design the highway facility under mixed traffic situations). For future research, a lot of extra information from various segments of two-lane highways with a variety in traffic volume and structure can be considered.

References

1. MoRT&H. (2015). *Annual report (2014–15)*. Government of India, New Delhi, India.
2. Ghosh, I., Chandra, S., & Boora, A. (2013). Operational performance parameters for two-lane roads: An assessment of methodological alternatives. *Procedia—Social and Behavioral Sciences, 104*, 440–448.
3. TRB. (2010). *Highway Capacity Manual*. Transportation Research Board, National Research Council, Washington, DC.
4. Al-Kaisy, A., & Karjala, S. (2010). Car-following interaction and the definition of free-moving vehicles on two-lane rural highways. *Journal of Transportation Engineering, 136*(10), 925–931.
5. Dixon, M., Sarepali, S., & Young, K. (2002). Field evaluation of highway capacity manual 2000 analysis procedures for two-lane highways. *Transportation Research Record, 1802*(1), 125–132.

6. Al-Kaisy, A., & Karjala, S. (2008). Indicators of performance on two-lane rural highways: Empirical investigation. *Transportation Research Record: Journal of the Transportation Research Board, 2071*(1), 87–97.
7. Van As, S. C., & Van Niekerk, A. (2004). The operational analysis of two-lane rural highways. In *23rd Southern African Transport Conference* (pp. 622–633), July, 2004.
8. Hashim, I. H., & Abdel-Wahed, T. A. (2011). Evaluation of performance parameters for rural two-lane roads in Egypt. *Alexandria Engineering Journal, Faculty of Engineering, Alexandria University, 50*(3), 245–255.
9. Moreno, A. T., Llorca, C., Sayed, T., & García, A. (2014). Field evaluation of traffic performance parameters for two-lane highways in Spain. In *93rd Transportation Research Board Annual Meeting* (pp. 1–17).
10. Al-Kaisy, A., & Freedman, Z. (2010). Estimating performance on two-lane highways. *Transportation Research Record: Journal of the Transportation Research Board, 2173*(1), 72–79.
11. Penmetsa, P., Ghosh, I., & Chandra, S. (2015). Evaluation of performance parameters for two-lane intercity highways under mixed traffic situations. *Journal of Transportation Engineering, 141*(10), 1–7.
12. Indian Road Congress. (1990). *Guidelines for capacity of roads in rural areas.* IRC (64:1990), New Delhi.
13. Al-Kaisy, A., & Durbin, C. (2008). Evaluating new methodologies for estimating performance on two-lane highways. *Canadian Journal of Civil Engineering*, 777–785.
14. Nokandeh, M., Ghosh, I., & Chandra, S. (2016). Determination of passenger-car units on two-lane intercity highways under mixed traffic situations. *Journal of Transportation Engineering, 142*(2), 548–556.
15. Saha, P., Sarkar, A. K., & Pal, M. (2015). Evaluation of speed–flow characteristics on two-lane highways with mixed traffic. *Transport, 30*(3), 19. Available at: http://www.tandfonline.com/doi/abs/10.3846/16484142.2015.1004369
16. Hashim, I. H. (2011). Analysis of speed characteristics for rural two-lane roads: A field study from Minoufiya Governorate, Egypt. *Ain Shams Engineering Journal, Faculty of Engineering, Ain Shams University, 2*(1), 43–52.
17. Fitzpatrick, K., Miaou, S. P., Brewer, M., Carlson, P., & Wooldridge, M. D. (2003). Exploration of the relationships between operating speed and roadway features on tangent sections. *Journal of Transportation Engineering, 131*(4), 261–269.
18. Miller, A. (1961). A queueing model for road traffic flow. *Journal of the Royal Statistical Society. Series B (Methodological), 23*(1), 64–90.
19. Gattis, B. J. L., & Low, S. T. (1999). Gap acceptance at a typical stop-controlled intersections. *Journal of Transportation Engineering, 125*(June), 201–207.
20. Chandra, S. (2004). Capacity estimation procedure for two-lane roads under mixed traffic situations. Paper No. 498. *Journal of Indian Roads Congress*, 139–170. Indian Road Congress, New Delhi.
21. Kouser, K., & Sunita. (2013). A comparative study of K means algorithm by different distance parameters. *International Journal of Innovative Research in Computer and Communication Engineering, 1*(9), 2443–2447.
22. Rousseeuw, P. J. (1987). Silhouettes: A graphical aid to the interpretation and validation of cluster analysis. *Journal of Computational and Applied Mathematics, 20*, 53–65.

Walkability Analysis of an Urban Area: Gender-Based and Combined Model Approach

Kavita Rani, Amardeep Boora, and Manoranjan Parida

Abstract Walkability analysis has becomes the point of interest of the researchers across the world due to its importance toward social life, economy, empirical quality, and health. It is the need to examine an urban area from the point of interest of health, economy, and safety to propose different walkability thresholds to analyze the walkability condition of a specific locality. Therefore, different study sites (i.e., 15 locations from three cities) were chosen from northern India to assess the walkability of urban transport network. Different models based on gender difference (i.e., male and female) and a combined model were developed using different techniques (i.e., factor analysis, stepwise regression method, and cluster analysis) to develop different walkability thresholds to make the job easier for sidewalk planners (i.e., to evaluate the walkability of an urban area).

Keywords Walkability · Health · Economic · Safety · Gender difference · Factor analysis · Stepwise regression · Clustering

1 Introduction

Walkability is a measure that examines how amenable a locality is to walk. In the past era, walkability was given least priorities by transport planners as it was considered as a minor mode of transport. Walkability is directly addressed to the level of a built environment regarding how much it is friendly to the pedestrian to walk at that particular area or facility. A neighborhood can be called walkable when people will be able to walk safely and easily on foot. A walkability assessment is an important tool for helping to evaluate the weaknesses of pedestrian networks of a particular community, become more physically active and healthy, and identify

K. Rani · M. Parida
Department of Civil Engineering, Indian Institute of Technology Roorkee, Roorkee 247667, India

A. Boora (✉)
Department of Civil Engineering, Jaypee University of Information Technology, Waknaghat, Himachal Pradesh 173234, India

A. K. Gupta et al. (eds.), *Advances in Construction Materials and Sustainable Environment*, Lecture Notes in Civil Engineering 196,
https://doi.org/10.1007/978-981-16-6557-8_9

ways to increase economic and environmental feasibility. Several studies have been carried out across the world focusing on the obesity which is directly related to the walkability. The neighborhood characteristics are associated with the frequency of walking for physical activity in older people [1]. Therefore, a walkable neighborhood increases the physical activity and helps in reducing unhealthy body weight. Before major revolution in the transportation facilities in the nineteenth century, all of the streets were designed in order to support the pedestrian's walkability [2]. Now, the scenario is changing progressively. Currently, it becomes the point of interest of the researchers across the world due to its importance toward social life, economy, empirical quality, and health. Researchers define walkability in different ways. Some researchers defined walkability as an important concern in sustainable urban design [3]. While another studies identified walkability as an important concern in urban planning [4, 5]. Earlier [6], it was concluded that the walkability improvements on a town's "Main Street" can surprise economic development (i.e., by saving travelers money otherwise spent on vehicular travel and saved money is available for spending that fuels the local economy). In another study [7], analysis was done for transit economic evaluation and proposed a model for evaluating non-motorized transportation. Walkable communities command the premium price for both residential and commercial property. Government also can benefit from sales tax revenue by providing a walkable environment or facilities in a particular area or locality without increasing the amount of land. A study carried out by the New York City Department of Transportation [8] included different indicators of economic vitality (sales tax receipts, commercial vacancies, number of visitors) for evaluating street redesigns by considering walking, cycling and public transit facilities, change traffic speeds or change vehicle parking conditions. Beside this, a walkable environment encourages people to walk which is beneficial for their health and ultimately reduces the healthcare cost. The older adults (age of 65 and over) living in more walkable neighborhoods had more transport activity and lower body mass index as they were found more physical activity relative to those living in less walkable neighborhoods [9]. Primary objective of the study is to know about the pedestrians' perception toward existing sidewalks in their localities. Other main objectives of the present study are to identify the principal parameters which affect the walkability of pedestrians of a city and proposing model to estimate the same in the field.

2 Background Studies

Different studies were conducted across the world to examine the walkability of a particular area. Researchers have broadly conducted different studies on walkability. Different methodologies were adopted in different studies to identify the main factors affecting the walkability. Earlier [10], correlation of walkability scores with household travel behavior using different methods namely walkability index, walk opportunity index, pedshed method, and walk score was examined. Among all the methods, pedshed method was identified to be the best walkability index when it comes to explaining the odds of walking to school, indicating that different walkability indices should be used when trying to understand the level to which the built environment encourages walking to various destinations. An attempt was also made in order to provide an urban design support system centered on pedestrian accessibility and walkability of places [11]. In this study, a planning decision support system termed as Walkability Explorer (WE) was proposed which was found appropriate to evaluate the walkability and assist urban design process focusing on accessibility and walkability of pedestrians. It was observed that the destinations and density have additional clarifying power when used together in order to examine other settings [12]. In another study [13], it was recommended to use Walk score as an alternative for estimating neighborhood density and access to facilities rather than as a global measure of neighborhood walkability. Later [14], association between physical and perceived walkability with neighborhood social environment was examined. Earlier, a study [15] was conducted to examine the neighborhood walkability and the walking behavior of Australian adults by utilizing geographic information system (GIS) method and different factors, namely dwelling density, street connectivity, land-use mix, and net retail area to calculate walkability index. However, the walkability index used in this study did not capture access to recreational destinations nor the quality of the pedestrian environment (e.g., sidewalk maintenance and aesthetics) due to which it was not able to explain walking for recreation. In a study [16], it was concluded that changes in walk mode share can be achieved after making improvement in walkability of the least walkable neighborhoods. It was also reported that the walkability generators performance has a profound negative effect on overall walkability performance compared with walkability catalysts performance [17]. Similarly, a study [18] utilized two indexes namely global walkability index and Asian index to measure walkability of a city. A study [7] was conducted in order to understand the walkability and the pedestrian's perception toward the development of pedestrian space. On the basis of results, it was recommended to revise the level of service standards of Highway Capacity Manual (HCM) to reflect better the convergence of other literature and research on what constitutes walkability or what contributes to pedestrian comfort and safety. It was also observed that good-pathway arrangement attracts the people to give preference to walk on pedestrian routes on the alternative mode of travel [19]. Beside this, researchers suggested that the environmental measures should measure separately (some objectively, and others through perceptions) [20]. In a previous study [21], it was observed that the appropriate land-use conditions and pedestrian facilities improvement programs in suburban areas can affect the mode choice and pedestrians travel significantly. Different attributes were identified which encourage the people to walk namely good weather condition, less risk from crime, good sidewalk condition, less car on the street, wide sidewalk, enough lighting, amenities/activities along the way, good traffic signal/signage for pedestrian, and slow movements of cars [22]. It was claimed that a substantial number of trips in Indian cities are made by foot, but pedestrian facilities are neglected and are not given ample focus [23]. Besides this, several studies [24–32] were conducted across the world that focused on the walkability analysis of an urban area.

Table 1 Description of the study sites

Study site	Location	Category	Footpath width (m)
Hisar	Bus Stand	Terminal	1.8
	Red Square Market	Commercial	1.8
	Govt. PG College	Institutional	3
	Sector 15	Residential	2.26
	Town Park	Recreational	1.81
Chandigarh	Sect. 17 Bus Stand	Terminal	1.9
	Shastri Market Sect. 22	Commercial	3.5
	PEC University	Institutional	2.2
	Sect. 21	Residential	1.95
	Rock Garden	Recreational	1.91
Jaipur	Sindhi Camp Bus Stand	Terminal	1.83
	Chaura Rasta	Commercial	1.8
	MNIT Jaipur	Institutional	1.86
	A. G. Colony	Residential	1.95
	Nehru Park	Recreational	3

3 Methodology of the Study

3.1 Site Selection and Data Collection

The present study is focused on the evaluation of pedestrian's perception toward existing sidewalk facilities in an urban area. The study area for this research is taken as Hisar and Chandigarh cities of Haryana state, Jaipur city of Rajasthan which are situated on level terrain in Northern India. The required data of the study area were collected through questionnaire survey (i.e., personal interview survey method). These study locations have different geometric conditions and different roadway characteristics, i.e., sidewalk width varying according to land uses such as in commercial area sidewalk width is greater than those in residential or terminal area. Details of survey locations in Hisar, Chandigarh, and Jaipur City are provided in Table 1. Pedestrian attraction points were selected for questionnaire survey so that higher number of respondents can take participation during the selected time period. The pedestrian perception of different places depends upon the land use, traffic condition, pedestrian infrastructure, personal security, comfort and convenience, accessibility, etc. The present research work is limited to the sidewalk only. According to walkability index score developed by Ministry of Urban Development of India (Ministry of Urban Development [33] by assessing pedestrian infrastructures of 30 cities, an average score of 0.52 was obtained out of 1 and Chandigarh has got the highest of 0.82. In the present study, information like pedestrian perceptions and socioeconomic characteristics of pedestrians were gathered by conducting questionnaire survey. A total of 704 pedestrians participated in the questionnaire survey. Data collection was performed on weekdays only at each of the study site from morning 8 a.m. to 6 p.m. except in institutional areas. In institutional areas, questionnaire survey was conducted in the morning during 8 a.m. to 10 a.m. and in evening from 4 p.m. to 6 p.m. as pedestrian flow was peak at these times. In the data cleaning process, 60 questionnaire forms were removed as these were found to be incomplete, while a total of 100 data samples was preserved for model validation process. Questionnaire was divided into two sections. First section included questions related to socio-demographic characteristics like age, gender, occupation, reason for not walking, and in the second section, respondents were asked to rate the subfactors defined under main factors that enhance walkability like safety from traffic, safety from crime, pedestrian convenience, sidewalk infrastructure, and accessibility. These variables were identified as independent variables, while another variable named as pedestrian's perceived walkability was identified as dependent variable. Traffic safety scale evaluates whether pedestrians are safe on the street (i.e., marking for pedestrian crossing, pedestrian signals, traffic control devices, underpass or foot-over bridge for crossing, etc.). Facilities provided for personal security, i.e., provision of lighting, closed-circuit television (CCTV) cameras, police patrolling, good visibility, etc., assesses the safety of pedestrians from being looted or get murdered by thieves. While comfort and convenience shows how much pleasant the environment is (i.e., how they feel means angry or happy from the facilities provided to them, e.g., cleanliness of sidewalk, street furniture, landscapes, public utility, etc.). Pedestrians infrastructure describes the basic facilities provided to the pedestrians for walking (i.e., footpath width, height, continuity of footpath, encroachment, etc.) and in the last, accessibility assesses the extent of easy access of the pedestrians to another area or destination (e.g., walkable distance to commercial area, to terminal, institutional building, etc.). A five-point Likert scale was used with "one" representing poor condition and "five" representing excellent condition of sidewalks. Table 2 presents the factors and subfactors which were taken into consideration for the study. Responses of pedestrians regarding importance and performance of existing sidewalks of Hisar and Chandigarh's pedestrians were compared using a five-point Likert scale. Initially, important rating was calculated by considering the pedestrians perception toward the variables used in the present study as shown in Table 2. It implies that pedestrian will prefer to walk if an improvement will be observed in these facilities. Weighted mean of the importance rating was calculated to understand the pedestrian's response precisely as shown in Table 3. On the basis of results, it was observed that pedestrians were found more concerned about the safety from traffic in Hisar and Jaipur city (i.e., more than 65%) as the guard rails were not provided on the footpaths. Beside this, there

Table 2 Description of study parameters

Factor	Variable ID	Variables
Safety from traffic (TRR)	TRR1	Traffic volume
	TRR2	Potential for vehicle conflict
	TRR3	Pedestrian signal
	TRR4	Traffic control devices
	TRR5	Traffic speed
	TRR6	Convenience for people crossing
	TRR7	Guard rail
	TRR8	Underpass/foot-over bridge
Safety from crime (PER)	PER1	Provision of lighting
	PER2	Outdoor lighting
	PER3	Police patrolling
	PER4	CCTV cameras
	PER5	Abandoned building or lot
	PER6	Good visibility
	PER7	Safety for walking
Pedestrians convenience (COM)	COM1	Cleanliness of sidewalk
	COM2	Street furniture
	COM3	landscapes
	COM4	Public utility functions
	COM5	Tactile flooring
	COM6	Curb cut
	COM7	Ramps
	COM8	Trees and shades
Sidewalk infrastructure (PIS)	PIS1	Footpath width
	PIS2	Footpath surface
	PIS3	Sidewalk maintenance
	PIS4	Continuity
	PIS5	Obstruction
	PIS6	Location of sidewalk
	PIS7	Encroachment
	PIS8	Footpath height
Accessibility (ACC)	ACC1	Pedestrian volume
	ACC2	Walkable distance to commercial area

(continued)

Table 2 (continued)

Factor	Variable ID	Variables
	ACC3	Walkable distance to bus stops
	ACC4	Walkable distance to institutional building
	ACC5	Walkable distance to mixed land uses
	ACC6	Another pedestrian access point
	ACC7	Two-way movement of pedestrian

Table 3 Weighted mean of importance rating for all the study sites

Parameters	Hisar city	Chandigarh city	Jaipur city
Safety from traffic	3.86	3.74	4.02
Safety from crime	3.36	3.82	3.30
Pedestrian's convenience	2.60	2.84	2.90
Sidewalk infrastructure	2.95	2.82	3.08
Accessibility	2.36	1.85	1.68

were no crossing facilities like foot-over bridge and under pass for pedestrians except at terminal areas. Safety from crime was another important parameter, which was followed by other parameters like sidewalk infrastructure, pedestrian's convenience, and accessibility as shown in Table 3. The main reason for feeling unsafe is due to the absence of CCTV cameras and very low frequency of police patrolling. Besides this, pedestrians complained about the absence of other facilities like provision of benches, public utility, cleanliness, etc. While in case of Chandigarh city, safety from crime was identified as the major concerns for the pedestrians during walking in the streets in early morning and late nights (i.e., due to inadequate lighting facilities) as it got the highest importance rating of 3.82 as shown in Table 3 and Fig. 1. However, on the major roads of the Chandigarh city, lighting arrangements were found up to the mark. Safety from traffic was identified as the other main concern (after safety from crime) having importance rating of 3.74 as there were no available foot-over bridge and under pass for pedestrians crossing at any of the locations except terminal area. Alternatively, pedestrians were also bothered about other parameters which could be ordered as convenience, sidewalk infrastructure, and accessibility. Later, a comparison was also made in weighted mean of satisfactory ratings observed for each subcriterion for all the study sites, namely Hisar, Chandigarh, and Jaipur City as presented in Table 4. From the point of safety from traffic, traffic volume, potential for

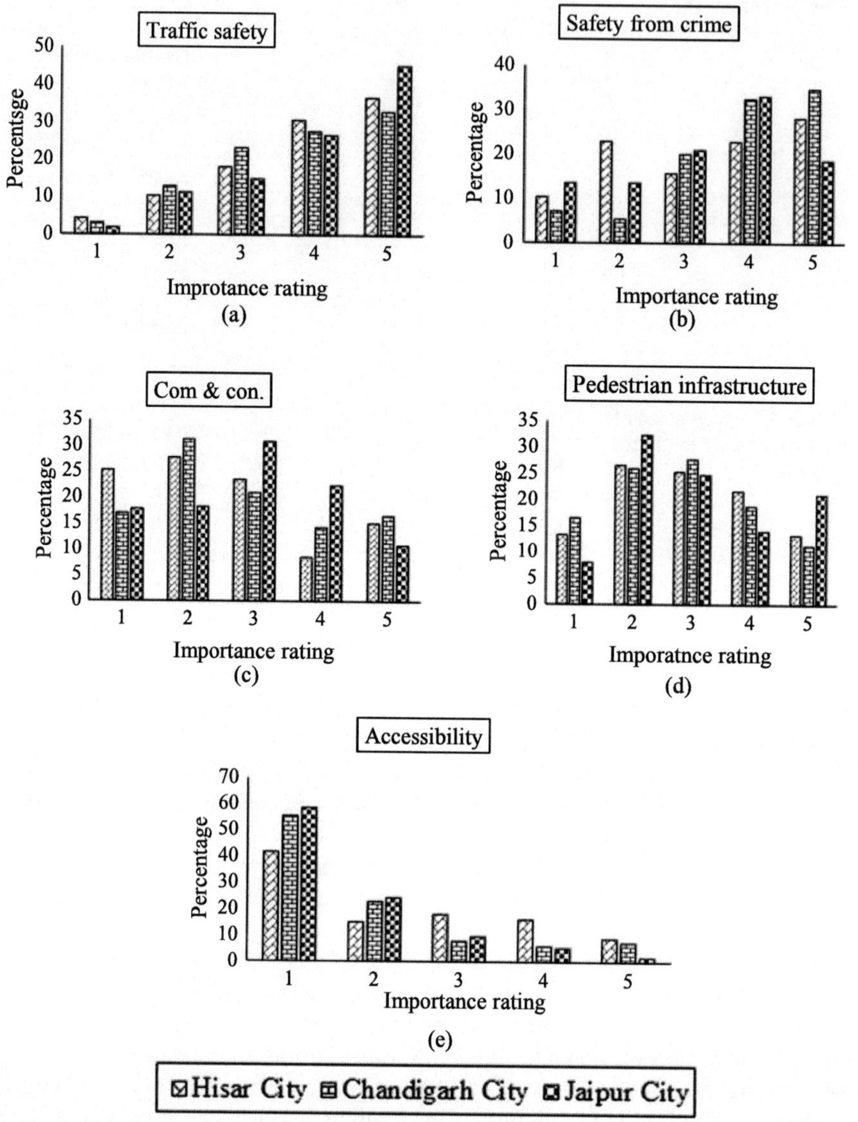

Fig. 1 Importance rating observed across all the study sites

vehicle conflict, traffic control devices, and traffic speed facilities were found satisfactory as per pedestrian perception at all the study sites. In case of personal security, provision of lighting, outdoor lighting, and good visibility were found satisfactory, while in case of pedestrian comfort and convenience, none of the facility was found up to the mark as observed in the field as well as from participant's point of view. However, people gave an average importance to the pedestrian facilities related to

Table 4 Satisfaction rating as per pedestrians perception

Parameters	Weighted mean		
	Hisar city	Chandigarh city	Jaipur city
Traffic volume (TRR1)	4.34	4.36	4.26
Potential for vehicle conflict (TRR2)	3.10	3.41	3.52
Pedestrian signal (TRR3)	1.00	1.50	1.92
Traffic control devices (TRR4)	2.70	2.65	2.68
Traffic speed (TRR5)	3.22	3.90	3.45
Convenience for people crossing (TRR6)	2.31	2.79	2.07
Guard Rail (TRR7)	1.00	1.31	1.39
Underpass/foot-over bridge (TRR8)	2.09	1.88	1.25
Provision of lighting (PER1)	3.41	3.88	3.78
Outdoor lighting (PER2)	3.36	3.86	3.75
Police patrolling (PER3)	2.48	2.42	2.60
CCTV cameras (PER4)	1.45	1.75	2.03
Abandoned building or lot (PER5)	1.92	2.04	1.61
Good visibility (PER6)	3.13	3.89	3.43
Safety for walking (PER7)	2.38	3.16	2.96
Cleanliness of sidewalk (COM1)	2.19	2.68	2.39
Street furniture (COM2)	1.16	1.47	1.62
Landscapes (COM3)	1.40	2.96	2.28
Public utility (COM4)	1.96	1.88	2.17
Tactile flooring (COM5)	1.96	2.18	2.57
Curb cut (COM6)	1.39	1.65	1.74
Ramps (COM7)	1.42	1.68	1.86
Trees and shades (COM8)	2.49	4.50	2.64
Footpath width (PIS1)	2.85	4.15	3.42
Footpath surface (PIS2)	2.58	2.65	2.89
Sidewalk maintenance (PIS3)	1.92	2.41	2.05
Continuity (PIS4)	2.08	2.99	2.59
Obstruction (PIS5)	2.58	1.90	2.32
Location of sidewalk (PIS6)	3.21	3.81	3.15
Encroachment (PIS7)	3.24	2.01	3.12
Footpath height (PIS8)	2.72	2.50	2.88
Pedestrian volume (ACC1)	4.31	4.07	4.08
Walkable distance to commercial area (ACC2)	3.21	3.63	3.68
Walkable distance to bus stops (ACC3)	2.04	3.64	3.33
Walkable distance to institutional building (ACC4)	3.14	3.82	3.19

(continued)

Table 4 (continued)

Parameters	Weighted mean		
	Hisar city	Chandigarh city	Jaipur city
Walkable distance to mixed land uses (ACC5)	3.18	3.63	3.44
Another pedestrian access point (ACC6)	1.90	2.14	2.18
Two-way movement of pedestrian (ACC7)	2.83	4.07	3.88

the pedestrian infrastructure (PIS). Beside these, in case of accessibility, which had the least priority to pedestrians, pedestrian volume, walkable distance to commercial area, walkable distance to institutional building, walkable distance to mixed land uses and two-way movement of pedestrian were performing satisfactory. On the basis of the pedestrians perception, it was concluded that improvement in the facilities as discussed above can result in increase in walkability of these cities.

4 Assessment of Walkability Models

An attempt was made to calibrate different walkability models in order to make the process of walkability analysis easier for sidewalk planner in the field. For the same, different models, namely based on gender difference and a combined model, were developed in the present study using different statistical techniques (i.e., factor analysis, stepwise regression analysis, etc.). Firstly, descriptive statistics was used to measure the central tendency of the data set. Subsequently, 38 parameters were factor analyzed for each model (i.e., based on gender difference, land use, and combine model) with the help of principal component analysis with orthogonal varimax rotation to reduce the number of variables and to group all the variables having high correlation.

In the factor analysis for female participant, a total of nine variables showing less correlation among all the variables, namely traffic speed, cleanliness of sidewalk, landscapes, continuity, obstruction, location of sidewalk, encroachment, footpath surface, and pedestrian volume was removed. In the model developed by considering only the female participants, nine factor solutions which explain a total of 69.08% variance in the whole dataset with eigenvalue greater than 1 were suggested as shown in Fig. 2a. While model developed by considering male participants alone, 11 factor solutions explaining a total of 69.75% of the variance in the whole dataset were suggested as shown in Fig. 2b. A total of seven variables showing less correlation among all the variables namely guard rail, abandoned building or lot, good visibility, pedestrian volume, another pedestrian access point, location of sidewalk and encroachment was removed.

In the primary phase of factor analysis, a number of variables were removed which showed correlation less than 0.40 [34]. It is to be noted that KMO value ranges from

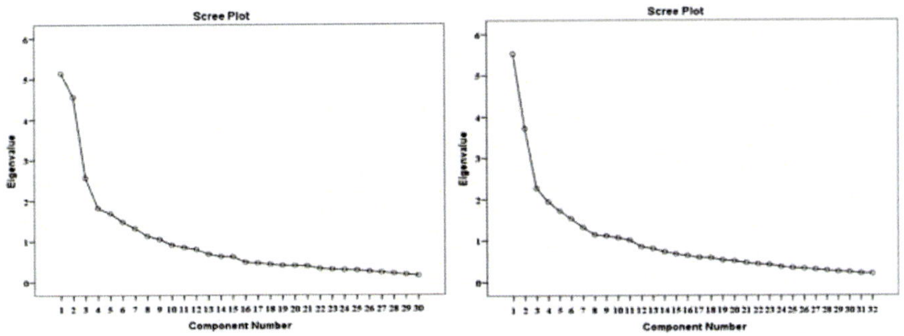

(a) Female (b) Male

Fig. 2 Scree plot for factor analysis based on gender difference

0 to 1 and the KMO value equal to or greater than 0.5 is recommended as satisfactory (i.e., minimum criteria) for factor analysis [35–37]. In a latest study [38], it is mentioned that the KMO statistic [39] was required to be above a minimum of 0.50. Consequently, suitability of factor analysis was evaluated with Kaiser Meyer Olkin (KMO) which exhibits the overall sampling adequacy of 0.759 (female) and 0.771 (male) for gender-based model. In case of combined model, KMO value of 0.785 was observed which explains 66.08% of the variance and suggested 9-factor solution. All the models are showing an appropriate KMO values which are greater than required KMO (i.e., KMO $\geq$ 0.5) as shown in Table 5. Bartlett test of sphericity also exhibited the overall significance of correlation matrix <0.000001. Beside this, Cronbach's alpha test was performed to examine the reliability and internal consistency of the variables for all the study areas. It is to note that a Cronbach's alpha coefficient value equal to or greater than 0.70 is acceptable [40]. The alpha coefficient for 34 parameters in case of gender difference-based model (male) was found 0.79, exhibiting good internal consistency among all the parameters. While in case of female participants based model (i.e., gender difference), alpha coefficient value of 0.775 was obtained which exhibited good internal consistency among 30 variables. A factor

Table 5 Summary of factor analysis

Analytic strategy	Gender difference-based model		Combined model
	Female	Male	
Cronbach's alpha	0.775	0.790	0.80
Kaiser measure of sampling adequacy (KMO)	0.76	0.75	0.79
Bartlett test of sphericity chi-square df significance	2957.48	4424.20	4989.316
	435	496	435
	0.000	0.000	0.000

analysis result for all the methods is tabulated in Table 5. Finally, a stepwise regression analysis was carried out to developed different models at 95% significance level. Stepwise regression systematically adds the most significant variable or removes the least significant variable during each step and by doing so the coefficients of best set of the variables that defines the suitable model will be finalized as the coefficients of regression analysis. Table 6 exhibits the result obtained from stepwise regression analysis for each type of model. All the parameters were found statistically significant at 95% confidence level except the underlined value. After analyzing all the models, combined model was found appropriate for the study. In stepwise regression model, independent variables were selected automatically through step-by-step iteration. From Table 6, it can be observed that both the model types (i.e., gender-based and combined model) are found significant at 95% confidence level. However, combined model is found more significant because of higher R^2 value in comparison to gender-based model (one variable named as PIS1 was found insignificant with female-based model), while COM7 and TRR4 were found negatively associated with walkability of an urban area as shown in Table 6. T-test results indicated that all independent variables are statistical significant. On the basis of graphical and statistical analysis,

Table 6 Summary of all developed models

Model	Category	Variables	Unstandardized coefficients (B)	Model R^2	St. error	t-test	p-value
Gender difference based	Male	(Constant)	0.888	0.24	0.291	3.05	0.002
		COM8	0.178		0.041	4.34	0.00
		PER2	0.191		0.051	3.78	0.00
		TRR8	0.106		0.041	2.57	0.10
		COM4	0.206		0.056	3.69	0.00
		TRR5	0.142		0.051	2.76	0.006
		COM7	−0.171		0.063	−2.6	0.007
		ACC5	0.092		0.045	2.03	0.042
	Female	(Constant)	0.788	0.27	0.323	2.440	0.015
		ACC3	0.170		0.044	3.834	0.000
		COM8	0.173		0.053	3.289	0.001
		TRR4	−0.186		0.053	3.500	0.001
		PIS8	0.155		0.071	2.165	0.031
		PIS1	0.145		0.072	1.998	0.047
Combined model		(Constant)	0.972	0.31	0.215	4.519	0.00
		COM8	0.237		0.031	7.579	0.00
		TRR4	0.204		0.034	5.982	0.00
		ACC2	0.127		0.034	3.61	0.00
		PER1	0.114		0.045	2.569	0.01
		ACC6	0.096		0.041	2.330	0.02

it was concluded that combined model will be helpful for evaluating the walkability condition of sidewalk in an urban area.

5 Conclusion and Future Scope

On the basis of the observations made during the study, it was concluded that safety from traffic is the principal parameters followed by safety from crime in Hisar as the pedestrians were mostly concerned (more than 65%) due to the absence of pedestrian signals, guard rails, foot-over bridge, and many more. Beside this, pedestrian's convenience, sidewalk infrastructure, and accessibility were other important factor. In case of Jaipur, similar type of trend was observed as found in case of Hisar while pedestrians of Chandigarh were found more concerned about safety from crime followed by safety from traffic, pedestrian's convenience, sidewalk infrastructure, and accessibility. Among all the subparameters which come under safety from traffic are namely traffic volume (TRR1), potential for vehicle conflict (TRR2) and traffic speeds (TRR5) were the primary concern for the pedestrians. On the basis of pedestrian's perception for all the study sites, traffic control devices (TRR4) were found up to the mark (i.e., satisfactory in operation) while people were found to cross the road with little hesitation (i.e., satisfactory rating <3 across all the sites). In case of personal security, provision of lighting (PER1), outdoor lighting (PER2), and good visibility (PER6) were found satisfactory, while frequent patrolling by police (i.e., satisfactory rating <3 which seems not good) can result in the improvement of walkability, especially in the night. Pedestrians were found unsatisfied from the facilities being provided to them for their comfort and convenience. Mixed responses (i.e., an average satisfactory rating) were observed regarding pedestrians infrastructure (PIS). However, encroachment (satisfactory rating >3 across all sites) was the main factor which affects the walkability most among all the pedestrians infrastructure-related parameters. Accessibility-related facilities were found having the least priority for the pedestrians at all the study sites. Pedestrian volume (satisfactory rating >4) was identified as the main parameter affecting walkability. Among all the walkability models in terms of walkability score, combined model was found most appropriate with high R^2 value and positively association of all the variable to the walkability in comparison to others. After combining all the data samples, different walkability score thresholds (i.e., from A to F) were proposed using clustering analysis by which a pedestrian or town planner can evaluate the walkability of an urban area very easily. For the future scope, more study sites from different regions of India can be selected for validation of the proposed methodology. Household survey and mobile method can also be used for the same. Study can be extended further by developing age-based models as the perception will vary from age to age (i.e., child to senior citizen). Same study can be done in the hilly regions also (i.e., after including other factors). Walkability score thresholds for each type of model can be developed with the help of different clustering techniques like fuzzy clustering, hierarchical clustering, and many more.

References

1. Shephard, R. J. (2008). Association of the built environment with physical activity and obesity in older persons. *Yearbook of Sports Medicine*, 97–99. https://doi.org/10.1016/s0162-0908(08)79167-0
2. Newman, P., & Kenworthy, J. (1999). *Sustainability and cities: Overcoming automobile dependence*. Island Press.
3. Shelton, T. (2008). Visualizing sustainability in urban conditions. *WIT Transactions on Ecology and the Environment, 113*, 253–262.
4. Rattan, A., Campese, A., & Eden, C. (2012). Modeling walkability. *Arc. User*, 30–33.
5. Stanford, C. B. (2003). Upright: The evolutionary key to becoming human. Houghton Mifflin.
6. Litman, T. A. (2009). Economic value of walkability. *Transportation Research Record: Journal of the Transportation Research Board, 1828*(10), 3–11.
7. Lo, R. H. (2009). Walkability: What is it? *Journal of Urbanism. International Research on Place Making and Urban Sustainability, 2*(2), 145–166.
8. NYCDOT. (2012). *Measuring the street: New metrics for 21st century streets*. Department of Transportation.
9. King, A. C., Sallis, J. F., Frank, L. D., Saelens, B. E., Cain, K., Conway, T. L., Chapman, J. E., Ahn, D. K., & Kerr, J. (2011) Aging in neighborhoods differing in walkability and income: Associations with physical activity and obesity in older adults. *Social Science and Medicine, 73*(10), 1525–1533. https://doi.org/10.1016/j.socscimed.2011.08.032
10. Manaugh, K., & El-Geneidy, A. (2011). Validating walkability indices: How do different households respond to the walkability of their neighborhood? *Transportation Research Part D: Transport and Environment, 16*(4), 309–315.
11. Blecic, I., Cecchini, A., & Trunfio, G. A. (2015). Towards a design support system for urban walkability. *Procedia Computer Science, 51*, 2157–2167.
12. Glazier, R. H., Creatore, M. I., Weyman, J. T., Fazli, G., Matheson, F. I., Gozdyra, P., Moineddin, R., Shriqui, V. K., & Booth, G. L. (2014). Density, destinations or both? A comparison of measures of walkability in relation to transportation behaviors, obesity and diabetes in Toronto, Canada. *PLoS ONE, 9*(1), e85295.
13. Carr, L. J., Dunsiger, S. I., & Marcus, B. H. (2010). Walk score™ as a global estimate of neighborhood walkability. *American Journal of Preventive Medicine, 39*(5), 460–463.
14. Jun, H. J., & Hur, M. (2015). The relationship between walkability and neighborhood social environment: The importance of physical and perceived walkability. *Applied Geography, 62*, 115–124.
15. Owen, N., Cerin, E., Leslie, E., Dutoit, L., Coffee, N., Frank, L. D., Bauman, A. E., Hugo, G., Saelens, B. E., & Sallis, J. F. (2007). Neighborhood walkability and the walking behavior of australian adults. *American Journal of Preventive Medicine, 33*(5), 387–395.
16. Weinberger, R., & Sweet, M. N. (2012). Integrating walkability into planning practice. *Transportation Research Record*, 2322, 20–30.
17. Al-Hagla, K. S. (2009). Evaluating new urbanism's walkability performance: A comprehensive approach to assessment in Saifi Village, Beirut, Lebanon. *Urban Design International, 14*(3), 139–151.
18. Yusuf, A., & Waheed, A. (2015) Measuring and evaluating urban walkability through walkability indexes: A case of Murree. *European Transport-Transporti Europei, 59*.
19. Luadsakul, C., & Ratanvaraha, V. (2013). The study of walkability index: A case study in Nakhon Ratchasima Province. *International Journal of Emerging Technology and Advanced Engineering, 3*(3), 471–476.
20. Smith, A. (2009). Contribution of perceptions in analysis of walking behavior. *Transportation Research Record: Journal of the Transportation Research Board, 2140*, 128–136.
21. Moudon, A. V., Lee, C., Cheadle, A. D., Garvin, C., Johnson, D., Schmid, T. L., Weathers, R. D., & Lin, L. (2006). Operational definitions of walkable neighborhood: theoretical and empirical insights. *Journal of Physical Activity and Health, 3*, S99–S117.

22. Ariffin, R. N. R., & Zahari, R. K. (2013). Perceptions of the urban walking environments. *Procedia-Social and Behavioral Sciences, 105*, 589–597.
23. Wilbur Smith Associates and Ministry of Urban Development. (2008). *Study on traffic and transportation-policies and strategies in urban areas in India.*
24. Datey, A. D. V., Patel, T., & Mahadevia, D. (2012). *Walking and cycling in Indian cities a struggle for reclaiming road edges.* Center for Urban Equality (CUE).
25. Southworth, M. (2005). Designing the walkable city. *Journal of Urban Planning and Development, 131*(4), 246–257.
26. Guo, Z. (2009). Does the pedestrian environment affect the utility of walking? A case of path choice in downtown Boston. *Transportation Research Part D: Transport and Environment, 14*(5), 343–352.
27. Kelly, C. E., Tight, M. R., Hodgson, F. C., & Page, M. W. (2011). A comparison of three methods for assessing the walkability of the Pedestrian environment. *Journal of Transport Geography, 19*(6), 1500–1508.
28. Gehrke, S. R. (2012). A review of walkability measures and the proposal of a standardized classification scheme. In *Transportation research board 91st annual meeting* (No. 12-0361).
29. Bhattacharyya, D. B., & Mitra, S. (2013). Making Siliguri a walkable city. *Procedia—Social and Behavioral Sciences, 96*, 2737–2744.
30. Gori, S., Nigro, M., & Petrelli, M. (2014). Walkability Indicators for Pedestrian-friendly design. *Transportation Research Record: Journal of the Transportation Research Board, 2464*, 38–45.
31. Hung, W. T., Manandhar, A., & Ranasinghege, S. A. (2010). A walkability survey in Hong Kong. In *Hong Kong: The 12th international conference on mobility and transport for elderly and disabled persons* (Vol. 1).
32. Ghani, F., Rachele, J. N., Washington, S., & Turrell, G. (2016). Gender and age differences in walking for transport and recreation: Are the relationships the same in all neighborhoods? *Preventive Medicine Reports, 4*, 75–80. https://doi.org/10.1016/j.pmedr.2016.05.001
33. Ministry of Urban Development (MOUD). (2008). *Studies on traffic and transportation policies and strategies in urban areas in India.* Final Report of Ministry of Urban Development, New Delhi, India.
34. Stevens, B. F. (1992). Price value perceptions of travelers. *Journal of Travel Research, 31*(2), 44–48.
35. Hidayat, N., Choocharukul, K., & Kishi, K. (2011). Pedestrian level of service model incorporating pedestrian perception for sidewalk with vendor activities. In *The 9th International Conference of Eastern Asia Society for Transportation Studies* (pp. 197–197).
36. Yong, A. G., & Pearce, S. (2013). A beginner's guide to factor analysis: Focusing on exploratory factor analysis. *Tutorials in Quantitative Methods for Psychology, 9*(2), 79–94.
37. Williams, B., Brown, T., & Onsman, A. (2010). Exploratory factor analysis: A five-step guide for novices. *Australasian Journal of Paramedicine, 8*(3).
38. Watkins, M. W. (2018). Exploratory factor analysis: A guide to best practice. *Journal of Black Psychology, 44*(3), 219–246.
39. Kaiser, H. F. (1974). An index of factorial simplicity. *Psychometrika, 39*, 31–36.
40. Choi, Y., Seo, M. J., & Oh, S. H. (2016). Walkability analysis of Busan's urban residential zones. *KSCE Journal of Civil Engineering, 20*(6), 2535–2547.

State of Art: Review for Sustainable Application of Waste Material in Rigid Pavement

Amardeep Boora and Ankit Dharma

Abstract Quite a lot of attempts have been made by the researchers throughout the world in order to specify the effects of byproducts (waste) of the different industries. It is to not that the large-scale production of the cement has been identified as the one of the major factor responsible for polluting the water and several diseases to the human beings. Therefore, it is necessary to take an initiative to utilize these byproducts as the replacement of cement (by looking at their chemical compositions). Present study is aiming to identify different possible byproduct to introduce as a replacement of cement as it may be partial replacement or fully replacement which will be helpful in maintain the balance between the society and the environment. Along with this, other objective of the current study is to distinguish the best method to utilize the byproduct of different industries in rigid pavement construction.

Keywords Cement · Byproduct · Rigid pavement · Health

1 Introduction

Availability of good infrastructure (i.e., roads, flyover, buildings which includes educational, commercial and residential, airports facilities, harbors, etc.) exhibits the improvement in the economy of a country which is not possible without utilizing the cement, i.e., in construction. It is a powdered material utilized as a binding substance for in the construction after mixing with other ingredients, i.e., water, sand and aggregates (i.e., highways or building). Along with the benefit of utilizing the cement, it has some serious concerns with respect to the environment and with the health of living beings due to the emission of several hazardous gases. Earlier [44], the utilization of the cement was found to be increased with the rate of 10% per annum. Recently [9], it was reported as the most disposable material after water. In a latest report [25], it was mentioned that the Indian companies are producing 0.83–43.8 million tons cement per year which makes the situation more critical. To preserve

A. Boora (✉) · A. Dharma
Department of Civil Engineering, Jaypee University of Information Technology, Waknaghat 173234, India

A. K. Gupta et al. (eds.), *Advances in Construction Materials and Sustainable Environment*, Lecture Notes in Civil Engineering 196,
https://doi.org/10.1007/978-981-16-6557-8_10

the natural resources and to maintain the sustainability of environment, there is a need to utilize different waste material upto maximum extent without compromising with required properties (i.e., strength, etc.). However, some guidelines have been provided by Indian standards [26] for utilizing the fly ash in road construction while no guidelines are provided for other waste materials. Therefore, it is the necessity of the situation to introduce another waste material or the combination or two or more as an alternative of the cement. Different emission standards (i.e., harmful objects or gases) have been notified for the cement plants in the latest report of Indian government [2]. For example, dust emission for existing cement plants is limited to 1.50 mg/Nm. While, in case of critically polluted areas, the limit was restricted to 100 mg/Nm. Figure 1 is exhibiting the main constituents of cement. It is to note that production of waste in India from different construction work and as well as from different companies was found to be varying from 0.10 to 5.14 million tons per year [66] as shown in Fig. 2. A past study [48] revealed that the minor changes in the

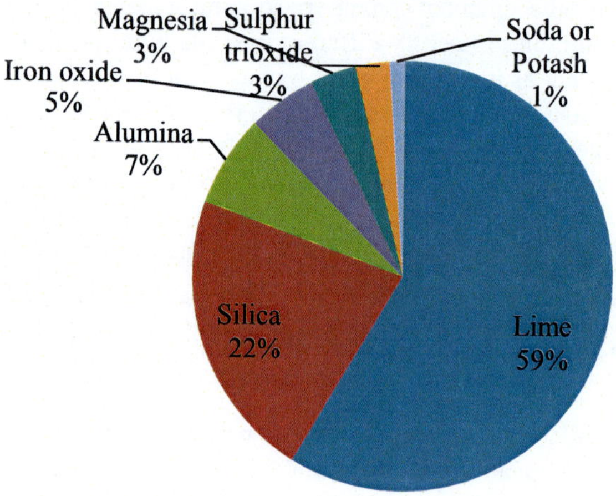

Fig. 1 Major constituents of cement

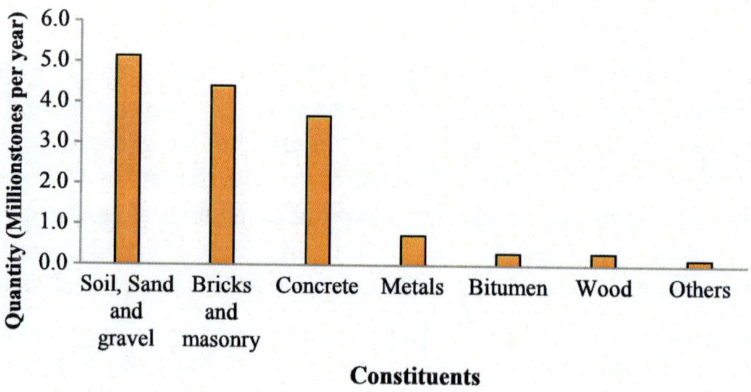

Fig. 2 Waste production rate in India

Table 1 List of bogus compounds of cement [48]

Compounds	Percentage (%)
Tricalcium silicate	40
Dicalcium silicate	30
Tricalcium aluminate	11
Tetracalcium aluminoferrite	12

constituents of the cement can be made by replacing it with other waste materials but it also affect the properties of the cement.

Table 1 is demonstrating diverse bogues mixes of concrete which are framed by blending of concrete with water. It is to note that the C3S (Tri calcium Silicate) is the principal compound for achieving the premature strength and the initial setting time of concrete, while C2S (dia calcium silicate) is responsible for increasing the later strength of cement which uses as a low heat cement for different construction purposes, i.e., bridge, piers, abutments, etc. On the other hand, C3A (Tri calcium aluminate) is responsible for controlling the setting time and shrinkage of cement, while contribution of C4AF (Tetra calcium alumino ferrate) was found very low toward the strength, control setting time, and impart color to cement. In 2014, effect of different gases on human health as well as on environment was examined [46]. Various diseases like tuberculosis, heart-attack, chronic bronchitis, asthma attacks, and many more are caused due to the mass production of the cement due to the emission of different hazardous gases. On the basis of the literature, cement was identified one of the most polluting industries (i.e., as the cement industries have their contribution more than 5% in the emission of greenhouse gasses). Many suspended solids and other substances found to be present in cement which shows a high Chemical Oxygen Demand (COD) are liable for sullying the water assets bringing about the loss of water species in light of the nearness of nitrate and the phosphors.

Siddiqui et al. [61] also found that the consistency of ground water is decreasing due to unnecessary extraction and the existence of absolute solidity, alkalinity, total hardness, and iron within its limits. The most significant effect of the gases emitted during cement manufacturing on the human health was reported by World Health Organization (WHO) report that the several non-treatable diseases are responsible for 82% of deaths and among which 10.7 % deaths are reported because of chronic respiratory diseases, asthma, and chronic obstructive pulmonary [23]. Another study [44] revealed that the workers residing in or near to the factories found suffering from several diseases like lung cancer in respiratory system (1%), chest tightness (49%), impairment of lung function, obstructive, and restrictive lung diseases, while it is responsible for getting affected the gastro intestinal system by causing oral cavity, mechanical trauma, mucosal inflammation, loss of tooth surface, periodontal diseases, dental caries, dental abrasion, liver diffuse, swelling, and proliferation of sinusoidal (hepatic) lining cells. A past study [55] concluded that the creation of wheat crop gets affected as its entire leaf covers with the cement residue with low chlorophyll content. It is to note that the wild life population also can be get affected due to

(a) (b) (c)

Fig. 3 **a** Leaves covers with cement dust. **b** Animals feed grass outside the cement mill. **c** Contaminated water resources

the establishment of these plants near to their livelihood because of eating the contaminated leaf or fruit. Figure 3 displays the various areas found to be contaminated by cement particles.

2 Background Studies

Recently, a review study was conducted [72] focusing on solid waste and nano-particles in order to reduce CO_2 emission. Based on the findings, it was accounted that the expansion of nano-particle of SiO_2 would improve the strength and durability of the structure. Zinc oxide (ZnO) nano-particles are responsible to decreases the pores in concrete structure which gives extra mechanical and flexural strengths to structure. A detailed discussion about the uses of different waste materials to the till date is provided below.

Recycled Concrete Aggregate (RCA)

Demolished concrete waste is a form of solid waste which has been used by different researchers across the world in landfills (i.e., in the form of aggregate). On the basis of the several study, it was observed that the utilization of recycled concrete aggregate can be done up to 100% as the replacement of fresh coarse aggregates. For an example, in a study [76] recycled concrete and demolished concrete waste was utilized as the substitute of natural aggregate (NA). For the same, nearly 50 samples were made by utilizing the waste materials as mentioned earlier in different proportion, i.e., 0%, 25%, 50% up to 100% (i.e., in the difference of 25%) having super plasticizer from 0 to 1.3%. While, 10% of silica fume was added as an alternative of cement content. This study indicated that the utilization of recycled concrete aggregate will compromise with the workability and the compressive strength of the concrete mix.

Later [24], NA were got replaced with 100% of recycled course aggregate. Different samples having different sizes were casted to know the bulk density, specific gravity, and moisture content of the aggregates. Compressive and splitting tensile strength was found to reduce with the use of recycled aggregate. In another study [32] recycled concrete made with limestone aggregate was utilized as the sub-base layer material which reduced the compressive, tensile, shear strengths, and modulus of elasticity of RCA up to 40%. Earlier [10], it was observed that the strength of concrete mix having recycled aggregate along with 5% cement can be increased by increasing the curing time period from 7 days to 28 days. Along with this, another study [12] stated that the concrete road recycled aggregates can be utilized in the construction of roller compacted concrete (RCC) as the performance was found upto the mark. Previously [14], utilization of recycled aggregate up to 25% as the replacement of fine aggregate was found by focusing the least requirement of strength and workability. In another study [49], some specifications were proposed by concerning the use of construction waste as the sub base material for rigid pavement. All the waste materials were showing the similar properties except shear resistance; i.e., RCA was demonstrating greater shear resistance in comparison to gravel and CSA. An earlier study [78] was conducting by focusing the bond strength between recycled concrete and steel bars. On the basis of the results, a decrement in the compressive strength up to 20% was observed in case of RCA in comparison to conventional one. Bond strength was found to be decreased from 50 to 100% in case of plane bars, while no decrement was observed in case of deformed bars. Another study [62] was focusing on the utilization on recycled aggregate (RA) as the replacement of virgin aggregate (VA). Beside this, several other studies [35, 52, 60, 70 and 74] were also utilized and concluded that the recycle concrete and demolition waste can be utilized in construction of rigid pavement.

Red Clay Brick Waste (RCBW)

Red clay brick waste is produced during the manufacturing process of bricks, as the bricks get out of order during the firing and breaks when transportation from one place to another. For the same purpose, behavior of alkaline activation of red clay brick waste (RCBW) and a mixture of Na_2SiO_3 and OPC was investigated [54]. Different samples were casted having different proportion of RCBW, i.e., four samples having variation in range of 5–20% while one sample was casted having 100% RCBW. Results were depicting the highest compressive strength for the samples having 10% of OPC with 28 days curing time period. Along with this, the compressive strength was found to be increased with the additional corroborates.

In another study [50], OMC was found on the higher side due to the utilization of 100% of RCA as the substitute of natural aggregates while maximum dry density was decreasing. A lower CBR value was observed in case of crushed clay brick comparatively consisting of RCA. Balbo et al. [11] stated that the sub-base layer of rigid pavement can be constructed by utilizing the crushed stone. Recently [71], the effect of recycled concrete powder (RCP) on mechanical properties of cement paste was examined. On the basis of the results, a decrement in compressive strength was reported with an increment of RCP more than 20%, while the flexural strength was found to be on higher side.

Fly Ash

Fly ash mostly procured from thermal power plants as the residue of burnt coal. By utilization of fly ash as the alternate of cement provides a cost-effective solution for green concrete structure. Previously [28], the durability of concrete was evaluated by incorporating the fly ash content. It was observed that the utilization of 45% fly ash with the dense graded aggregate increased the durability of the pavement as the specific gravity, permeability, water absorption, and voids in concrete mix get reduced. Another study also revealed that the 15% fly ash can be utilized for pavement construction in base layer as well as a cementitious material [45]. A similar study [1] was conducted to examine the performance of fly ash and bagasse ash as the road construction material. Fly ash (FA) and bagasse ash (BA) were mixed with soil to increase the bearing capacity of lower surface layer of pavement. Several samples were made with the addition of these wastes in different proportion (FA and BA), i.e., 5–35% by weight of soil. Results concluded that the 25% mixing of FA and BA increase the load bearing capacity of soil, while unconfined compressive strength of the soil was found to be increased with inclusion of FA and BA at 30% and 35%, respectively. In another study [75], performance of new type of fly ash named as fly ash microsphere (FAM) was examined in the construction of high-strength concrete. From the results, it was concluded that the 15% replacement of cement with fly ash can be done in order to get optimum results. It was also revealed that the 8% addition of ultrafine fly ash (UFFA) as the replacement of the cement in the concrete mix resulted in the increment in compressive strength, porosity, and durability due to the forming of extra C–S–H gel [63].

Bagasse Ash

Bagasse ash is the by-product of sugar industries generated during cogeneration process which is generally disposed off in landfills. The use of cogeneration plants, and subsequently the quantity of ash generated and disposed, has increased significantly in India. Earlier, a study [13] was conducted to define the fineness of bagasse ash in order to utilize as the replacement of cement in concrete mix. Replacement of cement was done with bagasse ash up to 20%, while the particle size of the bagasse ash was found to be less than 60 micron which do not allow to compromise with the compressive strength of the mix. A study [3] was conducted to identify the impact of bagasse ash on the strength or chloride resistivity of concrete. Different proportion of bagasse ash ranging from 5 to 30% was replaced with cement. While conducting different tests during the study, it was observed that the utilization of bagasse ash (up to 20%) as the replacement of OPC can be done without compromising the requisite properties (i.e., as obtained in case of the conventional mix) of the mix. The main reason for the same could be its higher surface area in comparison to cement and the large amount of silica content, Al_2O_3, Fe_2O_3, and CaO, though, density, specific gravity, and mean grain size were found lower than the cement. In 2015, bagasse ash was utilized as an alternative supplementary cementitious material in concrete mix [8]. On the basis of these findings, it was concluded that the 25% of substitution of cement with bagasse ash can be done. The main reason could be the presence of high composition of silica (SiO_2) which was found to be responsible for the formation of CSH gel resulted in the increment in the strength. Later, different properties of bagasse ash (0–20%) were examined to check its suitability as a construction material [20]. After addition of bagasse ash as the replacement of cement, an alkaline byproduct was found to be formed which is the responsible one to increase the durability of mix. Consequently, it was concluded that the replacement of cement with 20% of bagasse ash will not have any adverse effect on the mix (i.e., properties of the mix). In another study [47], partially replacement of cement was done with the sugarcane bagasse ash, i.e., 0–10% to prepare a mix of M20 and M15 grade of concrete. Result concluded that 5% substitution of cement with bagasse ash will increase the compressive strength up to 12%. Beside this, slump value was found to be more resulted in higher workability.

Ground Granulated Blast Furnace Slag (GGBS)

GGBS is the waste obtained from blast furnaces (i.e., quenching process). Problems, like face dehydration, permeability, thermal expansion, and cracking can be cured by using GGBS. Earlier, [39], the application of ground granulated blast furnace slag (GGBS) as the partial replacement of OPC cement was studied. On the basis of the results, it was stated that the 50% of GGBS can be used as the replacement of OPC without any changes in the strength. Beside this, a decrement in the voids of the mix was observed with the further addition of GGBS while strength gets increased. Consequently, GGBS was utilized in concrete mix for the construction purposes in marine environment [18]. A total of 16 cylindrical samples were made in which some were exposed to water containing NaCl and $CaCl_2$ while some to normal water. Results concluded that the permeability of concrete get decreased and chloride penetration also up to 36% after incorporating GGBS into it which increase its life in the sea water. In another study [31], it was revealed that the late stage of curing of GGBFS containing concrete mix resulted in the increase in the compressive strength as the pozzolanic reaction take place with the mix. In a recent study [43], the strength and permeability of green concrete (i.e., by introducing ground granulated blast furnace slag (GGBS) from 70 to 100% and rice husk ash (RHA) from 0 to 30% as the replacement of cement) was examined. Several tests were executed to measure compressive strength, split tensile strength, chloride permeability, and sorptivity of concrete after 90 days of curing. In order to get the microstructure property, scanning electron microscopy (SEM), energy dispersive spectroscopy (EDS), and x-ray diffraction (XRD) tests were also performed. On the basis of the findings, it was observed that the 15% RHA can be utilized as the replacement of cement having 85% of GGBS.

Ladle Furnace Slag

Ladle furnace slag (LFS) is a solid waste obtained during the steel making process. Earlier, a study [42] was conducted on utilization of LFS in construction. It was concluded that the ladle furnace slag is suitable for paving roads as a soil–cement mixture. Its low cost and time dependent properties, i.e., resistance to load and durability, also contribute to its potential use. Later, characterization of ladle furnace slag was done with the help of sieve size of 0.125–0.063 mm. A study [40] was conducted for utilizing the LFS as a binder by replacing the cement in order to make the sustainable environment. LFS is obtained in a slow cooling process and presents a large content of fine particles (20–35% particles having size less than 75 μm). Proportion of calcium oxide and calcium magnesium silicates was the 88% of the total mix. On the basis of the findings, LFS was found to be suitable as a substitute of cement instead of lime. Beside this, other studies [38, 53] were also made an attempt in the same direction, and it was concluded that LFS can be used in the rigid pavement.

Plastic Waste

Plastic becomes a part of almost all our activity. Naturally, plastic waste is non-biodegradable and takes very long time to degrade. Latest report of Central Pollution Control Board [16] estimated the total annual plastic waste generation in India around 3.3 million metric tons per year. Therefore, some researchers made some attempts to replace plastic as a fine aggregate in place of sand in conventional concrete. In 2014, recycled plastic was utilized as a construction material in India [29]. The coarse aggregates were partially substituted with diverse proportions of *E*-plastic by volume ranging from 10 to 50%. Findings of the study was revealing that the utilization of plastic waste up to 30% can be done without facing any decrement in the requisite compressive strength and other properties also. Later, polyethylene waste was utilized by mixing it with cement [27] in varying ratio, i.e., 15%, 20%, etc. An increment in the density and the compressive strength was observed with the 25–30% plastic waste utilization but the workability was found to be decreased with the plastic waste. Beside this, Appiah et al. [4] also incorporated plastic waste in the construction of flexible pavement to reduce the formation of potholes.

Recycled Asphalt Pavement (RAP)

RAP is defined as material which is detached from the pavement having bitumen and aggregates and obtained during the reconstruction of the pavement (i.e., during removal process of existing wearing course). In a previous study [36], the effect of RAP on number of base and subbase layer was observed. For this study, the calculation of layer coefficient, thickness, and drainage coefficient of each layer was carried out. A total of 9 samples of RAP/crushed-stone base and RAP/gravel-borrow subbase were made by mixing of RAP from 0 to 50% (by weight). Thereafter, 27 samples were tested for resilient modulus and 12 samples were made to check the hydraulic conductivity. As the result shows that the resilient modulus will be increased with the increment in the ratio of RAP. Disposing cost can be compensated by laying the RAP in the subbase layer of the rigid pavement. Structural number was found to be increased with the increment in RAP content. Another study [5] was carried out by focusing on the use of reclaimed asphalt pavement (RAP) in varying proportion, i.e., 100, 50, 30, and 15% and recycled concrete aggregate (RCA) blends in pavement sub bases. Results obtained from different tests revealed that the 15% of RCA blend with RAP is best suited to be used as a subbase layer material.

Later, the effect of extra binders mixed with reclaimed asphalt pavement for construction purposes was examined [6]. Different tests like durability, thermal resistance, stiffness and rutting of material, etc., were performed to check the suitability of recycled concrete as the replacement of natural aggregates. Different RAP binders also have to be added at 135 °C temperature. It was observed that 15% of RAP can be added easily without using any binder, while another study [77] revealed the rutting resistance of rubberized asphalt concrete pavement while mixing with the reclaimed asphalt pavement. Experiment was conducted in two stages. In 2010, the expansion characteristics of RAP and virgin aggregate (VA) were examined [19]. A total of 17 materials were selected to conduct the study. Total 7 RAP material namely surface binder, surface RAP, air-cooled blast furnace, gravel– crushed-stone RAP, steel slag RAP, stone mastic asphalt, and gravel RAP were compared with VA. All the samples were subjected for 7 days curing. It was observed while conducting the expansion test that the RAP materials settled down without showing any expansion. Expansions in samples were noted down after the compaction of specimens. Temperature and pH value of samples were measured as the amount of swelling or expansion depends upon the reactivity of the aggregates and the alkalinity of the solution. Later, a cost-benefit analysis of RAP in various highway applications was conducted [22]. If RAP is used in the pavement construction instead of freshly prepared concrete that will be beneficial from the environmental and economical point of view. For practice purpose, a road was constructed that have RAP in base and asphalt in surface layering heavy loaded truck traffic. Cost-benefit analysis was conducted at the several stages of the construction of road containing RAP. The value of B/C ratio obtained in the present study was found higher than 1 which reflects that the saving can be increased exponentially if large project is deal with RAP as a base layer. When we used RAP in the subbase layer of the pavement, it also reduced the dust particle of soil that is produced by the regular passage of vehicles. Sustainability of roads is also one of the major factors which promote the usage of RAP. Recently [57], decrement in CBR value was reported with the increment in RAP value. Another study [69] was conducted to examine the effect of RAP in the construction of road base and subbase. On the basis of the results it was observed that freshly prepared RAP is not suitable for base layer of the pavement but it can be used for the subbase layer of the rigid pavement.

Besides these, other waste materials were also utilized for the same purpose, i.e., steel slag, rice husk, silica fume, paper and cotton waste, wood and glass waste, marble powder, volcanic ash, e-waste, palm oil and ceramic waste, etc. Earlier [64], steel slag, fly ash, and phosphogypsum were used as the road base course materials. Recently [34], it was concluded that the steel slag powder can be utilized to manufacture high-strength concrete. In a recent study, [56] it was concluded that the cement block containing agricultural waste material (like rice husk, sawdust, peanut shell, rice straw, and coconut shell) in the proportion of 1:5:1 as mentioned above reflects the same properties as of normal mix of block, while water absorption capacity was found to decrease up to 30% due to the utilization of rice husk ash [79]. In a study [33], the effect of silica powder and recycled PVC waste on the shear strength of concrete was examined. A decrement in the shear strength (up to 21%) of concrete was observed after addition of 10% of silica as the replacement of cement in comparison to normal mix. Beside this, several other studies were also utilized the steel slag in concrete mix [37, 65]. Recently [7, 59], waste marble powder was also introduced as an alternative of cement and sand up to some extent, i.e., 15–20%. A review study [41] also concluded that the palm oil and ceramic waste can be used as a partial replacement of cement up to 20% and 30%, respectively. Consequently, in a recent study [68] with the combination of 5% POFA and 10% ceramic waste gives compressive strength up to 27.1 MPa of M30 grade concrete.

While, in another study [58], it was stated that the bone china ceramic waste can replace cement up to 60%. In 2018, wood waste was utilized along with magnesia-phosphate cement (MPC) into fiber reinforcement [73]. Extreme flexural and tensile strength was observed with 35 um diameter of wood fiber reinforcement having 12mm of fiber length and 2% dosage of PVA. In another study [67], waste glass was also utilized in construction purpose in order to make a sustainable environment. Earlier, another study [51] was conducted by focusing on the utilization of recycled paper waste and cotton in the construction of bricks. Study revealed that the 85% PW, 5% CW, and 10% of the cement can be used to make a suitable mix. While, in another study, it was concluded that the paper sludge fly ash could be a feasible replacement of cement; however, its high water demand limits the substitution level to 10% by mass of binder [21]. Recently, a study was conducted to check the suitability of volcanic ash as the replacement of cement into the concrete mixture [30].

3 Discussion and Conclusions

Present study gives an overview of the strategies which were used across the globe for the employment of waste materials in road construction. Earlier [66], different issues related to the generation and handling of wastes (i.e., produced from demolition of existing building) in India were studied. Globally, 2–3 billion tons of C&D waste is produced per year consisting 30–40% of concrete waste. Centre for Science and Environment (CSE) [15] reported 530 million ton (MT) of C&D waste production in India during 2013 which was found 44 times higher than official record and only 1% of it is recycling. Recycling and the reuse of the demolished concrete will help in build the sustainable environment which reduces the landfill cost also. The present study is summing up the existing practices related to the utilization of waste materials (i.e., fully or partially) in construction. Along with that, different experiences observed by the different researchers on the technical, environmental, and economic aspects of the varied applications of these wastes materials are also discussed in detail. This study revealed that the previously several studies were conducted by utilizing various waste materials as the replacement of cement but the definite proportion of the same cannot be defined till now. Along with this, no single research was found to be succeeding in introducing another material as a substitute of cement (i.e., 100% replacement) for the construction purposes. Though, some of the waste materials are utilizing for the same purpose (i.e., in flexible and rigid pavements) like rubber tires, plastics wastes, sewage sludge, coal fly ash, and many more. Among all of these waste materials, GGBS shows its potential to use as a partial substitute of cement with or without any further addition of superplasticizers, i.e., 85% and 50%, respectively, while other materials, namely baggash ash, and RHA were found to have the capability to utilize as the partial substitute of cement up to 25% and 15%, respectively. Besides this, Central Pollution Control Board (CPCB) [17] reported in his latest report that the India produced around 1014961 tons e-waste in 2019–20. Therefore, government should bring some tax policies to minimize waste generation and its unsafe disposal. Besides this, the main challenge is to produce environment friendly concrete mixes by considering the long term performance of rigid pavement which may be the concern of engineers and researchers. For the future scope, another study can be conducted by substituting the cement with the waste material, i.e., independently or in a combine form which will be an aid to the environment, human health and construction field also. Beside this, study can be conducted by focusing the adverse effects of different waste materials, i.e., contamination of ground water by leaching different chemicals. Therefore, an experimental analysis should be done to know the chemical composition of all the waste materials as well as of leachate in order to quantify the disadvantages of the same on human health and ecosystem.

References

1. Anupam, A. K., Kumar, P., Ransinchung, G. D., & Shah, Y. U. (2017). *Study on performance and efficacy of industrial waste materials in road construction: Fly ash and bagasse ash.* Airfield and Highway Pavements 2017. https://doi.org/10.1061/9780784480946.005
2. Annual Report on Environment and Air Pollution India. (2016–2017). *Industrial pollution control.* http://envfor.nic.in/sites/default/files/Environment%20AR%20English%202016-2017.pdf
3. Amin, N. (2011). Use of Bagasse Ash in Concrete and Its Impact on the Strength and Chloride Resistivity. *Journal of Materials in Civil Engineering, 23*(5), 717–720.
4. Appiah, J. K., Berko-Boateng, V. N., & Tagbor, T. A. (2017). Use of waste plastic materials for road construction in Ghana. *Case Studies in Construction Materials, 6*, 1–7. https://doi.org/10.1016/j.cscm.2016.11.001
5. Arulrajah, A., Piratheepan, J., & Disfani, M. M. (2014). Reclaimed asphalt pavement and recycled concrete aggregate blends in pavement subbases: Laboratory and field evaluation. *Journal of Materials in Civil Engineering, 26*(2), 349–357. https://doi.org/10.1061/(asce)mt.1943-5533.0000850
6. Al-Qadi, I. L., Elseifi, M., & Carpenter, S. H. (2007). *Reclaimed asphalt pavement—A literature review.* Illinois Center of Transportation.
7. Ashish, D. K. (2018). Feasibility of waste marble powder in concrete as partial substitution of cement and sand amalgam for sustainable growth. *Journal of Building Engineering, 15*, 236–242.
8. Bahurudeen, A., Kanraj, D., Dev, V. G., & Santhanam, M. (2015). Performance evaluation of sugarcane bagasse ash blended cement in concrete. *Cement and Concrete Composites, 59*, 77–88.
9. Bourtsalas, A. T., Zhanga, J., Castaldi, M. J., & Themelis, N. J. (2018). Use of non-recycled plastics and paper as alternative fuel in cement production. *Journal of Cleaner Production, 181*, 8–16.

10. Behiry, & , A. E. A. E.-M. (2013). Utilization of cement treated recycled concrete aggregates as base or subbase layer in Egypt. *Ain Shams Engineering Journal, 4*(4), 661–673.
11. Balbo, J. T. (1997). High quality cement treated crushed stones for concrete pavement bases. In *6th Purdue conference on concrete pavement* (Vol. 1, pp. 195–207). November 18–21, 1997. Indianapolis.
12. Courard, L., Michel, F., & Delhez, P. (2010). Use of concrete road recycled aggregates for roller compacted concrete. *Construction and Building Materials, 24*(3), 390–395.
13. Cordeiro, G. C., Filho, R. D. T., Tavares, L. M., & Fairbairn, E. M. R. (2009). Ultrafine grinding of sugar cane bagasse ash for application as pozzolanic admixture in concrete. *Cement and Concrete Research, 39*, 110–115.
14. Cuttell, G. D., Snyder, M. B., Vandenbossche, J. M., & Wade, M. J. (1997). Performance of rigid pavements containing recycled concrete aggregates. *Transportation Research Record. Journal of the Transportation Research Board, 1574*, 89–98.
15. Centre for Science and Environment. (2014). *Construction and demolition waste*. India.
16. Central Pollution Control Board. (2018–2019). *Annual report on implementation of plastic waste management rules*. Ministry of Environment, Forest and Climate Change, India.
17. Central Pollution Control Board. (2020). *E-waste management in India*. Ministry of Environment, Forest and Climate Change, India.
18. Das, S., Singh, G., Ahmed, A. A., Saha, S., & Karmakar, S. (2015). Ground granulated blast furnace slag (GGBS) based concrete exposed to artificial marine environment (AME) and sustainable retrofitting using glass fiber reinforced polymer (GFRP) sheets. *Procedia-Social and Behavioral Sciences, 195*, 2804–2812.
19. Deniz, D., Tutumluer, E., & Popovics, J. (2010). Evaluation of expansive characteristics of reclaimed asphalt pavement and virgin aggregate used as base materials. *Transportation Research Record: Journal of the Transportation Research Board, 2167*, 10–17.
20. Deepika, S., Anand, G., Bahurudeen, A., & Santhanam, M. (2017). Construction products with sugarcane bagasse ash binder. *Journal of Materials in Civil Engineering, 29*(10), 04017189.
21. Doudart de la Grée, G. C. H., Yu, Q. L., & Brouwers, H. J. H. (2018). Upgrading and evaluation of waste paper sludge ash in eco-lightweight cement composites. *Journal of Materials in Civil Engineering, 30*(3), 04018021. https://doi.org/10.1061/(asce)mt.1943-5533.0002
22. Franke, R., & Ksaibati, K. (2014). A methodology for cost-benefit analysis of recycled asphalt pavement (RAP) in various highway applications. *International Journal of Pavement Engineering, 16*(7), 660–666.
23. Gizaw, Z., Yifred, B., & Tadesse, T. (2015). Chronic respiratory symptoms and associated factors among cement factory workers in Dejen town, Amhara regional state, Ethiopia. *Multidisciplinary Respiratory Medicine, 11*(1), 13.
24. Huda, S. B., & Alam, M. S. (2014). Mechanical behavior of three generations of 100% repeated recycled coarse aggregate concrete. *Construction and Building Materials, 65*, 574–582.
25. Indian Minerals Yearbook. (2015). *(Part-III: mineral reviews) 54th edition cement*.
26. Indian Road Congress. (1976). Tentative guidelines of cement fly ash concrete for rigid pavement construction. In *IRC code of practices, 1976 vol (68)*. New delhi, India.
27. Jassim, A. K. (2017). Recycling of polyethylene waste to produce plastic cement. *Procedia Manufacturing, 8*, 635–642.
28. Jerath, S., & Hanson, N. (2007). Effect of fly ash content and aggregate gradation on the durability of concrete pavements. *Journal of Materials in Civil Engineering, 19*(5), 367–375.
29. Kumar, K. S., & Baskar, K. (2014). Recycling of E-plastic waste as a construction material in developing countries. *Journal of Material Cycles and Waste Management, 17*(4), 718–724.
30. Kupwade-Patil, K., Chin, S. H., Johnston, M. L., Maragh, J., Masic, A., & Büyüköztürk, O. (2018). Particle size effect of volcanic ash towards developing engineered Portland cements. *Journal of Materials in Civil Engineering, 30*(8), 04018190. https://doi.org/10.1061/(asce)mt.1943-5533.0002348
31. Kuo, W. T., Chen, S. H., Wang, H. Y., & Lin, J. C. (2013). A study on the mechanical and electricity properties of cement mortar added with GGBFS and piezoelectric powder. *Construction and Building Materials, 49*, 251–256.

32. Kuo, S. S., Mahgoub, H. S., & Nazef, A. (2002). Investigation of recycled concrete made with limestone aggregate for a base course in flexible pavement. *Transportation Research Record, 1787*, 99–108.
33. Kurup, A. R., & Senthil Kumar, K. (2017). Effect of recycled PVC fibers from electronic waste and silica powder on shear strength of concrete. *Journal of Hazardous, Toxic, and Radioactive Waste, 21*(3), 06017001.
34. Liu, J., & Guo, R. (2018). Applications of steel slag powder and steel slag aggregate in ultra-high performance concrete. *Advances in Civil Engineering.*
35. Malešev, M., Radonjanin, V., & Marinkovic, S. (2010). Recycled concrete as aggregate for structural concrete production. *Sustainability, 2*(5), 1204–1225.
36. MacGregor, J., Highter, W., & DeGroot, D. (1999). Structural numbers for reclaimed asphalt pavement base and subbase course mixes. *Transportation Research Record: Journal of the Transportation Research Board, 1687*, 22–28.
37. Maddalena, R., Roberts, J. J., & Hamilton, A. (2018). Can Portland cement be replaced by low-carbon alternative materials? A study on the thermal properties and carbon emissions of innovative cements. *Journal of Cleaner Production, 186*, 933–942.
38. Maghool, F., Arulrajah, A., Haghighi, H., & Horpibulsuk, S. (2017). The influence of a curing regime on the geotechnical properties of ladle furnace slag as used in pavement applications. *Geotechnical Frontiers*, 324–333.
39. Manjunatha, L. R., Anvekar, S. R., & Yogananda, M. V. (2014). Recent developments in the Indian concrete industry in the use of GGBS in concrete at RMC batching plants as partial replacement to OPC cement and its effects on concrete durability and sustainabilty in the Indian context. In *International congress on durability of concrete* (Vol. 1).
40. Marinho, A. L. B., Santos, C. M. M., Carvalho, J. M. F., Mendes, J. C., Brigolini, G. J., & Peixoto, R. A. F. (2017). Ladle furnace slag as binder for cement-based composites. *Journal of Materials in Civil Engineering, 29*(11), 04017207.
41. Mazenan, P. N., Khalid, F. S., Shahidan, S., & Shamsuddin, S. M. (2017). Review of palm oil fuel ash and ceramic waste in the production of concrete. *IOP Conference Series: Materials Science and Engineering, 271*(1), 012051.
42. Manso, J. M., Losanez, M., Polanco, J. A., & Gonzalez, J. J. (2005). Ladle furnace slag in construction. *Journal of Materials in Civil Engineering, 17*(5), 513–518.
43. Mehta, A., & Siddique, R. (2018). Sustainable geopolymer concrete using ground granulated blast furnace slag and rice husk ash: Strength and permeability properties. *Journal of Cleaner Production, 205*, 49–57.
44. Mehraj, S., & Bhat, G. (2013). Cement factories, air pollution and consequences. Department of Environmental Sciences & Centre of Research for Development. University of Kashmir, Jammu and Kashmir, India.
45. Mohammadinia, A., Arulrajah, A., Horpibulsuk, S., & Chinkulkijniwat, A. (2017). Effect of fly ash on properties of crushed brick and reclaimed asphalt in pavement base/subbase applications. *Journal of Hazardous Materials, 321*, 547–556.
46. Mishra, S., & Siddiqui, N. A. (2014). A review on environmental and health impacts of cement manufacturing emissions. *International Journal of Geology, Agriculture and Environmental Sciences, 3*, 26–31.
47. Modani, P. O., & Vyawahare, M. R. (2013). Utilization of bagasse ash as a partial replacement of fine aggregate in concrete. *Procedia Engineering, 51*, 25–29.
48. Mindess, S., & Young, J. F. (1981). *Concrete* (p. 671). Prentice-Hall, Inc.
49. Park, T. (2003). Application of construction and building debris as base and subbase materials in rigid pavement. *Journal of Transportation Engineering., 29*(5), 558–563.
50. Poon, C. S., & Chan, D. (2006). Feasible use of recycled concrete aggregates and crushed clay brick as unbound road subbase. *Construction and Building Materials, 20*(8), 578–585.
51. Rajput, D., Bhagade, S. S., Raut, S. P., Ralegaonakar, R. V., & Mandavgane, S. A. (2012). Reuse of cotton and recycle paper mill waste as building material. *Construction and Building Materials, 34*, 470–475.

52. Rao, A., Jha, K. N., & Misra, S. (2007). Use of aggregates from recycled construction and demolition waste in concrete. *Resources, Conservation and Recycling, 50*(1), 71–81.
53. Ra đenović, A., Malina, J., & Sofilić, T. (2013). Characterization of ladle furnace slag from carbon steel production as a potential adsorbent. *Advances in Materials Science and Engineering*, 1–6. https://doi.org/10.1155/2013/198240
54. Robayo, R. A., Mulford, A., Munera, J., & Mejía de Gutiérrez, R. (2016). Alternative cements based on alkali-activated red clay brick waste. *Construction and Building Materials, 128*, 163–169.
55. Sai, V. S., Mishra, M. P., & Mishra, G. P. (1987). Effect of cement dust pollution on trees and agricultural crops. *Asian Environment, 9*(1), 11–14.
56. Sathiparan, N., & De Zoysa, H. T. (2018). The effects of using agricultural waste as partial substitute for sand in cement blocks. *Journal of Building Engineering, 19*, 216–227.
57. Seferoğlu, A. G., Seferoğlu, M. T., & Akpınar, M. V. (2018). Investigation of the effect of recycled asphalt pavement material on permeability and bearing capacity in the base layer. In *Advances in civil engineering*.
58. Siddique, S., Shrivastava, S., & Chaudhary, S. (2018). Influence of ceramic waste as fine aggregate in concrete: Pozzolanic, XRD, FT-IR, and NMR Investigations. *Journal of Materials in Civil Engineering, 30*(9), 04018227.
59. Singh, M., Lamba, P., Srivastava, A., & Bhunia, D. (2017). Long term effect of partially replacing cement by waste marble slurry in concrete. In *Proceedings of the First Congress on Technical Advancement* (pp. 65–76).
60. Silva, R. V., De, B. J., & Dhir, R. K. (2014). Properties and composition of recycled aggregates from construction and demolition waste suitable for concrete production. *Construction and Building Materials, 65*, 201–217.
61. Siddiqui, N. A., Shukla, R. N., & Ziauddin, A. (2011). A impact of cement industry on the ground water quality—A case study. *Journal of Industrial Pollution Control, 27*(2), 1–5.
62. Singh, P. (2015). *Resilient modulus of recycled aggregates as road pavement materials*. Southern Illinois University at Carbondale.
63. Shaikh, F. U., & Supit, S. W. (2015). Compressive strength and durability properties of high volume fly ash (HVFA) concretes containing ultrafine fly ash (UFFA). *Construction and Building Materials, 82*, 192–205.
64. Shen, W., Zhou, M., Ma, W., Hu, J., & Cai, Z. (2009). Investigation on the application of steel slag–fly ash–phosphogypsum solidified material as road base material. *Journal of Hazardous Materials, 164*(1), 99–104.
65. Sobolev, K., Kozhukhova, M., Sideris, K., Menéndez, E., & Santhanam, M. (2018). Alternative supplementary cementitious materials. *RILEM State-of the-Art Reports, 25*, 233–282.
66. Srivastava, S., & Chini, A. (2009). Construction materials and C&D waste in India. In *CIB W115 construction material stewardship* (pp. 72–76). International Council of Building Research Studies and Documentation.
67. Sudharsan, N., Palanisamy, T., & Subhash, C. Y. (2018). Environmental sustainability of waste glass as a valuable construction material—A critical review. *Ecology, Environment and Conservation, 24*, 335–342.
68. Sheikh Khalid, F., Natasya Mazenan, P., Halim Abdul Ghani, A., Bazilah Azmi, N., Irwan Juki, M., Shahidan, S., & Haziman Wan Ibrahim, M. (2018). An utilization of palm fuel ash (POFA) and ceramic waste as cement materials replacement in concrete production. *International Journal of Engineering & Technology, 7*(3.9), 89. https://doi.org/10.14419/ijet.v7i3.9.15284
69. Taha, R., Al-Harthy, A., Al-Shamsi, K., Al-Zubeidi, M. (2002). Cement stabilization of reclaimed asphalt pavement aggregate for road bases and subbases. *Journal of Materials in Civil Engineering, 14*(3), 239–245.
70. Topcu, I. B., & Şengel, S. (2004). Properties of concretes produced with waste concrete aggregate. *Cement and Concrete Research, 34*(8), 1307–1312.
71. Topic, J., Prošek, Z., & Plachý, T. (2017). Influence of increasing amount of recycled concrete powder on mechanical properties of cement paste. *IOP Conference Series: Materials Science and Engineering, 236*(1).

72. Vishwakarma, V., & Ramachandran, D. (2018). Green concrete mix using solid waste and nanoparticles as alternatives—A review. *Construction and Building Materials, 162*, 96–103.
73. Wang, L., Iris, K., Tsang, D. C., Yu, K., Li, S., Poon, C. S., & Dai, J. G. (2018). Upcycling wood waste into fibre-reinforced magnesium phosphate cement particleboards. *Construction and Building Materials, 159*, 54–63.
74. Wan, D. S. L. Y., Aslani, F., & Ma, G. (2018). Lightweight self-compacting concrete incorporating perlite, scoria, and polystyrene aggregates. *Journal of Materials in Civil Engineering, 30*(8), 04018178.
75. Wang, Q., Wang, D., & Chen, H. (2017). The role of fly ash microsphere in the microstructure and macroscopic properties of high-strength concrete. *Cement and Concrete Composites, 83*, 125–137.
76. Wagih, A. M., El-Karmoty, H. Z., Ebid, M., Okba, S. H. (2013). Recycled construction and demolition concrete waste as aggregate for structural concrete. *HBRC Journal, 9*(3), 193–200.
77. Xiao, F., Amirkhanian, S., & Juang, C. H. (2007). Rutting resistance of rubberized asphalt concrete pavements containing reclaimed asphalt pavement mixtures. *Journal of Materials in Civil Engineering, 19*(6), 475–483.
78. Xiao, J., & Falkner, H. (2007). Bond behavior between recycled aggregate concrete and steel rebars. *Construction and Building Materials, 21*(2), 395–401.
79. Zabihi, S. M., Tavakoli, H., & Mohseni, E. (2018). Engineering and microstructural properties of fiber-reinforced rice husk-ash based geopolymer concrete. *Journal of Materials in Civil Engineering., 30*(8), 04018183.

Study on Cost Modeling and Economical Design of Superstructure

Peerzada Danish, Kamil Ashraf Bhat, S. Ganesh, and J. Anita Jessie

Abstract Due to the escalation of the prize of construction materials, Engineers are forced to reduce the construction cost in buildings by adopting various techniques. In this project, an attempt is made to reduce the cost of the buildings in the super structures, since 70% of the cost of the building has been spent on it. A comparison of detailed estimate for steel as well as concrete has been made. To make the building economical and structurally efficient, the traditional method of load-bearing walls has been avoided by adopting steel framed structure, since the cost of a reinforced concrete-framed structures is high because of the use of cement, sand, and aggregates. Most of the loads are not fully carried by concrete, and hence, the members are not efficient. For this reason, steel super structure is considered here. The optimization on the structural components has been done by adopting techniques like bracings, lightweight floor slabs, roof slabs, and partition walls which are analyzed and designed in ETABS. Generally, the bracings are the structural component that converts lateral load into axial force. Therefore, the moment on the structure has been reduced, and hence, the stability can be improved. As steel structures are prefabricated the assemblies of these systems require less time and environmental factors can also be avoided. The structure can be completed within the stipulated time based on the requirements of the client. Hence, the construction time can be reduced in addition to reduction in the cost of the building. The economical alternative among steel and reinforced concrete building can be analyzed by doing the cost estimation of the building.

P. Danish (✉)
Department of Civil Engineering, Government College of Engineering and Technology, Safapora, Ganderbal 193504, Jammu and Kashmir, India

P. Danish · K. A. Bhat · J. Anita Jessie
School of Civil Engineering, Vellore Institute of Technology, Vellore 632014, Tamil Nadu, India
e-mail: anitajessie.j2015@vit.ac.in

S. Ganesh
School of Civil Engineering, Lovely Professional University, Jalandhar 144411, Punjab, India
e-mail: ganesh.19421@lpu.co.in

A. K. Gupta et al. (eds.), *Advances in Construction Materials and Sustainable Environment*, Lecture Notes in Civil Engineering 196,
https://doi.org/10.1007/978-981-16-6557-8_11

Keywords Modeling · Cost optimization · ETABS · Reinforced concrete framed structure · Steel structure

1 Introduction

The construction industry is the most resource-intensive industry globally. The economics of the construction industry depends on the several factors like availability and scarcity of resources, local cost of materials, etc. [1]. Decreasing the economy of the cost of the construction can help in expanding the industry and improve the economic resilience [2]. Cost of construction plays an important role in construction projects. Even wood may be considered as the construction material for the high-rise buildings to reduce the cost [3]. Apart from the cost and economy, the time of construction is also of high importance [4, 5]. Steel buildings generally require lower time for construction [6, 7]. The cost of reinforced concrete and steel structure depends on the availability of the resources and cost of the materials of the area. Adopting the material with less cost can result in the reduced cost of the construction [8, 9]. Usually reinforced concrete structures are preferred for the small-scale construction, but reinforced concrete structures have more dead load due to which they are not preferred for high-rise structures. Increased dead load can also increase the seismic load of the structure which makes it prone to more structural damage. Steel structures can undergo more plastic deformation before failure and have more strength/weight ratio which makes steel structures more preferable for the regions with high seismicity. Steel structures have higher rigidity and ductility and are cost effective too and are thus used in variety of structures like high-rise buildings, bridges, towers, etc. The reinforced concrete buildings can be economical in cases where the buildings are small as concrete can increase weight of high-rise structures by a lot due to more dead load of concrete structures, but reinforced concrete structures can prove to be more economical in places where the price of steel may be more or in small-scale construction in which dead load is less. Another advantage of steel structure is that it can be erected in less time as most steel parts are cast in factories and not on site like in case of concrete structures. Due to these factors, one may take into consideration adopting steel structures, but for estimating the economic benefit of steel structures a detailed investigation needs to be done. The objective of this project is to determine the economical design of a super structure. It has been carried out by comparing concrete structure with steel structure. In this project, a normal $G + 2$ building has been taken. The design has been carried out with both concrete and steel structure [10, 11] of the normal $G + 2$ building, and then, the estimation cost for both concrete structure and steel structure has been done. The steel structure with bracing is also designed, and then, its cost is estimated [12, 13]. Generally, the steel structures are faster to build with less labor. The steel buildings are used for number of residential and commercial buildings because of their versatility and its structural stability. The idea of bracing is to reduce the cost of steel structure [14, 15]. By adopting the bracing technique, maximum moment can be resisted with less steel section than those required for steel structure without bracing [16]. Bracings can also be used as a diagrid along the circumference of the building which can reduce the steel required for the construction of the building [17]. In this paper, analysis and design of the two-story reinforced concrete structure and steel structure are done, and then, the cost of the buildings is estimated as per the price in Chennai (during the year 2012) and the more economical alternative is determined.

2 Methodology

The building is symmetrical about the local axes, and stability of structure is obtained by good geometry. Grouping of the columns is also done from the center line plan of the building. The building is modeled and designed in ETABS [18, 19]. For the modeling purposes, M20 grade of concrete and Fe415 steel are considered. For steel building, the standard I-sections are adapted, and for the bracing, the standard angle sections are adapted as per Indian Standards [19, 20]. The steel section used has the yield strength of 250 N/mm^2. The building as shown in Fig. 1 is modeled in ETABS as concrete and steel structure. The story height of 3.5 m is adopted for the building. The live load of 5 kN/m and the dead load in place of walls and wherever necessary are used [21, 22]. The earthquake load is also considered for the design. The building is then designed for the load combinations given in IS 1893 [23], and the optimum section is adopted for consideration in the estimation and costing. We get the details of the steel and concrete used for the concrete structure and the steel used for the steel structure from ETABS.

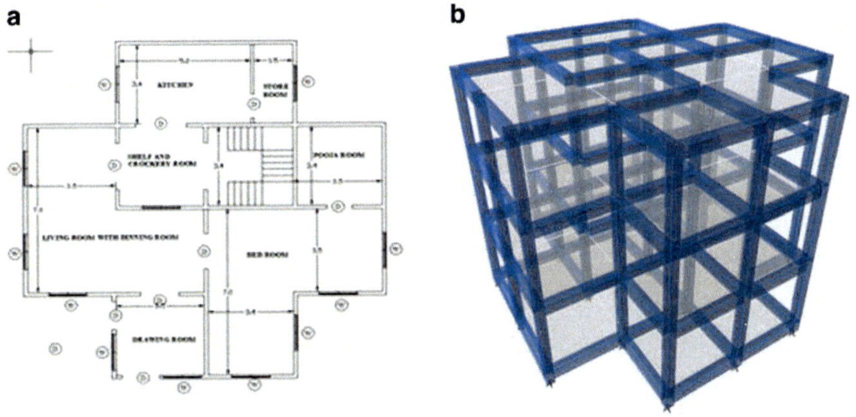

Fig. 1 **a** Sectional plan of the building **b** 3D model of the building in ETABS

Table 1 Material cost of the buildings

Particulars	Concrete structure	Steel structure (unbraced)	Steel structure (braced)
Weight of beams and columns (tons)	18.386	12.05	10.65
Weight of slabs (tons)	16.892	16.892	16.892
Material cost of beams and columns	Rs. 489,469	Rs. 554,300	Rs. 489,900
Material cost of slabs	Rs. 348,413	Rs. 348,413	Rs. 348,413
Total material cost	Rs. 837,882	Rs. 902,712	Rs. 838,312

3 Estimation and Costing

After the design and calculations, the cost estimation of the building is done. The weights and material costs of the beams and columns and slabs of the building and their material costs are calculated. The material costs of the concrete building are the costs of the steel reinforcement and the concrete utilized. In case of the steel building, the weight of the building is the weight of the steel utilized. In both cases of the concrete and steel buildings, the reinforced concrete slabs are used. The details of the weight of beams and columns and the slabs in the concrete, steel, and braced steel buildings and their material costs are given in Table 1.

The material cost of Rs. 45,000 for reinforcement and Rs. 46,000 for structural steel per metric ton is adopted, and the cost of Rs. 4500 per m^3 is adopted for concrete as per the cost of these materials in Chennai (during the year 2012).

Aside from the cost of materials, the cost of construction is also to be taken into consideration. The labor cost is given in Table 2.

The labor cost for the steel building and the braced steel building will be almost same and is given in Table 3.

The total costs of the reinforced concrete building are the sum of material and labor costs which is Rs. 1,040,777. The total cost of the normal steel building is Rs. 1,104,312.5, and the total cost of the steel building with the bracings is Rs. 1,039,912.5.

4 Results and Discussion

Analysis has been done on the building using the load combinations for all considerations of design. For similar loading, analysis and design have been done with ETABS model. During the analysis, there was considerable reduction in the bending moment and shear force of the steel structure and showed the inefficiency of concrete building to cater heave live loads. The deflection and moment in the steel structure were less as compared to that of concrete and are given in Table 4.

Table 2 Labor cost of concrete building

Labor cost	
Total rate for 1 day	Rs. 4563
Cost of curing for 7 days	Rs. 2800
Total cost for 1.5 m of column concrete	Rs. 7363
Cost of concrete for 3.5 m height	Rs. 17,200
Total rate for 1 day	Rs. 4563
For Slab	
Rate for slab concrete	Rs. 19,805
Cost of curing for 10 days	Rs. 3500
Total cost of labor for slab concrete	Rs. 23,305
Rate of labor for whole building	Rs. 121,515
For water service 5% of the total cost	Rs. 6080
For bar bending 40% of the total cost	Rs. 48,600
For shuttering 15% of the total cost	Rs. 18,200
For the other services 7% of the cost	Rs. 8500
Total cost	Rs. 202,895

Table 3 Labor cost of steel building

Cost for a beam column junction	Rs. 3200.00
Number of beam column junction	Rs. 21
Cost of one floor	Rs. 67,200
Labor cost for steel building	Rs. 201,600

Table 4 Comparison of cost of concrete and steel building

S. No.	Particulars	Concrete structure	Steel structure
1	Dead weight of building	35.278 tons	28.942 tons
2	Cost of building superstructure	837,882/-	902,712/-
3	Labor cost	202,895/-	201,600/-
4	Time for completion	6 months	4 months

Study was again carried out to reduce the structural cost of the steel building. For reducing the moment in the beams, bracing was used. The bracings took the load from the beam to the column effectively as axial force. It requires only a minimum cross section as there was zero moment action on it. The design was optimized in ETABS, and the amount of steel required was found out. The bracings were given on the outer faces of the building and at the interior columns and beams.

The bracings were attached on the beams at midpoints where moment was maximum, so the beam was split up into two, and the moment was taken out of the

Table 5 Comparison of cost of simple steel and braced steel building

S. No.	Particulars	Normal	Braced
1	Construction materials	28.942 tons	27.542 tons
2	Cost of building superstructure	899,994/-	780,127/-
3	Labor cost	201,600/-	201,600/-
4	Time for completion	6 months	4 months

junction by the bracings. This methodology of providing bracings reduced required steel of the buildings. The comparison between the braced and normal steel building is given in Table 5.

The total steel required for the braced structure was reduced by 1.4 tons (4.8%). Including bracings steel quantity reduced and shows that bracings were taking greater vertical load and in a reduced section. The cost of building was also reduced by 1.2 lakhs (13.3%). This shows that braced steel structure is very much adoptable for cost-effective construction methods, which also produced more serviceability and greater stability.

5 Conclusions

The sectional inefficiency of concrete structures is due to increased sections of concrete. The structure has to be provided with more strength to cater with its own self-weight. Comparing the self-weight of the structure steel braced and concrete structure, steel braced showed 22% reduction in dead weight of structure as compared to concrete structure. The total cost of the construction on the other hand was about 5.8% less for concrete structure as compared to the normal steel building but was 5.6% more than the braced steel building. It shows that the braced steel building can be more economical in Chennai. There are also more advantages of the steel building due to which steel building can be preferred such as higher yield capacity and less dead weight of the building which can increase the seismic performance of the building. Another factor that needs to be taken into consideration is the time of completion of the construction project. Steel buildings take less time for construction and can be adopted where there is less time available for construction or project needs to be completed as soon as possible. Due to these reasons, one may be more inclined to consider the braced steel structure for the building.

Steel brace structure is highly efficient structure showing good stability and economic values. As construction materials are getting scarce, steel can be used efficiently in structures with less wastage.

Acknowledgements The authors feel highly indebted to the Chancellor of Vellore Institute of Technology, Vellore, for the services provided to carry out the experiments.

References

1. Jones, A., Stegemann, J., & Sykes, J. P. (2016). Winslow, adoption of unconventional approaches in construction: The case of cross-laminated timber. *Construction and Building Materials, 125*, 690–702.
2. European Commission. (2019). *Final towards a circular economy: A zero waste programme for Europe COM 398 Final, 2014* (pp. 398). Available from: http://ec.europa.eu/transparency/regdoc/rep/1/2014/EN/1-2014-398-EN-F1-1.Pdf. Accessed February 25, 2019.
3. Mohammadi, J., & Ling, L. (2017). Can wood become an alternative material for tall building construction. *Practice Periodical on Structural Design and Construction*. https://doi.org/10.1061/(ASCE)SC.1943-5576.0000334
4. Ahmed, S., & Arocho, I. (2021). Analysis of cost comparison and effects of change orders during construction: Study of a mass timber and a concrete building project. *Journal of Building Engineering, 33*, 101856. https://doi.org/10.1016/j.jobe.2020.101856
5. Gagandeep. (2020). Time and cost comparison of reinforced cement concrete and steel structure. *Materials Today: Proceedings*. ISSN: 2214-7853. https://doi.org/10.1016/j.matpr.2020.08.672
6. John, S., & Buchanan, A. (2012). Cost and construction time for a 3-story post tensioned timber building compared with concrete and steel building. In *Proceedings of the 2012 World Conference on Timber Engineering*. Auckland, New Zealand.
7. Bockholta, M. T., Kristensen, J. H., Colli, M., Jensen, & Wæhrens, B. V. (2020). Exploring factors affecting the financial performance of end-of-life take-back program in a discrete manufacturing context. *Journal of Cleaner Production, 258*, 120916. https://doi.org/10.1016/j.jclepro.2020.120916
8. Zaveri, B. H., Gadhiya, J. A., & Dhameliya, H. K. (2016). A review on the comparative study of steel, RCC and composite building. *International Journal of Innovative Science Engineering and Technology, 5*.
9. Jirage, M., Sayagavi, V. G., & Gore, N. G. (2012). Comparative study of RCC and composite multi-storeyed building. *International Journal of Applied Science and Engineering (IJSEAS), 1*(6). ISSN: 2395-3470.
10. Krishna Raju, N. (2006). *Reinforced concrete design*. New Age International Pvt. Ltd.
11. Punmia, B. C. *RCC designs*. Laxmi Publications.
12. Dutta, B. N. (2002). *Estimating and costing in civil engineering* (25th ed.). UBS Publishers and Distributors Pvt. Ltd.
13. Chakraborti, M. (2006). *Estimation, costing, specific and valuation in civil engineering* (Kolkata ed.). M. Chakraborti Publisher.
14. Wang, M., Luo, D., & Hu, Z. (2020). Seismic performance of steel-enhanced damping concrete core walls with concealed steel plate bracings. *Engineering Structures, 213*. https://doi.org/10.1016/j.engstruct.2020.110564
15. Nassani, D. E., Hussein, A. K., & Mohammed, A. H. (2017). Comparative response assessment of steel frames with different bracing systems under seismic effect. *Structures, 11*, 229–242. https://doi.org/10.1016/j.istruc.2017.06.006
16. Safarizki, H. A., Kristiawan, S. A., & Basuki, A. (2013). Evaluation of the use of steel bracing to improve seismic performance of reinforced concrete building. *Procedia Engineering, 54*, 447–456. https://doi.org/10.1016/j.proeng.2013.03.040
17. Bhat, K. A., & Danish, P. (2021). Analyzing different configurations of variable angle diagrid structures. *Materials Today: Proceedings, 42*(2), 821–826. https://www.sciencedirect.com/science/article/pii/S2214785320389902
18. IS 456. (2000). *Indian Standard Code of practice for plain) and reinforced concrete* (4th rev.). Bureau of Indian Standards.
19. IS 800. (2007). *Code of practice for general construction in steel*. Bureau of Indian Standard.
20. IS 808. *Dimensions for hot rolled steel beam, column, channel and angle sections*. Bureau of Indian standards.

21. IS 875 (Part 1). (1987). *Code of practice for design loads for buildings and structures, dead loads*. Bureau of Indian Standards.
22. IS 875 (Part 2). (1987). *Code of practice for design loads for buildings and structures, live loads*. Bureau of Indian Standards.
23. IS 1893 (Part 1). (2002). *Criteria for earthquake resistant design of structures*. Bureau of Indian Standards.

Structural Assessment of an Overhead Storage Reservoir Using NDT: A Case Study

Ram Prakash and Sunita Bansal

Abstract The Indian mainland is considered among the world's most catastrophes inclined regions with almost 59% of total landmass prone to earthquakes, 8.5% to cyclones and 5% to floods. Also, the United Nation's Hyogo Framework requires pre-disaster mitigation of all hazards instead of post disaster reliefs. Thus, all Public Structures like Over Head Storage Reservoirs (OHSR) that predate mandatory earthquake resistant design formulation need structural assessment, especially the ones situated amidst densely inhabited areas. There is an accentuated need for residual life estimation followed by restoration and retrofitting of all utilitarian structures. Non-destructive tests (NDT) have been instituted to determine the in situ strength and durability levels of concrete. However, some of these tests give misleading results, and testing points are more or less left to the judgment of the assessor. In this study, a case study of a 40 year old OHSR is presented on which NDT like Rebound Hammer and Ultrasonic Pulse Velocity were conducted to estimate the in situ compressive strength. In addition, semi-destructive tests on the cores extracted were done to validate the strength values obtained from NDT tests. The rebound hammer gave higher compressive strength results as compared to core tests at the same point due to carbonation effects. This study will help the structural assessors and other stakeholders of Architecture Engineering and Construction Industry to estimate the structural capacity based on single or multiple non-destructive tests.

Keywords Over head storage reservoir · Shaft type staging · Structural assessment · NDT · Core tests

R. Prakash (✉) · S. Bansal
Department of Civil Engineering, Faculty of Engineering and Technology, Manav Rachna International Institute of Research and Studies, Faridabad, India

S. Bansal
e-mail: sunitabansal.fet@mriu.edu.in

A. K. Gupta et al. (eds.), *Advances in Construction Materials and Sustainable Environment*, Lecture Notes in Civil Engineering 196,
https://doi.org/10.1007/978-981-16-6557-8_12

1 Introduction

The Policy systems on disaster management depend on overall, proactive, multi-disaster, and innovation driven procedures to have a safe and disaster-resilient constructed facilities. This can be accomplished through a culture of counteraction, relief and readiness for disasters. The targets directing the arrangement plan overall incorporate "reconstruction to rebuild back better and construct disaster-resilient structures and habitats". Researching the impacts of seismic tremors has for some time been perceived as a fundamental headway to comprehend the characteristic dangers and its hazard to the general public. A fast evaluation of general structural capacity review and documentation of significant visual perceptions, other than crisis management and post disaster recovery exercises, recognizes the need of follow-up research. Be that as it may, long-term readiness requires extensive research on the distinguished issues with proposals for readiness. Basic structural assessment is at the cutting edge of basic and material research. Assessment frameworks empower engineers to accumulate information of structures and auxiliary components. Assessment can be extremely helpful for identifying of potential damages in structures and to recognize the reasons for their likelihood. The structural damage can be in the form of cracks which are viewed as increasingly harmful in the structure as they decrease the life and strength sturdiness of the structure. If the constructed facilities show distress signs, they need to be visually inspected and tested for in-situ compressive strength and reinforcement integrity. Any unwanted spalling, corrosion in structures, which have been designed with codal considerations and principles has caused genuine concerns to the engineers and researchers to survey the toughness of the structures.

One of the most important infrastructure element, "Overhead Water Tanks" of many types and configuration in RCC and steel have been used to store water for water supply schemes. Reinforced concrete water towers are considered advantageous over other types as they are not influenced by climatic changes, are sealed, give more noteworthy shape versatility, besides giving sufficient strength, resistance to cracking and adequate life. Standing structure is difficult to assess by customary strategies to assess structural adequacy and concrete compressive strength. Consequently for overhead structures, constructed years ago, an assessment productive framework for in situ quality and durability is diligently required. Non-destructive testing has the capability of turning into a successful means for quality appraisal and harm assessment. Non-destructive procedures are valuable for assessing the states of the structure by performing indirect evaluation of concrete properties. Distress of the overhead structure can occur because of the various factors during its proposed life cycle. Disintegration or ultimate failure can occur before the evaluated life pattern of the structure in the various conditions, prevalently brutal in hot and dry areas where high temperatures, humidity and saltiness exists. The life expectancy of the structure without regular upkeep relies on the corrosion rate of the reinforcement and subsequent spalling of concrete cover. As observed from previous disaster studies, many RCC overhead water tanks got profoundly harmed during serious seismic tremors, throughout the world. General perceptions bring up the reasons identified with the structural inadequacy essentially of the supporting framework, more pronounced for continuous RC cylinder shaft type support towers. Hence, the supporting tower arrangement of the raised tanks is more sensitive to seismic security than the tank itself. The majority of the harms seen in the earthquakes emerge due to inappropriate/unacceptable strength of supporting framework, ill-designed, and specifications of strengthening steel and defects during development of the supporting framework. Because of geographical topology, India is moderately high disaster-inclined zones on the planet. The nation is prone to numerous perils like quakes, floods, typhoons, and dry spells, and these disasters frequently cause huge harm to property and life. Post the "Hyogo Framework," India sees the need to move from a post debacle responsive strategy to a pre-calamity dynamic philosophy which involves mitigation and preparedness to avoid losses. The Government of India detailed the Disaster Management Act in 2005, formulated a National Policy on Disaster Management in 2009, and in 2015 got three general pacts including the "Sendai Framework for Disaster Risk Reduction," "Sustainable Development Goals 2015–2030," and the "Paris Agreement on Climate Change." The country has got a multihazard and multi-sectoral framework approach, which hence will make a resilient nation.

Structural elements of an overhead reservoir consist of tank portion with roof and roof beams, floor slab, staging portion, consisting of columns' staging or a continuous R/C shaft staging [1]. Most suitable type of staging with correct estimation of dead loads, imposed loads like live loads, snow loads, wind loads (both empty and full tank) seismic loads [2] (both empty and full tank) and appropriate load combinations have to be considered in the design. The motivation behind this research was the collapsed staging structure of two, not so old functional overhead water tanks in the Jabalpur earthquake in 1997. The barrel-shaped support structure experienced circumferential flexural cracks close to the base. Comparative damages to staging structures had been seen in past tremors and as of late in the Bhuj earthquake, 2001, which is characteristic in countless water tanks of varying capacities and as much as 100 km from the epicenter [3]. The Bihar Nepal Earthquake likewise caused harms and breakdown of various overhead water reservoirs in Bihar. Such behavior of essential infrastructure like water tanks is not adequate, as they are relied upon to stay utilitarian and safe to work considerably after the biggest possible quake. The safe water supplies are required to forestall flare-ups of illness and to monitor fires after a quake. The safety of existing overhead water tanks during extreme quakes is doubtful, particularly of those situated in zone IV and V. Thus, periodic inspection and structural assessment of reinforced concrete overhead water tanks are required. The primary aim of this study was identifying a distressed overhead tank in the region, conduct visual survey followed by NDT, critical examination of the NDT readings and thus decide on the course of rehabilitation. The tank under study (one of present 21 in the region) is more than 40 years old but functional and was showing signs of distress.

2 Literature Review

First and foremost, a literature survey of the design considerations of the overhead tanks shows the extensive changes in the seismic safety measures of raised tanks.

The seismic design considerations in the code apply only to overhead tanks and ground supported tanks are not covered. Also, resultant stresses due to sloshing mode of vibration in the event of an earthquake are not covered. In contrast to present global practice for seismic plan of tanks, there are numerous impediments in the arrangements of IS 1893:1984 [2], some of which have been talked about by Jain et al. [4]. In this manner, one finds that at present in India, there is no legitimate Code/Standard for seismic structure of fluid stockpiling tanks.

The "Guidelines for Seismic Design of Liquid Storage Tanks" formulated later, however, consider analysis of ground supported tanks and two degree of freedom system idealization instead of earlier single degree of freedom system. The lateral stiffness of tank staging is recommended to be increased by bracing beams. For raised tanks with shaft type staging, notwithstanding the impact of flexural distortion, the impact of shear stresses might be incorporated while ascertaining the lateral stiffness of staging.

The response reduction factor in seismic analysis is higher at 2.5 for frame type staging than RC shaft type for which it is estimated as 1.8, because it lacks redundancy and has poor ductility [5]. Ensuing assessment on ground acceleration response of some under study tanks shows that the sections of the frame-bolstered and the shell of the shaft staging are defenseless to tensile strain, especially at empty tank condition. Such an instance is additionally disturbed if the impact of soil–structure interaction (SSI) is overlooked in structural design [6]. Ghosh and Chowdhury [7] made a study of seismic structure of such raised water tanks with shaft staging against frame staging with reference to seismic codes. Rai [5] pursued seismic assessment and repair of shaft type R/C overhead tanks.

Tank's structural adequacy judgment and retrofitting choices are enormously upgraded with non-destructive testing results which may utilize rebound hammer (RH), sonic, and ultrasonic pulse velocity (UPV) estimations to give a thorough testing of the tank concrete condition. The UPV readings give concrete quality, distinguish areas with interior breaking and delaminations, and give extent of concrete splitting or cracks. Mulik and Balki [8] research accentuated on the mechanical properties of concrete utilized in common structures. Samples were casted to associate the concrete strength got by consolidated NDT strategy with cubical strength extricated by destructive techniques using data by testing samples according to prescribed methodology by IS codal provisions. The utilization of joined strategies expands the precision of the in situ concrete compressive strength. Binda et al. [9] recognized an experimental research after the floods happened in Italy that harmed the Heritage brick work structures. The primary target of the research was to check the impact of non-destructive systems in finding the presence of water and the drying procedure.

Latif et al. [10] recommended the advancement of a real-time checking framework utilizing micro-scale sensors and incorporates the approval of the framework through progressive corrosion and cover deterioration testing. The framework contained a remote terminal unit that incorporates a linear polarization strategy for corrosion location under the covering and a micro-strain gauge technique for observing stresses. The fifth revision of IS 1893 in 2002, and all other further studies on seismic stability of overhead tanks have not been applied on the overhead tanks in India constructed before 1990s, unless some structural assessment and subsequent rehabilitation measures are taken.

In the city under study, 21 Overhead tanks are being used since last 40–50 years. These overhead tanks are of various ages and because of a few reasons the structural wellbeing of these overhead tanks are ordinary whenever, and some are close to accomplish their life expectancy and some are yet to offer service. This research is centered taking into account evaluation of structural durability of chosen overhead tank. Research work will be valuable for the consultancies and the organizations for designing the overhead structures to get satisfactory strength and durability of the structure, to get ready for upkeep and maintenance and for supporting the life of overhead tanks.

3 Methodology

The overhead tank under study as shown in Fig. 2 was a circular, reinforced concrete (R/C) elevated water storage reservoir of capacity: 100 m^3 with a shaft-type R/C support of external Diameter 8.54 m. The function of the structure was to provide water under pressure, and the elevated tank is reported to have been constructed in 1970s. Thus, the structure was almost 50 years old. The tank had a single post broken R/C staircase located inside the shaft. The overhead tank was found to have visible signs of distress and surface cracks which led to the requirement of assessment, repair, and structural rehabilitation. An extensive test approach has been embraced to decide the impact of durability over the age for the Overhead water tank. This began with the assortment of visual perceptions information for the assessment followed by non-destructive methods. The non-destructive tests were performed on different components of the chosen structure. The structural assessment work of the tank consisted of the following steps (Fig. 1).

First and foremost, activity in the structural investigation was a systematic visual inspection so as to gather readily available information about the structure in question. Further visual observations were also recorded at the time of tests.

The purpose of the visual inspection was to observe the extent of deterioration by observing:

- Identification of cracks-minor/major and its pattern study
- Identifying loss in concrete cover or spalling
- Identifying scaling or corrosion of reinforcement or structural steel
- Locating areas of water seepage
- Reporting severity of damages and probable causes

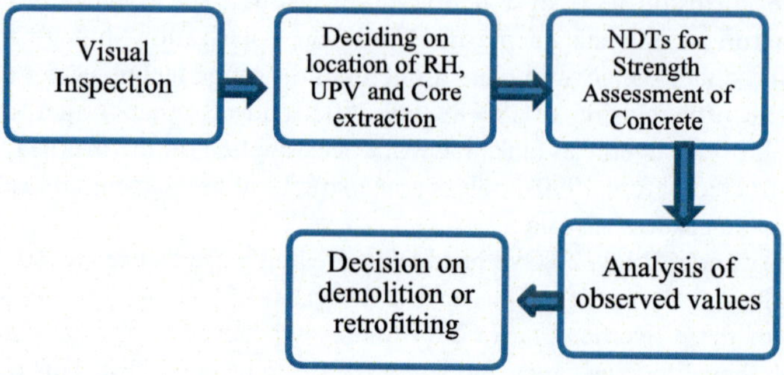

Fig. 1 Methodology of the study

Fig. 2 Completely deteriorated side of the tank

- Studying damages due to expansion joints
- Crack development at interface of concrete-structural steel.

The entire tank structure was thoroughly inspected visually. An unacceptable level of deterioration, including spalling and unsound concrete was observed in the bottom of shaft support and one side. The support structure of the tank has visible corroded bars, spalled, and biologically fouled concrete due to seepage.

In addition to the visual survey, the following tests were performed.

- Rebound hammer test
- Ultrasonic pulse velocity test
- Core test.

The RH test is one of the non-invasive tests regularly used to acquire data about the properties of solidified cementitious composites without harming the structure. This test could be utilized for evaluating the compressive strength of concrete with the assistance of appropriate relationships across rebound record and strength. It might be noticed that the RH test is not expected as a replacement for code specified compressive strength test led on 15 cm cubes at 7 and 28 days, but used when a gauge of hardened concrete is required in a previously built structure. This test was directed according to the Bureau of Indian Standards Code "IS 13311 (Part 2): 1992 (Reaffirmed 2004) Non-Destructive Testing of Concrete—Methods of Test—Part 2—Rebound Hammer" (Fig. 2).

The RH consists a weight controlled by spring of specific stiffness, slides on a plunger inside a cylindrical tubular element. When the plunger of the hammer is pressed against the structural element, the spring controlled mass bounces back and the level of such reflex relies on the "surface hardness" of concrete. The "surface hardness," and consequently the reading as rebound number on the graduated scale is correlated with the strength of concrete. Before doing a RH test, it is essential to clean and smoothen the concrete surface. It ought to be noticed that RH gives an idea of the concrete quality on the surface as it were. The primary factors that influence the readings are as age of concrete; size of aggregates; surface texture; carbonation; concrete mix characteristics; moisture content. Although the RH test is used for estimation of concrete strength, even in the best of situations, it is unlikely to estimate strengths with an accuracy of $+25\%$. The surface locations approximately 100 mm $\times$ 100 mm in area were prepared by rubbing with carborundum stone, to a perfectly smooth condition. RH readings were taken on the surface locations prepared. The average of the readings mentioned earlier thus represents the average rebound number at that location under consideration. The readings of RH were later correlated with the strength of concrete using the calibration graph. Carbonation affects the RH results; however, effect is likely to be similar in all the concrete location; hence, no correction for particular locations in this regard was considered necessary.

The dimensions of structural member do not affect the velocity of ultrasonic pulse; however, it relies upon the elastic properties of the material. The real pulse velocity acquired relies essentially on the materials and concrete mix. The guideline of the UPV test is based on the premise that higher velocities and subsequent lesser time for pulse reception is obtained for good quality concrete free from internal cracks, delaminations and is homogeneous. Lower UPVs are received when the nature of concrete is poor. In the instance of a break, defect inside the concrete mass which comes in the way of transmission of pulses, the pulse gets weakened and it goes around the flaw, in this way making the travel distance longer. Accordingly, pulses are received delayed, and the instrument shows lower values of velocities. USPV test comprises of sending ultrasonic pulses through concrete and estimating the travel time. The path length (for the most part the thickness of concrete part) divided by the travel time gives the velocity which is typically communicated in Km/sec and can be corresponded to concrete quality. USPV values can be interpreted to assess the condition of concrete with regards to its homogeneity, integrity, presence of voids, and relative quality of concrete within or between the members. Concrete quality can be appraised as "excellent," "good," "medium" or "doubtful." The meanings of the term "excellent," "good," "medium" and "doubtful" are based on ultrasonic pulse velocity measured at site and are as per the nomenclature of "IS 13311-1 [11]".

In this research, values have been deciphered to make a subjective evaluation with respect to homogeneity and structural integrity of concrete. 24 points were chosen for taking UPV test by surface probing method [11]. A transducer is held in contact with one surface of the concrete part under test. The subsequent transducer is held in contact with the surface at the very least separation of 150 mm. An electronic timing circuit empowers the estimation of travel time (T) required to cross a known path length (L). The pulse velocity is given by $V = L/T$. To guarantee that the ultrasonic pulses created at the transmitting transducer go into the concrete and are then recognized by the receiving transducer, it is fundamental that the concrete surface where the transducer is to be put is smoothened. The basic method for estimating the strength of concrete in a structure is to cut a core specimen utilizing a rotational diamond drill. In spite of the issues related with this sort of tests (for example, probability of sample damage, strength reliance on the cored area, presence of structural steel), they are accepted to be the most fitting for appraisal of mechanical properties of concrete. This strategy is especially suggested (for example during appraisal of structural adequacy). The specimens are taken from concrete structure components, starting with ferro-detector tests to find the presence of stirrups and longitudinal reinforcement. Besides, these location points are selected ensuring that the potential damage caused to the structure due to drilling is minimal. The drilling at locations are performed by a portable rotary drilling machine mostly in horizontal directions. The locations of coring were selected mostly at such locations where in situ RH readings, and UPV test readings are also available. These samples are then tested for compressive strength in Universal testing machine. This test gives us the actual accurate in situ compressive strength of concrete. The cores are visually examined for any delamination before undertaking uni-axial compression test. The estimated in situ cube compressive strength was calculated. The strength obtained from compression test is corrected for equivalent strength of cylinder with $l/d = 2$, using the correction factor corresponding to the l/d ratio of the tested core. The correction factor is obtained from the relevant figure in the code. The equivalent in situ cube compressive strength is then obtained by multiplying the equivalent cylinder strength by 1.25 as recommended in the code.

4 Results

In the present case, Schmidt's rebound hammer has been used for estimating the compressive strength of concrete. The rebound hammer results, ultrasonic pulse velocity readings, and core tests results are shown in Tables 1, 2, and 3, respectively.

Table 1 Rebound hammer readings

S. No.	Rebound hammer values						Average rebound no	Estimated cube compressive strength (N/mm^2)
	Point 1	Point 2	Point 3	Point 4	Point 5	Point 6		
1	34	37	38	38	32	33	35.333	34
2	36	35	31	38	37	34	35.167	34
3	36	30	39	35	40	39	36.500	38
4	34	34	34	34	33	37	34.333	30
5	38	32	33	36	36	30	34.167	30

Table 2 Ultrasonic pulse velocity readings

S. No.	Test method	Requirement as per IS:13311 [11]	Result	Concrete quality grading
1	Ultrasonic pulse velocity (km/s) IS 13311 (P-1) Indirect method or surface probing for a distance of 300 mm	Above 4.5 = Excellent 3.5–4.5 = Good 3.0–3.5 = Medium Below 3.0 = Doubtful	2.81	Doubtful
2			0.87	Doubtful
3			3.80	Medium
4			2.34	Doubtful
5			0.74	Doubtful
6			0.97	Doubtful
7			1.24	Doubtful
8			0.86	Doubtful
9			2.89	Doubtful
10			2.14	Doubtful
11			0.68	Doubtful
12			1.60	Doubtful
13			1.73	Doubtful
14			2.18	Doubtful
15			1.98	Doubtful
16			0.97	Doubtful
17			1.32	Doubtful
18			3.12	Doubtful
19			2.96	Doubtful
20			2.37	Doubtful
21			2.06	Doubtful
22			1.78	Doubtful
23			1.86	Doubtful
24			0.91	Doubtful

Table 3 Core—tests results

Core No.	Length (mm)	Diameter (mm)	*L*/*D*	Ultimate load (KN)	Cylinder strength (N/mm^2)	Correction factor	Corrected cylinder strength	Equivalent cube compressive strength (N/mm^2)
1	115	67	1.7164	39.6	11.2320	0.9695	10.8896	13.612
2	117	67	1.7463	67.3	19.0887	0.9727	18.5680	23.210
3	122	67	1.8209	53.8	15.2596	0.9807	14.9658	18.707
4	113	67	1.6866	74.5	21.1309	0.9663	20.4189	25.523
5	116.6	67	1.7403	42.5	12.0545	0.9721	11.7180	14.647
6	119.3	67	1.7806	58.0	16.4509	0.9764	16.0629	20.078
7	81.3	67	1.2134	55.9	15.8552	0.9154	14.5146	18.143
8	89	67	1.3284	9.7	2.7513	0.9278	2.5526	3.191
9	71	67	1.0597	59.8	16.9614	0.8989	15.2469	19.059
10	95.3	67	1.4224	33.3	9.4451	0.9379	8.8586	11.073
11	69	67	1.0299	64.6	18.3229	0.8957	16.4120	20.515

5 Data Interpretation and Inferences

A total (12 Core Tests and 30 RH Test in situ compressive strength tests were conducted on the shaft structure support. The average estimated cube compressive strength by RH test is 32 N/mm^2. The estimation of the strength of concrete by RH technique cannot be held to be extremely exact. The plausible precision of expectation of concrete strength in a structure is ±25% according to provision 8.1 of IS 13311 Part 2 [12].

12 specimens were sampled using a rotary diamond drill Cores of 75 mm in diameter were obtained by drilling from various available and permitted locations on the shaft support structure. On the basis of core tests conducted, the estimated average cube compressive strength was found to be 18.7 N/mm^2 (discarding cube values deviating by more than 15%). This value is lower than the minimum grade of concrete of 20 N/mm^2 as per the current code of practice—IS 456 [13], and 30 N/mm^2 required to construct tanks as per IS 3370 [14].

In total, 24 UPV Tests were performed. The average overall quality of concrete for the points tested was found to be "Doubtful." This fact was authenticated by the fact that some of the cores extracted were broken, and very few cores had lengths equal to the thickness of the shaft support. Surface examining is not so proficient as cross testing, on the grounds that the signal received at the transducer has an amplitude of just 2–3% of that created by cross probing, and the test outcomes are incredibly impacted by the surface layers of concrete which may have various properties from that of concrete inside the structural part. The aberrant velocity is constantly lower than the direct velocity by cross probing on a similar concrete component. This distinction may fluctuate from 5 to 20% relying to a great extent upon the nature of the concrete under test. For good quality concrete, a distinction of about 0.5 km/set may by and large be experienced. The results of Core tests, UPV test interpreted together with RH test results indicate low strength of concrete along with the presence of voids and cracks inside the structure.

6 Conclusion

The structure is old and shows visible signs of deterioration of concrete and corrosion of reinforcement. The location being in auto market, carbonation has increased surface hardness substantially, and therefore, RH test could not be used for accurate estimation of concrete strength. Results of the UPV tests suggested that the quality of concrete were doubtful at all locations. The estimated average cube compressive strength was found to be 18.7 N/mm^2 which is not adequate as per the BIS code regulations. Further, the tank is located in Faridabad, Haryana which lies in Zone IV as per the vulnerability atlas of India and is highly vulnerable to earthquakes. The cylindrical, R/C shaft-type staging for elevated tanks has insufficient redundancy, damping and seismic resistance typically present in beam-column support systems. The framed support structures, however, provide alternate load transfer paths which is not there in the continuous shaft supports. The seismic performance of such structures rests heavily on the ductility of the shaft support. One side of the cylindrical shaft of the tank has totally deteriorated and spalled concrete. Also, the reinforcing bars are completely corroded. Thus, the structure may not be able to withstand slight tremors also. The tank is situated amidst highly inhabited area and its collapse will result in great losses of property and humans. The tank should be demolished and a new one built with column-beam staging. The strengthening of the water tank, with considerations given to the results obtained from NDTs, retrofitting techniques such as braced system, carbon fiber jacketing construction in staging of the water tank may not be successful and cost effective.

References

1. BIS. (1985). *IS: 11682 criteria for design of RCC staging for overhead water tanks*. Bureau of Indian Standards.
2. BIS 1893 (Part 1). (2002). *Criteria for earthquake resistant design of structures, Part 1: General Provisions and Buildings*. Bureau of Indian Standards.
3. Rai, D. C. (2001). Performance of elevated tanks in Mw 7.7 Bhuj earthquake of January 26, 2001. In *International Conference on Seismic Hazard with Particular Reference to Bhuj Earthquake*.
4. Jain, S. K., & Sameer, S. U. (1993). A review of requirements of Indian codes for aseismic design of elevated water tanks. *Bridge and Structural Engineer, XXIII*(1).
5. Rai, D. C. (2002). Seismic retrofitting of R/C shaft support of elevated tanks. *Earthquake Spectra, Earthquake Engineering Research Institute, 18*(4), 745–760.

6. Dutta, S. C., Dutta, S., & Roy, R. (2009). Dynamic behavior of elevated tanks with soil–structure interaction. *Engineering Structures, 31*(11), 2617–2629.
7. Ghosh, P., & Chowdhury, P. R. (2017). A critical study of seismic behavior of R.C elevated water tanks on shafts type of staging system. *International Journal of Engineering Research and Science & Technology (IJERST), 6*(1). ISSN 2319-5991.
8. Mulik, T. F., & Balki, M. S. (1985). Seismically induced fluid forces on elevated tanks. *Journal of Technical Topics in Civil Engineering*, 1–15.
9. Binda, I., Cardani, C., & Zanzi, J. (2010). *The physiology of the grid: An open grid services architecture for distributed systems integration.* Technical report. Global Grid Forum.
10. Latif, J., Khan, Z. A., & Stokes, K. (2019). Structural monitoring system for proactive detection of corrosion and coating failure. *Sensors and Actuators A: Physical*, 111693.
11. BIS 13311 (Part 1). 1992. (Reaffirmed 2004). *Non-destructive testing of concrete-methods of test part 1—Ultrasonic pulse velocity.* Bureau of Indian Standards.
12. BIS 13311 (Part 2). (1992). (Reaffirmed 2004). *Non-destructive testing of concrete-methods of test part 2—Rebound hammer*. Bureau of Indian Standards.
13. BIS. (2000). *IS: 456 Indian Standard for plain and reinforced concrete.* Bureau of Indian Standards.
14. BIS 3370. (1967). *Code of practice for concrete structures for the storage of liquids.* Bureau of Indian Standards.

Fracture and Impact Studies on Steel Fibre and Wire Mesh Embedded Concrete

S. Kanchidurai, P. Jaishankar, R. Vidya, and Prakhash Neelamegam

Abstract Steel fibre and welded wire mesh could perform effectively with concrete. Usually, increasing the ductile properties by adding steel is the finest option. The effect of steel fibre and welded mesh on concrete is still unknown. Firstly, the fracture study was conducted with Edge Notched Disc Bend (ENDB) specimen of mode-I category. Mode-I (opening) fracture testing could determine the fracture toughness of the concrete model. Secondly, the impact resistance of nine 600 × 600 × 65 mm thin slab was cast and tested. A steel ball with a drop weight of 3.8 kg is used to determine the unreinforced concrete slab's impact strength. The thin concrete slab with added steel fibres on concrete follows 0.5, 1 and 1.5%, with and without welded mesh embedment. The experimental results showed that the combination of steel fibre and welded mesh on fibre mesh concrete (FMC) shows higher compressive strength, fracture toughness and impact resistance.

Keywords Drop weight impact · Steel fibre · Welded mesh · Slab · Concrete · Compressive strength · Crack resistance · Crack resistance ratio

1 Introduction

A concrete floor slab is a structural element consisting of a horizontal surface. It is generally classified as ground bearing or suspended. The slab directly resting on the ground is called ground bearing, or else suspended. A structure subjected to various types of loads in its lifetime in which impact and creep load is significant. Impact loading can affect the entire system or a specific part. Cracking is the

S. Kanchidurai (✉) · P. Jaishankar · R. Vidya · P. Neelamegam
School of Civil Engineering, SASTRA Deemed to be University, Thanjavur, India
e-mail: kanchidurai@civil.sastra.edu

P. Jaishankar
e-mail: jai@civil.sastra.edu

R. Vidya · P. Neelamegam
SRM University, Chennai, India

A. K. Gupta et al. (eds.), *Advances in Construction Materials and Sustainable Environment*, Lecture Notes in Civil Engineering 196,
https://doi.org/10.1007/978-981-16-6557-8_13

major problem faced by all airports in runway pavement. If the pavement is not good, it will affect the engine. Repairing the engine is not easy, and the cost will be very high. Hence, impact energy plays a vital role in runway pavement [1]. So, it is critical to understand the concrete members under impact loading. Concrete is a brittle material, but the fibre concrete is highly ductile that can bear more deformation without failure. So, it received more attention [2]. Steel fibre improved the impact of strength. By providing meshes, we can avoid shrinkage cracks—rectangular and square mesh used in construction, especially for slabs. Generally, lapping not needed for the welded mesh [3]. It has higher elasticity and tensile strength than conventional reinforcement—performed impact test, to find the amount of energy absorption by the material. The effects of impact loading in fibre and mesh reinforced concrete slab investigated. In recent years, the mesh plays a vital role as it has low density, great strength and corrosion resistance. The present study mainly deals with improving slabs' impact behaviour without increasing the fibre percentage but providing welded mesh and steel fibre (1%). Pan et al. [4] had discussed the glass fibre-reinforced polymer (GFRP) strengthened with engineered cementitious composites (ECC). Found that fibre could resist the impact without failure, the minor micro-cracks were noticed after the experimentation. Amira et al. [5] had probed strain-hardening cementitious composite (SHCC) was used RC slabs either at the top or bottom of the piece and achieved promising results. Mastali et al. [6] discussed functionally graded reinforced concrete used as reinforcement, and the calculated steel fibre also used for the slabs. For the first category with low-velocity impacts, the functional-graded reinforced concrete slabs made. For the second category, with high-velocity implications, the functionally graded reinforced concrete slabs were made. RC slab with steel fibre increased the number of blows. The results showed excellent performance for small penetration depth. Elavarasi et al. [7] investigated that different types of slabs by slurry fibre concrete SIFCON. It has been noted that the crack resistance of SIFCON specimen was higher than the control specimen. Also, the punching shear failures lowered in SIFCON slabs. Kumar et al. [8] had discussed fibre concrete slab specimen against impact behaviour. It used 243 kg free-falling hammer with different predefined height.

The results are highly reliable of increases in stiffness and decreases in deflection. Hassan [9] had discussed the patch test was done on the wired mesh reinforced cementitious slab. By increasing the volume fraction of wire mesh, it increased the punching of the specimen. Mansourian et al. [10] discussed reclaimed asphalt pavement (RAP) against impact testing. Rooholamini et al. [11] tested a three-point bending test employed on the notched beam in this study. Different types and different lengths of mono and hybrid fibres were used in fibre-reinforced roller-compacted concrete. Abadel et al. [12] had discussed carbon fibre-reinforced polymer (CFRP) used as reinforcement in reinforced concrete slabs. The specimens tested under projectile impacts. Mousavi et al. [13] had discussed the fibre-reinforced polymer bars and steel were as reinforcement in concrete slabs. Found that the maximum width of cracks and distribution of damages.

2 Experimental Investigation

2.1 Materials

2.1.1 Cement

The test specimens prepared with Ordinary Portland cement grade of 53 from manufacture. The specific gravity of cement was 3.17.

2.1.2 Fine Aggregate

River sand used as fine aggregate. The fineness modulus of river sand 2.29%. The specific gravity of fine aggregate found to be 2.64.

2.1.3 Steel Fibre

In the present study, the steel fibre size was 0.5 mm in diameter and 25–35 mm. The ductile steel has 250 MPa strength.

2.1.4 Mesh Reinforcements

Compared to conventional support and fibre and mesh reinforced concrete slab, a two-way slab reinforced at a spacing of 150 mm with HYSD deformed bars of 8 mm diameter. Welded mesh used 2.5 mm for the slab.

2.1.5 Coarse Aggregate

From the initial test for the 12.5 mm average size coarse aggregate, the fineness modulus of coarse aggregate found to be 7.29, and the impact value found to be 20.8%. The specific gravity of the coarse aggregate was 2.8 (Fig. 1).

2.2 Experimental Program

2.2.1 Compressive Strength and Impact Coupon Test

Maintained concrete mix M25 as 1:1:3 (cement 390 kg/m^3: sand 390 kg/m^3: 12.5 mm coarse aggregate 1170 kg/m^3) to prepare the specimen. The compressive strength was determined using a 100 × 100 × 100 mm cubic specimen. The compressive

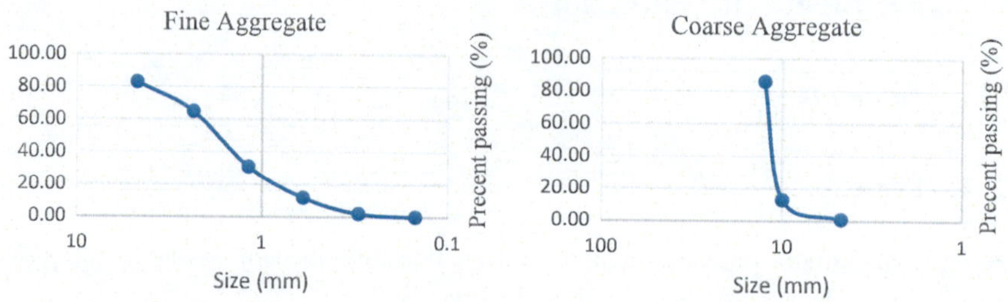

Fig. 1 Aggregates gradation curve

Fig. 2 Concrete cubes and cylindrical specimens against compressive strength test

strength tested on 21 cubic samples and the drop weight impact test performed on 21 impact cylindrical specimens with welded mesh, steel fibres containing 0.5, 0.75, 1% and combined steel fibre and welded mesh. The impact test specimen size 150 mm diameter and 50 mm thickness of the cylindrical specimen. Figure 2 shows the compression cube of welded mesh and impact cylinders of the control specimen.

2.2.2 Mode-I Fracture Test

The fracture toughness test performed as per the Irwin formula. The size of the specimen was 150 mm × 50 mm disc. 1.5 mm thickness edge notch were maintained for 20 mm depth. The test setup was shown in Fig. 3. Following Irvin equation was used to determine mode-I fracture toughness (stress intensity factor)

$$K_{1C} = 0.479 \times 6P_C S\sqrt{\pi} a/Rt^2 \quad (1)$$

where

K_{1c} = Fracture toughness, R = Radius of the ENDB specimen, t = Height of the ENDB, P_c = critical fracture load, s = span, a = notch depth.

Fig. 3 Edge notched disc bend (ENDB) specimen against Mode-I fracture test setup

2.2.3 Impact Resistance Test for Concrete Slabs

The impact test performed as per ACI committee 544.2R-89 (Eqs. 2, 3 and 4). The slabs were supported on four edges by the welded channel section, and a steel ball of 3.8 kg used to impact the slab on its centre at the height of 457 mm. Blows needed to initiate the crack, and the deterioration was recorded. Figure 4 shows the impact of testing specimen

$$\text{Energy absorption capacity} = N \times m\,g\,h \tag{2}$$

$$\text{Crack resistance} = \text{Energy absorption}/(L \times W \times D) \tag{3}$$

$$\text{Crack resistance ratio} = \text{Crack resistance/fck} \tag{4}$$

where

m = mass of a ball (3.8 kg), g = acceleration due to gravity (9.81 m/s^2), h = fall height (457 mm), N = number of blows, L = Total length of all cracks, W = width of the crack, D = depth of the crack and fck = compressive strength of concrete.

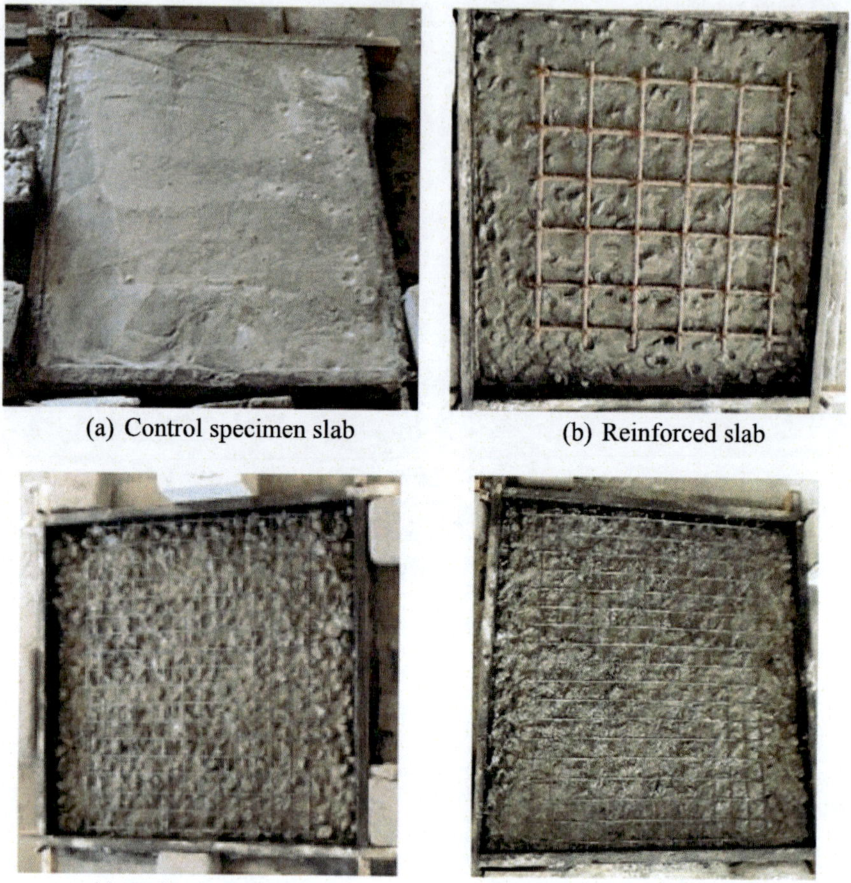

(a) Control specimen slab (b) Reinforced slab

(c) Welded mesh slab (d) Fiber and mesh reinforced slab

Fig. 4 Testing of mesh reinforcement slabs against impact testing

3 Results and Discussion

3.1 Compressive and Impact Strength of Concrete Cubes

For compressive strength of concrete, the addition of steel fibre is 0.5, 0.75 and 1% considered first set, also, welded mesh and steel fibre added together in another set. The compressive strength test observed that the addition of steel fibre could considerably increase strength and performance. The result shows that the welded mesh addition's compressive strength was 17.6% higher than the control specimen. The fibre and mesh reinforced concrete (FMRC) 27% compressive strength more elevated than the control specimen. Table 1 shows the compressive strength and impacts cylinders result.

Table 1 Avg. compressive strength and impact resistance of cube and ENDB specimens

Specimen No.	Type of mix	Compressive strength (N/mm^2)	Number of blows at failure (impact disc)
1	CS	28.65	8
2	WM	33.07	16
3	SF (0.5%)	29.13	13
4	SF (0.75%)	30.28	17
5	SF (1%)	32.56	21
6	FMRCS	36.08	26

CS control specimen, *WM* welded mesh, *SF* steel fibre, *FMRCS* fibre and mesh reinforced concrete specimen

Table 2 Calculation of fracture toughness on ENDB specimen

Specimen No.	Type of mix	Load in kN	Fracture toughness (MPa m$^{0.5}$)	Average (MPa m$^{0.5}$)
1	CS1	4.12	0.974	0.9955
2	CS2	4.30	1.017	
3	WMI	10.08	2.384	2.6315
4	WM2	12.17	2.879	

3.2 *Fracture Toughness*

Three-point bend fracture test was done to understand the compatibility and crack growth in mesh embedded specimens. The 50 mm thickness ENDB (Edge notch disc bend) specimens were tested with and without welded mesh embedment. The welded mesh embedment in concrete shows excellent performance against fracture [14–17]. Table 2 shows fracture toughness of ENDB specimen. Identifying fracture toughness basically by Irwin formula and not referred with any standard.

3.3 *Impact Test*

The fibre and mesh reinforced concrete slab had 600 × 600 × 65 mm (thickness) 0.1% volume fraction of hooked end steel fibre distributed to the mix. Welded mesh pieces size of 560 × 560 mm considered as reinforcements. The layer of mesh put on half of the thickness (centre of the depth). By seeing the results, the blows difference is 6 for the control specimen. But for the fibre and mesh reinforced concrete, the blows difference is 1047. We can see a tremendous increase in the FMRC slab. Nearly, 1041 gap of blows increased than the control specimen. Table 3 shows blows needed to initiate crack and ultimate crack. Figure 5 shows the impact resistance.

Table 3 Numbers of blow at crack initiation and deterioration

Specimen No.	Type of mix	Blows to initiate crack ($N1$)	Blows at deterioration ($N2$)	Blows difference ($N2$–$N1$)
1	CS	73	79	6
2	SF	153	247	94
3	FMRC	387	1434	1047

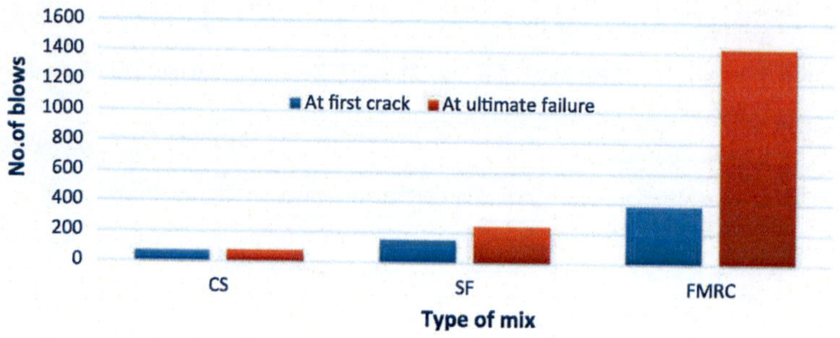

Fig. 5 Comparison between impact blows crack initiation and failure

3.4 *Energy Absorption Capacity*

The results showed that the control specimen reached rupture failure when the cracks propagated on the slab surface and fragmented gradually after the first crack. But fibre and mesh reinforced concrete slab was took many blows until the failure. It noticed that the FMRC absorbed more energy at the initiation of damage and deterioration than the control specimen. At the ultimate loss, the energy absorption capacity of control specimen, steel fibre, FMRC were 1.34 kJ, 4.20 kJ, 24.42 kJ, respectively [18, 19]. While comparing the result, SF absorbs more energy than the CS, whereas FMRC absorbs 17.19% more and more power than the SF. At the first crack initiation, the energy absorption capacity of CS, SF and FMRC were 1.24 kJ, 2.60 kJ,

Table 4 Energy absorption capacity of slabs

Specimen No.	Type of mix	Numbers of blows to initiate crack ($N1$)	Numbers of blows at deterioration ($N2$)	Energy absorption capacity ($N1$) kJ	Energy absorption capacity ($N2$) kJ
1	CS	73	79	1.24	1.34
2	SF	153	247	2.60	4.20
3	FMRC	387	1434	6.59	24.42

(a) Control specimen after 79 blows (b) Steel fiber slab after 247 blows (c) Fiber and mesh reinforced concrete slab after 1434 blows

Fig. 6 Failure pattern of steel fibre and mesh embedded concrete slabs

6.59 kJ, respectively. So, the energy absorption capacity of ultimate failure increased by the first crack initiation [20, 21]. Table 4 shows the calculation of energy absorption capacity for slabs. The comparison of control, SF and FMRC against energy absorption was specified in Fig. 6.

3.5 Crack Resistance

SF's crack resistance is increased by 21.42% than CS, whereas fibre and mesh reinforced concrete slab increased by 11.12% than CS. Finally, the crack resistance ratio of CS, SF and FMRC is 0.65, 2.02, 13.81, respectively. It observed that the crack resistance ratio of fibre and mesh reinforced concrete slab was 14.62% higher than SF. As per IS 456-2000, the max crack width of the reinforced slab was 0.3 mm. For the SF slab specimens, the maximum width is decreased than CS, whereas the full width of FMRC decreased than SF. The addition of welded mesh enhanced the bond strength. So, it reduced the shrinkage effect that causes the depreciation of crack width opening. Table 5 shows the crack resistance ratio of slabs.

Table 5 Crack measurement and resistance ratio of slabs

Specimen No.	Type of mix	At failure of slabs			At failure of slabs	
		Length of all cracks (mm)	Maximum width (mm)	Slab depth (mm)	CR (N/mm^2)	CRR
1	CS	2672	0.42	65	18.45	0.65
2	SF	2088	0.36	65	86.12	2.02
3	FMRC	1797	0.27	65	774.62	13.81

CS control specimen, *SF* steel fibre, *FMRC* fibre and mesh reinforced concrete slab, *CR* crack resistance, *CRR* crack resistance ratio

3.6 Failure Pattern

The figure shows that the failure pattern of CS, SF, FMRC slabs under impact loading. The level of damage decreased in FMRC compared to CS. The width of the crack in FMRC also decreased by CS. The FMRC slab could sustain nearly 1434 blows at deterioration. But while only adding steel fibre alone could sustain 247 impacts only. Control specimen failed by brittle and punching shear failure found in the FMRC slab. Figure 6a shows the control specimen after 79 blows. Figure 6b shows a steel fibre slab after 247 clashes. Figure 6c shows fibre and mesh reinforced concrete slab after 1434 blows.

4 Conclusion

The energy absorption capacity is more in the fibre and meshes reinforced concrete slabs than fibre-reinforced slab by 17.19%. The damage occurred more at the centre in all the slabs. The compressive strength results showed a drastic increase due to the addition of fibre along with mesh. The addition of 1% of steel fibre could increase the number of blows to reach the ultimate failure. We can see the noticeable increment in resistance for the fibre and mesh reinforced concrete slab (FMRC). Control specimen slab failed suddenly by brittleness than FMRC broken into three pieces. The crack resistance of reinforced steel fibre slab is increased by 21.42% than the control specimen.

The crack resistance of fibre and mesh reinforced concrete slab increased by 11.12% than the fibre-reinforced slab. The maximum crack width at ultimate failure was less due to the addition of steel fibre and welded mesh because steel fibres give better bonds and excellent resistance to impact. The crack resistance ratio of fibre and mesh reinforced concrete slab increases by 14.62% than fibre-reinforced steel slab. Control specimen slab failed by brittle, whereas fibre and mesh reinforced concrete slab by punching shear failures. The test results showed promising results in strength and energy absorption in the fibre and meshed reinforced concrete slab.

References

1. Guo, T., Xie, Y., & Weng, X. (2018). Evaluation of the bond strength of a novel concrete for rapid patch repair of pavements. *Construction and Building Matererial, 186*, 790–800.
2. Kaikea, A., Achoura, D., Duplan, F., & Rizzuti, L. (2014). Effect of mineral admixtures and steel fiber volume contents on the behavior of high performance fiber reinforced concrete. *Materials and Design, 63*, 493–499.
3. Sharaky, I. A., Mohamed, H. A., Torres, L., & Emara, M. R. (2020). Flexural behavior of rubberized concrete beams strengthened in shear using welded wire mesh. *Composite Structures, 247*, 112485.

4. Pan, Y., Wu, C., Cheng, X., Li, V. C., & He, L. (2020). Impact fatigue behaviour of GFRP mesh reinforced engineered cementitious composites for runway pavement. *Construction and Building Materials, 230*, 116898.
5. Elnagar, A. B., Afefy, H. M., Baraghith, A. T., & Mahmoud, M. H. (2019). Experimental and numerical investigations on the impact resistance of SHCC-strengthened RC slabs subjected to drop weight loading. *Construction and Building Materials, 229*, 116866.
6. Mastali, M., Ghasemi Naghibdehi, M., Naghipour, M., & Rabiee, S. M. (2015). Experimental assessment of functionally graded reinforced concrete (FGRC) slabs under drop weight and projectile impacts. *Construction and Building Materiasls, 95*, 296–311.
7. Elavarasi, D., & Saravana Raja Mohan, K. (2018). On low-velocity impact response of SIFCON slabs under drop hammer impact loading. *Construction and Building Materials, 160*, 127–135.
8. Kumara, V., Iqbala, M. A., & Mittal, A. K. (2017). Impact resistance of prestressed and reinforced concrete slabs under falling weight indenter. *Structural Integrity Procedia, 6*, 95–100.
9. Ibrahim, H. M. (2011). Experimental investigation of ultimate capacity of wired mesh-reinforced cementitious slabs. *Construction and Building Materials, 25*, 251–259.
10. Mansourian, A., Hashemi, S., Reza, M., & Aliha, M. (2018). Evaluation of pure and mixed modes (I/III) fracture toughness of Portland cement concrete mixtures containing reclaimed asphalt pavement. *Construction and Building Materials, 178*, 10–18.
11. Rooholamini, H., Hassani, A., & Aliha, M. R. M. (2018). Fracture properties of hybrid fibre-reinforced roller-compacted concrete in mode I with consideration of possible kinked crack. *Construction and Building Materials, 187*, 248.
12. Abadel, A., Abbas, H., Almusallam, J., Al-Salloum, Y., & Siddiqui, N. (2017). Local impact damage response of CFRP strengthened concrete slab. *Procedia Engineering, 173*, 85–92.
13. Mousavi, T., & Shafei, E. (2019). Impact response of hybrid FRP-steel reinforced concrete slabs. *Structures, 19*, 436–448.
14. Irwin, G. R. (1957). Analysis of stresses and strains near the end of a crack traversing a plate. *Journal of Applied Mechanics, 24*(3), 351–369.
15. Mousavi, S. R., Afshoon, I., Bayatpour, M. A., Amirhossein Davarpanah, T. Q., & Miri, M. (2021). Effect of waste glass and curing aging on fracture toughness of self-compacting mortars using ENDB specimen. *Construction and Building Materials, 282*, 122711.
16. Fuan, S., Ke, M., Kanghe, L., Kun, L., & Aliha, M. R. M. (2021). Influence of specimen geometry on mode I fracture toughness of asphalt concrete. *Construction and Building Materials, 276*, 122181.
17. Aliha, M. R. M., Razmi, A., & Mousav, A. (2018). Fracture study of concrete composites with synthetic fibers additiveunder modes I and III using ENDB specimen. *Construction and Building Materials, 190*, 612–622.
18. Saleem, M. A., Saleem, M. M., Ahmad, Z., & Hayat, S. (2021). Predicting compressive strength of concrete using impact modulus of toughness. *Case Studies in Construction Materials, 14*, e00518.
19. Sathurshan, M., Yapa, I., Thamboo, J., Jeyakaran, T., Navaratnam, S., Siddique, R., & Zhang, J. (2021). Untreated rice husk ash incorporated high strength self-compacting concrete: Properties and environmental impact assessments. *Environmental Challenges, 2*, 100015.
20. Alwesabi, E. A., Abu Bakar, B. H., Alshaikh, I. M. H., & Akil, H. M. (2021). Impact resistance of plain and rubberized concrete containing steel and polypropylene hybrid fiber. *Materials Today Communications, 25*, 101–640.
21. Vivas, J. C., Zerbino, R., Torrijos, M. C., & Giaccio, G. (2021). Effect of the fibre type on concrete impact resistance. *Construction and Building Materials, 264*, 120–200.

Physico-Mechanical and Thermal Properties of Lightweight Structural Concrete with Light Expanded Clay Aggregate for Energy-Efficient Buildings

Rajesh Kumar, Rajni Lakhani, and Ashok Kumar

Abstract Energy-efficient buildings are one of the very important issues recently, which consider both economic and environmental factors. In this paper, main focus was to develop cost effective and durable lightweight concrete having high thermal insulation properties using different grades of light expanded clay aggregate (LECA) as coarse aggregates. As a part of the study, optimization of LECA aggregates, ordinary Portland cement (OPC), and water/cement (w/c) ratio were done. The uniaxial compressive, flexural strength, and drying shrinkage of the developed lightweight aggregate concrete (LWAC) specimens have been determined along with thermal conductivity. The results revealed that compressive and flexural strength of the developed LWAC blocks was in the range of 16.7–23.9 MPa and 2.8–3.4 MPa, respectively. Thermal conductivity of the LWAC specimens were found in-between 0.23 and 0.30 W/m K. From the analysis, it was inferred that the LECA concrete shows good strength at lower density, lower thermal conductivity, and lower drying shrinkage that makes it a suitable material for building constructions.

Keywords Sustainable concrete · Light expanded clay aggregate · Lightweight concrete · Thermal conductivity

1 Introduction

Energy-efficient buildings are being designed to use energy as less as possible. Many developing countries are leading to construct green buildings. The increasing

R. Kumar (✉) · R. Lakhani · A. Kumar
Organic Building Materials (OBM) Group, CSIR-Central Building Research Institute, Roorkee, Uttarakhand 247 667, India
e-mail: rajeshkumar@cbri.res.in

R. Lakhani
e-mail: rlakhani@cbri.res.in

A. Kumar
e-mail: ashokkumar@cbri.res.in

A. K. Gupta et al. (eds.), *Advances in Construction Materials and Sustainable Environment*, Lecture Notes in Civil Engineering 196,
https://doi.org/10.1007/978-981-16-6557-8_14

demand of normal weight aggregates (NWAs) to develop normal weight concrete has extremely reduce natural stone deposits and thereby causing irreplaceable damage to our environment. In building infrastructure industry, annually about 1.2 billion tons of OPC and 7.5 billion tons of natural aggregates are being used for regular construction and development. To overcome the problem of natural stone deterioration and for making energy-efficient buildings, lightweight aggregates (LWAs) have been using recently. LWAs are a type of coarse/fine aggregates used for the manufacturing of LWAC products, and these products are used in different kinds of structural work [1]. For energy-efficient buildings, LECA is used as a LWA.

LECA aggregates are manufactured from low lime plastic clay. For aggregate preparation, clay is firstly heated and then dried. After drying, sintering is being done in specific kinds of rotary kilns at temperature of 1100–1300 °C. In this process, the gas, i.e., released after heating process, gets entrapped inside the pellets during cooling. LECA aggregates has rounded or sub rounded in shape with rough texture and incorporates many multi-separated interconnected voids of different sizes which make it lightweight. These aggregates are also available in yellow or black color due to varieties in chemical composition and process of manufacturing. It has pH value as 7 and inert in nature. It can be used as replacement of both coarse and fine aggregates. Due to the cellular structure of particular aggregate particles, the aggregates used in structural concrete are lightweight and the specific gravity and unit weight of LWAs is lower than that of NWAs. The maximum size grading designations of LWAs generally available are in 19 mm, 13 mm, or 10 mm. It is possible to make LWAC having compressive strength of 80 MPa. LWAs have an ability of more water absorption (up to 5–20%). The demand of water for concrete is strongly influenced by the surface texture, size, and shape of the LWA particles. Due to porous type structure of LWAs, the aggregate needs to be wet for 24 h before adding it into the mix. In the case of LWC, the bond between aggregate and the matrix is stronger than the normal concrete. LWAC provides the reduction in dead loads, improved thermal and fire properties of buildings. In LWAC excessive mixing should be avoided, because it tends to break up aggregate particles [2]. Zohrabi et al. [3] investigated properties of LWAC with LECA, metakaolin, polypropylene (PP), and steel fibers. The PP fibers imparted more effect on energy absorption capability of LWC rather than on strength properties. Addition of 1% steel fibers significantly increased the strength in LWAC. Grabois et al. [4] determined different properties of LWACs containing steel fibers. The 28-day compressive strength was found above 30 MPa for density of 1700–1900 kg/m^3. Results showed that the thermal conductivity was decreased by 60% by using coarse and fine LECA aggregates. Bogas and Cunha [5] investigated the physico-mechanical behavior of LWAC. It was found that the concrete incorporated with coarse LECA aggregate, having diameter of 4–8 mm showed thermal conductivity of 0.23 W/mK while, the tensile and compressive strength at 28 days were found as 0.5 and 5.3 MPa, respectively. Reddy et al. [6] studied the flexural strength of conventional concrete with partial to complete substitution of coarse aggregate by LECA. Nahhab and Ketab [7] studied the effects of maximum size of aggregate (MSA) of LECA, on the properties of self-compacting lightweight concrete (SCLWC). It was concluded that MSA of 10 mm provided best compressive and flexural strengths. Chung et al. [8]

studied the effect of different LWA (expanded glass, expanded clay, and foam glass) on durability characteristics of LWACs having density less than 1200 kg/m^3.

The main objective of the presented study was to develop eco-efficient LWAC using LECA (newly prepared by ***PERL TECH, Ahmedabad***) after optimization for their packing density along with optimization of aggregate and w/c ratio, SP %, slump, and cement content. After optimizing each of the above parameter, different kind of physico-mechanical and thermal properties were determined, and then one final optimized LWAC mix were suggested having desired attributes, as per ACI 213 recommendations.

2 Methodology

Raw materials used in this study were OPC, sand, and superplasticizer (SP). Ordinary Portland cement (OPC) with surface area of 280 m^2/kg and specific gravity of 3.15 was used. Mean particle sizes of FA, and cement are 19.67 μm, and 20.59 μm, respectively. According to Blaine's apparatus of specific surface area, the specific surface area of FA was 325 m^2/kg. Aggregates used in this study were light expanded clay aggregate (LECA) of different sizes 0–2, 2–8 and 8–15 mm. These aggregates were purchased from ***M/s. PERL TECH, Ahmedabad***, and used for the experimental work after physico-chemical characterization. All the characterization testing were done according to IS 2386 [9, 10] and IS 383 [11] as shown in Table 1. Table 2 presents the chemical composition of LECA. LECA as lightweight aggregates was selected for making LWAC having density ranging from 1200 to 1300 kg/m^3 with targeted compressive strength ranging from 17 to 20 MPa, as per Neville [12]. Apart from LECA, any other kind of aggregates such as expanded slag, sintered fly ash aggregates were not selected by keeping cost economics issues in mind.

Table 1 Physical properties of LECA

LECA size ranges (mm)	Properties				
	Specific gravity (g/cm^3)	Water absorption (%)	Crushing value (%)	Loose bulk density (kg/m^3)	Moisture content (%)
0–2	0.8	61	–	580	0.75
2–8	0.6	30	–	430	0.5
8–15	0.7	25	20.28	390	0.2

Table 2 Chemical properties of LECA

Oxide	CaO	SiO_2	Al_2O_3	Fe_2O_3	MgO	K_2O	LOI
Wt. (%)	2.14	61.73	21.11	3.29	3.41	2.01	1.29

(a) (b) (c)

Fig. 1 **a** Drum mixer; **b** mechanical vibration; **c** lightweight concrete blocks

LECA used in concrete mix were presoaked in water for 24 h so that the aggregate can absorb the maximum quantity of water as LECA absorbs 25–40% of water by volume. After 24 h, the aggregate were placed on a sieve for 2 h to dry off the water in order to reach in saturated surface dry condition. After deciding the optimization criteria for targets (i.e., optimization of cement, aggregates, w/c ratio and SP %), the raw materials were mixed properly in a drum mixer for 5 min (Fig. 1a). Casting was done mechanically using vibration process (Fig. 1b). Water with SP content is poured into the mixer to get proper mix. In the experimental investigation, the specimens were cast (Fig. 1c) and were used to analyze the properties of the concrete with all materials. Specimens were casted and tested after 7 and 28 days when curing process is over. Three types of curing conditions (moist, water dip and air curing) were taken for the best results. In this paper, three types of cases were taken to get the final mix proportion. Case 1 shows the aggregate and w/c ratio optimization, Case 2 shows the SP % and slump value optimization, and Case 3 shows the cement optimization.

Case 1: Optimization of LECA Aggregate for Improved Packing Density

In this study, experimental investigation was done on concrete using LECA as a coarse aggregate. LECA sizes 0–2, 2–8, and 8–15 mm were used in different percentages of total coarse aggregate, as shown in Table 3. Different mix proportions are prepared named as MA-1, MA-2, MA-3, MA-4, MA-5, and MA-6. For each batch of mixing, the quantity of cement and sand was same for all mix while w/c ratios and LECA % were different. w/c ratio were ranges from 0.35 to 0.60. In the experimental investigation, the concrete cubes were used to analyze the compressive strength of

Table 3 Different mix proportion of LWACs (OPC: 400 kg/m^3, Sand: 550 kg/m^3 and LECA: 325 kg/m^3)

Mix No.		MA-1	MA-2	MA-3	MA-4	MA-5	MA-6
w/c		0.35	0.40	0.45	0.50	0.55	0.60
LECA (%)	0–2 mm	20	25	**30**	35	40	0
	2–8 mm	40	25	**40**	45	20	0
	8–15 mm	40	50	**30**	20	40	100

LWAC using different % of aggregate and w/c ratio. 100 × 100 × 100 mm samples were cast, and the 7 and 28 days compressive strength test was conducted on the prepared samples. Table 3 shows the mix proportion to develop LWACs.

Case 2: Optimization of Superplasticizer (SP) % and w/c Ratio
After aggregate optimization; it was concluded (through bold texts in Table 3) that the MA3 mix having (0–2 mm: 30%, 2–8 mm: 40% and 8–15 mm: 30% by total volume of LECA) was better than other 5 mixes. Then after that, different concrete mixes were prepared using different % SP content and w/c ratio. Table 4 shows the mix design using SP content. Slump test was carried out using slump cone method as shown in Fig. 2.

SP in the LWAC mix was added to prevent segregation. Cement, sand, and aggregate content in mix were same. Mixes prepared were named as MS-1, MS-2, MS-3, MS-4, MS-5, and MS-6 on the basis of SP dosage. It was observed that SP improved the workability of concrete which is directly related to slump value of LWACs.

Table 4 Different mix proportion of LWACs (OPC: 400 kg/m^3, Sand: 550 kg/m^3 and LECA: 325 kg/m^3)

Mix No.	MS-1	MS-2	MS-3	MS-4	MS-5	MS-6
w/c	0.32	0.36	0.40	0.44	0.48	0.52
SP (%)	1.1	0.86	0.78	0.71	0.63	0.54
LECA (kg/m^3)	325 (0–2 mm: 30%, 2–8 mm: 40%, 8–15 mm: 30%)					

Fig. 2 Slump test

Table 5 Different mix proportion of LWACs (Sand: 550 kg/m^3 and LECA: 325 kg/m^3)

Mix No.	MC-1	MC-2	MC-3	MC-4	MC-5
OPC (kg/m^3)	300	325	350	375	400
w/c	0.40				
SP (%)	0.78				
LECA (kg/m^3)	325 (0–2 mm: 30%, 2–8 mm: 40%, 8–15 mm: 30%)				

Case 3: Optimization of Cement
After aggregate, SP and w/c ratio optimization, the next step was to optimize the cement for further studies. In this study, cement was optimize by taking cement content ranging between 300 and 400 kg/m^3, while the quantity of sand and aggregate were same for all mixes. The w/c ratio was taken as 0.40, while SP content 0.78% for all mixes, as shown in Case 2. Table 5 shows the mix proportion. The specimens were moist cured for 28 days to achieve the maximum strength. After curing period is over, the specimens were tested according to different IS codes.

3 Results and Discussion

3.1 *Slump*

A standard slump flow cone (height 300 mm, base and top diameter 200 and 100 mm respectively) was used for this test. In general, all mixes exhibit a good workability. Figure 3 shows that the slump value increases significantly as the w/c ratio increases. Here, the linear variation is depicted in the graph. In this paper, slump required for LWAC lies in the range of 5–42 mm. Adding SP content in mix increases the

Fig. 3 w/c ratio versus slump

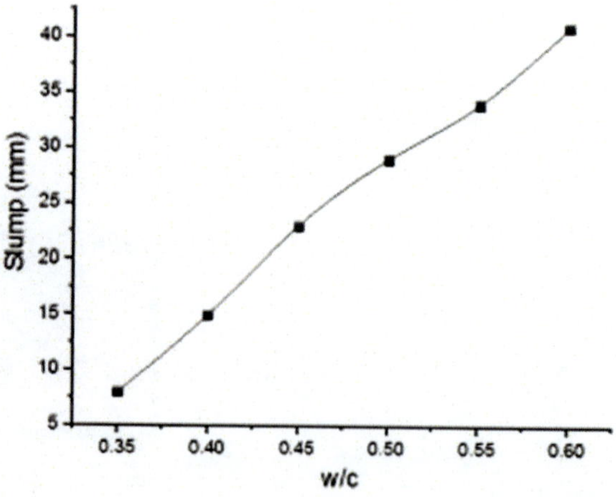

Fig. 4 Compressive strength test set up

workability of concrete. Mix MS-3 shows the slump within the permissible limits (12–15 mm).

3.2 Compressive Strength

The cast concrete specimens (100 mm cubes) were carried out at 7 and 28 days according to IS: 516-1959 (Fig. 4) [3]. Compressive strength is equal to compressive load divided by its cross sectional area. After analyzing the results LECA % were taken as 0–2, 2–8 and 8–15 mm are 30, 40, and 30%, respectively. Figure 5 summarizes the compressive strength of LWAC using different cement content. Here, the graph represents that the compressive strength of LWAC increases with increment of the cement content in the mix. MC-5 shows the maximum 28 days compressive strength (24.5 MPa) in which cement content was 400 kg/m^3. Hence, the mix MC-5 was taken as a final mix, and for further studies.

3.3 Flexural Strength

Flexural strength test was performed after 28 days according to IS 516 [13]. It is the ability of a beam (size of 100 × 100 × 500 mm, in this case) or slab to resist failure in bending. The load shall be applied without shock and increasing continuously at a rate of 180 kg/min for the 10.0 cm specimens and increased until the specimen fails (Fig. 6a). The result shows that the flexural strength of LWAC increases as the

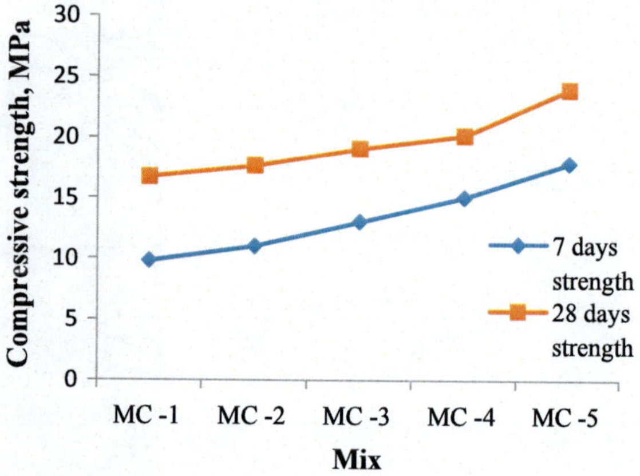

Fig. 5 Compressive strength

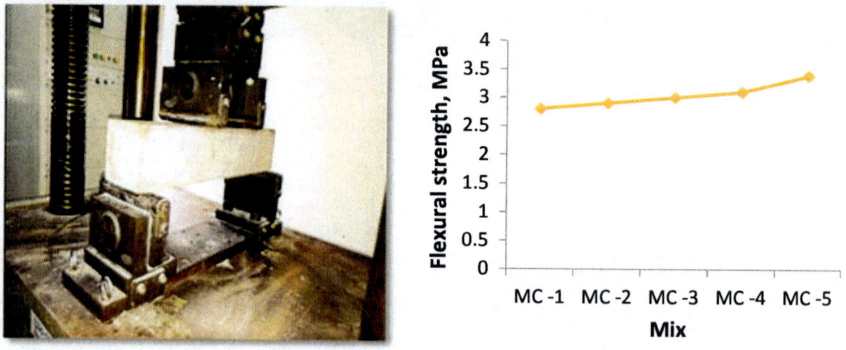

Fig. 6 Flexural strength: **a** Set up; **b** results

cement content in mix increases. LECA concrete shows flexural strength range of 2.8–3.4 MPa (Fig. 6b). In the above graph, it is clearly shown that the mix MC-5 shows the maximum flexural strength (3.25 MPa) out of all concrete mixes.

3.4 *Drying Shrinkage*

Due to hydration of cement, the drying shrinkage process begins when concrete is subjected to drying conditions. Drying shrinkage test was conducted according to IS: 4031 (part 10) [14]. Prism specimen 285 × 75 × 75 mm was casted to measure the drying shrinkage of different mixes (Fig. 7a). LWAC has no or lower drying shrinkage than the normal weight concrete. Results found that the M-5 concrete show minimum drying shrinkage (−0.03%) at 28 days as compared to other mix proportion (Fig. 7b).

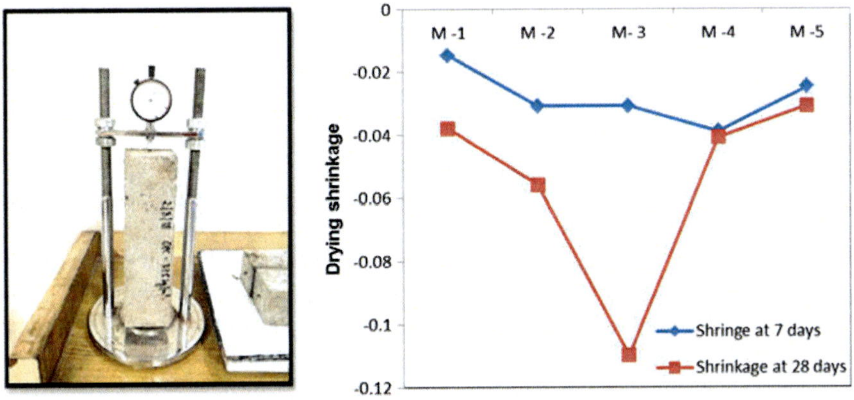

Fig. 7 Drying shrinkage: **a** set up; **b** results

As the rate of shrinkage decreases in LWAC which makes it durable material. During drying, the content of water decreases and shrinkage occurs slowly, of which water still remains into micropores.

3.5 *Thermal Conductivity*

Thermal conductivity of concrete is measure using guarded hot plate method as per IS: 3346 (Fig. 8). Specimen of dimensions 300 × 300 × 50 mm was casted for testing [15]. After 24 h, the specimens were de-molded, and moist curing of samples was done for 28 days. Before testing, samples were dried in ventilated oven at temperature of 100–121 °C to remove moisture present in the samples. In

Fig. 8 Thermal conductivity test set up

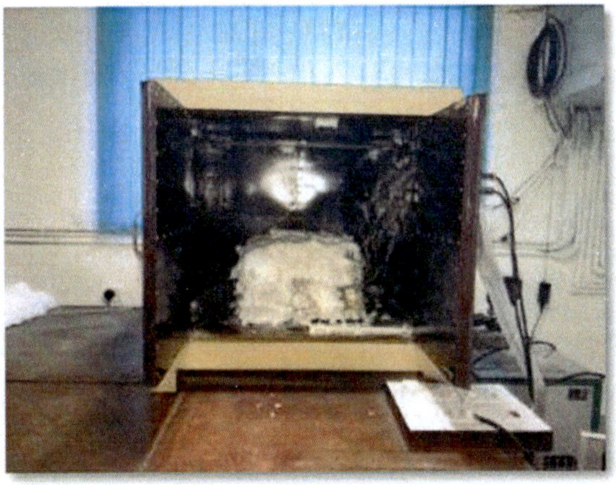

guarded hot plate method, mean temperature was taken 40 °C, while the maximum and minimum temperature was 60 and 20 °C, respectively. Results revealed that the thermal conductivity of the LWAC mix MC5 was obtained as 0.230 W/m K, which was lower than other mixes, i.e., MC1 (0.281 W/m K), MC2 (0.279 W/m K), MC3 (0.257 W/m K), and MC4 (0.243 W/m K). The thermal conductivities were decreased due to more amount of free water available which formed air entrapped voids and, thus, decreased thermal performance. Concrete prepared from normal weight aggregate (NWAs) (such as Basalt, Granite, Quartzite, Dolomite, etc.) shows higher thermal conductivity values [16]. Thermal conductivity of normal weight concrete was observed as 3.5, 3.2 and 2.65 W/m K for concrete containing Quartzite, Dolomite, and Granite as coarse aggregates. Lower the thermal conductivity better the material is a heat insulator. Thus, it shows that the LWAC mix MC5 shows better thermal insulations property.

4 Conclusions

In the presented study, the optimization of aggregate, cement, SP content were done, and different test was conducted according to different IS codes. Several mix design were prepared to get the final mix proportion. Based on the presented studies, the conclusions were:

1. MS-3 LWAC concrete mix provides the slump value within the permissible limits.
2. It is possible to produce LWAC having compressive strength between 16.7 and 23.9 MPa, while the flexural strength of 2.8–3.4 MPa.
3. As the cement content in the mix increases, the strength of concrete increases simultaneously.
4. The incorporation of LECA in the mixture increased thermal insulation property of LWAC which makes the building as energy-efficient envelope.
5. Drying shrinkage of LWAC was lower than the NWC. Initially, shrinkage of LWAC is higher than the NWC but after time it become constant.

Thus, it can be concluded that the LECA used as a concrete provide lower thermal conductivity, shrinkage, and comparable compressive strength than that of NWC.

Acknowledgements The authors are thankful to Director, CSIR-Central Building Research Institute, Roorkee (U.K.) for granting permission to publish this paper. The financial supports from 'Indo-US Science and technology Forum (IUSSTF), New Delhi' (Grant No. GAP:0806) and 'The Ministry of Environment, Forest and Climate Change, New Delhi, Government of India' (File Number: 354/2018/RE; Project No.: GAP0120) are gratefully acknowledged.

References

1. Alengaram, U. J., Muhit, B. A., & Jumaat, M. Z. (2013). Utilization of oil palm kernel shell as lightweight aggregate in concrete—A review. *Construction and Building Materials, 38*, 161–172. https://doi.org/10.1016/j.conbuildmat.2012.08.026
2. ACI Committee 213. (1999). *Guide for structural lightweight-aggregate concrete.* American Concrete Institute.
3. Zohrabi, M., Zohrabi, A., & Chermahini, A. G. (2015). Investigation of the mechanical properties of lightweight concrete containing LECA with Metakaoline Pozzolan using polypropylene and steel fibers. *Journal of Applied Environmental and Biological Sciences, 5*(12S), 11–15.
4. Grabois, T. M., Cordeiro, G. C., & Filho, R. D. T. (2016). Fresh and hardened-state properties of self-compacting lightweight concrete reinforced with steel fibers. *Construction and Building Materials, 104*, 284–292. https://doi.org/10.1016/j.conbuildmat.2015.12.060
5. Bogas, J. A., & Cunha, D. (2017). Non-structural lightweight concrete with volcanic scoria aggregates for lightweight fill in building's floors. *Construction and Building Materials, 135*, 151–163. https://doi.org/10.1016/j.conbuildmat.2016.12.213
6. Reddy, S. R., Swetha, N., & Desai, V. B. (2018). Flexural study on slab specimens with partial to fully replacement of natural coarse aggregate by light weight expandable clay aggregate (LECA). *International Journal of Technical Innovation in Modern Engineering and Science, 4*(4), 2935–2942.
7. Nahhab, A. H., & Ketab, A. K. (2020). Influence of content and maximum size of light expanded clay aggregate on the fresh, strength, and durability properties of self-compacting lightweight concrete reinforced with micro steel fibers. *Construction and Building Materials, 233*, 117922. https://doi.org/10.1016/j.conbuildmat.2019.117922
8. Chung, S., Sikora, P., Kim, D. J., El Madawy, M. E., & Abd Elrahman, M. (2021). Effect of different expanded aggregates on durability-related characteristics of lightweight aggregate concrete. *Materials Characterization, 173*, 110907. https://doi.org/10.1016/j.matchar.2021.110907
9. IS 2386 (I). (1963). *Methods of test for aggregates for concrete: Particle size and shape.* Bureau of Indian Standards.
10. IS 2386 (III). (1963). *Methods of test for aggregates for concrete: Specific gravity, density, voids, absorption and bulking.* Bureau of Indian Standards.
11. IS 383. (1970). *Specification for coarse and fine aggregates from natural sources for concrete.* Bureau of Indian Standards (revised 2002).
12. Neville, A. M. (1996). *Properties of concrete* (4th ed.). Wiley.
13. IS: 516. (1959). *Methods of tests for strength of concrete.* Bureau of Indian Standard.
14. IS: 4031 (Part IV). (1988). *Determination of consistency of standard cement paste.* Bureau of Indian Standard.
15. IS: 3346. (1980). *Method for the determination of thermal conductivity of thermal insulation materials (two slab, Guarded hot-plate method).* Bureau of Indian Standard.
16. Mehta, P. K., & Monteiro, P. J. M. (2006). *Concrete-microstructure, properties and materials* (3rd ed.). McGraw-Hill.

Agro-Industrial Wastes Incorporated Cement Stabilized Mud Composites for Roof and Wall Assembly in Energy Efficient Building Envelope

Rajesh Kumar, Rajni Lakhani, Bibhakar Kumar Singh, Mahesh Sharma, and S. K. Negi

Abstract Mud phuska is common type of course which is applied on top of the flat roofs in hot dry regions of the country. It acts as an insulating material and thus helps in providing the thermal comfort. The present experimental study aims to examine the potential use of different types of agriculture and industrial wastes, i.e., wheat straw, crumb rubber, and bagasse fiber in the development of the engineered mud phuska tiles. Four numbers of cubes were cast for computations of compressive strength, green-dry density, and thermal conductivity (k) values for each mix proportion. The experimental observations revealed that the mix proportion with ratio of crumb rubber and wheat straw in soil and cement mix up to 7.5% has achieved the desired value of compressive strength (2.65 MPa), flexural strength (1.85 MPa), and k value (0.25 W/m K). For demo testing, two model houses ($L \times B \times H = 430 \times 510 \times 330$ mm) were constructed (one conventional house and other mud phuska roof tile house). Temperature and humidity monitoring was done, and it was found that the waste incorporated mud phuska roof tile house was cooler as compare to conventional model house and have average temperature difference was up to 5–6 °C.

Keywords Mud phuska · Stabilized mud composites · Wheat straw · Baggase fibre · Waste tyre · Crumb rubber · Thermal conductivity

---

R. Kumar (✉) · R. Lakhani
Organic Building Materials Group, CSIR-Central Building Research Institute, Roorkee, Uttarakhand 247 667, India
e-mail: rajeshkumar@cbri.res.in

R. Lakhani
e-mail: rlakhani@cbri.res.in

B. K. Singh
Birla Institute of Technology, Mesra, Ranchi, Jharkhand 835215, India

M. Sharma
Department of Civil and Environmental Engineering, University of California, Davis, USA

S. K. Negi
CSIR-Central Building Research Institute, Roorkee, Uttarakhand 247 667, India

A. K. Gupta et al. (eds.), *Advances in Construction Materials and Sustainable Environment*, Lecture Notes in Civil Engineering 196,
https://doi.org/10.1007/978-981-16-6557-8_15

1 Introduction

One of the challenges of construction industry is building durability and sustainability. Therefore, it is required to make such building which is more economical at the same time more sustainable [1]. Nowadays, when population growth rate is so high and every people need a home. Therefore, making of such building which meets all the acceptance of customer is in demand. At the same time, we have to take care of our environment because construction rate is growing blindly, and due to this construction activities, the emission of CO_2 increases in our atmosphere. Many researches have conducted different studies to improve the technology of construction materials and to reduce the use of traditional aggregate. Presently in India, about 960 million tons (MT) of solid waste is being generated annually [2]. Of this 550 MT are agricultural sources, 290 MT are inorganic waste of industrial and mining sectors and 4.5 MT are hazardous in nature. After little improvement and recycling, these wastes show good engineering properties. Such as, in India, CSIR-CLRI, Chennai work on the field of recycling of solid leather waste, they want to use leather waste to make nanocomposite materials that is tough enough to make the body of a car, bike, and aircraft [3]. Agriculture waste viz. wheat straw and husk, bagaase fiber, jute fiber, groundnut shell, wooden mill waste, etc., show good thermal insulation when it is used with cement [4].

There are different types of solid waste where used for providing thermal insulation to the buildings. In India, many people use traditional technique for providing thermal comfort inside the house, brickbat coba, and mud phuska technique very common in India and use by civilians from many years. In mud phuska technique, agriculture waste like wheat straw and sugarcane waste use like a fibrous materials. Out of 550 MT of agricultural waste, about 110 MT of wheat straw and 141 MT of sugarcane waste are being produced annually in India [5]. But nowadays, due to advancement in technology, there are huge possibilities to use different types of solid waste such as rubber tyre and leather industry waste for providing thermal insulation to the building. In 2011, India produced 90,000 metric tons of reclaimed rubber from waste tyres [6]. The amount of waste generated from leather industry was noted as 5–6 MT [7].

Agriculture wastes include waste from crops and livestock. In developing countries, these wastes do not pose any serious problem as most of it is as used as raw materials in building construction. Many studies already showed that some agriculture wastes contain silica and showing good result when used with cement. India generate huge quantities of agriculture wastes per year [8]. Agricultural wastes can be of many types but in civil engineering applications mainly three types of waste are used most such as coconut waste, rice husk ash, and sugar cane waste. Various studies show that the use of coconut shells as replacement of conventional aggregate in construction is possible.

Mud is a versatile building material that has been used to make some extraordinary architectural marvels across the world [9]. In India, sun-baked brick of mud and straw is very popular to construct conventional rural houses. The use of mud as main building constituent is very beneficial and cost effective. Several studies have been conducted compressive strength tests on stabilized soil blocks/bricks with fibers and binders [10].

Mud phuska/composite—An insulating medium in roof finish of a layer of compacted soil underlying another layer of soil, mixed with building material such as fibrous reinforcing ingredients. Its use is cheap, reasonably durable, and adds enough thermal insulation [11]. After adding other fibrous reinforcing agent, its properties get improved and make it more durable as compared to bitumen roofing. Ashour et al. [12] developed stabilize soils with cement, gypsum, and straw fibers and found that the thermal performance was increased with increasing fiber content, while increasing with more cement and gypsum. Singh et al. [13] observed that for soil treated with 18% of tyre-scrap (size: 425–600 μm), unconfined compressive strength (UCS) of 1.75 kg/cm^2 was achieved. Tajabadipour et al. [14] used the scrap tire as the reinforcement element in the mechanically stabilized earth walls and found that scrap tire strip was capable of increasing pullout resistance by more than 2–2.5 times. Rajasekar and Srinivasan [15] investigated that for soil with 8% of shredded rubber tyre; california bearing ratio value of reached its optimum value as 13.1% at 2.5 mm penetrations and UCS values was increases as 0.41 MPa. Wang et al. [16] studied the influences of granular rubber on the erosion resistance of cement soil (CS) matrix. It was concluded that granular rubber effectively improved the erosion resistance of CS matrix. The existing mud phuska/composite treatment techniques (Bunker fill roof, Arch or vault roof, etc.) require lot of time, resource, and skilled labor. Also, due to shrinkage problem, many cracks are observed after drying which causes poor performance (thermal/durability) of Mud phuska roof tiles. Thus, proposed research work had following activities:

1. Development of precast mud phuska composite roof tile using different wastes
2. Remedial measures to overcome the problem of shrinkages in traditional mud-phuska techniques.

2 Materials and Methods

The aim of the experimental study was to develop the precast mud phuska tiles by using soil, wheat straw (WS), bagasse fiber and crumb rubber powder (recycled rubber from footwear waste and truck scrap tires), as primary raw materials (Fig. 1a–e). Crumb rubber waste is generated from leather waste. This solid waste shows some engineering properties that might be key to use these waste as future building components. The thermal conductivity value can be decreased by using crumb rubber, and hence, in civil engineering, it is use for construction of thermal insulation building. The use of leather wastes as insulator materials is able to provide an additional solution for reducing energy consumption and resolving problems in the handling of enormous wastes generated in tannery industries. The negative effect on mechanical strength is control by adding cement into the soil- waste matrix. Wheat straw was used in mud phuska from decades. Many possible composite products can

(a) Less cohesive soil (b) Wheat straw (c) Baggase fibre

(d) Rubber tyre waste (left); Footwear waste (right) (e) Crumb rubber powder

Fig. 1 Constituent materials for casting

be made using fibers including structural and non-structural materials. Sugarcane bagasse waste was used to perform the studies on the possible use of other natural fibers to enhance the properties of soil blocks which will add to knowledge regarding effective utilization of other alternative natural fibers in soil matrix.

The physical or chemical tests of the crumb rubber are given in Table 1. Properties of wheat straw and bagasse fibers are depicted in Table 2.

Soil used was stabilized in the view of plasticity index (PI) as per **IS: 2115–1980** recommendations [17], which gives specification for PI up to 10–15%. Therefore,

Table 1 Parameters of the crumb rubber

Classification	Volatility (%)	Sieving	D (μm)	Moisture content (%)
Sample used	0.33	99.37	365	0.47
Reference value	≤1.0	≥98.0	250	≤1.0

Table 2 Parameters of wheat straw and bagasse fibers

Property	Fiber form	Length, Avg. (mm)	Diameter, Avg. (mm)	Specific weight (g/cm^3)	In-situ moisture content (%)	Water absorption (%)
Wheat straw	Single	15	4	0.12	8.24	134
Bagasse fibers	Single	150	1.2	0.62	14.93	198

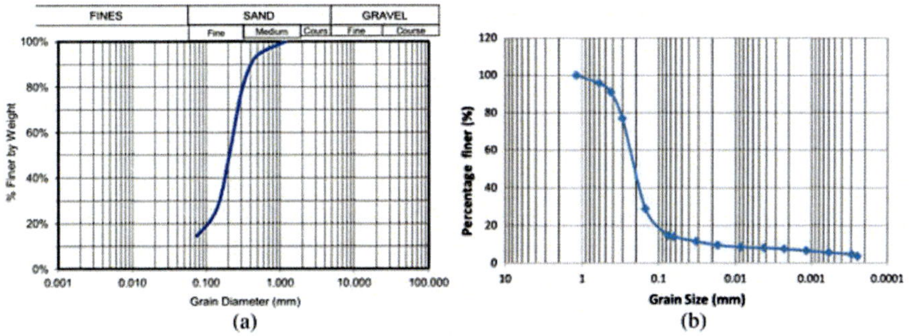

Fig. 2 Grain size analysis for local soil sample. **a** Sieve analysis; **b** hydrometer analysis

firstly soil was stabilized by adding OPC-43 cement ranging from 2 to 10%. After adding cement, water binder ratios were optimized for each mix proportions.

Particle size distribution of local soil is given in Fig. 2a. Hydrometer method was used to determine the particle size distribution of fine-grained soils passing 75 μm sieve (Fig. 2b).

Liquid limit (LL) of local soil was determined by Digital cone penetrometer as shown in Fig. 3.

Plastic limit (PL) was obtained as 13%, and thus, PI (LL-PL) calculated was 11%. After making three sample of different soil–cement ratios, it was concluded that 8% of cement was enough to stabilize less cohesive soil. Standard compaction test was also performed to understand the compaction characteristics of different soil with change in moisture content. Curves were drawn between moisture content (%) and dry density (gm/cc) to obtain the maximum dry density (MDD) and the optimum moisture content (OMC, %), as shown in Fig. 4a–b.

In total, 30 Nos. of Mix-proportions were made by replacement and addition of WS, crumb rubber and bagasse (Table 3). Further, for making such tile, the mix design was prepared by targeting *k*-value ($\leq$0.30 W/m K) and compressive strength (2.36–2.42 MPa) [18]. After mix design trails, it was observed that to develop stabilized mud composite for mix *A*, *B*, *C* and *D*; wastes (wheat straw/bagasse/crumb rubber) was used as 5.88% replacement of binder. To develop stabilized mud composite for

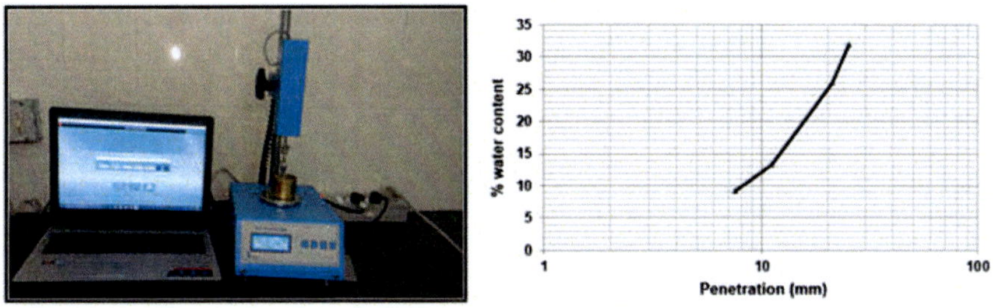

Fig. 3 Digital cone penetrometer for determination of liquid limit after adding 8% cement

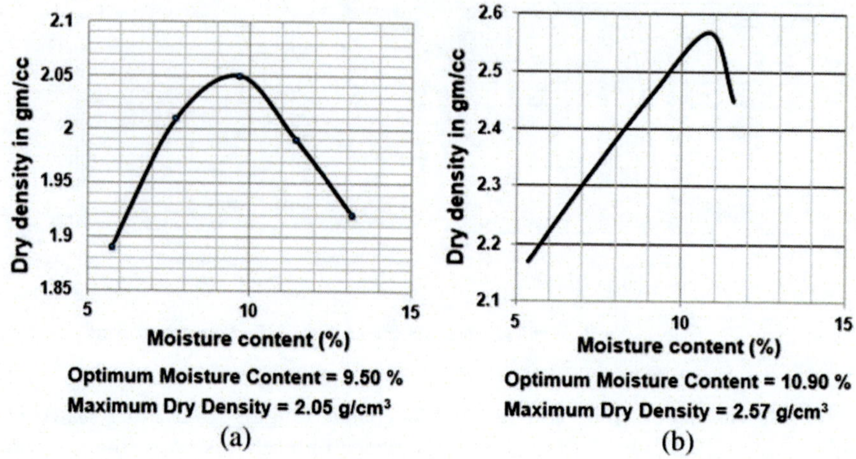

Fig. 4 Standard compaction test. **a** Soil with 0% cement; **b** soil with 8% cement

Table 3 Mixes for developing mud phuska composite

Mix designation	Raw materials
A	Soil + Ordinary Portland cement + Water
B	Soil + Ordinary Portland cement + Straw + Water
C	Soil + Ordinary Portland cement + Bagasse + Water
D	Soil + Ordinary Portland cement + Crumb rubber + Water
G_A	Soil + Ordinary Portland cement + Crumb rubber + Wheat straw + Water (when waste materials added into soil)
G_R	Soil + Ordinary Portland cement + Crumb rubber + Wheat straw + Water (when soil is replaced with waste materials)

Mix G_R, wheat straw was 5.88% of soil and cement along with 10, 20, and 30% replacement of soil by crumb rubber, while Mix G_A-stabilized mud composite was developed by adding wheat straw up to 5, 7.5 and 10% of soil and cement along with 5, 7.5, and 10% replacement of soil by crumb rubber (Table 3).

Four numbers of cubes were cast (Fig. 5) to check the compressive strength, green-dry density, and thermal conductivity (*k*) value for each mix proportion. Two types of curing, i.e., water curing and moist curing were implemented for both soil stabilized wall blocks and Mud phuska roof tile (Fig. 6a, b).

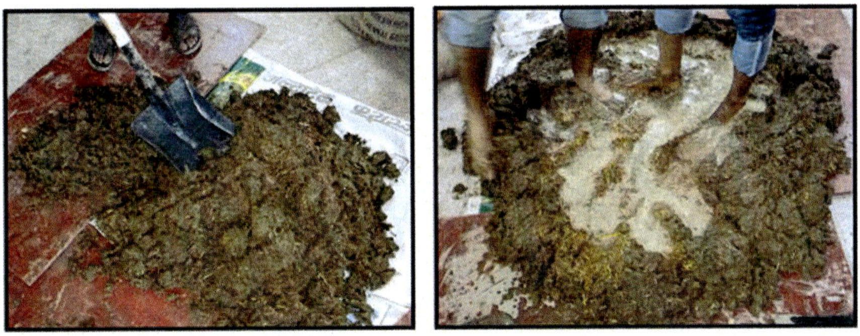

Fig. 5 Mixing of raw materials

(a) (b)

Fig. 6 Stabilized mud composites. **a** Wall blocks; **b** mud phuska roof tile

3 Results and Discussion

After curing the specimen for 7 days and 28 days, density, *k*-value, and compressive strength of specimens were determined. For Mix *A*, minimum density was observed as 1639 kg/m^3. While for optimized mix G_A, obtained density was 840 kg/m^3 which was about half than that of the control mix without cement. Due to high density of control mix, *k*-value was observed as 0.41 W/m K, while it was observed that the sample in which crumb rubber was added gives acceptable *k*-value of 0.29 W/m K but the specimens were not light weight, i.e., to targeted value. The mix proportion G_A; in which crumb rubber and wheat straw were added in soil, and cement was mixed up to 8%, has desired compressive strength (2.65 MPa) and *k*-value (0.25 W/m K) with dry density of 1020 kg/m^3. It was found that flexural strength was obtained as 1.85 MPa which was higher than the stabilized earth blocks developed using coal-ash and cassava peels (0.50 MPa) [19]. The compressive, flexural strength, and drying shrinkage were 0 improved due to addition of agricultural fibers. The fibers bonded the soil–cement matrix together, and thus, strength values were increased up to 15–20%. Drying shrinkage was reduced due to restrain provided by wheat straw fiber.

Addition of crumb rubber improved thermal properties due to insulation behavior of crumb rubber powder.

For small-scale field testing, two model houses ($L \times B \times H = 430 \times 510 \times 330$ mm) were made as shown in Fig. 7 (one conventional house and other precast mud composite house). To make conventional house, at top and bottom, 2 slabs of M30 Grade (Thickness: 70 mm) were cast, and walls were made by using ordinary bricks. Mud phuska house was made by giving treatment of mud phuska precast tile of 50 mm thickness as roof thermal insulation material with cement stabilized mud wall blocks; made by waste agricultural fibers (Fig. 6).

After construction of both the model houses, temperature and humidity monitoring were done by the use of electronic thermo-humidity meter and compared with each other to see the effect of using mud composite in place of conventional brick (Fig. 8). After analysis, it was found that conventional brick house becomes more heater during day time and temperature inside the house was more than the outside temperature. While, in the evening when outside temperature decreased slowly, the temperature in house were still more, and it takes significant time to cool down.

Figure 8b shows that the mud phuska as the thermal insulating medium makes house cooler as compared to the conventional model house. It was concluded further that the mud phuska roof tile house has significantly better thermal insulation properties as compared to conventional model house and average temperature difference was up to 4–6 °C. However, as per IS 2115, mud phuska layer by itself cannot provide adequate waterproof for the roof. Therefore, waterproofing layer or membrane is

(a) (b)

Fig. 7 Small-scale model house. **a** Conventional using bricks and concrete; **b** mud composite

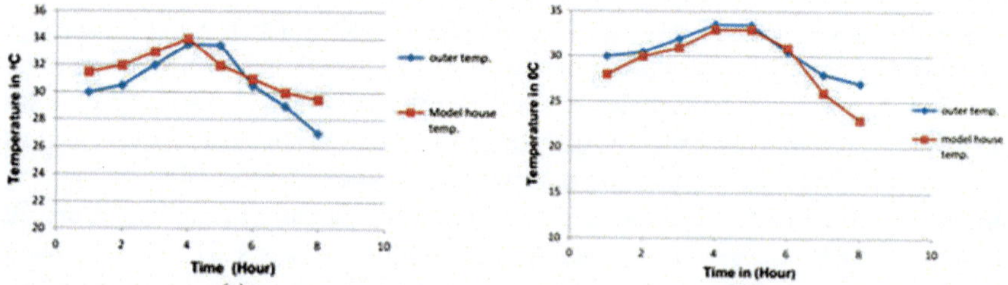

(a) (b)

Fig. 8 Temperature versus time graph of model house made of **a** bricks and concrete; **b** mud composite

Table 4 Desirable properties of the developed mud-phuska composite tiles

Properties of tiles	Observed values	Desired targets
Dimensions	300 × 300 × 50 mm	300 × 300 × 50 mm
Density (kg/m^3)	1020	900–1100
Final mix proportion • Cement • Wheat straw addition • Crumb rubber addition	• 8% of less cohesive soil • 7.5% of (OPC + Soil) • 10% of (OPC + Soil)	Less use of cementitious binder and Maximum utilization of waste
Thermal conductivity (W/m K)	0.25	≤0.30
Wet transverse strength (MPa)	1.85	≥1.50
Drying shrinkage (%)	0.071	≤0.08- 0.10

provided, to be effective in this respect. The slope of roof should be less than 1/40. The durability of the mud phuska layer depends on how the protective: outer surface, made up of the mud plaster or the brick tiles layer as the case may be, are maintained free of cracks.

The final mix proportion was used to make mud phuska tiles with the desirable properties, are shown in Table 4.

4 Conclusions

This technique will not only provide good thermal comfort to the dwellers, but also will attract less dead load from superstructure which in turn, will result into smaller cross-section of columns, beams and foundations, etc. In conclusion, these developed indigenous materials will have significantly better thermal properties as compared to contemporary building materials.

Till now, only physico-mechanical and thermal properties along with demo house testing have been conducted. Studies of micro-structural and durability parameters are in continuation and will be reported in future. Also, further experimental studies on full scale roof assembly will prove to be more valuable and will increase the know-how.

Acknowledgements The authors are thankful to Director, CSIR-Central Building Research Institute, Roorkee (Uttarakhand) for granting permission to publish this paper. The financial supports from Council of Scientific and Industrial Research (CSIR), New Delhi' (Project No. RSP-4051 (OLP-0393)) regarding CSIR-800 project, and 'The Ministry of Environment, Forest and Climate Change, New Delhi, Government of India' (File Number: 19-45/2018/RE; Project No.: GAP0090) are gratefully acknowledged. The assistance and support from Mr. Vikas Prabhakar is also appreciated.

References

1. Binici, H., Aksogan, O., Bodur, M. N., Akca, E., & Kapur, S. (2007). Thermal isolation and mechanical properties of fibre reinforced mud bricks as wall materials. *Construction and Building Materials, 21*(4), 901–906. https://doi.org/10.1016/j.conbuildmat.2005.11.004
2. Pappu, A., Saxena, M., & Asolekar, S. R. (2007). Solid wastes generation in India and their recycling potential in building materials. *Building and Environment, 42*(6), 2311–2320. https://doi.org/10.1016/j.buildenv.2006.04.015
3. Kanagaraj, J., Velappan, K. C., Babu, N. K., & Sadulla, S. (2006). Solid wastes generation in the leather industry and its utilization for cleaner environment. *ChemInform, 37*(49) (2006). https://doi.org/10.1002/chin.200649273
4. Al-Jabri, K. S., Hago, A. W., Taha, R., Alnuaimi, A. S., & Al-Saidy, A. H. (2009). Strength and insulating properties of building blocks made from waste materials. *Journal of Materials in Civil Engineering, 21*(5), 191–197. https://doi.org/10.1061/(asce)0899-1561(2009)21:5(191)
5. Devi, S., Gupta, C., Jat, S. L., Parmar, M.: Crop residue recycling for economic and environmental sustainability: The case of India. *Open Agric, 2*(1), 486–494 (2017). https://doi.org/10.1515/opag-2017-0053
6. Choudhary, V., & Choudhary, A. (2017). Use of tyre waste in concrete: A review. *International Research Journal of Engineering and Technology, 04*(10), 1924–1936.
7. ITC Sustainability Report 2016: https://www.itcportal.mobi/sustainability/sustainability-report-2016/environment/recycling-and-waste-management.aspx. Accessed on April 18, 2021.
8. https://dramarnathgiri.blogspot.com/2014/02/production-of-energy-from-agricultural.html. Accessed on March 05, 2021.
9. https://www.downtoearth.org.in/indepth/mud-housing-is-the-key-30237. Accessed on March 06, 2021.
10. Danso, H., Martinson, B., Ali, M., & Mant, C. (2014). Performance characteristics of enhanced soil blocks: A quantitative review. *Building Research and Information, 43*(2), 253–262. https://doi.org/10.1080/09613218.2014.933293
11. http://frontdesk.co.in/construction-materials/mud/. Accessed on 06 March 2021
12. Ashour, T., Korjenic, A., Korjenic, S., & Wu, W. (2015). Thermal conductivity of unfired earth bricks reinforced by agricultural wastes with cement and gypsum. *Energy and Buildings, 104*, 139–146. https://doi.org/10.1016/j.enbuild.2015.07.016
13. Singh, S., Dhiman, U., & Sharma, R. (2017). Soil stabilization using scrap rubber tyre. *International Research Journal of Engineering and Technology, 04*(05), 2952–2956.
14. Tajabadipour, M., Dehghani, M., Kalantari, B., & Lajevardi, S. (2019). Laboratory pullout investigation for evaluate feasibility use of scrap tire as reinforcement element in mechanically stabilized earth walls. *Journal of Cleaner Production, 237*, 117726 (2019). https://doi.org/10.1016/j.jclepro.2019.117726
15. Rajasekar, M., & Srinivasan, K. (2020). Experimental investigation on stabilization of clay soil using shredded rubber tyre. *Journal Seybold Report, 25*(9), 4281–4292.
16. Wang, F., Zhao, H., Ma, J., & Kang, T. (2021). Experimental study on the erosion resistance of rubberized cement-soil. *Soils and Foundations*. https://doi.org/10.1016/j.sandf.2021.02.003
17. IS 2115. (1980). *Code of practice for flat roof finish: Mud phuska*. Bureau of Indian Standards.
18. Lima, S. A., Varum, H., Sales, A., & Neto, V. F. (2012). Analysis of the mechanical properties of compressed earth block masonry using the sugarcane bagasse ash. *Construction and Building Materials, 35*, 829–837. https://doi.org/10.1016/j.conbuildmat.2012.04.127
19. Villamizar, M. C., Araque, V. S., Reyes, C. A., & Silva, R. S. (2012). Effect of the addition of coal-ash and cassava peels on the engineering properties of compressed earth blocks. *Construction and Building Materials, 36*, 276–286. https://doi.org/10.1016/j.conbuildmat.2012.04.056

Seismic Response of Reinforced Concrete Frames with Masonry Infills

Vinayak Sharma and Sushil Kumar Madan

Abstract Reinforced concrete (RC) frames with masonry infills are one of the most commonly used structural systems in the world, especially in the regions susceptible to earthquake. The infill walls are generally not considered as a structural element in seismic codal provisions. But it might be not profitable to neglect the role of infills. Moreover, the infilled masonry panels may considerably affect the stiffness and strength of RC frames. In the past few decades, great losses in lives and properties are witnessed in RC framed structures under the action of earthquakes, thus depicting the vulnerability of these structures. Over the past few years, extensive research has been carried out to study the behaviour of RC structures under seismic loading. Experimental investigations of RC frames with masonry infills are carried out to determine their complex seismic performance. However, these experimental tests are significantly expensive and are thus performed less in quantity. Hence, this study focuses on the analytical investigation on masonry infilled RC frame structures under seismic loading using computer programme ETABS. In this investigation, 3 eight storeyed planar frames are considered, namely bare frame (BF), fully infilled frame (IF), and open ground storey frame (OGSF). The masonry infill is modelled as an equivalent diagonal strut conforming to IS 1893:2016. The frame models are subjected to response spectrum analysis and time history analysis. The response of the structures was observed in terms of storey displacement, inter-storey drift, base shear, and various other parameters. All the analysis performed has shown that the inclusion of masonry infill in RC frames affect stiffness, strength, and displacement of these structures.

Keywords Reinforced concrete · Masonry infill · Earthquake · Response spectrum · Time history analysis

V. Sharma (✉) · S. K. Madan
Department of Civil Engineering, National Institute of Technology Kurukshetra, Kurukshetra, India

S. K. Madan
e-mail: skmadan@nitkkr.ac.in

A. K. Gupta et al. (eds.), *Advances in Construction Materials and Sustainable Environment*, Lecture Notes in Civil Engineering 196,
https://doi.org/10.1007/978-981-16-6557-8_16

1 Introduction

Concrete is the most commonly used building material in construction practices. In the past few decades, reinforced concrete (RC) structures infilled with masonry walls have been immensely used to design various types of buildings for people to reside and work in [1]. Masonry infill is generally not considered as a structural member but as a partition wall in reinforced concrete structures and steel frame buildings [2]. These structures have been mostly designed and build considering various codal provisions, commonly neglecting the contribution of the infill for the structural performance. It has been observed that the masonry infill affects stiffness, strength, and the displacement capacity of a building [3–5]. The inclusion of masonry panels may be beneficial or not for the seismic behaviour of the structure, depending on various factors like their plan and height distribution, condition of extremities, relative stiffness and strength between the infill panel and the frame members, and the mechanical and material properties of infill [6]. A comparative study of reinforced concrete frames infilled with and without masonry suggests that the inclusion of masonry infill causes a change in the approximated seismic performance of the RC framed structure due to the contribution of the masonry in the load transfer mechanism. It has been identified that masonry infilled frames have greater strength and rigidity in comparison with the bare frames, and their negligence has been the reason of failure of many multi-storied structures. Extensive research has been done throughout the world in the past fifty years, and many new techniques have been introduced for analysing the dynamic performance of RC building infilled with masonry.

Not considered as a structural element, infill wall is still usually joined with the main RC or steel frame and also lessen the deformation of the building. In this case, walls infilled with masonry constitute an important part of the vibrating structural system. The impact of masonry infilled walls is generally considered only by the involvement of extra loads and masses thoroughly distributed along the interfaces between the surrounding frame and the infill walls. This affects the performance of the infilled frame panel and also the entire building [7]. The earliest studies on reinforced concrete structures infilled with masonry walls began in the last 70 years. But mostly, these investigations were performed considering either small-scale frame models or frame designs that may not correctly represent existing buildings. Still, these researchers have identified a great amount of complex failure mechanisms that can be possibly caused by the frame-panel interaction [8]. To induce the performance of infilled steel frames subjected to racking loads, a mathematical model that can generate the hysteretic behaviour of RC frame elements accurately is required. It is also essential to define parameters that are indicative of the damage sustained by each frame element in a structure and also the overall damage in the structure [9]. The lateral stiffness of masonry infilled frames was studied by considering an equivalent diagonal strut to substitute the infill. Model experiments were conducted to check the theoretically derived equivalent diagonal strut. To investigate the vulnerability of RC frames with masonry infill, it is necessary to consider ways essential for the assessment scale pursued (kind of district, type of area, aggregate, single or multiple buildings). The introduction of walls infilled with masonry generally causes an overall improvement in the strength and stiffness of the RC framed structure. It was also reported that for the equivalent strut, the effective width lies around one-fourth of infill diagonal length for a square infill to one-eleventh of infill diagonal length for a side's ratio of five to one of an infill [10]. It was experimentally demonstrated that under seismic excitation, reinforced concrete floors serve as a diaphragm to disperse the forces of inertia that are produced by the dynamic action of earthquake to the resisting frame and wall elements [11]. It has been shown that the models with strong frames and strong panels exhibit greater performance as compared to those with weak frames and weak panels in respect to the load resistance and energy-dissipation capability [12].

The present study aims to study the behaviour of reinforced concrete planar frames under seismic loading. An important objective of the present study is to study the seismic response of eight storeyed planar frames. The respective frames are analysed to investigate their performance under linear dynamic analysis (Response Spectrum Method) and linear dynamic analysis (Time History Analysis) using ETABS Software.

2 Modelling of Masonry Infills

Based on the RC framed structures with masonry infill constructed in everyday life, some representative planar masonry infilled RC frame configurations were considered for investigating the seismic behaviour of these structures analytically. The analytical investigation was conducted on three 8 storeyed planar frames, namely, bare frame (BF), fully masonry infilled frame (IF), and open ground storey frame (OGSF). The masonry infills are modelled as equivalent diagonal single strut as per IS 1893:2016 [13]. The various frame configurations are depicted in Fig. 1. The masonry infilled RC frames with the given configurations were modelled in accordance with the IS 1893:2016 using the computer program ETABS [14]. The RC frames are considered to be located at Seismic Zone 5 ($Z = 0.36$). Only dead loads and seismic loads are considered in the analysis of the frames in the present work. Here, the response reduction factor (R) was specified as 5.0 to account for the inherent capability of the frame to undergo inelastic deformations without reduction in strength or stiffness. Importance factor (I) was taken as 1.0. Concrete having M25 grade and Fe415 HYSD rebar are considered for all the RC frame models used in this investigation.

All the columns and beams in the current investigation were modelled as frame elements with the beam-column joints assumed as rigid. Elastic properties of the given materials are considered as per IS 456:2000 [15]. M25 concrete grade and Fe415 HYSD Rebar were used to model all the frames.

For performing seismic analysis of infilled frames, the masonry infilled walls were modelled as equivalent diagonal strut. Thus, the depth of the masonry wall was

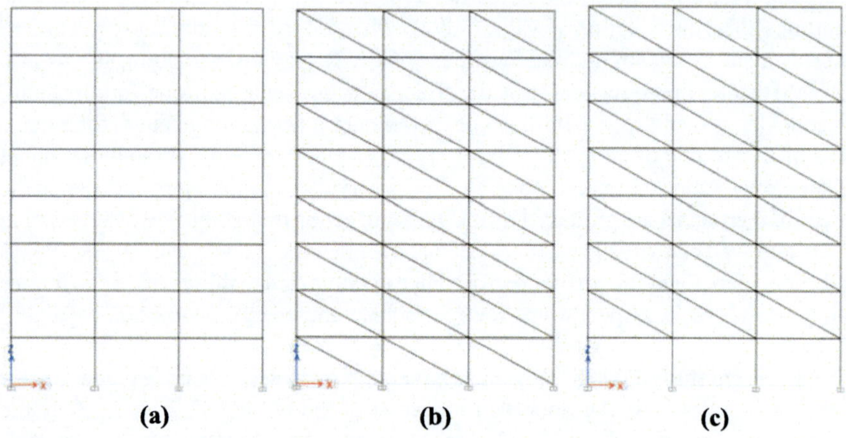

Fig. 1 RC frame configurations: **a** bare frame (BF), **b** fully masonry infilled frame (IF), **c** open ground storey frame (OGSF)

Table 1 Section properties

Property	Beam	Column	Masonry Wall
Width (mm)	350	450	230
Depth (mm)	450	450	775
Length (m)	6	3	6

calculated as per IS 1893:2016. The RC frame section properties are presented in Table 1.

3 Seismic Analysis of RC Frames

The seismic analysis of RC frames with and without infills is performed by 2 methods, namely response spectrum analysis and time history analysis. Thus, the seismic behaviour of given frames is assessed through dynamic analysis. The analytical investigation is performed on commercial software ETABS.

3.1 Response Spectrum Analysis

Response spectrum analysis is a method to approximate the structural response to dynamic excitation (earthquake). Response spectrum analysis is performed using multimode responses, where the free vibration modes are computed using Eigen vector or Ritz Vector analysis in ETABS. In the present study, response spectrum

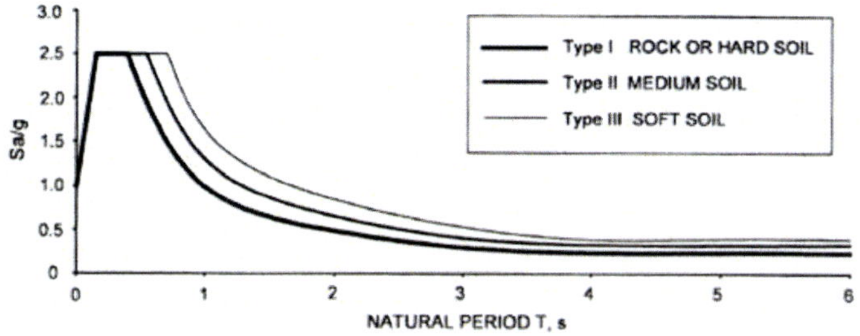

Fig. 2 Design response spectrum as per IS 1893:2016

data from IS 1893:2016 is considered for the analysis using Ritz vector method in the software (Fig. 2).

3.2 Time History Analysis

The time history analysis can be either performed as a linear or nonlinear analysis. Here, the linear time history analysis is conducted on RC frame models to analyse their seismic behaviour. In the present study, time history data of El Centro (1940) earthquake is considered. Here, the N-S component of the earthquake is taken into account (Fig. 3). It can be seen from the Fig. 3 that the peak ground acceleration (pga) of 0.349 g occurs at 2.12 s.

Fig. 3 Time history of El Centro earthquake—NS component (1940)

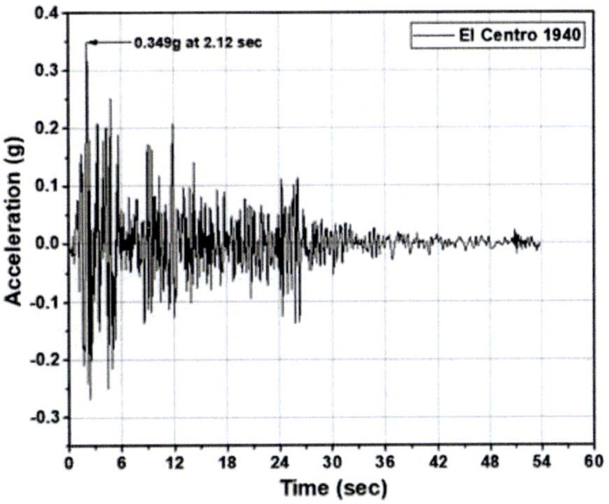

4 Results and Discussion

In this section, the results obtained from the analytical investigation on bare frame (BF), fully masonry infilled frame (IF), and open ground storey frame (OGSF) are presented. Furthermore, these results are discussed and interpreted so as to analyse the seismic behaviour of RC frames with and without infill.

4.1 *Structural Response from Response Spectrum Analysis*

Storey Displacement. The displacement of frames at various storey levels for response spectrum analysis is shown in Fig. 4a. The plotted curve shows that the top storey displacement is maximum for BF and minimum for IF, for the given loading conditions. It is observed that the infill acing as equivalent diagonal struts are responsible for the increase in the stiffness of storeys. Thus, the inclusion of infill reduces storey displacements. This result is in accordance with the past studies. Now, from Fig. 4a, it can be seen that the first storey displacement of OGSF is even greater than that of bare frame by nearly 28%. This is attributed to the decrease in stiffness of bottom storey of OGSF. To rectify this, various strengthening options for open ground storey frames need to be considered.

Inter-Storey Drift. As per IS 1893:2016, the storey drift in any storey should not be greater than 0.4% of the storey height. In the present 8 storeyed models, this maximum allowable drift comes out to be 0.012 as the storey height is 3 m in all the frames. The inter-storey drift at various storey levels for response spectrum analysis is shown in Fig. 4b. From the observed results, it is clear that all the modelled frames have storey drift value far lesser than the allowable drift. The plotted curve shows that the maximum value of inter-storey drift is for BF and minimum for IF, for

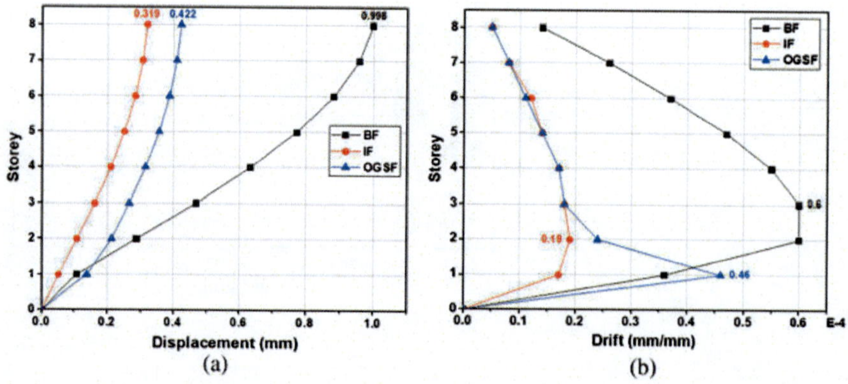

Fig. 4 Response from response spectrum analysis: **a** storey displacement, **b** inter-storey drift

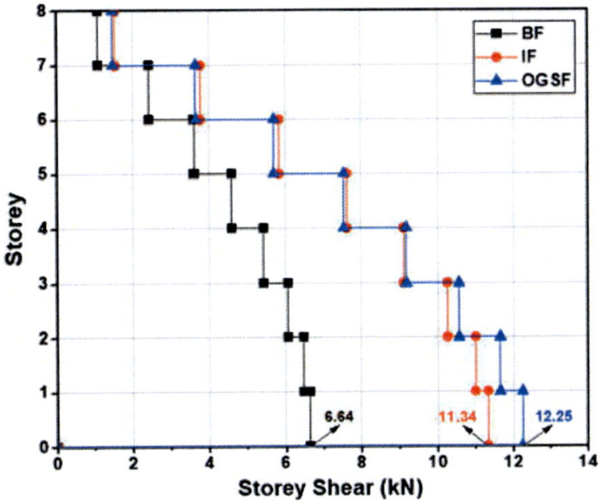

Fig. 5 Storey shears from response spectrum analysis

the given loading conditions. It was reported that the introduction of infill controls both the lateral displacements and storey drifts. However, no significant difference in storey drift of BF and OGSF was found in the upper stories. This depicts that the storey displacements for upper stories remain uniform in masonry infilled structures irrespective of absence of infill in the ground frame.

Storey Shear. The shear values of all the models at various storey levels for response spectrum analysis are shown in Fig. 5. The base shear is calculated as 6.64, 11.34, and 12.25 kN in BF, IF, and OGSF, respectively. Thus, the percentage increase in base shear is found to be around 71% for IF and 84% for OGSF, when compared with BF. The high increase in base shear in masonry infilled frames is attributed to their increased stiffness and strength as compared to frame without infill. It is also reported that there is no considerable difference in the shear values for the upper stories in RC frames with infill.

4.2 Structural Response from Time History Analysis

Base Shear. For the modelled frames, the variation of base shear with time is depicted in Table 2. It is determined that the base shear values are highest in OGSF model. The maximum absolute value of base shear is reported as 64.253, 74.351, and 87.318 kN.

Table 2 Base shear (kN) from time history analysis

Base shear (kN)	BF	IF	OGSF
Maximum	64.253	74.351	82.533
Minimum	– 60.74	– 68.251	– 87.318

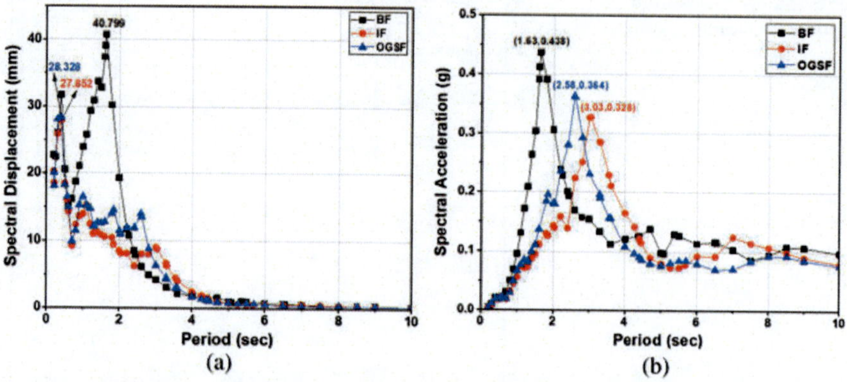

Fig. 6 Response of spectral parameters: **a** displacement, **b** acceleration

Table 3 Top storey spectral parameters response from time history analysis

Spectral parameter	BF	IF	OGSF
Displacement (mm)	40.799	27.852	28.328
Velocity (mm/s)	389.3	173.77	222.35
Acceleration (g)	0.438	0.364	0.328

When compared with bare frame, the percentage increase in the maximum base shear is around 28% for OGSF and 16% for IF, while for minimum base shear, it is 44% for OGSF and 12% for IF model. This validates the increase in results of base shear in masonry infilled frames from previous section.

Spectral Parameters. The response of spectral parameters of top storey in terms of spectral displacement and spectral acceleration is depicted in Fig. 6. Table 3 shows the maximum values of spectral parameters for all the frame models. Thus, it can be seen that out of the three spectral parameters, as compared to the bare frame, the reduction in displacement and velocity is more in fully infilled frame than open ground storey frame but the reduction in acceleration is more in OGSF than IF.

5 Conclusions

The present paper focuses on the investigation of reinforced concrete frames with masonry infill using ETABS software by performing dynamic seismic analysis. It has been observed that masonry infill has a considerable impact on the dynamic properties, stiffness, strength, and seismic behaviour of RC frame structures. The infill walls prevent the damages concentrated in top storey. The inclusion of infill increases stiffness and strength of the building, reducing the inter-storey drift of the structure. The infill acing as equivalent diagonal struts are responsible for the increase in the stiffness of storeys. The introduction of infill controls both the lateral displacements and storey drifts. When compared to bare frame, the top storey displacements are 68% lesser for fully infilled frame and 58% lesser for open ground storey frame. The first storey displacement of OGSF was reported to be greater than that of bare frame by nearly 28%. This is attributed to the decrease in stiffness of bottom storey of OGSF. To rectify this, various strengthening options for OGSF must be considered. Compared to bare frame, the percentage reduction in the spectral displacement is around 32% for IF and 30% for OGSF. Whereas, the value of spectral velocity reduces by 55% for IF and by 43% for OGSF, when compared with bare frame. Also, the spectral acceleration in IF is 17% lesser than BF and in OGSF, and it is 25% lesser as compared to BF. To compare the dynamic response of various structures, it is better to perform nonlinear analysis (time history and pushover analysis) on these frames. So, future investigations should focus more on nonlinear analysis of the reinforced concrete frame structures with masonry infill.

References

1. Mohamed, H. M. (2017). *Seismic risk assessment of reinforced concrete frames with masonry infill*. Ph.D. Thesis, Faculdade de Engenharia da Universidade do Porto.
2. Shing, P. B., & Mehrabi, A. B. (2002). Behaviour and analysis of masonry-infilled frames. *Progress in Structural Engineering and Materials, 4*(3), 320–331. https://doi.org/10.1002/pse.122
3. Mehrabi, A. B. (1994). *Behavior of masonry-infilled reinforced concrete frames subjected to lateral loadings*. Ph.D. Thesis, University of Colorado.
4. Bergami, A. V., & Nuti, C. (2015). Experimental tests and global modeling of masonry infilled frames. *Earthquakes and Structures, 9*(2), 281–303.
5. Shan, S., Li, S., Xu, S., & Xie, L. (2016). Experimental study on the progressive collapse performance of RC frames with infill walls. *Engineering Structures, 111*, 80–92. https://doi.org/10.1016/j.engstruct.2015.12.010
6. Furtado, A., & Teresa de Risi, M. (2020). Recent findings and open issues concerning the seismic behaviour of masonry infill walls in RC buildings. *Advances in Civil Engineering, 2020*, 1–20. https://doi.org/10.1155/2020/9261716
7. Ferraioli, M., & Lavino, A. (2020). Irregularity effects of masonry infills on nonlinear seismic behaviour of RC buildings. *Mathematical Problems in Engineering, 2020*, 1–18. https://doi.org/10.1155/2020/4086320
8. Polyakov, S. V. (1960). On the interaction between masonry filler walls and enclosing frame when loaded in the plane of the wall. *Translations in Earthquake Engineering, 2*(3), 36–42.
9. Holmes, M. (1961). Steel frames with brickwork and concrete infilling. *Proceedings of the Institution of Civil Engineers, 19*(4), 473–478.
10. Smith, B. S. (1962). Lateral stiffness of infilled RC frames. *Journal of Structural Division, Proceedings of ASCE, 88*, 183–199.
11. Smith, B. S., & Carter, C. (1969). A method of analysis for infilled RC frames. *Proceedings of the Institute of Civil Engineers, 44*, 31–48.
12. Mehrabi, A. B., Shing, P. B., Schuller, M. P., & Noland, J. L. (1996). *Experimental evaluation of masonry-infilled RC frames*.
13. Indian Standard, IS 1893:2016 (Part 1) Criteria for earthquake resistant design of structures. Bureau of Indian Standards.

14. CSI, Extended Three-Dimensional Analysis of Building System, ETABS v18.1.1 (2019) Computers and Structures Inc.
15. Indian Standard, IS 456 (2000) Indian standard for plain and reinforced concrete code of practice, Bureau of Indian Standards.

Analysis of Seismic Behavior of Buildings With and Without Shear Walls in Various Seismic Zones and Soil Types

Diptanshu Lal, Biplav Regmi, Haris Farooq Bhat, and S. Anbu Kumar

Abstract Earthquakes have a direct impact on all buildings. The impact varies as per the zone earthquakes occur in. Buildings behave differently to earthquake loading depending on the condition and types of soil—hard, medium, and soft soil. Since the soil properties of all these soils are quite different, seismic waves also behave in different ways when they pass through the layers. The foundation of buildings interacts with the soil beneath them during earthquakes. Thus, the ground structure and the impact of the earthquake it faces are directly influenced by soil type and the type of structure itself. Thus, multiple analyses of the same structure in various seismic zones are carried out by changing the soil type between hard, medium, and soft for every simulation. For different zones and different types of soil, responses like story drift and displacement are plotted. IS 1893:2016 'Criteria for Earthquake Resistant Design of Structures' provides the response spectrum criteria for simulations in these various soil types. This research has substantial benefits in the field of structural designing and geotechnical engineering.

Keywords Structural simulation · Numerical analysis · Seismic analysis · Soil-structure interaction

D. Lal (✉) · B. Regmi · H. F. Bhat · S. Anbu Kumar
Department of Civil Engineering, Delhi Technological University, Delhi 110042, India
e-mail: diptanshulal_2k17ce28@dtu.ac.in

B. Regmi
e-mail: biplavregmi_2k17ce21@dtu.ac.in

H. F. Bhat
e-mail: harisfarooqbhat_2k17ce30@dtu.ac.in

S. Anbu Kumar
e-mail: sanbukumar@dce.ac.in

A. K. Gupta et al. (eds.), *Advances in Construction Materials and Sustainable Environment*, Lecture Notes in Civil Engineering 196,
https://doi.org/10.1007/978-981-16-6557-8_17

1 Introduction

Earthquakes are considered a major cause of failure in the structures. The adverse effects caused due to earthquakes may result in swaying, cracking, and ultimately destruction of the structure. Despite their uncertainty and moderate intensity, earthquakes pose a threat to cause extensive property damage and associated financial losses. The mitigation of these hazards involves the collaboration of engineers across the world to provide and advance the knowledge of seismically resistant structures. In this context, various technological advancements have been made over the years on seismic vulnerability assessments for civil engineering structures. The major focus has been given to the case studies of earlier seismic damages and their correlation to the present context so that the situation can be improved over the years.

In most of the countries, the majority of structures, generally comprising the old buildings, are designed to resist generally the dead load and movable loads and this motivates an extensive assessment to be done on such structures. This paper analyzes the seismic response of various building types by running numerical simulations on a simplified analytical model. Various seismic zones across India pose varying effects on buildings and again by varying the soil types simultaneously can provide a detailed response study. Such numerical modeling can be of great help to extract the most suitable parameters for building seismically resilient structures and for obtaining optimal solutions for structural retrofitting, thereby saving human lives and property.

2 Literature Review

Poonam [1] discussed the responses of frames due to seismic excitation. Multiple studies have been carried out on the analysis of torsional response in multistoried structures using corresponding eccentricity [2] and using pushover analysis [3].

Spyrakos [4] discussed the properties of soil-structure interaction on the feedback by developing and calculating the frequency domain via a 4-degree freedom system. The expressions then were used to deduce the coefficients of damping and stiffness for rigid surfaces.

Henderson [5] studied and discussed the properties of base-isolated structures and came to the conclusion that when isolators are directly equivalent to soil flexibility, the handouts of SSI should not be disregarded.

Kelly [6] performed an experimental study relating to nuclear structures constructed on soft soils with isolated bases. The conclusion of the research implied the design of the isolator should be considered for displacement demands.

Pandey [7] explained the behavior of seismic waves with relation to soil-structure interaction on hill slopes. In his study, he found that with an increase in time period, the response reduction factor decreases but remains constant after achieving a definite value of the time period.

Anand [8] analyzed the seismic behavior in the presence and absence of shear walls in RCC structures under distinct soil conditions. When the soil type changes from one type to another, the values of horizontal displacement, axial force, base shear, and moment in the column increase for all types of structures. Therefore, we deduced that interaction between soil and structure should be kept in mind during the designing process.

Jennifer Priyanka [9] explained how tall buildings are affected by lateral forces with a distinct type of inconsistency. In this study, she concluded that in soft soil, there is more deflection in structures than in medium and hard soil. The stiffness irregularities in structures give more deflection and swaying than the structures with distinct irregularity.

Constantinou [10] put forward a method for the calculation of damping of seismically isolated structures, by keeping in mind the disappearance of the energy of the bearing and damping radiation in the soil.

Nowak [11] reviewed the available data base and shear model for reinforced concrete beams without shear reinforcement and select the most efficient model for design code for concrete structure.

Shah [12] presented statistical model for the prediction of shear strength of high strength reinforced concrete (HSRC) beams. By comparing the actual and predicted values of shear strength of beams, it shows that the proposed equation is conservatives for various longitudinal reinforcement ratios (ρ).

3 Methodology

Building analysis is conducted using ETABS [13], which is a general-purpose civil engineering program that designs by means of a realistic and intuitive object-based modeling environment that simplifies and increases the efficiency of the engineering process. All structures from basic to advanced, either of simple or complex geometry, can be modeled, evaluated, and enhanced.

The multistory structure was designed and examined under various soil types—hard, medium, and soft. Response spectrum method was used in the evaluation of all 18 structures (Table 1). The seismic analysis was performed as per IS 1893 (Part I): 2016 [14].

3.1 Load Combination

As per IS 1893 (Part I):2016, load combinations provided below are taken into account in the limit state design of prestressed and reinforced concrete structures:

1. 1.5 (DL + IL)
2. 1.2 (DL + IL ± EL)

Table 1 Building specifications for analysis

Building name	Type of soil	Seismic zone	Shear wall
BH3P	Hard	III	Present
BH4P		IV	
BH5P		V	
BM3P	Medium	III	Present
BM4P		IV	
BM5P		V	
BS3P	Soft	III	Present
BS4P		IV	
BS5P		V	
BH3A	Hard	III	Absent
BH4A		IV	
BH5A		V	
BM3A	Medium	III	Absent
BM4A		IV	
BM5A		V	
BS3A	Soft	III	Absent
BS4A		IV	
BS5A		V	

3. 1.5 (DL ± EL)
4. 0.9DL ± 1.5EL.

3.2 Building Parameter

- Number of stories: G + 7
- Grade of concrete:M25
- Grade of steel:HYSD 500
- Beam sizes:0.30 * 0.40 m
- Secondary beam size:0.25 * 0.30 m
- Column sizes:0.45*0.45 m
- Slab thickness:0.125 m
- Importance Factor:1.2
- Story height:3.3 m.

3.3 Loading Parameters (as Per IS: 875 (Part I)—2015 [15], IS: 875 (Part I)—2015 [16])

- Wall load on primary beam:10.24 kN/m
- Wall load on the secondary beam:5.12 kN/m
- Wall load for parapet wall:2.5 kN/m
- Floor finish:1.5 kN/m^2
- Live load:4 kN/m^2
- Roof live load:1.5 kN/m^2.

4 Response Spectrum Method

Response spectrum method is used for the determination of the response of the structure to nondeterministic and infrequent earthquake loads. It provides information about dynamic behavior by calculating pseudo acceleration, velocity, and displacement as a function of the structural period for a given amount of time and damping level. In this method, a graph is plotted between the maximum response and an undamped natural period for various damping values and may be expressed in terms of maximum absolute acceleration, maximum relative velocity, or maximum relative displacement. Since this method relates the structural type selection with dynamic performance, response spectrum analysis is very crucial for design decision-making. The structures with longer periods undergo larger displacement, whereas the structures with shorter periods undergo a greater acceleration. Thus, these structural performance objectives should be considered at the time of designing the structures.

5 Modal Pushover Analysis

Pushover analysis is a static nonlinear process to determine the seismic structural deformations. Since dynamic forces in structure moves to another component after one has yielded or failed, pushover analysis exploits this phenomenon to find a weak link in the structure thereby revising the model to a structure with changes caused by the weak link. But the consideration of fundamental mode only and the inappropriateness of inelastic theorem for conventional pushover analysis made the necessity of modal pushover analysis where the seismic response of a structure is calculated by pushing the structure in each mode to its modal target displacement, and the overall response is determined by combining each modal response by a modal combination rule where higher modes are distinctly considered.

6 Drift on the Structure

The total lateral displacement of the top of the building is termed 'drift,' while the relative lateral displacement between two consecutive stories is called the 'inter-story drift'. 'Drift index' relates to the lateral stiffness of the building and is the relative translational displacement between two consecutive floors divided by the story height as defined by Eqs. 1 and 2, respectively.

$$\text{Total Drift Index} = \text{Total drift/Building height} = \Delta/H \tag{1}$$

$$\text{Inter Story drift index} = \text{Inter story drift/story height} = \delta/h \tag{2}$$

7 Soil-Structure Interaction

Seismic waves cause a strong vibrating motion in the structure. The vibrating motion depends on the vibrational characteristics and hence on the layout of the building. The inertia of the structure must be overcome as well to react to the motion of the seismic waves. Stiffness properties of the soil play a vital role in altering the response of structures observed during earthquakes at the foundation level. This implies the importance of the physical properties of the foundation medium during earthquakes.

The two aspects of building foundation during earthquakes can be summarized as:

1. When a building is founded in a rigid structure, the response to earthquake motion is vastly different from that of a building found on deformable soil.
2. When earthquakes occur, the motions recorded at the foundation level of the buildings majorly vary from the motion which would have been recorded had there not been a building in that location.

8 Result and Discussion

The G + 7 residential buildings with and without shear walls are shown in Fig. 1 to understand the seismic forces acting on them. Comparisons for different seismic responses obtained were made for both types of building frames.

Table 1 shows the configuration of all the 18 building types analyzed in the numerical modeling software ETABS. The results obtained for the maximum deflection and drift are discussed below, which shows a systematic as well as a drastic variation for the varying soil types and seismic zones. As shown in Fig. 2, a variation of 34.448 mm can be observed for the variation in the seismic zone from III to V for the same G + 7 building structure without shear walls in the soft soil.

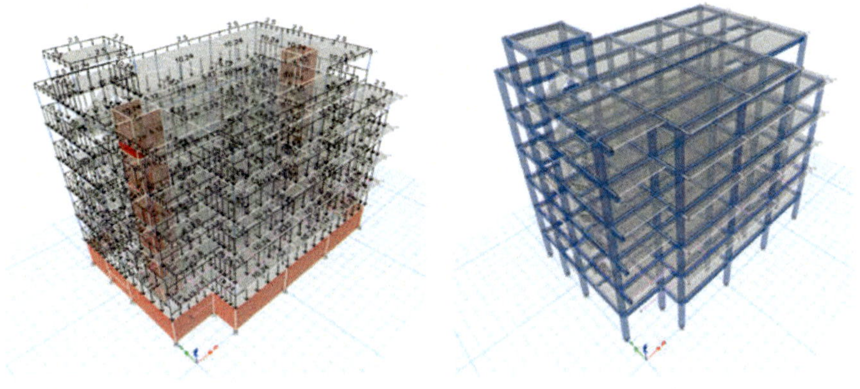

Fig. 1 Frame design and load assigns representation

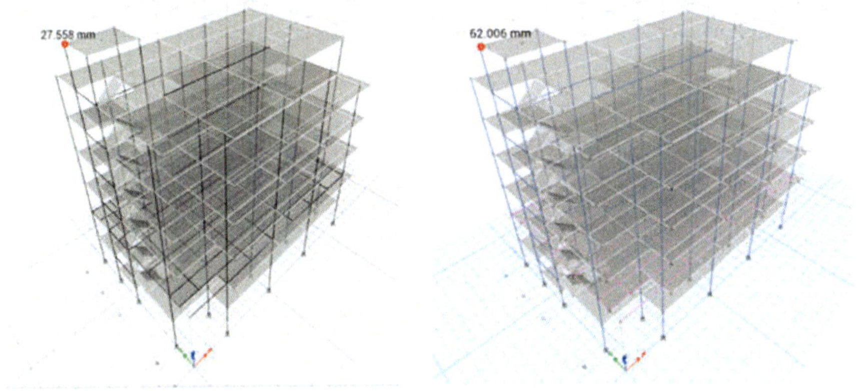

Fig. 2 Lateral deflection for the G + 7 building in seismic zones III (left) and V (right)

It can be deduced from Table 2 that the variation in the seismic zone and soil types vary the seismic load to a great extent and, hence, varies the lateral deflection of the structure. As seen in the building types BH3A and BS3A, the change in the soil type from hard to soft, while keeping the same seismic zone itself increased the lateral deflection by 66.99%. A similar pattern can be observed in the building types BH4A–BS4A and BH5A–BS5A. Similarly, for the change in soil type between the building structure BS4A and BM4A, the lateral deflection is seen to be increasing from 33.664 to 41.337 mm. While in the presence of shear walls, the lateral deflection of the structures is decreased by a lot. For the building types BH3P and BS3P, the lateral deflection seems to decrease from 10.479 mm to 6.275 mm. Hence, in the case with added shear walls, the deflection decreases by 40.12% on changing the soil type from soft to hard soil. The same pattern can be observed for the similar building types in the seismic zone IV and V (that is, for BS4P–BH4P and BS5P–BH5P). At the

Table 2 Lateral deflection values obtained for all the 18 building types

Seismic zone III						
FLOOR	BS3A	BS3P	BM3A	BM3P	BH3A	BH3P
BF	0	0	0	0	0	0
GF	3.551	0.084	2.891	0.069	2.126	0.05
IF	9.315	1.241	7.586	1.01	5.578	0.743
2F	15.107	3.059	12.303	2.491	9.046	1.832
3F	20.233	5.122	16.478	4.171	12.116	3.067
4F	24.141	7.15	19.66	5.823	14.456	4.281
5F	26.521	9.016	21.598	7.342	15.881	5.399
Stair cover	27.558	10.479	22.443	8.534	16.502	6.275
Seismic zone IV						
FLOOR	BS4A	BS4P	BM4A	BM4P	BH4A	BH4P
BF	0	0	0	0	0	0
GF	5.326	0.126	4.337	0.103	3.189	0.076
IF	13.973	1.861	11.379	1.516	8.367	1.114
2F	22.661	4.589	18.454	3.737	13.569	2.748
3F	30.35	7.683	24.716	6.257	18.174	4.601
4F	36.212	10.725	29.49	8.734	21.684	6.422
5F	39.781	13.524	32.397	11.014	23.821	8.098
Stair cover	41.337	15.718	33.664	12.8	24.753	9.412
Seismic zone V						
FLOOR	BS5A	BS5P	BM5A	BM5P	BH5A	BH5P
BF	0	0	0	0	0	0
GF	7.989	0.19	6.506	0.154	4.784	0.114
IF	20.96	2.792	17.069	2.273	12.551	1.672
2F	33.991	6.883	27.682	5.606	20.354	4.122
3F	45.525	11.525	37.074	9.385	27.261	6.901
4F	54.318	16.087	44.235	13.101	32.526	9.633
5F	59.672	20.286	48.595	16.521	35.732	12.147
Stair cover	62.006	23.577	50.496	19.2	37.129	14.118

same time, the effect of the variation in the seismic zone on the lateral deflection can also be observed from the table. As seen in BM3P, BM4P, and BM5P, the change in the seismic zone from III to IV and from IV to V has increased the lateral deflection by 4.26 mm and 6.40 mm, respectively, for medium soil type. A similar change is observed in building types BS3P, BS4P, and BS5P and BH3P, BH4P, and BH5P as well.

The bar diagram in Fig. 3 shows the lateral deflection in all stories of the building containing shear walls in medium soil type but different seismic zones. A systematic

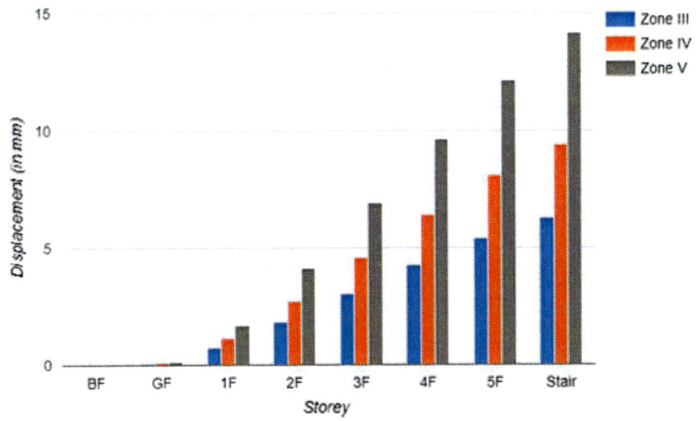

Fig. 3 Story-wise lateral deflection with varying seismic zones in buildings in medium soil type containing shear walls

variation is observed where the deflection is always the highest in seismic zone V irrespective of the story. Furthermore, the variation in lateral deflection increases with the increase in height of the building, i.e., the variation in the 5th floor for different seismic zones is higher than that on the 4th floor.

Similarly, the line graph in Fig. 4 has the same seismic zone (IV), but soil type varies. These buildings have shear walls. In this case, too, the height is proportional to the lateral displacement. Despite the soil types varying, we can observe and analyze the displacement being the largest in soft soil and the displacement being the least in hard soil. This value, when compared to that of buildings without shear walls in the

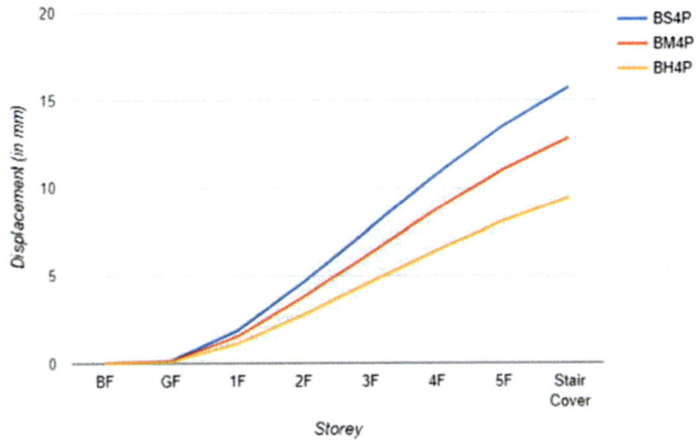

Fig. 4 Story-wise lateral deflection with varying soil types in buildings in seismic zone IV containing shear walls

	Drift
BH3P	0.00037
BH4P	0.000561
BH5P	0.000842
BM3P	0.000509
BM4P	0.000764
BM5P	0.00115
BS3P	0.000625
BS4P	0.000938
BS5P	0.00141
BH3A	0.00105
BH4A	0.00158
BH5A	0.00236
BM3A	0.00143
BM4A	0.00214
BM5A	0.00322
BS3A	0.00175
BS4A	0.00263
BS5A	0.00395

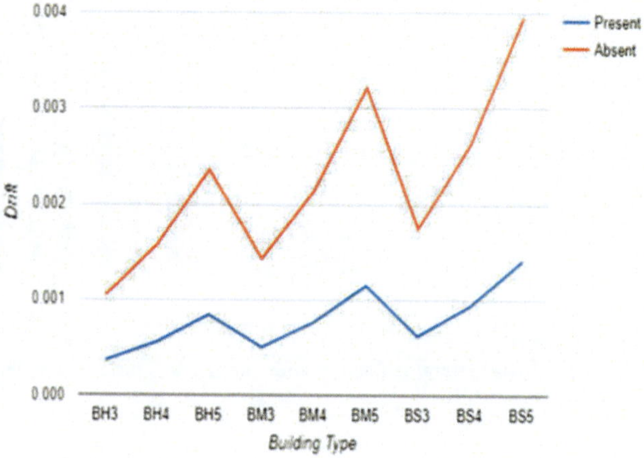

Fig. 5 Drift in buildings with and without shear walls in varying soil types and seismic zones

same seismic zones, shows a significant increase in lateral displacement in all floors and all soil types.

The drift represented in the line graph in Fig. 5 also follows a similar pattern to that of the lateral displacement mentioned above. The drift increases with the change in the type of soil, gradually from soft to hard, and removal of shear walls. The maximum drift can be observed in BS5A. The building, BS5P, has the same conditions and criteria; bar the presence of the shear wall. It shows the drift of 0.00141. The drift in BS5A, compared to this, is increased by 180.1%.

The comparison of the two Figs. 6 and 7, is of the same buildings in the same soil type—hard, but in different seismic zones. One of the buildings had a shear wall; the other one did not. As the graph shows, the maximum lateral displacement for both cases can be observed in the stair cover. This is true for all seismic zones too. The displacement of the stair cover of a building with shear walls decreases by an average of 61.97% in seismic zones III, IV, and V when compared to the displacement of stair cover in the building without the shear wall.

9 Conclusion

This research gives us a distinct idea about the importance of shear walls in multi-storied structures using only one kind of building. Despite having multiple variables to choose from, only one was varied at a time so that a direct comparison could be made to the same building in another seismic zone or another soil type.

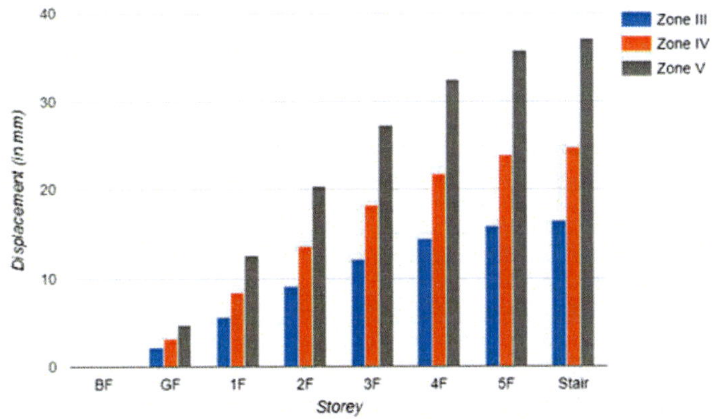

Fig. 6 Story-wise lateral deflection with varying seismic zones in buildings in hard soil type without shear walls

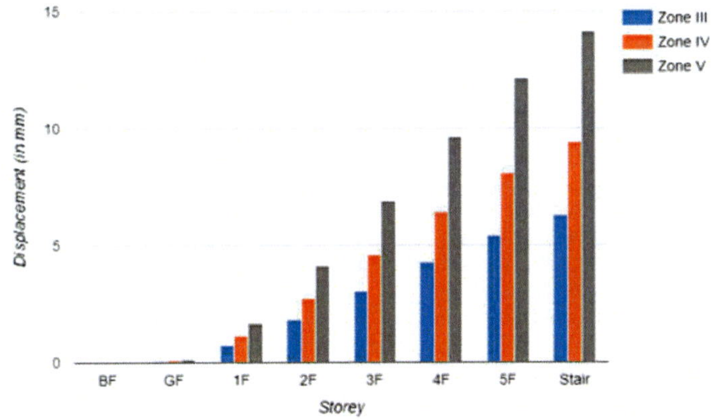

Fig. 7 Story-wise lateral deflection with varying seismic zones in buildings in hard soil type containing shear walls

In ETABS, all 18 buildings were analyzed with the configuration as shown in Table 1. The results show that the building with shear walls shows better structural stability in terms of lateral displacement and drift. Furthermore, these buildings behave better when built-in hard soil type in seismic zone III. The displacement is observed to be maximum (yet within the limits), when the structure is built on soft soil type in seismic zone V. This research also depicts the necessity to plan for seismic loads in areas susceptible to earthquakes. The data obtained from this study can be used as a reference for future planning in zones III, IV, and V. Furthermore, the data can be used for further research in similar or new terrains.

Acknowledgements The authors would like to thank the Department of Civil Engineering, Delhi Technological University for the opportunity and constant support.

References

1. Poonam, A. K. (2012). Study of response of structural irregular building frames to seismic excitation. *International Journal of Civil, Structural, Environmental and Infrastructure Engineering Research and Development*, 25–31.
2. Saffari, T. A. (2011). Evaluation of the torsional response of multistory buildings using equivalent static eccentricity. *ASCE Journal of Strucutral Engineering*, 862–868.
3. Chopra, C. A. (2004). Seismic response of vertically irregular frames: response history and modal pushover analyses. *ASCE Journal of Structural Engineering*, 1177–1185.
4. Spyrakos, C. A. (2002). Effect of soil-structure interaction on seismically isolated bridges. *Journal of Earthquake Engineering*, 391–429.
5. Henderson, M. N. (1989). Base isolated buildings with soil-structure interaction. *Earthquake Engineering Structural Dynamic*, 751–765.
6. Kelly, J. M. (1991). Shake table tests of long period isolation systems for nuclear facilities at soft soil sites. *UBC/EERC-91/03.*
7. Pandey, P. K. (2011). Seismic soil-structure interaction of buildings on hill slopes. *Journal of Civil and Structural Engineering.*
8. Anand, N. A. (2010). Seismic behavior of RCC shear wall under different soil conditions. In *Proceedings of the Indian Geotechnical Confrence on Geo-trends.*
9. Jennifer Priyanka, R. A. (2012). Effect of lateral force on tall buildings with different types of irregularities. In *Proceedings of the INCACMA Conference on Advances in Construction, Manufacturing and Automation Research.*
10. Constantinou, M. K. (1986). Effect of soil-structure interaction on damping and frequencies of base-isolated strucutres. In *3rd US National Conference on Earthquake Engineering* (pp. 671–681).
11. Nowak, P. A. (2008). *Shear resistances of reinforced concrete beams without web reinforcements.* Department of Civil Engineering, University of Nebraska.
12. Shah, S. A. (2009). Statistical model for the prediction of shear strength of high. *The Arabian Journal for Science and Engineering.*
13. ETABS. Integrated building design software, Computers and Structures, Inc. Berkeley, California, USA, Version 9, November 2005
14. Bureau of Indian Standards: IS 1893 Part (1):2016, Criteria of Earthquake Resistant Design of Structures: Part 1 General provisions on Buildings.
15. Bureau of Indian Standards: IS 875, part (1):1987, Code of practice for design loads (other than earthquake) for buildings and structures, New Delhi, India.
16. Bureau of Indian Standards: IS 875, part (2):1987, Code of practice for design loads (other than earthquake) for buildings and structures, New Delhi, India.

Review on the Durability Parameters of Self-compacting Concrete

Reshul Raj, Mayur Bhat, Achal Agrawal, and Narayan Chandak

Abstract The disposal of toxic environmental waste is one of the main challenges at present. The use of various kinds of waste as a substitute for aggregate or cement in concrete is a proven way to manage the waste. Various research experiments have been carried out using waste material in self-compacting concrete, resulting in improved strength and resilience, whether it is traditional concrete or self-compacting concrete. This paper highlights the improvements in the durability of self-compacting concrete through the use of different combinations of various waste materials. The study also explains the influence of mineral and chemical admixtures on self-compacting concrete. It also addresses the effect of waste materials, fiber content, and type of admixtures on the durability parameters such as Rapid Chloride Penetration Test (RCPT), sulfate attack, water penetration, and carbonation test on the self-compacting concrete.

Keywords Self-compacting concrete · Super-plasticizers · RCPT · Sulfate attack · Water penetration · Acid attack · Fibers

1 Introduction

Self-compacting concrete (SCC) is a front runner of concrete that does not require any vibratory concrete machine to be installed. Because of this concrete consistency, this type of concrete settles under its own weight and is commonly used in the field of construction. Particularly in congested formwork, where compaction is not fissile, it is a good alternative to conventional concrete and work with self-compacting concrete is preferred in this situation.

The use of waste materials from construction industries in making of self-compacting concrete can reduces tons of accumulated construction waste. The present review paper is a research on durability properties of self-compacting

R. Raj · M. Bhat · A. Agrawal (✉) · N. Chandak
Department of Civil Engineering, SVKM's NMIMS, MPSTME, Shirpur, Maharashtra 425405, India
e-mail: achal.agrawal@nmims.edu

A. K. Gupta et al. (eds.), *Advances in Construction Materials and Sustainable Environment*, Lecture Notes in Civil Engineering 196,
https://doi.org/10.1007/978-981-16-6557-8_18

concrete after partial replacement of natural aggregates and cementitious content with different mineral admixtures and waste aggregates.

Most research work have shown that partial replacement of cement and natural aggregates with waste mineral admixtures, and waste aggregates improves the durability properties of self-compacting concrete.

2 Review Significance

This paper gives a review on self-compacting concrete (SCC) mix made using various mineral admixtures and materials as partial replacement. A concrete which when poured from a certain height, compacts with its own weight due to gravity and does not require any external vibrations is known as self-compacting concrete. Different authors review papers and case studies are included in this paper, and it shows that effects of minerals admixtures and waste materials on the durability properties of SCC.

The main objective of this review is to find the change in durability of SCC on partial replacement of cement and aggregates with mineral admixtures and recycled materials.

3 Materials Used

Table 1 shows variety of material used by the authors in their research work, which gives a better understanding of different combination and effects of proportions of mineral admixtures and waste/recycled aggregates on the durability of SCC. A detailed review of 145 mixes is done in this study where 23 research studies are reported. Table 2 shows the mix design proportion used by different authors. The cement that was used widely for mix design was Ordinary Portland Cement. Mineral admixtures are also used in different combination with OPC. Fly ash, limestone powder, rice husk ash, pumice, zeolite, coal fly ash, silicon fillers, metakaolin, and waste asphalt filler was used in specific proportion with OPC.

Here, Fig. 1 shows the frequency of mineral admixtures used with the cement by different authors. It clearly shows that frequency of fly ash with cement and limestone powder with cement is used the most at 28 and 22%, respectively. Another frequently used material is metakaolin at 14%. The graph shows that fly ash, limestone powder, and metakaolin was used frequently in combination with OPC and other mineral admixtures by most authors.

In some SCC mixes, addition of silica fume and rice husk ash are used as mineral admixture in specific proportion. Pumice in SCC has superior resistance to harsh weather conditions like freezing and thawing. Pumice powder has pozzolanic properties, especially if it is used as combination with silica fume.

Table 1 Materials use by different authors

Reference	Year	SP type	Material used			
			Fine aggregates	Coarse aggregates	Cement	Mineral admixtures
Siddique [1]	2013	PCE	Bottom ash, Natural aggregate	Natural	OPC	Fly ash
Frazão et al. [2]	2015	Ether polycarboxylate (ViscoCrete 3005)	Fine river sand (< 1.19 mm), Natural aggregates	Coarse river sand (< 4.76 mm), Natural aggregates	CEM I 42.5 R Portland cement	Limestone filler, Steel fibers
Singh et al. [3]	2016	Polycarboxylate-based SP	Natural	RCA, NA	OPC grade 43	Fly ash, Metakaolin
Kapoor et al. [4]	2016	PCE	Natural	RCA, NA	OPC	Fly ash, Metakaolin, Silica fumes
Kannan et al. [5]	2014	SNP	Natural	Natural	OPC ASTM C 150 (Type1)	Rice husk Ash, Metakaolin
Samimi et al. [6]	2017	Polycarboxylate ether (PCE 180)	Natural	Natural	Portland cement type II	Limestone powder, Pumice, Zeolite
Kou et al. [7]	2009	SP ADVA-109	Recycled aggregates, Natural aggregates	Recycled aggregates, Natural aggregates	OPC	Fly ash
Persson [8]	2003	Polycarboxylic ether—brand Glenium 51	Natural	Natural	OPC	Limestone filler
Gupta et al. [9]	2020	Auramix-400	Copper slag	Natural	OPC 43 grade	Fly ash
Siddique [10]	2011	Polycarboxylicether-based SP	Natural	Natural	OPC (Grade 43)	Fly ash
Sharma et al. [11]	2017	Glenium SKY 8765 based on PCE	Copper slag	Natural	OPC of 43 grade ASTM Type I	Fly ash

(continued)

Table 1 (continued)

Reference	Year	SP type	Material used			
			Fine aggregates	Coarse aggregates	Cement	Mineral admixtures
Pereira-de-Oliveira et al. [12]	2014	Modified polycarboxylate-based SP	Natural	RCA, NA	Portland cement type CEM I 42.5R	Limestone powder
Chopra et al. [13]	2015	Conplast SP430 based on SNP	Natural	Natural	OPC grade 43	Rice husk ash
Assie et al. [14]	2006	Polycarboxylate modified superplasticizer	Natural	Natural	CEM II/A-LL 32.5 R and CEM I 52.5 N	Limestone filler
Valcuende et al. [15]	2010	Polycarboxylate with polyethylene condensate defoamedGlenium C303	Natural	Natural	CEM II/B-M (V-LL) 32.5 N and CEM II/B-M (V-LL) 42.5R	Limestone powder
Esquinas et al. [16]	2018	Glenium 303 based on PCE	Natural	Natural	OPC type CEM I 42.5 R/SR	Non-conforming fly ash, Siliceous filler
Esquinas et al. [17]	2018	Glenium 303 PCE	Natural	Natural	OPC CEM I 42.5 R/SR	Waste asphalt filler
Gill et al. [18]	2018	Conplast SP400	Natural	Natural	OPC	Rice husk ash, Metakaolin
Sasanipour et al. [19]	2019	PCE	RCA, NA	RCA, NA	OPC type II	Limestone powder, Silica fumes
Ofuyatan et al. [20]	2015	Conplast SP432MS	Natural	Natural	OPC type I	Palm oil fuel ash

(continued)

Table 1 (continued)

Reference	Year	SP type	Material used			
			Fine aggregates	Coarse aggregates	Cement	Mineral admixtures
Vaidevi et al. [21]	2020	Master glenium SKY 8233 based on PCE	NA, Marble waste	Natural	OPC grade 53	Fly ash
Tang et al. [22]	2018	Master glenium 51con 35%, BASF	Natural	NA, Artificial aggregates	OPC CEM I 42.5 N	Coal fly ash
Kapoor et al. [23]	2018	PCE	RCA, NA	RCA, NA	OPC grade 43	Fly ash

Table 2 Mix design proportions used by authors

Reference mix		Water (kg/m^3)	Total powder content (kg/m^3)	Aggregates						Super-plasticizer		Viscosity modifying agents	
				FA (kg/m^3)	CA (kg/m^3)	Replaced aggregates				Type	%	Type	(kg/m^3)
						Type	FA (kg/m^3)	CA (kg/m^3)	%				
				913	589	BA	0	0	0	PCE	2	N M	N M
	SCC2	260.9	550	821	589	BA	91	0	10	PCE	1.85	N M	N M
	SCC3	281.4	550	730	589	BA	183	0	20	PCE	1.9	N M	N M
	SCC4	300.5	550	639	589	BA	274	0	30	PCE	1.88	N M	N M
Frazão et al. [2]	SCC	127.8	766	N M	648	RS	198	722	N M	PCE	1.02	N M	N M
	SFRSCC	127.8	766	N M	640	RS	195	713	N M	PCE	1.02	N M	N M
Singh et al. [3]	CF-RCA0	277	615	846	646	RCA	0	0	0	PCE	0.28	N M	1.72
	CF-RCA25	277	615	846	484.5	RCA	0	150.5	25	PCE	0.29	N M	1.72
	CF-RCA50	277	615	846	323	RCA	0	301	50	PCE	0.31	N M	1.72
	CF-RCA75	277	615	846	161.5	RCA	0	451.5	75	PCE	0.35	N M	1.93
	CF-RCA100	277	615	846	0	RCA	0	602	100	PCE	0.35	N M	1.93
	CFM-RCA25	277	615	846	484.5	RCA	0	150.5	25	PCE	0.35	N M	1.72
	CFM-RCA50	277	615	846	323	RCA	0	301	50	PCE	0.41	N M	1.72
	CFM-RCA75	277	615	846	161.5	RCA	0	451.5	75	PCE	0.46	N M	1.8
	CFM-RCA100	277	615	846	0	RCA	0	602	100	PCE	0.46	N M	2.15
Kapoor et al. [4]	C-RO	277	615	846	602	RCA	0	0	0	PCE	0.8	N M	1.72
	C-R50	277	615	846	301	RCA	0	280	50	PCE	1	N M	2.58

(continued)

Table 2 (continued)

Reference mix		Water (kg/m³)	Total powder content (kg/m³)	Aggregates						Super-plasticizer		Viscosity modifying agents	
				FA (kg/m³)	CA (kg/m³)	Replaced aggregates				Type	%	Type	(kg/m³)
						Type	FA (kg/m³)	CA (kg/m³)	%				
	C-R100	277	615	846	0	RCA	0	560	100	PCE	1.2	N M	3.44
	C-SFR0	277	594.5	846	602	RCA	0	0	0	PCE	0.8	N M	1.72
	C-SFR50	277	594.5	846	301	RCA	0	280	50	PCE	1	N M	2.58
	C-SFR100	277	594.5	846	0	RCA	0	560	100	PCE	1.2	N M	3.44
	C-MKR0	277	604.5	846	302	RCA	0	0	0	PCE	0.8	N M	1.72
	C-MKR50	277	604.5	846	301	RCA	0	280	50	PCE	1	N M	2.58
	C-MKR100	277	604.5	846	0	RCA	0	560	100	PCE	1.2	N M	3.44
Persson [8]	KN	163	601	1169	363	N M	N M	N M	N M	PCE	0.49	N M	N M
	KOB	167	802	1007	371	N M	N M	N M	N M	PCE	0.51	N M	N M
	KN8	160	589	1145	355	N M	N M	N M	N M	PCE	0.54	N M	N M
	KO	165	609	1184	367	N M	N M	N M	N M	PCE	0.56	N M	N M
	KOT	163	603	1171	363	N M	N M	N M	N M	PCE	0.61	N M	N M
	SO	162	510	1208	402	N M	N M	N M	N M	PCE	0.59	N M	N M
	RO	168	431	861	862	N M	N M	N M	N M	PCE	1.7	N M	N M
Gupta et al. [9]	0CS-SCC	168	500	960	760	CS	0	0	0	PCE	1.2	N M	N M

(continued)

Table 2 (continued)

Reference mix		Water (kg/m³)	Total powder content (kg/m³)	Aggregates						Super-plasticizer		Viscosity modifying agents	
				FA (kg/m³)	CA (kg/m³)	Replaced aggregates				Type	%	Type	(kg/m³)
						Type	FA (kg/m³)	CA (kg/m³)	%				
	10CS-SCC	168	500	864	760	CS	96	0	10	PCE	1.2	N M	N M
	20CS-SCC	168	500	768	760	CS	192	0	20	PCE	1.2	N M	N M
	30CS-SCC	168	500	672	760	CS	288	0	30	PCE	1.2	N M	N M
	40CS-SCC	168	500	576	760	CS	384	0	40	PCE	1.2	N M	N M
	50CS-SCC	168	500	480	760	CS	480	0	50	PCE	1.2	N M	N M
	60CS-SCC	168	500	384	760	CS	576	0	60	PCE	1.2	N M	N M
Siddique [10]	SCC1	205	550	910	590	N M	N M	N M	N M	PCE	1.95	N M	N M
	SCC2	205	550	910	590	N M	N M	N M	N M	PCE	2	N M	N M
	SCC3	210	550	910	590	N M	N M	N M	N M	PCE	1.8	N M	N M
	SCC4	215	550	910	590	N M	N M	N M	N M	PCE	1.8	N M	N M
	SCC5	220	550	910	590	N M	N M	N M	N M	PCE	1.8	N M	N M
Sharma et al. [11]	OF-CS0	247.5	550	0	700	CS	0	0	0	PCE	0.8	N M	N M
	OF-CS20	247.5	550	0	700	CS	284.4	0	20	PCE	0.8	N M	N M
	OF-CS40	247.5	550	0	700	CS	568	0	40	PCE	0.6	N M	N M
	OF-CS60	247.5	550	0	700	CS	853.2	0	60	PCE	0.5	N M	N M

(continued)

Table 2 (continued)

Reference mix		Water (kg/m^3)	Total powder content (kg/m^3)	Aggregates						Super-plasticizer		Viscosity modifying agents	
				FA (kg/m^3)	CA (kg/m^3)	Replaced aggregates				Type	%	Type	(kg/m^3)
						Type	FA (kg/m^3)	CA (kg/m^3)	%				
	OF-CS80	247.5	550	0	700	CS	1137.6	0	80	PCE	0.4	N M	N M
	OF-CS100	247.5	550	0	700	CS	1422	0	100	PCE	0.4	N M	N M
Pereira-de-Oliveira et al. [12]	SCC	161.3	655.1	730.8	807.9	RA	0	0	0	PCE	0.52	N M	N M
	SCC20	163.2	655.1	730.8	646.3	RA	0	149.6	20	PCE	0.73	N M	N M
	SCC40	160.7	655.1	730.8	484.8	RA	0	299.3	40	PCE	0.7	N M	N M
	SCC100	162.4	655.1	730.8	0	RA	0	808.5	100	PCE	0.93	N M	N M
Chopra et al. [13]	0RHA	226	550	910	590	N M	N M	N M	N M	SNP	1	N M	N M
	10RHA	226	550	910	590	N M	N M	N M	N M	SNP	1	N M	N M
	15RHA	226	551	910	590	N M	N M	N M	N M	SNP	1	N M	N M
	20RHA	226	550	910	590	N M	N M	N M	N M	SNP	1	N M	N M
Assie et al. [14]	SCC 20	205	465	900	771	N M	N M	N M	N M	PCE	1.72	N M	N M
	SCC 40	191	490	888	791	N M	N M	N M	N M	PCE	2.57	N M	N M
	SCC 60	189	520	884	793	N M	N M	N M	N M	PCE	2.6	N M	N M
Valcuende et al. [15]	S-65-32	178.75	486.67	911.87	816.06	N M	N M	N M	N M	PCE	0.96	N M	N M
	S-55-32	178.75	522.6	902.06	797.94	N M	N M	N M	N M	PCE	1	N M	N M

(continued)

Table 2 (continued)

Reference mix		Water (kg/m³)	Total powder content (kg/m³)	Aggregates						Super-plasticizer		Viscosity modifying agents	
				FA (kg/m³)	CA (kg/m³)	Replaced aggregates				Type	%	Type	(kg/m³)
						Type	FA (kg/m³)	CA (kg/m³)	%				
	S-55-42	178.75	522.6	902.06	797.94	N M	N M	N M	N M	PCE	0.93	N M	N M
	S-45-42	180	580.09	879.89	769.43	N M	N M	N M	N M	PCE	1.03	N M	N M
Esquinas et al. [16]	SCC-1	176.93	410	938.1	807.65	N M	N M	N M	N M	PCE	2.25	N M	N M
	SCC-12	179.4	503.53	933.65	797.81	N M	N M	N M	N M	PCE	1.8	N M	N M
	SCC-2	182.75	482.94	942.98	789.07	N M	N M	N M	N M	PCE	1.8	N M	N M
Esquinas et al. [17]	SCC-SF	180.41	510.01	922.11	793.88	N M	N M	N M	N M	PCE	1.8	N M	N M
	SCC-RF	183.38	512.29	922.11	793.88	N M	N M	N M	N M	PCE	1.8	N M	N M
Gill et al. [18]	Control mix	211.2	480	900	670	N M	N M	N M	N M	SNP	1.5	N M	N M
	5MK10RHA	250.8	570	810	670	N M	N M	N M	N M	SNP	2	N M	N M
	10MK10RHA	250.8	570	810	670	N M	N M	N M	N M	SNP	2	N M	N M
	15MK10RHA	250.8	570	810	670	N M	N M	N M	N M	SNP	2	N M	N M
Ofuyatan et al. [20]	A1	9.76	18.97	20.54	42.78	N M	N M	N M	N M	SNP	0	N M	N M
	A2	9.76	18.97	20.54	42.78	N M	N M	N M	N M	SNP	0	N M	N M
	A3	9.76	25.41	20.54	42.78	N M	N M	N M	N M	SNP	2	N M	N M
	A4	9.76	24.46	20.54	42.78	N M	N M	N M	N M	SNP	2	N M	N M

(continued)

Table 2 (continued)

Reference mix		Water (kg/m^3)	Total powder content (kg/m^3)	Aggregates						Super-plasticizer		Viscosity modifying agents	
				FA (kg/m^3)	CA (kg/m^3)	Replaced aggregates				Type	%	Type	(kg/m^3)
						Type	FA (kg/m^3)	CA (kg/m^3)	%				
	A5	9.76	24.33	20.54	42.78	N M	N M	N M	N M	SNP	2	N M	N M
	A6	9.76	25.41	20.54	42.78	N M	N M	N M	N M	SNP	2	N M	N M
	A7	9.76	25.42	20.54	42.78	N M	N M	N M	N M	SNP	2	N M	N M
	A8	9.76	27.32	20.54	42.78	N M	N M	N M	N M	SNP	2	N M	N M
	Control SCC	193	640	738	847	MA	0	0	0	PCE	2	N M	N M
	SCCMFA25	193	640	553	847	MA	185	0	25	PCE	2	N M	N M
	SCCMFA50	193	640	369	847	MA	369	0	50	PCE	2	N M	N M
	SCCMFA100	193	640	0	847	MA	738	0	100	PCE	2	N M	N M
Tang et al. [22]	Mix 1 30%	139.5	489	572.9	816.09	AA	0	241.43	30	PCE	1.2	N M	N M
	Mix 1 60%	139.5	489	572.9	558.37	AA	0	483	60	PCE	1.2	N M	N M
	Mix 2 30%	139.5	489	572.9	816.09	AA	0	234.29	30	PCE	1.2	N M	N M
	Mix 2 60%	139.5	489	572.9	558.37	AA	0	468.71	60	PCE	1.2	N M	N M
	Mix 3 30%	139.5	489	572.9	816.09	AA	0	236.14	30	PCE	1.2	N M	N M
	Mix 3 60%	139.5	489	572.9	558.38	AA	0	472.14	60	PCE	1.2	N M	N M
Kapoor et al. [23]	C0F0	277	615	846	602	RCA	0	0	N M	PCE	0.8	N M	1.72

(continued)

Table 2 (continued)

Reference mix		Water (kg/m^3)	Total powder content (kg/m^3)	Aggregates						Super-plasticizer		Viscosity modifying agents	
				FA (kg/m^3)	CA (kg/m^3)	Replaced aggregates				Type	%	Type	(kg/m^3)
						Type	FA (kg/m^3)	CA (kg/m^3)	%				
	C50F0	277	615	846	301	RCA	0	278	N M	PCE	0.9	N M	2.15
	C50F25	277	615	635	301	RCA	193	278	N M	PCE	1	N M	2.15
	C50F50	277	615	423	301	RCA	386	278	N M	PCE	1	N M	2.15
	C100F0	277	615	846	0	RCA	0	556	N M	PCE	1.1	N M	2.58
	C100F25	277	615	635	0	RCA	193	556	N M	PCE	1.1	N M	3.01
	C100F50	277	615	423	0	RCA	386	556	N M	PCE	1.2	N M	3.01

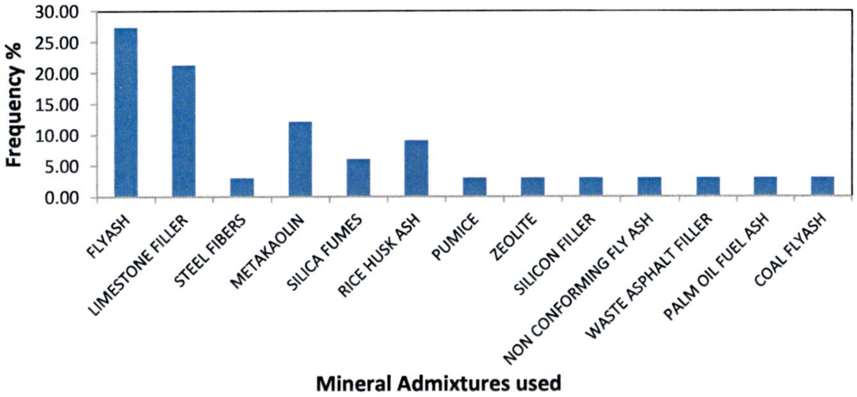

Fig. 1 Frequency of mineral admixtures used as partial replacement for cement

Figure 2 shows the frequency of materials used as partial replacement to natural aggregates by different authors. Recycled concrete aggregates were used frequently by authors in combination with natural aggregates. Other materials used as partial replacement to natural aggregates include bottom ash, copper slag, marble aggregate, and artificial aggregate.

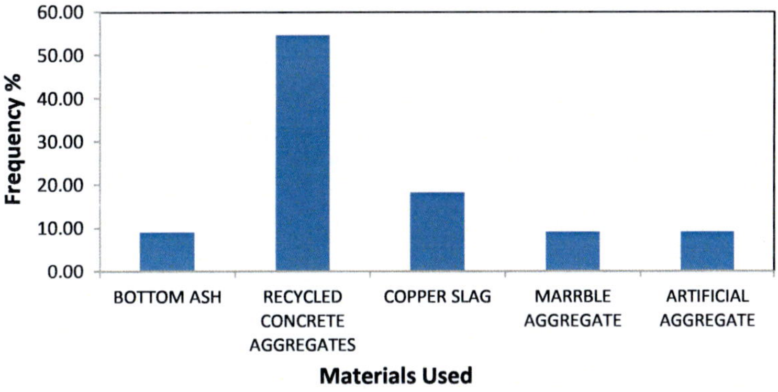

Fig. 2 Frequency of materials used as partial replacement for aggregates by 11 authors

4 Durability Test

4.1 Carbonation Depth Test

The process in which atmospheric carbon-dioxide (CO_2) reacts with hydrated cementitious mineral in the presence of moisture is known as carbonation. Calcium carbonate is formed when calcium hydroxide and carbon dioxide reacts together. Carbonation prevents corrosion of reinforcement as it reduces alkalinity and increases mechanical strength. Carbonation proves to be not good for concrete structure.

To check for carbonation depth, multiple specimens are prepared for different mix designs. After respective days of curing, cube specimens were split prior to testing, conditioning of specimen was done. Phenolphthalein was used as an indicator for carbonation, and then the depth of carbonation was measured. Figure 3 shows the result obtained by different authors for the carbonation test performed by on their mix samples.

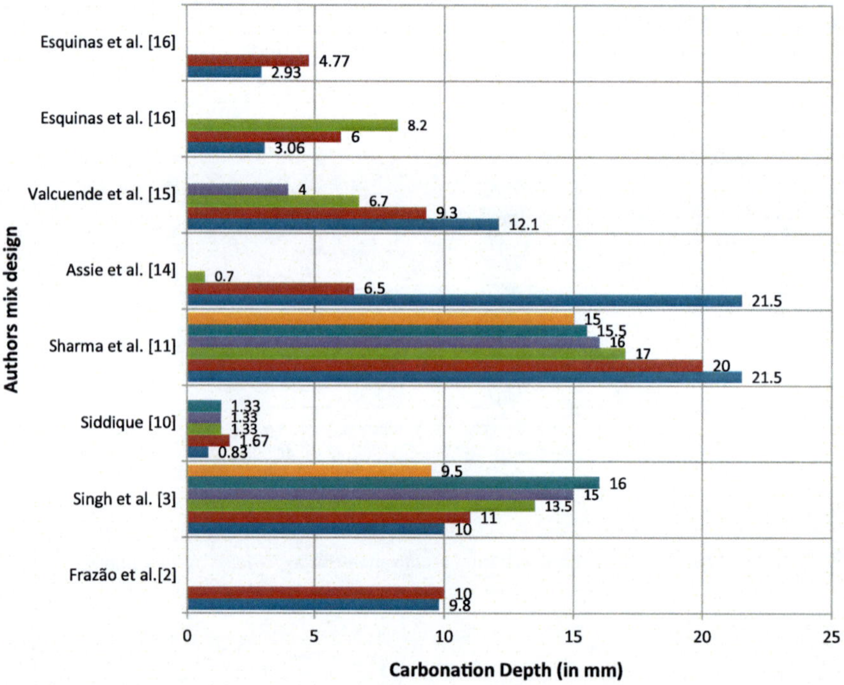

Fig. 3 Carbonation depth test (28 days curing)

4.2 *Rapid Chloride Penetration Test*

Chloride penetration leads to corrosion of reinforced steel and corrosion related damages to concrete, and it is a very common issue. This issue of durability has drawn attention toward it due to its frequent occurrence and high repair costs. Chlorides will even penetrate a crack free concrete by kind of mechanisms appreciate capillary absorption, fluid mechanics pressure, diffusion, etc. Diffusion is caused once the chloride concentration on the skin of the SCC concrete structure is higher than that on the inside.

For the purpose of Rapid Chloride Penetration Test (RCPT), cylindrical specimens of self-compacting concrete were selected (100 mm diameter, 50 mm thick). As per ASTM C1202, the tests were conducted at different intervals. The results obtained by different authors are shown in Fig. 4. Table 3 shows Chloride ion penetrability

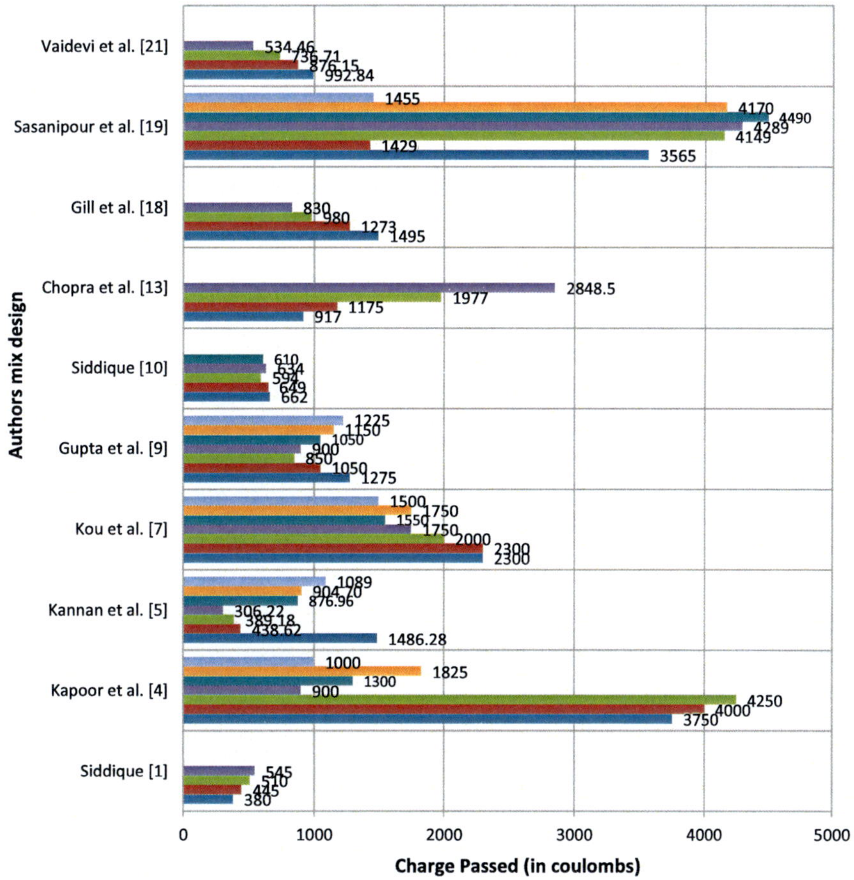

Fig. 4 Rapid chloride penetration test

Table 3 Chloride ion penetrability based on charge passed

Charge passed (Coulomb)	Chloride ion penetrability
> 4000	High
2000–4000	Moderate
1000–2000	Low
100–1000	Very low
< 100	Negligible

Source ASTM 1202-97

based on charge passed (ASTM 1202-97).

4.3 Sulfate Attack Test

According to ASTM C1012, the test for checking percentage mass change due to sulfate attack was performed. Cube specimens of size 150 mm × 150 mm × 150 mm were immersed in 5% Na_2SO_4 solution. Change in weight percentage of the specimens were noted down at an exposure period of 28, 56, 90, 120, and 365 days. Figure 5 shows the result for sulfate attack test performed by various authors on their respective mixes.

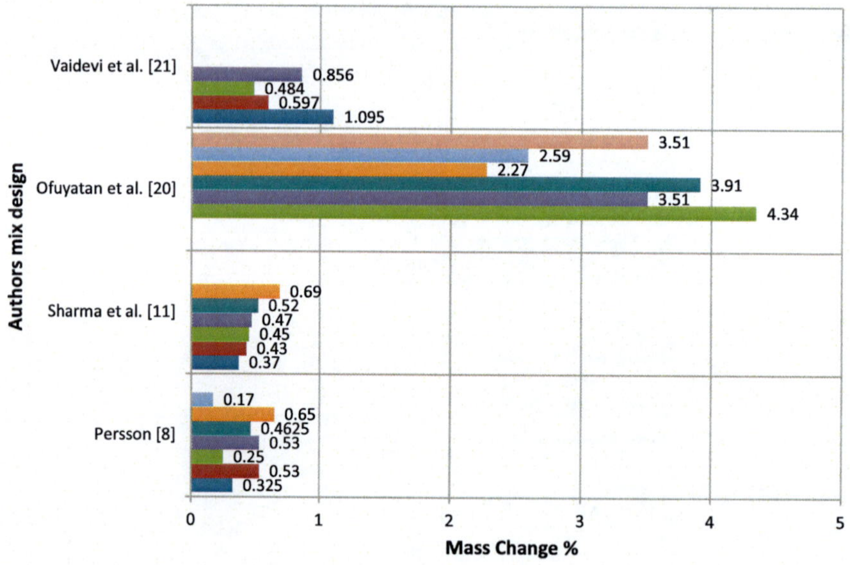

Fig. 5 Sulfate attack test (28 days)

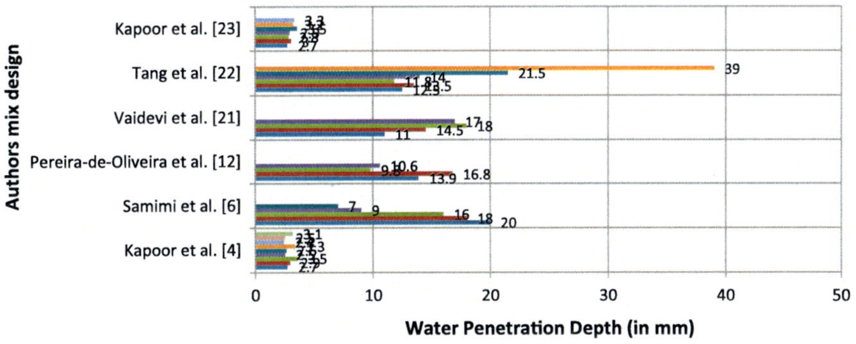

Fig. 6 Water penetration depth test (28 days)

4.4 Water Penetration Test

According to BSEN12390-8:2000, specimen of size 150 mm × 150 mm × 150 mm were selected for curing ages of 28, 56 and 120 days. A constant water pressure of 0.5 MPa was maintained over the specimen for 72 h. After 72 h, the specimens need to be removed from the apparatus and needs to split in two halves, perpendicular to the face on which the water pressure was applied. As soon as the split face had dried to such an extent that the water penetration front could be clearly seen, the waterfront was marked on the specimen. The reading for maximum water penetration depth was recorded to the nearest millimeter. The result of depth of penetration performed by different author son their mix is reported in Fig. 6.

5 Conclusion

By reviewing all the papers, it was concluded that addition of waste/recycled aggregates and mineral admixtures as partial replacement to cement and natural aggregates will improve the durability properties of self-compacting concrete. As shown in the tables and figures, it is clear that most suitable cement used is ordinary Portland cement in combination with various mineral admixtures such as fly ash, silica fumes, limestone powder, silica fumes, etc. The use of super plasticizer or high range water reducer to maintain the lower water to cement or water to powder ratio and achieve desired property to qualify as a SCC. Following inferences were derived from the study:

- Addition of recycle concrete aggregates reduces the durability of self-compacting concrete mix, but on addition of 30% of total binding material as powdered mineral admixtures such as fly ash, metakaolin, silica fumes or combination of the three, alongside recycled aggregates makes the specimen less vulnerable to chemical attacks;

- Addition of 10% metakaolin along with 20% fly ash or limestone powder to the mix increases the resistance of mix against carbonation by 17.5%, rapid chloride ion penetration by 60% and water penetration by 10%;
- Addition of 20% silica fumes in self-compacting concrete increases resistance against carbonation up to 70%, chloride ion penetration by up to 70% and water penetration by up to 10%;
- Addition of steel fibers in self-compacting concrete reduces resistance against carbonation attack by up to 20%.
- Major change in properties were observed when 30% fly ash was added as partial replacement to cement. Fly ash decreases the coulomb value of concrete in rapid chloride penetration test from 3000 coulombs to below 1000 coulombs which lies in very low range of chloride ion penetration according to ASTM 1202-97.
- Addition of 15% rice husk ash reduced the penetration of chloride ions by up to 80%. 15% was found to be optimum value of rice husk ash, increase or decrease in this value reduced the resistance of the mix. It was also found that every mix with rice husk ash lied in very low range of chloride ion penetration.

Abbreviations

NM	Not Mentioned
OPC	Ordinary Portland Cement
SP	Super-Plasticizer
FA	Fly Ash
BA	Bottom Ash
LF	Limestone Filler
FS	Fine River Sand
PC	Portland Cement
RCA	Recycled concrete aggregate
PCE	Polycarboxylate Ether
RHA	Rice Husk Ash
MK	Metakaolin
SNP	Sulphonated Naphthalene Polymer
LP	Limestone Powder
FO	Fatty Oils
SF	Silica Fumes
CS	Copper Slag
MA	Marble Aggregate
AA	Artificial Aggregate
RA	Recycled Aggregate

References

1. Siddique, R. (2013). Compressive strength, water absorption, sorptivity, abrasion resistance and permeability of self-compacting concrete containing coal bottom ash. *Construction and Building Materials, 47*, 1444–1450.
2. Frazão, C., Camões, A., Barros, J., & Gonçalves, D. (2015). Durability of steel fiber reinforced self-compacting concrete. *Construction and Building Materials, 80*, 155–166.
3. Singh, N., & Singh, S. P. (2016). Carbonation and electrical resistance of self compacting concrete made with recycled concrete aggregates and metakaolin. *Construction and Building Materials, 121*, 400–409.
4. Kapoor, K., Singh, S. P., & Singh, B. (2016). Durability of self-compacting concrete made with recycled concrete aggregates and mineral admixtures. *Construction and Building Materials, 128*, 67–76.
5. Kannan, V., & Ganesan, K. (2014). Chloride and chemical resistance of self compacting concrete containing rice husk ash and metakaolin. *Construction and Building Materials, 51*, 225–234.
6. Samimi, K., Kamali-Bernard, S., Maghsoudi, A. A., Maghsoudi, M., & Siad, H. (2017). Influence of pumice and zeolite on compressive strength, transport properties and resistance to chloride penetration of high strength self-compacting concretes. *Construction and Building Materials, 151*, 292–311.
7. Kou, S. C., & Poon, C. S. (2009). Properties of self-compacting concrete prepared with coarse and fine recycled concrete aggregates. *Cement and Concrete composites, 31*(9), 622–627.
8. Persson, B. (2003). Sulphate resistance of self-compacting concrete. *Cement and Concrete Research, 33*(12), 1933–1938.
9. Gupta, N., & Siddique, R. (2020). Durability characteristics of self-compacting concrete made with copper slag. *Construction and Building Materials, 247*, 118580.
10. Siddique, R. (2011). Properties of self-compacting concrete containing class F fly ash. *Materials and Design, 32*(3), 1501–1507.
11. Sharma, R., & Khan, R. A. (2017). Durability assessment of self compacting concrete incorporating copper slag as fine aggregates. *Construction and Building Materials, 155*, 617–629.
12. Pereira-de-Oliveira, L. A., Nepomuceno, M. C. S., Castro-Gomes, J. P., & Vila, M. D. F. C. (2014). Permeability properties of self-compacting concrete with coarse recycled aggregates. *Construction and Building Materials, 51*, 113–120.
13. Chopra, D., & Siddique, R. (2015). Strength, permeability and microstructure of self-compacting concrete containing rice husk ash. *Biosystems Engineering, 130*, 72–80.
14. Assie, S., Escadeillas, G., & Waller, V. (2007). Estimates of self-compacting concrete 'potential' durability. *Construction and Building Materials, 21*(10), 1909–1917.
15. Valcuende, M., & Parra, C. (2010). Natural carbonation of self-compacting concretes. *Construction and Building Materials, 24*(5), 848–853.
16. Esquinas, A. R., Alvarez, J. I., Jiménez, J. R., & Fernández, J. M. (2018). Durability of self-compacting concrete made from non-conforming fly ash from coal-fired power plants. *Construction and Building Materials, 189*, 993–1006.
17. Esquinas, A. R., Álvarez, J. I., Jiménez, J. R., Fernández, J. M., & De Brito, J. (2018). Durability of self-compacting concrete made with recovery filler from hot-mix asphalt plants. *Construction and Building Materials, 161*, 407–419.
18. Gill, A. S., & Siddique, R. (2018). Durability properties of self-compacting concrete incorporating metakaolin and rice husk ash. *Construction and Building Materials, 176*, 323–332.
19. Sasanipour, H., Aslani, F., &Taherinezhad, J. (2019). Effect of silica fume on durability of self-compacting concrete made with waste recycled concrete aggregates. *Construction and Building Materials, 227*, 116598.

20. Ofuyatan, O. M., Olutoge, F. A., & Olowofoyeku, O. A. (2015). Durability properties of palm oil fuel ash self compacting concrete. *Engineering, Technology and Applied Science Research, 5*(1), 753–756.
21. Vaidevi, C., Kala, T. F., & Kalaiyarrasi, A. R. R. (2020). Mechanical and durability properties of self-compacting concrete with marble fine aggregate. *Materials Today: Proceedings, 22*, 829–835.
22. Tang, P., & Brouwers, H. J. H. (2018). The durability and environmental properties of self-compacting concrete incorporating cold bonded lightweight aggregates produced from combined industrial solid wastes. *Construction and Building Materials, 167*, 271–285.
23. Kapoor, K., Singh, S. P., & Singh, B. (2018). Water permeation properties of self compacting concrete made with coarse and fine recycled concrete aggregates. *International Journal of Civil Engineering, 16*(1), 47–56.

An Approach Towards Zero-Waste Building Construction

Anil Soharu, Naveen BP, and Arjun Sil

Abstract Construction is one of the key parameters for social and economic growth for any nation, but it is also named as one of the major wastes generating process. However, it has sufficient potential to minimize or reduce any waste generation. Advance planning and monitoring start from the project till the end of construction can reduce waste generation up to a large extent. "Zero waste" is one of the most visionary approaches to overcome waste problems. In this research, the zero-waste approach is correlated for building construction work to assess the practical applications on a design and process throughout the life cycle. A state of zero waste can be reached by holistically implementing the revolutionary concept of prefabrication, demonstrability, circular economy, green rating tools and the principle of reducing, reuse and recycle. Further practices like smart waste audit, smart waste collection, high-value mixed waste processing, and the collaborative platform further help towards the prime target of zero waste construction.

Keywords Sustainability · Zero-waste designing · Life cycle assessment

1 Introduction

The construction industry is one of the major sectors and key contributor to gross domestic product (GDP) for any nation. Since the construction industry is linked with the development of nations, large engagement of man and machinery, and consumption of resources, it influences the environmental, economic and social elements of the human being. The construction sector accounted for 6% of the world gross domestic product in the year of 2018 and employed more than 100 million people worldwide as per the world economic forum [1–3]. Social and economic benefits are very important for any nation as it contributes to the overall development of the nation. However, it

A. Soharu (✉) · Naveen BP
Department of Civil Engineering, Amity University Haryana, Gurgaon, India

A. Sil
Department of Civil Engineering, National Institute of Technology, Silchar, India

A. K. Gupta et al. (eds.), *Advances in Construction Materials and Sustainable Environment*, Lecture Notes in Civil Engineering 196,
https://doi.org/10.1007/978-981-16-6557-8_19

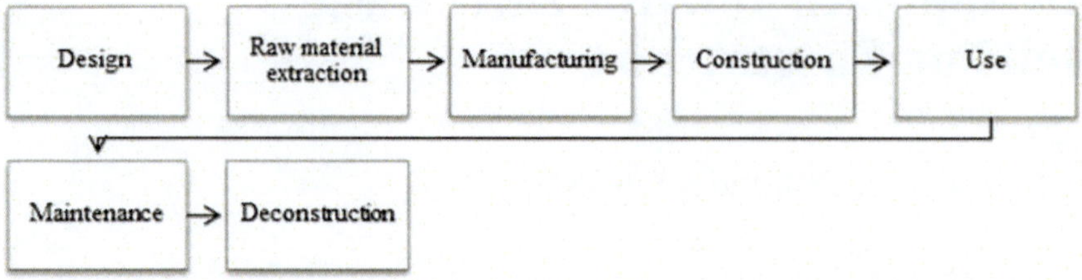

Fig. 1 Life span phases of a building

also creates risks towards the environment, because of which any construction must have equal importance towards social, economic and environmental factors. Building sustainable communities and achieving sustainable consumption and production are two of the 17 defined goals adopted for the 2030 global agenda towards sustainable development [4]. To achieve these goals, our present conventional construction techniques need to re-evaluate on a scale of social, economic and environmental weightage. We need to explore alternative construction techniques with proper selection of raw materials, building framework, the life span of the building, its functionality, flexibility during life span and end-of-life usage.

Most of the research is focussed on the environmental impact of the building during its operation usage in terms of high energy and waste generation, whereas on-site construction is not even modelled during designing to avoid waste generation during construction and its end-of-life usage. Phases of a building's life cycle that contribute to environmental impacts and energy use are portrayed in Fig. 1.

Lots of research has been done on the material phase for a building and its related impacts from extraction and manufacturing from raw material. Research has also been done focussing on end-of-life for buildings [5]. Much research indicated that the construction activities have a major impact on the environment but it is still underestimated [6]; however, few of the research also states that its impacts are negligible [7].

A brief chronology for the evolution of sustainable construction capturing key concepts of life cycle assessment, zero waste, circular economy, and cradle to cradle is summarized in Table 1.

During the year of 1713 word "sustainability" was first introduced, which got attention from all corners of the globe. In 1963, life cycle assessment (LCA) was presented by Harold Sminth at the work-energy conference. Paul Palmer coined the term Zero waste during the early 1970, Seattle introduced the Pay-As-You-Throw (PAYT) system in 1988 at the USA. 1989 major development took place as the California Integrated Waste Management Act was passed to achieve 25% waste diversion from landfill by 1995 and 50% by 2000. Further in 1995 Australia Canberra also passed the "No Waste by 2010" bill. The USA included Zero waste as guiding principles in North Carolina, Seattle, Washington, and Washington, DC. M. Braungart and W. McDonough introduced the revolutionary concept of Cradle to Cradle in 2002. Later in the year 2012, the zero-waste business council was established in the

Table 1 Evolution of sustainable construction

Year	Event
1713	Word Sustainability Coined in German, the original term was Nachhaltigkeit, meaning "sustained yield." It first appeared in a handbook of forestry published in 1713 [8]
1963	Harold Smith laid the foundations for LCA using the cumulative energy concept, at the World Energy Conference [9]
1970	Paul Palmer formed the company Zero Waste Systems Inc. before that the term "Zero Waste" had never been used publicly [10]
1976	The concept of an economy in loops was introduced by Walter Stahel and Genevieve Reday [11]
1981	The term cradle to cradle was first coined by Walter R. Stahel [11]
1987	The definition of sustainable development given in the Bruntland report considered as the start of the sustainability era [12]
1992	Earth Summit in Rio de Janeiro, sustainable development entered into the global stage [13]
1994	Definition of sustainable construction was given by Professor Charles J. Kibert; as "the creation and responsible management of a healthy built environment based on resource-efficient and ecological principles" [14]
1998	Zero waste was termed as one of the guiding principles in Washington, DC [15]
1999	Definitions and frameworks for sustainable construction practices were defined in Agenda 21 on Sustainable Construction [16]
2002	Book Cradle-to-Cradle was published [17]
2012	The Zero Waste Business Council was established in the USA [18]
2015	The Sustainable Development Goals (SDGs) were adopted by all United Nations Member States [19]
2019	LEED introduced the integrated design approach towards Zero Waste in Construction
2020	100+ Countries Supported to integrate the SDGs into national planning [19]

USA. In Year 2015, all United Nations Member States adopted 17 nos Sustainable Development Goals (SDGs) and by 2020, 100+ Countries Supported to integrate the SDGs into national planning.

1.1 Sustainable Construction

Sustainable construction is one of the pressing topics of attention in construction research and has been defined by Kibert [20] and Kashyap et al. [21] as the "creation of resource-efficient and ecological designed healthy built environment." Sustainable construction aims to follow [21]

- Maximum efficient utilization of resources
- Flexibility for future user demands
- Quality of life and customer satisfaction

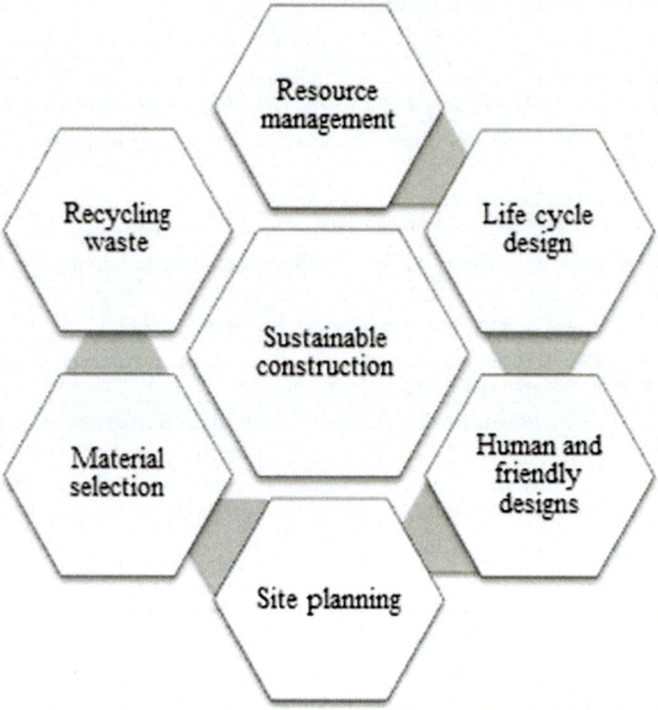

Fig. 2 Considerations for sustainable construction

- Desirable ecological and social environments.

Sustainable construction targets economic growth with stress on social and environmental uprightness connecting with a large number of processes associated with the whole life cycle of a project [20]. Key driving forces for sustainable construction stated in Fig. 2 [22, 23], and all the factors hold interrelation with each other and need to work in conjunction to achieve the desired goal for a sustainable built environment. An early thought of sustainability initiatives is the primary key for a sustainable building [24].

1.2 *Life Cycle Assessment (LCA)*

Life cycle assessment is a methodology, wherein environmental impact is assessed during every stage of any process, product or service. In the case of building construction, environmental impacts are evaluated starting from raw material extraction, followed by its processing, transportation, construction at the worksite, building operational usage and finally end of life for the building. Life cycle assessment is associated with the evaluation of environmental impacts both for the individual intermediate stages and complete process as a whole. Thus, results are used for further improvement of process, product, or services so that the ultimate goal towards reduction of environmental impact can be achieved. International Organization for Standardization divided the process of life cycle assessment into 4 phases [25, 26] as illustrated in Fig. 3.

- Phase 1—Study of goal and scope to clearly define the objective and boundaries for the required application.
- Phase 2—Life Cycle Inventory (LCI), flow model is developed using flow chart for all the activities involved in the required process, product or services.
- Phase 3—Life Cycle Impact Analysis (LCIA) is done for evaluating potential environmental impact based on inventory flow model defied in phase 2.
- Phase 4—Life Cycle Interpretation, results from the life cycle inventory phase and life cycle impact analysis phase is evaluated and interpreted and results are used to find the possible improvements, recommendations or limitations within a studied process, product, or services.

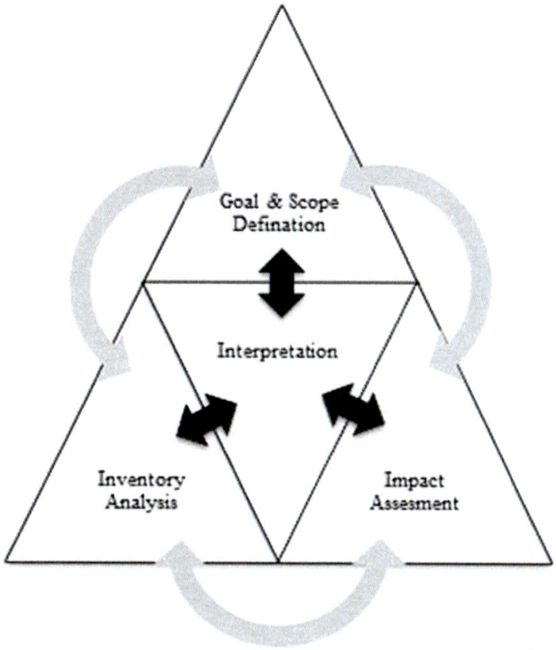

Fig. 3 Life cycle assessment phases

1.3 Zero Waste

In the year of 1973, the term "zero waste" was firstly introduced by Paul Palmer [27]. The concept of zero waste includes a variety of notions as indicated in Fig. 4, wherein a complete sequence of actions towards avoiding, reduce, reuse, recycle, recover, treat, dispose of is sorted.

The primary focus of the zero-waste design is waste reduction and reuse of material and then lastly recycle of materials. Materials in demolition waste are mostly not reused but dumped on a landfill [28]. Even when materials are "recycled," limitations exist. Materials are down-cycled or cannot be reused in their current form. This leads to increasingly lower usability of materials and ultimately to waste. For new uses, this generates a requirement for new materials to be extracted and produced. Wasting materials and energy should not be acceptable. Buildings should be designed for proper end-of-life solutions that allow for proper re-use and recycling. With an awareness of this problem in the design phase, the correct technologies and materials should be utilized to make buildings in such a way that they do not generate waste; zero-waste [29, 30].

Moreover, the resources of the earth are limited. With expected growth in population and prosperity, it will become unsustainable to dispose of materials in landfills or incinerators. When buildings are designed according to a zero-waste approach, this problem could hypothetically be solved.

2 Design Approach

The literature study is done by gathering and reading information relevant to the research. The information is then collected, ordered and summarized. More information is gathered by analysing reference projects and systems. The goal of the literature research is to establish requirements for a zero-waste design. These requirements are used as a benchmark in the design part of this research. A design choice can be made using these requirements and will give an objective method to test the design as presented in Fig. 5.

Fig. 4 Steps to achieve zero waste

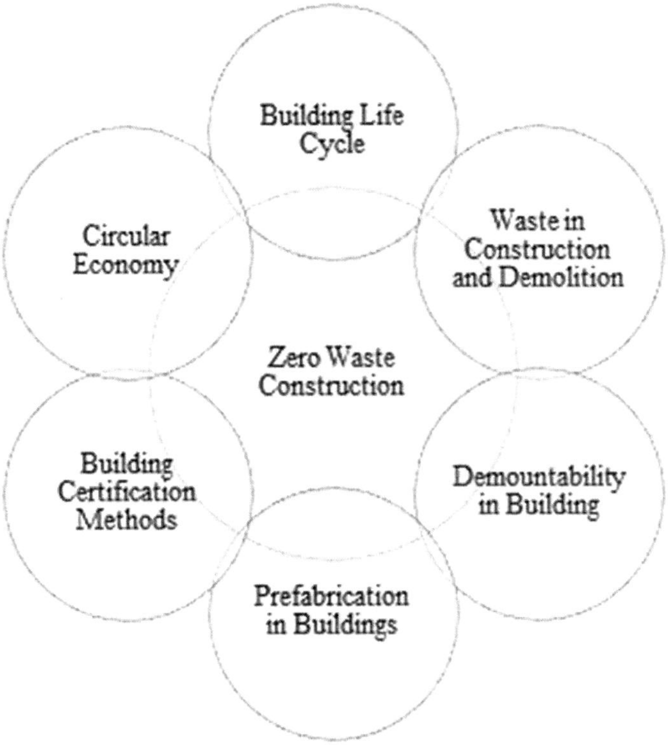

Fig. 5 Design approach for zero waste building construction

2.1 *Current Building Life and Lifecycle in Relation to Environmental Impacts*

The life of a building is mostly seen as a linear process. It starts with extracting materials, processing, manufacturing and then uses. It is disposed-off when the building is no longer needed back in the environment. This mostly leads to pollution, depletion of resources, habitat loss and energy consumption.

Buildings are commonly designed for a single type of use, and the end of life of a building is neglected. The design is optimized to reduce cost and time during construction and the use phase. At the end of life, a building is demolished and most of the materials end up in landfills, even when the materials do not complete their technical lifetime. This is a major waste of materials and thus energy to produce it.

To minimise the negative environmental impacts, a design should not only concern construction and service life but also the complete life cycle of the structure and its components. This study focuses on the important factors in buildings life-cycle and environmental impacts in their life-cycle phases. Three types of life can be distinguished in a building:

- Design working life—use for the intended purpose
- Technical service life—building performance
- Functional working life—meeting the demands of the users.

Buildings are subject to change every 30 years. Technically a building structure can last up to 100 years. The lifetime for building types is mostly restricted by functional working life. This leads to a relatively short life for building types like shops, businesses, educational facilities and industrial buildings of around 30 years. Residential buildings have a survival probability of 225 years.

Another important aspect is the different life expectancies of components in the building itself. Therefore, layers are distinguished within a building with different lifetimes, in order: structure, skin, services and space plan. Materials should be used accordingly as these also have different life expectancies.

The environmental impact of buildings is enormous. Currently, the operating energy of buildings is responsible for the largest part of the environmental impact. The important factors to judge the environmental impacts of buildings are energy, materials and waste. To make optimal use of the embodied energy in materials, a building should be designed for a closed cycle without the need for further processing. The best options are reuse, followed by recycling. Landfilling is the least environment friendly option. Deconstruction is a method that allows for recycling and reuse.

2.2 *Waste in Construction and Demolition*

Most buildings are designed with a life expectancy of a few decades with no consideration of what will happen when the building is demolished. Large amounts of waste are produced by the building industry which is associated with negative environmental, economic, and social impacts. The resources of the Earth are limited. With continuous growth around the world, it will become unsustainable to dispose of materials in landfills.

The building industry uses many available resources. The industry does not focus on what happens with the buildings when they are demolished. Most of the time this means that demolition leads to waste which is dumped in landfills. Some recycling and reuse take place, but this is mostly based on the economic value of the materials. In some countries, high recycling is achieved, but this recycling of materials mostly consists of down-cycling. This waste of resources and materials is unsustainable if the world becomes more prosperous and populated.

Waste is defined as a material that cannot be directly used for a purpose. This mostly contributed to the mixing or contamination of materials. A lot of energy, and therefore also money, must be invested to regain the original material. Often this is also not possible due to chemical reactions and bonds, such as concrete and silicon fillings, which cannot be recycled back into their original components. Construction waste originates mostly from packaging materials. Other sources are cutting and other types of adjustments on the building site or factory. 80% of the construction waste is caused by stone tablets, piles, concrete, sand-lime elements and roof tiles. The amount of demolition waste is considerably larger than construction waste. Demolition waste mostly consists of concrete and masonry materials. To achieve more sustainable resource use, the industry should aim for zero-waste. This is a strategy that aims for a closed loop of all materials. Materials and products should be reused or be decomposed to their natural state or be returned to the technical system.

Zero waste can be achieved by making use of the method of design for deconstruction. Deconstruction is an end-of-life approach in which the building is a system that is carefully and methodically disassembled. The disassembled parts can be reused in new buildings or recycled into other products. Down-cycling is avoided, and all materials should return to either the technical system or the bio-system. Deconstruction also eliminates mixed debris, allowing for direct reuse or recycling. The best materials for zero-waste design are recyclable materials and natural source materials. Steel and aluminium can be recycled completely and are also suitable for direct reuse. Timber products also offer great potential as they can be grown naturally and act as a carbon sink. It must however originate from sustainably managed forests. Concrete and masonry are very poor materials for zero-waste design and must be avoided.

2.3 Demountability in Buildings

The demonstrability of buildings allows for the deconstruction of a building, which is one of the methods to design for zero-waste buildings. To improve the environmental impact of the life cycle of a building, the perception and technical composition of buildings need changing. Deconstruction is an important element in sustainable construction as it brings attention to the building assembly and disassembly and the building materials and their connections. Designing for deconstruction has the following advantages over conventional design:

- Less generation of waste and landfill need
- Generation of reusable materials
- Less need for raw materials and less energy to process it
- Decreased difficulty in constructing
- The increased total efficiency of a building
- Less environmental impact.

Demountable buildings have been around for a long time, and their concepts have been studied in various designs. Several buildings have already been designed using the concept of demonstrability and show environmental potential. Several principles can be distinguished to design for deconstruction. Demountable construction is linked to building materials, connections, components, assembly and disassembly. Some materials are better suitable to be used when considering demountability. Concrete has several disadvantages in weight, size and uniqueness. The disadvantage of masonry lies in the required time associated with construction and the cleaning of mortar or flue off the components after deconstruction. Wood has a large potential, as this material could be reused with little modification, but screws and nails can create problems. Steel and other metals are suitable when connected properly, however, contamination, such as sprayed protection, may create difficulties.

2.4 Prefabrication in Buildings

When designing for deconstruction, prefabrication becomes a possible solution to the build-up of the building components. As prefabricated products are assembled on-site and commonly connected using principles related to deconstruction. Prefabrication allows for creating large volumes of products at low costs, and it improves the quality of the product compared to on-site construction. Prefabrication is the first step into industrialization, before mechanization, automation, robotics and reproduction. Building systems make use of prefabricated components, unrelated to specific designs but as general solutions. This allows for system building, a set of components and rules which are used to produce buildings. In the past, complete prefabricated systems have been tried but mostly failed due to the exclusivity of the system regarding components. Current day systems become more successful as these are more customizable to the client. The advantages of prefabrication compared to on-site construction are listed below:

- Cost-efficient in large numbers
- Less handicraft and skilled work required on building site
- Significant reduces building time
- Gravity is no restraining factor in build order
- Dry jointing solutions
- Controlled conditions inside the factory
- Higher quality compared to on-site construction
- Less waste
- Organizational advantages.

Different building systems can be distinguished in prefabrication. Light frame construction is based on lightweight structural members that form the frame of the walls and roof. This is mostly done in structural steel, reinforced concrete, or wood. Flat-pack is a system based on pre-assembled flat panels, stacked together on a trailer for transport. Modular building construction is based on the assembly of prefabricated large 3D-modules. Prefabrication can become more preferable in combination with mass customization. Components can be assembled in different combinations to achieve different results. Standardized and factory-produced products from the base for buildings are chosen by the customer.

2.5 Green Building Certification System

Green rating systems emphasizes sustainable building construction. All integrated parameters towards the environment, social and economic are picked up. Most of these rating tools have an predefine certain mandatory set of points which must be adhered to for getting the certain level of certification by these rating tools. Mandatory set credit points safeguard the ultimate interest of mankind for ensuring a better-built environment for our future generations to come [31].

A particular rating system is chosen based on its usage, coverage and popularity as per the below functions:

- Marker rating system
- Certification steps
- Cost parameters
- Usage region/area
- Composition of system
- Effectiveness of rating tools outside the originating country.

2.6 Circular Economy

The primary objective of the circular economy is to dismantle the relation between economic growth and environmental degradation and resource consumption through new production practices and technological developments, satisfying consumer needs in different, more sustainable ways.

2.7 Product Stewardship

Product stewardship is an approach that manages the impact of material or product used in an activity at its end-of-life scenario. The process involves all the stakeholder's responsibility in a project, which starts from raw material extraction, construction, usage and its disposing so that material is used in such an efficient way that it does not lead to any negative impact on the environment. Product stewardship focuses on the product, and all the stakeholder during the life span of the product need to take responsibility to ensure that product does not impact our environment at any stage of time.

2.8 Environmental Management System

The environmental management system is an approach to achieving the objective of a sustainable environment through its regular review, validation, and improvement processes.

3 Material and Methods

Materials form the base of the building thus selection of the right material with the right quantity and quality is particularly important in achieving the primary objective for a zero-waste building. The demands are aimed towards the sustainable use of materials. Only materials that have an indefinite source should be used, i.e., renewable materials such as wood or materials that can endlessly be reused. The materials should also not be dangerous or polluting to the environment, as this might inhibit reuse.

Another important demand is the identifiability of the materials in the design. It is necessary to know what component of material can be reused or recycled. Otherwise, correct reuse is not possible, and the possibility exists that the wrong method is used, possibly eventually causing waste. The last demand is to keep the amount of different used materials to a minimum. A larger diversity will require more work to separate the materials from each other. This requires more time and money and thus increases the likelihood of generating waste as wasting would be a simpler option than reusing or recycling.

As stated in Fig. 6, the first level of assessment is done on the material level. At this level, it is required to describe information for each homogeneous material that is used in the building such as aluminium, coating, wood and plastic. This information is used to track the material and have basic details about it and includes:

- The name of the material
- The function of the material
- Amount of material used (in kilograms)
- The percentage of this material that is from a recycled source
- The recycling percentage of the material that is achieved in intended recycling processes
- The intended use of the material after use in the building.

After the assessment of all materials in a component, an assessment can be made for that component. Components are defined as separate parts that can be seen as semi-manufactured goods, for example, a door handle, hollow core slab floor or corrugated sheets. Again, the assessment starts with a description. In this description, the following information is required:

- Name of the component
- The function of the component

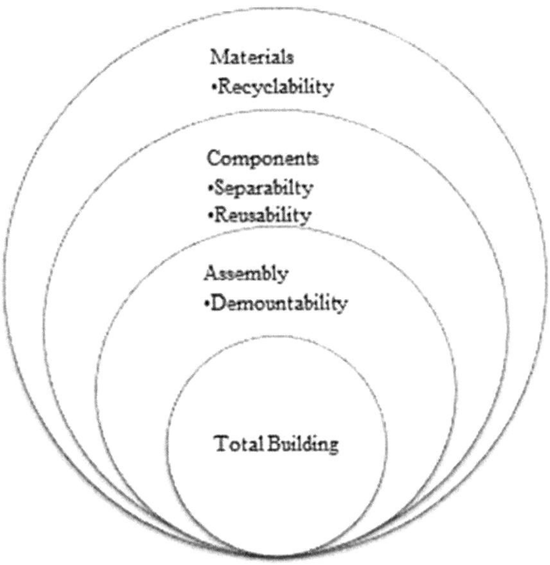

Fig. 6 Overview of assessment methods

- The number of instances of the component in the sub-assembly
- Weight of component (in kilograms)
- Producer (if applicable).

The percentages of the component from a recycled source are the intended method of recycling of component and intended component use of component post use in the sub-assembly.

The next layer in the hierarchy is the assembly level. Assemblies are defined as subsystems of a building that in theory could serve their main function without needing other sub-systems. Examples of assemblies are a curtain wall, a door or a ventilation system. The following descriptions of assemblies are required:

- Name of assembly
- Function
- Times used
- Total weight of assembly (in kilograms)
- Producer (if applicable)
- Intended method of assembly within building
- Intended method of disassembly within the building
- Intended tools needed for (dis-)assembly within the building
- Intended method of recycling of assembly
- Intended use post use in building.

Again, a positive or negative conclusion is drawn regarding the assemblies. The last layer in the hierarchy is the level of the total building. It includes the whole building and all its assemblies. At this level, the following aspects are to be described:

- The intended method of building construction
- The intended method of building deconstruction
- The intended tools needed for (dis)assembly of building.

This enables us to conclude the total building regarding zero-waste and is used in the final assessment of the building.

4 Cost Perspective

View of sustainability cannot be directly correlated only with economic benefits as we are dealing with a major issue of our depleting natural resources over the earth. However, still, the cost is one of the key driving factors for motivating the investors or developers for taking initiative towards the environment during the construction and operation of the building. A green rating building construction adds to 0.40–21.00% cost premium over a conventional building construction [32, 33], out of which almost 75% of cases cost premiums was within a range of 0.0–4.0% [34]. The expected payback period for a cost premium of 4% is calculated as 7 years and 12 years for a cost premium of 9% [34]. Further, in the case of zero waste construction building, more efforts are made to reduce the waste generation, and material product cycle is to be maintained so that at end of life of building most of the materials used in a building can be recovered and reused to the maximum extend. Thus, if we holistically review the overall economics for a complete life cycle for building, then a zero-waste building will be even more cost-effective in comparison to a conventional and green-rated building.

5 Barriers for Zero Waste Construction

In India though the concept of zero waste construction is very new but still many of its key approaches such as prefabrication and green building certification system are very popular in construction industries and is being implemented in most of recent industrial and governmental projects. Sardar Patel Stadium, Motera is one of the best examples for same.

However, in the current Indian scenario, before the start of any construction as per government bye-laws construction approval is to taken from the regulating authorities; however, no specific focus of effective waste management is captured in the bye-laws which are to be followed. Further, since the construction industry is very unorganized and spread all over the globe from small house construction to mega industrial projects thus a lack of recycling infrastructures across the locations, lack of skill, training and public awareness and old misconception existing in the complete system.

Zero waste construction is a necessity for us to conserve a healthy build environment for our future generations to come. However, lack of governmental policies, no legal framework, no strict regulation on landfills, lack of recycling infrastructures, pre-existing perception for the high cost of construction, lack of public awareness, missing training of workers and more time of construction is some of the barriers which retard the adoption of zero waste system in the present construction industry. In addition to the above some inherent barriers such as old and standard government building byelaws, lack of design standards for reducing waste generation and its reuse, further lead to ineffectiveness towards zero waste construction.

6 Discussion

Zero-waste approach is correlated with building construction work to assess the practical applications on a design and process throughout the life cycle. Building construction design approach has been evaluated using principles of building life-cycle, waste in construction and demolition, demountability design for buildings, construction techniques for prefabrication, green building certification system, use of circular economy for materials, follow product stewardship for construction material and environmental management system. Further for effective material management a building as a whole is studies in breakdown structures of assembled material to its independent components and then to basic raw material for its effective characteristics towards reusability and recyclability.

Sustainable development emphasises the integration of environmental, social and economic aspects as per Fig. 7.

Sustainable construction waste management must ensure a balance between economics, social and environmental and should not limit its focus only on environmental angle alone. Nevertheless, we should view the issue from the whole building life cycle outlook as:

- Design with a lack of future foresight reduces the building usage life as it fails to incorporate required changes in future and may lead to early demolition even before the life span of the building is over.
- Incorrect coordination between various design elements for building usage may lead to unnecessary demolition during construction.
- Wrong building operational usages reduce the life span of building and lead to demolition waste
- Lack of construction waste management at the site.
- No demolition plan for reuse of components of a building.

Buildings should be intended and made in such an approach that allows for the reuse and recycling of materials. The key problem of this study is how a building can be designed to achieve net-zero waste in overall phases of its construction and demolition and highlighted as per 3 questions as stated below:

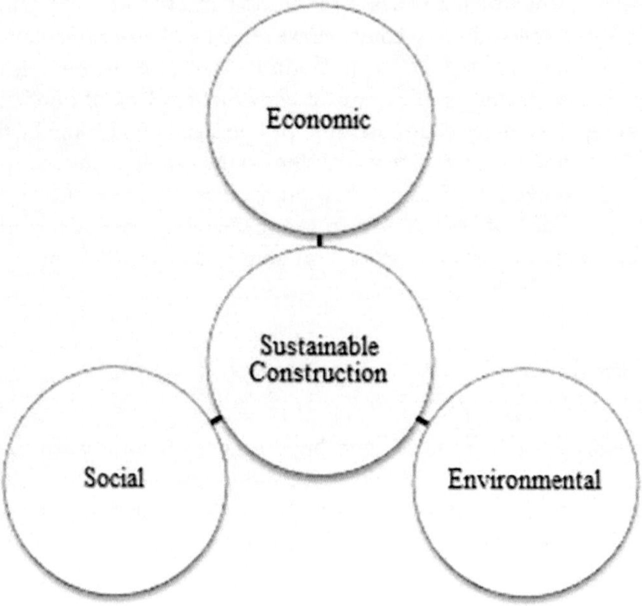

Fig. 7 Three major elements in sustainable development

1. What are the key factors which contribute to a zero-waste building design?
2. What are the functional requirements of a zero-waste building design?
3. What is a possible design for an industrial building according to zero-waste principles and requirements?

These questions are answered using a mixed approach of research and design for a hypothetical problem of new building construction.

- Life cycle assessment—Study will provide a complete outline of project to list down all the possible potential environmental concern it can have from start of design to end-of-life demolition.
- Circular economy for materials—Ensure reusability or recyclability of basic raw material for construction work.
- Product stewardship—manages the impact of material or product used in an activity at its end-of-life scenario
- Demountability in design—Enhance reusability of building components.
- Prefabrication—Construction techniques ensuring fast execution with better quality and waste control measure, as prefabrication will be done in controlled factory environment.
- Green building certification system—System check point for compliances of all above.

Building construction design approach has been evaluated using principles of building life-cycle, waste in construction and demolition, demountability design for buildings, construction techniques for prefabrication, green building certification system, use of circular economy for materials, follow product stewardship for construction material and environmental management system. Further for effective material management, a building as a whole is studies in breakdown structures of assembled material to its independent components and then to basic raw material for its effective characteristics towards reusability and recyclability.

7 Conclusions

Although construction is one of the key parameters for social and economic growth for any nation, it is also a major waste generating process. However, it also has sufficient potential to minimize or reduce any waste generation. Advance planning and monitoring start from the project till the end of construction can reduce waste generation up to a large extent. A state of zero waste can also be reached by holistically implementing the revolutionary concept of prefabrication, demonstrability, circular economy, green rating tools and the principle of reducing, reuse and recycle. Further practices like smart waste audit, smart waste collection, high-value mixed waste processing and the collaborative platform further help towards the prime target of zero waste construction. The structure targets to help stakeholders in the waste value chain to identify the tools that can be implemented to achieve sustainable zero waste in construction building. The research concludes three main points for a zero-waste design.

- No waste generation during any time of the life cycle of a building.
- Every material used in the building should remain in the material cycle during its life span.
- Reuse of materials, ensuring the required efficiency of the material.

In terms of zero waste construction design, the following was concluded:

- Using the right materials
- Maximizing the attractiveness of disassembly at the end of the life of the building by appropriate detailing and system choice.

The above conceptual framework derived from this paper motivates future research on the application of the zero-waste concept to the construction industry.

References

1. Deloitte. (2018). *Global powers of construction.* Communications, Brand and Business Development Department.
2. Mas, J. (2017). *The case of Latin America.* FIIC.
3. Eurostat, "Eurostat," Eurostat, 2017. [Online]. Available: https://ec.europa.eu/. Accessed May 05, 2020.

4. U. Nations. (2015). *UN sustainable development summit* [Online]. Available: https://sustainabledevelopment.un.org/. Accessed June 14, 2020.
5. Guggemos, A., & Horvath, A. (2003). Strategies of extended producer responsibility for buildings. *Infrastructure Systems, 9*(2), 65–74.
6. Soharu, A., & Malik, R. K. (2017). C & D waste management using life cycle assessment tool. *Open Access International Journal of Science and Engineering (OAIJSE), 2*(8), 5–8.
7. Junnila, S., & Horvath, A. (2003). Environmental sensitivity analysis of the life-cycle of an office building. In *2nd International Symposium Integrated Life Cycle Design of Materials and Structures ILCDES*, Kuopio.
8. Carlowitz, H. C. (1713). Sylvicultura oeconomica – Anweisung zur wilden, Leipzig.
9. Smith, H. (1963). Cumulative energy concept. in *World Energy Conference.*
10. Palmer, P. (1970). *Zero waste institute* [Online]. Available: https://zerowasteinstitute.org/?page_id=202. Accessed June 04, 2020.
11. Stahel, W. R., & Mulvey, G. R. (1981). *Jobs for tomorrow: The potential for substituting manpower for energy.* Vantage Press.
12. WCED. (1987). *Our common future: Report of the world commission on environment and development* [Online]. Available: http://www.un-documents.net/our-common-future.pdf. Accessed June 03, 2020.
13. E. Summit. (1992). *United Nations conference on environment and development (UNCED)*, Rio de Janeiro, Brazil.
14. Kibert, C. J. (1994). Sustainable construction: Proceedings of the first international conference of CIB TG 16. In *International Conference of CIB TG 16*, Florida.
15. Connett, P. (1998). *The zero waste solution.* Chelsea Green Publishing.
16. Plessis, C. (2002). Agenda 21 for sustainable construction in developing. In *International Council for Research and Innovation in Building and Construction.*
17. Braungart, M., & McDonough, W. (2002). *Cradle to cradle: Remaking the way we make things.* North Point Press.
18. Connett, P. S. B. (2001). *Citizen's agenda for zero waste: A strategy that avoids incinerators and eventually eliminates landfills.* GrassRoots Recycling Network [Online]. Available: http://archive.grrn.org/zerowaste/community/activist/. Accessed 15-06-2020.
19. Kibert, C. (2012). *Sustainable construction: Green building design and delivery.* Wiley.
20. Kashyap, M. Z.-A. N. K. M. (2003). Proposal for achieving sustainability in construction projects through concurrent engineering. In *The RICS Foundation Construction and Building Research Conference*, Wolverhampton.
21. Adetunji, I., Price, A. D. F., Fleming, P. R., & Kemp, P. (2003). Sustainability and the UK construction industry: A review. *Engineering Sustainability, 156*(4), 185–199.
22. Sev, A. (2009). How can the construction industry contribute to sustainable development? A conceptual framework. *Sustainable Development, 17*(3), 161–173.
23. Demaid, A., & Quintas, P. (2006). Knowledge across cultures in the construction industry: Sustainability, innovation and design. *Technovation, 26*, 603–610.
24. ISO-14040:2006, ISO 14040:2006. (2006). *Environmental management—Life cycle assessment—Principles and framework.* International Organisation for Standardisation.
25. ISO-14044:2006, ISO 14040:2006. (2006). *Environmental management—Life cycle assessment—Requirements and guidelines.* International Organisation for Standardisation.
26. Palmer, P. (1974). *The faux zero waste movement is spreading*, California.
27. Laefer, D. F., & Manke, D. F. (2008). Building reuse assessment for sustainable urban reconstruction. *Journal of Construction Engineering and Management, 134*(3), 217–227.
28. Baba, S. H., Skinder, B. M., Muzaffar, A. B. (2020). Zero waste: a sustainable approach for waste management. In *Innovative Waste Management Technologies for Sustainable Development* (pp. 134–155). IGI Global.
29. Liyanage, K., Waidyasekara, A., & Mallawarachchi, H. (2018). Enabling zero waste concept in the construction industry: A. In *The 7th World Construction Symposium 2018: Built Asset Sustainability: Rethinking Design, Construction and Operations*, Colombo.

30. Dwaikat, L., & Ali, K. (2016). Green buildings cost premium: A review of empirical evidence. *Energy Building, 110*, 396–403.
31. Soharu, A., Naveen, BP., & Sil, A. (2021). Effectiveness of existing green rating systems towards zero waste construction. *AIP Conference Proceedings*, https://doi.org/https://doi.org/10.1063/5.0066416.
32. Leskinen, N., Vimpari, J., & Junnila, S. (2020). A review of the impact of green building certification on the cash flows and values of commercial properties. *Sustainability, 12*(2729), 1–22.
33. Kats, G. (2010). *Greening our built world: Costs, benefits, and strategies.* Island Press.
34. Plebankiewicz, E., Juszczyk, M., & Kozik, R. (2019). Trends, costs, and benefits of green certification of office buildings: A polish perspective. *Sustainability, 11*(2359), 2–17.

Improvement of Hard Water Characteristics and Scale Formation Under the Effect of Pulsating Electromagnetic Field

Amrit Anand Dosar and Vivek Srivastava

Abstract Scale deposition in water pipe due to hard water circulation often leads to various technical and economical problems. The conventional chemical treatment methods use hazardous chemical which affects human health as well as water chemistry. This study shows the effect of physical water treatment methods like pulsating electromagnetic field on water characteristics and scale reduction under different turbulent flow conditions and pipe materials. The scale removal rate was analyzed by the formation of aragonite crystal in water pipes in place of calcite crystals after electromagnetic treatment. The morphology of aragonite and calcite crystals was analyzed by field emission scanning electron microscope on different pipe materials. The water flow rate was maintained at 3, 5, and 7 L/min. After electromagnetic treatment, the result shows that the scale removal rate increases from pipe wall and reduces the total dissolved solids, electrical conductivity, hardness, and alkalinity of water. These water characteristics are further decreasing on increasing the flow rate from 3 to 7 L/min. The reduction rate of these water characteristics was higher for the first 15 h of circulation time than the remaining 15 h. On investigating the effect of electromagnetic treatment on pipe material, it was obtained that the polyvinyl chloride pipe is much effective than the galvanized iron and copper pipes.

Keywords Electromagnetic water treatment · Pulsating magnetic field · Scaling · Hard water · Pipe material · Flow rate

1 Introduction

Scale deposition due to hard water in various household and industrial equipment causes significant economical and technical losses [1]. Scale reduces the heat transfer

A. A. Dosar (✉)
BBD University, Lucknow, India
e-mail: amritdosar@bbdu.ac.in

V. Srivastava
DIT University, Dehradun, India

A. K. Gupta et al. (eds.), *Advances in Construction Materials and Sustainable Environment*, Lecture Notes in Civil Engineering 196,
https://doi.org/10.1007/978-981-16-6557-8_20

from pipes in heat exchangers and blocking the water flow in pipes [2]. For scale removal, various conventional methods are generally used like the application of chemical product, the pre-precipitation of the scale with soda ash or lime and by ion exchange process. All these conventional methods are very efficient, but they can change the water chemistry and are much costly. So to avoid the chemical uses, physical methods were developed; pulsating electromagnetic treatment device method is the example of such method to prevent scale formation and water treatment [3]. Chemical products used in conventional methods were harmful to the atmosphere as well as human health [4, 5]. Electromagnetic treatment device helps in reducing the scale formation on the walls of various domestic and industrial equipment. In water system, electromagnetic water treatment (EMWT) has been used for several years as a preventing and controlling tool for scale deposition [6]. Various electromagnetic devices have been developed to get maximum anti-scale prevention efficiency in last few decades. It changes only the morphology of $CaCO_3$ scale without changing its chemical composition [7].

The alternating electromagnetic field hindered the precipitation of $CaCO_3$ and formed aragonite and veterite crystals as $CaCO_3$ precipitate [8]. Li et al. [9] analyzed using ultrasonic time domain reflectometer that the scale layer of deposited $CaCO_3$ is denser and thicker in nonmagnetic field treatment than with electromagnetic treatment. They also observed using SEM and XRD methods that after electromagnetic treatment, the formation of calcite crystals reducesand veterite and aragonite crystals increases. Piyadasa et al. [10] observed that the pulsed electromagnetic field treatment is efficient in scale removal and biofouling in reverse osmosis membrane. They also found that the treatment is also helpful in increasing the efficiency of heat exchanger. Simonic et al. [11] analyzed using SEM and XRD methods that initially calcite precipitates were formed in water, which converted in to nonadhesive aragonite form after magnetic treatment (0.6 T) that could be removed easily by turbulent flow of water.

The efficiency of MWT depends on the material of pipe, used for water circulation. Alimi et al. [12] investigated the effect of magnetic treatment on different pipe material (PTFE, tygon, copper, PVC, and stainless steel) by circulating hard water at 0.16 T magnetic fields. They found that the increase in total precipitation ratio was obtained, when the magnetic field was applied through pipes of nonconductive material.

The rate of flow of water affects the scale formation and water characteristics due to the Lorentz force. Shahryari et al. [13] investigated the effect of modulated electromagnetic field on fouling in double pipe heat exchanger. They observed that at different flow rates (0.5, 0.8, and 1.3 m/s), the fouling reduction rate changes. They also found that the electromagnetic treatment decreases the Ca^{2+} measurement in water. Latva et al. [14] examined using a pilot project which showed that MWT gives the best result at 2.3 m/s flow velocity of water and found that the magnetic field reduced calcium scaling by 15% from the pipe.

Magnetic field effects the water characteristics by changing its hardness, alkalinity, pH, TDS, electrical conductivity, etc. Mghaiouini et al. [15] investigated using experimental study that in a closed loop system at low speed of 0.18 m/s, and the electromagnetic treatment changes the conductivity, TDS, salinity, and temperature by 3.66%, 4.0%, 4.4%, and 0.76%, respectively. Khater et al. [16] analyzed the effect of the magnetic field (18 Gauss) on the physicochemical characteristics of water, as it decreases conductivity, pH, carbon dioxide, and alkalinity in water samples. The magnetic field also decreases the heavy metal concentrations like zinc, copper, cobalt, and aluminum in water. Bali et al. [17] analyzed the scaling effects on treated water using the magnetic method. The result shows that magnetic treatment increases pH and decreases the resistivity of the solution. Helal et al. [18] analyzed the effect of magnetic treatment at 1500 °C at 6480 Gauss magnetic field. They analyzed that the magnetic field reduces scale formation and changes the electrical conductivity of $CaCO_3$. The application of a magnetic field decreases the surface tension of water and increases its volume evaporated [19].

Electromagnetic field treatment method is also effective on biofouling. It reduces the bacterial count and diversities in biofilm. It also minimizes the mineral precipitates, carbonate, and silicate content in biofilm [20]. Othman et al. [21] studied the effect of magnetic water treatment under various conditions. They observed that MWT controls the growth rate of scale; hence, improves the life of water pipeline and its performance.

Out of all the above researches, very few literatures are available on the effect of the electromagnetic treatment on scale formation and characteristics of water flow through different pipe materials. The effect of flow rate and circulation time on the percentage change in water characteristics also needed to be explained.

The objective of this research was to experimentally analyze the effect of pulsating electromagnetic treatment on the formation of aragonite crystal in different materials of pipe and to analyze the change in water characteristics on increasing the flow rate and circulation time. The effect of pipe material on water characteristics also has to be examined.

2 Experimental Procedures

The experimental electromagnetic treatment device (EMTD) is shown in Fig. 1.

It consists of a solenoid coil by which magnetic field was generated in all the three pipe materials like polyvinyl chloride (PVC), galvanized iron (GI), and copper one after another. The coil is further connected to the relay switch to provide pulsating current to the solenoid coil. The water that has to be treated passed through the solenoid coil arranged at one end of the pipe fitting. The magnetic field generated by the coil was perpendicular to the flow of water. The hard water to be treated was circulated in pipe through solenoid coil by the centrifugal pump. The water was circulated in pipes at three different flow rates in a turbulent range (3, 5, and 7 L/min). The solenoid coil with relay switch was used to generate pulsating electromagnetic field. Hardened water was circulated in the different materials of pipe-like copper, GI, and PVC under magnetic and nonmagnetic condition (magnetized sample and unmagnetized sample) and different flow rates. Further, the analysis was done to

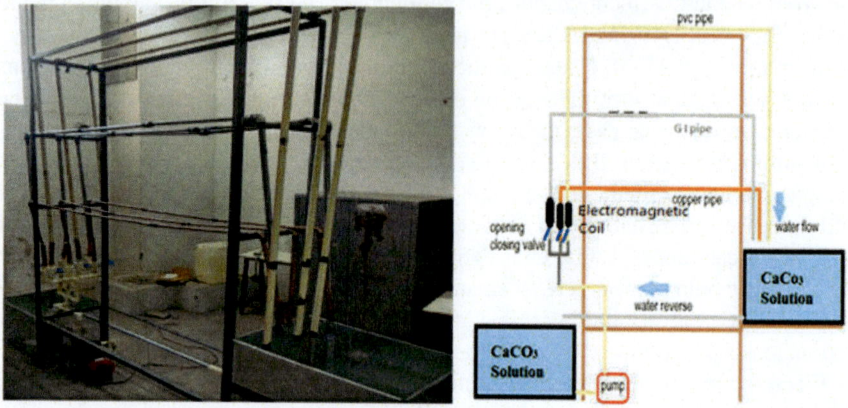

Fig. 1 Experimental setup and its corresponding layout

obtain the effect of scale formation on different pipe materials. The total dissolved solids, electrical conductivity, hardness, and alkalinity in the water samples were measured for all flow rate conditions.

The field emission scanning electron microscope (FESEM) model Gemini Zeiss Supra 35 VP is used to analyze the morphology of crystal $CaCO_3$ on pipe walls. The magnifications of this test were at 9000 X zoom for all materials of pipe and in both magnetic and nonmagnetic case.

The total dissolved solids (TDS), electrical conductivity (EC), hardness, and alkalinity of water were checked with a one-liter sample of magnetized and nonmagnetized water for all flow rates and different materials of pipes.

3 Result and Discussions

3.1 Effect of Magnetic Field on Scale Removal

A small number of deposited scales were formed on the inner surface of the pipe walls after magnetized and nonmagnetized treatment of water pipes of different materials which was analyzed by FESEM method. Figure 2 shows the size and morphology of the scale.

All these FESEM images show that the crystal structure on the inner surfaces of pipes of treated water is of orthorhombic or needle crystal, which means after magnetic treatment, aragonite crystals are formed which have low adhesive calcium crystal phase and can easily remove by the turbulent flow of water, while on untreated water pipes surface trigonal-rhombohedral shaped calcite crystals are formed which are usually associated with a hard scale and are difficult to remove.

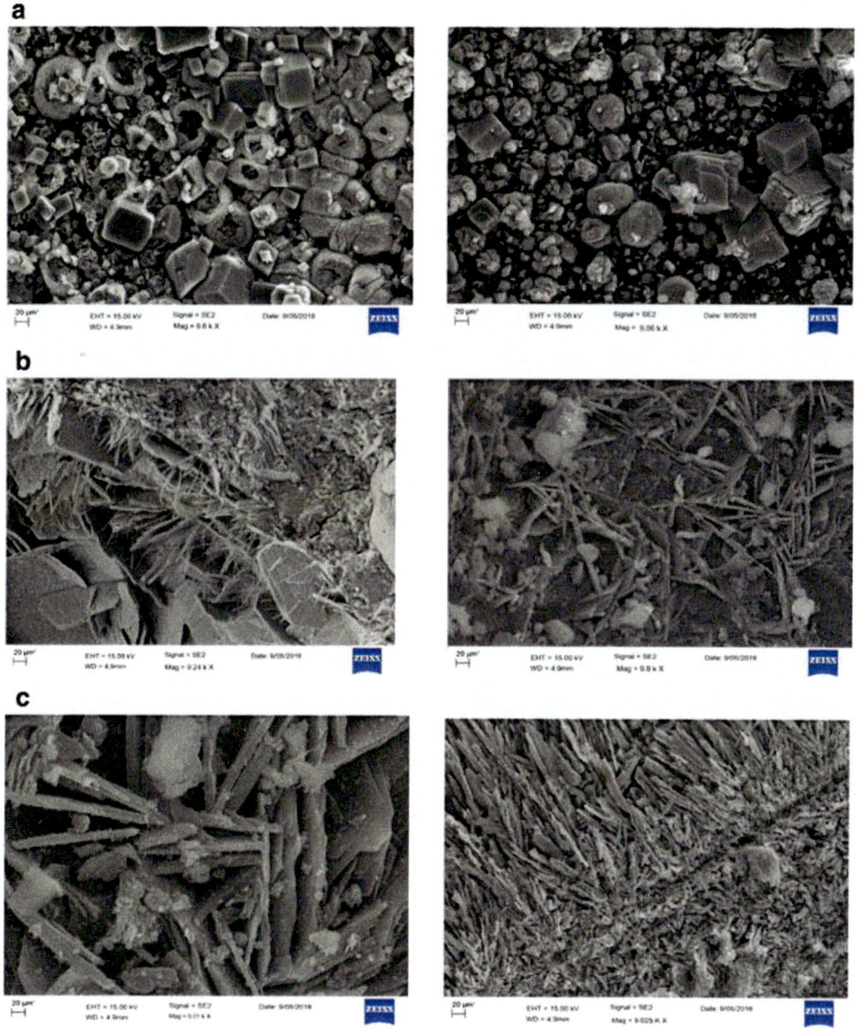

Fig. 2 **a–c** shows the magnetized and nonmagnetized FESEM images of copper, GI, and PVC pipes, respectively

After electromagnetic treatment, FESEM images show that due to the formation of aragonite crystals, the scale removal from pipe increases and less scale formed on the surface of pipes. The results are confirmed with an earlier study in which the experiment claimed that the aragonite crystals are formed due to the magnetization of the solution [22]. Lipus et al. [23] investigated the effect of scale formation in industrial water processing system with high (0.1–0.2 T) magnetic field. They found that electromagnetic field reduces the scale formation on at least 10 times less cost than the ion exchange process. Chang et al. [24] also analyzed the impact of the magnetic field on aragonite crystals. They observed that under the effect of magnetic field aragonite crystal grows and they grew faster at low pH, high level of supersaturation, and at activity ratio far from unity.

Hence, by the formation of aragonite crystals after electromagnetic treatment, the scale formation in pipes can be reduced and enhances the life of the pipe.

3.2 *Influence of Pipe Material*

For analyzing the effect of pipe material, the electromagnetic water treatment was done on three different materials of pipe-like PVC, GI, and copper pipe. The water was circulated for 30 h, and different water characteristics are analyzed.

Figure 3 shows the influence of pipe material on the TDS of water under the effect of the electromagnetic field. For the above study, the flow rate was fixed at 7 L/min and the only TDS of water was measured, because after study, the best result of water treatment was obtained at 7 L/min flow rate. The result shows that after magnetic treatment maximum reduction in TDS was observed for PVC pipe is 25.17%, while in GI and copper pipes, a small TDS reduction of 23.91 and 20.76% was observed, respectively. This shows that the efficiency of the magnetic field is different for different materials of pipe and it is maximum for the pipe with lesser conductivity. As known that the PVC pipe is nonconductive, so the treatment effect is maximum for this pipe material, while for conductive copper pipe treatment effect is minimum. Alimi et al. [12] who experimentally analyzed the effect of MF treatment on various pipe materials also verify the above results. They exposed the water to the magnetic field (0.16 T) with different flow rates (0.54–0.94 L/min) for 15 min and found that nonconductive pipe materials are much effective than others in respect to the homogeneous and total precipitation ratios. The type of material of pipe plays an important role in the formation of scale also [25, 26].

The rate of TDS reduction is different for different materials of pipe. In PVC and GI pipes, the maximum reduction is obtained in the first 15 h, while for copper pipe,

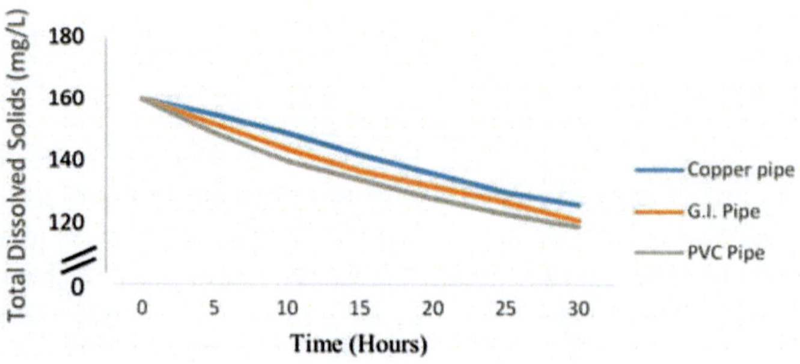

Fig. 3 Influence of pipe material on TDS of water

maximum reduction is obtained in the last 15 h because magnetic field effectiveness is higher for nonconductive and less for conductive material.

3.3 *Influence of Magnetic Field on the Characteristic of Water*

The electromagnetic field affects the total dissolved solids (TDS), electrical conductivity (EC), hardness, and alkalinity of water under different flow rates in turbulent ranges for PVC pipe. Water circulation time for the study was taken 30 h.

Figure 4a–d shows the effect of flow rate on TDS, EC, total hardness, and alkalinity of water after magnetic treatment. For the above study, PVC pipe material is used in all cases because PVC pipe is much efficient for magnetic water treatment than other pipe materials. The graphs are plotted between the characteristics of water and the magnetic circulation time. All the four figures show that on increasing the circulation time, the above characteristics of water goes on decreasing due to the magnetohydrodynamic (MHD) effect. In the magnetic water treatment process, the magnetohydrodynamic (MHD) mechanism is involved [27, 28].

Hence, the magnetic field reduces the TDS, EC, hardness, and alkalinity of treated water. The results are consistent with an earlier study in which Zhang et al. [29] experimentally analyzed the effect of the electromagnetic field on circulated water

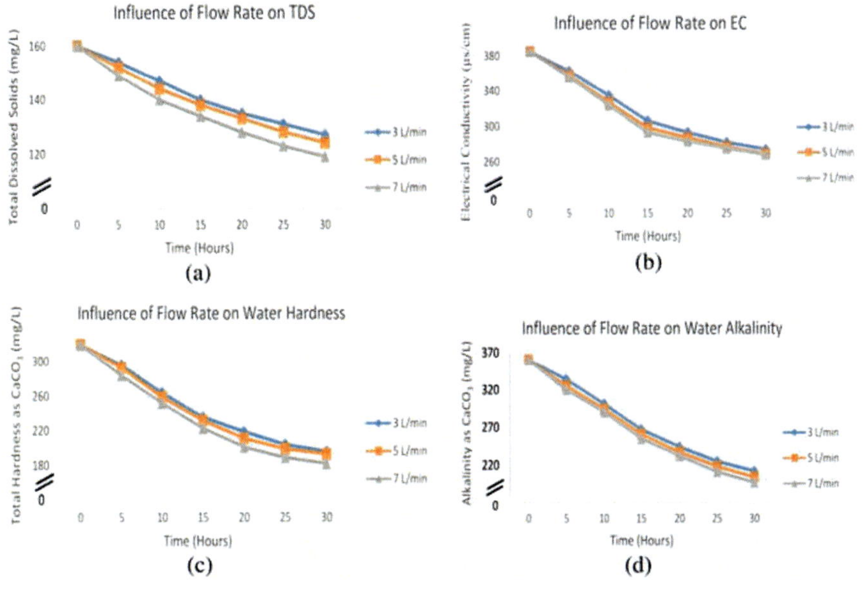

Fig. 4 Influence of flow rate on **a** total dissolved solids (TDS), **b** electrical conductivity (EC), **c** total hardness as $CaCO_3$, and **d** alkalinity as $CaCO_3$

Table 1 Water characteristics change rate for circulation time

Water characteristics	Percentage reduction	
	For the first 15 h	For the last 15 h
TDS	13.46	9.80
EC	21.45	10.69
Alkalinity	27.23	16.47
Total hardness	28.47	20.55

at constant flow rate. They found that after treatment physicochemical characteristics of water like TDS, turbidity, conductance, and hardness of water by 9.64%, 84.8%, 9.57%, and 17.11%, respectively. Kashef et al. [30] examined the effect of the magnetic field ($B = 1.45$ T $\pm$ 0.05) at open and dead-end flow conditions at a flow rate of 41.93 L/min and 52.16 L/min, respectively. They found that the MF reduces the electrical conductivity and TDS of water. Electrical conductivity analysis was done by Szczes et al. [31] who analyzed the effect of magnetic treatment by exposing water to weak static magnetic field generated from permanent magnets ($B = 15$ mT) at flow condition. They found that the conductivity of water decreases, which is inversely proportional to the flow rate. Khater et al. [16] also examined the effects of the magnetic treatment on water purification. Water was exposed to the weak magnetic field of 18 Gauss. They found that the magnetic field decreases the conductivity and alkalinity of water.

3.4 *Influence of Circulation Time*

It is also observed from Fig. 4 that the water characteristics change rate is higher for the first 15 h of the circulation time and comparatively became less for the remaining 15 h. Table 1 shows the percentage reduction of TDS, EC, alkalinity, and hardness of water after 30 h of magnetic water circulation in the PVC pipe.

The reduction in water characteristics is higher for initial hours because of MHD affects. This process is going on as the circulation time increases but most of the bonds break in the initial interval of time due to MHD effect and maximum physicochemical properties of the solution changed in that period.

3.5 *Influence of Flow Rate*

Figure 4a–d is plotted between water characteristics and magnetic circulation time for different flow rates (3, 5, and 7 L/min). The figure shows that the flow rate is inversely proportional to the above water characteristics. The average reduction in TDS, EC, hardness, and alkalinity of water on increasing the flow rates is 21.87%, 28.45%, 40.17%, and 42.94%, respectively, while at 7 L/min flow rate, the change in above water characteristics is 25.17%, 29.83%, 42.79%, and 44.85%, respectively. These changes are occurring due to the Lorentz force. The Lorentz force is defined by the following equation [1]:

$$|F_L| = q|v \times B| = qvB \sin\theta \quad (1)$$

The experiment shows that the flow is perpendicular to the magnetic field, and q and B are fixed. So on increasing the velocity of particle, the Lorentz force increases, which increases the magnetic field effect. The above statement can be justified by Saksono et al. [32] who related the effect of flow rate with Lorentz force. They stated that the increase of Lorentz force directly proportional to the increase of velocity of moving particles. Therefore, on increasing the flow rate, the magnetization effect also increases. Alimi et al. [4] also analyzed that the effectiveness of the magnetic field on $CaCO_3$ precipitation was increased with the flow rates increased.

4 Conclusion

In the present study, the hard water is circulated in different materials of pipe (PVC, GI, and copper) and at different flow rates (3, 5, and 7 L/min). After 30 h of circulation, it is concluded that

(1) The magnetic treatment enhances the aragonite crystals formation, which is less adhesive than calcite crystals.
(2) Magnetic field reduces the TDS, EC, hardness, and alkalinity of the water and it goes on decreasing, on increasing the circulation time of water.
(3) The reduction rate of water characteristics like TDS, EC, hardness, and alkalinity is higher for the first 15 h in comparison with the remaining 15 h of circulation time.
(4) Flow rate is inversely proportional to the water characteristics (TDS, EC, hardness, and alkalinity), means on increasing the flow rate from 3 to 7 L/min, and the TDS, EC, hardness, and alkalinity decreases at a faster rate.
(5) At 7 L/min, the TDS reduction rate in PVC pipe (25.17%) is higher than GI (23.91%) and copper (20.76%) pipe.
(6) For magnetic treatment, the PVC pipe is much effective than GI and copper pipe.

References

1. Hui, F., & Lédion, J. (2002). Evaluation methods for the scaling power of water. *Journal Europeen D Hydrologie, 33*(1), 55–74.
2. Gabrielli, C., Jaouhari, R., Maurin, G., & Keddam, M. (2001). Magnetic water treatment for scale prevention. *Water Research, 35*(13), 3249–3259.

3. Lin, L., Jiang, W., Xu, X., & Xu, P. (2020). A critical review of the application of electromagnetic fields for scaling control in water systems: Mechanisms, characterization, and operation. *npj Clean Water, 3*(1), 1–19.
4. Alimi, F., Tlili, M., Amor, M. B., Gabrielli, C., & Maurin, G. (2007). Influence of magnetic field on calcium carbonate precipitation. *Desalination, 206*(1–3), 163–168.
5. Joshy, N., & Meera, V. (2020). Scale control on pipe materials: A review. In *Green Buildings and Sustainable Engineering* (pp. 421–429). Springer.
6. Georgiou, D., Bendos, D., Kalis, M., & Koutis, C. (2018). Removal and/or prevention of limescale in plumbing tubes by a radio-frequency alternating electric field inductance device. *Journal of Water Process Engineering, 22*, 34–40.
7. Čolić, M., Chien, A., & Morse, D. (1998). Synergistic application of chemical and electromagnetic water treatment in corrosion and scale prevention. *Croatica Chemica Acta, 71*(4), 905–916.
8. Han, Y., Zhang, C., Zhu, L., Gao, Q., Wu, L., Zhang, Q., & Zhao, R. (2019). Effect of alternating electromagnetic field and ultrasonic on CaCO3 scale inhibitive performance of EDTMPS. *Journal of the Taiwan Institute of Chemical Engineers, 99*, 104–112.
9. Li, J., Liu, J., Yang, T., & Xiao, C. (2007). Quantitative study of the effect of electromagnetic field on scale deposition on nanofiltration membranes via UTDR. *Water Research, 41*(20), 4595–4610.
10. Piyadasa, C., Ridgway, H. F., Yeager, T. R., Stewart, M. B., Pelekani, C., Gray, S. R., & Orbell, J. D. (2017). The application of electromagnetic fields to the control of the scaling and biofouling of reverse osmosis membranes-A review. *Desalination, 418*, 19–34.
11. Simonič, M., & Urbancl, D. (2017). Alternating magnetic field influence on scaling in pump diffusers. *Journal of Cleaner Production, 156*, 445–450.
12. Alimi, F., Tlili, M. M., Amor, M. B., Maurin, G., & Gabrielli, C. (2009). Effect of magnetic water treatment on calcium carbonate precipitation: Influence of the pipe material. *Chemical Engineering and Processing: Process Intensification, 48*(8), 1327–1332.
13. Shahryari, A., & Pakshir, M. (2008). Influence of a modulated electromagnetic field on fouling in a double-pipe heat exchanger. *Journal of materials processing technology, 203*(1–3), 389–395.
14. Latva, M., Inkinen, J., Rämö, J., Kaunisto, T., Mäkinen, R., Ahonen, M., Matilainen, J., & Pehkonen, S. (2016). Studies on the magnetic water treatment in new pilot scale drinking water system and in old existing real-life water system. *Journal of Water Process Engineering, 9*, 215–224.
15. Mghaiouini, R., Elaoud, A., Garmim, T., Belghiti, M. E., Valette, E., Faure, C. H., et al. (2020). The electromagnetic memory of water at kinetic condition. *International Journal of Current Engineering and Technology*, 10.
16. Khater, Z., & Ibraheim, M. (2016). Some ecological studies on the impact of magnetic field on the tap water. *Egyptian Journal of Aquatic Biology and Fisheries, 20*(2), 51–60.
17. Bali, M., & Gueddari, M. (2018). The effect of magnetic treatment on the physico-chemical and microbiological characteristics of hard waters. *Separation Science and Technology, 53*(9), 1405–1411.
18. Al-Helal, A., Soames, A., Gubner, R., Iglauer, S., & Barifcani, A. (2018). Influence of magnetic fields on calcium carbonate scaling in aqueous solutions at 150° C and 1 bar. *Journal of Colloid and Interface Science, 509*, 472–484.
19. Amor, H. B., Elaoud, A., Salah, N. B., & Elmoueddeb, K. (2017). Effect of magnetic treatment on surface tension and water evaporation. *International Journal of Advance Industrial Engineering, 5*, 119–124.
20. Xiao, Y., Seo, Y., Lin, Y., Li, L., Muhammad, T., Ma, C., & Li, Y. (2020). Electromagnetic fields for biofouling mitigation in reclaimed water distribution systems. *Water Research, 173*, 115562.
21. Othman, A., Sohaili, J., & Supian, N. S. (2019). A review: Methodologies review of magnetic water treatment as green approach of water pipeline system. *Pertanika Journal of Science and Technology, 27*(1).

22. Tai, C. Y., Chang, M. C., Shieh, R. J., & Chen, T. G. (2008). Magnetic effects on crystal growth rate of calcite in a constant-composition environment. *Journal of Crystal Growth, 310*(15), 3690–3697.
23. Lipus, L. C., Ačko, B., & Hamler, A. (2011). Electromagnets for high-flow water processing. *Chemical Engineering and Processing: Process Intensification, 50*(9), 952–958.
24. Chang, M. C., & Tai, C. Y. (2010). Effect of the magnetic field on the growth rate of aragonite and the precipitation of CaCO3. *Chemical Engineering Journal, 164*(1), 1–9.
25. MacAdam, J., & Parsons, S. A. (2004). Calcium carbonate scale formation and control. *Reviews in Environmental Science and Bio/Technology, 3*(2), 159–169.
26. Doyle, J. D., Oldring, K., Churchley, J., & Parsons, S. A. (2002). Struvite formation and the fouling propensity of different materials. *Water Research, 36*(16), 3971–3978.
27. Busch, K. W., Gopalakrishnan, S., Busch, M. A., & Tombácz, E. (1996). Magnetohydrodynamic aggregation of cholesterol and polystyrene latex suspensions. *Journal of Colloid and Interface Science, 183*(2), 528–538.
28. Johan, S., Fadil, O., & Zularisham, A. (2004). Effect of magnetic fields on suspended particles in sewage. *Malaysian Journal of Science, 23*, 141–148.
29. Zhang, P., Cheng, S., & Guo, B. (2009). Effect of rotating-electromagnetic field on scaling in hard water. In *2009 International Conference on Energy and Environment Technology* (Vol. 1, pp. 614–617). IEEE.
30. El-Kashef, E., El-Shamy, A. M., Abdo, A., Gad, E. A., & Gado, A. A. (2019). Effect of magnetic treatment of potable water in looped and dead end water networks. *Egyptian Journal of Chemistry, 62*(8), 1467–1481.
31. Szcześ, A., Chibowski, E., Hołysz, L., & Rafalski, P. (2011). Effects of static magnetic field on water at kinetic condition. *Chemical Engineering and Processing: Process Intensification, 50*(1), 124–127.
32. Saksono, N., Yuliusman, Y., Bismo, S., Soemantojo, R., & Manaf, A. (2010). Effects of pH on calcium carbonate precipitation under magnetic field. *Makara Journal of Technology, 13*(2), 79–85.

Critical Overview of Reinforcing Sand Using Geocell for Shallow Foundation

K. Anusha Raj, Pragya Sinha, Sanjeev Kumar, and Davinder Singh

Abstract Geocells are the non-planar form of geosynthetics which is used to reinforce weaker soils. These are interconnected cells which forms like a mattress that confines the soil within the individual cells, enabling the soil–geocell system to bear higher loads with minimum settlement values. Several parameters that affect the performance of the geocell are—height of the geocell, pocket size of the cell, characteristics of the infill material, stiffness of the geocell, and location of placement. Researchers have conducted experimental as well as numerical analysis to understand the working mechanism of the geocells under static as well as dynamic loading. Finite element modeling has been used to simulate the behavior of these materials in various applications such as in embankments and foundations. It has been found that the inclusion of geocells can substantially improve the bearing capacity of soil to 7 times that of unreinforced soil and considerably reduce the settlement to optimum values. The objective of the paper is to identify the optimum values of various parameters of geocell which gives suitable results under static loading conditions. The optimum values for u/B and h/B were found to vary between 0.05–0.1 and 0.4–2, respectively. Moreover, with the addition of geogrid at the base of geocell mattress, further improvements in load-bearing capacity of the soil can be obtained.

Keywords Geocell · Static loading · Dynamic loading · Bearing capacity · Settlement

1 Introduction

Geocells are recent advancements made in the field of geotechnical engineering used for soil stabilization and finds their applications in slope stabilization, reinforcing pavements, and embankments. Geocells are three-dimensional structures made of

K. Anusha Raj (✉) · P. Sinha · S. Kumar · D. Singh
Dr BR Ambedkar National Institute of Technology Jalandhar, Jalandhar 144011, India

D. Singh
e-mail: singhdj@nitj.ac.in

A. K. Gupta et al. (eds.), *Advances in Construction Materials and Sustainable Environment*, Lecture Notes in Civil Engineering 196,
https://doi.org/10.1007/978-981-16-6557-8_21

planar forms of geosynthetics such as geotextiles or geogrids. They consist of honeycomb cells that provide lateral confinement to the soil and uniformly distribute the applied load to a wider area, thus improving its bearing capacity and reducing settlement to a permissible limit. There is a large improvement in the properties of the soil when it is reinforced with geocell, in place of planar forms of geosynthetic. Even when two or more layers are provided, the soil sandwiched in between the layers will squeeze out when it overcomes the frictional resistance, whereas geocell on the other hand gives sturdy confinement for the soil in between the cells enabling it to perform as a singular unit even at large footing settlements and thereby increasing the bearing capacity of the reinforced system. Figure 1 shows the basic working mechanism of the geocell [1]. The vertical confinement in geocells is achieved in two ways—due to the frictional resistance between the geocell and the geocell wall and the geocell acts as a mattress by restricting the movement of the soil beneath the geocell upwards to the loading area thus by reducing the heave.

Geocells made from geotextiles due to their low stiffness values and high flexibility are incapable of confining the soil which may result in lateral deformation. This causes geocell to fail even before attaining the failure strain. Geocells made of geotextiles are therefore not usually employed to improve the load-bearing capacity [3]. While comparing the efficiency of the geocell and geotextile as reinforcing material and it was seen that, for the same settlement value of 4%, the improvement in the bearing capacity ratio was 2.73 for geocell whereas it was 1.88 for geotextile [4]. It is proved that geocell fabricated from a geogrid shows a better performance in terms of the settlement, surface deformation also improved the bearing capacity improvement factors [5]. Though geocells are superior to the planar forms of geosynthetics, there is a minor setback in terms of the compatibility of the geocell with the soil.

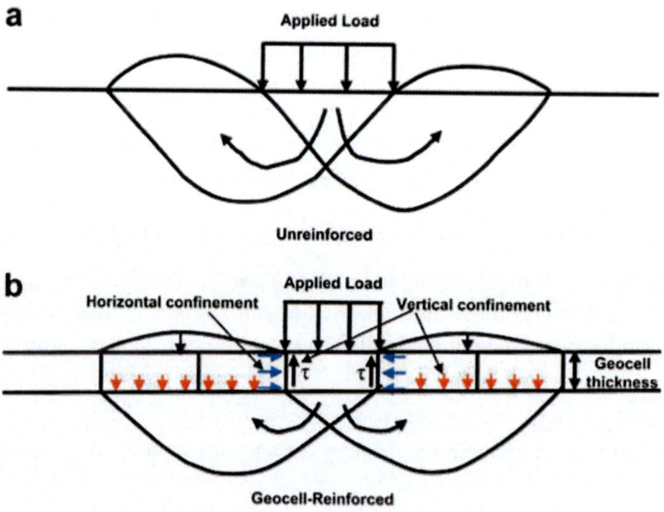

Fig. 1 Figure shows the mechanism of the geocell reinforcement [2]

Considering the case where there is an excessive settlement of the footing, strain mobilization occurs in the geocell mattress which causes the mattress to rupture. This happens because the geocell cannot withstand higher values of strain unlike the planar forms [3]. In a study, it was noticed that the ratio of the aperture size of the geogrids to the maximum aggregate size should be between 0.7 and 1.35 [2].

Nomenclature

u	depth of first layer of geocell from the base of footing
h	spacing between the layers of geocell
b	width of geocell
B	width of footing
BCR	bearing capacity ratio
q_u	bearing capacity of reinforced soil
q_0	bearing capacity of unreinforced soil
I_f	improvement factor
PRS	percentage reduction in settlement
Dr	depth of geocell layer
β	load dispersion angle
D_{50}	mean diameter of soil
D_{max}	maximum aggregate size
ϕ	angle of friction
Ψ	angle of dilation

2 Key Factors Influencing Efficiency

Various researchers have studied the effects of geocell–soil interactions under static loading both experimentally and numerically. From previous studies, the values for optimum placement of the cell, aperture size to be used, the height of the cell to be used, and many other parameters have been studied. Improvement in the bearing capacity of the soil is quantified by a dimensionless entity BCR, which is the ratio of the ultimate bearing capacity of the reinforced soil to that of unreinforced soil.

$$\mathrm{BCR} = \frac{q_u}{q_0}$$

BCR values are dependent on several factors—the type of geocell used, type of infill material used, placement depth, and width of the geocell—that affect the BCR of the reinforced system. Another parameter used to quantify the effectiveness of the load-carrying capacity is I_f which is the ratio of the bearing pressure of the reinforced soil at a value of the settlement to the same in unreinforced soil.

$$I_f = \frac{q_r}{q_0}$$

To understand the settlement reduction, another entity is used which is the percentage reduction in the settlement.

$$\text{PRS} = \frac{S_o - S_r}{S_o} \times 100$$

The optimum values for the selection and placement of the geocell and the infill materials are based upon the values produced when these parameters show the highest values.

2.1 *Effect of Characteristics of Geocell*

Various characteristics like height, pocket size, stiffness modulus, etc., of the geocell, greatly affect the soil–geocell interaction and thereby affect the reinforcing efficiency. Chevron pattern gives a better performance as a geocell mattress compared to diamond pattern [4]. A geocell mattress with smaller heights a shallow beam that deflects the under-loading, whereas a geocell mattress with larger heights behaves like a deep beam that evenly distributes the load to the underlying soil. Thick geocells, however, pose an issue of bending of the geocell wall which is not a desirable outcome, and therefore, geocell with optimum height should be employed for reinforcing purposes.

The aspect ratio of the geocell should be such that it effectively confines the soil within the cells and helps in the uniform load transfer. It is only sensible that an increase in the stiffness of the geocell would help to bear higher loads. Nevertheless, geocells with remarkably high stiffness will not be compatible with the soil and can also buckle and failure can occur very easily. The width of the geocell mattress to be laid is also an important factor for the practical application of the geocell. With the increase in the width of the mattress, the soil–geocell structure acts as a shallow beam, and there is a uniform distribution of the load applied from the above structures (Fig. 2).

Fig. 2 Vertical stress dispersion mechanism [6]

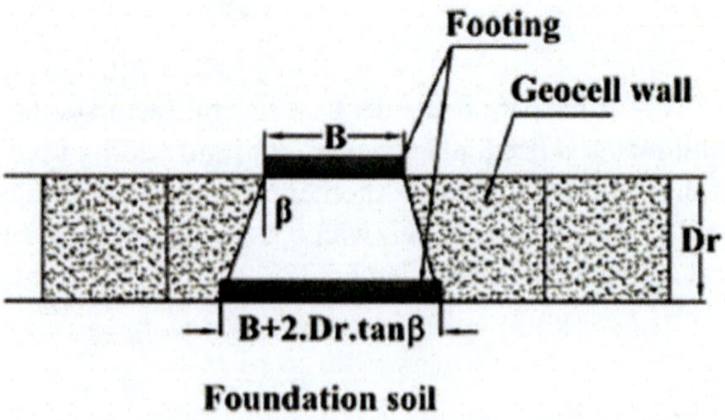

Table 1 Major characteristics of geocells

Characteristics of geocell	Notation	Optimum value	References
Depth of placement	u/B	0.05–0.1	[7] [8] [9]
Height	h/B	0.4–2	[10] [11]
Aspect ratio	a/D	1–1.67	[4] [12] [13]
Stiffness modulus	k_g	100–200 kN/m	[14]
Width of reinforcement	b/B	4–5	[4] [11]
Pocket size	d/D	0.8–1.0	[13]

Major parameters that affect the selection of the type of geocell to give optimum results are described in Table 1.

The optimum values can only be shown as a range because, the material used for geocells, type of footing used, type of soil is different in each study.

2.2 *Effect of the Soil*

It was seen that, to bear maximum BCR value, the ratio of the cell's size (b) to the mean diameter of the sand (D50) should be 15. In the case of larger backfill material, the backfill–geocell interaction gets disturbed due to rupture of the sand particle whereas, less soil interaction will be less when the ratio is less than 15. Both the cases result in lesser BCR value. Also, the ratio of cell's size and maximum aggregate size (Dmax) should be in the range 7–11. It is observed that for an optimum BCR value, the ratio of the size of the plate (B) and the mean diameter of the soil (D50) should be in the range of 13–27, with an approximate value of 20 [15].

Considering the infill materials, the preferred material is sand due to its soil particle interaction thus confinement effect is more seen in granular sand. It was also seen that other locally available materials can also be used as infill material [14]. This is pointed out that the locally dredged materials, or quarry waste, steel slag, etc., can be used as infill material [16]. The angle of friction (φ) and the dilatancy angle (ψ) also plays an important role in the reinforcing purpose. As the angle of internal friction increases, there is an increase in the confinement characteristics of the soil. The study showed that by increasing φ from 30° to 40°, the ultimate load-carrying capacity has improved by 1.15 times [17].

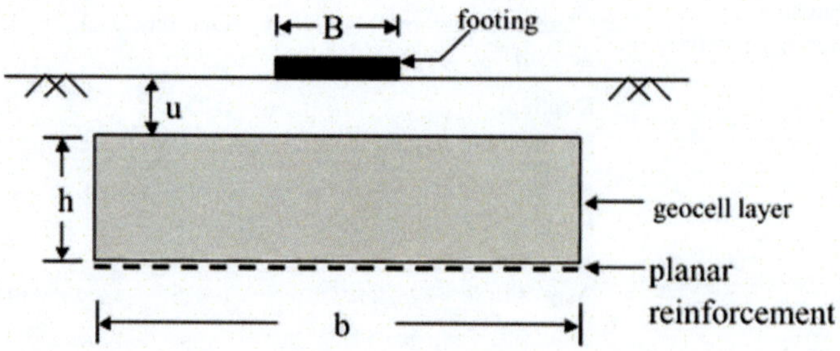

Fig. 3 Sectional view of geocell–basal planar reinforcement in the shallow foundation [10]

2.3 *Effect of Additional Planar Reinforcement*

For the further improvement of the geocell reinforced soil, additional planar reinforcement of either geotextile or geogrids is provided [10, 18]. Additional reinforcement can be either placed at the top or the bottom of the geocell mattress. Planar reinforcement placed above the mattress has little effect on the load-bearing capacity because of the pull-out behavior of the reinforcement above the mattress. Beneficial effects have been seen from the planar reinforcement placed in the bottom of the geocell mattress. This assembly is beneficial as, for a geocell mattress under load, there are chances of the infill materials seep downwards due to the footing penetration. This downward movement can be arrested by placing a planar reinforcement, preferably a geogrid.

The beneficial effect of this additional reinforcement is twofold—downward movement of the infill material can be avoided, and this reduces the lateral movement of the infill material, secondly, and geogrids will ensure a uniform distribution of the pressure to the underlying soil. The efficiency of this basal reinforcement is dependent on the height of the geocell. In the case of a circular footing with geocell reinforcement which is placed on a clay layer and infilled with sand which has a relative density of 70%, the bearing capacity of the reinforced system was found to be seven times the unreinforced case [11]. In the case of clayey soils, the addition of basal geogrid gives optimum values when $c_u \leq 15$ kPa for shorter geocells with $h/D \leq 1.05$ [18] (Fig. 3).

3 Numerical Analysis

Numerical analysis of the geocell is done using finite element analysis. Some of the software which uses the finite element method are—PLAXIS 3D, FLAC 3D, GEOFEM, and ABAQUS. These software are mainly used to model and simulate the conditions in the field and study the soil–geocell interactions (Fig. 4).

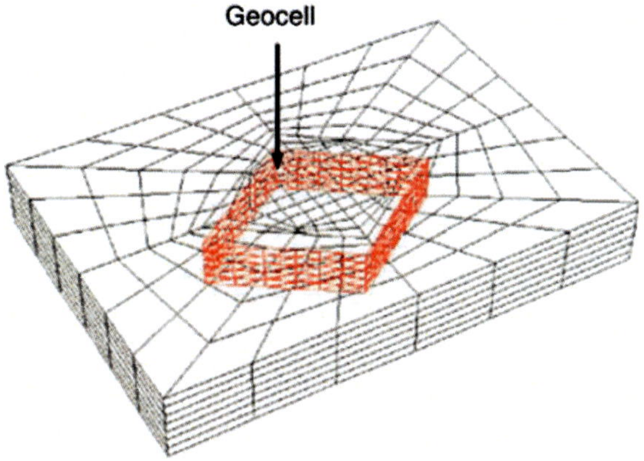

Fig. 4 Figure shows the 3D view of a single cell geocell in FLAC 3D [19]

Full model experiments are not always possible in laboratory experiments and so, due to the scale effect, the results produced in the laboratory need not translate to the field. The advantage of numerical analysis is that the in situ characteristics can be modeled also expected loading values can also be applied, and the results produced from the analysis can give the approximate expected values from the field. The interaction of the geocell and the soil can be thoroughly analyzed. These software packages help the modeling of foundations as well as, tunnels, anchor plates, and embankments can also be easily modeled and studied at different conditions. Analysis of dynamic loading conditions is easier by using the software. Various researchers have used numerical analysis to validate their experimental results (Table 2).

4 Discussion and Conclusion

From the study, it can be deduced that geocells are very effective in stabilizing weak soil. When the geocell is reinforced with additional planar reinforcement, it gives better results owing to the extra "membrane effect" by the basal geogrid, and the BCR value can even reach up to 7. That is the strength of the reinforced system is seven times stronger than the unreinforced soil. Several factors affect the efficiency of the geocell reinforced soil which when used under optimum conditions can give sturdy soil which can carry large loads with very small settlement values.

The numerical analysis with the application of the geocells can be simulated with the help of different software packages, and the results from it have been shown to agree with the experimental results. Software packages that are used for the modeling of the geocell reinforced structure do not limit themselves to the reinforcement of the footings but had also is helpful in slope stabilization, safe placement of anchor plates, and supporting a variety of other structures. Even with the large-scale applicability of the geocell reinforcement theoretically, and with the backup of numerical and experimental results, the field application of the geocells is limited.

Table 2 Review of related research work

References	Software used	Material model used for soil	Objective	Result
[20]	PLAXIS 3D	Hardening soil model	Uplift behavior of square shallow anchor plate	Optimum values Plate anchor size = 500 mm Geocell stiffness = 1000 Kn/m $u/B = 0$ $h/B = 0.75$
[21]	PLAXIS 3D	Mohr–Coulomb model	Analysis of three-dimensional shell foundation	Conical foundation showed (i) $I_f = 3.5$ at 14% settlement (ii) PRS = 80%
[22]	ABAQUS	Mohr–Coulomb model	Effect of geocell reinforcement in embankment	Smaller settlement and stress under embankment than without
[23]	ABAQUS	Mohr–Coulomb model	Contact pressure distribution in geocell reinforcement	Contact pressure on subgrade soil reduced to a great extent and soil remain undisturbed until large load values
[24]	ABAQUS	Mohr–Coulomb model	The factor of safety using strain reduction method when geocell is employed	The factor of safety found to be 3 $\beta =$ 45° and 75°
[25]	FLAC 3D	Mohr–Coulomb model	Machine foundation reinforced with geocell	61% displacement amplitude reduction
[7]	PLAXIS 2D	Mohr–Coulomb model	Machine foundation reinforcement in comparison with geogrids	Geocells effective than geogrids C_u increased by 2.7 times with a 40% reduction in peak particle velocity

References

1. Pokharel, S. K., Han, J., Leshchinsky, D., Parsons, R. L., & Halahmi, I. (2010). Investigation of factors influencing behavior of single geocell-reinforced bases under static loading. *Geotextiles and Geomembranes, 28*(6), 570–578. https://doi.org/10.1016/j.geotexmem.2010.06.002
2. Palmeira, E. M., & Góngora, I. A. G. (2016). Assessing the influence of some soil-reinforcement interaction parameters on the performance of a low fill on compressible subgrade. Part I: Fill performance and relevance of interaction parameters. *International Journal of Geosynthetics and Ground Engineering, 2*(1), 1–17. https://doi.org/10.1007/s40891-015-0041-3

3. Kargar, M., & Mir Mohammad Hosseini, S. M. (2018). Influence of reinforcement stiffness and strength on load-settlement response of geocell-reinforced sand bases. *European Journal of Environmental and Civil Engineering, 22*(5), 596–613. https://doi.org/10.1080/19648189.2016.1214181.
4. Tafreshi, S. N. M., & Dawson, A. R. (2010). Comparison of bearing capacity of a strip footing on sand with geocell and with planar forms of geotextile reinforcement. *Geotextiles and Geomembranes, 28*(1), 72–84. https://doi.org/10.1016/j.geotexmem.2009.09.003
5. Dash, S. K., Sireesh, S., & Sitharam, T. G. (2003). Behaviour of geocell-reinforced sand beds under circular footing. *Proceedings of the Institution of Civil Engineers—Ground Improvement, 7*(3), 111–115. https://doi.org/10.1680/grim.2003.7.3.111
6. Sitharam, T. G., & Hegde, A. (2013). Design and construction of geocell foundation to support the embankment on settled red mud. *Geotextiles and Geomembranes, 41*, 55–63. https://doi.org/10.1016/j.geotexmem.2013.08.005.
7. Beds, S. (2018). Numerical analysis of machine foundation resting on the geocell reinforced soil beds.
8. Dash, S. K., Krishnaswamy, N. R., & Rajagopal, K. (2001). Bearing capacity of strip footings supported on. *Geotextiles and Geomembranes, 19*(4), 235–256.
9. Oliaei, M., & Kouzegaran, S. (2017). Efficiency of cellular geosynthetics for foundation reinforcement. *Geotextiles and Geomembranes, 45*(2), 11–22. https://doi.org/10.1016/j.geotexmem.2016.11.001
10. Dash, S. K., Rajagopal, K., & Krishnaswamy, N. R. (2001). Strip footing on geocell reinforced sand beds with additional planar reinforcement. *Geotextiles and Geomembranes, 19*(8), 529–538. https://doi.org/10.1016/S0266-1144(01)00022-X
11. Dash, S. K., Sireesh, S., & Sitharam, T. G. (2003). Model studies on circular footing supported on geocell reinforced sand underlain by soft clay. *Geotextiles and Geomembranes, 21*(4), 197–219. https://doi.org/10.1016/S0266-1144(03)00017-7
12. Latha, G. M., & Murthy, V. S. (2007). Effects of reinforcement form on the behavior of geosynthetic reinforced sand. *Geotextiles and Geomembranes, 25*(1), 23–32. https://doi.org/10.1016/j.geotexmem.2006.09.002
13. Kumar, S., Sahu, A. K., & Naval, S. (2019). Performance of circular footing on expansive soil bed reinforced with geocells of chevron pattern. *Civil Engineering Journal, 5*(11), 2333–2348. https://doi.org/10.28991/cej-2019-03091415
14. Madhavi Latha, G., & Rajagopal, K. (2007). Parametric finite element analyses of geocell-supported embankments. *Canadian Geotechnical Journal, 44*(8), 917–927. https://doi.org/10.1139/T07-039.
15. Tavakoli Mehrjardi, G., Behrad, R., & Moghaddas Tafreshi, S. N. (2019). Scale effect on the behavior of geocell-reinforced soil. *Geotextiles and Geomembranes, 47*(2), 154–163. https://doi.org/10.1016/j.geotexmem.2018.12.003.
16. Pokharel, S. K., Han, J., Leshchinsky, D., & Parsons, R. L. (2018). Experimental evaluation of geocell-reinforced bases under repeated loading. *International Journal of Pavement Research and Technology, 11*(2), 114–127. https://doi.org/10.1016/j.ijprt.2017.03.007
17. Halder, K., & Chakraborty, D. (2020). Influence of soil spatial variability on the response of strip footing on geocell-reinforced slope. *Computers and Geotechnics, 122*, 103533. https://doi.org/10.1016/j.compgeo.2020.103533.
18. Biswas, A., & Krishna, A. M. (2018). Behaviour of geocell-geogrid reinforced foundations on clay subgrades of varying strengths. *International Journal of Physical Modelling in Geotechnics, 18*(6), 301–314. https://doi.org/10.1680/jphmg.17.00013
19. Han, J., Yang, X., Leshchinsky, D., & Parsons, R. L. (2008). Behavior of geocell-reinforced sand under a vertical load. *Transportation Research Record, 2045*, 95–101. https://doi.org/10.3141/2045-11
20. Yünkül, K., Usluoğulları, Ö. F., & Gürbüz, A. (2021). Numerical analysis of geocell reinforced square shallow horizontal plate anchor. *Geotechnical and Geological Engineering*. https://doi.org/10.1007/s10706-021-01679-1

21. Ari, A., & Misir, G. (2021). Three-dimensional numerical analysis of geocell reinforced shell foundations. *Geotextiles and Geomembranes*. https://doi.org/10.1016/j.geotexmem.2021.01.006.
22. Khalaj, O., Nejad, S. A., & Jenicek, S. (2020). The effect of geocell reinforced embankment construction on the behaviour of beneath soil layers using numerical analysis. *IOP Conference Series: Earth and Environmental Science, 609*(1). https://doi.org/10.1088/1755-1315/609/1/012015.
23. Dash, S. K., Saikia, R., & Nimbalkar, S. (2019). Contact pressure distribution on subgrade soil underlying geocell reinforced foundation beds. *Frontiers in Built Environment, 5*. https://doi.org/10.3389/fbuil.2019.00137.
24. Arvin, M. R., Zakeri, A., & Bahmani Shoorijeh, M. (2019). Using finite element strength reduction method for stability analysis of geocell-reinforced slopes. *Geotechnical and Geological Engineering, 37*(3), 1453–1467. https://doi.org/10.1007/s10706-018-0699-0.
25. Venkateswarlu, H., Ujjawal, K. N., & Hegde, A. (2018). Laboratory and numerical investigation of machine foundations reinforced with geogrids and geocells. *Geotextiles and Geomembranes, 46*(6), 882–896. https://doi.org/10.1016/j.geotexmem.2018.08.006

Road Bridges Across Cooum and Adyar Rivers in Chennai City—Need for Structural Health Monitoring

A. Rose Enid Teresa, S. Stella, M. Goutham Priya, P. Gajalakshmi, and J. Revathy

Abstract Bridges are the lifelines for public, and most of the road bridges in India were not designed for the vehicular loads and their speed that they are currently subjected to. As construction of new bridges involves Nation's huge economy, structural health monitoring of the existing bridges will help in terms of safety and economy. In Chennai City, there are many old road bridges, once functional, and are critical life lines, which are in need of SHM for the safer public usage. This paper expounds on the need for SHM on road bridges built across Cooum and Adyar Rivers in various zones, along with a case study on a RCC road bridge constructed across Cooum River. A live load testing was conducted on a single span of the bridge to study the flexural responses of the girders. The strain histories obtained from the analytically investigated model of the loaded bridge were compared with the field measured strain data, and necessary improvements were done on the model to measure the effective stiffness property of the girders. This investigation has helped in providing recommendations on retrofitting applications of the bridge components ensuring public safety and the continuous serviceability. Thus, SHM of the road bridges will increase the life of the structure and serving public in a longer run.

Keywords SHM · Bridges · Live load test · Non-destructive test

A. Rose Enid Teresa (✉) · M. Goutham Priya
Rajalakshmi Engineering College, Chennai, Tamilnadu, India

S. Stella
Karunya Institute of Technology and Sciences, Coimbatore, Tamilnadu, India

P. Gajalakshmi · J. Revathy
B.S. Abdur Rahman Crescent Institute of Science & Technology, Chennai, Tamilnadu, India
e-mail: gajalakshmi@bsauniv.ac.in

J. Revathy
e-mail: revathyj@crescent.education

A. K. Gupta et al. (eds.), *Advances in Construction Materials and Sustainable Environment*, Lecture Notes in Civil Engineering 196,
https://doi.org/10.1007/978-981-16-6557-8_22

1 Introduction

Roadways and railways are the most important means of transportation in India. Most of the bridges are designed to serve the public for a longer duration of time such as 100 years. After 1950, both traffic volume and ageing of the bridge structures which have increased by 9.1% per year and have played a significant role in their deterioration. Due to the growing population, the traffic congestion has reached a chronic stage in metropolitan cities such as Chennai, Bangalore, Mumbai, Delhi and Kolkata. Government has taken initiatives in constructing many new road and rail bridges to meet the needs of the growing population, and huge investment is spent on it. As huge economy of the nation is involved in these constructions, assessing the current structural health condition of the bridges is essential.

Structural health monitoring (SHM) targets at monitoring the current structural conditions and performance to prevent catastrophic failure. SHM includes traditional methods as well as advanced techniques for various infrastructures like building, bridges and towers. There remain numerous technical challenges, which involves the deployment of SHM systems, deciding the information required, developing strategy to deliver information and to interpret the information for decision-making [24]. According to highways policy note, Tamil Nadu State has 8639 major and minor bridges. About 8600 major and minor bridges in the state are in different stages of deterioration. For the first time, the sensor-based SHM system has been installed on Ekkattuthangal bridge along Adyar River on Jawaharhal Nehru Salai. The quality assurance and research wing of state highways have been entrusted with the task of collecting the data under Tamil Nadu Innovation Initiative (TANII) scheme (Express News Service, 2019).

Nationwide, there are initiatives taken by researchers in assessing the heath condition of the structures, listing few: Sathyanarayanan et al. [21] described the importance of remote structural monitoring (RSM) in SHM and the features developed at SERC, Chennai, India. Sharma and Mehta [3] dealt with the study of the inspection methods using high resolution digital image. Umesh et al. [23] predicted the location of the damage and its severity in an existing bridge structure by electro-mechanical impedance (EMI) technique proving its efficiency. Rose Enid Teresa et al. [19] conducted a live load testing on the reinforced cement concrete simply supported bridge to measure the flexure response using sensors and data acquisition system, and the reports were submitted to the authorities for decision-making.

Internationally, SHM has gained interest and there is a worldwide demand for its advances in information technologies and their implications in SHM for bridges and civil infrastructures [8–10, 16, 17, 22, 25]. Sensor arrays were instrumented to monitor the ambient vibration, strain and displacement of numerous structures and bridges worldwide [15].

Peters and Inaudi [18] employed fibre optic sensors for SHM. Isabela et al. [13] incorporated a tunable piezoelectric circuitry to measure the impedance, thus proposing an improved impedance-based damage identification method. Wyczalek and Wyczalek [12] used photogrammetric approach on selected bridge structures.

Jamali et al. [14] conducted an innovative vibration-based damage prediction methods on timber bridges and cable stayed bridges. Australian Network of Structural Health Monitoring (ANSHM) [2] was established in 2009 to promote and advance the field of SHM in Australia. The contributions include reliability-based load-carrying capacity assessment of bridges using SHM and non-linear analysis [14], innovative vibration-based damage identification methods with applications to cable-stayed, steel-truss or timber bridges [1, 2, 6].

A state-of-the-art review on various technical updates in SHM and the application of wireless sensor network (WSN) system and wireless smart sensor network (WSSN) system in SHM was given by Rose Enid Teresa et al. [20]. Kedare and Mundada [5] conducted a study employing electro mechanical impedance (EMI) technique to monitor the health of the component and damage occurred to automobile components on timely basis to avoid accidents.

This paper focusses on the study on the need of SHM of the road bridges across Cooum River (zone V, VIII, IX and XI in various wards) and road bridges across Adyar River (zone X and XIII in various wards) in Chennai City, Tamil Nadu, India. There remain numerous technical challenges, which involves the deployment of SHM systems, deciding the information required, developing strategy to deliver information and to interpret the information for decision-making. Most of the bridges in the specified zones/wards have become the life lines of transport for public nowadays, and the recommendations on retrofitting of the bridge components/structures from SHM report increases the durability of the structures.

2 Location of Bridges and Basis of Selection

This paper focusses on the study of the need of SHM on road bridges built across Cooum River (zone V, VIII, IX and XI in various wards) and road bridges across Adyar River (zone X and XIII in various wards) along with a case study. Listing few, bridges built across Cooum River in zone V, naming few, Binny Bridge was built in 1825, Anderson Bridge was built in 1829, Harris Bridge (now Adhithanar Bridge) was built in 1855, etc. in zone VIII, namely Anna Nagar Bridge, Naduvankarai bridge, in zone IX, Napier Bridge which was built in 1869, a bridge built across Coovum river in zone XI, connecting Maduravoyil road and chinna Nolambur, in zone XIII, bridge, namely Thiru Vi Ka Bridge, which was inaugurated in 1973 and damaged in November 1985 due to a rush of flood water. The images of the old bridges, namely Adyar Bridge, Kakkan Bridge and Napier Bridge, are shown in Figs. 1, 2 and 3.

Most of the bridges across the rivers were built before 100 years. It is the need of the hour to detect the damages or deteriorations in the bridge structures subjected to operational loading, accidental impacts, wind speed, ground motions, effects on the bridges due to ageing, natural calamities, such as cyclones, tsunami, flooding, changes in temperature and relative humidity, and to have an intensive study on the same.

Fig. 1 Adyar river bridge (courtesy: Google)

Fig. 2 Kakkan bridge (courtesy: Google)

Fig. 3 Napier bridge (courtesy: Google)

As per the HSCTC feasibility report, the existing width of most of the river bridges have reduced drastically due to encroachment. The high impact force of large quantum of flood flowing across the reduced river widths on the bridge piers has to be investigated too. The selection of SHM techniques for these bridges is based on the challenges involved in deployment of SHM systems in the bridge locations, information required from the SHM report and for decision-making. The bridges mentioned above are the life lines of most of the public and monitoring their health, thereby rehabilitating the structures ensures the benefits in terms of economy and public usage.

3 Need for SHM

SHM of the bridges is imperative to assess their structural health and integrity and it is intended to:

a. Prepare detailed report on the durability and criticality of the bridge components/structures.
b. Provide information to the stakeholders and the concerned authorities on the damage detection under environmental conditions, design, operation and management of bridges throughout the structures' lifespan.
c. Provide recommendations on retrofitting applications of the bridge components (if needed) for the political decision-making.
d. Ensure public safety and the continuous serviceability of bridges.

4 Case Study—SHM of a RCC Bridge Across Cooum River in Chennai City

A multispan simply supported bridge (Koyambedu Bridge in Chennai which is across Cooum River) of span length 129.7 m shown in Fig. 4 was taken for study. The widened portion which is isolated from the existing structure by an expansion joint was taken for the study. There are eight spans each of 16.21 m span length with longitudinal (four nos) and cross girders (five nos). The superstructure is simply supported by multi-column bents (four columns) over plain elastomeric bearing pads. The bridge was originally designed with M25 concrete and Fe415 steel.

Fig. 4 Photograph of the study bridge

4.1 Instrumentation and Testing Procedures

Strain gauges were mounted on the girders of a span (limited to a single span) in a non-destructive manner, and live load test was conducted to measure their flexural responses. Figures 5a, b and 6a, b show the arrangement of strain gauges in the longitudinal girders and the cross girder and the dimensions of the same.

A multi-axle truck was employed to conduct a live load testing on the strain gauge mounted span of the bridge. The truck was driven on the prescribed longitudinal path.

(a) (b)

Fig. 5 **a, b** Arrangements of strain gauges in the longitudinal girder

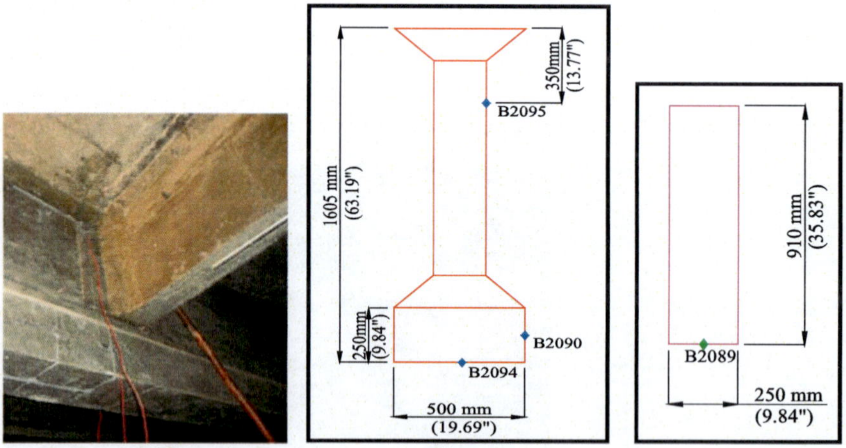

(a) (b) (c)

Fig. 6 **a–c** Arrangements of strain gauges in the girders

Fig. 7 Truck employed for bridge testing

Fig. 8 Auto clicker and reflector arrangement

The truck employed and the auto clicker-reflector arrangement are shown in Figs. 7 and 8.

4.2 Modelling and Analysis Using Wingen and WINSAC Programme

A two-dimensional grid model was modelled using WINGEN [11], a programme to define a planar bridge model. Simulation of the field load testing [4, 7, 11] was done in the 2D model. The strain histories recorded in the field were compared with the programme generated strain history by overlaying both. Calibration of the initial model is done until the strain histories superposed each other. Figures 9 and 10 show the truck dimension, axle weights and 2D finite element model of the bridge. The field measured strain histories were nonlinear for the longitudinal girder and linear for the cross girder.

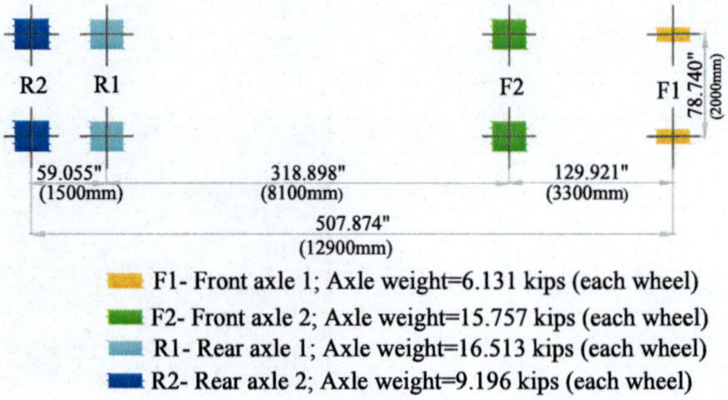

Fig. 9 Axle weights of the truck and truck path

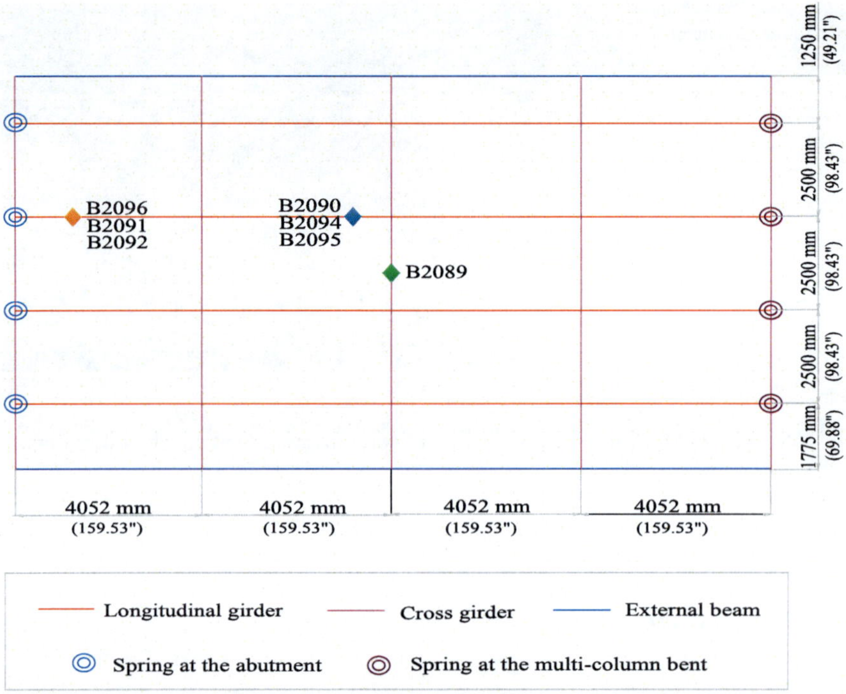

Fig. 10 2D finite element model using WINGEN

4.3 Test Results—Response History Plot

The model which was generated using the WINGEN and was analysed with the structural analysis programme WINSAC [11]. The WINGRF programme is a graphing utility, specifically designed for viewing field measured data from the structural testing system (STS). The field measured data were compared with the responses predicted by the structural analysis programme WINSAC. Data from the STS were imported directly into the WINGRF and viewed as load position-based response histories. The initial model was calibrated by modifying the longitudinal girder stiffness, until the results matched with the values measured in the field, as it indicated non-linear behaviour in the experimental investigation.

4.3.1 Inference Form the Strain History of the Transducers—Results and Discussion

a. B2090 (Mounted at the side of the bottom flange of the longitudinal girder near the midspan). B2090 measured a maximum strain value of 70 micro-strains (Fig. 11). This is due to the provision of lesser concrete cover in the longitudinal girder and due to the ageing effect. The field measured strain history matched well with the computed strain history.
b. B2094 (Mounted at the bottom of the bottom flange of the longitudinal girder near the midspan). Transducer measured a maximum strain value of 90 micro-strains (Fig. 12). The field measured and analytically computed strain history values matched very well.
c. B2091 (Mounted at the side of the bottom flange of the longitudinal girder near the abutment). Measured a maximum strain value of 40 micro-strains (Fig. 13). This is due to the provision of lesser concrete cover in the longitudinal girder and due to the ageing effect.
d. B2092 (Mounted at the bottom of the bottom flange of the longitudinal girder near the abutment). B2092 measured a maximum strain value of 27 micro-strains (Fig. 14). The strain histories indicated a sudden shift in the magnitude, indicating a non-linear behaviour in the member.

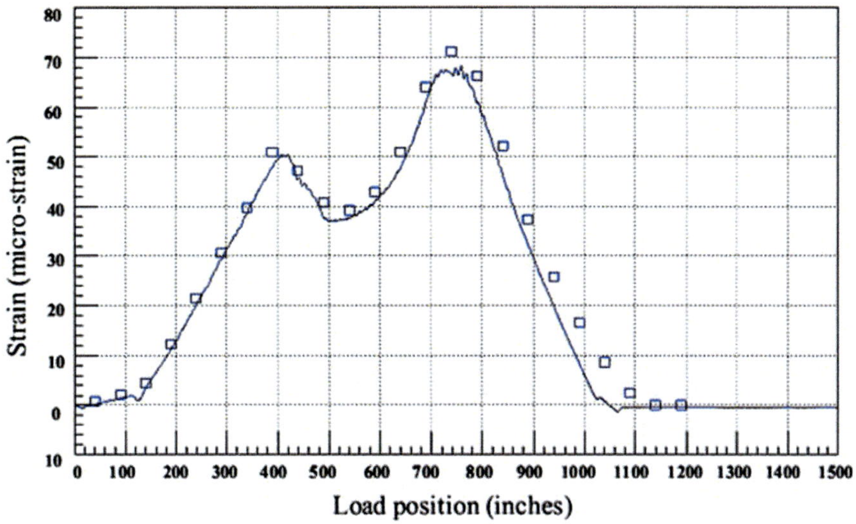

Fig. 11 B2090 strain history

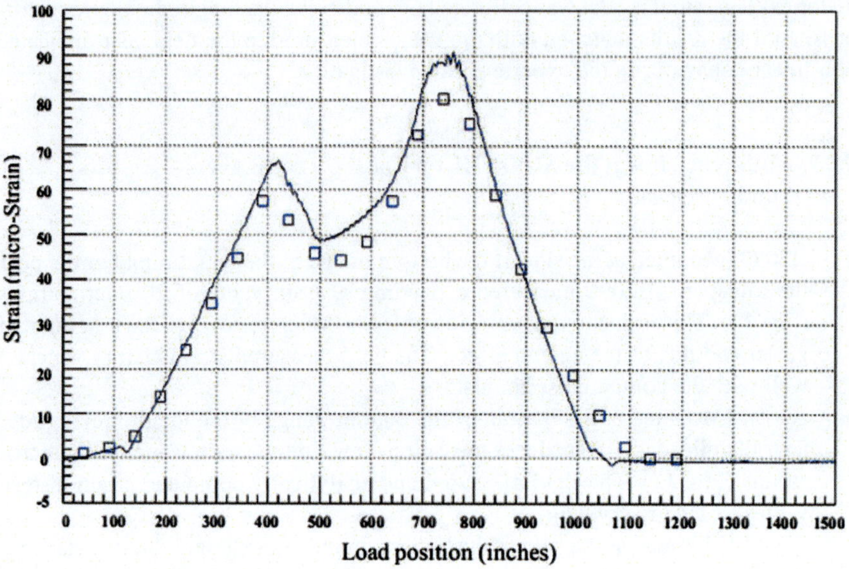

Fig. 12 B2094 strain history

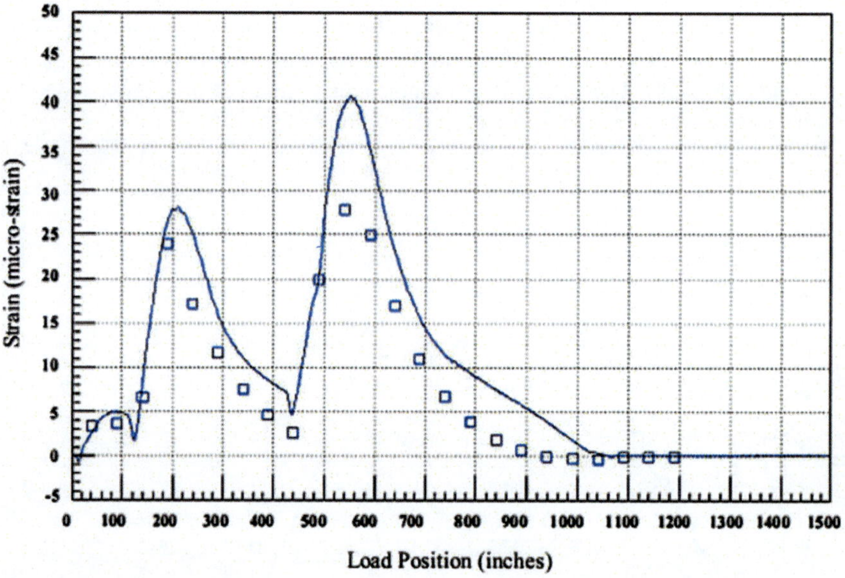

Fig. 13 B2091 strain history

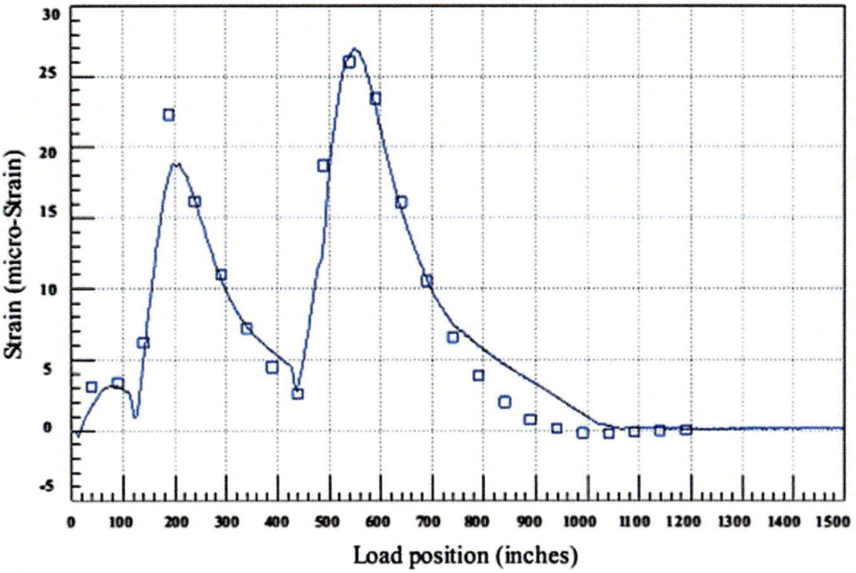

Fig. 14 B2092 strain history

e. B2095 (Mounted at the top of the longitudinal girder near the midspan). In the field measured data, there were a sudden shift in magnitude (sudden shift of the neutral axis) under the wheel load (Fig. 15). Since this type of behaviour cannot be modelled, gauges that display such irregular shapes were not included in the model correlation.
f. B2089 (At the bottom of the cross girder at the midspan). The strain values measured by the transducer exhibited a linear behaviour when the truck moved across the bridge (there was no sudden shift in the magnitude) (Fig. 16). When the rear axles moved over the position of the transducer, there was an increase in the strain as the rear axle carries a greater load, when compared with the front axle.

5 Conclusion

SHM of the bridges plays a vital role in assessing their behaviour in the current scenario. SHM includes traditional methods as well as advanced techniques which includes like impedance-based, non-destructive evaluation using vibration signature, limit strain measurement, data fusion method, inverse method, etc. for various infrastructures like building, bridges and towers. In Chennai City, there are many old road bridges, once functional, and are critical life lines, which needs health monitoring for the safer public usage. This paper focusses on the need for SHM on most important

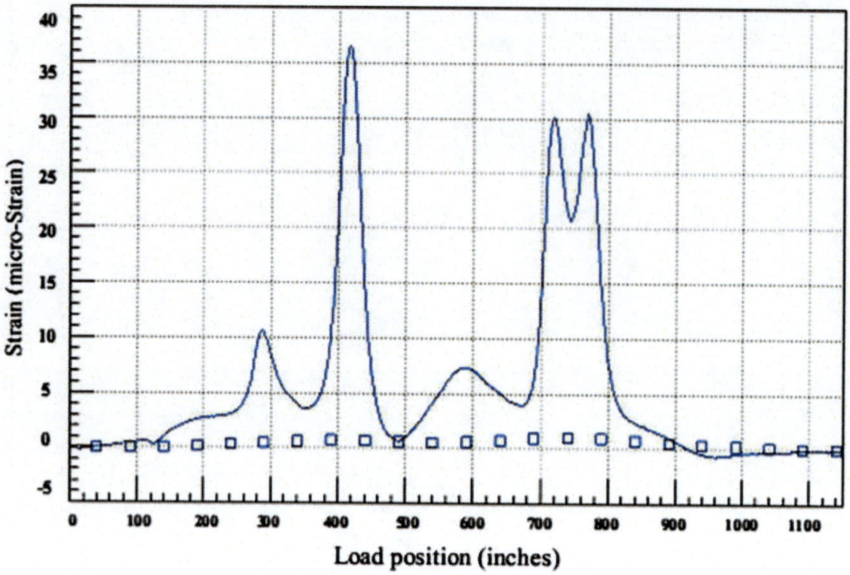

Fig. 15 B2095 strain history

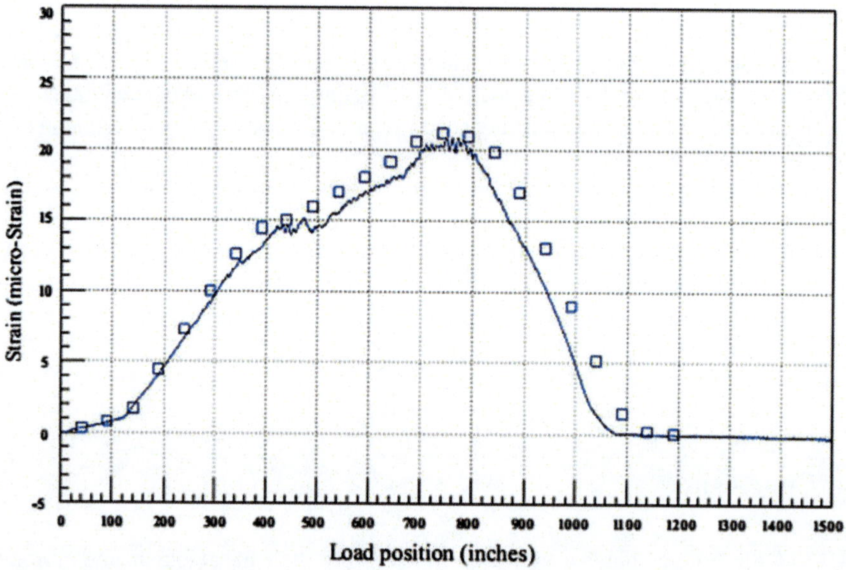

Fig. 16 B2089 strain history

road bridges which were built across Cooum and Adyar Rivers in various zones. This paper also includes a case study which was conducted on a RCC road bridge constructed across Cooum River. A live load testing was conducted on a single span of the bridge to study the flexural responses of the girders. Field measured strain history was nonlinear for the longitudinal girder and linear for the cross girder. The strain histories obtained from the analytically investigated model of the loaded bridge were compared with the field measured strain data, and necessary improvements were done on the model to measure the effective stiffness property of the girders. Thus, the effective stiffness (EIeff) property of the longitudinal girders was found to be 80% of the gross stiffness. An intense study with this finding on a single span has helped in providing recommendations on retrofitting applications ensuring public safety and the continuous serviceability of bridges. Thus, SHM of the road bridges will increase the life of the structure and serving public in a longer run.

References

1. Alamdari, M. M., Dang Khoa, N. L., & Wang, Y., et al. (2018). A multiway data analysis approach for structural health monitoring of a cable-stayed bridge. *Structural Health Monitoring, 2018.* https://doi.org/10.1177/1475921718790727.
2. Nguyen, A., Chan, T. H. T., & Zhu, X. (2019). Special issue: Real world application of SHM in Australia. *Structural Health Monitoring, 18*(1), 3–4.
3. Sharma, A., & Mehta, N. (2016). Structural health monitoring using image processing techniques—A review. *International Journal of Modern Computer Science (IJMCS), 4*(4). ISSN: 2320-7868 (Online).
4. Kinney, R. P., & River, E. (2013). *Field testing and load rating report: The MiloAdventist Academy's covered bridge.*
5. Kedare, A., & Mundada, K. (2020). Miniaturized device for SHM using electromechanical impedance technique. In *Inventive computation technologies.*
6. Dackermann, U., Smith, W. A., Alamdari, M. M., et al. (2018). Cepstrum-based damage identification in structures with progressive damage. *Structural Health Monitoring.* https://doi.org/10.1177/1475921718804730.
7. Commander, B., & Diaz-Alvarez, H. (2010). *Field testing and load rating of a short-span thermoplastic bridge system, bridge T-8519.*
8. Gastineau, A., Johnson, T., & Schultz, A. (2009). *Bridge health monitoring and inspections–A survey of methods.*
9. González, I. (2011). *Study and application of modern bridge monitoring techniques.* M.Sc. Thesis, Department of Civil and Architectural Engineering, the division of Structural Engineering and Bridges, KTH Royal Institute of Technology.
10. Guan, H., Karbhari, V. M., & Sikorsky, C. S. (2006). Web-based structural health monitoring of an FRP composite bridge. *Computer-Aided Civil and Infrastructure Engineering, 21*(1), 39–56.
11. *Integrated approach to load testing instruction manual.* (2003). Bridge Diagnostics, Inc.
12. Wyczalek, I., & Wyczalek, M. (2017). The potential of photogrammetric method of Measurement dynamic displacements of flexible bridges. *Archives of Institute of Civil Engineering.*
13. Isabela, et al. (2015). Electromechanical impedance-based structural health monitoring instrumentation system applied to aircraft structures and employing a multiplexed sensor array. *Journal of Aerospace Technology and Management, 7*(3), 294–306.

14. Jamali, S., Chan, T.H.T., Nguyen, A., et al. (2018). Reliability-based load-carrying capacity assessment of bridges using structural health monitoring and nonlinear analysis. *Structural Health Monitoring*. https://doi.org/10.1177/1475921718808462.
15. Ko, J. M., & Ni, Y. Q. (2005). Technology developments in structural health monitoring of large-scale bridges. *Engineering Structures, 27*, 1715–1725.
16. Lim. (2015). Development of self-powered wireless structural health monitoring (SHM) for wind turbine blades, a thesis on wireless sensor networks for structural health monitoring.
17. Mesquita, E., Arêde, A., Silva, R., Rocha, P., Gomes, A., Pinto, N., Antunes, P., & Varum, H. (2017). Structural health monitoring of the retrofitting process, characterization and reliability analysis of a masonry heritage construction. *Journal of Civil Structural Health Monitoring, 232*(7).
18. Peters, K., & Inaudi, D. (2014). Fiber optic sensors for assessing and monitoring civil infrastructures. *Sensor Technologies for Civil Infrastructures: Sensing Hardware and Data Collection Methods for Performance Assessment, 1*, 121.
19. Rose Enid Teresa, A., & Umarani, C. (2020). Experimental and analytical investigation of T-beam cum slab bridge. *International Journal of Bridge Engineering* (2). ESCI Indexed.
20. Rose Enid Teresa, A., Stella, S., & Jasper Daniel, J. (2018). Evolution of structural health monitoring system—The state-of-the-art review. *Journal of Applied Science and Computations*. 16.10089.JASC.V6I3.453459.150010623.
21. Sathyanarayanan, S., Ravisankar, K., Sreeshylam, P., & Murthy, S. G. N. (2010). Structural health monitoring development of a real-time remote structural monitoring scheme for civil infrastructural systems. *Structural Health Monitoring, 8*(30), 509–513.
22. Tomizuka, M., Yun, C.-B., & Giurgiutiu, V. (Eds.). (2006). Smart structures and materials 2006: Sensors and smart structures technologies for civil, mechanical, and aerospace systems. In Proceedings of the SPIE—Conference on Smart Structures and Material: Sensors and Smart Structures Technologies for Civil, Mechanical, and Aerospace Systems, San Diego, CA. SPIE.
23. Jagadale, U. T., Kharade, R. D., Nayakand, C. B., & Deulkar, W. D. (2020). Experimental investigation for damage evaluation of bridges using piezo-transducers. *Advances in computer methods and geomechanics*. Lecture notes in civil engineering (Vol. 56). https://doi.org/10.1007/978-981-15-0890-5_27319.
24. Vardanega, P. J., Webb, G. T., Fidler, P. R. A., & Middleton, C. R. (2016). Chapter 29: Bridge monitoring. *Innovative bridge design handbook*.
25. Wenzel, H. (2009). *Health monitoring of bridges*. Wiley.

Evaluating Factors Affecting Red Mud Interfacial Strength Using Binder Cement Kiln Dust and Polypropylene Fiber

Pankaj Sharma, Saurabh Rawat, and Ashok Kumar Gupta

Abstract Red mud (RM) and cement kiln dust (CKD) are major industrial waste products from aluminum and cement industries and generate disposal complexities. This paper focuses on the factors affecting red mud interfacial strength using binder cement kiln dust (CKD) and polypropylene fiber (PPF). Laboratory tests were conducted to investigate the dry density, bearing capacity, and shear strength of the material. Red mud was mixed with varying CKD and PPF percentages to achieve the optimal mix proposition. The maximum dry density of red mud barely was found at 14 kN/m^3 at optimum moisture contents of 23%. The optimal mix proposition was established as 79.5%RM + 20%CKD + 0.5%PPF, which reveals maximum dry density up to 17.3 kN/m^3. The percentage increase in CBR values and interfacial shear strength of materials was found significantly high for the optimal mix proposition. The shear strength parameter measured for the optimal mix using the consolidated drained triaxial shear test is found 45 kN/m^2 (cohesion (c)) and 43.74° (interfacial friction angle), respectively. Energy dispersive X-ray (EDX) analyzer tests were also performed to study the elemental composition and intermolecular reaction of materials. The EDX image shows that mix contains calcium, aluminum, and silica in the mix proposition. This optimal mix proposition can be used to strengthen subgrade layers of low-cost pavements, structural fill, or backfill material for retaining walls.

Keywords Triaxial shear tests · Interfacial strength · EDX

P. Sharma (✉) · S. Rawat · A. K. Gupta
Department of Civil Engineering, Jaypee University of Information Technology, Solan (H.P) 173234, India

S. Rawat
e-mail: saurabh.rawat@juit.ac.in

A. K. Gupta
e-mail: ashok.gupta@juit.ac.in

A. K. Gupta et al. (eds.), *Advances in Construction Materials and Sustainable Environment*, Lecture Notes in Civil Engineering 196,
https://doi.org/10.1007/978-981-16-6557-8_23

1 Introduction

Disposal of different industrial waste is a critical problem in the world. Industrial waste generates different environmental problems for the disposal areas like land degradation, poor air quality or air pollution, uneconomical disposal, etc. The use of different industrial waste as construction material becomes a way out to the established construction material [1, 2]. The consumption of industrial wastes in engineering fill and road construction has been of great interest throughout the globe. Scarcity of previously established construction material and reducing environmental impact renders industrial wastes as a substitute [3]. Fly ash, cement kiln dust, red mud, slag, and mine tailing are key industrial waste products that are a problem for society. Red mud (RM) is an incombustible residual part of the incinerated waste of the aluminum industry. Red mud is disposed of as slurry having 10–15% solid concentration with high pH and ionic value. The reason for collecting red mud as slurry is that this method is economical and easy for the elimination of ash. Samal et al. [4] and Ghosh [5] reported that disposal of red mud is a massive environmental problem. The generation of red mud in India is around nine million tons per year, while globally generation is reported more than 150 million tons which generate disposal complexities as reported by the Ministry of Mines in 2019 [6]. Similarly, cement kiln dust (CKD) is waste clinker dust which includes calcium oxide having pozzolanic properties. The ash generated aluminum and cement industry are quite high despite using it as much as possible, 40% ash has to be directly placed into landfills. These ashes contain various toxic substances which decrease the air quality index and cause various airborne diseases. These ashes also degrade the agricultural land and the crops of neighborhood areas of these industries. Moreover, it sometimes causes cancer risk to residents [3–7]. The dumping cement kiln dust requires a large landfill area for dumping having a high cost and significant environmental impact from a physical, chemical, and biological point of view. The utilization of such industrial waste in construction activities solves problems of recycling, dumping, and ecologically friendly [8–10]. The engineering properties of materials can be significantly improved by different types of fiber [11, 12]. The present paper examines the laboratory use of red mud in highway subgrade construction using binder cement kiln dust. Further, polypropylene fibers were added to the mix propositions to increase the tensile strength of the material. The properties being investigated in the current study are dry density, California bearing ratio, and shear strength of the mix. The study aims to find out the optimum mix proportions to utilize the same for large-scale embankment construction. The projected mix could be used in low-cost embankments construction, earth walls, land renovation, and backfill material in retaining walls.

2 Experimental Methodology

Samal et al. [4] reported that red mud is solid waste generated from the Bayer process of bauxite or mixed ore for the production of alumina. Cement kiln dust (CKD) is ash residues of underburn or overburn clinkers which are generated as a by-product in the cement industry during cement production [1, 2]. Figure 1 shows the grain size analysis of red mud and cement kiln dust used in the present study. The grain size of the curve of both red mud and cement kiln dust (CKD) shows that the material seems like poorly graded sand having fine to medium coarse particles (Fig. 1). As per IS: 2720–4 (1985) [13], the coefficient of uniformity (C_u) and coefficient of curvature (C_c) of red mud was measured at 1.69 and 1.13, respectively, while for cement kiln dust (CKD), the C_u and C_c were recorded 3 and 1.33, respectively.

Figure 2 shows the images of polypropylene fiber. The aspect ratio of fiber was used in the present study was equal to 600. The aspect ratio is length by diameter ratio, and in the present study, length of fiber is 12 mm while the diameter of polypropylene fiber was equal to 0.02 mm with a high melting point of 165 °C. Kumar and Gupta [11] and Kumar and Sharma [1] reported that fiber with an aspect ratio of 300–600 is used for laboratory testing. In which fiber with aspect ratio 600, it offers significantly

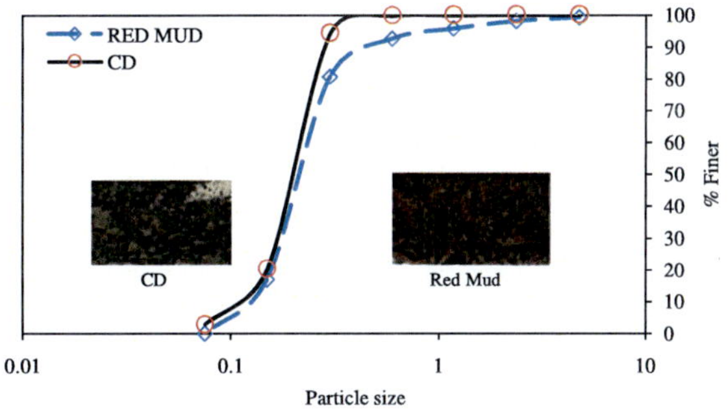

Fig. 1 Grain size analysis of CKD and red mud

Fig. 2 Polypropylene fibers

higher strength due to collective effect of adhesion or cohesion and friction among the fiber and dust particles.

For the sample preparations, the ovendry sample was first mixed manually with a known quantity of cement kiln dust and fiber in a mixing plate in dry condition. Then, the sample was transferred to the small-laboratory mixer where mixing was carried out at the strain rate of 1.25 mm/min in the dry state for a minimum of 5 min. Further, the calculated amount of water was added to the sample and again mixing was carried out in the laboratory mixer for 5 min. A similar type of procedure was also adopted by Kumar and Gupta [11] for the mixing of the sample containing fiber. For the field application of fiber, the volume of earthwork is required to calculate. Then, suggested percentage of fiber (i.e., 0.5% in the present study) of the total volume of earthwork was measured, which is then transferred to the mixer along with other mix propositions for uniform mixing purposes. In the present study, overall, 26 tests were carried out on different proportions to evaluate the factors affecting the interfacial strength of the mix proportions. As per IS 2720-7 (BIS 1983) [14], optimum moisture content (OMC) and maximum dry density (MDD) values were investigated of different mix proportions [10]. In addition, as per IS 2720-16 (BIS 1973), the California bearing penetration tests were conducted for evaluating the stability of subgrade materials [15]. Finally, as per IS 2720-13, the consolidated drained triaxial shear tests were conducted on the best performing mix [16]. The details of the testing program are given in Table 1.

Table 1 Experimental testing scheme

Combinations	Designation
1. Compaction test (total of 17 mixes) %Red Mud: %cement kiln dust: %PP fiber	100RM0CKD0PPF, 100RM0CKD, 90RM10CKD, 85RM15CKD, 80RM20CKD, 75RM25CKD, 70RM30CKD 89.5RM10CKD0.5PPF, 84.5RM15CKD0.5PPF, 79.5RM20CKD0.5PPF, 74.5RM25CKD0.5PPF, 69.5RM30CKD0.5PPF, 89RM10CKD1PPF, 84RM15CKD1PPF, 79RM20CKD1PPF, 74RM25CKD1PPF, 69RM30CKD1PPF
2. For CBR test (total of 6 mixes) %Red mud: %cement kiln dust: %PP fiber	100RM0CKD0PPF, 84.5RM15CKD0.5PPF, 79.5RM20CKD0.5PPF, 74.5RM25CKD0.5PPF, 69.5RM30CKD0.5PPF, 79RM20CKD1PPF
3. For triaxial shear test (total of 3 mix) %Red mud: %cement kiln dust: %PP fiber	100RM0CKD0PPF, 79.5RM20CKD0.5PPF, 79RM20CKD1PPF

RM Red mud; *CKD* Cement kiln dust; *PPF* Polypropylene fiber

3 Results and Discussion

3.1 Compaction Characteristics

To measure the OMC and MDD of the different mix, light compaction tests were conducted with varying proposition cement kiln dust and polypropylene fiber as shown in Table 1. Figure 3 shows the variation of dry density and moisture contents for different mix propositions of red mud and cement kiln dust. The peak dry density was recorded for mix proposition having 80% red mud (RM) and 20% cement kiln dust (CKD). The MDD and OMC of 80%RM + 20%CKD were obtained 16 kN/m^3 and 28%, respectively. Also, the MDD of red mud without adding cement kiln dust, the MDD and OMC were recorded 14 kN/m^3 and 23%, respectively. Further, with the replacement of 0.5% of polypropylene fiber to 79.5%RM + 20%CKD, the maximum dry density of mix increases up to 17.3 kN/m^3 at an optimum moisture content of 25% as shown in Fig. 4. The maximum dry density of red mud increases 14.55% with the addition of 20% cement dust and 0.5% polypropylene fiber. Further, in another trial, when fiber percentage increases up to 1%, the dry density again starts decreasing as shown in Fig. 5. Thus, the optimum mix was considered as 79.5%RM: 20%CKD: 0.5%PPF with a peak density of 17.25 kN/m^3 at 25% of water content. The density curve showing bulking nature of composition which depicts that material behaves

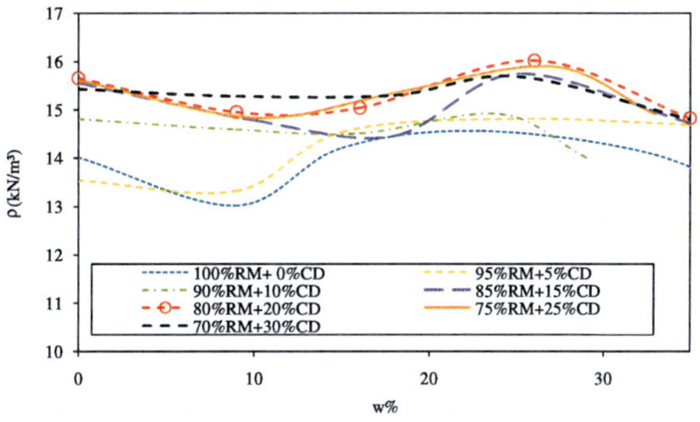

Fig. 3 MDD verses water content for different mix propositions of red mud and cement kiln dust

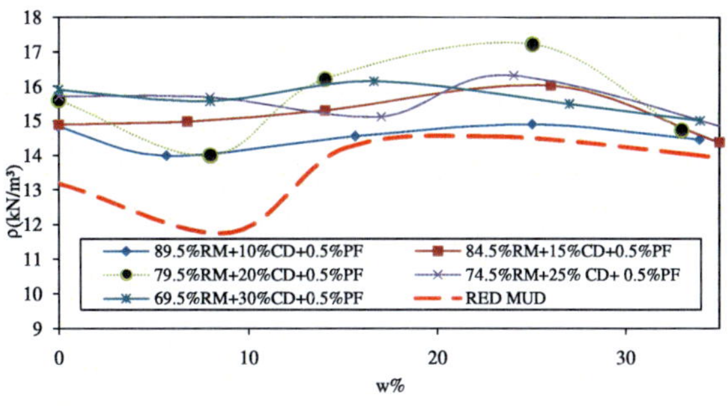

Fig. 4 MDD verses water content for different mix propositions of red mud and cement kiln dust, when PPF = 0.5%

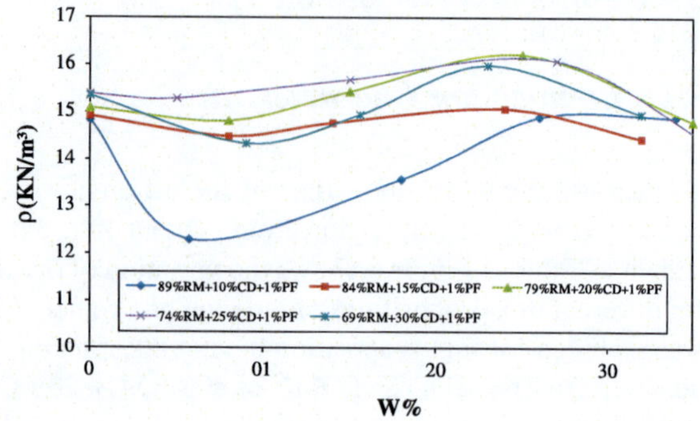

Fig. 5 MDD verses water content for different mix propositions of red mud and cement kiln dust, when PPF = 1%

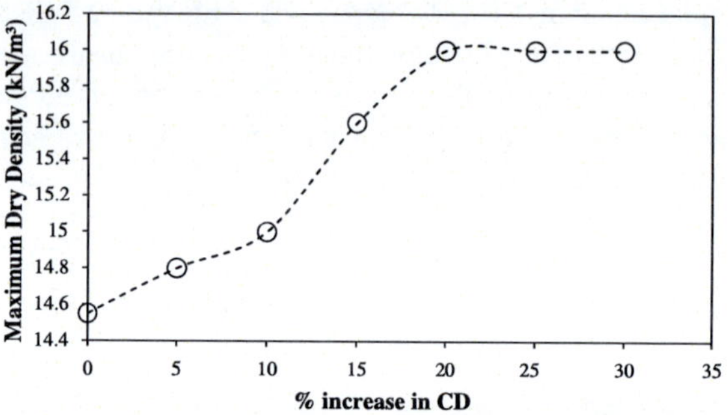

Fig. 6 MDD with % increases in cement kiln dust

like sandy soil. In addition, maximum dry density was found to increase with cement kiln dust and becomes constant after 20% of cement kiln dust as shown in Fig. 6.

3.2 California Bearing Ratio (CBR) Tests

The soaked CBR test represents the worst field condition of a subgrade. Thus, to simulate the worst field condition, the soaked CBR test was conducted on the different mix proportions. Evident from Fig. 7, the CBR value of red mud increases gradually from 16 to 28% with the addition of CKD and 0.5% PPF, which is found maximum at 79.5% red mud, 20% cement kiln dust, and 0.5% polypropylene fiber.

The CBR% start decreasing with an increase in fiber percentage up to 1%, these decreases in CBR value may be due to an increase in repulsive forces with increases in fiber percentage. Again, based on CBR value, the 79.5% RM: 20%CKD: 0.5%PPF mix was found to be optimum having peak CBR value.

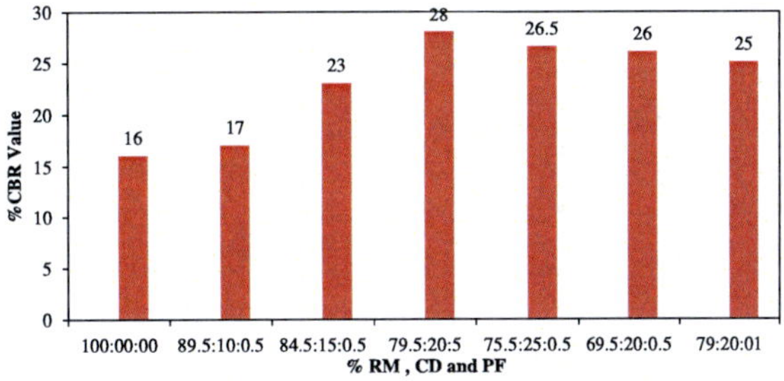

Fig. 7 CBR values for different mix proportions

3.3 Shear Strength Parameters

Consolidated drained triaxial shear tests were conducted on pure red mud, 79.5%RM + 20%CKD + 0.5%PPF, and 79%RM + 20%CKD + 1%PPF under different cell pressure to calculate the shear strength parameters of the material. The test was conducted at a controlled strain rate of 1 mm/min. Evident from Fig. 8, the optimum mix (79.5%RM + 20%CKD + 0.5%PPF) showing maximum shear strength comparative to pure red mud and 79%RM + 20%CKD + 1%PPF. The cohesion (c) and angle of internal friction (ϕ) values for red mud lone are 12.5 kN/m^2 and 24.22°, whereas, for 79.5% red mud + 20% CKD + 0.5% PPF, the values are 45 kN/m^2 and 43.74°, respectively, as shown in Table 2. The increase in interfacial strength is due

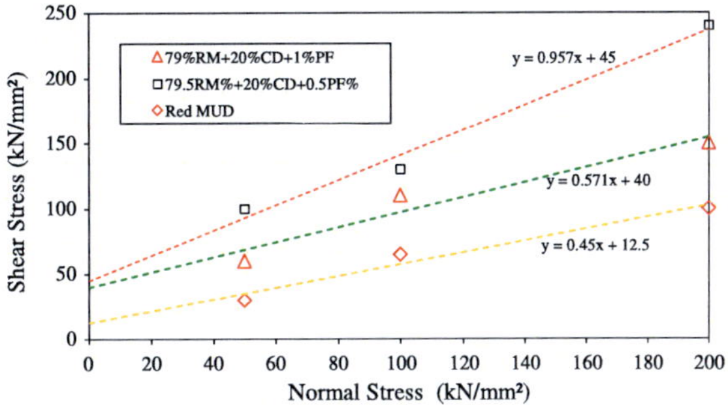

Fig. 8 Shear stress against normal stress

Table 2 Shear strength parameter

Sr. No.	Mix proportion	Cohesion, c (kN/m^2) or adhesion	Angle of internal friction, ϕ (degrees)
1	Red mud	12.5	24.22
2	**79.5%RM + 20%CKD + 0.5%PPF**	**45**	**43.7**
3	79.5%RM + 20%CKD + 1%PPF	40	29.6

to binder cement kiln dust and polypropylene fiber. The significant increases in cohesion (c) value are due to the water-retaining capacity of fiber, while cement kiln dust contributes to increasing the angle of internal friction (ϕ) owing to pozzolanic action which ultimately increases in the shear strength of the material. Further, with an increase in fiber up to 1%, the fiber starts expanding due to absorption of significant high water and thus the bond between material and fiber starts weakening resulting decrease in shear strength parameter. A similar kind of result has been published by a different author in the context of industrial waste and polypropylene fiber [1, 11, 17].

3.4 Energy Dispersive X-Ray (EDX) Analysis Studies

EDX was used to recognize the elemental composition of materials using the X-ray technique. Figure 9 showing the EDX analysis of cement kiln dust shows that it contains sulfur, iron, calcium, sodium, magnesium, alumina, and silica. Evident the increase in strength of different mixes using cement kiln dust is mostly due to calcium, alumina, and silica. Similarly, from Fig. 10, it is clear that red mud has to reach iron and silica content, thus when mix with cement kiln dust, it makes a good agreement with each other. Thus, due to calcium, alumina, and silica content, the dry density, CBR value, the shear strength of the mix increase significantly.

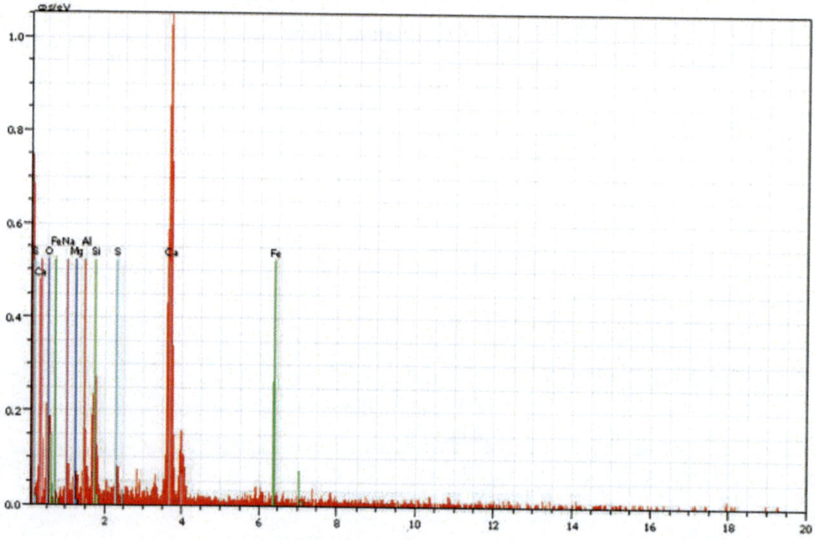

Fig. 9 Energy dispersive X-ray analysis of cement kiln dust

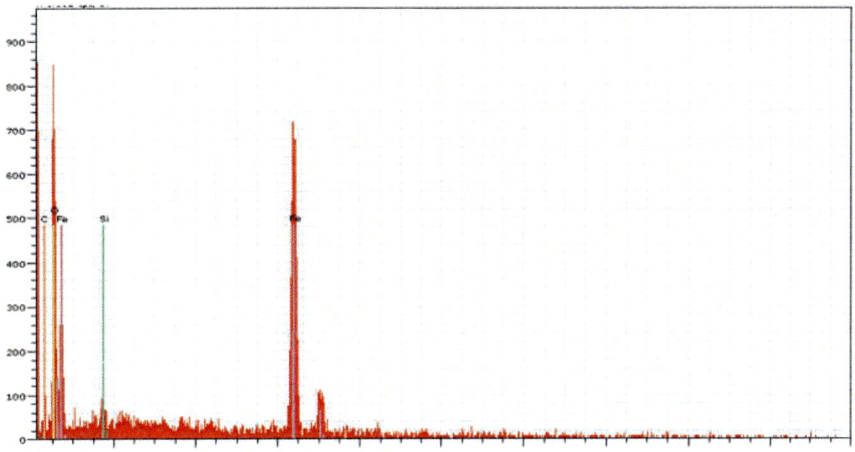

Fig. 10 Energy dispersive X-ray analysis of red mud

4 Conclusions

After a laboratory testing on red mud using binder cement kiln dust and polypropylene fiber, the following conclusions were made:

1. The MDD and OMC of pure red mud were recorded 14 kN/m^3 and 23%, which further increases up to 17.3 kN/m^3 and 25%, respectively, for the best mix (i.e., 79.5%RM + 20%CKD + 0.5%PPF). This revel that the industrial wastes CKD has binding properties which amplify the dry density substance.
2. The c and angle of ϕ values for red mud lone are 12.5 kN/m^2 and 24.22°, whereas, for 79.5% red mud + 20% CKD + 0.5% PPF, the values are 45 kN/m^2 and 43.74°, respectively. The increase in interfacial strength is due to binder cement kiln dust and polypropylene fiber. The significant increases in cohesion (c) value are due to the water-retaining capacity of fiber, while cement kiln dust contributes to the increasing angle of internal friction (ϕ) owing to pozzolanic action which ultimately increases in the shear strength of the material.
3. CBR value of red mud increases gradually from 16 to 28% with the addition of kiln dust and 0.5% polypropylene fiber, which is found maximum at 79.5% red mud, 20% cement kiln dust, and 0.5% polypropylene fiber.
4. Energy dispersive X-ray analysis of cement kiln dust and red mud reveals that cement kiln dust contains sulfur, iron, calcium, sodium, magnesium, alumina, and silica which help to increases the shear strength of the different mix. Thus, an optimal mix proposition can be used to strengthen subgrade layers of low-cost pavements, structural fill, or backfill material for retaining walls.

References

1. Kumar, S., & Sharma, P. (2018). Geotechnical properties of pond ash mixed with cement kiln dust and polypropylene fiber. J. Mater. Civ. Eng. **30**(8), 04018154-1-11. https://doi.org/10.1061/(ASCE)MT.1943-5533.0002334
2. Kumar, J. S., & Janewoo, U. (2016). Stabilization of expansive soil with cement kiln dust and RBI grade 81 at subgrade level. *Geotechnical and Geological Engineering, 34*(4), 1034–1046.
3. Muntohar, A. S., & Widianti, A. (2013). Engineering properties of silty soil stabilized with lime and rice husk ash and reinforced with waste plastic fiber. *Journal of Materials in Civil Engineering, 25*(9), 1260–1270. https://doi.org/10.1061/(ASCE)MT.1943-5533.0000659
4. Samal, S., Ray, A., & Bandopadhyay, A. (2013). Proposal for resources, utilization and processes of red mud in India—A review. *International Journal of Mineral Processing, 118*, 43–55. https://doi.org/10.1016/j.minpro.2012.11.001
5. Ghosh, P. K. (2009). *Utilization of red mud and pond ash for construction of embankments.* Dissertation, National Institute of Technology, Rourkela, India.
6. Workshop by Ministry of Mines on "Effective Utilization of Red Mud". Press Information Bureau Government of India Ministry of Mines (2019).
7. Sharma, P. (2017). *Utilization of pond ash mixed with cement kiln dust and polypropylene fiber for construction of embankment.* M.tech Dissertation, DAV Institute of Engineering and Technology, Jalandhar, India.
8. Reddy, P. S., Reddy, N. G., & Serjun, V. (2021). Properties and assessment of applications of Red Mud (Bauxite residue): Current status and research needs. *Waste Biomass Valor, 12*, 1185–1217. https://doi.org/10.1007/s12649-020-01089-z
9. Kang, X., Chang, K. K. T., & Ge, L. (2014). Chemically stabilized soft clays for road-base construction. *Journal of Materials in Civil Engineering, 27*(7), 1–9. https://doi.org/10.1061/(ASCE)MT.1943-5533.0001156
10. Kumar, S., & Sharma, P. (2019). Effect of cement kiln dust and RBI grade 81 on engineering properties of plastic clay recycled waste materials. *Lecture Notes in Civil Engineering, 32*. https://doi.org/10.1007/978-981-13-7017-5_5
11. Kumar, A., & Gupta, D. (2016). Behavior of cement-stabilized fiber reinforced pond ash, rice husk ash-soil mixtures. *Geotextiles and Geomembranes, 44*(3), 466–474. https://doi.org/10.1016/j.geotexmem.2015.07.010
12. Onyejekwe, S., & Ghataora, G. S. (2014). Effect of fiber inclusions on flexural strength of soils treated with nontraditional additives. *Journal of Materials in Civil Engineering, 26*(8), 04014039. https://doi.org/10.1061/(ASCE)MT.1943-5533.0000922
13. IS: 2720-4. (1985). *Grain size analysis.* Bureau of Indian Standard (BIS).
14. BIS (Bureau of Indian Standards). (1983). *Methods of test for soils: Determination of water content–dry density relation using light compaction.* IS 2720-7. BIS.
15. BIS (Bureau of Indian Standards). (1973). *Methods of test for soils: Laboratory determination of CBR.* IS 2720-16. BIS.
16. BIS (Bureau of Indian Standards). (1991). *Methods of test for soils: Determination of shear strength parameters by consolidated drained triaxial shear test.* IS 27290-12. BIS.
17. Sarkar, R., Abbas, S. M., & Shahu, J. T. (2012). Geotechnical behaviour of randomly oriented fiber reinforced pond ashes available in Delhi region. *International Journal of Earth Sciences and Engineering, 5*(1), 44–50.

Review on Thermal Energy Efficiency Using Gypsum Integrated Phase Change Materials in Buildings

Kavita Vaishnav, Gift Pon Lazarus, Sunita Bansal, and Yaman Hooda

Abstract Thermal performance of gypsum integrated with phase change materials in buildings plays a very important in conserving energy in a sustainable manner without any harmful effect over the environment. Some important parameters of gypsum integrated with PCMs are melting point, freezing point, latent heat, thermal conductivity, density specific heat and thermal efficiency. This paper presents a review about various techniques and practices for thermal energy storage and methods through which the building will become thermally efficient. Techniques used for temperature control are heating ventilation and air conditioning and latent heat thermal energy storage systems. LHTES includes use of phase change material and gypsum composites. Fabricated composites are more useful as it reduces the risk of fire and poor thermal conductivity. Microencapsulation of PCMS is more efficient way of integrating PCMs. The enthalpy values of fabricated composites lie between 55 and 73 J/g with good thermal stability and are more useful in TES applications. It was also investigated that most suitable low temperature for heating ventilation and air conditioning systems are between 21 and 23 °C. The temperature reduces from 5 to 9 °C when gypsum integrated with PCMs in wall boards with 20% replacement which helps in stabilizing indoor temperature as compare to only adding PCMs reduces 4 °C indoor temperature with low-compressive strength.

Keywords Thermal energy storage (TES) · Phase change materials (PCMs) · Gypsum · Energy saving

---

K. Vaishnav (✉) · G. P. Lazarus · S. Bansal · Y. Hooda
Department of Civil Engineering, Manav Rachna International Institute of Research and Studies, Faridabad, Haryana, India

G. P. Lazarus
e-mail: lazarusd.fet@mriu.edu.in

S. Bansal
e-mail: sunitabansal.fet@mriu.edu.in

Y. Hooda
e-mail: yaman.fet@mriu.edu.in

A. K. Gupta et al. (eds.), *Advances in Construction Materials and Sustainable Environment*, Lecture Notes in Civil Engineering 196,
https://doi.org/10.1007/978-981-16-6557-8_24

1 Introduction

In building sectors, requirement of energy demand is changing extremely over the time so researchers are also working energetically towards shortening energy swallowing. In buildings for controlling, the temperature by maintaining the temperature constant HVAC (Heating Ventilation and air conditioning) techniques play leading role in order to offer a comfort zone to occupants. For energy saving and decreasing temperature, the most feasible choice is latent heat thermal energy saving (LHTES) by the use of phase change material. It has been found that passive heating and cooling techniques by keeping the air temperature continuous and reducing the optimum energy quantity of HVAC systems a successful technique using phase change material in buildings for energy management to improve upgrade comfort conditions. In building thermal management, PCMs worked as a capable applier. The building engaged with PCMs reduces the external heat gain by consuming temperature during day time and losses during night time when the ambient temperature goes down the melting temperature to maintain thermal comfort [1]. Solar energy has been measured as 10,000 times more than the globe energy demand because solar energy is the energy source from the ancient time which is accessible to the whole world throughout the year to conserve the energy so that decrement could be possible in active energy utilizing by heating, ventilation and air conditioning (HVAC) process the solar plays an important role because this energy can be stored by a mechanism of phase change material when applied in the buildings. From the view of environmental conservation, social and economic aspects the integration of PCMs in walls, floors and ceilings are necessary in building construction so that it can satisfy the sustainable development and their dimensions. PCM acts as a passive thermal regulator for maintaining internal temperature constant by integrating with mortar. When PCMs undergo phase transition with ambient temperature heat gets stored and loses heat during phase change. It was investigated that high energy absorbing, temperature and density attenuating ability of phase change materials are good. Gypsum panel wall boards, mortar and concrete, are phase change materials which are generally composed with these materials [2]. According to the Government of India, Ministry of Urban Development (2015) the buildings expenditure for energy consumption needed to be lessened to build the cities of India alert and more reliable. The functionality of the building and the climate-responsive design needs to be composed for achieving an energy-sensible design. Assimilating climate-responsive features can reduce the expenditure of lighting and air-conditioning in the building design. Additionally, in tropical climates nowadays, there is a lack of deep research over the design strategies of high-rise buildings for energy consumption [3] (Table 1).

Table 1 Properties of sample with varying gypsum and water content [16]

Authors	Parameters	Tools and techniques	Methodology	Conclusions and remarks
David et al. [4]	High heat of fusion, in liquid and solid-state melting temperature, thermal conductivity, heat, density and heat capacity	Heating ventilation and air conditioning systems implemented to improve performance of building. Analytic hierarchy process and entropy works together for calculating the weighting criteria	The most useful method founded was TOPSIS use to help in determining material selection. Other methods used are COPRAS-G method, TOPSIS method, VIKOR method and spearman's correlation coefficient	It was investigated that for sensible thermal energy storage CES selector software can be used and their value lies between 150 and 200 °C. In cold places PCMs behave very good both in day and night
Awni and Alkhazaleh [5]	Important factors in this study were mass loss, peak temperature, modulus and stress and failure, melting or freezing temperature and time interval	Techniques used for preparing samples are direct incorporation, immersion and encapsulation	Flexural strength of samples are calculated using three-point bending test. The room temperature for further experiment taken as 20 °C. During experiments mass loss% and residue % was also checked at a temperature of 350 °C	Direct incorporation technique was founded the most beneficial and economical method in these experiments. It was founded that temperature at which the phase transits are ± 4 °C. The value of latent heat during phase transition was noticed to be ± 0.65 J/g. It was noticed that medium low temperature range of gypsum wall boards were 5–40 °C
Nadiya et al. [6]	Thermal cycles are important parameter of this study and some other factors include thermo-gravimetric and thermal conductivity	For determining the values of thermal conductivity and thermo-gravimetric values differential scanning calorimetry analysis used	Thermo-physical properties are measured then thermal cycles are performed in a room for testing purpose	In cold energy thermal storage generally 90 thermal cycles are useful. Gypsum integrated PCMs use to reduce the heating and cooling requirement of buildings by 25%

(continued)

Table 1 (continued)

Authors	Parameters	Tools and techniques	Methodology	Conclusions and remarks
Gwang et al. [7]	Thermal conductivity, specific heat, density and chemical properties are important factors in this study	There are various techniques for analysis used as thermal conductivity analysis, thermal durability analysis, dynamic heat transfer analysis and FTIR peak intensity analysis	Chemical properties are investigated of gypsum integrated PCMs composite by vacuum impregnation method then thermal properties are analyzed	The bending strength calculated for gypsum integrated PCMs are 138 N. The most important thing for the composition is moisture content of it is required to lie below 16%. Gypsum integrated PCMs shows increase in the thermal conductivity by 149 and 130%
Borreguero et al. [8]	Average apparent heat capacity, microencapsulation efficiency, energy storage capacity and density are important parameter of this study	Modulated Differential scanning calorimetry technique was used in it for calculating melting point of paraffin and apparent heat capacity	Microencapsulation form of PCMs synthesis performed then analytical methods are used for determining physical properties, and finally the temperature profile checked	When paraffin wax was integrated with PCM its value of Mass ratio 0.5 then value of obtained energy storage capacity was 40 J/g and resulting in microencapsulation efficiency value 70%
Zhu and Wang [9]	The properties of optic variable wall factors are low or high reflectivity, indoor air temperature, maximum and average degree comfort, discomfort hours, thermal comfort degree and Discomfort degree hours	There are various techniques used like heat transfer process, thermal comfort assessment, acceptable indoor air temperature	The optic variable wall was used as a sample for maintaining thermal comfort. This optic variable wall then analyzed and their values then compared with indoor temperature discomfort hours of absorptive and reflective wall	The indoor air temperature values of walls based on reflective and absorptive behaviour as for reflective wall panel its minimum value is less than 1.2 °C and for absorptive wall panel the maximum indoor temperature was founded 1.5–5.1 °C. It was noticed that maximum value of discomfort degree is measured around 16 °C and value of degree hours of discomfort of wall in horizontal and vertical hours as 41–100 and 21–51%

(continued)

Table 1 (continued)

Authors	Parameters	Tools and techniques	Methodology	Conclusions and remarks
Zhang and Yang [10]	Direct and diffuse solar radiation, air temperature, air velocity and air relative humidity	Tools of experimental instruments are diffusometer, temperature humidity recorder& universal anemometer pyranometer and data acquisition	Various measurements are taken by TESTO 175 for relative humidity, air temperature. After this air velocity also recorded with WFWZY-1	If air temperature range lies between 35 and 65 °C than its accuracy will be equal to ± 0.5 °C. If the air relative humidity range lies between 0 to 100% accuracy will be around ± 3%. Air velocity range recorded for accuracy 0.05 m/s is ± 0.05–30 m/s
Wu [11]	Important factors are unit weight, thermal expansion, tensile strength and compressive strength	Experimental investigation includes various techniques or tests are longitudinal shear test, unit shear strength test. With ultimate load values lower bound theorem also used in it	A sample as a wall was prepared with gypsum integrated PCMs and reinforced with glass fibre after this it was tested for flexural strength and shear strength	The compressive strength and tensile strength obtained after experiments were 170 and 40 kN/m. It has fire resistance property also but its value is less than 3.1 h. These types of walls used as conventional reinforced concrete walls with thermal properties
Wang and Lei [12]	Ventilation rate, average room temperature and average water temperature	A numerical model was used so that the rate of steady state mass flow can be recorded	The thermal cycles are analyzed at different levels to know the temperature variation with respect to the temperature of rooms. Its value lies nearly 4.9%	The ventilation rate of building increased by 5% when water wall was used with 52% of transmission of solar energy. The average room temperature also effects through it and leads to 4% increment. As the thickness of gypsum integrated wall panel reduces the average room temperature and rate of ventilation system increasing by 7.5%

(continued)

Table 1 (continued)

Authors	Parameters	Tools and techniques	Methodology	Conclusions and remarks
Ouakarrouch et al. [13]	Some important factors are thickness, conductivity, density and specific heat	Guarded hot plate method used for used for measuring thermal effusively, Thermal characterization methods for diffusivity and volumetric thermal capacity	A composition of gypsum plaster integrated with chicken feathers than thermo-physical properties determined	Greenhouse gas emission deceases when gypsum composed with chicken feathers in plaster form. The composite of gypsum incorporated with chicken feather leads to reduction in heating consumption rate. In summer it starts reducing heating by 25% but in winters it decreased heating consumption by 30%

2 Phase Change Material

In modern time, use of phase change materials is termed as smart materials in the field of building constructions. For maintaining thermal comfort by decreasing energy consumption PCMs are used. There are two different excellent properties of PCMs one side is latent heat storage capacity and on other hand is their positive impact over the environment due to these reasons thanks to PCMs. Many combinations are prepared with the help of PCMs as incorporated in concrete, gypsum and other building materials [14]. Heating and cooling loads of a building are established by building envelopes as they are directly exposed to the radiations of sun. Due to direct solar radiation, it is necessarily required to use methods or solutions of low carbon and needs to be energy-efficient. As the PCM keeps heat into latent form so PCMs are extra functional as related with other materials for collecting the solar energy and larger heat collecting absorbency per unit space over other envelope materials. A factor of fifteen PCM has frequently altered their phase from liquid to solid and from solid to liquid which helps in increasing the thermal collecting absorbency of using PCMs in building materials like concrete, mortar and plaster. Directly including PCMs into the building materials is tough. Microencapsulation: (MPCM) is a technique for storing energy. During this process minute size molecules of solid, liquid and gases are enclosed in an inner shell. Thermal energy storage systems are of mainly three types, i.e. latent, sensible and thermo-chemical heat storage systems. In this process towards reaction, the energy is stocked, and during reverse the energy is loosed. It has been reported the PCM is the most feasible way to manage the heat storage characterization of MPCM. This method is very useful for achieving building energy enhancement: For analyzing composite walls thermal stability and mechanical behaviour, composite Portland cement and MPCM prepared. It has been concluded that wall composite with 55% of encapsulation efficiency measured and also results in good thermal reliability [15]. Latent heat storage plays a leading role in storing thermal energy comparing to other techniques. PCMs consume or lose heat energy from the environment during phase transition. There are various advantages of PCMs such as massive latent heat absorbing ability, outstanding chemical status.

PCMs are materials which undergo phase transition from one phase to another either through gaining or by losing heat around at the same temperature. Solid–liquid–solid (melting and solidification cycle), liquid–vapour–liquid (Evaporation/boiling and condensation cycle) are the most common phase transition process, whilst concerning to PCM, solid–liquid–solid is of most interested phase transition taken for thermal energy storage. Phase transition involves complex phenomenon and also requires high-specific volumes for storing and releasing energy [1]. The advantages of organic PCM includes: (1) Huge Temperature Range, (2) High Chemical Stability, (3) Lesser effect during Phase Transformation, (4) Non-toxic in nature and good compatibility with conventional materials. Whereas the advantages of inorganic PCM includes: (1) Good Thermal Conductivity, (2) Lesser volume change comparatively, (3) Easy availability and (4) Non-Flammable in nature.

3 Selection of PCM

PCMs are selected on the concept of their requirement or can be selected on the need of application of PCMs building inside space thermal managing like cooling and heating energy requires maintaining low temperature and power generation system requires high temperature. In case of low temperature, the amount of thermal energy storage is 50 °C and in high-temperature energy storage system 150 °C. It is found that after having so much research by the researchers there is still moving towards better solution by them in the form of PCMs that can be applied to building environmental circumstances. Some major points are considered for selecting PCMs like melting temperature requires in functional temperature limit, phase change nature requires high enthalpy of fusion (200 kJ/kg), specific heat capacity must be more than the 2.5 kJ/kg K. Thermal conductivity needs to be maintained at high level of 0.6 W/m C in the phases, thermal cycles can be completed more than 5000 times for completing cycles for charging and discharging, super cooling has the property of not solidifying completely when it goes below the freezing point, change in volume is minimal during phase transition if larger the volume it may leads to the requirement of large container size, congruent melting includes homogeneity when PCMs transforms its phase, vapour pressure should be low so that containment problem can be avoided. Non-corrosive property of PCMs makes it eco-friendly, economic or availability of PCM makes it reliable in to all works and PCMs must be non-flammable in nature so that fire accident problem can be reduced [16]. Inorganic PCMs were

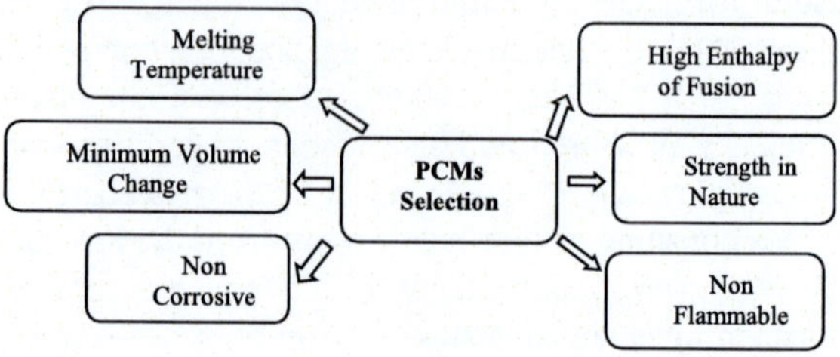

Fig. 1 PCM selection concept

tested from last two decades with or without composition with other materials so that their applications in building materials could be investigate. These inorganic materials were fatty acids, long chain alcohols, paraffin's, and polyglycols and the compositions of all these. The only reason for testing these inorganic PCMs was thermal energy storage characteristics. When organic PCMs are directly integrated with the materials then it was found that compressive strength decreases because PCMs in organic are hydrophobic nature and due to this reason organic PCMs lead to reduction in compressive strength [17] (Fig. 1).

4 Properties of Gypsum at Different Percentage of Water Content

Composition made by gypsum without integrating phase change materials for determining the properties of gypsum at different water percentage on adding to get variation [18] (Table 2).

Table 2 Properties of sample with varying gypsum and water content [18]

Sample	Gypsum content (%)	Water content (%)	Density (kg/m^3)	Thermal conductivity (W/(m K))	Specific heat capacity (J/(kg K))
A1	62.1	31.6	1193	0.485	926
A2	65.2	34.8	1100	0.512	809
A3	63.2	36.8	1323	0.414	884
A4	62.0	41.6	1273	0.512	855
A5	58.4	46.6	992	0.350	856
A6	53.4	45.4	1070	0.325	844
A7	56.6	43.0	809	0.387	837
A8	57.6	42.4	769	0.343	948
A9	54.1	45.9	985	0.251	987
A10	50.7	49.3	934	0.251	937
A11	48.2	51.8	871	0.291	868

4.1 Application of Gypsum Integrated with PCMs

As the continuous advancement in the technologies demand for internal thermal comfort is increasing day by day due to this reason energy consumption rate increases rapidly for decreasing, this energy consumption PCMs are added to the building materials like in floors, roofs, concrete and wall boards with incorporation of gypsum material in various forms to improve the mechanical properties and overcoming the drawbacks of PCMs.

Applications for building Envelopes: For decreasing, the peak temperature load also energy demand buildings PCMs are integrated in bricks, concrete, wall board and plasters. In buildings interior walls, ceilings, and exterior walls PCMs are incorporated with G/C (Gypsum/Cement) board for serving as thermal energy storage units. Organic PCMs have latent heat and low vapours pressure in melting properties so that it can be opted to the ventilation systems of the structures. G/C materials are not structural materials; these are only used for interior purpose only [7].

In buildings for maintaining, the thermal comfort and indoor temperature gypsum and PCM are composed together in microcapsules form and used to prepare several of building materials like wall board, cement, mortar and brick. These composites are applied to the various components of the building are: (1) Building walls can be plastered with the mortar prepared by adding MPCMs and gypsum for enhancing energy storage. Gypsum blocks are another form incorporated with microcapsule content. These blocks are used in building walls for altering the external temperature profile of the building wall. The temperature experienced by the walls decreases with respect to time when the number of microcapsules increases in the gypsum blocks. This combination leads to the increase in thermal energy storage capacity for maintaining thermal comfort in the building, (2) In building roofs, concreting work can be done by adding gypsum and PCM incorporation to it as by storing and releasing energy, (3) Interior walls of the buildings prepared by gypsum wall boards and incorporation of PCM for maintain the indoor thermal comfort, (4) False ceilings prepared in the buildings can be insulated by the help of gypsum board coffering and applying plaster prepared with the help of PCM and gypsum composition and (5) The structures or buildings which are already constructed then gypsum and PCM plays a very important role because those buildings can be converted in to structures for working on thermal energy saving and the inside conditions of buildings can be maintained by thermal insulation. But when the PCM amount increased in gypsum board composition then there is a decrease in thermal conductivity. Using MPCM is very useful as the latent temperature of MPCM is 170 kJ/kg and melting temperature of MPCM is 25 °C [15].

4.2 Prefabricated Structures

Gypsum and PCM are composed together in manufacturing the prefabricated structures. One of most important examples of gypsum and PCM composites are prefabricated gypsum wall panels. As pre cast structures have fewer problems during its manufacturing process because in precast structures standard procedure and dimensions are followed for all elements. Gypsum and PCMs can be used in concrete in mortar manufacturing, but this process is more difficult and needs more carefulness because these are prepared directly on site and are also not a controlled process.

5 Ventilation System

It is a type of heat recovery system which contains rotary thermal wheels, fixed plate heat exchangers, heat pipes and phase change materials integrated with gypsum used in various forms as per their requirement according to topography. Gypsum-cement board are use full in those areas where ventilation is required. As the gypsum cement board stores and releases thermal energy from atmospheric air temperature depends upon the climatic nature.

- G/C board acts as a PCM holder during phase transformation it prevents PCM from leakage
- G/C board can be opted to the structures where less allowable shrinkage and less deformation is mandatory.
- G/C boards are applied to the fire-resistant structures as g/c board is more stable and can be easily nailed or screwed.
- The structures with requirement of reducing peak temperature and time racking effect can also be fulfilled by application of G/C board.

These boards are prepared by integration of cement and gypsum. These materials are inorganic in nature. Generally, gypsum is chemically referred to as calcium sulphate dehydrate; it is prepared by obtaining a stable quantity of water to the calcium sulphate hemihydrates in powder form. The component mixed with gypsum governs the property of gypsum [18].

Computational techniques for disguising actual temperature disbursement it was gypsum plaster boards having the outstanding property of saving energy and the more beneficial in both the cases of new and old buildings but during this process, it was observed that there is no proportion maintained between the efficiency and number of PCMs applied.

6 Microencapsulation in Building Materials

Microencapsulation is one of the best techniques by which sensitive materials can be prevented from the adverse effect of the environment and helps in easy and safe active materials and their handling. It also improves thermal properties and helps in decreasing reactivity. Micro capsules are available in spherical shape with a very less diameter lying in between 10 and 100 μm. As the dehydrating property of gypsum, at a temperature of 120 °C and making it fire resistant, gypsum plaster boards prevents the adverse effect of fire in the building [19]. There are various methods used for making core and shells are polymerization emulsion, polymerization mini-emulsion also some other techniques like interfacial polymerization. Inorganic materials in the form of polymers and PCMs are employed as cores and the shells but it is found that encapsulated phase change materials be a major role in magnifying the thermal conductivity of PCM [9]. Paraffin-based binary composites (RT 42 and RT 28) considered as core and shell material $CaCO_3$ is engaged. In formation of $CaCO_3$ shell, $CaCl_2$ imparts a major role in which $CaCO_3$ is used for the purpose of predecessor. The thermal conductivity received by EPCM is 2–3 times more than the RT 28 and RT 42 which are pure form. It was found that if there is more amount of $CaCO_3$ then it's helping in excellent effective manner to increase the thermal conductivity [20]. In shells of EPCMs mostly polymers are used for the thermal conductivity of polymers are very low so these polymers are needed to be modified for the purpose of increasing the thermal conductivity. RT21 utilized as a core to synthesize EPCM and used polymethyl methacrylate (PMMA) as a shell. The excellent thermal conductivity was noticed when the particle size lies 9.4 μm and more than it. The concentration of nitrate was increased it also increased the apparent thermal conductivity of EPCM. These are directly comparable with respect to the coverage over EPCMs surface by the silver coating [21]. Nano-alumina content is added to the EPCMs it results in improving the thermal conductivity. The emulsion polymerization methodology was preferred in which paraffin wax core and shell of poly (methyl methacrylate-co-methyl acrylate) was used for composing EPCMs. Although integration more amount of nano-alumina concentration may cause decrease in the phase change enthalpy [22]. PCM panels were composed by $CaCl_2.6H_2O$/expanded graphite integrated with PCM. All this experiment was carried out for checking application of PCMs in buildings. After the experiments, it was observed that best results of PCM panel thickness was obtained by 8–10 mm range. It was observed that temperature fluctuations also decrease due to presence of PCM panels because the test room temperature range is lesser than the temperature range of conventional rooms [23]. Novel double-layer radian floor for determining the temperature field of the floor system. In these, PCMs are used which is favourable to the electric peak-conservation. It was deduced that when radiant floor prepared by double layer it satisfies the demand of heat. But when the temperature of thermal storage PCMs increases, it leads to the decrease in the energy consumption rate during the heat energy storage process [24]. Building energy saving techniques for environmental prospective and for saving energy are the most effective

Table 3 Measured properties of samples with PCM microcapsules [18]

Sample	Density (kg/m^3)	Thermal conductivity (W/(mK))	Specific heat capacity (J/(kg K))
C1	1121	0.398	910
C2	905	0.286	960
C3	970	0.276	1033
C4	985	0.395	934
C5	970	0.320	933
C6	850	0.356	1157
C7	881	0.296	962
C8	880	0.249	1046
C9	929	0.289	945
C10	904	0.302	1028
C11	893	0.293	1050
C12	851	0.253	1150
C13	893	0.246	1116
C14	800	0.230	972
C15	725	0.183	1225

and sustainable technologies. To modifying the fossil fuel consumptions in buildings, it was noticed that one of the most effective ways to solve this problem are heat recovery systems. This recovery system helps in reducing the effect of greenhouse gases over the environmental air and also helps in conservation of energy [25] (Table 3).

7 Discussions and Conclusion

It was observed that integrating additives which are having high-thermal conductivity was composed with PCMs so as to achieve increased thermal conductivity this is most successful technique. In terms of prevention control from leakage when PCMs transits from one phase to another phase in melting some materials like polyethylene having high density were used. These materials were speaking or supporting type and also used to increase the stability of PCMs. Expanded graphite and metal foam having the property of absorbing PCMs thoroughly and helps in supporting the composition of PCM. These graphite and metal foam were mainly used in pore structures. Carbon-based and metal-based additives are used. In carbon-based additives some marvellous properties are present. These excellent properties are low density and have excellent stability when these are applied. Nano-additives are very frequent nowadays after the development of nano-technology. As additives have various advantages on the other hand additives also have some disadvantages like unequal distribution of additives which may cause harmful effects like it destroys the temperature uniformity during energy storage of system. Secondly, when additives are added to the PCMs, it leads to increase in the weight of the PCMs that causes the use of PCMs composed with additives to a limited extent only. Other disadvantages are aggregation, precipitation and if denser additives applied it decreases the use of PCMs application in various fields. Integration of additives may cause a decrease in capacity of storing latent heat. PCMs undergo various problems during phase transition like phase segregation, leakage in melting stage and corrosiveness are major issues. Therefore, for solving these problems, PCMs are enclosed in capsules. Through this PCMs can be prevented from environmental effect. This is helping in extending the utilization of PCMs for a long period. Encapsulation has more benefits like it provides a huge increase in the thermal contact area and in thermal energy storage systems it increases the rate of heat transfer. When PCMs transits phase from one phase to another phase capsules help in managing the volume change of PCM.

Increasing thermal conductivity of PCMs in this review paper is gypsum, graphite and carbon-based additives. Amongst both these methods gypsum-based additives are more beneficial in properties of density and stability as compared to the other additives. There are various applications of PCMs as in buildings, cooling systems, and in solar energy systems. In addition to the methods explained in the review paper, some new advanced materials and methods are needed to be investigated so that the demand of thermal energy storage can be full-filled. As PCMs have various applications in different fields because PCMs have some outstanding features like PCMs maintain constant temperature throughout the phase transition process and possess large latent heat. A suitable temperature during phase transition needs to be maintained in PCMs, whilst selecting a PCM for application. High-latent heat storage capacity is one of the main reasons for application of PCMs in recovery heat systems and solar energy systems so that the requirement of demand and supply can be adjusted. To maintain optimal environmental temperature, PCMs are applied to the buildings because PCMs have the property of constant rate temperature throughout the process due to this reason PCMs are useful in buildings, cooling systems for maintaining thermal comfort and in textiles. PCMs play a very important role in saving natural resources. In the energy development systems, PCMs are essential materials and will give great contributions in the future advanced developments.

Thermal conductivity of PCMs can be measured for a range of temperature when the range of temperature lies between 5 and 35 °C then the material density and thermal conductivity can be used because both are independent of temperature [18]. Secondly for human, thermal comfort requires average temperature 18–28 °C. Thermal conductivity and material density of gypsum are measured at room temperature 23 °C. It was investigated through various experiments that microencapsulated PCMs are more efficient in terms of compressive strength as compared to composite PCMs but in other aspects like reaction, thermal properties, workability and hydration composite PCMs and microencapsulated PCMs has similar effects over these properties. Many times, composite PCMs showing cracks after absorbing [26]. When PCMs are used in building envelope then it is required to fabricate in a panel as compared to direct incorporation of PCMs in concrete because direct incorporation of PCM causes various problems like risk of fire, poor thermal conductivity and causing low-mechanical strength of material which are mixed with PCMs are fabricated in the panels a vast amount of PCM is applied in the structures without having any harmful effect or any strength loss of mechanical properties [27].

References

1. Arivazhagan, R., & Prakash, S. (2020). Performance analysis of concrete block integrated with PCM for thermal management. *Materials Today: Proceedings, 22*, 370–374.
2. Rao, V. V., & Parameshwaran, R. (2018). PCM-mortar based construction materials for energy efficient buildings. A review on research trends. *Energy and Buildings, 158*, 95–122.
3. Bano, F. (2018). Evaluation of energy-efficient design strategies: Comparison of the thermal performance of energy-efficient office buildings in composite climate, India. *Solar Energy, 176*, 506–519.
4. David Beltran, R., & Martinez-Gomez, J. (2019). Analysis of phase change materials (PCM) for building wall boards on the effect of environment. *Journal of Building Engineering, 24*. https://doi.org/10.1016/j.jobe.2019.02.018.
5. Alkhazaleh, A. H. (2020). Preparation and characterization of isopropyl palmitate/expanded perlite and isopropyl palmitate/nanoclay composites as form-stable thermal energy storage materials for buildings. *Journal of Energy Storage*. https://doi.org/10.1016/j.est.2020.101679.
6. Philip, N., Veerkumar, C., & Sreekumar, A. (2020). Lauryl alcohol and stearyl alcohol eutectic for cold thermal energy storage in buildings: Preparation, thermophysical studies and performance analysis. *The Journal of Energy Storage, 31*, 101600. https://doi.org/10.1016/j.est.2020.101600.
7. Jeong, S. G., Wi, S., Chang, S. J., Lee, J., & Kim, S. (2019). An experimental study on applying organic PCMs to gypsum-cement board for improving thermal performance of buildings in different climates. *Energy and Buildings, 190*, 183–194.
8. Borreguero, A. M., Luz Sanchez, M., & Luis Valverde, J. (2010). Improvement of thermal behaviour of gypsum blocks by the incorporation of microcapsules containing PCMs obtained by suspension polymerization with an optimal core/coating mass ratio. *Applied Thermal Engineering, 30*(10), 1164–1169.
9. Zhu, Y., Wang, C. (2019). Smart utilization of solar energy with optic variable wall (OVW) for thermal comfort. *Energy and Buildings, 202*, 109376.
10. Zhang, H., & Yang, X. (2020). The CPMV for assessing indoor thermal comfort and thermal acceptability under global solar radiation in transparent envelope buildings. *Energy and Buildings, 225*, 110306.
11. Wu, Y.-F. (2009). The structural behaviour and design methodology for a new building system consisting of glass fibre reinforced gypsum panel. *Construction and Building Materials, 23*(8), 2905–2913. https://doi.org/10.1016/j.conbuildmat.2009.02.026.
12. Wang, H., & Lei, C. (2020). A numerical investigation of combined solar chimney and water wall for building ventilation and thermal comfort. *Building and Environment, 171*, 106616.
13. Ouakarrouch, M., El Azhary, K., Laaroussi, N. (2020). Thermal performance and environmental analysis of a new composite building material based on gypsum plaster and chicken feathers waste. *Thermal Science and Engineering Progress, 19*, 100642. https://doi.org/10.1016/j.tsep.2020.100642.
14. Eddhahak-Ouni, A., & Drissi, S. (2014). Experimental and multi-scale analysis of the thermal properties of Portland cement concretes embedded with microencapsulated Phase Change Materials (PCMs).
15. Rathore, P. K. S. (2020). Potential of microencapsulated PCM for energy savings in buildings: A critical review. *Sustainable Cities and Society, 53*, 3–9.

16. Rathore, P. K. (2019). Potential of macroencapsulated PCM for thermal energy storage in buildings: A comprehensive review.
17. Karaipekli, A. (2016). Development and thermal performance of pumice/organic PCM/gypsum composite plasters for thermal energy storage in buildings.
18. Toppi, T., & Mazzarella, L. (2013). Gypsum based composite materials with micro-encapsulated PCM: Experimental correlations for thermal properties estimation on the basis of the composition. *Energy and Buildings, 57*, 227–236.
19. Mohaine, S. (2016). Development and thermal performance of pumice/organic PCM/gypsum composite plasters for thermal energy storage in buildings. *Energy and Buildings, 131*, 99–112.
20. Wang, T. Y., & Wang, S. (2016). Microencapsulation of phase change materials with binary cores and calcium carbonate shell for thermal energy storage. *Applied Energy, 171*, 113–119.
21. Sharif, M. K. A., & Al-Abidi, A. A. (2015). Review of the application of phase change material for heating and domestic hot water systems. *Renewable and Sustainable Energy Reviews, 42*, 557–568.
22. Jiang, X., & Luo, R. (2015). Synthesis characterization and thermal properties of paraffin microcapsules modified with nano-Al_2O_3. *Applied Energy, 137*, 731–737.
23. Ye, R. D., & Lin, W. (2017). Experimental and numerical investigations on the thermal performance of building plane containing $CaCl_2.6H_2O$/expanded graphite composite phase change material. *Applied Energy, 193*, 325–335.
24. Xia, Y., & Zhang, X. (2016). Experimental research on a double-layer radiant floor system with phase change material under heating mode. *Applied Thermal Engineering, 96*, 600–606.
25. Cuce, P. M., & Rifflat, S. (2015). A comprehensive review of heat recovery systems for building applications. *Renewable and Sustainable Energy Reviews, 47*, 665–682.
26. Drissi, S., Ling, T. C., Mo, K. H., & Eddhahak, A. (2019). A review of microencapsulated and composite phase change materials: Alteration of strength and thermal properties of cement-based materials. *Renewable and Sustainable Energy Reviews, 110*, 467–484.
27. Al-Yasiri, Q., & Szabó, M. (2020). Incorporation of phase change materials into building envelope for thermal comfort and energy saving: A comprehensive analysis. *Journal of Building Engineering, 36*, 102122.

Coastline Protection Using a Rubble Mound Seawall: A Case Study

Prakhar Joshi, Prashant, Pritesh Goyal, and Pradeep K. Goyal

Abstract The coastal areas are very strategic and have significant economic importance. However, there are plenty of natural disasters coastal regions are facing, like a tsunami, cyclone, storm surge, etc. Wave action sediment transport can alter the coastline by causing silting, sedimentation, abrasion, and erosion. As per the study, Ramakrishna Beach in Visakhapatnam in state Andhra Pradesh, located on the east coast of India, is one of the most eroded regions. In this study, a rubble mound sea wall is designed to protect the Ramakrishna beach's coastline against the brutal forces of the waves. A rubble mound sea wall consists of three layers which are armor, secondary layer, and core. For designing, Engineering Manual EM 1110-2-1614 was used in this study. This will help to protect the coastline, and it is crucial in relation to coastal disaster mitigation.

Keywords Natural disaster · Environment · Coastline erosion · Rubble mound seawall · Mitigation

1 Introduction

Sea walls act as the defense system of the masses when it comes to fight the natural hazard that nature poses in the coastal regions, i.e., the waves which due to their regular interaction with the coastline erode and deteriorate it. The seawall acts as a protection to the coastline as well as to the structures that are on or near the coastline. There are various types of seawall like vertical seawall, circular seawall, concave seawall, convex seawall, mound seawall. All having their different functions to perform, but the vertical seawall is widely used due to its simplicity, convenience, and economy.

P. Joshi (✉) · Prashant · P. Goyal · P. K. Goyal
Department of Civil Engineering, Delhi Technological University, Delhi, India

P. K. Goyal
e-mail: pkgoyal@dtu.ac.in

A. K. Gupta et al. (eds.), *Advances in Construction Materials and Sustainable Environment*, Lecture Notes in Civil Engineering 196,
https://doi.org/10.1007/978-981-16-6557-8_25

There is an urgent need to pay attention to the coastal areas and to the coastline in specific as due to global warming the temperature has gone up by 0.08 °C per decade from 1981 and the sea level has been rising faster than ever seen in the history of mankind which puts our coast at a risk, which further is accelerated by the wave action that erodes the beaches.

This paper discusses the case of such a beach in Visakhapatnam, Andhra Pradesh. It is a beach whose erosion issue has been raised in the Parliament and the Press Information Bureau (PIB), and outlook newspaper press release supports the same.

There are three major types of seawalls which are generally used in coastal areas which are

(i) Vertical seawall; (ii) Curved seawall; (iii) Mound seawall.

They are described as below:

Vertical Seawall—They stand perpendicular to the approaching waves and receive the wrath of waves head on. The damage they receive is the maximum of the three walls. They are cheaper and most convenient to construct with certain drawbacks.

Curved Seawall—They stand at an angle to the approaching waves and receive the wrath of waves at an angle at which they are constructed. The damage they receive is reduced as compared to the vertical wall as the horizontal component of force of wave only contributes to the damage. They are complex, and their maintenance is difficult, but the water reflection coefficient of this wall is the maximum of the all three.

Mound Seawall—They are trapezium in shape with multiple layer which were absent in the above two. This type of wall is made by anchoring the stones into a specific position. It is a cheaper and environment friendly option proving to be a sustainable option that also cares of the marine life. It can be designed as porous and non-porous, thus giving us a greater edge above other walls.

Hanbin et al. [1] studied wave characteristics in front of an emerging or submerged ocean wall in the form of NewFLUME straight 2D numbers. Wave processes and reflection coefficients in front of the ocean wall were compared with the experimental data, in which case the direct distribution of high-velocity velocity and vertical velocity was compared with wavelength theory. Esteban et al. [2] summarized that there are no mechanisms in place for the development of weapon units in the fight against tsunamis. Both the results of laboratory tests and breakwater were studied in the field where they were analyzed according to well-established formulas. Ryu et al. [3] analyzed the failure process and processes for the destruction of rubble mound structures under unusual wave attacks.

Kato et al. [4] conducted a Two-phase simulation to calculate the pressure in the trapezoid-shaped coastal armor made with concrete throughout the tsunami overflow. The result was crosschecked with a large-scale model experiment where the pressure on the bottom of the coastal armor is measured. The simulation revealed that tension in the dike body intensified with overflow depth at the dike's seaward slope toe and that sheet piles at the dike's seaward slope toe mitigated pressure increase. Chun et al. [5] looked into the ability of perforated wall harbor structures to regulate waves during the design stage. They measured reflection and transmission coefficients, wave pressure exerted on perforated walls, and the dimensions of the mound supporting the upper, perforated wall. Yajnheswaran et al. [6] discussed about the berthing structures built in ports and harbors. Diaphragm walls and anchor rods were provided to support the formation of open spaces on seawater. The walls of the diaphragm were loaded due to the ground layer on one side of the building. Tamrin et al. [7] experimented and found out that perforated concrete blocks act as coral reef and acted as a hindrance in path of waves, thus reducing the wave height. Thus, it helps to make the coastal protection into an ecofriendly and economical option. These researches bring about that various study has been done in the developed countries, and this can be done in India as well.

This paper proposes the seawall to protect the coastline of the coastal disasters that nature poses every moment in the form of wave erosion.

2 Site Selection

For the purpose of the study, a site has been selected as Ramakrishna Beach, Visakhapatnam, Andhra Pradesh, India. The latitude and longitude of the site are 17.7142° N and 83.3237° E, respectively. The wall is measured to be 2900 m by the help of Google Maps measuring tool. The proposed seawall on the Ramakrishna beach is shown in Fig. 1.

The site, i.e., Ramakrishna Beach, has been selected as it has been eroding since 1990, and till 2014, it had lost 55 m of the beach due to constant erosion due to the lack of sand from the south side. Shri Y. S. Chowdary (MoS Science and Technology and Earth Sciences) said, "20 m beach was lost in the last two years while answering in the Parliament in 2015."

The Visakhapatnam Port Trust (VPT) has been trying to control soil erosion by regularly feeding Rama Krishna Beach with sand found in the outer harbor. In 2013 and 2014, hurricanes "Phailin" and "Hudhud" also accelerated the erosion of

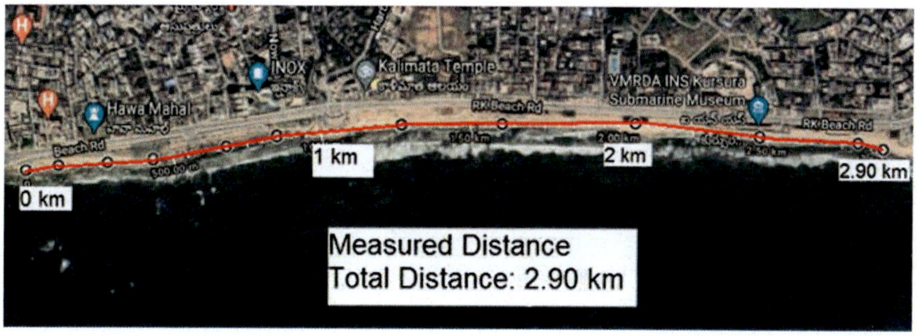

Fig. 1 Length of proposed seawall. *Source* Google Maps

Rama Krishna Beach. The population in this area is dense, and the minister was also concerned about this area's people and the importance of the road as this is a major road that connects to the port so to preserve it seawall is a necessity in a long run. On average, about 40% of India's ocean suffers from various erosions. The data of the last 25 years (1990–2014) has shown that 25–33% of the Andhra Pradesh coastline is experiencing erosion of various amounts.

A workshop on beach erosion in 2017 was conducted in Visakhapatnam where discussion on how the near shore currents were eroding the Ramakrishna beach and also affecting the biodiversity which was badly held as reported. The rubble mound wall is best suited to protect the biodiversity as it is made with stones and no use of any factory material is there so the biodiversity can regenerate with helping them to grow without damaging marine life [8].

The images of Ramakrishna Beach are shown in Fig. 2a–e for the year 2005, 2012, 2015, 2017, and 2020, respectively. The line marked in red color shows the coastline (as the base year of 2005) and is superimposed on the successive images,

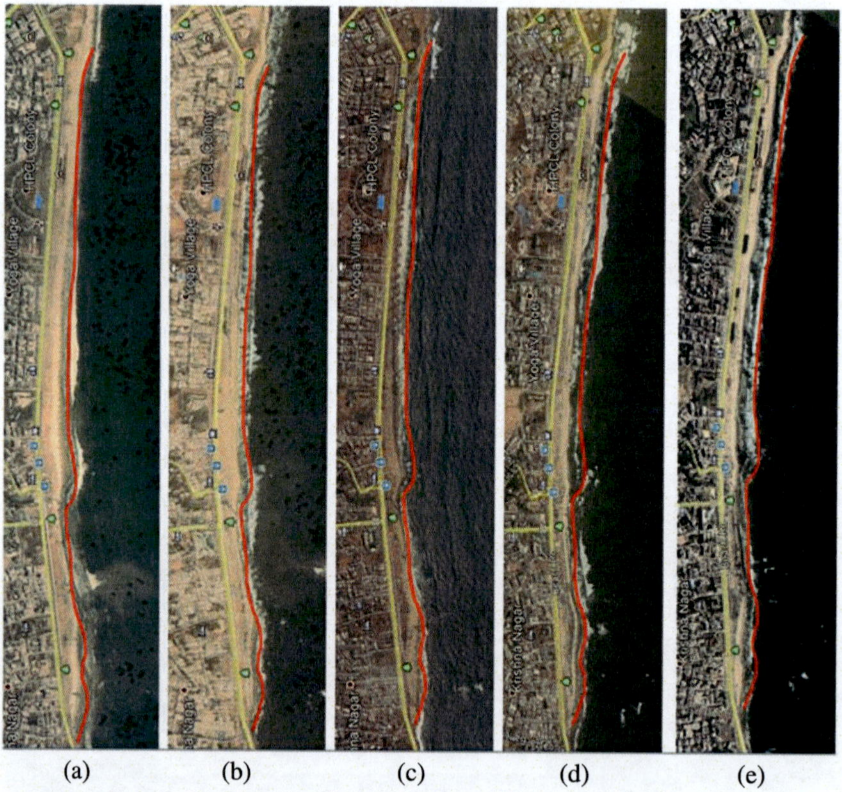

(a) (b) (c) (d) (e)

Fig. 2 Images of Ramakrishna Beach for the year 2005, 2012, 2015, 2017, and 2020 thus showing the erosion of the beach in coming years. This beach is selected as gradual erosion of beach can be seen from satellite images.

The stability constant (K_D) is defined as the cube of stability number, i.e., derived from Van der Meer formula and divided by the value of cot θ (as in section 2-16 of EM 1110-2-1614). The values of stability constant (K_D) are given in Table 1. The rough angular stone is chosen where placement of the stones is at random and slope is predetermined at 1.5, so the value of stability constant (K_D) comes out as 2.0 as highlighted in the table in yellow box.

The layer coefficient (K_Δ) has been found out using multiple experiments and has constant values as given in table below. The values of layer coefficient (K_Δ) are given in Table 2. The rough angular quarry stone is chosen where placement of the stones is at random and number of stones is taken as 2, so the value of layer coefficient (K_Δ) comes out as 1.0 as highlighted in the table in yellow box.

The average significant wave height at Ramakrishna Beach, Visakhapatnam, came out to be 1.516 m. This image is taken on February 24, 2021, at 5:30 pm for reference. This data is further used for designing of the rubble mound seawall.

Table 1 Values of stability constant (K_D)

Armor unit	n1	Placement	Slope (cot q)	K_D
Quarry stone				
Smooth rounded	2	Random	1.5–3.0	1.2
Smooth rounded	> 3	Random	1.5–3.0	1.6
Rough angular	1	Random	1.5–3.0	Do not use
Rough angular	2	Random	1.5–3.0	2
Rough angular	> 3	Random	1.5–3.0	2.2
Rough angular	2	Special[2]	1.5–3.0	7.0–20.0
Graded riprap	24	Random	2.0–6.0	2.2

Source EM 1110-2-1614

Table 2 Values of layer coefficient K_Δ

Armor unit	n	Placement	K_Δ	P (%)
Quarry stone (smooth)	2	Random	1	38
Quarry stone (rough)	2	Random	1	37
Quarry stone (rough)	≥ 3	Random	1	40
Graded riprap	2[a]	Random	N/A	37
Tetrapod	2	Random	1.04	50
Tribar	2	Random	1.02	54
Tribar	1	Uniform	1.13	47
Dolos	2	Random	0.94	56

Source EM 1110-2-1614

3 Designing of Rubble Mound Sea Wall

For designing of rubble mound seawall, the specifications for design are given in Table 3. Following procedure is followed as given below (Fig. 3).

3.1 Armor Design

Armor design consists of (i) Weight of armor unit; (ii) Crest width of armor layer; (iii) Thickness of armor layer.

Weight of Armor Uni

Weight of armor unit is calculated using the formula given as in Eq. 1.

Table 3 Specifications for design

Design conditions	Values
Specific gravity of armor	$S_r = 165$ lb/ft^3; $S_r = 165 * 16.0185 = 2643$ kg/m^3
Specific gravity of sea water	$S_w = 64$ lb/ft^3; $S_w = 64 * 16.0185 = 1025.18$ kg/m^3
Slope	1–1.5
Stability coefficient (K_D)	2.0 (as using rough quarry stone) (as per Table 1)
Significant wave height (H_s)	1.516 m
Wave height (H)	$1.516 + 1 = 2.516$ m
Depth of water (d)	2.516 m

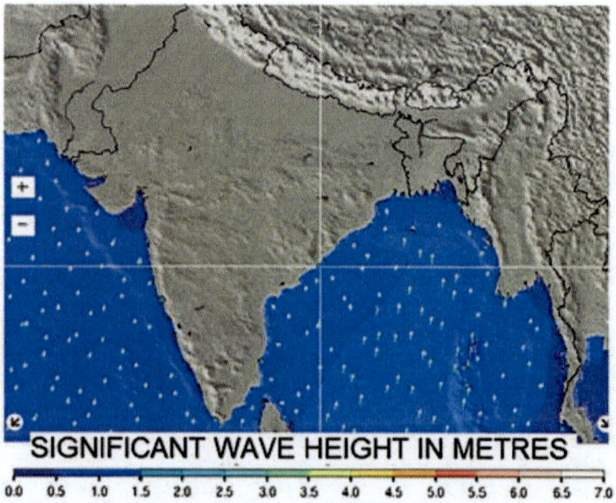

Fig. 3 Significant wave height [9]. *Source* https://www.surf-forecast.com/breaks/Vizag-East-Point/

$$W_{ar} = (S_r * H^3) / [K_D * \{(S_r/S_w) - 1\}^3 * \mathrm{Cot}\theta] \quad (1)$$

(as in section 2-16 equation 2-15 of EM 1110-2-1614) [10]

where

S_r = specific gravity of armor,
H = wave height,
K_D = stability constant taken from Table 1,
S_w = specific gravity of seawater, and
Cot θ = 1/slope.

After substituting all the values of the given variables mentioned above, we get the value of weight of armor as:

$$W_{ar} = (2643 * (2.516)^3)/[2 * \{(2643/1025.18) - 1\}^3 * 1.5] = 3570.40\,\mathrm{kg}$$

Thickness of armor layer

Thickness of armor layer is calculated using the formula given as in Eq. 2

$$t_{ar} = n * K_\Delta * (W_{ar}/S_r)^{1/3} \quad (2)$$

(as in section 2-17 equation 2-22 of EM 1110-2-1614) [10]

where

n = number of stones which is taken as 2,
K_Δ = constant as shown from Table 2,
W_a = weight of the armor layer, and
S_r = specific gravity of armor.

After substituting all the values of the given variables mentioned above, we get the value of thickness of armor as:

$$t_{ar} = 2 * 1.00 * (3570.40/2643)^{1/3} = 2.21\,\mathrm{m}$$

Crest width of armor layer

Crest width of armor layer is calculated using the formula given as in Eq. 3

$$B_{ar} = n * K_\Delta * (W_{ar}/S_r)^{1/3} \quad (3)$$

(as per equation VI-5-116 of EM 1110-2-1100 (Part VI)) [11]

where

n = number of stones which is taken as 3,
K_Δ = constant as shown from Table 2,
W_{ar} = weight of the armor layer, and
S_r = specific gravity of armor.

After substituting all the values of the given variables mentioned above, we get the value of crest width of armor as:

$$B_{ar} = 3 * 1.00 * (3570.40/2643)^{1/3} = 3.316\,m$$

3.2 Under Layer Design

Weight of the under layer is the one tenth to one fifteenth of weight of the armor layer unit. We have already calculated the weight of armor as 3570.40 kg from Eq. 1. Hence, the weight of under layer varies from 357.04 to 238.02 kg. As per the engineering manual, thickness of the under layer is equal to the thickness of the armor which comes out to be 2.21 m (as in section 2-20 of EM 1110-2-1614).

3.3 Toe Berm Design

Weight of toe berm is the one tenth to one fifteenth of weight of the armor layer unit. We have already calculated the weight of armor as 3570.40 kg from Eq. 1. Hence, the weight of toe berm varies from 357.04 to 238.02 kg. As per the engineering manual, the width of the toe berm is equal to twice the significant wave height (H_s) which comes out to be 3.032 m. Depth of toe berm is calculated as four by tenth of the depth of water which comes out to be 1.064 m (as in section 2-19 of EM 1110-2-1614).

3.4 Core Design

Weight of core layer is calculated as one hundredth to four hundredth of the weight of armor layer unit (this is the value which lies in for the core in designing of breakwaters so using the same as well). We already calculated the weight of armor as 3570.40 kg from Eq. 1. Hence, the weight of core varies from 35.704 to 8.926 kg. So, the thickness of the core layer is equal to the depth of the toe berm which comes out to be 1.064 m.

3.5 Filter Layer Design

Weight of the filter layer is equal to one thousandth to six thousandth of the weight of the armor layer unit (this is the value which lies in for the core in designing of breakwaters so using the same as well). We already calculated the weight of armor

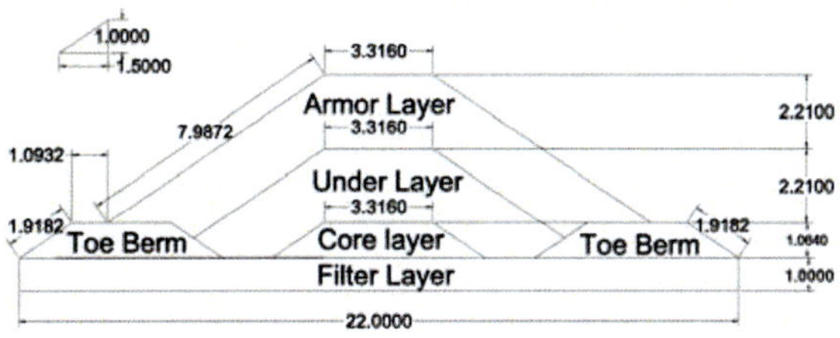

Fig. 4 Drawing of designed seawall (all dimensions in meters)

Table 4 Weight of components in a seawall

Layer	Area (m^2) (as per AutoCAD design)	Length (m)	Unit weight (kg)	Total weight (t)
Armor layer	30.0598	2900	3570.40	311,243.978
Under layer	20.6373	2900	357.04	21,368.19
Core	5.2264	2900	35.704	541.1498
Toe berm	2 * 4.9242 = 9.8484	2900	357.04	10,197.1909
Filter layer	22	2900	3.5704	227.7915

as 3570.40 kg from Eq. 1. Hence, the weight of filter layer varies from 3.5704 to 0.5905 kg. The height of the filter layer is taken as 1 m.

Structure height is the sum of thickness of armor layer, thickness of under layer, depth of toe berm, and the thickness of filter layer which comes out as 6.484 m (2.21 + 2.21 + 1.064 + 1 = 6.484).

All the components of the seawall are drawn in the AutoCAD and are shown in Fig. 4.

Various components of the rubble mound seawall are given in the table below where area of each component is calculated using the AutoCAD as shown in Fig. 4; the length of the seawall is taken as 2900 m from Fig. 1, and the unit weight of each component is calculated above, so the total weight of each component is given in Table 4. The cumulative weight of the seawall comes out to be 343,578.3 t.

4 Conclusion

The seawall is a structure constructed to prevent coastal erosion and mitigate ocean originated disasters such as storm surge and flooding as the region is densely populated due to its proximity to the Visakhapatnam port and is of much economic importance. It will also protect the strategic road alongside the beach which connects interiors of Visakhapatnam to the Visakhapatnam Port. As a case study, a seawall is designed for the Ramakrishna Beach with the height of 6.484 m and total weight of 343,578.3 tons which will almost resolve the problem of beach erosion. This design is made using the standard procedure of the engineering manuals, based on the literature, and it will be validated at later stage. This is the case study specifically for the Indian coastal region affected from coastal erosion. If the proposed seawall is constructed along the Ramakrishna Beach, we can prevent the erosion of coastline by 10 m (approx.) annually. This seawall is also ecofriendly as there is no use of factory materials and will help in conservation of the marine life.

References

1. Hanbin, G. U., Pengzhi, L. I. N., Yanbao, L. I., Taiwen, H. S. U., & Jianlue, H. S. U. (2003). Wave characteristics in front of vertical sea-walls. In *International Conference on Estuaries and Coasts*, November 9–11.
2. Esteban, M., Morikubo, I., Shibayama, T., & Munoz, R. A. (2012). Stability of rubble mound breakwaters against solitary waves. *Coastal Engineering*.
3. Ryu, C.-R., & Kim, H.-J. (1994). Failure of rubble mound structures due to the storm duration and the irregularity of ocean waves, Chapter 110. *Coastal Engineering* (pp. 1526–1540).
4. Kato, F., Saito, M., Himeno, K., & Suwa, Y. (2014). Computed tomography imaging inside the beach by matching of refractive index. *Journal of Japan Society of Civil Engineers, Series B2 (Coastal Engineering), 70*(2), I_971–I_975.
5. Chun, I., Min, Y., Ha, D., & Park, K.-s. (2013). A general purpose numerical tool for determining wave control capability of perforated coastal structures. A general purpose numerical tool for determining wave control capability. *Journal of Coastal Research* (65), 189–194.
6. Yajnheswaran, Ranjan, B., & Subba Rao. (2014). 2d and 3d analysis of a diaphragm wall type berthing structure under static loading. *International Journal of Scientific Engineering and Technology*, 371–376. ISSN: 2277-1581, Issue Special.
7. Tamrin, S. P., Parung, H., & Thaha, A. (2014). Experimental study of perforated concrete block breakwater. *International Journal of Engineering & Technology IJET-IJENS, 14*(03), 6–10.
8. https://www.outlookindia.com/newsscroll/researchers-discuss-beach-erosion-in-vizag/985431
9. https://www.surf-forecast.com/breaks/Vizag-East-Point/
10. Engineering Manual 1110-2-1614. Design of coastal revetments, seawall and bulkheads.
11. Engineering Manual 1110-2-1100. Coastal Engineering Manual.

Prediction of Compressive Strength of Rubberized Concrete Using Ordinary Least Squares Regression Model

Prabhat Kala, Shivam Upadhya, Pradhyumna Asthana, and Pradeep K. Goyal

Abstract Rubberized concrete (popularly known as Rubcrete) is a type of concrete in which crumb rubber replaces some amount of fine sand particles as fine aggregates. In this paper, ordinary least squares (OLS) regression was deployed to check its reliability in terms of predicting the compressive strength of rubcrete. To prepare the training data set for the model, concrete samples were prepared with percentage of rubber as a component of fine aggregate varying from 0 to 40% in increments of 5 percentage points. Each sample mix was tested for its compressive strength after intervals of 7, 14 and 28 days, respectively. Upon training the model with relevant input parameters, it predicted the compressive strength of rubcrete with a fair degree of accuracy. The score and mean squared error (MSE) were evaluated for the OLS model to find its general performance. For the model, a score of 0.959 and mean squared error of 1.745 indicated that the model was efficient and reliable.

Keywords Concrete · Rubberized concrete · Compressive strength · Regression

1 Introduction

The scarcity of natural fine aggregate, i.e. sand has led to the development of better alternatives which can lead to sustainability in the construction industry. One of such alternatives is crumb rubber which solves the problems of depleting resource of sand and the ever-increasing rubber waste. Partial substitution of mineral aggregate

P. Kala · S. Upadhya (✉) · P. Asthana · P. K. Goyal
Department of Civil Engineering, Delhi Technological University, Delhi, India
e-mail: shivamupadhya_2k17ce93@dtu.ac.in

P. Kala
e-mail: prabhatkala_2k17ce68@dtu.ac.in

P. Asthana
e-mail: pradhyumnaasthana_2k17ce69@dtu.ac.in

P. K. Goyal
e-mail: pkgoyal@dtu.ac.in

A. K. Gupta et al. (eds.), *Advances in Construction Materials and Sustainable Environment*, Lecture Notes in Civil Engineering 196,
https://doi.org/10.1007/978-981-16-6557-8_26

(fine and coarse aggregate) with recycled tyres crumb rubber has been seen as an appealing solution as it leads to the formation of concrete that has many improved properties. Lavagna et al. [1] showed that there is a two-fold environmental benefit associated with the usage of crumb rubber as fine aggregate in concrete with reduction in landfilling and burning for tyre disposal and doing away with excavation processes in quarries or along river courses.

Ganjian et al. [2] discussed that rubcrete is a term that is used for rubberised concrete in which a partial volume of mineral aggregates is replaced with an equal volume of crumb rubber or rubber chips. It finds use in the construction industry for its useful properties as well, aside from reducing environmental impact. Benazzouk et al. [3] found that there is significant potential in the use of waste rubber as fine aggregates for developing lightweight concrete. Also, water propagation is avoided due to the presence of rubber, thus offering protection to the steel reinforcement against corrosion. Khaloo et al. [4] observed that rubcrete is an effective absorber of sound and shockwave energy. Also fresh rubcrete had lower unit weight and desirable workability as compared to normal concrete. Hernandez-Olivares et al. [5] carried out studies to show that addition of crumb rubber to concrete slabs improved fire resistance. Liu et al. [6] observed that due to the rubber particles having low modulus of elasticity compared to mineral aggregates, rubcrete exhibited an obvious deformation performance and better toughness. Jalal et al. [7], Mohammed et al. [8] and Su et al. [9] observed that despite the various advantages of rubcrete, reduction in strength of concrete with increase in rubber content limits its use. Still, it can find its application when significant strength of hardened concrete is not a necessity. Miller and Tehrani [10] found that for dynamic applications of lightweight concrete, rubcrete is a good choice as the artificial aggregates provide better ductility and toughness to the concrete. Youssf et al. [11] concluded that pre-treatment methods for rubber used in rubcrete is not worth the time and cost involved. They result in the lowering of workability of concrete mix without improving the compressive strength to a great extent. Raj et al. [12] concluded that rubcrete samples showed lesser decrease in compressive strength than ordinary concrete samples when exposed to saline water and acidic environment for a duration of 90 days.

Since past research has shown that rubcrete is a material that may be increasingly used in the future, knowing the relationship of rubber content in concrete and its corresponding compressive strength is vital to create optimum design mixes. Since sampling concrete mixes and performing destructive tests to obtain the characteristic compressive strength takes a significant amount of time, machine learning has been employed in the research in order to predict the compressive strength of rubcrete samples for various durations.

This paper presents an approach that enables the prediction of in-situ compressive strength of concrete with a fair level of accuracy. Using a sample data set obtained by experimentation with rubcrete, an ML (machine learning) model has been created for the same by the researchers. The feasibility of such a model was checked by computing errors or losses induced and also by comparing the predictions made with the help of the model with the training data set as well as the validation data set.

2 Experimental Investigations

Indian standard specifications were followed to create cubical concrete samples of 15 cm × 15 cm × 15 cm size of M40 grade (M40 denotes a concrete mix that has a minimum compressive strength of 40 N/mm^2 at the end of 28 days when 15 cm cube specimens are tested). Based on the Indian Standard IS: 10262-2009 (BIS 2009), the mix proportions for M40 grade concrete were finalized [13]. These are given in Table 1. Compressive tests of concrete cubes made by replacing fixed percentages of fine aggregates with crumb rubber in the concrete mix for M40 grade were conducted.

2.1 *Compressive Test on Cubic Samples*

The compressive test was then performed on the 7th, 14th and 28th days from the commencement of curing by performing compressive strength tests on ACTM-5000 (Automatic Compression Testing Machine with capacity up to 5000 kN).

Table 2 shows the experimental results of compression tests conducted on concrete cube specimens on intervals of 7, 14 and 28 days.

After obtaining the experimental data, the compressive strength of concrete mix by increasing the rubber content was plotted to observe the drop in the compressive strength.

Figure 1 shows the results of compressive strength conducted on concrete cubes after intervals of 7, 14 and 28 days.

Table 1 Mix proportions

Mix designation	Cement (kg/m^3)	FA (kg/m^3)	CA (kg/m^3)	Water (kg/m^3)	Crumb rubber (kg/m^3)
M40R0	350	896	1140	140	0
M40R5	350	851.2	1140	140	44.8
M40R10	350	806.4	1141.2	140	89.6
M40R15	350	761.6	1140.6	140.2	134.4
M40R20	350	716.9	1139.7	140.4	179.1
M40R25	349.6	671.9	1140.5	140.5	224.1
M40R30	349.3	627.2	1139.2	140.7	268.8
M40R35	348.9	582.5	1140.9	140.8	313.5
M40R40	349.8	537.6	1140.1	141	358.4

Note In the designation, M stands for mix, 40 denotes concrete with 40 N/mm^2 characteristic compressive strength which is observed after a period of 28 days. RX represents the crumb rubber content as X% where X is the percentage of total fine aggregate in the mix which is replaced by crumb rubber

Table 2 Results of the OLS regression model for determining accuracy

Mix designation	Number of days	Experimental value (E)	Predicted value (P)	Difference $D = \|E - P\|$	Error percentage[a]
M40R0	7	26.8	30.1	3.3	12.3
M40R0	14	34.5	32.3	2.2	6.4
M40R0	28	38.3	36.8	1.5	3.9
M40R5	7	25.0	27.0	2.0	8.0
M40R5	14	31.9	29.2	2.7	8.5
M40R5	28	35.2	33.6	1.6	4.5
M40R10	7	21.7	23.9	2.2	10.1
M40R10	14	27.2	26.1	1.1	4.0
M40R10	28	30.6	30.5	0.1	0.3
M40R15	7	19.3	20.8	1.5	7.2
M40R15	14	25.3	23.0	2.3	9.1
M40R15	28	27.2	27.4	0.2	0.7
M40R20	7	15.1	17.7	2.6	17.2
M40R20	14	20.5	19.9	0.6	2.9
M40R20	28	23.4	24.3	0.9	3.8
M40R25	7	13.5	14.6	1.1	8.1
M40R25	14	18.0	16.8	1.2	6.7
M40R25	28	19.5	21.2	1.7	8.7
M40R30	7	11.8	11.5	0.3	2.5
M40R30	14	15.8	13.7	2.1	13.3
M40R30	28	16.7	18.1	1.4	8.4
M40R35	7	9.1	8.4	0.7	7.7
M40R35	14	11.0	10.6	0.4	3.6
M40R35	28	12.4	15.0	2.6	21.0
M40R40	7	6.9	5.2	1.7	24.6
M40R40	14	9.7	7.5	2.2	22.7
M40R40	28	10.6	11.9	1.3	12.3

[a] Error percentage = [(Difference)/(Experimental value)] * 100

3 OLS Regression Model: An Overview

Regression is a method of developing relationships between one (or multiple) input parameters and one (or multiple) output parameters. Uyanik et al. [14] claimed regression as a method of developing relationships between one (or multiple) input parameters and one (or multiple) output parameters. In statistics, ordinary least squares (OLS) is a method which is used to establish the best possible equation of a line

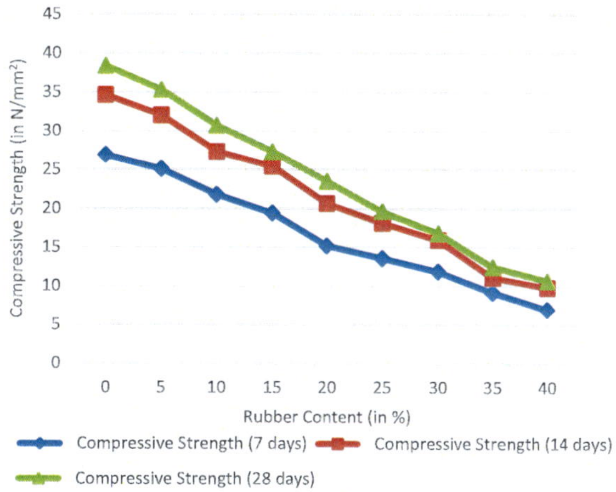

Fig. 1 Results of compressive strength tests conducted on concrete cubes after intervals of 7, 14 and 28 days

(known as line of best fit) between a dependent variable and one or more independent variables. Thus, the OLS model performs iterations by adjusting the coefficients associated with the input parameters so as to arrive at an optimal value of the individual coefficients which results in the minimum possible value of the mean square error (loss). This corresponds to the equation of line of best fit.

3.1 Training Data

For the creation and calibration of the OLS regression model, training data needs to be developed. Thus, the experimental data recorded was compiled in an excel sheet in CSV (comma-separated values) format with the columns being as follows:

- Rubber percentage
- Number of days
- Rubcrete strength.

3.2 Results of OLS Regression Algorithm

After the initial preparation of the model using the employed training dataset, the latter was cross-checked with the predictions for the same input parameters and hence differences in compressive strengths as obtained through experimentation and by the model's predictions were evaluated. These results are given in Table 2.

To find the model's general performance, its score and MSE (mean squared error) were evaluated. Ideally, a high score and low MSE are deemed to be ideal for a good model. These are given as follows:

Fig. 2 Plot of line of best fit

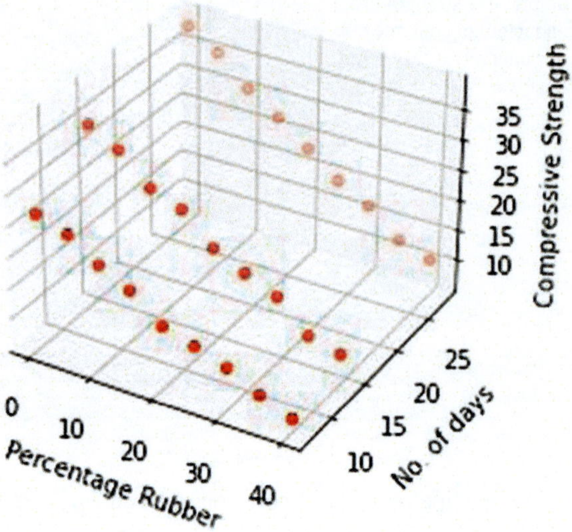

- Score: 0.959462
- MSE: 1.744713

The regression coefficients and intercept for the line of best fit were also calculated using the model. These are given below:

Regression coefficients:

- For rubber content: −0.6218
- For no. of days: 0.3161
- Intercept: 27.9022

The line of best fit has been depicted by a 3-D scatter plot as shown in Fig. 2.

The computed regression coefficients indicate that with the increase in rubber content, the compressive strength decreases, and with increase in number of days after which samples are tested, the compressive strength increases. It is practically advisable to use the model, with the number of days not exceeding 28 days, as the sample attains 99% of its full compressive strength by the end of 28 days.

Visual representation of the feasibility of the model is provided through Figs. 3, 4 and 5 in which experimental compressive strength and predicted compressive strength are pitted against each other on a graph of compressive strength vs rubber content for 28, 14 and 7 days, respectively. The graph obtained through experimental procedure nearly coincides with the line of best fit of the regression model, further augmenting the success of the research.

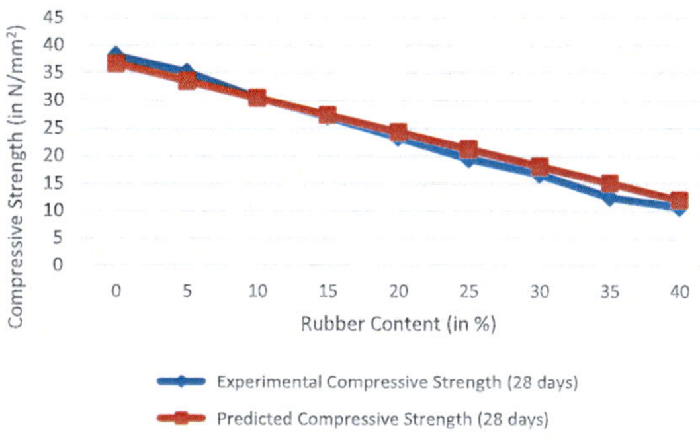

Fig. 3 Comparison of experimental results and OLS regression model results for compressive strength at 28 days

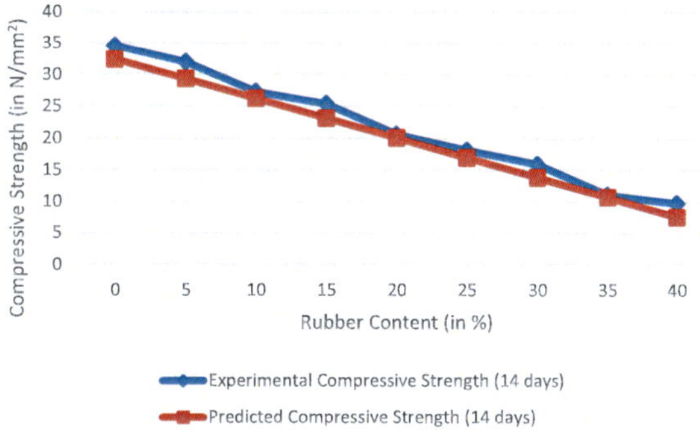

Fig. 4 Comparison of experimental results and OLS regression model results for compressive strength at 14 days

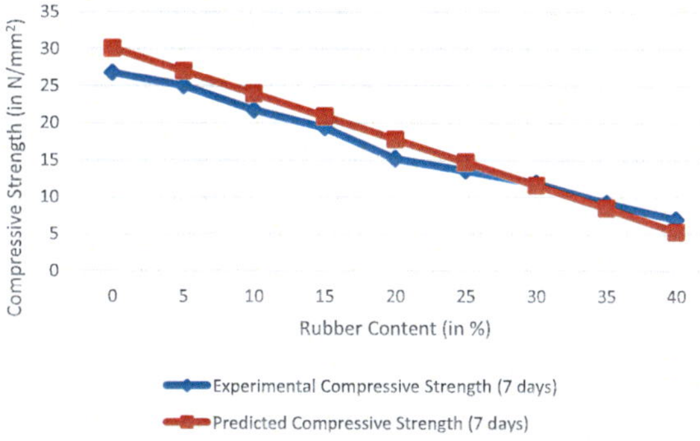

Fig. 5 Comparison of experimental results and OLS regression model results for compressive strength at 7 days

4 Conclusion

This research paper aimed at providing an insight into the strength characteristics of the innovative material called rubcrete and further attempted at creating a machine learning model for the same in order to predict the compressive strength of the resulting sample from the design mix itself. Major conclusions that can be drawn from the research are presented as follows:

- Whilst it is well established that the compressive strength of a rubcrete mix is less than that of a conventional concrete mix, rubcrete as a novel composite material can be used in the construction industry when the required final strength is not high. Since high-strength performance is not expected out of a rubcrete mix, appropriate modifications can be done to the design mix (for normal concrete) to incorporate the different strength trends as observed through experimentation.
- It takes 28 days to find the compressive strength of a single sample of concrete mix. Thus, it becomes tedious to carry out the compressive strength tests for all the different samples. Machine learning can prove to be extremely efficient in such cases as it can help in predicting results with a high degree of accuracy in a time efficient manner. Obtaining an estimate for the compressive strength of a concrete mix using the machine learning algorithm can help speed up construction activities which would otherwise take longer due to the delay in computing the compressive strength results.
- The experimental and regression results were close enough for the model to be ascertained as viable and practicable for future use. The ML model further makes the process of sampling obsolete to a certain extent.

References

1. Lavagna, L., Nisticò, R., Sarasso, M., & Pavese, M. (2020). An analytical mini-review on the compression strength of rubberized concrete as a function of the amount of recycled tires crumb rubber. *Materials, 13*(5), 1234. https://doi.org/10.3390/ma13051234
2. Ganjian, E., Khorami, M., & Maghsoudi, A. A. (2009). Scrap-tyre rubber replacement for aggregate and filler in concrete. *Construction and Building Materials, 23*(5), 1828–1836. https://doi.org/10.1016/j.conbuildmat.2008.09.020
3. Benazzouk, A., Douzane, O., Langlet, T., Mezreb, K., Roucoult, J. M., & Quéneudec, M. (2007). Physico-mechanical properties and water absorption of cement composite containing shredded rubber wastes. *Cement & Concrete Composites, 29*, 732–740. https://doi.org/10.1016/j.cemconcomp.2007.07.001
4. Khaloo, A. R., Dehestani, M., & Rahamatabadi, P. (2008). Mechanical properties of concrete containing a high volume of tire-rubber particles. *Waste Management, 28*(12), 2472–2482. https://doi.org/10.1016/j.wasman.2008.01.015
5. Hernandez-Olivares, F., & Barluenga, G. (2004). Fire performance of recycled rubber-filled high-strength concrete. *Cement and Concrete Research, 34*(1), 109–117. https://doi.org/10.1016/s0008-8846(03)00253-9
6. Liu, H., Wang, X., Jiao, Y., & Sha, T. (2016). Experimental investigation of the mechanical and durability properties of crumb rubber concrete. *Materials, 9*(3), 172. https://doi.org/10.3390/ma9030172
7. Jalal, M., Nassir, N., & Jalal, H. (2019). Waste tire rubber and pozzolans in concrete: A trade-off between cleaner production and mechanical properties in a greener concrete. *Journal of Cleaner Production, 238*, 117882. https://doi.org/10.1016/j.jclepro.2019.117882.
8. Mohammed, B. S., Anwar Hossain, K. M., Eng Swee, J. T., Wong, G., & Abdullahi, M. (2012). Properties of crumb rubber hollow concrete block. *Journal of Cleaner Production, 23*(1), 57–67. https://doi.org/10.1016/j.jclepro.2011.10.035.

9. Su, H., Yang, J., Ling, T.-C., Ghataora, G. S., & Dirar, S. (2015). Properties of concrete prepared with waste tyre rubber particles of uniform and varying sizes. *Journal of Cleaner Production, 91*, 288–296. https://doi.org/10.1016/j.jclepro.2014.12.022
10. Miller, N. M., & Tehrani, F. M. (2017). Mechanical properties of rubberized lightweight aggregate concrete. *Construction and Building Materials, 147*, 264–271. https://doi.org/10.1016/j.conbuildmat.2017.04.155
11. Youssf, O., Hassanli, R., Mills, J. E., Skinner, W., Ma, X., Zhuge, Y., Roychand, R., & Gravina, R. (2019). Influence of mixing procedures, rubber treatment, and fibre additives on rubcrete performance. *Journal of Composites Science, 3*(2), 41. https://doi.org/10.3390/jcs3020041.
12. Raj, A., Nagarajan, P., & Shashikala, A. P. (2020). Behaviour of fibre-reinforced rubcrete beams subjected to impact loading. *Journal of the Institution of Engineers (India): Series A, 101*(4), 597–617. https://doi.org/10.1007/s40030-020-00470-4.
13. BIS (Bureau of Indian Standards). (2009). *Concrete mix proportioning—Guidelines. IS: 10262-2009*. BIS.
14. Uyanık, G. K., & Güler, N. (2013). A study on multiple linear regression analysis. *Procedia - Social and Behavioral Sciences, 106*, 234–240. https://doi.org/10.1016/j.sbspro.2013.12.027

Effect of Soil Fill on the Load Distribution Characteristics of RC Skew Box Culverts for Road Under Bridge Design

Shimol Philip, R. Rakendu, and Rajesh Lal

Abstract Box culvert for road under bridge (RUB) design is widely used for providing overhead passages for trains and highways through the box culvert. The varying topsoil fill height may positively or negatively affect the structural integrity of box culverts. The height of backfill over the culvert can influence vertical earth pressure and internal structural forces. The current work aimed to investigate the effect of soil filling on the load distribution characteristics of RC skew box culvert for RUB design. The RUB structures considered in this work are associated with a complex set of loadings such as self-weight of the structure, superimposed dead loads (SIDL) from railway components and soil fill, lateral earth pressures, surcharge from dead and live loads, high- and low-intensity live loads from different dynamic effects such as axle movement, replacement of existing tracks, curved tracks, wind forces, and longitudinal forces. The height of soil fill above the top slab was varied from 0.5 to 10 m with increments of 0.5 m, and corresponding loading intensities were calculated. A detailed comparative study on the varying dispersion of live and dead loads through soil fill for different fill heights was also carried out. The results show that the live-load effect does diminish with increasing fill depth. At fill depths beyond about 6.5 m, the live-load pressures are less than 10% of the dead-load pressure, and beyond this fill height, live-load effects can be neglected. Moreover, the intensity of lateral earth pressure was observed significantly higher than that of vertical soil fill loads.

Keywords Box culvert · Road under bridge (RUB) design · Skew box culverts · Soil fill · Earth pressure

S. Philip (✉) · R. Rakendu
Department of Civil Engineering, Saintgits College of Engineering (Autonomous), Kottayam, Kerala, India

R. Rakendu
e-mail: rakendu.r@saintgits.org

R. Lal
Kerala Water Authority, Kollam, Kerala, India

A. K. Gupta et al. (eds.), *Advances in Construction Materials and Sustainable Environment*, Lecture Notes in Civil Engineering 196,
https://doi.org/10.1007/978-981-16-6557-8_27

1 Introduction

Box culverts are essential infrastructure for transportation with great economic relevance [1]. According to IRC 5-1998, slab or box culverts are generally used over major bridges for crossing obstacles with a span of less than or equal to 6 m [2]. The increasing population and traffic problems on the Indian railways paved the way for constructing road under bridge (RUB) and road over bridge (ROB) structures. Since cities are densely populated, land acquisition for the construction of ROB structures is difficult, and most engineers prefer RUB structures. The design of the RCC box culvert for road under bridge (RUB) is based on the Indian Railway Standards (IRS) Bridge Rule.

The major forces acting on box culverts include dead loads, live loads, earth pressure, water pressure, and surcharge from live and dead loads. Typically, the dead loads acting on the box culvert for the RUB design include superimposed dead loads (SIDL) from the sleepers, rails, ballast, footpath, and soil filling, as well as the self-weight of the culvert. The culvert installation type, side fill compaction, and culvert foundation significantly affect the dead load contribution to the culvert [3]. Live load refers to the various dynamic effects from traffic across the culvert and the general environmental factors such as wind, rainfall, and temperature fluctuations. The live loads acting on the railway tracks are distributed through the sleepers, ballast, rails, and soil fill before it approaches the box culvert. The distribution of these loads has a major impact on the magnitude of loads that will approach the culvert. If the amount of load distribution is wider than what is presumed in the design calculations, the magnitude of live load stress is smaller than expected, and its effect on the box culvert is reduced [3]. The effect of the live load may be negligible compared to the influence of the dead load at certain fill depths, and therefore need not be included in the design of box culverts.

Orton et al. [3] determined the effects of live load on RC box culverts under soil fills with different thicknesses and found that the live load effect of culverts diminishes with increasing fill height. The calculated live load pressure was less than 10% of the dead load pressure for a fill height of beyond 1.82 m. Similarly, Abdel-Karim et al. [4] studied experimentally the pressure distribution in concrete box culverts covered with different amounts of soil fill by driving a test truck across the culvert. The research revealed that peak stress in the soil was about 300 psf with 2 ft of fill and decreased to about 100 psf at 8 ft of fill. They observed that the effect of the live load decreases beyond 8 feet of filling. Hence, the live load effect was ignored when the live load effect was less than 5 percent of the total load effect. Kim and Yoo [5] conducted a study by varying fill heights from 15.2 to 61 m for embankment installation conditions and showed that the load on a structure was affected by installation conditions in addition to the height of the soil backfill. Also, they concluded that the soil structure interaction factor increases with an increase in backfill height. Sharma and Sinha [6] studied the structural design of box culverts made of RCC with and without cushion and have shown that the live load effect from braking force and impact from traffic should be considered for box culverts without providing soil cushions.

In summary, previous research has indicated that increasing soil fill resulted in a reduction of the live load effect and a corresponding increase in dead loads from soil fill. However, most of the studies on the dispersion of live and dead loads for different soil fill heights were performed on single-celled box culvert for road over bridge (ROB) design where the path of the natural stream through box culvert crosses roads, railways flyovers, etc., above box structure. The current work aims to determine the effects of soil fill heights on the dead load and the live load distribution characteristics of a reinforced concrete box culvert for road under bridge (RUB) design. For a RUB design, railway traffic is passing above the culvert, and the highway is going through the structure. Therefore, this study was conducted by considering complex combinations of live load effects from railway traffic as well as other dead loads. Live load effects on box culverts usually reduce with an increase in soil filling. Moreover, when a significant fill is placed above the crown of the culvert, the influence of the live load is negligible compared to the effect of the dead load.

2 Methodology

2.1 Description of the Box Culvert

The site chosen for the current work is a road under bridge (RUB) built between the Faridabad to Baramulla of the WDFC project. The Western Dedicated Freight Corridor (WDFC) is a broad gauge corridor under construction that connects Mumbai and Delhi. This project was intended to provide the conveyance for fertilizers, iron, cement, salt, steel, and food grains. The WDFC project covers a length of 1504 km of electric track with dual-lane from Jawaharlal Nehru Port (JNPT) to Dadri, crossing Vadodara, Ahmedabad, Palanpur, Phulera, and Rewari. The design variables required for the work, such as geometrical features of box culvert, and railway track sections, backfilling soil properties, details of steel, and the concrete properties, were collected from this site.

Geometry

A schematic representation of the detailed dimensions of a two-celled box culvert obtained from the site under consideration is shown in Fig. 1.

The major geometric parameters are the clear horizontal opening, vertical opening, effective width, effective height, skew angle, depth of haunches, wall and slab thicknesses. The clear dimensions of a single cell are 10.87 m $\times$ 5.60 m, and the center-to-center distances or effective dimensions are 11.750 m $\times$ 6.70 m. In this work, the railway track is above the culvert, and a highway passes through the culvert at a skew angle of 15°. Moreover, the corners of the box culvert have been provided with slight bends called haunches for restricting the concentration of stresses, and the depth and

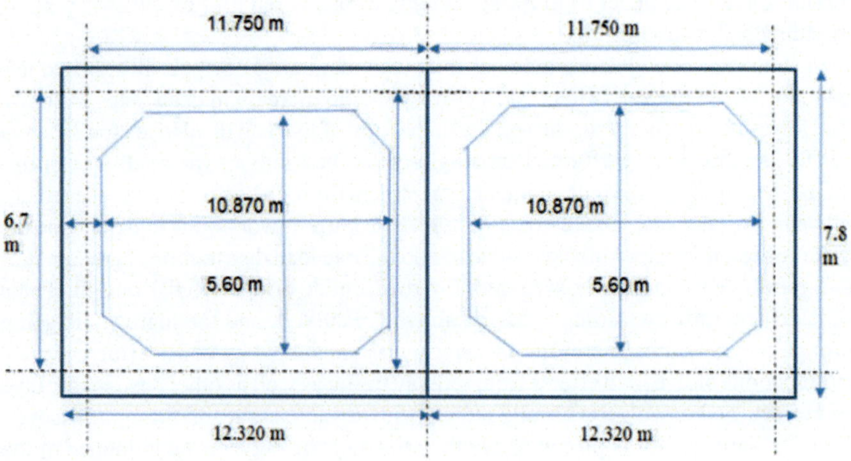

Fig. 1 Dimensions of a two-celled box culvert

width of haunches are 150 mm each. The two cells of the box culverts have a fixed slab thickness and an external wall thickness which is equal to 1100 mm. The two cells were separated by an intermediate wall of thickness 600 mm.

Material properties
The material properties used in modeling of the culverts were the standard grade of concrete with a compressive strength of 35 MPa (5075 psi), steel grade with a yield strength value of 500 MPa, and the modulus of elasticity of 200,000 MPa. The density of the concrete was 25 kN/m^3 and that of the underlying soil and ballast in the railway track was 20 kN/m^3 and 20.40 kN/m^3, respectively. The soil was considered at rest condition with a lateral pressure coefficient K_o of 0.5.

2.2 *Loading*

Dead loads, super-imposed dead loads, live loads, earth pressure, and surcharge due to the dead and live loads were the various loads considered in this study to investigate the load distribution in RCC box culverts. The culvert was subjected to super-imposed dead loads (SIDL) from topsoil fill, footpath, ballast, sleepers, and rails. The culvert self-weight is not a variable, and therefore, it is not included in the comparative analysis of dead load and the live load. However, the deadweight of the culvert should be a part of the load combination of the soil and live loadings in the final analysis and design of the culverts. The height of soil fill above the top slab was varied from 0.5 to 10 m (32.80 ft), with increments of 0.5 m (2 ft). The super-imposed dead loads from rails, sleepers, and ballast have been calculated by considering the corresponding dispersion widths as per clause 2.3.4.2(b) of the IRS

Bridge rule [7]. The dead and live loads from the railway footpath were calculated based on the dimensions of the kerb, handrail, slab, and ballast retainer of the footpath. The lateral earth pressure acting on the box culvert was determined based on clause 5.7.1 of IRS Foundation and substructure code [8].

The live Equivalent Uniformly Distributed Load (EUDL) was calculated as described in the IRS Bridge Rules [7]. For the maximum bending moment at the center and maximum shear force at the support section of the loaded beam of the effective length of the beam, EUDL has been calculated corresponding to maximum moment and maximum shear produced due to all five combinations of trains given in Appendix XXVI and XXVII as per IRS DFC 32.5 t axle loading. Following Appendix XVII of the IRS Bridge Rule [7], a design approach for estimating live load intensity from derailment loads of box culvert with guard rails was determined. The surcharge loads from both live and dead loads were calculated by considering two cases; when the depth of section "h" < (L - B) or box length is greater than dispersed width and vice versa. The live load surcharge for unit length was taken from Table 3 (for DFC loading (32.5 t axle load) of IRS substructures and foundation of bridges [8]. According to tables 1 and 2 of IS 875 part 3 [9], the design speed for calculating live load effects from wind was determined by considering three criteria: risk level, terrain roughness, height, and size of the structure, and local topography. The live load intensity from longitudinal loads due to traction and braking forces was calculated as per Appendix XXVIII of Bridge rule for IRS DFC loading 32.5 t axle loads. The load due to Plasser's Quick Relay System (PQRS) for broad gauge (BG) shall be considered for reduced coefficient of dynamic augment for a maximum speed of 20 kmph as per Appendix X of IRS Bridge rule for the most unfavorable position. For the curved track, provision must be made for the centrifugal motion of the moving load, and all tracks in the structure were considered to be occupied. According to clause 2.5.3(b) of the IRS Bridge rule, the horizontal load due to centrifugal force has been assumed to act at a height of 3000 mm for "DFC loading (32.5 t axle load)" for broad gauge above rail level. Table 1 shows the intensity of loading for different soil fill heights considered in this study to investigate the load distribution in concrete box culverts.

3 Results and Discussion

3.1 Effect of Fill Height on the Dispersion of Superimposed Dead Loads

The soil fill, railway components, and railway footpath above the culvert and self-weight of the structure were the contributors of dead loads in RUB box culverts considered in the current work. From Table 1 and Fig. 2, it was evident that the superimposed dead loads (SIDL) from soil fill increased with an increase in soil fill height. However, SIDL from the footpath and railway components decreases with an

Table 1 Loads acting on box culvert for RUB design

Type of loading	Standard codes	Intensity of load for different soil fill heights (kN/m)																			
		0.5	1	1.5	2	2.5	3	3.5	4	4.5	5	5.5	6	6.5	7	7.5	8	8.5	9	9.5	10
Dead load due to self weight	IRS Bridge Rule	97.5	97.5	97.5	97.5	97.5	97.5	97.5	97.5	97.5	97.5	97.5	97.5	97.5	97.5	97.5	97.5	97.5	97.5	97.5	97.5
SIDL from rail, sleeper and ballast	IRS Bridge Rule	9.51	8.85	8.29	7.8	7.39	6.98	6.63	6.31	6.02	5.76	5.52	5.29	5.09	4.9	4.72	4.56	4.41	4.27	4.13	4.01
SIDL from footpath	IRS Bridge Rule	0.722	0.67	0.62	0.59	0.57	0.52	0.50	0.47	0.45	0.43	0.41	0.40	0.38	0 37	0.35	0.34	0.33	0.32	0.31	0.30
SIDL from top soil fill	IRS Bridge Rule	12	22	32	42	52	62	72	82	92	102	122	132	142	152	162	172	182	192	202	202
Total SIDL on box culvert	IRS Bridge Rule	22.23	31.52	40.91	50.39	59.96	69.51	79.13	88.79	98.48	108.19	117.94	127.7	137.48	147.27	157.08	166.91	176.74	186.59	196.45	206.31
Active earth pressure	IRS Foundation and substructure code	101.68	108.16	114.64	121.11	127.59	134.06	140.54	147.02	153.5	159.97	166.45	172.92	179.4	185.88	192.36	198.83	205.31	211.78	218.26	224.74
Earth pressure at rest	IRS Foundation and substructure code	117	124.46	131.91	139.36	146.81	154.27	161.72	169.17	176.62	184.08	191.53	198.98	206.43	213.89	221.34	228.79	236.24	243.7	251.15	258.6
Live load (LL) from footpath	IRS Bridge Rule	0.32	0.3	0.28	0.27	0.26	0.24	0.23	0.22	0.21	0.2	0.19	0.18	0.17	0.17	0.16	0.16	0.15	0.15	0.14	0.14
LL intensity from dynamic effects for bending moment	IRS Bridge Rule	30.53	21.77	20.4	19.2	18.62	17.17	16.31	15.52	14.81	14.17	13.58	13.03	12.52	12.06	11.62	11.23	10/85	10.5	10.17	9.86

(continued)

Table 1 (continued)

Type of loading	Standard codes	Intensity of load for different soil fill heights (kN/m)																			
		0.5	1	1.5	2	2.5	3	3.5	4	4.5	5	5.5	6	6.5	7	7.5	8	8.5	9	9.5	10
Derailment load	IRS Bridge Rule	30.13	29.04	28.02	27.07	26.19	25,36	24.58	23.84	23.15	22.5	21.89	21.3	20.75	20.23	19.73	18.8	18.37	17.96	17.56	17.56
Wind load	IS 875 part 3 and IRS Bridge Rule	6.33	5.89	5.52	5.2	5.04	4.65	4.41	4.2	4.01	3.83	3.68	3.53	3.39	3.26	3.15	3.04	2.94	2.84	2.75	2.67
Longitudinal Force	IRS Bridge Rule	4.63	4.31	4.04	3.8	3.69	3.4	3.23	3.07	2.93	2.8	2.69	2.58	2.48	2.39	2.3	2.22	2.15	2.08	2.01	1.95
PQRS load	IRS Bridge Rule and IRS CBC	43.07	37.66	34.38	31.66	29.36	27.41	26.07	25.13	24.28	23.5	22.78	22.11	21.49	20.91	20.36	19.84	19.36	18.9	18.46	18.04
LL surcharge (at rest condition)	IRS Foundation and substructure code	11.76	10.94	10.25	9.64	9.11	8.63	8.25	8.05	7.86	7.67	7.49	7.33	7.16	7.01	6.86	6.72	6.59	6.46	6.33	6.21
LL surcharge (active condition)	IRS Foundation and substructure code	7.26	6.75	6.32	5.95	5.62	5.32	5.09	4.97	4.85	4.73	4.62	4.52	4.42	4.33	4.32	4.15	4.06	3.98	3.91	3.83
DL surcharge (at rest condition)	IRS Foundation and substructure code	4.75	4.42	4.14	3.9	3.68	3.49	3.31	3.15	3.01	2.9	2.84	2.78	2.72	2.66	2.61	2.56	2.51	2.47	2.42	2.38
DL surcharge (active condition)	IRS Foundation and substructure code	2.93	2.73	2.56	2.41	2.27	2.15	2.04	1.95	1.86	1.79	1.75	1.71	1.68	1.64	1.61	1.58	1.55	1.52	1.49	1.47

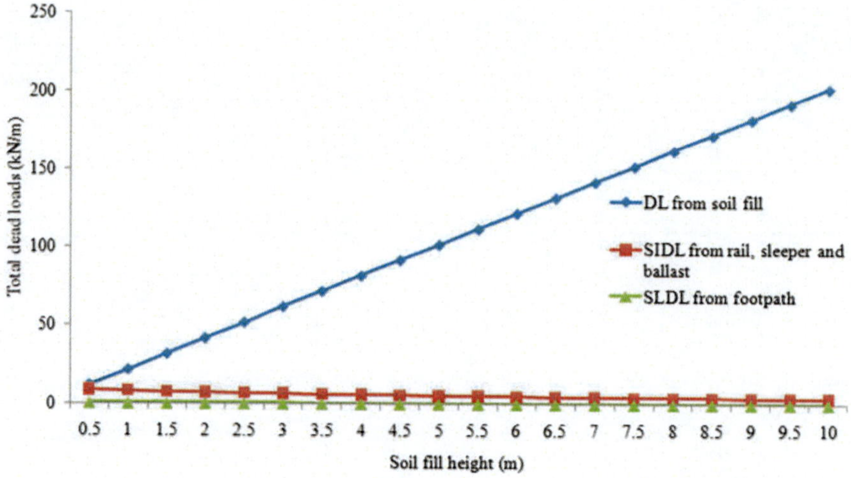

Fig. 2 Variation of dead loads from soil, sleeper, rail, ballast, and footpath with fill height

increase in fill height. The magnitude of SIDL from soil fill and railway components was comparable up to a fill height of 0.5 m. After 0.5 m soil fills height, there is a large increase in dead load from the soil, but the SIDL from railway components and footpath decreased with increased soil fill after 0.5 m. This is due to the dispersion of applied loads in a direction perpendicular to the direction of applied loads through soil fill. The increased dead loads can lead to large-scale deflections on the top slabs and sidewalls of the box culvert, and therefore, thickened slabs are required to resist the large loads.

3.2 *Effect of Soil Fill on the Dispersion of Live Loads*

Figure 3 shows a comparison of the intensity of total SIDL from the soil, footpath, and railway components with the major live load effects on the RUB box culvert from various dynamic movements on the railway track. The RUB box structures were subjected to complex live loads such as force produced by the centrifugal action of the partially unbalanced weight of the reciprocating parts of the locomotive, impacts from derailed trains, centrifugal action of moving loads on curved track, and replacement of existing rails. From Table 1 and Fig. 3, it is seen that the intensity of live loads from dynamic effects is relatively lower than the total SIDL except at 0.5 m fill height, and the intensity of live loads diminishes with increased fill height beyond 0.5 m. Moreover, beyond the soil filling height of 0.5 m, there was a large increase in total SIDL. Generally, the PQRS and derailment loads are examined for design purposes only since these are not always applicable during train movement. The dynamic effects from the axle movement of rolling stock are considered as the

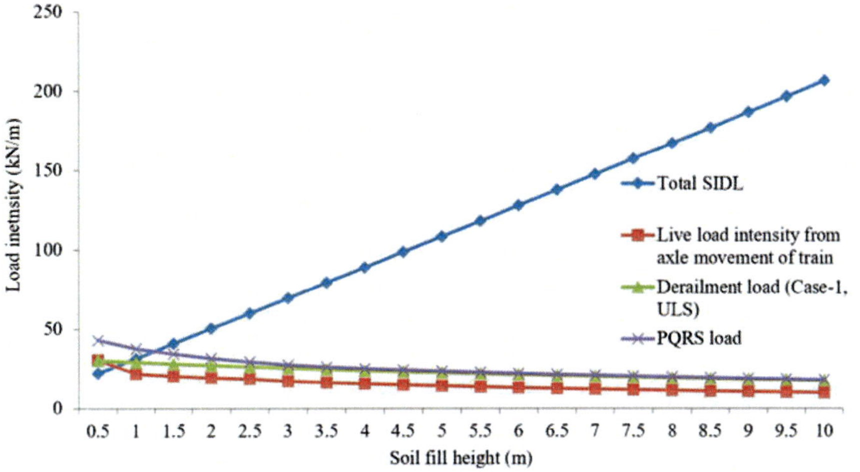

Fig. 3 Variation of high-intensity live loads from dynamic effects with different fill heights

significant live load from the railway track. The magnitude of all the dynamic loads follows a similar value reduction trend with increased fill height up to around 7 m. Thereafter, reduction in live loads with soil fill height was comparatively minimal. The live load applied to the structure is distributed along the transverse direction. According to clause 2.3.4.2 of IRS Bridge Rules, the load is assumed to be distributed from sleepers to the ballast and then spread by soil fill at a slope of not more than half horizontal to one vertical and is generally known as the length of dispersion. The moving loads greatly affected the length of dispersion. The reduction in live load effects on box culvert with increment in fill height is due to the dispersion of loads through the soil fill, and the live load is nearly negligible compared to dead load from the soil. Similar observations have been shown by Orton et al. [3]. According to Gilliland [10], live load effects below 10% of the dead load can be ignored. Therefore, a filling height of 6.5 m can be considered as the optimal height, and beyond that, the live load effects are found to be less than 10% of the total SIDL.

Figure 4 shows the dispersion characteristics of the comparatively low-intensity live loads acting on the box structures, such as loads from wind forces, longitudinal forces, and the footpath. It can be seen from the calculation that this live load intensity decreases drastically with the increase in fill height up to 7 m fill height, and the reduction rate decreases gradually after 7 m and then remains almost constant at higher fill heights. Moreover, the intensity of wind loads was comparatively higher than that from the longitudinal forces for all the soil fill heights. The difference between wind loads and longitudinal forces is maximum at lower fill heights (i.e., at 0.5 m), and the difference in magnitude of these loads decreases as fill height increases. The contribution of live loads from the railway footpath is negligible compared to wind loads and longitudinal forces. The live load from the footpath

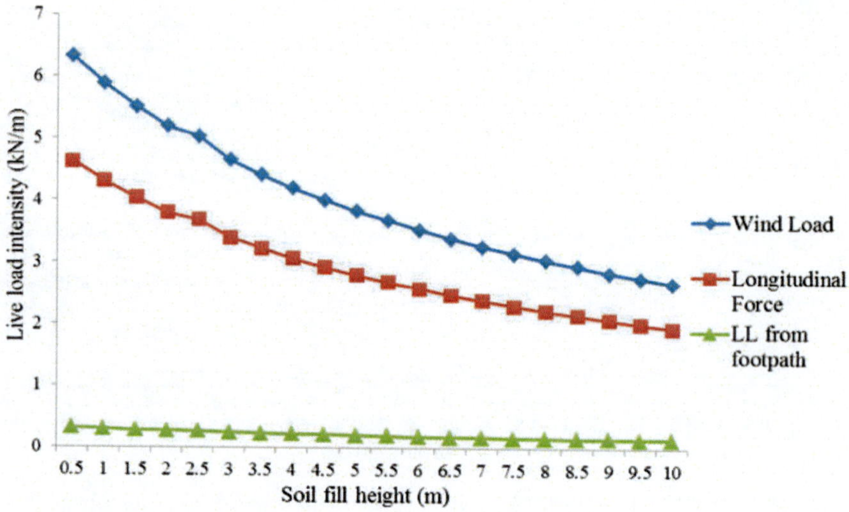

Fig. 4 Variation of low intensity live loads with different soil fills heights

also decreases with an increase in soil fill height up to 7 m and then remains almost constant at higher fill heights.

3.3 Influence of Varying Fill Height on Lateral Loads

The effect of soil fill height on the dispersion characteristics of earth pressure and total SIDL are plotted in Fig. 5. From Fig. 5, it can be observed that the lateral earth pressure acting on the box culvert increases with the increase in soil fill height. A similar observation of increased lateral earth pressure on the sidewall of box culvert with the increase in the ratio of the height of soil fill to culvert width (*H*/*B*) was stated by Tao et al. [5]. Here, an equal increment in earth pressure has been observed at each fill height. Also, the intensity of lateral earth pressures at rest and active conditions was higher than that of vertical loads from topsoil fills. However, the horizontal earth pressure due to surcharge from dead and live loads (Fig. 6) decreases with increased fill height, and a drastic reduction has not been observed after 5 m of fill height. Besides, the surcharge load is almost constant beyond 5 m soil filling height. It is clear from Table 1 that the surcharge due to the live and dead load is considerably small compared to the total SIDL.

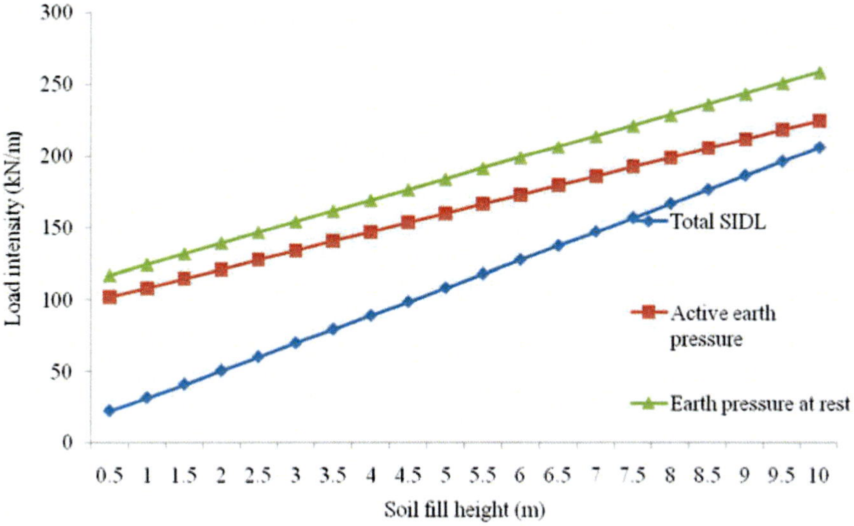

Fig. 5 Effect of soil fill height on earth pressure at rest and active conditions

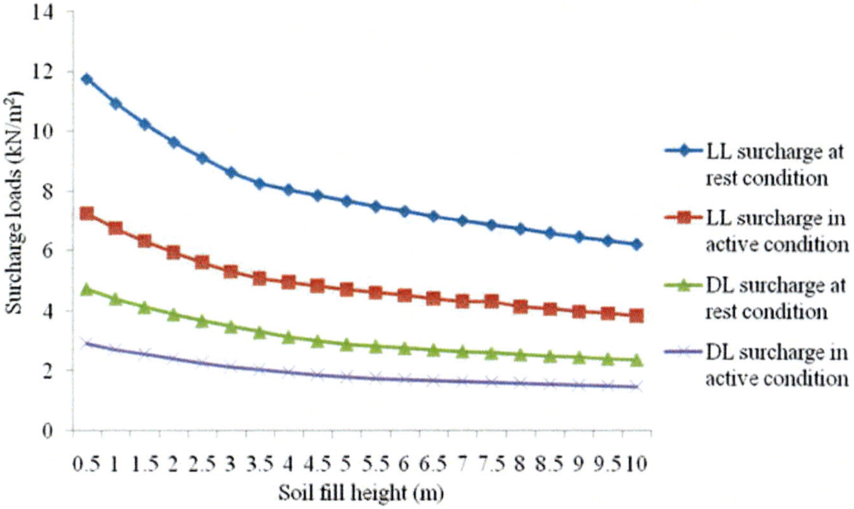

Fig. 6 Effects of increasing soil fill height on surcharge from dead and live loads

4 Conclusions and Future Outlook

The earth cushions or soil fill over the RUB box culverts play a crucial role in the effective dispersion of live loads from railway tracks, and increased fill height can also lead to structural instability. Hence, the soil fill height of buried structures allowing the transport of high-speed trains is a major parameter that influences the design of RUB box culverts. The current work examined the effect of soil fill on the load distribution characteristics of RC skew box culverts for road under bridge (RUB) design. As a part of this work, the soil fill height has been varied from 0.5 to 10 m, and the corresponding live loads and dead loads were determined. A comparative study on the effects of live and dead loads on a two-celled RCC skew box culvert for different soil filling heights has also been conducted. From the study conducted on the influence of soil fill heights for RCC skew box culvert, the following conclusions can be reached:

- The magnitude of SIDL from soil fill and railway components was comparable up to a fill height of 0.5 m. After 0.5 m soil fills height, there is a large increase in dead load from the soil, but the SIDL from railway components and footpath decreased with increased soil fill after 0.5 m.
- The effective dispersion of live load intensity through soil fill above the box culvert in the transverse direction increased with an increase in fill height. The magnitudes of live load from dynamic movements in the railway tracks were higher than the total SIDL at 0.5 m fill height. But after 0.5 m, the total SIDL increased considerably, and live loads decreased rapidly.
- With increased fill heights, the magnitude of PQRS and derailment loads decreased, which is not generally considered for design purposes.
- The intensity of live loads from wind decreases significantly with the fill heights. The intensity of live load from wind is comparatively higher than the longitudinal forces from braking, and traction effects for all the fill heights studied.
- The intensity of lateral earth pressures at rest and active conditions was higher than that of vertical loads from topsoil fills. However, the horizontal earth pressure due to surcharge from dead and live loads decreases with increased fill height, and a drastic reduction has not been observed after 5 m of fill height. Besides, the surcharge load is almost constant beyond 5 m soil filling height.
- A fill height of 6.5 m can be considered as the optimum height beyond which live load effects were observed to be lower than 10% of total SIDL.

The ultimate contribution of this work is to provide more accurate load distribution features to RCC box culverts for RUB design. The paper concludes by comprehensively listing the magnitude of all the dead and live loads at different soil filling heights, and these load values along with the key findings of the current work will be helpful to the researchers in this area. Experimental works can be performed with advanced instruments to determine the load dispersion characterizes of RUB box culverts, and these experimental data can be verified with the theoretical data. Also, a detailed comparative study on the possible lateral and vertical loads acting on box culverts and their contribution to the structural stability of box sections can be investigated. Moreover, the impacts of compressibility characteristics of the poor safe bearing capacity (SBC) and good SBC soils on the overall design criteria of box culverts are yet another future scope of this research. Further investigation into the effects of a skew angle greater than 30° and provision of haunches at the corners on the internal forces of the box culvert is also recommended.

References

1. Abuhajar, O., El Naggar, M. H., & Newson, T. (2015). Static soil culvert interaction the effect of box culvert geometric configurations and soil properties. *Computers and Geotechnics, 69*, 219–235.
2. Patel, R., Jamle, S. (2019). Analysis and design of box culvert: A manual approach. *International Journal of Advanced Engineering Research and Science (IJAERS), 6*, 286–291.
3. Orton, L., Loehr, J. E., Boeckmann, A., & Havens, G. (2015). Live load effect in reinforced concrete box culverts under soil fills. *Journal of Bridge Engineering, 20*(11), Article ID 04015003.
4. Abdel-Karim, A. M., Tadros, M. K., & Benak, J. V. (1990). *Live load distribution on concrete box culverts*. Transportation Research Board.
5. Kim, K., & Yoo, C. (2005). Design loading on deeply buried box culverts. *Journal of Geotechnical and Geoenvironmental Engineering, 131*, 20–27.
6. Sinha, B. N., & Sharma, R. P. (2009). RCC box culvert—Methodology and designs including computer method. *Journal of Indian Road Congress*, 189–219.
7. Indian Railway Standard, Rules specifying the loads for the design of super structure and sub structure of bridges and for the assessment of the strength of the existing bridges. IRS Bridge rule – 2014. Lucknow, India
8. Indian Railway Standard, Code of practice for the design of sub-structures and foundations of bridges. IRS Substructures and Foundation of Bridges – 2013. Lucknow, India
9. Indian Standard, Code of practice for design loads (other than earthquake) for buildings and structures. IS 875.3 part 3 – 1987. New Delhi, India
10. Gilliand, M. K. (1986). *Cost effective concrete box culvert design*. Master's thesis, University of Nebraska, Lincoln, NE.

Chloride Ion Penetration of GGBS-Based Geopolymer Concrete with Different Molarities of NaOH

N. Sailaja, M. Naveen, S. K. Amir Basha, B. Sarath Chandra Kumar, C. Ravi Kumar Reddy, Y. Himath Kumar, and J. D. Chaitanya Kumar

Abstract Geopolymer concrete is an inventive development material that is delivered by the compound activity of inorganic particles like sodium hydroxide and sodium silicate. The cementations material utilized in the current investigation is ground granulated blast-furnace slag (GGBS). The sodium hydroxide arrangements having the molarities of 12, 14, and 16 M are utilized. The proportion of the alkali activators utilized is 1:2.5. The specimens are presented to the NaCl arrangement. The specimens are drilled at different depths ranging from 5 to 25 mm. The powder is gathered from the specimens and is titrated against $AgNO_3$ to decide the level of chloride. From the test results, it was concluded that the days increases the percentage of chloride ion penetration decreases.

Keywords GGBS · Sodium hydroxide · Sodium silicate · Geopolymer · Alkali solutions · NaCl exposed

1 Introduction

The demand for concrete as a construction material is increasing exponentially, as is the demand for regular Portland cement. Its effects on the environment are well known [1]. To meet these challenges, cement is partially replaced by other compounds that have binding capacity [2, 3].

Davidovits proposed in 1978 that binders could be shaped by polymerizing alkaline fluid with aluminon-silicate materials such as fly ash, GGBS, and so on, and that these binders could then be used as geopolymers. Geopolymer is a green material, since it emits less carbon [2].

N. Sailaja · M. Naveen · S. K. Amir Basha · B. Sarath Chandra Kumar (✉) · C. Ravi Kumar Reddy
Department of Civil Engineering, Kallam Haranadhareddy Institute of Technology, Chodavaram, Guntur, Andhra Pradesh, India

Y. Himath Kumar · J. D. Chaitanya Kumar
Department of Civil Engineering, Koneru Lakshmaiah Education Foundation, Guntur, Andhra Pradesh, India

A. K. Gupta et al. (eds.), *Advances in Construction Materials and Sustainable Environment*, Lecture Notes in Civil Engineering 196,
https://doi.org/10.1007/978-981-16-6557-8_28

Geopolymer concrete is made by blending NaOH and Na_2SiO_3 with aluminon-silicate materials. In present day times, it has gotten an exceptionally basic subject to explore. The essential justification its acknowledgment is that the utilization of geopolymers can completely supplant Portland concrete as the cementitious material [4, 5]. Geopolymers are more alluring because of their essential segment, which is generally an assembling result wealthy in silica and alumina [4]. The soluble base arrangements utilized in this investigation to predicament the blend are one part of sodium hydroxide and 2.5 parts of sodium silicate [6–11].

GGBS is the cementitious substance utilized. The cubes are cast and presented to NaCl arrangement, which contains 3.5% NaCl by weight. The cubes are drenched in the liquid for 28, 56 and 90 days [6–11]. The primary objective of the present study is to evaluate the variation of chloride content under different molarities of NaOH.

1.1 Geopolymer Technology

Davidovits described a wide scope of geopolymer materials for the arrangement of inorganic atoms through chains and organizations. Thermally enacted normal materials and other assembling side-effects, for example, GGBS and Metakaolin are utilized to give a stock of silicon and alumina, which is broken down in an AAS and polymerizes into atomic fastens and organizations to make the solidified cover. This concrete is otherwise called a soluble base enacted concretes.

Geopolymer is an inorganic polymer with an amorphous microstructure. Under basic conditions, the synthetic response utilized in the polymerization interaction on Si–Al minerals made out of Si–O–Al–O bonds. The polymerization response is exothermic, requiring atmospheric pressure at temperatures below 100 °C.

1.2 Chloride Ion Penetration

Reinforced concrete buildings are subjected to harsh conditions and are often required to survive for extended periods of time with little or no rehabilitation or maintenance (often 100 years or more). To do this, a long-lasting system must be built [4]. One of the most common type of environmental assault on reinforced concrete bridges is chloride entrance, which causes weakening of the reinforcing steel and, as a result, a decrease in the structure's efficiency, serviceability, and esthetics. This can result in the foundation being repaired or replaced too soon [1].

A typical way of avoiding such corrosion is to use relatively impenetrable concrete to prevent chlorides from entering the structure to the level of the reinforcing steel bar [2]. The limit of chloride particles to penetrate concrete should then be resolved for development and quality control purposes. The passage of chloride particles into the solid, then again, is a lethargic activity. It can't be estimated exactly in a time span accommodating as a quality control measure. Therefore, to assess chloride entrance,

Table 1 Physical properties of GGBS

Parameter	GGBS (%)	IS: 12089-1987
CaO	37.34	–
Al_2O_3	14.42	–
Fe_2O_3	1.11	–
SiO_2	37.73	–
MgO	8.71	Max. 17%
MnO	0.02	Max. 5.5%
Sulphide sulphur	0.39	Max. 2%
Loss of ignition	1.41	–
Insoluble residue	1.59	Max. 5%
Glass content (%)	92	Max. 85%

a test strategy that speeds up the interaction is required, permitting the assurance of dispersion esteems in a sensible measure of time [3].

2 Materials Used

The materials and preparation of GPC were adopted from Sarath Chandra Kumar et al. [12–19].

2.1 GGBS

The GGBS is a byproduct of iron assembly that when applied to concrete, increases its utility, consistency, and sturdiness. The physical properties of GGBS can be found in Table 1 (Fig. 1).

2.2 Sodium Hydroxide (NaOH)

It is an inorganic compound which is highly soluble in water. 12, 14, and 16 M were included in this experiment. The solution is made 24 h ahead of time for the casting.

2.3 Sodium Silicate (Na_2SiO_3)

That is sometimes referred to as a water glass or a liquid glass.

Fig. 1 Ground granulated blast furnace slag

3 Experimental Procedure

3.1 Casting of Cubes

The cubes were 150 mm × 150 mm × 150 mm in size. To remove the voids, the cubes are mounted on the vibrator. The specimens are demolded after 24 h and exposed to NaCl solution up to 90 days, respectively (Fig. 2).

Fig. 2 Casting of cubes

Fig. 3 Cubes exposed to NaCl solution

3.2 *Exposed to NaCl Solutions*

The 3.5% NaCl is dissolve in water. The cubes are immersed in the solution. Cubes with different moralities are immersed in NaCl and allowed to cure [7–11] (Fig. 3).

3.3 *Drilling of Cubes*

The cubes are air-dried for 24 h before being drilled at different depths. The powder removed from the cubes is put away and utilized for titration [7–11] (Fig. 4).

3.4 *Titration Process*

The cubes were drilled at different depths. To achieve the results, the extracted powder is titrated against an $AgNO_3$ solution [7–11] (Fig. 5).

Experimental procedures for compressive strength identification, chemical analysis and calculation of chloride ion penetration are adopted from Raziya Tasneem et al. [19].

Fig. 4 Drilling of cubes

Fig. 5 Titration process

4 Results and Discussion

4.1 Compressive Strength

The cubes are subjected to NaCl for 28 days before being tested under the compressive testing machine for 56 days, and the findings are shown in Fig. 6 [12–18].

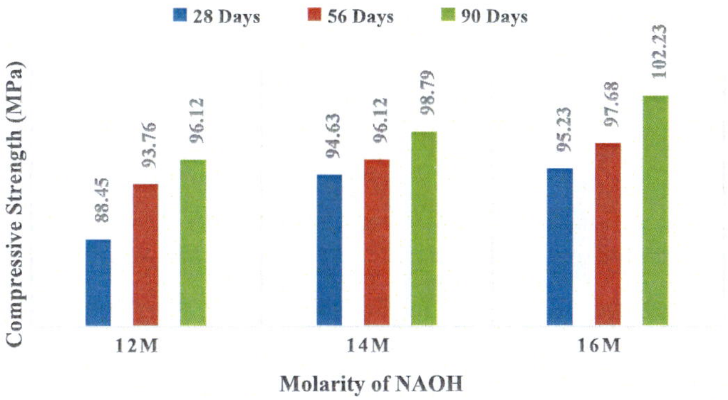

Fig. 6 Compressive strength of GPC cubes

4.2 *Cubes Exposed to NaCl Solution*

The level of chloride particle infiltration is given by the powder taken from the specimen that is titrated as expressed in the above compound examination. The charts portraying chloride particle infiltration versus depth of the entrance are found in Figs. 7, 8 and 9.

GGBS-based geopolymer concrete made with 12 M NaOH gains a higher percentage of chloride ion concentration after 90 days in NaCl. This is due to the fact that concrete has a lot of pores when it's dry.

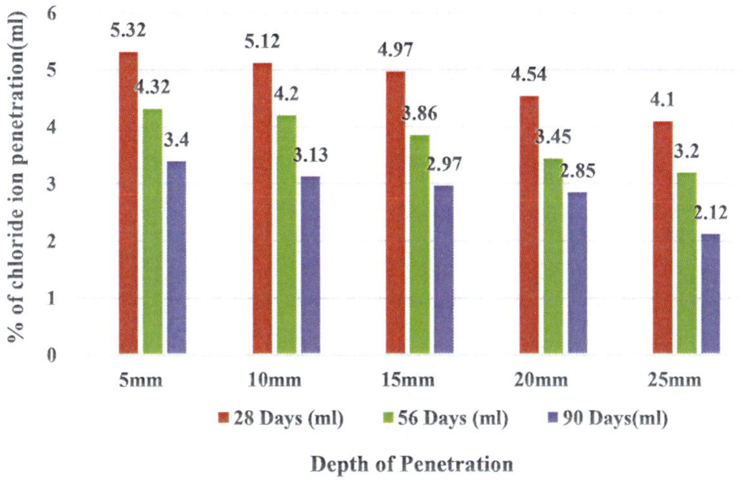

Fig. 7 Variation of chloride ion penetration for 12 M NaOH

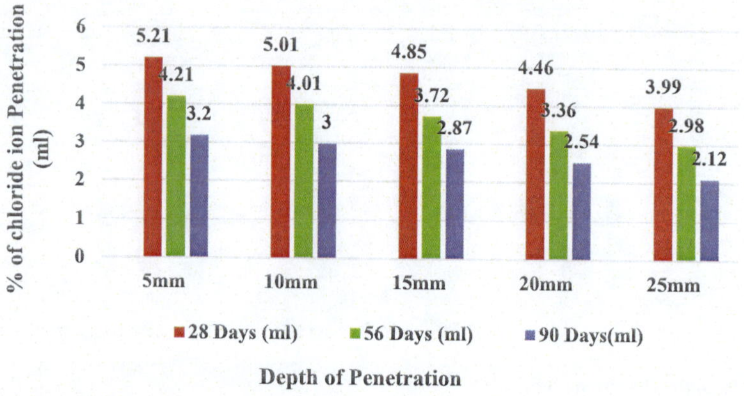

Fig. 8 Variation of chloride ion penetration for 14 M NaOH

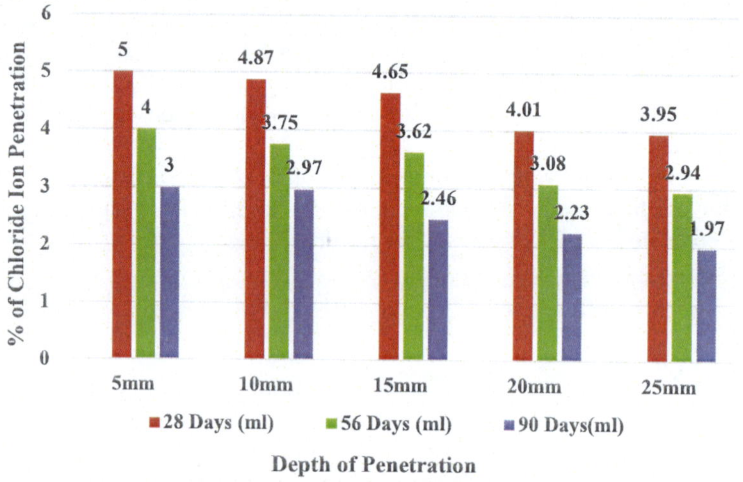

Fig. 9 Variation of chloride ion penetration for 16 M NaOH

The consequence of a higher alkali ratio in the mixture, expressed by soluble Na_2SiO_3, stopped concrete from corrosion. The use of soluble silica is thought to reduce pore size due to the creation of a stronger geopolymeric gel.

Higher NaOH concentrations in GGBS-based geopolymer concrete clearly resulted in higher chloride binding capability. It also aided in the exchange of chloride ions on the surface of concrete.

5 Conclusion

The following results are taken from the tests.

Usage of fine materials like GGBS decreases chloride ion into concrete. If penetration of chloride ions is less, then it will take more time to get corroded. From the results, one can understand the amount of chloride ion penetration in relation to both time and depth.

GGBS are efficient to resist the penetration of chloride ions. Moreover, they will also resist the penetration of the sulfate, carbonates, and other material which are toxic to the concrete.

From the results, it is clear that the chloride ion penetration in the NaCl cured cubes decreases with increase in the depth of penetration. The penetration of chloride ion decreases as the days of curing increases which substantially increases the life span of structure.

References

1. Reed, M., Lokuge, W., & Karunasena, W. (2014). Fibre-reinforced geopolymer concrete with ambient curing for in situ applications. *Journal of Materials Science, 49*(12), 4297–4304.
2. Krishnan, L., et al. (2014). Geopolymer concrete an eco-friendly construction material. *Magnesium, 1*, 1.
3. Lloyd, N., & Rangan, V. (2010). Geopolymer concrete with fly ash. In *Proceedings of the Second International Conference on sustainable construction Materials and Technologies, 2010*. UWM Center for By-Products Utilization.
4. Kayali, O. (2016). Sustainability of fibre composite concrete construction. *Sustainability of construction materials* (pp. 539–566). Elsevier.
5. Aswani, E., & Karthi, L. (2017). A literature review on fiber reinforced geopolymer concrete. *International Journal of Scientific and Engineering Research, 8*(2), 408–411.
6. Davidovits, J. (1989). Geopolymers and geopolymeric new materials. *Journal of Thermal Analysis and Calorimetry, 35*, 429–441.
7. Hema Durga Rajeswari, Ch., & Kameswara Rao, B. (2019). Service life prediction of high-performance concrete incorporated with GGBS and silica fume. *International Journal of Recent Technology and Engineering, 7*(6C2), 448–455.
8. Rajesh, G., & Kameswara Rao, B. (2019). Service life prediction of high-performance concrete with respect to chloride ion penetration by incorporated with GGBS. *International Journal of Recent Technology and Engineering, 7*(6C2), 478–483.
9. Daya Rani, B., & Kameswara Rao, B. (2019). Service life prediction of high-performance concrete with respect to chloride ion penetration by incorporated with fly ash and silica fume. *International Journal of Recent Technology and Engineering, 7*(6C2), 484–489.
10. Prudhvi sai, E. V., & Kameswara Rao, B. (2019). Service life prediction of high-performance concrete with respect to chloride ion penetration by incorporated with fly ash. *International Journal of Recent Technology and Engineering, 7*(6C2), 496–501.
11. Suwito, A., & Xi, Y. (2004). Service life of reinforced concrete structures with corrosion damage due to chloride attack. In *Life-cycle performance of deteriorating structures: assessment, design and management* (pp 207–218).
12. Ashok, K., Kameswara Rao, B., & Sarath Chandra Kumar, B. (2021). Experimental investigation on nano alumina based concrete. *ARPN Journal of Engineering and Applied Sciences, 16*(01), 76–87. ISSN: 18196608.

13. Ashok, K., & Sarath Chandra Kumar, B. (2020). Influence of nano particles in concrete—a review. *International Journal of Advanced Research in Engineering and Technology, 11*(06), 248–263.
14. Srinivasa Rao, G., & Sarath Chandra Kumar, B. (2019). Experimental investigation of GGBS based geopolymer concrete with steel fibers. *International Journal of Recent Technology and Engineering (IJRTE), 07*(6C2), 49–55.
15. Sarath Chandra Kumar, B., Karuppusamy, S., & Ramesh, K. (2019). Correlation between compressive strength and split tensile strength of GGBS and MK based geopolymer concrete using regression analysis. *Journal of Mechanics of Continua and Mathematical Sciences, 14*(1), 21–36.
16. Sarath Chandra Kumar, B., Ramesh, K., & Poluraju, P. (2017). An experimental investigation on flexural behavior of GGBS and metakaolin based geopolymer concrete. *ARPN Journal of Engineering and Applied Sciences, 12*(07), 2052–2062.
17. Sarath Chandra Kumar, B., & Ramesh, K. (2017). Durability studies of GGBS and metakaolin based geopolymer concrete. *International Journal of Civil Engineering and Technology (IJCIET), 8*(01), 17–28.
18. Sarath Chandra Kumar, B., & Ramesh, K. (2016). Experimental study on strength properties of metakaolin and GGBS based geopolymer concrete. *ARPN Journal of Engineering and Applied Sciences, 11*(21), 12414–12422.
19. Raziya Tasneem, A., Sarath Chandra Kumar, B., & Himath Kumar, Y. (2020). Service life prediction of geopolymer concrete with respective to chloride ion penetration. In *3rd International Conference on Advances in Mechanical Engineering*, SRM Institute of Science and Technology, Kattankulathur, Tamilnadu, India. IOP Conference Series: Materials Science and Engineering (Vol. 912, p. 062033).

Application of RSM in the Optimization of GGBS and Metakaoline Based Geopolymer Concrete

Ch. Pavan Kalyan, D. Anil Kumar, K. Saloman Raju, B. Sarath Chandra Kumar, C. Ravi Kumar Reddy, Y. Himath Kumar, and J. D. Chaitanya Kumar

Abstract Waste products like fly ash, ground granulated blast furnace slag (GGBS), and Metakoaline are used to make geopolymer cement concrete (MK). GGBS and MK are used as cementitious ingredients in this analysis to prepare geopolymer concrete (GPC). In a steel plant, GGBS is generated as waste. In this analysis, the RSM Method is used to optimize the mechanical properties of GGBS and MK-based GPC with 10 M concentrations NaOH solution is optimized using Analysis of Variance (ANOVA). The findings showed that the mechanical properties of the GGBS and MK have been improved. RSM investigates the relationships between a number of explanatory variables and one or more reaction variables. Based on the RSM performance, optimization mixtures were created and experimentally tested with less than 2% error. Finally, it was found that the utilization of GGBS, and MK based GPC is replaced with ordinary Portland cement concrete.

Keywords GGBS · Metakaoline · ANOVA · RSM · Optimization

1 Introduction

Concrete, apart from water, is the most often used building material. Ordinary Portland Cement (OPC) is traditionally the most commonly used binder in the concrete manufacturing. However, the method of producing Portland cement emits a significant volume of greenhouse gas into the atmosphere. One ton of OPC requires 2.8 t of raw materials such as lime, diesel, and other materials, so it is well understood that cement manufacture depletes a large amount of natural resources. As a result,

Ch. Pavan Kalyan · D. Anil Kumar · K. Saloman Raju · B. Sarath Chandra Kumar (✉) · C. Ravi Kumar Reddy
Department of Civil Engineering, Kallam Haranadhareddy Institute of Technology, Chodavaram, Guntur, Andhra Pradesh, India

Y. Himath Kumar · J. D. Chaitanya Kumar
Department of Civil Engineering, Koneru Lakshmaiah Education Foundation, Guntur, Andhra Pradesh, India

A. K. Gupta et al. (eds.), *Advances in Construction Materials and Sustainable Environment*, Lecture Notes in Civil Engineering 196,
https://doi.org/10.1007/978-981-16-6557-8_29

Table 1 Mix proportions

Materials used	Cementitious materials	Fine aggregate	Coarse aggregate	Sodium hydroxide	Sodium silicate
Quantity of materials in kg/m^3	414	660	1136	53	133

Table 2 Combinations of GGBS and MK

Mix ID	M1	M2	M3	M4	M5	M6	M7	M8	M9	M10	M11
MK (%)	100	90	80	70	60	50	40	30	20	10	0
GGBS (%)	0	10	20	30	40	50	60	70	80	90	100

1 t of CO_2 is emitted in the de-carbonation of lime in the manufacture of one ton of cement.

There is currently a great deal of uncertainty about the production of Portland cement substitutes. As a result, attempts are being made to produce alternative cementitious materials for use in concrete production. Davidovits introduced a new substance in 1978 that serves as an alternative binder to cement. For the reaction, a geologically dependent raw material of alkaline liquid is used, and the material is known as a geopolymer [1–6].

1.1 Development and Selection of RSM Models

Response surface methodology (RSM) is a mathematical research approach in which each response is interconnected with a variety of variables to assess the consequences, association, and interaction between the variables and responses. The Design-Expert 11 program was used in the current study for process variable design, mathematical simulation, statistical analysis, and optimization. ANOVA can be used to determine the association and relationship between the strengths of GPC. In this case, the Quadratic model consists of 33 experimental runs for the two variables, GGBS and Metakaoline, with values ranging from 0 to 100%.

2 Experimental Program

2.1 Mix Design of Geopolymer Concrete

Mix design was adopted from Sarath Chandra Kumar et al. [7]. Table 1 shows the blend proportion for geopolymer concrete, and Table 2 shows the various variations of GGBS and Metakaolin investigated.

2.2 GPC Preparation

The preparation of GPC was adopted from Sarath Chandra Kumar et al. [7] and the compressive, flexural, and split tensile tests are conducted for the 10 M GPC, testing of the specimens was shown in Figs. 1, 2 and 3. The test results have been plotted in Table 4. The mix proportions are listed in Table 1 (Table 3) [8–14].

GPC Cubes | Ambient Curing of GPC Cubes | Compression Testing of GPC Cubes

Fig. 1 Compressive strength of GPC

Fig. 2 Split tensile strength of GPC

Fig. 3 Flexural strength of GP

Table 3 Mix proportioning the GGBS and MK

Mix ID	M1	M2	M3	M4	M5	M6	M7	M8	M9	M10	M11
MK (kg/m^3)	413.8	371.7	330.4	289.1	247.8	206.5	165.2	123.7	82.6	42.3	0
GGBS (kg/m^3)	0	42.3	82.6	123.7	165.2	206.5	247.8	289.1	330.4	371.7	413.8

Table 4 Test results of the 10 M GPC at 28 days

Mix ID	Compressive strength (MPa)	Split tensile strength (MPa)	Flexural strength (MPa)
M1	19.64	0.6	2.07
M2	24.06	0.88	2.12
M3	25.14	0.37	2.36
M4	28.35	1.00	2.91
M5	30.51	0.86	3.37
M6	35.23	1.41	3.92
M7	38.34	1.77	4.10
M8	47.35	1.71	4.20
M9	49.94	1.61	4.61
M10	55.50	2.71	5.72
M11	66.03	3.43	6.65

3 Results and Discussion

In this section, the effect of all variables was interacted and optimized according to mechanical properties. The specimens were tested at the age of 28 days after curing at

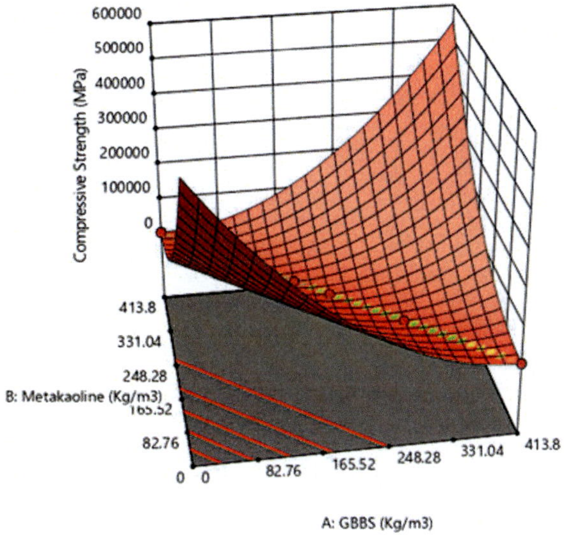

Fig. 4 RSM analysis 3D surface diagram for compressive strength

the room temperature. The mechanical properties of all the mixtures were increased with the age. The test results for the 10 M GGBS at 28 days are listed in Table 4.

3.1 Compressive Strength

The equation gives the relationship between the GGBS and MK which is useful for the best optimized proportion of GPC (Figs. 4 and 5).

$$\text{Compressive Strength (MPa)} = 5.54993\text{E}+05 - 2681.51545 * \text{A} - 2679.44560 * \text{B} + 6.47324 * \text{A} * \text{B} + 3.23939 * \text{A}^2 + 3.23413 * \text{B}^2 \quad (1)$$

where “A” indicates for GGBS content in kg/m^3, “B” for Metakaoline content in kg/m^3.

In the above optimization, target compressive strength is 65 MPa achieved at 290 kg of GGBS and 126 kg of Metakaoline.

3.2 Split Tensile Strength

$$\text{Split Tensile Strength (MPa)} = -31799.97343 + 153.14279 * \text{A} + 154.26130 * \text{B} - 0.371442 * \text{A} * \text{B} - 0.184355 * \text{A}^2 - 0.187074 * \text{B}^2 \quad (2)$$

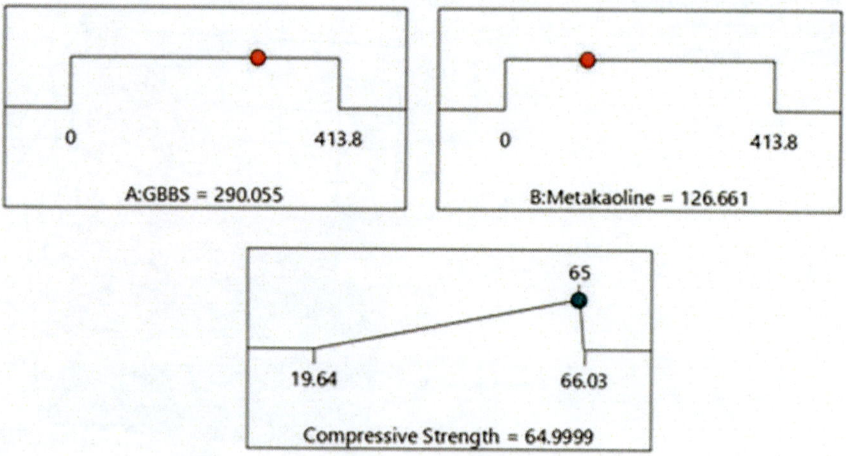

Fig. 5 Graphical ramp views for optimized for compressive strength

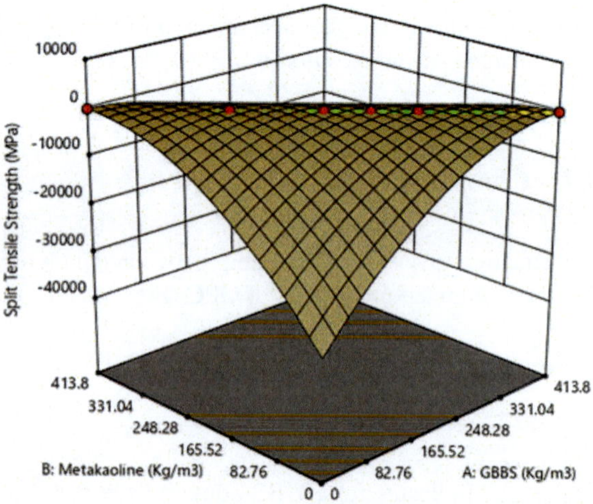

Fig. 6 RSM analysis 3D surface diagram for split tensile strength

where "A" indicates for GGBS content in kg/m^3, "B" for Metakaoline content in kg/m^3. In the above optimization, target Split Tensile strength is 3.4 MPa achieved at 401.1 kg of GGBS and 14.6 kg of Metakaoline (Figs. 6 and 7).

3.3 Flexural Strength

$$\text{Flexural Strength (MPa)} = -1.37953\text{E} + 05 + 666.50194 * \text{A} + 667.19993 * \text{B} - 1.61170\,\text{A} * \text{B} - 0.804995 * \text{A}^2 - 0.806707 * \text{B}^2 \quad (3)$$

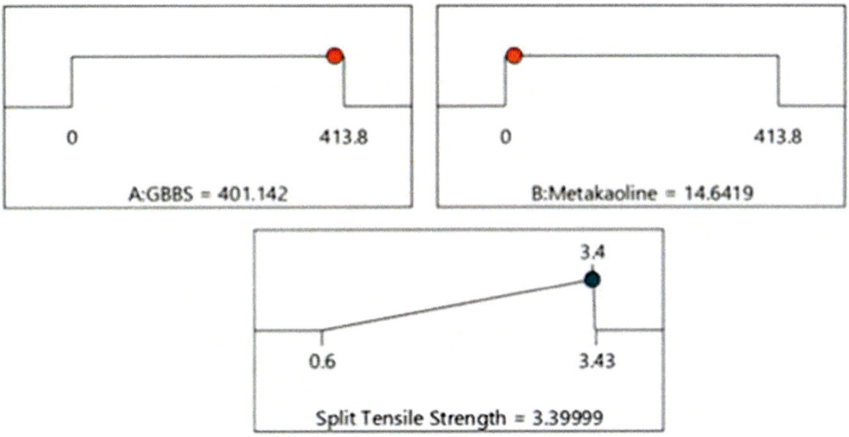

Fig. 7 Graphical ramp views for optimized for split tensile strength

where "A" indicates for GGBS content in kg/m^3, "B" for Metakaoline content in kg/m^3 (Figs. 8 and 9).

In the above optimization, target flexural strength is 6.65 MPa achieved at 413 kg of GGBS and 0.17 kg of Metakaoline.

Fig. 8 RSM analysis 3D surface diagram for flexural strength

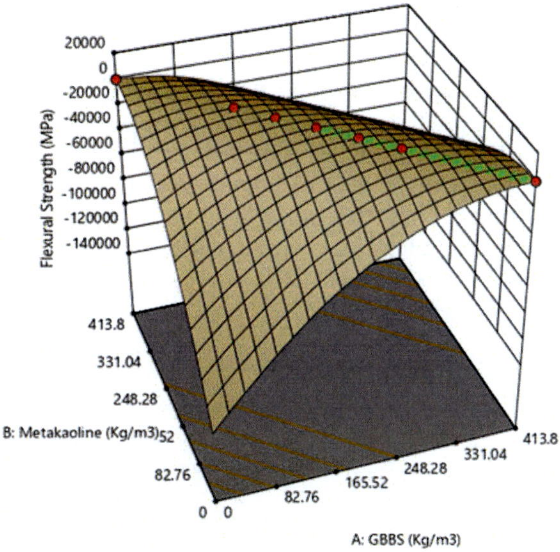

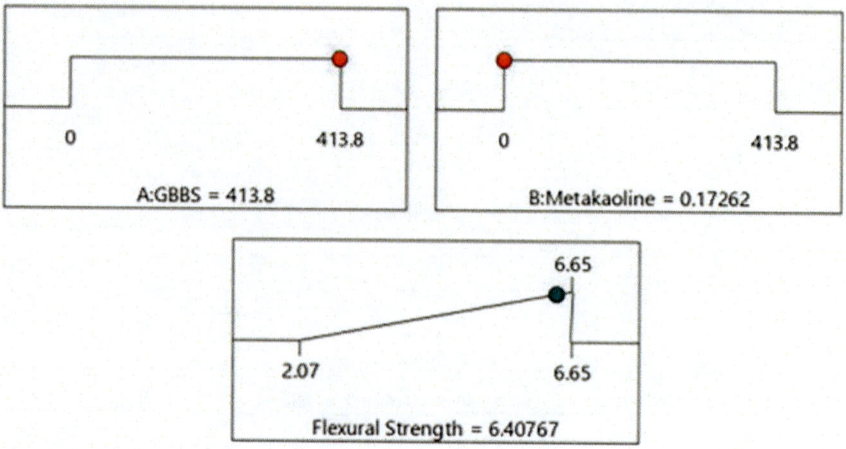

Fig. 9 Graphical ramp views for optimized for flexural strength

4 Conclusion

The effect of GGBS and MK quantity on GPC's 28-day compressive power, split tensile strength, and flexural strength was investigated. The experiments were constructed using RSM's Quadratic process, and the preparation parameters of GGBS and MK-based GPC were thoroughly examined. The findings revealed that the order of the variables in terms of their effect on power. The RSM and experimental findings were used to develop a concrete 28-day strength prediction model. Using the response surface model, the preparation parameters were designed to increase the 28-day resistance of geopolymer concrete.

The optimum preparation parameters are quantity of GGBS and Metakaoline. The maximum compressive strength is the predicted as 65 MPa achieved at 290 kg of GGBS and 126 kg. The maximum split tensile strength is the predicted as 3.4 MPa achieved at 401.1 kg of GGBS and 14.6 kg of MK. The maximum flexural strength is the predicted as 6.65 MPa achieved at 413 kg of GGBS and 0.17 kg of MK. The model equations were developed, and the results were validated using ANOVA by observing the model significance value exceeding 90%.

References

1. Davidovits, J. (1994). Properties of geopolymer cements. In *First International Conference on Alkaline Cements and Concretes*. SRIBM, Kiev State Technical University, Kiev, Ukraine (pp. 131–149).
2. Daniel, K., Sanjayan, J., & Sagoe-Crentsil, K. (2006). The behaviour of geopolymer paste and concrete at elevated temperatures. In *International Conference on Pozzolan, Concrete and Geopolymer*, Khon Kaen, Thailand (pp. 105–118).

3. Davidovits, J., Comrie, D. C., Paterson, J. H., & Ritcey, D. J. (1990). Geopolymeric concretes for environmental protection. *ACI Concrete International Journal, 12*(7), 30–40.
4. Van Jaarsveld, J. G. S., van Deventer, J. S. J., & Lukey, G. C. (2002). The effect of composition and temperature on the properties of fly ash and kaolinite-based geopolymers. *Chemical Engineering Journal, 89*, 63–73.
5. Douglas, E., Bilodeau, A., & Brandstetr, J. (1991). Alkali activated ground granulated blast-furnace slag concrete: Preliminary investigation. *Cement and Concrete Research, 21*, 101–108.
6. Temuujin, J., van Riessen, A., & Williams, R. (2009). Influence of calcium compounds on the mechanical properties of fly ash geopolymerpastes. *Journal of Hazardous Materials, 167*, 82–88.
7. Sarath Chandra Kumar, B., & Ramesh, K. (2016). Experimental study on strength properties of metakaolin and GGBS based geopolymer concrete. *ARPN Journal of Engineering and Applied Sciences, 11*(21), 12414–12422.
8. Joseph. (1999). Chemistry of geopolymeric systems, terminology. In C. James (Ed.), *Geopolymer International Conference*, France (pp. 9–40).
9. Purdon, A. O. (1940). The action of alkalis on blast-furnace slag. *Journal of the Society of Chemical Industry, 59*, 191–202.
10. Srinivas Rao, G., & Sarath Chandra Kumar, B. (2019). Experimental investigation of GGBS based geopolymer concrete with steel fibers. *International Journal of Recent Technology and Engineering, 7*(6C2), 49–55.
11. Saad, A.-R., & Nildem, T. (2018). Performance of self-compacting geopolymer concrete with and without GGBFS and steel fiber. *Advances in Concrete Construction, 4*(6), 323–344.
12. Patel, A. J., Patel, D. D., Patel, D. H., & Patva, V. A. (2014). Self-compacting concrete with use of waste material. *International Journal for Technological Research in Engineering, 1*(9), 1–10.
13. Memon, A. A., Nuruddin, M. F., Demie, S., & Shafiq, N. (2012). Effect of superplasticizer and extra water on workability and compressive strength of self-compacting geopolymer concrete. *Research Journal of Applied Sciences, Engineering and Technology, 4*(5), 407–414.
14. Fadhil Nuruddin, M., Samuel Demie, M., Ahmed, F., & Shafiq, N. (2011). Effect of superplasticizer and NaOH molarity on workability, compressive strength and microstructure properties of self-compacting geopolymer concrete. *International Journal of Civil and Environmental Engineering, 3*(2), 122–129.

Evaluation of Use of Plastic and Rubber in Road Construction

Neeraj Kumar, Nikhil, Ashutosh Kumar, and A. R. Kongan

Abstract Plastic is a widely used product, which we can find everywhere in our day-to-day products. Amount of plastic trash flowing into ocean is expected to triple by 2040 to 29 million metric tons [1]. Cumulative global plastic production rose from 2 million tons to 7.82 billion tones for the period 1950–2015. Plastic waste generated in India in 2010 is 4.49 million tones. Manufacturing and packing industry produce most of plastic waste. As disposal of plastic waste is a huge challenge, alternate method must be devised such as in road construction. Use of plastic waste has shown improvement in mix, reduction in usage of natural aggregates, reduction in void spaces, no leaching of plastics, no effect of UV radiation and overall life of road. This paper presents changes in Marshall stability and flow value of different samples with addition of pure bitumen, plastic and plastic-rubber case, the latter giving better results with 50% increased strength. The bulk density of all the samples is found to be in the range of 2.25–2.35.

Keywords Plastic road · Rubber road · Flexible pavements · Green road · Plastic waste · Rubber · Marshall stability

1 Introduction

Waste plastic production is increasing on a day-to-day basis, as the production is massive and it isn't disposed off correctly all over the world. Ministry of Petroleum and Natural Gas, India reports that by 2022, annual per capita plastic consumption would be 20 kg. According to a Central Pollution Control Board (CPCB) report, plastic constitutes 8% of total solid waste. Only 60% of total plastic waste is currently recycled.

According to CPCB estimation, collection efficiency is 80.28% in 2014, out of which 28.4% [2] was treated and remaining were disposed in landfills or open dumps. Due to overgrowing demand and projected increase in plastic waste production, it

N. Kumar (✉) · Nikhil · A. Kumar · A. R. Kongan
Delhi Technological University, New Delhi, India

A. K. Gupta et al. (eds.), *Advances in Construction Materials and Sustainable Environment*, Lecture Notes in Civil Engineering 196,
https://doi.org/10.1007/978-981-16-6557-8_30

has become a necessity to find the optimal use of plastic waste. Previous studies and uses of plastic in road construction have shown improvement in road quality and its performance in stress condition. Use of plastic as a binder has shown reduction in voids and helps reduce moisture absorption as compared to conventional methods. Bitumen of low grade has poor resistance toward water and adding of plastic improve its properties.

Bhageerathy and Alex [3] studied usage of bio-medical plastic waste in road construction and concluded plastic modified mix Marshall Stability Value is 51% more than the normal mix.

Dr. Vasudevan [4] in his studies found out plastic coated aggregates decrease the porosity, dampness and improves the strength. His tests showed better Marshall stability value and increase in stripping value suggesting better bond of plastic with aggregate and bitumen.

Prasad and Sowmya [5] found out in his studies that along with waste plastic, rubber can be used in decreasing the bitumen content and increasing the strength of mix. At different quantities of plastic and rubber studies were conducted and 6% plastic by weight was found to be optimum value. They also discovered that PET containers showed better result, PET being one of the most used plastic types; it can be easily utilized in road construction purposes.

Rajasekaran and Vasudevan [6] explained the plastic coated aggregates shows many favorable design aspects along with improving the quality of flexible pavements along with aggregate quality. This technology helps in utilization of plastic from domestic and packing materials.

Separate use of plastic and rubber in road construction has shown improvement in different properties and increased strength of road. Our work aims to highlight how using both rubber and plastic improves the road quality and strength.

2 Methodology

With reference to IRC: SP:98—2013, Guidelines for the use of waste plastic in hot bituminous mixes (Dry Process). In wearing courses, different samples with varied quantities of plastic and 10 mm coarse aggregate were prepared.

Primary tests such as water absorption, specific gravity, gradation–sieve analysis and moisture content were done to ascertain the quality of coarse and fine aggregate which will be used as per IS 2386 Part I and Part III—1963. As per IRC: 111—2009, specification for dense graded bituminous mixes, bitumen of viscosity grade—30 (VG 30) was used.

Melt Flow Index as per ASTM D 1238—2010 for plastic was found to be 11 g/10 min.

Sample was prepared following Fig. 1 and with reference to IRC SP:98—2013 [7] and as per MORTH guidelines for Bituminous Concrete [8]. The Marshall tests were conducted as per ASTM D1559 by preparing compacted spherical specimens of diameter 101.6 mm and height 63.5 mm. Sample is kept in water bath for 30 min at

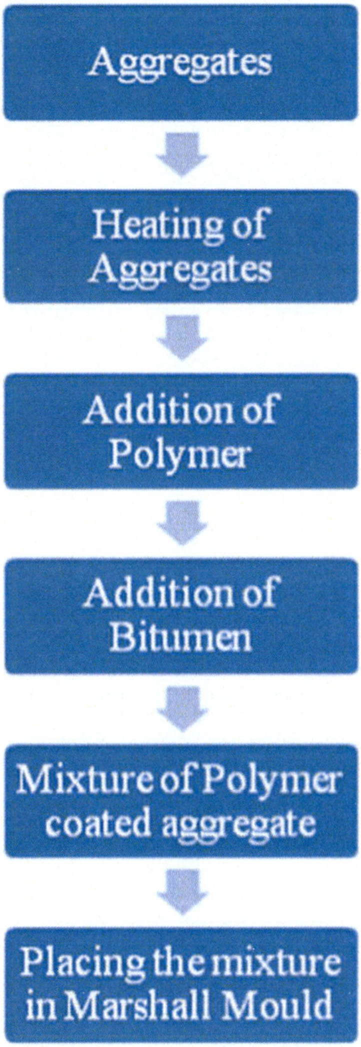

Fig. 1 Flow diagram for sample preparation

60 °C and then load was applied perpendicular to the axis of the cylindrical specimen through a testing head consisting of a pair of cylindrical segment, at a constant rate of deformation of 51 mm per min. 36 Samples were tested with varying quantities of bitumen, plastic, rubber and plastic–rubber mix. 3 Samples each of pure bitumen, 8% plastic, 8% rubber and 4% plastic–4% rubber mix with 5, 5.5 and 6% Bitumen were prepared and tested and compared with each other.

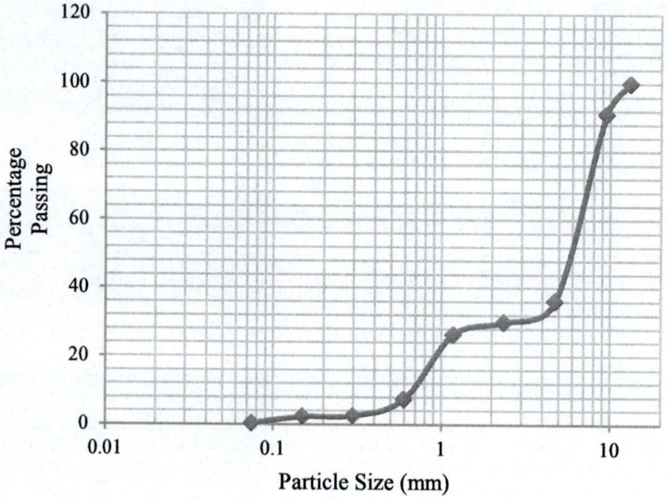

Fig. 2 Gradation curve for coarse aggregate as per IS 2386-1 (1963)

3 Experimental Tests

3.1 Gradation of Fine and Coarse Aggregates

See Figs. 2 and 3.

3.2 Penetration, Softening Point, Specific Gravity, Ductility and Flash Point Test on Bitumen

Bitumen used was tested for its properties and fitness, and results are tabulated in Table 1.

3.3 Bulk Density of Pure Bitumen, 8% Plastic Sample, 8% Rubber Sample and 4% Rubber–4% Plastic Composite Sample

$$G_{\text{bcm}} = \frac{W_a}{W_s - W_w} \tag{1}$$

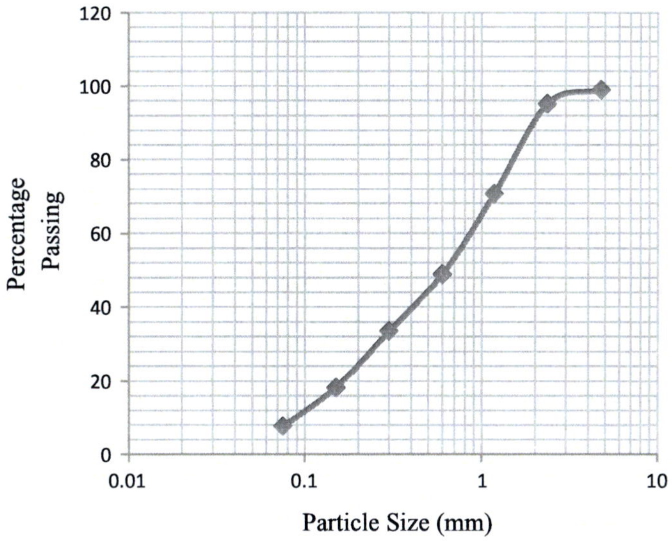

Fig. 3 Gradation curve for fine aggregate as per IS 2386-1 (1963)

Table 1 Bitumen test results

S. No.	Description	Test result	Requirement as per IS:73—1992
1	Penetration at 25°, 100 g, 5 s, 1/10 mm	64	60–70
2	Softening point, °C	52	45–55
3	Specific gravity	1.32	Min. 0.99
4	Ductility at 27 °C, cm	92	Min. 75
5	Flash point °C	210	Min. 175

where, W_a = weight of sample in air; W_s = weight of SSD sample; W_w = weight of sample in water. Bulk Density of all the samples is found to be in the range of 2.25–2.35 and is plotted in Fig. 4.

4 Quantities of Materials Used

Quantities of different materials have been found using MORTH Table for Bituminous Concrete and required modifications were made for composite samples and quantities used has been listed in Table 2.

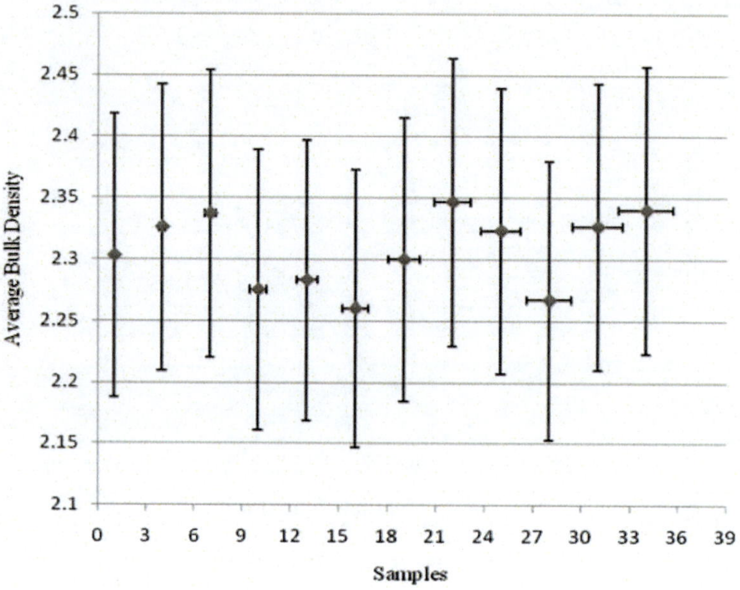

Fig. 4 Average bulk density of sample

5 Marshall Stability Test for Stability and Flow Value

Marshall stability and flow value were obtained for the samples prepared by keeping the sample in water bath at 60 °C for 30 min. Marshall stability of the mix is defined as the maximum load carried by the sample at a standard test temperature of 60 °C. The flow value is the deformation that the test specimen undergoes during loading up to maximum load, expressed in mm unit.

It is evident from Table 3 that from sample number 10, Marshall stability and flow value have increased considerably for sample with plastic content.

Sample number 28–30 have shown slight better results and improvement in Marshall stability values indicating better strength.

6 Results and Discussions

Average bulk density of all the samples is found to be in the range of 2.25–2.35. The Marshall stability value has significantly improved on addition of plastic and plastic–rubber by 50% with respect to pure bitumen samples, thus indicating we can built better roads in less resources using waste such as plastic and rubber in construction of flexible pavements. Sample No. 28–30 have shown great results with respect to addition of bitumen and other cases. 8% plastic and 8% rubber samples

Table 2 Quantities of different materials used in sample preparation

S. No.	Coarse aggregate (g)	Fine aggregate (g)	Lime (g)	Bitumen (g)	Plastic (g)	Rubber (g)
1	592.8	513	34.2	60	0	0
2						
3						
4	589.7	510.3	34	66		
5						
6						
7	586.6	507.6	33.8	72		
8						
9						
10	592.8	513	34.2	55.2	4.8	
11						
12						
13	589.7	510.3	34	60.72	5.28	
14						
15						
16	586.6	507.6	33.8	66.24	5.76	
17						
18						
19	592.8	513	34.2	55.2	0	4.8
20						
21						
22	589.7	510.3	34	60.72	0	5.28
23						
24						
25	586.6	507.6	33.8	66.24	0	5.76
26						
27						
28	592.8	513	34.2	55.2	2.4	2.4
29						
30						
31	589.7	510.3	34	60.72	2.64	2.64
32						
33						
34	586.6	507.6	33.8	66.24	2.88	2.88
35						
36						

Table 3 Marshall stability and flow values

S. No.	Specification		Marshal stability test	
			Stability (kN)	Flow value (mm)
1	Without plastic	5% Bitumen	19.024	2.05
2			18.65	2.02
3			19.48	2.09
4		5.5% Bitumen	17.25	3.18
5			15.55	3.57
6			17.18	3.09
7		6% Bitumen	12.45	3.46
8			14.83	3.9
9			15.55	3.78
10	With 8% plastic	5% Bitumen	28.52	3.08
11			26.5	3.04
12			27.4	3.2
13		5.5% Bitumen	24.97	3.34
14			24.95	3.35
15			23.87	3.32
16		6% Bitumen	26.81	3.32
17			22.9	3.29
18			25.52	3.31
19	With 8% rubber	5% Bitumen	25.56	3.06
20			25.14	3.25
21			24.2	3.12
22		5.5% Bitumen	23.66	3.3
23			24.23	3.34
24			24.69	3.35
25		6% Bitumen	25.55	3.32
26			25.9	3.31
27			26.36	3.31
28	With 4% plastic and 4% rubber	5% Bitumen	29.9	3.35
29			30	3.35
30			30.52	3.2
31		5.5% Bitumen	28.52	3.08
32			27.95	3.05
33			27.87	3.2
34		6% Bitumen	30.3	3.29
35			30.87	3.31
36			30.55	3.33

have shown similar results with respect to Marshall stability value. 4% rubber–4% plastic composite samples have shown increased in stability value by 50% and 15% with respect to pure bitumen and 8% plastic samples, respectively.

7 Conclusions

Coating of polymers such as plastic, rubber and plastic-rubber had shown improved Marshall stability value indication better strength and use of less bitumen thus saving on natural resources and cost saving for laying road. In this dry process, more polymers can be added than hot polymer process and can be carried out in-situ thereby reducing cost and time.

The roads made with these polymers show better road surface condition over a period of time and heavy traffic. Addition of 4% plastic and 4% rubber together has shown the best results in Marshall stability test.

Acknowledgements The author would like to thank the Department of Civil Engineering, Delhi Technological University, New Delhi, India for providing resources and facilities that were needed for the study.

References

1. National Geographic. (n.d.). https://www.nationalgeographic.com/science/2020/07/plastic-trash-in-seas-will-nearly-triple-by-2040-if-nothing-done/
2. *Fact sheet on plastic waste in India.* (n.d.). https://www.teriin.org/sites/default/files/files/factsheet.pdf
3. Bhageerathy, K. P., & Alex, A. P. (2014). Use of biomedical plastic waste in bituminous road construction. *International Journal of Engineering and Advanced Technology (IJEAT).*
4. Vasudevan, R. (2006). Utilization of waste plastics for flexible pavements. *Indian Road Congress, 34*, 105–111.
5. Prasad, A. R., & Sowmya, N. J. (2015). Bitumen modification with waste plastic and crumb rubber. *International Journal of Engineering Research and Technology (IJERT).*
6. Rajasekaran, S., & Vasudevan, R. (2013). Reuse of waste plastics coated aggregates bitumen mix composite for road application—Green method. *American Journal of Engineering Research (AJER)*, 01–13.
7. IRC SP:98 Guidelines for the use of waste plastic in hot bituminous mixes (dry process) in wearing courses (2013).
8. *MORTH Specification.* (2009), Section 500.

Evaluation of Cloth Bag and Gunny Bag as Potential Reinforcing Materials for Pond Ash

Sujit Kumar Pradhan, Anwesha Rath, and Goutam Kumar Pothal

Abstract The quantity of pond ash produced by the thermal power plants have improved in the latest years. The unutilized ash prompts an over increasing ponding region for putting away ash and correlated environmental problems for the society surrounding the power plants. Pond ash can possibly be utilized as a fill-up substance in retaining walls, embankment and structural land filling, etc., because it is a non-plastic cohesionless material. But the strength of the consolidated pond ash fills retained moderately by reinforcing it suitably. Reinforcing soil can be considered as one of the standardized techniques for improving the strength parameters of soil. Keeping this in mind along with the cost-effectiveness cloth bags and gunny bag as geotextile reinforcement in this study. The impact of water content, degree of compaction, cloth bag and gunny bag as a reinforcement, etc., on several geotechnical properties are concentrated during the work. The strength parameters of the cloth bag reinforced geotextile are observed from a sequence of tests including proctor test, direct shear test and triaxial shear test. As the number of layers of reinforcement increases, there was an evident increase in the deviator stress of specimen. Even for the same confining pressure increase in the number of cloth bags and gunny bag geotextile reinforcing layer increases the deviator stress of specimen. Increase in the spacing between the layers of cloth bags reinforcement layers cause reduction in the deviator stress and shear strength parameters.

Keywords Pond ash · Cloth bag · Shear behavior · Reinforcement

1 Introduction

The demolition of the unconsumed waste obtained through processing of items from factories has been an extraordinary concern today. The ash formed by blazing of coal is arranged by blending it with water, and specific synthetic compound to become

S. K. Pradhan (✉) · A. Rath · G. K. Pothal
Department of Civil Engineering, Indira Gandhi Institute of Technology Sarang, Dhenkanal, Odisha 759146, India
e-mail: sujitpradhan@igitsarang.ac.in

A. K. Gupta et al. (eds.), *Advances in Construction Materials and Sustainable Environment*, Lecture Notes in Civil Engineering 196,
https://doi.org/10.1007/978-981-16-6557-8_31

eco-friendly known as ash ponds. The removal system utilized for unloading ash is thick slurry removal that utilizes definite chemical process that contributes rapid consolidation of ash slurry when it is dumped. The disposal of ash can have adverse impact on the environment, and relocation; water resource pollution; air pollution and may cause human health hazards. The design of waste removal zones is ineffective with regard to savings and land use. Amassing ash in swallow ponds by fencing of natural low lands, brings about inadequate utilization of land regions for storage of high-volume pond ash. Thus, disposal of these wastes is a problematic issue and needs to be checked for the sustainability of the environment. So effective utilization of these wastes as a substitute for natural ingredients would benefit the construction industry in many ways. The components influencing pond ash is the function of various characteristics such as source from which coal is obtained, loading and firing condition, degree of pulverization, boiler unit, handling and storage methods. Many researchers utilized geotextile as a reinforcing materials for poor graded soil and many waste material such as fly ash, pond ash, etc. Using geotextile in pond ash dykes may be useful for implementing it as a reinforcement material in pavement embankment and fills [1]. Geomembrane improve the hydraulic stability of the costal structure in short term durability performance [2]. Addition of one layer of geocomposite, enhanced the strength parameters of the poor quality soil. With increasing the number of reinforcement layers improve the strength as compared to one layer of geocomposite [3]. Geosynthetic reinforcements will expand pavement performance which substitute to improve weak poor quality of soil [4]. There is greater resistance to penetration offered by the geotextile observed by [5]. Use of geotextile showed good performance as subgrade stabilization [6]. Improvement in bearing capacity with increase in footing size along with increase in settlement rate by using gunny bags as geotextile and sand as a soil media to improve soil bearing capacity [7]. Sand was included in pond ash bed which was again reinforced with jute geotextile at varying depths and observed improvement in the bearing capacity with increasing in nos. of layer [8]. The purpose of this study is to evaluate the impact of cloth bag and gunny bag as a reinforcing materials for pond ash.

2 Materials

Pond ash utilized for this study was collected from NTPC, Talcher, Odisha, India. It was analyzed as unified soil classification system which is slit of low compressibility. The sample was dried at 100–110° and screened through 2 mm diameter sieve. Cloth bag and Gunny bags collected locally. Table 1 shows the physical properties of pond ash, cloth bag and gunny bag.

In this research, cloth bags and gunny bags as two types of geotextiles were investigated (Fig. 1).

Table 1 Physical properties of pond ash, cloth bag and gunny bag

Properties	Pond ash	Cloth bag	Gunny bag
Color	Gray		
Shape	Sub-rounded		
Uniformity coefficient	1.93		
Coefficient of curvature	1.5		
Plasticity index	Non-plastic		
Mass per unit area (g/m^2)		450	1200
Thickness (mm)		2	5

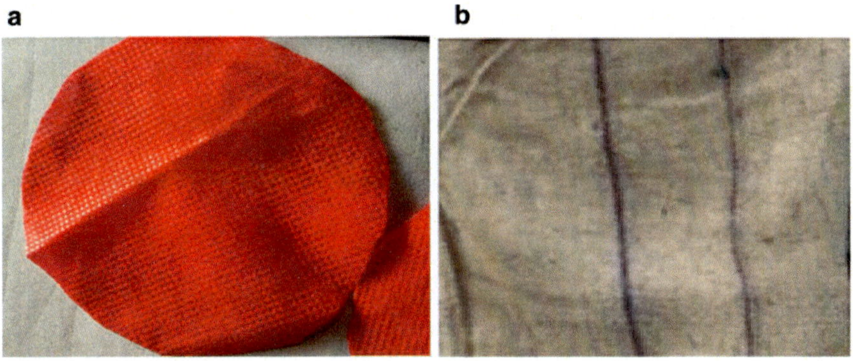

Fig. 1 **a** Cloth bag **b** Gunny bag

3 Methodology/Experimental Program

Line diagram for unreinforced pond ash, 1 layer, 2 layer, 3 layer reinforcement presented in Fig. 2.

Table 2 shows the various experiments carried out on the samples.

The optimum moisture content (OMC) and maximum dry density (MDD) were performed (Fig. 3) as per IS code mentioned above and found to 23% and 1.272 gm/cc, respectively.

Triaxial shear test

The triaxial test was performed on a cylindrical sample having height to diameter ratio of 2. From the three principal stresses, two of them were loaded by the water pressure in the confining cell, and the third one is loaded by a loading ram along the top of the cell.

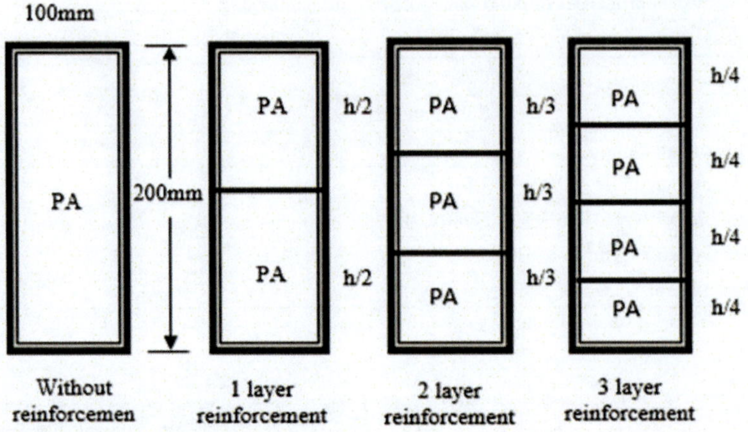

Fig. 2 Line diagram for placement of reinforcement

Table 2 Code adopted for different tests

Test	Code
Particle size distribution	IS: 2720 (Part-4):1985 [9]
Compaction	IS: 2720 (Part-7):1980 [10]
Triaxial shear	IS: 2720 (Part-11):1993 [11]
Plasticity index	IS: 2720 (Part-5):1985 [12]

Fig. 3 Compaction curve of pond ash

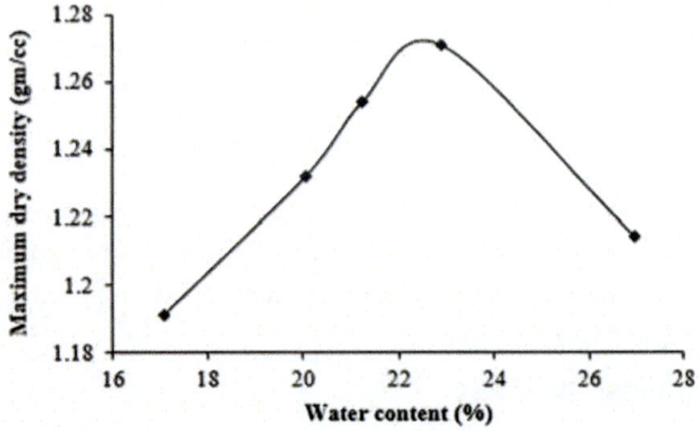

4 Results and Discussion

4.1 *Impact of Cloth Bag Geotextile Reinforcement on Shear Strength Parameters*

The shear strength parameters observed are cohesion of unreinforced pond ash, adhesion between the pond ash and non-woven cloth bag carry geotextile reinforcement, angle of internal friction of unreinforced pond ash and friction angle of non-woven cloth bag geotextile reinforced pond ash. It was seen that shear strength parameters improves with raising in the nos. of layer. Increasing space in between the layers and decrease in the number of layers of non-woven carry bag geotextile, it was observed that there was a decrease in deviator stresses. This may happen due to initial compression of the reinforcement due to applied pressure before the tensile force is mobilized in the reinforcement. Unlike standard UU test the samples exhibited higher value of because the samples were not saturated. Also, there was an enhancing the shear strength parameters with reduce in spacing with increment in number of layers. Enhancement of shear strength is because of the interference of geotextile with the potential shearing plane and redistribution of the stress. Figure 4 shows the effect of reinforcement in the shear strength specifications.

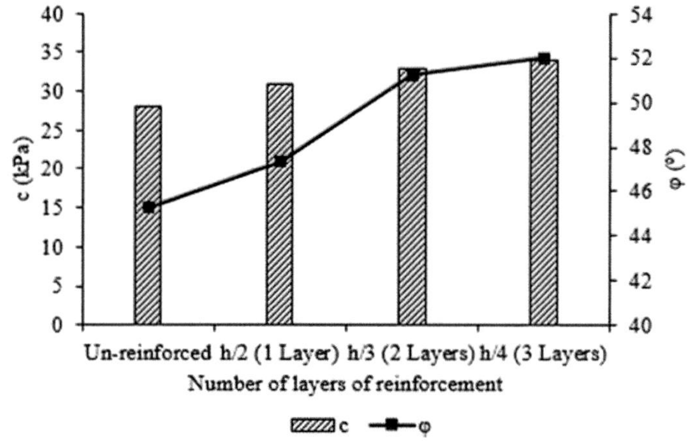

Fig. 4 Effect of cloth bag reinforcement in the shear strength parameters

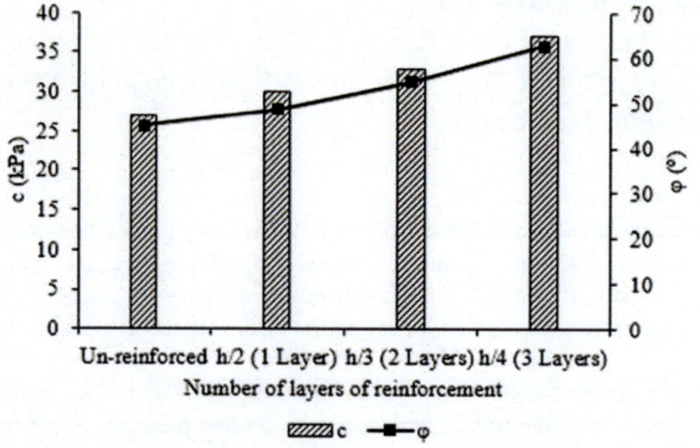

Fig. 5 Effect of cloth bag reinforcement in the shear strength parameters

4.2 *Impact of Gunny Bag Geotextile Reinforcement on Shear Strength Parameters*

With improvement of the number of layers and decline in spacing brought about an increment in the shear strength parameters. The unsaturated samples exhibited higher value of φ in the UU test. It is due to the redistribution of stress and the interference of geotextile with the potential shearing plane. As the deviator stress increased with increase in number of layers, c and φ increased. Figure 5 summarizes the influence of gunny bag geotextile on shear strength parameters.

4.3 *Effect of Cloth Bag Geotextile Reinforcement on Deviator Stress*

The deviator stress versus axial strain curve for pond ash was obtained by changing the number of reinforcing layers at a confining pressure of 100 kPa, 200 kPa and 300 kPa, respectively. It was observed that accumulating the number of layers of reinforcement leads to increase in the peak of the deviator stress (Figs. 6 and 7). Lateral expansion of the sample is restricted by the tensile property of the reinforcement. As the number of reinforcement layers are increased at different levels of the sample, the sample is restricted from lateral expansion there by the deviator stress increased.

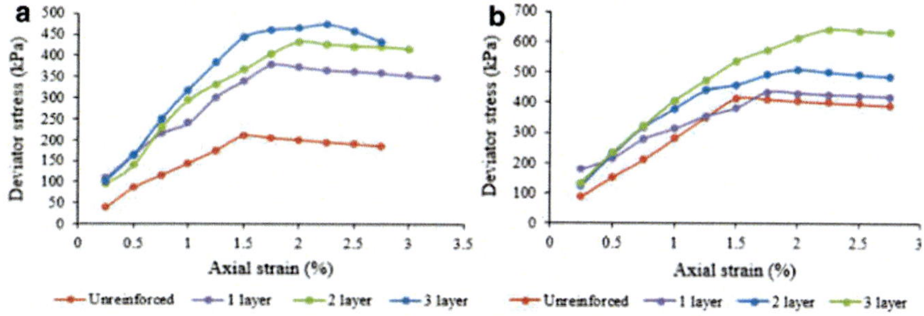

Fig. 6 Deviator stress versus axial strain **a** 100 kPa **b** 200 kPa confining pressure

Fig. 7 Deviator stress versus axial strain 300 kPa confining pressure

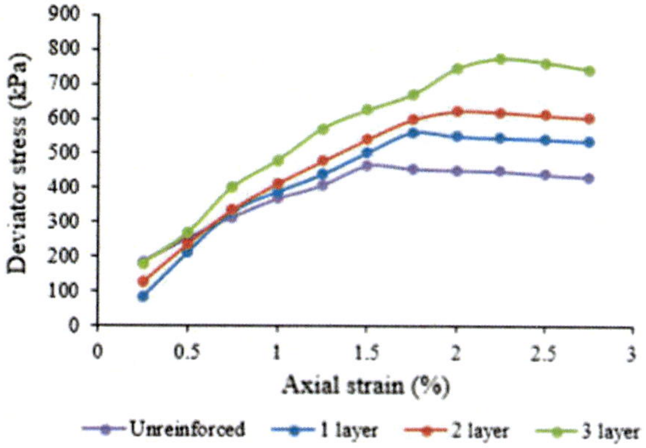

4.4 *Impact of Gunny Bag Geotextile Reinforcement on Deviator Stress*

With increase in the number of reinforcement layers (gunny bag) prompts to enhance the peak of the deviator stress versus axial strain plot. Furthermore, it was observed that on increasing the number of layers of reinforcement leads to increase in the peak of the deviator stress (Figs. 8 and 9).

5 Conclusion

Increasing confining pressure leads to deviator stress at failure enhances and also shear strength improves. Pond ash attains most of its shear strength from internal friction and indicates some amount of cohesion.

- Deviator stress enhances with the increase in number of layers of reinforcement for the same confining pressure.

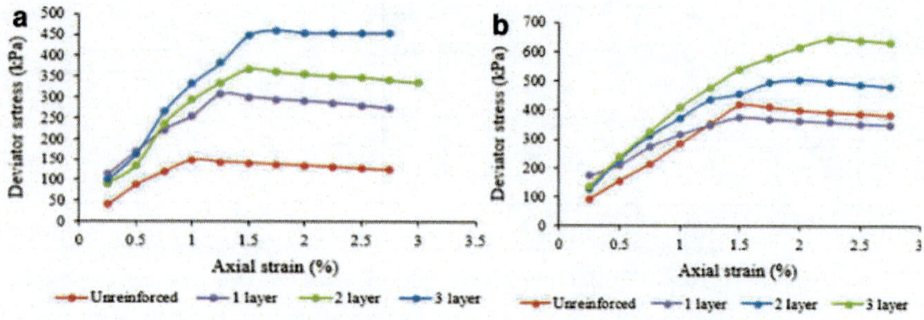

Fig. 8 Deviator stress versus axial strain **a** 100 kPa **b** 200 kPa confining pressure

Fig. 9 Deviator stress versus axial strain 300 kPa confining pressure

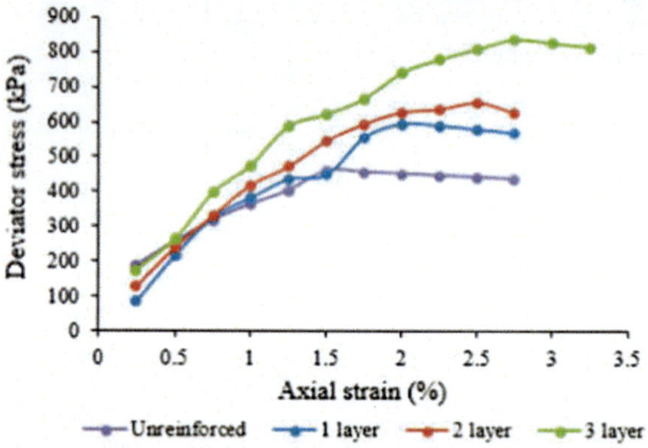

- The shear strength parameters of pond ash reinforced with non-woven cloth bag geotextile increases with the increase in the number of layers of reinforcement.
- Deviator stress and shear strength reduces with increase in spacing between the layers of reinforcement.

References

1. Vijayasri, T., Raychowdhury, P., & Patra, N. R. (2020). Dynamic behavior of a geotextile reinforced pond ash embankment. *Journal of Earthquake Engineering, 24*(11), 1803–1828. https://doi.org/10.1080/13632469.2018.1483848
2. Oyegbile, B. O., & Oyegbile, B. A. (2017). Application of geosynthetic membrane in soil stabilization and coastal defence structures. *International Journal of Sustainable Built Environment, 6*(2), 636–662. https://doi.org/10.1016/j.ijsbe.2017.04.001
3. Carlos, D. M., Pinho-Lopes, M., & Lopes, M. L. (2016). Effect of geosynthetic reinforcement inclusion on the strength parameters and bearing ratio of a fine soil. *Journal of Procedia Engineering, 143*, 34–41. https://doi.org/10.1016/j.proeng.2016.06.005
4. Kaur, M., & Aggarwal, S. K. (2020). Weak sub-grade soil reinforced with Geogrid material—A review. *IOP Conference Series: Materials Science and Engineering, 955*, 012108. https://doi.

org/10.1088/1757-899X/955/1/012108
5. Singh, M., Trivedi, A., & Shukla, S. K. (2020). Influence of geosynthetic reinforcement on unpaved roads based on CBR, and static and dynamic cone penetration tests. *International Journal of Geosynthetics and Ground Engineering, 6* (13). https://doi.org/10.1007/s40891-020-00196-0
6. Cuelho, E. V., & Perkins, S. W. (2017). Geosynthetic subgrade stabilization—Field testing and design method calibration. *Journal of Transportation Geotechniques, 10*, 22–34. https://doi.org/10.1016/j.trgeo.2016.10.002
7. Panigrahi, B., & Pradhan, P. K. (2019). Improvement of bearing capacity of soil by using natural geotextile. *International Journal of Geo-Engineering, 10*, 9. https://doi.org/10.1186/s40703-019-0105-7
8. Sharma, A., & Singh, K. (2020). Bearing capacity of sand admixed pond ash reinforced with natural fiber. *Journal of Natural Fibers*. https://doi.org/10.1080/15440478.2020.1848699
9. IS: 2720: Part 4. (1985). Methods of test for soils—Grain size analysis (Second revision). Bureau of Indian Standards, New Delhi.
10. IS: 2720: Part 7. (1980). Methods of test for soils—Determination of water content-dry density relation using light compaction. Bureau of Indian Standards, New Delhi.
11. IS: 2720: Part 11. (1993). Methods of test for soils—Determination of the shear strength parameters of a specimen tested in unconsolidated undrained triaxial compression without the measurement of pore water pressure. Bureau of Indian Standards, New Delhi.
12. IS: 2720: Part 5. (1985). Methods of test for soils—Determination of liquid limit and plastic limit (Second revision). Bureau of Indian Standards, New Delhi.

The Effect of Sisal Fiber on Mechanical Strength of Concrete M20 Grade

Celso Januário Baúque, Ankit Thakur, and Bhartesh

Abstract The material of construction made by mix of cement and fine aggregate and coarse aggregate and water is a concrete, and can be considered one of the structure made by human that are mostly used in the field of construction industry. This work is aiming to study or find the effect of sisal fiber on mechanical strength of concrete M20 grade in different mix proportions, as in 0, 1, 2 and 3% of sisal fiber to replace by the volume of concrete mix. Tests were carried out on compressive strength, split tensile strength and flexural strength in 3, 7 and 28 days of concrete curing. The results of incorporation of sisal fiber in concrete were found that the properties on mechanical strength were affected. According to the results, sisal fiber affected the workability of concrete. The compressive strength was found decreasing when the amount of sisal fiber increased. The results of tensile strength was found increasing up to 1% of sisal fiber by the volume and after was verified decreasing, and the flexural strength showed early increasing by 2% of sisal but at 28 days registered that 1% of sisal fiber was increasing the flexural strength.

Keywords Concrete · Sisal fiber · Mechanical strength

1 Introduction

Concrete has been the principal mix used in the construction industry. The days of today the challenging is to improve the properties of the concrete, and then the uses of natural fibers have been normal in research field of construction and materials. For this issue of improvement of properties, in the practice of concrete, was found that there some poor issues due to the degradation of natural fiber of cement, and to

C. J. Baúque
Department of Civil Engineering, Instituto Superior Politécnico de Songo, ISPS, 2304 Songo, Cahora Bassa, Tete Province, Mozambique

A. Thakur (✉) · Bhartesh (✉)
Department of Civil Engineering, Alakh Prakash Goyal Shimla University, Mehli, Shimla, Himachal Pradesh 171013, India

A. K. Gupta et al. (eds.), *Advances in Construction Materials and Sustainable Environment*, Lecture Notes in Civil Engineering 196,
https://doi.org/10.1007/978-981-16-6557-8_32

alleviate this effect in matrix can be used some techniques to improve or to modify the properties of concrete [1].

The aim of this work was to study the incorporation of natural fiber in concrete, to get the optimal performance on mechanical strength of M20 grade of concrete, knowing that sisal has been considered a one of the strongest natural fiber in the world [1]. So the purpose of this investigation is to discover the optimal strength of concrete M20 grade when is mixed with natural sisal fiber using proportion 0, 1.0, 2.0 and 3.0% using IS: 456 (2000).

When is incorporated sisal fiber in the concrete mix have been found that does not increase the compressive strength of concrete but found reduced, despite that the tensile strength was increased [2, 3]. Was reported the mix of sisal fiber and concrete, and was found that the concrete got a superior effect on mechanical properties, also was more economic when compared with another concrete mixed by other fiber [4].

The investigation of using concrete mixed with sisal fiber to improve the properties on mechanical strength also was found decreasing the compressive when increase the amount of sisal, but increasing the ductility and the energy dissipation [5].

Ramakrishna and Sundararajan [6] investigating the behaviors of tensile strength concluded that sisal fiber was found lower than the control concrete and were similarity in the results of compressive and flexural [6].

1.1 Sisal Fiber

Sisal fiber can be extracted from the leaves of sisal plant *Agava Sisalana.* The sisal plant can be considered that is from Mexico [7–9]. Also is written that Brazil is the largest producer of Sisal fiber even in India and in Mozambique the sisal is produced but in smaller quantities [9]. This fiber has a great commercial application in the World [7–9], see Fig. 1a–c that shows the sisal plant.

Along of the length of the sisal fiber can be found a lot of different resistance, then the sisal can vary from the lower part, intermediate part and the tip part. Was found that the sisal fiber to the properties in tensile strength is low [9, 11], modulus of elasticity and percentage of elongation showed that the resistance in the region in basal part are low, while the deformation is high [11], being that the intermediate

Fig. 1 Sisal fiber **a** Sisal plant, **b** Processed sisal fiber and **c** Sisal fiber image, according to Uppal et al. [10]

Table 1 Properties of sisal fibers

Fiber	Properties	References	Fiber	Properties	References
Fiber length	30 mm	2	Cellulose (%)	47–78	8
Fiber diameter	0.10–0.13 mm	2	Lignin (%)	10–24	8
Aspect ratio	230–300	2	Hemicellulose (%)	7–11	8
Tensile strength	371.28 MPa	2			
Tensile modulus	12.43223 GPa	2			
Shape	Straight	2			
Color	Creamy white	2			
Density	0.113 g/cm^3	2			
Water absorption	43.58%	2			
Specific gravity	0.73	2			

Okeola et al. [2] and Saxena et al. [9]

part is resistant and hard, and the apical part presents moderate strength and stiffness [9, 11].

Was found the each leaf having 3–4% of fiber, and 8% of dry matter and also was found 87.25% of water. The same investigation was found 3% of weight of fiber considered 600 g of leaf, and the shape of fiber was cylindrical [9] the difference of properties and structure can be verified due to the different factors, method of extraction also the age of plant even the sources [12]. Many authors found that sisal fiber consist mainly of cellulose, hemicellulose and lignin [9, 11–13]. Some of the properties are in Table 1.

2 Materials and Methods

2.1 Cement and Water

Were used water, fine and coarse aggregates, cement of 43 grade according to IS 8112.1989 [14], in Table 2 is listed the properties of cement according to [14–16], following to the practice the water used in concrete mix design was normal and had a 6–8 of pH [15, 16], the properties of cement can be seen in Table 2.

2.2 Fine Aggregates

This experiment was using the natural sand to fill up the space of voids of coarse aggregates. The fine aggregates used had a size between 4.75 and 0.75 mm, as is written in IS 383 1970. The percentage passing by weight of sand was found in

Table 2 Properties of cement in term of physical and chemical

Content	Requirement
Physical properties	
Minimum compressive strength	
3 day	23
7 day	33
28 day	43
Fineness	
Minimum specific surface (m^2/kg)	225
Soundness, expansion	
Le Chatelier test (mm), Max	10.0
Setting time (min)	
Initial, min	30
Final, max	600
Chemical properties	
Loss on ignition (%), max	5.0
Magnesium MgO (%), max	6.0
Total sulphuric anhydride (SO_3)	3.5

Baúque et al. [16]

grading zone II, and all fine aggregate was in prescribed size from the code [17] see Fig. 2.

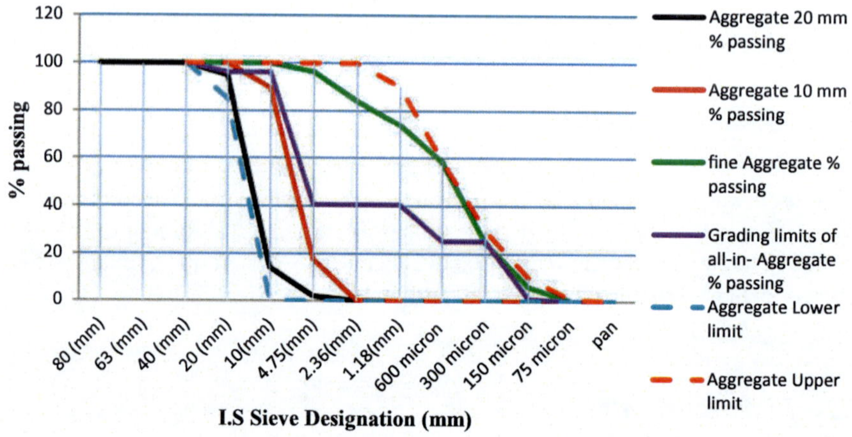

Fig. 2 Gradation of aggregate

Table 3 Code of mix

Code of mix	M0	M1	M2	M3
Proportion of the mix (%)	0.0	1.0	2.0	3.0

2.3 Coarse Aggregates

The size of 10 and 20 mm was considered in this experiment then the aggregate was found retained according to the IS 383 1970 [17]. See Fig. 2, the coarse aggregates considered in this experimental were natural and was without organic dross.

2.4 Sisal Fiber

The sisal fiber in concrete mix design actually was to discover his improvement in mechanical properties and his influence in technology of concrete. This work was used the sisal fiber of 30 mm of length also was using the intermediate part.

2.5 Concrete Mix

The mix design considered was M20 grade of concrete 1:1.5:3:0.5 for the mix proportions of Cement: Fine Aggregate, Coarse Aggregate and Water Cement ratio. The concrete was mixed with sisal fiber, where the sisal fiber was added in the matrix of concrete in proportion of 0, 1, 2 and 3% in volume, and the code mix of the proportions used is in Table 3.

2.6 Specimens to Perform the Mechanical Properties Tests in Laboratory

To perform a compression test was used a specimen with metal structure and also cubical specimen with size of 150 mm × 150 mm × 150 mm according to the description in IS: 516-1959 [18]. For the tensile strength, test was using the cylindrical specimen having the size of 150 mm in diameter 300 mm long as described in IS: 5816-1999 [19], to perform flexural strength were used the standard size with 150 mm × 150 mm × 700 mm, according to IS: 516-1959. And the cube specimens were metal structure [18].

Table 4 Concrete mix proportions for compressive strength test for Sisal Fiber

Component w/c = 0.5	0.0% S.F	1% S.F	2% S.F	3% S.F
	Content	Kg/m^3		
Cement (kg)	16.20	16.04	15.88	15.71
Fine aggregate (kg)	24.30	24.06	23.81	23.57
Coarse aggregate (kg)	48.60	48.11	47.63	47.14
Sisal fiber (kg)	0.00	0.41	0.81	1.22

2.7 *Measurement of Workability*

The test was performed in the apparatus consisting in a metallic mold which is in a form of a frustum of a cone. The dimensions of the specimen in the bottom had diameter of 20 cm, the top diameter with 10 cm and having a height of 30 cm. Was added the water–cement ratio 0.50 in concrete mix and then after the mixing, was started to conduct the test finding the reduction value of slump, and also the mold was being filled in four layers, all the procedures followed was found in IS: 1199-1959 [20].

2.8 *Compressive Strength Test*

To perform the compression test, was calculated the mix proportions 1:1.5:3:0.5 according to the IS 456 (2000) for M20 grade concrete [21], and was found that for the concrete mix control was 16.20 kg of cement and in the corporation of sisal fiber was found in amount of volume replacement of 0, 1, 2 and 3% of volume of concrete in Table 4.

2.9 *Splitting Tensile Strength Test*

Also the tensile strength was performed and the mix design concrete in proportions 1:1.5:3:0.5 according to the IS 456 (2000) for M20 grade concrete [21], the concrete mix control had of 25.45 kg of cement and the incorporation of sisal fiber was for 0.0, 1, 2 and 3% of replacement by the volume of concrete in Table 5.

2.10 *Flexural Strength of Concrete*

To perform the flexural strength, the concrete mix control was found amount of cement in 76.6 kg/ in proportions of 1:1.5:3:0.5 for M20 grade concrete and the sisal

Table 5 Concrete mix proportions for splitting tensile strength test for Sisal Fiber

Component w/c = 0.5	0.0% S.F	1% S.F	2% S.F	3% S.F
	Content	Kg/m^3		
Cement (kg)	25.45	25.19	24.94	24.68
Fine aggregate (kg)	38.17	37.79	37.41	37.03
Coarse aggregate (kg)	76.34	75.58	74.81	74.05
Sisal fiber (kg)	0.00	0.64	1.27	1.91

Table 6 Concrete mix proportions for flexural strength test for Sisal Fiber

Component w/c = 0.5	0.0% S.F	1% S.F	2% S.F	3% S.F
	Content	Kg/m^3		
Cement (kg)	75.60	74.84	74.09	73.33
Fine aggregate (kg)	113.40	112.27	111.13	110.00
Coarse aggregate (kg)	226.80	224.53	222.26	220.00
Sisal fiber (kg)	0.00	1.89	3.78	5.67

fiber was found in proportions of 0, 1, 2 and 3% to making a replacement by the volume of concrete in Table 6.

To analysis the results obtained after testing, were used a statistical program of variance analysis system called SISVAR. The analysis was taken the minimum significance value of 5%, considering the Scott—Knott test. To verify the effect of sisal fiber in concrete, the result should be found in ($p < 0.05$) and if sisal fiber had no effect on concrete the value should be ($p > 0.05$).

3 Results and Discussion

3.1 Slump of Concrete Mix

Using the water–cement ratio of 50% for concrete mixed with sisal fiber the results of slump test was found that reduction in slump of reference concrete was 50.00 mm, and with addition of sisal fiber there was an tendency of reduction in slump up to 30.30 mm due to the existing an reinforce of fresh concrete with fibers of sisal Fig. 3.

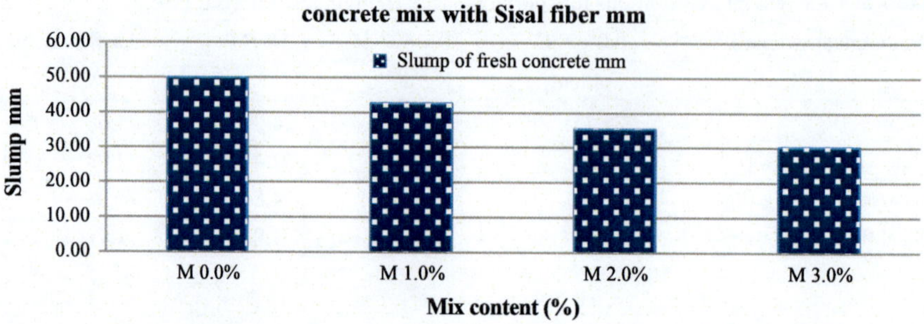

Fig. 3 Slump of concrete mix

3.2 *Mechanical Strength of Concrete Mix*

Compressive strength of concrete mixed with sisal fiber

The compressive strength of concrete with sisal fiber in 3 days of curing was found that the reference concrete had a higher compressive strength compared to other mixes, it can be seen in figure that was found having different results, the control concrete 0%, followed by the 1% of sisal fiber and also, followed by the concrete of 2% of sisal fiber and finally of 3% according to the Scott–Knott test at a nominal level of 5% significance, there was a highly significant effect ($p = 0.0000$) of mix proportions.

In 7 days of concrete curing, the control concrete had a higher compressive strength, then after of 1% of sisal fiber and, followed by of 2 and 3% that was equal and lower compressive, the results were according to the Scott–Knott test at a nominal level of 5% significance. There was a highly significant effect ($p = 0.0000$) in the concrete of mix proportions tested.

The results of compressive strength in 28 days, statistically the results showed that the reference concrete was higher than other mixes, followed by the concrete with 1% of sisal fiber, and the 2 and 3% were found equal and lower using the Scott–Knott test at a nominal level of 5% significance. In terms of mix proportions to the concrete was verified that there highly significant effect ($p = 0.0000$) as shown in Fig. 4.

The results reported that increasing the amount of sisal was decreasing the compressive in concrete, in this case was found that the compressive strength also was found affected ($p < 0.05$) by the incorporation of sisal fiber in all ages of curing tested, this could be caused by the various factors like cylindrical shape of the sisal fiber, but also it can be considered the factor of absorbing water that can modify the weight also and other properties in the strength of the concrete.

Another study related that there some impact of sisal fiber on concrete. Despite that was for partially replaced cement with shell ash, in the report was found that using sisal fiber of 40 mm the compressive strength was increasing [22].

In the study of Okeola et al. [2], the decreasing was verified since were added sisal in concrete mix [2], another study found the similar result, and the sisal fiber in

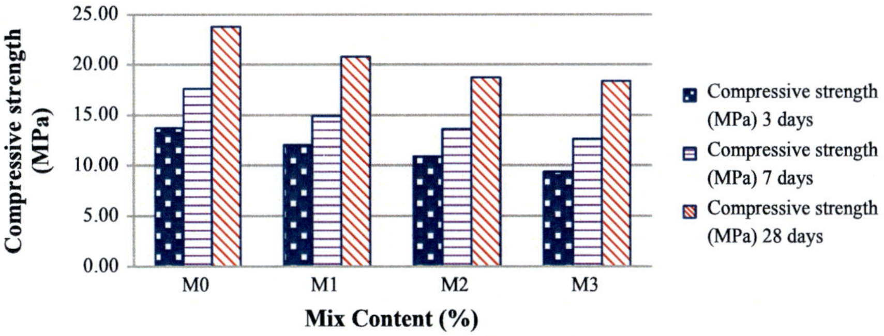

Fig. 4 Compressive strength of concrete with sisal fiber

concrete was decreasing the compressive strength [23]. The risk that can be considered in concrete mixed with fiber is shown on its reducing of workability, things that can influence negatively on compressive strength of concrete [7].

Tensile strength of Concrete mixed with sisal fiber

The results reveal that the mix proportions of concrete is affected, the tensile strength was improved up to 1% of replacement of the volume of concrete, and after this 1%, the tensile strength decreased, due to the surface and shape of sisal into concrete mix, the tensile strength also was found affected ($p < 0.05$) by the incorporation of sisal fiber in all ages of curing tested.

In 3 days of concrete curing, the concrete with 1% of sisal fiber was higher than other mixes the control concrete and the 2% of sisal fiber on concrete, respectively, were equal, and the 3% of the sisal fiber was less in the evaluation according to Scott–Knott at a nominal level of 5% significance, this result also means that the mix proportions was highly significant effect ($p = 0.0001$).

Statistically in 7 days of concrete curing, the tensile strength was found that 1% of sisal fiber in concrete was bigger than others, followed by concrete control, and after the third was the 2% of concrete with sisal and lowest were 3% of mix proportions in Scott–Knott test at a nominal level of 5% significance. In terms of mix proportions tested, there was a highly significant effect ($p = 0.0000$).

The tensile strength in 28 days of concrete curing, reported that 1% of sisal fiber was higher than others, followed by the concrete control, then the third was the 2% of concrete with sisal and lowest was 3% of mix proportions according in Scott–Knott test at a nominal level of 5% significance. In terms of mix proportions, there was a highly significant effect ($p = 0.0000$) as can be seen in Fig. 5.

The tensile strength in this experiment was found higher at 1% with 3.72 MPa, when found that sisal fiber has ability to transfer the ductility in concrete improving the tensile strength and the cracking achievement. [23]. But after 1% of sisal fiber mixed with concrete that transference start to reduce, and it can be attributed due to the amount of surface of sisal fiber concentrated highly in concrete then reducing the tensile strength. The tensile strength was found increasing, was observed also that

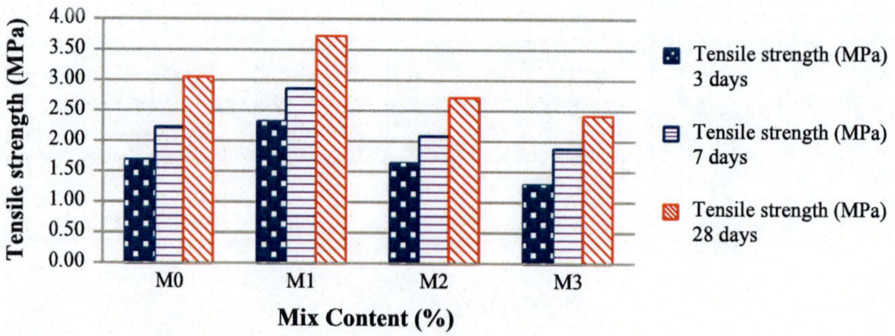

Fig. 5 Tensile strength of concrete with sisal fiber

sisal in concrete mix was increasing till 1% [2], these results also was verified in the current study.

Flexural strength of Concrete mixed with sisal fiber

The flexural strength of concrete with sisal fiber in 3 days of curing statistically the reference concrete and others that contains sisal fiber was equal according to the Scott–Knott test at a nominal level of 5% significance. In terms of treatment also in the mix proportions, there was no significant effect ($p = 0.4048$).

In 7 days of concrete curing the flexural strength of concrete with 2% of sisal fiber was higher than others and was, followed by the concrete with 3 and 1% of sisal that was equal and the reference control was the lowest according to the Scott–Knott test at a nominal level of 5% significance the mix proportions tested, there was a highly significant effect ($p = 0.0006$).

In 28 days of concrete curing, was found concrete with 1% of sisal fiber higher than others and, followed by the concrete with 2% of sisal fiber, after the 3% and the reference concrete was the lowest according to the Scott–Knott test at a nominal level of 5% significance in terms of mix proportions, there was a highly significant effect ($p = 0.0000$). See Fig. 6.

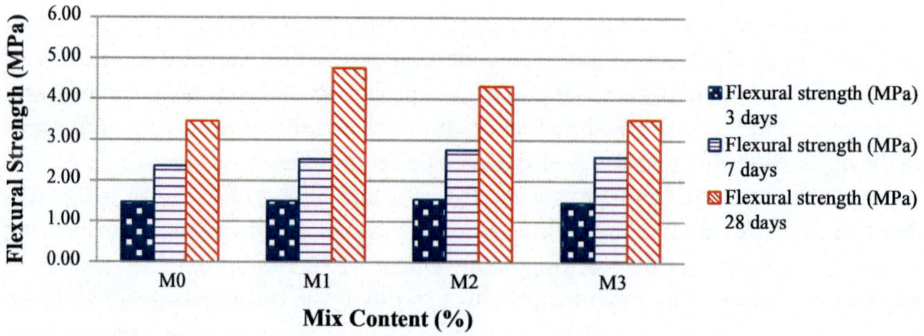

Fig. 6 Flexural strength of concrete with sisal fiber

According to the results was found that the incorporation of sisal fiber in concrete increase up to 1% of sisal fiber in volume, and after 1% the flexural strength was decreasing, generally the flexural strength also was affected ($p < 0.05$) by the incorporation of sisal fiber in all ages of curing tested. The variation of flexural strength can be also attributed to the high concentration of the surface of sisal fiber in concrete, then after 1% it start to lose his properties with concrete.

4 Conclusion

The results of the investigation about the incorporation of sisal fiber in concrete were found that the mechanical strength was affected, and also there was found an optimal replacement in volume of sisal fiber in to concrete.

According to the results of slump test, was concluded that the incorporation of sisal fiber in concrete affect the workability of fresh concrete, and is concluded that the reference concrete show a high reduction in 50.0 mm, while the concrete mixed with 3% of sisal fiber the reduction were found lower at 30.0 mm due to the reaction of fiber in reducing the workability and absorbing water. And also were found that the 1 and 2% of replacement were 42.7 mm and 35.54, respectively. Can be concluded also that when the sisal fiber is mixed with concrete will be there some reduction of workability on fresh concrete.

The reference concrete shows the better compressive strength at all days or ages of curing concrete, and were found decreasing 12.46% at 28 days using 1% of sisal fiber, for the mix of 2% of sisal also was decreasing from 23.76 to 18.74 MPa, and for the mix of 3% of sisal in concrete was found decreasing about 22.64% of the compressive strength, so with this results, is concluded that the compressive strength at all ages of concrete mixed with sisal fiber is decreased. Using the Scott–Knott test at a nominal level of 5% significance, the compressive strength was found affected or highly significant ($p < 0.05$) at all ages or days.

The results of tensile strength at 3, 7 and 28 days of curing were improving or increasing up to 1% of sisal fiber replacement of the volume. Then generally, the strength according to the Scott–Knott test, the concrete was affected ($p < 0.05$). And after that there was decreasing of tensile. It can be concluded also that in 28 days of the curing concrete mixed with 1% of sisal fiber the improvement is 21.97% having 3.72 MPa of tensile strength, compared to the normal concrete with 3.05 MPa. For the 2% and 3% of sisal fiber in concrete are concluded that the tensile decrease in order of 10.82% and 20.66%, respectively. And the 1% of sisal fiber can be considered as optimal replacement to improve the tensile strength.

The flexural strength of concrete with sisal fiber at 3 days was found not affected with sisal fiber ($p > 0.05$) in the Scott–Knott test, and in 7 days of curing was found increasing up to 2% of sisal in replacement by volume of concrete. Thus, at 28 days of curing was found that the concrete with 1% of sisal fiber was increased up to 37.86% of tensile strength (4.77 MPa) compared to the reference concrete with 3.46 MPa, also at 28 days, to the 2% and 3% of sisal fiber mixed with concrete was found increasing compared to the reference concrete in 4.32 MPa and 3.51 MPa, respectively. Thus, the concrete with 1% was concluded as optimal to the flexural strength and also according to the Scott–Knott test there was highly significance effect ($p < 0.05$).

References

1. Siddique, R. (2008). *Waste materials and by-products in concrete. Engineering materials*. Springer. ISBN: 978-3-540-74293-7.
2. Okeola, A., Abuodha, S., & Mwero, J. (2018). Experimental investigation of the physical and mechanical properties of sisal fiber-reinforced concrete. *Fibers, 6*(3), 53. https://doi.org/10.3390/fib6030053
3. Bulut, H. A., & Şahin, R. (2017). A study on mechanical properties of polymer concrete containing electronic plastic waste. *Composite Structures*. https://doi.org/10.1016/j.compstruct.2017.06.058
4. Hossam A. E., Abou Haloub, M. A., & Rustom, R. N. (2019). Effect of new mixing method of glass powder as cement replacement on mechanical behavior of concrete. *Construction and Building Materials, 203*, 75–82. https://doi.org/10.1016/j.conbuildmat.2019.01.077
5. Ismail, Z. Z., & AL-Hashmi, E. A. (2009). Recycling of waste glass as a partial replacement for fine aggregate in concrete. *Waste Management, 29*(2), 655–659. https://doi.org/10.1016/j.wasman.2008.08.012
6. Ramakrishna, G., & Sundararajan, T. (2019). Long-term strength and durability evaluation of sisal fiber composites. *Durability and Life Prediction in Biocomposites, Fibre-Reinforced Composites and Hybrid Composites*. https://doi.org/10.1016/B978-0-08-102290-0.00010-6
7. Thakare, A. A., & Suryawanshi, S. R. (2018). Structural properties of concrete using sisal fiber. *Journal of Advances and Scholarly Researches in Allied Education, Ignited Minds Journals, XV* (2), 364–369. ISSN 2230-7540; Multidisciplinary Academic Research. https://doi.org/10.29070/15/56848
8. Carvalho, R. R. C., Neto, G. C., & Candido, V. S. (2016). Utilização de fibra de Sisal como Agente De Reforço Em Matriz Cimentícia para Fabricação de blocos de concreto, 60o Congresso Brazileiro de Cerâmica 15 a 18 de Maio de 2016, Águas de Lindóia, São Paulo, pp. 1619–1627
9. Saxena, M., Pappu, A., Haque, R., & Sharma, A. (2011). Sisal fiber based polymer composites and their applications. *Cellulose Fibers: Bio- and Nano-Polymer Composites*, 589–659. https://doi.org/10.1007/978-3-642-17370-7_22
10. Uppal, N., Pappu, A., Patidar, R., & Gowri, V. S. (2019). Synthesis and characterization of short sisal fibre polyester composites. *Bulletin of Materials Science, 42* (3). https://doi.org/10.1007/s12034-019-1792-6
11. Martin, A. R., Martins, M. A., Mattoso, L. H. C., & Silva, O. R. R. F. (2009). Caracterização química e estrutural de fibra de sisal da variedade Agave sisalana. *Polímeros, 19*(1), 40–46. https://doi.org/10.1590/s0104-14282009000100011
12. Soto Izquierdo, I., Soto Izquierdo, O., Ramalho, M. A., & Taliercio, A. (2017). Sisal fiber reinforced hollow concrete blocks for structural applications: Testing and modeling. *Construction and Building Materials, 151*, 98–112. https://doi.org/10.1016/j.conbuildmat.2017.06.072
13. Li, Y., Mai, Y.-W., & Ye, L. (2000). Sisal fiber and its composites: A review of recent developments. *Composites Science and Technology, 60*, 2037–2055, PII: S0266-3538(00)00101-9
14. Santhakumar, A. R. (2007). *Concrete technology*. First Edition published, Oxford University Press, ISBN-13: 978-0-19-567153-7, New Delhi 110001, India.
15. Shetty, M. S. (2013). *Concrete technology theory and practice*. Revised Edition, S Chand And Company Limited (AN ISO 9001: 2008 Company), ISBN: 978-81-219-0003-4, Ram Nagar, New Delhi 110055.

16. Baúque, C. J., Thakur, A., Bhartesh, & Sharma, R. (2020). The use of partial glass powder as cement replacement in concrete M20 grade on mechanical strength. *International Journal of Recent Technology and Engineering (IJRTE), 8* (5), 4476–4481. ISSN: 2277-3878. https://doi.org/10.35940/ijrte.E6699.018520
17. IS 383. (1970). Specification for coarse and fine aggregates from natural sources for concrete [CED 2: Cement and Concrete].
18. IS 516. (1959). Indian standard method of tests for strength of concrete [CED 2: Cement and Concrete].
19. IS 5816. (1999). Method of test splitting tensile strength of concrete [CED 2: Cement and Concrete].
20. IS 1199. (1959). Methods of sampling and analysis of concrete [CED 2: Cement and Concrete].
21. IS 456. (2000). Indian standard plain and reinforced concrete—Code of practice [CED 2: Cement and Concrete].
22. Afolayan, J. O., Wilson, U. N., & Zaphaniah, B. (2019). Effect of sisal fibre on partially replaced cement with Periwinkles Shell Ash (PSA) concrete. *Journal of Applied Sciences and Environmental Management, 23*(4), 715. https://doi.org/10.4314/jasem.v23i4.22
23. Okeola, A. A., Abuodha S. O., & Mwero, J. (2018). The effect of specimen shape on the mechanical properties of Sissal fiber—Reinforced concrete (Research Article). *The Open Civil Engineering Journal, 12*, 368–382. https://doi.org/10.2174/1874149501812010368

Waste Plastic Management via Pyrolysis as Sustainable Route

Sahil Chauhan, Subhankar Basu, Sk Aakash Hossain, and Arasavilli Srija

Abstract Plastics gained popularity worldwide due to their broad spectrum of advantages, for which they have been used in many industries. Today, 6300 MT of plastics are being manufactured in USA out of which 79% is disposed in landfills and the same for India is 3360043 tonnes per annum out of which 1,344,017.2 TPA is littered. Plastics have made human lives comfortable, but they also have disturbed the environment. Different procedures to treat plastic wastes have been discussed with a special emphasis on pyrolysis technique in terms of its types (slow pyrolysis, fast pyrolysis, flash pyrolysis). The types of plastics that have been reviewed are polyethylene terephthalate (PET), high-density polyethylene (HDPE), low-density polyethylene (LDPE), polypropylene (PP), polystyrene (PS), poly vinyl chloride (PVC). The final products of pyrolysis are gas, liquid and solid residue or char. From literature, it has been observed that the oil yield of PET with and without catalyst during pyrolysis is 34.4 wt% (600 °C), 11.4 wt% (900 °C) and 52.9 wt% (600 °C), 19.3 wt% (900 °C), respectively. The same for HDPE and LDPE was 94 wt% (5% catalyst) and 93.2% (10% catalyst), respectively. Such product yields for rest of the types have also been given in this chapter. The applications of plastics type in civil engineering have been tabulated and finally a thermo gravimetric analysis (TGA) has been highlighted to assess the pyrolysis products. The TGA analysis is discussed for a better understanding of the optimum temperature to degrade the feedstock into simple compounds.

Keywords Plastic waste · Pyrolysis · Utilization · Recovered energy · Product yield · Sustainability

S. Chauhan (✉) · S. Basu · A. Srija
National Institute of Foundry and Forge Technology, Ranchi, Jharkhand, India

S. A. Hossain
Jadavpur University, Kolkata, West Bengal, India

A. K. Gupta et al. (eds.), *Advances in Construction Materials and Sustainable Environment*, Lecture Notes in Civil Engineering 196,
https://doi.org/10.1007/978-981-16-6557-8_33

1 Introduction

India stands second most populated country in the world, with a net population of 1.3 billion (2011) and resulting in 17.7% of the world's total population. In addition to the trending scenario of rising population, rapid industrialization and urbanization are major contributors to critical solid waste management problems. With the changing lifestyle of people, there is continuous and rising demand for modern materials such as plastics. Plastics are considered as one of such commodities which have made human life versatile and easy-going because of their tremendous advantageous features such as durability, non-corrosiveness, flexibility, cost-effectiveness, lightweight and non-degradable nature [1]. The dependency of the population (urban as well as rural) on plastics due to huge requirement in various areas such as for packaging materials, toys, electronic gadgets, shopping and garbage bags, containers, plastic bottles and related applications as shown below in (Fig. 1) causes an indirect effect on environmental health.

The industrial evolution of polymers, as well as the unmatched popularity of plastics, has made the manufacturing production rate increasing day by day, both on an Indian and global perspective. The recent data reveals that in the year 2015 in US nearly 6300 MT of plastic waste was generated, out of which 9% was recycled, 12% was incinerated, and 79% was disposed in landfills [3]. According to the report of the Central Pollution Control Board in 2018–19, nearly, 3,360,043 tonnes per annum (TPA) of plastic waste is generated in India [4]. Out of this, only up to 60%, i.e. 2,016,025.8 TPA of plastic waste is recycled, and the rest around 1,344,017.2 TPA still remains under littered condition. So ultimately, the milestone is how to deal with the disposal of such waste since directly or indirectly, it is creating a detrimental impact on environmental ecology, marine life, as well as human or public health because of its disposal issue. One of the most crucial problems, i.e. municipal solid waste (MSW) is the key producer of acute as well as chronic environmental, public health and economic impacts in the developing nations [5]. Therefore, it must be

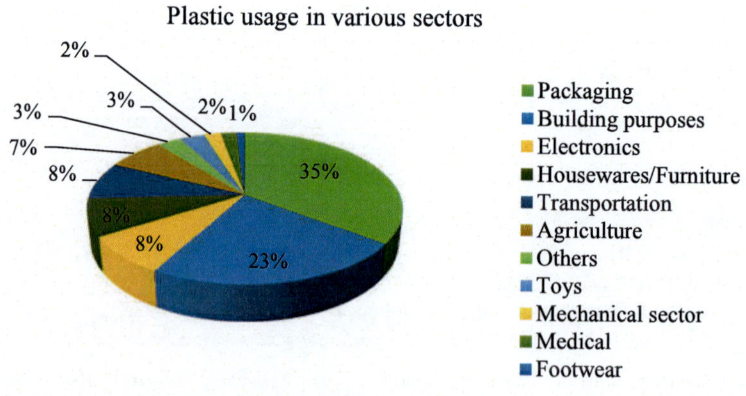

Fig. 1 Usage of plastics as a commodity in various fields [2]

taken into account that the issue of plastic waste needs proper or efficient treatment in terms of a sustainable and clean environment. So, the rudimentary idea of this chapter is to highlight the management of waste plastics in an eco-friendly manner. The appropriate option for handling such type of waste is pyrolysis since it involves the conversion of plastic waste into value-added products with the added advantage of the complete elimination of waste plastics in the form of oil, residues and char. Furthermore, the purpose of this chapter is to address the treatment routes available for waste plastics, categories of waste plastics with a focus on pyrolysis process and few civil engineering applications of waste plastics.

2 Available Treatment Routes/Methods for Waste Plastics

Several studies revealed there are multiple methods available for the treatment of waste plastics, but every method has its own disadvantages and advantages. Ideally, waste plastics should be minimized at the initial source level only in order to withstand waste management practices. But in actual, it is seldom practiced since the waste management pyramid is being followed entirely opposite, i.e. most common scenario being ultimate disposal. As each country has its own rules for proper management of such waste and these set guidelines must be abiding by the citizens of that particular country. To support such waste issues, the Government of India has also implemented and mandated some regulations for waste plastics called plastic waste management rules 2016 [6]. Apart from these rules, there are some methods available in order to deal with the problem of plastic waste, which includes mechanical recycling, incineration or combustion, landfilling, microbial degradation. Mechanical recycling seems to be a green operation, but in reality, it is time taking technique due to the collection, sorting, transportation and processing of plastic waste for turning into serviceable product and the recovery of energy is done to reduce the volume of organic and inorganic fractions of waste materials via incineration/combustion but usage of this type of technique leads to emission of several air pollutants such as dioxins and furans, SO_x, NO_x which are obviously dangerous gases the environmental surroundings [7]. Landfilling is the non-ideal solution since we have a scarcity of space for landfills, and most importantly, plastics take millions of years to degrade into the land. Microbial degradation needs proper design to efficiently degrade the waste plastics. So, we can call it a time taking process since it involves the cumulative performance of mechanical, chemical, thermochemical and biotechnological recycling techniques with microbial and fungal degradation and also needs proper control over process conditions. Such various demerits are observed in the conventional techniques, which lead to the operation of some other process that proves to be energy-oriented as well as ecologically sustainable. Most of the researchers have examined the way for converting waste into energy through an efficient process called pyrolysis in a sustainable way in order to meet the peak energy demand as well as to minimize the plastic waste problem throughout the nation. The proper transformation of waste plastics has a great potential to change it as a form of recovered energy since the plastics are known to be synthesized by a petrochemical source. This is mainly due to the high calorific nature of the plastics. Therefore, pyrolysis is proven to be one of the attractive pathways to ensure the minimization of waste. This process essentially needs oxygen-free atmosphere in order to thermally degrade the long-chain polymeric structures into short-chain polymeric compounds, i.e. from complex molecules to simpler ones or less complex molecules through an external source of heat. Pyrolysis is an endothermic process and requires heat from an external source; thus, it is often known as destructive distillation. Upon heating in the absence of oxygen, most of the substances get converted into three major byproducts, namely oil, gas and char as shown in the following basic reaction:

$$\textbf{Pyrolysis Feedstock} \rightarrow \textbf{Gas} + \textbf{Liquid} + \textbf{Solid residue (char)}$$

These generated products have various applications in industries that are mostly petrochemical refineries. This is why; the pyrolysis has been gaining attention in recent years. The typical sequence of processes that pyrolysis undergoes is represented in (Fig. 2), and it also shows that there are three basic groups of classification of pyrolysis depending upon the operational conditions (temperature, heating rate, residence time). Slow pyrolysis is the pyrolysis process where the rate of heating is comparatively lower to that of fast pyrolysis, or it is a conventional system for heating the reactor at a slow pace which finally produces lower liquid and gaseous products. This type of pyrolysis is mainly concerned to produce charcoal; for this reason, and it is also known as carbonization. Fast pyrolysis works at the increased level of heating, and this rapid heating ensures better performance to generate products in the form of liquids and gases. The mechanism of heating allows the favourable condition for the pyrolytic conversion of unstable feedstock into liquid products, therefore, escaping the formation of coke. Flash pyrolysis is considered as the extended reform of fast pyrolysis with reference to heating rates. It involves minimization of secondary cracking, favours extreme high temperature (450- 1000 °C) and ensures very short

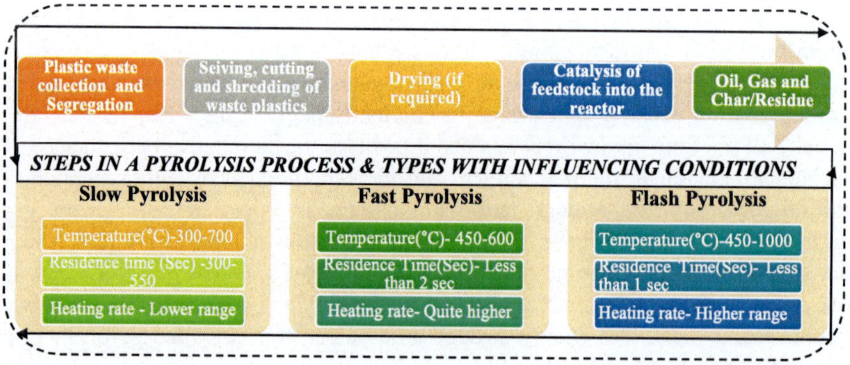

Fig. 2 Sequential flow sheet of pyrolysis process and types of pyrolysis with essential influencing condition

residence time for vapours inside the reactor. In the case of flash pyrolysis, the particle size distribution of the feedstock is very less. To make pyrolysis process occur, it should follow some process conditions which are mandatory to produce products with the required yield and quality. The concerning issues for the treatment of plastic waste and peak energy demand can be dealt together by generating fuel out of plastics [8].

3 Categories of Plastics with Focus on Pyrolysis as Effective Treatment Approach

The types of plastics with its unique identification code, chemical structure and widely used applications are summarized in Table 1. Each of these types is discussed

Table 1 Category of plastics with their identification codes, chemical formula and wide applications [9]

Category of plastics	Identification code	Chemical formula	Wide applications
Polyethylene terephthalate (PET, PETE)	01 PET	$-(C_{10}H_8O_4)_n-$	Clear soft drinks and beverage bottles, food packaging, microwaveable food trays, fruit juice containers
High-density polyethylene (HDPE)	02 PE-HD	$-(CH_2–CH_2)_n-$	Bottles (especially for food products, detergents and cosmetics), industrial wrapping and film, sheets
Polyvinyl chloride (PVC)	03 PVC	$-(CH_2CHCl)_n-$	Bottles, packaging films, credit cards, water containers, water pipes, mouth wash bottles
Low-density polyethylene (LDPE)	04 PE-LD	$-(CH_2\text{-}CH_2)_n-$	Bread and shopping bags, carpet, clothing, furniture, flexible containers
Polypropylene (PP)	05 PP	$-[CH_2\text{-}CH(CH_3)]_n-$	Potato chip bags, straw, microwave dishes, lunch boxes, cattles, garden furniture
Polystyrene (PS)	06 PS	$-[CH_2\text{-}CH(C_6H_5)]_n-$	CD cases, plastic cutlery, low-cost brittle toys, video cases

in the following subsections in the sequence of their identification codes. The applications of these waste plastics, their types and performance improvement in civil engineering are shown in Table 2.

3.1 Polyethylene Terephthalate (PET)

PET is a daily applied thermoplastic polymer that comes under the polyester domain of polymers. These types of plastics are well known for their chemical resistivity as well dimensional stability. PET area usually employed for mineral water bottles, juice containers, aerated drink bottles, packaging of food products, and there are several other modes of usage of this type of plastics in X-ray and photographic films, insulation of electrical parts and magnetic tapes [10]. The basic features of PET, such as thermal stability and slow degradable nature, give rise to crucial deterioration of human health as well as environmental surroundings [11, 12]. Waste plastics in the form of PET contribute remarkable waste generation to the surroundings, and it is extremely difficult to efficiently dispose them off. The ultimate challenge to deal with the PET waste is to collect them efficiently at a faster pace with optimum cost involved in the transportation. Asadi and Miskolczi [13] have conducted the co-pyrolysis of PET and PE using Ni loaded catalysts and the 50% feedstock of each plastic was allowed to feed into horizontal tubular reactor in the temperature range of 600–900 °C. In the experiment, the nitrogen purging was done to create sweeping atmosphere. The pyro-oil yield without using catalyst was 34.4 wt% at 600 °C and 11.4 wt% at 900 °C, respectively, whereas feedstock with catalyst increased the oil yield to 52.9 wt% at 600 °C and 19.3 wt% at 900 °C. The gaseous product without catalyst obtained was 21.4 wt% and 48.3 wt% at 600 °C and 900 °C respectively. The implication of catalyst in the feedstock increased the gas yields in the range of 42.4–69.2 wt%. It was, therefore, observed that both gas and oil yields were affected by temperature and catalysts.

3.2 High-Density Polyethylene (HDPE)

These kinds of polymers are cost-effective and high-strength thermoplastics with straight chain line structure and usually exhibit very low degree of branching. HDPE shows excellent toughness even at very low temperatures. Moreover, this type of plastics are most commonly applied in the manufacturing of trays, food packaging caps, huge containers utilized in industries, garbage containers, toys, oil cans, detergent bottles and so on. As far as the waste HDPE concerned, it has great potential to be utilized in pyrolysis process for fuel production by employing proper process conditions. Various researchers have done studies based on their set operational process parameters and obtained variation in product yield as well as quality of the

Table 2 Civil engineering applications with plastics type, percentage and its key role in performance improvement

S.No	Applications	Plastics type	Percentage of plastics	Performance improvement	References
1.	Modified bitumen	PET	5, 10, 15% by wt. of bitumen	Increase in strength, fatigue life and pavement performance	[21]
		Plastic bottles, cups, polymers etc	4, 6, 8 and 10% by wt. of bitumen	Optimum plastic content is 8% Here, Marshall stability is high	[22]
		PE	6, 8, 10, 12 and 14% by wt. of bitumen	Optimum plastic content is 12% Here, Marshall stability is high	[23]
		PET bottles	1, 3 and 5% by wt. of aggregate	Optimum plastic content is 1% by wt. of aggregate and 15% by wt. of bitumen. Here, Marshall stability is high	[24]
2.	Plastic aggregate in concrete	LDPE	Sand is substituted with LDPE by 0, 5, 10, 20, 30, 40, 50 and 60%	Optimum LDPE is 50–60% at which high performance is seen	[25]
		PET bottles	Fine aggregate is replaced with PET by 0.5, 1, 2, 4 and 6%	Optimum PET is 2% because after 2% the compressive strength decreased	[26]
		PET	Fine aggregate is replaced with PET by 5, 10 and 15%	5% PET yields optimum compressive strength	[27]
3.	Soil stabilization	HDPE	0, 1, 2 and 4%	Optimum California bearing ratio is observed at 4%	[28]
		PET	0, 0.2, 0.6, 1%	Optimum PET is at 0.6% which resists fatigue cracking and rutting	[29]

(continued)

Table 2 (continued)

S.No	Applications	Plastics type	Percentage of plastics	Performance improvement	References
4.	Plastic bricks	PET	0, 1, 3 and 7%	Optimum compressive strength is observed for 1% PET with particle size < 6.3 mm	[30]
5.	Cement	PE	15, 20, 25, 30, 35, 40, 50 and 60%	At 25, 30 and 35% compressive strength is high	[31]
6.	Geosynthetics	PET	0, 0.25, 0.5, 1, 2 and 4%	Optimum California bearing ratio is observed at 2%	[32]

experimented feedstock. Singh et al. [14] explored the thermo-catalytic degradation of waste HDPE via pyrolysis using $CuCO_3$ as catalyst (0, 2, 5, 8%) in a fixed bed reactor at a temperature range of 23–390 °C. The heating rate was maintained as 20 °C/min with a total residence time of 4 h 20 min. The corresponding liquid, gaseous and solid yield were obtained at specified feed to catalyst ratio with maximum oil yield of 94 wt% using 5% catalyst and maximum gaseous yield of 14.67 wt% using pure waste HDPE as feedstock. They analyzed that the conversion of waste HDPE plastic with $CuCO_3$ as the catalyst using a pyrolysis catalytic technique gave various hydrocarbon fractions.

3.3 *Polyvinyl Chloride (PVC)*

One of the major thermoplastic material found in nature with wide variety of applications but the mode of manufacturing of PVC differs a lot from that of other thermoplastics since it is not only derived from oil but also requires salt for its formation. The monomer of vinyl chloride is formed by combining ethylene from natural gas or oil with chlorine produced from the electrolysis of salty water. The versatile nature of the PVC to get accommodated with various other admixtures is attractive to consumers. The broad applications of PVC include the electrical insulation, automobile sectors, building construction, blood bags, tubing, piping, siding, debit cards, wind shielding of various components etc. Past literature reveals that there is very scant research done on PVC composed waste, this is mainly due to the generation of toxic HCl fumes inside the furnace when exposed to heated environment. It has been observed that dealing with the high cost of treatment for PVC contained waste; many industrial and petrochemical sectors find it to be disadvantageous. But now as the time is progressing the researchers are gaining interest to tackle this problem and manage the pyrolysis of PVC effectively. Miskolczi et al. [15] explored the pyrolysis of PVC—containing mixed plastic waste for recovery of hydrocarbons in a horizontal tube reactor at 530 °C. They determined the effect of the PVC concentration on the characteristics of products so obtained. Yield of light oils were achieved in the range of 36.9–59.6% which was solely dependent on composition of feedstock. They found that the highest concentration of PVC was achieved in gases (HCl being dominant compound), and therefore, they concluded, more the PVC content in feedstock, the higher is the chlorine contamination in the products.

3.4 Low-Density Polyethylene (LDPE)

LDPE is popularly used polymer in day-to-day life in the form of plastic bags as it is flexible and light enough to carry providing easiness to human mankind. These types of plastics are basically branched oriented due to the presence of weak intermolecular forces which imparts less tensile strength and more durability. LDPE stands second in terms of its contribution to waste in MSW after polypropylene. LDPE possesses less temperature resistance as well as less crystalline structure, thus can be easily shaped. LDPE exhibits waterproof property and good electrical insulation. These types of polymers are used as light packaging materials, wash bottles in laboratories, used as fine layer protection against corrosion, covers for hardware. Soliman et al. [16] investigated thermal and catalytic pyrolysis of waste LDPE with a temperature in between (550–650 °C) in a semi-pilot scale autoclave reactor. The heating rate was ensured as 10 °C/min. They determined and average yield of 70% as liquid and 1.5% as solid residue. They observed insignificant change in the liquid product but there was significant reduction in the temperature for getting first drop of the oil during experimental runs with catalytic pyrolysis. Abadi et al. [17] conducted thermochemical pyrolysis of LDPE using FCC catalyst in a stirred semi-batch reactor under atmospheric pressure. The reactor was heated at a rate of 25 °C/min upto a degradation temperature between 420 and 510 °C. They ensured the improved mass and heat transfer by employing stirred mechanism. At pyrolysis temperature of 450 °C, the obtained condensable product was highest, i.e. 93.2 wt% when the catalyst to polymer ratio was 10% and non-condensable product and coke were maximum i.e., 13.1 and 7.2 wt% at catalyst to polymer ratio of 60%. They inferred this temperature as economically suitable in terms of cost-effectiveness and liquid production.

3.5 Polypropylene (PP)

PP is a rigid or tough thermoplastic derived from propene or propylene monomer, and it is saturated long chain polymer having linear hydrocarbon structure. As a plastics commodity it has least density and enough hardness which makes this type of polymers suitable for automotive industry, plastic industries, consumer goods, etc. PP shows excellent resistance to environmental stress cracking, but it is one of the flammable materials. These types of plastics have higher softening points, i.e. do not melt at least below 160 °C. These polymers also exhibit great resistance against fatigue, heat and chemicals. This shows that requirements of PP on daily basis remain all time at its peak level. So, in order to deal with the demand of PP by consumer pyrolysis can open doors to extract energy out of it and use that valuable recovered energy further. There are many investigators who have studied the degradation pattern at various process conditions needed for pyrolysis. Hakeem et al. [18] studied the catalytic pyrolysis of waste polypropylene into liquid fuel using kaolin as catalyst under a temperature of 450 °C and catalyst to feed ratio of 1:3. The experimentation was carried out in a horizontal chemical vapour deposition (CVD) reactor with a heating rate of 30 °C/min and residence time of 30 min. The product yield in the thermal run were 67.48 wt% (liquid), 8.85 wt% (gaseous) and 23.67 wt% (solid residue), whereas product yield in case of catalytic run (cat: feed as 1:3) were 79.85 wt% (liquid), 1.48 wt% (gaseous) and 18.67 wt% (solid residue). The observations after conducting both the runs enabled them to suggest that catalytic runs generated liquid products which were almost comparable with conventional fuels such as diesel oil. Heydariaraghi et al. [19] investigated the thermal degradation of PP in a stirred batch reactor with a temperature of 450 °C and heating rate of 8 °C/min. They obtained the maximum liquid yield as 88 wt % as compared to lower product yield of gases and residue with total product conversion of 99 wt%.

3.6 Polystyrene (PS)

Polystyrene is produced by high temperature reaction of petroleum hydrocarbons via polymerization. It is one of the common thermoplastics made from togetherness of propylene monomers. The polymeric structure typically consists of phenyl group attached at particular localized carbon atom. Being easily available and considerably inexpensive, these plastics are widely used in various fields such as textiles, consumer products packaging, medical, constructional applications, electronic gadgets etc. Due to diversified applications, the PS have several characteristics which includes elasticity, toughness and chemical resistance. On the other hand, the continuous usage of PS is also becoming one of the severe problems as it leads to waste generation in municipal solid waste each year and being directly dumped into the landfills. The challenging task to treat the PS waste can find its way through pyrolysis process and therefore converting into valuable fuel. Such investigations are carried out by various researchers. Prathiba et al. [20] explored their studies on slow pyrolysis of polystyrene waste using activated carbon as catalyst in a semi-batch reactor. They maintained the reactor temperature up to 500 °C. From their research, the oil yield achieved was around 82.5 wt%, gaseous yield as 8.72 wt% and residue yield as 8.78 wt% having total product conversion of around 90.24 wt% at a temperature of 434 °C (Table 2).

4 Role of Pyrolysis Parameters and Analytical Tool for Pyrolysis of Waste Plastics

There are several operational parameters responsible to predict the pyrolysis behaviour of variety of waste plastics and they affect the obtained valuable products such as pyrolytic oil, carbonaceous char and gases via pyrolysis. However, the generated oily products need to be reformed by distillation and further upgraded to be used for transportation purposes. Various parameters or process conditions include constituents of feedstock, degradation/process temperature, reactor type, heating rate, pressure, residence time and catalyst contact mode which play vital role during the overall pyrolysis reaction. The cracking of waste plastics via pyrolysis is responsible for substantial change in any of these operational factors. The two most critical components are process temperature and heating rate and functions directly to yield distributed products in the form of oil, char and gases. The higher range of both of these factors critically disintegrates the fraction of waste plastics quickly into low-molecular weight compounds as gaseous yield and lower ranges indicates the production of more liquid yield from the pyrolysis process [33]. Next essential parameter is type of reactor and residence time. The calorific value (CV) of gaseous components is critically affected if the type of reactors are not carefully examined such as fluidized bed reactors uses fluidizing gaseous medium which reduces the CV of gaseous fractions [34]. Fixed bed reactors are easier to operate and maintain and impose least effect on the calorific value of pyrolysis products. Residence time inside the reactor also affect the quality as well as yield of obtained products. All of the aforementioned parameters show significant role during the pyrolysis of waste plastics and therefore affects the waste management aspect of environmental engineering, a specialized branch of civil engineering. Therefore, before choosing the pyrolysis route for waste management of plastics, decent analysis should be taken into account with respect to critical factors of pyrolysis process.

Data analysis for the recovered products is another important aspect which provides some quantified information about the pyrolysis operational conditions. In agreement to this, sophisticated analytical instrumentations differentiated by make and model are used to reveal the information with regard to these parameters of pyrolysis feedstock such as waste plastics. One of the most popular analytical equipment to determine the sample mass variation with degradation temperature as a key parameter for the duration of pyrolysis reaction is thermo gravimetric analysis (TGA). During TG analysis, the feedstock like waste plastics is washed, dried and grinded in smaller particles. Initially, the amount of waste plastic samples is usually taken in few milligrams (5–10 mg). This amount may correspond to variation in different heating rates, pressure, catalyst effect or any other factor which influences degradation pattern of pyrolysis conditions. The optimum temperature required for degradation or dissociation of waste plastics can be known using TG analysis. The effect of heating rates with temperature can be inferred by plotting the TGA curve. Similar investigation revealed the TG analysis of different kinds of plastics and combined waste plastics at 5, 10 and 20 °C per min [34]. Their work showed that higher heating rates shifts the curve towards right which indicated the lag in degradation of waste plastics, whereas the lower runs of heating rates provided sufficient residence time inside the reactor to degrade the contents of waste plastics into its product components with narrower temperature range. Moreover, they explored that the combination of waste plastics as feedstock if allowed to degrade via pyrolysis, are dissociated into valuable components at around 300°C earlier than single type of waste plastics (350 °C). Essentially, the TGA plot for all the waste plastic fractions reaches to a stabilized state once the pyrolytic reaction comes to end. It signifies complete dissociation of all components of waste plastics during the primary and secondary reactions of pyrolysis process and provides the clear information to adopt optimum degradation temperature corresponding to employed heating rate of the process.

5 Conclusion and Outlook

In the present global scenario, plastics have become an in-separable commodity in human lives. Plastics are in demand for their durability, non-corrosiveness and cost affectivity. Plastics have been established as the core material used in various industrial sectors. The problem arises when these plastics and their products are not well managed and are littered as wastes. Thus, it is imperative to deduce certain efficient treatment methods for treating plastic wastes in terms of sustainability and making the environment a cleaner one. There are various methods such as landfill, microbial degradation, mechanical recycling, incineration or combustion. But, each of these methods have some issues which make them difficult for a widespread application. Therefore, pyrolysis has been suggested as the appropriate technique as it converts the waste into recovered energy thus meeting the required energy demand. The pyrolysis process takes place in environments that are devoid of oxygen to thermally degrade the long polymeric chain structures or complex structures into shorter chains. The mechanism of heating during pyrolysis causes the favourable formation of stable liquid products from unstable feedstock because of which the formation of coke is inhibited.

Literature described in the aforementioned sections give realistic insight about the product yield of pyrolysis products of each type of waste plastics by varying different operational parameters. For example, PET gave a pyro-oil yield of 34.4 and 11.4 wt% at 600 and 900 °C, whilst using catalyst the same increased to 52.9 and 19.3 wt% at the same temperature. The gas yield from the feedstock was increased from 21.3 to 48.3 wt% to the range of 42.4–69.2 wt% in the presence of catalyst at the above-mentioned temperature. HDPE gave a liquid yield of 94 wt% using 5% $CuCO_3$ catalyst and the gaseous yield was 14.67% using pure HDPE waste. The feedstock from PVC yielded light oils in the range of 36.9–59.6% depending on the composition of the feedstock. LDPE was pyrolyzed at 450 °C in presence of FCC catalyst giving the highest liquid yield product of 93.2 wt% at 10% of catalyst to polymer ratio. The product yield for PP at 450 °C for gas, liquid and solid residue was changed from 67.48% to 79.85%, 8.85% to 1.48% and 23.67% to 18.67%, respectively, in the presence of kaolin catalyst to feed ratio of 1:3. It was observed that the increase of temperature up to 450 °C caused a decrement of liquid yield and solid residue. On the other hand, it increased the yield of gas thereby increasing the total conversion of products up to 99.8 wt%. During pyrolysis at 434 °C, PS produced an oil yield of 82.5%, gaseous yield as 8.72% and solid residue of 8.78% with a total product conversion of about 90.24%.

Plastic wastes have found a broad spectrum of good applications in civil engineering. They have been successfully implemented in the preparation of lightweight concrete. Plastic wastes have increased the binding efficiency of bitumen during the construction of roads where they have been used in small quantities (5–10%), increasing the strength, durability and stability of roads. In geotechnical engineering, these plastic wastes have been used in place of costly additives in order to increase the engineering properties of soil, such as maximum dry density (M.D.D.), California Bearing Ratio (C.B.R.) and optimum moisture content. (O.M.C.) Plastic wastes have been used in the preparation of geosynthetics such as geocells and geogrids. The geocell plastic waste had a significant improvement in the bearing strength of soil in comparison to the untreated natural soil. Waste plastic bottles have substituted geogrid as a reinforcing agent where it has been seen that it reduced the cost by 60% and giving a strength increment of 15% in soil. Shredded plastics of size less than 6.3 mm have been used in compressed earth brick using 1% by weight which gave an increment of compressive strength by 244.4%, giving the least corrosion level in the bricks as well as on the outer surface of the walls constructed by these reinforced bricks. Thermoplastics such as polystyrenes, mixed plastics and polycarbonates are useful materials when used in varying proportions by weight (0–10%) along with sand, fly ash and OPC for manufacturing eco-friendly bricks. These bricks had reduced thermal conductivity and sufficient increment in compressive strength. Role of operational parameters of pyrolysis and subsequent data analysis gives realistic picture to achieve healthy environment and open doors for scaling up this technology. Thus, all these applications as well as waste plastics conversion to recover energy and valuable products via pyrolysis can lead to a sustainable environment by reducing the potential contamination level by plastics and its wastes.

References

1. Cleetus, C., Thomas, S., & Varghese, S. (2013). Synthesis of petroleum-based fuel from waste plastics and performance analysis in a CI engine. *Journal of Energy* 1–10. https://doi.org/10.1155/2013/608797
2. Nkwachukwu, O., Chima, C., Ikenna, A., & Albert, L. (2013). Focus on potential environmental issues on plastic world towards a sustainable plastic recycling in developing countries. *International Journal of Industrial Chemistry, 4*, 34. https://doi.org/10.1186/2228-5547-4-34
3. Geyer, R., Jambeck, J. R., & Law, K. L. (2017). Production, use, and fate of all plastics ever made—Supplementary information. *Science Advances, 3*, 19–24. https://doi.org/10.1126/sciadv.1700782
4. CPCB: Annual Report PWM Rules. 17 (2019).
5. Nizami, A. (2016). Trends and sustainability criteria for biofuels .pdf.

6. MoEF&CC: [Published In the Gazette of India, Part-II, Section-3, Sub-section (ii)] Ministry of Environment, Forest and Climate Change NOTIFICATION New Delhi, the 29. 317, 1–15 (2016)
7. Hopewell, J., Dvorak, R., & Kosior, E. (2009). Plastics recycling: Challenges and opportunities. *Philosophical Transactions of the Royal Society B: Biological Science, 364*, 2115–2126. https://doi.org/10.1098/rstb.2008.0311
8. Kunwar, B., Moser, B. R., Chandrasekaran, S. R., Rajagopalan, N., & Sharma, B. K. (2016). Catalytic and thermal depolymerization of low value post-consumer high density polyethylene plastic. *Energy, 111*, 884–892. https://doi.org/10.1016/j.energy.2016.06.024
9. Sharma, S., & Mallubhotla, S.: Plastic waste management practices. *Zero Waste*, 105–113. https://doi.org/10.1201/9780429059247-7
10. Anuar Sharuddin, S. D., Abnisa, F., Wan Daud, W. M. A., & Aroua, M. K. (2016). A review on pyrolysis of plastic wastes. *Energy Conversion and Management, 115*, 308–326. https://doi.org/10.1016/j.enconman.2016.02.037
11. Park, C., Kim, S., Kwon, Y., Jeong, C., Cho, Y., Lee, C. G., Jung, S., Choi, K. Y., & Lee, J. (2020). Pyrolysis of polyethylene terephthalate over carbon-supported Pd catalyst. *Catalysts, 10*, 1–12. https://doi.org/10.3390/catal10050496
12. Hurley, R., Woodward, J., & Rothwell, J. J. (2018). Microplastic contamination of river beds significantly reduced by catchment-wide flooding. *Nature Geoscience, 11*, 251–257. https://doi.org/10.1038/s41561-018-0080-1
13. Al-Asadi, M., & Miskolczi, N. (2018). Pyrolysis of polyethylene terephthalate containing real waste plastics using Ni loaded zeolite catalysts. *IOP Conference Series: Earth Environmental Science, 154*. https://doi.org/10.1088/1755-1315/154/1/012021
14. Singh, M. V., Kumar, S., & Sarker, M. (2018). Waste HD-PE plastic, deformation into liquid hydrocarbon fuel using pyrolysis-catalytic cracking with a $CuCO_3$ catalyst. *Sustain. Energy Fuels, 2*, 1057–1068. https://doi.org/10.1039/c8se00040a
15. Miskolczi, N., Bartha, L., & Angyal, A. (2009). Pyrolysis of polyvinyl chloride (PVC)-containing mixed plastic wastes for recovery of hydrocarbons. *Energy and Fuels, 23*, 2743–2749. https://doi.org/10.1021/ef8011245
16. Soliman, A., Farag, H. A., Nassef, E., Amer, A., & ElTaweel, Y. (2020). Pyrolysis of low-density polyethylene waste plastics using mixtures of catalysts. *Journal of Material Cycles and Waste Management*. https://doi.org/10.1007/s10163-020-01028-z
17. Abbas-abadi, M. S., Mcdonald, A. G., Nekoomanesh, M., & Yeganeh, H. (2015). Using different process parameters in a stirred reactor 2, 39–47
18. Hakeem, I. G., Aberuagba, F., & Musa, U. (2018). Catalytic pyrolysis of waste polypropylene using Ahoko kaolin from Nigeria. *Appl. Petrochemical Res., 8*, 203–210. https://doi.org/10.1007/s13203-018-0207-8
19. Heydariaraghi, M., Ghorbanian, S., Hallajisani, A., & Salehpour, A. (2016). Fuel properties of the oils produced from the pyrolysis of commonly-used polymers: Effect of fractionating column. *Journal of Analytical and Applied Pyrolysis, 121*, 307–317. https://doi.org/10.1016/j.jaap.2016.08.010
20. Prathiba, R., Shruthi, M., & Miranda, L. R. (2018). Pyrolysis of polystyrene waste in the presence of activated carbon in conventional and microwave heating using modified thermocouple. *Waste Management, 76*, 528–536. https://doi.org/10.1016/j.wasman.2018.03.029
21. Bhatnagar, A., Kumar, G., Pal, G. S., Singh, P. P., & Gupta, V. (2019). To enhance the strength of bituminous roads using 4958–4961.
22. Singh, M., Khan, S., Sharma, S., Tiwari, H., & Dohare, P. D. (2019). Improvising pavement strength using plastic waste 4628–4631.
23. Rajput, P. S., & Yadav, R. K. (2016). Use of plastic waste in bituminous road construction. *International Journal of Science and Technology and Engineering, 2*, 509–513.
24. Badejo, A. A., Adekunle, A. A., Adekoya, O. O., Ndambuki, J. M., Kupolati, K. W., Bada, B. S., & Omole, D. O. (2017). Plastic waste as strength modifiers in asphalt for a sustainable environment. *African Journal of Science, Technology, Innovation and Devlopment, 9*, 173–177. https://doi.org/10.1080/20421338.2017.1302681

25. Ohemeng, E. A., & Ekolu, S. O. (2019). Strength prediction model for cement mortar made with waste LDPE plastic as fine aggregate. *Journal of Sustainable Cement-Based Materials, 8*, 228–243. https://doi.org/10.1080/21650373.2019.1625826
26. Ramadevi, K. K., & Manju, R. (2012). Experimental Investigation on the Properties of Concrete With Plastic PET (Bottle) Fibres as Fine Aggregates. *Journal of Emerging Technology and Advanced Engineering, 2*, 42–46.
27. Rahmani, E., Dehestani, M., Beygi, M. H. A., Allahyari, H., & Nikbin, I. M. (2013). On the mechanical properties of concrete containing waste PET particles. *Construction and Building Materials, 47*, 1302–1308. https://doi.org/10.1016/j.conbuildmat.2013.06.041
28. Salim, N. M., Al-Soudany, K. Y. H., & Ahmed, A. A. (2018). The impact of using recycled plastic fibres on the geotechnical properties of soft Iraqi soils. *IOP Conference Series: Materials Science and Engineering, 433*. https://doi.org/10.1088/1757-899X/433/1/012017
29. Hafez, M., Mousa, R., Awed, A., & El-Badawy, S. (2019). Soil reinforcement using recycled plastic waste for sustainable pavements. Springer International Publishing.
30. Akinwumi, I. I., Domo-Spiff, A. H., & Salami, A. (2019). Marine plastic pollution and affordable housing challenge: Shredded waste plastic stabilized soil for producing compressed earth bricks. *Case Studies in Construction Materials, 11*, e00241. https://doi.org/10.1016/j.cscm.2019.e00241
31. Jassim, A. K. (2017). Recycling of polyethylene waste to produce plastic cement. *Procedia Manufacturing, 8*, 635–642. https://doi.org/10.1016/j.promfg.2017.02.081
32. Shah, A., Modha, H. (2020). *Improving the soil subgrade with plastic waste reinforcement—An experimental study*. Springer Singapore.
33. Demirbas, A. (2004). Pyrolysis of municipal plastic wastes for recovery of gasoline-range hydrocarbons. *Journal of Analytical and Applied Pyrolysis, 72*, 97–102. https://doi.org/10.1016/j.jaap.2004.03.001
34. Singh, R. K., & Ruj, B. (2016). Time and temperature depended fuel gas generation from pyrolysis of real world municipal plastic waste. *Fuel, 174*, 164–171. https://doi.org/10.1016/j.fuel.2016.01.049

Seismic Fragility of Buildings Subjected to Pounding Effects with Soil–Structure Interaction

Rajan L. Wankhade, Ajinkya Sawarkar, Ayush Chandwani, Shahaji Chavan, Pratik Malkar, and Gaurav Sawarkar

Abstract In this paper, seismic analysis is performed for multi-storey buildings with soil structure–interaction considering pounding effect. Parameters including displacement, shear force, bending moment, and drift ratios are compared for various models. The nonlinear dynamic analysis, i.e. most accurate with time history, is performed on these building structures to compare their behaviour during different strong EQ ground motions. M20-grade concrete and Fe500 steel are used as material of RCC. The structure was assessed again and again for such properties of SSI and pounding. Further corrections are made to these frames so that it meets the minimum requirement possible to make it stiff and sufficient strong to sustain in strong EQ motion. It is observed that the building frames with different storey height exhibits 8% and 23% higher defection than building frames with same storey height at top and bottom levels, respectively. Reduction in moment is observed to be 41.63% for building frames spaced faraway that of closely spaced frames. Hence, the results shown here are for the worst situation possible under the action of SSI effect and pounding.

Keywords Structural pounding · Different building configurations · Earthquakes · Soil–structure interaction

1 Introduction

Earthquakes, amongst the existing natural disaster, have the ability to cause larger damages to the building components and entire structure. Technological tools need to be sharpened and developed accordingly to meet the requirements of analysing the structures under EQ forces as earthquake forces are most random and versatile. Elastic linear design criteria fails for structure subjected to strong ground motion. Thus, seismic response cannot be accurately predicted by linear design characteristics and its methods. Damage loss occurred to structural and non-structural elements

R. L. Wankhade (✉) · A. Sawarkar · A. Chandwani · S. Chavan · P. Malkar · G. Sawarkar
Government College of Engineering, Nagpur, Maharashtra, India

A. K. Gupta et al. (eds.), *Advances in Construction Materials and Sustainable Environment*, Lecture Notes in Civil Engineering 196,
https://doi.org/10.1007/978-981-16-6557-8_34

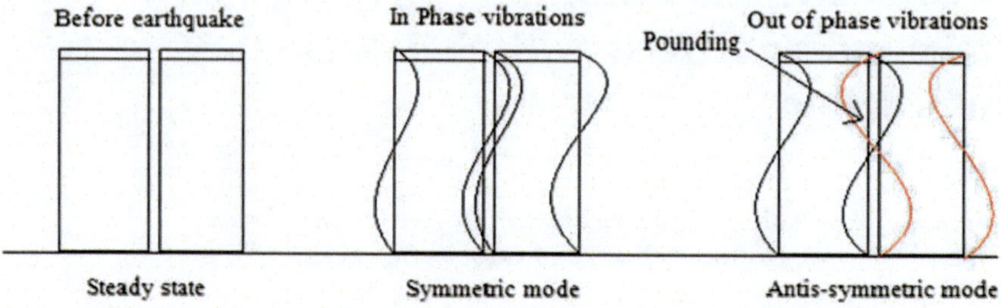

Fig. 1 Different modes in which building vibrates during strong EQ motion

cannot be captured during EQ motion and under the action of predominant EQ forces. Hence, traditional analysis and design approaches fail over such variety of forces. Elastic or nonlinear analysis gives more accurate and sometimes exact results under variety of loading. EQ with magnitude six causes greater damage to structural as well as non-structural elements. Soil–structure interaction plays vital role in strong ground motion and forces causing due to it. Pounding effect occurs if buildings are placed near to each other subjected to symmetric and anti-symmetric vibration modes. Due to pounding, adjacent buildings can have severe damage if it vibrates in out of phase. If insufficient gap is not available in two adjacent buildings, then energy dissipation and further damages are predominant. Pounding occurs when the adjacent buildings start vibrating out of phase during the strong EQ ground motion. Pounding is considered to be one of the important causes of severe EQ damages and collapse of structure. Figure 1 shows different modes in which building vibrates during strong EQ motion.

Some seismic codes state to provide a sufficient enough separation gap between adjacent structures to make it ineffective for pounding effect. Soil–structure interaction can be added by considering the soil properties at the foundation level. The phenomenon of strong nonlinearities is not captured by many of seismic design codes. Pounding between closely spaced structures is considered as a very complex phenomenon making the analysis and design complicated. We aimed to analyse models which are prepared for defining the structural response of closely and far placed structures during a strong earthquake. It is found that displacements and forces with performance-based seismic design of structural frames are available [1, 2]. Comparative response of SDF systems considering near-fault and far-fault earthquake ground motions by considering spectral regions is estimated [3]. Xue [4] proposed a displacement-based seismic design method of in-elastic frames. It also performed nonlinear modelling and further analysis of earthquake-induced vibration with pounding of buildings [5]. Sayani [6] provided assessment for response of minimally compliant low-rise base isolated and conventional steel MRF framed buildings in their work. Performance-based design to estimation of forces for framed buildings with EQ loading is also studied [7]. Wankhade and Bajoria [9] worked on vibration characteristics of structural members including beams. Further, performance-based analysis and revised design of building frames with EQ loading are given and found

Fig. 2 Building frames resting on soil and rock for SSI effect

economical [8, 10]. Yesane et al. [11] proposed a review on soil–structure interaction for structural components. The performance characteristics of RC MR frames using different rubber bearing base isolation techniques are effective in earthquake resistance [12]. Provision of vibration analysis of some beams and plated members for sensing and actuating applications under dynamic loading is found useful [13–21]. Nonlinear pounding of multi-span and simply supported bridges composed of beam excited with strong ground motions are further studied [22].

2 Soil–Structure Interaction

Soil–structure interaction (SSI) shown in Fig. 2 is the phenomenon in which the response of the soil governs the motion of the structure and the motion of the structure governs the response of the soil vice-versa. In such case, the structural motion and the ground motion are independent from each other. Traditional concept of structural design disregards the SSI effects, which is acceptable only for light structures on relatively stiff soil. Hence, it can be accepted that the presence of the soil below foundation affects the response of the structure. Also, structures resting on rock have different behaviours than that of resting on soft and medium soil.

3 Description of Structures and Modelling Assumptions

The building considered under study and analysis is made from an original plan. Two floor and three floor building are used for analysis using SSI and pounding effect. The floor to floor height of both the models is taken as 2.8 m. The basement height is considered different to create different cases. Hence, basement height is kept as 2.8 m and 3.8 m for different cases. The material used are M20 concrete and Fe500 steel bars as reinforcement. Different checks like column PMM interaction ratio, beam/column capacity ratio, column/beam capacity ratio, and drift storey check were performed on the building to check the stability of structure, and wherever necessary, the size of column is increased to pass the checks. 350 × 350 mm and 400 × 400 mm column, 230 × 300 mm beam and 130 mm slab is used in the building, and the slab used for staircase is 125 mm thick. For soil interaction, spring is used in place of soil and stiffness of string is used to depict different types of soil (Table 1).

In above, A_b area of the foundation under consideration, B and L 0.5 * width and 0.5 * length of a foundation. I_{bx}, I_{by}, and I_{bz} moment of inertia of the foundation area with respect to longitudinal (x), lateral (y), and vertical axes (z), respectively (Fig. 3) (Table 2).

For the purpose of this study, the following models shown in Figs. 4 and 5 are developed. The model is created and modified depending upon the data to be implemented. Analysis is then executed, and optimized design is taken into consideration. Time–history displacements and moments are provided further.

The Above cases of structural pounding are considered as shown in Fig. 6. In each case, four different types of soil are considered so like these 16 cases are there.

Table 1 Different spring constants calculated using spring stiffness formulae given by George Gazetas

Degrees of freedom	Stiffness of equivalent soil spring
Translation X (longitudinal direction)	$[2GL/(1-\nu)](0.73+1.54X^{0.75})$ with $X = A_b/4L^2$
Translation Y (lateral direction)	$[2GL/(2-\nu)](2+2.50X^{0.85})$ with $X = A_b/4L^2$
Translation Z (Vertical direction)	$[2GL/(2-\nu)](2+2.50X^{0.85}) - [0.2/(0.75-\nu)]GL[1-(B/L)]$ with $X = A_b/4L^2$
Rotation about X (about longitudinal)	$[G/(1-\nu)]I_{bx}{}^{0.75}(L/B)^{0.25}[2.4+0.5(B/L)]$
Rotation about Y (about lateral)	$[G/(1-\nu)]I_{by}{}^{0.75}(L/B)^{0.15}$
Rotation about Z (Torsion)	$3.5GI_{bz}{}^{0.75}(B/L)^{0.4}(I_{bz}/B^4)^{0.2}$

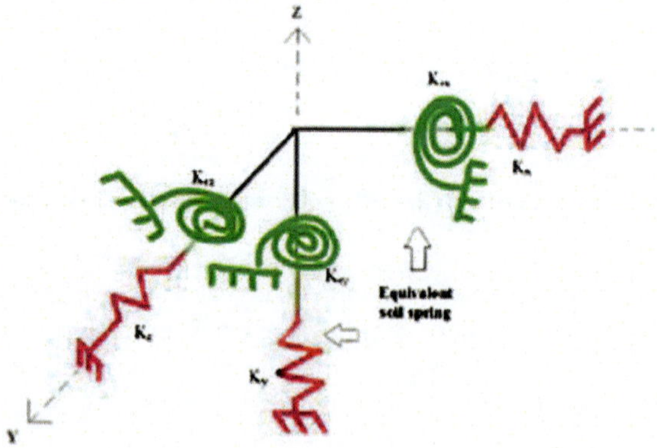

Fig. 3 Equivalent spring stiffness (translational and rotational)

Table 2 Soil properties calculated and adopted for analysis

No	Strata	Modulus of elasticity (kN/m^2)	Poisson ratio (M)	Shear modulus (G)	Unit weight (kN/m^3)
1	Soft soil	15,000	0.45	10,875	16
2	Medium soil	50,000	0.4	35,000	16
3	Hard soil	120,000	0.4	84,000	18
4	Rocky basalt	15,000,000	0.3	9,750,000	18
5	Fixed	–	–	–	–

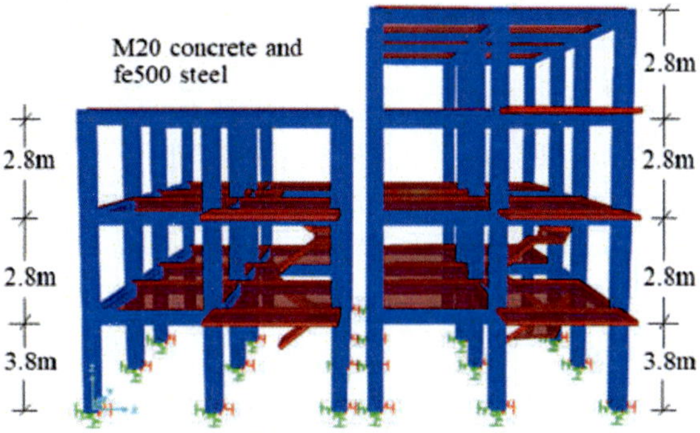

Fig. 4 Buildings used for analysis

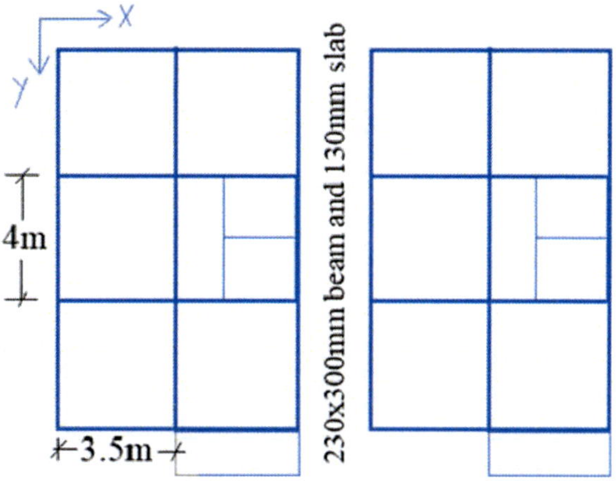

Fig. 5 Top view of building

4 Seismic Analysis

Seismic analysis plays an important role while studying structural damping and SSI effect. Responses of building frames under dynamic load are governed by time history analysis. On the basis of external load and ground motion, these characteristics

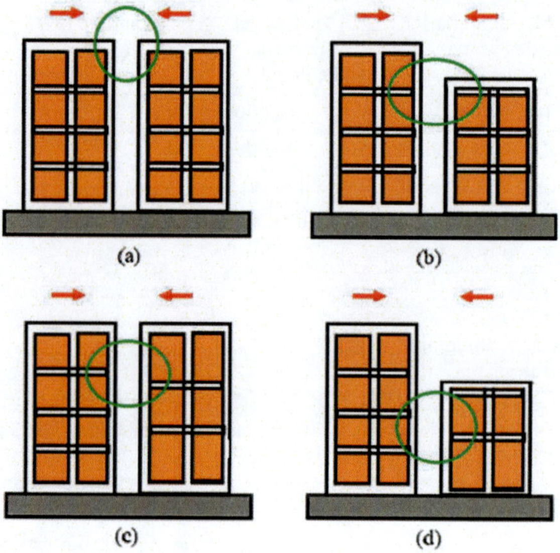

Fig. 6 Different cases for pounding **a** Same storey Ht and same Ht of building **b** Same storey Ht and different Ht of building **c** Different storey Ht and same Ht of building **d** Different storey Ht and different Ht of building

can be considered. Based upon the type of external loading and sustainability of structure, the seismic analysis can be classified as: (1) linear static analysis, (2) nonlinear static analysis, (3) linear dynamic analysis, and (4) nonlinear dynamic analysis. For a structure with sufficient small height and regular geometry, linear static analysis is employed. Further, response spectrum method is also used for time domain under linear dynamics. While considering the variety of forces and strong ground motion, time history analysis is preferred. Nonlinear dynamic analysis considered P-delta effect incorporating strong ground motion at the base. Hence, elasto-plastic deformations can also be treated in the analysis.

4.1 Nonlinear Time History Analysis

Nonlinear time history analysis falls under nonlinear dynamic analysis which is the most realistic and accurate analysis method available. The old data of earthquake is collected, and using this data, earthquake loading is applied on structure model incorporating elements with inelastic force–deformation relationship and p-delta effect. When different ground motions are applied to the structure, which generate complete response histories for any quantity of interest like displacement, moments, etc., they also discussed about performance-based design like Wankhade [8–11] paper; PBD is used for the analysis of the structure subjected to seismic loading, and the advancement of the computational methods and computer technology expands the capability and applicability of PBD. How the use of PBD is made for seismic analysis in very simple language is explained. In PBD, nonlinear analysis is performed on

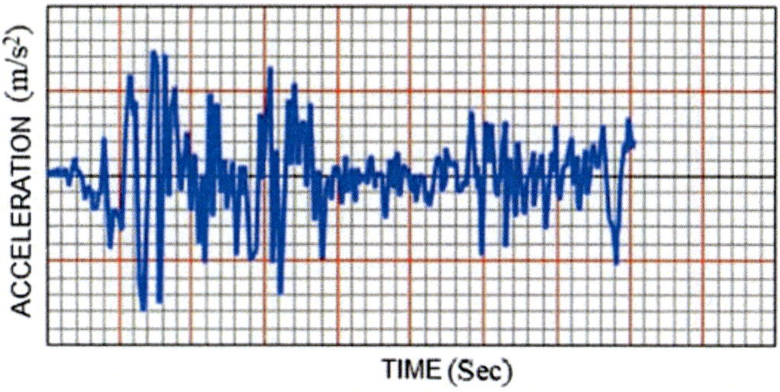

Fig. 7 El Centro time function (time vs. acceleration graph)

the structure. Hong-Yu et al. [22] used gap elements and explained its use. Gap element lets the software know about the space between different structure and helps in ponding analysis.

The use of time history analysis is done for finding maximum displacements, moments, base shear, and drift ratios. In SAP2000, there are inbuilt time functions, so ELECENTRO time function is used, and it has magnitude of earthquake up to 0.2763 g and there is high variation of magnitude also; we can see its graph in Fig. 7.

5 Conclusion

In the present study, pounding effect is considered for adjacent frames having sufficient gap between them. Soil–structure interaction effect is also considered for these frames which are considered in the analysis of different models. Time history analysis is performed for setting the results in terms of displacement and moments. Hence, different combinations of the results are provided in Table 3 and Figs. 8, 9 and 10. The displacement of the joints in $+ x$-direction is shown and compared for the respective joint. Moment at support is also given in tabular form when the buildings are separated from each other with specified distances. It is observed that the moment at the support of the frames which are closely spaced to each other is increasing as the distance between these frames is increasing. Whereas the moment at the support for the frames which are very far from each other the end supports there the moment is decreasing as the distance between the frames are increasing. If we consider maximum displacement, the displacement increases as the spacing between the frames increased. It is observed that the building frames with different storey heights exhibit 8% and 23% higher defection than building frames with same storey height at top and bottom levels, respectively. When the frames are kept close to each other, they hit each other as due to insufficient space between them, and therefore, the moment at the support close to each other has less moment as compared to the frames spaced far apart. Reduction in moment is observed to be 41.63% for

Table 3 Maximum displacements in X-Dir (mm) when buildings are separated at a distance 2.5 inch apart and maximum moments at left and right support of two frames

	Same storey Ht and same Ht of building		Same storey Ht and different Ht of building		Different storey Ht and same Ht of building		Different storey Ht and different Ht of building	
Max. displacements in *X*-Dir (mm)	Frame 1	Frame 2	Frame 1	Frame 2	Frame 1	Frame 2	Frame 1	Frame 2
Level 5	–	–	–		52.974	–	–	–
Level 4	46.996	46.996	–	46.732	47.318	44.238	–	50.803
Level 3	39.208	39.208	37.329	37.329	37.743	37.743	43.106	43.106
Level 2	26.381	26.381	24.543	24.543	25.775	25.775	30.471	30.471
Level 1	11.159	11.159	9.696	9.696	11.413	11.413	13.756	13.756
Separation gap between adjacent buildings	Moments at support (kN mm)							
1 m	46,083.6	97,285.5	71,039.9	76,060	35,714.2	41,114.1	37,217.7	125,537
3 m	42,224.9	945,162	70,417.7	72,005	34,129.2	39,701.4	33,331.5	120,653
5 m	38,869.1	91,337.9	68,914.2	68,105	31,651.6	37,630.8	31,233.4	115,667
Far away	36,281.0	78,236.0	15,776.3	39,293	205,672	25,654.0	26,277.2	97,429.3

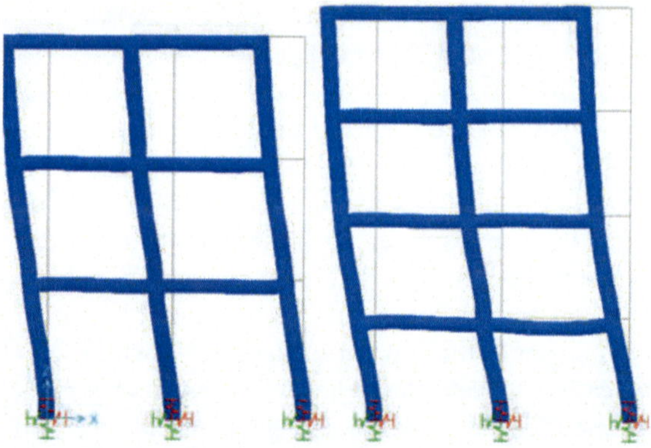

Fig. 8 Deformed shape of adjacent building frames with equivalent spring at base

building frames spaced faraway that of closely spaced frames. Soil-–structure interaction is also studied, and it is noted that as the soil was getting stiffer the SSI effect became less significant, and as a result, the structure maximum drift decreased.

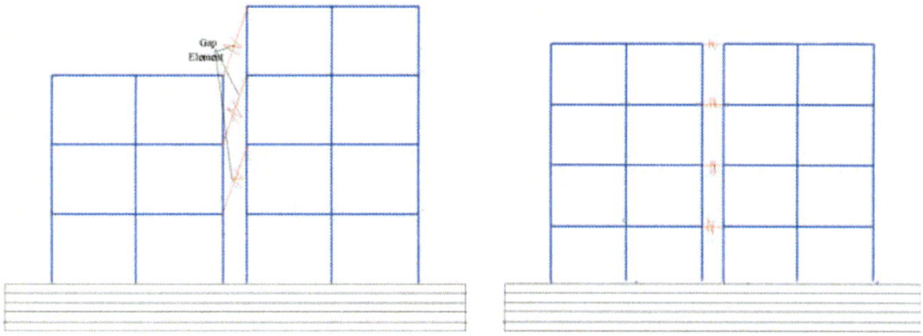

Fig. 9 G + 2 and G + 3 frame having gap element for pounding effect with hard soil as base

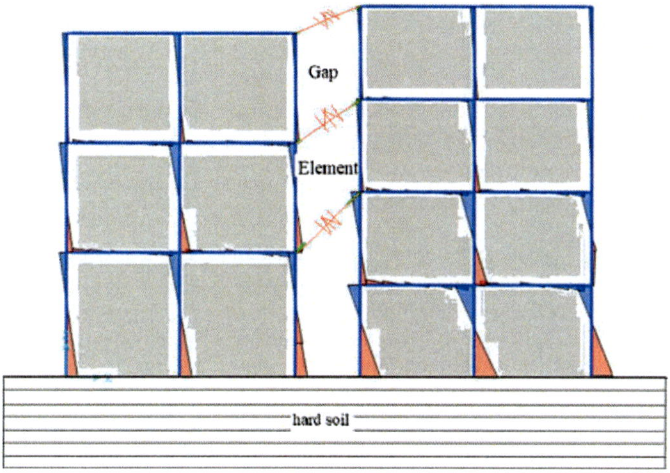

Fig. 10 Moment for frames having gap element for pounding effect with hard soil as base

References

1. Whittaker, A., Constantnou, M., & Tsopelas, P. (1998). Displacement estimates for performance based seismic design. *Journal of Structural Division, ASCE, 124* (8), 905–912
2. Priestley, M. J. N. (2000). Performance based seismic design. *Bulletin of the New Zealand Society for Earthquake Engineering, 33*, 325–346.
3. Chopra, A. K., & Chintanapakdee, C. (2001). Comparing response of SDF systems to near-fault and far-fault earthquake motions in the context of spectral regions. *Earthquake Engineering and Structural Dynamics, 30*, 1769–1789.
4. Xue, Q. (2001). A direct displacement based seismic design procedure of inelastic structures. *Engineering Structures, 23*(11), 1453–1460.
5. Jankowski, R. (2004). Non-linear modelling of earthquake induced pounding of buildings' published on ICTAM, August 15–21, 2004, Warsaw, Poland
6. Sayani, P. J., Erduran, E., & Ryan, K. L. (2011). Comparative response assessment of minimally compliant low-rise base isolated and conventional steel moment resisting frame buildings. *Journal of Structural Engineering. ASCE, 137*(10), 1118–1131.

7. Wankhade, R. L. (2014). Performance based design and estimation of forces for building frames with earthquake loading. In *Paper 141, International conference on recent trends and challenges in civil engineering*, December 12–14, 2014, MNNIT Allahabad, India.
8. Wankhade, R. L., & Landage, A. B. (2014). Static analysis for fixed base and base isolated building frame. In *Proceedings of national conference on advances in civil and structural engineering (NCACSE-2014).*
9. Wankhade, R. L., & Bajoria, K. M. (2015). Vibration of cantilever piezo laminated beam with extension and shear mode piezo actuators. In *Proceeding of SPIE active and passive smart structures and integrated systems*, Vol. 9431, p. 943122-1.
10. Wankhade, R. L., & Landage, A. B. (2016). Performance based analysis and design of building frames with earthquake loading. *International Journal of Engineering Research, 5*(1), 106–110.
11. Yesane, P. M., Ghugal, Y. M., & Wankhade, R. L. (2016). Study on soil–structure interaction: A review. *Int. J. Eng. Res., 5*(3), 737–741.
12. Wankhade, R. L. (2017). Performance analysis of RC moment resisting frames using different rubber bearing base isolation techniques. In *International conference on innovations in concrete for infrastructure challenges Nagpur*, India October 6–7.
13. Bendine, K., & Wankhade, R. L. (2016). Vibration control of FGM piezoelectric plate based on LQR genetic search. *Open Journal of Civil Engineering, 6*, 1–7.
14. Wankhade, R. L., & Bajoria, K. M. (2016). Shape control and vibration analysis of piezolaminated plates subjected to electro-mechanical loading, Open. *Journal of Civil Engineering, 6*, 335–345.
15. Wankhade, R. L., & Bajoria, K. M. (2017). Numerical optimization of piezolaminated beams under static and dynamic excitations. *Journal of Science: Advanced Materials and Devices, 2*(2), 255–262.
16. Bendine, K., & Wankhade, R. L. (2017). Optimal shape control of piezolaminated beams with different boundary condition and loading using genetic algorithm. *International Journal of Advanced Structural Engineering, 9*(4), 375–384.
17. Wankhade, R. L., & Bajoria, K. M. (2019). Vibration analysis of piezolaminated plates for sensing and actuating applications under dynamic excitation. *International Journal of Structural Stability and Dynamics, 19* (10), 1950121.
18. Wankhade, R. L., & Niyogi, S. B. (2020). Buckling analysis of symmetric laminated composite plates for various thickness ratios and modes. *Innovative Infrastructure Solution, 5*, 65. https://doi.org/10.1007/s41062-020-00317-8
19. Wankhade, R. L., & Bajoria, K. M. (2021). Vibration attenuation and dynamic control of piezolaminated plates for sensing and actuating applications. *Archive of Applied Mechanics, 91*(1), 411–426.
20. Wankhade, R. L., Niyogi, S., Gajbhiye, P. (2021). Buckling analysis of laminated composites considering the effect of orthotropic material. *IOP Science, Journal of Physics: Conference Series, 1706* (1) 012188.
21. Kouider, B., & Wankhade, R. L. (2021). Piezoelectric energy harvesting from a curved plate subjected to time-dependent loads using finite elements. *IOP Science, Journal of Physics: Conference Series, 1706* (1), 012008.
22. Jia, H.-Y., Lan, X.-L., Luo, N., Yang, J., Zheng, S.-X., Zhang, C. (2019). Nonlinear pounding analysis of multispan and simply supported beam bridges subjected to strong ground motions. *Shock and Vibration, 2019* (8759428) (Hindawi Publication). https://doi.org/10.1155/2019/8759428

Sustainable Bituminous Pavement: A Study on Low-Density Polymer Modified Bituminous Binder

Vishnu Vijayan, Jeevan Mathew Tharayil, R. Rakhil Krishna, Jiji Saji, Divya S. Shaji, and G. Lakshmi

Abstract Plastics, especially low-density polyethylene (LDPE), is now being the most effective pollutant on Earth due to its non-degradability and increased demand. The reduced reusability and recyclability of these plastics makes more hurdle in proper disposal and treatment of these wastes. Thus, this problem can be managed only by implementing an alternative method of recycling these polymeric compounds for other beneficial purposes. The aim of this study is to suggest an ecofriendly approach of disposing these wastes by using it as bitumen modifiers. The effect of LDPE on bitumen has been analyzed by different wt% of LDPE from 2 to 8 wt% and was compared with pristine bitumen. Characterization tests mainly viscosity test, standard penetration and softening point, Fourier transform infrared spectroscopy, thermogravimetric analysis, and rheological analysis were done in order to examine the effect of LDPE addition to bitumen. The results show rise in thermal stability and deformation resistance for modified bitumen. However, a rise in polymer content above an optimum limit is not preferable. LDPE addition of 4% shows a better performance and is suggestable for employing in road sectors.

Keywords Low-density polyethylene · Modified bitumen · Eco-friendly · Road pavement

1 Introduction

Owing to the use in many aspects of our lives, plastics produce a significant quantity of waste on Earth such as packaging, building and manufacturing, automobile, electrical, and computer uses. Plastics are polymers derived primarily from fossil fuels by-products and made primarily by variations of carbon, hydrogen, and oxygen [1]. Though these possess extreme temperature tolerance, resistant to UV irradiation, and often non-biodegradable, these are able to continue polluting the whole Earth for centuries. Plastics appear to detach into smaller, meso-, and micro-structured

V. Vijayan (✉) · J. M. Tharayil · R. R. Krishna · J. Saji · D. S. Shaji · G. Lakshmi
Department of Civil Engineering, Baselios Mathews II College of Engineering, Kollam, Kerala, India

A. K. Gupta et al. (eds.), *Advances in Construction Materials and Sustainable Environment*, Lecture Notes in Civil Engineering 196,
https://doi.org/10.1007/978-981-16-6557-8_35

nature that have unique and important effects on the environment that may have a detrimental impact on the health of humans and animals involved with the structure. Bioaccumulation of these to the food chain causes severe impacts on humans [2]. The asphalt industry is extensively involved in polymer improvements in bitumen to boost the properties of bentonite, impacting breaking, breakdown, and rolling pavements in thermal tolerance. Polymers are well known for their usage to significantly improve the performance of bitumen. Polymers are capable of increasing the elasticity and the stability, therewith increasing the durability, versatility, and deformation resistance for asphalt mixtures and reducing their temperature sensitivity [3].

Polymer alteration of bitumen has been shown to enhance different properties of bitumen including high-temperature rigidity, moisture tolerance, fatigue life, and the ability to lower temperature stresses. Bitumen modifiers have three types of polymers, according to the chemistry and properties of their materials: plastomers, elastomers, and reactive polymers [4]. Each serves different functions like increase in stiffness can be attained by addition of plastomers, whereas fatigue properties can be raised using elastomers. Majority of plastomers effect the thermal properties of binder, especially reduces the thermal susceptibility of bitumen [5]. Plastomers not only just decreases the stiffness parameter but also provides deformation resistance to a greater extent. The major plastomers which we usually deal with are polyethylene and are waste materials after use. Polyethylene has many forms out of which high-density polyethylene (HDPE) and low-density polyethylene (LDPE) are more available. LDPE is commonly the non-reusable waste plastic which has a melting temperature of around 125 °C. For common bitumen binders, the temperature adopted is above 170 °C which makes LDPE to consider as bitumen modifier as its melting temperature is less than that of bitumen [6].

Hinislioglua and Agar worked on various concentration of HDPE additions in asphalt concrete HDPE concentrations on four to eight wt. % in asphalt concrete 20. The temperature variations have been done at 145–160 °C with mixing duration ranging from 5 to 30 min. The results show high-stability and high-marshal quotient, and out of those, the best results were obtained for 4%/135 °C/30 min composition. Moreover, this mix was also highly resistant to permanent deformation (rutting) [7]. Deformation resistance was also notable for this mixture. In a study done by Zoorab et al. [8] has done incorporation of polypropylene and low-density polyethylene in bituminous concrete mixture this shows an increase in properties like durability and resistance to deformation. The incorporation of recycled plastic in pavement design will not only increase the basic pavement properties but also give economic as well as environmental advantages [8]. Nearly, every industry uses plastics and its use is very likely to grow as plastics technologies are developed, causing plastic waste to increase in turn. Because they are found to modify the ecosystem's functionality it is important that instead of being left freely in nature or landfilled, to take into account preventive or reduced plastic waste and to restore this ecologically hazardous waste. As the present waste management concentrates on 3R principle, more changes have to be introduced [9]. Novel strategies are to be established before these turns out to be non-manageable crisis. Shredded plastic-reinforced hot mix asphalt is a research field for such a goal that is still in its early stages. It can be used as strengthening components for pavement asphalt of concrete. In the past literatures, polyethylene waste was typically applied with dry process to the asphalt mixture or used as an aggregate for improving the HMA output in the asphalted mixture (aggregated substitution) [10]. The literatures has revealed an increase in permanent resistance to deformation, Marshall stability, stiffness, and tiredness in asphalt mixtures, while a reduction in moisture damage was also observed while using shredded plastics in pavements. The past works also highlights a reduction specific gravity for these asphalt mixtures.

Even though there are a few investigations that have been done on using polyethylene as binder modifiers, the incorporation of recycled low-density polyethylene into bitumen was less explored one. Thus, the aim of this work is to investigate the effect of recycled LDPE in modifying bitumen properties. LDPE with different wt% (2, 4, 6, and 8%) were studied, and physical, thermal, chemical, and rheological properties of every samples were analyzed and are compared with pristine hot mix asphalt.

2 Materials

2.1 Fabrication of Shredded Polyethene

The polyethene waste was collected from the dumping yard at Kollam, Kerala. The collected plastic waste includes, low-density and high-density polyethylene derivatives and a few of PET low-grade bottles. The initial steps were sorting of the plastics, cleaning, and then drying. The polyethene bags were shredded into flakes so as to get a size range of 2.36 to 4.75 mm by maintaining an aspect ratio of 2:1 shown in Fig. 1. The tensile strength analysis of the flakes has been done with 10 mm/min of elongation. The results obtained were on an average of 0.31KN, thereby providing an elongation rate of 25%.

Fig. 1 Shredded LDPE

Table 1 Mixing of polymer and bitumen blend

Sample	LDPE (wt%)	Bitumen (wt%)
Pristine bitumen	0	100
LDPE 1	2	98
LDPE 2	4	96
LDPE 3	6	94
LDPE 4	8	92

2.2 *Incorporation of Polyethylene to Bitumen*

For preparing bitumen—polyethylene blend—the bitumen was initially heated to a temperature of 170 °C. Then, the shredded low-density polyethylene flakes were added to the bitumen, thereby obtaining bitumen polyethylene blend. The LDPE additive was incorporated to pristine bitumen at 2, 4, 6, and 8% weight ratios and is described in Table 1. Uniform polymer–bitumen blended mixture was obtained using shear mixer at a rate of 4000 rotations per minute. The mixing was done for an average of 2 h maintaining the temperature of 180 °C.

3 Methodology and Characterization

3.1 *Viscosity*

Rotational viscosity tests were conducted according to ASTM D-4402 at 135 °C. The tests were done for pristine as well as modified binders. Brookfield DV III rheometer was employed for testing.

3.2 *Penetration and Softening Point*

The softening point was measured using ring and ball analysis in accordance with ASTM D36. The penetration test was conducted by following ASTM D-5 [11]. Both tests were done for pristine and modified binders.

3.3 *FTIR Analysis*

FTIR tests were done for identifying the functional groups present in both pristine and modified bitumen. Fourier-transform infrared spectroscopy (FTIR in transmission mode, Cary 630, Agilent Technologies).

3.4 TGA Analysis

Thermogravimetric analysis (TGA) of nanoparticle was carried out under a constant flow of nitrogen at a heating rate of 10 °C/min on the Hitachi, STA7200 TGA instrument.

3.5 Rheological Analysis

The rheological tests were done using dynamic shears rheometer (DSR) (Physica MCR302, Anton Paar). Frequency sweep test was conducted for bitumen at linear viscoelastic region at 60 °C, and the frequency range adopted was 0.1–10 Hz. For this, DSR plates with a radius of 12.5 mm having a thickness of 1 mm. The storage and loss modulus were obtained as output for respective frequency. The shear modulus and phase angle were calculated according to [12]

$$|\mathrm{G}| = \sqrt{G_1^2 + G_2^2} \tag{1}$$

$$\delta = tan^{-1}(G_2/G_1) \tag{2}$$

where G_1 is storage modulus; G_2 is loss modulus; G is complex shear modulus; δ is the phase angle.

4 Results and Discussion

4.1 Viscosity

Viscosity is considered to be a basic physical property of bitumen which gives the flow resistance of fluid. The rotational viscosity of bitumen and bitumen polyethylene blend has been tested for analyzing the resistance to shear stress. The results showing the effect of varying concentration of LDPE are given in Fig. 2, and from the results, it depicts that by increasing LDPE concentration, there shows a significant rise in viscosity in comparison with pristine bitumen [12]. At 130 °C, the viscosity was about 0.59 Pa.s for normal pristine bitumen. The variation in polyethylene content was done for 2%, 4%, 6%, and 8 wt% which rises the viscosity to 0.75, 1.21, and 2.98. The obtained results are on agreement with previous literatures [13]. The rise in viscosity also results in rising the temperature stability of the bitumen polyethene blend. The enhancement in viscosity with rise in polyethene concentration is due to the presence of asphaltenes, which results in the formation of complex internal

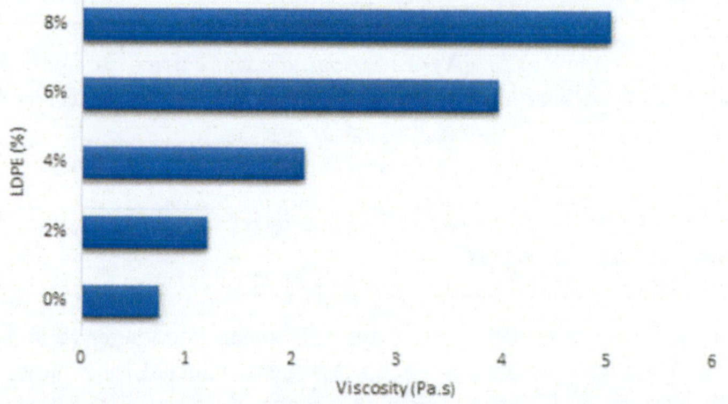

Fig. 2 Viscosity of LDPE–bitumen samples

structure. But increased viscosity may effects workability of the mix for 8 wt% sample which has effects on mixing and workability.

4.2 Determination of Penetration and Softening Point

The results of penetration and softening point was according to the expectation that these properties get good results with addition of LDPE and are given in Figs. 3 and 4. The penetration test depicts the flow and deformation properties of the LDPE–bitumen mix. The penetration decreased with rise in the polymer concentration. Pristine bitumen shows penetration value of 55.1 mm, while with rise in concentration of polymer to 2wt%, 4wt%, 6wt%, and 8 wt%, the penetration results were 52.3 mm,

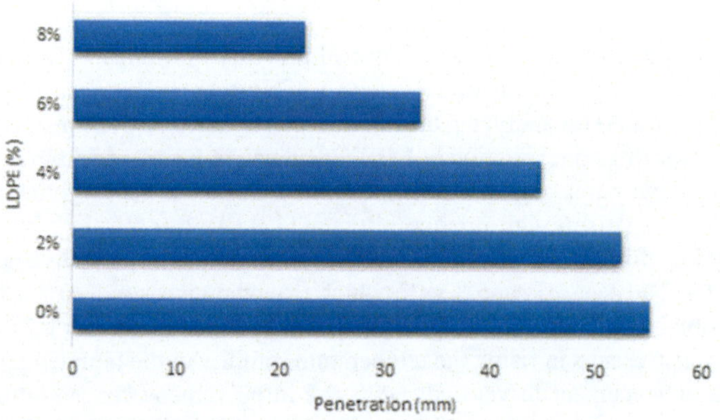

Fig. 3 Penetration of LDPE–bitumen samples

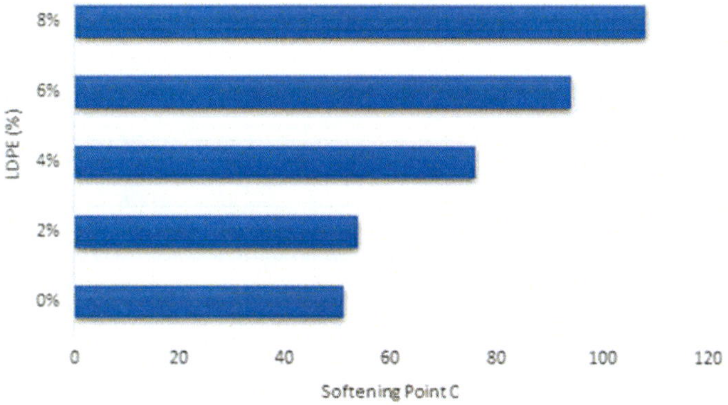

Fig. 4 Softening point of LDPE–bitumen samples

44.7 mm, 33.2 mm, and 22.2 mm, respectively. The reduction in the penetration of the bitumen polymer blend shows the increased hardening of the bitumen with addition of polymer [14]. The effect of polymer in physical properties of bitumen is evident from reduction in penetration results. The reduced penetration results show the ability of modified bitumen to resist temperature effects and rutting [15].

Rise in softening point values was observed with rise in polymer concentration. The softening point of the pristine bitumen was 51 °C which gets increased to 54, 76, 94, and 108 °C with 2, 4, 6, and 8 wt% of polyethene additions. The complexation in internal structure caused by addition of polymer results in the rise in softening point of modified mix. From the raised softening point is evident that the modified mix has better resistance to deformation at varying range of temperature. These blended mixtures will be strong enough to resist rutting [16].

4.3 FTIR

The FTIR analysis of both pristine and modified asphalt samples was done and is given in Fig. 5. It is evident that both pristine and modified bitumen have major similar peaks with only few new peaks for modified bitumen. The peaks at pristine bitumen at 2918.3 cm^{-1} show the presence of CH stretching vibrations, thus confirming the presence of organic hydrocarbons in the bitumen. These peaks are also seen in polymer blended bitumen samples at 2916.21 cm^{-1}. C=C attraction was found to be observed at 1598 cm^{-1}. The addition of LDPE is confirmed by the peaks in range of 1200–750 cm^{-1}. The SO_2 peaks at 1329 cm^{-1} confirm the addition of polymer of bitumen matrix [17].

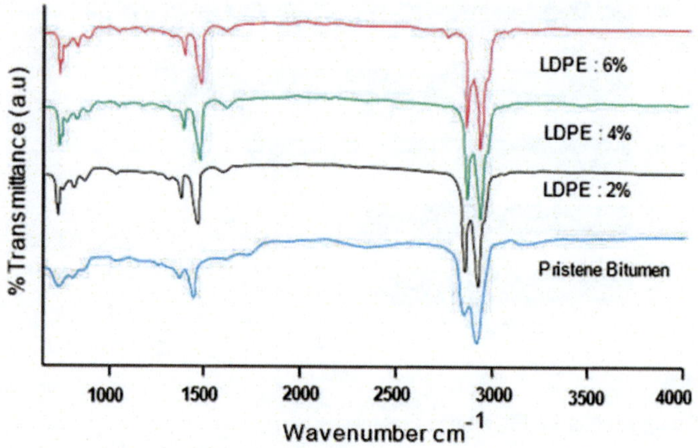

Fig. 5 FTIR spectra of LDPE–bitumen samples

4.4 *Rutting and Fatigue Properties*

The rutting and fatigue properties were analyzed by conducting DSR test of the LDPE-added bitumen mixes. The test was done at a wide range of temperatures as 64 and 70 °C. The rutting was analyzed in reference to ASTM D2872. The result of DSR test is given in Table 2. The shear modulus (*G*) and phase angles were analyzed by the DSR tests. *G* indicates that the shear modulus depicts the stiffness of asphalt mix samples. The elasticity at varying temperature was determined as phase angles (δ). As phase angles decreases, the elasticity of the bitumen mix rises which in turn increase the resulting rutting tolerance. From the results, the modified bitumen binders has reduced δ value for both aged and unaged RTFO samples [18]. Ideally, the bitumen possesses a specific stiffness value to relieve rutting. From Table 2, it is evident that a reduction in *G* value can be noted, thus causing reduction of rutting parameter ($G/\sin\delta$). Referring the pavement criterions, AASHTO: MP1, the value of rutting parameter should be greater than 2.2 kPa; for aged RTFO sample, it must be 1 kPa. From the results, the values of the modified bitumen are favorable, and thus concludes its better tolerance against rutting. G Sin θ δ is considered as the fatigue parameter and must be less than 5000 kPa for PAV-pressurized aged vessel bitumen samples [19]. The results are under the higher limit and fatigue prominent at temperature less than 20 °C, thus concluding that the polymer bitumen mixture has enough rutting and fatigue resistance.

4.5 *TGA*

Thermal gravimetric analysis was done to test the thermal stability of the prepared samples. The decomposition of the samples with respect to temperature was identified

Table 2 Dynamic shear test rheometer result

Bitumen	Temperature	Parameter	Pristine bitumen	2% (w/w) LDPE	4% (w/w) LDPE	6% (w/w) LDPE	8% (w/w) LDPE
Original (Unaged)	64 °C	*G** (kPa)	1.4195	1.3564	1.1969	1.1521	1.1103
		δ (∘)	88.07	86.94	86.18	85.16	84.58
		*G**/sin *δ* (kPa)	1.4203	1.3583	1.1996	1.0132	0.8331
	70 °C	*G** (kPa)	0.6421	0.62465	0.57001	0.53112	0.51231
		δ (∘)	88.73	87.99	87.33	86.45	85.88
		*G**/sin *δ* (kPa)	0.6394	0.6249	0.5821	0.5433	0.5216
RTFOT residue	64 °C	*G** (kPa)	3.0898	3.0333	3.0292	3.0272	3.0188
		δ (∘)	86.24	85.79	85.36	84.88	84.27
		*G**/sin *δ* (kPa)	3.0965	3.0395	3.0292	3.0181	3.0012
	70 °C	*G** (kPa)	1.5342	1.3664	1.3499	1.3488	1.3411
		δ (∘)	87.58	87.19	86.88	86.55	86.32
		*G**/sin *δ* (kPa)	1.5692	1.3612	1.3558	1.3496	1.3443

by TGA analysis. The respective weight loss at varying temperature is provided with TGA curve in Fig. 6. From the results, it is seen that the weight loss was evidently high during a temperature range of 380–500 °C [20]. The loss in weight at this temperature

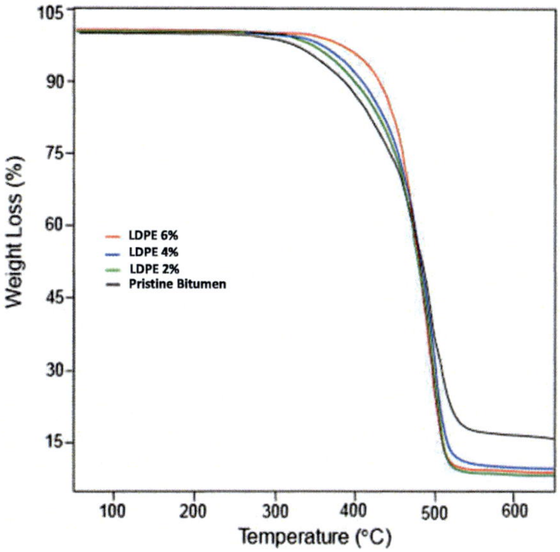

Fig. 6 TGA analysis of LDPE–bitumen samples

range was because of decomposition of bituminous compounds in mixture. These bituminous compounds include the aromatics and saturates in asphalt [21]. The pristine bitumen shows a higher decomposition than modified samples. The reduced decomposition of modified samples is due to its reduced evaporation, and thus makes it clear that these modified samples have a greater stability over varying temperature ranges.

5 Conclusion

The performance of the pristine asphalt has been intensified by the polyethene addition. The reusing of polyethene bags as an asphalt additive was done in this work which in turn reduces the plastic contamination of nature. The characteristic analysis shows that there is a notable enhancement in the overall properties of the pristine bitumen. As expected, the physical properties like viscosity and softening point get raised with increase in concentration of polymer. The penetration and softening point results give emphasis that the stiffness gets raised with rise in polymer concentration. The chemical composition was analyzed using FTIR, and it seems not to have drastic changes in absorbance peaks. The temperature stability was in higher side for modified bitumen mixes. From the DSR test, the rise in fatigue and rutting resistance was confirmed as these modified binders have much more rutting and fatigue resistance than pristine bitumen. Discussing about the limitation, the viscosity seems to affect the workability at higher concentration of polymer. A drastic rise in viscosity was observed for LDPE 3 and LDPE 4 mixes which are of polymer 6 and 8 wt%; this heavy rise in viscosity will curb the workability of the mixture. Thus, an additive concentration of 4% can be suggested so as to get an optimum workability along with increased overall performance. As a conclusion, it can be summarized that incorporation of additive gives favorable effects on rheological as well as overall properties of bitumen. Thus, using polyethene waste as bitumen modifier has a great potential in road pavements as well as for a clean environment.

References

1. Carson, H. S., Colbert, S. L., Kaylor, M. J., & McDermid, K. J. (2011). Small plastic debris changes water movement and heat transfer through beach sediments. *Marine Pollution Bulletin, 62*. https://doi.org/10.1016/j.marpolbul.2011.05.032
2. Katsanevakis, S., Verriopoulos, G., Nicolaidou, A., & Thessalou-Legaki, M. (2007). Effect of marine litter on the benthic megafauna of coastal soft bottoms: A manipulative field experiment. *Marine Pollution Bulletin, 54*. https://doi.org/10.1016/j.marpolbul.2006.12.016
3. Gama, D. A., Yan, Y., Rodrigues, J. K. G., Roque, R. (2018). Optimizing the use of reactive terpolymer, polyphosphoric acid and high-density polyethylene to achieve asphalt binders with superior performance. *Construction and Building Materials, 169*. https://doi.org/10.1016/j.conbuildmat.2018.02.206

4. García-Travé, G., Tauste, R., Moreno-Navarro, F., Sol-Sánchez, M., & Rubio-Gámez, M. C. (2016). Use of reclaimed geomembranes for modification of mechanical performance of bituminous binders. *Journal of Materials in Civil Engineering, 28*. https://doi.org/10.1061/(asce)mt.1943-5533.0001507
5. Brasileiro, L., Moreno-Navarro, F., Tauste-Martínez, R., Matos, J., Rubio-Gámez, M., & del, C. (2019). Reclaimed polymers as asphalt binder modifiers for more sustainable roads: A review. *Sustainability (Switzerland), 11*. https://doi.org/10.3390/su11030646
6. Costa, L. M. B., Silva, H. M. R. D., Peralta, J., & Oliveira, J. R. M. (2019). Using waste polymers as a reliable alternative for asphalt binder modification—Performance and morphological assessment. *Construction and Building Materials, 198*. https://doi.org/10.1016/j.conbuildmat.2018.11.279
7. Hinislioglu, S., & Agar, E. (2004). Use of waste high density polyethylene as bitumen modifier in asphalt concrete mix. *Materials Letters, 58*. https://doi.org/10.1016/S0167-577X(03)00458-0
8. Zoorob, S. E., & Suparma, L. B. (2000). Laboratory design and investigation of the properties of continuously graded Asphaltic concrete containing recycled plastics aggregate replacement (Plastiphalt). *Cement and Concrete Composites, 22*. https://doi.org/10.1016/S0958-9465(00)00026-3
9. Cobut, A., Blanchet, P., & Beauregard, R. (2015). The environmental footprint of interior wood doors in non-residential buildings—Part 1: Life cycle assessment. *Journal of Cleaner Production, 109*. https://doi.org/10.1016/j.jclepro.2015.04.079
10. Baghaee Moghaddam, T., Soltani, M., & Karim, M. R. (2014). Evaluation of permanent deformation characteristics of unmodified and Polyethylene Terephthalate modified asphalt mixtures using dynamic creep test. *Materials and Design, 53*. https://doi.org/10.1016/j.matdes.2013.07.015
11. Read, J., & Whiteoak, D. (2003). *The Shell bitumen handbook*.
12. Oliveira, J. R. M., Silva, H. M. R. D., Abreu, L. P. F., & Fernandes, S. R. M. (2013). Use of a warm mix asphalt additive to reduce the production temperatures and to improve the performance of asphalt rubber mixtures. *Journal of Cleaner Production, 41*. https://doi.org/10.1016/jjclepro.2012.09.047
13. Sybilski, D. (1997). New simplified equation for the computation of absolute viscosity of polymer-bitumens. *Materials and Structures/Materiaux et Constructions, 30*. https://doi.org/10.1007/bf02486391
14. Cong, P., Xun, P., Xing, M., & Chen, S. (2013). Investigation of asphalt binder containing various crumb rubbers and asphalts. *Construction and Building Materials, 40*. https://doi.org/10.1016/j.conbuildmat.2012.11.063
15. Yu, J., Zeng, X., Wu, S., Wang, L., & Liu, G. (2007). Preparation and properties of montmorillonite modified asphalts. *Materials Science and Engineering A, 447*. https://doi.org/10.1016/j.msea.2006.10.037
16. Bala, N., Kamaruddin, I., & Napiah, M. (2017). The influence of polymer on rheological and thermo oxidative aging properties of modified bitumen binders. *Jurnal Teknologi, 79*. https://doi.org/10.11113/jt.v79.10180
17. Gürü, M., Çubuk, M. K., Arslan, D., Farzanian, S. A., Bilici, I. (2014). An approach to the usage of polyethylene terephthalate (PET) waste as roadway pavement material. *Journal of Hazardous Materials, 279*. https://doi.org/10.1016/j.jhazmat.2014.07.018
18. Presti, D. L., Izquierdo, M. A., & Jiménez del Barco Carrión, A. (2018). Toward storage-stable high-content recycled tyre rubber modified bitumen. *Construction and Building Materials, 172*. https://doi.org/10.1016/j.conbuildmat.2018.03.226
19. Nizamuddin, S., Jamal, M., Gravina, R., & Giustozzi, F. (2020). Recycled plastic as bitumen modifier: The role of recycled linear low-density polyethylene in the modification of physical, chemical and rheological properties of bitumen. *Journal of Cleaner Production, 266*. https://doi.org/10.1016/j.jclepro.2020.121988
20. Farahani, H. Z., Palassi, M., & Sadeghpour Galooyak, S. (2017). Thermal analysis of bitumen modified with LDPE and CR. *Petroleum Science and Technology, 35*. https://doi.org/10.1080/10916466.2017.1319385

21. Zhang, F., & Hu, C. (2016). The research for crumb rubber/waste plastic compound modified asphalt. *Journal of Thermal Analysis and Calorimetry, 124*. https://doi.org/10.1007/s10973-015-5198-4

Identification of Parking Sites in the Kukatpally Region Using GIS and AHP

Ramu Penki, T. Srinivasa Rao, G. Vinod Naik, and Rapaka Aparna

Abstract Finding a parking space in urban areas has become very difficult due to less number of available parking spaces. The traffic flow in urban areas is very high compared to rural areas. Kukatpally area is one of the busiest places in Hyderabad city, where finding a parking space is a major concern for vehicle users. Lack of parking spaces leads vehicle users to park their vehicles at inappropriate locations such as the shoulders of the carriageways, no parking zones, etc., obstructing the ongoing traffic and pedestrians. Identification of suitable parking sites helps vehicle users in parking their vehicles at appropriate locations. In this study, a geographical information system (GIS) based multi-criteria decision making method (MCDM), i.e. analytical hierarchical process (AHP) is used for the selection of suitable sites for vehicle parking. AHP method is used for ranking and weighing different criteria which affect the selection of parking site. Various factors considered are distances from commercial centres, administrative centres, educational centres, hospitals, cultural and recreational centres, roads. The selected factors are ranked and weighed using AHP. Thematic maps are created for each criterion in the GIS environment using spatial analysis. Using the weights obtained from the AHP, the maps are reclassified and are used in overlay analysis to obtain the parking suitability index of the entire study area, i.e. Kukatpally region. Based on the value of obtained suitability index parking facilities can be planned in line with the local government regulations. The suitability index is found to be in the range of 0–60% which indicates the lack of highly suitable parking spaces in the study region.

Keywords GIS · AHP · Parking site · MCDM

R. Penki (✉)
GMRIT, Srikakulam, AP, India
e-mail: ramu.p@gmrit.edu.in

T. Srinivasa Rao · G. Vinod Naik
VNR VJIET, Hyderabad, India

R. Aparna
Sree Dattha Institute of Engineering and Science, Hyderabad, India

A. K. Gupta et al. (eds.), *Advances in Construction Materials and Sustainable Environment*, Lecture Notes in Civil Engineering 196,
https://doi.org/10.1007/978-981-16-6557-8_36

1 Introduction

Public parking facilities play an important role in planning and urban transportation system. In a year on average, a car runs on the roads only for 400 h out of 8760 h, whilst the rest of the time it will be in a parked position [1]. All vehicle users like to park their vehicles as close as possible to their destination to reduce their walking distance [2]. In urban areas and major metropolitan cities where a lot of trip attraction centres (TAC) such as administrative centres, commercial places, etc., are present, parking demand is normally very high and vehicle parking may become a major problem if facilities are not provided to park the vehicle [1, 3]. Sometimes vehicle users may have to travel long distances to find a parking place close to the TAC's which will increase the walking distance and also lead to more fuel consumption which further leads to vehicle operating cost and air pollution that damages the ecosystem [4–6]. Lack of parking facilities causes inconvenience to vehicle users leading them to inappropriate parking of the vehicles, which eventually causes other problems such as traffic jams and accidents. The urban planners and traffic engineers should plan for proper public parking facilities in concern with the increasing vehicular count and parking demand [1]. Suitable site selection for designing a parking facility is very important to overcome the problems. Proper site selection for public parking increases the traffic fluency and reduces the accident rate by avoiding on-street parking or parking at inappropriate places. Site selection is a process of deciding between different alternatives considering various criteria. Site selection for public parking through the traditional method is a time taking process and lacks consideration of all criteria in spatial context [7]. In this study, the public parking site selection for Kukatpally region is carried out using multi-criteria decision making method, i.e. analytical hierarchical process (AHP) and geographic information system (GIS). AHP and similar techniques such as fuzzy AHP, TOPSIS AHP are used successfully in several site selection problems [9–12]. Studies [9, 11] demonstrate that implementation of GIS and AHP simplifies the complexity involved in public parking facilities and improves the overall efficiency of the process.

2 Study Area

Kukatpally is a part of Hyderabad city in the state of Telangana, India. It is located in the northwestern part of Hyderabad city at the latitudes and longitudes of 17°29′0″ N and 78°25′0″ E and an elevation of 505 m. Kuktapally is one of the busiest areas of Hyderabad city with a density of 33,076 persons per km^2. It also hosts the residential colony named Kukatpally Housing Board colony (KPHB) which is a major residential area of Hyderabad. Kukatpally area is connected to the nearest IT hub known as Hi-tech city and there are many numbers of TACs such as commercial centres, administrative centres, cultural and recreational centres, etc., present in this area which attracts a large number of people who travel through different modes such as government buses, auto-rickshaws, cabs, private vehicles, bikes, light rail transits, etc. With the high concentration of TAC's at one place and lack of proper parking facilities, the vehicles are parked on roads near to the TACs. This creates many traffic-related problems like parking on the edges of the carriageways, etc., in this area.

3 Methodology

Step 1 Delineation of the study area.

A base map representing the Kukatpally area is collected from the Google Earth imagery. ArcGIS software is used for geo-referencing and digitization of the study area. WGS 1984 coordinate reference system is used for the preparation of the boundary map of Kukatpally.

Step 2 Selection of criteria affecting the parking sites.

Based on the literature review, several criteria considered for the selection of parking site are categorized as the distance from roads and distance from TACs. The TAC's are differentiated as administrative centres, commercial centres, cultural and recreational centre, educational centre and hospitals.

Administrative centres such as government offices, banks, etc., are considered as one of the important criteria for site selection as many people visit these facilities regularly from longer distances also. Parking facility is very much important here as people generally drive vehicles to these places. Commercial centres such as shopping malls, theatres, etc., at a location are also considered as a criterion for providing parking space. Commercial centres which are mostly located by the roadside cause lot of inconvenience if a proper parking facility is not provided within a short distance as vehicle parking can happen on kerbs of roads if parking not provided. In the same way, the cultural and recreational centres are provided with the proper site for parking. Hospitals are considered as one of the important criteria because it is very essential to any hospital to provide a parking space to the people who visit the hospitals and most of the people are patients who cannot spare time to wait for their vehicles to get parked. Educational centres are considered as the important criteria as they attract more number of vehicles driven by students and employees which needs sufficient space for parking. Roads are also considered as a criterion for parking site selection because there are many small commercial developments happens near to the roads, apart from this ongoing traffic also requires some parking facility [8]. Hence, it is very essential to provide a parking site near the main roads.

Parking sites are best suitable if provided in the vicinity of the TACs. Hence, based on distance from TAC's different sites can be ranked. The distances are grouped into different ranges in case of each criterion and are described as follows. The sites which are nearest to the TAC's are most preferable, while distance sites are less preferable.

Table 1 Representation of criteria and sub-criteria

Criteria	Sub-criteria
Commercial centres	$C1 = 0$–100 m, $C2 = 100$–225 m, $C2 = 225$–350 m, $C3 = >$ 350 m
Administrative centres	$A1 = 0$–150 m, $A2 = 150$–225 m, $A3 = 225$–350 m, $A4 = >$ 350 m
Cultural and recreational centres	CR1 = 0–200 m, CR2 = > 200 m
Hospitals	$H1 = 0$–150 m, $H2 = 150$–250 m, $H3 = > 250$ m
Educational centres	$E_1 = 0$–250 m, $E_2 = 250$–350 m, $E_3 = > 350$ m

In case of commercial centres to the parking sites, the distances considered are (0–100 m), (100–225 m), (225–350 m) and > 350 m, and similarly for cultural and recreational centres (0–200 m), > 200 m, for administrative centres (0–150m), (150–225 m), (225–350 m), > 350 m, for educational centres (0–200 m), (200–350 m), > 350 m, for hospitals (0–150 m), (150–250 m), > 250 m and for roads the distances considered are (0–100 m), (100–250 m), (250–350 m) and > 350 m. The presence of TAC's and roads in the vicinity are considered as the main criteria for site selection and the distance of a site from these criteria are considered as sub-criteria. Table 1 summarizes various criteria and sub-criteria considered in the current study.

Step 3 Calculation of weights using AHP

In the present study AHP, which is an MCDM technique is used for calculating the weights of various criteria and sub-criteria. AHP was introduced by Thomas saaty in 1980, it is an MCDM method used to solve complex decision-making problems involving multiple criteria. AHP method makes use of the pairwise comparison for ranking and rating the criteria. The criteria are ranked on a scale of 1–9 during the process of pairwise comparison where each numerical value represents its importance as shown in Table 2.

Figure 1 represents the flow chart of methodology consisting of various criteria used in the AHP process and overlay analysis. As all criteria are grouped into 2 categories, i.e. TAC's and roads, they are mutually compared and weights are calculated as shown in Table 6. Table 4 indicates the details of ratings assigned and calculated weights for the main criteria. In the case of roads, weights are calculated for sites

Table 2 AHP numerical values judgement

Importance of criteria to another one	Numerical value
Huge importance	9
Significant importance	7
High importance	5
Slight importance	3
Equal importance	1
The importance between former spaces	2, 4, 6, 8

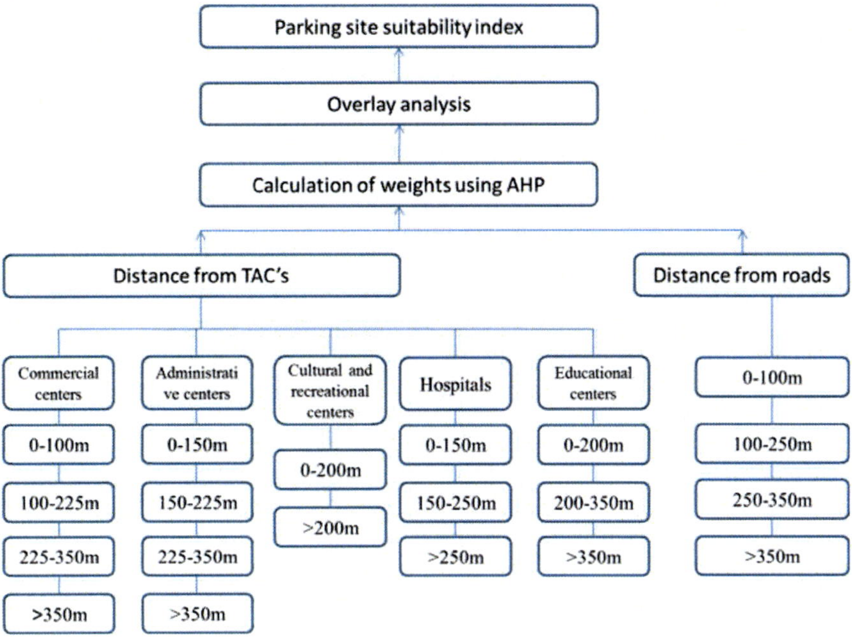

Fig. 1 Flow chart representing AHP methodology

based on their distance to the nearest road as given in Table 5 Weights for sub-criteria are also calculated using the same approach; the obtained results are tabulated in Table 3.

Step 4 Parking site selection

For all the criteria mentioned in Table 1 proximity analysis is performed using the Euclidian distance method to select the final parking site. The proximity analysis results of all the criteria are prepared as thematic maps containing distance zones using ArcGIS. The weights calculated for each distance zone of criteria are assigned to the thematic maps using reclassification analysis. The reclassification results are shown in Fig. 2 All the thematic layers are combined using Eq. 1.

$$0.14 * \mathrm{R} + 0.86 * \mathrm{TAC} \tag{1}$$

The level 1 criteria represent Roads (R) and trip attraction centres (TAC's), whereas the criteria under the TAC's, i.e. commercial centres (C), administrative centres (A), cultural and recreational centres (CR), hospitals (H) and educational centres (E) represents level 2 criteria. Thematic maps of Level 2 criteria are joined through overlay analysis using Eq. 2.

$$\mathrm{TAC} = 0.49 * \mathrm{C} + 0.09 * \mathrm{CR} + 0.23 * \mathrm{H} + 0.13 * \mathrm{A} + 0.06 * \mathrm{E} \tag{2}$$

Table 3 Representation of weights of sub-criteria

Criteria	Sub-criteria (distance in m)	Weights
Commercial centres	0–100	0.57
	100–225	0.25
	225–350	0.13
	350 <	0.05
Administrative centres	0–150	0.53
	150–225	0.28
	225–350	0.13
	350 <	0.05
Educational centres	0–200	0.61
	200–350	0.30
	350 <	0.09
Cultural and recreational centre	0–200	0.83
	200 <	0.17
Hospitals	0–150	0.62
	150–250	0.28
	250 <	0.10
Roads	0–100	0.52
	100–250	0.24
	250–350	0.14
	350 <	0.09

Table 4 Pairwise comparison and calculated weights of TACs

	D1	D2	D3	D4	D5	Calculated weights
D1	1	6	3	5	6	0.49
D2	1/6	1	1/3	1/3	2	0.09
D3	1/3	3	1	4	3	0.23
D4	1/5	3	1/4	1	3	0.13
D5	1/6	½	1/3	1/3	1	0.06

Table 5 Pairwise comparison and calculated weights of roads

Roads	0–100	100–250	250–350	350 <	Calculated weights
0–100	0.55	0.62	0.53	0.40	0.52
100–250	0.18	0.21	0.27	0.30	0.24
250–350	0.14	0.10	0.13	0.20	0.14
350 <	0.14	0.07	0.07	0.10	0.09

Table 6 Pairwise comparison of TAC's and roads

	Distance from roads	Distance from trip attraction centres	Calculated weights
Distance from roads	1	1/6	0.14
Distance from trip attraction centres	6	1	0.86

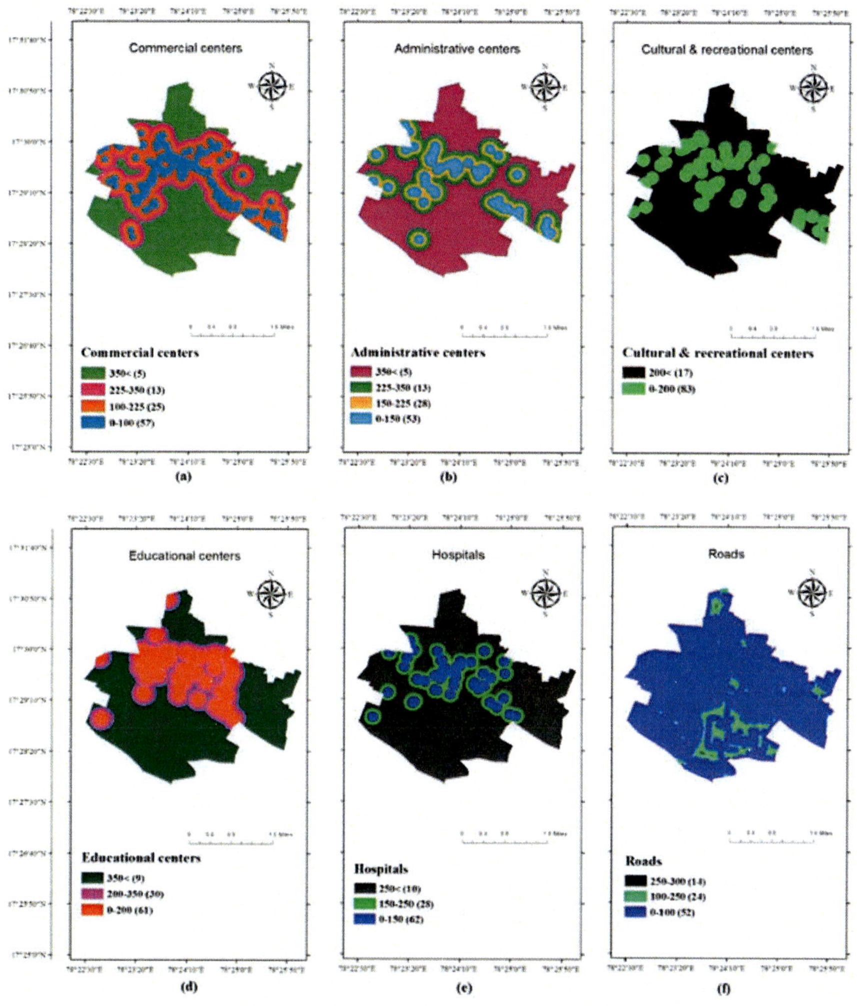

Fig. 2 **a** Distance from commercial centres. **b** Distance from administrative centres. **c** Distance from cultural and recreational centres. **d** Distance from educational centres. **e** Distance from hospitals. **f** Distance from roads

4 Results and Discussion

The study reveals that the combined usage of GIS and AHP is an effective method for solving problems related to parking site identification. The result shows a map representing the suitability index on a scale of 0–100 for parking site selection in the Kukatpally area. The maximum value of the suitability index is found to be 59. This shows the unavailability of highly suitable parking spaces in the study area. The map is divided into three zones of suitability, i.e. (0–30) not suitable or less suitable, (30–45) as moderately suitable, and (> 45) as highly suitable. The suitability index of various locations is represented in Fig. 3. According to the results obtained, very less area is available under the highly suitable region. This indicates a lack of suitable

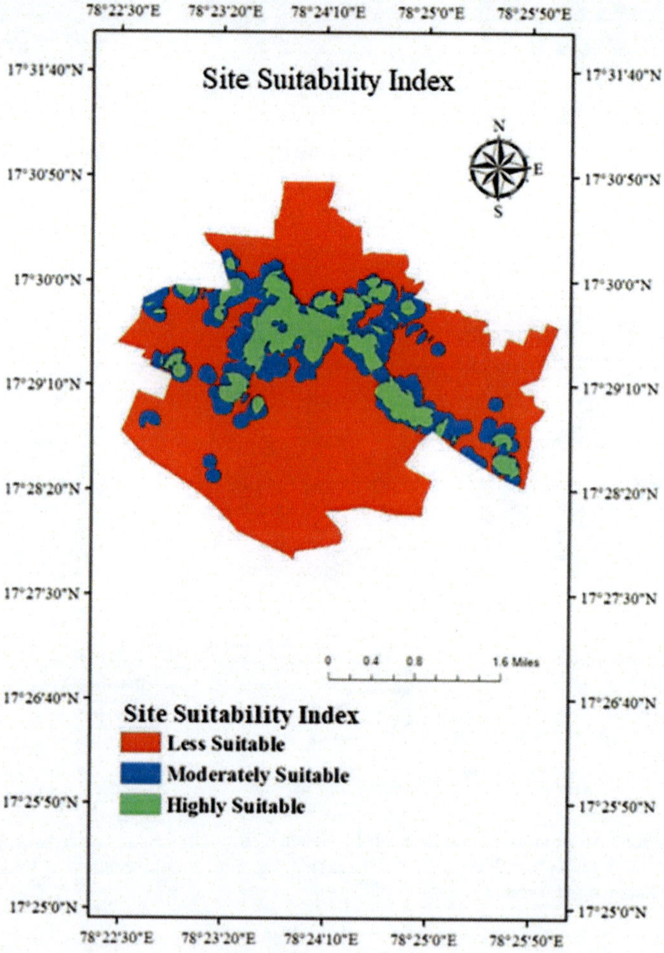

Fig. 3 Result of a study showing suitability index

areas for providing parking facilities. So further we can perform a study to find the free spaces available in the highly suitable area for establishing the new public parking.

5 Conclusions

Conventional methods of site selection for parking site facility often results in poor results due to omission of certain criteria which plays a key role in site selection for public parking. In this study, GIS and AHP methodology are used for selecting an appropriate public parking site by considering several criteria which are highly important for public parking site selection. The conclusions derived from this study are as follows.

- It was noticed that there were no existing parking sites in the study area, and this leads to on-street parking of vehicles which causes obstruction to the ongoing traffic in high traffic areas and may also cause accidents.
- The maximum value of the parking site suitability index was found to be ~ 60 on a scale of 100 and was found at very few locations in the study area. This shows the level of segregated development in the key areas where the demand for parking is expected to be high. This further resulted in less availability of area for planning a highly suitable parking facility in the study area.
- The study inherently also represents where parking demand is high, from the final suitability index map we can observe the parking demand is centralized, and it also indicates less than 10% of the existing area demands for highly efficient parking facilities.
- The metrics chosen for suitability index in this study are < 30 less suitable, 30–45 moderately suitable and above 45 highly suitable. However, it should be noted that if zone wise suitability selection based on how different facilities exist in zone will yield a different results and can offer more reliable result compared to this study.
- The results from the study are limited only to the areas where commercial development has taken place and parking facilities are needed. In the portion of the study area where less development and fewer TAC's are present no parking facilities are shown in the result.
- As mentioned above exclusion of zonal development analysis for criteria evaluation and weightage selection is the limitation of the study and can be considered in further studies for evaluating and understanding the effectiveness of this methodology.

It is also observed that the AHP methodology is effective in planning parking facilities as appropriate locations are determined in a technical manner considering all the effective criteria.

References

1. Subramani, T. (2012). Study on existing parking condition on major roads in Salem and suggestions for improvement. *IOSR Journal of Engineering, 02*, 704–710.
2. Desai, K. N., & Vaidya, V. (2017). Parking study on major corridor of urban area: a case study on Ahmedabad City. *International Journal for Scientific Research and Development, 5* (02).
3. Brierley, J. (1972). *Parking of motor vehicles*. Applied Science Publishers Ltd.
4. Klappenecker, A., Lee, H., & Welch, J. L. (2014). Finding available parking spaces made easy. *Ad Hoc Networks, 12*, 243–249.
5. Aliniai, K., Yarahmadi, A., Zarin, J. Z., Yarahmadi, H., & Lak, S. B. (2015). Parking lot site selection: an opening gate towards sustainable GIS-based urban traffic management. *Journal of the Indian Society of Remote Sensing, 43*, 801–813.
6. Pandhe, A., & March, A. (2012). Parking availability influences on travel mode: Melbourne CBD offices. *Australian Planner, 49*, 161–171.
7. Public Parking Site Selection Using GIS. https://www.scribd.com/document/59517854/Public-Parking-Site-Selection-Using-GIS.
8. Moradijoz, M., Moghaddam, M. P., Haghifam, M., & Alishahi, E. (2013). A multi-objective optimization problem for allocating parking lots in a distribution network. *International Journal of Electrical Power & Energy Systems, 46*, 115–122.
9. Kazazi Darani, S. et al. (2018). Parking lot site selection using a fuzzy AHP-Topsis framework in Tuyserkan, Iran. *Journal of Urban Planning and Development, 144* (3), 04018022.
10. Prasertsri, N., & Sangpradid, S. (2020). Parking site selection for light rail stations in Muaeng District, Khon Kaen, Thailand. *Symmetry, 12*(6), 1055.
11. Baseri, M., Ahmadi, R., Malekabadi, M., & Gandomkar, A. (2012). Site selection of public parking in Isfahan City, using AHP model. *International Journal of Humanities and Social Sciences, 6* (4), 547–550.
12. Arsovski, S. et al. (2017). Model for selection of the best location based on fuzzy AHP and Hurwitz methods. *Mathematical Problems in Engineering, 2017*.

Minimization of Risks in Highway Projects Using Buffer

Rahul Garg and Saurabh Rawat

Abstract This paper aims at identifying the highway construction delays factors and rectifies those using buffers to time and money. A buffer sizing methodology based on fuzzy risk assessment is used to minimize risk factors causing delays to the road construction activities. The input of the questionnaire is subjected to frequency and severity indexing which are then used as input for the buffer size model. A total of 25 Fuzzy if–then rules are then applied using the fuzzy model. Data reliability is checked using SPSS in which the value of Cronbach's Alpha coefficient comes out to be 0.814. The finding identifies the top 10 delay factors in highway construction, with 'preparation of natural surface' requiring the maximum buffer size of 40.64% and 'approval from authorities' showing a small buffer size requirement of 10.89%. The project manager can identify the activities requiring immediate gearing up among simultaneous activities.

Keywords Time buffer · Fuzzy logic · Highway construction · Construction management · Importance index

1 Introduction

Timely completion of a highway project is considered to be a great success which saves time, money, and quality. Sometimes due to inadequate information available and the vulnerability of activities to various risk factors projects can also delay their completion. Risk causes higher project costs and schedule delays.

A highway project must complete under less construction cost, time, and satisfy quality. The selection of the type of road projects depends on various factors including geography; however, road projects require high investment cost and long durations for construction, the risk associated with failure of sub-grade soil and skilled labor requirement for construction work make them a reason for risk analysis [1, 2]. It is evident that buffers are used in construction and only causes of time buffer in

R. Garg (✉) · S. Rawat
Department of Civil Engineering, Jaypee University of Information Technology, Solan, Himachal Pradesh 173234, India

A. K. Gupta et al. (eds.), *Advances in Construction Materials and Sustainable Environment*, Lecture Notes in Civil Engineering 196,
https://doi.org/10.1007/978-981-16-6557-8_37

construction project task durations are determined. The critical chain method and PERT (Program Evaluation and Review Technique) can be used to adUncertainties, while scheduling the activities of the project. Farag [3] developed a Fuzzy logic model to note the actual variation in the highway project and calculate buffer values for the project. Various techniques have been used with the buffer determination process so that project can be complete on time. Ghaziani [4] used Fuzzy network analysis to calculate delay both at the project and activity level. Park et al. [5] concluded that to cope with the problem of delayed project schedule reliability buffering can be used and can complete the project in a shorter duration. Although statistical analysis can be used for calculations of buffer but it is not easy to find statistical data and calculate buffer on the basis of data. Russell et al. [6] had researched causes of time buffer and duration variations in which reason for adding time buffer to project task duration is determined. Moreover, to identify the structure of time buffer factors, the author used factor analysis. The most widely used scheduling method in projects is the critical path method (CPM) because it is a deterministic approach and does not incorporate uncertainty [7–10]. Moreover, impact and frequency of risk evaluated at activity level using linguistic terms and Fuzzy scheduling calculations were performed by software, namely MS PROJECT™.

Gunduz et al. [11] identified and categorize delay factors in construction projects and using the theory of fuzzy set delay assessment model was developed. The research used Mamdani-style fuzzy rules which are well suited to human input [12]. For describing the entire model in terms of linguistic variable Fuzzy if–then rules forms were used. The primary phase of risk assessment is the identification of risk factors. Samantra et al. [13] provides a Fuzzy risk assessment model for metropolitan construction projects in which risk was categorized as a function of two parameters (a) possibility of occurrence and (b) impact. The risk of a construction project can be easily modeled by a developed structure. Project scheduling is a complex process in the construction project. Trapezoidal Fuzzy numbers pose several advantages over triangular due to their nature of determining a specific value in a better manner [14, 15]. Fuzzy theory is being employed to evaluate the uncertainty in a fast manner. Whenever an unforeseen event occurs in a project then buffer acts as a cushion to the project manager. Buffer can be added in the form of resources like time, money, and labor, etc. Buffer can be determined for an individual activity or can be assigned for a complete project and often associated with the scheduling of the project. In this paper, time buffer is used as extra time in project management to create some flexibility so that a project can be completed on time.

1.1 Objectives of the Study

The main objectives of this study are:

- To determine time buffers for uncertainties during highway construction.

- To develop the Fuzzy logic model for the interrelationship between buffer size and uncertainties.

2 Methodology

Firstly, the causes of delay which affect the activity duration of road project are identified by a literature review and also from the expert opinion. After the determination of delay causes the next step is to determine activities in the road construction projects. The activities of road construction projects were determined from the previously completed project and also by taking interviews with experts. Because road projects can be of various types and based upon material specifications activities are defined. After finding road activities the next step is to create a questionnaire for taking a survey to know the importance of these delay causes. For this five- point Likert scale ranging from very low to very high for severity and frequency, no happen to always is adopted. The questionnaire was provided with variables to rank risk factors based on their knowledge and opinions. Questionnaires were distributed to various companies to obtain the viewpoint of professionals within the construction industry. The target respondents were Site Engineers, Design Engineers, Project Manager, Contractor, and Supervisor. Respondents should have experience of more than five years in the road construction projects. To check the consistency of collected data, a reliability test is performed. Reliability is checked using SPSS. The value of Cronbach's Alpha coefficient range from 0 to 1. If the coefficient value comes out to be greater than 0.7, then the data is said to be reliable [16–18].

After collecting survey data, the next step is to calculate the frequency and severity values for delay causes. An activity network is created and an activity interrelationship is taken into consideration. In road construction projects next activity can be started when its predecessor activity is half completed. In this paper, buffers are calculated for a particular activity rather than calculating a buffer for the whole project. For admeasure, the vastness of risk fuzzy risk assessment is used. For this purpose, fuzzy logic toolbox is used in MATLAB.

The frequency of occurrence and severity value of delay factors are the input values for the buffer size model. After determining the frequency and severity of delay cause the average value of both will put in the model and the output will be the size of the buffer for delay cause which is in terms of percentage of activity duration. After adding frequency and severity values, linguistic variable, membership function, and Fuzzy if-then rules are determined. For the membership function, linear functions were used for both input and output and can be changed easily. A linear function is used here as a membership function because it is most widely used. 25 fuzzy if-then rules were found which shows the interrelationship between the input and output variable. Based upon these rules, size of the buffer for road activities is determined in the fuzzy logic toolbox in MATLAB.

Now in the final part size of the buffer is calculated with the fuzzy assessment. For this purpose, fuzzy logic toolbox for MATLAB is used. The frequency of occurrence

Table 1 25 fuzzy rules for the model

Rule	Frequency values	Connection	Severity value	Result indicator	Buffer values
1	Very rarely	AND	Very low	THEN	Very small
2	Very rarely	AND	Very low	THEN	Very small
3	Very rarely	AND	Very low	THEN	Small
4	Very rarely	AND	Very low	THEN	Small
5	Very rarely	AND	Very low	THEN	Small
6	Rarely	AND	Low	THEN	Very small
7	Rarely	AND	Low	THEN	Small
8	Rarely	AND	Low	THEN	Medium
9	Rarely	AND	Low	THEN	Medium
10	Rarely	AND	Low	THEN	Medium
11	Sometimes	AND	Moderate	THEN	Small
12	Sometimes	AND	Moderate	THEN	Medium
13	Sometimes	AND	Moderate	THEN	Medium
14	Sometimes	AND	Moderate	THEN	Large
15	Sometimes	AND	Moderate	THEN	Large
16	Often	AND	High	THEN	Small
17	Often	AND	High	THEN	Medium
18	Often	AND	High	THEN	Large
19	Often	AND	High	THEN	Large
20	Often	AND	High	THEN	Very large
21	Always	AND	Very high	THEN	Small
22	Always	AND	Very high	THEN	Medium
23	Always	AND	Very high	THEN	Large
24	Always	AND	Very high	THEN	Very large
25	Always	AND	Very high	THEN	Very large

and severity of delay cause for road projects is the input values for the buffer model. After putting these values of frequency and severity as input the output will be the size of the buffer in terms of percentage of activity duration. The values of frequency and severity are taken from 204 surveys and the mean value of it is taken for analysis. By using the Linguistic variable, membership function, and 25 fuzzy if-then rules in the Fuzzy logic toolbox, the buffer size is calculated. The fuzzy logic toolbox does not limit its number of inputs and outputs allowed. The first input is the frequency of occurrence, and the second input is the severity of occurrence for the buffer size model. The output of the buffer size model will be the size of the buffer.

Fuzzy if-then rules are shown in Table 1, Fuzzy rules are defined in model because Fuzzy models can manipulate, interpret, and utilize data that are ambiguous and uncertain. It can be used for risk assessment in the construction project where less information is available and probabilistic assessment can't be done. By applying 25 fuzzy rules, the size of buffer for a particular activity can be calculated because sometimes it is not easy to understand noisy and unclear data. 25 fuzzy rules applied to the model which shows the interrelationship between buffer size and uncertainties.

After calculating the buffer for each cause of delay in road construction, the buffer for activity can be calculated. This degree of importance is provided for each delay cause by multiplying frequency and severity which shows the contribution of each delay cause to overall activity. The formula for importance index is represented in Eq. (1).

$$I_K = F_K \times S_K \tag{1}$$

where F_k represents the frequency of occurrence of delay cause/risk factor and S_k represents the severity of delay cause.

I_k Importance value of delay cause.

The buffer size for each activity is calculated by multiplying the size of the buffer of delay caused by their importance value and then adds the multiplied values. The first input is the frequency of occurrence, and the second input is the severity of occurrence for the buffer size model. The output of the buffer size model will be the size of the buffer.

3 Data Analysis and Results

Different causes of delay were identified by a detailed literature review and were divided based on the responsible group. 210 responses were collected from a survey out of which 204 were selected and statistically evaluated. The value of the Cronbach alpha coefficient was found to be 0.814 which means the internal consistency of data is good. Respondents had experience more than 5 years were considered qualified for survey. 17% of respondents held the position of Project Manager within a Construction Company, 60% of them are engineers, 14% are contractors, and 9% are supervisors. The maximum part of responses falls under the category of experts. The distribution of the company's profile includes 66% of respondents work as contractors, 14% designer, 11% consultant, and 9% as a client. The top 5 risk factors according to the survey are shown by a bar chart in Fig. 1.

3.1 Hypothetical Project

To show the results of this methodology, a hypothetical Bitumen road project is considered. The activities in this road project are considered by expert opinion and are consider occurring in every bitumen road project. A hypothetical project is considered to show how much buffer the activity will be needed for road projects considering

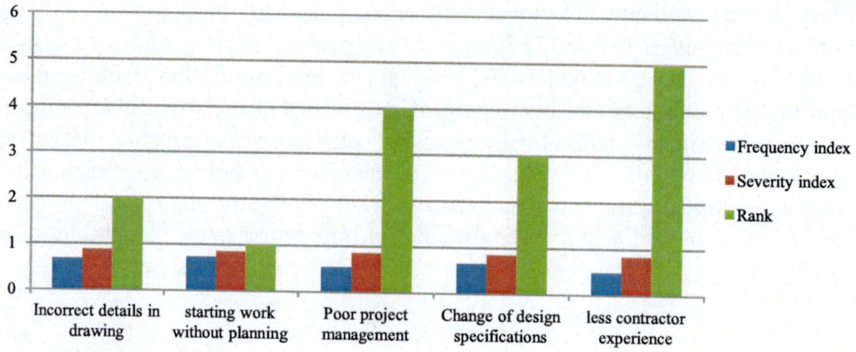

Fig. 1 Bar chart analysis of top five risk factors

risk associated with it. Buffer is added in the road activities in terms of percentage of activity duration. These road activities can be further change depending upon the requirement/specifications. The activities which are considered in the highway project are given as:

- Obtaining approvals from authority.
- Expropriate.
- Mobilization of construction material, machinery, and preparation of the site.
- Preparation of the natural surface.
- Preparation of sub-grade.
- Preparation of sub-base.
- Preparation of base course.
- Preparation of surface course.

To show how this methodology work, the first activity of this example project is considered an example. The project activities are considered so that buffer in terms of extra time can be calculated for the activities and project can be calculated within the estimated time. In the first step, the values of frequency and severity of occurrence of delay causes are considered which affects the duration of the first activity. Then, these values are entered in the model Table 2. Value of frequency and severity range from 0 to 1. In the next steps size of the buffer for the delay, causes are determined by putting these values in the model. The frequency and severity values in the form of input are entered in ruler view in MATLAB and by adding two inputs output will be the size of buffer for that particular risk. Then, buffer size for each activity is calculated by using buffer size and the importance value of delay causes and adding the multiplied values. All 25 fuzzy if-then rules have been tested and every part of the rule has fulfilled to get an output called aggregation process.

Fig. 2 shows the calculated value of buffer for frequency value of 0.721 and severity value of 0.737. The buffer size of 8.35% is obtained for the design error. After calculating buffer for each risk factor in MATLAB, buffer for each activity is calculated.

Table 2 Frequency and severity values for activity 1

Activity	Causes of delay	Frequency	Severity
Obtaining approval from authorities	Sudden change in government laws	0.4764	0.6382
	Getting permissions from government	0.433	0.7039
	Design error	0.7215	0.7373
	Not following contract conditions	0.5431	0.652
	Difference between project design specification and codes	0.5608	0.4794

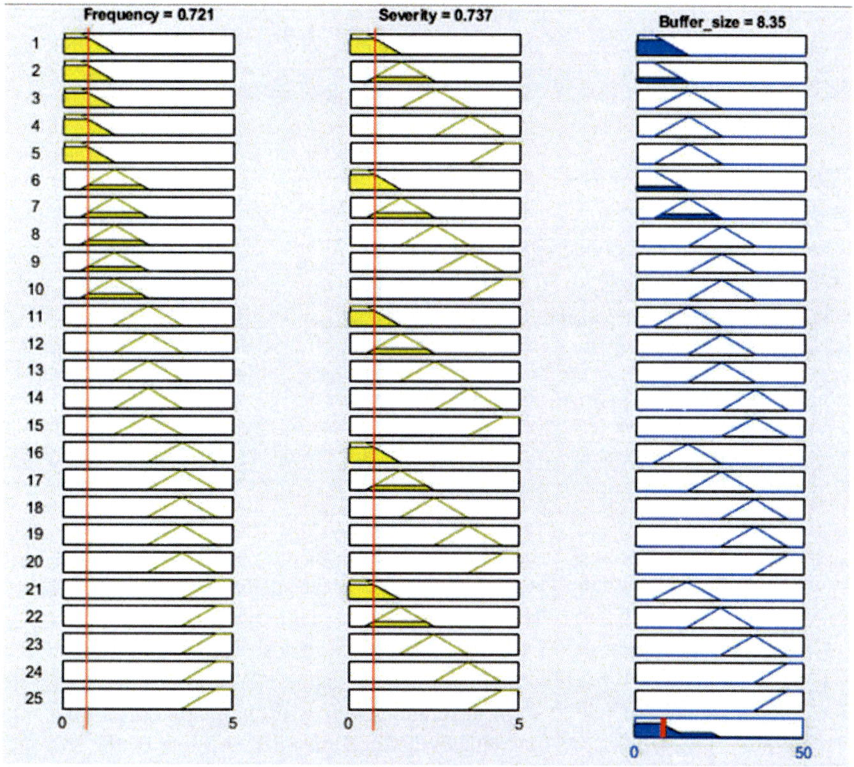

Fig. 2 Buffer sizes for input value 0.721, 0.737 from rule viewer in MATLAB

Fig. 3 shows the top 10 risk factors which have a higher buffer value. The buffer size for each activity is calculated and is shown in Table 3.

From Table 3, it can be seen that each activity has a different buffer size. In Table 3, preparation of natural surface has a bigger buffer size as compared to other activities, so the project team needs to minimize the impact of delay causes so that a

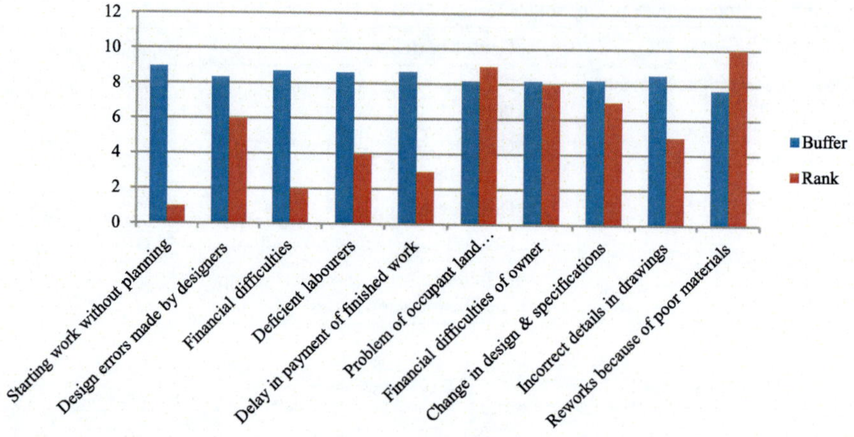

Fig. 3 Bar chart analysis of top 10 risk factors with the highest buffer value

Table 3 Buffer size for various activities

Activity name	Buffer size (%)
Obtaining approvals from authority	10.89
Expropriate	16.13
Mobilization of construction material, machinery and preparation of site	19.22
Preparation of natural surface	40.64
Preparation of sub grade	36.36
Preparation of sub base	35.79
Preparation of base course	29.24
Preparation of surface course	17.49

faster schedule can be created. In the preparation of the surface course buffer comes out to be less as compared to other layers and is reasonable because it is the top layer and the chance of failure or other risks is very less. It also shows that the activity of a higher buffer size is highly vulnerable to risk factors. 25 fuzzy rules applied to the model in MATLAB which shows the interrelationship between buffer size and uncertainties. MATLAB gives the value of buffer for a particular risk factor. Results clearly show that the higher the size of the activity buffer, the bigger will be the problem. So the project staff has to minimize the negative effect of delay cause, so that project can be completed on time without any delay.

The results clearly shows that by calculating buffer for the road activities a project can be completed within estimated time and budget. Project team will know which activity will need gearing up among other activities before the starting of project.

4 Conclusion

This study provides a methodology for a road construction project to eliminate the negative impact of delay caused in terms of time, money, and quality. This paper developed a buffer model for highway construction projects that can minimize the risk of causing delay to various construction activities. The value of Cronbach's Alpha coefficient comes out to be 0.814 and showed that the findings of the collected survey are reliable. The input of the questionnaire was subjected to frequency and severity indexing which was then used as input for the buffer size model. A total of 25 fuzzy if-then rules were applied using the fuzzy model. Fuzzy rules were applied to the model because sometimes it is not easy to manipulate ambiguous data, and fuzzy is mostly used where probabilistic assessment can't be done.

This paper identified the top 10 delay factors with 'preparation of natural surface' required maximum buffer of 40.64% and minimum requirement of 10.89% for 'approval from authorities.' Buffer size varies for each construction activity of the highway project. The buffers are calculated in terms of percentage of activity duration and by adding these in the total activity duration project can be completed in time. By calculating the buffer, the project team can identify the activities requiring immediate gearing up among simultaneous activities before the commencement of the project. The result of this study can be used for risk management, controlling, and scheduling highway projects.

References

1. Elawi, G. S. A., Algahtany, M., & Kashiwagi, D. (2016). Owners perspective of factors contributing to project delay: Case studies of road and bridge projects in Saudi Arabia. *Procedia Engineering, 145*, 1402–1409.
2. Sharma, S., & Gupta, A. K. (2021). Analysis of factors affecting cost and time overruns in construction projects. In: S. Kumar Shukla, S. N. Raman, B. Bhattacharjee, & J. Bhattacharjee (Eds.), *Advances in geotechnics and structural engineering. Lecture notes in civil engineering*, Vol. 143. Springer, Singapore. https://doi.org/10.1007/978-981-33-6969-6_6
3. Farag, M. A. M. (2010). An integration of a buffering assessment model using fuzzy logic with lean management for improving highway construction process.
4. Ghaziani, A. (2012). *A fuzzy delay assessment tool for construction projects* (Doctoral dissertation, M.S. Thesis, Civil Engineering, Middle East Technical University, Ankara).
5. Park, M., & Pena Mora, F. (2004). Reliability buffering for construction projects. *Journal of Construction Engineering and Management, 130*, 626–637.
6. Russell, M. M., Hsiang, S. M., Liu, M., & Wambeke, B. (2014). Causes of time buffer and duration variation in construction project tasks: Comparison of perception to reality. *Journal of Construction Engineering and Management, 140*, 04014016.
7. Budayan, C., Dikmen, I., Birgonul, M. T., & Ghaziani, A. (2018). A computerized method for delay risk assessment based on fuzzy set theory using MS Project. *KSCE Journal of Civil Engineering, 22*, 2714–2725.
8. Hasan, R., Suliman, S. M., & Malki, Y. A. (2014). An investigation into the delays in road projects in Bahrain. *International Journal of Research Engineering Sciences, 2*, 38–47.

9. Susmitha, R. S., Raja, K. H., & Asadi, S. S. (2018). A statistical approach for ranking of factors impacting the timeline of residential construction projects using important index method (IMPI). *International Journal of Civil Engineering and Technology, 9*, 1075–1083.
10. Vasishta, N., Chandra, S., & Asadi, S. S. (2018). Analysis of risk assessment in construction of highway projects using relative index method. *9*, 1–6.
11. Gunduz, M., Nielsen, Y., & Ozdemir, M. (2015). Fuzzy assessment model to estimate the probability of delay in Turkish construction projects. *Journal of Management in Engineering, 31*, 04014055.
12. Guettouche, M. S. (2013). Modeling and risk assessment of landslides using fuzzy logic. Application on the slopes of the Algerian Tell (Algeria). *Arabian Journal of Geosciences, 6*, 3163–3173.
13. Samantra, C., Datta, S., & Mahapatra, S. S. (2017). Fuzzy based risk assessment module for metropolitan construction project: An empirical study. *Engineering Applications of Artificial Intelligence, 65*, 449–464.
14. Zhang, J., Song, X., & Díaz, E. (2017). Critical chain project buffer sizing based on resource constraints. *International Journal of Production Research, 55*, 671–683.
15. Fayek, A. R. (2020). Fuzzy logic and Fuzzy hybrid techniques for construction engineering and management. *Journal of Construction Engineering and Management, 146*, 04020064.
16. Aziz, R. F., & Abdel-Hakam, A. A. (2016). Exploring delay causes of road construction projects in Egypt. *Alexandria Engineering Journal, 55*, 1515–1539.
17. Kazaz, A., Ulubeyli, S., & Tuncbilekli, N. A. (2012). Causes of delays in construction projects in Turkey. *Journal of Civil Engineering and Management, 18*, 426–435.
18. Kaliba, C., Muya, M., & Mumba, K. (2009). Cost escalation and schedule delays in road construction projects in Zambia. *International Journal of Project Management, 27*, 522–531.

Trend Modeling for Air Quality—An Approach

M. Goutham Priya and S. Jayalakshmi

Abstract Introduction of deleterious materials in the atmosphere is termed as air pollution. It causes waning changes in the environment distressing human life, flora, and fauna. Owing to boost in industries worldwide, there has been inflating mount in the quantity of air pollutants released into the atmosphere. Therefore, a study becomes necessary to swot up the trend in emission of the pollutant and its variation so as to predict the future. In this study, a review of various methods of evaluating the trend in air quality is discussed. Overall trend can be evaluated in two ways, viz simply by generating descriptive statistics or a deep analysis by hypothesis testing. For air quality monitoring, air quality index (AQI), fuzzy logic (ARIMA), and exceedance factor (EF) are certain factors with which the trend can be easily evaluated. Statistical trend modeling proved to be the robust method to study the trend of a phenomenon. Hence, in this paper, the role of different parametric tests like regression—simple, multiple, logistic, ordinal, multinominal, discriminant in analyzing trend in air pollutants are discussed with the help of case study and compared with the nonparametric tests like Mann–Kendall and Sen slope estimator. It was found that for trend modeling in air pollution, parametric and nonparametric tests give out that same accuracy, but the latter is fit for monotonic trend and also has good reproducibility of the results in presence of an outlier.

Keywords Air pollution · Trend modeling · Time series · Statistics

1 Introduction

Presentation of negative substances, for example, particulates, gases, and organic particles into the Earth's air, either from regular or anthropogenic source perhaps that influences environmental change, human well-being, nourishment crops, or the

M. Goutham Priya (✉)
Rajalakshmi Engineering College, Chennai, Tamil Nadu, India

S. Jayalakshmi
Institute of Remote Sensing, College of Engineering, Chennai, Tamil Nadu, India

A. K. Gupta et al. (eds.), *Advances in Construction Materials and Sustainable Environment*, Lecture Notes in Civil Engineering 196,
https://doi.org/10.1007/978-981-16-6557-8_38

common or fabricated condition is named as air contamination or air pollution. Due to economic vacillation, for better living, most of the world's population are in cities. One of the most exigent tasks of city planning is to plan the various amenities without having adverse effect on climate through poor air quality. The governing sources of air pollution are vehicular emission, biomass burning, etc. The major pollutants which causes dreadful effect on environment are NO_x, SO_x, ozone, RSPM, and CO. The world is bearing into the worst epoch of air pollution where 93% (WHO Report 2018) of the children breathe air with pollution levels exceeding their limits.

Multicity trend studies of air pollutants show the day by day disparity in air pollution levels. These fluctuations are caused by the changes in the country's economy, traffic, technology improvements, industrial activities, and many other factors. Therefore, a trend report is essential which summarizes the long-term changes in emission of air pollutants which gives comprehensive analysis on emission assessment and involve in forecast methodologies for a scrupulous period of time of study. As a consequence, methodological issues concerning time series analysis of air pollution have engrossed the thought of the scientific society and critics have raised concerns about the competency of existing models. This paper is focusing the different methods of trend modeling for studying the trend of various air pollutants by identifying, assessing, and illustrating.

2 Time Series Data and Trend Modeling

Repeated measurements of a distinct and precise quantity measured and compiled over a period of time are known as time series data. Time series can be classified into two types: viz stock series and flow series [1, 2]. A stock series is a measure of definite attributes at a point in time, whereas a flow series is measure of activity over a given period. Any time series can include three different types of movements namely recurring, drifting, and asymmetrical. Standard disparity in annual, monthly, weekly, and data is termed as recurring effect [3]. Any disparity which occurs steadily in the data which explicitly has seasonal variation is termed as drifting effect. The drifting effect seems to happen every year as it is seasonal. An irregular effect is the movement that occurred at a specific point in time, but is unrelated to a season or cycle. Therefore, a time series is used to smooth out the irregularities and to remove the serrated effect which in turn plots the trend over a long period of time for which the data is available [4].

Any change in the environmental parameters over a period of time can be identified and quantified by trend modeling. A trend is typically represented by a graph through a line or curve which depicts the trend in the gradual change of the quantity through a series of observations done on the quantity in a particular period of time for which the trend is required to study its pattern. It usually studies that the past data identifies and analyzes the trend, which is used to predict the future data. Time series data plot is the most useful tool for visualizing trend/change. The variable of interest is plotted against time. A trend line can be included to the data plot. There is much commercial software for visualizing trends some of which are Excel, Mathernatica, MATLAB, MINITAB, SAS, SPlus, SPSS, Systat, etc.

A considerable amount of literatures has been published by several researchers on trend modeling for a range of environmental parameters like soil [5], water [6–9], air, temperature [10–12], rainfall [11], etc., using different trend models. However, the purpose of this paper is to review the recent researches in trend on air pollutants.

Starting with trend of an air pollutant can be considered by simply analyzing their individual daily, monthly, seasonal, or yearly concentrations [13]. A detailed trend analysis can also be carried out by employing hypothesis testing, both parametric and nonparametric tests like F-test or T-test [14]. Community multiscale air quality (CMAQ) is a multiscale chemical transport model which can also be used to study trends in air quality motivated by meteorological simulations [15]. The validation of the model can be done by comparing the model output to several ground-based measurements. The model successfully reproduced the practical decreasing trends of SO_2, NO_2, and 8 h O_3 maxima in western countries. On the other hand, the declining trend in NO_3^- in USA was not detected by the model. Also, the model had uncertainties in NH3 emissions, and in China, the model did not incarcerate the trend of SO_2 and NO_2. In eastern Asia, the model produced a trustworthy trend of pollution index from 2005 to 2010 (5 years). Trend analysis can be done in two steps: viz detection and estimation. The outputs can be compared with linear regression method and found that both the methods give the output in the same accuracy, but the performance of Sen estimator is even more enhanced in presence of outliers than linear regression method [16].

Forecasting of air quality can be done on the basis of air quality index (AQI). Quiet few metro cities around the world use AQI to study the air quality which uses the principal component regression technique. AQI was estimated for a period of 7 years from 2000 to 2006. Forecasting is done by multiple linear regression and principal component regression model for the parameters NO_2, SO_2, RSPM, and SPM for four different seasons, summer, monsoon, post-monsoon, and winter, and the results were compared. PCM was found to better in forecasting AQI with the help of AQI of the previous day and the meteorological parameters [17]. Due to the inability of the conventional models to predict the future trend of air pollutants, a novel method SMOreg was employed for the trend prediction of Navi Mumbai such that it will help to envisage environment data and extort significant patterns from the environment data, so that it will allow the policy makers to take the counteractive measures [7].

Trend prediction using neural networks are much more accurate, as the classification is done using fuzzy logic. A prediction system for studying the trend in stock pricing in Taiwan using autoregressive integrated moving average model (ARIMA model) was developed. Using neural networks, the trend can also be detected using only the secondary difference data and even if the raw data is not available. Unlike conservative systems, on the contrast, the neural networks can be trained reliably and steadfastly from the available data. It is said that the neural networks can model any systems without any proper rules. Moreover, neural networks may be well thought-out as flexible nonlinear parameterized models where the parameters may be tailored according to the presented data [1].

Trend data analysis can also be done by scheming the exceedance factor (EF) as used by Central Pollution Control Board (CPCB) to indicate the severity of the pollution in a study area. This factor has been used to study the air quality in Bangalore from 2005 to 2011 with respect to each pollutant as compared to the standard values which results in a comprehensive work [18]. Frequency of count of exceedance of individual pollutants can also be considered to learn the trend of a pollutants in an area along with ratios of the pollutants and correlation among them to have an enhanced perceptive [19]. Recent trend studies have also been done on particulate matter (PM10 and PM2.5) using Mann–Kendall test for linear trend. The results can be validated using any other models such as positive matrix factorization (PMF) or multi-exponential fit of the data used [20].

Recently, the statistical and non-statistical methods are being widely used and have kindled the interest of the environmentalist to study the trend in air quality. The different methods of modeling of trend are moving average smoothing method, exponential weighted smoothing method, and statistical method [16]. But, as far as time series data as air pollution data is concerned, in general, there are three different ways in which they can be handled, viz smoothing methods, statistical trend analysis, and explicit trend models (Fig. 1) which are explained as follows.

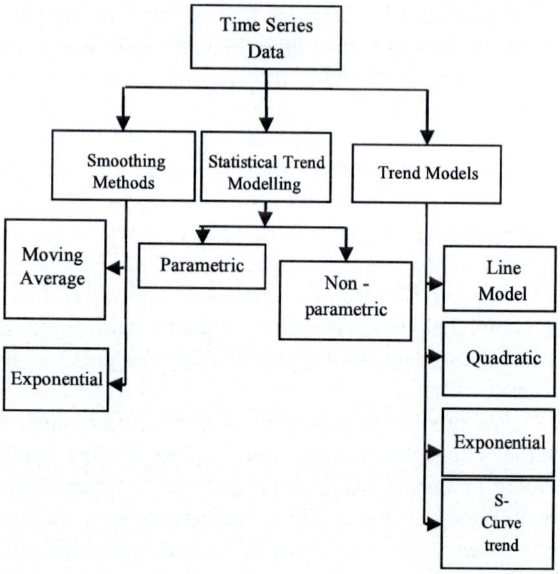

Fig. 1 Different methods of modeling time series data

3 Smoothing Methods

Moving Average

Moving averages are developed based on an average of observations, which tends to even out short-range abnormality in the data series. They are valuable if the data series remains fairly stable over time. To felicitate the studying of short-term trend, the technique consign identical weightages to the time series data and determine the overall trend in the given data [16, 21].

As the new data comes into picture, it averages the new data and the old data from which the previous average was calculated moves on. Thus, the moving average technique locks in the time period for averaging even when the data gets replaced frequently. Equation (1) gives the expression for moving average.

$$\text{Moving average} = \sum (\text{data})/n \tag{1}$$

Also, there are different methods available under moving average, viz centered moving average and weighted moving average to overcome its disadvantages like inability to capture peaks and troughs of the series. However, the weighted moving average determines the weighted average of the data in the window assigning weights from 0 to 1. The weights are assigned on the basis of the distance from the center data. Auto-regressive integrated moving average model was employed to study the trend of the pollutants CO, NO_2, SO_2, PM10, and PM2.5 in Varanasi, India. The trend was positive for PM10 and PM2.5, whereas negative trend was identified in CO, NO_2, and SO_2. An attempt was made to predict the pollutants in the year 2030 based on their trend, and it was possible only when the magnitude of trend is known [22].

Exponential Smoothing

Exponential smoothing provides an efficient method for forecasting, particularly when we have hardly any observations to furnish for conducting the forecast procedure. This method is suitable for series that move indiscriminately above and below a constant mean. The simple model of exponential smoothing is given in Equation (2)

$$F_{t+1} = F_t + \alpha(y_t - F_t) \tag{2}$$

where

F Forecast value for period,
y_t Actual value for period t,
α Smoothing constant.

However, if the series presents a trend or seasonal pattern, some modification is required. If an additional parameter, a 'linear-trend' is added to single exponential

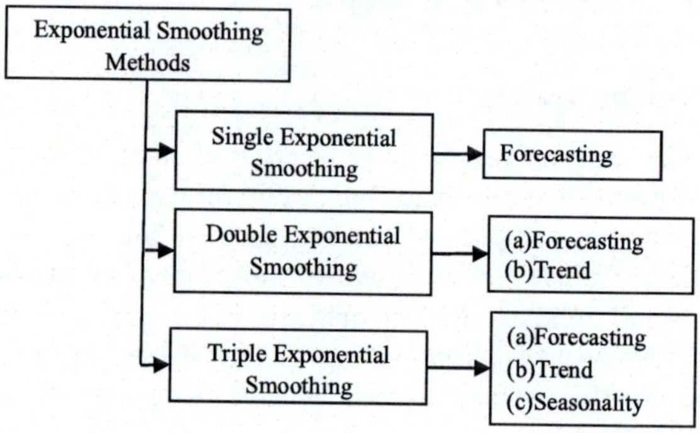

Fig. 2 Exponential smoothing methods—types and parameters

smoothing, it becomes double exponential smoothing [16, 20, 23, 24]. This model consists of two parameters one for the level of the series and another for trend. It is also called exponential smoothing with trend. The various methods involved in exponential smoothing is depicted in Fig. 2.

Triple exponential smoothing or winter's model has three parameters which includes forecasting and trend from the previous methods along with that a parameter of seasonality is also added for better accuracy and interpretation.

4 Statistical Trend Modeling

In time series, changes can happen in either of the two ways: steadily or abruptly. An abrupt change can lead to more complex form [25]. Whether the changes happen either steadily or abruptly, it affects the statistical descriptive of the time series data such as mean, median, and standard deviation. Selecting a statistic is very tricky and may depend on a number of factors such as normal distribution of data, redundant values and outliers, missing values, unexpected changes, and serial correlation [26–28]. Issues to consider while trend modeling are consistent data quality, time frame and number of samples, seasonality, type of data distribution, similar variance across data sets, minimal missing observations, exogenous variables, and differences in detection limits [11, 29–31]. Statistical trend modeling can be broadly classified into two types: parametric and nonparametric.

In parametric tests, the data are distributed normally where the parameter is regressed against time or any other spatial parameter. These tests are rarely appropriate for environmental data without resampling the data. Parametric tests are usually powerful than nonparametric as they quantify the changes in the data. It is a simple and straightforward method where the following assumptions are made about normality of the data, i.e., all the values appear or clustered near the mean. The more distant a value from the mean, the less likely it is to happen. Another assumption is about the "homoscedasticity" which means homogeneity of variance of the data or data are similar to each other or have identical straits of difference between each sample of a set to its mean [11, 31]. The data should facilitate the application central limit theorem for which a large sample is required, minimum 30 samples per parameter. It should have finite level of variance, and mean of all the samples of a parameter should be equal to the mean of the parameter itself, (e.g., linear regression).

Linear Regression

The expression for the linear regression is given by Eq. (3). This technique depicts the relationship between two parameters by fitting a linear equation. It considers one of the parameter to be explanatory variable and the other parameter to be dependent variable.

$$Y = c + mX, \tag{3}$$

where X —explanatory variable; Y—dependent variable; m—slope of the line; c—intercept.

There are different ways of recognizing the variables. Regressand, criterion, prognostic, and endogenous are the other names for the dependent variable [14, 21, 32]. In the same way, regressors, exogenous, and predictor variables are the other names of the independent variable. According to the number of dependent and independent variables, there are different types of regression, and they are summarized in Table 1.

Linear regression analysis, Mann–Kendall, Sen slope estimator, and Spearman's rho were compared for studying the trend of the pollutants SO_2 and PM in Aegean region of Turkey. It was realized that to carry out linear regression, there must be exiting linear trend in the data and also missing data are not entertained. On the contrary, Mann–Kendall and Sen slope estimator test include auto-correlated data with missing values [33].

Table 1 Types of linear regression and its characteristics

Type of regression	Dependent variable	Independent variable
Simple linear regression	1 (Interval or Ratio)	1 (interval or ratio or dichotomous)
Multiple linear regression	1 (Interval or Ratio)	2 + (interval or ratio or dichotomous)
Logistic regression	1 Binary	2 + (nominal or ratio or dichotomous)
Ordinal regression	1 Ordinal	1 + (nominal or dichotomous)
Multinomial regression	1 nominal	1 + (interval or ratio)
Discriminant regression	1 nominal	1 + (interval or ratio)

Mann–Kendall Test

To detect monotonic trend in time series data such as environment, climate, or meteorology, Mann–Kendall test, a statistical nonparametric test, is commonly used. Equations (4) and (5) give the tau value and the principle of statistic of the test. The trend in the time series dataset will be represented either as positive trend or negative trend as in the case represented in Eq. (6).

$$\tau = \frac{S}{n(n-1)/2} \tag{4}$$

where S = Mann–Kendall principle of statistic

$$S = \sum_{a=1}^{n-1} \sum_{b=k+1}^{n} \text{Sign}(X_b - X_a) \tag{5}$$

$$\text{sign}(X_j - X_i) = \begin{cases} 1 & \text{if } (X_b - X_a) > 0 \\ 0 & \text{if } (X_b - X_a) = 0 \\ -1 & \text{if } (X_a - X_b < 0 \end{cases} \tag{6}$$

where n—Number of data pairs; $X_1, X_2, \ldots, X_n$—Data points; X_a, X_b—data points at time a and b, respectively, such that ($b > a$).

If $\tau = 0$, no trend exist.

If $\tau = 0$, trend exist.

If τ takes up negative value or value less than "0", then the parameters X and Y are said to be negatively correlated. If one parameter increases, then the other decreases. Similarly, if τ takes up positive value or value greater than "0", then the parameters X and Y are positively correlated. Increasing one parameter will also increase the other. If τ is equal to "0", then no trend exist between the parameters.

Nonparametric tests are run on data that are not dependent on data distribution. They detect trend but do not quantify them. These tests are mostly applied on meteorological parameters where the data are randomly distributed. Mann–Kendall test, seasonal Mann–Kendall test, multivariate MK test, correlated seasonal MK test, partial MK test, partial correlation trend test, and Cox and Stuart trend test are some of the nonparametric tests conducted for the detection of trend in random data. Out of which to detect trend, Mann–Kendall test is mostly used [16, 34, 35]. The magnitude of trend can be estimated from Sen's slope estimator method or seasonal Sen's slope estimator method which has proven to be more successful [8, 16, 20, 34–36].

Sen Estimator method

The simplest method to determine the true slope or the magnitude of the trend detected in Mann–Kendall test is Sen estimator method. This method is applicable only when there is a significant trend.

Table 2 Comparison of linear regression and Mann–Kendall tests

Test characteristics	Name of the test	
	Linear regression	Mann–Kendall
Parametric/nonparametric	Parametric	Nonparametric
Data-driven/rank-based	Data-driven	Rank-based
Sensitivity to outliers	Sensitive	Less sensitive

$$T_i = \frac{x_j - x_k}{j - k} \tag{7}$$

where $i = 1, 2 \ldots n$

The median Q of these n values of T_i gives Sen slope which is calculated as given in Eq. (8).

$$Q = \begin{cases} \frac{\text{TN}+1}{2} & if\, N \text{ is Odd} \\ \frac{1}{2}\left[\frac{\text{TN}}{2} + \frac{\text{TN}+2}{2}\right] & \text{if } N \text{ is Even} \end{cases} \tag{8}$$

The linear trend is given in the form of Equation (9).

$$Y = Q^{*}t + B \tag{9}$$

where

Q Sen slope,
B Intercept, and
T Time.

The intercept (constant) value is calculated as depicted in Eq. (10)

$$B = \text{Median}\left[X_i - \left(Q^{*}t_i\right)\right] \tag{10}$$

The differences between the parametric and nonparametric test are summarized in the following Table 2.

5 Observations

The product of statistical trend analysis would facilitate to formulate efficient air pollution management policies as well as to broaden appropriate measures to safeguard the air pollution standards [37]. The trend analysis usually depends on the variability of the data due to various factors such as error in measurement due to nature and anthropogenic, seasonal and diurnal cycles and even an actual trend in the data. The Mann Kendall trend analysis was compared with an exceedance factor (EF) which is the ratio between the annual mean of a particular pollutant to the annual

Table 3 Zonation of EF

	Values	Air quality
Exceedance factor	> 1.5	Critical
	1–1.5	High
	0.5–1	Moderate
	< 1	Low

standard of the pollutant [38]. The study concluded that even after the determination of the EF, the trend in the EF has to be determined to study the trend of a pollutant in detail [39]. The pollutants studied were RSPM and PM. The classification of the air quality based on EF is given in Table 3.

The long-term trend in pollutants such as PM10, CO, and NO_x was studied in Malaysia with HYSPIT model. It was found that PM10 varied largely with large coefficient variation; on contrast, the pollutants CO and NO_x varied less [40]. The reason behind the variation is the trans-boundary pollution which occurs due to the biomass burning from Indonesia. The model failed to conclude the other parameters related to trans-boundary pollution which can be very well suited for the parametric test as described [41]. The conclusion of extreme value trend analysis would help to forecast the pattern and ruggedness of the future forceful events. This would assist to expand efficient safety defensive measures. Before trending, verification and documentation of all the assumptions should be done which have been tested; failing to do so may lead to erroneous results.

Multivariate analysis was carried out on various pollutants from the year 2010 to 2017 (7 years) in Madrid, Europe. The multivariate analysis like correlation analysis, principal component analysis, etc., was carried out on the study area and compared. The main disadvantage that the researcher faced is that these methods do not accept missing data and the data should form normal curve fit [42]. The major advantage of statistical trend modeling is that it gives magnitude of trend. It magnitude of the trend is represented as a number which is, justifiable and the disadvantage is that it can go erroneous and it is difficult to outline the error.

The easiest trend modeling to conduct is ordinary least square analysis which is usually not appropriate for environmental studies like air pollution as it requires data resampling. Trend can also be modeled using artificial neural networks (ANN) [43]. The generalized regression neural network (GRNN) model is an exclusively accumulated with radial basis function network (RBFN). It consists of an input layer and a hidden layer which is implied with nonlinear regression with feed forward neural network. On a supplementary approach, ARIMA models play a vital role in time series forecasting [44].

The two most largely used approaches for the trend modeling is the exponential smoothening and ARIMA models [45]. While the earlier deals with trend and seasonality of data, the latter deals an extra with autocorrelations. The performance of MK and Sen slope estimator was incorporated for the Swat river basin in Pakistan, and it showed consistent significance when compared with other trend analysis techniques [46].

Table 4 Comparison of parametric and nonparametric test

Property	Parametric test	Nonparametric test
Distribution	Normal Non-normal (central limit theorem)	Distribution free
Central value	Group mean	Group median
Variances (dispersion)	Unequal variances among the group	Equal variances among the group
Nature	Statistical power	Robust
Outliers	Affected	Not affected
Data	Continuous	Random
Example for correlation	Pearson's	Spearman's

The more robust method is nonparametric Mann–Kendall or seasonal Mann–Kendall test which gives out seasonal pattern which cannot be accomplished by any of the other methods, but it is highly demanding and computationally tedious. If there is no trend, then the data is incompetent and insufficient to detect trend. The magnitude of the trend can also be found by using Sen slope estimator [47]. The advantages and disadvantages of the parametric and nonparametric test are given in Table 4.

6 Conclusion

In this review paper, the different methods of trend analysis for time series data such as smoothing techniques, statistical trend modeling, and simple trend models to analyze the trend in air quality are discussed. It is concluded that the method of statistical trend modeling is open for research in environmental fields like climatology, meteorology, air pollution, etc. Therefore, the various statistical techniques, viz parametric and nonparametric, are discussed with the help of literatures. All the aspects of the statistical trend modeling are compared to give out a clear picture. It is found that parametric and nonparametric gives output of the same accuracy, but estimation of the magnitude of the trend of the time series data can be determined by nonparametric techniques like Sen slope estimator. Like Sen slope estimator, linear regression also gives out the same accuracy, but the latter is found to be more efficient in presence of outliers.

References

1. Wang, J.-H., & Leu, J.-Y. Stock market trend prediction using ARIMA-based neural networks.

In *Proceedings of international conference on neural networks (ICNN'96)*, Vol. 4, pp. 2160–2165.
2. Devi, B. U., Sundar, D., & Alli, P. (2013). An effective time series analysis for stock trend prediction using ARIMA model for Nifty Midcap-50. *International Journal of Data Mining & Knowledge Management Process, 3* (1), 65–78.
3. Tsai, W. T., & Lin, Y. Q. (2021). Trend analysis of air quality index (AQI) and greenhouse gas (GHG) emissions in Taiwan and their regulatory countermeasures. *Environments—MDPI, 8* (4).
4. Etuk, E. H., & Mohamed, T. M. (2014, July). Time series analysis of monthly rainfall data for the Gadaref rainfall station, Sudan, by SARIMA methods. *International Journal of Science Research in Knowledge*, 320–327.
5. Conor, J. A., Farhat, S. K., & Vanderford, M. (2012, November). *GSI Mann-Kendall toolkit for constituent trend analysis. User's manual.* GSI (Groundwater Services Inc.)
6. Yu, Y. S., Zou, S., & Whittemore, D. (1993). Non-parametric trend analysis of water quality data of rivers in Kansas. *Journal of Hydrology, 150*(1), 61–80.
7. Mumbai, N. *Trend analysis and prediction of air and water pollutants using regression algorithm SMOreg* (pp. 487–492).
8. Mustapha, A. (2013, January). Detecting surface water quality trends using Mann-Kendall Tests and Sen'S slope estimates. *International Journal of Agriculture Innovations and Research*, 108–114.
9. Helsel, D. R. (1993). Chapter 12—Trend analysis. Statistical methods in water resources. *Studies Environmental Science*, 323–355.
10. Hu, Y., Maskey, S., & Uhlenbrook, S. (2012). Trends in temperature and rainfall extremes in the Yellow River source region, China. *Climate Change, 110*(1–2), 403–429.
11. Jain, S. K., & Kumar, V. (2012). Trend analysis of rainfall and temperature data for India. *Current Science, 102*(1), 37–49.
12. Hedegaard, G. B. et al. (2008). Impacts of climate change on air pollution levels in the northern hemisphere with special focus on Europe and the Arctic. *NATO Science for Peace and Security Series C Environmental Security, Part F3* (240), 568–576.
13. Assessment, A. Q. (2010). *Trend analysis of air pollutants* (pp. 56–59).
14. Kindzierski, W. B., Chelme-Ayala, P., & Gamal El-Din, M. (2009). *Ambient air quality data summary and trend analysis.*
15. Xing, J., et al. (2015). Observations and modeling of air quality trends over 1990–2010 across the Northern Hemisphere: China, the United States and Europe. *Atmospheric Chemistry and Physics, 15*(5), 2723–2747.
16. Yadav, G., Mishra, N., Prashanthi, K., & Chaturvedi, S. (2015). *Air pollution trend analysis using Sen estimator method-A survey* (Vol. 4, No. 3, pp. 71–76).
17. Kumar, A., & Goyal, P. (2013). Forecasting of air quality index in Delhi using neural network based on principal component analysis. *Pure and Applied Geophysics, 170*(4), 711–722.
18. Trodd, N., & Santos, G. N. (2014). *Air pollution in Bangalore, India : A six-year trend and health implication analysis air pollution in Bangalore, India : A six—Year trend and health*, August 2015.
19. Biswas, J., Upadhyay, E., Nayak, M., & Yadav, A. K. (2011). An analysis of ambient air quality conditions over Delhi, India from 2004 to 2009. *Atmospheric and Climate Sciences, 01*(04), 214–224.
20. Pandolfi, M., et al. (2016). Trends analysis of PM source contributions and chemical tracers in NE Spain during 2004–2014: A multi-exponential approach. *Atmospheric Chemistry and Physics, 16*(18), 11787–11805.
21. Mills, T. C., & Patterson, K. D. (2015). Modelling the trend: The historical origins of some modern methods and ideas. *Journal of Economic Surveys, 29*(3), 527–548.
22. Jaiswal, A., Samuel, C., & Kadabgaon, V. M. (2018). Statistical trend analysis and forecast modeling of air pollutants. *Global Journal of Environmental Science and Management, 4*(4), 427–438.

23. Bostani, A., Salahedin, M., Rahman, M. M., & Khojasteh, D. N. (2017). Spatial mapping of soil properties using geostatistical methods in the Ghazvin plains of Iran. *Modern Applied Science, 11*(10), 23.
24. Jafari-samimi, A., Shirazi, B., & Fazlollahtabar (2007). A comparison between time series, exponential smoothing, and neural network methods to forecast. *Iranian Economic Review, 12* (19), 19–35.
25. Rai, R., Rajput, M., Agrawal, M., & Agrawal, S. B. (2011). Gaseous air pollutants: A review on current and future trends of emissions and impact on agriculture. *Journal of Scientific Research, 55*(771), 1.
26. Bayazit, M., & Önöz, B. (2007). To prewhiten or not to prewhiten in trend analysis? *Hydrological Sciences Journal, 52*(4), 611–624.
27. Daneshvar Vousoughi, F., Dinpashoh, Y., Aalami, M. T., & Jhajharia, D. (2013). Trend analysis of groundwater using non-parametric methods (case study: Ardabil plain). *Stochastic Environmental Research and Risk Assessment, 27* (2), 547–559.
28. Yue, S., & Wang, C. Y. (2002). Applicability of prewhitening to eliminate the influence of serial correlation on the Mann-Kendall test. *Water Resources Research, 38* (6), 4-1–4-7.
29. Pohlert, T. (2016). Non-parametric trend tests and change-point detection. *R Package*, 26.
30. Park, S. (2015). Time-series analysis of satellite-measured vegetation phenology and aerosol optical thickness over the Korean peninsula. *International Archives of the Photogrammetry, Remote Sensing and Spatial Information Science—ISPRS, 40* (7W3), 231–235.
31. Morell, O. & Fried, R. (2009). On nonparametric tests for trend detection in seasonal time series. In *Statistical Inference, Econometric Analysis and Matrix Algebra* (pp. 19–39).
32. Calculation, T. (1995). Linear regression analysis for STARDEX. *Cru.Uea.Ac.Uk.*
33. Cukurluoglu, S., & Bacanli, U. (2018). Trend analysis of the sulfur dioxide and particulate matter concentrations in the aegean region, Turkey. *International Journal of Engineering Science, 7*(9), 64–74.
34. Rahman, A., & Begum, M. (2013). *Application of non parametric test for trend detection of rainfall in the largest island of Bangladesh* (Vol. 2, No. 2, pp. 40–44).
35. Mondal, A., Kundu, S., & Mukhopadhyay, A. (2012). Rainfall trend analysis by Mann-Kendall test: A case study of north-eastern part of Cuttack District, Orissa. *International Journal of Geology, Earth and Environmental Sciences, 2*(1), 2277–208170.
36. Nury, A. H., & Hasan, K. (2016). Eer-21-1-58.Pdf (Vol. 21, No. 1, pp. 58–68).
37. Jain, S., & Mandowara, V. L. (2019). Study on particulate matter pollution in Jaipur City. *International Journal of Applied Engineering Research, 14*(3), 637–645.
38. Latif, M. T., Abidin, E. Z., & Praveena, S. M. (2015). The assessment of ambient air pollution trend in Klang Valley. *World Environmental, 5*(1), 1–11.
39. Thakur, A. (2017). Study of ambient air quality trends and analysis of contributing factors in Bangalore, India. *Oriental Journal of Chemistry, 33*(2), 1051–1056.
40. Wang, L., Wang, J., Tan, X., & Fang, C. (2020). Analysis of NOx pollution characteristics in the atmospheric environment in Changchun city. *Atmosphere (Basel), 11* (1).
41. Sentian, J., Herman, F., Yih, C. Y., & Hian Wui, J. C. (2019). Long-term air pollution trend analysis in Malaysia. *International Journal of Environmental Impacts. Mitigation Recovery, 2* (4), 309–324.
42. Núñez-Alonso, D., Pérez-Arribas, L. V., Manzoor, S., & Căceres, J. O. (2019). Statistical tools for air pollution assessment: Multivariate and spatial analysis studies in the Madrid region. *Journal of Analytical Methods in Chemistry, 2019*, 1–9.
43. Lang, P. E., Carslaw, D. C., & Moller, S. J. (2019, April). A trend analysis approach for air quality network data. *Atmospheric Environment X, 2*, 100030.
44. Jamaati, H., Attarchi, M., Hassani, S., Farid, E., Seyedmehdi, S. M., & Pormehr, P. S. (2018). Investigating air quality status and air pollutant trends over the metropolitan area of Tehran, Iran over the past decade between 2005 and 2014. *Environmental Health and Toxicology, 33* (2), e2018010.
45. Shikwambana, L., Mhangara, P., & Mbatha, N. (2020, April). Trend analysis and first time observations of sulphur dioxide and nitrogen dioxide in South Africa using

TROPOMI/Sentinel-5 P data. *International Journal of Applied Earth Observation and Geoinformation, 91*, 102130.

46. Ahmad, I., Tang, D., Wang, T., Wang M., & Wagan, B. (2015). Precipitation trends over time using Mann-Kendall and spearman's Rho tests in swat river basin, Pakistan. *Advances in Meteorology.*
47. Olstrup, H., Forsberg, B., Orru, H., Nguyen, H., Molnár, P., & Johansson, C. (2018). Trends in air pollutants and health impacts in three Swedish cities over the past three decades. *Atmospheric Chemistry and Physics, 18*(21), 15705–15723.

Traffic Analysis on Intersection Using PTV Vissim

Pranjal Sharma, Ashok Kumar Gupta, and Akash Bhardwaj

Abstract As India is in its developing stage and the traffic on the other side in India is very heterogeneous or mixed in its nature and the average growth rate of vehicles in India is about 8%. With the increase rate of urbanization in India it will lead to the considerable traffic and travel growth on the roads which will result in vehicular delays, long queues and traffic congestion. So, in this paper with the help of traffic simulation software, i.e. VISSIM, three simulation of an unsignalized intersection {Dadour and Una-Jahu, Nerchowk Rd. (NH-21),H.P} will be analyzed and will compare them on the basis of vehicular delays and long queues. These three simulation will be analyzed on the basis of real world traffic data which is less from the expectations due to the pandemic covid-19, theoretical traffic data (increase in real data by 30%) and theoretical traffic data {with traffic signals as theoretical data follows warrant 1 (Min. Vehicular Volume) shown in IRC:93:1985}. Result showed that with increase in vehicular data there was not so much variation in vehicular delays, whereas there was an increase in long queues or queue stops and whilst third simulation (with traffic lights) is done it shows that it overcomes the queue stops of the intersection.

Keywords Vissim · Intersection · Simulation

1 Introduction

Road intersections are perilous component of a road section. They are normally the most important bottleneck to smooth flow of traffic and a major crash (accident) point. In both urban and rural areas, the general principles of design are same. The difference lies only in the sight distance available, design speed, restriction on available land and the availability of pedestrians and cyclists in large volume in urban areas. With the increase rate of urbanization in India, it will lead to the considerable traffic and travel growth on the roads which will result in vehicular delays, long

P. Sharma (✉) · A. K. Gupta · A. Bhardwaj
Department of Civil Engineering, Jaypee University of Information Technology, Solan, H.P. 173234, India

A. K. Gupta et al. (eds.), *Advances in Construction Materials and Sustainable Environment*, Lecture Notes in Civil Engineering 196,
https://doi.org/10.1007/978-981-16-6557-8_39

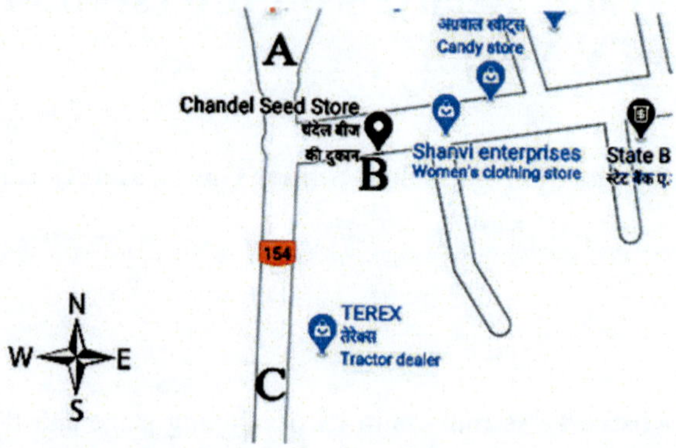

Fig. 1 Dadour and Una-Jahu, Nerchowk Rd. (NH-21), H.P

queues and traffic congestion [1]. Due to the increase in traffic and increasing technology, traffic engineers are mostly using microscopic simulation models in recent years. These simulation purposes are to analyse and optimize traffic flows without disturbing the existing traffic, or put people out of the level of risk. Unsignalized intersection {Dadour and Una-Jahu, Nerchowk Rd. (NH-21)} is adopted for the simulation because there is always misjudgment of gaps in traffic which leads to accidents, the area is prone to accidents and have inappropriate intersection traffic control. The intersection is in the shape of T shown in Fig. 1.

PTV VISSIM is a microscopic multipurpose traffic simulator based on the behaviour of traffic. There are many more simulation software's but VISSIM gives more accurate result as compared to other software's. Mostly, the software is adapted for traffic engineers. A scientific or mathematical model is required in any traffic simulator to represent the transportation supply framework simulating the organizational and technical angles of the physical transportation supply. Second, a demand model has to be created to model the demand of persons and vehicles. Travelling on the supply system, unlike macroscopic transport models, traffic control must be modelled exceptionally detailed depending on the supply and demand [2]. Lee et al. [3] aim's to investigate effectiveness and possible causes of countermeasures by using comparison sites or before and after studies and in all cases it shows a reduction in red light area (RLA) frequency. The largest reduction is done by RLR camera (− 60%) in crossing conflicts and in the other case where yellow signal interval is increased shows reduction of 12.8% of rear-ends conflicts. Paul et al. [4] proposes a methodology to calibrate VISSIM model under heterogeneous traffic condition for unsignalized intersection. Result shows that the optimum values and calibrated parameters are obtained by reducing the error between the field gap time and the simulated. By using gap data, the calibrated model is validated for other time period which is not implemented in the calibration and the observed error is less than 5%. So, it shows that the proposed model can be applicable for whole day traffic of the site as well as the site having similar traffic and road conditions.

Malim et al. [5] aim was to model the traffic flow at persiaran kayangan and persiaran permas intersection, section 7, Shah Alam by using any simulation software. The model was introduced to find out the suitable timing of green lights of traffic that minimize the average time at the intersection and overcome the congestion of traffic. Result showed that best green time of lights at the intersection was 75, 130, 100 and 120 s and the lowest average time at intersection was 55.65 seconds. Zhandong et al. [6]) uses control variables and dichotomy to obtain the thresholds as vehicles queuing starts at off-ramp using simulation in VISSIM software. Different control mode critical traffic volume threshold can be obtained. Experiment site is taken as shoushan intersection of south second ring expressway in Fuzhou. The simulation comparison analyses are introduced with different control modes on account of the different stages of the comprehensive zoning map. The result shows that the optimal control methods are at free flow stage using non-control mode, during the stage of congestion and amble level. Second, uses induction control and for the stage of highest congestion main road vehicles have priority or close off-ramp.

2 Objectives of the Study

- To identify the problems occurring with the increasing traffic on an intersection {Dadour and Una-Jahu, Nerchowk rd. (NH-21), H.P} through VISSIM.
- To compare and overcome the identified problems using PTV VISSIM.

3 Methodology

3.1 First Simulation (with Real Data)

Step 1: To replicate any model in VISSIM basic requirement is data which is collected from the intersection {Dadour and Una-Jahu, Nerchowk Rd. (NH-21)} by using videography method as it gives a permanent record of volume count, data can be cross checked and quality can be ensured (Table 1) and for the speed of vehicle a stretch near intersection is taken about 120 m in distance and vehicle took 10 sec. to complete the distance. Therefore, using formula (1), speed of vehicle comes out to be 12 m/s or 43.2–45 km/h.

$$\text{Speed} = \text{Distance}/\text{time} \tag{1}$$

This data is less as per expectations due to the Pandemic (Covid-19).

Step 2: Second stepin VISSIM is to create link and connectors which are the basic elements to design any road or intersection and model should always designed to the scale by tracing manually from the smart map in VISSIM.

Table 1 Volume of vehicle

S. No	Time A.M/P.M	(NH 21) Major road Veh./h	Dadour minor road Veh./h
1	9:00–10:00 A.M.	640	144
2	10:00–11:00	597	205
3	11:00–12:00	780	180
4	12:00–1:00	778	190
5	1:00–2:00	670	210
6	2:00–3:00	728	201
7	3:00–4:00	688	185
8	4:00–5:00	780	180
9	5:00–6:00 P.M.	798	207
		Average = 720 approx	Average = 190 approx

Step 3: As vehicles will turn on the intersection so there speed should be less, so for that case provide reduce speed area and use this option from the Network objects bar on left side of VISSIM and then select area where reduction of the speed of vehicle is required and set speed to 0-10 km per hr. as per the needs.

Step 4: Select input vehicle option from the bar, here add vehicles on the lane from the data which is already collected and shown in Table 1. After vehicle input routing of vehicle is also necessary. It is also done from the bar with option vehicle routing. With the help of this option select the route of vehicles by filling the dialogue box with relative flow data gathered using videography methodology shown in Fig. 2.

Step 5: After putting all data which is required for the VISSIM simulation find out the delay vehicles and queue length of vehicle by running the simulation.

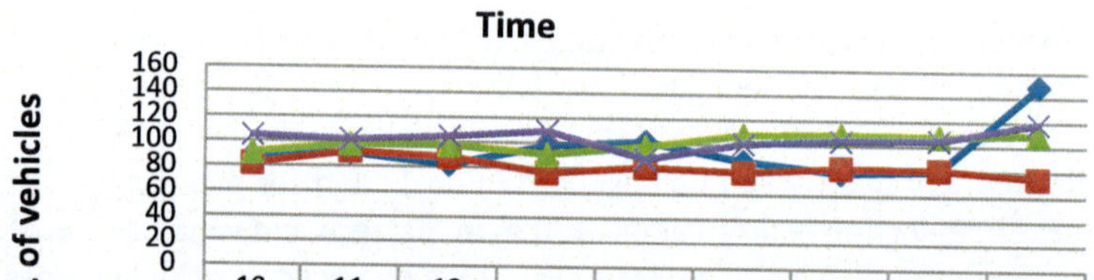

	10 A.M	11 A.M	12 P.M	1 P.M	2 P.M	3 P.M	4 P.M	5 P.M	6 P.M
	9 A.M	10 A.M	11 A.M	12 P.M	1 P.M	2 P.M	3 P.M	4 P.M	5 P.M
North-East	88	91	82	98	102	87	78	80	149
South-East	82	92	88	76	81	78	82	81	75
East-South	91	98	98	91	98	108	109	108	110
East-North	105	102	105	110	88	101	104	105	119

Fig. 2 Relative flow of vehicles

3.2 Second Simulation (Theoretical Data Increase of 30% in Real Data)

In first simulation, the vehicular volume data used was less due to the pandemic (covid-19) but in second simulation that value is increased by 30% so that it can follow warrant 1 (Min. Vehicular Volume) shown in IRC:93:1985 and will also help in third simulation. Rest all the steps are same as shown in first simulation process.

3.3 Third Simulation (Theoretical Data Increase of 30% in Real Data with Traffic Lights)

Data will be same as in second simulation but here as data follows warrant 1 (Min. Vehicular Volume) showed in IRC: 93:1985. So, design traffic lights for the unsignalized intersection, islands are not possible here because land area is not that much available. First, find out the green time for three phases as this intersection which is divided into three phases, so signal design timing includes:

Data: T-shape unsignalized intersection, Total 3 phases major street width (Road A and Road C) = 7m (2 lane) Minor street width (Road B) = 7m (2 lane)

Volume of Road A = 940 veh./h. Volume of Road B = 250 veh./h. Volume of Road C = 940 veh./h.

Pedestrian walking speed = 1.2m/s (from IS: 93:1985)

Step 1: Pedestrian crossing time

$$\begin{aligned}\text{For Road A} &= \text{Width of lane/Pedestrian walking speed} + 7\\ &= 3.5/1.2 + 7 = 9.91 \sim 10\,\text{s}.\end{aligned} \tag{2}$$

For Road B = Road C = 10 s.

Step 2: Minimum Green time for traffic

As minimum pedestrian crossing time for Road A is 10 s means min. green time for road B traffic is 10 s and same for road A & Road C.

$$\begin{aligned}\text{Green time Road A\&C} &= (\text{volume of A/volume of B})^{*}\text{Min.green time road A}\\ &= (940/250)^{*}10 = 37.6 - 38\text{ s}\end{aligned} \tag{3}$$

Adding clearance amber and initial amber of 2 s for each phase (total 3 phases). So, total cycle time = (2+10+2) + (2+38+2) + (2+38+2) = 98 s According to the IS code cycle time is always in multiple of 5 so new cycle time = 100 ss. The extra 2 s can be apportioned to the green time of Road B.

Step 3: Check for clearing the vehicles arrived during green phases

According to IRC: 93:1985, it is assumed that on each approach first vehicle will take 6 s to start after the stop on signal and all other vehicles behind that will constant headway of 2 s each.

No. of vehicles per lane per cycle of 100 s for Road A and C = 940/100 = 9.4–10 Minimum green time for Road A and C on above assumptions = 6 + (10 − 1) 2 = 24 s. Therefore, 24 s is less than the calculated green time for A and C which is 38 s. So, it is safe.

No. of vehicles per lane per cycle of 100 s for Road B = 250/100 = 2.5–3 Minimum green time for Road B on above assumptions = 6 + (3 − 1) 2 = 10 s.

Therefore, 10 s is less than the calculated green time for B which is 12 s. So, it is safe.

Step 4: Check for optimum cycle length by Webster's Equation 4:

$$C_0 = 1.5L + 5/(1 - Y_i) \tag{4}$$

where, C_0 = optimum cycle length; L = Loss of time in one cycle = $2N + R$, R is generally 16 s. N = No. of phases; R = Total red time; Y_i = YA + YB+YC.

To find Green signal use formula (5) as shown below:

$$G_\Delta = \frac{Y_a(C_0 - L)}{Y_i} \tag{5}$$

where, G_Δ = Green time for phase 1 and similar for other phases. Y_a = Na/Sa; Na = Vehicle accommodated on Road A in 1 h/lane; Sa = Saturated volume of vehicles accumulated on road A per hour per lane given in IRC: 93:1985.

Optimum cycle length comes out to be 290 s which is not suitable for the signal design because it will delay vehicles so for simulation Minimum 100 s is adopted so; Table 2 values will be followed for the third simulation traffic signal timing.

Table 2 Signal timing

Signal timing	Initial amber (s)	Green time (s)	Clearance amber (s)	Red time	Cycle length (s)
Major street A and C	2	38	2	58	100
Minor street B	2	12	2	84	100

Table 3 Delay measurement

TIMEINT	STOPDELAY (s)	STOPS (no.)	VEHDELAY (s)	VEHS
0–3600	0	0	5.66	12
0–3600	0	0	8.25	46
0–3600	0.52	0.46	5.88	13
0–3600	0.4	0.45	6.04	20
0–3600	0.06	0.06	14.28	54
0–3600	0.06	0.06	13.41	17

Table 4 Queue counters

SIMRUN	TIMEINT	QCOUNTER	QLEN (m)	QLENMAX (m)	QSTOPS (no.)
1	0–3600	1	0.88	25.56	13
1	0–3600	2	23.56	46.05	67
1	0–3600	3	8.18	55.23	44

4 Results

4.1 First Simulation (With Real Data)

After running first simulation output of delay measurements are shown in Table 3 and queue stops shown in Table 4.

4.2 Second Simulation (With Theoretical Data)

After running the simulation output of delay measurements are shown in Table 5 and queue stops shown in Table 6.

Table 5 Delay measurement

TIMEINT	STOPDELAY (s)	STOPS (no.)	VEHDELAY (s)	VEHS
0–3600	0.21	0.13	13.3	16
0–3600	0.12	0.09	14.79	58
0–3600	0.36	0.33	6.71	18
0–3600	0.23	0.36	6.71	22
0–3600	0.07	0.05	15.97	57
0–3600	0.07	0.11	13.91	19

Table 6 Queue counters

SIMRUN	TIMEINT	QCOUNTER	QLEN (m)	QLENMAX (m)	QSTOPS (no.)
1	0–3600	1	2.34	27.99	26
1	0–3600	2	29.31	46.06	78
1	0–3600	3	27.35	60.82	71

Table 7 Delay measurement

TIMEINT	STOPDELAY (s)	STOPS (no.)	VEHDELAY (s)	VEHS
0–3600	15.8	0.44	19.36	16
0–3600	11.31	0.34	14.26	59
0–3600	22.41	0.71	28.32	17
0–3600	35.9	0.91	43.44	22
0–3600	12.2	0.33	14.9	72
0–3600	10	0.24	12.14	21

Table 8 Queue counters

SIMRUN	TIMEINT	QCOUNTER	QLEN (m)	QLENMAX (m)	QSTOPS (no.)
1	0–3600	1	1.58	26.71	24
1	0–3600	2	15.61	46.14	36
1	0–3600	3	16.35	55.31	46

4.3 Third Simulation (With Theoretical Data and Traffic Lights)

After running the simulation output of delay measurements are shown in Table 7 and queue stops shown in Table 8.

Results shows that with increase in vehicular data it didn't show so much variation in vehicular delays, whereas there was an increase in long queues or queue stops, whilst comparing simulation 1st and 2nd data and when third simulation (with traffic lights) is done it shows that not so much variation in delay measurement but it overcomes the queue stops, whilst comparing 2nd and 3rd simulation data.

5 Conclusion

The objective of the study was to identify the problems occurring with the increase in traffic on an unsignalized intersection and to overcome those problems through VISSIM. As, the traffic in India is heterogeneous and mixed in its nature, travel growth on the road will lead to vehicular delays and long queues. To study these two parameters, three simulations were done and whilst comparing first two simulation results, it shows that with the increase in traffic volume there was not so much variation in vehicular delays but the no. of queue stop increases as in 1st simulation queue stops on each counter were 13, 67, 44, respectively, whereas in 2nd simulation result, queue stops increases to 26, 78, 71, respectively, and when 3rd simulation is done with traffic signal implementation as it follows warrant 1 {(Min. Vehicular Volume) for which cycle length of 100 s is taken for traffic signal and shows that there was again not so much variation in vehicular delays but it reduces the no. of queue stops to 24, 36, 46, respectively. So, it represents that with the implementation of traffic signal on intersection {Dadour and Una-Jahu, Nerchowk Rd. (NH-21), H.P} queues stop parameter can be reduced.

References

1. Arkatkar, S., Mitra, S., & Mathew, T. (2019). Department of Civil Engineering, Sardar Vallabhbhai National Institute of Technology, Surat, India, Civil Engineering Department, Indian Institute of Technology, Kharagpur, India, Department of Civil Engineering, Indian Institute of Technology Bombay, Mumbai, India.
2. Fellendorf, M., & Vortisch, P. (2010). Microscopic traffic flow simulator VISSIM. In Fundamentals of traffic simulation (pp. 63–93). Springer, New York, NY.
3. Lee, C., So, J., & Ma, J. (2018). Evaluation of countermeasures for red light running by traffic simulator–based surrogate safety measures. *Traffic injury prevention, 19*(1), 1–8.
4. Paul, M., Charan, V., Soni, V., & Ghosh, I. (2018). Calibration methodology of microsimulation model for unsignalized intersection under heterogeneous traffic conditions. *Urbanization challenges in emerging economies: Energy and water infrastructure; transportation infrastructure; and planning and financing* (pp. 618–627). American Society of Civil Engineers.
5. Malim, M. R., Halim, F. A., & Abd Rahman, S. S. (2019). Optimizing traffic flow at a signalized intersection using simulation. *Malaysian Journal of Computing, 4* (2), 261–269.
6. Zhandong, Z., Shaohui, C., Yanquan, Y., Aixiu, H., & Xinyi, Z. (2016). VISSIM simulation based expressway exit control modes research. *Procedia Engineering, 137*, 738–746.
7. Congress, I. R. (1994). *Guidelines for the design of at grade intersections in rural and urban areas.* IRC SP, 41-1994.
8. IRC 93:1985. (1985). Guidelines for design of signalized road intersections. Indian Roads Congress, New Delhi, India.
9. Bloomberg, L., & Dale, J. (2000). Comparison of VISSIM and CORSIM traffic simulation models on a congested network. *Transportation Research Record, 1727*(1), 52–60.
10. Lownes, N. E., & Machemehl, R. B. (2006, December). VISSIM: A multi-parameter sensitivity analysis. In *Proceedings of the 2006 Winter Simulation Conference* (pp. 1406–1413). IEEE.

Parametric Strength of Sustainable Concrete Using Fly Ash, GGBS and Recycled Aggregates as Per Taguchi's Approach

Yaman Hooda, Sunita Bansal, and Anjali Gupta

Abstract With the advancement in the building industry, the construction rate of concrete structures is increasing which directly impacts the increasing demand of the production and usage of concrete. One of the main points to be noted that concrete uses a lot of naturally available resources; making it as a non–eco-friendly material. Thus, the ingredients of the concrete should be such that they do not cause any harm to the environment, by partially replacing the main ingredients with the industry by–waste products so as to make it a "sustainable" concrete. This study focuses on the partially replacement of fine aggregates, coarse aggregates and cement with GGBS, recycled concrete aggregates and fly ash, respectively, with different proportions so as to determine the effect on its mechanical strength properties. The percentage replacement is obtained by considering Taguchi's Approach, which uses a set of orthogonal arrays to determine the different trial mixes for the determination of the strength of "sustainable concrete" with varying proportions of fly ash (FA), GGBS and recycled concrete aggregates. The strength in compression, tension and flexure of sustainable concrete is measured, depicting that by partially replacing the main ingredients of the concrete, the mechanical strength is almost increased by nearly 75, 60 and 35% at 7 days and 40, 60 and 30% at 28 days, respectively, in compression, tension and flexure. The results than traditional concrete mix, and thus making the by—products more useful and environment a bit safer.

Keywords Sustainable concrete · FA · GGBS · Recycled concrete aggregates · Parametric strength

---

Y. Hooda (✉) · S. Bansal · A. Gupta
Department of Civil Engineering, Manav Rachna International Institute of Research and Studies, Faridabad, Haryana, India
e-mail: yaman.fet@mriu.edu.in

S. Bansal
e-mail: sunitabansal.fet@mriu.edu.in

A. Gupta
e-mail: anjaligupta.fet@mriu.edu.in

A. K. Gupta et al. (eds.), *Advances in Construction Materials and Sustainable Environment*, Lecture Notes in Civil Engineering 196,
https://doi.org/10.1007/978-981-16-6557-8_40

1 Introduction

Concrete is known to be as a construction material, used extensively; manufactured by blending its basic ingredients of fine aggregates, coarse aggregates, cement and water. These constituents are added either in a desired pre-fixed proportion or one has to determine if the mix has to be prepared for the high-grade concrete. When all these constituents are mixed, they allowed to set and cure so as to make a hard–solid mass material, called concrete. The hard-solid mass like structure of concrete is due to the chemical reaction and heat produced during the process of mixing of constituents, between the cement, aggregates and water. Since this reaction continues for a lengthier period, the concrete mix gain its strength with age. Therefore, concrete is considered to be a composite material where the finer particles tend to fill the voids of the coarser particles to acquire strength, thus cement filled the voids of fine aggregates (or sand) and fine aggregates filled the voids of coarser aggregates. The cement–water paste which is formed during the process of mixing, also fill the voids of aggregates, in addition to coat the coarse aggregates and fine aggregates, binding the coarse and fine aggregates together and thus, making a compact concrete mass in which the aggregates are fixed together in a dense form. Thereby, it is well understood that the strength of the concrete mix directly depends on its constituents, and their respective physical, chemical and mechanical properties. As the technology advances, researches showed that properties of the concrete mix can be improved by the addition of some special constituents. Taking the example of FRC, in which different types of fibers are added to the concrete mix to enhance its properties, as the fibers restrict the cracks which are present in the concrete mix at micro-level, resulting in greater strength, durability and stiffness. Since the fibers are added as a reinforcement, the resulting concrete is called fiber-reinforced concrete. If the constituents of the concrete mix are replaced by some industrial by-products that concrete mix is considered as a green concrete or sustainable concrete. In sustainable concrete, one or all the constituents of the concrete mix is replaced by the waste products from the industry, so that the consumption of basic constituents of the concrete is less and the waste product can be used effectively. Some of the notable advantages of the green concrete are of using a great part of the recycled aggregates, as they may be environmental harmful if not employed correctly, which in turn, makes the structure more durable and low requirements in terms of maintenance.

For the purpose of the modification of the properties of concrete in a better way, constituents of concrete mix are replaced partially, which are available in the market as by-products or waste products. The by-products which are available for the replacement of cement, exhibit an important property through which they can pay the role of cement clinker, without negotiating the cement role in the construction industry. Moreover, such by-products minimize the process of clinker production, which in turn have an advantage over the conventional cement, in saving raw materials, fuel and emission. The examples of the by-products available into the market as a partial replacement are "silica fumes, GGBS, RHA, FA, Sugarcane Bagasse, Wood Sawdust and many more." Similarly, for fine aggregates (sand), the partially replaced materials are majorly derived from the industrial waste, and thus, they are considered to be as a great concern since they are considered to be hazardous for the human health and ecosystem. As a result, these waste products are effectively utilized and incorporated. Crumb rubber, Quarry dust, Copper Slag, FA, RHA are some of the considerable replaced materials for sand. Also, since as the rate of construction is increasing, rate of demolition of old structures is also increasing, thereby the finer demolished waste can also be used as a better replacement for sand in terms of strength, stiffness and durability. But, while considering partial replacement of coarse aggregates is considered, there is no such proper replacement for them is found out. Commonly used partial replaced materials for traditional coarse aggregates are recycled aggregates, recycled concrete aggregates, coconut shells, tile waters, marble wastes, high-density polymers and crushed rubber. In this study, the cement, fine aggregates and coarse aggregates are partially replaced by FA, GGBS and recycled concrete aggregates in different proportions in different concrete mix samples by considering Taguchi's Approach and then, the comments be made on which combination of the by-products is going to prove the optimum for obtaining the better mechanical strength of the concrete sample.

2 Literature Review

The study conducted by Kurda et al. [1] showed, while considering the proportion of fly ash upto 45%, concrete shows an interesting result in the strength in compression with FA at 90 days is same for 28 days. Contrary, as the proportion levels increased, the result doesn't show positive response. Thus, this study suggests that the strength class of a FA concrete should be decided based on the age of 90 days instead of 28 days. For the incorporation of FA above 36%, RCA concrete with great ion penetration rate and low permeability is obtained. With the addition of RCA into the concrete mix, the same experienced the phenomenon of shrinkage, which doesn't observe when FA was added to the concrete mix.

The durability of recycled aggregate concrete is mainly estimated by RA-blended adhered mortar. Guo et al. [2] observed that the inclusion of RA with a fly ash or blast furnace slag coating improves the pore structure of RA and, also can react with existing calcium hydroxide to produce additional secondary C–S–H products to strengthen the layer. Improved recycled aggregates with carbon dioxide may also react with hydration products to produce calcium carbonate for the strengthening of the micro-pore structure. Therefore, it had been concluded that for producing more durable concrete, it is recommended to use pozzolanic materials and treatment with carbon dioxide, prior to be used in concrete. Based on his study, Siavash Mahvash et al. [3] concluded that (1) Finer FA gives better strength than the corresponding activated coarser FA, (2) The gain in strength depends upon the curing time of the concrete mix and (3) Class F FA samples gained strength 3 times more than the sample of Class C FA. Also, he concluded that if the cement-based binders are used, the mechanical strength of the concrete is increased and having a better consistency.

In his study, Saha [4] focused the effect of FA on the various characteristics of the concrete, followed by the conclusions: (1) By adding class F fly ash, the compressive strength of the concrete mix after 28 days of curing was lesser and decreased sharply with the addition of FA content, (2) The addition of FA decreased the hydration rate, thus results in the drying shrinkage of FA concrete, (3) The FA-induced concrete has less-chloride permeability at 28 and 180 days of curing. This concept was attributed to the low inter-connecting voids and alkali binding associated with FA concrete and 5. The density of binder matrix has been improved by the inclusion of FA with the pozzolanic reaction.

Li et al. [5] noted that the large-sized coarse aggregates (LRCA) were used as a partial replacement. The experimental study focused the effect of LRCA on the various properties of the concrete, having following findings: (1) The LRCA bond was good. While performing the prism compressive test, a good amount of LRCA got fractured at cracking surface. The same was observed, while determining the tensile strength, (2) The difference in the magnitude of strength in compression for LRCA—induced concrete and normal concrete was marginal. The compressive strength in increased by 40 percent when incorporate with LRCA as compared to normal mix, (3) While considering the ratio of compressive strength in axial direction to experimental cube magnitudes, the value drops by 12%, and 4. While determining the split tensile strength, the error of less than 10% was observed, while comparing the magnitude of tensile strength, mathematically and experimentally.

On the basis of his studies, Hefni et al. [6] concluded that: (1) Addition of FA as fractional replacement of cement declines the strength of mix at initial ages. A noticeable upgradation was noted between 60 to 180 days, (2) The addition of FA majorly boosted strength of concrete in all dimensions after getting exposure of the elevated temperatures, (3) Aqueous sodium silicate activator was found out to be a great activating agent for enhancing strength in compression, split tensile and flexural and 4. Stimulation of FA has a significant consequence on the concrete mix during initial investigation.

Manjunath et al. [7], the workability of mortar and the setting time of fly ash are inspected with the substitutions. 50 and 25% of cement (C) are replaced with FA; and 100, 75, 50 and 25% of Sand (S) by GGBS. In this investigation, they concluded that with an increase in FA, consistency and setting time (both initial and final) got increased, and effectively using of CaCl2 decreases setting time of pastes.

An attempt is made by Vivek [8] to investigate characteristics of geo-polymer concrete with low Ca-FA, getting replaced by 5 varying proportions of slag. Na2Si03 (105 kg/m^3) and NaOH-8M mixture were employed as alkaline solution. They concluded that the magnitude of water absorption is less and compressive strength of around 75% can be achieved in just initial 4.5 setting hours.

"Basil Johny et al. [9] investigated the characteristics of sustainable concrete with replacement of RCA and slag. GGBS was replaced for 60, 50 and 40% with cement and optimal fraction was observed. For the mixes prepared by replacing 50% cement with slag and 50% coarse aggregate, it satisfies the strength criteria required for an M30 mix. Berndt [10] noted the properties of sustainable concrete containing fly ash, slag and recycled concrete aggregates by replacing cement by a percentage of fly ash or slag and natural aggregates by recycled concrete aggregates. The mixes containing 50% slag gave best overall performance."

Shakir Ahmed et al. [11] observed the strength of concrete with percentage replacement in natural coarse aggregate with recycled concrete aggregate for M20 mix concrete. The strength of concrete decreases as the percentage of RCA increases. Dixit et al. [12] did an experimental research on the effect of the partial replacement of cement, fine aggregates and coarse aggregates with GGBS, fly Ash and recycled aggregates, depicting an increase in strength of the concrete samples in both tension and compression. The strength with the partial replacement is increased by 60% in compression and 75% in tension.

3 Concrete and Its Components

3.1 Cement

The most important component which is required in the production of concrete is cement. The cement generally consists of oxides of several metal such as calcium, silicon, etc., which has their own functions toward the properties of the cement. The following table depicts the average composition of the oxides present in the cement structure. OPC cement is considered for the concrete sample production. The grade of the OPC used is OPC 43. Also, as per IS 4031 [13], various important experiments were performed to determine the properties of the cement as given in Fig. 1 (Table 1).

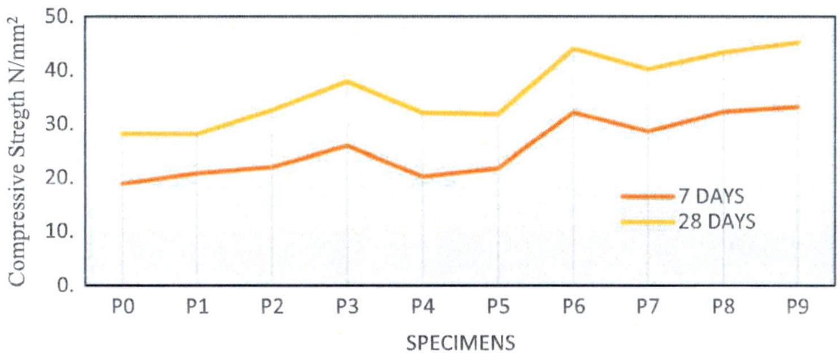

Fig. 1 Comparative strength in compression at 7 and 28 days

Table 1 Properties of cement

Properties	Values obtained
Fineness of cement	5%
Specific gravity	3.14
Standard consistency	39%
Initial setting time	205 min
Final setting time	385 min
Average compressive strength at 28 days	50.56 N/mm^2

Table 2 Properties of aggregates

Properties	Fine aggregates	Coarse aggregates
Fineness modulus	2.05	7.50
Water absorption (%)	1.20	3.44
Specific gravity	2.50	2.30
Bulking	22%	0.395
Flakiness index	–	35%
Elongation index	–	7.7%

3.2 *Aggregates*

The next group of constituents of cement consists of aggregates. There are two different types of aggregates—"Fine Aggregates and Coarse Aggregates." The aggregate size less than 4.75 mm are considered to be as "fine," otherwise considered to be as "coarse". Both type of aggregates is used in the production of the concrete, and thus have different roles. The amount (proportion) of the aggregates used in the concrete production depends upon the Mix Design. Similar to the cement, several experiments have to performed before using them in the concrete production as per IS 2386 [14] (Table 2):

3.3 *Fly Ash*

Fly ash (FA) is the by-waste originated from the electric power generation, considered to most widely used by-product in the concrete. The common structure of FA contains considerable amount of silicon dioxide, present in both amorphous state and crystalline state, calcium oxide and aluminum oxide, as given in the following Table 3. FA is available in to different forms: Class C FA and Class F FA, both have different functions and used as per the application. Since we have to use the FA to act as a binder, we will consider Class F FA. With the addition of FA, the concrete mix attains more strength as compared to the conventional concrete mix. And also, since it is a waste product and using as a partial replacement to cement, it will have a

Table 3 Chemical composition of FA

S. No.	Component	Amount (%)
1	Silica	62
2	Lime	0.8
3	Iron oxide	3.44
4	Alumina	4.55
5	Magnesia	0.6
6	Sulfur trioxide	0.5
7	Alkalies	2.0

Table 4 Composition of GGBS

S. No.	Component	Amount (%)
1	Silica	44
2	Lime	1.47
3	Iron oxide	1.25
4	Alumina	12
5	Magnesia	1.15
6	Alkalies	1.7

positive impact on the environment in two-folds: waste generated is using effectively and since cement is less used, there will be a net reduction in the emission of the greenhouse as well as other harmful gases.

3.4 Glass Granulated Blast Furnace Slag (GGBS)

It is derived from quenching process of "blast furnace slag" of molten iron in water or steam, so as to get a granular—glassy product, which again being processed after drying and converted into finer-texture particles. Since, GGBS is also considered to be as a by-product or waste product to the environment, and this is also used as a partial replaced material to the constituents of the concrete (both cement and sand). Here, GGBS is taken as a partially replaced material for sand, and the properties are determined for mixing the same into the concrete sample. The composition of GGBS is enlisted in Table 4.

3.5 Recycled Aggregates

RCA or specifically known as "Recycled Concrete aggregates" are basically crushed aggregates which can be easily available from the demolition site. Nowadays, some of the concrete manufacturing organizations are giving special importance toward the building of construction and demolition plant, which will only focus on such kind of demolished material, which after processed to some extent, can be used again into the construction industry. The basic purpose behind using the recycled concrete aggregates is their strength and durability. In the previous studies, the mechanical strength of the concrete sample got increased after the partial replacement (or fully sometimes) of the coarse aggregates with recycled concrete aggregates. Recycled concrete aggregates show better values as compared to normal coarse aggregates, in regarding to permeability and water absorption.

4 Concrete Mix Design

Concrete mix design is a stipulated procedure for the determination of the various proportions of the constituents of the concrete. After performing the concrete mix design, the various properties of the concrete mix can be determined, both in the fresh and hardened state. The fresh properties basically determine the workability, whereas the hardened properties include the determination of the mechanical strength (Table 5). In this study, the concrete mix proportioning is carried out for M 25 grade, as per the latest IS—10,262: 2019 [15]. As per IS—10,262: 2019, fck = 25 N/mm^2, Nominal maximum size of aggregate = 20 mm, cement content = 330 kg/m^3, Sand Content = 735 kg/m^3, Coarse Aggregates = 1480 kg/m^3 and W/C Ratio = 0.45. The actual quantities of the constituents for determination of the strength of one standard cube (150 mm × 150 mm × 150 mm) can be carries out as:

Volume of concrete required for 1 cube (assuming 25% wastage) 0.153 × 1.25 = 0.004218 m^3, thus Cement = 1.450 kg, Sand = 3.157 kg, Coarse Aggregates = 5.715 kg and Water = 0.640 kg. Also, for the determination of the strength in the tension, the volume of the concrete required for one cylindrical mold can be given as: Cement = 1.750 kg, Sand = 4.055 kg, Coarse Aggregates = 7.100 kg and Water = 0.755 kg. The results of the various properties of M25 grade mix can be noted as:

Table 5 Fresh and hardened properties of M25 concrete mix sample

S. No.	Property	Characteristic	Result
1	Fresh state	Slump value (mm)	159
2	Fresh state	Compaction factor	0.91
3	Hardened state	Compressive strength at 7 days (N/mm^2)	18.90
4	Hardened state	Compressive strength at 28 days (N/mm^2)	28.115
5	Hardened state	Split tensile strength at 7 days (N/mm^2)	1.855
6	Hardened state	Split tensile strength at 28 days (N/mm^2)	2.475
7	Hardened state	Flexural strength at 7 days (N/mm^2)	3.04
8	Hardened state	Flexural strength at 28 days (N/mm^2)	3.70

5 Methodology and Observations

The experimental approach adopted for this study is Taguchi's Approach. This method was invented by Dr. Taguchi, in which the problem can be solved by considering Orthogonal Arrays. With the help of this methodology, experimental specimens can be reduced by considering all the optimal situations and control parameters. Also, the array approach provides the best result due to the optimization of the given problem [16, 17]. In this study, fly ash, GGBS and recycled coarse aggregates are partially replaced as cement, fine aggregates and coarse aggregates, in different proportions; which can be carried out by Taguchi's Approach. The replacement proportions considered are 10, 20 and 30%; the representation can be done as given in Table 6.

With the consideration of Taguchi's Approach, we will have nine different specimens (P1–P9) from the above considered proportions which will yield optimum results. After considering the specimens with different proportions of the partially replaced constituents and performing the experiment, the strength in compression, tension and flexural at 7 and 28 days can be observed as follows (Table 7):

Table 6 Proportioning considered

Representation	Partially replaced materials	Proportion 1	Proportion 2	Proportion 3
X	Fly ash	10	20	30
Y	GGBS	10	20	30
Z	Recycled concrete aggregates	10	20	30

Table 7 Observation on strength (All magnitude in N/mm^2)

Specimen	Trial mix	7 days compressive strength	28 days compressive strength	7 days tensile strength	28 days tensile strength	7 days flexural strength	28 days flexural strength
PO	–	18.9	28.115	1.855	2.475	3.04	3.71
P1	X1–Y1–Z1	20.85	28.05	2.005	2.1	3.2	3.71
P2	X1–Y2–Z2	22	32.55	2.4	2.7	3.28	4
P3	X1–Y3–Z3	25.85	37.85	2.9	3.55	3.56	4.31
P4	X2–Y1–Z2	20.23	32.05	3.15	2.6	3.15	3.95
P5	X2–Y2–Z3	21.75	31.79	3.48	3.15	3.26	3.95
P6	X2–Y3–Z1	32.1	44	4.9	3.25	3.97	4.65
P7	X3—Y1–Z3	28.5	15	3.25	3.4	3.74	4.4
P8	X3–Y2–Z1	32.25	43.3	3.88	3.27	3.98	4.6
P9	X3–Y3–Z2	33.18	45.18	4.15	3.95	4.03	4.75

6 Results and Discussion

After conducting the experiments related to determining the mechanical strength of the concrete mix of grade M25, without and with partial replacement materials, it had been observed that usage of these by-products to the environment found productive.

Considering the strength in compression, it had been found that the strength of the sustainable concrete is maximum in case of the specimen which consists of the partial replaced proportion of 30% of FA with cement, 30% of fine aggregates with GGBS and 20% of coarse aggregates with recycled concrete aggregates. The percent increment with respect to normal concrete sample is 75.55% in case of 7 days and 37.77% in case of 28 days of curing. The graphical representation of the strength obtained after performing the compressive test is given as Fig. 1.

On the other hand, considering the strength in tension at 7 days, it had been found that the strength of the sustainable concrete is maximum in case of the specimen which consists of the partial replaced proportion of 20% of FA with cement, 30% of fine aggregates with GGBS and 10% of coarse aggregates with recycled concrete aggregates and at 28 days, the maximum strength was observed with the combination of partial replaced proportion of 30% of FA with cement, 30% of fine aggregates with GGBS and 20% of coarse aggregates with recycled concrete aggregates The percent increment with respect to normal concrete sample is 62.14% in case of 7 days and AT 28 days, the percentage was found 59.60. The graphical representation of the strength obtained after performing the split tensile test is given as Fig. 2.

While considering the strength in flexure, it had been found that the strength of the sustainable concrete is maximum in case of the specimen which consists of the partial replaced proportion of 30% of FA with cement, 30% of fine aggregates with GGBS and 20% of coarse aggregates with recycled concrete. The percent increment with respect to normal concrete sample is 32.56% in case of 7 days and nearly 28% in case of 28 days of curing. The graphical representation of the strength obtained after performing the split tensile test is given as Fig. 3.

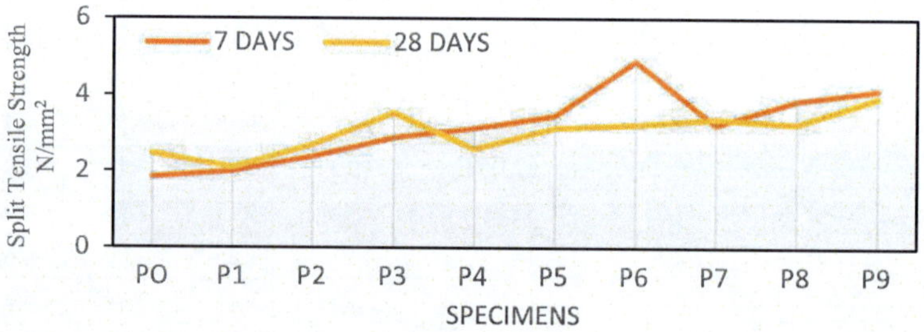

Fig. 2 Concrete strength in tension at 7 and 28 days

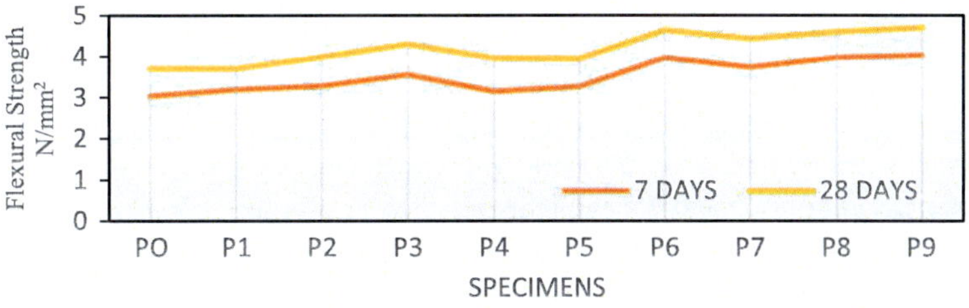

Fig. 3 Flexure strength at 7 and 28 days

7 Conclusion

Considering the M25 grade concrete, as per the latest revision of IS 10262: 2019, the proportions of the various concrete constituents are determined, and then, they were partially replaced with the waste or by-products, which are harmful to the environment as well as human beings, in different proportion as per Taguchi's Approach, the study noted that:

1. With the partial replacement of the cement, sand and coarse aggregates with FA, GGBS and Recycled concrete aggregates, respectively, the strength in both compression as well as tension is increased.
2. For strength in compression, the optimum specimen found out in case of "S9," where the specimen is prepared by 30% partially replaced cement and fine aggregates with FA and GGBS, and 20% partially replaced coarse aggregates with recycled concrete aggregates. The percent increase in the strength in compression is found to be 75.55% (7 Days) and 37.77% (28 days).
3. For strength in tension. the optimum specimen found out in case "S6" where the specimen is prepared by 20% partially replaced cement with FA, 30% with of fine aggregates with GGBS, and 10% partially replaced coarse aggregates with recycled concrete aggregates at 7 days and "S9" where the specimen is prepared by 30% partially replaced cement and fine aggregates with FA and GGBS, and 20% partially replaced coarse aggregates with recycled concrete aggregates at 28 days. Percent increase in the strength in compression is found to be 62.14% (7 Days) and 59.60% (28 days) of curing.
4. For strength in flexure, the optimum specimen found out in case of "P9," where the specimen is prepared by 30% partially replaced cement and fine aggregates with FA and GGBS, and 20% partially replaced coarse aggregates with recycled concrete aggregates. The percent increase in the strength in compression is found to be 32.56% (7 Days) and 27.89% (28 days).
5. While considering the strength, the specimen "S9"-30% partially replaced cement and sand and 20% partially replaced coarse aggregates with cement, fine aggregates and coarse aggregates, is found to be the ideal combination, followed by the combination of partially replaced materials with proportions of 20, 30 and 10%, respectively.

Previous studies [6, 8, 12] showed that the partial replacement of cement, fine aggregates and coarse aggregates with GGBS, fly ash and recycled aggregates by using the Taguchi's Approach, depicts strength increment in concrete samples in both tension and compression. The strength with the partial replacement is increased by 60% in compression and 75% in tension. Also, from this study, we can conclude that a sustainable concrete can be produced with the help of waste or by-products, which are harmful to the environment. With this approach, we will minimize the use of cement, through which a enormous volume of CO_2 is generated during the concrete making process when mixed with water, and also, the injurious impacts on the environment if the stated waste products aren't used in an effective way. When we combine concrete with these materials, the results are positive in the mechanical strength, and thus we can blend these waste materials with the concrete and finds application in concrete industry, thus making a green and sustainable environment.

References

1. Rawaz, K., de Brito, J., & Silvestre, J. D. (2017). Combined influence of recycled concrete aggregates and high contents of fly ash on concrete properties. *Construction and Building Materials 157*, 554–572.
2. G. Hui, Caijun, S., Xuemao, G., Jianping, Z., Yahong, Tung-Chai, L., Haibo, Z., & Yuli, W. (2018). Durability of recycled aggregate concrete: A review. *Cement and Concrete Composites, 89*, 251e259.
3. Siavash, M., López-Querol, S., & Bahadori-Jahromi, A. (2017). Effect of class F fly ash on fine sand compaction through soil stabilization. *Heliyon, 3*, e00274
4. Ashish Kumer, S. (2018). Effect of class F fly ash on the durability properties of concrete. *Sustainable Environment Research, 28*, 25e31.
5. Tan, L., Jianzhuang, X., Cimian, Z., & Zheng, Z. (2016). Experimental study on mechanical behaviors of concrete with large-size recycled coarse aggregate. *Construction and Building Materials, 120*, 321–328.
6. Yasmin, H., Abd El Zaher, Y., & Wahab, W. A. (2018). Influence of activation of fly ash on the mechanical properties of concrete. *Construction and Building Materials, 172*, 728–734.
7. Manjunath, S. N., Sivapullaiah, P. V., & Kumar, M. P. (2015). Implication of partial replacement of cement with FA and sand by GBS on setting time and workability of mortar. *International Journal of Research in Engineering and Technology (IJRET), 4*(1), 47–52.
8. Vignesh, P., & Vivek, K. (2015). An experimental investigation on strength parameters of flyash based geopolymer concrete with GGBS. *International Research Journal of Engineering and Technology (IRJET), 2* (02).
9. Johny, B., George M. V., & John, E. (2014). Study of properties of sustainable concrete using slag and recycled concrete aggregates. *International Journal of Engineering Research and Technology, 3* (09).
10. Berndt, M. L. (2009). Properties of sustainable concrete containing fly ash, slag and recycled concrete aggregate. *Construction and building materials, 23*(7), 2606–2613.
11. Ahmed, M. S., & Vidyadhara, H. S. (2013). Experimental study on strength behaviour of recycled aggregate concrete. *International Journal of Engineering Research Technology, 2*, 76–82.
12. Dixit, A., & Hooda, Y. (2019, June). Experimental evaluation on compressive and tensile behavior of concrete utilising GGBS, fly ash and recycled aggregates. *International Journal of Engineering and Advanced Technology (IJEAT)*. ISSN: 2249-8958, 8 (5).

13. IS 4031: 1988. Methods of physical tests for hydraulic cement.
14. IS 2386: 1963. Methods of test for aggregates for concrete.
15. IS 10262: 2019. Concrete mix proportioning—Guidelines.
16. Ghani, J. A., Jamaluddin, H., Rahman, M. N., & Deros, B. M. (2013). Philosophy of Taguchi approach and method in design of experiment. *Asian Journal of Scientific Research, 6* (1), 27–37, 19.
17. Antony, J., Warwood, S., Fernandes, K., & Rowlands, H. (2001). Process optimization using Taguchi methods of experimental design. *Work Study, 50*(2), 51–58.

Numerical Analysis on Voided Slab with Different Reinforcement on ANSYS 2020R1

Nikita Jain and Asif Hussain

Abstract The urbanization and increasing demand for infrastructure are increasing the use of reinforced concrete as a primary construction material. The conventional design methods provide large member sizes to fulfill structural requirements. These large dimensions of structural elements like slab, beam, column, foundation, etc., affect the quantity of concrete and steel used for construction; this, in turn, alarms us for the need to optimize the use of concrete and steel to reduce structure's dead weight. The present analytical and numerical study tries to evaluate the results. This study involves a typical bi-axial voided slab. It eliminates the concrete from the middle of the slab. The results observed showed a maximum reduction of 17% in weight; maximum increase of 20% was observed in total deformation; maximum reduction of 77% in equivalent strain, and maximum reduction of 72.4% is equivalent stress. The results so obtained fulfilled the criteria of strength and serviceability.

Keywords Bi-axial voided slab · Spherical void former · Rectangular void former · Cylindrical void former · Square void former · Trapezium void former · ANSYS 2020 R1 · Composite reinforcement bar · Steel

1 Introduction

This paper describes types of hollow slab technologies that have appeared in the last few decades worldwide. The bi-axial voided slabs are slabs in which the excess of concrete, i.e., the concrete which neither takes part in compression nor tension, is replaced, creating voids in the slab's core. It was introduced in the late 1950s but was used in one spanning slabs and supported by fixed walls, but later researchers came up with the plan that this methodology will be more advantageous for two-way concrete slab to having same capabilities but relatively less weight as the excess concrete is being replaced by the voids created. In any construction structure, the central part is the slab and also the most extensive element consuming concrete [2].

N. Jain · A. Hussain (✉)
Department of Civil Engineering, Jamia Millia Islamia, New Delhi 110025, India

A. K. Gupta et al. (eds.), *Advances in Construction Materials and Sustainable Environment*, Lecture Notes in Civil Engineering 196,
https://doi.org/10.1007/978-981-16-6557-8_41

Table 1 Material properties

Material	Properties	Value
Concrete (M20)	Modulus of elasticity (MPa)	2500
	Compressive strength (MPa)	20
	Poisson's ratio	0.2
	Density (kg/m^3)	2400
Conventional reinforcement bar (Steel)	Modulus of elasticity (MPa)	200,000
	Tensile strength (MPa)	500
	Poisson's ratio	0.3
	Density (kg/m^3)	8050
HDPE	Modulus of elasticity (MPa)	1030
	Density (kg/m^3)	950
	Poisson's ratio	0.4
Composite reinforcement bar (Glass fiber)	Modulus of elasticity (MPa)	55,000
	Tensile strength (MPa)	1250
	Poisson's ratio	0.4
	Density (kg/m^3)	2040

Steel being the most used ductile material in slabs, but nowadays vast research is going on for overcoming the only drawback of steel, i.e., self-weight. Composite reinforcement bar showed promising results with huge weight reduction. Properties of material used are given in Table 1.

Research done by various researchers in last five years have been discussed below:

Sagadevan and Rao [6]: Presented paper during a conference where the researchers did an analytical and experimental investigation on three specimen samples. One was a standard slab with a 150 mm depth, the second was voided slab with a 90 mm diameter spherical voids spaced at a 160 mm center to center with 150 mm depth, and therefore, the third was voided slab with 120 mm diameter spherical voids spaced at 210 mm center to center with 250 mm depth. In his investigation, the researcher found both experimental and analytical that both the slabs, i.e., conventional slab and voided slab, showed an equivalent flexural behavior and cracks in X form, the same as the yield pattern. The last word, the load-carrying capacity of the slabs, was also an equivalent. The initial stiffness of voided slab was 37% of the traditional slab.

Paul and George [5]: Studied comparison of voided and solid flat slab system. It analyzed square and rectangular slab of thickness 28 cm on ANSYS. The response spectrum and structural spectrum showed a similar pattern in deformation and stress of voided slab over the conventional slab. The ultimate load capacity of VBS is very close to that of a conventional slab. Providing the void former near-neutral axis does not subsequently affect the deflection behavior of slabs under seismic effects, and it also leads to the saving of material and cost.

Fanella et al. [3]: This paper shows that the test results will not determine the dynamic behavior of a bi-axial voided slab to verify a FEM model. The numerical model was made on FEM code, and therefore, the mode shapes obtained by experimental and numerical analysis were compared. Nondestructive testing was conducted on the plan by use of seven accelerometer; it was placed on the slab and was the applied with 0.5 Hz of acceleration at intervals. The FEM code created was tested in NI LabView software. The four natural frequencies were recorded and compared with numerical records. The great agreement was recorded in both numerical and experimental results that suggest that the measurements were done properly. It has also appeared that connections between the slabs and walls did not perform important role.

Bhade and Barelikar [1]: Studied different arrangements of HDPE balls sandwiched in reinforced concrete slabs. It verified the load-carrying capacity, deflection, and weight reduction tested under a single-point loading system. There were three specimens; first was conventional slab made of M30 grade of concrete; and the other two were bubble deck slab. In the Type I bubble deck slab, the reinforcement mesh was on both the faces, but the HDPE balls were placed. In alternative square and Type II bubble deck slab, the reinforcement mesh was on both sides, but the HDPE balls were placed at every square. The load-bearing capacity was highest by bubble deck slab Type II around 325 kN. Bubble deck slab Type II has improved elasticity of slabs, for example, the conventional sample deflects lesser than the solid slab. The number of voids created affects the elasticity of the bubble deck system. Reduction in self-weight being the important factor, and the weight was reduced up to 25%.

Onchuru and Opiyo [4]: Presented the comparison of different characteristics such as stiffness reduction, weight reduction, bending strength for different loads, and different diameter of balls. They concluded that using voided slabs save 40% time and also reduced the weight by 30–50%. It also stated that voided slabs are environmental friendly as it reduces the emission of CO_2 on casting of concrete.

2 Objectives

(1) To analyze the slab with conventional steel reinforcement and composite reinforcement bar with void former using ANSYS 2020 R1 for deflection, equivalent stress, equivalent strain, and weight.

(2) To compare the results obtained from models with different void former having conventional steel reinforcement and composite reinforcement bar (voids in shear zone).

3 Methodology and Analysis

To achieve the objectives stated above, the following steps were taken:

(1) Slab were modeled in a CAD software named as CREO with dimension of slab—300 mm × 300 mm × 100 mm.
(2) The voids were created at time of modeling of different shapes, i.e., sphere, cylinder, rectangle, square, and trapezium.
(3) The voids so created were assumed to be made of high-density polyethylene (HDPE).
(4) The reinforcements were placed in two meshes, one at bottom and one at top.
(5) The slabs were analyzed twice using ANSYS 2020 R1; Firstly with conventional reinforcement (Steel) and secondly with composite reinforcement.
(6) The results observed were tabulated for weight, total deformation, equivalent strain, and equivalent stress.
(7) Comparison of results for slab with conventional reinforcement and slab with composite reinforcement in different types of void former.

4 Results

The slab were analyzed on ANSYS 2020 R1. The results obtained are given in Table 2.

4.1 Graphical Comparison of Results Tabulated

The results obtained were compared graphically. The slabs were represented with S; where: Slab 1—Conventional slab with no void former and steel; Slab 2—Conventional slab with no void former and composite reinforcement bar; Slab 3—Slab with spherical voids and steel; Slab 4—Slab with spherical voids and composite reinforcement bar; Slab 5—Slab with cylindrical voids and steel; Slab 6—Slab with cylindrical voids and composite reinforcement bar; Slab 7—Slab with rectangular voids and steel; Slab 8—Slab with rectangular voids and composite reinforcement bar; Slab 9—Slab with square voids and steel; Slab 10—Slab with square voids and composite reinforcement bar; Slab 11—Slab with trapezium voids and steel; Slab 12—Slab with trapezium voids and composite reinforcement bar.

Figure 1 shows the comparison of weight. The weight of conventional slab was 2 kg; reduction in weight was observed due to voids and due to composite reinforcement bar. The maximum reduction in weight was observed for Slab 10 (Slab with square void former and composite reinforcement bar), i.e., 1.75 kg which is 17% when compared with conventional slab.

Figure 2 gives the comparison of total deformation. The total deformation observed for conventional slab was 0.0029 m; increase in total deformation was

Table 2 Results

Void former	Point load applied	Weight (kg)	Maximum total deformation (m)	Maximum equivalent strain (m/m)	Maximum equivalent stress (Pa)
Conventional slab with no void former and steel	1,000,000 N	2.00	0.0029	0.96	17.7×10^9
Conventional slab with no void former and composite reinforcement bar	1,000,000 N	1.90	0.0029	0.22	4.94×10^9
Slab with spherical voids and steel	1,000,000 N	1.88	0.0029	0.37	21.6×10^9
Slab with spherical voids and composite reinforcement bar	1,000,000 N	1.78	0.0029	0.31	8.29×10^9
Slab with cylindrical voids and steel	1,000,000 N	1.87	0.0031	1.06	22.7×10^9
Slab with cylindrical voids and composite reinforcement bar	1,000,000 N	1.80	0.0032	0.4	7.64×10^9
Slab with rectangular voids and steel	1,000,000 N	1.85	0.0034	0.97	19.1×10^9
Slab with rectangular voids and composite reinforcement bar	1,000,000 N	1.76	0.0035	0.33	8.18×10^9
Slab with square voids and steel	1,000,000 N	1.84	0.0035	0.94	18.1×10^9
Slab with square voids and composite reinforcement bar	1,000,000 N	1.75	0.0033	0.45	8.62×10^9
Slab with trapezium voids and steel	1,000,000 N	1.86	0.0032	0.62	20.9×10^9

(continued)

Table 2 (continued)

Void former	Point load applied	Weight (kg)	Maximum total deformation (m)	Maximum equivalent strain (m/m)	Maximum equivalent stress (Pa)
Slab with trapezium voids and composite reinforcement bar	1,000,000 N	1.78	0.0030	0.31	7.55×10^9

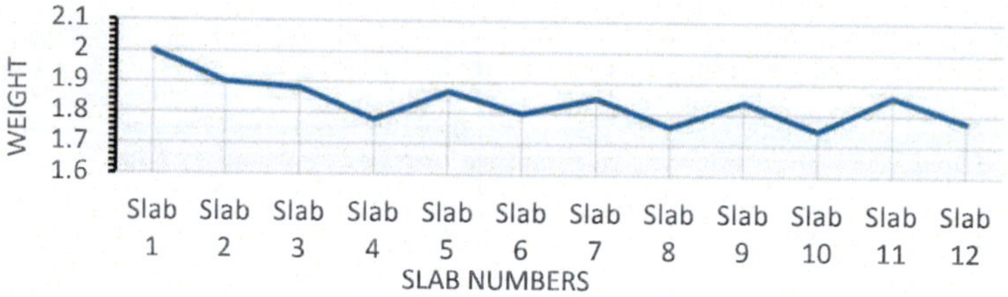

Fig. 1 Comparison of slab for weight

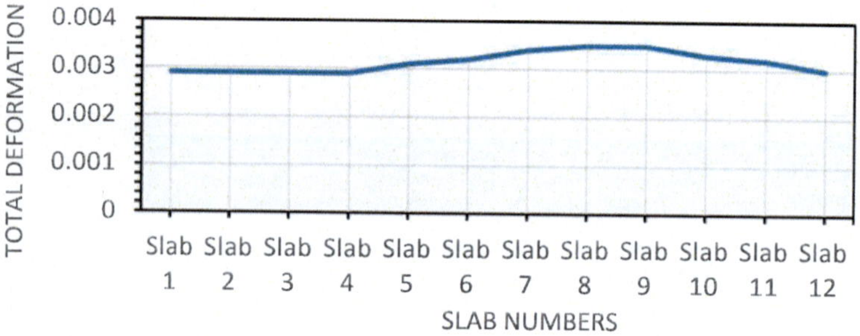

Fig. 2 Comparison of slab for total deformation

observed due to voids and due to composite reinforcement bar. There was no change observed for spherical void former. The minimum increase in total deformation was observed for Slab 12 (Slab with trapezium voids and composite reinforcement bar), i.e., 0.0030 m, which is 3.4% when compared with conventional slab. Maximum increase was observed for Slab 8 (Slab with rectangular voids and composite reinforcement bar), i.e., 0.0035 m, i.e., 20% when compared with conventional slab.

Figure 3 gives the comparison of equivalent strain. The equivalent strain observed for conventional slab was 0.96; increase and decrease in equivalent strain were observed due to voids and due to composite reinforcement bar. The maximum increase in equivalent strain was observed for Slab 5 (Slab with cylindrical voids and steel), i.e., 1.06 which is 10% when compared with conventional slab. Maximum

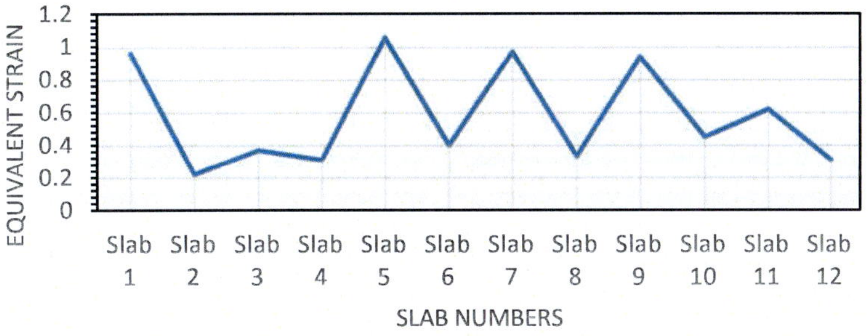

Fig. 3 Comparison of slab for equivalent strain

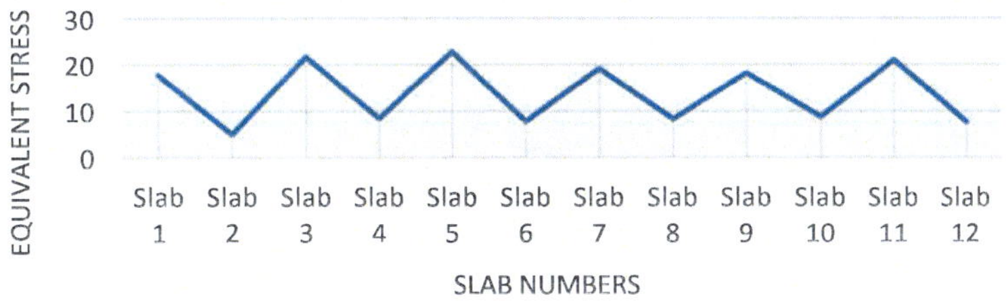

Fig. 4 Comparison of slab for equivalent stress

reduction 0.22 was observed for Slab 2 (Conventional slab with no void former and composite reinforcement bar), i.e., 0.22, i.e., 77% when compared with conventional slab.

Figure 4 is comparison of equivalent stress. The equivalent stress observed for conventional slab was 17.7×10^9 Pa; increase and decrease in equivalent stress was observed due to voids and due to composite reinforcement bar. The maximum increase in equivalent stress was observed for Slab 5 (Slab with cylindrical voids and steel), i.e., 22.7×10^9 Pa which is 28% when compared with conventional slab. Maximum reduction 0.22 was observed for Slab 2 (Conventional slab with no void former and composite reinforcement bar), i.e., 0.22, i.e., 72.4% when compared with conventional slab.

4.2 Figures for Results Evaluated

4.2.1 Total Deformation

As seen above, total deformation did not changed for most cases, but it increase minimum by 3.4–20%. The results evaluated can be verified with Fig. 5a–c as given below:

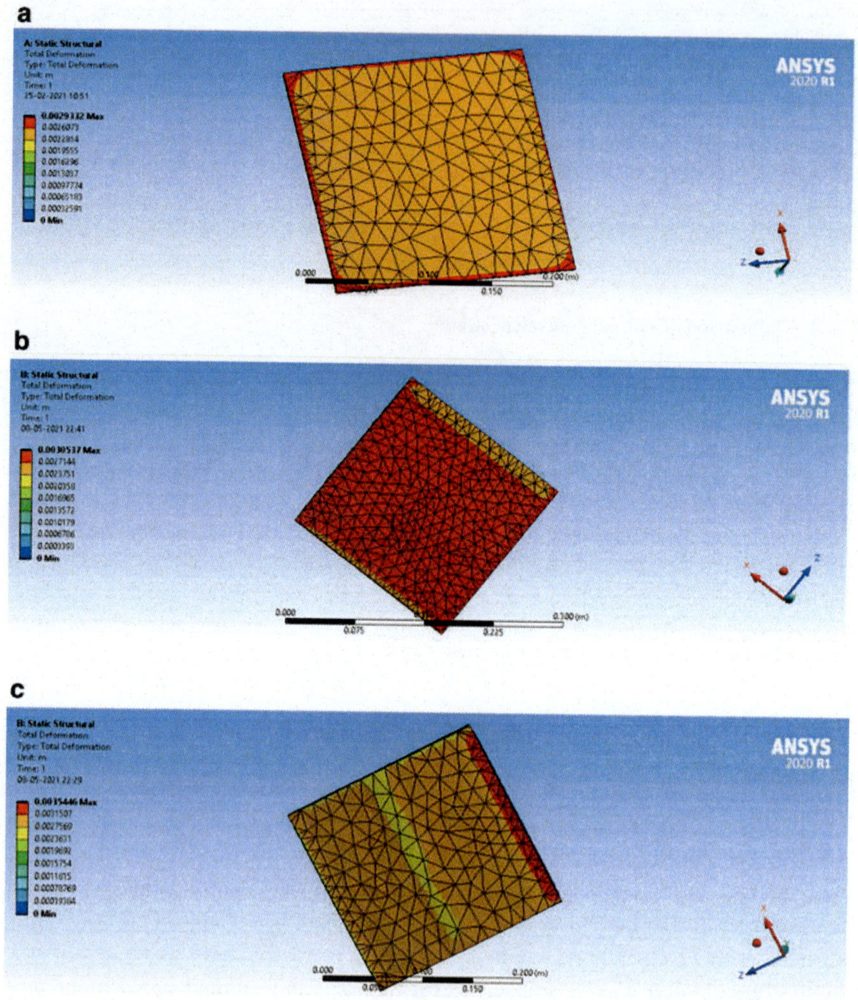

Fig. 5 **a** Total deformation for Slab 1 (conventional slab with no void former and steel), **b** total deformation for Slab 12 (slab with trapezium voids and composite reinforcement bar), **c** total deformation for Slab 8 (slab with rectangular voids and composite reinforcement bar)

4.2.2 Equivalent Strain

As evaluated above, equivalent strain showed considerable increase and decrease when compared with conventional slab. The maximum increase was of 10%, but maximum decrease was 77%. The results evaluated are validated from Figs. 6a–c as given below:

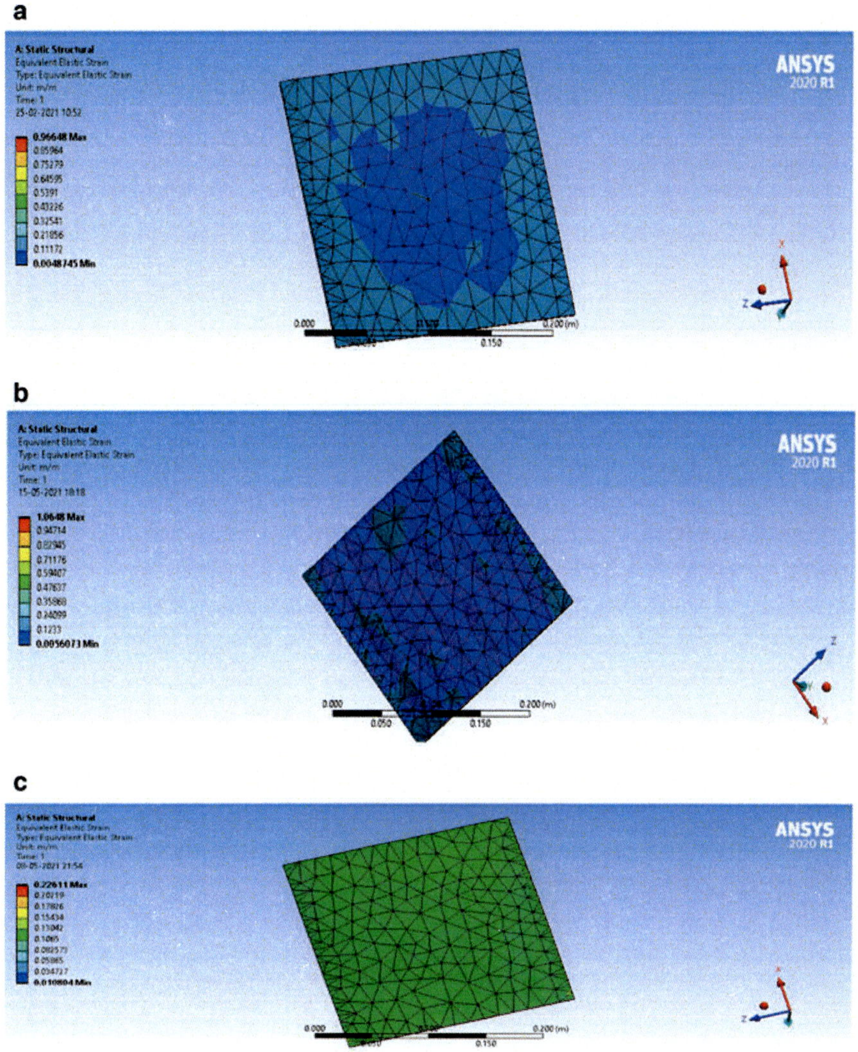

Fig. 6 **a** Equivalent strain for Slab 1 (conventional slab with no void former and steel), **b** Equivalent strain for Slab 5 (slab with cylindrical voids and steel), **c** Equivalent strain for Slab 2 (conventional slab with no void former and composite reinforcement bar)

4.2.3 Equivalent Stress

As evaluated earlier, considerable increase and decrease was observed in results for equivalent stress when compared to conventional slab, and the results are validated from Figs. 7a–c. The maximum increase was of 28%, but maximum decrease was 72.4%.

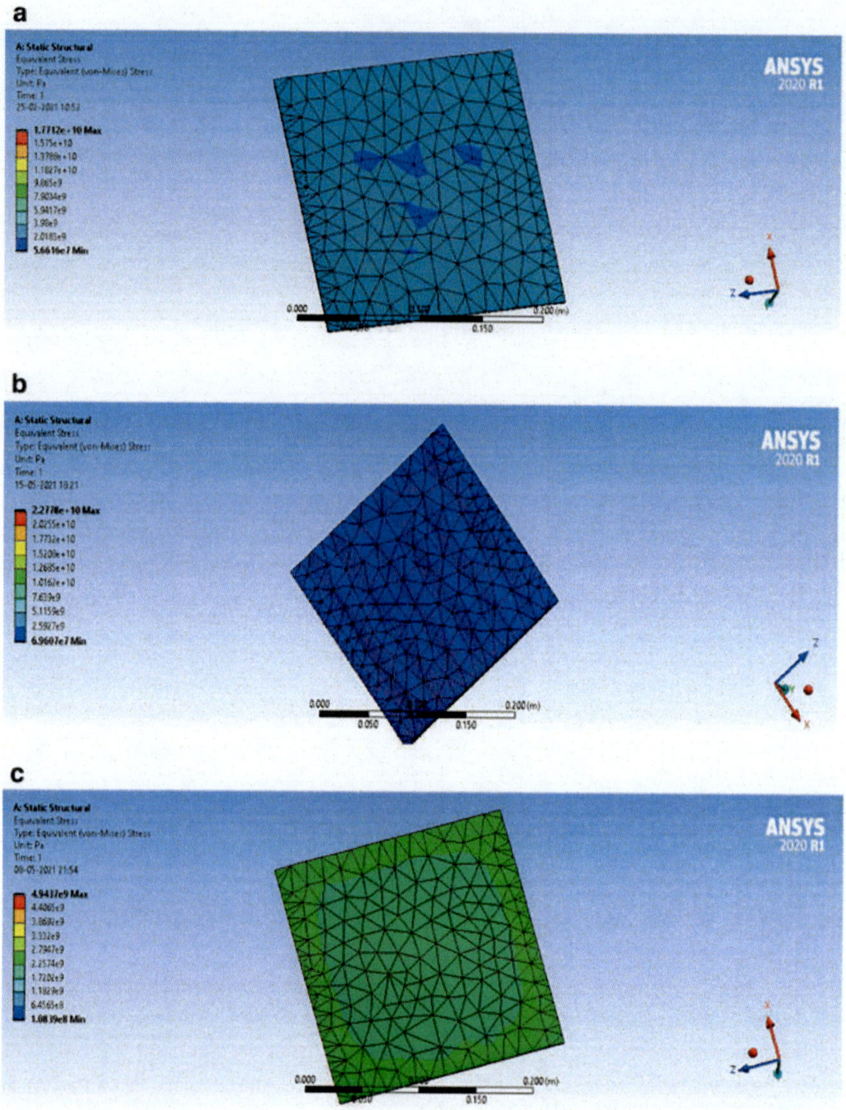

Fig. 7 **a** Equivalent stress for Slab 1 (conventional slab with no void former and steel), **b** equivalent stress for Slab 5 (slab with cylindrical voids and steel), **c** equivalent stress for Slab 2 (conventional slab with no void former and composite reinforcement bar)

5 Conclusion

From the above results, we can conclude that:

(a) The weight reduction achieved with composite reinforcement bar ranged between 5 and 18%.
(b) The total deformation showed no change for slab with no voids and slab with spherical voids, but it increased by 20% in slab with cylindrical and rectangular voids.
(c) The equivalent strain maximum increase was 10% and maximum decrease was 77%. Maximum reduction in equivalent strain was found in slab with no void and maximum increase slab with cylindrical void and steel.
(d) The equivalent stress maximum increase was 28% and maximum decrease was 72.4%. Maximum reduction in equivalent strain was found in slab with no void and maximum increase slab with cylindrical void and steel.

So, from the above results, we can conclude that the voided slab with composite reinforcement bar showed satisfactory results for weight, total deformation, equivalent strain, and equivalent stress. There was decrease in equivalent stress and strain in composite reinforcement bar to reduce stress and to strengthen the structure.

References

1. Bhade, B. G., & Barelikar, S. (2016). An experimental study on two way Bubble deck slab system with spherical elliptical hollow balls. *International Journal of Recent Scientific Research*, 11621–11626.
2. Bindea, M., Moldovan, D., & Kiss, Z. (2013). Flat slabs with spherical voids. *Part I: Prescriptions for Flexural and Shear Design, Acta Technica Napocensis: Civil Engineering & Architecture, 56*(1), 67–73.
3. Fanella, D. A., Mahamid, M., & Mota, M. (2017). Flat plate–voided concrete slab systems: Design, serviceability, fire resistance, and construction. *Practice Periodical on Structural Design and Construction.*
4. Onchuru, M. W., & Opiyo, M. O. (2015). Comparative structural analysis of Biaxial voided slabs and solid slab.
5. Paul, V. J., & George, M. (2018). Parametric investigation on the seismic response on voided & solid flat plate system. *International Journal of Innovative Science, Engineering and Technology*, 255–259.
6. Sagadevan, R., & Rao, B. N. (2019). Experimental and analytical investigations on two-way flexural capacity of bi-axial voided slab. In *Proceedings of National Conference on Advances in Structural Technologies (CoAST-2019).* https://doi.org/10.1016/j.istruc.2019.03.013.

Laboratory-Based Study of Flexible Facing in Soil Nailed Slope

Mohammad Farhad Ayazi, Samrity Jalota, and Amanpreet Tangri

Abstract One of the biggest mishaps occurring in construction industries leading to loss of lives and loss of property is slope instability caused due to various reasons such as adverse climatic conditions, earthquake, lower bearing capacity of soil, high rainfall intensity, and seepage. A recent in-situ technique developed in order to stabilize such slopes is "soil nailing" which incorporates positioning of a vertical reinforcing bar called steel nail into the existing soil slope for the sake of improvising the mechanical properties of the soil. In the present study, a physical model is prepared comprising of a hydraulic jack for load application, a multimeter to record the strain produced corresponding to the respective stress, sensors adhered to the nails, a perplex sheet to observe values of horizontal and vertical deflections, and six number of soil nails arranged in rectangular pattern. In addition to a slope considered without any facing material, four other flexible facing materials were also considered, namely geotextile fabric, coir mat, bi-axial geogrid, and geomembrane. The results are computed and compared in order to depict a clear idea of the strength properties of the flexible materials used. The comparison shows that coir mat withstands a maximum stress of 1.476 N/mm^2, while the slope without any facing material takes minimum load of 0.738 N/mm^2. Thus, it can be established that provision of flexible facing materials in soil nailing technique further aids in slope stabilization.

Keywords Reinforcement of slopes · Soil nailing · Flexible facing · Physical model · Slope stabilization

M. F. Ayazi · S. Jalota
Chandigarh University, Mohali, Punjab 140413, India

A. Tangri (✉)
Department of Civil Engineering, Chandigarh University, Mohali, Punjab 140413, India
e-mail: amanpreet.civil@cumail.in

A. K. Gupta et al. (eds.), *Advances in Construction Materials and Sustainable Environment*, Lecture Notes in Civil Engineering 196,
https://doi.org/10.1007/978-981-16-6557-8_42

1 Introduction

Slope stability is a subject matter of key concern in the construction industries as it eventually becomes hazardous causing hindrance to human activities, social, as well as economic losses, and often leading to loss of human lives. In India, one such emerging remedial measure in vogue to stabilize and support the existing soil slopes is "soil nailing." It is grabbing a lot of popularity and attention due to cost and time effectiveness, reliability, and in all superiority to the other conventional slope supporting methods.

The design guidelines for soil nailing in India are drafted by Indian Road Congress (IRC) adopting the guidelines provided by Federal Highway Administration (FHWA).

Soil nailing walls mainly comprise of soil nail, facing component, and drainage component.

Soil nail is a vertical inclusion of specified diameter placed into the slope for its passive reinforcement and to resist bending, shear, and tensile forces. A soil nail can be hollow or solid as per the requirement. Based on the technique of installation of the nail, they are categorized into four types: Drilled and grouted soil nail, self-drilling soil nail, driven soil nails, and jet grouted soil nails.

Facing is a very important component of soil nail structures as it provides support to the exposed soil mid excavation and nail fixture. It also provides continuity throughout and furnishes esthetic finish together with erosion control at excavation face of wall.

The function of spacing is to guard the surface of soil nails slopes which is exposed by holding the soil between the nails and thus providing continuity throughout in the soil nailed structures.

2 Literature Review

Viswanadham and Rotte [1]: A laboratory-based physical model of soil nailed slope was prepared with horizontal slope and 10° inclination of the soil nail in order to find out the effect of stiffness of the facing materials used on the performance of the soil slopes. The impact of seepage pressure was the parameter of utmost concern and was induced into the system with the help of displacement and pore water pressure transducers with increasing ground water table. Different facing materials were used for facing of slope and also a slope without any facing material was experimented for comparison. The results obtained from centrifuge physical tests were compared to the finite element method which was found to be in good agreement. The results showed that the slope without any facing material undergoes maximum settlement, whereas settlement reduces significantly for slopes provided with facing materials.

Azzam and Basha [2]: A soil nailed slope situated in clayey soil was reinforced with soil nails, and tests were conducted on it in order to establish the stress–strain relationship and to find out the characteristic strength of the clayey soil post the reinforcement. The main aim of this study was to determine the effect of soil nailing technique in cohesive soils on the shear strength and settlement properties. Various parameters including nail radius, depth of nail embedment, and no. of total vertical inclusions were varied to study their impact on slope stability. It was observed that in order to attain an optimum reduction in failure of slope due to shear and to reduce the settlement, a sufficient number of soil nails must be employed along with sufficient embedded depth. The tests also confirmed that the application of soil nailing adds to shear strength and reduces the horizontal settlement of soil slopes.

Wu and Fang [3]: A soil nailing system in expansive soils was considered, and for the calculation of various parameters such as bond strength of soil nailing system, nail withdraw strength, and strength of unsaturated soils, simplified calculation formulas were given to examine the stability of the soil nailed slopes in expansive soils. In the calculations, the various varying parameters such as soil nail inclination, position of installation, and length of soil nail were used for calculations. It was found out that the used methods and the procedures of calculations proved okay in expansive soils for analysis of soil nailing stability and for optimization of its design. Also, a reference was provided for optimized design for slopes in the expansive soils.

Kumar et al. [4]: A physical model of soil nailed slope was prepared in the laboratory with the use of four different facing materials, i.e., hexagonal geonet, Tuflex geonet, hexagonal drainage geonet, and bi-axial geogrid along with one rigid facing. The load was applied onto the model with the help of the pressure gauge. The results showed that out of the four flexible materials used, bi-axial geogrid showed the best performance as it took maximum stress of 78 kg/cm^2. While on comparing results of rigid and flexible facing, it was found that rigid facing took a maximum stress of 98 kg/cm^2 and hence performed better than flexible facing.

Loghu Prasad et al. [5]: This paper compares the results of the soil nailed slopes and un-nailed soil slopes of 2H: 1 V in less cohesive soils at 10° and 20° inclinations. This paper also aims at studying the behavior of the soil slopes on application of soil nailing technique. The results proved that soil nailing technique is economic, time saving and offers better strength to soil slopes in less cohesive soils as compared to other conventional methods of stabilization of soil, and also shows that application of shotcreting on the soil slope helps in preventing shear failures of soil slope.

Mangnejo et al. [6]: In this study, attempt was made to study out the various parameters of soil nailing such as various soil nail inclinations, i.e., 20°, 25°, 30°, 35°, and 40° on the varying diameter of the soil nails, i.e., 25 and 40 mm in order to find out their impact on the stability of the existing soil slope in clay using limit equilibrium method and to encourage use of soil nailing system for stability of existing clay slopes. The results show that the existing slope was instable as it had lower FOS. Thus, it was soil nailed with 25 and 40 mm soil nails in three rows and at various angles. Out of which, 40 mm dia soil nails proved to be more effective than the 25 mm dia soil nails as they improved the FOS and thus the stability of the slope significantly.

3 Materials and Methodology

The soil is mixed with water to a OMC of 12% as determined by the laboratory tests and is filled in the physical model in three layers of 150 mm each. Each layer is compacted thoroughly 25 number of times, and nails are installed at desired spacing as per the guidelines of US CODE FHWA-NHI-14-007, shown in Fig. 2 keeping a slope angle of 60°.

A physical model of dimensions 500 mm * 450 mm * 400 mm is prepared in the laboratory with the help of plywood and Perspex sheet. A hydraulic jack with pressure gauge having 5-ton capacity is used for load application. The flex sensor is attached to every nail having 320 mm length and 10 mm dia to detect the strain produced on loading and a multimeter for recording the values of strain is used.

The horizontal and vertical displacements of soil slope are measured after application of every load with the help of ruler as it is easily visible due to installation of Perspex sheet.

The stress is applied until the failure of the slope which is identified by the cracks' development on the soil slope, giving the maximum values of strain and displacements of the slope.

3.1 *Soil Properties*

Various tests were conducted in the laboratory in order to identify the type of the soil sample and to determine the various index properties (Table 1) of soil sample to be used for the physical model.

The value of Cu determined from particle size distribution graph in sieve analysis is 5.069, and value of Cc is equal to 1, and therefore, the soil is identified as well-graded sandy soil.

Table 1 Properties of soil

S. No.	Soil properties	Values
1	Water content (%), w	13.64
2	Specific gravity, G	2.5
3	Liquid limit (%)	14
4	Optimum moisture content (OMC) (%)	12
5	Maximum dry density (kg/m^3)	1983

Table 2 Properties of flexible facing Materials

Property	Geomembrane sheet	Biaxial geogrid	Geotextile fabrics	Coir mat
Thickness	1.5 mm	1.5 mm	1 mm	7 mm
Material	Polymer	Polymer	Polyethylene	Coconut fibers
Tensile strength	15 KN/m	8 KN/m	2 KN/m	20 KN/m

Tensile Strength testing: ASTM D6693[(1)], ASTM D6637[(2)], ASTM D4632[(3)], ASTM D6818[(4)]

3.2 *Facing Materials*

In order to conduct analogical study and to evaluate the behavior of the various flexible facing materials under similar loading conditions, four types of flexible materials have been used. Table 2 depicts the flexible facing materials employed and their characteristics:

3.3 *Scaling and Description of Physical Model*

To embrace the behavior of reinforced slopes toward the various flexible facing materials, five small-scale laboratory testing of soil nailed slopes reinforced with six nails has been conducted. For scaling purpose, the various parameters associated with the soil, the slopes and the soil nail have been established. A 1: N small scale has been used by scaling down the various dimensions related to soil, slope, and soil nails by a N factor for simulating the actual conditions in the model. For proper simulation of loading conditions, a hydraulic jack has been used, and for simulation of vertical and horizontal displacements, a Perspex sheet has been used on the side walls.

In the present study, a physical model is prepared in laboratory conforming to US CODE FHWA-NHI-14-007, which measures 500 mm*450 mm*400 mm (Fig. 1).

Components of physical model:

I. *Hydraulic jack*: It is used with pressure gauge for load application on the soil slope.
II. *Multimeter*: It is used to record the strain produced corresponding to the stress applied by hydraulic jack.
III. *Sensors*: The flex sensors are adhered to the nails as they transmit the values of strain produced to the multimeter and thus assist strain measurement.
IV. *Perplex acrylic sheet*: This transparent sheet aids in observing the values of horizontal and vertical deflections
V. *Soil nails*: Six number of soil nails of 320 mm length and 10 mm diameter arranged in rectangular pattern are provided (Fig. 2).

Fig. 1 Physical model

4 Results

4.1 Without Facing

The slope without facing when subjected to loading, withstands a maximum stress of 0.738 N/mm^2, whereas the maximum horizontal displacement and maximum vertical displacement undergone by the slope are 2.3 cm and 3.1 cm, respectively, as depicted by the graphs in Figs. 3 and 4, respectively.

Fig. 2 Nails arrangement

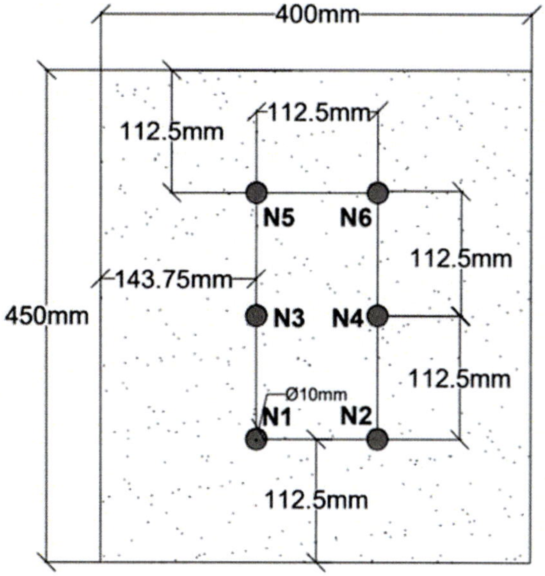

Fig. 3 Stress versus strain graph for slope without facing

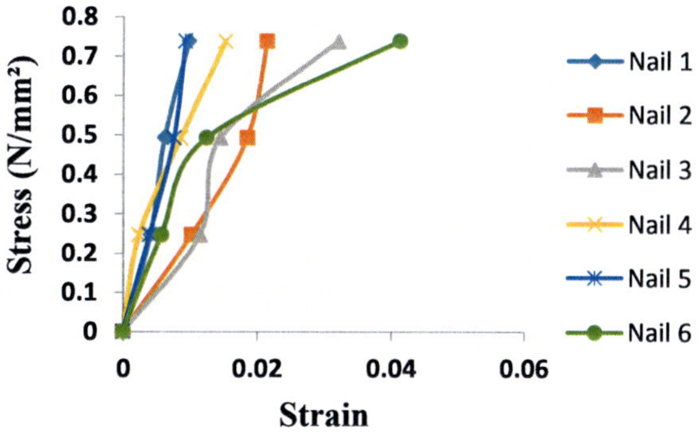

Fig. 4 Stress versus ΔV and ΔH graph for slope without facing facing

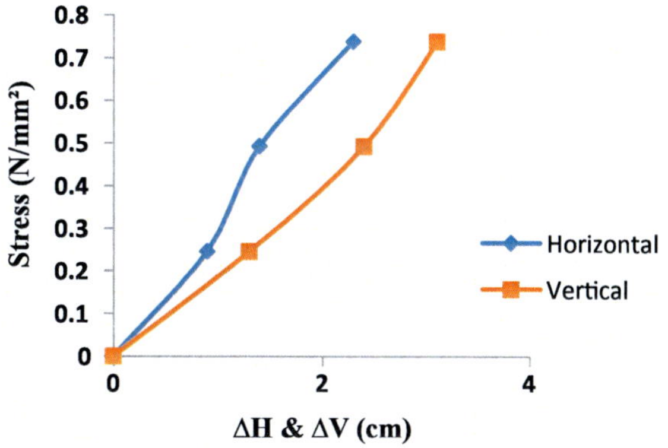

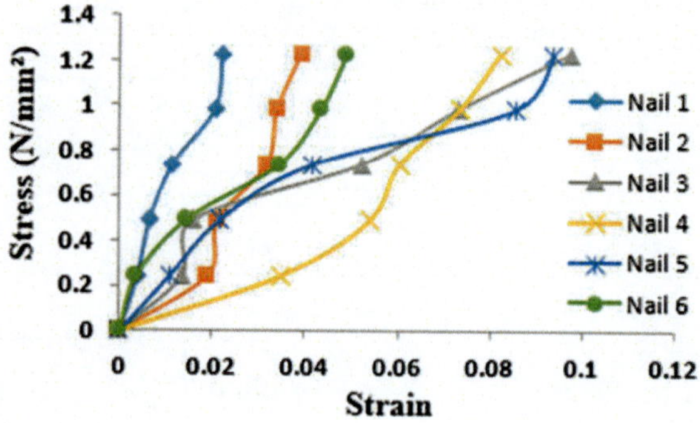

Fig. 5 Stress versus strain graph for slope with geomembrane sheet facing

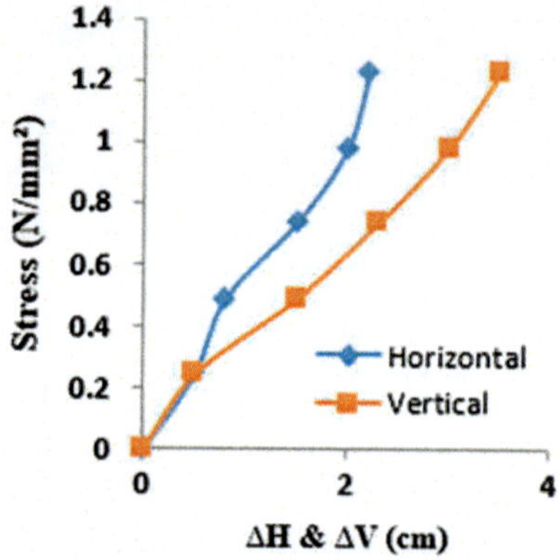

Fig. 6 Stress versus ΔV and ΔH graph for slope with geomembrane sheet facing

4.2 Geomembrane Sheet

The slope with geomembrane sheet facing when subjected to loading withstands a maximum stress of 1.23 N/mm^2, whereas the maximum horizontal displacement and maximum vertical displacement undergone by the slope are 2.2 cm and 3.5 cm, respectively, as depicted by the graphs in Figs. 5 and 6, respectively.

4.3 Biaxial Geogrid

The slope with bi-axial geogrid facing when subjected to loading withstands a maximum stress of 1.107 N/mm^2, whereas the maximum horizontal displacement and vertical displacement undergone by the slope are 1.8 cm and 2.4 cm as depicted by graphs in Figs. 7 and 8, respectively.

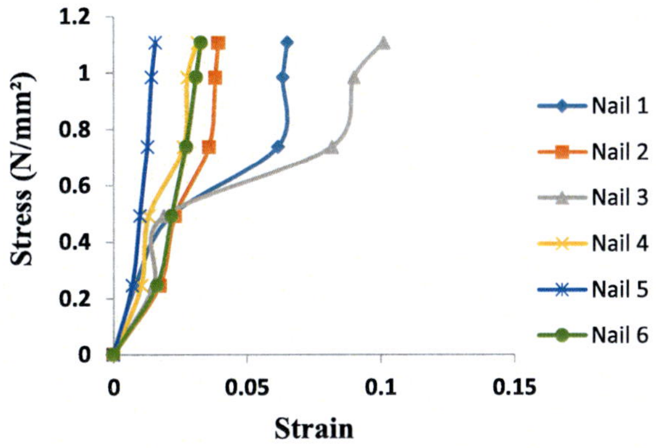

Fig. 7 Stress versus strain graph for slope with bi-axial geogrid facing

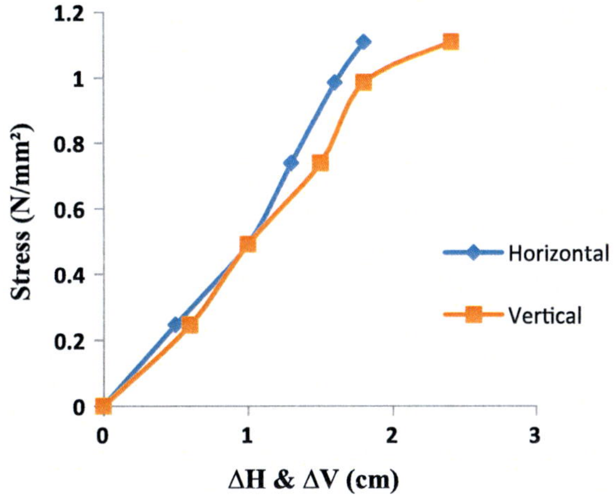

Fig. 8 Stress versus ΔV and ΔH graph for slope with bi-axial geogrid facing

4.4 Geotextile Fabric

The slope with geotextile fabric facing when subjected to loading withstands a maximum stress of 1.107 N/mm^2, whereas the maximum horizontal displacement and vertical displacement undergone by the slope are 1.5 cm and 2 cm as depicted by graphs in Figs. 9 and 10, respectively.

4.5 Coir Mat

The slope with coir mat facing when subjected to loading, withstands a maximum stress of 1.476 N/mm^2, whereas maximum horizontal displacement and vertical

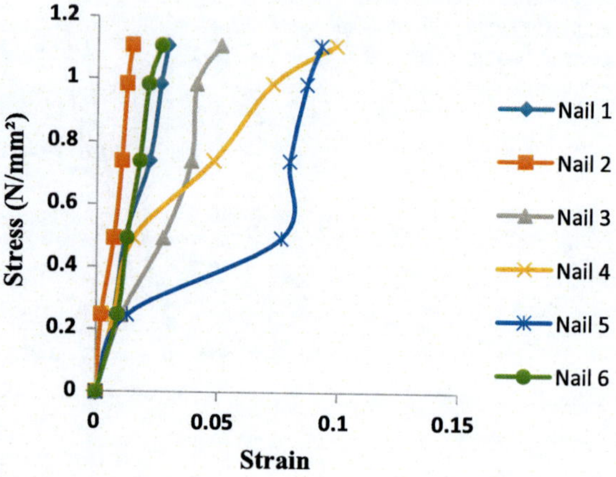

Fig. 9 Stress versus strain graph for slope with geotextile fabric facing

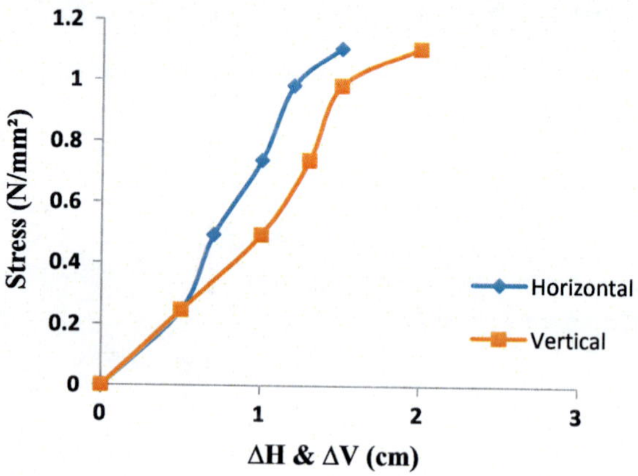

Fig. 10 Stress versus ΔV and ΔH graph for slope with geotextile fabric facing

displacement undergone by the slope are 1.5 cm and 2.8 cm, respectively, as depicted by the graphs in Figs. 11 and 12, respectively.

5 Discussion

5.1 Vertical Displacement

On comparing the vertical displacement under the application of load in the slopes aided with facing materials along with the slope not provided with any facing material, it is observed that the slope without any facing material fails at lowest stress value of 0.738 N/mm^2 showing the maximum vertical displacement of 3.1 cm as depicted in Fig. 3.

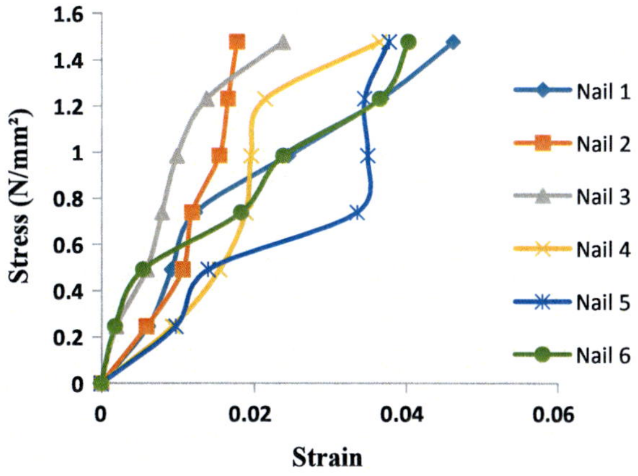

Fig. 11 Stress versus strain graph for slope with coir mat facing

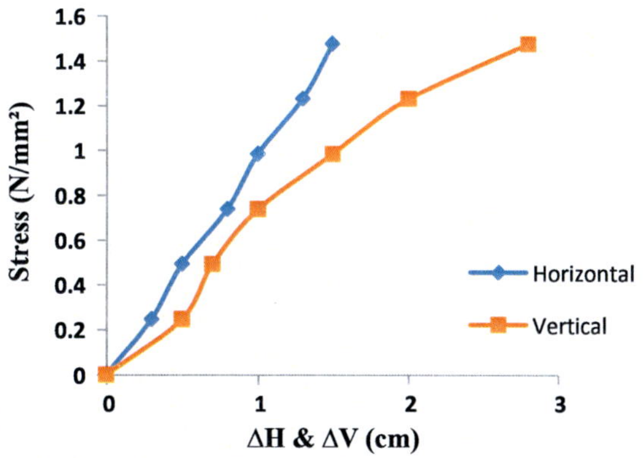

Fig. 12 Stress versus ΔV and ΔH graph for slope with coir mat facing

In case of the slopes provided with facing materials, the maximum vertical displacement is shown by geomembrane of 3 cm at a stress of 1.23 N/mm^2, while the minimum vertical displacement of 2.4 cm is shown by bi-axial geogrid at a stress of 1.107 N/mm^2 as depicted in Figs. 6 and 8, respectively.

5.2 *Horizontal Displacement*

On comparing the horizontal displacement under the application of load in the slopes aided with facing materials along with the slope not provided with any facing material, it is observed that the slope without any facing material fails at lowest stress value of 0.738 N/mm^2 showing the maximum horizontal displacement of 2.3 cm as depicted in Fig. 3.

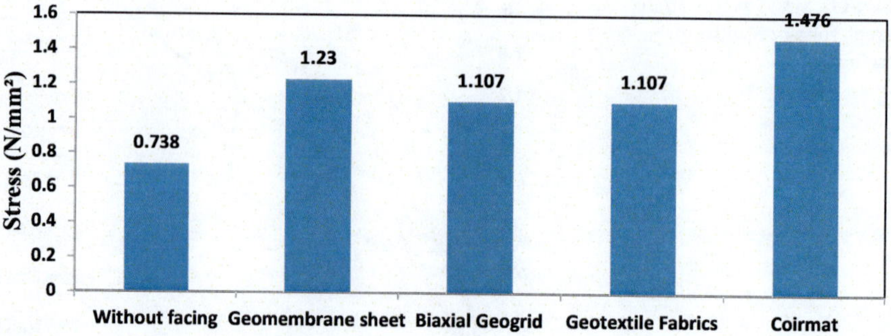

Fig. 13 Comparison b/w stress taken by several facing materials

In case of the slopes provided with facing materials, the maximum horizontal displacement is shown by geomembrane of 2.2 cm at a stress of 1.23 N/mm^2, while the minimum horizontal displacement of 1.5 cm is shown by coir mat at a stress of 1.476 N/mm^2 as depicted in Figs. 6 and 12, respectively.

5.3 *Stress*

Figure 13 depicts the comparison of slopes with the flexing facing materials and the slope without facing material in terms of maximum stress withstood by them with stress values on y-axis and different flexible facing materials used on x-axis. Clearly, the coir mat takes maximum stress of 1.476 N/mm^2 as it is a very ductile material, and also it has very good tensile strength properties. On contrary, the minimum stress of 0.738 N/mm^2 is taken by the slope without any facing material winding up that addition of flexible facing material to soil nailed slopes increases their strengths and assists in reduction of displacement of soil slope in horizontal and vertical direction.

6 Conclusion

(1) The slope without any facing material fails earliest at a stress of 0.738 N/mm^2 as compared to the slopes with flexible facing materials, and thus, the provision of flexible facing material to the soil nailed slopes increases its strength and reduces displacement significantly.

(2) Coir mat proves to be the best flexible facing material among all the other flexible materials used as it shows highest strength of 1.476 N/mm^2 with very low amount of displacements, i.e., 1.5 cm and 2.8 cm in horizontal and vertical directions, respectively.

(3) The addition of coir mat increases the strength of the slope without facing material by almost 50% which is a considerable amount, i.e., the maximum stress of 0.738 N/mm^2 taken by the soil nailed slope without any facing increased to 1.476 N/mm^2 when coir mat is used as a flexible facing in soil nailed slopes.
(4) Flexible facing can be a good remedial solution for increasing the stability of soil nailed slopes by almost 50%.
(5) Flexible facing can be a good substitute to rigid facing in terms of economy where long-term strength is not required or where structure is of non-critical type.

References

1. Viswanadham, B. V. S., & Rotte, V. M. (2015). Effect of facing type on the behavior of soil-nailed slopes: Centrifuge and numerical study. *Discovery, 46*(215), 214–223.
2. Azzam, W. R., & Basha, A. (2017). Utilization of soil nailing technique to increase shear strength of cohesive soil and reduce settlement. *Journal of Rock Mechanic and Geotechnical Engineering*, 1104–1111.
3. Wu, K., & Fang, J. (2018). A study on the method of stability calculation of soil nailing expansive soil slope. In *2018 International Conference on Civil, Architecture and Disaster Prevention, IOP Conf. Series: Earth and Environmental Science 218* (pp. 1–11). https://doi.org/10.1088/1755-1315/218/1/012028
4. Kumar, S., Tangri, A., & Rana, A. S. (2019). Model testing of reinforced soil slope under surcharge loading with adaptable confronting. *International Journal of Innovative Technology and Exploring Engineering (IJITEE), 8*, 802–806. ISSN: 2278-3075
5. Loghu Prasath, S., Malini, P., Mohanchandru, B., Nataraj, N., & Mohanraj, M. (2019). Study on the behavior of slope using soil nailing and shotcreting. *International Research Journal of Engineering and Technology(IRJET), 06*, 1971–1974.
6. Mangnejo, D. A., Oad, S. J., Kalhoro, S. A., Ahmed, S., Laghari, F. H., & Siyal, Z. A. (2019). Numerical analysis of soil slope stabilization by soil nailing technique. *Engineering, Technology and Applied Science Research (ETASR) Journal*, 4469–4473. https://doi.org/10.48084/etasr.2859

Summer Vs Winter Air Pollutants Variation for Year 2019 and Lockdown Effect for CRRI Mathura Road Station Delhi

Shahbaz Ahsan, Shashi Tiwari, S. M. Huzaifa Abbadullah, Deepak Nader, and Gaurav Kumar

Abstract Delhi is confronting extreme air contamination due to fast financial development and speed up urbanization. $PM_{2.5}$, PM_{10}, CO and O_3 pollutant data at CRRI Mathura Road monitoring station are downloaded for analyzed for year 2019 and 2020. The concentrations of $PM_{2.5}$ and PM_{10} are increased from summer to winter, though $PM_{2.5}$ with high jump (doubled) in concentration. Whereas the O3 level is more during summer months. However, CO level is almost same throughout the year. Extreme episode has shown during Diwali festival in the year 2019 and its effect remained in the atmosphere for almost one week. $PM_{2.5}$ was discovered to be the influential species among the contaminations influencing the air quality during winter, recommending that sources like biomass burning and fire crackers should be banned and high attention needed to improve the air quality during winter. The data analyzed herein supplied is a support for source apportionment modeling of pollutants and their health effects.

Keywords $PM_{2.5}$ · PM_{10} · CO · O_3 · CRRI Mathura road

S. Ahsan (✉) · S. M. H. Abbadullah · D. Nader · G. Kumar
Department of Civil Engineering, Manav Rachna International Institute of Research and Studies, Faridabad 121004, India
e-mail: shahbaz_ahsan1819@manavrachna.net

S. M. H. Abbadullah
e-mail: syed_mdhuzaifa2018@manavrachna.edu.in

D. Nader
e-mail: deepak_nader1819@manavrachna.net

G. Kumar
e-mail: gaurav4_kumar2017@manavrachna.net

S. Tiwari
Department of civil engineering, Institute of Engineering and Technology, Lucknow 226021, Uttar Pradesh, India
e-mail: stiwari.ced.cf@ietlucknow.ac.in

A. K. Gupta et al. (eds.), *Advances in Construction Materials and Sustainable Environment*, Lecture Notes in Civil Engineering 196,
https://doi.org/10.1007/978-981-16-6557-8_43

1 Introduction

Healthy air is a combination of different gases like nitrogen, oxygen, argon, carbon dioxide and modest quantities of different gases in a fixed extent. However, in the event that structure of air modifies using any and all means; it is known as air contamination which can prompt impact on climate, human wellbeing, and other living animals. Gaseous pollutants like SO_x, NO_x, O_3, CO, etc., and particulate matter PM_{10} and $PM_{2.5}$ are the pollutants which cause pollution. Particulate matter is in the environment through both natural and anthropogenic sources. The anthropogenic sources like vehicular discharge, mechanical burning cycle, business and private ignition and development enterprises. While gaseous toxin contamination fundamental source is consuming of carbonaceous fills, outflow from IC motors, Hydrocarbons and NO_x after responding with daylight brings about O_3 development [1]. A major effect due to these pollutants is on health and environment. Human wellbeing impacts incorporates respiratory issues, liver fibrosis, lung/liver disease, heart stroke, bone issues because of PM [1], anoxemic prompting different cardiovascular issues. Babies, pregnant ladies and old individuals are at higher danger respiratory issues, asthma, bronchitis and so on. Due to gaseous pollutants especially due to carbon mono-oxide (CO) and ozone (O_3) [1]. Effects on human health are very severe and need attention to take preventions. World Health Organization announced in 2014, Delhi is the most contaminated city of the world [2]. Exposure to ambient fine particulate matter ($PM_{2.5}$) has been identified as the leading health risk in India [3] with a staggering health [2] and economic [4] burden. In this study, we have collected data for CPCB CRRI Mathura road, Delhi. It has shown that pollution level varies significantly with, time and climatic conditions. We discovered that the effects of air pollutions such $PM_{2.5}$, PM_{10}, CO and OZONE are present in the air in different ranges, in different times (in the festive season like Diwali) pollute the air of DELHI NCR which result in high levels of pollutant concentration. Whereas in the lockdown phase when vehicles and every industry were shut the pollutant level falls to extremely low.

2 Method and Data

The data of different air pollutants ($PM_{2.5}$, PM_{10}, CO and OZONE) collected from Central pollution control board (CPCB). The data available on website for $PM_{2.5}$ and PM_{10} and two vaporous contaminations (CO and O3) at CRRI Mathura road of Delhi during the period of April 2019 to June 2020 downloaded to quantify the pollutant levels for this station. Central pollution control board (CPCB) has 40 air quality monitoring stations for year 2020 at different locations. The data shown here has been collected for central road research institute (Mathura Road, Delhi) monitoring stations as shown in Fig. 1.

The latitude, longitude and elevation of CRRI, Mathura Road monitoring station are 28.5512005, 77.2735737 and 194, respectively.

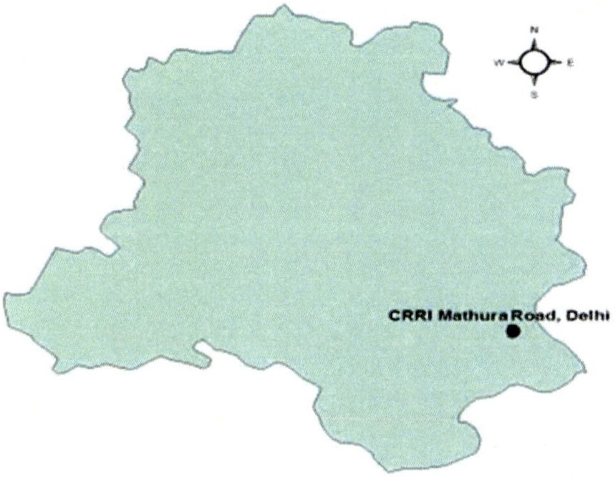

Fig. 1 Location of CRRI Mathura Road on Delhi shape file

The monitoring systems installed to determine the concentrations of the particulate matter PM_{10}, $PM_{2.5}$ and gaseous pollutants including SO_2, NO_2, CO and O_3 at each site. The permissible concentration for particulate matter and gaseous pollutants are shown in Table 1.

We downloaded the daily data from CPCB website and averaged it monthly to see monthly variation. We carried out another round of quality check prior to our main analysis and only considered the days where more than 20 h of valid data are recorded. Extreme episode analysis is also done to see days having high concentration and reasons behind using daily data.

Table 1 National ambient air quality standards (2009) (Pollution & Board, 2012)

Pollutants	Time	Ambient air concentration		Measurement methods
		Residential, industrial, and other areas	Ecologically sensitive area	
PM_{10}, $\mu g/m^3$	Annual	60	60	– Gravimetric – TEOM – Beta attenuation
	24 h	100	100	
$PM_{2.5}$, $\mu g/m^3$	Annual	40	40	– Gravimetric – TEOM – Beta attenuation
	24 h	60	60	
Ozone (O_3), $\mu g/m^3$	8 h	100	100	– UV Photometric – Chemiluminescence – Chemical method
	1 h	180	180	
Carbon monoxide (CO), mg/m^3	8 h	02	02	– Non dispersive infrared (NDIR) – Spectroscopy
	1 h	04	04	

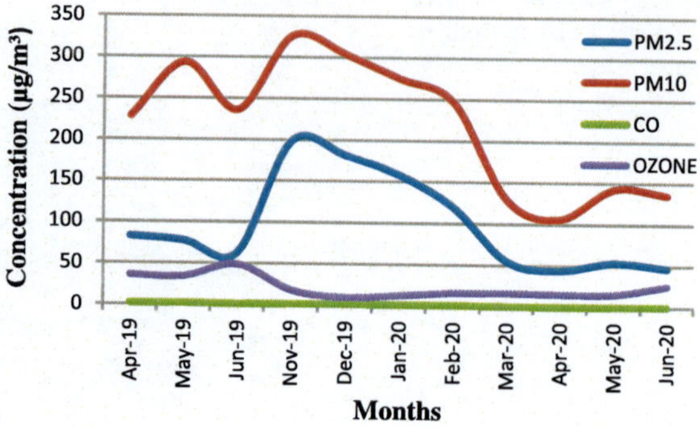

Fig. 2 Monthly variation during summer and winter months for year 2019

3 Result and Analysis

3.1 Monthly Variation of Pollutions

We compared the monthly data, for four air pollutants (i.e., $PM_{2.5}$, PM_{10}, CO, OZONE) for winter season and summer season months. Figure 2 shows the fluctuation in different parameters of air pollution in every month from winter season to summer season. During year 2019 summer months, $PM_{2.5}$ values are in the range of 40–80 $\mu g/m^3$, while it rises to 200 $\mu g/m^3$ during winter season. However, PM_{10} concentration is less during summer season compared to winter. But as we see a large hike for $PM_{2.5}$ pollutant from summer to winter season, is not the case for PM_{10}. On the other hand, interestingly, the ozone concentration is more during summer season. While CO concentration is very less about zero during both the seasons.

3.2 Effect of Crackers Burnt on Diwali

In 2019, 27th October was celebrated for Diwali Festival. On the auspicious occasion people do burn crackers. However, Government had banned crackers strictly for year 2019. Though we have seen pollution rise during Diwali and its effect for almost one week due to non-dispersion of pollutants. Figure 2 shows the effect of crackers burnt on Diwali. After 1 day of Diwali in October 2019, Delhi air quality rose sharply to severe health condition. Delhi showed $PM_{2.5}$ levels of up to 521.77 micrograms per cubic meter ($\mu g/m^3$).

3.3 Decreases in Air Pollution During Lockdown Due to Pandemic Coronavirus in 2020

A massive decrease in air pollution has been observed in the months April, May and June 2020 as compared to 2019. During this time, India was on complete lockdown due to COVID-19. All the industries were shut down and transportation was completely restricted due to which there was a sudden decrease of air pollutants in the atmosphere. The below figures show the difference in range of air pollutants (i.e., $PM_{2.5}$, PM_{10}, CO and OZONE) between the year 2019 and 2020. Figures 3 and 4 show the monthly variation in pollution without and with lockdown due to COVID-19 for year 2019 and 2020, respectively.

During 2019, the PM_{10} concentration varies between 225 to 300 $\mu g/m^{3}$, while in 2020 it remained between 110 and 150. Which is almost 50% sharp reduction of year 2019. Similarly, $PM_{2.5}$ values remained between 40 and 50 $\mu g/m^3$ for year 2020, which is under permissible limit specified by CPCB as shown in Table 1. Interestingly, Ozone concentration is also reduced by 50% from 2019 to 2020 showing values between 18 and 25 $\mu g/m^3$. However, pollutant ozone concentration was very less in 2019 and 2020 and has no significant effect for Delhi NCR. (Fig. 5)

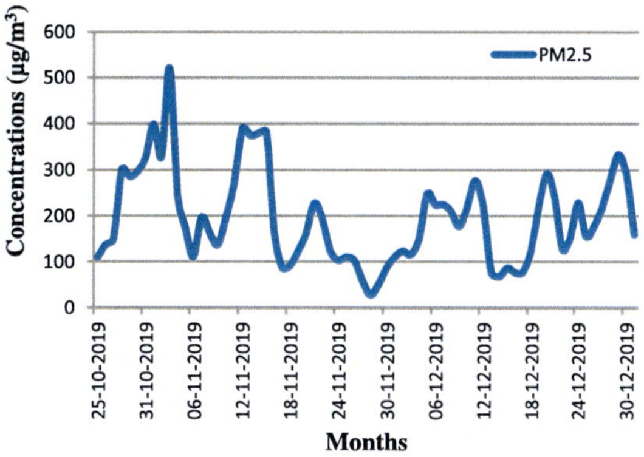

Fig. 3 Daily variation for $PM_{2.5}$ concentration after Diwali celebration in 2019

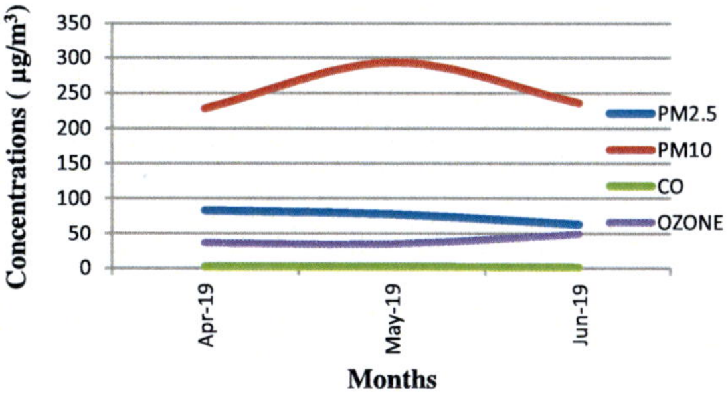

Fig. 4 Monthly variation of pollutants during summer season in 2019

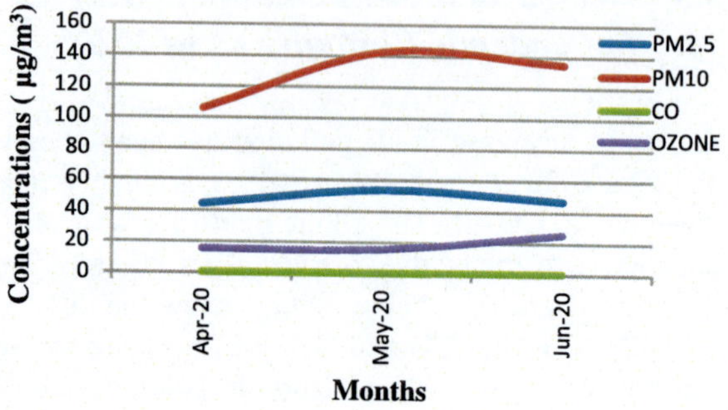

Fig. 5 Monthly variation of pollutants during summer season in 2020

Table. 2 Pollutant variation for summer and winter season 2019

$PM_{2.5}$		PM_{10}		CO		Ozone	
Summer	Winter	Summer	Winter	Summer	Winter	Summer	Winter
74	180	253	301	1.7	1.5	40	14

3.4 Winter Versus Summer Concentration

Table 2 shows the variation of pollutants for summer and winter months with concentration values in µg/m^3. $PM_{2.5}$ concentration is increased from summer season to winter season by 1.4%, while PM_{10} values increased by 0.2% only. The CO concentrations have not shown much variation seasonally. However, O_3 Concentration in more during summer season and decreased by 0.7% during winter.

4 Conclusions

During winter season due to atmospheric stability and less boundary layer height, pollutants do not disperse quickly and remained high for various days. PM_{10} concentration is almost high during both the season. The winter months $PM_{2.5}$ concentration is double or more than the values during summer season. The ozone concentration is high during summer season. However, CO concentration is remained same during both the season. After Diwali celebration for year 2019, peaks are shown from 29th October to 5th Nov 2019. It remained in the atmosphere for almost 7 days in the atmosphere. From the literature, it is evident that per 10 µg/m^3 increase in $PM_{2.5}$ concentrations, there is 3.1% increase in COPD hospitalization [5]. So it is very important to minimize the concentration especially during winter season. Biomass burning and crackers should be strictly banned. To overcome and for betterment pollutant emission control strategies is highly demanded.

Acknowledgements We acknowledge central pollution control board for providing pollutants data through website.

References

1. Central Pollution Control Board (CPCB). (2015). Air pollution in Delhi—An analytical study. Envis Centre Cpcb, pp. 1–20.
2. Balakrishnan, K., Dey, S., Gupta, T., Dhaliwal, R. S., Brauer, M., Cohen, A. J., Stanaway, J. D., Beig, G., Joshi, T. K., Aggarwal, A. N., Sabde, Y., Sadhu, H., Frostad, J., Causey, K., Godwin, W., Shukla, D. K., Kumar, G.A., Varghese, C. M., Muraleedharan, P., & Dandona, L. (2019). The impact of air pollution on deaths, disease burden, and life expectancy across the states of India: The Global Burden of Disease Study 2017.
3. *The Lancet Planetary Health, 3*(1), e26–e39. https://doi.org/10.1016/S2542-5196(18)30261-4
4. Dandona, L., Dandona, R., Kumar, G. A., Shukla, D. K., Paul, V. K., Balakrishnan, K., Prabhakaran, D., Tandon, N., Salvi, S., Dash, A. P., Nandakumar, A., Patel, V., Agarwal, S. K., Gupta, P. C., Dhaliwal, R. S., Mathur, P., Laxmaiah, A., Dhillon, P. K., Dey, S., & Swaminathan, S. (2017). Nations within a nation: Variations in epidemiological transition across the states of India, 1990–2016 in the Global Burden of Disease Study. *The Lancet, 390*(10111), 2437–2460. https://doi.org/10.1016/S0140-6736(17)32804-0
5. Pandey, A., Brauer, M., Cropper, M. L., Balakrishnan, K., Mathur, P., Dey, S., Turkgulu, B., Kumar, G. A., Khare, M., Beig, G., Gupta, T., Krishnankutty, R. P., Causey, K., Cohen, A. J., Bhargava, S., Aggarwal, A. N., Agrawal, A., Awasthi, S., Bennitt, F., & Dandona, L. (2021). Health and economic impact of air pollution in the states of India: The Global Burden of Disease Study 2019. *The Lancet Planetary Health, 5*(1), e25–e38. https://doi.org/10.1016/S2542-5196(20)30298-9
6. Li, M. H., Fan, L. C., Mao, B., Yang, J. W., Choi, A. M. K., Cao, W. J., & Xu, J. F. (2016). Short-term exposure to ambient fine particulate matter increases hospitalizations and mortality in COPD: A systematic review and meta- analysis. *Chest, 149*(2), 447–458. https://doi.org/10.1378/chest.15-0513

Use of Fly Ash—A Resourceful Byproduct in Road Embankment: A Review

Deepak Kumar Sahay and Sunita Bansal

Abstract Fly ash is an acidic, fine powdery waste by-product material produced during combustion of coal in thermal power plants (TPP) for generation of electricity. Safe disposal of it is one of toughest challenge being faced by mankind. Considerable research has been done since worldwide use of coal began in 1920 for generation of electricity. Coal continues to be used as a major source of fuel for power generation worldwide. Thermal power is the largest source of electricity of India and coal-based one hundred ninety-seven (197) TPP account for around 72% of total installed power generation capacity. Indian coal has high ash content in-between 30 and 45%. India being third largest worldwide producer of coal, generated 226.13 Mt (Million tons) of fly ash in 2019–20. Exhaustive research was conducted since its inception to explore safe disposal and properties of this environmental pollutant. During the year 2019–20, as per CEA, GOI, reuse of fly ash touched 83.05% while 16.95% could not be utilized. Though fly ash, being a valuable waste material, its disposal through conventional method may lead to environmental, radiation, and magnetic issues on air, water, soil, man, animal, and aquatic life.

Keywords Fly ash · Sulfate resistance · Sustainability · Soil stabilization

1 Introduction

Since 1920 with the wide use of coal for generation of power from TPPs, safe disposal, and utility of fly ash, being environmental pollutant, hazardous, acidic by product, posed a Herculean task worldwide. For disposal of the total fly ash produced in India, initially, the only option available was dumping it in low-lying areas or filling of abandoned mines and collieries. Because of hazardous and environmental issues

D. K. Sahay (✉) · S. Bansal
Manav Rachna International Institute of Research and Studies (MRIIRS), Faridabad, Haryana, India

S. Bansal
e-mail: sunitabansal.fet@mriu.edu.in

A. K. Gupta et al. (eds.), *Advances in Construction Materials and Sustainable Environment*, Lecture Notes in Civil Engineering 196,
https://doi.org/10.1007/978-981-16-6557-8_44

attached with fly ash disposal, a wide range of research were taken up in India and worldwide.

Coal and lignite continue to be used as a major source of fuel for power generation worldwide. Thermal power is the largest source of electricity of India and coal-based one hundred ninety-seven (197) thermal power plants (TPP) account for around 72% of installed capacity for power generation. The different usages of fly ash depend upon coal's chemical composition, reactivity, and mineralogy. Further Indian coal is rich in ash content, between 30 and 45%. India is third largest worldwide producer of coal and in the year 2019–20 generated 226.13 Mt (Million tons) of fly ash. A significant 19.08%, i.e., 1.055 Mt, of fly ash was utilized in the roads, embankments, and fly overs construction. Around 30% quantity produced is utilized in partial replacement of OPC and hydraulic plaster, with improved setting time and chemical attacks, while another 40% is disposed as landfills. Fly ash reuses are in partial replacement of OPC and sand, road embankment fills, in production of Grout and flowable fill grout, soil and waste stabilization, production of cement clinkers, mines reclamation, road subbase construction material, as mineral filler in asphaltic concrete, for brick production, loose application on rivers, roads and parking lots to melt ice, production of geo-polymers, tiles for roof, floor, ceiling, Insulated panels etc.

2 Literature Review

2.1 Coal

Basically coal can be divided into four categories with varying chemical composition, heat value, ash content and geological origin and process of handling in wet, conditioned or dry format (Table 1). They are: (i) Anthracite, (ii) bituminous, (iii) sub-bituminous, and (iv) lignite.

Table 1 Chemical composition of coal

S. No.	Chemical composition (in wt. %)	Lignite	Bituminous	Sub-bituminous
1	SiO_2	15–45	20–60	40–60
2	Al_2O_3	10–25	5–35	20–30
3	Fe_2O_3	4–15	10–40	4–10
4	CaO	15–40	1–12	5–30
5	MgO	3–10	0–5	1–6
6	SO_3	0–10	0–4	0–2
7	Na_2O	0–6	0–4	0–2
8	K_2O	0–4	0–3	0–4
9	LOI	0–5	0–15	0–3

Fly ash is also defined on the basis of type of coal used in the boiler. They are chemically of either calcareous and silicious and also known as Class C and F class. Anthracite coal is generally not used in burning in utility boilers, so quantity of this type of fly ash is very less available. ASTM C618-05 [1].

2.2 CEA Annual Report 2019–20 [2]

Stressed need for great efforts to achieve 100% utilization of fly ash from present 83% (187.81 Mt). Non-utilization of fly ash in India has created hazardous environment both at the point of deposit and at construction point. At present, utilization of fly ash varied 3–57%, yet it averages only 16% of the total world's ash quantity. During the year 2019–20, as per CEA annual report, 25.60% of fly ash produced was used in cement manufacture and replacement, while 15.5% is utilized in land fill, 9.80% for ash dyke, 9.46% for manufacture of tiles and bricks, 9.27% in construction of embankment for roads and flyovers, and 4.69% in mines filling, 0.74% in concrete works, for agriculture work, it was 0.06%, for other utilization, it was 7.92%, while 16.95% could not be utilized.

As per MORTH annual report 2019–20 [3], India stands as having second largest of road network worldwide with 5.898 million kilometer of National Highways, Expressways, State Highways, Major District Roads, Other District Roads, and Village Roads of which National Highway and Expressway make a length of 1,32,500 Kms, while State Highways and other roads make it to 1,56,694 and 56,08,477 km, respectively. The year 2009 amendment notification of MOEF, GOI (03/11/2009) [4] laid down a target for achieving 100% utilization of fly.

ash in:

(a) Five (5) phases for all coal- and lignite-based TPP stations commissioned before 3/11/2009 for achieving 100% utilization within 5 years of commissioning and
(b) Four (4) phases for all TPPs commissioned after 3/11/2009 within 4 years (Fig. 1).

The latest amendment by MOEF, GOI, (28/04/2016) [5] makes TPPs responsible to bear the cost of transportation for supplying fly ash up to 100 KMs and from sharing the transportation cost 50:50 beyond 100 km up to 300 km. It also made use of mandatory for all works including buildings, roads, etc., under State and Central Government. Ministry of Surface Transport, GOI, has been working on plans to further improve the connectivity, increase speed of traffic, and upgrading quality of road parameters. From Table 1, it is clear that fly ash of sub-bituminous coal and lignite have higher content of CaO and lower content of LOI than that of bituminous one, while it contains higher percentage of SO_3. As per ASTM C618, if content of:

$$SiO_2 + Al_2O_3 + Fe_2O_3 > 70(\text{wt.\%})\text{then fly ash is called as Class F and if it is :}$$

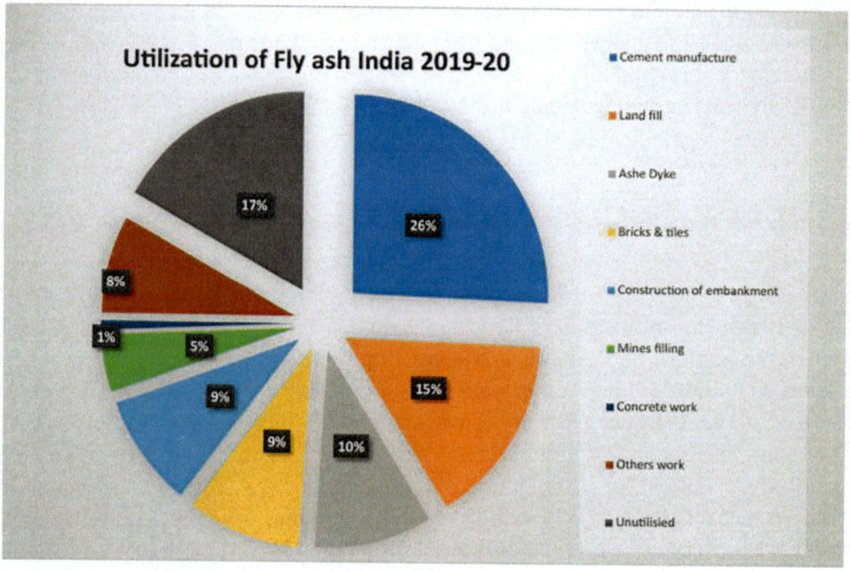

Fig. 1 Fly ash utilization

$$50(\text{wt.}\%) < SiO_2 + Al_2O_3 + Fe_2O_3 < 70(\text{wt.}\%), \text{ then known as Class C.}$$

Generally, the total of sodium and potassium and SO_3 is higher in Class C fly ash than Class F. End use of fly ash is generally dependent on the required property for a particular utilization. It fly ash is to be used as structural fill, then one has to look for its ability to be compacted in lower unit weight layers which is a particle size distribution function and for post-compaction fly ash with high calcium may be beneficial. For base course of highways, strengthening may need low calcium fly ash if there is enough of time.

2.3 *Micro Structure Study of Fly Ash*

Hemmings and Berry [6] studied fly ash using SEM and termed morphologically to be a heterogeneous material. Fly ash is composed of lightweight hollow and spherical, particles called cenospheres, large glassy particles plerospheres and some part-melted spherical particles clathroshpheres indicating occurrence of intensive chemical for shorter duration within particles in the furnace zone with smoother outer surfaces due to low calcium content.

Table 2 Chemical composition of Indian and USA fly ash

S. No.	Chemical composition (in %)	Indian fly ash			U.S. Fly ash
		DF	Range	Mean	
1	SiO_2	60.12	53–63	59.3	28–59
2	Al_2O_3	30.16	27–37	26.6	7–38
3	Fe_2O_3	6.36	3.3–6.1	5.5	4–42
4	CaO	1.00	0–3	1.5	0–13
5	MgO	0.53	0–0.8	1.0	–
6	Na_2O	0.06	0.1–0.4	0.1	–
7	K_2O	0.007	0–0.9	1.07	–
8	SO_3	0.01	–	0.2	–

2.4 *Properties of Fly Ash*

As defined by Ahmaruzzaman [7] as follow:

i. **Physical properties**:

Fly ash is alkaline abrasive generally gray in color by nature consists of glossy fine powder either hollow or solid particles spherical in shape and the angular particles, are composed of carbonaceous material. Bituminous fly ash's particle size distribution is almost similar to that of silt finer than 75 μm, but sub-bituminous fly ash is coarser than bituminous fly ash. Its sp.gr. is in between 2.1 and 3 sp. surface area range between 170 and 1000 m^2/kg. Depending on quantity of iron and unburnt carbon, its color can be from generally red/tan/grey to black [8], and pH varies from 1.2 to 12.5 [9]. Havanagi [10] reported Sp. Gravity of Dadri Plant of NTPC fly ash as 2.2, specific surface area 3520 cm^2/g, and loss on ignition (%) as 0.4.

ii. **Chemical properties**:

Table 2 gives a comparison picture of range of fly ash in India and USA. The Dadri Fly ash data as per Gayathri [11]. Range by Sridharan [12] from tests on fly ash from 13 TPPs of India, USA data from Yudharbir and Honjo [13].

It also contains small percentage of B, Cu, Mn, Mo, and Zn. Its various geotechnical properties make it suitable for use in a number of Civil Engineering applications.

2.5 *Construction of Road Embankments*

Raymond [14] and Joshi et al. [15] concluded from various geotechnical tests conducted on trial embankment done in UK and USA that fly ash can be used as a filler material on soft ground due to its lighter weight. Fly ash as filler material, therefore, gives a better choice than by soils in construction of road embankment with RE walls techniques thus avoiding costly and slow paced, requiring precious more space-retaining wall construction in urban areas. The other advantage of using ash fill in embankments is reduction of shear stress and lateral deformations of geo-textiles with steeper sides.

Martin et al. [16] summarized from that UCS substantially increased during first week of curing followed by nominal increase after third week and then decreased after 8 weeks, on samples compacted at OMC and cured up to 8 weeks, while in the case of soaked samples, after 2 weeks, the UCS was not as fast as un-soaked case but after 8 weeks, UCS was more for soaked ones. It was also observed that there was a drop of 40% strength of saturated sample compared with unsaturated case. Studies were conducted on quality of water for use with fly ash for embankment construction on highways by Mathew et al. [17] in Maryland brought out that the ground water data indicated that the leachate from fly ash had no discernible impact on ground water quality.

Baykal et al. [18] studied embankment constructions on highways, in colder regions with fly ash. Due to availability in abundance and low cost of by products and mineral waste, reduction in cost of construction can be achieved. Fly ash, a residue of coal from TPP, has been widely used in the construction works. Its compaction of is very crucial to water requirement which may create complication in embankment construction over 1–2% OMC. This study reports a unique way for working out quantity of water for fly ash mix to increase formation of cementitious mineral without letting down compactibility. In an investigation, fly ash compaction was done at OMC, and in another test, fly ash was compacted with 10% more (by weight) crushed ice and set up cured from 1 to 90 days at a temperature of 21 °C. UCS and tensile tests (splitting) done after end of curing. Using powder XRD analysis and ESEM forming of cementitious mix was observed with micro fabric changes during curing timing. Added extra snow in it increased the UCS and tensile strength (split) were noticed onward 14 days of curing. Increase in shear strength can be explained from ESEM micrographs and X-ray diffractograms. The better performance of test result by using snow was encouraging. The new methodology can be used for allowing construction of embankments and other structures even during winter in colder climate, wherein no construction work is taken up during winter.

The restriction put by Indian code (IRC: SP 58—2001) [19] was also examined by Basu and Sarasati [20] and concluded that with achievable compressive strength more than 53 N/mm^2 in 28 days by using high-strength cement and very high lime reactivity (more than 6 N/mm^2) of fly ash and cement reactivity more than 90%, it is possible to achieve up to 70% replacement of cement in HVFAC using super plasticizers. IS: 12330— (1988) [21] & IS: 12269—(2013) [22] are generally used for embankment work. IS: 6909–1990 [23] is a blend of 70% granulated blast furnace slag, 20% calcium sulfate and 7% 33 grade ordinary portland cement and can be used in places below 40 °C. Iron oxide present in supersulfated portland cement combines with C_3A and is limited to 3.5% due to its susceptibility to sulfate action. Yoon et al. [24] after study of a demonstration embankment built with sixty to forty ratio (by

Table 3 '*R*' sulfate resistance factor versus concrete expansion

Sample	Expansion (in %)				$R = \frac{(CaO-5)}{Fe_2O_3}$
	Wetting and drying test		10% soak test		
	700 days	1200 days	700 days	1200 days	
M 6498	0.02	0.04	0.00	0.02	0.17
M-6754	0.02		0.03	0.07	1.06
M 6730	0.02	–	0.03	0.06	1.67
No ash	0.05	0.15	0.05	0.10	–
M 6510	0.08	0.18	0.02	0.05	2.33
M 6577	0.07	0.09	0.05	0.05	2.57

weight) of fly ash and bottom ash mixture, confirmed result as per field and laboratory test and its viability in comparison with conventional other filler materials.

Price and Peterson (1961) [25], investigated the concern of increased sulfate resistant of fly ash concrete, made with Katchewan lignite. Researchers with worked with lignite fly ash from other TPP and found contradicting results. Dunstan [26] proposed a sulfate resistance factor, $R = \frac{(CaO-5)}{Fe_2O_3}$ (oxide analysis %). If R works out to be less than 1.5, fly ash improves it. Dustan also reported that presence of very active high calcium in fly ash close to field of gehlenite which plays important role in sulfate attack and was correlated to above value of *R* (Table 3).

A number of research conducted study on admixture fly ash and concrete reported a number of positive points to addition of fly ash to OPC. It reduces water demand thus reducing water cement ratio, an improved particle size, and thereby, reducing air entrainment whose effect minimizes if allowed to cure, concrete's performance stands satisfactorily. It was reported an increase to corrosion resistance due to reaction forming stable cement with $Ca(OH)_2$ with corrosive fluid. Ravina and Mehta (1986) [27] reported that a replacement of 35 to 50% cement by brought a reduction of 5–7% water demand for a particular slump. Mukherjee and Vesmawala [28] demonstrated that fly ash is a lightweight material as compared to soils. Compacted ash demonstrated higher strength (shear) than to soils used as filler. In the cases of soils of weaker bearing capacity, fly ash gives advantage. Permeability of fly ash in un-compacted state is better. Compacting fly ash is easier as compared to soils, as moisture-density curve is more even. Thus, fly ash is a better fill material than soils in road embankment or approach embankment construction, backfilling, etc., and is environment friendly. The water resisting property of supersulfated cement gives it edge over OPC in hydraulic engineering work, RCC pipes for ground water, concrete work in soil containing sulfate, sewer system carrying industrial effluents, etc. SRPC is suitable for mass concrete due to its low heat of hydration, which is maximum limit of 60 and 70 cal/g Jain and Garg [29]. The following two cases of completed fly ash road embankments are discussed below:

Reinforced fly ash RE wall Okhla Bridge in New Delhi (1996): The approach road embankment close to NH 2 is about 60 m with height varying from 5.3 to 7.3 m

Table 4 Materials properties of Okhla fly over

Properties	Backfill soil	Bottom ash	Pond ash
L.L ((%)	26	NP	NP
P.I. (%)	NP	NP	NP
Silt and clay size < 0.0075, (%)	50	14	78
Sand size -0.0075–2.0 mm	50	86	22
MDD (kN/m^3)	19.6	10.1	10.4
OMC (%)	10	37	35
Cohesion, kPa	10	-	-
Angle of friction	32	37	31

was successfully completed and opened for traffic.. Ponded ash and bottom ash from Badarpur Thermal Power station were transported to site, a lead distance of 6 Kms. Geogrids both mono- and bi- were used in construction of embankment. During rains, it was observed that unlike construction with soil as fill material, and in this case, work could be resumed within four hours of interruption. For compaction of fill, both static (8 Tons) and vibratory rollers were used. A MDD of 95% of modified proctor test was achieved, Malik et al. [30] (Table 4)

Approach Road Embankment of 2nd Nizamuddin Bridge: (1995) is located downstream the one existing bridge and id about 1.9 km long and 6–9 m high. Filling of low lying areas were done with fly ash from Indraprastha TPP about 2 km away. Fly ash was laid in compacted layers of 250 mm with static 8 tons and vibratory rollers to achieve 95% of modified proctor MDD (Table 5)

During mining and washing 10–15% coal is generated as coal gangue and this gangue occupy land and damages environment as well as ground water. It also causes landslides and debris flow, etc. Fly ash generated in TPP is stored in silos or in ash pond. Over the time, a large amount of fly ash accumulates on ground surface [32, 33] which occupies large piece of land which could been used for otherwise and

Table 5 Materials properties of Nizamuddin approach embankment [31]

Material properties	Soil	Fly ash	Gravel
Sp. gravity	2.6–2.7	1.80–2.0	2.65–2.72
LL (%)	26–30	NP	–
P.I. (%)	5–8	NP	–
MDD (in kN/m^3)	18.1–20.1	11.2–12.8	20.2–21.6
OMC (%)	8–11	24–30	7–8
Coefficient of permeability (cm/s)	10^{-6}	10^{-4}	–
Cohesion, c (kPa)	0–10	0	0–12
Angel of friction, φ (°)	30–32	35–42	30–33

Table 6 Specific mass ratio of gangue and fly ash

Material	Specific ratio (mass ratio)								
Gangue	1.0	1.0	1.0	1.0	1.0	0.9	0.7	0.5	0.3
Fly ash	0.3	0.5	0.7	0.9	1.0	1.0	1.0	1.0	1.0

is threatens land, water bodies, and soil with migration and leaching. Li et al. [34]. Sekine and Sunaga [35] made a full size embankment made with fly ash and studied the stability and protocols. Shekhovtsova et al. [36] studied the of fly ash reaction with cement system that was alkali activated and developed a system to find out the reaction of low calcium fly ash in a activated alkaline system to find out the strength of fly ash binder. Fly ash and 50 mm particle size gangue was selected to be used in land reclaiming in areas of coal mines and road bed backfill also. SEM images showed gangue as dense particles with uneven surfaces filled with small particles of fly ash. The mixed fly ash had an irregular structure but was highly plastic indicating that it can be used to fill gaps between gangue particles. The mass ratio tried was (Table 6).

2.6 Environmental Concern

Following environmental consideration must be addressed for use of fly ash in construction of road embankment, pavement and as filling material: (i) Erosion due to wind, (ii) Erosion of surface due to water, (iii) Leaching due to surface water runoff, and (iv) Leaching due to percolation of runoff water into ground water.

The embankment area as such should be made impervious or made of low permeable materials and area should be sealed suitably to avoid percolation of water.

Using the modified oedometer and lead solution, to cross through the compacted sample and analyzed the readings. Pandian et al. [37, 38] that fly ash rich in calcium retain more lead ions compared to fly ash low in calcium content. Leaching studies were done by Pandian and Balasubramonian [39] with Jar method, zinc concentration increased for first 3 days to decrease slightly later with increase of value of pH for Neyveli fly ash. Pandian et al., concluded that Neyveli fly ash has higher retention of lead than other types of fly ash depending upon following factors: (i) High specific surface area (SSA), (ii) lower capacity of cation exchange, (iii) type of fly ash, (iv) reaction time duration, and (v) pH value of solution.

3 Future Scope for Research

Fly ash has a great potential in Environmental and Civil Engineering applications and interesting alternative to replace old practices in construction work. The reduction in total cost will help make construction cost effective. High volumes of fly ash (60–75% by weight) with up to 12% carbon content, can be used from sources not meeting RMC specifications, in autoclaved cellular concrete. However, further research in this area is needed. Considering fly ash to be a resourceful material, further studies need to be carried out for facilitating the use of fly ash for innovative applications. The effect of fly ash on human being, animal, plant life, and its magnetic effect must be studied as fly ash is going to cover major aspects of our planet life.

4 Conclusion

Even being an environment pollutant, fly ash plays a very important raw material in various applications and enhances a number of properties. It also brings a reduction in cost of different processes. The benefit as such provided by fly ash cannot be overlooked and further research is the demand of hour for taking these benefits.

References

1. ASTM standard specification for coal fly ash and raw or calcined natural pozzolan for use in concrete (C 618- 05). In: Annual book of ASTM standards, concrete and aggregates, vol. 04.02. American Society for Testing Materials, 2005CEA, Ministry of Power, GOI, Annual report 2019–20, Nov. 2020, New Delhi, India (2021) (pp. 1–2 & 13–14)
2. CEA, Ministry of Power, GOI, Annual report 2019–20, Nov. 2020, New Delhi, India (2021) (pp. 1–2 & 13–14)
3. MORTH, Ministry of Road Transport & Highways, GOI, Annual report 2019–20, New Delhi, India.
4. MOEF, GOI, (03/11/2009) & CEA, GOI, Annual report 2019–20, Nov.2020, pp 1, New Delhi, India.
5. MOEF, GOI, (28/04/2016) & CEA Annual report 2019–20, GOI, Ministry of Power, Nov. 2020, New Delhi, India (p. 1).
6. Hemmings, R. T., & Berry, E. E. (1987). On the glass in coal fly ashes: Recent advances. In Proceedings of Materials Research Society Symposium on Fly Ash and Coal Conversion By-Products: Characterization, Utilization and Disposal IV, Boston, Massachusetts, U.S.A (Vol. 113, pp. 3–8).
7. Ahmaruzzaman, M. (2010). A review on the utilization of fly ash. *Progress in Energy and Combustion Science, 36*, 327–363. https://doi.org/10.1016/j.pecs.2009.11.003
8. Fisher, G. L., et al. (1978). Physical and morphological studies of size-classified coal fly ash. *Environmental Science and Technology, 12*, 447–451.
9. Kolbe, J. L., et. al. (2011). Use of alkaline coal ash for reclamation of a former strip mine. In *Proceedings of 2011, World of Coal Ash (WOCA) Conference, Denver, CO, USA*, May 9–12, 2011.
10. Havanagi, V. G. (1999). *Geotechnical characterization, strength and erosion aspects of fly ash-soil mixtures* [Ph.D. thesis, Department of Civil Engineering, Indian Institute of Technology, New Delhi, India].
11. Gayathri, V.: *Geotechnical characterization, strength and erosion aspects of fly ash-soil mixtures* [Ph. D. thesis, Department. of Civil Engineering, Indian Institute of Technology, New Delhi, India].

12. Sridharan, A. (2001). Physical, chemical and engineering properties of Indian coal ashes. In *National Seminar on Utilisation of fly ash in Water Resources Sector, Central Soil and Materials Research Station, New Delhi* (pp. 15–28).
13. Yudharbir, Honjo. (1991). Application of geotechnical engineering to environmental control. In Proceedings of IX Regional Conference on Soil Mechanics and Foundation Engineering, Bangkok, Thailand (Vol. 2, pp. 431–469).
14. Raymond, S. (1961). Pulverised fuel ash as embankment material. *Proceedings of Institute of Civil Engineers, 19*, 515–536.
15. Joshi, R. C., et al. (1975). New and Conventional engineering uses of fly ash. *Journal of Transportation Engineering ASCE, 101*(TE4), 791–806.
16. Martin, P. J., Browning, J. S., & Biehl, J. F. (1975). Properties and use of fly ash for Embankments. *Journal of Energy Engineering ASCE, 116*(2), 71–76.
17. Mathew, W. E., Keating, W. R., & Hodges, K. W. (1999). Evaluation of water quality conditions associated with the use of coal combustion products for highway embankments. In *International Ash Utilization Symposium and Centre for Applied Energy Research, University of Kentucky*, and paper # 31.
18. Baykal, G., Edinçliler, A., & Saygılı, A. (2004). Highway embankment constructions using fly as in cold regions. *Resources Conservation and Recycling, 42*(3), 209–222.
19. IRC: SP 58. (2001). Guidelines for the use of flu ash in road Embankments, Indian Road Congress, Special Publication 58, New Delhi, India
20. Basu, P. C., & Saraswati, S. (2006). High volume fly ash concrete with indian ingredients. *The Indian Concrete Journal*, 37–48.
21. IS: 12330. (1988). Specification for sulphate resisting Portland cement, Bureau of India Standards, New Delhi, India.
22. IS:12269. (2013). 53 Grade Ordinary Portland Cement - Specification , BIS, New Delhi, India.
23. IS: 6909. (1990). Supersulphated Portland Cement-Specification, Bureau of India Standards, New Delhi, India.
24. Yoon, S., Balunaini, U., Yildirim, I., Prezzi, M., & Siddiki, N. (2009). Construction of an embankment with a fly and bottom ash mixture: Field performance study. *Journal of Materials in Civil Engineering* (2009).
25. Price G. C., & Peterson R. (1961). *Performance of concrete.* University of Toronto Press.
26. Dunstan, Jr., E. R. (1980). *Proceeding of the American Concrete Institute, Las Vegas* (p. 30).
27. Ravina, D., & Mehta, P. K. (1986). Properties of fresh concrete containing large amounts of fly ash. *Cement and Concrete Research, 1986*(16), 227–238.
28. Mukherjee, P. S., & Vesmawala, G.: (2013). Exploring fly ash utilization in construction of highways in India. *IOSR, Journal of Mechanical and Civil Engineering (IOSR-JMCE), 8*(4), 23–32.
29. Jain, N., & Garg, M. (2015). Formulation of sulphate resistant super sulphated cement using fluorogypsum and granulated blast furnace slag. *IOSR, Journal of Mechanical and Civil Engineering (IOSR-JMCE), 12*(3), 153–159. e-ISSN: 2278-1684, p-ISSN: 2320-334X.
30. Malik, F. A., et. al. (2014). Management of fly ash in construction of National Highway Projects. *International Journal of Advanced Engineering Research and Studies.* E-ISSN 2249-8974/III/IV.
31. Vittal, G. (2001). Bulk utilisation of fly ash in construction of road embankments in India. *Coal Ash India, A Newsletter of the GEP Project, New Delhi, India, 2*, 6–7.
32. Capasso, I., et al. (2019). Reuse of mining waste as aggregates in fly ash-based geopolymers. *Journal of Cleaner Production, 220*, 65–73. https://doi.org/10.1016/j.jclepro.2019.02.164
33. Deng, X., et al. (2017). Experimental characterization of the influence of solid component on the rheological and mechanical properties of cemented paste backfill. *International Journal of Mineral Processing, 168*, 116–125. https://doi.org/10.1016/j.minpro.2017.09.019
34. Li, M., et al. (2020). *International Journal of Rock Mechanics and Mine Sciences, 125*, 1–10 (2020). https://doi.org/10.1016/j.ijrmms.2019.104170
35. Sekine, E., & Sunaga, M. (1991). Study on utilization of fly ash as embankment materials in railway. *Quality Report RTRI, 32*, 244–250.

36. Shekhovtsova, J., et al. (2018). Estimation of fly ash and reactivity for use in alkali- activated cements—A step towards sustainable building material and waste utilization. *Journal of Cleaner Production, 178*, 22–33. https://doi.org/10.1016/j.jclepro.2017.12.270
37. Pandian et al. (1995). Fly ash-lime system for the retention of lead ions. In *Proceedings of Indian Geotechnical Conference, Bangalore, India* (Vol. 1, pp. 219–222).
38. Pandian, N. S., Sridharan, A., & Rajasekhar, C. (2001). Heavy metal retention behaviour of fly ashes. *Indian Geotechnical Journal, 31*(3), 309–321.
39. Pandian & Balasubramonian, S. (2000). Studies on leaching behaviour of Indian fly ashes. In *Proceedings of Indian Geotechnical Conference, Indore, India* (pp. 453–456)

Application of Industrial Wastes for Soil Strength Improvement

S. Muthu Lakshmi, S. Geetha, M. Selvakumar, S. Revathy, and K. M. Shri Varshini

Abstract Industrial wastes like ground granulated blast furnace slag (GGBFS) and rice husk (RH) ash were utilized to improve the strength of subgrade soil available locally in Vembakkam region of Thiruvannamalai district in Tamil Nadu. To improve the bonding between soil and industrial wastes, 4% quick lime was mixed with the soil along with the industrial wastes GGBFS and RH ash. Soil was identified as clayey sand (SC) based on its index properties. GGBFS was varied in proportions of 20–40% by weight of SC and RH ash was added in quantities of 10–50% by weight of SC, to decide the best dosage of GGBFS and RH ash to be mixed with SC so as to achieve maximum soil strength after curing for 3 days. Based on the unconfined compressive strength (UCCS) test and soaked CBR test, ideal percentage of GGBFS and RH ash was found to be 30% and 20%, respectively. For SC with 4% lime, 30% GGBFS, and 20% RH ash, UCCS was observed to increase by 3.36 times that of virgin soil and soaked CBR value was found to increase by 9.23 times that of virgin soil after 3 days curing period.

Keywords Clayey sand · Quick lime · Ground granulated blast furnace slag · Rice husk ash · Unconfined compressive strength · Soaked CBR strength

---

S. M. Lakshmi (✉) · S. Geetha · M. Selvakumar · S. Revathy · K. M. S. Varshini
Department of Civil Engineering, Rajalakshmi Engineering College, Thandalam, Chennai, India
e-mail: muthulakshmi.s@rajalakshmi.edu.in

S. Geetha
e-mail: geetha.s@rajalakshmi.edu.in

M. Selvakumar
e-mail: selvakumar.m@rajalakshmi.edu.in

S. Revathy
e-mail: revathy.s.2014.civil@rajalakshmi.edu.in

K. M. S. Varshini
e-mail: shrivarshini.km.2014.civil@rajalakshmi.edu.in

A. K. Gupta et al. (eds.), *Advances in Construction Materials and Sustainable Environment*, Lecture Notes in Civil Engineering 196,
https://doi.org/10.1007/978-981-16-6557-8_45

1 Introduction

Industrialization transformed the economies of nations, but one of the drawbacks of industrialization is environmental pollution that has harmfully affected the humans and other living species. Lots of industrial waste by-products are generated that are disposed off in water bodies, landfills, or any open areas available for disposal. This has led to widespread pollution in the form of air, water, land pollution, environmental degradation, etc. Dumping of wastes in open landfills has led to soil contamination and wastage of valuable land area. This issues can be resolved either by reducing the quantity of industrial wastes that is being generated or by utilizing these wastes as resources in the production of other useful products. Recently, lot of research is being carried out in utilizing these industrial wastes in construction industry as a resource for producing concrete and also as an additive in soil treatment. Gobinath [1] observed that precise cementitious mineral blend from industrial wastes improved the geotechnical properties of soil. In this paper, industrial wastes like ground granulated blast furnace slag (GGBFS) and rice husk (RH) ash were used as an additive with locally available soil and studied their influence on the strength of soil. By improving the soil strength, load-bearing capability of soil is improved which in turn has positive effect on design parameters thus reducing the cost of construction. As per Saravanan [2], GGBFS, silica fume, and RH ash are the basic waste materials which enhances the soil properties and cuts the cost of pavements. Mahesh [3] suggested that treatment of black cotton soil by GGBFS not only stabilized the soil but also resolved the problem of waste disposal. Pozzolanic behavior of RH ash allows its use in the construction industry, thus providing another solution to its dumping [4]. Thomas [5] suggested that application of RH ash as stabilizer is a reasonable green technology and also provides expansive clays the required strength for pavement construction.

GGBFS is a discarded by-product produced during manufacture of pig iron and steel and nearly 2–4 tons of wastes are produced for every ton of steel manufactured. GGBFS contains calcium, magnesium, manganese, aluminum silicates, and oxides in different combinations, and due to its pozzolanic properties, it is used in manufacturing cement [6]. Specific gravity of GGBFS was 2.82, and the range of particle size was found to be between 75 μm and 2 μm. Michael [7] observed that GGBFS improved the properties of lime-stabilized soil better than fly ash. Lang [8] used lime and carbide slag to activate GGBFS for stabilizing dredged sludge instead of Portland cement.

Rice husk obtained from paddy is burnt as fuel and RH ash contains about 85%–90% amorphous silica which is appropriate as cementitious material. A paddy mill produces approximately 22% husk which when burned in boilers produces about 25% RH ash. Specific gravity of RHA was 2.04 and particle size was found to be in the range of 75 μm–2 μm. Jittin [4] suggested that addition of 10–20% RH ash enhanced the engineering characteristics of construction products. Emmanuel [9] utilized RH ash and cement kiln dust for enhancing the mechanical strength of weak soil. Argaw [10] treated expansive soil with steel slag, RH ash, quick lime, and observed that stiffness of treated soil improved by 58–78% as compared to clay. Younes [11] found that strength and post peak strength of soil with cement, lime, RH ash significantly enhanced with increase in binder content.

In the current work, GGBFS and RH ash were added in varying percentages along with 4% quick lime to the local subgrade soil identified as clayey sand (SC), to evaluate the improvement in its strength after curing for 3 days. 4% quick lime was added along with GGBFS and RH ash to improve the bonding between SC and the industrial wastes. 4% lime was decided as the ideal quantity to be added to SC based on the earlier experimental investigations conducted on SC soil. Rahul [12] suggested that GGBFS-lime combination is a better binder than GGBFS for the treatment of black cotton soil. As RH ash contains small amounts of calcium, it is combined with lime to be used as stabilizing agent [13]. Marta [14] suggested that optimum lime quantity depends on characteristics of soil to be treated. Adding lime increased pH of soil thus stimulating pozzolanic reactions and forming cementitious products which tend to stabilize the soil [15, 16].

2 Experimental Investigation

Soil utilized in the project was obtained in disturbed state from Vembakkam in Thiruvannamalai district of Tamil Nadu. Ground surface was cleared of vegetation and the soil was taken from below a depth of half a meter from the ground surface. Based on the engineering characteristics of soil, it was identified as clayey sand (SC). Specific gravity of soil was 2.6 [17]. As per [18], maximum dry density of 2.01 gm/cc at optimum moisture content (OMC) of 11% was observed, and based on this, soil sample for unconfined compressive strength (UCCS) tests & soaked CBR tests were prepared at relative compaction of 97%. Industrial wastes like GGBFS and RH ash were added in varying percentages to SC along with 4% quick lime to determine their influence on UCCS and soaked CBR value of SC after 3 days curing. Table 1 shows the percentage of additives such as quick lime, GGBFS and RH ash that were added to SC to prepare different soil specimens for UCCS test and soaked CBR test. Soil samples were prepared by first mixing the soil and the additives in dry state and then water corresponding to OMC was added to form a uniform wet mix. For UCCS test, this wet mix of soil with additives was kept covered with polythene sheets for a period of 3 days to avoid escape of moisture from the mix. After 3 days curing, soil specimen for UCCS tests were prepared and tested as per Indian Standard [19]. In case of soaked CBR test, wet mix of soil with additives was compacted in CBR mold using dynamic compaction, and then, soaked in water for a period of 3 days after which the CBR test was conducted as per Indian Standard [20].

Table 1 Quantity of quick lime, GGBFS, and RH ash added to SC to prepare different soil samples

Soil specimen	Quick lime (%)	GGBFS (%)	RH ash (%)
Virgin soil	0	0	0
SC + 4% Lime	4	0	0
SC + 4% Lime + 20% GGBFS	4	20	0
SC + 4% Lime + 25% GGBFS	4	25	0
SC + 4% Lime + 30% GGBFS	4	30	0
SC + 4% Lime + 35% GGBFS	4	35	0
SC + 4% Lime + 40% GGBFS	4	40	0
SC + 4% Lime + 10% RH ash	4	0	10
SC + 4% Lime + 20% RH ash	4	0	20
SC + 4% Lime + 30% RH ash	4	0	30
SC + 4% Lime + 40% RH ash	4	0	40
SC + 4% Lime + 50% RH ash	4	0	50
SC + 4% Lime + 30% GGBFS + 20% RH ash	4	30	20

2.1 *SC with Industrial Wastes*

GGBFS was added in percentages of 20%, 25%, 30%, 35%, and 40% by weight of SC with 4% lime. UCCS tests and soaked CBR tests were completed on different soil samples, and Fig. 1 shows the UCCS in kPa and soaked CBR strength (%) of SC with varying content of GGBFS after 3 days curing.

Maximum UCCS of 83.1 kPa and maximum soaked CBR of 22.2% were obtained for SC with 4% lime and 30% GGBFS. Thus, ideal percentage of GGBFS was found to be 30% based on the maximum soil strength attained. RH ash was added in percentages of 10%, 20%, 30%, 40%, and 50% by weight of SC with 4% lime. Figure 2 shows the UCCS in kPa and soaked CBR strength (%) of SC with variable quantity of RH ash after curing for 3 days.

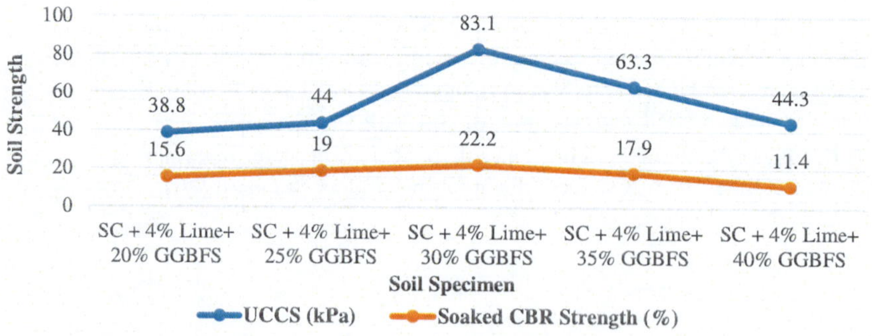

Fig. 1 UCCS and soaked CBR strength of SC with variable content of GGBFS

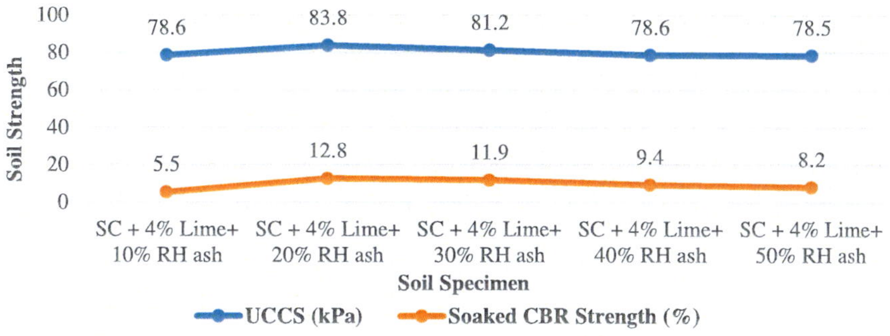

Fig. 2 UCCS and soaked CBR strength of SC with variable content of RH ash

Peak UCCS of 83.8 kPa and maximum soaked CBR of 12.8% was obtained for SC with 4% lime and 20% RH ash. Thus, ideal quantity of RH ash was found to be 20% for which ultimate soil strength was observed.

2.2 *UCCS of SC with Optimum Quantity of Industrial Wastes*

UCCS test was carried out on SC with 4% lime and optimum content of GGBFS and RH ash after 3 days curing. 30% GGBFS and 20% RH ash were added along with 4% lime to SC, and the observed UCCS was compared with that of virgin soil, to determine the effect of industrial wastes on UCCS. Figure 3 shows the stress–strain curves of different soil samples after curing for 3 days.

As per Fig. 3, SC with 4% lime, 30% GGBFS and 20% RH ash failed at a lesser strain value of 0.0015 at maximum stress of 132.4 kPa. It showed a brittle failure

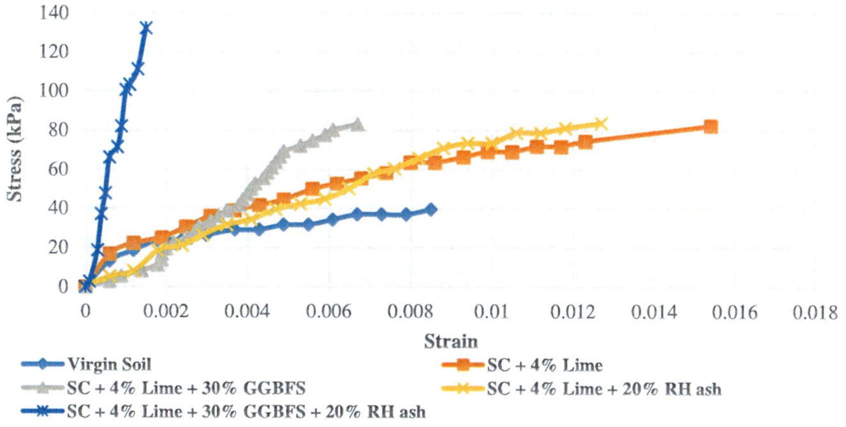

Fig. 3 Stress–strain curves obtained from UCCS test

Table 2 UCCS of soil specimens

Soil specimen	UCCS (kPa)	Increase in UCCS w.r.t. virgin soil
Virgin soil	39.4	–
SC + 4% Lime	82.4	2.09
SC + 4% Lime + 30% GGBFS	83.1	2.11
SC + 4% Lime + 20% RH ash	83.8	2.13
SC + 4% Lime + 30% GGBFS + 20% RH ash	132.4	3.36

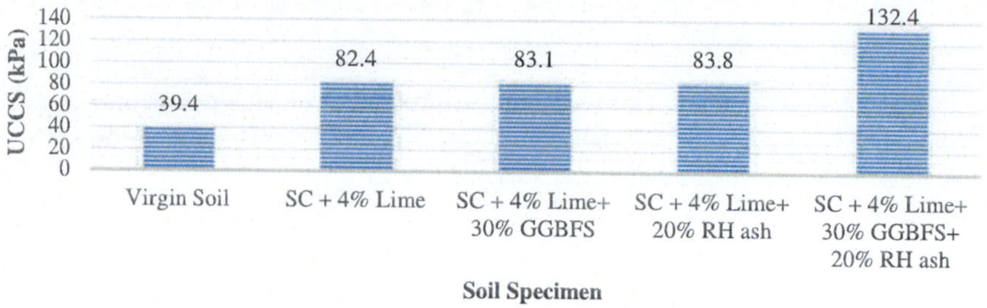

Fig. 4 UCCS of SC with optimum content of lime, GGBFS, and RH ash

as compared to other soil specimens which showed a ductile failure at higher strain values. Table 2 shows the UCCS of different soil specimens and the increase in UCCS compared to that of virgin soil. Figure 4 shows the UCCS of SC with optimum content of lime, GGBFS, and RH ash after 3 days curing.

For SC with 4% lime, for SC with 4% lime, 30% GGBFS and for SC with 4% lime, 20% RH ash, nearly same UCCS was achieved in the range of 82.4 kPa to 83.8 kPa, and the rise in UCCS was approximately 2.1 times that of virgin soil. Peak UCCS of 132.4 kPa was achieved for SC with 4% lime, 30% GGBFS, and 20% RH ash, and the rise in UCCS of 3.36 times that of virgin soil was witnessed. This rise in UCCS may be attributed to the formation of calcium silicate hydrates (CSH) and calcium aluminate hydrates (CAH) due to the pozzolanic reaction that occurs between silica, alumina, and calcium found in soil, GGBFS, and RH ash. Abhinaba [21] observed that reaction products accountable for soil strength increase are similar to that attained in cement stabilization. Lei Lang [8] suggested that in carbide slag—GGBFS stabilized dredged sludge, CSH, and CAH were the chief hydration products, and ettringite was also noticed. Jinrong [22] suggested that RH ash has high pozzolanic activity which fills the pores of the soil thus forming new cementing material with lime/cement and hardens soil.

2.3 *Soaked CBR Strength of SC with Optimum Quantity of Industrial Wastes*

Soaked CBR test was conducted on SC with 4% lime and optimum content of GGBFS and RH ash after curing for 3 days. 30% GGBFS and 20% RH ash were added along with 4% lime to SC, and the observed soaked CBR strength was compared with that of virgin soil, to determine the effect of industrial wastes on CBR strength. Figure 5 shows the load-penetration curves of different soil specimens after 3 days curing.

As per Fig. 5, load–penetration curves of virgin soil and SC with 4% lime showed a similar trend. Table 3 shows the soaked CBR values of different soil specimens and also the increase in soaked CBR compared to that of virgin soil. Figure 6 displays the soaked CBR values of SC with optimum content of lime, GGBFS, and RH ash after curing for 3 days.

Addition of 30% GGBFS to SC with 4% lime was found to improve the soaked CBR strength by 7.4 times that of virgin soil. For SC with 4% lime and 20% RH ash, soaked CBR strength was found to increase only by 4.27 times that of virgin soil. Thus, it can be inferred that in terms of soaked CBR strength, addition of 30%

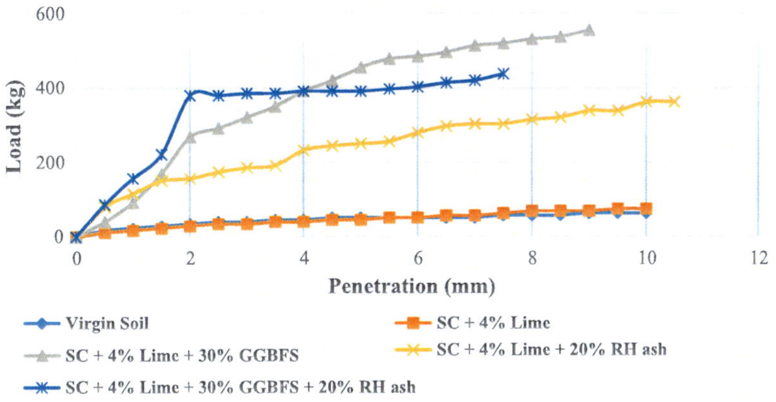

Fig. 5 Load–penetration curves obtained from soaked CBR test

Table 3 Soaked CBR values of soil samples

Soil specimen	Soaked CBR (%)	Increase in soaked CBR w.r.t virgin soil
Virgin soil	3	–
SC + 4% Lime	2.6	0.87
SC + 4% Lime + 30% GGBFS	22.2	7.4
SC + 4% Lime + 20% RH ash	12.8	4.27
SC + 4% Lime + 30% GGBFS + 20% RH ash	27.7	9.23

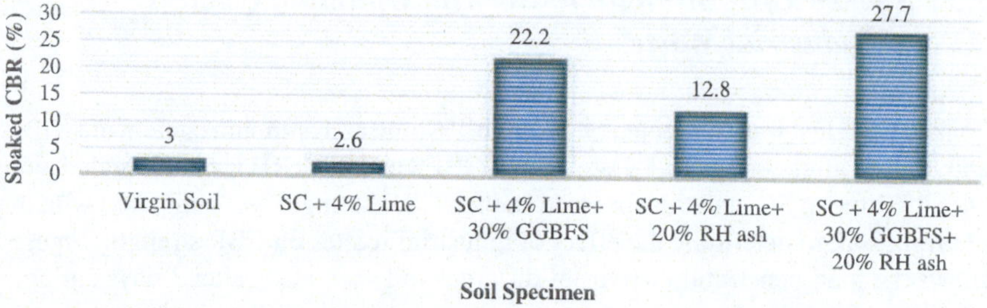

Fig. 6 Soaked CBR value of SC with optimum content of lime, GGBFS, and RH ash

GGBFS to SC was more beneficial compared to that of addition of 20% RH ash to SC. Also maximum soaked CBR value of 27.7% was achieved for SC with 4% lime, 30% GGBFS and 20% RH ash and the rise in soaked CBR value of 9.23 times that of virgin soil was achieved. This maximum increase in soaked CBR strength may be attributed to the pozzolanic reaction that formed CSH & CAH which improved the soil strength. Marta and Jacques [23, 24] observed that pozzolanic reaction progressed with curing period. Thus, it can be concluded that addition of industrial wastes like GGBFS and RH ash along with lime to SC in ideal amounts, improved the shear strength and CBR strength of SC subgrade to large extent, and this technique of soil strength improvement can be utilized to enhance the strength of in-situ soil comprising of SC. This will have a positive outcome on the design of foundations resting on SC and also reduce the thickness of flexible pavement thus decreasing the cost of construction. Another advantage of using GGBFS and RH ash as additives with SC would be decreasing the quantity of industrial wastes that otherwise would have been dumped in landfills thus reducing the magnitude of environmental pollution. As per Yaolin [25], both eco-friendly and cost-effective benefits can be expected by utilizing lime-GGBFS for stabilizing soft clay.

3 Conclusion

Ideal quantity of GGBFS and RH ash to be added to SC along with 4% quick lime was 30% and 20%, respectively. For SC with 4% lime, for SC with 4% lime, 30% GGBFS and for SC with 4% lime, 20% RH ash rise in UCCS was found to be approximately 2.1 times that of virgin soil. Maximum UCCS of 132.4 kPa was achieved for SC with 4% lime, 30% GGBFS, and 20% RH ash, and the rise in UCCS of 3.36 times that of virgin soil was witnessed. For SC with 4% lime, 30% GGBFS and for SC with 4% lime, 20% RH ash, soaked CBR strength was found to increase by 7.4 times and 4.27 times that of virgin soil respectively. Thus, addition of 30% GGBFS to SC was more beneficial compared to that of addition of 20% RH ash to SC. Maximum soaked CBR of 27.7% was achieved for SC with 4% lime, 30% GGBFS, and 20% RH ash, and the rise in soaked CBR value of 9.23 times that of virgin soil was achieved.

References

1. Gobinath, R., Raja, G., Prasath, E., et al. (2020). Studies on strength characteristics of black cotton soil by using novel SiO_2 combination as a stabilizing agent. *Materials Today: Proceedings, 27*, 657–663. https://doi.org/10.1016/j.matpr.2020.01.597
2. Saravanan, R., Udhayakumar, T., & Dinesh, S., et al. (2017). Effect of addition of GGBS and lime in soil stabilisation for stabilising local village roads in Thanjavur region. *IOP Conference Series: Earth and Environmental Science, 80*(1).https://doi.org/10.1088/1755-1315/80/1/012060/meta
3. Vastrad, M., Karthik, M., Dhanavandi, V., & Shilpa, M. S. (2020). Stabilization of black cotton soil by using GGBS, lime and nano-silica. *International Journal of Research in Engineering, Science and Management, 3*(9), 1–7. https://doi.org/10.47607/ijresm.2020.273
4. Jittin, V., Bahurudeen, A., & Ajinkya, S. D. (2020). Utilisation of rice husk ash for cleaner production of different construction products. *Journal of Cleaner Production*, 121578. https://doi.org/10.1016/j.jclepro.2020.121578
5. Karatai, T. R., Kaluli, J. W., Kabubo, C., & Thiongo, G. (2017). Soil stabilization using rice husk ash and natural lime as an alternative to cutting and filling in road construction. *Journal of Construction Engineering and Management, 143*(5), 04016127. https://doi.org/10.1061/(ASCE)CO.1943-7862.0001235
6. Indian Minerals Yearbook (Part II: Minerals and Alloys). (2018). 57th Edition, Slag—Iron and Steel, Government of India, Ministry of Mines, Indian Bureau of Mines.
7. McCarthy, M. J., Csetenyi, L. J., Sachdeva, A., & Dhir, R. K. (2014). Engineering and durability properties of fly ash treated lime-stabilised sulphate-bearing soils. *Engineering geology, 174*, 139–148. https://doi.org/10.1016/j.enggeo.2014.03.001
8. Lang, L., Chen, B., & Li, N. (2021). Utilization of lime/carbide slag-activated ground granulated blast-furnace slag for dredged sludge stabilization. *Marine Georesources & Geotechnology*, 1–11. https://doi.org/10.1080/1064119X.2020.1741050
9. Adeyanju, E., Okeke, C.A., Akinwumi, I., & Busari, A. (2020). Subgrade stabilization using rice husk ash-based geopolymer (GRHA) and cement kiln dust (CKD). *Case Studies in Construction Materials, 13*, e00388. https://doi.org/10.1016/j.cscm.2020.e00388
10. Ashango, A. A., & Patra, N. R. (2016). Behavior of expansive soil treated with steel slag, rice husk ash, and lime. *Journal of Materials in Civil Engineering, 28*(7), 06016008. https://doi.org/10.1061/(ASCE)MT.1943-5533.0001547
11. Bagheri, Y., Ahmad, F., & Ismail, M. A. M. (2014). Strength and mechanical behavior of soil–cement–lime–rice husk ash (soil–CLR) mixture. *Materials and structures, 47*(1), 55–66. https://doi.org/10.1617/s11527-013-0044-2
12. Pai, R. R., & Patel, S. (2019). Effect of GGBS and lime on the strength characteristics of black cotton soil. In *Ground improvement techniques and geosynthetics* (pp. 319–328). Springer. https://doi.org/10.1007/978-981-13-0559-7_36
13. Zaika, Y., & Suryo, E. A. (2020). The durability of lime and rice husk ash improved expansive soil. *International Journal of Geomate*, 171–178. https://doi.org/10.21660/2020.65.5539
14. Di Sante, M., Fratalocchi, E., Mazzieri, F., & Pasqualini, E. (2020). Prediction of shear strength parameters in soil–lime mixture design—part 1: quicklime. *Proceedings of the Institution of Civil Engineers - Ground Improvement, 173*(2), 93–104. https://doi.org/10.1680/jgrim.17.00076
15. Bessaim, M. M., Bessaim, A., Missoum, H., & Bendani, K. (2018). Effect of quick lime on physicochemical properties of clay soil. In *MATEC Web of Conferences 149*: 02065. EDP Sciences. https://doi.org/10.1051/matecconf/201814902065
16. Mehta, K. S., Sonecha, R. J., Daxini, P. D., et al. (2014). Analysis of engineering properties of black cotton soil & stabilization using by lime. *Journal of Engineering Research and Application, 4*(5), 25–32.
17. IS: 2720 (Part III/Sec 2)—1980, Determination of specific gravity - fine, medium and coarse grained soils.

18. IS:2720 (Part VII)—1980, Determination of water content - dry density relation using light compaction.
19. IS:2720 (Part 10)—1991, Determination of unconfined compressive strength.
20. IS:2720 (Part 16) RH ash which was added in quantities 1987, Laboratory determination of CBR.
21. Paul, A., & Hussain, M. (2020). Sustainable use of GGBS and RHA as a partial replacement of cement in the stabilization of Indian Peat. *International Journal of Geosynthetics and Ground Engineering, 6*(4), 1–15. https://doi.org/10.1007/s40891-020-0185-7
22. Ma, J., Su, Y., Liu, Y., & Tao, X. (2020). Strength and microfabric of expansive soil improved with rice husk ash and lime. *Advances in Civil Engineering*. https://doi.org/10.1155/2020/9646205
23. Di Sante, M., Fratalocchi, E., Mazzieri, F., & Brianzoni, V. (2015). Influence of delayed compaction on the compressibility and hydraulic conductivity of soil–lime mixtures. *Engineering Geology, 185*, 131–138. https://doi.org/10.1016/j.enggeo.2014.12.005
24. Locat, J., Bérubé, M. A., & Choquette, M. (1990). Laboratory investigations on the lime stabilization of sensitive clays: Shear strength development. *Canadian Geotechnical Journal, 27*(3), 294–304. https://doi.org/10.1139/t90-040
25. Yi, L., Gu, L., & Liu, S. (2015). Microstructural and mechanical properties of marine soft clay stabilized by lime-activated ground granulated blastfurnace slag. *Applied Clay Science, 103*, 71–76. https://doi.org/10.1016/j.clay.2014.11.005

Physical and Mechanical Characteristics of Cement Mortar with Coal Bottom Ash as Fine Aggregate Under Elevated Temperature

Abhishek Srivastava, S. K. Singh, and Rajesh Kumar

Abstract This experimental study aims at evaluating the performance of coal bottom ash (CBA) mortar after exposure to elevated temperatures. Two mortar mixes were prepared in which one was conventional natural sand (NS) based mortar and other mix contains CBA as 100% replacement of NS. The test specimens were cured for 28 days at ambient condition and then was heat treated at different elevated temperature (200, 400, 600 and 800 °C) using an electric controlled high-temperature furnace. The effect of elevated temperature on the mass loss, ultrasonic pulse velocity (UPV) and compressive strength of mortar specimens was evaluated after the high-temperature treatment. In addition, empirical equation derived from regression analysis was proposed to predict the residual mechanical strength of mortar incorporating CBA as fine aggregate at different elevated temperature. The test results signify that both mass and UPV decreases with increases in temperature, however, residual compressive strength of CBA mortar was better or comparable to NS mortar up to 400 °C, but above 400 °C; strength reduction rate in CBA mortar was higher than NS mortar.

Keywords Coal bottom ash · Fine aggregate · Elevated temperature · Mortar · Compressive strength · Mass loss · UPV

1 Introduction

The concrete and masonry structures are made up of constituent like concrete, mortar, bricks, reinforcement, etc. During their service life, these structures get accidently exposed to very high temperature which results in loss of integrity and stability

A. Srivastava (✉) · S. K. Singh · R. Kumar
CSIR-Central Building Research Institute, Roorkee, Uttrakhand 247667, India

S. K. Singh
e-mail: sksingh@cbri.res.in

R. Kumar
e-mail: rajeshkumar@cbri.res.in

A. K. Gupta et al. (eds.), *Advances in Construction Materials and Sustainable Environment*, Lecture Notes in Civil Engineering 196,
https://doi.org/10.1007/978-981-16-6557-8_46

of structure. Exposure to elevated temperature makes the structure dilapidated and sometimes may leads to collapse [1]. Therefore, understanding the behavior of constituents (concrete, mortar, etc.) of structure at an elevated temperature is very important in order to predict the response of structure and to minimize the detrimental impact of high temperature.

The natural sand is one of the key components of construction industry used for manufacturing the concrete, mortar and plasters. Due to shortage of good quality sand and high demand of it has led the rapid extraction. Rapid extraction of natural sand from river beds has led the serious environmental issues [2, 3]. Loss of bio-diversities, land, coastal erosion, and river shores sliding, lowering of river beds, etc., are some of the detrimental impact of rapid extraction of sand from river beds [4, 5]. Therefore, researchers have shifted their attention toward finding the alternative sources of sand. Recently, utilization of industrial waste in concrete and mortar as alternative to natural sand is gaining importance due to its various environmental benefits. CBA is produced from thermal power plant during burning of coal, accumulating on the land sites and causing various environmental and health issues. In India, about 407–524 million tons of coal is burned annually in different thermal power plants, generating approximately 105–173 million tons of coal ash which constitutes 30–35 million tons of CBA alone [6–8]. Presence of hazardous metals in CBA may contaminate the surface and sub-surface water bodies lying near the landfills [7]. Furthermore, exposure to CBA causes pulmonary vasculitis and hypertension in living organisms due to alteration in cardio-pulmonary organs [8]. Hence, recycling of CBA instead of dumping into the landfills is the need of an hour for its safe disposal.

Utilization of CBA as “Alternative sand” is a promising solution of meeting the growing demand of sand and reducing the negative impact occurring due to over dredging of natural sand. The performance of CBA mortar at ambient temperature is a well studied by various researchers worldwide [9–12]. However, no study has been reported in the literature investigating the behavior of CBA mortar at elevated temperature. To address this research gap, this study investigates the behavior of CBA mortar under elevated temperature and compares it with natural sand (NS) based mortar. Experimental investigation involves preparation of NS and CBA based mortar mixes of binder: fine aggregate ratio of 1:4. The mortar specimens were exposed to 200, 400, 600 and 800 °C and then examine for mass loss, UPV and compressive strength.

2 Experimental Program

2.1 *Materials and Mix Proportions*

In the present study, OPC43 cement conforming to IS 8112 [13] is used as binder in mortar. The natural sand lying in the zone II of IS 383 [14] was obtained from local supplier, while the CBA was collected from NTPC, Dadri. Figure 1 shows the

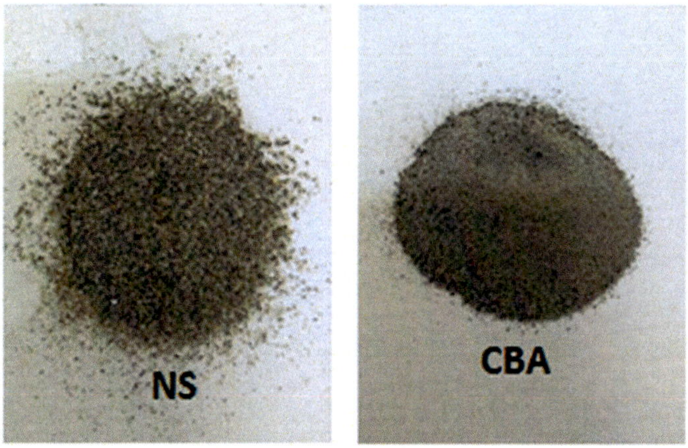

Fig. 1 Pictorial representation of NS and CBA

Table 1 Physical property of NS and CBA

Property	NS	CBA
Specific gravity (G_s)	2.56	1.87
Fineness modulus (F_m)	2.46	2.51

pictorial representation of NS and CBA used in the study. The physical characterization of NS and CBA was examining in accordance with IS 2386 (Part-III) [15] and is presented in Table 1. SEM images shown in Fig. 2 depicts that CBA has porous structure and rough surface texture compare to the NS which is solid as well as has smooth surface.

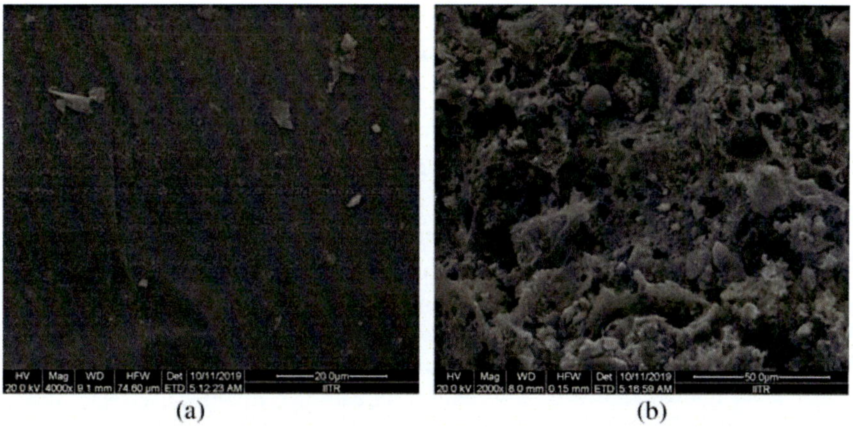

Fig. 2 SEM micrograph of **a** NS and **b** CBA surface

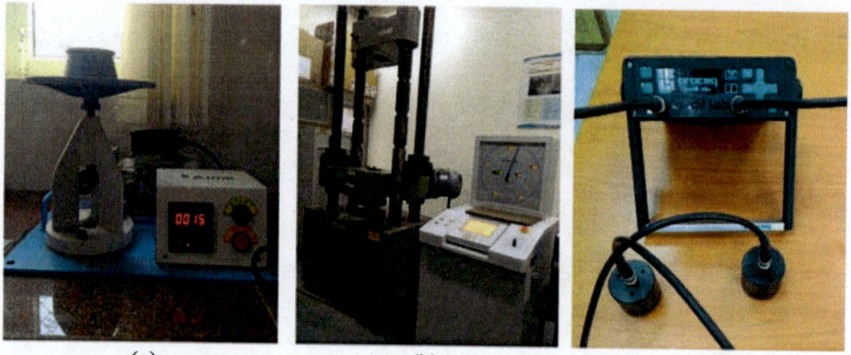

(a) (b) (c)

Fig. 3 Pictorial representation of **a** flow table, **b** UTM machine and **c** UPV measuring apparatus

Table 2 Mix proportions, kg/m^3

Mix code	Cement	NS	CBA	Water	Plasticizers
NS	385	1790	–	270	1.9
CBA	385	–	1510	400	1.9

For preparing the mortar mixes, the materials were mixed as per the procedure conforming to ASTM C305 [16] in fine: binder ratio of 1:4 (volumetrically) using the standard mortar mixer. The quantity of water in the mix was made such that the obtained mortar mix has the standard flowability of 110–115% according to IS 2250:1981 [17]. The flowability of the mortar mixes was determined using electrically operated standard flow table apparatus (Aimil, AIM 411-1) as shown in Fig. 3a and as per the procedure given in ASTMC1437 [18]. Table 2 presents the NS and CBA mortar mixture proportions. NS mortar contains conventional NS while CBA mortar contains 100% CBA as replacement of NS for fine aggregate.

2.2 Sample Preparation and Testing Procedure

For compressive strength and UPV test, 50 mm cubes as per ASTM C109 [19] and 40 × 40 × 160 mm bars [20] were prepared, respectively, for each mortar types. The mortar specimens were demolded from the casting molds after 24 h and then put for curing for 28 days. After 28 days, specimens were taken out, dried and heated at temperature of 200, 400, 600 and 800 °C. Heating procedure involves exposing the mortar specimen to high temperature using Carbolite Gero (HTF 1800) high-temperature furnace. Each heating cycle includes three steps: (i) Exposing the specimens to a temperature rise at constant rate of 5 °C/min up to desired temperature (ii) then peak temperature is kept constant for 1 h and (iii) finally, specimens were allowed to cool down naturally to room temperature at approximate rate of 2 °C/min. After heating exposure, mass loss, UPV and compressive strength were evaluated.

The mass loss was calculated by deducting the mass of mortar specimen before and after the heat exposure. Compressive strength of cubes was estimated as per the ASTM C109 [19] using 1000 kN UTM machine as shown in Fig. 3b. For UPV test, UPV testing device, of Proceq Pundit Lab shown in Fig. 3c, is used for measuring the velocity of ultrasonic pulses through mortar specimens after thermal exposure.

3 Results and Discussions

3.1 *Workability*

The flow of both the mortar type was made constant at 110–115% in accordance with IS 2250:1981 [17] and the amount of water required to attain the standard flow or workability was determined in respect of water to cement ratio (w/c). Figure 4 presents the variation in w/c ratio of NS and CBA mortar samples. From Fig. 4, it can be noted that w/c ratio of mortar mixes containing CBA as fine aggregate was higher than the NS mortar mixes. The higher water demand of CBA mortar for certain standard flowability is attributed due to its porous structure and rough surface texture [12, 21] as can also be observed in Fig. 2.

3.2 *Mass Loss*

Mass loss of NS and CBA mortar specimens when exposed to elevated temperature is shown in Fig. 5. With rise in temperature, all the mortar specimens show increase in mass loss. The specimen's subjected to 200 °C shows mass loss of 2.01–4.35%, due to elimination of free water present in the voids of the specimens through evaporation [22, 23]. CBA mortar specimen's shows higher reduction in weight relative to NS mortar, which is attributed to more water content in CBA mortar mix as shown in Fig. 4. Further increment in temperature up to 400 °C causes drastic rise in mass loss. This trend is attributed due to commencement of dehydration of calcium hydroxide (CH), C-S–H gel and elimination of chemical water at 400 °C [24].

Beyond 400 °C, the mass loss in mortar specimens of NS and CBA is because of CH and C–S–H gel decomposition [25]. Between 600 and 800 °C, the mass loss was 7.44–8.71% and 9.83–11.2% in NS and CBA mortar, respectively.

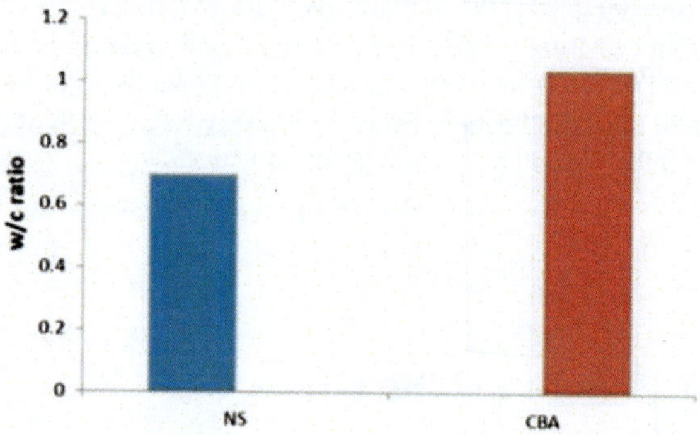

Fig. 4 Water-cement ratio of NS and CBA mortar mix

3.3 *Ultrasonic Pulse Velocity (UPV)*

The UPV values of NS and CBA mortar specimens at different elevated temperature are shown in Fig. 6. UPV values of both type of mortar specimens decreases with increase in temperature. A minimal fall in UPV values was observed at 200 °C. The exposure of specimens to 400 °C temperature exhibited a reduction of 9.81% and 20.04% in UPV values for NS and CBA mortar mixes, respectively. The chemical water present in specimens gets eliminated when expose to elevated temperature of 400 °C [26], which results into empty air voids. The velocity of ultrasonic pulses reduces, while passing through these air voids [27].

A sharp reduction in UPV values was noted between 600 and 800 °C as presented in Fig. 6. For instance, UPV values of specimens exposed to temperature 600–800 °C reduces to 27.78–55.17%. This decrement in UPV values is attributed due to

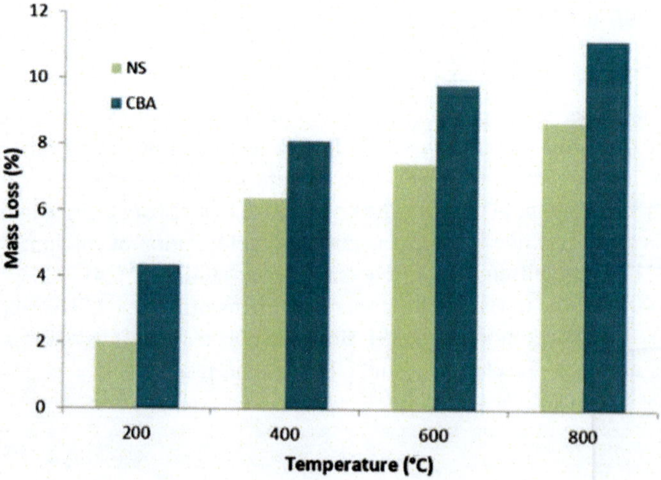

Fig. 5 Variation in mass loss of specimens at elevated temperature

increase in porosity caused by more internal cracks, dehydration and decomposition of compounds like C–S–H and CH [24, 28].

3.4 Residual Compressive Strength

Figure 7 depicts the ratio of residual compressive strength of NS mortar and CBA mortar at different elevated temperature to their strength at 27 °C (room temperature). It can be noted that as the exposure temperature increases, the compressive strength ($f_T/f_{27°C}$) of both NS and CBA mortar continuously decreases. At 200 °C, the NS and CBA mortar specimens exhibit 76% and 81% of its initial strength. Further exposure to 400 °C leads to reduction in specimen's strength. In this case, NS and CBA mortar samples exhibit 72% and 66% of their initial strength, respectively, which is comparable as can be seen in Fig. 7.

Heating of specimens at higher temperature 600 °C, more prominently to 800 °C, leads to significant reduction in residual compressive strength. CBA mortar specimens possess 42% and 30% of their initial strength hat 600 °C and 800 °C respectively, lower than strength retention by NS mortar specimens (62% and 47%). Lower residual strength of mortar specimens containing CBA is attributed due to their high porosity caused by more water content than NS mortar. Lateral tension zones are formed around these pores when tested for compressive strength, leading to local failure which results in lower strength values [10].

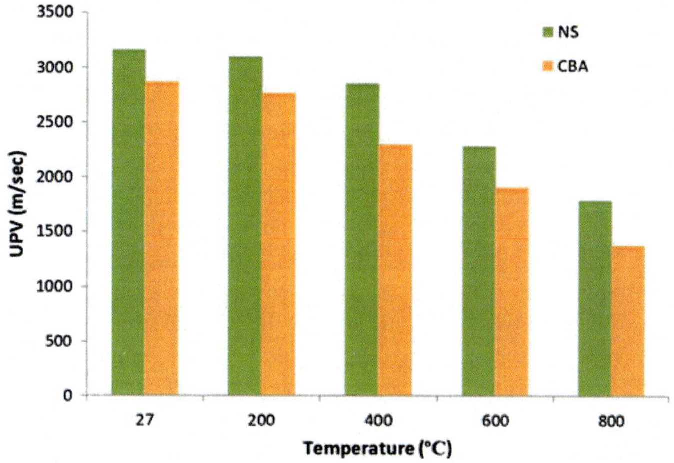

Fig. 6 Variation in UPV of specimens at elevated temperature

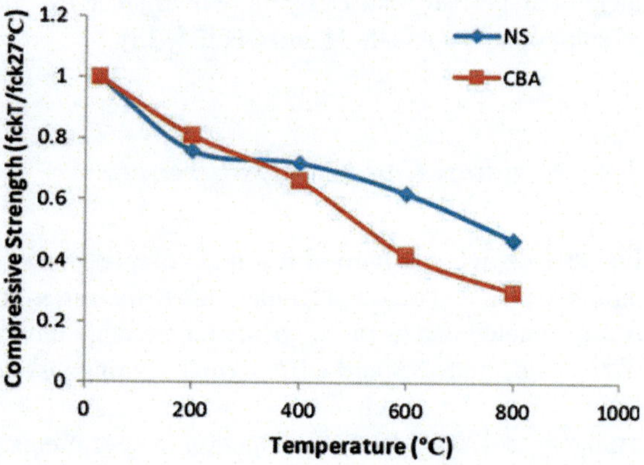

Fig. 7 Variation in residual compressive strength of NS and CBA mortar specimens

4 Correlation Among Residual Compressive Strength and Elevated Exposure Temperature

The experimental test results are used to develop the empirical equations for estimating the residual compressive strength ($f_{\text{ckt}}/f_{\text{ck27°C}}$) of CBA mortar at different elevated temperature (200 °C < T < 800 °C). All the experimental test results were merged and by using regression analysis, empirical equation is proposed in which residual strength (α_t) is the dependent variable, while the exposure temperature was the independent or predictor variable. The obtained Eq. (1) is as follows:

$$\alpha_t = \begin{cases} 1, & 27\ °\text{C} \\ 1 - 0.00092T, & 200\ °\text{C} \le T \le 800\ ° \end{cases} \tag{1}$$

where, α_t is $f_{\text{ckT}}/f_{\text{ck27°}}$ T represents the exposure temperature. The coefficient of determination (R^2) for the developed Eq. (1) was 99% (close to 100%), thus specifying the acceptable confidence levels. To further establish the reliability of the proposed Eq. (1), the predicted strength was compared with the experimental test results as presented in Fig. 8. From Fig. 8, it can be noted that most of the data points are close to similarity line indicating that proposed empirical equation can be applied in the post fire analytical studies for determining the residual strength of CBA mortar exposed up to 800 °C.

5 Conclusion

The present study aimed to evaluate the effect of elevated temperature on the performance of CBA mortar. From the results obtained, following specific conclusions are formulated:

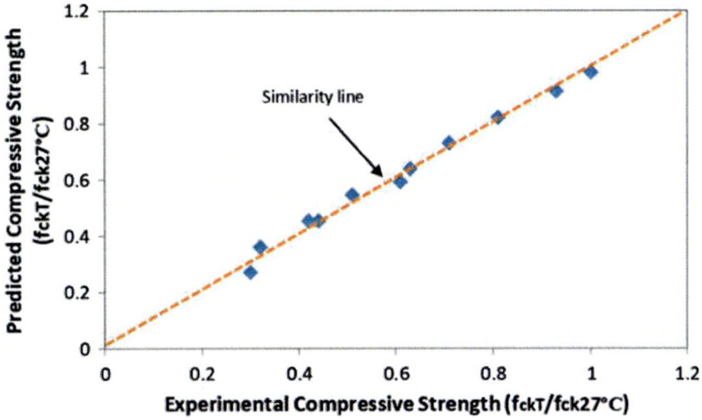

Fig. 8 Comparison between predicted and experimental strength data

- The water demand of CBA mortar was 48.57% higher than NS based mortar for attaining standard flow due to porous and rough surface texture of CBA compare to NS.
- The mass loss in mortar specimens was increased with increase in exposure temperature. However, CBA mortar shows 27.07–45.27% higher reduction in mass compare to NS mortar specimens due to the higher water content.
- The residual compressive strength of CBA mortar is 6.57% higher or comparable to NS mortar up to 400 °C exposure temperature. Beyond 400 °C, strength reduction in CBA mortar was higher than the NS mortar specimens.
- The UPV values for CBA mortar specimens were 9.20–28.15% lower as compare to NS mortar specimens at different elevated temperature due to its higher porosity.
- An empirical equation is proposed having high coefficient of determination ($R^2 >$ 0.99) to predict the residual compressive strength of CBA mortar being exposed to high temperature.

Acknowledgements The authors acknowledge the financial support received from the NBCC, New Delhi (Grant: GAP 0018).

References

1. Alonso, C., & Fernandez, L. (2004). Dehydration and rehydration processes of cement paste exposed to high temperature environments. *Journal of Materials Science, 39*(9), 3015–3024.
2. Wenger, A. S., Harvey, E., Wilson, S., Rawson, C., Newman, S. J., Clarke, D., & Erftemeijer, P. L. (2017). A critical analysis of the direct effects of dredging on fish. *Fish and Fisheries, 18*(5), 967–985.
3. Srivastava, A., & Singh, S. K. (2020). Utilization of alternative sand for preparation of sustainable mortar: A review. *Journal of Cleaner Production, 253*, 119706.

4. Gavriletea, M. D. (2017). Environmental impacts of sand exploitation. Analysis of sand market. *Sustainability, 9*(7), 1118.
5. Kim, H. K., Jeon, J. H., & Lee, H. K. (2012). Flow, water absorption, and mechanical characteristics of normal-and high-strength mortar incorporating fine bottom ash aggregates. *Construction and Building Materials, 26*(1), 249–256.
6. Singh, M., & Siddique, R. (2014). Compressive strength, drying shrinkage and chemical resistance of concrete incorporating coal bottom ash as partial or total replacement of sand. *Construction and Building Materials, 68*, 39–48.
7. Singh, M., & Siddique, R. (2016). Effect of coal bottom ash as partial replacement of sand on workability and strength properties of concrete. *Journal of Cleaner Production, 112*, 620–630.
8. Singh, N., Mithulraj, M., & Arya, S. (2018). Influence of coal bottom ash as fine aggregates replacement on various properties of concretes: A review. *Resources, Conservation and Recycling, 138*, 257–271.
9. Baite, E., Messan, A., Hannawi, K., Tsobnang, F., & Prince, W. (2016). Physical and transfer properties of mortar containing coal bottom ash aggregates from Tefereyre (Niger). *Construction and Building Materials, 125*, 919–926.
10. Ramadoss, P., & Sundararajan, T. (2014). Utilization of lignite-based bottom ash as partial replacement of fine aggregate in masonry mortar. *Arabian Journal for Science and Engineering, 39*(2), 737–745.
11. Srivastava, A., Singh, S. K., & Sharma, C. S. (2021). Correlation between ultrasonic pulse velocity (UPV) and compressive strength of coal bottom ash mortar. *Journal of The Institution of Engineers (India): Series A*, 1–13.
12. Torkittikul, P., Nochaiya, T., Wongkeo, W., & Chaipanich, A. (2017). Utilization of coal bottom ash to improve thermal insulation of construction material. *Journal of Material Cycles and Waste Management, 19*(1), 305–317.
13. IS: 8112, Ordinary Portland cement, 43 Grade-Specification, Bureau of Indian Standards, New Delhi, India, 2015.
14. IS: 383, Coarse and fine aggregate for concrete-Specification, Bureau of Indian Standards, New Delhi, India, 2016
15. IS 2386(Part-III)-1963, Indian standard code of practice, Methods of test for aggregates for concrete, Part-III specific gravity, density, voids, absorption and bulking, Bureau of Indian standards, New Delhi, 2002.
16. Standard, A. S. T. M. C305. Standard Practice for Mechanical Mixing of Hydraulic, Cement Pastes and Mortars of Plastic Consistency. Annual book of ASTM standards, 2006.
17. IS: 2250, Code of Practice for Preparation and Use of Masonry Mortars, Bureau of Indian Standards, New Delhi, India (1981)
18. Standard, A. S. T. M. C1437. Standard Test Method for Flow of Hydraulic Cement Mortar, 2007.
19. Standard, ASTM C109-standard test method for compressive strength of hydraulic cement mortars. ASTM International, West Conshohocken, PA, 2008.
20. Gupta, L. K., & Vyas, A. K. (2018). Impact on mechanical properties of cement sand mortar containing waste granite powder. *Construction and Building Materials, 191*, 155–164.
21. Singh, M., & Siddique, R. (2014). Strength properties and microstructural properties of concrete containing coal bottom ash as partial replacement of fine aggregate. *Construction and Building Materials, 50*, 246–256. https://doi.org/10.1016/j.conbuildmat.2013.09.026
22. Guo, Y. C., Zhang, J. H., Chen, G. M., & Xie, Z. H. (2014). Compressive behaviour of concrete structures incorporating recycled concrete aggregates, rubber crumb and reinforced with steel fibre, subjected to elevated temperatures. *Journal of Cleaner Production, 72*, 193–203.
23. Noumowé, A., Lefevre, A., & Duval, R. (2002). Porosité supplémentaire consécutive à la fusion de fibres de polypropylène dans un béton à hautes performances. *Revue française de génie civil, 6*(2), 301–313.
24. Mohamad, S. A., Al-Hamd, R. K. S., & Khaled, T. T. (2020). Investigating the effect of elevated temperatures on the properties of mortar produced with volcanic ash. *Innovative Infrastructure Solutions, 5*(1), 1–11.

25. Savva, A., Manita, P., & Sideris, K. K. (2005). Influence of elevated temperatures on the mechanical properties of blended cement concretes prepared with limestone and siliceous aggregates. *Cement and Concrete Composites, 27*(2), 239–248.
26. Nematzadeh, M., & Mousavimehr, M. (2019). Residual Compressive Stress-Strain relationship for hybrid recycled pet–crumb rubber aggregate concrete after exposure to elevated temperatures. *Journal of Materials in Civil Engineering, 31*(8), 04019136.
27. IS 13311. (1992). Method of Non-destructive Testing of Concrete–Part 1: Ultrasonic Pulse Velocity, 1992.
28. Demirel, B., & Keleştemur, O. (2010). Effect of elevated temperature on the mechanical properties of concrete produced with finely ground pumice and silica fume. *Fire Safety Journal, 45*, 385–391.

Prediction of Air Pollution Due to Mobile Sources Using Line Source Models

M. Selvakumar, S. Geetha, and S. Muthu Lakshmi

Abstract Industrialisation has resulted in environmental pollution which leads to health hazards. The major pollution problem is caused in air, water, and land. Monitoring air pollution involves costly equipments. Hence, researchers have developed many models to predict air pollution by considering certain parameters. This work deals with monitoring air pollutants like particulate matter and CO and validating it with the use of the available line source models and finding out which models suits the best. The area selected for the study was Chennai-Sriperambudhur National Highway. The line source models used in the study are Delhi finite line source model (DFLSM) and General Finite Line source model (GFLSM) model. The data that were arrived by these models were validated through real-time monitoring data. The duration of monitoring was 12 h (7 am to 7 pm), and the readings were observed for a month. Change in the pollutant concentration during peak hour traffic and normal hours were also observed and the data required for modelling like traffic count, wind velocity, and wind direction were collected. It was observed that the value of total suspended particulate matter varied from 100 to 152 μg/m^3. Respirable particulate matter was observed to vary from 16 to 70 μg/m^3. The concentration of CO was observed to be from 3 to 15 ppm. The concentrations of air pollution are highly dependent on the climatic conditions and varied based on the meteorological parameters.

Keywords Line source · Pollutants · Traffic · Vehicle count · Meteorological parameters

---

M. Selvakumar · S. Geetha (✉) · S. Muthu Lakshmi
Civil Engineering Department, Rajalakshmi Engineering College, Thandalam, Chennai, Tamil Nadu, India
e-mail: geetha.s@rajalakshmi.edu.in

M. Selvakumar
e-mail: selvakumar.m@rajalakshmi.edu.in

S. Muthu Lakshmi
e-mail: muthulakshmi.s@rajalakshmi.edu.in

A. K. Gupta et al. (eds.), *Advances in Construction Materials and Sustainable Environment*, Lecture Notes in Civil Engineering 196,
https://doi.org/10.1007/978-981-16-6557-8_47

1 Introduction

Road transport is an integral part of urbanisation for any country. In a developing country like India, which is now focussing on Industrial revolution, the use for road transport is a major source. In spite of its advantage in contributing the economic development of a nation, road transport contributes to the major air pollution resulting in many health hazards if it exceeds its threshold level [1, 2]. The emission of air pollution from vehicles increases the pollutant concentration in the air despite the other point sources like industries [3–5]. Monitoring air pollution can be done by establishing monitoring stations in many locations which involves huge cost. Hence, the use of mathematical modelling has increased in the past few years which has been proven to be efficient and close to the monitored data by many researchers across the globe [6–11]. The crucial data points for modelling air quality by any mathematical model involve a traffic survey, emission inventory, and meteorological data [12]. It has proven by many studies that the use of mathematical models correlates well with the observed pollution concentration [13]. Apart from Gaussian-based Line source mathematical models [14, 15], there are many other models using remote sensing and GIS techniques for air pollution monitoring [16, 17]. The pollutants that cause air pollution are broadly of five types: Particulate matter which can be total suspended particulate matter (TSPM) or respirable particulate matter (RPM), carbon monoxide (CO), sulphur dioxide (SO_2), and nitrous oxide (NO_x). Each pollutant has a permissible threshold level above which they are considered to have fatal health effects on humans exposed or those inhaling these pollutants. Meteorological factors and daily variation in the weather also plays a major role in the concentration of the pollutants that are suspended in air [18–20]. Many other simulation models based on dispersion of pollutants and the chemistry of pollutant dispersion have also been used to explain the pollution due to vehicular emission [21, 22]. The parameters like cloud cover, temperature, humidity, and wind speed enable the dispersion and dilution of the pollutants. Also regulation in traffic and proper road network design is an integral part in mitigating air pollution. This will help us to take necessary preventive and control measures. The objective of this study is to find the air pollutant concentration due to vehicular emission and to find a suitable prediction model from the various line source models that are available.

2 Methodology

2.1 Study Area Characteristics

The study area in this work is a 22 km road on the national highway NH-48 (Bangalore Highway) connecting Sriperambudur to Chennai bypass, which has a sipcot and Toll plaza included in the study area. The major class of vehicles plying in this area is the container trucks which import cars and goods from the Sipcot. Many popular companies like Hyundai, Iljin, Wabco, Saint Gobain, and many Engineering colleges are also located in this area. The average number of vehicles was estimated to be 75,000 per day. There are no proper traffic signals at the two junctions in the stretch of road considered in this study. The study area selected for this study is on national highway NH-48, and the location of the monitoring stations is given in Fig. 1.

Fig. 1 Details of study area and location of monitoring stations

2.2 Methodology

This study was carried out in five monitoring stations which were approximately 4.5–5 km apart. Particulate matter (PM 10 and PM2.5) was monitored using a sensor-based smart air quality monitor (Airveda) which is commercially available. For measurement of carbon monoxide, CO analyser (Ecotech) was used. The range of the analyser was 0–200 ppm with 0.05 ppm sensitivity. The analyser is a microprocessor controlled combined with gas filter correlation and non-dispersive infrared spectrophotometry. The sample flow rate is 1 slpm. Pollutant concentration was collected daily from all the stations during June to December in order to study the variation of pollution due to weather conditions. Variation in wind speed, wind direction, temperature, and humidity were also noted. Traffic survey was done to find the number of vehicles passing through. The concentration of the pollutants was measured at 3 m height from the ground level by placing the equipment in first floor of the buildings near the road side. Also, readings were taken very close to the road at a height of 1 m from the ground level. Delhi finite line source model was used in this study by the previous study [20]. Monitored data was compared with the modelled data, and the accuracy of the model was determined by R^2 value.

3 Results and Discussion

The stations were selected near the traffic signals where there is a convergence of 2–4 roads meeting at a point. Station S-5 is near the toll gate where several vehicles are in idling condition as they wait to clear the toll charges and pass the toll gate. The other three stations are intermediate stations with considerably less traffic. The stations were selected based on the congestion of vehicles at a place due to the traffic signal and also perpendicular road crossings where there were no proper signalling systems.

Metrology plays a major role in air pollution. It decides the concentration of pollutants prevalent at a point. The major metrological parameters that significantly affect the air pollutant dispersion are temperature, humidity, wind speed, and wind direction. Wind data represented by the wind rose diagram as given in Fig. 2 shows that the maximum wind speed was 16 kmph along the north direction. Temperature and humidity were also collected during the monitoring period. To facilitate the modelling of air pollutant concentration using line source models, wind speed and wind direction are the major parameters that are needed. Apart from that vehicle count was also taken in all the stations during the monitoring period. It was found that 12% of the vehicles were containers and trucks, 22% were buses, 26% were cars, and 40% were motor cycles.

The pollutants monitored using the monitoring equipment are plotted in Figs 3, 4 and 5. It shows that the pollution concentrations in the two stations S-1 and S-5 are the maximum. The reason for this is that the first station was taken near a junction,

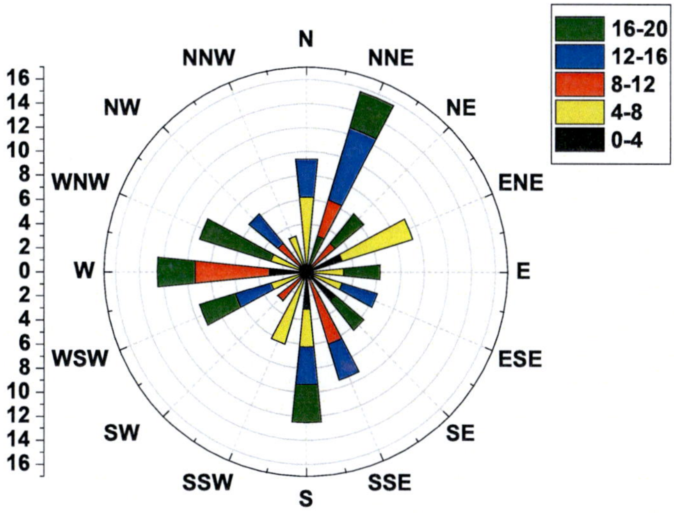

Fig. 2 Wind-rose diagram showing wind direction and wind velocity

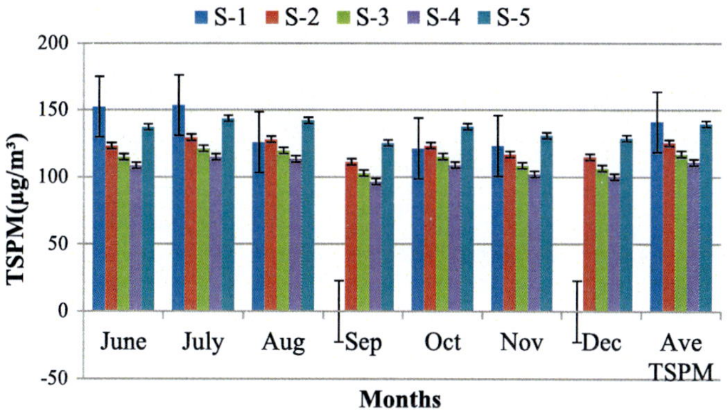

Fig. 3 Observed TSPM in 5 stations

places where four roads meet at a point, and no proper traffic signal has been used here. So, the unorganised traffic flow in perpendicular directions resulted in traffic congestions often with a long queue of vehicles to cross the junction. The maximum total suspended particulate matter (TSPM) that was recorded was 153 $\mu g/m^3$. The least was 100 $\mu g/m^3$. With reference to respirable particulate matter (RPM) and total suspended particulate matter (TSPM), the concentration was observed to be more during summer season. This could be attributed to the dry ground that facilitates suspension of dust particles easily in the air. The maximum RPM recorded was 70 $\mu g/m^3$, and the minimum value was 15 $\mu g/m^3$. The peak hour traffic was at 8 am and at 4 pm. This stretch of road had a cluster of industrial sipcots and many

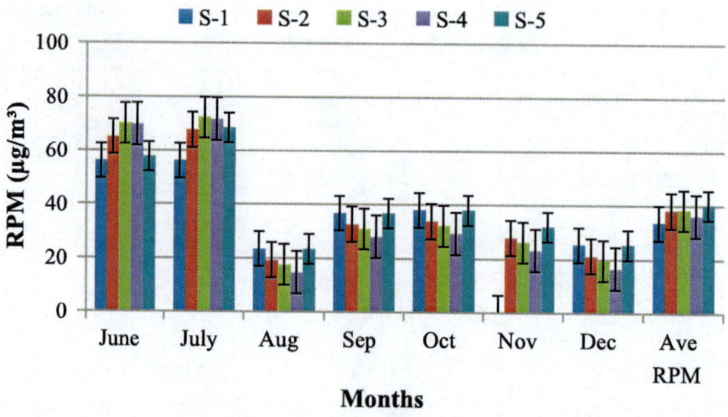

Fig. 4 Obsered RPM in 5 stations

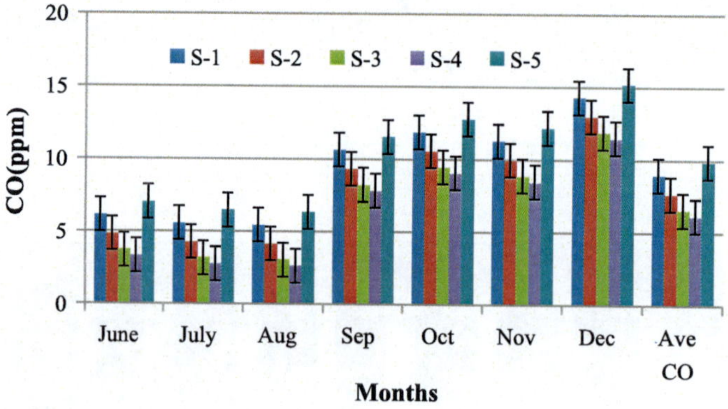

Fig. 5 Observed CO in 5 stations

educational institutions. It was observed that at peak hours, the number of buses plying in this road was high there were 7 Engineering colleges located in this stretch and each college had around 50 to 70 buses. Also apart from Industrial Sipcots which had factories like Hyundai, WABCO, ILJIN, and many more there were many parking areas to park the containers which transport the goods of these industries. The main reason for traffic congestion was found that these trucks blocked the traffic when taking U turns to reach the parking lots during the peak hour. This particular issue resulted in traffic congestion increasing the pollutant concentration at these junctions.

Similarly, the carbon monoxide (CO) concentration was higher during the winter months of Nov and Dec. The maximum CO that was recorded was 16 ppm, and the minimum was 3 ppm. The values of these pollutants when compared with the standard allowable pollutant concentrations (NAAQS) was found to exceed the permissible level of 9 ppm for CO, 150 $\mu g/m^3$ for TSPM and 35 $\mu g/m^3$ for RPM. The source of RPM could be the dusty roads apart from the vehicular emission as due to heavy traffic more vehicles are prone to ride along the shoulders which are not black topped.

3.1 General Finite Line Source Model (GFLSM)

This GFLSM was developed by Luhar and Patil in 1986 [23] which involves co-ordinate transformation between the wind system (x_1, y_1, z_1) and line source system (x, y, z) which is calculated from the geometry of the road. Midpoint of the line source is taken as the origin for both the systems, and also Z co-ordinate is same for both the systems. This model was has been proved to be very effective when finite length of road segments are considered for the study. Stability class was taken as neutral as given by Luhar [23] as it has been explained that along a finite length of road the stability is neutral. The dispersion parameter as given in Eq. (1) is based on the wind-road orientation angle and the source distance.

Pollutant concentrations ($\mu g/m^3$) are calculated as given in the following equation:

$$C = \frac{Q}{\sqrt{2\pi}\sigma_Z u \sin\theta}\left[\exp\left\{-\frac{1}{2}\left(\frac{z-h_0}{\sigma_z}\right)^2\right\} + \exp\left\{-\frac{1}{2}\left(\frac{z+h_o}{\sigma_z}\right)^2\right\}\right] \times \left[\operatorname{erf}\left(\frac{\sin\theta(L/2-y) - x\cos\theta}{\sqrt{2}\sigma_y}\right) + \operatorname{erf}\left(\frac{\sin\theta(L/2+y) + x\cos\theta}{\sqrt{2}\sigma_y}\right)\right] \quad (1)$$

where Q—is emission rate per unit length, g/m s; u—the mean ambient wind speed, m/s; σ_y and σ_z are horizontal, vertical dispersion parameter, respectively, m; x—the distance from the receptor to the line source, m; y—receptor distance from the roadway centre line along the line source, m; z—height of the receptor relative to the ground, 1.8 m; h_0 is line source height (m) taken as 0.5 m; H—the plume centre height relative to the ground, m; L—length of the source, m; θ—angle between the ambient wind and the road; and erf() is the error function.

The observed and predicted data using the model are validated using scatter plots as shown in Fig. 6, 7 and 8. The results show good correlation. The emission data as given in Table 1 by Goel [24] have been used in this study for both the models.

3.2 Delhi Finite Line Source Model (DFLSM)

Another model termed as e Delhi finite Line source model (DFLSM) devised by Khare and Sharma [25] as given in Eq. (2) is derived by removing the error function of GFLSM given in the previous section.

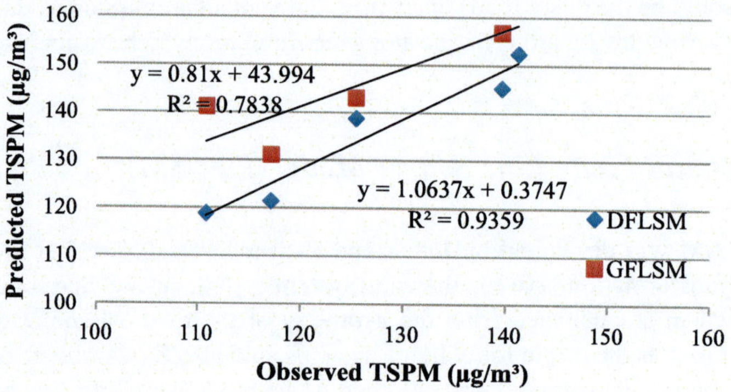

Fig. 6 Observed versus predicted TSPM in 5 stations

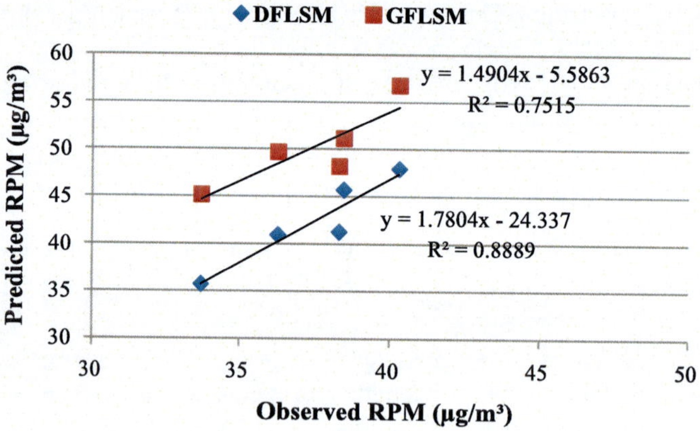

Fig. 7 Observed versus predicted RPM in 5 stations

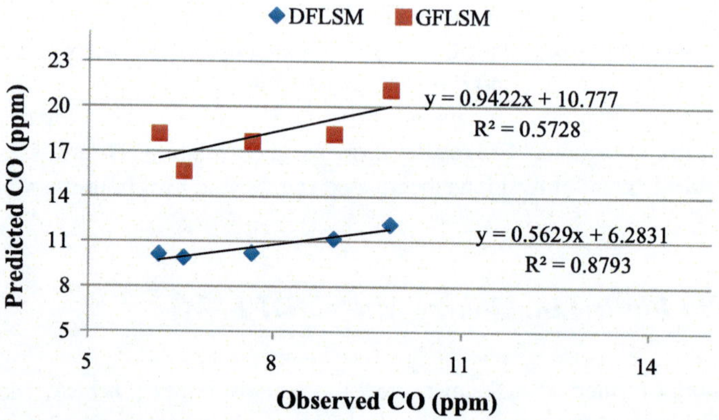

Fig. 8 Observed versus predicted CO in 5 stations

Table. 1 Average emission factor (Q) for different vehicles

Vehicle type	PM 2.5 (gm/km)	CO (gm/km)
LMV	0.461	52.22
HMV	6.975	42.39
3 wheelers	0.712	7.42
2 wheelers	0.371	3.42

$$C = \frac{Q}{2\sqrt{2}\sigma_Z u_e}\left[\exp\left\{-\frac{1}{2}\left(\frac{z-h_o}{\sigma_Z}\right)^2\right\} + \exp\left\{-\frac{1}{2}\left(\frac{z+h_o}{\sigma_Z}\right)^2\right\}\right] \tag{2}$$

where Q is emission rate per unit length, g/m s; u—the mean ambient wind speed, m/s; σ_y and σ_z are horizontal, vertical dispersion parameter, m; x—distance from the receptor to the line source, m; y—Receptor distance from the roadway centre line along the line source, m; z—height of the receptor relative to the ground, 1.5 m; h_0 is line source height (m) taken as 0.5 m.

The modelled data with reference to the parameters as given in Eq. (2) has been compared with the predicted data and a scatter plot as given in Figs. 6, 7 and 8 are given in order to validate the results by means of R^2 value. It has been seen that the experimental values and the modelled data correlated well except with CO which had a higher variation. The values predicted by GFLSM were high compared to that of DFLSM. This could be attributed to the parallel wind along the road side as explained by Luhar. Also the metrological data used in the models are those taken from the nearby metrology department and onsite wind speed and direction is not usually measured, which again have variations prevailing with the predicted data compared to the observed one. Moreover, the emission data was taken as an average value which could vary depending upon the vehicle condition and the fuel used in the vehicle. The modelled data was higher for CO than the monitored data. In the scatter plots given for the two models used in the study, it is observed that DFLSM fits well with the predicted value for all the three pollutant concentration, as given by the R^2 value. Gaussian-based models have been very useful in predicting the air pollution in many areas as studies have been reported by many researchers. When it is not possible to have a monitoring station and there is a need to analyse the air pollution in a particular road, these types of models can be used to get an overview of the prevalent pollutant concentrations.

4 Conclusion

The following conclusions can be arrived from the study that has been carried out:

- Improper traffic policies like absence of signal near the junctions can increase the pollutant concentrations. Hence, traffic management can be effectively planned by providing automatic signal systems which works based on the volume of traffic.

- In places where it is not possible to have equipment to find the pollutant concentration, mathematical modelling using line source models can be used to find the pollutant concentration.
- The concentration of pollutants for CO was high during the winter months which could be due to the high humidity of the air and difficulty in dispersion of the pollutants, and for particulate matter, it was high during the summer months which are due to more dust in dry weather. Planting of trees in the vicinity of the roads can minimise the air pollution to a certain extent.
- DFLSM finds good correlation with the observed data and can be used for modelling the pollution data at any point from the wind data and emission factors.

References

1. Wang, X. (2008). *Spatial analysis of long-term exposure to air pollution and cardiorespiratory mortalities in Brisbane, Australia* [Dissertation, University of Technology, Queensland].
2. Gan, W. Q., Koehoorn, M., Davies, H. W., Demers, P. A., & Tamburic, L. (2011). Long-term exposure to traffic-related air pollution and the risk of coronary heart disease hospitalization and mortality. *Environmental Health Perspectives, 119*(4), 501–507.
3. Ahn, K., & Rakha, H. (2008). The effects of route choice decisions on vehicle energy consumption and emissions. *Transportation Research Part D: Transport and Environment, 13*(3), 151–167.
4. Panis, L. I., Broekx, S., & Liu, R. (2006). Modelling instantaneous traffic emission and the influence of traffic speed limits. *Science of the Total Environment, 371*, 270–285.
5. Chen, C. H., Huang, C., Jing, Q. G., Wang, H. K., Pan, H. S., Li, L., Zhao, J., & Dai, Y. (2007). On-road emission characteristics of heavy-duty diesel vehicles in Shanghai. *Atmospheric Environment, 41*(26), 5334–5344.
6. Bell, M., Morgenstern, R., & Harrington, W. (2011). Quantifying the human health benefits of air pollution policies: Review of recent studies and new directions in accountability research. *Environmental Science & Policy, 14*(4), 357–368.
7. Zhang, K., & Batterman, S. (2010). Near-road air pollutant concentrations of CO and PM2.5: A comparison of MOBILE6.2/CALINE4 and generalized additive models. *Atmospheric Environment, 14*, 1740–1748.
8. Modig, L., Sunesson, A. L., & Levin, J. O. (2004). Can NO2 be used to indicate ambient and personal levels of benzene and 1,3-butadiene in air? *Journal of Environmental Monitoring, 6*, 957–962.
9. Schnitzhofer, R., Beauchamp, J., Dunkl, J., Wisthaler, A., Weber, A., & Hansel, A. (2008). Long term measurements of CO, NO, NO_2, benzene, toluene and PM10 at a motorway location in an Austrian valley. *Atmospheric Environment, 42*, 1012–1024.
10. Beckerman, B., Jerrett, M., Brook, J. R., Verma, D. K., Arain, M. A., & Finkelstein, M. M. (2008). Correlation of nitrogen dioxide with other traffic pollutants near a major Expressway. *Atmospheric Environment, 42*(2), 275–290.
11. Mohan, M., Bhati, S., Sreenivas, A., & Marrapu, P. (2011). Performance evaluation of AERMOD and ADMS-urban for total suspended particulate matter concentrations in megacity Delhi. *Aerosol Air Qual Res., 11*(7), 883–894.
12. EPA. (2012). *Dispersion modelling home page*. [Cited 20 December 2019] http://www.epa.gov/scram001/dispersionindex.htm.
13. Wheeler, A. J., Smith-Doiron, M., Xu, X., Gilbert, N. L., & Brook, J. R. (2008). Intra-urban variability of air pollution in Windsor, Ontario—Measurement and modeling for human exposure assessment. *Environmental Research, 106*, 7–16.

14. Lin, J., & Niemeier, D. A. (2003). Regional driving characteristics, regional driving cycles. *Transportation Research Part D, 8*, 361–381.
15. Sivacoumar, R., & Thanasekara, K. (1998). Estimation of vehicular Pollution from transport sector for Madras City. *Research Journal of Chemistry and Environment, 2*(3), 45–52.
16. Ko, Y.-W., & Cho, C.-H. Characterization of large fleets of vehicle exhaust emissions in middle Taiwan by remote sensing. *Science of the Total Environmenr, 354*, 75–82.
17. Muncaster, G. M., Hamilton, R. S., & Revitt, D. M. (1996). Remote sensing of carbon monoxide vehicle emissions. *Science of the Total Environment, 189*, 149–153.
18. Sillman, S. (2003). *Tropospheric ozone and photochemical smog* (pp. 407–431). Academic press.
19. Levy, J. I., Bennett, D. H., Melly, S. J., & Spengler, J. D. (2003). Influence of traffic patterns on particulate matter and polycyclic aromatic hydrocarbon concentrations in Roxbury, Massachusetts. *Journal of Exposure Science and Environmental Epidemiology., 13*, 364–371.
20. Madhavan, S., Meenambal, T.(2010). Prediction of CO using finite line source model for Coimbatore city. *International Journal of Applied Environmental Science, 5*(3), 449–455.
21. Woodwarda, H., Stettlerb, M., Pavlidisc, D., Aristodemoud, E., ApSimona, H., & Painc, C. (2019). A large eddy simulation of the dispersion of traffic emissions by moving vehicles at an intersection. *Atmospheric Environment, 215*, 1–16.
22. Kovacs, A., Leelossy, A., Tettamanti, T., Esztergar-Kiss, D., Meszaros, R., & Lagzi, I. (2021). Coupling traffic originated urban air pollution estimation with an atmospheric chemistry model. *Urban Climate, 31*, 1–10.
23. Luhar, A., & Patil, S. (1986). A general finite line source model for vehicular pollution prediction. *Atmospheric Environment, 23*, 555–562.
24. Goel, R., & Guttikunda, S. K. (2015). Evolution of on-road vehicle exhaust emissions in Delhi. *Atmospheric Environment, 105*, 78–90.
25. Khare, M., & Sharma, P. (1999). Performance evaluation of general finite length source model for Delhi traffic conditions. *Transportation Research Part D: Transport and Environment, 4*, 65–70.

Estimation of Methane Generation from Municipal Solid Waste of Mohali Landfill Site

Rishi Rana, Abhinav Choudhary, Karma Yangzom, and Kaushal Kumar

Abstract Municipal solid wastes (MSW) are mostly decomposable solid waste, which under aerobic and anaerobic decomposition produces green house gases and almost 50% of the green house gases consists of methane gas. The landfills used for waste disposal act as a source of methane (CH_4) gas. As methane has higher green house potential as compared to carbon dioxide (CO_2), it can significantly damage the environment as it becomes an important contributor of global warming, and it may lead to degradation of ecosystem. Therefore, to reduce the harmful effect of the methane, it is important to measure the methane generation from all the sources is important part to plan a contingency plan to tackle methane utilization in more appropriate ways which will ultimately lead to sustainable use of methane. Methodology includes (a) Study of waste generation and segregation in Mohali City, (b) Physical characterization of MSW, (c) Methane generation techniques adopted, and (d) Assessment of methane generation. Methane gas generated from the waste disposable into the city taken into consideration, i.e., Mohali city, we have taken two methods (i). The default method (DM) (given by IPCC) and (ii) Modified triangular method (MTM) for assessing the amount of methane generated. In this paper, the estimated amount of methane is 25.89 and 16.2 Gg respectively from the two methods, viz. DM and MTM.

Also, the deviations associated with the two methods of methane generation for calculating quantity of methane has also been discussed in this paper. Latest technologies for harnessing methane from landfills can significantly minimize the amount of methane released in the environment which can be judiciously utilize to produce energy, encourage strategic marketing of compost, and digestate.

Keywords Methane generation · Municipal solid waste management (MSWM) · Landfill · Default method · Modified triangular method · Global warming

R. Rana (✉) · A. Choudhary · K. Yangzom · K. Kumar
Civil Engineering Department, Jaypee University of Information Technology, Waknaghat, Solan, Himachal Pradesh 173234, India
e-mail: rishi.rana@juit.ac.in

A. K. Gupta et al. (eds.), *Advances in Construction Materials and Sustainable Environment*, Lecture Notes in Civil Engineering 196,
https://doi.org/10.1007/978-981-16-6557-8_48

1 Introduction

With the increase in MSW generation over years, India is facing various environmental challenges. With the heaps of MSW coming out in various cities on regular basis some of them even look like an artificial hill. Apart from destroying the aesthetic view of the city, it is causing health hazards in the city to such an extent that communities living in the vicinity of these landfill/disposal sites have to fled the area. These MSW landfills results in the increase of microbial pathogens which are polluting air as well as water [1]. Organic components of MSW play a major role in the emission of green house gases, which leads to global warming and is a responsible for climate change and is ultimately treat to all livings creature in this planet [3, 4, 8–10, 12].

Methane, whose concentration has increased by around 2 percent in last few decades, is a green house gas has global warming potential of 20 times more than that of standard carbon dioxide (CO_2).WSELF dumping of MWS is a major source of methane estimated at 3–19% of methane emissions [5].

Generation of methane gas has been assessed and quantified using may disposal models and methods such as Zero-order decay, First-order decay (FOD), IPCC's Default Method (DM), Modified triangular (MTM), Multi-order decay models, as well as software such as Land GEM can also be used to test potential methane production in a landfill [6, 10].

2 Generation of Waste in Mohali City

Due to population growth and development activities, magnitude of waste created has amplified many times with each passing year. The amount of waste generated in the year 2017 was 54.48 Gg [8], for expected growth in population in the year 2020, the quantity of waste will be 56.27 Gg, and following similar trend, it is expected that the waste generated in the next upcoming years from 2015 to 2030 will be approximately 900 Gg. It has been estimated that about 90% of the total waste generated in the city is disposed of in an open landfill and only 10% or is less is utilized in reuse and recycling [7].

Table 1 Calculation of waste (in kg/day) per capita generation

S. No.	Year	Waste per capita (kg/day)	Comments/remarks
1	2015	0.458	Population basis
2	2016	0.461	Population basis
3	2017	0.512	Calculated from the total waste generation Mohali City population
	Avg	0.481	

Generation of waste in Mohali city per capita for 16 years (i.e., from 2015 to 2030) is computed and shown in Table 1 per capita generation of MSW generation using the projected population of Mohali city.

Individual disposal of waste in 2015 and 2016 is considered to be the allocation for a specific population and for 2017; waste per capita is estimated based on the overall amount of MSW in Mohali on daily basis and as per the 2011 Govt. of India Census report [2, 9].

An average waste of 0.481 kg/day is used in conjunction with the population growth expected to compute the amount of per capita waste generated in the various years into consideration. The population of the city taken in this has been calculated using a geometric growth method [2]. Total city waste disposal per annum and the number of MSW that has reached the landfill site per year is listed in Table 2.

The amount of MSW achieved at the landfill annually is calculated by accounting for 90% of the waste collected at the disposal land every year [7]. As per Table 2, we note that the total MSW value for the forecasted duration of sixteen years is approx.

Table 2 Waste generation per year and per capita waste generated in Mohali from the forecasted population

S. No.	Forecasted year	Estimated population[a]	Quantity of waste ($\times 10^3$ kg/day)	Quantity of waste ($\times 10^3$ kg/year)	Quantity of waste ($\times 10^6$ kg/year)	Quantity of waste reaching landfills($\times 10^3$ kg/day)
1	2015	246,559	113.42	41,397.26	41.40	37.26
2	2016	271,117	124.71	45,520.54	45.52	40.97
3	2017	292,587	149.22	54,465.07	54.46	49.01
4	2018	315,794	151.58	55,327.11	55.33	49.80
5	2019	318,498	152.88	55,800.85	55.80	50.22
6	2020	321,225	154.19	56,278.62	56.28	50.65
7	2021	323,976	155.51	56,760.60	56.76	51.08
8	2022	326,750	156.84	57,246.60	57.25	51.53
9	2023	329,547	158.18	57,736.63	57.74	52.00
10	2024	332,369	159.54	58,231.05	58.23	52.41
11	2025	335,215	160.90	58,729.67	58.73	52.86
12	2026	338,085	162.28	59,232.49	59.23	53.31
13	2027	340,980	163.67	59,739.70	59.74	53.77
14	2028	343,900	165.07	60,251.28	60.25	54.23
15	2029	346,844	166.48	60,767.07	60.78	54.68
16	2030	349,814	167.91	61,287.41	61.29	55.16
	Total				898.78	808.94

[a] Estimated as per census of 2011

Table 3 Physical characterization of MSW in Mohali city [13]

S. No.	Components	MIG	LIG	HIG	Avg.
1	Unit wt. (kg/m^3)	430 ± 8.7	550 ± 8.8	395 ± 2.9	464.9
2	Paper/cardboard	4.5 ± 2.9	4.6 ± 2.9	7.8 ± 1.8	5.3
3	Compostable	53.9 ± 2.56	54.9 ± 1.4	46 ± 6.1	46.7
4	Plastic/polythene	5.7 ± 1.7	4.7 ± 0.2	8.0 ± 1.44	6.6
5	Leather/rubber	0.7 ± 0.04	0.5 ± 0.02	1.9 ± 0.03	1.21
6	Metals	0.1 ± 0.03	0.1 ± 1.7	0.1 ± 1.6	0.6
7	Glass	1.0 ± 2.0	0.4 ± 3.1	1.9 ± 4.10	1.4
8	Inert substances	25 ± 1.58	32.2 ± 7	30.1 ± 9.6	28.5
9	Misc	9.3 ± 5.29	2.6 ± 1.7	4.2 ± 3.89	9.1
10	Sum total	100	100	100	100

900 Gg; total MSW generated for landfill is approximately 810 Gg for duration of sixteen years.

2.1 *Characterization (Physical) of MSW in the Mohali City*

The tangible properties related to MSW are useful in the management systems of MSW. Characterization of MSW is done by its designation as organic / vegetable, paper, and plastic, wood and leaves, iron and metal, textiles, etc. Details of the physical description are provided in Table 3.

It can be seen from the table that the waste produced is mostly made from a low-cost group and non-fertilizer waste such as paper and plastic is mostly found in the high-income group. We can also see that 46.7% of the waste is the total amount of degradable waste produced, and the mineral composition is very small.

2.2 *Assessment Methods for Methane Generation*

The quantity of methane generated from this landfill has been calculated using various techniques and with the help of software. Techniques for methane estimation used are as follows:

2.2.1 Default Methodology of Methane Estimation

The IPCC's (DM) method is widely used methods to assess methane gas emissions from dumpsites/landfills. Its suitability is in the event that data in MSW is limited.

Assumptions related to Methane gas calculation using Default Method:

- Waste generation is assumed to be uniform every year in the city selected.
- Potential amount of methane generated is assumed to be emitted from that year's waste disposal.
- Out of all the emission, fraction of methane gas is taken constants value of **0.5** as per IPCC consideration.

Methane generation associated with the default method is based on Eq. (1): [5]

$$CH_4\left(\frac{Gg}{year}\right) = \left(MSW_T \times MSW_F \times MCF \times DOC \times DOC_F \times F \times \frac{16}{12} - R\right) \times (1 - OX) \quad (1)$$

where CH_4 = Quantity of methane generated per year in Gg/year; 1 Gg/year (Giga gram) = 1000 tons/year; MSW_T = total quantity of MSW produced per year in Gg. which is equal to 54.76 Gg per year; MSW_F = the part MSW reaches the disposal site (assuming 90%), which is equal to 49.375 Gg per year; MCF (Methane correction factor) = 0.41 (for depth of landfill up to 3 m);

DOC = Decomposable organic carbon, computed from Eq. 2 given as:

$$DOC = 0.4A + 0.17B + 0.15C + 0.3D \quad (2)$$

where A (cardboard/paper) = 0.053 (From Table 3); B (leaves, woods etc.) = 0.474 (From Table 3); C (vegetable waste) = 0.467 (From Table 3);

$D = 0$; *The value of DOC is computed using* Eq. 2, DOC = 0.1721

- $\boldsymbol{DOC_F}$ (dissimilated fill gas) = $0.014*T + 0.28 = 0.77$

Where T = 35 °C, is temperature of a landfill at which under anaerobic condition methane will be generated.

- F (fraction of methane gas) = 0.5 (assumed)
- R (recovered methane gas) = 0 (as Mohali hasn't applied any recovery technique up to the year 2019)
- OX (Oxidation factor) = 0 (default value)

2.2.2 Modified Triangular Method (MTM) of Methane Generation

It works in the absence of detailed data, the amount of methane gas emissions. This approach believed that CH_4 removal would only begin when waste disposed is a year old. The first stage assumed that methane gas would be released at the beginning of 2nd years after garbage dumping and would strengthen successively until the 6th year and after this at second stage, and it would start and decrease methane emissions from year 7 and then zero to 16 years. The triangular form of methane generation is illustrated in Fig. 1.

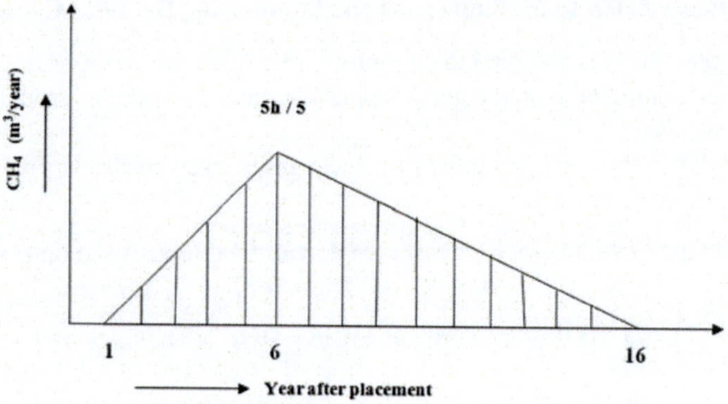

Fig. 1 Triangular shape of gas generation

The quantity of methane generated is computed using default method along with density of methane (0.714 kg/m^3). By comparing the vol. of CH_4 to the area inside the triangle, the abscissa of triangle (m^3) is estimated. Using abscissa of triangle of 6th year, the abscissa of other triangles is calculated. The abscissa of triangle of 6th year is represented as methane emission. The area inside the triangle will give the emission trend of methane on yearly basis.

3 Results and Discussion

3.1 Methane Generation by IPCC's Default Method

Using Eq. (1), quantity of methane gas generated from the MSW disposal site in Mohali city has been computed in Table 4. From Table 4, following observations can be made;

- For 2020: 50.66 Gg/year MSW reached landfill and 1.62 Gg/year is methane potential.
- Estimation for 2030: 55.16 Gg/year waste will reach landfill and 1.768 Gg/year is the estimated methane gas emission potential.
- The emission coefficient of 0.032 is calculated from DM. The overall MSW reaching disposal site is computed to be 809.01 Gg/year and that of total methane emission is 25.888 Gg/year for the duration of 16 years (2015–2030). The estimated emission of methane in various years is shown in Fig. 2.

The graph showed that methane gas emissions from the landfill site increased according to the calculated data from the default method. The DM method usually provides a limited amount of methane emissions compared to the other method [6].

Table 4 Methane generation by default method over various years

S. No. (1)	Forecast year (2)	Quantity of waste reached dumpsite/landfill site (Gg per year) (3)	CH_4 (Gg/year) = 0.032 * 3) (4)
1.	2015	37.25	1.192
2.	2016	40.98	1.311
3.	2017	49.02	1.568
4.	2018	49.82	1.594
5.	2019	50.23	1.607
6.	2020	50.66	1.621
7.	2021	51.07	1.634
8.	2022	51.54	1.649
9.	2023	52.01	1.664
10.	2024	52.42	1.677
11.	2025	52.85	1.691
12.	2026	53.33	1.706
13.	2027	53.79	1.721
14.	2028	54.22	1.735
15.	2029	54.67	1.749
16.	2030	55.15	1.768
	Sum total	809.01	25.888

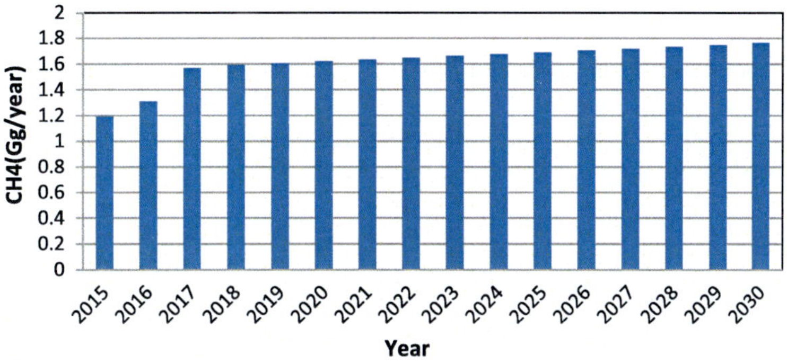

Fig. 2 Methane generation trend

3.2 Methane Release by MTM

A methane emission coefficient from DM is used in MTM to calculate the amount of methane gas emissions. Compared to DM, this method estimates the appropriate

Table 5 Methane generation from MSW with modified triangular method (MTM) technique

S.No. (1)	Year (2)	CH_4 (Gg/year) (3)	CH_4 (m^3) = (3) × 10^6/(0.714 kg/m^3) (4)	Height of triangle (m^3) = 2 × (4)/15 (5)	Ht. of triangle, CH_4 (m^3) by MTM (6)	Ht. of triangle = 0.714 × 10^{-6} × (6) (7)	CH_4 (Gg) by MTM (area of triangle) (8)
1	2015	1.19	1,666,666.67	222,222.22	0	0	
2	2016	1.31	1,834,733.89	244,631.18	60,504.20	0.43	0.22
3	2017	1.57	2,198,879.55	293,183.94	121,008.40	0.86	0.64
4	2018	1.59	2,226,890.76	296,918.77	181,512.61	1.29	1.07
5	2019	1.61	2,254,901.96	300,653.59	242,016.81	1.73	1.53
6	2020	1.62	2,268,907.56	**302,521.01**	302,521.01	2.16	1.94
7	2021	1.63	2,282,913.16	304,388.42	272,268.91	1.94	2.05
8	2022	1.65	2,310,924.37	308,123.25	242,016.81	1.73	1.84
9	2023	1.66	2,324,929.97	309,990.66	211,764.71	1.51	1.62
10	2024	1.68	2,352,941.18	313,725.49	181,512.61	1.29	1.39
11	2025	1.69	2,366,946.78	315,592.90	151,260.51	1.08	1.20
12	2026	1.71	2,394,957.98	319,327.73	121,008.40	0.86	0.98
13	2027	1.72	2,408,963.58	321,195.14	90,756.30	0.65	0.76
14	2028	1.74	2,436,974.79	324,929.97	60,504.20	0.43	0.52
15	2029	1.75	2,450,980.39	326,797.38	30,252.10	0.22	0.35
16	2030	1.77	2,478,991.60	330,532.21	0	0	0.09
	Sum						16.21

amount of methane gas emissions. Calculation of methane extraction by MTM method is shown in Table 5.

The graphical representation of total methane estimated by MTM is as shown in Figs. 3 and 4.

4 Comparison of Potential Methane Gas Generation Between DM and MTM

See Fig. 5.

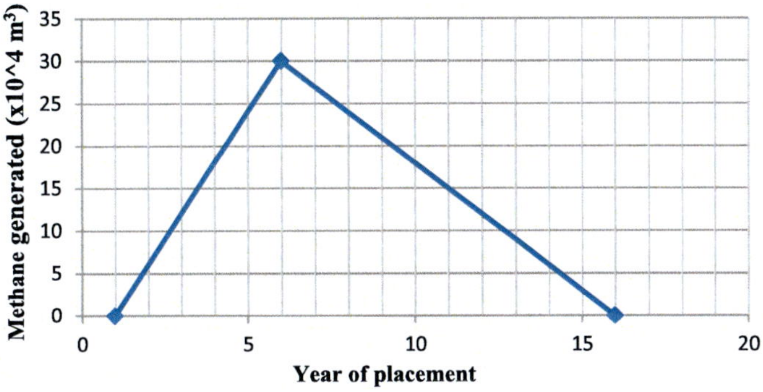

Fig. 3 Triangular form for gas production

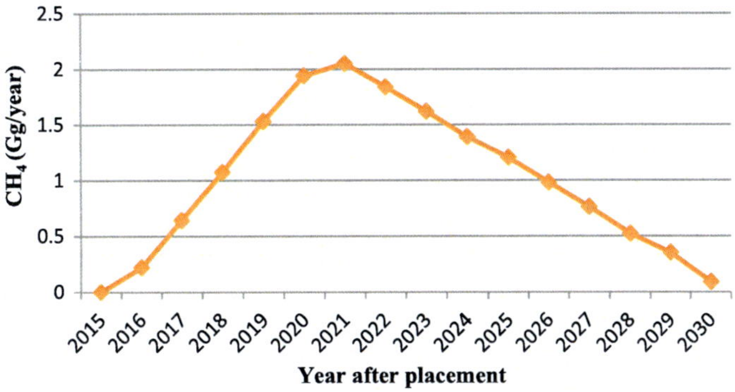

Fig. 4 Emission profile of landfill gas (methane)

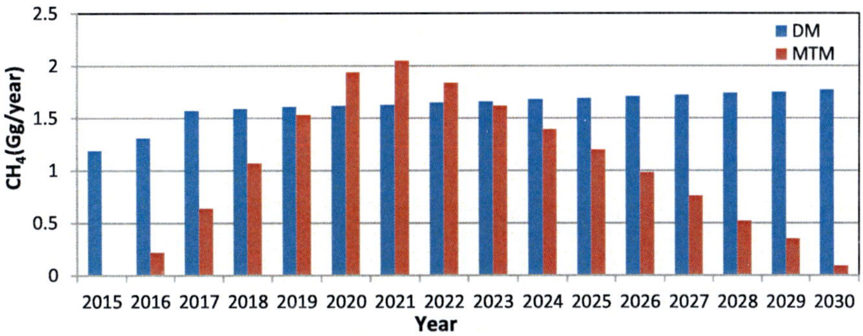

Fig. 5 Comparison of methane generation by DM and MTM

Table 6 Unpredictability of various constants as per IPCC [6]

S. No.	Parameters	Uncertainty range	Value range
1.	Degradable organic carbon (DOC) = 0.21 (maximum default value)	−50% to 30%	0.105–0.252
2.	Fraction of degradable organic carbon dissimilated (DOC_F) = 0.77	−30% to 0%	0.539–0.770
3.	Methane correction factor (MCF) = 0.4	−30% to 30%	0.280–0.520
4.	Fraction of CH_4 in landfill gas (F) = 0.5	0% to 20%	0.500–0.600

5 Variability Related to Estimation of Methane Gas Generation

Dubiety in the estimation of methane gas from a variety of sources and is mainly associated with automatic emissions where the characteristics of waste, methane gas fraction, and other values is similar to taking regular years of irregular waste disposal. The default parameters used are compiled by the IPCC and are given in Table 6 and using those percentages of uncertainty percentages can be adjusted.

The overall volume of the methane generation from any site varies upon many parameters and one of the important factors is type of model/methods used for methane estimation.

6 Conclusions

- An estimate of 810 Gg of MSW reaching landfills/dumpsites in Mohali city in next 16 years which poses a threat to environment be it soil pollution, groundwater pollution or air pollution. Apart from different types of pollution caused by landfills, anaerobic decomposition of waste produces methane gas which has very high global warming potential, and the assessment of methane gas generation had been done for the city.
- Estimation of methane emission at the MSW landfill site in Mohali city was conducted using two different methods to measure methane extraction method default and modified triangular method.
- The potential methane gas estimation by both the methods is DM-25.88 Gg and MTM-16.25 Gg for the duration from year 2015 to 2030.
- The estimated quantity of methane gas emitted by default methodology is higher to that based on modified triangular method. DM takes assumption that emission coefficient is always present every year which can sometimes lead to incorrect calculations of methane emission.
- Among various methods/models for methane generation, the default methodology and the modified triangular method are most effective and are widely used where

there is a lack of detailed data or information related to a particular of waste disposal.

- Modern techniques for recovering methane from landfills and sewage can significantly reduce the amount of methane. Another option to effectively utilize this untapped resource is to produce energy, encourage strategic marketing of compost and digestate.

References

1. Bhuvanendran, R., & Shimi, S. L. (2010). Municipal solid waste management: Projection of methane emission from landfill disposal system in Chandigarh.
2. Gawatre, D., Kandgule, M., & Kharat, S. (2016). Comparative study of population forecasting methods. *IOSR Journal of Mechanical and Civil Engineering, 13*(04), 16–19.
3. Holbrook, S. J., Schmitt, R. J., & Stwphens, J. S., Jr. (1997). Changes in an assemblage of temperate reef fishes associated with a climate shift. *Ecological Applications, 7*, 1299–1310.
4. Hughes, L. (2000). Biological consequences of global warming: Is the signal already apparent? *Trends in Ecology & Evolution, 15*(2), 56–61.
5. IPCC. (1996). IPCC Guidelines for National Greenhouse Gas Inventions: Reference manual. *National Physical Laboratory, New Delhi, India, 1996*, 6–15.
6. IPCC. (2002). CH4 emissions from solid waste disposal. Background Papers—IPCC Expert Meetings on Good Practice Guidance and Uncertainty Management in National Greenhouse Gas Inventories (pp. 419–439).
7. Mohali Municipal Corporation, Mohali Developmental Plan and Program Report, Mohali Developmental Report (2013).
8. Mooney, H. A. (1991). Biological response to climate change: An agenda for research. *Ecological Applications, 1*, 112–117.
9. Myers, S. S. (2009). Global environmental change: The threat to human health. Worldwatch Institute. *Worldwatch Report, 181*.
10. Oonk, H., & Boom, T. (1995). Validation of landfill gas formation models. *Studies in Environmental Science., 65*, 597–602. https://doi.org/10.1016/S0166-1116(06)80251-7
11. Jain, P., Wally, J., Townsend, T. G., Krause, M., Tolaymat, T.: Greenhouse gas reporting data improves understanding of regional climate impact on landfill methane production and collection. *PloS One*.
12. Srivastava, A. N., & Chakma, S. (2020). Quantification of landfill gas generation and energy recovery estimation from the municipal solid waste landfill sites of Delhi, India. *Energy Sources, Part A: Recovery*.
13. Rana, R., Ganguly, R., & Gupta, A. (2017). Physico-chemical characterization of municipal solid waste from Tricity region of Northern India: A case study. *Journal of Material Cycles and Waste Management, 20*(1), 678–689.

Evaluation of Water Quality Index to Assess the Impact of River Pollution on Vembanad Lake—A Ramsar Site

Rohan Nair, K. V. Hariprasad, S. Ashwin Shenoi, M. P. Amrutesh, Kiran V. Gireesh, Gevargis Muramthookil Thomas, and S. N. Jyothi

Abstract Evaluation of the degree of pollution in the water bodies is becoming a matter of prime concern globally, as the number of countries becoming water stressed is on the rise. A single number that adequately subsumes the pertinence of various physicochemical parameters on the quality of water like pH, dissolved oxygen, alkalinity, dissolved solids, hardness, coliform content, etc., is the water quality index. In the study, a generic Python code was developed to promptly compute the water quality index, using a method proposed by the Council of Ministers of the Environment of Canada, for any water body, by inputting 24 physicochemical parameters for four consecutive years. Vembanad Lake is one among the Wetlands of International Importance in Kerala and is listed as a Ramsar site. The lake receives drainage from ten rivers and is rich in biodiversity. The wetland is a major source of livelihood for the people in various districts of the state over which the lake is spread. The levels of pollution in the rivers draining into the Ramsar site have been analyzed by determining its water quality index, using the Python code. It has been observed that some of the river stations of River Periyar that drain into the lake are having marginal and poor water quality, due to high levels of fecal and total coliform. As a result of the pollution from the river and few other reasons like tourism and agricultural practices, the water quality index of Vembanad Lake is 60 and falls in the marginal category.

Keywords Vembanad Lake · Water quality index · Water pollution · Ramsar site · Kerala rivers

---

R. Nair · K. V. Hariprasad · S. A. Shenoi · M. P. Amrutesh · K. V. Gireesh
Department of Computer Science and Engineering, Amrita Vishwa Vidyapeetham, Amritapuri, India

G. M. Thomas · S. N. Jyothi (✉)
Department of Mechanical Engineering, Amrita Vishwa Vidyapeetham, Amritapuri, India
e-mail: jyothisn@am.amrita.edu

A. K. Gupta et al. (eds.), *Advances in Construction Materials and Sustainable Environment*, Lecture Notes in Civil Engineering 196,
https://doi.org/10.1007/978-981-16-6557-8_49

1 Introduction

Anthropogenic activities have resulted in the introduction of various substances into water bodies. When the concentration of such substances exceeds the tolerable limits, the water bodies get polluted and become unfit for domestic consumption [1]. Rising levels of pollutants in the water bodies of India are a matter that has still not received the deserved attention [2]. Quality of water depends on various physicochemical parameters, and the tolerable limits for these parameters are set by various organizations like the World Health Organization (WHO) [3], Indian Council for Medical Research (ICMR) [4], and Bureau of Indian Standards (BIS) [5]. Water quality index incorporates all these physicochemical parameters into a single number and provides a method for assessing the degree of contamination of a water body [6–8]. The Canadian Council of Ministers of the Environment water quality index (CCME WQI) is evaluated using a method that was proposed by the British Columbia Ministry of Environment, Lands, and Parks, which was subsequently revised by Alberta Environment [9, 10]. The CCME WQI value ranges between 0 and 100, with 100 indicating excellent water quality and 0 being the worst, as shown in Table 1 [9, 10].

Kerala is one of the southernmost states in the Indian subcontinent, located between 8° 17′ N–12° 48′ N and 74° 27′ E–77° 38′ E. Around 40% of the Western Ghats, a global biodiversity hotspot, is situated in Kerala where 41 rivers originate. There are three freshwater and seven major estuarine lakes in Kerala. A Ramsar site is an ecosystem designated as wetland of international importance under the Ramsar Convention. The Wetlands Convention is an intergovernmental treaty that establishes a structure for wetland protection and resource management. Wetland ecosystems are among the most complex and active ecosystems on the planet. Vembanad Lake in Kerala is an estuarine lake that has been considered as one of the most important wetland ecosystems in India [11].

The evaluation of the water quality of Vembanad Lake and the ten rivers, namely Keecheri, Puzhakkal, Karuvannur, Chalakkudy, Periyar, Muvattupuzha, Pamba, Meenachil, Manimala, and Achenkovil, from which the lake receives drainage has been performed [12, 13]. The CCME WQI was evaluated, for this purpose, using several physicochemical parameters over a span of four years (2014–2017). A generic Python code was developed as part of this study, for calculating the CCME WQI, for any water body, which will generate a single valued output, the significance of which

Table 1 Water quality categories

CCME WQI	Category
0–44	Poor
45–64	Marginal
65–79	Fair
80–94	Good
95–100	Excellent

can be understood by a common man. The study has helped to shed some light on some of the reasons for the deterioration in the water quality of the Vembanad Lake.

2 Methodology

The CCME WQI has been evaluated for each river station, using the 24 physicochemical parameters, as shown in Table 2, using the data collected from the Kerala State Pollution Control Board (KSPCB) [14]. The permissible limits for the parameters have been selected based on the standards set by BIS and ICMR [4, 5]. However, due to the lack of available data, only the parameters such as nitrate, dissolved oxygen, ammonia, pH, biochemical oxygen demand, and fecal coliform and total coliform were used to calculate the CCME WQI for Vembanad Lake.

Table 2 Physicochemical parameters and their limits

S. No.	Parameter	Permissible limit
1	pH	6.5–8.5
2	Electrical conductivity (μ S/cm)	300
3	Turbidity (NTU)	1
4	Total alkalinity (mg/L)	200
5	Chlorides (mg/L)	250
6	Hardness (mg/L)	200
7	Calcium (as $CaCO_3$ mg/L)	75
8	Magnesium (as $CaCO_3$ mg/L)	30
9	Sulfates (mg/L)	200
10	Total dissolved solids (mg/L)	500
11	Boron (mg/L)	0.5
12	Biochemical oxygen demand (mg/L)	5
13	Fluoride (mg/L)	1
14	Dissolved oxygen (mg/L)	5
15	Nitrate (mg/L)	45
16	Total coliform (MPN/100 ml)	500
17	Fecal coliform (MPN/100 ml)	500
18	Cadmium (mg/L)	0.003
19	Copper (mg/L)	0.05
20	Lead (mg/L)	0.01
21	Chromium (mg/L)	0.05
22	Nickel (mg/L)	0.02
23	Zinc (mg/L)	5
24	Iron (mg/L)	0.3

The CCME WQI is calculated in a three-step process [9, 10]:
The scope (F_1) is calculated using Eq. (1):

$$F_1 = \frac{\text{Total number of failed parameters}}{\text{Total number of parameters}} * 100 \tag{1}$$

The frequency (F_2) is calculated from Eq. 2 as

$$F_2 = \frac{\text{Total number of failed tests}}{\text{Total number of tests}} * 100 \tag{2}$$

The amplitude (F_3) is calculated in a three-step process:
The first step in calculating F_3 is to find the excursion (Eq. 3), in the case, when the test value must not exceed the standard, excursion is calculated as

$$\text{excursion}_i = \frac{\text{FailedTestValue}_i}{\text{Objective}_j} - 1 \tag{3}$$

In cases, when the test values must not fall below, the stipulated standard using Eq. 4:

$$\text{excursion}_i = \frac{\text{Objective}_j}{\text{FailedTestValue}_i} - 1 \tag{4}$$

The normalized sum of excursions (*NSEs*) are calculated by Eq. 5:

$$\text{nse} = \frac{\sum_{i=1}^{n} \text{excursion}_i}{\text{Total number of tests}} \tag{5}$$

Finally, amplitude (F_3) is obtained by Eq. 6:

$$F_3 = \frac{\text{nse}}{0.01 * \text{nse} + 0.01} \tag{6}$$

CCME WQI can be calculated using Eq. 7:

$$\text{CCMEWQI} = 100 - \frac{\sqrt{F_1^2 + F_2^2 + F_3^2}}{1.732} \tag{7}$$

In this study, the calculation of CCME WQI has been done by developing a Python program. All the data, including the different parameters of different water bodies, have been stored in the program database, and a formula is coded which calculates scope (F_1), frequency (F_2), and amplitude (F_3) separately and finally CCME WQI. The time complexity of the program is linear, i.e., $O(n)$. Programming is done without using any complex feature to make it user friendly.

3 Results and Discussion

The validation of the Python program was done by matching the analytically calculated CCME water quality index. Figure 1 shows the screenshot of the program output, and Fig. 2 depicts the results of the code matching the analytical calculation and displays the equation used for calibration.

A bubble plot has been prepared to map the CCME WQI values of all the ten rivers and lake, chosen for study, to their corresponding geographical locations in the state of Kerala. Figure 3 illustrates the bubble plot, where the bubble size represents the value of the CCME WQI.

The CCME WQI of the river stations of rivers Keecheri, Puzhakkal, Karavannur, Chalakkudy, Muvattupuzha, Meenachil, and Manimala has been illustrated in Fig. 4. It was observed that most of these rivers have good water quality and have relatively lower levels of pollution. Most of the parameters are within the permissible limits. Figure 5 illustrates the calculated CCME WQI of all the river stations of the rivers Pamba and Achenkovil and both these rivers show good water quality. However, increasing urbanization along the riverbanks and poorer compliance with pollution control norms can result in deterioration of the water quality.

It was observed that some of the river stations of Periyar had only marginal to poor CCME WQI values, mostly in the highly industrialized areas of Eloor and Pathalam,

Fig. 1 Screenshot of program output

Fig. 2 Validation of code with analytical results

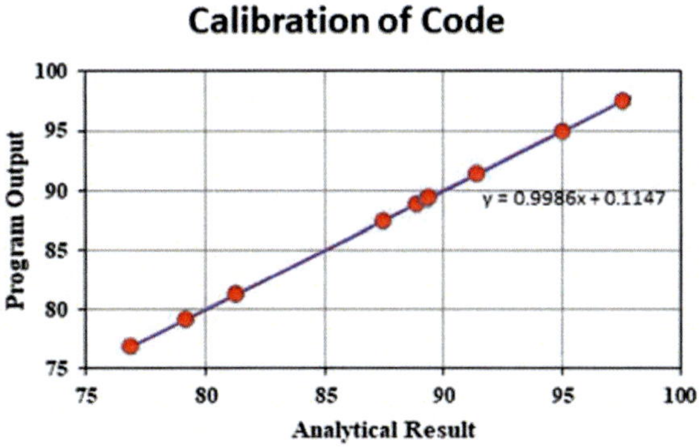

Fig. 3 CCME WQI bubble plot surfaced on Kerala map

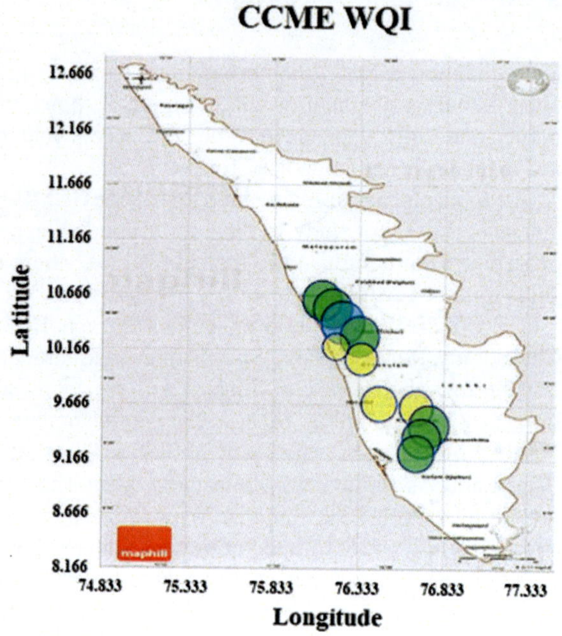

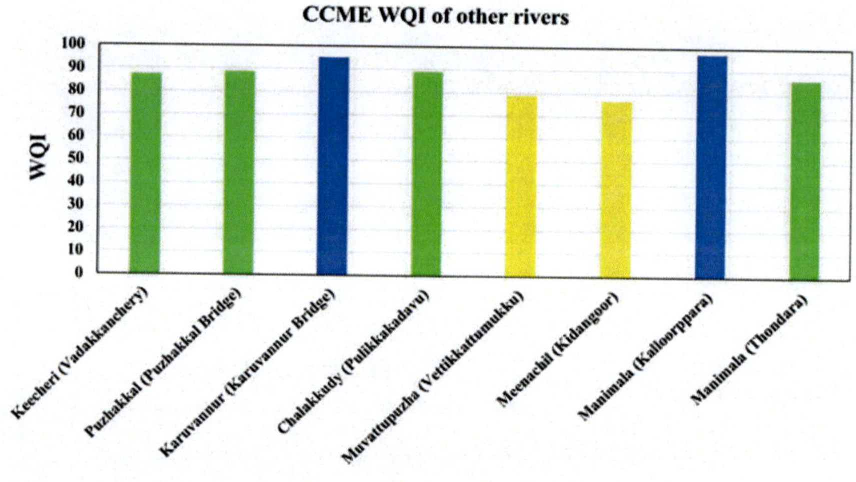

Fig. 4 CCME WQI of river stations of other rivers draining into Vembanad Lake

as shown in Fig. 6. A study of the data shows high levels of concentration of fecal coliform and total coliform around these river stations. Unprocessed wastes from septic tanks are one of the most common sources for coliform bacteria [6]. Effluents from process industries, along the banks of the river, if released without conforming

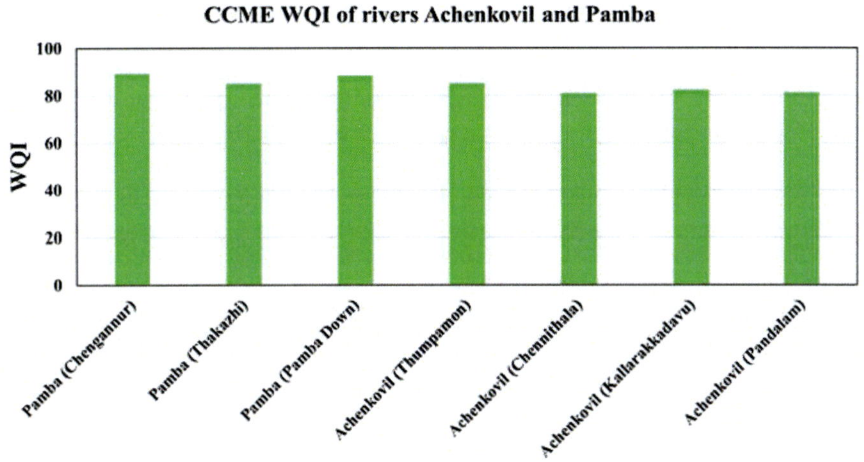

Fig. 5 CCME WQI of river stations of Achenkovil and Pamba rivers

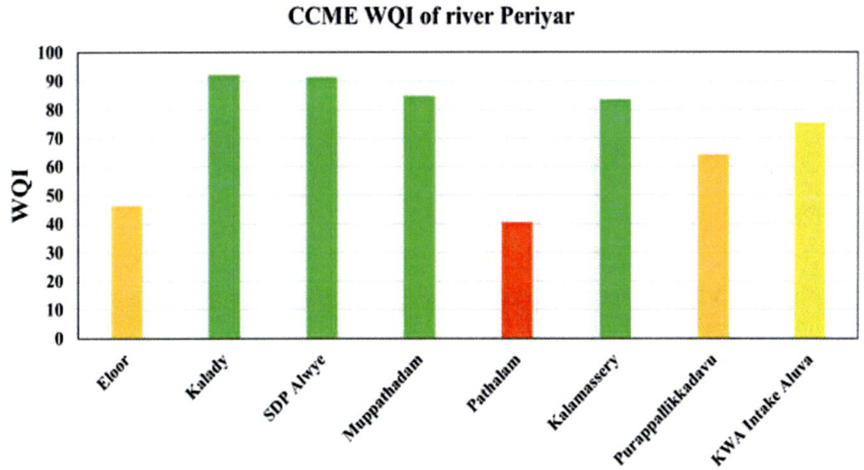

Fig. 6 CCME WQI of river stations of Periyar River

to the regulations, could also be another reason for the degrading water quality of the river.

Among all the rivers that drain into the Vembanad Lake, the higher pollution levels in the River Periyar could have caused the reduction the water quality of the lake. It has been observed that more than two million m^3 of urban sewage enters the Vembanad Lake, on a daily basis, from the city of Kochi. Through the extensive network of canals in Kochi and the rivers that lead to the lake, a considerable volume of unprocessed wastes from hospitals and slaughterhouses also enters the lake. The Pathalam bund, which is a temporary barrage, constructed every year on the Eloor branch of the

River Periyar, helps to prevent the salinity ingress from the Vembanad backwaters. However, this results in the pollution reaching toxic levels due to stagnation of large quantities of wastewater that are being released into this branch. The stagnation occurs due to the improper flow hindering the natural flushing out of the contaminants from the water.

The tourism industry has been flourishing recently in Kochi, Alappuzha, and Kumarakom areas around the lake. This has in turn resulted in a drastic growth in the number of resorts, hotels, and houseboats in the area. These have also become a major source for unprocessed sewage, released into the lake, and raising the levels of fecal coliform and total coliform.

The coir industries, found in and around the Vembanad Lake, follow traditional retting to extract coir fiber from the husks of coconut. This process puts additional immense pressure on the lake system. Large-scale agriculture and aquaculture are very common in the drainage basins, and huge amounts of fertilizers are applied to enhance the productivity of the paddy cultivations in the Kuttanad region. Flood irrigation practices and increased precipitation cause excess nutrients to enter the lake system, along with the eroded topsoil, resulting in eutrophication and depletion of dissolved oxygen levels. Finally, the Thanneermukkom bund, which was built in order to protect the Kuttanad farmlands from the salinity ingress has also contributed to altering the pollution dispersion capabilities of the lake, thereby contributing to degradation of the wetland.

4 Conclusion

As a result of the high population density in Kerala, greater number of people directly depend on the lakes and rivers for their daily domestic consumption. Evaluation of CCME WQI helps to quantify the levels of pollution in these rivers and lakes by placing them into various water quality categories. The evaluation of CCME WQI requires at least four parameters that are tested minimum four times, the accuracy increases with a greater number of parameters. As a part of this study, a total of 24 parameters were considered for the years 2014–2017 to get a more accurate value for the CCME WQI for the ten rivers that drain into the Vembanad Lake. The calculated CCME WQI for the river station Pathalam is 40.5 making it fall in the poor water quality category, while the calculated CCME WQI for the river stations Eloor and Purappallikkadavu is 46 and 63.9, respectively, placing them in marginal water quality category. The study also shows that majority of the rivers, draining into the lake, possess good water quality as their CCME WQI values are in the range of 81–92, which led to the conclusion that the Periyar River is one of the major contributors to the rising pollution levels of Vembanad Lake.

The marginal water quality of the river stations in Periyar River can be attributed to the increasing industrialization and urbanization along its riverbanks, causing the levels of conductivity, chloride, calcium, magnesium, sulfates, dissolved oxygen, lead, iron, and coliform to fall out of the permissible limits. The Vembanad Lake has marginal quality of water, with a CCME WQI of 60, due to the parameters dissolved oxygen, fecal and total coliform falling outside permissible limits. Other notable reasons are the rapid growth of the tourism industry around Vembanad Lake, poor agricultural practices, and large volumes of unprocessed sewage flowing into the lake. Vembanad wetland, a location of great ecological importance, plays an integral role in Kerala's agriculture and aquaculture, and hence, it is a necessity to take appropriate measures to bring down the pollution levels in the lake. Detailed field survey would be needed to identify other diffuse sources of pollution causing the degradation of water quality in the lake. Furthermore, the quality of other major rivers that flow into the lake needs to be maintained, since a further decline in the lake's water quality can eventually lead to the loss of its status as a Ramsar site.

References

1. Ramesh, M. V., et al. (2017). Water quality monitoring and waste management using IoT. In IEEE Global Humanitarian Technology Conference (GHTC) (pp. 1–7). https://doi.org/10.1109/GHTC.2017.8239311.
2. Vijayakumar, A., & Mahesh, A. S. (2019). Quality assessment of ground water on small dataset. *International Journal of Innovative Technology and Exploring Engineering (IJITEE), 8*(5).
3. World Health Organization (W.H.O.). (1998). *Guideline for drinking water quality. Health criteria and other supporting information* (2nd ed., pp. 231–270)
4. ICMR. (1975). *Manual of Standards of Quality for drinking water supplies.* Indian Council of Medical Research, Special Report No. 44:27.
5. BIS. (2012). Indian Standard Drinking Water—Specification (second revision), Bureau of Indian Standards, New Delhi.
6. Jyothi, S. N., Thomas, G. M., Rohith Raj, R. V., Masetti, A., Tammana, A., Motheram, M., & Gutlapalli, N. C. (2020). Assessment of water quality Index and study of the impact of pollution on the rivers of Kerala. *Materials Today: Proceedings.* ISSN 2214-7853. https://doi.org/10.1016/j.matpr.2020.09.084
7. Brown, R. M., McCleiland, N. J., Deininger, R. A., & O'Connor, M. F. (1972) A water quality index—Crossing the psychological barrier. In *Proceedings of the International Conference on Water Pollution Research, Jerusalem* (pp. 787–797).
8. Horton, R. K. (1965). An index number system for rating water quality. *Journal of Water Pollution Control Federation, 37*(3), 300–306.
9. CCME. (2001). Canadian water quality guidelines for the protection of aquatic life: CCME Water Quality Index 1.0, User's Manual. In *Canadian Environmental quality guidelines (1999).* Canadian Council of Ministers of the Environment, Winnipeg, Manitoba.
10. Canadian Council of Ministers of the Environment. www.ccme.ca/publications/ceqg_rcqe.html
11. Sajeev, S., Sekar, S., Kumar, B., Senapathi, V., Chung, S.Y., & Gopalakrishnan, G. (2020). Variations of water quality deterioration based on GIS techniques in surface and groundwater resources in and around Vembanad Lake, Kerala, India. *Geochemistry, 80*(4), Supplement, 2020, 125626. https://doi.org/10.1016/j.chemer.2020.125626
12. Prasad, G., & Ramesh, M. V. (2019). Spatio-temporal analysis of land use/land cover changes in an ecologically fragile area Alappuzha district, southern Kerala, India. Natural Resources Research (pp. 1–12). https://doi.org/10.1007/s11053-018-9419-y
13. Prasad, G., Vinod, P. G., & John, S. E. (2018). Delineation of ground water potential zones using GIS and remote sensing—A case study from midland region of Vamanapuram river basin, Kerala, India. In *AIP Conference Proceedings 1952.* https://doi.org/10.1063/1.5031990
14. Kerala State Pollution Control Board. (2014–2017). Water and air quality directory.

A Scientometric Analysis on Bio-Bitumen

Ramu Penki, Banna Madhavi, K. Akhilesh Patnaik, and A Sri Divya

Abstract Bitumen is a result of oil creation and is utilized to tie the surfaces of cleared streets. A few analysts, worried about its ecological effects and future inventory if petrol refining eases back down, have begun to create options, i.e., bio-bitumen. Bio-bitumen is an alternative for the fossil-based bitumen which stalls the blended family canister waste to create a dark, exceptionally thick liquid-like bitumen—a material ordinarily utilized in street surfacing. An all-encompassing method for conducting the first scientometric analysis on Bio-bitumen is discussed in this research. The investigation accepted a comprehensive survey idea utilizing scientometric examination and science planning innovation, and extensive conversation to feature the most powerful distribution sources, most utilized catchphrases, most dynamic analysts and foundations, just as writing with the most noteworthy effect on the field of Bio-bitumen, to inspect the present status of the workmanship research center, and to distinguish the ebb and flow research gaps. The study gives a brief note on 1998–2018 data from the Scopus database on relevant bibliographic data. The exploration gaps distinguished were in the regions of Bio-bitumen properties. In this study by giving a scopus file to the Vos-viewer software, we will know the following results such as maximum no. of papers published in 2019 and followed by 2020, degree of collaboration 0.97, highest multiple authore during the period 2019–2020, etc., more than 90% of the papers are of multiple authors. Further long-haul contemplates are needed in these territories to give a premise to an administrative structure for the appropriation of Bop-bitumen. This examination will assist scientists with understanding the latest thing in Bio-bitumen, opening more space for additional exploration just as fill in as a wellspring of data for strategy creators, diary editors, experts, and exploration organizations.

Keywords Scientometric analysis · Bio-bitumen · Fossil-based bitumen · VOS software

R. Penki · B. Madhavi (✉) · K. Akhilesh Patnaik · A. Sri Divya
GMRIT, Srikakulam, Andhra Pradesh, India

R. Penki
e-mail: ramu.p@gmrit.edu.in

A. K. Gupta et al. (eds.), *Advances in Construction Materials and Sustainable Environment*, Lecture Notes in Civil Engineering 196,
https://doi.org/10.1007/978-981-16-6557-8_50

1 Introduction

Scientometrics is a scientific discipline concerned with assessing and analyzing scholarly literature. Bibliometrics have a sub-field called scientometrics. The assessment of the influence of academic journals and review papers, as well as the interpretation of scientific citations, as well as their use in policies and strategy contexts, are all major research topics [1]. Scientometrics are a field that studies the structure and development of science by analyzing scientific publications. The methods used to analyze different aspects of a project that are subjective journal using bibliometric/scientometric/informetric techniques [2]. Bibliometrics/scientometrics study includes studies on the dispersal and development of the literature, as well as the productivity of authors, document obsolescence, distribution of scientific literature by region, language, and other topics that aid in the monitoring of research growth and trend [2]. In this paper by using Scopus file in Vos-viewer, we extract the information. The literature review is a study or, more precisely, a survey that uses academic resources to analyze previously published information on a topic or research question. As a consequence, in order to write the literature review, you must first gather information. You must be an authority on the research subject. The findings and conclusions will be published and made available to the general public, especially scientists working in the field [3]. Determine the need for further analysis (justifying your research) study the correlation among research focused on their exposure to the subject and the contributions of other works. Make a case for why more research is needed by comparing your own findings to the existing literature [4]. We could go on and on about how the literature reviews feed new research, which feeds the literature reviews, and so on. The truth is that one exerts a force on the other, which is why science, as a global discipline, is continuously developing and evolving. Writing the literature review will help you advance in your career as a scientist by distinguishing you from the expert elite in your field of interest [4]. The literature review establishes context, informs methodology, promotes creativity, prevents duplication of effort, and ensures that professional standards are met. Iterative literature reviews take time and should be done during the proposed study [4]. Justifies the necessity of doing specific research in a specific area. Allows some new analysis technique to be used instead of the old ones by assisting in the correct collection of data. Allows the manuscript's readers to respond to the following questions in order to improve the manuscript's chances of being published [4]. It demonstrates to a Ph.D. committee that the student has read a significant amount of statistical literature in order to show that the student is familiar with a broad variety of theory and methodology analysis relevant to the proposed research subject [4]. The literature review can be viewed as a goal in and of itself, either to educate practise or to provide a thorough understanding of what is known about a subject. Depending on your goal, the procedure for performing this form of the literature review varies. If you want to develop your clinical practise, you should focus your literature review on an issue you came across in your practice [3]. When undertaking qualitative research, some texts advise against conducting a systematic literature review because it may lead the researcher to see "what someone else think they saw" instead of examining with new perspectives [3]. There are several different types of literature reviews, which are Narrative or Traditional literature reviews, Scoping Reviews, Systematic Quantitative Literature Review, Cochrane Reviews, and Campbell Collaboration [5]. A narrative or conventional literature review is a thorough, objective and vital examination of the existing state of knowledge on a particular subject. They are important step in the investigation because they help to create a theoretical structure and a research emphasis or background [6]. A scoping analysis is a newer approach to evidence synthesis that differs from systematic analyses in terms of meaning and objectives. A scoping review's aim is to provide a broad overview of the available research evidence without providing a definitive response to a specific research query [7]. Systematic quantitative literature review is a smart and effective method for conducting literature reviews, particularly for research students and others who are interested in new fields. It connects the dots between conventional narrative study and meta-analysis [8]. A Cochrane Review is a comparative study of research in health care and health policy conducted in the Cochrane Database of Systematic Reviews. The Campbell Alliance is a worldwide scientific research platform that provides exceptional evidence analogs, synopses, and policy briefs. [8].

The bio-bitumen is subjected to a systematic scientometric study in this article. Bitumen/asphalt is commonly used around the world, but production is declining, so bio-bitumen is produced from bio-degradable materials such as wood, agriculture and herbal wastes, industrial and municipal solid wastes, animal wastes, aquatic wastes, and so on, using various processes. The majority of bituminous binders used in pavement materials come from fossil fuels, namely crude petroleum. Bio-binders can now be produced using bio-renewable materials, which has both technological and economic advantages. Bio-binders can be used in three ways to reduce the need for crude petroleum–derived bituminous binders: direct alternative binder (100 percent replacement), bitumen extender (25–7% replacement), and bitumen extender (25–7% replacement) [9]. By examining the rheological properties, the feasibility of producing bio-binders from oakwood bio-oils for use as a direct substitute (100% replacement) is demonstrated [10].

The scientometric analysis is used to analyze the scientific knowledge domain objectively, whereas the critical review is used to suggest research themes and difficulties based on the scientometric results. [12] The monitoring of science, the evaluation of writers' scientific contributions, papers, or individual works, as well as the study of the distribution mechanism of scientific information are all common uses of scientometric [13]. In such approaches, researchers have developed methodological principles for gathering information produced. Researchers have used various approaches such as citation analysis, data visualization, and co-word assessment in their correspondence, and content analysis, as well as text-mining, to accomplish these goals. Many bibliometrics studies may concentrate on authorship or assess the contribution of authors. [13]

2 Methodology

The flowchart of the methodology adopted for the study is given in Fig. 1.

2.1 Co-authorship Network: Authorship Pattern

Table 1 clearly shows that more than 90% of the articles are co-authored by multiple writers, with only a handful by single authors. Four writers contributed 74 papers out of 280, accounting for 26.43 percent of the total. With 21.07% and 17.85%

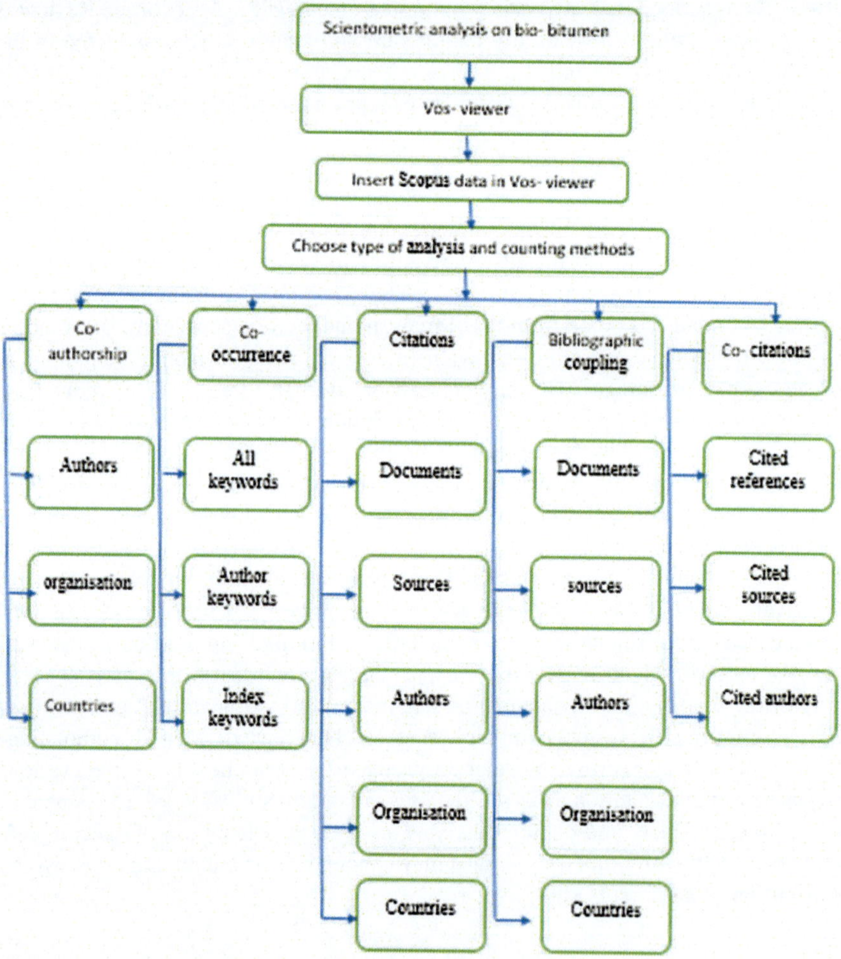

Fig. 1 Flow chart representing scientometric methodology

Table 1 Years wise distribution of Papers

#Authors	No. of papers	% Percentage	#Authors	No. of papers	% Percentage
Single	6	2.14	Seven	15	5.35
Two	28	2.14	Eight	6	2.14
Three	50	17.85	Nine	6	2.14
Four	74	26.43	Eleven	2	0.71
Five	59	21.07	Fifteen	1	0.35
Six	33	11.78	Total	280	

respectively, 5 authors and 3 authors came in second and third. Just 2.14 percent of papers are written by a single author, i.e., only 6 papers are authored by a single person.

2.2 *Co-authorship Network: Authorship Pattern Year Wise*

Table 2 shows the data for Authorship Trend year-by-year, i.e., year and single, double, triple, four, five, six, seven, eight, nine, ten, eleven, and fifteen in respective

Table 2 Authorship trends by year

Year	1	2	3	4	5	6	7	8	9	10	11
1998							1				
2002			1								
2004					1						
2008	1		1	1							
2009		2	2								
2010		4	2	2	1						
2011				2	1		1				1
2012		3	2	1	2						
2013		3	6	1	3						
2014		3	4	11							
2015		1	4	8	3	4					
2016		3	2	6	6		4				
2017	1	1	4	6	7	4	2	1	2		1
2018		1	8	9	11	3	2	1	2		
2019	3	6	8	14	8	11	1	2	1	1	
2020	2	4	6	9	12	9	3	1	1		
2021				4	4	2	1	1			

years. There are just a few years of single-authored articles, and the rest are multi-authored papers. Single writers are present in the years 2008, 2017, 2019, and 2020. In comparison with previous years, 2019 has a higher number of double authors.

2.3 *Co-authorship Network: Author Productivity*

Table 3 contains details on author productivity. Between 1998 and 2021, the year 1998 has the highest authorship efficiency, while the years 2013 and 2014 have the lowest, followed by 2010, 2009, and 2012. Per year, an average of 12.2 papers are shared. To measure authorship productivity, use the formulas 1 and 2 as given below.

$$\text{Average Authors per Paper} = \text{No. of Authors}/\text{No. of Papers} \tag{1}$$

$$\text{Productivity per Author} = \text{No. of Papers}/\text{No. of Authors} \tag{2}$$

2.4 *Co-authorship Network: Degree of Collaboration*

The following formula, suggested by Subramanyam, was used to assess the strength of Collaboration (DC) (Eq. 3).

$$\text{DC} = \text{Nm}/(\text{Nm} + \text{Ns}) \tag{3}$$

where DC = Degree of Collaboration; Nm = Number of Multiple Authored Papers; Ns = Number of Single Authored Papers. (Table 4).

2.5 *Co-authorship Network: The Pattern of Co-authorship (CAI)*

The following equation, introduced by Garg and Padhi, was used to examine the Pattern of Co-Authorship (CAI) as given by Eq. 4.

$$\text{CAI} = \left(N_{ij}/N_{io}\right)/\left(N_{oj}/N_{oo}\right) \tag{4}$$

where N_{ij} = No. of papers having authors in block *i*; N_{io} = Total output of block *i*; N_{oj} = No. of papers having *j* authors for all books; N_{oo} = Total no. of papers for all authors and all blocks (Table 5).

Table 3 Author productivity

Year	Total number of papers	Total number of Authors	AAPP	Productivity per author	Year	Total no. of papers	Total no. of authors	AAPP	Productivity per author
1998	2	7	7	0.14	2014	19	32	1.77	0.56
2002	1	3	3	0.33	2015	20	93	4.65	0.215
2004	1	5	5	0.2	2016	21	94	4.94	0.202
2008	3	7	2.66	0.42	2017	29	102	3.64	0.274
2009	4	9	2.5	0.44	2018	37	155	4.3	7.33
2010	9	20	2.22	0.45	2019	52	207	3.98	0.25
2011	5	31	7.75	0.12	2020	45	114	2.53	0.395
2012	8	22	2.75	0.36	2021	12	60	5	0.2
2013	13	24	1.71	0.58	Total	281	985	65.6	12.566

Table 4 Degree of collaboration

Year	Single	Multiple	Degree of collaboration	Year	Single	Multiple	Degree of collaboration
1998		1	1	2014		19	1
2002		1	1	2015		20	1
2004		1	1	2016		21	1
2008	1	2	0.66	2017		29	1
2009		4	1	2018		37	1
2010		9	1	2019	3	48	0.94
2011		5	1	2020	2	43	0.95
2012		8	1	2021		12	1
2013		13	1	Total		280	0.97

Table 5 Pattern od co-authorship

Blocks	Single	CAI	Double	CAI	> 2	CAI	Total
1998–2003	0	0	1	292.7	2	75.54	3
2004–2009	1	501.7	2	219.5	5	70.82	8
2010–2016	0	0	17	157.14	78	93.03	95
2017–2021	6	253.3	12	60	163	105.53	175
Total	7	755	32	729.34	248	344.97	281

2.6 *Country-Wise Documents, Citations, and Total Link Strength*

The number of documents submitted to countries is listed in Table 6, followed by citations and total relation power. The United States is the most productive country in terms of contribution, citations, and total relation power. China, France, the United Kingdom, Malaysia, and other countries are ranked second, third, and fourth, respectively. This information is obtained from Vos-viewer by giving it a Scopus file. This Table 6 shows the relationship between countries, documents, citations, and total link strength [11]. (Table 6; Figs. 2 and 3).

2.7 *Source of Documents and Link Strength*

Figure 4 shows the effect of a diary's absolute attachment intensity to other friends verified on diaries, the number of reports distributed, and its comparison consider. With a full attachment power of 14,052, a distribution absolute of 658, and a comparison search of 658, Construction and Building Materials is the highest in terms of

Table 6 Country document network

ID	Country	Document	Citations	Total link strength
51	United States	89	2099	6964
6	China	63	1075	5809
12	France	35	448	1690
50	United-kingdom	14	305	974
29	Malaysia	17	257	2297
22	Japan	8	183	189
21	Italy	19	139	1829
44	Sweden	9	125	1815
42	South Korea	10	122	317
13	Germany	10	110	1078
31	Netherlands	4	101	327
16	India	8	90	519
43	Spain	7	90	353
2	Australia	4	69	845
37	Portugal	5	55	587
46	Taiwan	2	31	514
35	Pakistan	2	21	740
34	Norway	2	15	459
7	Colombia	2	1	263
18	Iran	2	0	423

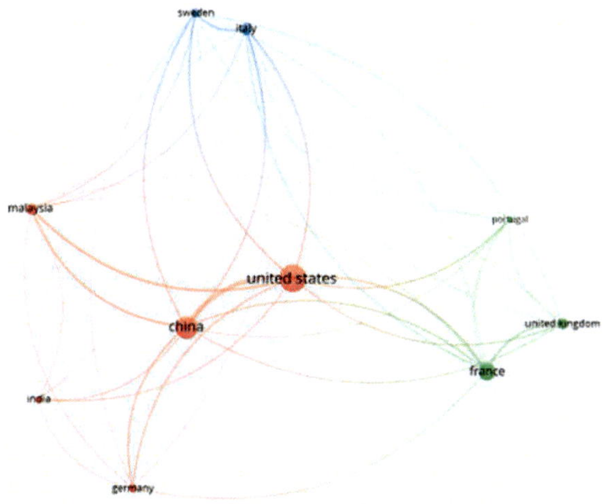

Fig. 2 Visualization of countries

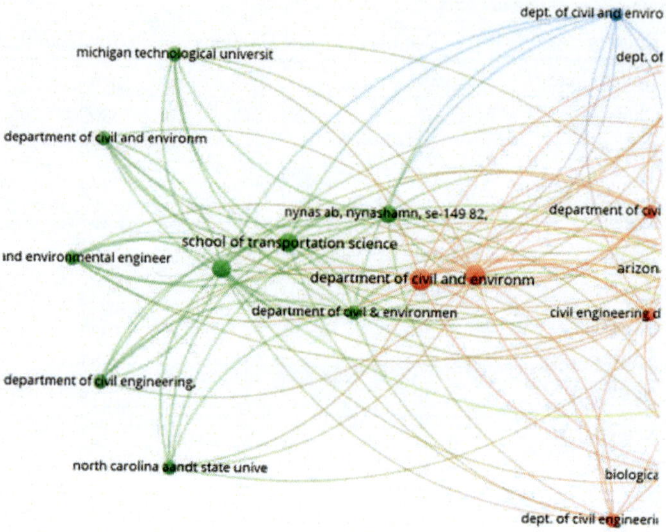

Fig. 3 Visualization of organizations

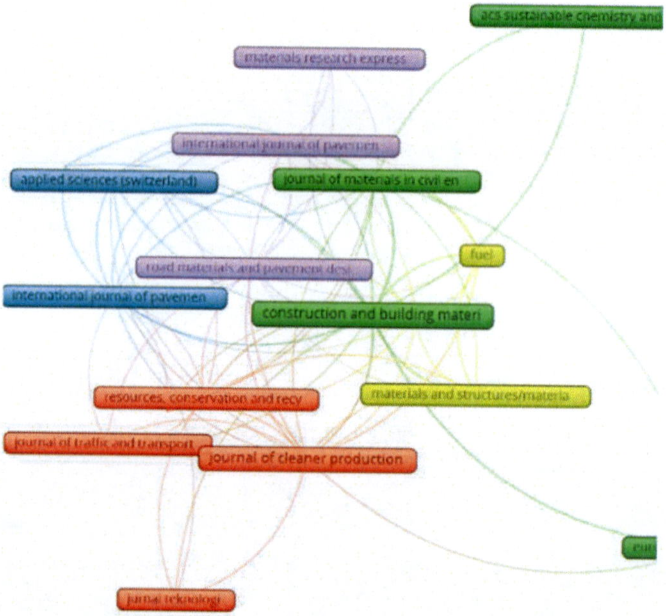

Fig. 4 Visualization of sources

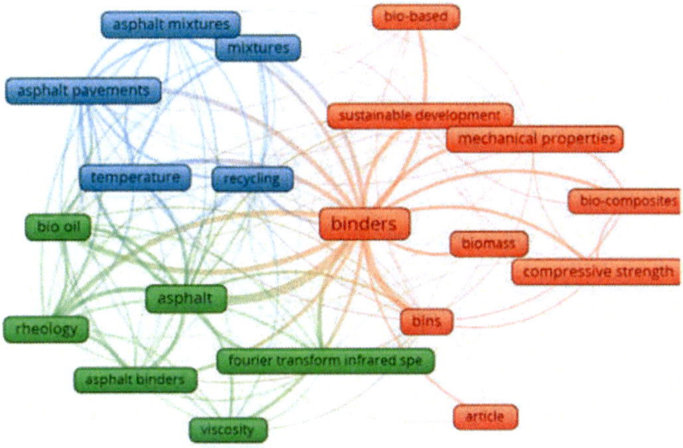

Fig. 5 Visualization of keywords

contact. Construction and Building Materials is the key diary in Fig. 5, with distributions on geopolymer concrete since it has a bigger reputation. Figure 6 depicts the archives by source for each year that had geopolymer concrete distributions between 2008 and 2020 (March). The diagram, created with Scopus analyzer, depicts the fact that Construction and Building Materials [11].

2.8 Research Keywords and Link Strength

The keywords reflect the substance of research and suggest the research domain's core field. The most frequently used keywords in the study papers used for this analysis are shown in the table below. The top three most commonly used keywords are binders, asphalt, and asphalt pavement. The figure depicts the visualization of the most frequently occurring keywords and their connectivity to one another based on their correlation strength. However, it is also noted that bio-bitumen, bio-oil, rheology, and asphalt binder are the main research areas. Keyword occurrences and relation strength of key works are listed in the table. The term "binder" appears many times and has a connection strength of 475 [11].

3 Conclusions

According to the findings, the majority of articles are written by multiple writers, with just a few written by single authors. This report gives a brief data for scientific literature, the idea of scientometric research and also the discussions of findings.

Other bodies like documents, most cited authors, most influenced institutions and countries involved in the study of bio-bitumen. And also visualized together with the citations and the link strength. This paper also reveals the results of 100% replacement of bio-bitumen. The formula by Subramanian is used to calculate degree of collaboration that gives a very high degree of collaboration. The global average in the paper and the increasing trajectory of co-authored papers are reflected in the average co-authorship index for all authors.

- The maximum no of papers published in 2019 and minimum in 2004 followed by no papers published in years 1999, 2000, 2001, 2003, 2005, 2006, 2007.
- During this time span, 20,192,020 the number of research papers contributed by multiple authors was at an all-time high.
- The degree of collaboration was 0.97
- It was found that the average value for CAI was around 281 of total.
- The relation between countries, authors and documents are gathered from Vos-viewer.
- The word binder has a high link strength of 475.
- Construction and Building materials is the highest source for maximum number of authors and papers.

References

1. Scientometrics, https://en.wikipedia.org/wiki/Scientometrics
2. Rajendran, P, Jayashankar, R, Elango B. : Scientometric analysis of contributions to journal of scientific and industrial research. International journal of digital library services. Vol 1., Issue 2, 2011.
3. The importance of Literature Review in Research Writing, https://scientificpublishing.webshop.elsevier.com/researchprocess/importance-literature-review-research-writing/
4. Literature Review and Focusing the Research, https://www.sagepub.com/sites/default/files/upmbinaries/29986_Chapter3.pdf.
5. Organizing Your Social Sciences Research Paper: 5. The Literature Review, https://libguides.usc.edu/writingguide/literaturereview.
6. Byrne, J.A.: Improving the peer review of narrative literature reviews, https://researchintegrityjournal.biomedcentral.com/articles/https://doi.org/10.1186/s41073-016-0019-2.
7. Cochrane Reviews | Cochrane Library, https://www.cochranelibrary.com/.
8. Saran, A.: What is a systematic review?, https://www.campbellcollaboration.org/what-is-a-systematic-review.html.
9. Raouf, M. A., & Williams, R. C. (2010). Temperature and Shear Susceptibility of a Nonpetroleum Binder as a Pavement Material. *Transportation Research Record: Journal of the Transportation Research Board., 2180*, 9–18.
10. Zahoor, M., Nizamuddin, S., Madapusi, S., & Giustozzi, F. (2021). Recycling asphalt using waste bio-oil: A review of the production processes, properties and future perspectives. *Process Safety and Environmental Protection., 147*, 1135–1159.
11. Zakka, W.P., Lim, N.H.A.S., Khun, M.C.: A scientometric review of geopolymer concrete. Journal of Cleaner Production. 280, 124353 (2021).
12. Explore scientific, technical, and medical research on ScienceDirect, https://www.sciencedirect.com/.

13. What are Scientometric and Bibliometrics? https://provalisresearch.com/solutions-2/applications/scientometrics-bibliometrics-software/

A Scientometric Analysis on Aggregate Blending

Kota Komal Kumar and Ramu Penki

Abstract An all-encompassing approach for the first scientometric analysis on aggregate blending or proportions of aggregates is described in this study. The investigation accepted a comprehensive survey idea utilizing scientometric examination and science planning innovation, and extensive conversation to feature the most powerful distribution sources, most dynamic analysts, and foundations, effect on the field of Aggregate blending, to inspect the present status of-the-workmanship research center, and flow research gaps. The study gives a brief note on 19,732,021 related bibliographic data taken from the Scopus database. This examination will assist scientists, policymakers, professionals, research institutions with understanding the latest technology in aggregate blending and opening more space for additional exploration.

Keywords Scientometric analysis · Aggregate blending · VOS viewer · Dry blend · Bibliometrics

1 Introduction

Aggregate blending is the method of evaluating the proper proportions of aggregates for the blend to meet the gradation specifications [1]. The preliminary step in dry mix design is aggregate blending. The task of determining individual properties and proportioning the constituent materials is tedious. The coarse aggregate structure and degree are also essential to the design. Dry blend configuration aims to determine the proportions of various aggregate sizes to create a blend with the greatest thickness [2]. The dry blend configuration involves three main elements, namely aggregate selection, aggregate gradation, and aggregate blending. When it comes to the aggregate collection, different sizes of aggregates are taken and measured. Flaky and gravel aggregates give the mix more consistency than rounded aggregates, coarse aggregate is described as material retained on a 2.36 mm sieve. Fine aggregates are aggregates that passing 2.36 mm and are retained on 75 microns sieve [3].

K. K. Kumar · R. Penki (✉)
GMR Institute of Technology, Rajam, Srikakulam, Andhra Pradesh, India

A. K. Gupta et al. (eds.), *Advances in Construction Materials and Sustainable Environment*, Lecture Notes in Civil Engineering 196,
https://doi.org/10.1007/978-981-16-6557-8_51

There are several approaches for mixing aggregates that can be divided into three categories: (a) Graphical method (b) Trial and Error method and (c) Optimization techniques. The use of traditional methods will become complex with an increase in the number of different sizes of aggregates and will require higher computational time and sometimes uneconomical solutions. Optimization methods are the current trend and most researches were doing on this technique [4]. The current study compares the different approaches available in the state of art of Technology and summaries the various approaches in graphical format for easy interpretation and identification of further research needs in this area.

Scientometrics is a branch of research that analyses scientific publications to explore the structure and growth of science. The term has increased in usage and appreciation in recent decades, especially since Tibor Braun founded the dedicated Journal of Scientometrics in 1978 [5]. Mulchenko first suggested the concept of scientometrics as "A Quantitative Study of Research on The Progress of Science" [6]. The term scientometrics, which started in Russia, is also used interchangeably with the term bibliometrics. Scientometrics is the science of measuring science, which includes counting items to the processing and use of knowledge and drawing conclusions from the counts [5]. The results of this study are expected to give researchers and practitioners a thorough understanding of the current state and research trends in aggregate blending research in the construction industry, as well as encourage further research in this field [7].

2 Methodology

The keyword aggregate blending was used to extract related documents and data from the Scopus database. As a result, all documents related to aggregate blending were displayed and downloaded. Science mapping is performed for this data. Science mapping emphasizes the systematic and productive dimensions of scientific study, as well as the relationships between disciplines, sectors, writers, and individual publications [8].

In this paper, VOSviewer [9] was used for the analysis along with Scopus analyser. VOSviewer is a free open-source software for creating and reviewing scientometric maps. Unlike other bibliometric mapping applications. Data was collected from Scopus, tools were selected, data mining, data processing, and analysis, visualization and presentation, description, discussion of observations, gap detection, limitation, and conclusion were all used in the research [10].

A flowchart containing the methodology of current study is given in Fig. 1. Bibilometric analysis in the VOSviewer software has been considered. A total of 309 articles were identified from the literature search performed in the Scopus database. The data is analyzed with multiple analysis types and results for four major types are presented.

The literature samples indicate that research into aggregate mixing has turned up a notch for over a decade. However, research into computational strategies for

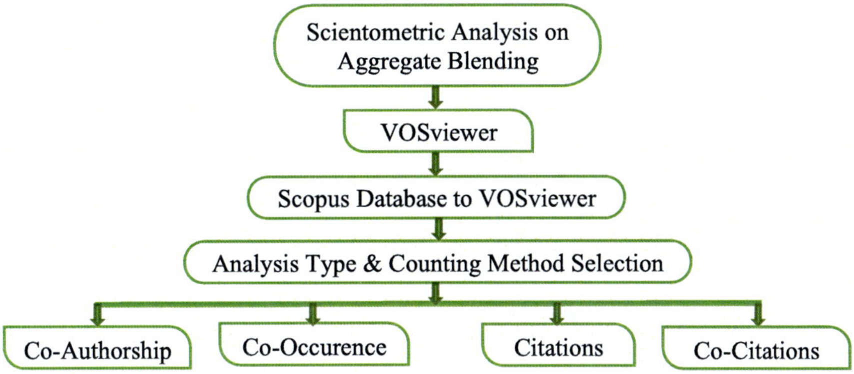

Fig. 1 Methodolgy

aggregate blending is still in its early stages compared to other techniques that have gained widespread popularity and have informed many policy decisions on aggregate blending. Figure 2 depicts the forms of documents contained in the Scopus database as reviewed by the Scopus analyser; journal articles, conference papers, book chapter, review papers, and notes. This research mainly focuses on these 309 articles. In the results it is shown that the co-authorship, co-occurrence, citation, co-citation analysis.

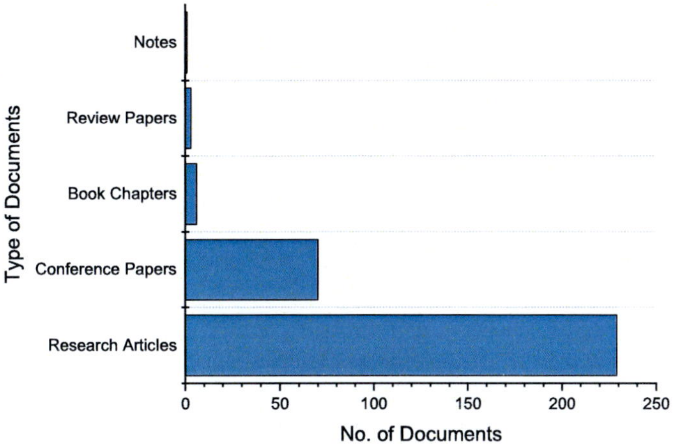

Fig. 2 Types of documents collected versus number of documents

3 Results

3.1 Co-authorship Analysis

According to Zhong [6], with the advancement of information technology systems and an increase in academic exchanges, research collaboration has grown. The researcher is an information carrier. Understanding the current state of research and locating influential authors requires identifying collaboration relationships among researchers. Co-authorship means a temporal and collegial association, which brings it closer to social network study [11]. Figure 3 depicts documents published year wise. Table 1 shows the Total douments per year and number of reaserch papers and

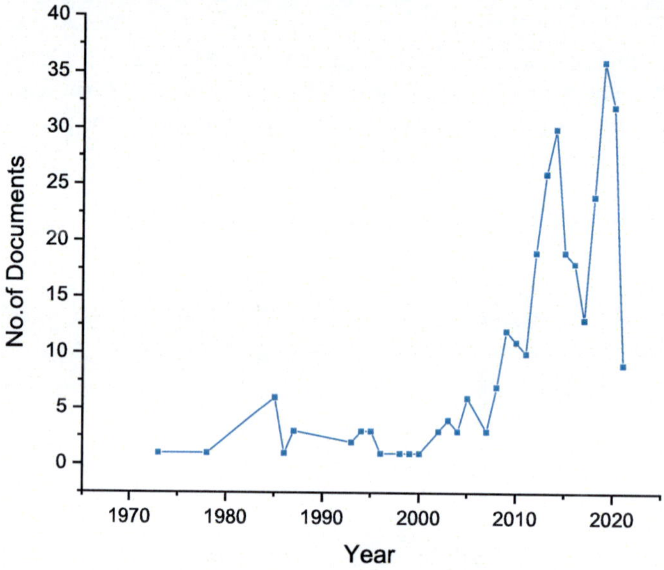

Fig. 3 Number of document versus year of publication

Table 1 Year wise distribution of documents

Year	Total documents	Research papers	% research papers
2021	9	7	77.77
2020	32	29	90.62
2019	36	20	55.55
2018	24	21	87.50
2017	13	12	92.30
1985	6	5	83.33

Note Total documents with less than 5 per year not listed

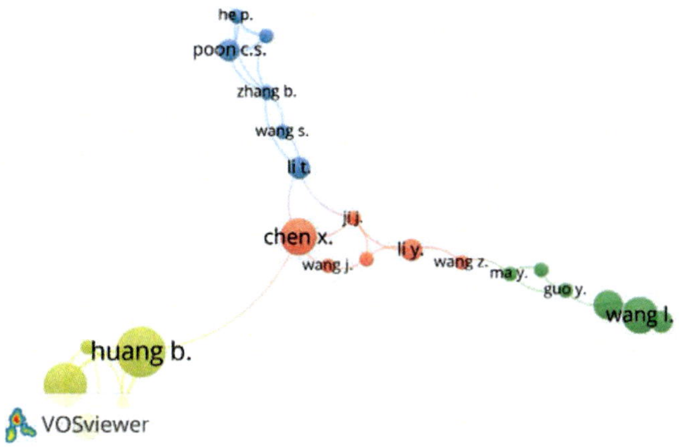

Fig. 4 Visualization of co-authorship and authors network

percentage of research paper per year. It is evident that year 2019 had high number of documents.

3.1.1 Co-authorship and Authorship Network

Evaluations and exchanges are the subjects of the author unable to actively associate papers, journals, or institutions with one another. The network of co-authors is the key for the Analysis of different types of relationship co-occurrences [6]. A co-authorship network is shown in Fig. 4 where each node represents an author and the edges mean collaborative actions.

3.1.2 Authorship Pattern Year Wise

Table 2 contains information on authorship patterns by year. Years 2015, 1987 and 1985 has the most single-authored contributions. In terms of double-authored contributions year 2018 has the most number of publishings. Year 2019 has the most multi-authored (upto and more than six authors) contributions in this area of research.

3.1.3 Author Productivity

The data about author productivity has presented in Table 3. Average authors per paper is found to be 3.13. The average productivity per author is 0.41 during the year 19,732,021. The years 2018, 2016 and 2014 have relatively equal productivity per

Table 2 Authorship pattern year wise

Year	No. of authors							Year	No. of authors						
	1	2	3	4	5	6	> 6		1	2	3	4	5	6	> 6
2021	0	3	0	2	1	2	1	2004	1	0	0	2	0	0	0
2020	0	2	13	4	5	5	3	2003	2	0	0	0	1	0	0
2019	2	2	8	12	6	2	5	2002	0	1	1	1	0	0	0
2018	1	6	7	4	1	3	2	2000	0	0	1	0	0	0	0
2017	0	0	1	3	5	3	1	1999	0	1	0	0	0	0	0
2016	1	2	3	7	3	2	0	1998	0	0	0	0	1	0	0
2015	3	4	6	1	2	0	3	1996	0	1	0	0	0	0	0
2014	0	4	12	6	5	1	2	1995	1	2	0	0	0	0	0
2013	2	6	6	6	3	2	1	1994	0	1	2	0	0	0	0
2012	2	3	5	8	0	0	0	1993	0	1	1	0	0	0	0
2011	2	2	1	5	0	0	0	1987	3	0	0	0	0	0	0
2010	0	0	5	3	1	0	2	1986	0	1	0	0	0	0	0
2009	0	2	6	1	1	1	1	1985	3	2	1	0	0	0	0
2008	0	3	0	3	0	1	0	1978	0	0	1	0	0	0	0
2007	1	0	1	1	0	0	0	1973	1	0	0	0	0	0	0
2005	2	1	2	1	0	0	0	Total	27	50	83	70	35	22	21

Table 3 Author productivity

Year	Total docs	Total number of authors	AAPP	Productivity per author
2021	9	38	4.22	0.24
2020	32	152	4.75	0.21
2019	36	158	4.39	0.23
2018	24	92	3.83	0.26
2017	13	65	5.00	0.20
2016	18	69	3.83	0.26
2015	19	65	3.42	0.29
2014	30	115	3.83	0.26
2013	26	91	3.50	0.29
2012	19	67	3.53	0.28
2011	10	29	2.90	0.34
2010	11	50	4.55	0.22
2009	12	44	3.67	0.27
Total*	309*	1152*	3.04*	0.41*

Note The data with Total documents less than 9 are not presented author when compared to the average author paper productivity has been calculated with the following formula [12].

1. Average Authors Per Paper (AAPP) = No. of Authors/No. of Papers
2. Productivity Per Author = No. of Papers/No. of Authors.

3.1.4 Degree of Collaboration

To determine the strength of Collaboration (DC), the formula as given by Eq. 1 and suggested in [7] has been employed.

$$DC = \frac{N_m}{N_m + N_s} \tag{1}$$

where DC = Degree of Collaboration; N_m = Number of Multiple Authored Papers; N_s = Number of Single Authored Papers.

Degree of Collaboration (DC) of authors is presented in Table 4. The degree of collaboration ranges from 0.33 to 1.00. The average degree of collaboration is 0.83 during the period and it brings out clearly that there exists a higher level of collaboration in the journal.

Table 4 Degree of collaboration

Year	Single	Multiple	DC	Year	Single	Multiple	DC
2021	0	9	1.00	2004	1	2	0.67
2020	0	32	1.00	2003	2	1	0.33
2019	2	35	0.95	2002	0	3	1.00
2018	1	23	0.96	2000	0	1	1.00
2017	0	13	1.00	1999	0	1	1.00
2016	1	17	0.94	1998	0	1	1.00
2015	3	16	0.84	1996	0	1	1.00
2014	0	30	1.00	1995	1	2	0.67
2013	2	24	0.92	1994	0	3	1.00
2012	2	16	0.89	1993	0	2	1.00
2011	2	8	0.80	1987	3	0	0.00
2010	0	11	1.00	1986	0	1	1.00
2009	0	12	1.00	1985	3	3	0.50
2008	0	7	1.00	1978	0	1	1.00
2007	1	2	0.67	1973	1	0	0.00
2005	2	4	0.67	Total	27	281	0.83

3.2 *Co-occurrence, Keyword*

Keywords are the explicit explanations of the material of the research paper. The hot topics can be detected in the field of information over a certain period via the keyword co-occurrence network. The network of growth will over time expose the development of the field of study [13]. A network of co-occurring keywords was generated with VOSviewer containing as shown in Fig. 5. Table 5 shows the top 10 frequently used keywords were "blending" (occurrences = 289) followed by "aggregate" (occurrences = 213) and mixtures (occurrences = 55), etc.

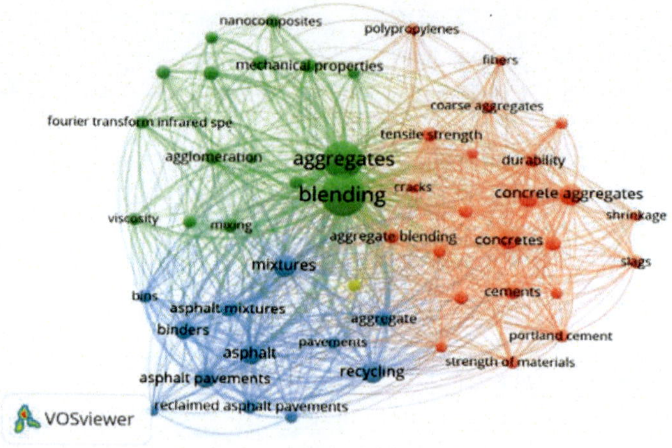

Fig. 5 Visualization of co-occurrence's and keywords

Table 5 Co-occurrence of research keywords

S. No.	Keyword	Occurrences	Total link strength
1	Blending	289	776
2	Aggregates	213	637
3	Mixtures	55	260
4	Concrete aggregates	54	173
5	Recycling	51	254
6	Asphalt	47	228
7	Concretes	43	182
8	Asphalt pavements	40	223
9	Compressive strength	38	149
10	Binders	37	196

3.3 Citation Analysis

Citation analysis provides the source publications or documents which enables researchers to understand which source has highest number of publications.

3.3.1 Source-Citation Network

The top source journals for aggregate blending research in the construction industry are listed in Table 6. Construction and building materials published 19 documents, average citation (17.74%) related to aggregate blending and occupied the top position. This indicates that the journal had more authority in this field and was well-known by scholars. The percentage of citations in top 5 sources have been pictorially represented in Fig. 6. The pictorial representation also reveals that the Construction and building materials source has more number of citations.

Table 6 Source-citation network

S. No.	Source	Documents	Citations	Avg. citation
1	Journal of Materials in Civil Engineering	13	351	27.00
2	Construction and Building Materials	19	337	17.74
3	Cement and Concrete Research	6	322	53.67
4	Journal of Cleaner Production	5	219	43.80
5	Macromolecules	3	163	54.33

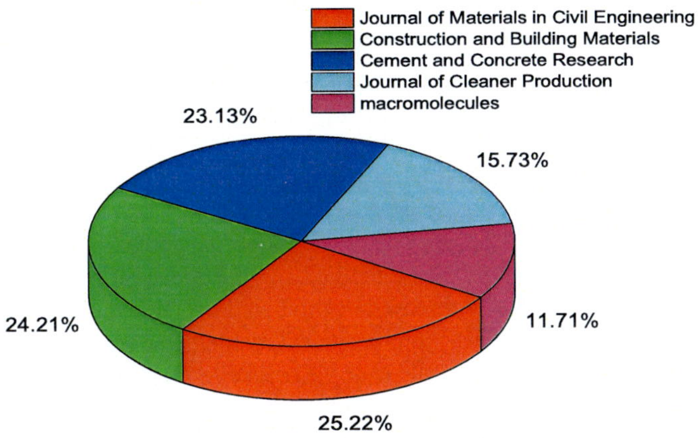

Fig. 6 Pictorial representation of citations in top 5 sources

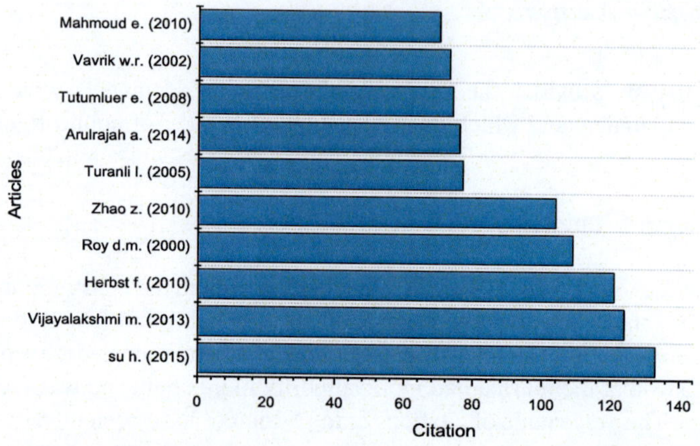

Fig. 7 Top 10 articles with high citations

Table 7 Articles with the highest citation

S. No.	Document	Citations	Links
1	SU H. (2015)	133	0
2	Vijayalakshmi M. (2013)	124	1
3	Herbst F. (2010)	121	0
4	Roy D. M. (2000)	109	0
5	Zhao Z. (2010)	104	0
6	Turanli L. (2005)	77	0
7	Arulrajah A. (2014)	76	0
8	Tutumluer E. (2008)	74	0
9	Vavrik W. R. (2002)	73	0
10	Mahmoud E. (2010)	70	0

3.3.2 Articles with the Highest Citation

Citation frequency is often used as a key metric to reveal the most influential articles when reviewing "cited journal" in the field of aggregate blending techniques. The scientometric methodology and research method are close to what was done previously with the "cited author" [14]. Bar graphical method of representation is given in Fig. 7 with highest citation in articles of different authors. A top 10 articles with highest citation are shown in Table 7.

3.3.3 Organization-Citation Network

This analysis revelas the Top organizations which are published with more documents containing highest citations. Oraganisations with no of citations is given in Table

Table 8 Organization-citation network

S. No.	Organization	Documents	Citations	Average citation
1	University of Alaska Fairbanks, United States	2	68	34
2	University of Tennessee, United States	2	66	33
3	University of Tennessee, Knoxville, United States	1	18	18
4	Virginia Tech, United States	1	14	14
5	Dept. of Civil Engineering, Ryerson University of Canada	1	14	14

8. The University of Alaska Fairbanks and University of Tennessee published two documents with an average citation of 34 and 33. Among these top organizations Virginia Tech, United States and Dept. of Civil Engineering, Ryerson University of Canada had the lowest average citation.

3.3.4 Countries-Citation Network

Table 9 shows the Countries where most of the research was conducted on this subject. From all countries, the United States with documents-89 (citations-1298) followed by China (citations-740) and India (citations-244). Among all countries, United States and Canada has high link strength. Hong Kong has higher number of average citations. A bar graphical representation of top coutries with highest citation is shown in Fig. 8.

Table 9 Countries-citation network

S. No.	Country	Documents	Citations	Average citations	Total link strength
1	United States	89	1298	14.58	19
2	China	81	740	9.14	3
3	India	22	244	11.09	11
4	Canada	14	190	13.57	19
5	United Kingdom	11	306	27.82	2
6	Brazil	10	117	11.70	1
7	France	9	123	13.67	2
8	Turkey	9	136	15.11	6
9	Hong Kong	7	349	49.86	8
10	Spain	7	96	13.71	2

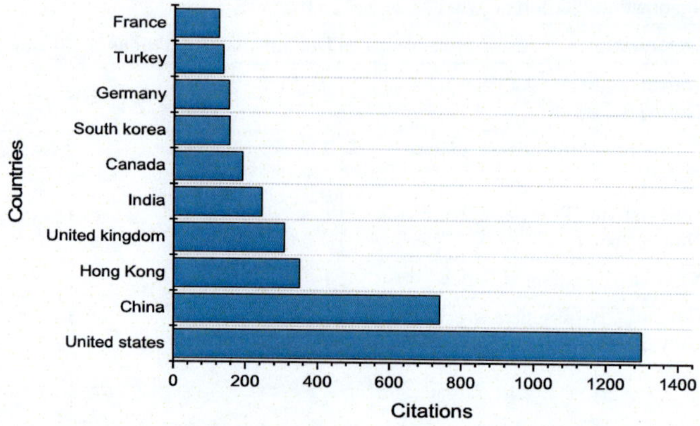

Fig. 8 Top 10 countries versus citations count

3.4 *Co-citation Analysis*

Co-citation analysis results are presented as below.

3.4.1 Co-citation Author Network

Author co-citation analysis (ACA), first suggested by [12], is a widely used technique among information domains. In this article, the author sequence is considered in ACA, and a sequence-based ACA approach is proposed. This suggested approach acknowledges distinct contributions of co-authors affecting the influence of ACA by applying varying weight values to authors of different sequences [14].

Table 10 shows that the top authors with more co-citations among all authors. Figure 9 containing the visualization of author and co-citation network.

Table 10 Author and co-citation network

S. No.	Author	Citations	Total link strength
1	Huang, B	93	630
2	Shu, X	85	619
3	Zhang, Y	63	499
4	Liu, J	44	607
5	Wang, Y	44	297
6	Wang, L	43	386
7	Li, Y	42	494
8	Li, G	37	432
9	Li, J	37	233
10	Liu, Y	36	246

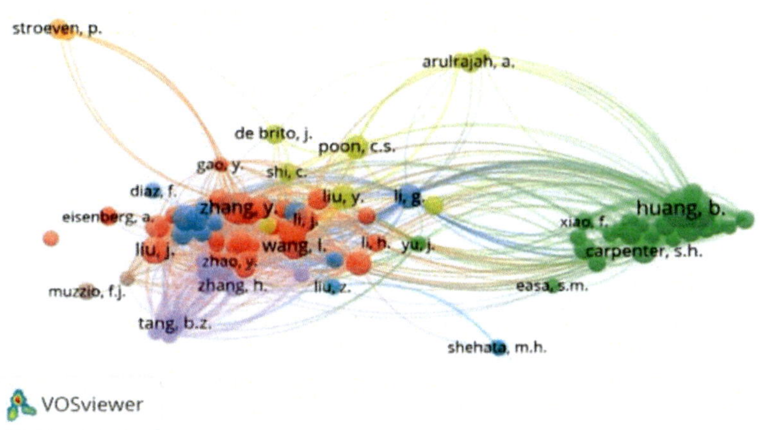

Fig. 9 Visualization of co-citation–author network

4 Conclusions

This study aimed to perform a scientometric analysis on the blending of aggregate using VOSviewer and Scopus data analyzer. The use of aggregate blending methods has proven beneficial in the pavement construction sector. This present study used co-author analysis, co-occurrence analysis, citation analysis, and co-citation analysis to visualize the state of art research publishings. The Scopus index was used to compile the data in this area of research from 1973. The current study provides researchers and clinicians with a thorough understanding of the proportion of aggregates research, as well as recognize current themes to help guide prospective studies.

It was observed that the research articles are more in number among all publications (~ 74.11%). Co-authorship and author network analysis reveals that multiple author documents with three authors or more in this area of research. Productivity analysis shows that highest productivity per author is during the year 2017. Research keywords "blending" and "aggregates" were more co-occurrence keywords and their total link strength is 776 and 637, respectively. Construction and Building Materials, Journal of Materials in Civil Engineering has highest number of publications.

Article citation analysis reveals that SU H. publishings has the highest citations (133) and Huang B., Shu X. has second highest co-citations. United states and China are the top countries with highest publications of average citation of 14.58 and 9.14.

The keywords and their frequency of usage were established in this analysis, with aggregate blending being the most commonly used. In addition, the study discover that there are gaps in the modern techniques in aggregate blending that require further studies.

References

1. Kikuchi, S., Kronprasert, N., & Easa, S. M. (2012). Aggregate blending using fuzzy optimization. *Journal of Construction Engineering and Management, 138*(12), 1411–1420. https://doi.org/10.1061/(asce)co.1943-7862.0000557
2. Soujanya, C., Priya, G. A., Ramu, P., & Kumar, C. N. (2019). A novel technique to design optimum bituminous mix designs based on R studio and autograph with integration. *Materials Today: Proceedings, 18*, 4566–4579. https://doi.org/10.1016/j.matpr.2019.07.431
3. Government of India Ministry of India. (2013). *Road transport and highways specifications for road and bridge works*. India Roads Congress
4. Ramu, P., Sarika, P., Kumar, V. P., & Sravana, P. (2016). Analytical method for asphalt concrete job mix formula design. *International Research Journal of Engineering and Technology, 03*(10), 927–932. https://doi.org/10.13140/RG.2.2.19075.81447
5. Ramy, A., Floody, J., Ragab, M. A. F., & Arisha, A. (2018). A scientometric analysis of knowledge management research and practice literature: 2003–2015. *Knowledge Management Research and Practice, 16*(1), 66–77. https://doi.org/10.1080/14778238.2017.1405776
6. Zhong, B., Wu, H., Li, H., Sepasgozar, S., Luo, H., & He, L. (2019). Automation in construction a scientometric analysis and critical review of construction related ontology. *Automation in Construction, 101*, 17–31. https://doi.org/10.1016/j.autcon.2018.12.013
7. Rajendran, P., Jeyshankar, R., & Elango, B. (2011). Scientometric analysis of contributions to journal of scientific and industrial research. *International Journal of Digital Library Services, 1*(2), 79–89.
8. Xu, Y., Zeng, J., Chen, W., Jin, R., Li, B., & Pan, Z. (2018). A holistic review of cement composites reinforced with graphene oxide. *Construction and Building Materials, 171*, 291–302. https://doi.org/10.1016/j.conbuildmat.2018.03.147
9. Van Eck, N. J., & Waltman, L. (2010).Software survey: VOSviewer, a computer program for bibliometric mapping. *Scientometrics, 84*(2), 523–538. https://doi.org/10.1007/s11192-009-0146-3
10. Zakka, W. P., Abdul Shukor Lim, N. H., & Chau Khun, M. (2021). A scientometric review of geopolymer concrete. *Journal of Cleaner Production, 280*, 1–49. https://doi.org/10.1016/j.jclepro.2020.124353
11. Liu, X., Bollen, J., Nelson, M. L., & Van De Sompel, H. (2005). Co-authorship networks in the digital library research community. *Information Processing and Management, 41*(6), 1462–1480. https://doi.org/10.1016/j.ipm.2005.03.012
12. Yoshikane, F., Nozawa, T., Shibui, S., & Suzuki, T. (2007). An analysis of the connection between researchers' productivity and their coauthors' past attributions, including the importance in collaboration networks. In *Proceedings of ISSI 2007—11th International Conference of the International Society for Scientometrics and Informetrics*, 18700244 (pp. 783–791).
13. He, Q., Wang, G., Luo, L., Shi, Q., Xie, J., & Meng, X. (2017). Mapping the managerial areas of building information modeling (BIM) using scientometric analysis. *International Journal of Project Management, 35*(4), 670–685. https://doi.org/10.1016/j.ijproman.2016.08.001
14. Bu, Y., Wang, B., Chinchilla-Rodríguez, Z., Sugimoto, C. R., Huang, Y., & Huang, W. B. (2020). Considering author sequence in all-author co-citation analysis. *Information Processing and Management, 57*(6), 1–13. https://doi.org/10.1016/j.ipm.2020.102300

Contextual Analysis on Antenatal and Postnatal Effects of 2018 Flood in Ernakulam, Kerala

Athira B. Menon, Devi priya, Krishna Rajeev, Geena Prasad, and Gevargis Muramthookil Thomas

Abstract Floods in the past years have immensely affected the life of the common man in India. A case study conducted about the dominant explanation for Kerala flood and landslide catastrophe in the year 2018, where Ernakulam district of Southern Kerala, India was distinctly drawn into concern. In 2018, Kerala went through the largest flood since 1924. It relies upon extensive review of data collected by survey and statistics provided. The study seeks to find out the hurdles faced by the people during this hazard, the reverberation of the hazard in day-to-day life and the reasons emanated like unfitting drainage systems, modern trends in house construction with strong compound walls and tiled courtyards, considerably low ground water level, lack of alarming system, etc., which will lead to the flood. To find a permanent solution for this, we collected data through various ways including, conducted a survey by visiting a flood prone village in Ernakulam district (Kaduval), Consulted Dr Dineshan retired senior scientist (KSCSTE), experiments, etc. Finally, the study propounds natural methods like water trench method, a method through which we can increase water penetration to soil so that to increase ground water level, tiling reduction which will reduce water stagnation as well as production of harmful gases into the atmosphere. Technical methods proposed include development of an android application in the future, this app will collect information on disaster prone areas and their chances of occurrence which will minimize the disaster effects and death toll if developed.

Keywords Flood · Kerala · Disaster management · Hazard

A. B. Menon (✉) · Devi priya · K. Rajeev
Department of Computer Science Engineering, Amrita Vishwa Vidyapeetham, Amritapuri, India

G. Prasad · G. M. Thomas
Department of Mechanical Engineering, Amrita Vishwa Vidyapeetham, Amritapuri, India
e-mail: geena@am.amrita.edu

G. M. Thomas
e-mail: gevargismt@am.amrita.edu

A. K. Gupta et al. (eds.), *Advances in Construction Materials and Sustainable Environment*, Lecture Notes in Civil Engineering 196,
https://doi.org/10.1007/978-981-16-6557-8_52

1 Introduction

Floods are one of the deadliest disasters in India. The frequency of flooding and severe precipitation increased significantly as a result of climatic changes. Many areas of India that were previously thought to be secure are now the most vulnerable to floods and other natural disasters.

Kerala, also known as "God's own country," is one of the tourist places in India, with a beautiful land in the south with one half surrounded by the Indian Ocean and another shared with Tamil Nadu. Kerala is famous for its attractive and green habitat. It has wide varieties of flora where some of the commonly found plants are palm, black pepper bamboo, wild cardamom, vetiver grass. In addition, the state has rich fauna too, holding 90% of vertebrate fauna, enriched by the Western Ghats, one of the world's biodiversity hotspots. Kerala is close to the equator, with a delightful climate throughout the year. The temperature in the state normally ranges from 28 to 32 °C [1]. The soil is of different types—specifically different varieties of coastal alluvium, mixed alluvium, acid saline, kari, laterite, red, hill, black cotton, and forest soils [2]. For the last 3 years, Kerala received an unexpected amount of rainfall. The state experienced the fear of flood at its peak in the year 2018 and later in its consecutive years. Kerala received 2990.3 mm instead of the expected 2925 mm in the year 2020 [1, 3]. Usually floods due to rainfall were constrained to few parts of the state every year mainly like upper Kuttanad of Alappuzha district, but the 2018 flood made this district very dangerous [4].

Ernakulam, a district in Kerala, was legally established on 1st of April 1958 comprising Kanayannoor, Kochi, Kunnathunadu, Aluva and Paroor taluks. Periyar river originates from cardamom hills of Western Ghats and spreads throughout the entire district. Idamalayar dam generates electricity and is an influent of Periyar river streams. Presence of conservation areas helped to conserve flora and fauna. Mangalavanam, is one among such ecologically preserved areas for migratory birds and mangroves, which helps in decreasing the pollution [5]. Ernakulam is bounded by Thrissur on North, Idukki on East, Kottayam, and Alappuzha to the South. The terrifying flood occurred in Ernakulam too in 2018. It affected thousands of lives and damaged the infrastructure and livelihood. In 2018, an unprecedented rainfall occurred in Kerala during the months of June, July, and August [1, 4] and made Ernakulam's infrastructure to be rebuilt. Many houses were washed off, people had to move to refugee camps. Rivers like Periyar, Pamba, Achankovil changed their courses of flow and they made it through the hinterland.

2 Methodology

The study (Fig. 1) begins with Kerala setting the context to Ernakulam district followed by reviewing Kerala's geographical structure, climate change, temperature, landslides, topography, waterways, drainage system, etc. This is done with the

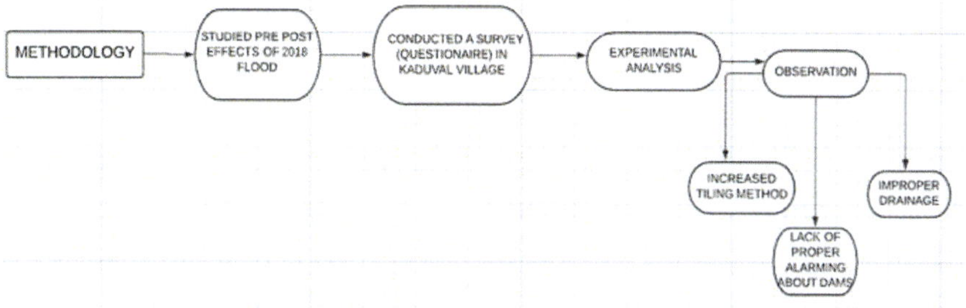

Fig. 1 Flowchart of methodology

purpose of understanding how flood disasters are occurring in Kerala. The review throws light to 2018 flood and landslides, where average rainfall and most affected areas are studied along with appraising the possible voids which lead to this disaster, turned out to be lack of proper alarming system, unsustainable construction and thereby studying how these effects are increasing the death toll. The framework of Ernakulam's floods and landslides has been outlined as a major part of the study. The required data for the study consists of seeking public perception about the disaster through a survey done at Kaduval village in Ernakulam District. The incentive is to suggest possible solutions which are put forth as technical and non technical solutions which include and thus revitalize the state from this frequent disaster.

2.1 Survey

The primary survey is done in altogether 50 houses from Perumbavoor, Parapuram, Kaduval village and is assumed to be representatives of other flood affected areas of Ernakulam and Kerala. These villagers experienced flood issues from the year 2018 which they have never experienced before. Field works were done in the month of January, 2021. The way villager's lives were muddled by change in usual weather have been outlined in detail and issues have been brought forward as a result of this exercise. The questionnaire for the survey (Fig. 3) was prepared through due thoroughness and after a pilot survey done in the neighborhood. The final questionnaire had questions which enabled both collection of data and observations needed for the research. The collected data includes the area of the forecourt of the village, total area occupied, methods of groundwater management and water scarcity problems faced. The observations drawn out include the supposed reasons of disaster, causes of infiltrations which restricts normal flow of water and extended to observing, calculating, and comparing levels of groundwater between the area where trees and plants are present and the area which is tiled or concreted (Fig. 2). The survey recorded the villagers' ideas and suggestions for water conservation and proper drainage systems

Fig. 2 Year versus no of courtyard graph

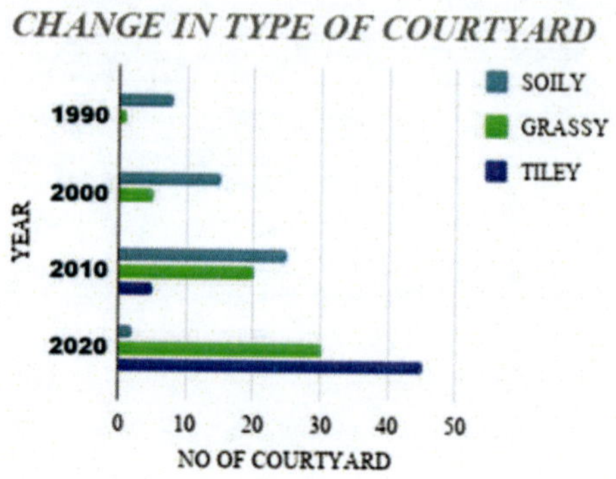

Fig. 3 Survey questionnaire

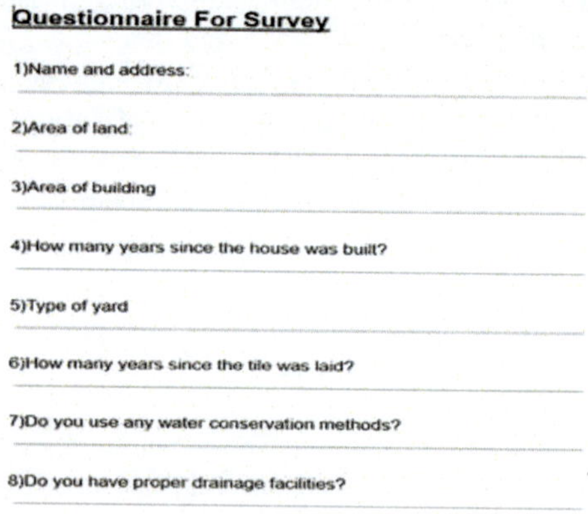

Questionnaire For Survey

1)Name and address:

2)Area of land:

3)Area of building

4)How many years since the house was built?

5)Type of yard

6)How many years since the tile was laid?

7)Do you use any water conservation methods?

8)Do you have proper drainage facilities?

to restore the village's prominence and resolve flood issues in the city and throughout Kerala.

2.2 *Experimental Analysis*

A simple experiment is done to learn how inadequate it is to pave tiles and concrete floors. This experiment was conducted by taking three containers filled with soil- container A contained only soil in it, container B contained soil with grassing on the top and tiling was done on top of the soil in container C. Three beakers were kept below each container and a piping was connected between the containers and beakers. Two liters of water was poured into each of the containers and observations recorded (Table 1).

According to the observations, the grass paved container B holds more water. Container A let the water penetrates through it and the beaker collected the same amount of water poured. On the other hand, the tiled container C made it hard for the water to penetrate to the ground. To infer from the experiment, there is a need

Table 1 Quantity of water penetration in different containers (2L)

Containers	Quantity of water penetrating through soil (mL)	Quantity of running water (mL)
A	1000	1000
B	1400	600
C	300	1700

for the water trenching method as concrete or tiled areas does not let running water seep to the ground causing unwanted flooding of the area.

3 Flood

Flood is overflowing with water without limits. It is also inferred as an inflow of tide or backflow of a river, which occurs at the place where rivers meet [6]. Floods are regularly brought about by hefty precipitation, fast snowmelt or a tempest flood from a typhoon or wave in waterfront zones [7]. Floods can cause far and wide annihilation, bringing about death toll and harms to individual property and basic general wellbeing framework. Between 19,982,017, floods influenced over 2 billion individuals around the world [8]. Individuals who live in floodplains or non-safe structures, or need cautioning frameworks and attention to flooding peril, are generally helpless against floods [9, 10].

There are 3 kinds of floods [11]:

- Flash floods: that brought about by fast and exorbitant rainfall. Maximum water levels of waterways, streams, channels, or streets might be surpassed.
- River floods: caused when reliable downpour or snow-melting increases the water level rapidly in a waterway causing it to surpass the limit.
- Beach front floods: brought about by storm floods related to hurricanes and tsunamis.

3.1 2018 Kerala Flood

Very few areas in Kerala were affected by floods before the year 2018 and even the fatalities were comparatively less in number. After the 2018 flood, this disaster began to happen every consecutive year.

The State was struck by the biggest era of natural attacks from nature in the year of 2018. This disaster struck hard at the livelihood of people without discriminating between rich or poor. All the districts were constantly under the alert. The roads were destroyed by the flood, communication lines disrupted though relief and rescue organizations operated diligently [8, 11].

Usually, Kerala receives moderate rainfall in the month of June–September. During the month of June, the expected rainfall was 649.8 mm, whereas the actual rainfall received in 2018 was 749.6 mm, which leads to a departure of 15%. In July, the actual rainfall was 857.4 mm continued to be greater than normal rainfall of 726.1 mm. During the first half of August the expected rainfall was 287.6 mm, the amount received increased thrice 758.6 mm, leading to 168% departure. This shows how drastically the water level increased in the State. Throughout the monsoon in 2018, the actual rainfall was 2346.6 mm, which was twice the normal 1649.5 mm with a total of 42% departure [2, 12].

On 18th June 2018 in Ernakulam, the area received the actual rainfall levels of 727.6 mm, whereas 696.4 mm was the normal rainfall and had a departure of 4.48%. In the next month, the actual rainfall was 956.51 mm but expected to be 670.2 mm rainfall with a departure of 42.72%. In the month of August, 401.8 mm was the normal rainfall but got a level of 410.3 mm and had a departure of 2.37% [13].

In Aluva, in June 696.4 mm was the normal rainfall but got a level of 680.7 mm and had a departure of – 2.25%. In the month of July, 670.2 mm was the normal rainfall but got a level of 803.4 mm and had a departure of 19.87%. On 18th August, 401.3 mm was the normal rainfall but got a level of 374.4 mm and had a departure of – 6.7% [14].

In Piravom area, during 18th June, the normal rainfall was 696.4 mm while measured rainfall was 1178.1 mm with a departure of 69.7%. In July, the usual rainfall was 670.2 mm and got more than twice actual, i.e., 1601.1 mm in 2018. In August, 401.3 mm was the regular rainfall but the reported level was 720.2 mm and had a departure of 79.47% [12].

In Perumbavoor, on June 18th, the normal rainfall was 696.4 mm while actual received was 831.2 mm with a departure of 19.36%. In July on the day of 18th, the normal and actual rainfall was 670.2 mm and 968.7 mm with a departure of 44.54%. The next month on the 18th the normal rainfall was 401.3 mm but it was 664.4 mm with a departure of 65.56% [2, 13].

In CIAL Kochi, on the day 18th June the normal rainfall was 696.4 mm, but it was 764 mm with a departure of 9.71%. The next month the actual rainfall was 956.8 mm with normal expected 803.4 mm and departure of 42.76%. During the 18th of August, the actual rainfall was 146.4 mm and normal was 401.3 mm with a departure – 63.52%, which turned out to be deficient [12, 13].

Figure 4 shows the rainfall analysis in different places in Ernakulam in the month of (a) June (b) July and (c) August. From all the figures, Piravom is the most affected place. Aluva is least affected in Ernakulam in June and July. The rainfall has increased proportionally comparing June.

4 Result and Discussion

The 2018 Kerala flood was an eye opener in many ways for Keralites. Flood occurred mainly due to lack of preparedness, profit before safety, carelessness of government

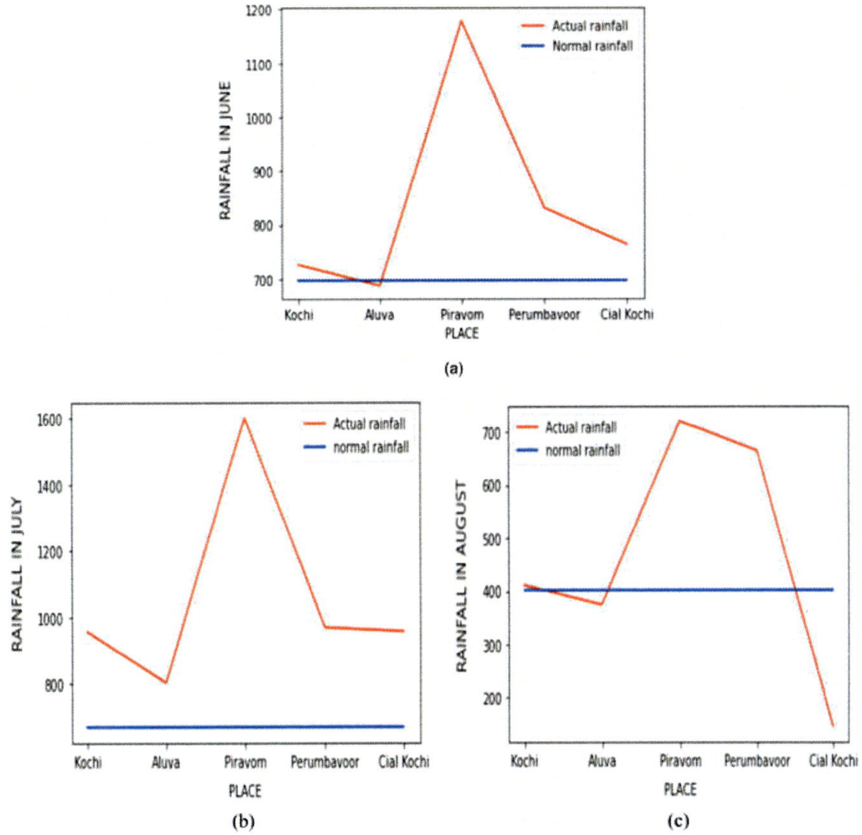

Fig. 4 Location versus rainfall (mm) in **a** June, **b** July, **c** August

officials, etc. The officials tried to produce the maximum amount of electricity at the peak time when water levels in dams were near to their maximum capacity. The study proposes what must be done for the self-awareness and safety of people, in order to avoid future hazards. The 2018 flood had been an unexpected hazard. Focusing on Ernakulam district, the district itself is not a flood prone area but the 2018 flood will always remain as a fearful experience in the minds of the localities. The event changed the life of many in varied ways. From, economic to health perspective and lifestyle has changed. Many lost their housing and living situations leading to complete loss of ability to fulfill their daily needs. Among the common contagious diseases during the period was leptospirosis, which claimed the lives of around 500 people.

4.1 Natural Methods

Natural methods focus on different natural techniques through which flood can be reduced and it can start from implementing small measures. This includes maintaining a proper drainage. We should allow the rainwater to drain or to seep through the soil so as to avoid flood as well as drought. Water trenching is one of the best natural methods. Reducing tiling and concreting roads and increasing grassing are other alternatives. These measures can be implemented at Panchayat level asking landowners to not tiling or concreting in a fixed portion of land so as to allow rainwater to seep through soil and increase ground water level.

4.2 Technical Methods

A proposal is made to develop a flood prediction application or server which includes all the details about a region (weather, temperature, dams, etc.). This will help to give live updates to people providing alerts and they can take necessary steps. Machine learning algorithms can be used for flood prediction. The data can be collected from the websites [15, 16], which updates the water level in area wise. The target variables are extremely flooded, severe flooded, and normal flood. The area can be distinctly classified based on the water level. The top 5 areas with largest precipitation fall under the "Extremely flooded category." With the respective categories, red, yellow and green alerts can be given. Warning signals can be provided to the people living in the red zone areas. The application's algorithm grasps specific keywords when people in red alert areas post messages in social media and send it to Non-Governmental Organizations. With this the people living in the red alert areas will be tracked and be aware of the situation.

4.2.1 Floodguard Application

Flood guard, is the proposed app which provides a wide range of information of flood to the subscribers. If it is developed, it will help the subscribers to know whether they are safe from flood. Whoever has downloaded the app will receive a message allowing them to know in which zone they are, that is when the zone is critical to flood the app sends a message "YOU ARE IN THE RED ZONE." Other than this the features included in the app are:

- Water level in dams
- Water level of nearby rivers/water bodies
- Groundwater level
- Remedies to rejuvenate the flood prone areas.

4.2.2 Mode of Collecting DATA that Can Be Incorporated

- Water levels of rivers and dams can be accessed from the Government of India website [12, 15, 16].
- Groundwater level data can be collected and recorded using a groundwater sensor through application of IoT devices. This can be done in a small area in Kaduval village on an experimental scale. So, we can use soil moisture sensors to detect and record the soil moisture level and thus predict the flood—if the particular area's recorded soil moisture levels are more than a specific value then that zone is viable to some disruption during abnormal conditions.
- Data related to rejuvenation includes the drainage system and ecosystem balance of that area. This data can either be collected through supervision or from other online sources available.

4.2.3 Working of Flood Guard Application

The data received from different sources are stored in the cloud database, through which subscribers can access the information of that area that is the past records of flood and rain [15]. The data stored in the database is accessed and monitored by the system and these datasets can be pre-set accordingly to various ranges of values. Once the recently added data exceeds the pre-set values the area is classified as yellow, orange and red zone and respective alert messages will be sent to the subscribers.

Figure 5 illustrates a simple mode of collecting data using a soil sensor which calculates the moisture of the soil and sends it to the server. The application installed in the phone will display the necessary output based on results from the server side. An alert will be sent when the moisture level is higher than usual. Groundwater sensors are used to know the water level. The reforms needed to maintain the necessary water levels are decided accordingly.

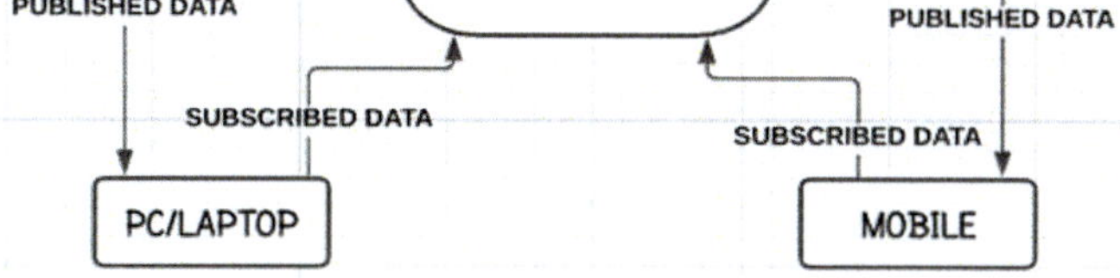

Fig. 5 OSI model

5 Conclusion

The recent consecutive flood inside Kerala has created a major impact on the livelihood of people. Inspired by the experiment data analysis on Kaduval Village, the pre and post impacts caused by flood on the village have been marked down. A survey was conducted based on how floods affected the people. Analyzing the data collected, we came to the following conclusion:

1. From 2013 onwards Perumbavoor, Kaduval village area is facing the threat of flood. During 1980s, the number of houses in this area was much less. As the number of houses, compound walls and tiling increased. This obstructed normal flow of water and prevented the percolation of rain water into soil leading to flooding. It also reduces the level of underground water.
2. Experiments conducted to test the water penetrating capacity of various surfaces proved that 90% of water seeps down to soil through grass land, 70% seeps down through plain soil and 0% seeps down through tiled land. 1 cm rain brings 930 L water from 1000 ft^2 area and for 1 acre area it becomes 40,463 L. As the intensity of rain increases, the volume of water will also increase. As there are no chances for water to flow away or seep down into soil, the level of water increases enormously when it reaches an 18 ft wide road.

Recent events of increased fatalities, it is required to overcome the death rates and reduce the casualties. Hence, it is advisable to devise a solution of making people aware from time to time by creating an app, which will provide necessary information including whether they are in a flood prone area and how critical a situation they are in. Natural methods like maintaining a proper drainage system and letting the rainwater to pass to ground level by planting grass instead of tiling or concreting the floor. Rather than constructing strong compound walls, adopt fencing methods which don't obstruct the normal flow of water. Usage of ponds, rivers and paddy fields for construction purposes can be avoided and laws can be put forth to provide one fourth of land area for water conservation purposes and rainwater reservoir. All these solutions will aid for proper disaster management.

References

1. Balasubramanian, A. (2017). *Kerala—At a glance*. https://doi.org/10.13140/RG.2.2.19375.43680
2. Mishra, V., Aaadhar, S., Shah, H., Kumar, R., Pattanaik, D. R., & Tiwari, A. D. (2018). The Kerala food of 2018: Combined impact of extreme rainfall and reservoir storage. *Hydrology and Earth System Science*, 1–13. https://doi.org/10.5194/hess-2018-480
3. Sabu, S., & Joby, N. E. (2018). Kerala floods—A model of rescue and rehabilitation using information technology and social media based crowdsourcing. *IEEE India Info, 13*(3).
4. Prasad, G., & Ramesh, M. V. (2019). Spatio-temporal analysis of land use/land cover changes in an ecologically fragile area Alappuzha district, southern Kerala, India. *Natural Resources Research*, 1–12. https://doi.org/10.1007/s11053-018-9419-y

5. Azeez, P. A., Bhupathy, S., Rajasekaran, A., & Arun, P. R. (2004). Ecodevelopment plan for the Mangalavanam mangrove area, Ernakulam, Kerala. SACON Technical report No. 62. https://doi.org/10.13140/RG.2.1.5099.4082
6. Tingsanchali, T. (2012). Urban flood disaster management. *Procedia Engineering, 32*, 25–37. https://doi.org/10.1016/j.proeng.2012.01.1233
7. Aiswarya Jyothi, K. G., & Meera, B. (2019). An apocalyptic analysis of the narratives on 1924 flood in Kerala. *OPUS: International Journal of English Language, Literature in Humanities, 7*(6).
8. Prasad, G., & John, S. E. (2018). Delineation of groundwater potential zones using GIS and remote sensing a case study from midland region of Vamanapuram river basin, Kerala, India. In *AIP Conference Proceedings* (Vol. 1952, pp. 020028). AIP Publishing.
9. Singh, S. (2019). Analytical study of Kerala floods. *IJRASET, 7*(VI), 45–98. ISSN: 2321-9653. https://www.ijraset.com/fileserve.php?FID=23709
10. Nair, B. B., & Rao, S. (2017). Flood water depth estimation-A survey. In *2016 IEEE International Conference on Computational Intelligence and Computing Research*, ICCIC 2016.
11. Prasad, G., Thomas, G. M., & Ramesh, M. V. (2020). Trace metal analysis of pre-flood and post-flood drinking water at Alappuzha District, Southern Kerala, India. In *Materials Today: Proceedings*. https://doi.org/10.1016/j.matpr.2020.10.958
12. Shaharban, D. V., & Rathnakaran, A. (n.d.). Disaster prevention and management in the era of climate change with special reference to Kerala flood 2018. *Journal of Composition Theory*. https://doi.org/10.29322/ijsrp.8.3.p7548
13. Vishnu, C. L., Sajinkumar, K. S., Oommen, T., Coffman, R. A., Thrivikramji, K. P., Rani, V. R., & Keerthy, S. (2019). Satellite-based assessment of the August 2018 flood in parts of Kerala, India. *Geomatics, Natural Hazards and Risk, 10*(1), 758–767. https://doi.org/10.1080/19475705.2018.1543212
14. Prasad, G., Kentilitisca, J. Y., Ramesh, M. V., & Suresh, N. (2020). An overview of natural organic matter. In A. Kumar, M. Paprzycki, & V. Gunjan (Eds.), *ICDSMLA 2019. Lecture Notes in Electrical Engineering* (Vol. 601). Springer. https://doi.org/10.1007/978-981-15-1420-3_150
15. https://sdma.kerala.gov.in/dam-water-level/
16. Sherpa, S. F., Shirzaei, M., Ojha, C., Werth, S., & Hostache, R. (2020). Probabilistic mapping of August 2018 flood of Kerala, India, using space-borne synthetic aperture radar. *IEEE Journal of Selected Topics in Applied Earth Observations and Remote Sensing, 13*, 896–913. https://doi.org/10.1109/jstars.2020.2970337

Mechanical Properties of Concrete Containing Plastic Fiber

Shubham Sharma and Amardeep Boora

Abstract India is globally ranked 11th as a generator of plastic waste. According to CPCB, India produces about 3.3 million metric tons of plastic every year out of which only 9% is being recycled. In this paper, waste plastic sheets (LDPE) are shredded to form plastic fiber (PF). These PFs are added to concrete with a varying volume percentage of 0–10, concerning the total volume of the aggregates. A total of 40 samples are prepared to have 150 × 150 × 150 mm dimensions and varying percentage volume of PF as 0, 2, 5, 7, and 10. Other than these 20 samples are prepared for flexural strength test beam with 450 mm length and 150 mm, width and height. The mix design prepared is for M25 grade concrete. Different tests are done on the samples for strength and water absorption. The results from various tests conclude that the samples with PF exhibit higher flexural strength and a higher water absorption rate as compared to the conventional concrete samples. A decrease in compressive strength is observed due to the increased void ratio of concrete. The flexural strength of plastic concrete is much better because the PF act as a reinforcing material to the plastic concrete. The higher absorption rate of the cubes is due to the greater void formation in the samples allowing the water to penetrate easily in the test cubes. This PF concrete can be used in low-loading structures like partition walls, walkways, parking lots, etc.

Keywords Concrete · Plastic fiber · Void ratio · Flexural strength · Water absorption

1 Introduction

Cement concrete is only next to the water in terms of the amount of material used on our planet. Over hundreds of years, concrete has become the material of choice for constructing residential and commercial buildings, infrastructural facilities such as highways, dams and bridges, canals, ports, and other important facilities. The

S. Sharma (✉) · A. Boora
Department of Civil Engineering, Jaypee University of Information Technology, Solan, Himachal Pradesh 173234, India

A. K. Gupta et al. (eds.), *Advances in Construction Materials and Sustainable Environment*, Lecture Notes in Civil Engineering 196,
https://doi.org/10.1007/978-981-16-6557-8_53

popularity of concrete owes to its economy, ability to be cast into any shape, ability to be fabricated practically anywhere, and last but not the least, its inherent durability. Concrete is made by mixing different aggregates in various proportions. This proportion decides the grade of concrete which also decides the compressive strength of the concrete.

On the other hand, plastic is also a widely used product on earth. It does not decompose in nature easily, and the recycling process of plastic is very energy consuming and creates pollution to a higher extent. Reddy [1] used low-density polyethylene waste (LDPE) and fine aggregates to manufacture paver bricks. As a result, the water absorption was reduced by 2%. Melting point reduced due to the presence of plastic aggregates. Eneh [2] studied the plastic waste and the result showed that the plastic could be used in the construction of building blocks, manufacturing damp proof membrane, drainage pipes, and a lot more. Gu [3] used plastic fibers and plastic aggregates and partially replaced them with fine and coarse aggregates to make lightweight concrete which has good strength as compared to conventional concrete due to plastic fibers. Choi et al. [4] cut PET bottles in the range of 5–15 mm and mixed them with granulated blast-furnace slag as a coating on PET aggregates and was mixed with Portland cement. Reduction in compressive strength was noted after the replacement of plastic aggregates increased more than 20%.

This present study aims to develop high-strength concrete by using PF waste and conventional aggregates that are used to make concrete. The plastic fiber is to be added with varying proportions of 2, 5, 7, and 10% of the total aggregate volume [5]. Ambient air curing of cubes is done and is kept for 3, 7, 28, and 90 days of curing period. Each sample is tested for compressive strength, flexural strength, water permeability, and ultrasonic pulse velocity.

2 Objectives of the Study

- Development of plastic fiber reinforced concrete of desirable strength (M25).
- Strength and durability study under various tests.
- Property study by non-destructive testing.

3 Preparation of Test Cubes

3.1 Materials

Waste LDPE plastic sheets (used or food packing) are used to make plastic fiber. The plastic fiber (PF) was obtained by shredding the waste LDPE sheets. The length of the fiber is reduced according to the width to get a reduced aspect ratio. Plastic fibber with a higher aspect ratio would result in lump formation in the concrete mix. To

Table 1 Properties of plastic fiber

Size of PF	Varying length of 0.15–1.75 mm and width of 0.12–1.25 mm
Aspect ratio	1.25–1.4
Density	910–940 kg/m^3
Melting point	105–115 °C
Compressive strength	Very poor
Tensile strength (psi) ASTM D638	1400
Water absorption (%) ASTM D570	0.10

Table 2 Properties of aggregates used in mortar formation

The specific gravity of fine aggregates	2.52 (10 μm)
The specific gravity of coarse aggregates	2.64 (20 mm)
Compaction factor	0.9
Specific gravity of cement	3.10 (33 grade PPC)

ensure, a uniform mix in concrete aspect ratio is reduced. The PF is tough, flexible, and is a good sound and dampness insulator. The plastic fiber will exhibit similar properties as LDPE sheets. The properties of plastic fiber (PF) areas in Table 1 as per ISO 1183 [6] and the properties of various aggregates used to form mortar are shown in Table 2 as per IS 383 [7].

3.2 *Mix Design*

The mix design is prepared for M25 grade concrete as per the IS codes [IS 10262:1982, IS 456:2000, IS 269:2016, IS 383:1970] [7–10]

For all mixes, the ratio obtained is 1:0.97:2.18

Thus, the values of various aggregates for 1 cement bag

Cement = 50 kg

Water content = 18.5 L

Fine aggregates = 48.5 kg

Coarse aggregates = 109 kg.

The total amount of aggregates is 207.5 kg; hence, the PF volume to be added is given in Table 3.

Table 3 Quantity of PF to add to the mix according to varying percentages

PF (%)	2	5	7	10
PF volume (kg)	4.15	10.37	14.52	20.75

3.3 *Casting of Cubes*

The PF is dry mixed with cement until the PF has a good coat of dry cement. It is done to ensure the proper and even mixing of PF with the concrete mix prepared. Laboratory concrete mixer is used to prepare concrete is used and fresh concrete of M25 grade is prepared. First, the cement, coarse aggregates, and fine aggregates are mixed in a dry condition for 3–4 min, and then, the water was added and is mixed for about 6–8 min for the proper bonding of all the aggregates. After this, the calculated amount of cement coated PF is added to the concrete mix prepared and is mixed well ensuring an even mix of the PF.

The cube molds are kept on the vibrating machine to remove any air entrapped bubbles during the pouring of fresh concrete into molds. The sizes of cubes to be cast are 150 × 150 × 150 mm. For the flexural strength test the dimensions of the beam cast is, 150 × 150 × 450 mm. The cubes are left for ambient curing for the period of 3, 7, 28, and 90 days (Fig. 1).

(a) (b)

Fig. 1 **a** Cubes with PF giving uneven finish, **b** cubes without PF giving an even finish

4 Experimental Study

4.1 Compressive Strength Test as Per IS 516 [11]

The test is conducted for each specimen with dimensions 150 × 150 × 150 mm. A total of 20 samples are tested with varying PF volume and different curing periods. A digital compressive testing machine is used to determine the compressive strength values of the cubes.

4.2 Flexural Strength Test as Per IS 516 [11]

UTC 5533 having 100 kN of loading capacity was used in the flexural strength test. The test is conducted on a 150 × 150 mm concrete beam with a length of 450 mm. This test is done to check the resistance failure while bending. The beam is placed in the instrument. The specimen rest on the edge support and even loading are applied from the top of the beam. The loading is increased gradually till the specimen fails.

4.3 Water Absorption Test as Per ASTM C140 [12]

After the complete curing of samples, weight of each sample is taken (W1). The samples are kept in the oven at 105 °C for 72 h. After 72 h, the cube samples are taken and are emerged in the tank for 24 h. The cubes are taken out and are surface dried, and weight is taken again ($W2$). The water absorption percentage is calculated using Eq. 1.

$$\mathrm{WA}\% = [(W2 - W1) \div W1] * 100 \tag{1}$$

4.4 Ultrasonic Pulse Velocity Test as Per IS-13311 Part 2[13]

It is non-destructive testing of concrete that is widely used to check the quality of concrete. This test was conducted on the blocks that were used to do the water absorption test and flexural strength test. The transducers were placed on the opposite faces (one on the top face and another on the bottom face) and 3 readings are taken from each cube, and the pulse velocity is calculated using Eq. 2

$$\mathrm{PV} = w/T \tag{2}$$

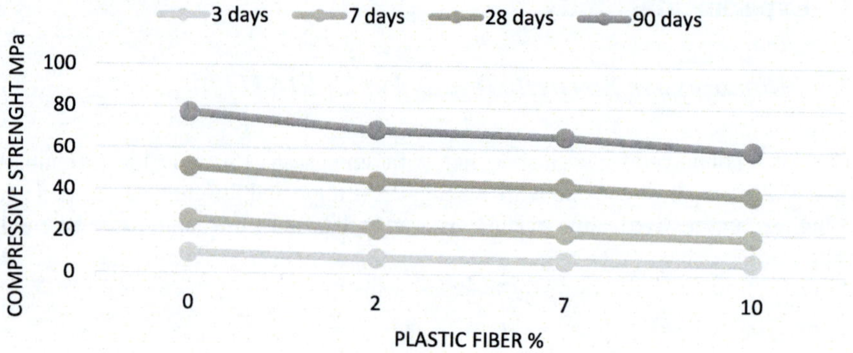

Fig. 2 Variation in compressive strength with the varying volume of PF at different curing time

where PV = pulse velocity
w = width/thickness of sample
T = Time taken by pulse to go through.

5 Result and Discussion

All the tests are done according to the Indian Standard codes IS 516 [11] and IS-13311 [13].

5.1 *Result for Compressive Strength Test*

A decrease in compressive strength of the cubes containing PF (Fig. 2) as the void formation is more due to the fibers, and cracking tendency is more in samples containing PF. When the volume of PF increases more than 7%, the compressive strength reduces and gets lower than the compressive strength of the sample containing 2% of PFs.

5.2 *Result for Flexural Strength Test*

An increase in flexural strength is noticed as shown in Fig. 3. It is due to the PF added as they act as minor reinforcements and resists the bending of the beam against loading. All the samples with PF present have high flexural strength as compared to the samples with no PFs present.

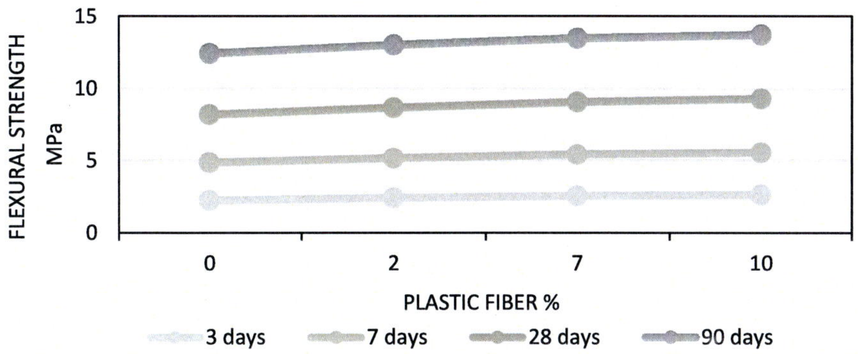

Fig. 3 Variation in flexural strength with the varying volume of PF at different curing time

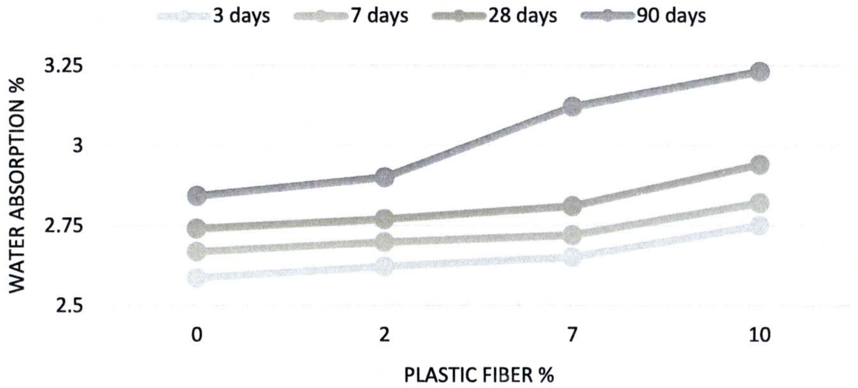

Fig. 4 Variation in water absorption percentage with varying PF % and curing time

5.3 *Result for Water Absorption Test*

Water absorption percentage also increased as seen from Fig. 4. It is due to the increased porosity of the concrete cubes that have PF in them. The void formation is high in concrete having a curing period of 90 days.

5.4 *Result for Ultrasonic Pulse Velocity Test*

A decrease in pulse velocity is noticed (Table 4) when the test is performed on the cubes from the water absorption test. It is known that the pulse velocity decrease by 2–5% in the concrete containing moisture content.

Table 4 Ultrasonic pulse velocity test result

PF (%)	Ultrasonic pulse velocity (km/s)							
	Samples from the water absorption test				Samples from flexural strength test			
Curing days (%)	3 days	7 days	28 days	90 days	3 days	7 days	28 days	90 days
0	3.90	4.14	4.37	4.45	4.05	4.19	4.40	4.52
2	3.69	3.76	4.05	4.32	3.74	3.82	4.17	4.43
7	3.45	3.58	3.84	3.95	3.67	3.78	3.92	4.21
10	3.32	3.45	3.69	3.81	3.55	3.62	3.81	3.95

Decreased pulse velocity is also observed in the concrete containing PF. It's because plastic fiber that is present in the cubes hiders the traveling of an electric pulse generated by the transducers from one face to another. The quality of concrete is still good as the value of UPV reaches 4 as we test the 90-day curing sample with PF 7%. The pulse velocity remained less than 4 when the PF percentage gets more than 7%.

6 Conclusion

There is an increase in flexural strength. This is because of the PFs present in concrete. The PFs take the load and equally divides the load and work as reinforcement increasing flexural strength.

Compressive strength was reduced due to the high void formation. These voids are the major reason for the failure of concrete. The same is with the water absorption test results. The voids increase due to which water penetration increases and concrete soak more water.

Value more than 4.3 in UPV test donates good concrete quality. We can conclude from the test results that there is a decrease in the pulse velocity when tested on moist samples. The decrease is up to 2–5%. Replacement till 7% results in UPV value to be 4.2 which is quite equivalent to 4.3; hence, the concrete quality is good and is ok for construction.

Hence, the PF can improve various properties of concrete if added in a limited ratio. If this ratio increases more than 7%, the various strengths get decreased. This PF concrete can be used in the construction of partition walls, parking lot floorings, footpath flooring, and other low-loading areas.

References

1. Reddy, C. S. K., Chandra Kumar, B. S., & Asadi, S. S. (2019). Utilization of low density

polyethylene waste in the manufacturing of paver brick. *International Journal of Recent Technology Engineering, 7*(6C2), 62–67.
2. Eneh, A. E. O. (2015). Application of recycled plastics and its composites in the built environment. *International Journal of Management Innovation in Technology and Engineering, 3*(3), 9–16.
3. Gu, L., & Ozbakkaloglu, T. (2016). Use of recycled plastics in concrete: A critical review. *Waste Management, 51*, 19–42. https://doi.org/10.1016/j.wasman.2016.03.005
4. Choi, Y. W., Moon, D. J., Chung, J. S., & Cho, S. K. (2005). Effects of waste PET bottles aggregate on the properties of concrete. *Cement and Concrete Research, 35*(4), 776–781. https://doi.org/10.1016/j.cemconres.2004.05.014
5. Sharma, R., & Bansal, P. P. (2016). Use of different forms of waste plastic in concrete—A review. *Journal of Cleaner Production, 112*, 473–482. https://doi.org/10.1016/j.jclepro.2015.08.042
6. Association, F. S. (2019). Plastics. Methods for determining the density of non-cellular plastics. *Sfs-En Iso 1183–12019.*
7. B. of I. S. (BIS). (1970). IS 383: 1970 *Specification for coarse and fine aggregates from natural sources for concrete* (pp. 1–24). Indian Standard.
8. BIS:10262. (2009). *Indian standard guidelines for concrete mix design proportioning*. Bureau of Indian Standards.
9. IS 456. (2000). *Concrete, plain and reinforced* (pp. 1–114). Bureau of Indian Standards.
10. B. of I. Standards. (2013). *IS 269: 2013 Ordinary portland cement specification.* Bureau of Indian Standards.
11. IS 516:2014. (2004). *Method of tests for strength of concrete.* IS 516-1959 (Reaffirmed 2004).
12. A. C140. (1955). Water absorption. *Lancet, 265*(6860), 393–394. https://doi.org/10.1016/S0140-6736(55)91288-5
13. IS: 13311 (Part 2). (1992). *Non-destruct we testing of concrete-methods of test part 2 rebound hammer*. Bureau of Indian Standards [Online].

Structural Health Monitoring Through the Application of Piezoelectric Sensors – State of the Art Review

Aishwarya Thakur and Saurav

Abstract Electromechanical impedance-based (EMI) method is generally utilized for health monitoring. Structural health monitoring (SHM) could show a significant part in keeping up safety of "in-working structures." The EMI method is generally acknowledged as an inexpensive and extremely sensitive procedure for SHM and non-destructive evaluation (NDE) of a variation of engineering organizations. The utilization of piezoelectric sensors (PZT) is one of SHM's intelligent detecting technologies. The piezoceramic materials act simultaneously as an actuator and sensor for SHM. This paper reviews online strength increment monitoring of early age concrete is identified with the service of EMI discovering method. RMSD values for different patches were compared at various frequency ranges (300,400 kHz). Different conductance signature values were also compared and found that with increase in duration RMSD value increases, which provided conclusive evidence in the health monitoring process of a structure by utilizing PZT sensor. Upon connecting PZT sensors in series conductance value subsequently increases from 2.01 at Day 7 to 7.31 after 21 Days of monitoring the beam at different curing region. The root mean square deviation index (RMSD) was additionally used for automatically specifying the change in the impedance of the integrated PZT sensor below temperature and load. Also, it was reviewed that impedance analyzer at range from 40 Hz to 1000 kHz applied at different piezoelectric sensors and showed that integrated PZT sensor can efficiently determine the result in quantification and indication of structural damage; this benefits the accurate assessment of structural damage. Thus, from the study, we can conclude that the proposed strategy for damage recognition is very successful and sensitive to concrete.

Keywords Non-destructive evaluation · Piezoceramic · Concrete · Sensor · EMI technique · Root mean square deviation

A. Thakur (✉) · Saurav
Civil Engineering Department, Jaypee University of Information Technology, Solan, Himachal Pradesh, India
e-mail: 206602@juitsolan.in

Saurav
e-mail: saurav@juit.ac.in

A. K. Gupta et al. (eds.), *Advances in Construction Materials and Sustainable Environment*, Lecture Notes in Civil Engineering 196,
https://doi.org/10.1007/978-981-16-6557-8_54

1 Introduction

Civil infrastructure needs timely inspection to assess its reliability and strength for the future advantage of human culture, wellbeing of life, and productivity. Its economic and safe presentation represents a great advantage for the civilization, since it stabilizes financial administration and security. In over-all, their presentation contains a great deal of uncertainties since they are frequently subject to disturbing natural tragedies and heavy use. To conquer these difficulties, numerous independent research associations and entities are devising new methods for large-scale presentation evaluation and clarification of the behavior of structures, which is called SHM. Structural health monitoring is a trending topic today to monitor or certify the complete performance of the organization [1].

An effective SHM system for civil engineering is able to identify the location and quantity of faults (rust, cracks, etc.) in real time, so the building can be fixed and strengthened on schedule to confirm integrity and structural wellbeing [2]. Today, various structural health finding techniques are enforced to different civil structures or their segments. Like as the standard static deformation test, vibration documents techniques and non-destructive test (NDT) techniques: ultrasound, impedance, acoustic emission, X-rays, pulse radar and infrared thermal imaging. But maximum of the methods are qualitative, and real-time detection is difficult. Smart materials, such as PZT materials, cement-based smart composites, fiber optic sensors, and magneto-restrictive materials, offer a new technique for long-term and real-time checking of the health of civil structures [3].

In the middle of the several smart materials, piezoelectric materials, which are mostly signified by piezoelectric ceramics, have the benefits of integrating detection and guidance that makes them fit for SHM [4]. The word "piezo" comes from a Greek word that implies pressure. The phenomenon of piezoelectricity was found in 1880 by Pierre and Paul-Jacques Curie [5]. Piezoelectric Lead Zirconate Titanate (PZT) is an actuator worked framework. It is a solitary PZT that behaves together as an actuator and as a sensor when the PZT head is connected or implanted inside the frame. Among the several smart materials, piezoelectric materials, which are mostly signified through piezoelectric ceramics, have the benefits of integrated sensing and guidance integration, creating them appropriate for SHM. Simultaneously, piezoelectric materials have good linear ratio and quick response, and maximum of piezoelectric materials has low power utilization, minimal cost and easy.

2 Literature Remark

This article includes a brief review of the literature on the various methods available to monitor the structural health of the RC structure.

Bharathi Priya et al. [6] investigate an embedded concrete beam with six PZT covers glued into slabs of expanding thickness by exposing the beam to variance curing. A new process called "sequential detection" is utilized, where patches are associated in sequence, and their admission marks are noted from 10 to 1000 kHz for various frequency varieties from day 2 to day 28. Since the various reaches, 300–400 kHz is ideal with a higher amount of peaks (higher modal density) noticed.

Methodology: RMSD among the signatures was the most suitable index of damage to characterize structural damage.

Expression utilized as given in Eq. 1

$$\text{RMSD} = \left[\frac{\{\sum(G_i^1 - G_i^0)\}_{\text{sub - range}}}{\{\sum G_i^{02}\}_{\text{full - range}}} \right]^{0.5} \tag{1}$$

where G_i^1 –conductance at the point of the ith dimension in Day of interest; G_i^0– conductance at the point of the ith dimension in Day 2 (baseline).

Patches 5, 3, and 7 situated on the good cure side show higher root mean square deviation (RMSD) values than patches 4 and 6 situated on the poor cure side. RMSD values were calculated from 300 to 400 kHz in patch4 and patch7.

Experimental data: The RMSD values calculated from patch 4 and patch 7 were compared. Patch 4 and patch 7 were linked in series; it is showed in Table 1.

The calculated values are also compared to the RMSD calculated from the patch 4 and patch 7 individual readings. It is showed in Table 2.

Table 1 Percentage change in RMSD values of patches4 and 7 connected in series

Patch	EMSD (%)		
	Day 7	Day 14	Day 21
4 (360–375 kHz)	1.10	2.52	2.01
7 (340–355 kHz)	1.23	5.54	7.31

Table 2 Percentage change in RMSD values of recorded individually from patches 4 and 7

Patch	RMSD (%)		
	Day 7	Day 14	Day 21
4 (360–375 kHz)	5.03	4.10	6.90
7 (340–355 kHz)	2.28	8.09	10.68

From the results, it was clear that the good cure side has a higher RMSD value than the bad curing side. Furthermore, the RMSD of the sequential detection patches was found to follow a similar pattern to the RMSD determined by the information of the individual patch. The results and perceptions from sequential detection and separable record show that the serial detection process utilizing PZT is an efficient, simple, and rapid technique that has surprising and promising utilizes in the field of SHM.

Demi Ai et al. [7] studied that the proposed model was proved through the experimental test outcome of a produced integrated PZT sensor. The inserted PZT sensors were subjectively compared along those fixed to the surface in the investigation of sensing a concrete beam broken shown through means of the subsequent cracking of the concrete covers. As a final point, the efficiency of the integrated PZT sensor was investigated for quantifying structural destruction utilizing the slope-based RMSD index. In addition, a new modifiable RMSD reference index was additionally planned to estimate the impact effect of dual kinds of PZT sensors.

Methodology: The impedance analyzer scanned a PZT patch along measurements of $10 \times 10 \times 0.5$ mm in a free vibration state as of 40 Hz to 1000 kHz, as presented in Fig. 1. A PZT patch along a size of $26 \times 12 \times 0.5$ mm was also measured for comparison in the following experiment. The sizes and the properties of the material used for PZT sensors are additionally presented in Table 3.

Experimental data: A slope situated RMSD list is planned to measure the efficiency of measured destruction values, which (i) displays an extra direct upward trend in destruction quantification; (ii) it can eliminate the value of the suspiciously incorrect damage index of the single PZT sensor, for the benefit of an accurate measurement of the structural damage.

It may also be that the RMSD values of the PZT sensors attached to the surface are almost twice that of the integrated sensors, the slope of the RMSD PZT 2 and PZT 3 values attached to the surface are 0.574 and 0.621 correspondingly, which are additionally greater than PZT integrated sensors. It was originated that the integrated PZT sensor can efficiently filter the impression impact by quantifying and

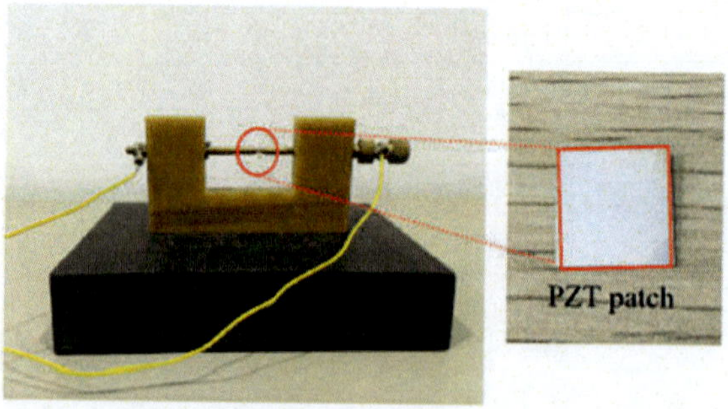

Fig. 1 Experimental arrangement to calculate the PZT patch in a free vibration state

Table 3 Properties of materials and dimensions of PZT sensors

Symbol	Name	Value
l_2	Length	10/25 mm
h_2	Width	10/12 mm
h_2	Thickness	0.5 mm
ε^{33}	Dielectric constant	1.53E−8 facad/m
${\gamma_{33}}^{\varepsilon}$	Young's modulus	67 GPa
δ	Dielectric loss factor	0.012
η	Mechanical loss factor	0.003
ρ	Density	7860 kg/m^3
d_{31}, d_{33}	Piezoelectric constant	– 24, 41E−10 m/volt
γ_{13}, γ_{33}	Poisson's ratio	0.38, 1.05

indicating structural damage, benefiting from an accurate assessment of structural damage (Figs. 2 and 3).

Bilgunde et al. [8] mathematical study to arrange the in-situ soundness of piezoelectric sensors utilized for SHM of huge automotive structures, civil, and aeronautical. The system planned in these work activities to demonstrate in-homogeneities in the

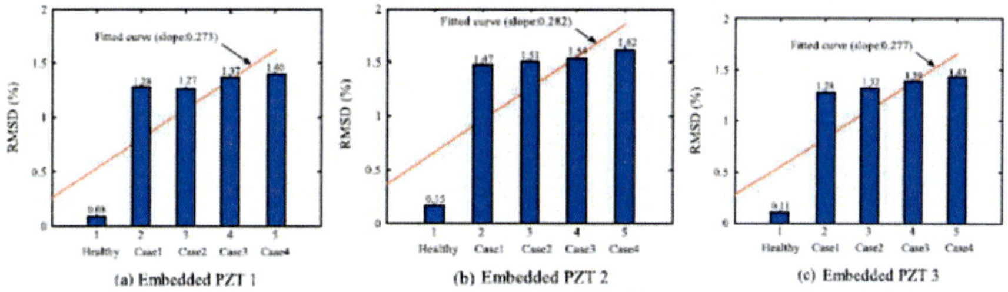

Fig. 2 RMSD values based on the slope of the inserted PZT sensor 1–3

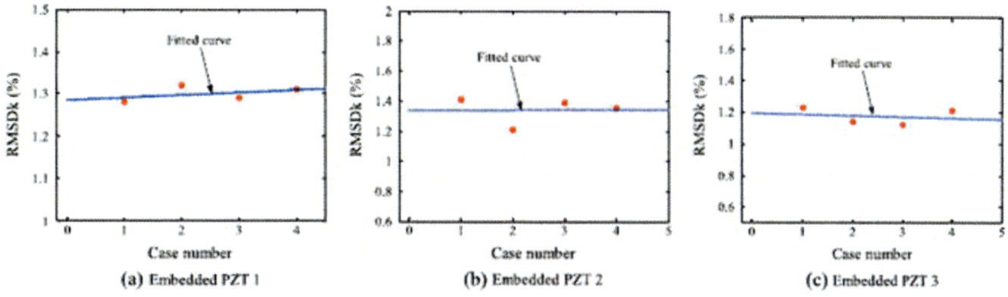

Fig. 3 RMSDk values determined via three inserted PZT sensors in four destruction cases

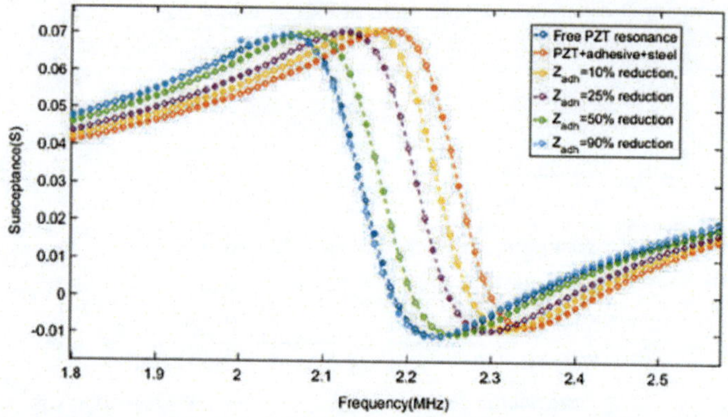

Fig. 4 Decrease in the value of the resonant frequency shown as a function of percentage of decrease in Z_{adh}.. In case of 90% decrease, the assemblage starts vibrating at a frequency which equal to the resonant frequency of free PZT-5Adisc

glue along which the sensor is usually attached toward the SHM building. The failure of the connection state was found to justify a discount in resonant frequency of structure and, ultimately, methodologies of the resonance appearances of a PZT material below non-traction limit conditions. These movements in the resonance spectrum are moreover measured utilizing the RMSD-based destruction index. As the outcomes, it was found that EMI technique can be used to monitor the heath of a structure by using an attached sensor.

Experimental data: As per temperature rises, it has been analytically seen that storing modulus of epoxy decreases. The decrease of Z_{adh} reasons an equivalent reduction of the resonant frequency along respect to the base spectrum (presented in Fig. 4). At the point when Z_{adh} is diminished through 90%, comparable to the original binding state, the resonance characteristics are close to the resonance characteristics of a PZT disk.

Changes in the resonance spectrum due to the variation in binding state can also utilize the RMSD-based destruction index. As the percentage of Zadh reduction rises, the destruction rate of the adhesive as well rises. With a 90% decrease in Zadh, the destruction index is 95.7, while free resonance index is 97.80, representing that boundary condition in connected PZT sensor is close to the situation of the situations limit without traction. In this work, we have studied that the decrease of the equal acoustic impedance reasons a decrease of the resonant frequency of the attached group, fundamentally imminent the resonance characteristics of the PZT element however it is in the boundary conditions without traction.

Xu et al. [9] studied in this paper a type of inserted impedance-based PZT sensor that was manufactured utilizing a combination of powdered cement and epoxy as the packing layer. The effects of charge and temperature on the conductance and impedance for the sensor were examined. In the outcomes, a clear temperature dependency is display as the reference point of the conductance spectra keeps on changing by rising the resonance and the temperature heights. Also, RMSD were moreover used here to instinctively visualize the changing impedance for an integrated sensor that is PZT under the effects of load and temperature.

Methodology: The temperature survey was carried out by installing the PZT sensor in a 40 mm × 40 mm × 40 mm mortar sample. The mortar sample with internal the PZT sensor was placed in the temperature examination chamber (Model MPC-710) next restoration on behalf of 28 days. The temperature variety of the experiment is between 2040 °C at a heating degree of 2 °C/min and temperature holding period is one hour each 10 °C. An analyzer was utilized impedance (Agilent 4294A model) to examine electrical impedance spectra of sensor (Fig. 5). The compression load test was performed utilizing a pressure testing instrument, and a growing load with a stage load of 10 kN was enforced to the mortar testers till damage.

Experimental data: The variety of resonance peaks shows that here is a clear temperature dependency of the conductance spectra of integrated PZT sensor (Fig. 6).

The destruction index is generally utilized to visibly specify the characterization of structural destruction. The RMSD index was designed dependent on the calculation given below using Eq. 2

$$\mathrm{RMSD} = \sqrt{\frac{\sum_{i=1}^{n}\left[|G_i| - |G_i^0|\right]^2}{\sum_{i=1}^{n}\left[|G_i^0|\right]^2}} \tag{2}$$

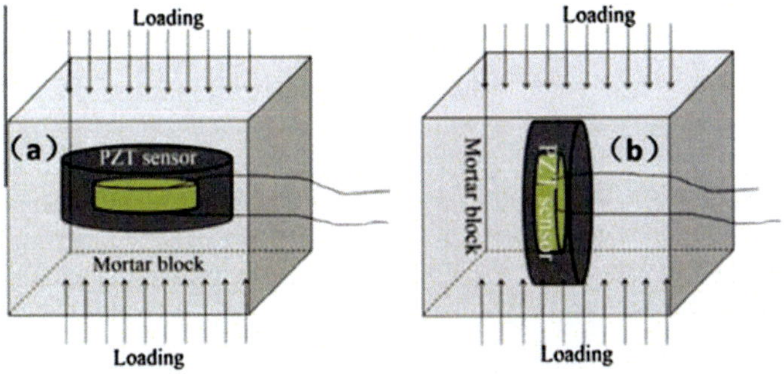

Fig. 5 Pressure being enforced on a mortar sample, **a** alongside the direction of the thick end, **b** beside the direction of flat end of the assigned sensor

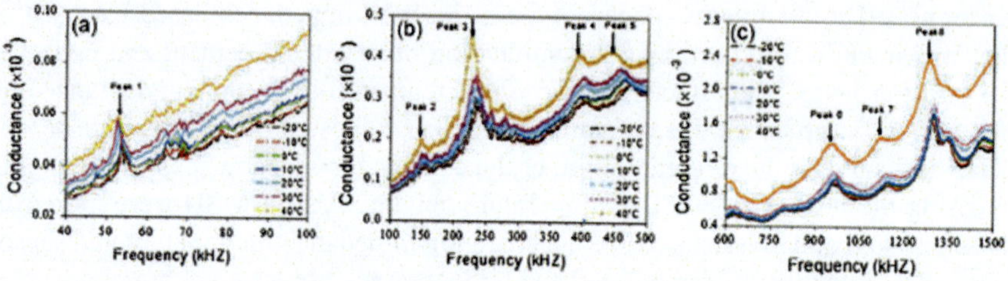

Fig. 6 Conductance spectra versus frequency **a** 40–100 kHz, **b** 100–500 kHz and **c** 600–1500 kHz on various temperature

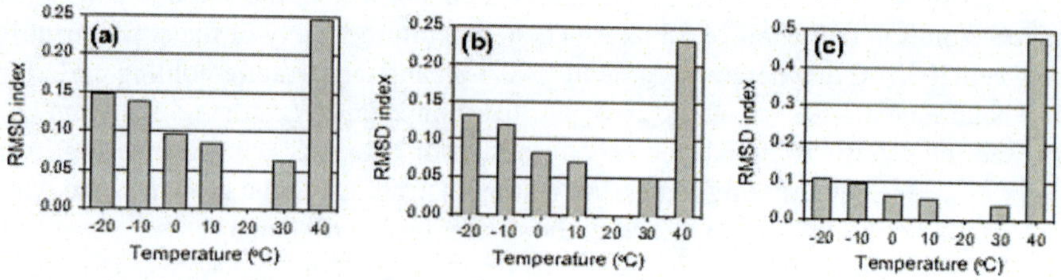

Fig. 7 The RMSD index of the PZT sensor at different temperature in frequency range of **a** 40–100 kHz, **b** 100–500 kHz, **c** 600–1500 kHz

The greater temperature that diverges from the baseline, the greater the dissimilarity in RMSD (Fig. 7). This also obviously specifies the temperature dependence of the EMI method.

The outcomes displayed the sensitivity of the sensor to the primary load at the time the external load was similar to the thickness of the sensor.

Yang et al. [10] this paper deals with the analysis of damage in the RCC structure through the comparative analysis of the sensitivity of the structural mechanical impedance and EM admittance toward damage in a concrete structure. An experiment forced a two-story concrete borderline equipped with 5 PZT sensors. The 5 PZT sensors were constantly associated with one of the segments on the leading flooring of the frame structure. Utilizing impedance analyzer, it is feasible to get conductance and susceptibility signatures from PZT sensors. Prior to destruction and next every of the two destruction stages, these 5 PZT sensors were checked to register PZT EM entry marks. The spaces of the 5PZT sensors as of the primary crack appeared in Fig. 8c were recorded at the 3rd part of Table 4.

Methodology: The loads to be tested were enforced along the base of horizontal motion along changed frequencies. The examination was carry out in a number of stages, and the applied base motion sequences are listed in Table 5.

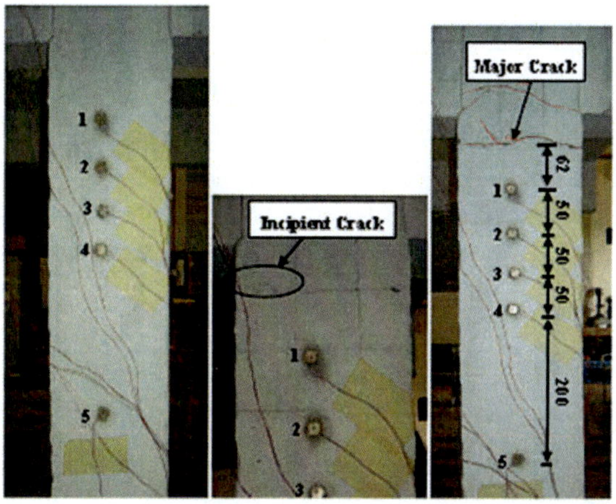

(a) Before damage (b) Damage Phase 1 (c) Damage Phase 2

Fig. 8 State of the column prior and later destruction (unit mm)

Table 4 Positions of PZT sensors and their distance to primary fissure

Patch Number	Distance from joint O along the column as shown in Fig. 2 (mm)	Distance of the PZT patch from the main crack (mm)
1	100	62
2	150	112
3	200	162
4	250	212
5	450	412

Table 5 Test sequences of the shake table

Phase No	Name	Description	Purpose
1	RND-A	Random, PGA = 0.02 g	Structural identification
2	CHL-	Chile, PGA = 023 g,	Moderate shaking
3	RND-B	Random, PGA = 0.02 g	Structural identification
4	CHL-	Chile, PGA = 0.46 g,	Sever shaking
5	RND-C	Random, PGA = 0.02 g	Structural identification

The 5PZT sensors were scanned prior to loading the base to record reference signatures for health conditions.

Experimental data: The 5PZT sensors were instrumented at positions 62 mm, 112 mm, 162 mm, 212 mm, and 412 mm from the focal crack, respectively, (marked in Fig. 8c). Curve fitting was performed by plotting Figs. 9 and 10. For every destruction stage, the values of RMSD for the *G* and *x* signatures show reducing tendency as space rises (Table 6).

For both stages of destruction, the highest RMSD values for *G* and *x* are examine in the PZT patch1, which is the closest space to the major break, and the lowermost values in the PZT patch5. This reflects that as the PZT patch destruction space rises, PZT patch's sensitivity to destruction reductions. In addition to Figs. 9 and 10, we can additionally see that the RMSD values on behalf of destruction stage 2 were

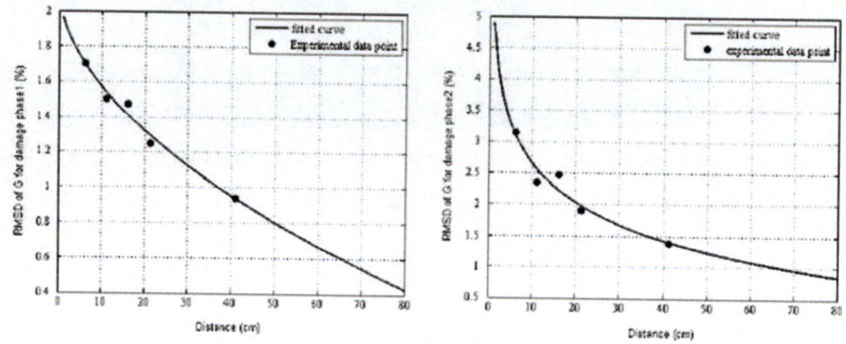

Fig. 9 RMSD values of *G* used for 2 destruction stages with respect to the distance of PZT to the crack

Table 6 Damage rates for entirely PZT sensors at two phase of destruction

PZT pitch no	RMSD of *G*	RMSD of *B*	RMSD of *x*	RMSD of *y*
Damage phase 1				
1	1.700904	0.079885	15.73118	0.077741
2	1.502574	0.106806	13.34286	0.098003
3	1.473301	0.086287	9.605244	0.063772
4	1.25146	0.11261	7.5918	0.095865
5	0.940053	0.116923	4.747043	0.117208
Damage phase 2				
1	3.139675	0.061062	25.89481	0.049555
2	2.343378	0.073566	20.84959	0.064063
3	2.46278	0.058494	18.08281	0.047549
4	1.898626	0.071459	10.93107	0.054598
5	1.390455	0.056117	8.161555	0.045829

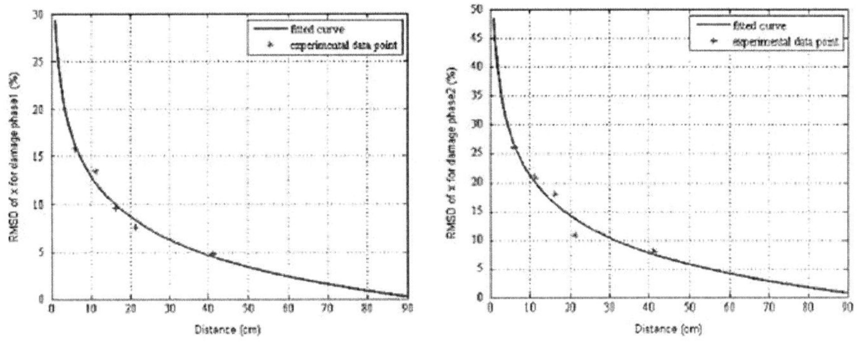

Fig. 10 RMSD values of x for two stages of destruction with respect to the PZT distance from the crack

entirely higher as compare to that of destruction stage 1. So, when the destruction is extreme, the sensitive of the PZT sensor also increases.

The results showed that the rate of damage causes an expansion of the damping and a reduction of mass and stiffness at a given point that the model is particularly sensitive to extreme destruction. Furthermore, it was concluded that the PZT sensor along a greater space for destruction is a smaller amount sensitive to destruction.

3 Summary of Data Analysis

Author name	Description	Formulation
C. Bharathi Priya et al.	The technique of localized changes in electromechanical impedance (EMI) along a piezoelectric sensor assembly (PZT), to sense destruction, is accelerated in a series-connected LCR circuit in a large-scale sensor distribution situation. In this research, an embedded concrete beam with six PZT covers glued into slabs of expanding thickness by exposing the beam to variance curing. RMSD among the signatures was the most suitable index of damage to characterize structural damage G_i^1–conductance at the point of the *i*th dimension in G_i^0–conductance at the point of the *i*th dimension in	Index of damage to monitor the structural damage, $\text{RMSD} = \left[\frac{\{\sum(G_i^1 - G_i^0)\}_{\text{sub-range}}}{\{\sum G_i^{0.2}\}_{\text{full-range}}}\right]^{0.5}$

(continued)

(continued)

Author name	Description	Formulation
Demi Ai et al.	The report coverers sensitivity in case of a new type integrated active PZT sensor in sensing structural effect destruction by theoretical and experimental studies. The planned model was demonstrated through the result of the experimental test of an integrated PZT sensor produced The RMSD as a scalar destruction metric has been generally acknowledged to assess the severity of destruction through guessing the quantity of variation among two signatures. G_i^{o} is the conductance calculated through the PZT sensor previous to destruction to the structure, which is regularly set as a baseline, and G_i^{1} is the relating conductance later damage to the ith calculating point. Unlike conventional RMSD, which uses fixed undamaged information as a baseline, RMSDk utilizes data from the previous state as a baseline; therefore, it is changed progressively when the next RMSD value is calculated. Along these lines, absolute variations can be discriminated in every case. Through integrating along the description of the actual admission sign, impact the corresponding effect can be measured utilizing RMSDk. G_i^k is the conductance calculated at the ith calculation, as a base line in previous condition, and G_i^k is the equivalent conductance after the variations	RMSD BASED on slope, $\mathrm{RMSD}\ (\%) = \sqrt{\frac{\sum_{i=1}^{N}\left(G_i^1 - G_i^0\right)^2}{\sum_{i=1}^{N}(G_i^0)^2}} \times 100$ RMSD via three PZT sensors in four destruction cases, $\mathrm{RMSDk}\ (\%) = \sqrt{\frac{\sum_{i-1}^{N}\left(G_i^k - G_i^{k-1}\right)^2}{\sum_{i-1}^{N}\left(G_i^{k-1}\right)^2}} \times 100\ k = 1, 2, \ldots, n$

(continued)

(continued)

Author name	Description	Formulation
Prathamesh N. Bilgunde et al.	Mathematical study to organize the in-situ soundness of PZT sensors utilized for SHM of huge automotive structures, civil, and aeronautical The sensitivity of the material whose characteristics depends upon temperature remains constant w.r.t. to the resonance spectrum is described through metric indices. One of those metrics consists of RMSD. ω_I is the beginning frequency and ω_F which is the completion frequency. $Z_{E,E}$ equal to electrical admittance rate of the set PZT + epoxy + steel; while $Z_{E,T}$ is the admission signature for the equivalent variations in the connection state denoted through the same acoustic impedance Z_{adh}	RMSD for adhesive $\text{RMSD} = \sum_{n=\omega_I}^{n=\omega_F} \sqrt{\frac{[Z_{E,E}(p) - Z_{E,T}(P)]^2}{Z_{E,E}^2(p)}}$
Xu Dongyu	A type of impedance-based integrated PZT sensor was made utilizing a mix of powdered cement and epoxy as a fill coating. In this paper a type of inserted impedance-based PZT sensor that was manufactured utilizing a combination of powdered cement and epoxy as the packing layer The destruction index is generally utilized to visibly specify the characterization of structural destruction. The RMSD index was designed dependent on the coming calculation. $\|G_i\|$ is the amplitude of the conductance of PZT sensor at a various temperature. $\left\|G_i^0\right\|$ is the amplitude of the conductance of PZT sensor at a mention temperature of 20 °C. n is the numeral of frequency sample points in the conductance spectra	RMSD under load at different frequencies, $\text{RMSD} = \sqrt{\frac{\sum_{i=1}^{n} [\|G_i\| - \|G_i^0\|]^2}{\sum_{i=1}^{n} [\|G_i^0\|]^2}}$

(continued)

(continued)

Author name	Description	Formulation
Yaowen Yang et al	Electromechanical piezoelectric ceramic based on PZT impedance technique for SHM has been effectively enforced in various engineering systems. Comparative examination on the sensitivity of electromagnetic admittance and structural mechanical impedance toward destruction in a concrete design. The destruction index is generally utilized to visibly specify the characterization of structural destruction In the past work, a productive statistical algorithm, in view of frequency-to-frequency comparison, was accessible as RMSD. $G^0{}_1$ is the conductance of the PZT calculated in strong situations of the structure and $G^1{}_i$ is the value after the equivalent destruction at the ith measure point	RMSD for two destruction phases vs distance, $\text{RMSD}\ (\%) = \sqrt{\frac{\sum_{i=1}^{N}\left(G_i^1 - G_i^0\right)^2}{\sum_{i=1}^{N}\left(G_i^0\right)^2}} \times 100$

4 Conclusions

This review was carried out to find the utilization of EMI method to carry out the health monitoring of structural. PZT sensor applied on the structure is more efficient in non–destructive evolution technique. The particular benefit of the EMI strategy over the other SHM strategies is its high sensitivity to early harms. The technique has been effectively enforced to different structures alternating from aerospace to civil structures and is ideal for following and web monitoring of damages in different designs or structures. The following conclusions have been made from the research reviews.

- RMSD was the most suitable damage to monitor the damage in structure. Conductance variance is most likely factor to find out the information at different PZT Patches.
- Also, it can be concluded that RMSD calculated with values obtained using impedance analyzer is efficient to find out the destructive values.
- Also, it was reviewed through research papers that impression effect is filtered more efficiently when integrated at PZT than, when integrated on surface. Thus, help to find the accurate assessment of structural damage.
- It was concluded that conductance factor depends upon temperature. Destructive index was also increased with increase in temperature and RMSD index was utilized to specify the characterization of structural destruction. RMSD was moreover used to instinctively visualize the change in impedance in case of integrated PZT sensor at different load and temperature.
- It was also reviewed that for damage detection method, the higher RMSD values sensed by a PZT sensor indicated a greater sensitivity of the PZT sensor to structural destruction.

References

1. Raghavan, A. (2007). *Guided-wave structural health monitoring*. Available online. https://deepblue.lib.Umich.edu/bitstream/handle/2027.42/77498/Raghavan_PhD_thesis_GWSHM.pdf?sequence=1. Accessed on March 15, 2020.
2. Rahim, N. A., & Ahmad, Z. (2017). Graphical user interface application in matlabtm environment for water and air quality process monitoring. *Chemical Engineering Transactions, 56*, 97–102. https://doi.org/10.3303/CET1756017
3. Tong, F., Dong, J., & Fan, Y. (2016). Application of piezoelectric smart materials to structural damage detection technology. *Journal of Liaoning University of Technology*, 26–38.
4. Zhu, X., & Hao, H. (2012). Development of an integrated structural health monitoring system for bridge structures in operational conditions. *Frontiers of Structural and Civil Engineering, 6*(3), 321–333. https://doi.org/10.1007/s11709-012-0161-y
5. Na, W. S., & Baek, J. (2018). A review of the piezoelectric electromechanical impedance based structural health monitoring technique for engineering structures. *Sensors, 18*(115), 1307.
6. Priyaa, C. B., Gopalakrishnana, N., & Raoa, A. R. M. (2015). Impedance based structural health monitoring using serially connected piezoelectric sensors *4*(1), 38–45.

7. Ai, D., Zhu, H., & Luo, H. (2016). Sensitivity of embedded active PZT sensor for concrete structural impact damage detection. *Construction and Building Materials, 111*, 348–357.
8. Bilgunde, P. N., & Bond, L. J. (2018). In-situ health monitoring of piezoelectric sensors using electromechanical impedance: a numerical perspective. In *AIP Conference Proceedings* (Vol. 1949, No. 1, p. 230011). AIP Publishing LLC.
9. Xu, D., Banerjee, S., Wang, Y., Huang, S., & Cheng, X. (2015). Temperature and loading effects of embedded smart piezoelectric sensor for health monitoring of concrete structures. *Construction and Building Materials, 76*, 187–193.
10. Yang, Y., Hu, Y., & Lu, Y. (2008). Sensitivity of PZT impedance sensors for damage detection of concrete structures. *Sensors, 8*(1), 327–346.

The Inhibitive Effect of Vitamin-C on the Corrosive Performance of Mild Steel in Ground Granulated Blast Furnace Slag-Based Concrete

Imran Qasim and Khushpreet Singh

Abstract Corrosion has been the main problem in the concrete structures due to which the pressure is formed in the concrete which points to the cracking and ultimately spalling of the concrete. The inhibitive behavior of ascorbic acid with the combination of ground granulated blast furnace slag (GGBS), as an organic inhibitor on the corrosive behavior of Mild Steel, was studied at the concentration of 1M NaCl with 5% of the ascorbic acid (AA). Linear Polarization Resistance (LPR) was used for the measurement of the corrosion rate and it was observed that the ascorbic acid revealed some resistance when combined with the GGBS. The results depicted that the combination of both the AA and GGBS was much effective in keeping the resistance to corrosion higher than the control sample. Potentiodynamic Polarization results showed that ascorbic acid is a mixed type inhibitor.

Keywords Anodic polarization · Chloride · Inhibitor · Oxidation

1 Introduction

Corrosion of reinforcement is one of the major concerns for the construction world in degrading the concrete structures. Corrosion nowadays is affecting the economy of the country. The effect of corrosion on the GDP of the country is huge as much as up to 5% of the total GDP of the country. According to the data enlisted in the report submitted by the NACE, it states, in India, the losses due to corrosion sums up to the expenditure of around 25,000 crores rupees per year. According to the report of Economic Times of India, India loses Rs. 2 trillion (USD40 billion) annually to the corrosion of infrastructure, industrial equipment, and other important installations [1]. Steel in the concrete is covered up or protected by the formation of the passive oxide layer around it, which makes it stable due to the alkaline nature of the concrete.

I. Qasim (✉) · K. Singh
Chandigarh University, Mohali 140301, India

K. Singh
e-mail: khushpreet.civil@cumail.in

A. K. Gupta et al. (eds.), *Advances in Construction Materials and Sustainable Environment*, Lecture Notes in Civil Engineering 196,
https://doi.org/10.1007/978-981-16-6557-8_55

Once the foreign matter attacks this passive layer, it breaks which tends to onset of the corrosion on the reinforcement. The onset of the corrosion process leads to the various physical changes in the reinforcement bar like the expansion of the cross-sectional area of the reinforcement due to the presence of the high amount of the corrosion residues. This process results in the spalling or crack formation in the concrete and ultimately leads to the instability of the structure. The most common cause of the onset of corrosion is the presence of chlorine in the soil or marine areas. Some of the approaches which are already being followed by the construction firms to repair the cracks and another sort of damages caused by the corrosion. Induction of new repair concrete at cracked places can be too costly and time taking work. The use of the inhibitors now has emerged as the demand in the industry whether surface applied or pre-mixed which leads to the proper maintenance of the concrete structures. In the last few years, this method of using the inhibitors has proven to be beneficial and economical but few exceptions are there on the use of the type of inhibitors. The choice of inhibitors has been a debatable topic for some time as some of the inhibitors have environmental impacts as well. The ascorbic acid used in the previous researches has proved to be the exhibiting inhibition properties [2–5]. Various other inhibitors have been used by the researchers to find out the effects of the ascorbic acid on the material ingredients of the concrete.

2 Materials

2.1 Ascorbic Acid

Ascorbic Acid (AA) (Table 1) is a chemical that is soluble in nature and organic acid, also known as vitamin-C or Hexuronic Acid. It is present in dissociated molecular form insoluble solutions having low pH. Ascorbic acid is also present as an antioxidant for use in various varieties of food and drink. The use of ascorbic acid has been adopted for some years by the researchers as a corrosion inhibitor for various

Table 1 Physical properties of ascorbic acid

S. No.	Physical properties	Property value
1	Molar Mass (g/mol)	176.12
2	pH	3
3	Appearance	Powdered form
4	Color	White
5	Melting Point	190 °C
6	Nature	Water-soluble
7	Molecular formula	$C_6H_8O_6$
8	Density (g/cm^3)	1.65

Source en.wikipedia.org

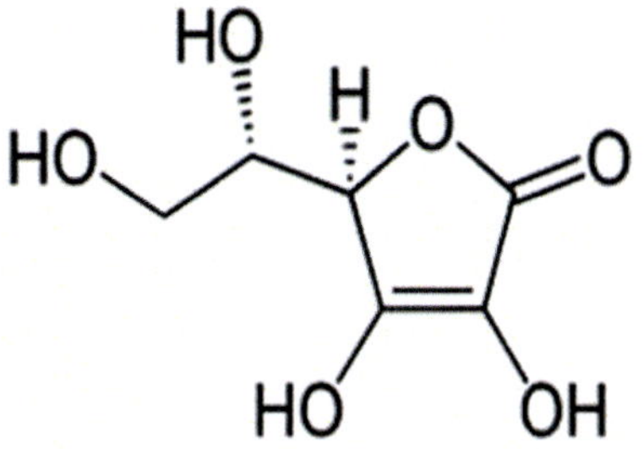

Fig. 1 Chemical structure of ascorbic acid. *Source* en.wikipedia.org

steel types. Corrosion inhibitors help inhibit the corrosion by adsorbing the inhibitor constituents on the metal surface.

The nitrogen atoms in the form of NH_2 or NH and the oxygen atoms in the form of OH^-, then compounds start acting as corrosion inhibitors due to the ubiquity of the active sites in the inhibitor molecular structure (Fig. 1) [2]. The interaction of the metal with the corrosive solution paves way for the inhibitor to play its role by adsorbing on the surface. The adsorption on the surface starts making a change in the mechanism of the electrochemical process. Some of the polar working groups contain reaction centers that are useful in stabilizing the process of adsorption [6].

The inhibitor outcome of ascorbic acid is linked to its adsorption upon the electrode surface. When ascorbic acid is present in concentrations of more than 200 ppm, it forms complex reactions consisting of soluble iron chelates. Ascorbic acid serves as a good corrosion inhibitor concerning mild steel with the addition of 2.34% Cr [7]. Concrete protects rebar physically by a concrete cover and chemically by the ubiquity of thin passive film that forms in the high alkaline environment of concrete. The formation of the passive layer is the indication that the product is resisting the corrosion but this protective layer splits down either by the carbonic acid existing in the environment or by the presence of chloride ions [8]. Carbon dioxide present in the air reacts with cement hydration products such as $Ca(OH)_2$, C–S–H gel, and many more, thus leading to the formation of $CaCO_3$ which results in lowering the pH of the pore solution [9, 10]. Due to the lowering of pH from 13.5 to nearly pH 8, results in the breakdown of the protecting passive layer that protects the rebar from corrosion [8]. The immersion time plays an important role in studying the corrosion, as in previous literature it was found that the corrosion rate declined as the time of immersion is increased [3].

2.2 Ground Granulated Blast Furnace Slag (GGBS)

GGBS is the by-product obtained after quenching the iron slag at the time of processing. It is used as the replacement material which tends to block the pores by forming a C–S–H bond, after reacting with cement. This leads to less passage of carbon or chloride particles to move in and thereby reducing the corrosion [11]. The present paper encompasses the effect of the ascorbic acid on concrete resistance to corrosion besides the ground granulated blast furnace slag (GGBS) in the presence of a chloride environment.

3 Experimental Program

In the laboratory, the concrete specimens of size 300 × 300 × 150 mm with a water-cement ratio of 0.45 were cast at room temperature. The specimens were cast with adding the 5% amount of inhibitors by weight of cement. After casting, the samples were de-molded after 24 h and were subjected to 14 days curing. Specimens were reinforced by one horizontally placed and fixed steel bar in each of the specimens used for corrosion testing.

The specimens which were cast featured recess on the top surface to allow ponding of chloride mix above an area of 220 mm^2. The cover for the concrete was 15 mm. The length of the reinforcement bars used in the experiment was 350 mm with a diameter of 10 mm. Out of the length 350, 25 mm on each side was covered by the insulated plastic tapes to protect the bar from getting corroded from outside moisture. The cement used was Pozzolona Portland Cement. The mix proportion of concrete is given in Table 2. The inhibitor used in the experiment was ascorbic acid, which was mixed in concrete as a corrosion inhibitor. The specimens used in this experiment are named accordingly as A_1, B_1, C_1, and D_1. The number bestows the sample number and the letter shows the category of the sample, where the A_1 sample is a control sample having PPC-based concrete, the other sample B_1 is the sample containing the ascorbic acid along with the PPC. Furthermore, C_1 is of the GGBS containing sample besides ascorbic acid, and D_1 is the sample having only GGBS as additive. The chloride cycles involved weekly ponding (58.5 g/L NaCl solution). The samples were tested after a two-week ponding cycle and the effect of chloride on corrosion was studied and compared to the control sample. All of these samples including the control sample were subjected to the same chloride conditions for four-week ponding cycles and then the results were correlated with control samples. The surface of the specimens was wetted by placing a wet sponge on the surface shown in Fig. 2a, before corrosion measurement to obtain low resistance for concrete during the data record. The corrosion apparatus used for the measurement was ACM Corrosion Analyzer (CORE/205), is shown in Fig. 2b, by which the corrosion potential, E_{corr}

Table 2 Mixture proportion

Materials	(kg/m^3)
Cement	393
Coarse aggregate	1191
Fine aggregate	603
Water/cement ratio	0.45

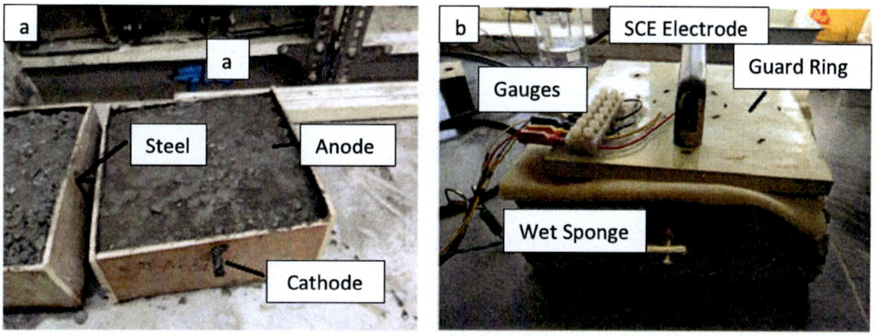

Fig. 2 **a** Figure showing the specimens formed of size 300 mm × 300 mm × 150 mm, **b** figure showing the specimen with ACM corrosion analyzer setup

with Calomel reference electrode, corrosion current density (I_{corr}), and the resistance of the concrete (R) was measured in the same time.

4 Results and Discussion

4.1 Weight Loss Measurement

The bars were polished and degreased at the time of casting before putting them into the concrete. After 30 days of exposure to the chloride conditions within the concrete, the reinforcement bars were taken out and cleaned with the Clarks Solution, and then with distilled water, air-dried for reducing any moisture left, and then weighed. The values were noted down and then compared with the values obtained at last. However, there was not much of the weight loss recorded but efficiencies were recorded on the weight loss basis. Weight loss was calculated by the following Eq. (1):

$$mpy = w \times 534/a \times t \times d \quad (1)$$

where w denotes the weight loss in grams, a denotes the area of the specimen in square inches, d denotes the density of the specimen in g/cm^3 and t is time in hours (Table 3).

4.2 Potentiodynamic Polarization

After ponding, the results of E_{corr}, I_{corr}, for the specimens were analyzed and the measurements are shown in graphical form in Fig. 3a, b. After the application of chloride, there was a small drop in E_{corr} values for all of the specimens except the

Table 3 Gravimetric test

Sample	Weight loss (mgm)	Corrosion rate (mmpy)	Inhibition efficiency (%)
Control with PPC (*A*1)	118	0.142	–
Control + AA (*B*1)	18.2	0.023	83
Control + AA + GGBS (*C*1)	11.3	0.014	90
Control + GGBS (*D*1)	19.6	0.024	83

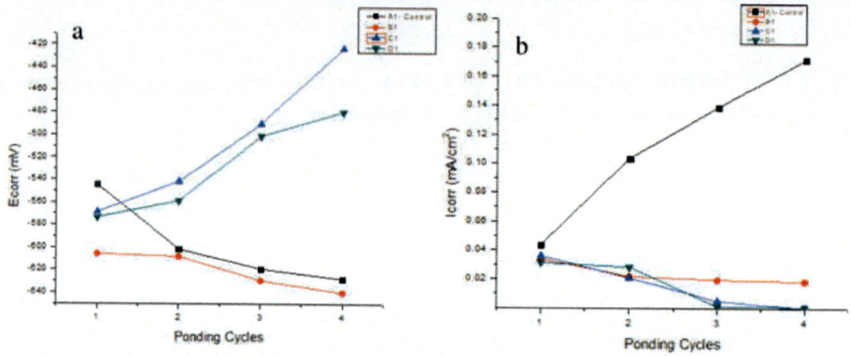

Fig. 3 **a, b** Comparison of E_{corr} and I_{corr} values with the ponding cycles of specimens

specimen C_1 having ascorbic acid and GGBS varying approximately between − 568 to − 423 mV. However, those values are corresponding to the corrosion potential of the steel in a chloride environment. The E_{corr} value of the control sample was reduced with the increase in the ponding cycles of the concrete from − 544 to − 628 mV. The E_{corr} values were recorded higher for the specimen C_1 having ascorbic acid and GGBS contained in it ranging from − 568 to − 423 mV as the chloride volume was increased with each ponding cycle. While the specimens such as A_1 and B_1 showed very low E_{corr} values, i.e., more negative as compared to the initial readings. However, the B_1 specimen showed the same range as taken initially and after the four cycles from − 605 to − 639 mV. I_{corr} values kept changing with the E_{corr} values which were at initial reading less than 1 mA/cm^2 and had increased with the application of chloride. After the initial increase in the values, the I_{corr} of the control sample increased from 0.043900 to 0.172000 mA/cm^2 and the kept on hanging between 0.043900 and 0.172000 mA/cm^2. Furthermore, the specimen with higher E_{corr} value had slightly lower I_{corr} than compared with other samples *A* and *B* and decreased from 0.036110 to 0.000093 mA/cm^2 as shown in Table 4. The modified stern-geary equations were used to calculate the polarization resistance (R_p) [12], in Eqs. (2)–(4):

Table 4 Potentiodynamic polarization test

Specimens	Immersion time	E_{corr} (mV)	b_a (mV/dec)	b_c (mV/dec)	I_{corr}	C.R (mA/cm^2)	I.E (%)
Control with PPC (A1)	First Readings	– 544.63	144	128	0.043900	0.4105	–
	After 4 Weeks	– 628.12	377	113	0.172000	0.0104	–
Control + AA (B1)	First Readings	– 605.44	158	115	0.034115	0.3194	22
	After 4 Week	– 639.91	171	100	0.018800	0.0245	94.2
Control + AA + GGBS (C1)	First Readings	– 568.6	140	109	0.036110	0.2011	51.7
	After 4 Weeks	– 423.64	174	117	0.000093	0.0159	96.1
Control + GGBS (D1)	First Readings	– 573.6	150	126	0.031200	0.2110	28.9
	After 4 Weeks	– 479.93	183	121	0.000682	0.0311	92.4

$$R_p = \Delta E/\Delta I * \Delta E \rightarrow 0 \quad (2)$$

$$I_{corr} = ß/R_p \quad (3)$$

$$I_{corr} = \beta_a \cdot \beta_c/2 \cdot 3\ (\beta_a \cdot \beta_c) \quad (4)$$

B is Stern–Geary constant and β_a and β_c (Volts/dec) are Tafel constants, respectively. The value of β should be determined empirically. Whereas ΔE and ΔI are the potential densities and current densities changing, respectively, in cm^2, I_{corr} is the corrosion current density in Eq. (4). From Eq. (2), E–I plots were used to measure the slope of polarization resistance. From measured R_p values, corrosion current densities were calculated using proportionality constants ß as in Eq. (5), as done in the literature [12].

After the initial increase in the values, the I_{corr} of the control sample increased from 0.043900 μA/cm^2 to 0.172000 μA/cm^2 and they kept fluctuating between 0.043900 and 0.172000 μA/cm^2. Furthermore, the specimen with more E_{corr} value had slightly lower I_{corr} as compared to the other samples and decreased from 0.036110 to 0.000093 μA/cm^2 as shown in Table 4. The other samples of the group remained at the stable state but the trend of changing vales kept in the sample C_1, which was having ascorbic acid and GGBS as the additives added. As the corrosion current is reduced by some margin, it means the effect of ascorbic acid is good in decreasing the corrosion rate.

Table 5 Anodic polarization test

Medium	Immersion time	I_p (μA/cm^2)	E_p (mV)	E_{pt} (mV)
Control with PPC (*A*1)	First readings	100	– 580	– 255
	At 4 weeks	100	– 642	– 449
Control + AA (*B*1)	First readings	100	– 554	+ 107
	At 4 weeks	0.002	– 379	+ 308
Control + AA + GGBS (*C*1)	First readings	100	– 400	+ 118
	At 4 weeks	0.2	– 264	+ 700
Control + GGBS (*D*1)	First readings	100	– 494	+ 171
	At 4 weeks	2.1	– 389	+ 543

4.3 *Anodic Polarization Test*

The passivation potential (E_p), passivation current (I_p), and the pitting potential (E_{pt}) were measured for steel in the chloride environment of NaCl as a control sample for other inhibitor mediums as reported in Table 5. From Table 5, it can be stated that in the case of the control sample, passivity has been highly reduced because of chloride ions. So the reduction in passivity measured in initial data is signified by the passivation and current potentials ($E_p = -580$ mV, $I_P = 100$ μA/cm^2) and after period of 30 days ($E_p = -642$ mV, $I_p = 100$ μA/cm^2). The pitting potential was perceived to be – 243 mV initially and – 429 mV after an exposure time of 30 days to the chloride environment. Initially, the inhibitor medium reveals a higher passivation current with more negative passivation potential and small positive pitting potential but later exposing it to chloride, the inhibitive medium gave very low passivation current (0.002–2.1 μA/cm^2). On the other hand, less negative passivation potential (– 260 to – 390 mV) and the pitting potential was recorded as more positive (+ 400 to + 700 mV). This data clearly shows that exposure time plays an important purpose in the formation of the strong passive film on the metal surface even in the chloride environment.

4.4 *Chloride Testing*

The chloride testing was carried on the specimens in the four-week cycle with the ponding of 58.5 mg/L of NaCl as per IS 6925:1973 (1) [13]. Figure 4a, b, show that chloride content is calculated by percentage cement mass with the depth. From the literature, it was found that the in NaCl solution, the ascorbic acid has maximum inhibition at the 200 ppm [2].

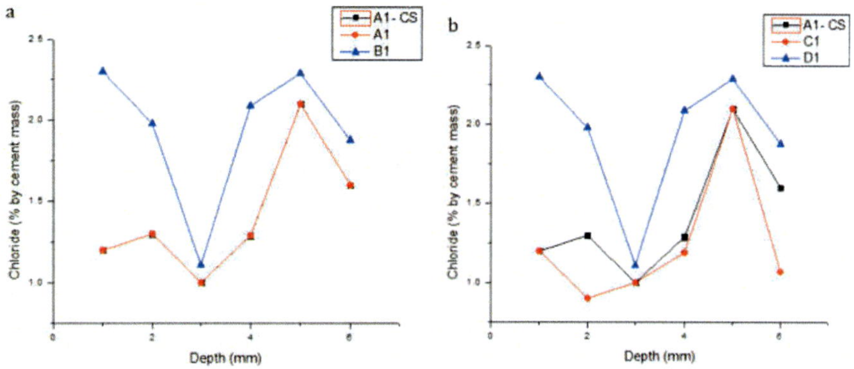

Fig. 4 **a**, **b** Profiles of Chloride for A_1, B_1, C_1 and D_1 specimens

Table 6 Corrosion values as per ASTM C876 for calomel electrode [14]

Half-cell potential	Corrosion risk
> – 126 mV	Low corrosion risk (10%)
– 126 to – 276 mV	Intermediate corrosion risk
< – 276 mV	High risk (> 90%)
< – 426 mV	Severe

4.5 Half-Cell Potential

The Half-Cell potential test was performed as per the ASTM C876 [14], which indicated the values for the Calomel electrode. As per Table 6, the values of our readings obtained after testing were not up to mark but the graph shows a decline in values was observed. For specimen C_1, the values of half-cell were recorded less as compared to the other specimens (Fig. 5).

5 Conclusion

The use of ascorbic acid as an inhibitor has been effective in combination with the GGBS in resisting corrosion. It kept the I_{corr} at 1 mA/cm^2 which can be considered as the origin content for the onset of corrosion, although the chloride content was kept the same as for the other specimens. As the chloride attacks the concrete there was a higher corrosion rate observed in some stances where the inhibitor became unproductive. It may also be noted that the resistivity recorded higher for some specimens may be due to the blocked pores within the specimens.

- The inhibition efficiencies calculated by the method of gravimetric weight loss method gave 83, 90, and 83% for $A1$, $B1$, $C1$, and $D1$ samples, respectively.

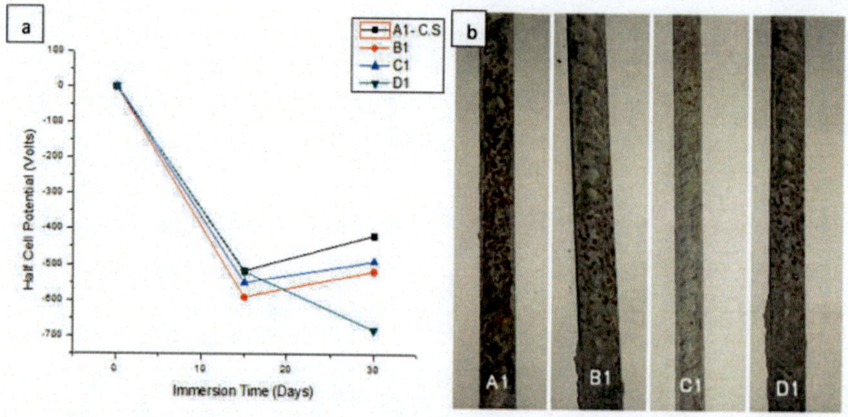

Fig. 5 **a** Half-cell potential measurements for all specimens immersed for 30 days in 1M NaCl, **b** photo of reinforcement bars after taking out of concrete. (From Left: A1, B1, C1, and D1)

- The inhibition efficiency calculated from potentiodynamic polarization data depicted 94, 96, and 92% for *B*1, *C*1, and *D*1 samples, respectively, at the end of 30 days exposure.
- Anodic Polarization data gave a clear picture that under the chloride contaminated concrete environment, the samples containing GGBS and the ascorbic acid recorded less passivation current, and hence, the stability of the passivity of the film is guaranteed, whereas in the other samples the passivation current was recorded high that depicted the passive film was rigorously destroyed by the presence of chloride ions.

The combination of the ascorbic acid as an organic inhibitor and the ground granulated blast furnace slag (GGBS) proved to be the good resistor of the corrosion onto the steel reinforcement in the chloride environment.

Conflict of Interest There is no conflict of interest.

References

1. India loses Rs 2 trillion annually to corrosion of infrastructure: Government—The Economic Times. https://economictimes.indiatimes.com/news/economy/finance/india-loses-rs-2-trillion-annually-to-corrosion-of-infrastructure-government/articleshow/48792499.cms?from=mdr
2. Sekine, I. (1988). The corrosion inhibition of mild steel by ascorbic and folic acids the development of new corrosion inhibitors of the non-toxic type which do not contain heavy metals and organic phosphates, has recently been regarded as important, and such organic, *28*, 987–1001
3. Fuadi, A. (2019). Investigation of ascorbic acid as environment-friendly corrosion inhibitor of low carbon steel in marine environment investigation of ascorbic acid as environment-friendly

corrosion inhibitor of low carbon steel in marine environment. https://doi.org/10.1088/1757-899X/536/1/012108
4. Fuchs-Godec, R., Pavlovic, M. G., & Tomic, M. V. (2015). The inhibitive effect of vitamin-C on the corrosive performance of steel in HCl solutions—Part II. *International Journal of Electrochemical Science, 10*, 10502–10512.
5. Brnardic, I. (2008). The inhibition activity of ascorbic acid towards corrosion of steel in alkaline media containing chloride ions *50*, 2705–2709. https://doi.org/10.1016/j.corsci.2008.06.018
6. Maduabuchi, Q. (2014). Ascorbic acid as corrosion inhibitor for Q235 mild steel in acidic environments. *Journal of Industrial and Engineering Chemistry*. https://doi.org/10.1016/j.jiec.2014.11.029
7. Ferreira, E. S., Giacomelli, C., Giacomelli, F. C., & Spinelli, A. (2004). Evaluation of the inhibitor effect of l-ascorbic acid on the corrosion of mild steel *83*, 129–134. https://doi.org/10.1016/j.matchemphys.2003.09.020
8. Kaur, K., Goyal, S., Bhattacharje, B., Oh, T., Kim, J., Lee, C., & Park, S. (2017). Electrochemical impedance spectroscopy to study the carbonation behavior of concrete treated with corrosion inhibitors *14*, 172–182. https://doi.org/10.3151/jact.15.738
9. Sisomphon, K., & Franke, L. (2007). Carbonation rates of concretes containing high volume of pozzolanic materials *37*, 1647–1653. https://doi.org/10.1016/j.cemconres.2007.08.014
10. Likhanova, N. V., Arellanes-lozada, P., Olivares-xometl, O., Hernández-cocoletzi, H., Lijanova, I. V., Arriola-morales, J., & Castellanos-aguila, J. E. (2019). Effect of organic anions on ionic liquids as corrosion inhibitors of steel in sulfuric acid solution. *Journal of Molecular Liquids., 279*, 267–278. https://doi.org/10.1016/j.molliq.2019.01.126
11. Cheng, A., Huang, R., Wu, J. K., & Chen, C. H. (2005). Influence of GGBS on durability and corrosion behavior of reinforced concrete. *Materials Chemistry and Physics., 93*, 404–411. https://doi.org/10.1016/j.matchemphys.2005.03.043
12. Feliu, V., Gonzalez, J. A., Andrade, C., & Feliu, S. (1998). [Part I], Equivalent circuit for modeling the steel-concrete interface complications in applying the stern-geary equation to corrosion rate determinations. *Corrosion Science, 39*(6), 0995–1006.
13. Standards, B. of I.: IS.6925.1973 Chloride Code (1973)
14. American Society for Testing and Materials. (2009). ASTM C 876/09—Standard test method for half-cell potentials of uncoated reinforcing steel in concrete. *Annual Book for ASTM Standards, American Society for Testing and Materials., 91*, 1–6.

Hydrodynamic Modeling for Identifying Flood Vulnerability Zones in Mahi Lower Sub-basin

R. Rathod Krina and Sudhanshu Dixit

Abstract Floods can cause widespread devastation, resulting in loss of life and damages to personal property and critical public health infrastructure. River flooding is the most common type of flooding in many parts of the world. It occurs when a water body exceeds its capacity to hold water and usually happens due to prolonged heavy rainfall. Mahi River located in Gujarat State of India, being the third major west-flowing river was inundated in August 2006. This present study aims to develop mathematical model using HEC-RAS 6.0 version to analyze one-dimensional and two-dimensional unsteady flow of water in Mahi lower sub-basin. The river stretch for hydrodynamic modeling starts from Khanpur gauge station located 139 km downstream of Kadana Dam at Mahisagar District of Gujarat to the Mujpur Village near the mouth of Mahi River in Gulf of Khambhat. The simulated results in 1D modeling show that approximately 64.38% of the total cross section have their bank station in overtopped condition for discharge equal or more than 31,061.91 cumec. It has been observed that right bank of river indicating eastern side of Mahi River near Vadodara City is more prone to water spill for higher water levels in River. 2D modeling results identified vulnerable areas of Kherda, Rajupura, and Bhanpura located between Anand and Vadodara cities for flood event 2006 with the highest water surface elevation of 30.95 m. Model outputs can be used as on-structural method of flood vulnerability assessment by government agencies to reduce flood damage.

Keywords Hydrodynamic modeling · HEC-RAS · Flood vulnerability assessment

1 Introduction

A flood is an overflow of water that submerges land which is typically meant to be dry. Major causes of floods in India include inadequate capacity within riverbanks to contain high flows, riverbank erosion and silting of riverbeds [1]; it can be prevented

R. Rathod Krina (✉) · S. Dixit
Water Resources Engineering-Civil Engineering Department, L.D. College of Engineering, Ahmedabad, Gujarat, India

A. K. Gupta et al. (eds.), *Advances in Construction Materials and Sustainable Environment*, Lecture Notes in Civil Engineering 196,
https://doi.org/10.1007/978-981-16-6557-8_56

if rivers are managed properly, especially in densely populated and flat areas [2]. Effective flood warning systems can help take timely action during natural calamities and may save lives [3].

Between 1998 and 2017, floods affected more than 2 billion people worldwide [4]. Floods within the Indian subcontinent have increased radically over the previous couple of decades including 2019 Kerala floods, 2020 Assam floods and Brahmaputra floods [5]. People that sleep in floodplains and lack of warning systems are most susceptible to floods. Mitigation strategies can significantly reduce the effects of floods, give time for people to migrate to safer locations and stock up essential utility items for future [3]. When dealing with flood mitigation measures, we have two types of mitigation measures: structural measures and non-structural measures [6]. Non-structural measures include flood forecasting and warning system, floodproofing, disaster prevention, preparedness, and response mechanisms, wherein flood vulnerability assessment plays a significant role in decision making [7]. Physical modeling is the most accepted and frequently used technique for food hazard assessment [8–10]. Based on the modeling approach, flood models are classified as 1D and 2D. 1D models such as HEC-RAS, MIKE 11, and 2D models such as LISFLOOD-FP, FLOW2D, MIKE 21, and HEC-RAS 2D [6]. Flood risk map was prepared using HEC-RAS 1D and Arc-GIS in Dhaka, Bangladesh [10], DEM of 90 m grid size was resampled to 30 m resolution, where high-resolution DEM of 5 m can give better results. Floodplain maps in Kabul River Pakistan were created using HEC-RAS [9] and compared using MODIS satellite image where calibration of manning's n value can generate accurate simulation. In this present study, HEC-RAS version 6.0 is used for simulation of unsteady flow condition both in one dimension and in two dimension.

HEC-RAS performs 1D and 2D computations using the St. Venant equations of conservation of mass and conservation of momentum [11]. Whereas 1D models solve the St. Venant equations along one dimension, a 2D model solves the St. Venant equations along two dimensions [12]. Moreover, 1D flood models assume that water remains inside the floodplain and does not consider any lateral flow [13], 2D hydrodynamic models consider a variation in flow both in longitudinal and in transverse directions of the river channel [14]. In HEC-RAS past version, HEC-GeoRAS software was needed to create flood inundation maps but in the version 6.0, HEC-RAS has an integrated geospatial capabilities part referred to as "RAS mapper" that permits the creation of flood inundation maps directly in HEC-RAS [15]. In the present study, hydrodynamic modeling is performed for identifying flood vulnerability zones in Mahi lower sub-basin under unsteady condition. The catchment of the Mahi River is mostly hilly and its shape is double fanned, which gives rise to high-intensity flash floods [16]. The total catchment area of Mahi River is 34,842 km^2 of which 34% lies in Gujarat. In the past 35 years of flow data highest discharge of 31,061.91 cumec was observed on 12th August 2006 at Khanpur gauge station with the highest rainfall recorded since 1971 in Mahi lower sub-basin of 1600 mm and was selected for simulation of flood event results. 37.4 km of river reach starting from Khanpur gauge station located 139 km downstream of Kadana dam at Mahisagar District of Gujarat to the Mujpur Village near the mouth of Mahi River in Gulf of

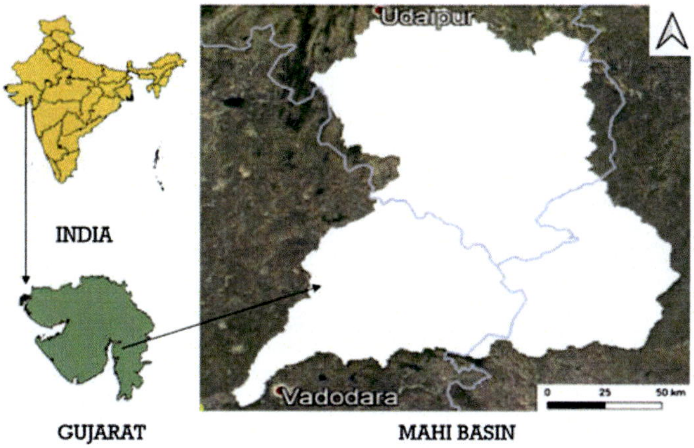

Fig. 1 Study area map

Khambhat was divided into equal cross-sectional cut lines of 3000 m width and 500 m interval and modeled for unsteady flow analysis [17].

2 Study Area

The current work focuses on Mahi lower sub-basin (Fig. 1) which is situated in south Gujarat. It originates in the northern slopes of Vindhyas Mountain ranges at an elevation of 500 m above mean sea level at about 22°35′ N and 74°15′ E near the village Sardarpur in Dhar District of Madhya Pradesh. Its total length is 583 km, traversing 167 km in Madhya Pradesh, 174 km in Rajasthan and remaining 242 km in Gujarat. Mahi is joined by the Bhadar, the Anas, the Panam, and the Meshri.

Mahi basin consists of two major sub-basins: Mahi upper sub-basin and Mahi lower sub-basin. Mahi upper sub-basin covers total area of about 24,959 km^2, accounting for 65.11% of total basin area and consisting of 41 watersheds, whereas the Mahi lower sub-basin covers an area of about 13,377 km^2 accounting for 34.89% of total basin area and consisting of 22 watersheds with size range of 372–873 km^2 [18].

3 Data Collection

The data has been collected as per Table 1.

Table 1 Summary of data collection and its application

Data	Frequency	Source	Application
Hydrologic data	Hourly discharge and WSE	State Water Data Center, CWC	Simulation of 1 D/2D unsteady flow in HEC-RAS
River geometry data, cross sections, LOB, ROB	Every 500 m interval	CWC, RAS Mapper in HEC-RAS	Simulation of 1D unsteady flow model
DEM of Mahi watershed	SRTM (30 m grid)	USGS earth explorer website	2D unsteady flow modeling

4 Model Development in HEC-RAS

Major new features have been added to HEC-RAS 6.0 since version 5.0.7. containing many additional features/upgrades such as HEC-RAS mapper editing tools, breach time series plot, 3D graphics/animation, terrain modification tools, 1D finite volume solver, non-Newtonian fluids option for 1D and 2D, and many more, among which HEC-RAS mapper is very important to effectively create river geometry and view analysis results. Methodology adopted for unsteady modeling is presented in Fig. 2.

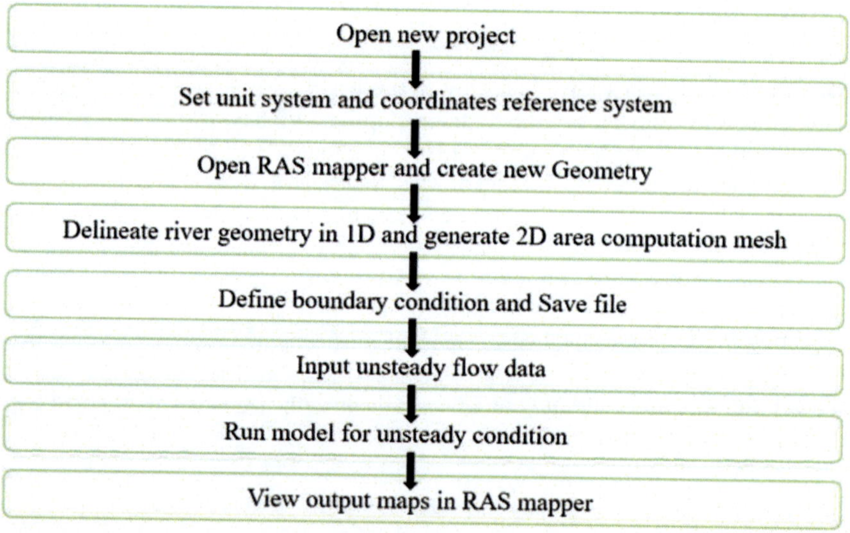

Fig. 2 Flowchart methodology of model development

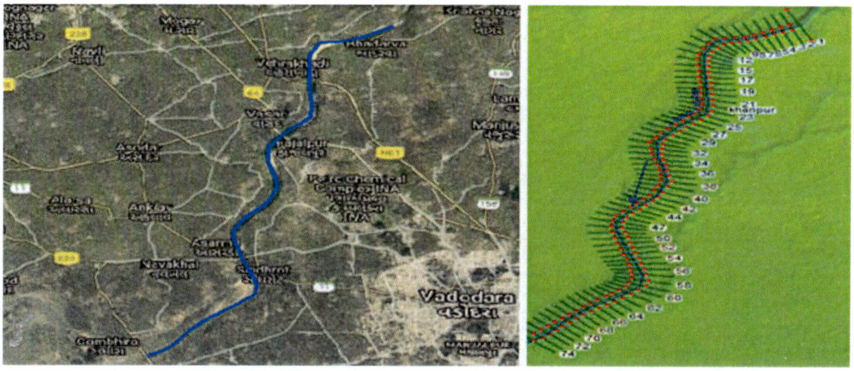

Fig. 3 River delineation in RAS mapper

4.1 Delineation of River Geometry

For unsteady flow analysis in HEC-RAS, start a new project and set SI units as default option. Projections of software and digital elevation model of given study area should be same to compare output maps with base map of Google satellite imagery. In the present study, Mahi estuarine area of Gujarat falls within WGS 84, UTM zone 43 boundaries.

For Geometry creation of 1D and 2D analysis, georeferenced DEM is required. DEM is converted into DTM by terrain creation tool in RAS mapper (Fig. 3). One-dimensional analysis requires delineation of river centerline (blue color), bank station (red points) and flow paths as shown in Fig. 3. Cross sections of 500 m equal interval [17] and width of 3000 m were generated perpendicular to river lines (green color). Total 74 numbers of cross section cut lines covering 37.4 km of river reach were delineated by edit geometry tool in RAS mapper.

In two-dimensional analysis, 2D area mesh as shown in Fig. 4 was generated with cell number of 29,667, using cell size of 100 m in terms of DX and DY. Computation points generated were able to calculate hydraulic property table for each grid cell.

4.2 Boundary Condition and Flow Data

Boundary conditions are required at the upstream and downstream ends of the 1D as well as 2D flow area. One can enter a flow hydrograph, stage hydrograph, rating curve or normal depth. In this present study of Mahi lower sub-basin, Khanpur gauge station is taken as upstream boundary and Mujpur Village is taken as downstream boundary for simulation of unsteady flow data. Flow hydrograph of 72 h duration and 3 h interval with maximum discharge of 31,061.91 cumec at 12 August 2006 derived from discharge data obtained from SWDC Gandhinagar is input for upstream

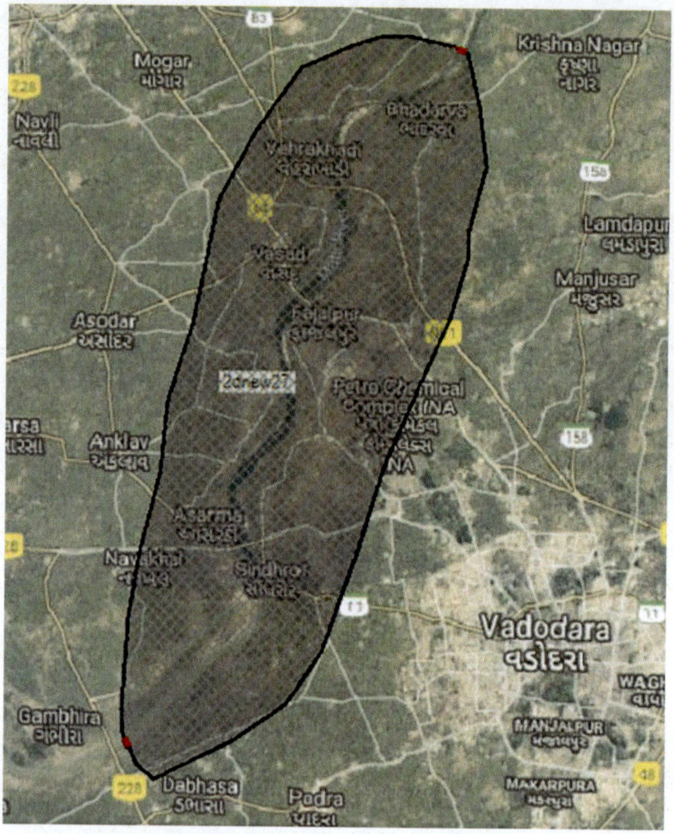

Fig. 4 2D mesh generated with geometry feature in RAS mapper

boundary condition and normal depth of 0.0018 m, which is obtained by calculating energy slope of water surface elevation is taken as downstream boundary condition for 1D and 2D HEC-RAS modeling.

Model is run for 30 s computation interval under run unsteady flow data tool in RAS mapper, and output maps are generated in terms of depth, velocity, and water surface elevation.

5 Results and Discussion

The simulated results in one-dimensional analysis show that out of total 74 cross sections, 47 numbers of cross section have left and right banks higher water surface elevation than ground level of their respective banks. It indicated that approximately 64.38% of total XS have possibilities of water spilling and inundation for discharge equal or more than 31,061.91 cumec. It has been observed that right bank of river indicating eastern side of Mahi River near Vadodara City is more prone to water spill for higher water levels in river than the left bank of city located on west side of river near Anand City.

Some of the cross sections having only either bank station inundated are classified as left bank critical and right bank critical, respectively. Cross sections having both the banks under inundation are highly vulnerable and presented in Table 2.

The cross sections having bank elevation lesser than relative water surface elevation on the corresponding bank are marked as unsafe and have probability of water spill. Some of them are presented in Figs. 5 and 6.

As shown in Fig. 7, behavior of water surface elevation in Mahi River under 2D modeling for flood event of the year 2006 has been studied which indicates gradual increase in water levels and starts from 11 August 2006, 10:00 h (green color). Maximum WSE recorded in river of 30.8 m (orange color) at 20:00 h on the same day with discharge of 17,287.95 cumecs at Khanpur gauge station. The right and left banks start spilling in area surrounding Poicha Temple near Vadodara City, between 11 August 2006 20:00 h to 12 August 2006 16:00 h. Highest water surface elevation reached up to 30.95 m at 10:00 h (orange color) on August 12th, inundating low-lying areas of Kherda, Rajupura near Poicha temple, and Bhanpura located between Anand and Vadodara cities near Vasad Village.

Time series plot of water surface elevation for entire simulation duration is presented in Fig. 8. Average water surface elevation lies between 20 and 31 m, and average terrain profile elevation of river is 13.22 m which indicates that most parts of river are inundated for this flood event.

Table 2 Critical bank stations under one-dimensional unsteady flow analysis

Flood event for unsteady flow	Left bank critical	Both right and left bank critical	Right bank critical
AUGUST 2006 Max Discharge of 31,061.91 cumec	CS-17, CS-20,CS-31, CS-48,CS-51, CS-52,CS-53	CS-1, CS-2, CS-4, CS-5, CS-7, CS-8, CS-9, CS-10, CS-11, CS-12, CS-13, CS-14, CS-15, CS-18, CS19, CS-21, CS-22, CS-23, CS-29, CS-30, CS-32, CS-33, CS-34, CS-35, CS-36, CS-37, CS-39, CS-40, CS-41, CS-46, CS-47, CS-48, CS-49, CS-50, CS-54, CS-55, CS-56, CS-57, CS-59, CS-60, CS-61, CS-62, CS-64, CS-66, CS-67, CS-68, CS-69, CS-70, CS-71, CS-73	CS-3, CS-6, CS-24, CS-25, CS-26, CS-28, CS-38, CS-42, CS-43, CS-44, CS-45, CS-58, CS-63, CS-65, CS-72

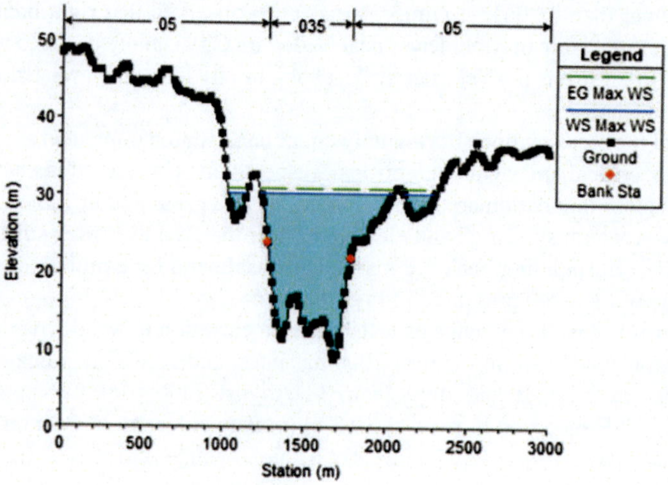

Fig. 5 Cross section of Khanpur gauge station

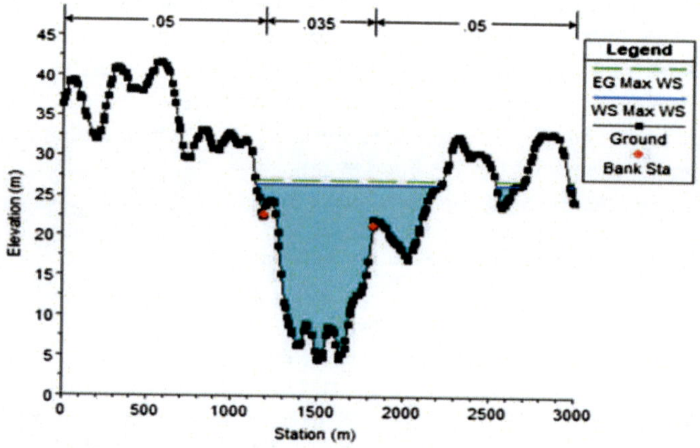

Fig. 6 Cross section of downstream side of Vasad Bridge

The results obtained can be verified through remote sensing [19] and geospatial techniques such as Sentinel 1 satellite imagery or MODIS satellite imagery [20, 21]. This study is limited to data collected through SWDC and CWC; however, one can verify actual cross section data through field visit and hydrologic data can be computed by rainfall–runoff analysis in HEC-HMS which converts precipitation excess into overland flow and channel runoff [22].

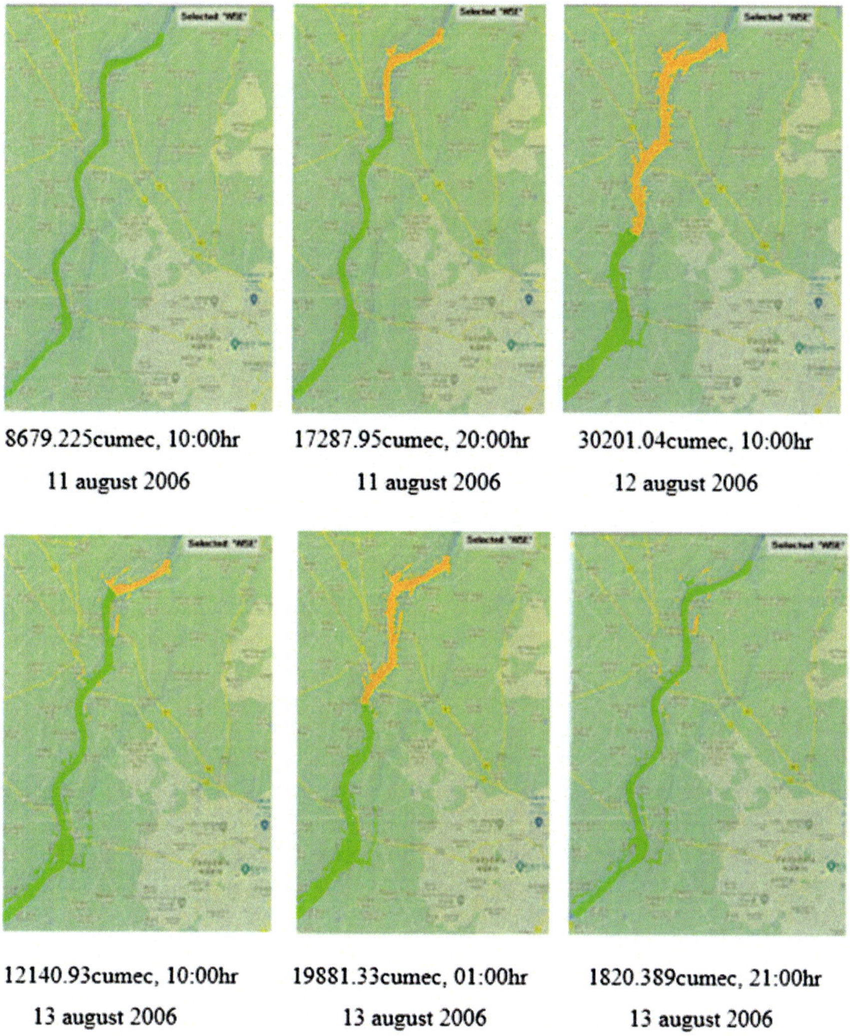

Fig. 7 Simulated flood water levels in Mahi River for flood event August 2006

6 Conclusions

The Mahi River channel of 37.4 km spread over upper reach of Khanpur gauge station located 139 km downstream of Kadana Dam at Mahisagar District of Gujarat to the Mujpur Village near mouth of Mahi River in Gulf of Khambhat and was modeled through HEC-RAS 6.0 version for unsteady flow condition in 1D and 2D. Model was calibrated and validated for flood event August 2006. Simulated results for 1D analysis show that 64.38% of total cross-sectional area are under inundation for

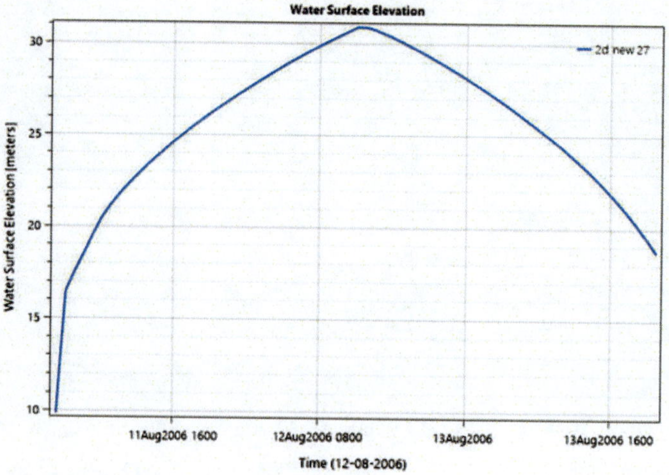

Fig. 8 Time series plot of WSE for August 2006 flood event

discharge equal or more than 31,061.91 cumec. Whereas 2D analysis results show that low-lying areas of Kherda, Rajupura, and Bhanpura located between Anand and Vadodara cities are identified as vulnerable zones for August 2006 event with the highest water surface elevation of 30.95 m. It has been observed that right bank of river indicating eastern side of Mahi River near Vadodara City is more prone to water spill for higher water levels in river. Considering the case of this research, it is concluded that unsteady analysis for 1D and 2D in HEC-RAS 6.0 can be satisfactorily used for flood vulnerability assessment in Mahi lower sub-basin. Based on this study, it is proposed that the capacity of all cross sections along river stretch should increase such that it can accommodate maximum discharge of 31,061.91 cumec without overtopping. River channel can be modified and guided along vulnerable area of river. Scope of the study includes sedimentation analysis in conjunction with unsteady flow analysis to better understand flow variation in river. This research work can be helpful in studying behavior of river and identifying vulnerable area under maximum flood event for government agencies to take preventive measures for sustainable development.

References

1. Mohit, P. M., Sahil, M., & Subhankar, K. (2020). Flood management in India: A focussed review on the current status and future challenges. *International Journal of Disaster Risk Reduction, 49.*
2. Das, T. (2014). River bank erosion induced human displacement and its consequences. *Living Reviews in Landscape Research, 8.*
3. Patrick, G. (2007). The effectiveness of early warning systems for the reduction of flood disasters. In *Some experiences from cyclone induced floods in Zimbabwe* (Vol. 9).

4. WHO. (2017). https://www.who.int/health-topics/floods
5. Ray, K., Pandey, P., Dimri, A. P., & Kishore, K. (2019). On the recent floods in India. *Current Science, 117*. https://doi.org/10.18520/cs/v117/i2/204-218
6. Farooq, M., Khattak, M. S., & Shafique, M. (2019). Flood hazard assessment and mapping of River Swat using HEC-RAS 2D model and high-resolution 12-m TanDEM-X DEM (WorldDEM). *Natural Hazards, 97*, 477–492.
7. Deepak, S., Rajan, G., & Jairaj, P. G. (2020). Geospatial approach for assessment of vulnerability to flood in local self governments. *Geoenvironment Disasters, 7*, 35.
8. Bahadar, I., Shafique, M., Khan, T., Tabassum, I., & Ali, M. (2015). Flood hazard assessment using hydro-dynamic model and GIS/RS tools: A case study of Babuzai-Kabal Tehsil Swat Basin. *Journal of Himalayan Earth Sciences, 48*(2), 129–138.
9. Khattak, M. S., Anwar, F., Saeed, T., Sharif, M., Sheraz, K., & Ahmed, A. (2016). Floodplain mapping using HEC-RAS and ArcGIS: A case study of Kabul River. *Arabian Journal for Science and Engineering, 41*, 1375–1390.
10. Masood, M., & Takeuchi, K. (2012). Assessment of flood hazard, vulnerability and risk of mid-eastern Dhaka using DEM and 1D hydrodynamic model. *Natural Hazards, 62*, 757–770.
11. Brunner, G. W. (2020). HEC-RAS users manual. CEIWR-HEC.
12. Brunner, G. W. (2020). HEC-RAS Hydraulic reference manual version 6.0 Beta. http://www.hec.usace.army.mil/
13. Patel, D. P., Ramirez, J. A., Srivastava, P. K., Bray, M., & Han, D. (2017). Assessment of flood inundation mapping of Surat city by coupled 1D/2D hydrodynamic modeling: A case application of the new HEC-RAS 5. *Natural Hazards, 89*, 93–130.
14. Brunner, G. W. (2020). HEC-RAS_6.0 2D modelling user manual, CEIWR-HEC, December 2020.
15. USACE. (2020). HEC-RAS river analysis system release notes. http://www.hec.usace.army.mil/, December 2020.
16. Anant, P. D., Dhruvesh, P. P., & Prakash, I. (2016). Flood modelling using Hec-Ras and geo-informatics technology in lower reaches of Shetrunji River, Gujarat, India. In *National Conference on Water Resources & Flood Management with special reference to Flood Modelling*, Surat, October 2016.
17. Geravand, F., Seiyed, M. H., & Behzad, A. A. (2020). Influence of river cross-section data resolution on flood inundation modeling: Case study of Kashkan river basin in western Iran. *Journal of Hydrology, 584*, 124743. ISSN 0022-1694.
18. G. o. India and Ministry of Water Resources. (2014). Mahi Basin Report, Version 2.0. www.india.wris.nrsc.gov.in, March 2014.
19. Sumit, D. (2019). Geospatial mapping of flood susceptibility and hydro-geomorphic response to the floods in Ulhas basin, India. *Remote Sensing Applications: Society and Environment, 14*, 60–74.
20. Ezzine, A., Saidi, S., Hermassi, T., Kammessi, I., Fadila, D., & Rajhi, H. (2020). Flood mapping using hydraulic modeling and Sentinel-1 image: Case study of Medjerda Basin, northern Tunisia. *The Egyptian Journal of Remote Sensing and Space Science, 23*(3), 303–310.
21. Nharo, T., Makurira, H., & Gumindoga, W. (2019). Mapping floods in the middle Zambezi Basin using earth observation and hydrological modeling techniques. *Physics and Chemistry of the Earth, 114*, 10278. ISSN 1474-7065.
22. Knebl, M. R., Yang, Z. L., Hutchison, K., & Maidment, D. R. (2005). Regional scale flood modeling using NEXRAD rainfall, GIS, and HEC-HMS/RAS: A case study for the San Antonio River Basin Summer 2002 storm event. *Journal of Environmental Management, 75*(4), 325–336, 2005.
23. Timbadiya, V., Patel, P. L., & Porey, P. D. (2011). Calibration of HEC-RAS model on prediction of flood for Lower Tapi River, India. *Journal of Water Resource and Protection, 3*(11).

Influence of Distinctive Parameters on Fundamental Time Period of the Building

Shubam Sharma and Aditya Kumar Tiwary

Abstract The first mode of vibration of a building is critical, and as per Indian Standard, it is a function of building height and base dimension only. The fundamental time period of the building is accountable for the recital of the building during seismic conditions as it is connected to the energy dissipation, acceleration, and fundamentally base shear of the building during ground motion acceleration. In this paper, an analytical study is contemplated to find out the unusual criterion that sequel the time period of the building. Parameters contemplate are story height, number of storys, seismic zone, type of soil, number of bays, size of column, and base dimension which are taken into consideration. The study is reckoned due to fact that the current code commentary is not ample to palliate all the variables which accomplish the fundamental time period of the building. The structure considered for investigation is a reinforced concrete frame without infill walls and the Indian Standards are followed for analysis. Modal analysis is executed over the different models by varying the different parameters which are taken into scrutiny for analysis, and the consequence of every single parameter is examined on the fundamental time period of the building. Analysis results revealed that seismic zone and soil type have no impact on the fundamental period of the structure. It was observed that the height of story, bay length, story number, and bay number are the chief factors which increase the fundamental time period by 159%, 132%, 119%, and 5.5%, respectively. But the size of column reduces the fundamental period by 14.5%.

Keywords Time period · Seismic zone · Story height · Response spectrum · Dynamic analysis · Modal analysis

S. Sharma (✉) · A. K. Tiwary
Department of Civil Engineering, Chandigarh University, Mohali 140413, India

A. K. Gupta et al. (eds.), *Advances in Construction Materials and Sustainable Environment*, Lecture Notes in Civil Engineering 196,
https://doi.org/10.1007/978-981-16-6557-8_57

1 Introduction

A reinforced concrete structure is a compound assemblage of elements, whichever structural or else non-structural. The non-structural elements, for instance, walls with infills, are associated with the structural assembly. Though several past studies [1–7] explored the effect of non-structural assemblies on the overall stiffness of the building, its strength, and damping ratio. Currently, the fundamental period of structures can be governed by an exact modal analysis eigenvalue analysis or Rayleigh's method, using the convenience of the fast-processing computer system.

The fundamental period governed by these approaches is typically slower compared with the period achieved by current code equations. Though the structural codes deliver empirical estimates based on the observed period of existing buildings during seismic activity, codal equations relate fundamental period with base dimension, type of building, and height of the building, but some critical parameters like the number of storys, seismic zone, type of soil, size of the column, and bays dimension need to be investigated. There is a dire need to perform an examination of the effect of respective parameters on the fundamental time period of the building as it directly influences the collapse condition or modes of the building. Minimal work was carried out in the literature [1–6, 8–12] regarding effect of mentioned parameter on fundamental time period. Also, the variation of fundamental period of the building reliably affects the acceleration, which is a significant parameter while designing tall structures.

In this study, the effect of non-structural components is ignored by succeeding the design philosophy, the strength of the structure is responsible for the building frame assembly namely beams and columns, and not by the non-structural members.

The software-based linear modal analysis was carried out in the study using ETABS [13], and the effect of certain parameters are examined. Parameters discussed in the study are adopted to achieve a close examination of how the structure responds when certain parameters are varied, such parameters include story height, number of storys, number of bays, bays dimension, seismic zone, soil type, size of the column.

As the effect of total building, height is incorporated in the IS 1893:2016 [14], but the effect of story height, size of the column, bay number, and length of each bay need to be carried out. The comparative study of variation of fundamental time period of the building due to addition of the number of bays and variation of bay length need to be examined to study what exactly effect the fundamental period.

Also, the story height and addition of the number of storys both increase the structure height, and thus, the fundamental period of the structure increases. The present code commentary states that the structural period depends upon the height of the building, and there is no information regarding the variation by the effect of column size, bay length, bay number, story height, and total number of storys.

Furthermore, the study also focused to investigate whether there is any variation occurs in the fundamental period of the building when the structure is constructed over different soil conditions and seismic zones.

2 Seismic Analysis

Thirty-six seismic models having a bare frame (base frame for comparison) Fig. 1 with the plan in Fig. 2 is studied under different parameters, and the influence of

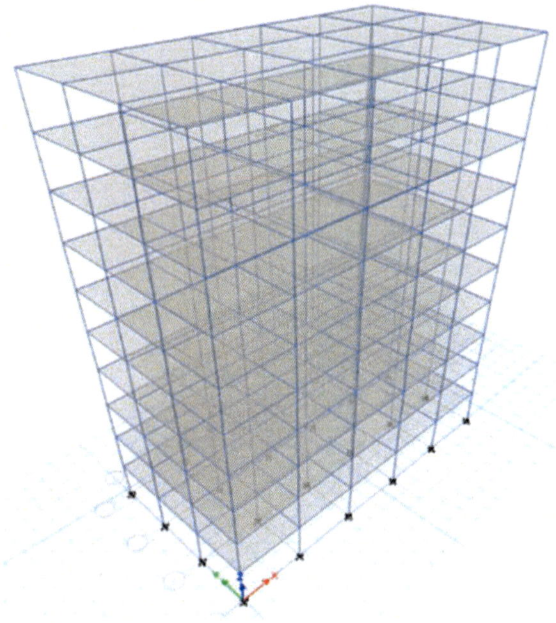

Fig. 1 Bare frame (base frame)

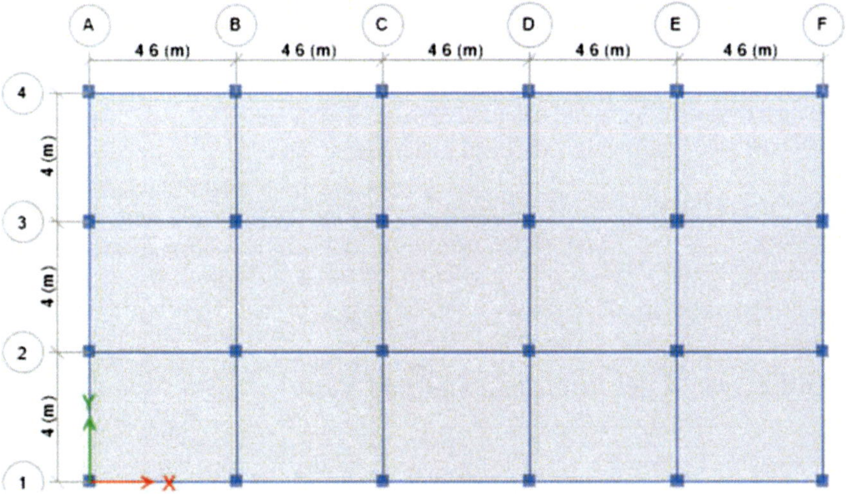

Fig. 2 Plan of the building

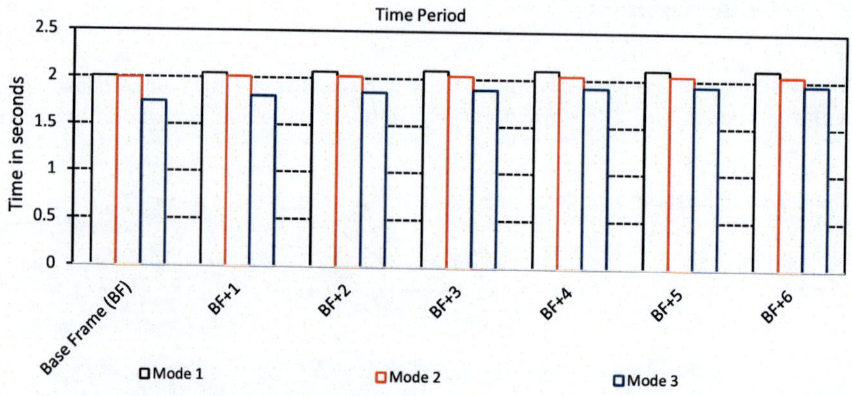

Fig. 3 Influence of number of bays on fundamental time period of the building

every single considered parameters on the fundamental time period of the building is examined. The seismic design adopted for analysis is as per IS 1893:2016 [14], and loading condition as per [15, 16] and the first three modes of modal analysis are examined as they are the critical mode of vibration. The base dimension of the benchmark model is 23 m * 12 m and having 5 bays in direction-X and 3 bays in direction-Y and having a story height of 3.2 m. The size of the column and beams conforming to IS 456:2000 [17] and IS 13920:2016 [18]. The fundamental mode of vibration shows a period of 2.00 s followed by the second and third modes of 1.99 s and 1.74 s.

3 Influence of Number of Bays

The consequence of the bay number on fundamental time period of the building is examined by providing an increment of one bay in each direction up to six bays, and the incremented bay length is identical to the base frame.

From Fig. 3, it is revealed that as the number of bays increased the fundamental time period of the building hiked. Contrasting the first mode of vibration, the incrementation of one bay increases the period by 1.84% followed by 2.95%, 3.89%, 4.48%, 4.98%, and 5.38%.

4 Influence of Seismic Zone and Soil Type

To investigate the impact of the seismic zone on the fundamental period of the building, four seismic zones, namely Zones-II, III, IV, and V, are considered for analysis. Parallelly, the soil type adopted for analysis as per IS 1893:2016 [14] as

Type-1 (Rocky), Type-2 (Medium), and Type-3 (Soft). The modal analysis results reveal that both seismic zone and soil type have no impact on the fundamental time period of the building as from the modal analysis, there is a variation of 0% observed for both the parameters and fundamental time period of the building remain constant as 2.00 s for the first mode, 1.99 s for the second mode, and 1.74 s for the third mode of vibration in all the cases which is similar to the base frame.

5 Influence of Bays Dimension

Examination of the effect of size of the bay on fundamental time period of the building is studied by providing an increment to the existing bay length by 20%, 40%, 60%, 80%, and 100% in both the direction. The influence of different bay lengths is shown in Fig. 4.

The first mode of vibration results indicates that incrementing 20%, 40%, 60%, 80%, and 100% bay length increases the fundamental time period by 22.13%, 46.11%, 72.23%, 100.74%, and 131.95%, respectively. From these results, it can be accomplished that the effect of bay length is substantial in terms of the fundamental time period of the building and is an important factor while designing tall structures.

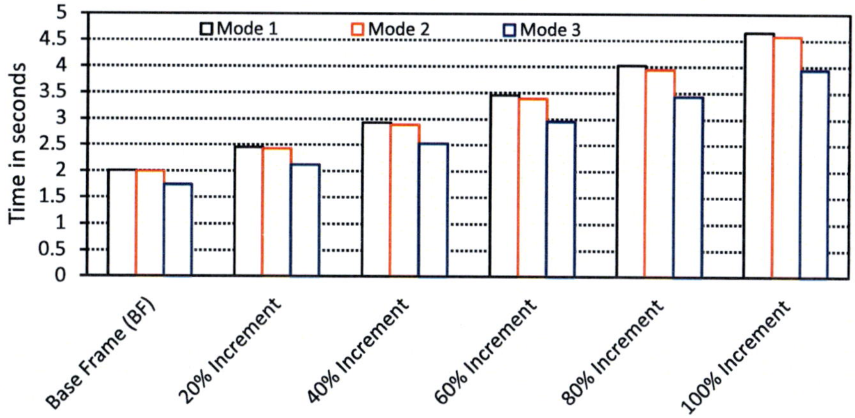

Fig. 4 Influence of the size of bays on fundamental time period of the building

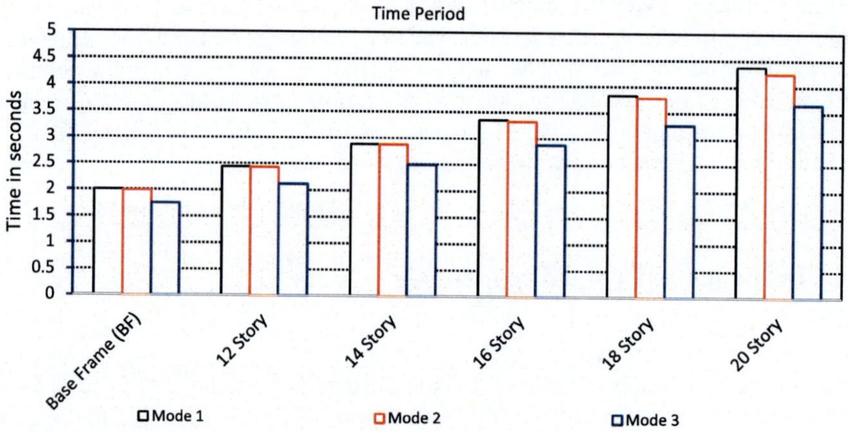

Fig. 5 Effect of distinctive storys on the fundamental period of building

6 Effect of Number of Storys

To reveal the effect of the number of storys on fundamental period of building, an iteration having incrementation of two storys to the base frame is adopted for analysis up to 20 storys. The analysis results are mentioned in Fig. 5.

Figure 5 directly indicates that as the number of storys augments the fundamental period of the building alters and a variation of 21.53%, 43.76%, 67.04%, 91.17%, and 118.34% is noticed. The variation noticed is significant, and it explicitly mentioned that the number of storys plays an imperative role in the modal response of the building as the incrementation in the fundamental period is exponential.

7 Influence of Height of Story

The effect of story height is investigated by providing an increment to the story height of base frame by 20%, 40%, 60%, 80%, and 100%. Figure 6 reveals that as

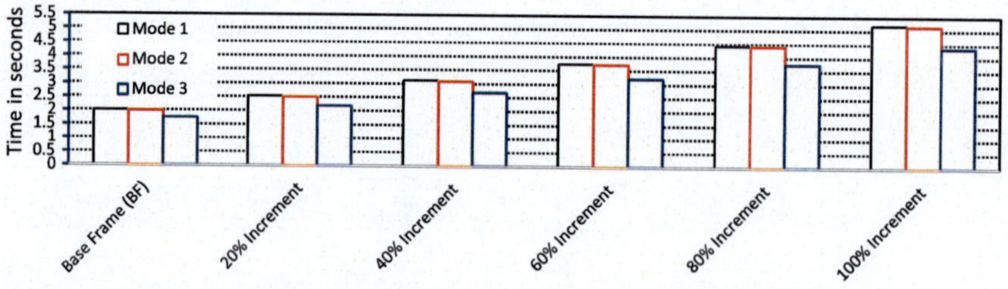

Fig. 6 Influence of story height on the fundamental time period of the building

each story is incremented by an iteration rate of 20% up to 100% results in 26.12%, 55.33%, 86.44%, 120.83%, and 158.72% intensification in the fundamental period of the building.

Before these results, another fact which revealed from the analysis is that on increasing the story height there is a negligible difference between the first and second modes of vibration and difference only noticed in the third mode of vibration, comparing these results with the effect by the number of bays, bays dimension, and height of the building, it is a different response which is significant and strongly recommended to be considered for design aspect.

8 Influence of the Size of Column

From seismic design philosophy, columns are designed to play the major role of ductility in the structure during ground motion acceleration. As the member of the building directly influence the overall performance of building under ground motion acceleration, and from IS 1893:2016, the stiffness modification factor of the column is 70% and for the beam is 35%. Thus, the effect of columns on the fundamental period of the structure needs to be assessed before the beam. The investigation is carried out by increasing the size of the column with an iteration of 20% up to 100% to the original dimension which is 400 mm * 400 mm. The response of the building is revealed in Fig. 7.

From Fig. 7, it is cleared that the increase in the size of the column decreases the fundamental period of the building. It can be explained in the manner as the size of the member increases, its stiffness increases, and thus, the fundamental period drops continuously with the incrementation. The rate of reduction in fundamental period noticed is −7.27%, −10.76%, −12.26%, −13.75%, and −14.75%, respectively.

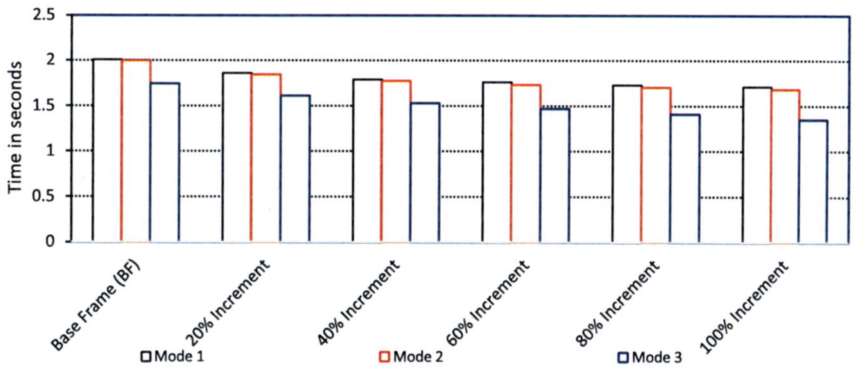

Fig. 7 Influence of the size of columns on the fundamental time period of the building

9 Results

The influence of numerous parameters is investigated comprehensively, and the effect of each parameter on the first mode of vibration or fundamental mode is shown in Table 1.

Contrasting Table 1, the chief factor which affects the fundamental period of the structure is the story height and then the bay length. Furthermore, the number of storys increases the fundamental period of the building, but if we contrast over the effect of size of the column, it reduces the fundamental time period of the building. Additionally, seismic zone and type of soil lay negligible effect on the fundamental period of the structure which is a positive indication from the design aspect as the ground motion acceleration has less time period in rocky soil compared to soft soil. Thus, it is revealed from the study that type of soil lay no effect on building response; therefore, the tall structure should recommend on rocky soil and small structures on soft soil or medium soil to prevent the state of resonance.

Another novel result report indicates that on comparing the vibrational mode 1, mode 2, and mode 3, it is revealed from Figs. 3, 4, 5, 6, and 7 the bay number, base dimension, story height, and the number of storys upsurges the fundamental period of the building but when noticed precisely height of the story is a most important parameter from all of these and novel result revealed that in case of increment in height of story the first and second modes of vibration becomes almost similar which helps the designer to make the structure translational irrespective of rotational in the first two-mode of vibration.

The size of the column decreases the fundamental period of the building due to an increase in stiffness of the structure, a decrease in the fundamental period results in a higher frequency to be transferred to the structure which is a critical parameter that causes the structure to collapse without any movement because of reduction in dissipation of energy. Thus, it is recommended not to decrease the fundamental period of the small building too much to avoid structural collapse.

Comparing the consequence of the bay number and bay length, it is revealed from the result that the number of bays has less impact on the time period of the building as compared to bay length. This behavior can be justified by the concept of building stiffness. Elaborating, as when we increase the bay length the increase in the stiffness of the structure be contingent on the size of the beam only which have a modification factor of 35% only as per IS 1893:2016, and thus, stiffness increases very less but when the number of bays increased the number of columns is increased in the structure which defines the stiffness varied mostly by columns and the columns have a stiffness modification factor of 70%, and thus, the stiffness increased is much higher as compared to the previous case.

Table 1 Effect of certain parameters on the fundamental period of the building

Parameter	Incrementation (period in seconds)						Story number (period in seconds)						Number of bays incremented (period in seconds)						
	0%	20%	40%	60%	80%	100%	10	12	14	16	18	20	0	1	2	3	4	5	6
Bay length	2.00	2.45	2.93	3.45	4.02	4.65	–	–	–	–	–	–	–	–	–	–	–	–	–
Bay number	–	–	–	–	–	–	–	–	–	–	–	–	2.00	2.04	2.06	2.08	2.09	2.10	2.11
Size of column	2.00	1.86	1.79	1.76	1.73	1.71	–	–	–	–	–	–	–	–	–	–	–	–	–
Story number	–	–	–	–	–	–	2.00	2.43	2.88	3.35	3.83	4.38	–	–	–	–	–	–	–
Height of story	2.00	2.53	3.11	4.43	3.74	5.19	–	–	–	–	–	–	–	–	–	–	–	–	–

10 Conclusion

The study intends to reveal the effect of distinctive parameters on the fundamental period of the building because the condition of resonance occurred when fundamental frequency matched with the frequency of ground motion acceleration. The effect of distinctive parameters is well-studied, and the following conclusion can be drawn:

- The current code commentary (IS 1893:2016) states that fundamental period is a function of building height and base dimension but from the results, it is indicated that base dimension and height play a significant role, but stiffness factor and height of story need to be considered for the analysis to avoid the state of resonance.
- The analysis results reflected that there is a need to revise the current code commentary and the significant parameters detected in this study need to be encompassed in further revisions.
- Comparative results reveal that increasing the base dimension without adding the number of columns (increasing the number of bays) has a significant impact on the time period of the building. Thus, for the small structures, it is recommended to avoid the addition of the number of columns and strongly endorsed to increase the bay length.
- As the height of story, bay length, story number, and bay number increase the fundamental time period by 159%, 132%, 119%, and 5.5%, respectively. But the size of column reduces the fundamental period by 14.5%. Thus, it is recommended to increase the size of column instead of addition of column for design purpose. Moreover, addition of column makes the structure more uneconomical instead of increase in size of column.
- It is found from the analysis that increasing the height of the story causes the first and second modes of vibration nearly the same, which results in the first and second modes of vibration to generate almost equal frequency which is a positive outcome for the regular structural engineer and designers as it helps them to make the structure response translational irrespective of rotational.
- There is no effect of the seismic zone as well type of soil is noticed on the fundamental period of the building which is significant because if seismic zone and kind of soil alters the fundamental period, then it is very challenging to study the real response during strong ground motion acceleration. As the time period of the ground motion acceleration is very less in rocky soil and vice versa for soft soil, therefore it is recommended to construct tall structures on rocky soil and small structure on soft soil because the fundamental period of the tall structure is high and vice versa for a small structure, and thus, the probability of the state of resonance will be minimized.
- It is recommended not to provide the large size of columns in small structures as it reduces the fundamental period of the building, and thus, the chances of failure will increase due to an increase in the probability of the state of resonance.
- Study strongly recommended that designers should not decrease fundamental time period of small buildings by extreme values as it results in predominancy of frequency, which results in collapse of building. Thus, small structures should

not be constructed with high stiffness. Similarly, for the tall structures the design approach is opposite of small structures. The tall structures must be designed with reasonable stiffness in order to save the structure from collapse due to drift and sway.

References

1. Powell, G. H. (2008). Displacement-based seismic design of structures. *Earthquake Spectra, 24*(2), 555–557.
2. Adham, S., Avanessian, V., Hart, G., Anderson, R., & Elmlinger, J. (1990). Shear wall resistance of lightgage steel stud wall systems. *Earthquake Spectra, 6*, 1–14.
3. Freeman, S. (1977). Racking tests of high-rise building partitions. *Journal of the Structural Division, 103*, 1673–1685.
4. Rihal, S. (1993). Research studies into seismic resistance and design of non-structural building components. Non-structural components. Design and detailing. SEAONC seminar proceedings.
5. Villaverde, R. (2006). Simple method to estimate the seismic nonlinear response of nonstructural components in buildings. *Engineering Structures, 28*, 1450–1461.
6. Cavaleri, L., & Papia, M. (2003). A new dynamic identification technique: Application to the evaluation of the equivalent strut for infilled frames. *Engineering Structures, 25*(7), 889–901.
7. Stewart, J. P., Chiou, S.-J., Bray, J. D., Graves, R. W., Somerville, P. G., & Abrahamson, N. A. (2002). Ground motion evaluation procedures for performance-based design. *Soil Dynamics and Earthquake Engineering, 22*(9–12), 765–772.
8. Takewaki, I. (2011). *Building control with passive dampers: Optimal performance-based design for earthquakes*. Wiley.
9. Wen, Y. (2001). Reliability and performance-based design. *Structural Safety, 23*(4), 407–428.
10. Ozkul, T. A., Kurtbeyoglu, A., Borekci, M., Zengin, B., & Kocak, A. (2019). Effect of shear wall on seismic performance of RC frame buildings. *Engineering Failure Analysis, 100*, 60–75. https://doi.org/10.1016/j.engfailanal.2019.02.032
11. Bush, T. D., Jones, E. A., & Jirsa, J. O. (1991). Behavior of RC frame strengthened using structural steel bracing. *Journal of the Structural Engineering (United States), 117*, 1115–1126. https://doi.org/10.1061/(ASCE)0733-9445(1991)117:4(1115)
12. Maheri, M. R., & Sahebi, A. (1997). Use of steel bracing in reinforced concrete frames. *Engineering Structures, 19*, 1018–1024. https://doi.org/10.1016/S0141-0296(97)00041-2
13. Computers and Structures CSI-ETABSv18. Integrated Solutions for building.
14. Bureau of Indian Standard (BIS). (2016). IS 1893:2016—Indian Standard Criteria for Earthquake Resistant Design of Structures , Part 1—General Provisions and Buildings. Bur Indian Stand 1893.
15. Bureau of Indian Standard (BIS). (2002). Code of Practice for Design loads (Other than Earthquake) for Buildings and Structures 875.
16. Code of Practice for Design Loads (Other Than Earthquake) for Buildings and Structures. (1997). Code of Practice For. Public Work 875.
17. Bureau of Indian Standard (BIS). (2016). *IS 456–2000 Plain and reinforced concrete- code of practice* (p. 2000). New Delhi.
18. Bureau of Indian Standard (BIS) (2016) IS 13920:2016 Ductile Design and Detailing of Reinforced Concrete Structures Subjected to Seismic Forces 13920.
19. Kadid, A., & Yahiaoui, D. (2011). Seismic assessment of braced RC frames. *Procedia Engineering, 14*, 2899–2905. https://doi.org/10.1016/j.proeng.2011.07.365
20. Jain, A. K. (1986). Seismic response of RC frames with steel braces. *Journal of the Structural Engineering, I*, 2138–2148.

21. Tahamouliroudsari, M., Entezari, A., & Hadidi, M. (2017). Experimental assessment of retrofitted RC frames with different steel braces. *Structures, 11*, 206–217. https://doi.org/10.1016/j.istruc.2017.06.003
22. Viswanath, K. G., Prakash, K. B., & Desai, A. (2010). Seismic analysis of steel braced reinforced concrete frames. *International Journal of Civil and Structural Engineering, 1*, 114–122.
23. Navya, G., & Agarwal, P. (2016). Seismic retrofitting of structures by steel bracings. *Procedia Engineering, 144*, 1364–1372. https://doi.org/10.1016/j.proeng.2016.05.166
24. Jose, A., & Pincheira, J. O. J. (1985). Seismic response of RC frames with steel brace. *Journal of the Structural Engineering (United States), 111*, 2138–2148. https://doi.org/10.1061/(ASCE)0733-9445(1985)111:10(2138)
25. Kose, M. M. (2009). Parameters affecting the fundamental period of RC buildings with infill walls. *Engineering Structures, 31*, 93–102. https://doi.org/10.1016/j.engstruct.2008.07.017
26. Hosseini, M., Hashemi, B., & Safi, Z. (2017). Seismic design evaluation of reinforced concrete buildings for near-source earthquakes by using nonlinear time history analyses. *Procedia Engineering, 199*, 176–181. https://doi.org/10.1016/j.proeng.2017.09.225
27. Buckle, I. G., & Mayes, R. L. (1990). Seismic isolation: History, application, and performance—A world view. *Earthquake Spectra, 6*(2), 161–201.
28. Priestley, M. (2000). Performance based seismic design. *Bulletin of the New Zealand Society for Earthquake Engineering, 33*(3), 325–346.
29. Higashino, M., & Okamoto, S. (2006). *Response control and seismic isolation of buildings*. Routledge.
30. Taniguchi, T., Der Kiureghian, A., & Melkumyan, M. (2008). Effect of tuned mass damper on displacement demand of base-isolated structures. *Engineering Structures, 30*(12), 3478–3488.
31. Sharma, S., & Tiwary, A. K. (2020). Analysis of a building under composite structural system: A review. *Materials Today: Proceedings, 33*, 1650–1659. https://doi.org/10.1016/j.matpr.2020.06.376
32. Whittaker, A., Constantinou, M., & Tsopelas, P. (1998). Displacement estimates for performance- based seismic design. *Journal of the Structural Engineering. American Society of Civil Engineers, 124*(8), 905–912.

Experimental Study of Fiber-Reinforced Concrete Prepared with Recycled Coarse Aggregate Bagasse Ash and Polypropylene Fiber

Harish Kumar and Aditya Kumar Tiwary

Abstract As the concern of environment leads to the management of waste, on the same hand problem of carbon dioxide is the reason behind the ozone layer depletion gives the researcher leads to the conclusion that the environmental waste can be easily utilized in the concrete with not much effect to its strength. This paper deals with the experimental properties of concrete which are prepared by recycled aggregates and bagasse ash in the replacement of coarse aggregate and fine aggregate and polypropylene fiber are added to strengthen the split tensile and flexural properties. Various ratios of recycled aggregates are used in the concrete, and optimum is found which is further used for the investigation of optimum ratio of replacement of fine aggregate with bagasse ash the addition of polypropylene fiber is done with different ratios. This replacement and addition of material give a 12% increment in the compressive strength of concrete, 22% increment in split tensile, and 15% in flexural strength of the concrete. Cost is decreased by 13.5% and water absorption is not disturbed by the addition of polypropylene fiber.

Keywords Composite concrete · Mechanical properties of concrete · Fiber · Compressive strength · Tensile strength

1 Introduction

Concrete is a very important matter in today's day-to-day life. An industrialization process cannot be possible without the use of concrete. As day-by-day industrialization is increasing its peak the demand for concrete also increasing day by day. In 2019–2020, the concrete demand was increased by 12%. This concrete produces many harmful gases like CO_2 and SO_2 to minimize this harmful environmental gas

H. Kumar (✉) · A. K. Tiwary
Department of Civil Engineering, Chandigarh University, Mohali, Punjab 140413, India

A. K. Tiwary
e-mail: Aditya.civil@cumail.in

A. K. Gupta et al. (eds.), *Advances in Construction Materials and Sustainable Environment*, Lecture Notes in Civil Engineering 196,
https://doi.org/10.1007/978-981-16-6557-8_58

production. So, researchers came up with the idea of utilizing the environmental waste in concrete, which reduces the risk of the formation of gases.

Bagasse ash: The most and widely consumed material is sugar. In 2019, it is reported that more than 300 million metric tons of sugar are being produced which leaves around 10 million metric tons of bagasse ash. Ganesan et al. [1] states that Jagadesh et al. [2] the density of this bagasse ash is too low that the only way to discard it is to dispose of it in the landfill which creates the land pollution problem. We can easily utilize this ash in concrete because it has pozzolanic action. Modani and Vyawahare [3] studies about sugarcane bagasse ash and stated that due to the pozzolanic action of SCBA, the rate of gain of strength is higher in the period of 28 days when compared to the 7-day rate. The purest form of bagasse ash can replace the fine aggregate by up to 30%. Inbasekar et al. [4] studies about the behavior of bagasse ash and steel fibers in reinforced cement concrete. The bagasse ash can be utilized in two ways (1) in the wet state and (2) in the wet state. If we use dry bagasse ash then the workability decreases when the percentage of replacement increases but when we use wet bagasse ash then the workability increases because of the water content of the ash.

Recycled aggregate: The demand for industrialization also makes an increment in the demand for fine aggregate and coarse aggregate. Qureshi et al. [5] studies the effect of replacement of coarse aggregate by recycled coarse aggregate. The reduction of cost is Rs. 130 but there is a decrease in the flexural strength of the specimen. Ravi et al. [6] states that the full replacement of coarse aggregate is possible in certain conditions of admixtures and water content. Yadav et al. [7] states about the use of recycled aggregates in the high strength of concrete which gives 20% replacement of coarse aggregate.

Polypropylene fiber: Polypropylene fiber is the by-product of the petroleum industry which can be converted into stereoregular polymer which has industrial values polypropylene fibers were first proposed as an admixture to concrete in 1965 for the development of resistant structures for the US Corps of Engineers. The fiber has along these lines been improved further, and at present, it is utilized either as short discontinuous fibrillated material for the creation of fiber strengthened concrete. Polypropylene fibers are new-age synthetic fibers. These have the fourth-largest volume after polyester, acrylics, and polyamides. Around 4 million tons of polypropylene fiber are production which is done on earth every year. The advantages of using polypropylene fibers are they have a high melting point (165 °C). They do not react with water because their water requirement is zero, increase the toughness and durability of concrete help in reducing drying shrinkage and improve concrete tensile strength [8–10]. [11] states that the use of polypropylene fiber used in the concrete should be restricted to the use of 0.2% by the weight of cement because of low density, the futuristic properties of concrete.

2 Methodology

For this experimental study, the concrete grade of M35 is being prepared with mixed proportions from is 10262:2019. The materials and their various physical properties are discussed in Table 1.

The specific gravity of materials like cement and bagasse ash is determined by specific gravity bottle test as per IS2720 part 3 and for RCA, CA and FA IS2686-1983 are used. For polypropylene fiber, the references of supplier are taken. The ratio of M35 grade concrete is taken as 1:1.6:2.9 with a water–cement ratio of 0.45. Further, this study is being divided into three parts:

Part 1 Replacement of coarse aggregate with recycled coarse aggregate.
Part 2 Replacement of fine aggregate with bagasse ash.
Part 3 Addition of polypropylene fiber.

The proportions of the replacement and addition are given in Table 2.

For this study cubes (150 × 150 × 150), beam (100 × 100 × 500) and columns (150 × 300) are prepared (Figs. 1 and 2). This study is done to study the mechanical properties like compression, split tensile, flexural strength, water absorption, and cost analysis of concrete.

The cost analysis of concrete is done in reference to the price taken from the Schedule of Rates by the Public Works Department by Punjab Government except for the price of Cement and PPF, these prices are taken from the market. The prices of the material are taken from the market suppliers Guru Building and Engineering Material Kharar Mohali, Punjab, which are given in Table 3.

Table 1 Materials and their properties

Materials	Properties	Result
Cement (OPC43)	Fineness	97%
	Specific gravity	3.15
	Setting time	29 min and 598 min
Fine aggregate	Fineness modulus	3.12
	Specific gravity	2.67
Coarse aggregate	Fineness modulus	8.21
	Specific gravity	2.6
Bagasse ash (sugar factory waste)	Fineness	99%
	Specific gravity	1.92
Recycled coarse aggregate (C&D waste)	Specific gravity	3.4
Polypropylene fiber	Specific gravity	1.81
	Sample length	25 mm

Table 2 Proportions of additives

Coarse aggregate replacement with recycled coarse aggregate in % weight of coarse aggregate	Fine aggregate replacement with bagasse ash in % weight of fine aggregate	Addition of polypropylene fiber in % weight of cement
0	5	0.1
10	10	0.2
20	15	0.3
30	20	0.4
40	25	0.5
50	30	0.6

Fig. 1 Casted specimen

3 Results and Discussion

When recycled aggregates are introduced in the concrete experimentation process, RCA increases the water demand of the concrete which decreases the workability of concrete due to the presence of fine particulates on the surface of aggregate it showed some pozzolanic action and help in changing the strength of concrete (Table 4).

Fig. 2 Testing of specimen

Table 3 Prices of material

Material	Price
Cement	8 Rs per Kg
Fine aggregate	3 Rs per Kg
Coarse aggregate	1.5 Rs per Kg
Polypropylene fiber	2000 per Kg

Due to fine particles, compressive and split tensile strength increase up to the replacement for 20% after that workability starts decreasing RCA usage, not negligible affect the flexural strength. 20% replacement gives a rise of 6.19% and 5.5% in

Table 4 Result for replacement of CA with RCA

Ratios	Compressive strength	Split tensile strength	Flexural strength
CA	N/mm^2	N/mm^2	N/mm^2
0%	35.60	3.38	4.18
10%	36.20	3.44	4.21
20%	37.80	3.59	4.30
30%	36.70	3.49	4.24
40%	35.20	3.34	4.15
50%	33.60	3.19	4.06

compressive and split tensile strength as shown in Figs. 3 and 4, whereas for flexure, it is only 3% as shown in Fig. 5.

The effect on compressive strength, split tensile strength, and flexural strength by replacement of Fa with SCBA is summarized in Table 5 (Fig. 6).

The replacement of FA with SCBA gives a good increase in the flexural and split tensile because the particles of bagasse ash are much finer than cement which fills the voids of cement gel and contributes to increase the strength. But after some limit, i.e., 20%, it goes on decreasing because there is lack of cement action as shown in Figs. 7 and 8.

Addition of polypropylene fiber does not affect the workability of concrete but the effect is seen in the flexural and split tensile strength of concrete due to the presence of fiber it acts as reinforcements so it increases the flexural and split tensile strength of concrete. The negligible but negative effect is seen in the concrete because the

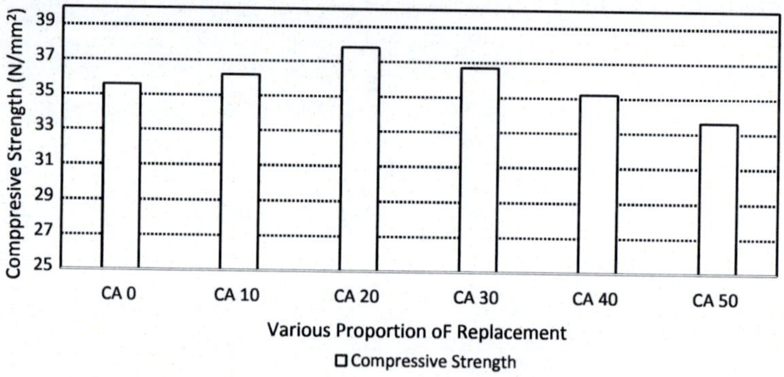

Fig. 3 Compressive strength result for the replacement of CA with RCA

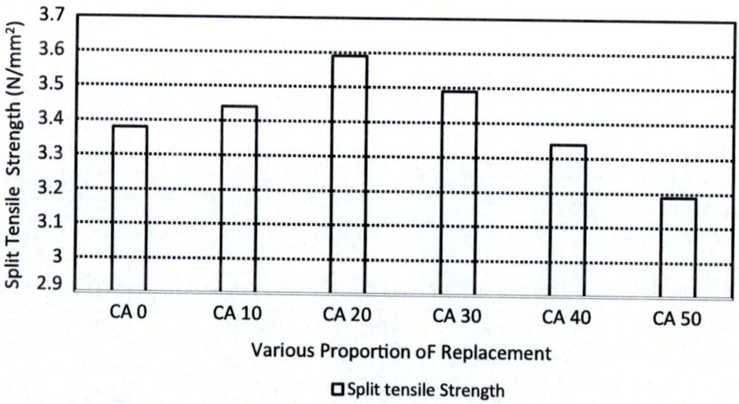

Fig. 4 Split tensile strength result for the replacement of CA with RCA

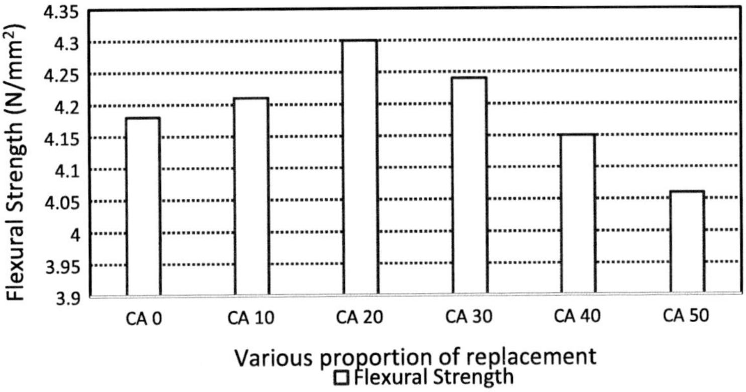

Fig. 5 Flexural strength result for the replacement of CA with RCA

Table 5 Result for the replacement of FA with SCBA

Ratios		Compressive strength	Split tensile strength	Flexural strength
CA (%)	FA (%)	N/mm^2	N/mm^2	N/mm^2
20	5	38.20	3.63	4.33
20	10	38.60	3.67	4.35
20	15	39.20	3.72	4.38
20	20	39.65	3.77	4.41
20	25	38.40	3.65	4.34
20	30	37.20	3.53	4.27

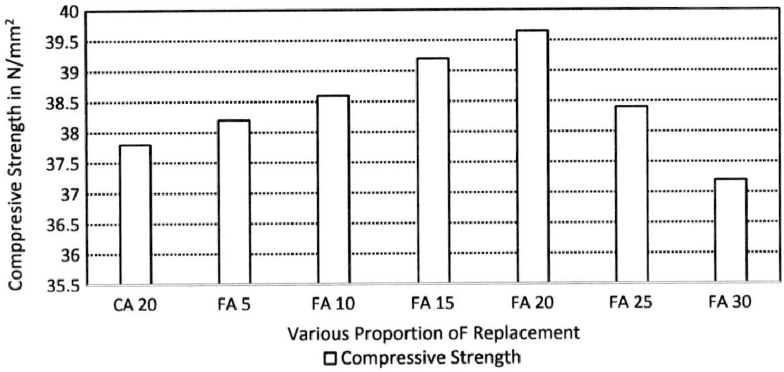

Fig. 6 Compressive strength result for the replacement of FA with SCBA

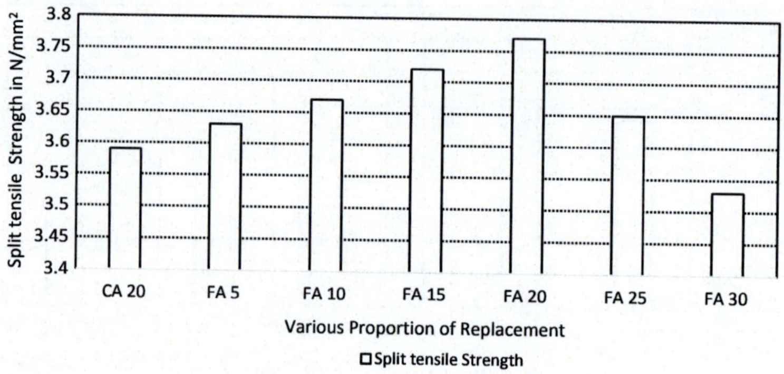

Fig. 7 Split tensile strength result for the replacement of FA with SCBA

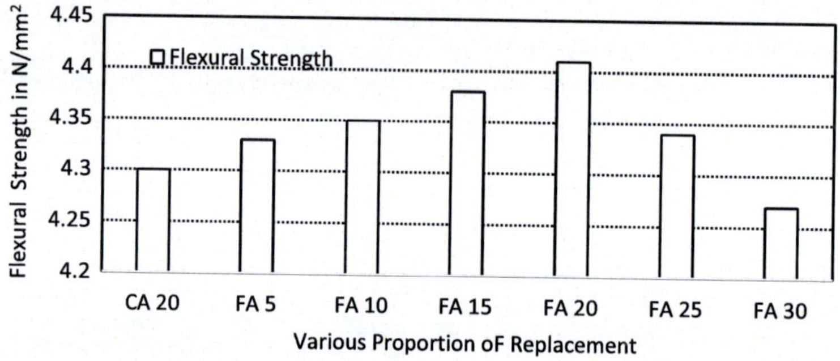

Fig. 8 Flexural strength result for replacement of FA with SCBA

low density of fibers increases the volume of concrete which reduces the concrete content and also reduces the load-bearing capacity of concrete.

The variation of compressive strength, split tensile strength, and flexural strength with varying percentage of CA, FA, and PPF is given in Table 6.

Table 6 Result for the addition of PPF

Ratios			Compressive strength	Split tensile strength	Flexural strength
CA (%)	FA (%)	PPF (%)	N/mm^2	N/mm^2	N/mm^2
20	20	0.1	39.46	3.70	4.50
20	20	0.2	39.25	3.90	4.90
20	20	0.3	39.09	4.20	5.20
20	20	0.4	38.8	4.50	5.60
20	20	0.5	38.3	4.7	5.7
20	20	0.6	38.3	5	5.9

Fibers do not affect the compressive strength of concrete directly but addition of fiber can result in the lack of cement gel in the concrete so fibers are used only up to the specific limit the other hand, it is recommended polypropylene fiber should not be used more than 100gm per bag of cement because fiber undergoes shrinkage phenomenon due to the heat of hydration of cement and external heat in contact. The strength goes on increasing as the fiber in concrete increases, but the effect will be seen at later stages (Figs. 9, 10, and 11).

Polypropylene fibers are by-product of petroleum industries which are type of waste which can be used to gain the tensile strength of concrete because the concrete is weak in tension using fiber to an extent can help concrete to be good in tension properties. In overall experimentation process, the compressive strength of concrete is due to increase in pozzolanic action by bagasse ash and additional silicious action provided by attached residue on the RCA which helps in the incrementation of the compressive strength of concrete.

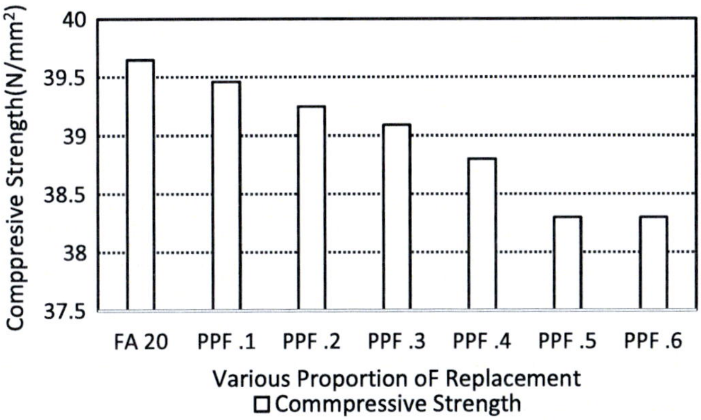

Fig. 9 Compressive strength result for addition of PPF

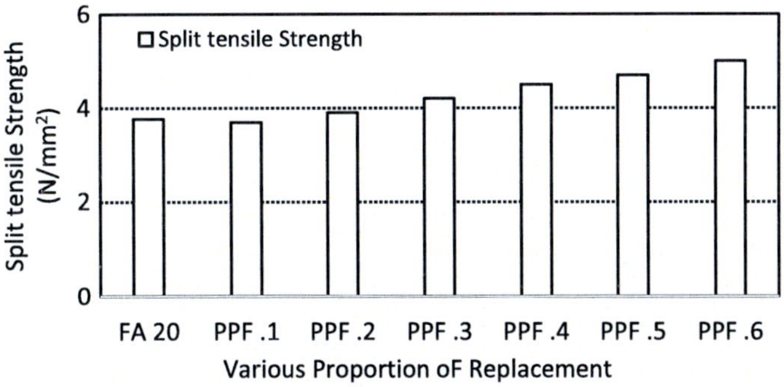

Fig. 10 Split tensile strength result for the addition of PPF

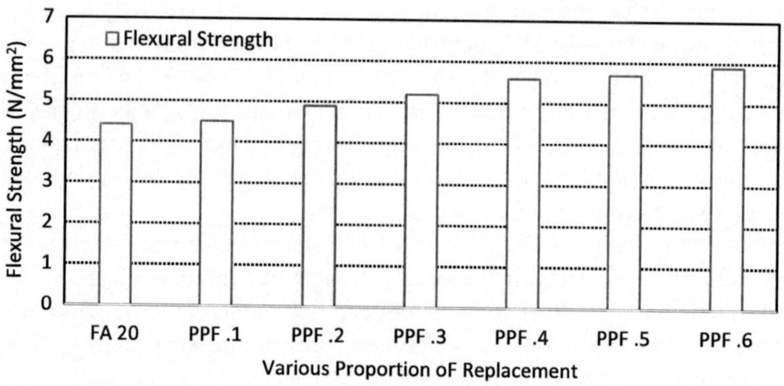

Fig. 11 Flexural strength result for addition of PPF

Table 7 Cost analysis of proposed concrete

S. No.	Concrete	Price (Rs.)	Comparison in cost (with conventional)	Water absorption (%)
1	Conventional	28.8	0	8
2	Modified (20% RCA)	20.5	10.2% decrease	11
3	Modified (20%RCA + 20%BA)	18.2	21.3% decrease	15
4	Modified (20%RCA + 20%BA + 0.2%PPF)	26.01	13.5% decrease	15

Cost Analysis and Water absorption: There is a cost of concrete decreases when the main constituents are replaced by the additive like ash and recycled aggregate but increase in cost due to polypropylene fiber because RCA and bagasse ash are a by-product of the industries. The price variation of concrete per cube of dimension 150 mm × 150 mm × 150 mm is given in Table 7.

The water absorption of the concrete is taken as the percentage of water absorbed by concrete in 24 h at room temperature after the curing period of 28 days.

4 Conclusions

Coarse aggregates in the concrete grade of M35 can be replaced by 20%, by giving an increment of 6% in compressive strength. Bagasse ash in concrete can be added in the concrete by 20% whose pozzolanic action increases the split tensile and compressive strength by approx. 3%. Using bagasse ash in the concrete specimen by 20% reduces the heat of hydration which results in less water demand for curing. Polypropylene fiber can be added in concrete up to 5% with no change in properties of concrete but due to the deformation under heat and pressure, it is only be used as 100 gm per bag of cement, i.e., 0.2 by weight of cement. The cost of modified concrete is 13% more than the cost of conventional concrete due to polypropylene fiber, and the water absorption does not change when polypropylene fiber is added to the concrete.

References

1. Ganesan, K., Rajagopal, K., & Thangavel, K. (2007). Evaluation of bagasse ash as supplementary cementitious material. *Cement and Concrete Composites, 29*(6), 515–524. https://doi.org/10.1016/j.cemconcomp.2007.03.001
2. Jagadesh, P., Ramachandramurthy, A., & Murugesan, R. (2018). Overview on properties of sugarcane bagasse ash (SCBA) as Pozzolan. *Indian Journal Geo-Marine Sciences, 47*(October), 1934–1945.
3. Modani, P. O., & Vyawahare, M. R. (2013). Utilization of bagasse ash as a partial replacement of fine aggregate in concrete. *Procedia Engineering, 51*(NUiCONE 2012), 25–29, 2013. https://doi.org/10.1016/j.proeng.2013.01.007
4. Inbasekar, M., Hariprasath, P., & Senthilkumar, D. (2016). Study on potential utilization of sugarcane bagasse ash in steel fiber reinforced concrete. *International Journal of Engineering Sciences & Research Technology, 5*(4), 43–50. https://doi.org/10.5281/zenodo.48824
5. Qureshi, L. A., Munir, M. J., Mubeen, T., & Tasaddiq, R. M. (2016). Effect of using recycled concrete as coarse aggregate on tensile and flexural strength of concrete. In *Fourth International Conference on Sustainable Construction Materials and Technologies.*
6. Patel, R., Vyas, C. M., & Bhatt, D. R. (2013). Experimental investigation for recycled coarse aggregate. *International Journal of Civil, Structural, Environmental and Infrastructural Engineering Research and Development, 3*(2), 35–42.
7. Yadav, A. K., Prasad, B., Kumar, V., & Mohan, M. (2017). Properties of polypropylene fiber reinforced concrete and structural behaviour of beam. *International Journal Innovative Research Science Engineering Technology, 6*(6), 10200–10210. https://doi.org/10.15680/IJIRSET.2017.0606011
8. Danso, H., Martinson, D. B., Ali, M., & Williams, J. B., Physical , Mechanical and Durability Properties of Soil Building Blocks Reinforced with Natural Fibres.
9. Prasad, J. R., Anuradha, V., & Gridharan, P. (2016). Experimental investigation on fibre reinforced concrete using waste materials. *International Journal of Engineering Trends and Technology (IJETT), 34*(3), 100–105. https://doi.org/10.14445/22315381/ijett-v34p221
10. Hanumesh, B., Harish, B., & Venkata Ramana, N. (2018). Influence of polypropylene fibres on recycled aggregate concrete. *Materials Today: Proceedings, 5*(1), 1147–1155. https://doi.org/10.1016/j.matpr.2017.11.195
11. Eidan, J., Rasoolan, I., Rezaeian, A., & Poorveis, D. (2019). Residual mechanical properties of polypropylene fiber-reinforced concrete after heating. *Construction and Building Materials, 198*, 195–206. https://doi.org/10.1016/j.conbuildmat.2018.11.209

Effect of Shear Walls on Tall Buildings with Different Corner Configuration Subjected to Wind Loads

Saransh Mahajan, Vikramaditya Yadav, Rahul Raj, and Ritu Raj

Abstract Many structures which are constructed at present times tend to be wind-sensitive because of their shape, slenderness, flexibility, and size. This leads to demand for more economical wind resistant designs of tall buildings. For the stability of high-rise structure, lateral forces like wind forces and earthquake forces play a critical role; however, in case of high-rise structure, wind loads are more critical than earthquake forces. Lateral load resisting systems like shear walls are used in construction to resist the effect of lateral forces on the building. This paper is focused on studying the effect of shear walls on the deflection of a tall building exposed to the wind load in lateral direction and observing the variance in the deflection that will occur on the three columns (windward, leeward, and central) of a high-rise building with and without using the shear walls in different scenarios using Bentley STAAD Pro software. Prototype buildings are assumed to be made up of RCC beams and columns that are designed on a software constitutes of 36 story (G + 35) having a height of 3.5 m of each floors, and response analysis of these prototype buildings has been done for different columns of building exposed to 0° wind incidence.

Keywords STAAD pro · Tall building · Wind loads · Square-shaped structure · Square-shaped structure with triangular corners · Square-shaped structure with recessed corners · Shear walls

1 Introduction

Since the ancient times, humans always admired tall structures although there is a growing need in present times due to exponential growth in the population for the construction of high-rise building and expansion in horizontal direction is not more a practical solution specially in the urban areas. Mainly, the construction of high-rise

S. Mahajan (✉) · V. Yadav · R. Raj · R. Raj
Delhi Technological University, New Delhi, India

R. Raj
e-mail: rituraj@dtu.ac.in

A. K. Gupta et al. (eds.), *Advances in Construction Materials and Sustainable Environment*, Lecture Notes in Civil Engineering 196,
https://doi.org/10.1007/978-981-16-6557-8_59

building is focused on four criteria mainly human comfort, stability, strength, and serviceability. So, there is a need for the use of shear walls in the tall building to provide resistance against lateral loads like wind and earthquake loads. RCC shear walls starts from the foundation and extends up to the roof level of the structure and are generally provided along the length and width of a building. In the designing of a tall structure, the use of shear walls is to provide strength and increase the stiffness of structure which will ultimately curb the damage occurring to the structure due to lateral loads by reducing the lateral sway of the building; shear walls are provided in a structure in symmetrical manner for distributing the inelastic deformation in a uniform way over the plan of a building rather than concentrating on particular walls.

Although the wind loads are estimated using the standard code of practice that are being used around the world, but the information provided is very minimal and given only for standard shapes like square plan or rectangular plan building exposed to wind. Various literature and researchers have come up with the analysis and effect of shear walls on the deflection due to wind loads subjected to a high-rise structure.

Various standard codes of practice for the design of building and estimation of design loads such as wind load, imposed load, and dead load are used across the world for a RCC building although these codes only provide minimal information for the design of high-rise structure that is only for standard and conventional shapes like square plan or rectangular plan, but these information are not exhaustive in nature [1–8]. Murali Krishna and Dr. Arunakanthi in 2014 used Etabs software to perform analysis for various different locations of shear walls in an unsymmetrical high-rise building in order to find the location which is effective, efficient, and ideal for shear walls by comparing results on various parameters like bending moment, torsion, and shear force [9]. Esmaili et al. in 2008 conducted study on one of tallest RCC building having 56 storys which was located in high seismic zone in which the shear wall system was used under lateral and gravity loads under irregular openings and the behavior of structural elements were recorded [10]. Amin and Ahuja conducted experiments on wind tunnel models of "L" shaped and "T" shaped building for calculating wind pressure distributions of different faces, and these experiments were conducted on two buildings of same plan area and height but having different cross-sectional shapes with wind direction varying from 0° to 180° at an interval of 150° [11]. Kwok, in 1988, conducted wind tunnel model tests in order to analyze the effect of shape of building on the wind-induced responses on a tall building having rectangular cross section which were performed in a long-wind and cross-wind directions with the help of chamfered, slotted, and horizontal corners [12]. Mukherjee et al., conducted a case study using k-ε method for N plan shape high-rise building for the determination of wind pressure coefficients for wind angle of incidence varying from 0° to 180° having an interval of 30° [13]. Sheng et al., investigate unsteady characteristics of a wind, and its complete information is like profile, intensity, fluctuation for the global, and local wind loads for a high-rise building structure using wind tunnel test at a scale of 1:300 for a tall building with a well-defined atmospheric layer [14]. Ahlawat and Ahuja, in 2015, performed experiments in order to investigate wind loads generated in interference and isolated condition, and the experiments were performed on models of tall buildings having "Y" plan shape in open-circuit wind tunnel [15]. Chakraborty et al., in the study carried out wind tunnel experiment for a "+" shaped tall building where the surfaces are exposed to wind angle for 0° and 45° and mean wind pressure coefficient was calculated using [16]. Nagar et al., in the paper studied the wind forces on unconventional plan shaped tall building although wind tunnel test also performed for "H" plan and square plan shaped tall building and for these models interference factors are calculated [17]. Yuan-Lung Lo et al., in 2019, performed experiments to find out the wind forced acting on two principal buildings with the aim to measure overturning moments in both cross-wind and along-wind by using dummy building located very closely to the principal building by using high-frequency balancer [18].

2 Methodology

The analysis of buildings is performed with the help of STAAD Pro, a general purpose civil engineering software which is widely used in order to model, design, and analyze various types of structures like bridges, buildings, towers, transportation, industrial, and utility structures. It helps to simplify and increase the efficiency of the engineering process.

2.1 Wind Flow

Design wind Pressure: The speed of wind increases with height and is approximately zero at ground level and increases to maximum in the atmospheric boundary layer at a height called the gradient height as given by Eq. 1. Above the mean ground level, the wind pressure can be calculated at any height in the structure by using the relationship between wind pressure and the wind speed.

$$P_z = 0.6 \times C_p \times V_z^2 \tag{1}$$

where P_z, V_z, and C_p are wind pressure (N/m^2), design wind speed (m/s) at a height z above the mean level, and pressure coefficient, respectively.

Design wind speed,

$$V_z = V_b * k_1 * k_2 * k_3 * k_4 \tag{2}$$

where V_b is the basic wind speed (m/s) given by Eq. 2, varies with zones of the country, and k_1, k_2, k_3, and k_4 are risk coefficient, terrain roughness and height factor, topography factor, and importance factor for the cyclonic region, respectively.

Force coefficient method: On any specific building, the wind load is calculated as per the formula given in Eq. 3.

$$F = C_f \times A_e \times p_z \quad (3)$$

where C_f, A_e, and p_z are force coefficient for the building, effective area normal to wind direction, and design pressure of the wind, respectively.

Here, three types of prototype buildings are constructed which are having same base area as well as same height, but the cross section varies due to different corner configurations. In all the buildings, shear walls are applied in two ways, and their effects are analyzed.

2.2 *Shear Walls*

Shear walls are vertical elements which are constructed to resist in-plane forces mostly wind and seismic loads and are hence constructed in areas with high wind and seismic activity. Shear walls are large structures that are usually built-in high-rise structures that are large. For construction of shear walls, materials such as concrete and masonry are used.

Shear walls case-1 In this case of shear wall, the shear walls are applied on the corners of the building such that they cover the start and end portion of all the external faces of the building forming L-shapes at corners when seen in the plan view.

Shear walls case-2 In this case of shear wall, the shear walls are applied on the central columns of the building such that it forms a square when seen in the plan view.

2.3 *Details of Prototype Buildings*

2.3.1 Types of Buildings

1. Square-shaped structure (Fig. 1)
2. Square-shaped structure with triangular corners (Fig. 2)
3. Square-shaped structure with recessed corners (Fig. 3)

2.3.2 Building Parameters

- Height = 126 m
- No. of floors = 36 (G+35)
- Concrete grade = M-25
- Steel reinforce = Fe-415
- Basic wind speed (Delhi) = 47 m/s
- Live load considered = 2.0 kN/m^2

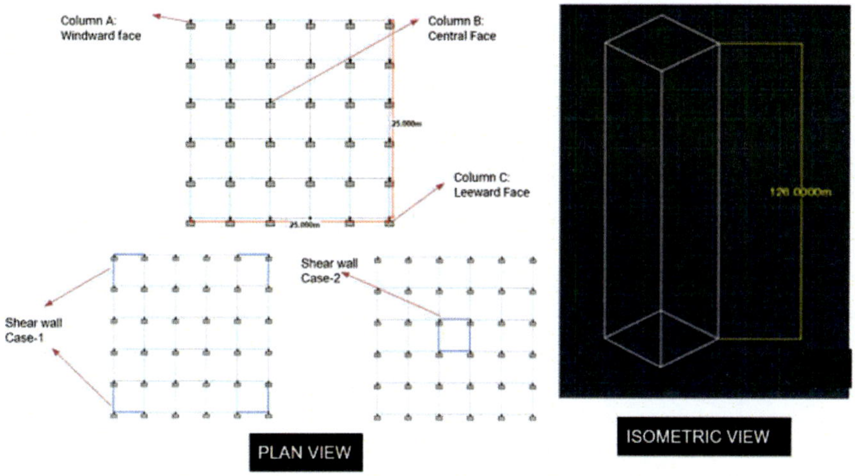

Fig. 1 Plan view of building, shear wall case-1, shear wall case-2, and isometric view of square-shaped building

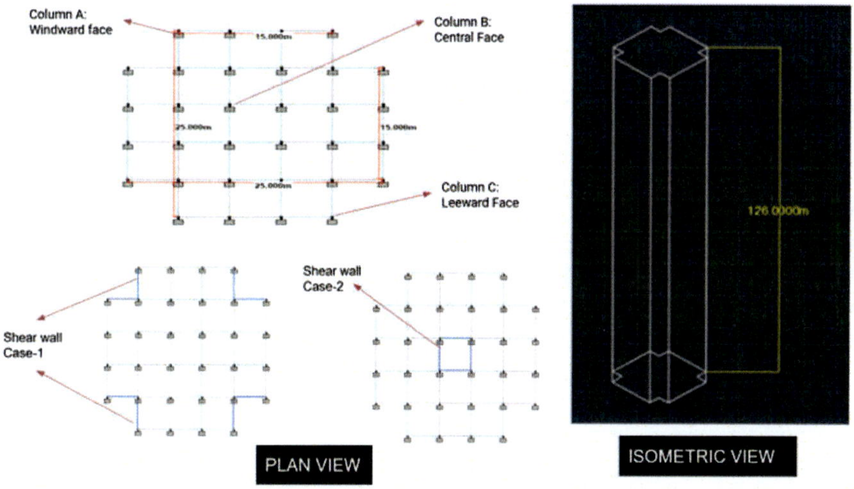

Fig. 2 Plan view of building, shear wall case-1, shear wall case-2, and isometric view of square-shaped building with triangular edges

- Floor finish = 1.0 kN/m^2
- Beam dimension = 450 × 450 mm
- Column dimension = 550 × 450 mm
- Main wall thickness = 230 mm
- Partition wall thickness = 150 mm
- Height of parapet wall = 1 m

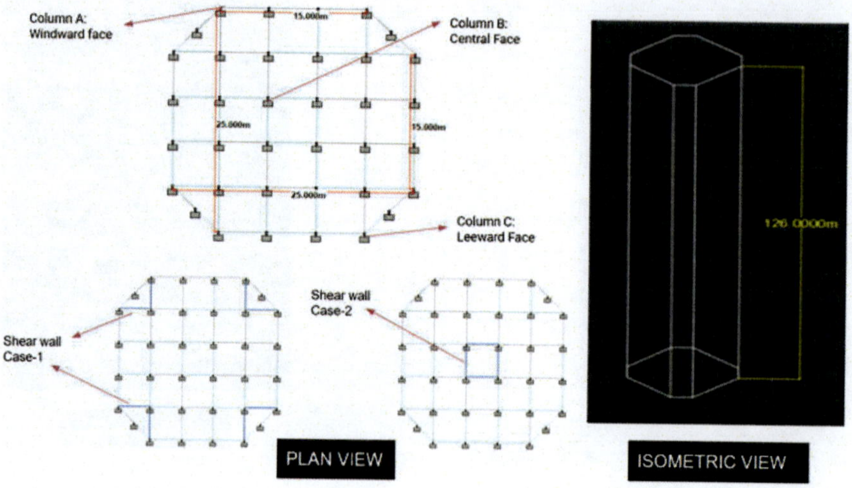

Fig. 3 Plan view of building, shear wall case-1, shear wall case-2, and isometric view of square-shaped building with recessed corners

- Slab thickness = 150 mm.

2.3.3 Loading Parameters (as per IS: 875 (Part I)—1987 [12], IS: 875 (Part II)—1987 [13])

- Self-weight = automatically generated
- Main wall load = 14.03 kN/m
- Partition wall load = 6.1 kN/m
- Parapet wall load = 4.6 kN/m
- Floor finish = 1 kN/m^2
- Floor live load = 2 kN/m^2.

3 Result and Discussion

3.1 *Square-Shaped Building*

The values of deflection increase as the height increases in all the three columns which are shown in Figs. 4, 5, and 6. When shear walls are applied, the values of deflection are found to reduce. Shear walls applied in first case gives better results than the shear walls applied in second case as they reduce deflection value by about 45%, while shear wall case-2 reduces the value by 34%. All the faces show similar

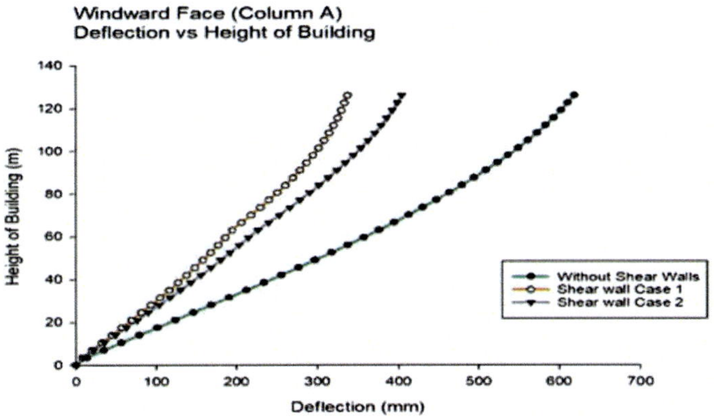

Fig. 4 Effect of shear walls on deflection at 0° incidence angle of wind on windward face (Column A) of square-shaped building

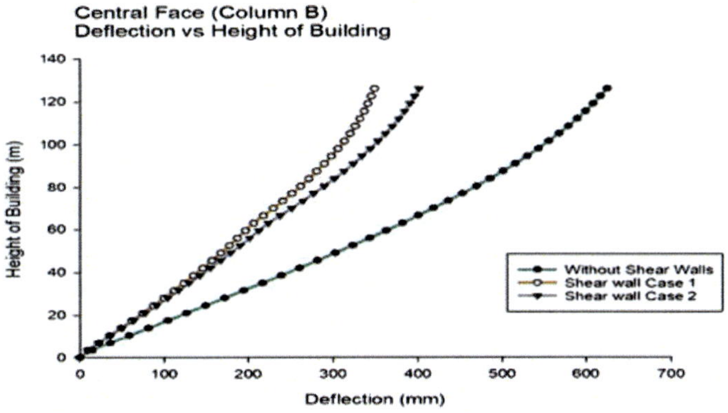

Fig. 5 Effect of shear walls on deflection at 0° incidence angle of wind on central face (Column B) of square-shaped building

values with slight difference in values with leeward face with maximum reduction first cases while central face in second case of shear walls.

3.2 Square-Shaped Building with Triangular Edges

The square-shaped building with triangular edges shows similar pattern (Figs. 7, 8, and 9) of increasing deflection with increasing height in all the three faces. Here, the shear walls in first case reduce the deflection by 21% with maximum in leeward face by slight margin, while shear walls in second case reduce deflection by 16% with slightly maximum reduction in central face.

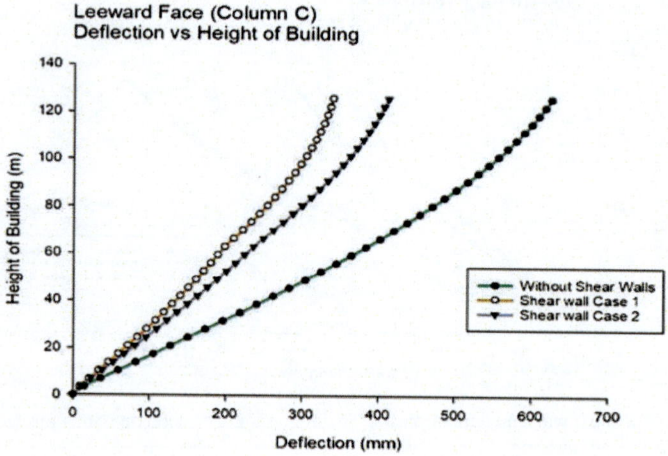

Fig. 6 Effect of shear walls on deflection at 0° incidence angle of wind on leeward face (Column C) of square-shaped building

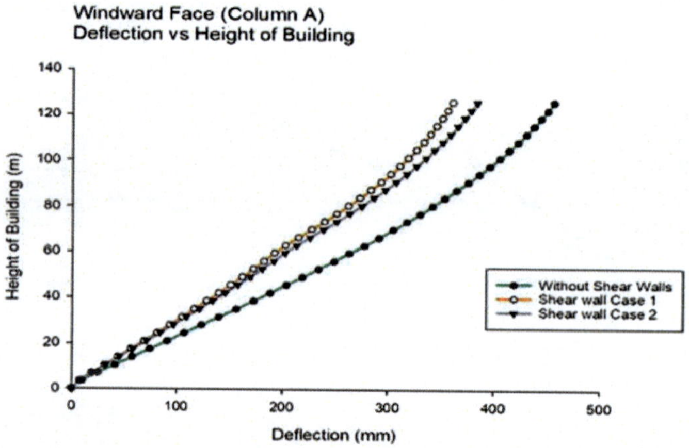

Fig. 7 Effect of shear walls on deflection at 0° incidence angle of wind on windward face (Column A) of square-shaped building with triangular edges

3.3 *Square-Shaped Building with Recessed Corners*

From Figs. 10, 11, and 12, it is found to increase in all three columns of square-shaped building with recessed corners with increasing height. The shear walls applied in first case helps to reduce the deflection by around 30% in leeward face and slightly less in other faces, while shear walls applied in second case helps to reduce the values of deflection by 18% in central face and slightly less in other faces.

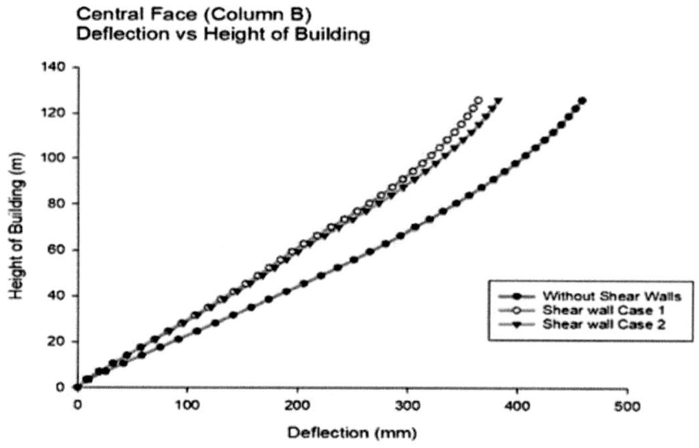

Fig. 8 Effect of shear walls on deflection at 0° incidence angle of wind on central face (Column B) of square-shaped building with triangular edges

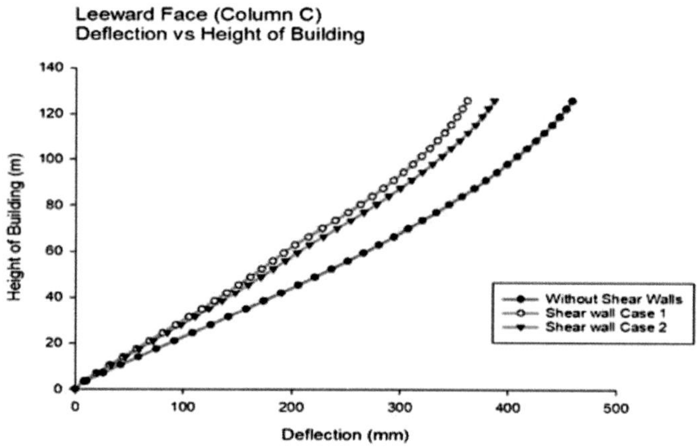

Fig. 9 Effect of shear walls on deflection at 0° incidence angle of wind on leeward face (Column C) of square-shaped building with triangular edges

4 Conclusion

1. The nature of graphs is found to be same in all three columns of all the buildings. The nature increases in deflection with increase in height of building.
2. All the three columns of any particular building show nearly same amount of reduction when shear walls of any one type are applied.
3. In square buildings, the first case of shear walls reduce deflection by about 45% while the second case of shear walls by about 34%.

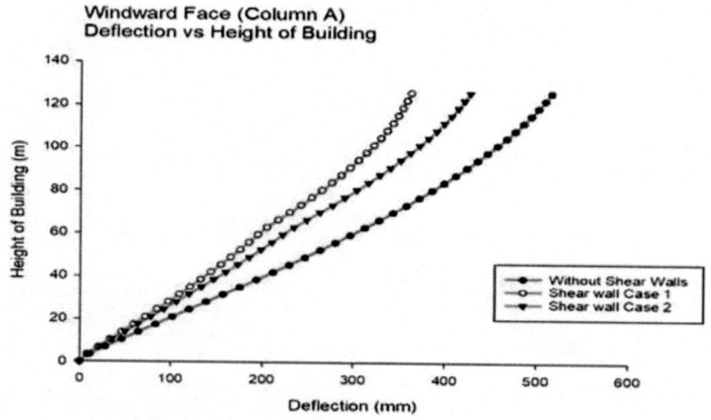

Fig. 10 Effect of shear walls on deflection at 0° incidence angle of wind on windward face (Column A) of square-shaped building with recessed corners

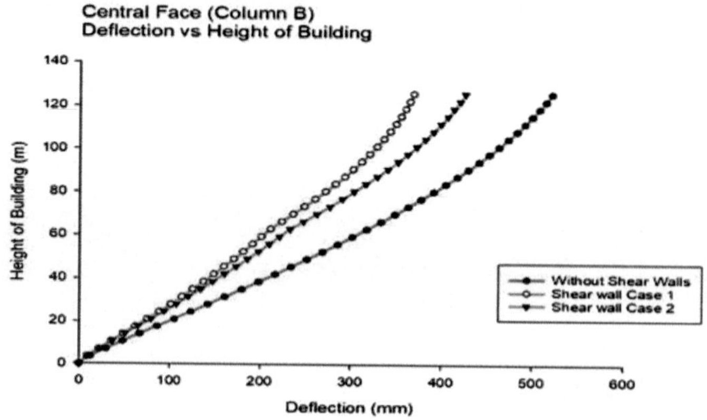

Fig. 11 Effect of shear walls on deflection at 0° incidence angle of wind on central face (Column B) of square-shaped building with recessed corners

4. In square buildings with triangular edges, the first case of shear walls reduces deflection by about 21% while the second case of shear walls by about 16%.
5. In square buildings with recessed corners, the first case of shear walls reduces deflection by about 30% while the second case of shear walls by about 18%.
6. The maximum reduction in deflection is seen in square building in both cases of shear walls in comparison with other two buildings.
7. Shear walls applied in first case (on the corners) are more effective in reducing deflection than the shear walls applied in second case (central columns) in all three buildings.

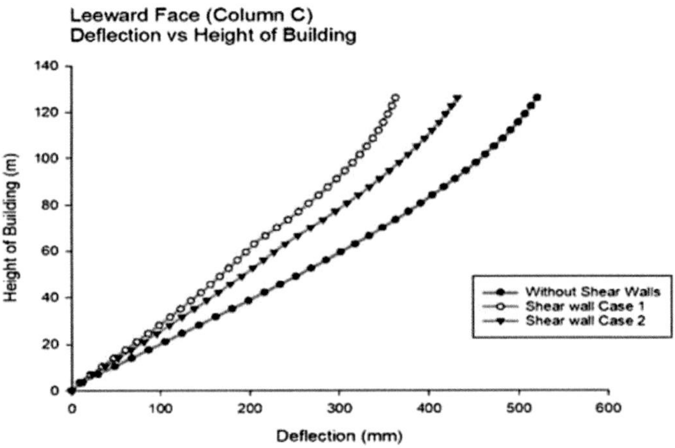

Fig. 12 Effect of shear walls on deflection at 0° incidence angle of wind on leeward face (Column C) of square-shaped building with recessed corners

Acknowledgements The authors gratefully acknowledge the technical support provided by the esteemed Delhi Technological University.

References

1. AS/NZS: 1170.2. (2002). Structural design actions, Part-2: Wind action.
2. ASCE: 7-02. (2002). Minimum design loads for buildings and other structures.
3. BS: 63699. (1995). Loading for buildings: Part 2-code of practice for wind loads.
4. EN: 1991-1-4. (2005). Euro code 1: Actions on structures wind actions.
5. IS:875-Part-3. (1987). Code of practice for design loads (other than earthquake loads) for buildings and structures-Wind loads.
6. IS:456. (2000). Plain and Reinforced Concrete—Code of Practice.
7. IS:875-Part-1. (1987). Code of Practice for Design Loads (other than Earthquake Loads) for Buildings and Structures—Dead Loads.
8. IS:875-Part-2. (1987). Code of Practice for Design Loads (other than Earthquake Loads) for Buildings and Structures—Imposed Load.
9. Krishna, M., & Arunakanthi, D. E. (2014). optimum location of different shapes of shear walls in unsymmetrical high rise buildings. *International Journal of Engineering Research & Technology (IJERT), 3*(9).
10. Esmaili, O. E. (2014). Study of structural RC shear wall system in a 56-Story RC tall building. In *14th World Conference on Earthquake Engineering.*
11. Amin, J. A., & Ahuja, A. K. (2008). Experimental study of wind pressures on irregular plan shape buildings. In *BBAA VI International Colloquium on: Bluff Bodies Aerodynamics and Application* (pp. 20–24).
12. Kwok, K. (1988). Effect of building shape on wind-induced response of tall building. *Journal of Wind Engineering and Industrial Aerodynamics, 28*, 381–390.
13. Mukherjee, A. B. (2017). Wind pressure and velocity pattern around 'N' plan shape tall building—A case study. *Asian Journal of Civil Engineering, 182*, 6399–6401.

14. Sheng, R., Perret, L., Calmet, I., Demouge, F., & Guilhot, J. (2018). Wind tunnel study of wind effects on a high-rise building at a scale of 1:300. *Journal of Wind Engineering and Industrial Aerodynamics, 174*, 391–403.
15. Ravinder Ahlawat, A. K. (2015). Wind loads on 'Y' plan shape tall building. *International Journal of Engineering and Applied Sciences (IJEAS), 2*(4).
16. Nagar, S. K., Raj, R., & Dev, N. (2014). Wind load on irregular plan shaped tall building—A case study. *Wind and Structures, 19*, 59–73.
17. Nagar, S. K., Raj, R., & Dev, N. (2020). Experimental study of wind-induced pressures on tall buildings of different shapes. *Wind and Structures*, 31, 431–443.
18. Yuan-Lung Lo, Y. C. (2019). Estimation of wind-induced response of high-rise buildings immersed in interfered flow. *International Journal of Applied Science and Engineering, 22*, 429–448.

Analytical Investigation of Moment Resisting Frame Structure—A Case Study on Performance-Based Capacity Spectrum Method

Ajay Singh Thakur and Tanmay Gupta

Abstract This paper has only two primary concerns (a) to properly adapt the performance-based methods with the Indian Seismic code; (b) to analyse the structure for its ductile behaviours. This case study addresses these concern by analysing the moment resisting frame having a six-storey with three bays model for elastic and nonlinear static procedures. The results support the code-based design but only for service-level earthquake and design-based earthquake, but for maximum considered, an earthquake, the performance-based design seems to be a perfect fit rather than a code-based approach. Results seem to be fit for prescribed target displacement, but to satisfy the global needs, the model needs strengthening which can be sorted with the hinge results of beams and columns. Ductile behaviour can be noticed through the pushover curve after the first yield. Change in size in frame members after certain floor needs to be designed carefully as a result the hinges can be seen extending beyond collapse prevention for the displacement of 159 mm in Step 27 in this paper. Pushover, hinge result, performance point and coefficient displacement results are presented and discussed.

Keywords Moment resisting frame · Performance level · Design-based earthquake · Maximum considered earthquake

1 Introduction

India is classified into IV zones under the category of the high seismic zone, moderate seismic zone, and low seismic zone. The zone factor for each zone corresponds to the maximum considered earthquake of that zone. The value of the zone factor is damped to get the design-based earthquake by dividing the value of zone factor by

A. S. Thakur · T. Gupta (✉)
Jaypee University of Information Technology, Waknaghat, India
e-mail: tanmay.gupta@juitsolan.in

A. S. Thakur
e-mail: 206605@juitsolan.in

A. K. Gupta et al. (eds.), *Advances in Construction Materials and Sustainable Environment*, Lecture Notes in Civil Engineering 196,
https://doi.org/10.1007/978-981-16-6557-8_60

denominator 2 given in Indian Seismic code IS1893:2016 [1]. The seismic ground motion is characterised by intensity, duration which is a function of magnitude, the distance between focus and epicentre, and characteristics of the propagation of waves; the waves show impulsive and complex behaviour due to which the structure responds by introducing the inertia forces as an opposing force to horizontal seismic force which is a type of surface wave. The fundamental natural period also plays an important role in calculating the average response acceleration coefficients for different soil conditions. The short period of vibration generates large inertia forces within a structure which may cause large stiffness degradation in members so to damp the seismic forces it is feasible to elongate the period which is done by using advanced techniques like base isolation, but these techniques are not cost effective for everyone. The partial factor of safety, redundancy and ductility help in extrapolating the design base shear to ultimate elastic force; thus, in Indian code, the nonlinearity is incorporated by response reduction factor denoted by *R*. By following the ductile detailing guidelines, we ignore to analyse the ductility part as the response beyond elastic is not a simple extrapolation. For analysing the response of structure beyond the elastic stage, the nonlinear method or dynamic method of time history is followed by analysing the structure on finite element software which is time consuming and needs an experienced designer to analyse the results of nonlinear time-history analyses [2]. Pushover analysis is used as an alternative for time-history analysis which saves computational time and can analyse the structure as per various international guidelines. Performance-based design is a design approach that eases out the process of designing for the owner to engineer by giving the flexibility to the owner to discuss the performance goals with the designer and engineer which helps to set up multiple performance objectives which is a function of performance level which expresses the extent of damage in structure due to hazard event [3, 4]. It is necessary to introduce PBD to meet the performance requirement in emerging society [5, 6]. In this case study, we will analyse the structure by cooperating the performance design approach with user-defined functions to reflect the values based on Indian Seismic codes.

2 State of Development

The performance-based design approach had been in development since the end of the '60s, and due to the unexpected economic loss in Loma Prieta Earthquake (1989) and followed by Northridge Earthquake (1994), the recommendation for the performance based was developed by the SEAOC committee under Vision 2000 [7]. The capacity spectrum method was one of the earliest methods to be applied in PBD for assessing the seismic vulnerability of structures. It is a graphical method that plots the relation between base shear and displacement. This method was used in the early '70s as a pilot project at the Puget Sound Naval Shipyard [8] supporting its correlation between ground motion and building performance by stating the hinge states from IO to CP and beyond due to which it serve as a verification design document in Tri-services [9]. The pushover curve which is a plot of base shear and displacement gives the capacity of the structure. With some modification over time, the pushover curve was converted to an acceleration displacement response spectrum which merges the base shear and displacement with (a) spectral acceleration versus spectral displacement representation, (b) spectral acceleration versus time and (c) spectral displacement versus time. N2 method and coefficient displacement method are the two advancements after the capacity spectrum method under performance-based design methodology [10]. The method for converting the base shear and displacement into spectral acceleration was covered using the relation given in ATC40 [11]. Spectral acceleration and displacement can be calculated through Eqs. 1 and 2.

$$S_a = \frac{\frac{V_B}{W}}{\left(\frac{M_K}{M}\right)} \quad (1)$$

$$S_a = \frac{\Delta\text{rooftop}}{P_K \Phi_{K\text{rooftop}}} \quad (2)$$

V_B represents the base shear, W is the seismic weight of the structure, Δ is the displacement of the rooftop, P_K is the modal participation factor and $\Phi_{k\ \text{rooftop}}$ is the modal amplitude at the rooftop.

S_a versus S_d will give the capacity curve, and for the demand curve, the response spectrum graph having S_a and T is also converted by relation given in Eq. 3 and [12].

$$S_d = \frac{T^2}{4\pi^2} S_a \quad (3)$$

T was represented by radial lines in the transformed plot. The response spectrum curve gives the demand curve. The intersection of the capacity curve and demand curve gives the performance point. The damping of spectra can vary from 5 to 20% or more. The pushover analysis was not introduced in Indian Seismic Code, but ATC 40 procedure can be adapted by using the value of C_a and C_v parameters concerning response spectra used in the Indian Seismic code for a different type of soil [13]. A case study model S6B3 was used from an original paper [14]. The CSM graphs will be discussed later in the paper.

3 Case Study

The model S6B3 consist of six storey and three bay with a spacing of 4 m in each direction. The building is an office building with an importance factor of 1.5 on medium soil having a floor height of 3 m with a mean live load of 4kN/m^2. Lateral design load will be calculated from the equivalent static method [1, 11]. ATC40 guidelines will be followed and adapted in the Indian code for finding the performance point. The 4% of height was selected as target displacement [1, 11]. M25 grade and Fe 415 grade steel were used. Beams of size 300 × 400 mm were used. The column

Table 1 Column size detail for S6B3

Floor	Exterior column	Internal column
1–3	450 × 450	530 × 530
4–6	350 × 350	450 × 450

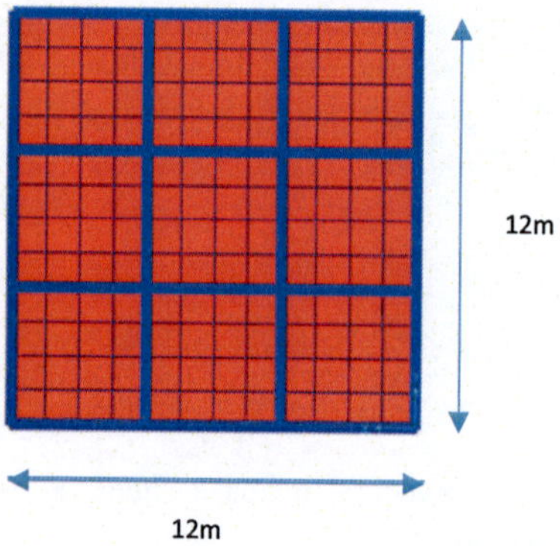

Fig. 1 Plan of S6

was designed separately for 1–3 floors and 4–6 floors with external and internal columns as given in Table 1. In this case study, we will use the modified value of C_a and C_v as per medium soil and Indian response spectra for 5% damping and check the structure for 0.004H permissible limit for Indian code and verify the structure hinges for collapse prevention parameter 0.04H. Property modifiers for beam and column as per Indian seismic codes were applied. It is very important after gravity load to add a new load case as gravity with nonlinear behaviours which will be force controlled, and after that push load cases can be used which will start from unstressed gravity load case. The plan of six storeys three-bay (S6B3) is shown in Fig. 1.

In Sap2000, pre-target displacement of 0.04H can be changed with the user-calculated target displacement. The push cases will be deformation controlled. Hinges can be user defined or predefined based on the designer. In this study, hinges were defined at 5 and 95% at both i and j node of the beam and column while for beams M3 hinge were assigned and for column P-M2-M3, hinges were used as a relationship to describe the post-yield backbone behavior of beam and column [15].

4 Results and Discussion

Pushover curve—Graph of base shear and displacement for the X- and Y-axis was discussed in Fig. 3. The target displacement was revised from 0.004H to 0.04H to consider more displacement effect as a result we can get the results for larger displacement and larger rotations in hinges. Figure 2 states the hinge result of the bottom node of frame object 214 hinges 2 (214H2).

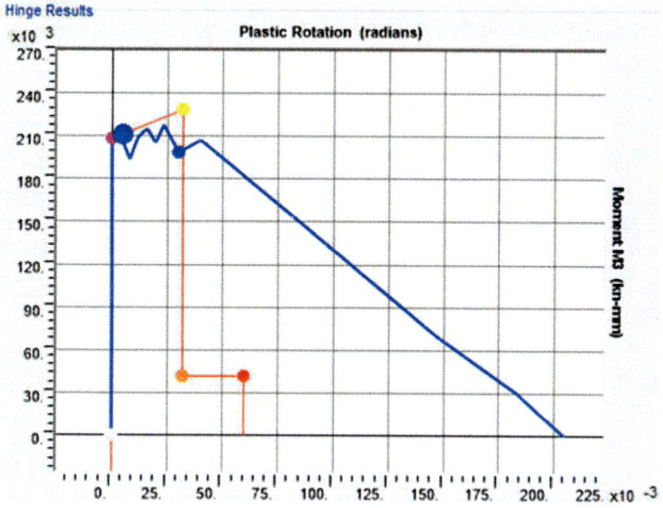

Fig. 2 214H2 P-M2-M3 hinge result

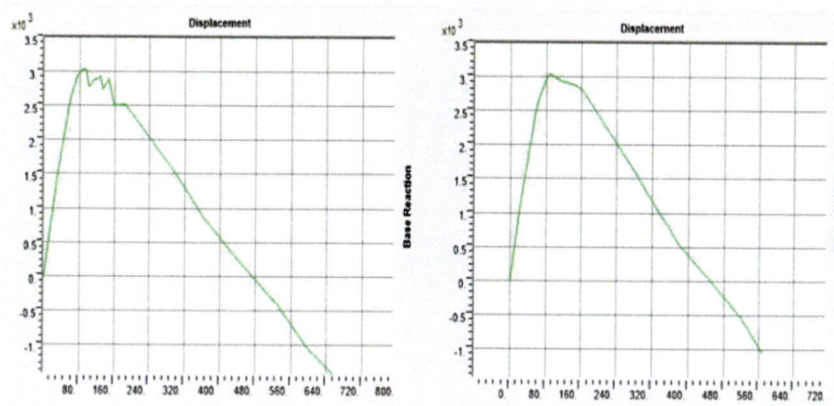

Fig. 3 Pushover curve X-axis and Y-axis

P-M2-M3 hinges were assigned for a column with M3 as hinge degree of freedom resulting in plastic rotation of 0.0304 radians at Step 27 of Push X stating the degradation of fibres after CP point. Beam hinges were found to be between IO-LS state, change in a property of column in fourth storey result in loss of stiffness in columns.

Pushover results also consider the hinge state and results for various push steps which will be discussed after covering the capacity spectrum graph. The top node of the structure (Node81) was pushed up to target displacement which results in loss of stiffness in members while exposing the ductile capacity of the structure. The S6B3 model performs satisfactorily till the permissible limit given by IS codes and follows the strong column weak beam methodology. At Step 11, base shear was 2949 kN and displacement was 76.46 mm with only four hinges which change their state from IO to LS as given in Table 2. According to IS code, the structure performance was elastic, but for global results, performance point must be checked thoroughly.

Capacity spectrum method shows the performance point by superpositioning the pushover curve given by green colour to the ADRS representation of damped responded spectrum by 5, 10, 15, 20% indicated by four red lines known as the demand curve as shown in Fig. 4. The S_a and S_d values correspond to the 5% damped response spectra for X and Y push [16]. The performance point for X (2694.003 kN, 63.779 mm) and Y (2677.617 kN, 64.827 mm) was given in the form of base shear and displacement (V, D point). C_a value was taken as 2.5(Z/2) and C_v value 1.39/T(Z/2) for medium soil for Indian Code [13]. T is the modal fundamental period of building corresponding to mode 1. Z is the zone factor 0.36. The 1.36/T value was referred from the value of the average response spectrum for medium soil when the period was between 0.55 and 4 s.

FEMA 356 Coefficient method [17] in Fig. 5 shows the bilinearisation of the pushover curve which represents the maximum displacement and force as a performance point for the structure. For Push X and Push Y, the value of performance point was (2297.31 kN, 210.587 mm) and (2533 kN, 158 mm). These values are discussed below in hinge results as shown in Fig. 6.

In Step 27, the hinges in the column of four storeys change their state to IO-LS at one end, and the far end column reached beyond C point in which the column degrades its strength abruptly with very minute rotation and displacement. The maximum performance point values approximately correspond to Step 27 to 29 value. The hinge state for the column is shown in Fig. 1. Step 35 consists of the response due to ultimate displacement which results in the collapse of the column at four storey; hence, the moment redistribution due to storey 4 columns was already at its peak from Step 27. The base shear can be checked for displacement and its hinge state from Table 2.

5 Conclusion

For designing an earthquake-resistant important structure, performance-based design is a perfect fit, as in this case the structure was design only for elastic base shear with the use of Indian seismic code and later on verified for a global response with nonlinear static analysis which shows the structure ductility performance, with the response of hinges generated at the nodes of beam and column in the relative distance of 0.5 and 0.95. These methods with the help of hinges show the potential beams, the column for retrofitting in the existing structure and redesigning in case of a new building to achieve the proper performance. Only selected frame members to be properly modelled as in our case label 195, 214, 217, and 199 need to be retrofitted to impart additional strength and ductility. The hinge results give a good indication of

Table 2 Pushover capacity curve

Pushover capacity curve												
Load case	Step	Displacement	Base force	A-B	B-IO	IO-LS	LS-CP	CP-C	C-D	D-E	E	Total
	Unitless	mm	kN									
P X	0	0.025704	0	480	0	0	0	0	0	0	0	480
P X	1	31.812695	1520.154	479	1	0	0	0	0	0	0	480
P X	2	56.653549	2511.808	396	84	0	0	0	0	0	0	480
P X	3	64.814425	2720.484	377	103	0	0	0	0	0	0	480
P X	4	64.814429	2718.852	377	103	0	0	0	0	0	0	480
P X	5	64.814436	2718.85	377	103	0	0	0	0	0	0	480
P X	6	72.03646	2881.369	351	129	0	0	0	0	0	0	480
P X	7	72.038725	2881.462	351	129	0	0	0	0	0	0	480
P X	8	73.261773	2903.57	344	136	0	0	0	0	0	0	480
P X	9	73.261809	2903.571	344	136	0	0	0	0	0	0	480
P X	10	73.269009	2896.84	344	136	0	0	0	0	0	0	480
P X	11	76.36627	2949.383	338	138	4	0	0	0	0	0	480
P X	12	76.374821	2949.762	338	138	4	0	0	0	0	0	480
P X	13	76.382021	2948.005	338	138	4	0	0	0	0	0	480
P X	14	85.379513	3018.694	322	142	16	0	0	0	0	0	480
P X	15	85.386713	3018.843	322	142	16	0	0	0	0	0	480
P X	16	86.182507	3027.189	322	142	16	0	0	0	0	0	480
P X	17	87.774095	3037.745	318	146	16	0	0	0	0	0	480
P X	18	89.365683	3041.413	316	148	16	0	0	0	0	0	480

(continued)

Table 2 (continued)

Pushover capacity curve												
Load case	Step	Displacement	Base force	A-B	B-IO	IO-LS	LS-CP	CP-C	C-D	D-E	E	Total
	Unitless	mm	kN									
P X	19	92.548858	3031.362	316	148	16	0	0	0	0	0	480
P X	20	95.732034	3011.489	316	146	18	0	0	0	0	0	480
P X	21	102.098375	2782.71	316	116	48	0	0	0	0	0	480
P X	22	114.831405	2894.847	316	116	48	0	0	0	0	0	480
P X	23	121.197756	2895.837	316	116	48	0	0	0	0	0	480
P X	24	127.564108	2934.963	316	116	48	0	0	0	0	0	480
P X	25	133.924557	2748.697	316	116	48	0	0	0	0	0	480
P X	26	146.657271	2881.442	316	116	48	0	0	0	0	0	480
P X	27	159.398875	2507.931	316	116	42	0	0	6	0	0	480
P X	28	184.867597	2523.08	316	116	16	0	0	32	0	0	480
P X	29	292.82261	1575.692	316	116	16	0	0	32	0	0	480
P X	30	364.823046	849.084	316	116	16	0	0	32	0	0	480
P X	31	436.839233	270.955	316	116	16	0	0	32	0	0	480
P X	32	526.681203	−398.161	316	116	16	0	0	24	0	8	480
P X	33	598.681604	−1044.15	316	116	16	0	0	0	1	31	480
P X	34	652.545648	−1418.557	316	116	16	0	0	0	0	32	480
P X	35	657.053206	−1471.126	316	116	16	0	0	0	0	32	480

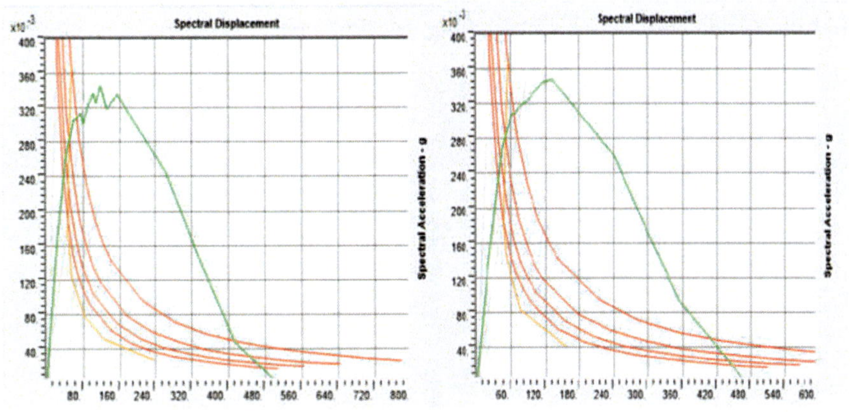

Fig. 4 ATC40 capacity spectrum for Push X and Push Y

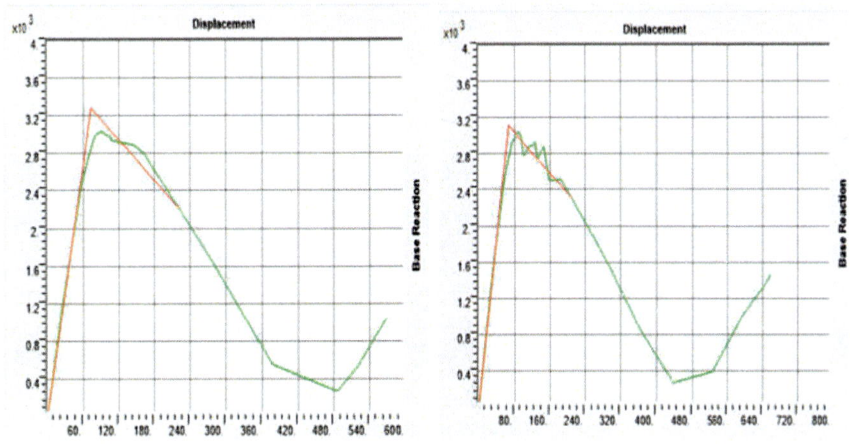

Fig. 5 FEMA356 coefficient method for Push X and Push Y

moment and rotation with each increment of push. As the hinge behaviour is deformation controlled and up to Step 26 shows satisfactory ductile behaviour while with the adaptation of C_a and C_v values, the performance point shifts its self-according to the Indian response spectra. The structure was not analysed for the nonlinear time-history method, but in the case, it should be assigned with more than seven ground motions to get proper supporting results.

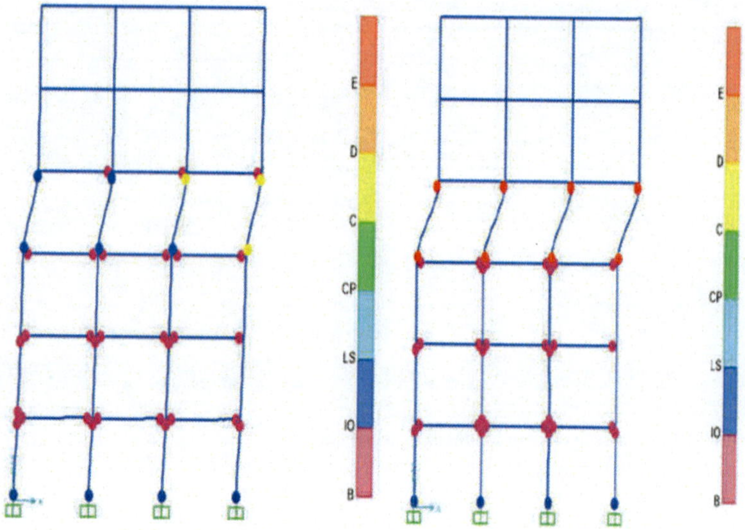

Fig. 6 Push X hinge formation at Step 27 and 35

References

1. IS1893 (Part 1)-2016. Indian Standard criteria for earthquake resistant design of structures, Part 1: General provision and buildings. Bureau of Indian Standards, New Delhi.
2. Dokainish, M. A., & Subbaraj, K. (1989). A survey of direct time-integration methods in computational structural dynamics. I. Explicit methods. *Computers & Structures, 32*(6), 1371–1386.
3. ASCE 41: American Society of Civil Engineers. Seismic rehabilitation of existing building; 2007 Reston, Virginia.
4. Zammeruddin, M. (2016). Review on recent development in the performance-based seismic design of reinforced concrete structures. *Structures, 6*, 119–133.
5. Dalal Sejal, P., Vasanwala, S. A., & Desai, A. K. (2011). Performance-based seismic design of structure: A review. *International Journal of Civil and Structural Engineering* (4), 795.
6. Wang, Y. (2021). Study on performance-based design theory. *IOP Conference Series: Earth and Environmental Science, 669*(2021), 012020. https://doi.org/10.1088/1755-1315/669/1/012020
7. SEAOC Vision 2000 Committee. (1995). Performance-based seismic engineering, Report prepared by Structural Engineers Association of California, Sacramento, California.
8. Sigmund, F. A. (2004). Review of the development of the capacity spectrum method. *ISET Journal of Earthquake Technology, Paper No. 438, 41*(1), 1–13.
9. Army. (1986). Seismic design guidelines for essential buildings. Departments of the Army (TM 5- 809–10–1), Navy (NAVFAC P355.1), and the Air Force (AFM 88-3, Chapter 13, Section A), Washington, D.C., U.S.A.
10. Priestley, M. J. (2000). Performance-based seismic design. In: *12 World Conference on Earthquake Engineering*, Auckland.
11. ATC 40 Seismic evaluation and retrofit of existing concrete buildings. Redwood City (CA): Applied Technical Council: 1996.
12. FEMA 356 Prestandard and Commentary for the Seismic Rehabilitation of Buildings. Federal Emergency Management Agency, Washington, DC, USA

13. Leslie, R. (2013). The pushover analysis explained in its simplicity. In: *Proceedings of 2nd National Conference.*
14. Zammeruddin, M., Sangle, K. K. (2021). Performance-based seismic assessment of reinforced concrete moment resisting frame. *Journal of King Saud University—Engineering Science, 33*, 153–165.
15. Al-jassim, S. A. B. J., & Husssain, M. A. (2018). Pushover analysis of G + 5 Reinforced concrete building in Basrah. *International Journal of Innovations in Engineering and Technology (IJIET), 11*(August 2018). ISSN: 2319-1058.
16. Fajfar, P. (1998). Capacity spectrum method based on inelastic demand spectra. Report EE-3/98, IKPIR, Ljubljana, Slovenia.
17. FEMA 440 Improvement in nonlinear static seismic analysis procedures. Washington (DC): Federal Emergency Management Agency; 2005.

Study of Slope Stability Using Flexible Facing

Amanpreet Tangri and Saurabh Rawat

Abstract Soil nailing is an improvement technique used to overhaul the nature of ground that ensures the security of new or existing slopes. Soil nailing is used for the stability of natural soil which are of weak nature. This technique is widely used in engineering because of certain advantages, like it is easy in construction, low cost, short construction period and mature construction techniques. In this present study, a series of lab tests were conducted on a physical model using six nails in a rectangular pattern. The strain was measured using the flex sensor using digital multimeter on the application of load by the hydraulic jack. In this study, sandwiched wire mesh facing material was used with and without jute gunny bag. These both were tested for maximum stress that they can bear safely. Alongside, the displacement in horizontal and vertical direction were also measured with the scale during the application of the load. The maximum stress taken by sandwiched wire mesh was 0.492 N/mm^2, and when sandwiched, wire mesh was used with the jute gunny bag was 0.780 N/mm^2. So, it was concluded that sandwiched wire mesh along with jute gunny bag has taken maximum stresses as compared to other condition in which just sandwiched wire mesh was used.

Keywords Flexible facing · Physical model · Sandwiched wire mesh · Strain · Stress

1 Introduction

The development of hilly and far-off areas, the dam's construction and hydroelectric projects in hard terrain, all needs stabilized ground in one way or another way. In the past, many stabilization methods have been developed to combat these situations [1]. Stabilization of the Earth such as mechanical and chemical land stabilization, drilling, earthworks, crushed stone and ground gravel, are useful in providing structural and

A. Tangri · S. Rawat (✉)
Department of Civil Engineering, Jaypee University of Information Technology, Waknaghat, Solan, H.P., India
e-mail: saurabh.rawat@juit.ac.in

A. K. Gupta et al. (eds.), *Advances in Construction Materials and Sustainable Environment*, Lecture Notes in Civil Engineering 196,
https://doi.org/10.1007/978-981-16-6557-8_61

service-oriented stability [2, 3]. In this paper, the detailed description of one such technique utilized for ground stabilization known as soil nailing is provided.

Soil nailing technology is employed to offer reinforcement for natural and steep slopes on site. If unstable existing grounds require support and nearby buildings are sensitive to deformation, the utilization of soil nailing provides stability through temporary construction or permanent support structures and take appropriate measures to decrease ground movement [4, 5]. Soil nailing walls are a widely used technique for keeping vertical cuts on any slope that is even more vertical in the ground with vertical slopes and normal angles. A large part of the cost of landing is related to the construction of the reinforced concrete surface. Soil facing mechanism helps to provide the stability between soil and inserted nails. In this manner, to get better stylish look of the nail structure, confronting assumes a significant job [6].

2 Literature Review

Soil nail advancement was first used in France to collect an unending holding divider cut in sensitive shake. The assignment, endeavoured in 1961, was the place steel nails were used to sustain a holding divider. The technique included presenting high-thickness, grouted soil nails into a 60-feet-high divider and facing it with reinforced framework. Mohammad Abubakar et al.: A case study was done in order to study causes of failure and to evaluate remedial measures on a fort built in eighteenth century and renovated ever since several times a portion of which collapsed due to failure of retained slope during earthquake. The studies suggested various possible reasons of instability of the slopes, out of which the seepage of water was the main reason along with the earthquake. The best remedial measure suggested was the reconstruction of the wall along with the reconstruction of the damaged slope using the soil nailing technique along with grouting [1]. Tao-Yang et al.: Using the three-dimensional rotational failure mechanism, the FOS of soil-nailed slopes subjected to seismic forces was evaluated applying the stress-reduction method and kinematic approach of limit analysis to replace the traditional evaluation using 2-D limit equilibrium method. The results were compared to validate the results obtained by the proposed approach. The various parameters of the soil nails were studied and both the tensile and pullout failures of soil nail were taken into consideration. This new approach proved to be an efficient design tool to evaluate the FOS of soil reinforced slope and also provides a set of chart to assess the slope safety rapidly [7]. Anant K. Kokane et al.: For the analysis of nail forces, force distribution in vertical and horizontal in nailed vertical cuts subjected to seismic forces, this paper aims at providing novel closed form solutions using a limit equilibrium concept based on pseudo-dynamic approach. It was concluded that the increase in the amplitude of initial vertical base acceleration increases the demand of the tensile force. The bottom-most nail had the largest nail force when subjected to seismic forces [8].

Nelson NS Chou et al.: A failure investigation was conducted on an earthquake hit site using finite element software PLAXIS. For the reconstruction of the slope, the conventional measures were replaced by the use of the MSE slope with the wrap around facing and geoweb with vegetation. Completed and well-designed after some time, the project has withstood several hazards and proved to be successful in terms of economy, safety and eco-friendly [6]. Nozomu Kotake and Eisuke Sato: In order to reduce the impact on the environment, reclaimed plastics have been used for construction of large flexible-bearing face plates. This study aims at reflecting light on the influences of bending stiffness and dimensions of bearing capacity and deformations. A number of model laboratory tests were conducted on rigid as well as flexible footings, and their results were compared. The lab experiments showed that with the use of the flexible footings, the ultimate bearing capacity reduced to 10–20% as compared to rigid footing which was not very large [9]. Mahesh Sharma et al.: This study gives methodology for analysis and design of helical nailed walls under static and seismic conditions. Laboratory tests were performed to investigate their stability analysis and to develop the equation of pullout capacity in dry cohesion less soils considering various parameters of the soil nail. The results were obtained and compared to the already existing results to validate this approach, and then equation which satisfied the boundary conditions was proposed to analyse the static and seismic conditions of the helical soil-nailed walls. This study suggested that the FOS values from the present method are lower than the existing methods [10].

3 Materials and Methodology

Initially, the soil is collected from Chandigarh University campus, and then various properties were studied on the basis of code (IS2720). The properties such as moisture content, specific gravity, compaction characteristics and unconfined compression strength have been analysed and shown in Table 1.

From the particle size distribution results as given in Fig. 1, the soil was categorized as well-graded sand. In the present study, materials used were sandwiched wire mesh as facing material with and without jute gunny bag. These both were tested for maximum stress that they can bear safely. The testing was done using a physical model.

Table 1 Properties of soil

Properties	Values
Water content (%)	11
Specific gravity	2.56
Optimum moisture content (OMC) (%)	7.3
Maximum dry density (MDD) (g/cc)	1.98
Coefficient of uniformity	6
Coefficient of curvature	1.04
Unconfined compressive strength (kg/cm^2)	0.118

Fig. 1 Sieve analysis of soil

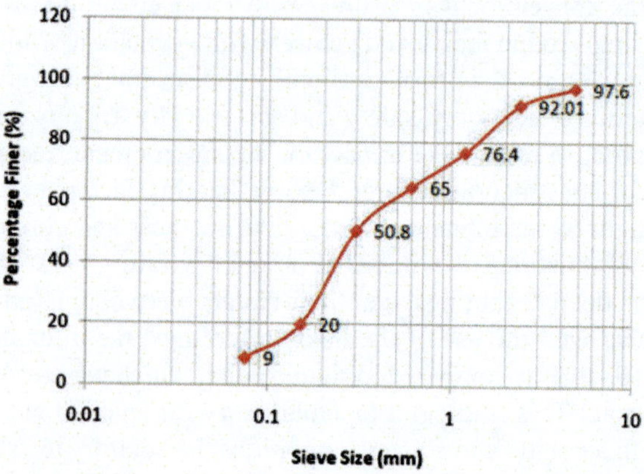

Physical Model

A physical model was prepared with the Perspex sheet (acrylic sheet) of thickness 8 mm. The dimensions of the model such as length, width and height were taken as 60 cm, 40 cm and 50 cm, respectively. The height of slope was considered as 45 cm. The nails used in present study were of 10 mm diameter and length 31.5 cm. The nail length is considered in accordance with the slope height as the length is considered 0.7*H*- *H*. So, in present study, it is taken as 0.7*H* ($0.7 * 45 = 31.5$ cm). The slope of 60° was prepared in a physical model as the flexible materials can be used maximum up to 70°. The scale effect on soil nails in granular soils is evaluated on the basis of following parameters, i.e. the effect of mean particle size on shaft resistance. For no scale effect on shaft resistance, the ratio of minimum shaft diameter (d) to mean grain size of soil (D50) is given by d/D50 should be greater than 30 [11], and in the present case, it is $10/0.3 = 33.3$. The scale factor used for model preparation was by considering $N = 30$ [12]. The various components used in the physical model are described below.

Pressure hydraulic jack

Pressure hydraulic jack of capacity 5 tonnes is used and placed in between the Perspex sheet. During the research, hydraulic jack is placed in between the Perspex sheet of 8 mm. The load was applied using hydraulic jack, and readings of change in resistance for all the nails were taken by using multimeter. Stress was calculated by dividing the load given on the dial gauge by the area. Further, the values of change in resistance were used for the calculation of strain.

Flex sensor

The flexible sensor is just like a variable resistor. As the nails bend, the resistance of the flexible sensor increases. A multimeter is attached to the switch to measure the resistance sensed by flex sensor, and then strain is calculated by using the formula as given in Eq. 1:

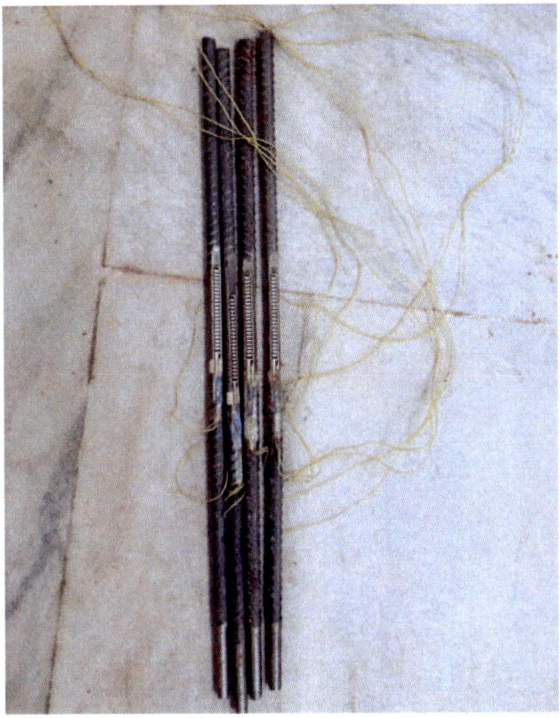

Fig. 2 Steel bars with sensor

$$\text{Strain} = \frac{\Delta R/R}{\sum_c} \quad (1)$$

Here, $\sum_c$ stands for gauge factor; its value lies between 2 and 2.5 and for 200 KΩ, the value of gauge factor is 2.1. Copper wire is attached with each nail and resistance is measured using digital multimeter. The flex sensor is used at the centre of nail and is pasted on the steel bar by using glue (Fig. 2). When the load is applied, the steel bars would be undergoing strain due to which there is change in the resistance value on multimeter, so by applying above formula, strain is calculated for each nail for a specific load. A total of six instrumented nails has been used with horizontal and vertical spacing as shown in Fig. 3.

4 Results and Discussions

Facing material is required to hinder deformation thus limiting the mobilization of shear stress along nails. Flexible facing for soil nails is designed to provide the necessary restrains to slope between the bearing plates as well as erosion control. As depicted in Fig. 4, the sandwiched triaxial mesh was used in two conditions, i.e. with

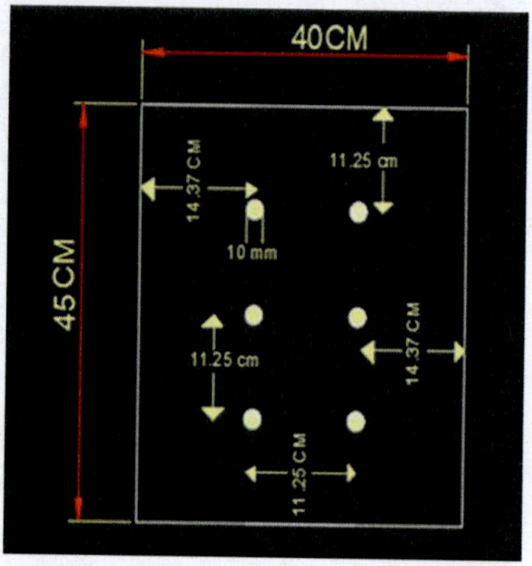

Fig. 3 Rectangular arrangement

Fig. 4 Sandwiched wire mesh facing slope

jute gunny bag and without jute gunny bag. The stress strain behaviour (Table 2) as well as the deformation in vertical and horizontal direction as obtained from model testing is shown in Table 3. The vertical and horizontal deformation were measured using the scale corresponding to each stress.

Sandwiched triaxial mesh facing

The stress–strain behaviour of sandwiched wire mesh along with jute gunny bag is given in Table 4. For all six nails, the stress is increasing with increase in strain. The

Table 2 Stress–strain values of testing with sandwiched wire mesh using 6 nails

Stress (N/mm^2)	Strain nail 1	Strain nail 2	Strain nail 3	Strain nail 4	Strain nail 5	Strain nail 6
0.123	0.00125	0.00485	0.00172	0.00474	0.00492	0.0068
0.246	0.00751	0.00809	0.01215	0.00949	0.00821	0.01133
0.369	0.01253	0.01133	0.01344	0.01423	0.01477	0.01587
0.492	0.01503	0.02591	0.01646	0.02164	0.0197	0.02267

Table 3 Values of horizontal and vertical displacement of testing with sandwiched wire mesh using six nails

Stress (N/mm^2)	ΔH (cm)	ΔV (cm)
0.123	0.3	2
0.246	0.5	2.5
0.369	1.2	3.7
0.492	1.8	4.2

Table 4 Stress–strain values of testing with sandwiched wire mesh and jute gunny bag using six nails

Stress (N/mm^2)	Strain nail 1	Strain nail 2	Strain nail 3	Strain nail 4	Strain nail 5	Strain nail 6
0	0	0	0	0	0	0
0.123	0.00342	0.00464	0.00175	0.0036	0.00634	0.0066
0.246	0.01027	0.01161	0.00702	0.00727	0.01058	0.0099
0.369	0.01884	0.01626	0.01581	0.0109	0.02328	0.01498
0.492	0.02398	0.02555	0.02108	0.01454	0.03174	0.01998
0.615	0.02911	0.03484	0.02987	0.01575	0.03597	0.02331
0.78	0.03768	0.04181	0.03338	0.0278	0.05079	0.02664

Table 5 Values of horizontal and vertical displacement of testing with sandwiched wire mesh and Jute gunny bag using six nails

Stress (N/mm^2)	ΔH (cm)	ΔV (cm)
0	0	0
0.123	0.3	0.6
0.246	0.7	1.2
0.369	1	1.5
0.492	1.4	2.1
0.615	2.2	2.7
0.78	2.7	3.4

maximum strain is found for nail number 5. This signifies that maximum deformation occurs near the toe of the nailed slope.

From Table 5, a maximum vertical deformation of 3.4 cm is observed for stress of 0.78 N/mm^2. Similarly, for the same magnitude of stress, horizontal displacement of 2.7 cm is also maximum. This further validates the occurrence of maximum strain values near the toe of the nailed slope.

Discussions

From the comparative results as obtained from Tables 2 and 4, it is found that maximum stress has been taken by the combinations of two materials, i.e. sandwiched wire mesh and jute gunny bag. This is due to the reason that when we are using two materials instead of one, so obviously, it would be having more stiffness as compared to just the sandwiched wire mesh. So, in the second case, more stresses are absorbed and due to this, the slope could bear more load.

Similarly, the comparative displacement in both the horizontal and vertical direction from Tables 3 and 5, it is concluded that both the horizontal and vertical displacement are more in the second case before the failure of material on the application of load. This is because sandwiched wire mesh along with jute gunny bag has more stiffness and both the flexional and axial stiffness of facing control the deformation.

5 Conclusions

- In the present study, the experiments were performed on physical model using different materials, and it can be concluded that use of flexible facing in soil nail walls will lead to an increase in the soil slopes load handling and slope stability
- In the present study, sandwiched wire mesh along with jute gunny bag has taken up higher stresses as compared to other materials for rectangular nail arrangement
- The stresses taken by combination of sandwiched wire mesh and jute gunny bag is 36.92% as compared to the stresses taken by sandwiched wire mesh alone
- Maximum deformation was observed in the second case when
- This research can be the beginning of learning guides for a construction technique that even reflects to be cost-effective and less environmental impact.

References

1. Naveed, M. A., Ali, Z., Qadir, A., Latif, U. N., Hamid, S., & Sarwar, U. (2020). Geotechnical forensic investigation of a slope failure on silty clay soil—A case study. *Frontiers of Structural and Civil Engineering, 14*(2), 501–517. https://doi.org/10.1007/s11709-020-0610-y
2. Prakasam, C., Aravinth, R., Nagarajan, B., & Kanwar, V. S. (2020). Site-specific geological and geotechnical investigation of a debris landslide along unstable road cut slopes in the Himalayan region, India. *Geomatics, Natural Hazards and Risk, 11*(1), 1827–1848. https://doi.org/10.1080/19475705.2020.1813812
3. Meng, C. C., & Tan, Y.-C. (2006). Soil nail design: A Malaysian perspective. In *International Conference on Slopes* (pp. 379–400).

4. Sharma, P., Rawat, S., & Gupta, A. K. (2021). Laboratory investigation of pullout behavior of hollow and solid shaft helical nail in frictional soil. *Acta Geotechnica, 16*(4), 1205–1230. https://doi.org/10.1007/s11440-020-01069-6
5. Hu, H. Y., Zhang, Y. C., & Chen, C. Y. (2020). Discussion on the displacement and stability of the soil nailing. *IOP Conference Series: Earth and Environmental Science, 510*(5). https://doi.org/10.1088/1755-1315/510/5/052012
6. Chou, N. N. S., Liu, T.-Y., Chen, P.-H., Fan, C.-C., & Zhang, J. (2020). Failure Investigation and sustainable renovation for slope at National Chi Nan University in Taiwan. *Journal of Performance of Constructed Facilities, 34*(5), 04020085. https://doi.org/10.1061/(asce)cf.1943-5509.0001459
7. Yang, T., Zou, J. F., & Pan, Q. J. (2020). Three-dimensional seismic stability of slopes reinforced by soil nails. *Computer and Geotechnics, 127*(June), 103768. https://doi.org/10.1016/j.compgeo.2020.103768
8. Kokane, A. K., Sawant, V. A., & Sahoo, J. P. (2020). Seismic stability analysis of nailed vertical cut using modified pseudo-dynamic method. *Soil Dynamics and Earthquake Engineering, 137*(May), 106294. https://doi.org/10.1016/j.soildyn.2020.106294
9. Kotake, N., & Sato, E. (2020). *Bearing capacity of a flexible plastic plate for soil nailing*, vol. 21, no. 1.
10. "Analysis of helical soil nailed walls under static and seismic conditions," pp. 1–49.
11. Schiavon, J. A., Tsuha, C. D. H. C., & Thorel, L. (2016). Scale effect in centrifuge tests of helical anchors in sand. *International Journal Physical Modelling in Geotechnics, 16*(4), 185–196. https://doi.org/10.1680/jphmg.15.00047
12. Rotte, V. M., & Viswanadham, B. V. S. (2013). Influence of nail inclination and facing material type on soil-nailed slopes. *Proceedings of the Institution of Civil Engineers—Ground Improvement, 166*(2), 86–107. https://doi.org/10.1680/grim.11.00026

Use of Waste Polymers in a Plastic Bricks as Sustainable Building and Construction Materials

Prajwal Madghe, Himanshu Berad, Abhijeet Roy, Nishant Vaidya, Nivesh Sakharwade, and Rajan L. Wankhade

Abstract This paper presents a brick composition comprising of thermoplastic material. We adopted low-density polyethylene (LDPE) and polyvinyl chloride (PVC) having cement-containing material for preparation of specimens of bricks. We used fly ash and waste plastic for preparing a specified form of brick comprising thermoplastic material and a cement-containing material. A proper mix design is composed with low-density polyethylene (LDPE), cement and fly ash in a better proportion to gain the desired strength. Plastic is a product which is easily available and became a hazardous material nowadays. Plastic is daily used, and after its use, it is simply thrown away, which creates hazards. Plastic is non-biodegradable hence can be re utilized. The reuse of plastic waste in construction and building materials is the recent application. We, in this work, prepared a brick with different compositions containing LDPE and PVC. Different strength and properties are then checked by means of various test on bricks to get desired optimum component. Results are presented further for optimum content of brick ingredients including plastic, cement and fly ash.

Keywords Low-density polyethylene (LDPE) · Polyvinyl chloride (PVC) · Cement · Fly ash

1 Introduction

It is observed that approximately 40 million tons of the municipal solid waste is collected in India per year. It has an increasing a rate of 1.5–2% for every year. Therefore, it became a need to dispose this plastic solid waste properly, to recycle and reuse it. It is possible to reutilize the plastic waste by properly treating it.

P. Madghe · H. Berad · A. Roy · N. Vaidya · N. Sakharwade
Department of Civil Engineering, Government College of Engineering, Nagpur, MH, India

R. L. Wankhade (✉)
Applied Mechanics Department, Government College of Engineering, Nagpur, MH, India

A. K. Gupta et al. (eds.), *Advances in Construction Materials and Sustainable Environment*, Lecture Notes in Civil Engineering 196,
https://doi.org/10.1007/978-981-16-6557-8_62

Some construction industries are trying to use plastic waste as a byproduct to be mixed with cementitious materials. Plastic is considered to be an engineering material having many good characteristics including versatility, lightness, hardness, resistant to chemicals, water and impact. It contains synthetic or semi-synthetic organic compounds that are malleable and can be molded into solid objects. After molten, plastic can be deformed into any desired shape and size. Jha et al. [1] studied behavior of plastic waste in fiber-reinforced concrete with application to pavement. Manhal and Farah [2] found out strength and behavior of concrete-containing waste plastic. Appiah et al. [3] stated the use of waste plastic materials for road construction in Ghana. Kamaruddin et al. [4] provided comprehensive a review on potential use of plastic waste as construction materials. Frigione [5] conducted recycling of PET bottles as fine aggregate in concrete. Osmani and Papp [6] performed an assessment of the compressive strength of glass-reinforced plastic waste-filled concrete for potential applications in construction. Ferreira et al. [7] showed influence of curing conditions on the mechanical performance of concrete-containing recycled plastic aggregate. Wankhade et al. [8–13] performed some studies on smart structures including beams and plates. Jha et al. [1] examined behavior of plastic waste fiber-reinforced industrial wastes in pavement applications. Wankhade [14] applied performance-based design and estimation of forces for building frames with earthquake loading. Wankhade and Landage [15] examined static analysis for fixed-base and base-isolated building frame. Wankhade et al. [16–19] used smart materials in construction purpose and did numerical optimization with piezo-laminated beam and plates. Yesane et al. [21] studied on soil–structure interaction. Yin et al. [22] studied fiber preparation and mechanical properties of recycled polypropylene for reinforcing concrete. Vanitha et al. [23] also suggested utilization of waste plastics as a partial replacement of coarse aggregate in concrete blocks. Yin et al. [24] demonstrated use of macroplastic fibers in concrete with a review. Kamaruddin [25] et al. gives potential use of plastic waste as construction materials with recent progress and future prospect. Wankhade et al. [26–36] did numerical optimization of piezo-laminated beams under static and dynamic excitations with buckling of plates. Bogas et al. [37] studied mechanical characterization of concrete produced with recycled lightweight expanded clay aggregate concrete. Binici and Aksogan [38] suggested eco-friendly insulation material production with waste olive seeds, ground PVC and wood chips. Shanmugapriya and Santhi [39] examined strength and chloride permeable properties of concrete with high-density polyethylene wastes.

Hence, in the present work, we aimed to prepare a brick with different composition of LDPE, PVC, cement and fly ash. The optimum content is found out to get desired results. These results are then presented in the graphical form.

2 Overview of Brick Making

A brick is made from concrete by a specified process. Concrete is formed by adding cement composition and water together. This pourable mixture of concrete may be poured into a brick mold. After hardens, it forms a brick.

In the same way, polyvinyl chloride (PVC) and low-density polyethylene (LDPE) can be mold in different shapes. PVC from plastic bottles may be recycled to produce further PVC bottles. Low-density polyethylene (LDPE) is a quite common plastic used in different utensils and products. Figure 1 shows a flowchart for preparation of plastic brick.

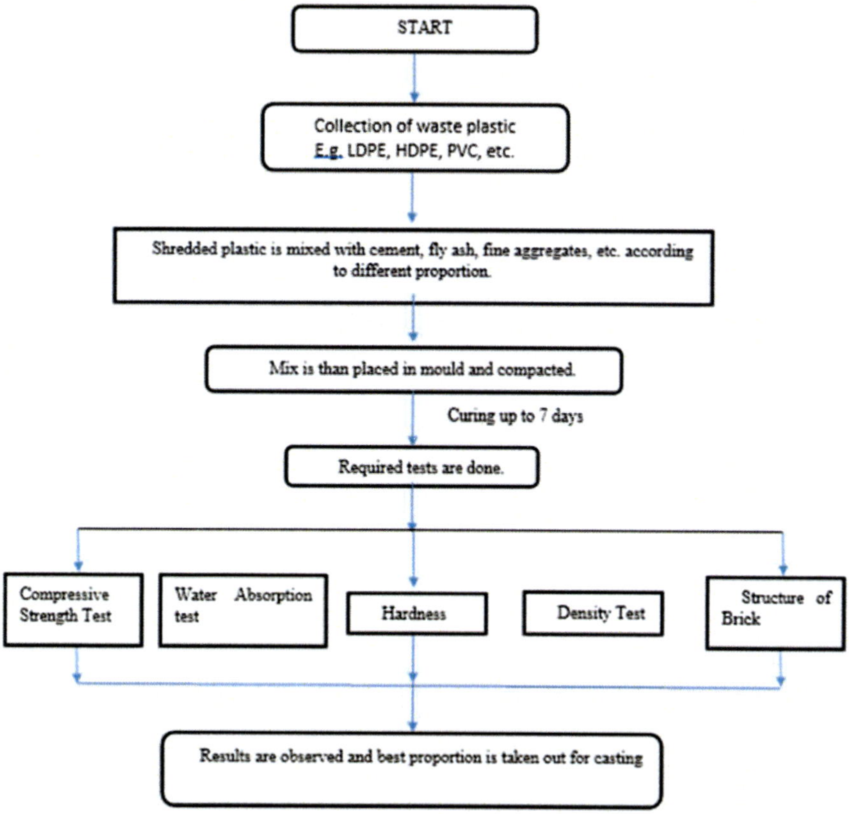

Fig. 1 Flow chart for preparation of brick sample

Table 1 Mix proportion for prepared for optimization of bricks

S. No.	Cement (kg)	Waste plastic (kg)	Fine aggregates (kg)	Fly ash (kg)	Total material (kg)
1	15%—0.525	20%—0.70	20%—0.7	45%—1.575	3.5
2	25%—0.875	35%—1.125	20%—0.7	20%—0.7	3.5
3	20%—0.8	30%—1.2	30%—1.2	20%—0.8	4
4	11%—0.385	30%—1.05	30%—1.05	29%—1.01	3.5
5	13%—0.39	30%—0.9	30%—0.9	27%—0.81	3
6	12%—0.48	25%—1	25%—1	38%—1.52	4
7	10%—0.4	20%—0.8	30%—1.2	40%—1.6	4
8	8%—0.32	40%—1.6	30%—1.2	40%—1.6	4

3 Mix Proportions for the Optimization of Brick

Mix proportion is prepared considering cement, waste plastic, fine aggregates and fly ash to form composite materials for the brick. Hence, Table 1 shows mix proportion for prepared for constructed bricks.

To get the optimum composition of plastic bricks, various mixes are prepared consists of plastic (PVC and LDPE) are made and tested to get the optimum results. The composition of plastic bricks are mentioned in the above table. For composition No. 4 and composition No. 5 LDPE (10%) + HDPE (15%) is used. And for other remaining composition, PVC is used. The above mix proportion are placed in mold and then cured for one week, and then the following tests are done to get the optimum result.

4 Results and Discussion

Different tests are carried out on the prepared brick samples. The tests include compressive strength test and density. To get the following results, samples of bricks of each composition are tested for every test mentioned above, and the average of results is considered. The outcomes of the tests are mentioned below.

For normal cement brick: Normal brick taken is fly ash brick (cement masonry unit (CMU)) (excluding plastic) of size 19 × 9 × 7 cm size. Mix proportions are placed in mold and then cured for one week, and then the tests (compressive strength, water absorption and density of brick) are done. In which, grade of cement used is OPC 43. We referred IS456:2000 and IS:8112-1989. Hence, we observed experimentally following results.

1. Compressive strength varies from 9.8 N/mm^2.
2. Water absorption is 14–15%.
3. Density of bricks is 1.8–1.9 g/cc.

Fig. 2 Shredded plastic in powdered (agglo) form with sieve and grinder

For bricks with polymer waste:

We used shredded plastic in powdered (agglo) form as shown in Fig. 2. Mixing procedure is done with help of shovels and hand. Aggregates used were fine aggregate (churi). Fly ash properties are given in end of document. We referred IS456:2000, IS:8112-1989, IS: 13767-1993 (burnt clay fly ash building bricks—specification) IS: 12894-2002 (Pulverized fuel ash-lime bricks—specification) (Fig. 3).

Physical and chemical properties of fly ash mixed with cement and waste polymer:

- The fineness of the fly ash is to be checked in dry sieving. The fly ash sample is sieved in 75 micron sieve and the percentage of retained on the 75 micron sieve is calculated.
- The specific gravity of fly ash is 1.90 for a sub-bituminous
- A dark gray to black color is typically attributed to an elevated un-burned content.
- Grade I type of Fly Ash as per IS Codes (IS 3812-1981) is derived from bituminous coal having fractions $SiO_2 + Al_2O_3 + Fe_2O_3$ greater than 70%.

Hence, Table 2, Figs. 4 and 5 shows comparison of compressive strength, water absorption and density test for different composition of bricks with above materials.

Table 2 Strength and characteristics for different composition of brick

Composition S. No.	Compressive strength (N/mm^2)	Water absorption (%)	Density of bricks (gm/cc)
1	10.5	9.8	1.58
2	13.9	10.5	1.49
3	11.3	8.9	1.66
4	7.6	9.9	1.5
5	9.9	10.2	1.49
6	13.25	12	1.6
7	12.01	10	1.65
8	11.15	9	1.63

Fig. 3 Bricks prepared from shredded plastic with cement and fly ash

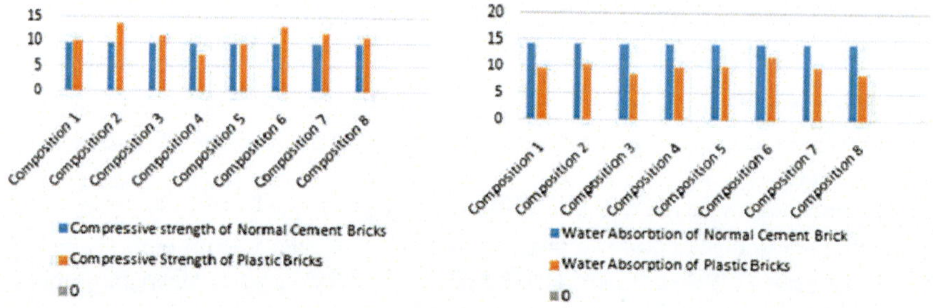

Fig. 4 Compressive strength and water absorption

5 Conclusion

In the present work, we obtained the optimum composition of plastic bricks, in which shredded plastic in powdered form is mixed with cement and fly ash. Total eight

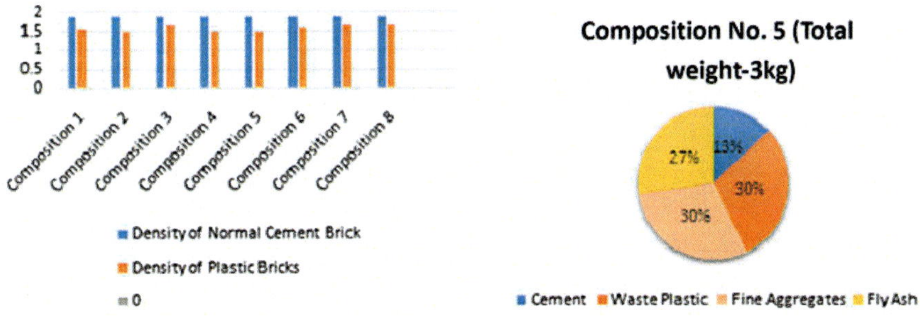

Fig. 5 Density and optimum composition

different compositions are tested for bricks out of which composition No. 5 gives desired results in terms of strength of bricks. It also has less cement in mix proportion. This brick composition provides less density and water absorption, provide higher compressive strength, and it is cost effective as compared to normal bricks. Hence, waste plastic (polymers) enhances the properties of bricks as it added in brick mix proportion to be useful as advanced building and construction material. Such recycled and reuse of waste plastic is treated as effective way for waste management. We performed several trials with different composition of bricks. It is observed that the maximum compressive strength sustained by the polypropylene/rubber composite brick is 13.25 N/mm^2 followed by LDPE/rubber composite brick with 12.01 N/mm^2 which is much higher than the fly ash brick which sustained only 9.82 N/mm^2. By implementation, this type of bricks in construction and industrial sector not only the plastic waste will be reduced but also new techniques of reusing it and making it as asset will discovered. By using this, deterioration of natural soil is also reduced. Suggested bricks in the present work are economic and cost effective.

References

1. Jha, J. N., Choudhary, A. K., Gill, K. S., & Shukla, S. K. (2014). Behavior of plastiwaste fiber-reinforced industrial wastes in pavement applications. *International Journal of Geotechnical Engineering, 8*(3), 277–286. https://doi.org/10.1179/1939787914Y.0000000044
2. Manhal, A. J., & Farah, P. (2016). Strength and behavior of concrete contains waste plastic. *Journal of Ecosystem Ecography, 6*(2). https://doi.org/10.4172/2157-7625.1000186
3. Appiah, J. K., Berko-Boateng, V. N., & Tagbor, T. A. (2017). Use of waste plastic materials for road construction in Ghana. *Case Studies in Construction Materials., 6*, 1–7.
4. Kamaruddin, M. A., Abdullah, M. M. A., Zawawi, M. H., & Zainol, M. R. R. A. (2017). Potential use of plastic waste as construction materials, recent progress and future prospect. *IOP Conference Series: Materials Science and Engineering, 267*(2017), 012011. https://doi.org/10.1088/1757-899X/267/1/012011
5. Frigione, M. (2010). Recycling of PET bottles as fine aggregate in concrete. *Waste management, 30*(6), 1101–1106.
6. Osmani, M., & Pappu, A. (2010). An assessment of the compressive strength of glass reinforced plastic waste filled concrete for potential applications in construction. *Concrete Research*

Letters, 1(1), 1–5.

7. Ferreira, L., de Brito, J., & Saikia, N. (2012). Influence of curing conditions on the mechanical performance of concrete containing recycled plastic aggregate. *Construction and Building Materials, 36*, 196–204.
8. Wankhade, R. L. (2011). Geometric nonlinear analysis of skew plates using finite element method. *International Journal of Advanced Engineering Technology, 2*(2), 154–163.
9. Bajoria, K. M., & Wankhade, R. L. (2012). Free vibration of simply supported piezolaminated composite plates using finite element method. *Advanced Materials Research, 587*, 52–56.
10. Wankhade, R. L., & Bajoria, K. M. (2012). Stability of simply supported smart piezolaminated composite plates using finite element method. In *Proceeding of the International Conference on Advances in Aeronautical and Mechanical Engineering-AME* (Vol. 1, pp. 14–19).
11. Deshpande, S. S., & Wankhade, R. L. (2013). Analysis of thick beams using first order shear deformation theory. In *International Conference on Structural Engineering and Mechanics*, December 20–22, 2013, Rourkela, India.
12. Wankhade, R. L., & Bajoria, K. M. (2013). Free vibration and stability analysis of piezolaminated plates using finite element method. *Smart Materials and Structures, 22*(125040), 1–10.
13. Wankhade, R. L., & Bajoria, K. M. (2013). Buckling analysis of piezolaminated plates using higher order shear deformation theory. *International Journal of Composite Materials, 3*, 92–99.
14. Wankhade, R. L. (2014). Performance based design and estimation of forces for building frames with earthquake loading. In *Paper 141, International Conference on Recent Trends and Challenges in Civil Engineering*, December 12–14, 2014, MNNIT Allahabad, India.
15. Wankhade, R. L., & Landage, A. B. (2014). Static analysis for fixed base and base isolated building frame. In *Proceedings of National Conference on Advances in Civil and Structural Engineering (NCACSE-2014).*
16. Wankhade, R. L., & Bajoria, K. M. (2015). Vibration of cantilever piezolaminated beam with extension and shear mode piezo actuators. In: *Active and Passive Smart Structures and Integrated Systems, Proceedings of SPIE* (Vol. 9431, p. 943122-1).
17. Wankhade, R. L., & Landage, A. B. (2016). Performance based analysis and design of building frames with earthquake loading. *International Journal of Engineering Research, 5*(1), 106–110.
18. Yesane, P. M., Ghugal, Y. M., & Wankhade, R. L. (2016). Study on soil–structure interaction: A review. *International Journal of Engineering Research, 5*(3), 737–741.
19. Wankhade, R. L. (2017). Performance analysis of Rc moment resisting frames using different rubber bearing base isolation techniques. In *International Conference on Innovations in Concrete For Infrastructure Challenges*, Nagpur, India, October 6–7, 2017
20. Bendine, K., & Wankhade, R. L. (2016). Vibration control of FGM piezoelectric plate based on LQR genetic search. *Open Journal of Civil Engineering, 6*, 1–7.
21. Wankhade, R. L., & Bajoria, K. M. (2016). Shape control and vibration analysis of piezolaminated plates subjected to electro-mechanical loading. *Open Journal of Civil Engineering, 6*, 335–345.
22. Yin, S., et al. (2015). Fiber preparation and mechanical properties of recycled polypropylene for reinforcing concrete. *Journal of Applied Polymer Science, 132*(16).
23. Vanitha, S., Natarajan, V., & Praba, M. (2015). Utilisation of waste plastics as a partial replacement of coarse aggregate in concrete blocks. *Indian Journal of Science and Technology, 8*(12).
24. Yin, S., et al. (2015). Use of macro plastic fibres in concrete: A review. *Construction and Building Materials, 93*, 180–188.
25. Kamaruddin, M. A., Abdullah, M. M. A., Zawawi, M. H., & Zainol, M.R.R.A. (2017). Potential use of plastic waste as construction materials: Recent progress and future prospect. *IOP Conference Series: Materials Science and Engineering, 267*, 012011. https://doi.org/10.1088/1757-899X/267/1/012011
26. Wankhade, R. L. (2014). Buckling of laminated composite plate subjected to axial compressive loading. In *International Conference on Recent Trends and Challenges in Civil Engineering*, December 12–14, 2014, MNNIT Allahabad, India.

27. Wankhade, R. L. (2014). Stability of composite laminates subjected to compressive in-plane loading. In *International Conference on Advances in Civil and Mechanical Engineering Systems*, at Government College of engineering, Amravati.
28. Patil, D. B., Wankhade, R. L., & Deshpande, P. K. (2015). Geometric nonlinear analysis of laminated composite plates using finite element method. *International Journal of Emerging Science and Engineering (IJESE), 3*(9), 1–5.
29. Bajoria, K. M., & Wankhade, R. L. (2015). Vibration of cantilever piezolaminated beam with extension and shear mode piezo actuators. *Proceedins of SPIE, 9431*(943122), 1–6.
30. Wankhade, R. L., & Bajoria, K. M. (2017). Numerical optimization of piezolaminated beams under static and dynamic excitations. *Journal of Science: Advanced Materials and Devices, 2*(2), 255–262.
31. Bendine, K., & Wankhade, R. L. (2017). Optimal shape control of piezolaminated beams with different boundary condition and loading using genetic algorithm. *International Journal of Advanced Structural Engineering, 9*(4), 375–384.
32. Wankhade, R. L., & Bajoria, K. M. (2019). Vibration analysis of piezolaminated plates for sensing and actuating applications under dynamic excitation. *International Journal of Structural Stability and Dynamics, 19*(10), 1950121.
33. Wankhade, R. L., & Niyogi, S. B. (2020). Buckling analysis of symmetric laminated composite plates for various thickness ratios and modes. *Innovative Infrastructure Solutions, 5*, 65. https://doi.org/10.1007/s41062-020-00317-8
34. Wankhade, R. L., & Bajoria, K. M. (2021). Vibration attenuation and dynamic control of piezolaminated plates for sensing and actuating applications. *Archive of Applied Mechanics, 91*(1), 411–426.
35. Wankhade, R. L., Niyogi, S., & Gajbhiye, P. (2021). Buckling analysis of laminated composites considering the effect of orthotropic material. *IOP Science, Journal of Physics: Conference Series, 1706*(1), 012188.
36. Bendine, K., & Wankhade, R. L. (2021). Piezoelectric energy harvesting from a curved plate subjected to time-dependent loads using finite elements. *IOP Science, Journal of Physics: Conference Series, 1706*(1), 012008.
37. Bogas, J. A., de Brito, J., & Figueiredo, J. M. (2015). Mechanical characterization of concrete produced with recycled lightweight expanded clay aggregate concrete. *Journal of Cleaner Production, 89*, 187–195.
38. Binici, H., & Aksogan, O. (2016). Eco-friendly insulation material production with waste olive seeds, ground PVC and wood chips. *Journal of Building Engineering, 5*, 260–266.
39. Shanmugapriya, M., & Santhi, H. (2017). Strength and chloride permeable properties of concrete with high density polyethylene wastes. *International Journal of Chemical Sciences, 15*(1).

Determination of the Probability of Collapse for Existing Building Using Rapid Visual Screening as Tool

Salil Jha and Shilpa Pal

Abstract Delhi is rapidly urbanizing city without considering the standard construction practices. As it is situated in seismic zone IV (as Per IS 1893:2016), it is a soft target for any severe seismic event. And, in order to safe guard structures against seismic excitation, a rapid performance evaluation strategy is the need of hour. Various developing countries including India have etched numerous guidelines to strategize the assessment of existing structures, but due to their complex in nature, they have failed to inherit the idea and eventually not been implemented. Adopting the various fundamentals of practices in various guidelines such as FEMA and ATC, this paper aims to simplify the application of RVS in purview of various Indian Standard Codes, in order to achieve a better probabilistic approach in estimation of probable life building. To support the study with realistic data, four different types of existing structures have been taken as case study, i.e., High-Rise RCC Building, Mid-Rise RCC Building, RCC plus Masonry Combined Building, and Load-Bearing Masonry Building. RCC plus Masonry Combined Building is found to have the least likelihood of being collapsed under MCE ground shaking, i.e., 0.63%, whereas the Mid-Rise RCC Building shows the highest likelihood of being collapsed under MCE ground shaking, i.e., 100%. However, the High-Rise RCC Building and Load-Bearing Masonry Building have been found to have fairly high likelihood of being collapsed under MCE ground shaking, i.e., 63%. From the study, it can be concluded that the RVS guidelines of FEMA P-154 (2015) can be used satisfactorily including permissible limits as per Indian standard for preliminary investigation, based on which a fair decision regarding the necessity of detailed technical evaluation can be done.

Keywords Vulnerability of building · Collapse probability of building · Case study on collapse probability

S. Jha (✉) · S. Pal
Department of Civil Engineering, Delhi Technological University, New Delhi 110042, India

A. K. Gupta et al. (eds.), *Advances in Construction Materials and Sustainable Environment*, Lecture Notes in Civil Engineering 196,
https://doi.org/10.1007/978-981-16-6557-8_63

1 Introduction

Rapid visual screening is a sidewalk survey of any building and is designed to be conducted at the streets mostly, where entry inside building may or may not be possible for the investigation of lateral load-resisting system and structural materials. Hence, in such cases, a building can rendered to be “Structurally Deficit,” whereas on the contrary it might be “Structurally Sound.” In this, a basic score pertaining to the structural hazard is assigned based on the lateral load-resisting system and seismicity region [2]. On the basis of various existing irregularities, different score modifiers are applied to basic score to arrive at final score. Eventually, the final score is related to collapse probability of existing building.

As the name suggests, its objective is to rapidly score the condition of existing structures, without carrying out any detailed structural analysis. It has been designed in a way that it identifies, groups, and ranks the building stocks that may be vulnerable, under risk targeted maximum considered earthquake (MCE_R) [3]. But, IS 1893:2016 (latest) is based upon MCE only; hence, MCE will be considered for evaluation. However, the screened building stock is then categorized as either seismic resistant or seismic vulnerable.

Rapid visual screening is a two stage method, i.e., level 1 rapid visual screening and level 2 rapid visual screening. For each building, level 1 screening is performed, whereas level 2 is an optional screening, done in order to assess the risk in a more refined form [3].

If a building has RVS score less than the cutoff value, then it is highly vulnerable and must be evaluated by some person of expertise, who would be responsible for the detailed analysis and suggestions on retrofitting [6].

2 Methodology of RVS to Calculate the Collapse Probability

Methodology of the rapid visual screening hovers around the aspects affecting the seismic vulnerability of any building such as seismic susceptibility of soil type available in the area of survey, irregularity in building, and presence of non-structural hazards [1].

In the entire process of Level 1 data collection, duration of 30 min at most is expected at one building site usually. The data collection sheet (FEMA P-154, 2015) pertaining to the pertinent seismicity of the area is used and is filled meticulously by correlating the existing condition with Indian standard codes, to reach a fair conclusion eventually.

Level 2 screening adds cost as it requires a structural engineer or someone having equivalent expertise, and the concerned person may have high hourly compensation. Level 2 screening may be done on high-priority structures, taking into account the expertise and cost involved [4].

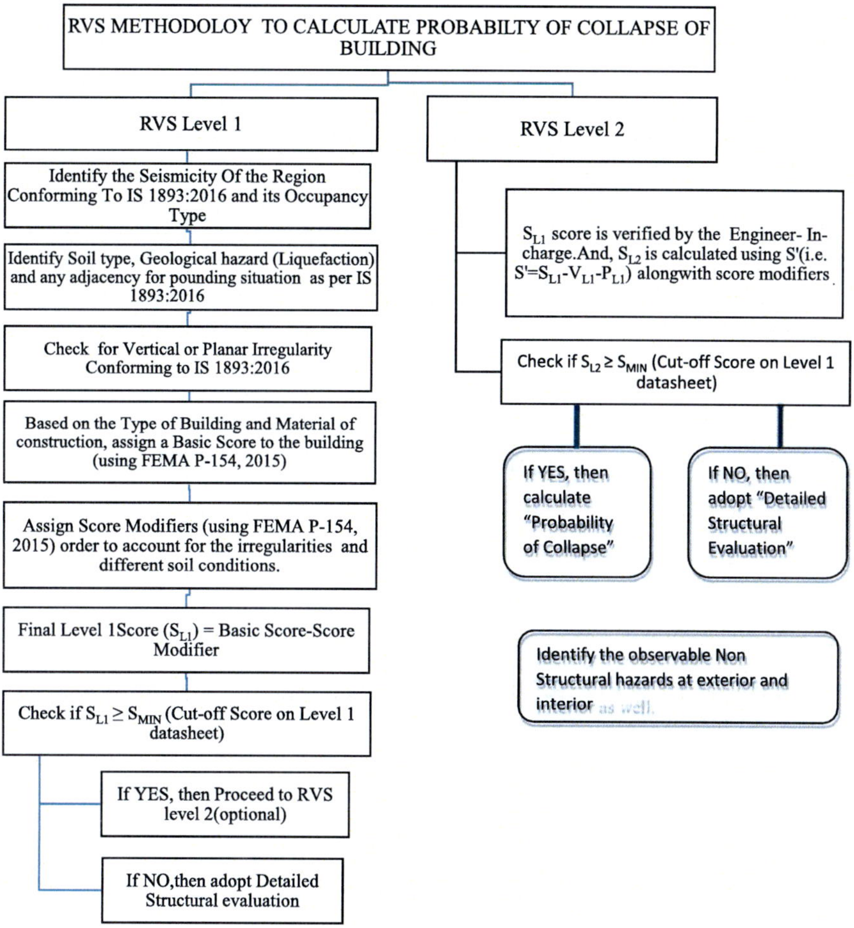

Fig. 1 Methodology of RVS for calculation of collapse probability [3]

3 Case Studies on Collapse Probability of Different Building Types

There are seventeen different types of buildings as per the material of construction, as per FEMA P-154 (2015). But, we have considered only four different types of buildings that are prevalent in the construction practices. Here in this case study, an attempt has been made to elucidate the procedure in a simplified and self-explanatory way. The different types of buildings are

(a) High-Rise RCC Framed Building
(b) Mid-Rise RCC Framed Building
(c) Combined RCC and Masonry Building

(d) Load-Bearing Masonry Building.

3.1 *Case Study on High-Rise RCC Framed Building*

Problem Statement A Basement + G + 11 storied residential building situated in Dwarka, Delhi, has five towers A, B, C, D, and E, where towers A, B, D, and E are identical in plan and form. Assess the structure for its collapse probability.

Comment

(A) As the buildings fall in zone IV as per IS 1893:2016, hence, it can be categorized as situated in region of high seismicity as per FEMA P-154, 2015.

(B) Soil type at the site was found to be of "Medium Soil" category (As per IS: 1893, 2016), which corroborates to soil type D/Stiff Soil (As per FEMA P-154, 2015). Moreover, geological hazard such as liquefaction is not present. Although, there are five blocks of building, but all of them are separated far enough, i.e., 4 m. Hence, chances of pounding has been ruled out.

(C) Vertical irregularity:- All the stories are spaced equally, vertical irregularity such as soft story, mass irregularity, vertical geometric irregularity, in-plane discontinuity, and weak story conforming to Table 6 of IS 1893:2016 [7], has not been observed during survey in the building blocks A, B, C, D, and E.

Plan irregularity: The floor plan of blocks A, B, D, and E was found to be same, and it has been found that the foresaid blocks have re-entrant corners, whereas plan irregularities such as excessive cutouts and out of plane offsets in vertical elements have been missing.

The floor plan of block C was found to have re-entrant corners, whereas plan irregularities such as excessive cutouts and out of plane offsets in vertical elements have been missing.

(D) Based on the type of building and its material of construction, the building can be categorized as concrete frames with unreinforced masonry infill walls (C3) as described In FEMA P-154, 2015. Hence, the basic score rendered for such a building in high seismicity zone is 1.2.

(E) As discussed above, only plan irregularity in terms of re-entrant corners [7] is present in all of the building blocks. Hence, score modifier for plan irregularity is −0.5. As per the data obtained, the building is 17 year old which means it falls within the scope of post-benchmark [5]. But, the score modifier for post-benchmark is N.A. for the given category of building (C3).

(F) Final level 1 score $(S_{L1}) = 1.2 + (-0.5) = 0.7 > S_{\min}$ (0.3). At this stage, the screener can stop the survey if the RVS level 1 score is twice of $S_{\min}$ or the cutoff valve. However, we now proceed toward RVS level 2. As RVS level 2 is a detailed assessment of entire building based on technical discretion of screener, hence, the screener must have strong technical background.

(G) The first step in RVS level 2 is to adjust the baseline score; therefore, $S' = S_{L1} - V_{L1} - P_{L1}$

where

S_{L1} = Final level 1 score.
V_{L1} = Score modifier due to vertical irregularity.
P_{L1} = Score modifier due to plan irregularity.

$S' = 0.7 - 0 - 0.5 = 0.2$

Now, apply score modifier on the basis of irregularity present in the building. However, in this building stock, only re-entrant corners are present, to which a negative score modifier of -0.4 is to be assigned. Moreover, the structure can be observed as a redundant one, and therefore, a positive score modifier of 0.3 can be assigned.

Eventually, final level 2 score $S_{L2} = 0.2 - 0.4 + 0.3 = 0.1 < S_{min}$ (i.e. 0.3).

Therefore, $S_{L2} = S_{min} = 0.3$

(I) Extent of review = All sides

Interior = Entered.
Drawing reviewed = No.
Level 2 screening performed = Yes.
Pounding potential = Ignore pounding as level 2 screening is performed.
Non-structural hazard = Absent.

Therefore, probability of collapse = 10^{-S}.

Probability of collapse = $10^{-0.3}$
Probability of collapse = 0.5
Probability of collapse = 50%

Hence, it can be inferred that the buildings have 1 in 2 chance of being collapse under MCE ground motion.

(J) Action required

Detailed structural evaluation is required as RVS level 2 score is less than cutoff.

3.2 *Case Study on Mid-Rise RCC Framed Building*

Problem Statement A Basement + G + 1 storied Commercial Building situated at Mathura Road, Delhi is an RCC OMRF Structure. Assess the Structure for its collapse Probability.

Comment

(A) As the buildings fall in zone IV as per IS 1893:2016, hence, it can be categorized as situated in region of HIGH seismicity as per FEMA P-154, 2015.

(B) Soil type at the site was found to be of "Medium Soil" category (As per IS: 1893, 2016), which corroborates to Soil type D/Stiff Soil (As per FEMA P-154, 2015). Moreover, geological hazard such as liquefaction is not present. However, the building is separated far enough from adjacent buildings. Hence, chances of pounding have been ruled out.

(C) Vertical irregularity: Ground story is 6.1 m high, whereas other stories are 3.1 m high. At ground story, various heavy instruments and apparatus for the calibration and testing of commercial machines are kept, due to which there is mass irregularity in the building along soft story. Vertical irregularity such as vertical geometric irregularity, in-plane discontinuity and weak story conforming to Table 6 of IS 1893:2016 [7], has not been observed during survey in the building block A, B, C, D, and E.

Plan irregularity: The floor plan at each story is found to be same, and it has been found that the building has re-entrant corners, whereas plan irregularities such as excessive cutouts and out of plane offsets in vertical elements have been missing.

(D) Based on the type of building and its material of construction, the building can be categorized as concrete frames with unreinforced masonry infill walls (C3) as described in FEMA P-154, 2015. Hence, the basic score rendered for such a building in high seismicity zone is 1.2.

(E) As discussed above, severe vertical irregularity along with plan irregularity in terms of re-entrant corners is present in the building. Hence, score modifier for severe vertical irregularity is −0.7 and for plan irregularity is −0.5. As per the data obtained, the building is 14 year old which means it falls within the scope of post-benchmark [5]. But, the score modifier for post-benchmark is N.A. for the given category of building (C3).

(F) Final level 1 score (S_{L1}) $= 1.2 + (-0.7) + (-0.5) = 0 < S_{\min}$ (0.3). At this stage, the screener must recommend the building for detailed structural evaluation.

(H) Extent of review = All sides

Interior = Entered.

Drawing reviewed = No.

Level 2 screening performed = No.

Pounding potential = No.

Non-structural hazard = Absent.

(I) Action required

Detailed structural evaluation is required as RVS level 1 score is less than cutoff.

3.3 *Case Study on RCC Plus Masonry (Combined) Building*

Problem Statement A Basement + G + 2 storied structure situated in Udyog Vihar, Gurgaon, Delhi NCR is an RCC Plus Masonry (Combined) Structure having glass gladding as a facade. Assess the structure for its collapse probability.

Comment

(A) As the buildings fall in zone IV as per IS 1893:2016, hence, it can be categorized as situated in region of HIGH seismicity as per FEMA P-154, 2015.

(B) Soil type at the site was found to be of "Soft Soil" category (As per IS: 1893, 2016), which corroborates to "Soil Type E/Soft Clay Soil" (As per FEMA P-154, 2015). Moreover, geological hazard such as liquefaction is not present. As adjacent buildings are separated far enough, hence, chances of pounding has been ruled out.

(C) Vertical irregularity: All the stories are spaced equally except basement as it has more height, vertical irregularity such as mass irregularity, vertical geometric irregularity, in-plane discontinuity and weak story conforming to Table 6 of IS 1893:2016 [7], has not been observed during survey, except soft story at basement level.

Plan Irregularity: The floor plan at each story was found to be same and plan irregularities such as excessive cutouts, re-entrant corners, and out of plane offsets in vertical elements have not been observed.

(D) Based on the type of building and its material of construction, the building can be categorized as concrete frames with unreinforced masonry infill walls (C1) as described In FEMA P-154, 2015. Hence, the basic score rendered for such a building in high seismicity zone is 1.5.

(E) As discussed above, only moderate vertical irregularity is present at the basement. Hence, score modifier for moderate vertical irregularity is −0.5. As per the data obtained, the building is 10 year old which means it falls within the scope of post-benchmark [5]. And, the score modifier for post-benchmark is 1.9 for the given category of building (C1).

(F) Final level 1 score (S_{L1}) $= 1.5 + (-0.5) + 1.9 = 2.9 > S_{\min}$ (0.3). At this stage, the screener can stop the survey if the RVS level 1 score is twice of $S_{\min}$ or the cutoff valve. However, we now proceed toward RVS level 2. As RVS level 2 is a detailed assessment of entire building based on technical discretion of screener, hence, the screener must have strong technical background.

(G) The first step in RVS level 2 is to adjust the baseline score; therefore, $S' = S_{L1} - V_{L1} - P_{L1}$

where

S_{L1} = Final level 1 score.
V_{L1} = Score modifier due to vertical irregularity.
P_{L1} = Score modifier due to plan irregularity.

$$S' = 2.9 - 0.5 - 0.0 = 2.4$$

(H) Now, apply score modifier on the basis of irregularity present in the building. However, in this building, only soft story is present at basement level, to which a negative score modifier of −0.5 is to be assigned. Moreover, the structure can be observed as a redundant one; therefore, a positive score modifier of 0.3 can be assigned.

Eventually, final level 2 score $S_{L2} = 2.4–0.5 + 0.3 = 2.2 > S_{min}$ (i.e., 0.3). Therefore, $S_{L2} = 2.2$

(I) Extent of review = All sides

Interior = Entered.
Drawing reviewed = NO.
Level 2 screening performed = YES.
Pounding potential = Ignore pounding as level 2 screening is performed.
Non-structural hazard = Glass cladding is present on all the faces.
Therefore, probability of collapse = 10^{-S}.

Probability of collapse = $10^{-2.2}$
Probability of collapse = 0.0063.
Probability of collapse = 0.63%

Hence, it can be inferred that the buildings have 1 in 158 chance of being collapse under MCE ground motion.

(J) Action required

Detailed structural evaluation is not required as RVS level 2 score is much greater than cutoff value.

However, non-structural hazards identified must be evaluated in detail.

3.4 Case Study on Load-Bearing Masonry Structure

Problem Statement A G + 3 storied hostel building situated in Shahbad Daulatpur, Delhi, is a Load-Bearing Masonry Structure. Assess the structure for its collapse probability.

Comment

(A) As the buildings fall in zone IV as per IS 1893:2016, hence, it can be categorized as situated in region of high seismicity as per FEMA P-154, 2015.

(B) Soil type at the site was found to be of "Medium Soil" category (As per IS: 1893, 2016), which corroborates to Soil type D/Stiff Soil (As per FEMA P-154, 2015). Moreover, geological hazard such as liquefaction is not present. As there is enough space between the buildings, therefore, pounding potential is ruled out.

(C) Vertical irregularity: All the stories are spaced equally, and therefore, vertical irregularity such as soft story, mass irregularity, vertical geometric irregularity, in-plane discontinuity, and weak story conforming to Table 6 of IS 1893:2016 [7], has not been observed during survey.

Plan irregularity: The floor plan at each story was found to be same and plan irregularities such as excessive cutouts, and out of plane offsets in vertical elements have not been observed. At certain locations, re-entrant corners can be seen.

(D) Based on the type of building and its material of construction, the building can be categorized as unreinforced masonry (URM) as described in FEMA P-154, 2015. Hence, the basic score rendered for such a building in high seismicity zone is 1.0.

(E) As discussed above, only plan irregularity is present at all floor in form of re-entrant corners. Hence, score modifier for plan irregularity is −0.4.

(F) Final level 1 score (S_{L1}) = 1.0 + (−0.4) = 0.6 > $S_{\min}$ (0.2). At this stage, the screener can stop the survey if the RVS level 1 score is twice of $S_{\min}$ or the cutoff valve. However, we now proceed toward RVS level 2. As RVS level 2 is a detailed assessment of entire building based on technical discretion of screener, hence, the screener must have strong technical background.

(G) The first step in RVS level 2 is to adjust the baseline score, and therefore, $S' = S_{L1} - V_{L1} - P_{L1}$

where

S_{L1} = Final level 1 score.
V_{L1} = Score modifier due to vertical irregularity.
P_{L1} = Score modifier due to plan irregularity.
$S' = 0.6 - 0.0 - 0.4 = 0.2$

(H) Eventually, final level 2 score $S_{L2} = 0.2 + (-0.4) + (0.3) = 0.1 < S_{\min}$ (i.e., 0.2).
Therefore, $S_{L2} = S_{\min}$(i.e., 0.2).

(I) Extent of review = All sides

Interior = Entered.
Drawing reviewed = No.
Level 2 screening performed = Yes.
Pounding potential = Ignore pounding as level 2 screening is performed.
Non-structural hazard = NONE.
Therefore, Probability of collapse = 10^{-S}.

Probability of collapse = $10^{-0.2}$
Probability of collapse = 0.63.
Probability of collapse = 63%

Hence, it can be inferred that the buildings have 1 in 1.58 chance of being collapse under MCE ground motion.

(J) Action required
Detailed structural evaluation is required as RVS level 2 score is less than cutoff value.

4 Results and Discussions

Each of the building, based on its characteristic and configuration, yielded different result. A comparison based on the collapse probability can be drawn among different building type, as to which building can perform satisfactorily under MCE ground shaking. Speaking of the performance of building, as per the case study, it can be clearly seen that although the buildings at first sight seemed fine. But after RVS, various irregularities either in plan or in vertical direction, summed as a negative aspect. These negative aspects, along with the various deficiencies present in the building, cause collapse of the building. It is also possible that there could be instances when the S_{L2} will be higher S_{L1}, and it can be accounted by detailed screening of building attributes and less conservative score modifier. Hence, a less conservative result will lead to an approximately exact assessment of structure.

As per our case study, a High-Rise RCC Building has been found as vulnerable, as its collapse probability is equal to 63%. And, on this basis, it can be referred to detailed structural evaluation. Similarly, a Mid-Rise RCC Building has collapse probability of 100%, rendering it dangerous for operations and requires immediate action.

On the other hand, a RCC Plus Masonry Combined Building has been found to have collapse probability less one percent, which hints the benefits of a regular structure free from deficiencies. However, a Load-Bearing Masonry Building has been found to have collapse probability of 63%, which is again vulnerable.

4.1 RVS Score and Its Interpretation

RVS score can be correlated to the damage grade, which a building could suffer in the event of seismic excitation.

FEMA P-154 has described various RVS scores and their corresponding damage grade as discussed below,

(a) If a building is having RVS score less than 0.3, then it can be said to have very high likelihood of Damage Grade 4, whereas Damage Grade 5 can also be observed.
(b) If a building is having 0.3 < RVS score < 0.7, then it can be said to have very high likelihood of Damage Grade 3, whereas Damage Grade 4 can also be observed.
(c) If a building is having 0.7 < RVS score < 2.0, then it can be said to have very high likelihood of Damage Grade 2, whereas Damage Grade 3 can also be observed.
(d) If a building is having 2 < RVS score < 3, then it can be said to have very high likelihood of Damage Grade 1, whereas Damage Grade 2 can also be observed.
(e) If a building is having RVS score > 3, then it can be said to have very high likelihood of Damage Grade 1.

Table 1 Building type and their collapse probability

S. No.	Building type	Photograph of building	RVS score	Collapse probability
1	High-Rise RCC Building		0.3	1 in 2, i.e., 50%
2	Mid-Rise RCC Building		0	1 in 1, i.e., 100%
3	RCC Plus Masonry Combined Building		2.2	1 in 158, i.e., 0.63%
4	Load-Bearing Masonry Building		0.2	1 in 1.58, i.e., 63%

5 Conclusions

In the study, four different types of buildings are selected, i.e., High-Rise RCC Building, Mid-Rise RCC Building, RCC Plus Masonry Combined Building, and Load-Bearing Masonry Building have been evaluated for calculating the RVS score and its corresponding probability of collapse under MCE. As per the results, it can be inferred that the probability of collapse of RCC Plus Masonry Combined Building under MCE ground shaking is least, i.e., 0.63%. The absence of plan irregularity and vertical irregularity in the building has resulted into least chance of collapse under MCE ground shaking. Hence, it has not been recommended for detailed evaluation.

Similarly, accounting to the presence of irregularities in High-Rise RCC Building and Load-Bearing Masonry Building, they have 63% chances of being collapsed under MCE ground shaking which is fairly high. Hence, the buildings have been recommended for further detailed evaluation.

And, due to a severe vertical irregularity present in the Mid-Rise RCC Building, it has resulted into 100% likelihood of being collapsed under MCE ground shaking. Therefore, it has been further recommended for detailed evaluation.

Based upon the study, it can be concluded that RVS guidelines of FEMA P-154 (2015) can be used as a tool for performing preliminary investigation of existing buildings.

References

1. American Society of Civil Engineers. (2003). *Seismic evaluation of existing buildings*, SEI/ASCE Standard No. 31–03 (444 p.). American Society of Civil Engineers, Reston, VA.
2. Federal Emergency Management Agency. (1998). *Handbook for the seismic evaluation of buildings—A prestandard.* FEMA310. http://www.degenkolb.com/0_0_Misc/0_1_FEMADocuments/fema310/prestnd.html
3. Federal Emergency Management Agency. (2015). *Rapid visual screening of buildings for potential seismic hazards: A handbook.* FEMA 154, 3rd edn.
4. Arya, A. S. (2011). *Rapid structural and non-structural assessment of school and hospital buildings in SAARC countries.* SAARC Disaster Management Centre, New Delhi.
5. IS 1893: 1962. (1962). *Criteria for earthquake resistance design of structure.* Indian Standard, Bureau of Indian Standards, New Delhi.
6. IS 4326: 1967. (1967). *Earthquake resistance design and construction of buildings, code of practice.* Indian Standard, Bureau of Indian Standards, New Delhi.
7. IS 1893: 2016. (2016). *Criteria for earthquake resistance design of structure part 1 general provisions and buildings.* Indian Standard, Bureau of Indian Standards, New Delhi.
8. Jain, S. K., Mitra, K., & Shah, M. (2010). A proposed rapid visual screening procedure for seismic evaluation of RC frame buildings in India. *Earthquake Spectra, 26*(3), 709–729.
9. Sinha, R., & Goyal, A. (2004). *A national policy for seismic vulnerability assessment of buildings and procedure for rapid visual screening of buildings for potential seismic vulnerability.*

Prioritizing Buildings for Seismic Retrofit on the Basis of RVS Score

Salil Jha and Shilpa Pal

Abstract As per the studies conducted in past, it has been stipulated that it is not the earthquakes which kill humans, instead it is the poor and substandard practices involved in the construction of building which leads to its failure in the seismic event. Although, various IS codes for the design of building are available, but due to the rapid urbanization, they have been overlooked. Therefore, such deficient buildings need to be screened at earlier stage in order to address their vulnerability. Various screening guidelines have been issued by different agencies, but FEMA P-154 (2015) supplements the screening in the most comprehensive manner by scoring the screened building for its various attributes like RVS score has been used to calculate risk of earthquake causing the collapse of building [1]. Two different types of structures have been taken, i.e. RCC G + 3 building and load-bearing masonry building for calculating the risk score and its associated vulnerability using the technique as per FEMA for any seismic activity in the next 50 years. It has been found that the load-bearing masonry building has 3.11% chance of being confronted by an earthquake that can cause collapse. And the probability of an earthquake causing the RCC G + 3 building to collapse over the next 50 years has been calculated to be 1.57%. The risk score is an indicator of the degree of fatality of the building, and it has been calculated that load-bearing masonry building is 100 times more fatal than the newly constructed load-bearing masonry building. Therefore, there is a need to detail structural evaluation for retrofitting the load-bearing masonry building to be safe under any seismic activity. Hence, as per the study, the methodology will help in prioritizing the building for detailed structural evaluation and retrofitting recommendations.

Keywords Retrofitting · Probability of collapse causing earthquake · Rapid visual screening

S. Jha (✉) · S. Pal
Department Of Civil Engineering, Delhi Technological University, New Delhi 110042, India

A. K. Gupta et al. (eds.), *Advances in Construction Materials and Sustainable Environment*, Lecture Notes in Civil Engineering 196,
https://doi.org/10.1007/978-981-16-6557-8_64

1 Introduction

RVS being a sidewalk survey, it offers a systematic approach to pinpoint those structures which could be potential threat in the event of seismic hazards [2]. A variety of structures are included in it ranging from low-rise to high-rise, which makes it an inclusive approach. Based on the RVS score and their corresponding cut-off values. Buildings are stated either "Structurally Deficit "or "Structurally fit", after which the former is further subjected to the detailed investigation [3]. A variety of screening methods are put forward by different agencies, in which the focus has been laid upon the detailed evaluation. But the advantage of the RVS methodology provided by FEMA P-154 (2015) is that it scores the existing building to be screened, as per the irregularities present in the building and the prevailing site condition. As per the study, RVS score is a measure of the collapse probability for any given building subjected to MCE ground shaking [2]. But RVS score does not cater for levels of shaking either less or greater than MCE ground shaking [4]. Hence, the RVS methodology provided by FEMA P-154 (2015) assists in distinguishing among the different building types based upon their RVS scores. Luco et al. (2007) put forward a design methodology, in which, if the collapse probability of building was 10% at a given risk targeted maximum considered earthquake [5], then the same building would have the collapse probability of 1% in next fifty years given any level of shaking [4]. On the same note, RVS score is correlated with the risk score, once the rapid visual screening is complete for a building [4]. The risk score calculated is then correlated with the probability of at least one earthquake occurs during *t* years that is strong enough to cause collapse [4]. Eventually, the probable fatality based on risk score and probability of collapse-causing earthquake is considered for the decision of detailed technical evaluation and subsequently seismic retrofit.

2 Methodology for the Calculation of Probability of Earthquake Leading to Collapse

Risk score when compared to RVS score is significantly unique in a way that it gives the likelihood of collapse at any level of shaking [6]. However, to evaluate the risk score of a building, it is required to have the RVS score of that building as shown in Fig. 1. As per our study, it has been found that FEMA P-154 does not shed any light on the score modifiers for cracks present in the building. Therefore, as per the discretion, it has been proposed to reduce the RVS score by 25% in the seismic zone IV and V to account for cracks or any defects present in the building which may go unnoticed during the survey. As per FEMA P-154 (2015), it has been found that only if a structure is retrofitted globally, then a score of 1.4 is added to the RVS score. Eventually, the risk score for that building for the future surveys will be increased by numerical value of 1.4 and correspondingly the probability of collapse causing earthquake for next 50 years will be reduced by significant percentage. Whereas, any

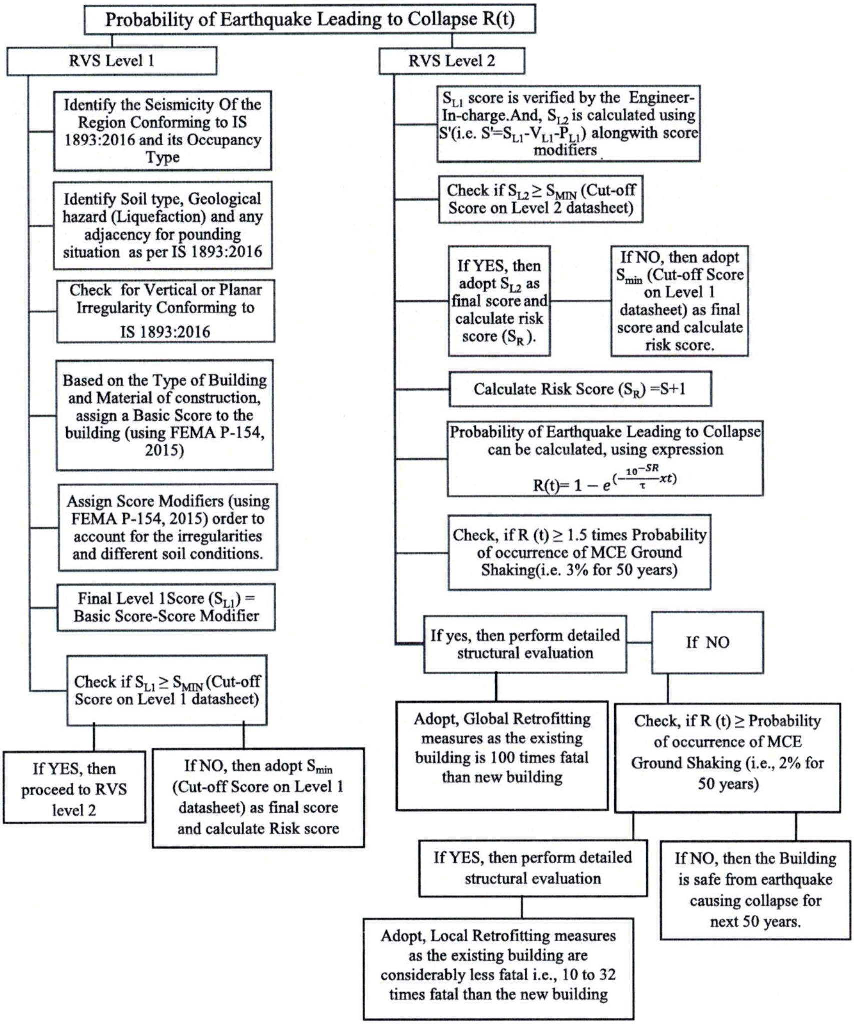

Fig. 1 Methodology of RVS for calculation of probability of collapse causing earthquake [4]

of the score modifier does not account for local retrofitting in RVS score. Therefore, a global retrofitting measure to be adopted must be supported by the cost–Benefit ratio. Although, as a general thumb rule, if the retrofitting is less than thirty percent of cost of reconstruction, retrofitting must be adopted [3].

3 Case Study on Retrofitting of Buildings

As per our study, based on the type of construction and different material used, seventeen types of building have been identified in FEMA P-154 (2015). But, to correlate the actual observation on site and risk score, we have adopted two different types of building. One of the structures is RCC G + 3 storied building whereas other is load-bearing masonry hostel building, and both are situated in Delhi. The case study has been carried out in a detailed manner to promote better understanding and discretion.

3.1 *Case Study on RCC G + 3 Storey Framed Building*

Problem Statement An RCC OMRF building having G + 3 storey situated in Delhi. It has masonry infills. It was constructed in the year of 1984 and has been under periodical repair since then. It has not gone through any major upgrades ever since. Comment on its probability of at least one collapse causing earthquake and retrofitting are summarized in Table 1.

3.2 *Case Study on Load Bearing Masonry Hostel Building*

Problem Statement A masonry hostel building having G + 3 storey situated in Delhi has been under periodical repair, and it has not gone through any major upgrades. Comment on its probability of at least one collapse causing earthquake and retrofitting given in Table 2.

4 Results and Discussion

As per the case study, the RCC G + 3 building has risk score of 2.4 and the probability of earthquake causing its collapse $R(t)$ is equal to 1.57%, which is less than probability of MCE (as per IS 1893:2016); hence, the building can be rendered as "SAFE". Moreover, as per the site survey done, the building was found to have sound structural system (Table 3). Although, there have been some patch work done to spalling of concrete slabs. But otherwise, building was in "Satisfactory condition".

Similarly, a load-bearing masonry building was surveyed, and its risk score is equal to1.5. The probability of earthquake causing collapse $R(t)$ is equal to 3.11%, which is greater than the probability of MCE (as per IS 1893:2016); hence, the building can be rendered as "UNSAFE". As per the actual site condition, building

Table 1 Comments on the probability of at least one collapse causing earthquake and retrofitting

S. No.	Rapid Visual Screening Procedure	RVS Score corresponding to the level of seismicity (using datasheet of FEMA P-154, 2015)	Remarks
I.	RVS Level 1		
A.	Seismicity of the region conforming to IS 1893:2016 and its occupancy type	Zone IV (IS 1893:2016)	
		HIGH seismicity (FEMA P-154, 2015)	
		Commercial occupancy	
B.	Soil type as per IS 1893:2016	Medium (IS 1893:2016)	
		Type D/Stiff Soil (FEMA P-154, 2015)	
	Geological hazard (Liquefaction)	None	
	Chances of Pounding	Adjacent buildings are situated far apart. Hence, no chances of pounding	
C.	Planar Irregularity Conforming to IS 1893:2016 (P_{L1})	Plan irregularities such as re-entrant corners, excessive cutouts and out-of-plane offsets in vertical elements are not observed	
D.	Vertical Irregularity Conforming to IS 1893:2016 (V_{L1})	Vertical irregularity such as soft storey, mass irregularity, vertical geometric irregularity, in-plane discontinuity and weak storey are not observed	

(continued)

Table 1 (continued)

S. No.	Rapid Visual Screening Procedure		RVS Score corresponding to the level of seismicity (using datasheet of FEMA P-154, 2015)	Remarks
E.	Type of Building (using FEMA P-154, 2015)		Building is identified as concrete frames with unreinforced masonry infill walls (C3)	
F.	Basic Score to the building (using FEMA P-154, 2015)		1.2	
G.	Score Modifier			
	i.	Planar Irregularity (P_{L1})	0.0	Not Observed
	ii.	Vertical Irregularity (V_{L1})	0.0	Not Observed
	iii.	Pre Code modifier	−0.1	Built before 2002[a]
	iv.	Post-Benchmark modifier	0.0	
H.	Final Level 1Score (S_{L1}) S_{L1} = Basic Score − Score Modifier (1)		1.1	> S_{min} (i.e., 0.3)
II.	RVS Level 2			
A.	Adjusted Baseline Score, $S'\left(\text{i.e. } SS' = S_{L1} - V_{L1} - P_{L1}\right)$ (2)		$1.1 - 0 - 0 = 1.1$	
B.	Score Modifier			
	i.	Planar Irregularity (P_{L2})	0.0	Not Observed
	ii.	Vertical Irregularity (V_{L2})	0.0	Not Observed
	iii.	Redundancy(R_1)	0.3	Space frame with fixed supports
	iv.	Retrofit(R_2)	0.0	Not Observed
C.	Final Level 2 Score S_{L2}, $\left(S_{L2} = S' - V_{L2} - P_{L2} + R_1 + R_2\right)$ (3)		1.4	> S_{min} (i.e., 0.3)

(continued)

Table 1 (continued)

S. No.	Rapid Visual Screening Procedure	RVS Score corresponding to the level of seismicity (using datasheet of FEMA P-154, 2015)	Remarks
D.	Risk Score (S_R) $= S + 1$ (4)	2.4	
	To account for unnoticed cracks of past earthquake, which could be catastrophic in the events of Earthquakes, the Risk Score must be reduced by 25%	$0.75 \times 2.4 = 1.8$	
E.	Probability of at least one earthquake occurs during t years that is strong enough to cause collapse, $R(t) = 1 - e^{\left(-\frac{10^{-SR}}{\tau} \times t\right)}$ (5) Where τ = design life in years (50 years) t = number of years (adopted 50 years)	$R(t) = 1 - e^{(-\frac{10^{-1.8}}{50} x50)}$ $R(t) = 1.57\%$	Less than the probability of MCE (i.e., 2% for 50 years)
F.	Recommendations		
	i. As probability of at least one earthquake occurs during 50 years that is strong enough to cause collapse is significantly less, when compared to probability of occurrence of MCE ground shaking. Hence, structure is prioritized as safe		
	ii. No need for detailed technical evaluation as the existing building is at thirty two times the fatality risk of a new building, which is considerably less. Hence, No requirement of retrofitting		

[a] Year 2002 has been taken as the benchmark year because IS 1893 was revised in this year. A lot of philosophical changes were observed such as four zones in India, realistic values of acceleration etc.

Table 2 Comment on the probability of at least one collapse causing earthquake and retrofitting

S. No.	Rapid visual screening procedure	RVS Score corresponding to the level of seismicity (using datasheet of FEMA P-154, 2015)	Remarks
I.	RVS Level 1		
A.	Seismicity of the region conforming to IS 1893:2016 and its occupancy type	Zone IV (IS 1893:2016)	
		HIGH seismicity (FEMA P-154, 2015)	
		Commercial occupancy	
B.	Soil type as per IS 1893:2016	Medium (IS 1893:2016)	
		Type D/Stiff Soil (FEMA P-154, 2015)	
	Geological hazard (Liquefaction)	None	
	Chances of pounding	Adjacent buildings are situated far apart. Hence, no chances of pounding	
C.	Planar Irregularity Conforming to IS 1893:2016 (P_{L1})	Plan irregularities such as re-entrant corners, excessive cutouts and out-of-plane offsets in vertical elements are not observed	
D.	Vertical Irregularity Conforming to IS 1893:2016 (V_{L1})	Vertical irregularity such as soft storey, mass irregularity, vertical geometric irregularity, in-plane discontinuity and weak storey are not observed	

(continued)

Table 2 (continued)

S. No.	Rapid visual screening procedure		RVS Score corresponding to the level of seismicity (using datasheet of FEMA P-154, 2015)	Remarks
E.	Type of Building (using FEMA P-154, 2015)		Building is identified as unreinforced masonry (URM)	
F.	Basic Score to the building (using FEMA P-154, 2015)		1.2	
G.	Score Modifier			
	i.	Planar Irregularity (P_{L1})	0.0	Not observed
	ii.	Vertical Irregularity (V_{L1})	0.0	Not observed
	iii.	Pre Code modifier	0.0	Year of construction is unknown
	iv.	Post-Benchmark modifier	0.0	Year of construction is unknown
H.	Final Level 1Score (S_{L1}) S_{L1} = Basic Score − Score Modifier (1)		1.0	> S_{min} (i.e., 0.2)
II.	RVS Level 2			
A.	Adjusted Baseline Score, $S'\left(\text{i.e. } SS' = S_{L1} - V_{L1} - P_{L1}\right)$ (2)		1.0 − 0 − 0 = 1.0	
B.	Score Modifier			
	i.	Planar Irregularity (P_{L2})	0.0	Not observed
	ii.	Vertical Irregularity (V_{L2})	0.0	Not observed
	iii.	Redundancy (R_1)	0.0	Not observed
	iv.	Retrofit (R_2)	0.0	Not observed

(continued)

Table 2 (continued)

S. No.	Rapid visual screening procedure	RVS Score corresponding to the level of seismicity (using datasheet of FEMA P-154, 2015)	Remarks
C.	Final Level 2 Score S_{L2}, $(S_{L2} = S' - V_{L2} - P_{L2} + R_1 + R_2)$ (3)	1.0	$> S_{\min}$ (i.e., 0.2)
D.	Risk Score $(S_R) = S + 1$ (4)	2.0	
	To account for unnoticed cracks of past earthquake, which could be catastrophic in the events of Earthquakes, the Risk Score must be reduced by 25%	$0.75 \times 2.0 = 1.5$	
E.	Probability of at least one earthquake occurs during t years that is strong enough to cause collapse, $R(t) = 1 - e^{\left(-\frac{10^{-SR}}{\tau} \times t\right)}$ (5) Where τ = design life in years (50 years) t = number of years (adopted 50 years)	$R(t) = 1 - e^{(-\frac{10^{-1.5}}{50} \times 50)}$ $R(t) = 3.11\%$	Greater than the 1.5 times the probability of MCE (i.e. 2% for 50 years)
F.	Recommendations		
	i. As probability of at least one earthquake occurs during 50 years that is strong enough to cause collapse is significantly high, when compared to probability of occurrence of MCE ground shaking. Hence, structure is prioritized as unsafe		
	ii. A detailed technical evaluation of the existing building must be carried out, as the fatality risk of the existing building is 100 times the fatality risk of new building. Hence, building must be retrofitted globally as per IS 4326:2013		

Table 3 Building type and their collapse probability

S. No.	Building type	Photograph of building	Risk score	Probability of earthquake causing collapse $R(t)$	Remarks
1	RCC G + 3 Building		2.4	1.57% i.e., 1 in 63	1. As R (t) is less than probability of MCE; therefore, building is prioritized as SAFE 2. The retrofitting is not required as building is considerably less fatal
2	Load Bearing Masonry Building		1.5	3.11% i.e., 1 in 32	1. As R (t) is greater than probability of MCE; therefore, building is prioritized as UNSAFE 2. The detailed technical evaluation for the building is required and global retrofitting must be done as the existing building is 100 times fatal than new building

has suffered a lot of damages in term of spalling and corrosion. However, the building is kept serviceable by periodical maintenance.

4.1 Risk Score and Its Inference

As per the study, fatality risk of existing building can be correlated to the new building, based on risk score. FEMA P-155 (2015) has described various risk score and their corresponding fatality risk as discussed below,

(a) A risk score of 1.5 implies that the fatality risk of an existing building is a hundred times the fatality risk of a new building.
(b) A risk score of 2.0 implies that the fatality risk of an existing building is thirty-two times the fatality risk of a new building.
(c) A risk score of 2.5 implies that the fatality risk of an existing building is ten times the fatality risk of a new building.
(d) A risk score of 3.0 implies that the fatality risk of an existing building is three times the fatality risk of a new building.
(e) A risk score of 3.5 implies that the fatality risk of an existing building is equal to the fatality risk of a new building.
(f) A risk score of 4.0 implies that the fatality risk of an existing building is one-third of the fatality risk of a new building.
(g) A risk score of 4.5 implies that the fatality risk of an existing building is one-tenth of the fatality risk of a new building.

4.2 Limitation of the Present Study

The present study involves the feasibility check of RVS for the prioritization of buildings for seismic retrofit. In this study, the evaluation of buildings has been confined to only preliminary stage, based on which the decision for detailed technical evaluation for retrofitting of buildings has been presented. Detailed technical evaluation for retrofitting of the building is not a part of this study.

5 Conclusion

In this study, two different types of structures have been taken, i.e. RCC G + 3 building and load-bearing masonry building for calculating the risk score. As per the case study performed, it has been found that the probability of earthquake causing collapse for RCC G + 3 building is 1.57% for next 50 years, which is less than the probability of MCE ground shaking (i.e. 2% for 50 years). As the fatality risk of existing building rounds off to only ten times fatality of new building; therefore, the existing RCC G + 3 building can be rendered as "SAFE" based on preliminary study.

Similarly, for the load-bearing masonry structure, the study suggests that the probability of earthquake causing collapse is equal to 3.11% for next 50 years, which is greater than the probability of MCE ground shaking (i.e. 2% for 50 years). For that matter, the fatality risk of existing building rounds off to hundred times fatality of new building. Therefore, load-bearing masonry building is rendered as "UNSAFE" based on the preliminary study.

Hence, based on RVS score, buildings can be prioritized for seismic retrofit.

References

1. Federal Emergency Management Agency. (2015) Rapid visual screening of buildings for potential seismic hazards: A handbook. FEMA 154, 3rd edn.
2. Federal Emergency Management Agency. (2015). Rapid visual screening of buildings for potential seismic hazards: Supporting documentation, FEMA report 155, 3rd edn.
3. IS 13935: 2009. *Seismic evaluation, repair and strengthening of masonry buildings—Guidelines, code of practice.* New Delhi: Indian Standard, Bureau of Indian Standards.
4. Federal Emergency Management Agency. (2005). BCA EQ Washington State seismic hazard data, Excel software on the Mitigation BCA Toolkit CD (January 2005).
5. Luco, N., Ellingwood, B. R., Hamburger, R. O., Hooper, J. D., Kimball, J. K., & Kircher, C. A. (2007). Risk-targeted versus current seismic design maps for the conterminous United States. (pp. 163–175).
6. Wang, Y., & Goettel, K. A. (2006). Prioritization of seismic retrofits of school buildings in Oregon using an enhanced rapid visual screening methodology. In *Proceedings of the 8th U.S. National Conference on Earthquake Engineering*, April 18–22, 2006, San Francisco.
7. Arya, A. S. (2011). *Rapid structural and non-structural assessment of school and hospital buildings in SAARC countries.* New Delhi: SAARC Disaster Management Centre.

Utilization of Rice Straw Ash as a Replacement of Cement and Fine Aggregate in Mortar Mixes

Mohammad Ihtesham Hadizai and Aditya Kumar Tiwary

Abstract This study is conducted to utilize the rice straw ash which is a waste material that provides for 20% of the global rice production of 649.7 million tons per year. The rice straw ash that was used was locally purchased. In this paper, the rice straw ash (RSA) was replaced with cement, fine aggregate, and both of them, the mix ratio was taken 1:2; cement ash was replaced with RSA 0%, 5%, 15%, and 25%; and then fine aggregate was replaced with RSA 0%, 5%, 15%, and 25% after that both cement and fine aggregate were replaced by RSA (5/5%, 15/15%, 25/25%). The compressive strength test was done on all samples after the age of 7, 14, and 28 days, the density test was done on the samples after the age of 28 days to see the density of the samples, water absorption test was done after the age of 28 days, and the porosity test was also done after the age of 28 days to see the volume of voids in the samples. The result showed that mortar with replacement of up to 15% of fine aggregate with RSA showed better result than the nominal and other mixes in respect to the compressive strength of mortar, and 5% replacement of RSA with cement showed a promise able result as compared to the others in respect to water absorption and porosity tests.

Keywords Rice straw ash · Mortar mix · Density · Porosity · Water absorption

1 Introduction

Mortar is a functional paste that hardens and is used to bind building blocks such as rocks, bricks, and concrete blocks, to fill and seal uneven spaces between them, to uniformly distribute their weight, and to add decorative colors or shapes to masonry walls [1, 2]. The global use of natural sand in mortar production has increased significantly in the last two decades, and the majority of promotional nations have

M. I. Hadizai (✉) · A. K. Tiwary
Civil Engineering Department, Chandigarh University, Mohali, Punjab, India

A. K. Tiwary
e-mail: aditya.civil@cumail.in

A. K. Gupta et al. (eds.), *Advances in Construction Materials and Sustainable Environment*, Lecture Notes in Civil Engineering 196,
https://doi.org/10.1007/978-981-16-6557-8_65

struggled to produce natural sand as fine aggregate in mortars [1]. Many countries are experiencing a scarcity of good natural sand for building. In other words, the main environmental threats and issues are thought to be fly ash (FA), bottom ash (BA), and agricultural losses. With the improvement and increase of these losses, the situation becomes even worse [3]. Many researches have been undertaken and are currently being conducted to resolve the scarcity of natural resources. The development and use of blended cements are growing rapidly [4, 5]. Pozzolans from industrial and agricultural by-products such as fly ash and rice husk ash are receiving more attention now since their uses generally improve the properties of the blended cement concrete, the cost, and the reduction of negative environmental effects; rice husk is an agricultural waste that provides for 20% of the global rice production of 649.7 million tons per year [6]. When used as a fuel, the partially burned husk from milling plants contributes to emissions, and attempts are being made to resolve this problem by using this material as a supplementary cementing material. These days, management and disposal of agricultural waste is a big problem. Rice husk is among the agricultural wastes. Paddy fields produce an average of 120 million tons of rice husk each year [7]. It would be a brilliant idea to use fine aggregate as a replacement for natural mortar materials. The use of agricultural waste materials as a substitute for natural materials in construction mortars has three benefits: Protect natural resources from depletion, removing waste products, and preserving land for various purposes [8]. The use of rice straw ash as a fine aggregate in mortar will help to minimize environmental impact and building material costs [8]. Traditional cementing materials are in short supply in many countries. In recent years, there have been significant attempts around the world to use natural and waste materials in concrete [9]. Rice husks are one of these materials, and the ash created by controlled burning, if sufficiently ground, can be used as a cement substitute material in concrete [9].

2 Literature Review

Chouhan Harshvardhan Singh et al. have assessed the impact of dimensional limestone squander on the mechanical properties like compressive strength, flexural strength, glue and bond strength, water retention, thickness, drying shrinkage, and microstructure analysis (SEM, XRD, and FTIR). In the test, dimensional limestone squander is utilized in various rates 0%, 20%, 40%, and 60% as fine total in concrete mortar. The outcome shows that an expansion in the DLSW and DLCS content up 20 to 60% expanded the compressive and flexural strength and decreasing the water–concrete proportion while expanding DLSW and DLSC up to 40%. Based on the above test outcome, we inferred that DLSW and DLCS can be utilized in mortar and solid creation. It will help in diminishing both ecological waste and save common assets [6].

Arunah et al., by these test works show to examine the physical properties (specific gravity, SEM, and XRD) and synthetic properties (XRF) of utilized material; in this analysis, utilized rice straw debris and miniature silica in distinction rate as 5%, 10%, 15%, 20%, 25%, 30%, and 2.5%, 5%, 7.5%, and 10% by the heaviness of common Portland concrete. This examination investigates the compressive strength and flexural strength conduct of the quality asphalt concrete. Numerous tests have been done the outcome show beginning and last setting time expanded while expanding the rice straw debris content [9].

Kabeer Ahmed Syed et al., have done an examination on mortar with the utilization of marble powder as fine total in mortar to diminish the employments of common material, lessen the expense of the venture, and have eco-accommodating climate. The pre-owned waste marble powder which is known as waste material and it is created from cutting and delivery of marble tiles. The accompanying substitution has been done from (0 to 100%) marble powder to waterway sand. What is more, the outcome shows subsequent to testing functionality, drying shrinkage, compressive strength, bond and cement qualities, thickness, water assimilation, and dynamic Young's modulus. Results show that mortar blends in with 20% replacement of stream sand by marble powder can be utilized for stone work and delivering purposes. The utilization of 20% of marble powder would empower significant saving water in and waterway sand in development ventures and furthermore improve the substance properties of the mortar-like compressive strength, new mass thickness, malleable security strength, and abatement the water ingestion [8, 10, 11].

Dabai et al., have done the examination on six mortar blocks which supplanted concrete by rice straw debris (0%, 2%, 4%, 6%, 8%, and 10%) by weight of concrete to check compressive strength. The consequence of compressive strength tests shows that at 2% supplanting with RSA at the time of relieving following 2, 7, and 28 days was discovered to be 15.77, 34.73, and 48.53 N/mm^2 and expanded with period of restoring, however, diminished at 10% trade for 7 days (18.06 N/mm^2) and 28 days (27.23 N/mm^2) with increment in RSA content for all blends. What is more, it shows that RSA can be utilized at 2%, 4%, and 6% to supplant concrete at 2 years old to 28 days period of restoring, and increment the underlying and last setting time when increment rice straw debris [1].

In Surajit et al., these trial works show to break down the actual properties (X-beam diffraction, compressive strength, flexural strength, normal consistency, workability, and water porousness of the solid. In this, analysis has utilized ground rice straw debris in various rates like 5%, 10%, and 15% by the heaviness of concrete. The test outcome shows the compressive strength and flexural strength steadily increment up to 10% of the fastener by rice straw debris contrasted with control concrete, and furthermore, the porousness of rice straw debris mortar relies upon the period of mortar and concrete substitution proportion. When all said is done, porousness diminishes with augmenting in the compressive strength and period of mortar. In this investigation, locally accessible rice straw debris has used to lessen both the ecological effect and the development cost. Locally, rice straw has been singed off the temperature of 600 °C to make rice straw debris. It has over 76% silica content [7].

In Revathi et al., these exploratory works show to dissect the mechanical properties like compressive strength, thickness, water retention, and sorptivity of the mortar.

Table 1 Physical properties of OPC

Physical properties	Ordinary Portland cement (OPC)
Specific gravity	2.79
Color	Gray
Surface area m^2/g	308

In this, analysis uses groundnut husk debris in various rates 0%, 10%, 20%, 30%, 40%, half, and 60% by fractional substitution of the characteristic waterway sand. The outcome shows thickness esteems decline with an expansion in groundnut husk debris content. Groundnut husk debris increases the compressive strength of the mortar, the sorptivity, and water assimilation of mortar expanded while expanding the level of groundnut husk debris content [4].

In this study, rice straw ash (RSA) was replaced with cement, fine aggregate, and both of them, using a 1:2 mix ratio. Cement ash was replaced with RSA 0%, 5%, 15%, and 25%, while fine aggregate was replaced with RSA 0%, 5%, 15%, and 25%. The compressive strength test was performed on all samples after 7, 14, and 28 days, the density test was performed on the samples after 28 days to determine density, the water absorption test was performed after 28 days, and the porosity test was performed after 28 days to determine the volume of voids in the samples.

3 Materials and Methods

3.1 Cement

The cement that was used in this research was ordinary Portland cement (OPC), Grade (43) confirming IS: 8112-2013 for the all the samples. The specific gravity test was done for the particular cement under normal conditions.

Physical and chemical properties of cement: The physical and chemical properties of cement were tested for the research in the laboratory, and the results were obtained as given in Tables 1 and 2.

3.2 Rice Straw Ash

Rice straw ash (Fig. 1) is a form of agricultural waste that can be found in rice-producing nations. The ash of the rice straw was purchased from the farmers around Kharar district of Mohali and was kept in normal temperature after for mixing; it was sieved by the sieve no 1.5 mm.

Table 2 Chemical properties of OPC

Chemical properties	Ordinary Portland cement
SiO_2	21.5
Al_2O_3	5.2
Fe_2O_3	3.5
CaO	65.4
MgO	1.2
SO_3	2.7
Na_2O	0.2
K_2O	0.2

Fig. 1 RSA before mixing in mortar

Physical and Chemical properties of RSA: The physical and chemical properties of rice straw ash were obtained from the tests that was conducted in the laboratory; the results obtained are given in Table 3.

3.3 Fine Aggregate

Fine aggregate that is used in this particular research was locally purchased and was dried in the oven, and it was sieved by sieve 4.75 mm confirming IS-3831970. The specific gravity test was done for the fine aggregate.

Physical Properties of Fine Aggregate: The physical properties for fine aggregate were obtained under different tests in the laboratory; the results obtained are given in Table 4.

Table 3 Physical and chemical properties of RSA

S. No.	Parameters	Values
1	Fineness passing 45 microns	96%
2	Specific gravity	2.06%
3	Specific surface (m^2/kg)	27,400
4	Silicon dioxide (SiO_2)	87.20%
5	Aluminum oxide (Al_2O_3)	0.15%
6	Ferric oxide (Fe_2O_3)	0.16%
7	Calcium oxide (CaO)	0.55%
8	Magnesium oxide (MgO)	0.35%
9	Sulfur oxide (SO_3)	0.24%
10	Carbon (C)	5.91%
11	Loss on ignition	5.44%
12	Pozzolanic activity	84%
13	Partial size	7

Table 4 Physical properties of fine aggregate

Description	Fine aggregate
Specific gravity	2.6
Water absorption	1.57%
Fineness modules	3.1
Bulk density	1450 kg/m^3

3.4 Water

Water plays an important role in concrete and mortar mixes; it initiates the response between cement, pozzolans, and aggregates. It needs special inspection before mixing it to the mixes. In this research, locally available water was used.

4 Mix Proportions

For the mortar, the mix ratio of 1:2 was chosen. In the particular mortar mix cement, fine aggregate and rice straw ash were used. Fine aggregate and cement were replaced by rice straw ash by (5%, 15%, and 25%) and (5%, 15%, and 25%), respectively in different samples, and after that cement and fine aggregate were replaced by rice straw ash at the same time by (5% F.A) and (5% C). For all the mixes, water cement ratio was kept (0.45). Tables 5, 6, and 7 show the different mixed proportions that were used.

Table 5 Mix proportion 1:2 (FA replaced by RSA)

S. no	ID (mix)	W/C ratio	Cement (kg/m^3)	Fine aggregate (kg/m^3)	RSA (kg/m^3)	Water (kg/m^3)
M-1	FA_{100} RSA_0	0.45	638.4	1420.4	0	287.2
M-5	FA_{95} RSA_5	0.45	638.4	1349.38	71.02	287.2
M-6	FA_{85} RSA_{15}	0.45	638.4	1207.38	213.02	287.2
M-7	FA_{75} RSA_{25}	0.45	638.4	1065.3	355.1	287.2

Table 6 Mix proportion 1:2 (Cement replaced by RSA)

S. No.	ID (mix)	W/C ratio	Cement (kg/m^3)	Fine aggregate (kg/m^3)	RSA (kg/m^3)	Water (kg/m^3)
M-1	C_{100} RSA_0	0.45	638.4	1420.4	0	287.2
M-2	C_{95} RSA_5	0.45	606.48	1420.4	31.92	272.9
M-3	C_{85} RSA_{15}	0.45	542.64	1420.4	95.76	244.18
M-4	C_{75} RSA_{25}	0.45	478.8	1420.4	159.6	215.46

Table 7 Mix proportions 1:2 (FA and cement were replaced by RSA)

S. No.	ID (mix)	W/C ratio	Cement (kg/m^3)	Fine aggregate (kg/m^3)	RSA (kg/m^3)	Water (kg/m^3)
M-1	FA_{100} C_{100} RSA_0	0.45	638.4	1420.4	0	287.2
M-8	FA_{95} C_{95} RSA_{10}	0.45	606.48	1349.38	102.9	272.9
M-9	FA_{85} C_{85} RSA_{30}	0.45	542.64	1207.38	308.78	244.18
M-10	FA_{75} C_{75} RSA_{50}	0.45	478.8	1065.3	514.7	215.46

5 Preparation of Test Specimen

The cubes that were cast for doing compressive strength, porosity test, water absorption, and density was (70.6 mm × 70.6 mm × 70.6 mm) size. Total number of cubes that were casted are 144, and from these, 90 cubes were casted for average compressive strength test to be carried out after 7, 14, and 28 days, respectively, and the remaining 54 were casted for the average porosity, density, and water absorption tests.

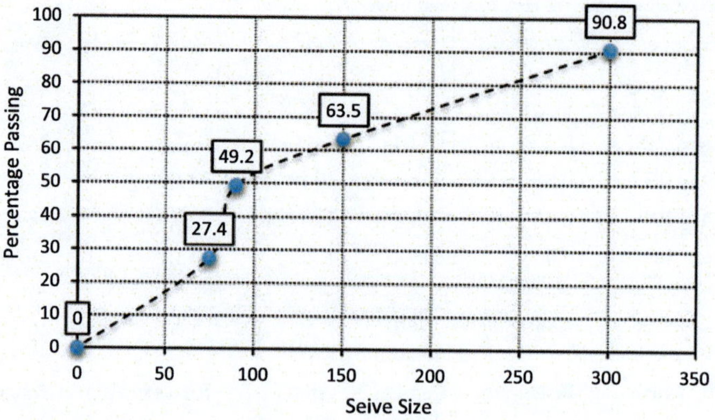

Fig. 2 Sieve analysis of rice straw ash

6 Result and Discussion

6.1 *Sieve Analysis*

Sieve analysis was done on rice straw ash and fine aggregate, the size of the sieve used is 300 μm, 150 μm, 90 μm, and 75 μm. The analysis was done by the sieve shaker, shaker was on for 20 min, and after that, the sieve were removed and the weight of the RSA that were remaining on the sieve was taken for the analysis. Result is shown in Fig. 2.

6.2 *Density*

The density for the specimen was calculated for different mixes as per data given in Tables 8, 9, and 10. As the result shows, the density for samples with nominal mortar has a vast difference as compare to the mix with RSA as a replacement to the fine aggregate and cement, and both of them in one mix.

6.3 *Water Absorption*

The test was done on the samples after 28 days of curing which are shown in the tabular form below. For the water absorption test (Eq. 1), the specimens are dried in an oven for a specified time and temperature and then placed in a desiccator to cool. Immediately upon cooling, the specimens are weighed. The material is then emerged in water at agreed upon conditions, often 23 °C for 24 h or until equilibrium.

Table 8 Density for RSA replacement for FA

Mix ID	Specimen no	Ratio 1:2	
		Weight of mortar cubes (Kg/m^3)	Average density (Kg/m^3)
$FA_{100}RSA_0$	1	0.865	1972.1
	2	0.870	
	3	0.868	
$FA_{95}RSA_5$	1	0.676	1530.3
	2	0.671	
	3	0.673	
$FA_{85}RSA_{15}$	1	0.622	1413.6
	2	0.619	
	3	0.625	
$FA_{75}RSA_{25}$	1	0.560	1275.7
	2	0.563	
	3	0.561	

Table 9 Density for RSA replacement for FA

Mix ID	Specimen no	Ratio 1:2	
		Weight of mortar	Average density
$Cement_{100}$ RSA_0	1	0.865	1972.1
	2	0.870	
	3	0.868	
$Cement_{95}$ RSA_5	1	0.698	1577.2
	2	0.694	
	3	0.690	
$Cement_{85}$ RSA_{15}	1	0.654	1477.2
	2	0.656	
	3	0.640	
$Cement_{75}$ RSA_{25}	1	0.616	1400.73
	2	0.620	
	3	0.613	

Specimens are removed, patted dry with a lint free cloth, and weighed. As the result shows, the water absorption of the specimen of nominal mix and the RSA replacement of FA up to 15% have a lower average range of water absorption, but the samples that contain RSA as replacement for cement and cement plus FA have a higher range of water absorption as given in Tables 11, 12 and 13.

Table 10 Density for RSA replacement for both FA and cement

Mix ID	No. of specimen	Ratio 1:2	
		Weight of mortar	Average density
$Cement_{100}$ FA_{100} RSA_0	1	0.865	1972.1
	2	0.870	
	3	0.868	
$Cement_{95}$ FA_{95} RSA_{10}	1	0.660	1494.69
	2	0.658	
	3	0.655	
$Cement_{85}$ FA_{85} RSA_{30}	1	0.583	1325.75
	2	0.586	
	3	0.581	
$Cement_{25}$ FA_{25} RSA_{50}	1	0.382	871.9
	2	0.385	
	3	0.384	

Table 11 WA for RSA replacement of cement

Percentage of RSA	Wet weight of sample	Dry weight of sample	Average water absorption in (%)
$Cement_{100}$ RSA_0	0.735	0.670	10.51
	0.768	0.686	
	0.740	0.672	
$Cement_{95}$ RSA_5	0.698	0.660	5.55
	0.690	0.652	
	0.694	0.660	
$Cement_{85}$ RSA_{15}	0.634	0.522	22.36
	0.656	0.530	
	0.640	0.525	
$Cement_{75}$ RSA_{25}	0.608	0.460	30.84
	0.592	0.450	
	0.572	0.444	

$$W = \frac{w1 - w2}{w1} * 100 \tag{1}$$

Table 12 WA for RSA replacement of FA

Percentage of RSA replacement of FA	Wet weight of sample	Dry weight of sample	Average water absorption (%)
FA_{100} RSA_0	0.735	0.670	10.51
	0.768	0.686	
	0.740	0.672	
FA_{95} RSA_5	0.656	0.544	20.15
	0.652	0.538	
	0.660	0.556	
FA_{85} RSA_{15}	0.541	0.430	25.42
	0.542	0.436	
	0.540	0.428	
FA_{75} RSA_{25}	0.455	0.330	37.4
	0.450	0.328	
	0.457	0.333	

Table 13 WA for RSA replacement for both FA and cement

Percentage of RSA replacement of (FA + C)	Wet weight of sample	Dry wet of sample	Average water absorption (%)
FA_{100} C_{100} RSA_0	0.735	0.670	10.51
	0.768	0.686	
	0.740	0.672	
FA_{95} C_{95} RSA_{10}	0.660	0.582	13.26
	0.658	0.580	
	0.654	0.579	
FA_{85} C_{85} RSA_{30}	0.530	0.383	37.7
	0.534	0.390	
	0.531	0.385	
FA_{75} C_{75} RSA_{50}	0.440	0.298	48.03
	0.437	0.295	
	0.441	0.296	

6.4 *Compressive Strength*

For the samples, compressive test was tested as per (IS-2550-1981) under the compression test machine after 7, 14, and 28 days, respectively, as given in Tables 14, 15 and 16. As per the results that are given in the table, conventional samples have best results in respect to the compressive strength, while the samples that contain RSA as a replacement for fine aggregate have good result compared to RSA as a replacement for cement and RSA replacement for both cement and fine aggregate.

Table 14 ACS for RSA replacement for cement

Percentage of RSA replacement of C	Average compressive strength		
	7 days	14 days	28 days
$Cement_{100}$ RSA_0	22.4	26.3	33.6
$Cement_{95}$ RSA_5	17.8	20.1	24.8
$Cement_{85}$ RSA_{15}	13.3	18.6	21.3
$Cement_{75}$ RSA_{25}	9.7	14.5	18.2

Table 15 ACS for RSA replacement of FA

Percentage of RSA replacement of FA	Average Compressive Strength		
	7 days	14 days	28 days
FA_{100} RSA_0	22.4	26.3	33.6
FA_{95} RSA_5	25.6	29.4	36.5
FA_{85} RSA_{15}	23.5	27.1	34.8
FA_{75} RSA_{25}	20.3	23.2	31.3

Table 16 ACS for RSA replacement for both FA and cement

Percentage of RSA replacement of FA and cement	Average compressive strength		
	7 days	14 days	28 days
FA_{100} C_{100} RSA_0	22.4	26.3	33.6
FA_{95} C_{95} RSA_{10}	19.3	22.6	27.6
FA_{85} C_{85} RSA_{30}	11.2	14.3	18.2
FA_{75} C_{75} RSA_{50}	5.3	7.2	10.8

6.4.1 Compressive Strength After 7 Days of Curing

Compressive strength test for the mortar after 7 days of curing was carried out under compression test machine as per IS-2250-1981. The values are given in the table form above. The result of the tests after 7 days as shown in Fig. 3 depicts that the result for conventional concrete in (M-1) is (22.4 MPa). However, the compressive strength result for the M-5 is (25.6 MPa) that shows 12.5% increase. M-6 has a result of (23.5 MPa) and shows a 4.6% of increase from the nominal sample. M-7 has a result of (20.3 MPa) that has decreased 10.3%. The specimen containing RSA as a replacement for cement M-2 had a result of (17.8 MPa), M-3 has a result of (13.3 MPa), M-4 is (9.7 MPa) which declares 25.8%, 68.4%, and 43.4% decrease as compare to the conventional mortar, respectively. The result for specimen containing RSA as a replacement for both FA and cement is that M-8 is (19.3 MPa), M-9 is (11.2 MPa), and M-10 is (5.3 MPa) that shows an amount of 17.1%, 50%, and 76.3% of reduction, respectively, in the compressive strength of mortar as to the nominal mortar.

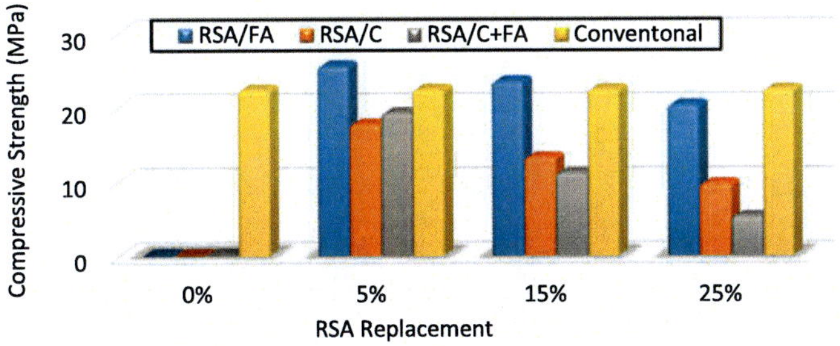

Fig. 3 Average compressive strength after 7 days of curing

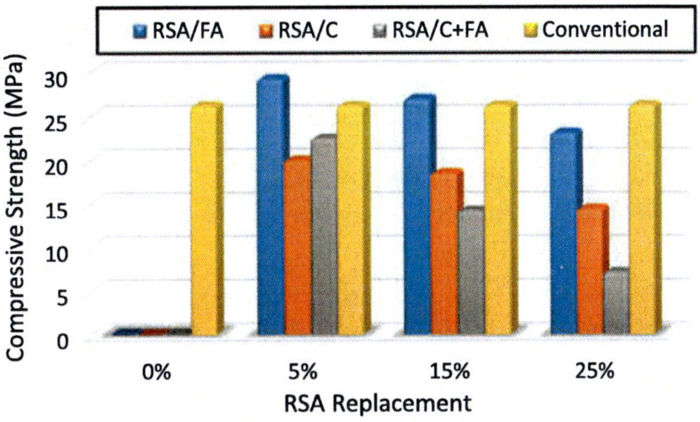

Fig. 4 Average compressive strength after 14 days of curing

6.4.2 Compressive Strength After 14 Days of Curing

The results for the compressive strength of the mortar mixes for 14 days of curing was shown as per IS-2250–1981 in the table form above. Figure 4 shows that nominal mix of mortar (M-1) has a strength of (26.3 MPa). The mixes that have RSA replacement with FA: M-5 has a strength of (29.4 MPa), while M-6 is (27.1 MPa) that show 12.5% and 3.04% increase in respect to the compressive strength, while M-7 has a strength of (23.2 MPa) which has 11.7% decrease from nominal mix. RSA replacement with cement results is such as: M-2 is (20.1 MPa), M-3 is (18.6 MPa) and M-4 the compressive test result is (14.5 MPa) that declares 23.5%, 29.27%, and 44.86% decrease in respect to the compressive strength. And the result of RSA replacement by FA and cement shows that M-8 is (22.6 MPa), M-9 is (14.3 MPa), and M-10 is (7.2 MPa) which figure out 14%, 45.62%, and 72.6% of decrease in compressive strength of the specimen.

6.4.3 Compressive Strength After 28 Days of Curing

As per IS-2250-1981 and the compressive test, result for the specimens after 28 days is shown in Fig. 5. The average compressive strength for the conventional mortar

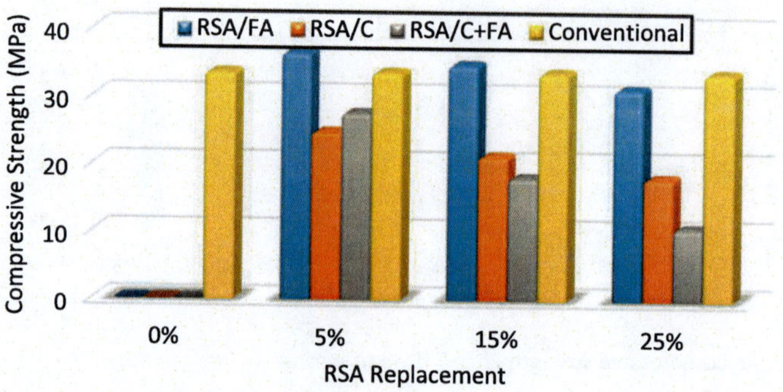

Fig. 5 Average compressive strength after 28 days of curing

sample M-1 is (33.6 MPa). However, the test result for the specimens containing RSA as replacement for FA: M-5 is (36.5 MPa) and M-6 is (34.8 MPa) which shows 8.6% and 3.5% increase, respectively. M-7 is (31.3 MPa) that sort out 6.8% of decrease in respect to compressive test. The mix that contains RSA as replacement for Cement M-2 is (24.8 MPa), M-3 is (21.3 MPa) and M-4 is (18.2 MPa) which declares 26.2%, 36, 6% and 54.1% of decrease, respectively, as compare to nominal mix. Replacement of RSA with FA and cement shows the following result in respect to the compressive strength test that M-8 is (27.6 MPa), M-9 is (18.2 MPa), and M-10 is (10.8 MPa); as per results, the compressive strength in these particular mixes has decreased 17.8%, 24.8%, and 67.85%, respectively.

6.5 Porosity

The porosity test for the samples was conducted after 28 days of curing. From the results (Fig. 5), it can be figured out that nominal mortar has the best result as compare to the other mixes, whereas the RSA as replacement for fine aggregate up to 15% has remarkable porosity result. But the samples that contain RSA as replacement for cement and has poor porosity result and the RSA replacement for both fine aggregate and cement up to 15% have good result as compare to RSA as replacement to the cement. The porosity was calculated using equation given by Demirel Bahar [12], Guneyisi Erhan et al. [13]. This property is used to determine effective porosity of the rock. In this method, the weight of a dry sample is measured, and then, the sample is immersed under vacuum in water or any other fluid that rock has the tendency to imbibe. After enough time, up to several days, the saturated sample is weighted. (Fig. 6).

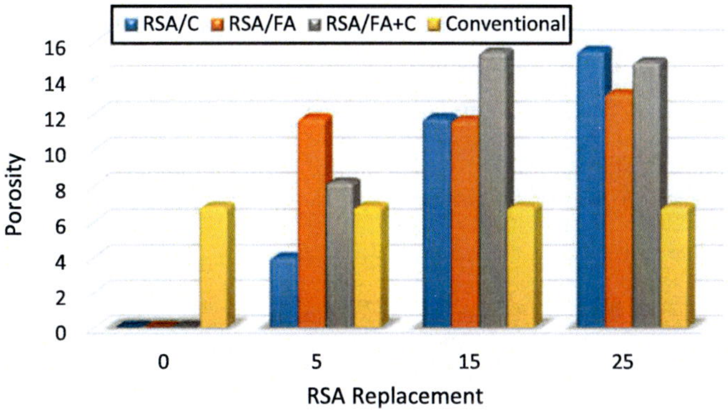

Fig. 6 Average porosity after 28 days of curing

7 Conclusion

The utilization of rice straw ash as replacement for fine aggregate, cement, and both of them was done; compressive strength, porosity, density, and water absorption tests were done on the samples after 7, 14, and 28 days, respectively, from the following results that were found.

- The replacement of fine aggregate up to 15% had a positive effect on the compressive strength of the mortar, but the increase in percentage of RSA from 15% had an inverse effect on the compressive strength of mortar.
- The replacement of cement with RSA had an inverse effect on the compressive strength of the mortar with any percentage (5%, 15%, and 25%).
- The replacement of (FA + C) with RSA up to 5% had increased the compressive strength of the mortar as compare to RSA as replacement for cement mortar but had less compressive strength result as compare to conventional mortar.
- RSA replacement for cement up to 5% showed 47% less water absorption as compare to the conventional mortar, but in other mixes, the percentage of water absorption was high as compare to the nominal mortar.
- Form the density point of view, the nominal mortar had a vast and great density as compare to the mortar mixes with RSA as replacement for fine aggregate or RSA as replacement for cement or mix of them.
- The porosity result for RSA replacement for 5% of cement was 41.5% less than conventional mortar, and the rest of the mixes show a high porosity result as compare to nominal mix of mortar.
- Finally, it can be pointed that mortar with replacement of up to 15% of fine aggregate with RSA showed better result than the nominal and other mixes in respect to the compressive strength of mortar, and 5% replacement of RSA with cement showed a promisable result as compare to the others in respect to water absorption and porosity tests.

References

1. Naveed, S. K., & Sharma, T. (2020). Utilization of rice straw ash as fine aggregate in mortar. *International Research Journal of Engineering and Technology (IRJET), 10*(3), 7809–7818.
2. Corinaldesi, V., Moriconi, G., & Naik, T. R. (2010). Characterization of marble powder for its use in mortar and concrete. *Construction and Building Materials, 24*, 113–117.
3. Tomar, A., Sharma, T., & Singh, S. (2020). Strength properties and durability of clay soil treated with mixture of nano silica and Polypropylene fiber. *Materials Today: Proceedings.*
4. Soil, C. (2019). Effect of polypropylene fiber on properties of bagasse ash-cement stabilized. July.
5. Kristulovic, P., Kamenic, N., & Popovic, K. (1994). A new approach in evaluation of filler effect in cement. *Cement and Concrete Research, 24*(4), 721–727.
6. Pandey, A., & Kumar, B. (2019). Preliminary study of cement paste admixed with rice straw ash, microsilica & rice straw ash-microsilica composite. January, 2019.
7. Nithyambigai, G. (2015). Effect of rice husk ash in concrete as cement and fine aggregate. *International Journal of Engineering Research & Technology, 4*(05), 934–936.
8. El Damatty, A. A., & Hussain, I. (n.d.). An economical solution for the environmental problem resulting from the disposal of rice.
9. Dabai, M. U., & Muhammad, M. (2017). Studies on the effect of rice straw ash as admixture of ordinary Portland cement mortar. *IJISET—International Journal of Innovative Science, Engineering & Technology, 4*(7), 112–119.
10. Karasahin, M., & Terzi, S. (2007). Evaluation of marble dust in the mixture of asphaltic concrete. *Construction and Building Materials, 21*(3), 616–620.
11. Binici, H., Kaplan, H., & Yilmaz, S. (2007). Influence of marble and limestone dusts as additives on some mechanical properties of concrete. *Scientific Research and Essays, 2*(9), 372–379.
12. Demirel, B. (2010). The effect of the using waste marble dust as fine sand on the mechanical properties of the concrete. *International Journal of the Physical Sciences, 5*, 1372–1380.
13. Guneyisi, E., Gesoglu, M., & Mermerdas, K. (2007). Improving strength, drying shrinkage and pore structure of concrete using Metakaolin. *Materials and Structures, 12*, 10–26.

Analytical Behavior of Concrete-Filled Single-Skin and Double-Skin Tube Columns Subjected to Axial Loading

Sakshi Bhatia and Aditya Kumar Tiwary

Abstract Columns made of concrete-filled steel tubes (CFSTs) have been found to improve the weight, stiffness, and ductility of concrete members. However, under axial tension, bending, or torsion, the central portion of the concrete in CFST columns is not completely used. This report presents the numerical results of the study on concrete-filled steel tube and concrete-filled double steel tube columns. The commercial platform ABAQUS is used to do a numerical analysis. Computational techniques are used, and the findings are checked against a literature-based experimental program. The numerical model will diagram the load deflection reaction of the CFST and CFDST specimens. The experimental results in the literature are compared to the column strength expected using finite element analysis and the Eurocode-4 and American Institute of Steel Construction specifications. The *D*/*t* ratio has a major effect on the compressive behavior of CFST specimens, according to the results. Concrete cores may also be confiningly confined by CFDST columns, particularly when both the outer and inner tube *D*/*t* ratios are small. These findings suggest that a smaller hollow steel tube will comfortably cover the central concrete to create double-skinned concrete-filled steel tube (CFDST) columns. The numerical investigation's findings are compared in terms of load–deformation trend and ultimate load strength.

Keyword Concrete-filled steel tube column (CFST) · Concrete-filled double steel tube column (CFDST) · Axial compression · FE modeling

1 Introduction

The CFST columns have outstanding seismic-safe underlying properties like high strength, high flexibility, and huge energy ingestion potential because of the upside of composite activity of the two materials. CFST sections have been the subject of a few logical and computational tests [1–10]. The axial load behavior of CFST sections has

S. Bhatia (✉) · A. K. Tiwary
Chandigarh University, Mohali, India

A. K. Gupta et al. (eds.), *Advances in Construction Materials and Sustainable Environment*, Lecture Notes in Civil Engineering 196,
https://doi.org/10.1007/978-981-16-6557-8_66

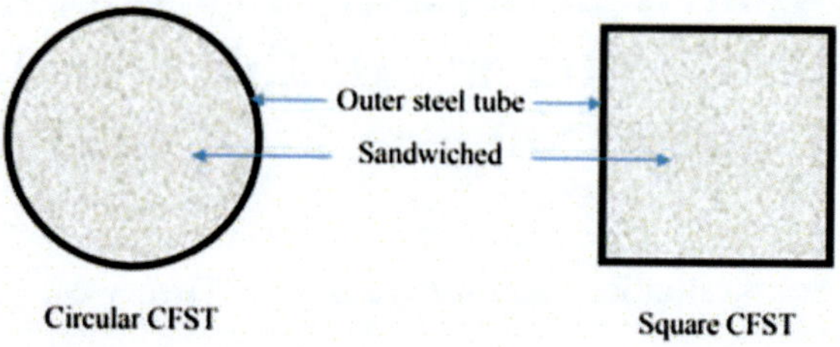

Fig. 1 Concrete-filled steel tube column (CFST)

been broadly explored. Many researchers [11–19] performed tests on concrete-filled steel tube sections (Fig. 1). These examinations were led utilizing round, square, and rectangular empty parts on concrete-filled carbon gentle steel and high-strength steel tube segments. The authors [11] led a test investigation on axially filled small CFST columns having *D*/*T* ratios of varying proportions from 17 to 50. FEA of CFST sections having square parts subjected to loading and chloride corrosion was also introduced [20].

Two concentric steel tubes of concrete filled between them make up a concrete-filled double skin tube (CFDST) column with a circular cross section (Fig. 2). CFDST segments are additionally more earthquake resistant [21–25] and lighter than CFT segments [26], just as being more fireproof. The overall strength of a CFDST segment is influenced by the concrete compressive strength, confined strain of concrete, tube's yield strength, and diameter-to-thickness proportions of internal tube as well as external tube. Circular cross sections are considered to have the strongest confinement effect [8] as well as better offshore loading resistance [27, 28]. Accordingly, CFDST segments with circular cross-sections are investigated when exposed to axial compressive loads. This paper proposes concrete and steel content constitutive models. At that point, utilizing the finite element program Abaqus [11], nonlinear examinations are completed and contrasted with experiment results from Tao et al. [23] and Zhao et al. [25]. At last, the impacts of the diameter-to-thickness proportion of outer tube (D_o/t_o) and diameter-to-thickness proportion (D_i/t_i) of internal tube on the confined strain are examined.

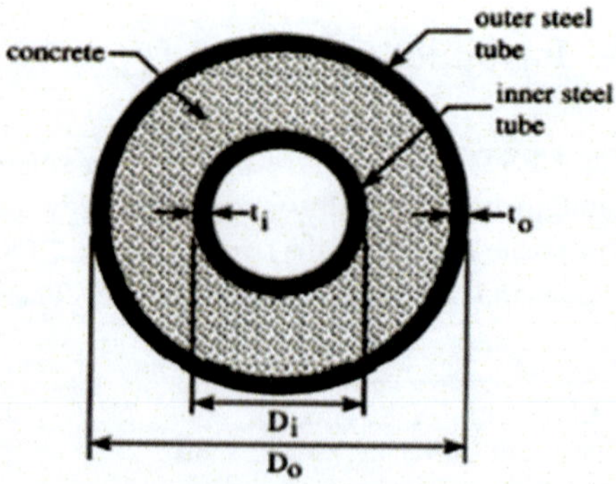

Fig. 2 Concrete-filled double steel tube column (CFDST)

2 Numerical Modelling of CFST and CFDST Column

Johnson Cook model is used for determining the external steel tube's inelastic behavior, while the concrete's inelastic behavior was determined using the concrete weakened plasticity model (CDP) accessible in ABAQUS/CAE [29]. Table 1 shows the geometry used to build representations of CFST columns. For representing the concrete's inelastic behavior, the model was based on the concept of isotropic damaged elasticity. Concrete's compressive strength was measured at 36.7 MPa.

2.1 Element Types

The in-filled concrete simulation was carried out using an eight-node linear solid element (C3D8R) of compressed integration and three degrees of freedom at each node in ABAQUS. As per the previous studies, the components utilized for the CFST's steel section were typically modeled with the help of four-node doubly-curved shell element (S4R) to capture the damage caused by compression and local buckling [30–33].

2.2 Loading and Boundary Conditions

For modeling purposes, the two closures were thought to be pivoted. Relocation in the x and y headings (U_x and U_y) is limited at the two closures, while interpretation and rotational levels of opportunity in the x, y, and z bearings were considered unhindered. A vertical compressive axial load was applied simultaneously on the top surfaces of steel and concrete with the help of displacement control option available in ABAQUS library [31, 32].

2.3 Interface Modeling

The rigid contact is visible among the two surfaces. Energy can only be transferred between the two surfaces if they are still in contact, causing the surfaces to separate due to the tensile force. As a result, the proper modeling of combined effort of concrete and steel tube is successfully attained to evaluate the nature of the CFST and CFDST columns [32].

Table 1 Geometric details and material properties of simulated specimens

Specimens	Outer tube dimensions (mm)	Inner tube dimensions (mm)	Height (mm)	D_o/T_o	D_i/T_i	f_y (MPa)	f_c (MPa)	Area (mm^2)		Tested by
								Steel (As)	Conc (Ac)	
CFST-1	140 × 3	–	600	47	–	285	36.7	1291	14,734	Tao et al. [23]
CFST-2	140 × 6	–	600	23	–	313	36.7	2526	14,095	Schneider [4]
CFST-3	180 × 3	–	600	60	–	276	36.7	1668	24,593	Tao et al. [23]
CFST-4	180 × 6	–	600	30	–	396	36.7	3280	23,767	Tao et al. [23]
CFDST-1	140 × 3	48 × 3	550	47	16	285	36.7	2091	15,652	Schneider [4]
CFDST-2	140 × 3	88 × 3	550	47	29	285	36.7	2468	10,212	Schneider [4]
CFDST-3	180 × 3	48 × 3	550	60	16	276	36.7	1714	21,958	Tao et al. [23]
CFDST-4	180 × 3	88 × 3	550	60	29	276	36.7	2091	17,688	Tao et al. [23]

2.4 Material Modeling

Steel: The ideal tri-linear stress–strain profile for steel was employed to make it simple. First, the steel's elasticity grows in a slopy manner to the yield point along with the modulus. The plastic nature is the second part that remains steady under yield stress [32]. Figure 3 shows the idealized tri-linear stress–strain profile of steel and can be given by Eq. 1.

$$\varepsilon_t = 10\varepsilon_y \tag{1}$$

Concrete: Modulus of elasticity, E_c is determined with the help of Eqs. 2 to 14 provided by ACI (1999).

$$E_c = 4700\sqrt{F_{nn}} \tag{2}$$

$$F_{nn} = F_n + K_1 f_1 \tag{3}$$

$$\in_{nn} = \in_n' \left(1 + K_2 \frac{F_1}{F_2}\right) \tag{4}$$

$$\frac{F_1}{F_2} = 0.043646 - 0.000832\left(\frac{D}{T}\right) \quad \text{CASE 1.} \quad 21.7 \le D/T \le 47 \tag{5}$$

$$n\frac{F_1}{F_2} = 0.006241 - 0.0000357\left(\frac{D}{T}\right) \quad \text{CASE 2.} \quad 47 \le D/T \le 150 \tag{6}$$

Non-associated plastic flow rule is taken into account by the CDP model. The CDP model's plastic potential function is the Druger–Prager hyperbolic function,

$$G = \sqrt{(F_C - mf_t \tan\beta)^2 + (\overline{q})^2} - \overline{p}\tan\beta - \sigma \tag{7}$$

Fig. 3 Idealized tri-linear stress–strain curve of steel

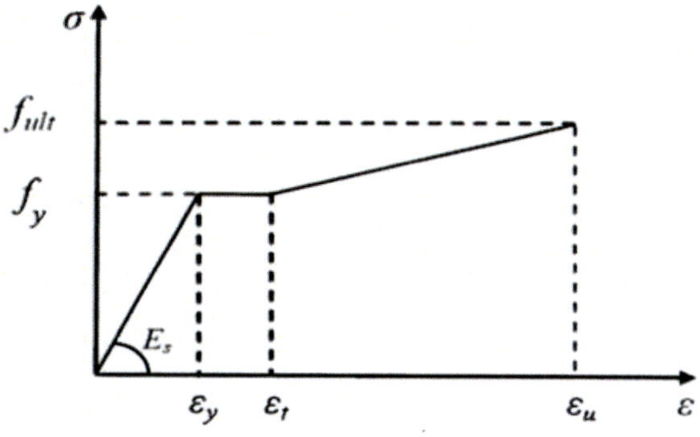

$$F'_B/F'_C = 1 \cdot 57 F_c'^{-0.09} \tag{8}$$

$$K_c = 0.71 F_c'^{-0.025} \tag{9}$$

$$E_c = \left(5000\sqrt{F'_c}\right)(\text{MPa}) \tag{10}$$

Dilation angle w.r.t lateral strain and axial strain relationship:

$$\beta = \tan^{-}\left[\frac{6(\nu_{c0} - \nu_{ce})}{\frac{3E_c\varepsilon_{cp}}{f'_c} + 2(\nu_{c0} - \nu_{ce}) - 3}\right] \tag{11}$$

$$\varepsilon_{cp} = \mu_a \frac{f_c'^{0.75}}{E_C}\left(\frac{\rho_C}{2320}\right)^{\varphi} \tag{12}$$

The CDP model damage parameters were as:

$$D_c\left(\tilde{\varepsilon}_c^{in}\right) = \frac{\tilde{\varepsilon}_c^{in}}{\tilde{\varepsilon}_c^{in} + \frac{2\sigma_c}{E_c}} < 0.8 \tag{13}$$

$$D_t\left(\tilde{\varepsilon}_t^{cr}\right) = \frac{\tilde{\varepsilon}_t^{cr}}{\tilde{\varepsilon}_t^{in} + \frac{5\sigma_t}{E_c}} < 0.8 \tag{14}$$

where $\tilde{\varepsilon}_c^{in} = \varepsilon_c - \sigma_c/E_c$ is the inelastic compressive strain; $\tilde{\varepsilon}_t^{cr} = \varepsilon_t - \sigma_t/E_c$ the cracking strain.

3 Specification for Design of CFST and CFDST

3.1 Eurocode-4

Eurocode-4 consists of the design guidelines for composite columns. The ultimate axial load capacity of the CFDST and CFST columns is given by Eqs. 15 to 20:

$$N_c = A_s f_y + A_c f_{ck} \tag{15}$$

N_{cr} = Composite column's Euler buckling strength and is given in Eq. (16),

$$N_{cr} = \frac{\pi^2(EI_{eff})}{l^2} \tag{16}$$

Further,

$$EI_{\text{eff}} = E_s I_s + 0.81 E_{\text{cm}} I_c \tag{17}$$

$$E_{\text{eff}} = \frac{E_{\text{cm}}}{Y_f} \tag{18}$$

Here, Y_f= safety factor and is taken as 1.35

$$EI_{\text{eff}} = E_s I_s + 0.6 E_{\text{cm}} I_c \tag{19}$$

$$N_c = \eta_2 A_s f_y + A_c f_{\text{ck}} \left(1 + \eta_1 \frac{t f_y}{d f_{\text{ck}}} \right) \tag{20}$$

3.2 AISC-LRDF

The code provides a design process for composite structures. As per the design mechanism of LRFD, the composite materials used in composite construction are believed to behave as a single unit to prevent bending, i.e., monolithically. The code also specifies that the minimum steel amount in composite elements must be greater than 4%.

i.e.,($\rho_{Sr} > 4\%$)

Equation (21) governs the value of steel tube thickness for resisting local buckling,

$$\frac{D}{T} = 0.15 \frac{E}{f_y} \tag{21}$$

Equation (22) gives the ultimate load capacity

$$P_n = A_s f_y + \phi A_c f_{\text{ck}} \tag{22}$$

For circular CFST and CFDST, ultimate load capacity can be determined using Eq. 23,

$$P_n = A_s f_y + 0.95 A_c f_{\text{ck}} \tag{23}$$

4 Comparison of Results and Discussion

The CFST and CFDST columns were simulated using a numerical model. The disparity between the experimental data and the ABAQUS peak load was used

Table 2 Experimental outcomes compared to numerical and theoretical results

Specimens	Experimental load, P_{exp} (kN)	Numerical load, P_n (kN)	Theoretically calculated load	
			Eurocode-4, N_{EC} (kN)	AISC-LRDF, N_{AISC} (kN)
CFST-1	881	795	908	882
CFST-2	1290	1265	1307	1281
CFST-3	1342	1315	1362	1318
CFST-4	1837	1823	1971	1928
CFDST-1	1422	1405	1470	1442
CFDST-2	1305	1228	1378	1359
CFDST-3	2068	1925	2078	2039
CFDST-4	1933	1827	1976	1944

to assess the numerical model's accuracy. Table 1 shows the CFST and CFDST columns' dimensions and other details. Table 2 compares the results outcomes with the approximate capacities of the CFST and CFDST specimens.

4.1 Validation of the Numerical Model

The experimental and ABAQUS load–deformation behavior for CFST specimens is shown in Figs. 5, 6, 7, and 8, while the experimental and ABAQUS load–deformation behavior for CFDST specimens is shown in Figs. 9, 10, 11, and 12. Table 2 shows that the experimental and numerical peak load values are in strong alignment. The numerical loading potential of CFST and CFDST columns varies within 10% of the experimental values (Fig. 4).

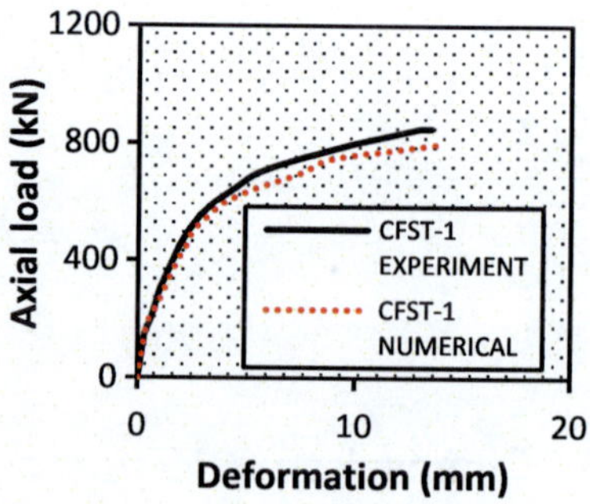

Fig. 4 Load–deformation curve for CFST-1 specimen

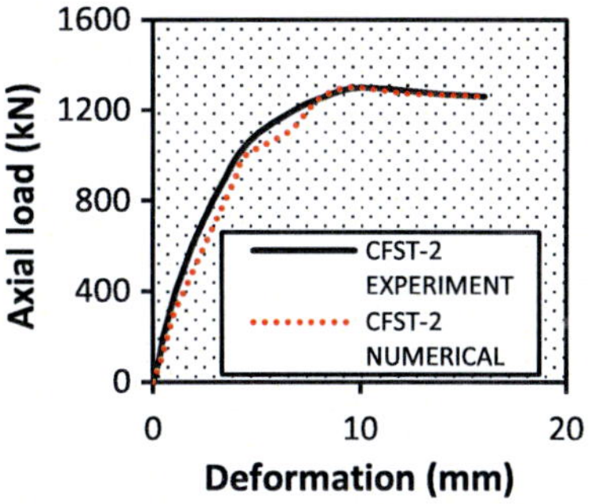

Fig. 5 Load–deformation curve for CFST-2 specimen

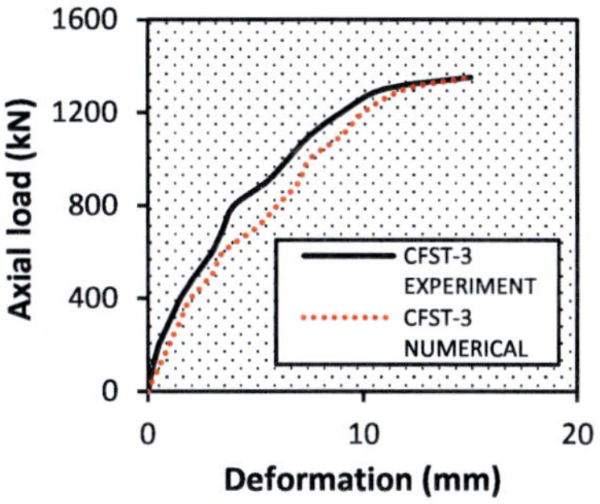

Fig. 6 Load–deformation curve for CFST-3 specimen

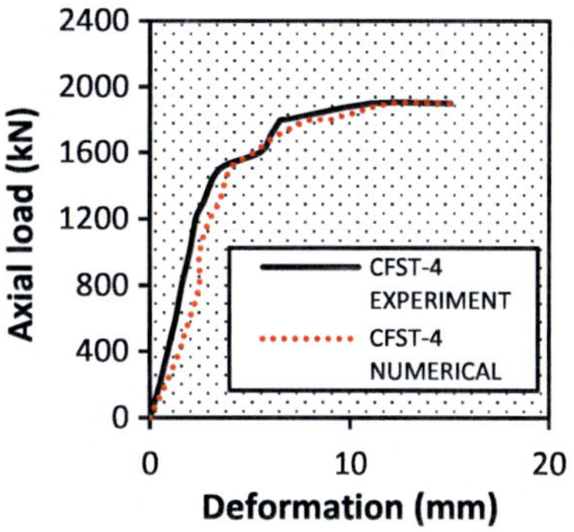

Fig. 7 Load–deformation curve for CFST-4 specimen

4.2 Verification of Results Through Design Codes

Table 2 shows detailed load-carrying capability measurements of composite columns with Eurocode-4 and AISC–LRFD. The axial load capability of CFST and CFDST columns is overestimated up to 7% according to Eurocode-4. In the AISC–LRFD code, the mean value of the approximation of axial strength potential Pexp/NAISC is 1.02. According to the AISC–LRFD specification, the estimated load for CFST and CFDST columns can be as high as 5%. Using code-based techniques and theoretical

Fig. 8 Load–deformation curve for CFDST-1 specimen

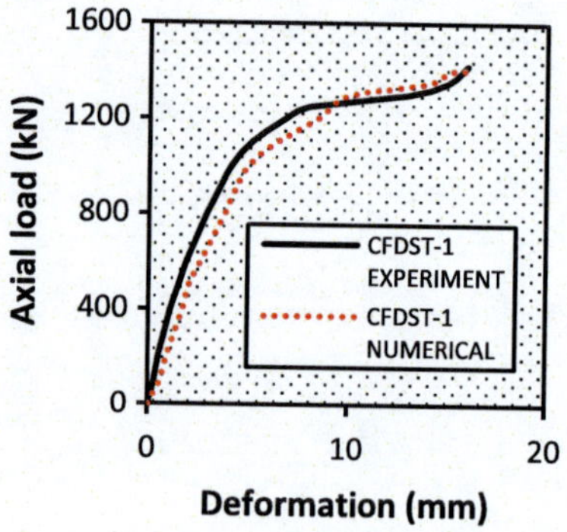

Fig. 9 Load–deformation curve for CFDST-2 specimen

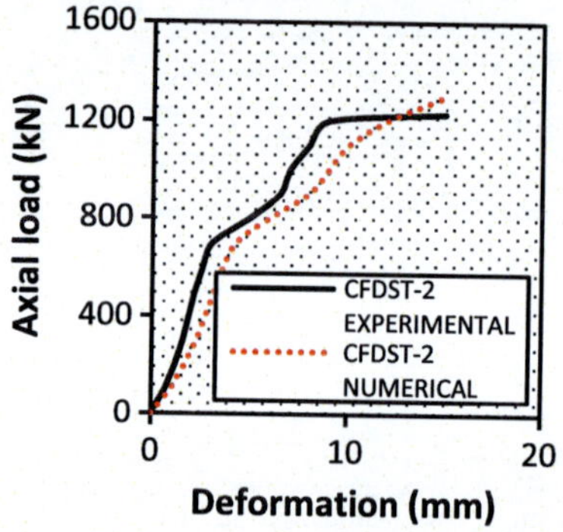

Fig. 10 Load–deformation curve for CFDST-3 specimen

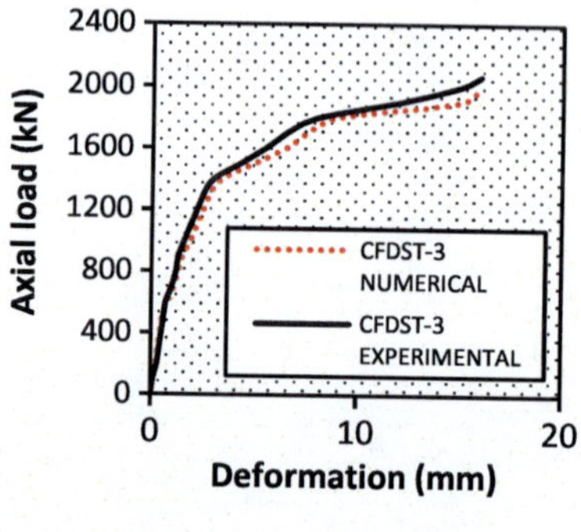

equations, a finite element model was predicted for the axial potential of the circular concrete filled single skin and concrete filled double skin tube columns (see Fig. 12).

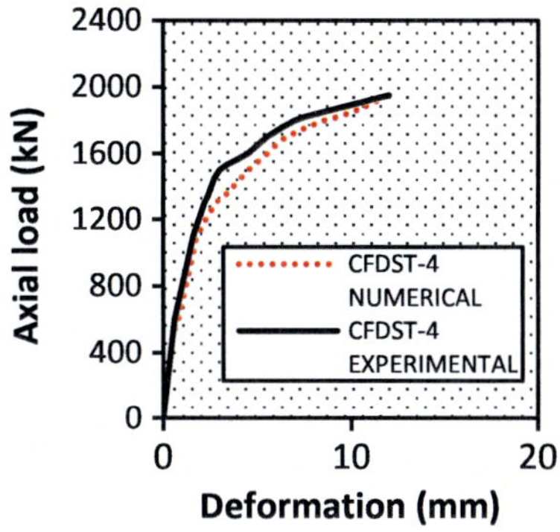

Fig. 11 Load–deformation curve for CFDST-4 specimen

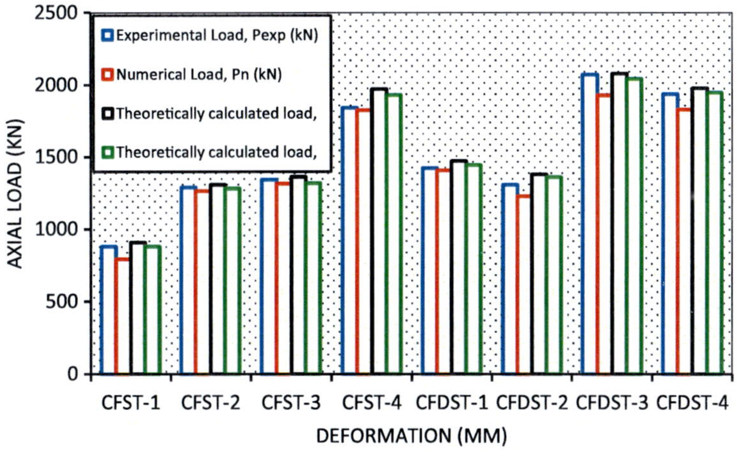

Fig. 12 Experimental and numerical results against theoretical results

5 Conclusion

For the numerical simulation of CFST and CFDST columns, ABAQUS was used to create a nonlinear FEM based model. The computational modeling using finite element modeling has led to the following conclusions:

- The theoretical model's results are compared to published experimental evidence to verify its load–deformation behavior. The experimental and model findings for CFST and CFDST columns agree well in terms of axial loading capacity, with a variance of less than 10%. The developed FE model can be used in more parametric studies to reveal additional variables that affect the CFST and CFDST behavior.
- The load-bearing strength of CFST and CFDST columns is greatly influenced by variations in steel tube thickness and diameter. The CFDST columns' axial

load capacity was reduced by 15% after the diameter of the inner steel shaft was reduced.

- The axial strength of CFST and CFDST columns is calculated using Codal parameters and formulas. In this study, the experimental findings are compared to the numerical analysis and theoretical results. Based on the findings, it is concluded that Eurocode-4 and AISC, both international codes, have conservative axial capability results for both CFST and CFDST columns, ranging from 2 to 5%, while Eurocode-4 gave a range of 3–7%.
- The impact of the *D*/*t* ratio of the columns on the projections' deviations from the findings can be attributed. There is a decrease in confinement effect when the *D*/*t* ratio is high.

References

1. Ge, H. B., & Usami, T. (1992). Strength of concrete-filled thin-walled steel box columns: Experiment. *Journal of Structural Engineering, 118*(1), 3036–3054.
2. Boyd, F. P., Cofer, W. F., & McLean, D. (1995). Seismic performance of steel-encased concrete column under flexural loading. *ACI Structural Journal, 92*(3), 355–365.
3. Bradford, M. A. (1996). Design strength of slender concrete-filled rectangular steel tubes. *ACI Structural Journal, 93*(2), 229–235.
4. Schneider, S. P. (1998). Axial loaded concrete-filled steel tubes. *Journal of Structural Engineering, 124*(10), 1125–1138.
5. Roeder, C. W., Cameron, B., & Brown, C. B. (1999). Composite action in concrete filled tubes. *Journal of Structural Engineering, 125*(5), 477–484.
6. Huang, C. S., Yeh, Y. K., Liu, G. Y., Hu, H. T., Tsai, K. C., & Weng, Y. T. (2002). Axial load behavior of stiffened concrete-filled steel columns. *Journal of Structural Engineering, 128*(9), 1222–1230.
7. Elchalakani, M., Zhao, X. L., & Grzebieta, R. H. (2002). Plastic mechanism analysis of circular tubes under pure bending. *International Journal of Mechanical Sciences, 44*(6), 1117–1143.
8. Hu, H. T., Huang, C. S., Wu, M. H., & Wu, Y. M. (2003). Nonlinear analysis of axially loaded CFT columns with confinement effect. *Journal of Structural Engineering, 129*(10), 1322–1329.
9. Hu, H. T., Huang, C. S., & Chen, Z. L. (2005). Finite element analysis of CFT columns subjected to combined axial force and bending moment. *Journal of Constructional Steel Research, 61*(12), 1692–1712.
10. Lu, F. W., Li, S. P., & Sun, G. (2007). A study on the behavior of eccentrically compressed square concrete-filled steel tube columns. *Journal of Constructional Steel Research, 63*(7), 941–948.
11. Schneider, S. P. (1998). Axially loaded concrete-filled steel tubes. *Journal of Structural Engineering, 124*, 1125–1138.
12. Uy, B. (2003). High-strength steel–concrete composite columns for buildings. *Structures and Buildings, 156*, 3–14.
13. Huang, C. S., Yeh, Y. K., Liu, G. Y., Hu, H. T., Wang, S. H., & Wu, M. H. (2002). Axial load behavior of stiffened concrete filled steel columns. *Journal of Structural Engineering, 128*(9), 1222–1230.
14. Han, L. H., & Yao, G. H. (2003). Influence of concrete compaction on the strength of concrete-filled steel RHS columns. *Journal of Constructional Steel Research, 59*, 751–767.
15. Mursi, M., & Uy, B. (2003). Strength of concrete filled steel box columns incorporating interaction buckling. *Journal of Structural Engineering, 129*(5), 626–639.

16. Liu, D., Gho, W. M., & Yuan, J. (2003). Ultimate capacity of high-strength rectangular concrete-filled steel hollow section stub columns. *Journal of Constructional Steel Research, 59*(12), 1499–1515.
17. Uy, B. (2001). Strength of short concrete filled high strength steel box columns. *Journal of Constructional Steel Research, 57*(2), 113–134.
18. Sakino, K., Nakahara, H., Morino, S., & Nishiyama, I. (2004). Behavior of centrally loaded concrete-filled steel-tube short columns. *Journal of Structural Engineering, 130*(2), 180–188.
19. Giakoumelis, G., & Lam, D. (2004). Axial capacity of circular concrete filled tube columns. *Journal of Constructional Steel Research, 60*, 1049–1068.
20. Hou, C., Han, L. H., & Zhao, X. L. (2013). Full-range analysis on square CFST stub columns and beams under loading and chloride corrosion. *Thin-Walled Structures, 68*, 50–64.
21. Lin, M. L., & Tsai, K. C. (2001). Behavior of double-skinned composite steel tubular columns subjected to combined axial and flexural loads. In *Proceedings of the First International Conference on Steel and Composite Structures* (pp. 1145–1152), Pusan, Korea, June 14–16.
22. Han, L. H., Huang, H., Tao, Z., & Zhao, X. L. (2006). Concrete-filled double skin steel tubular (CFDST) columns subjected to cyclic bending. *Engineering Structures, 29*(12), 1698–1714.
23. Tao, Z., Han, L. H., & Zhao, X. L. (2004). Behaviour of concrete-filled double skin (CHS inner and CHS outer) steel tubular stub columns and beam-columns. *Journal of Constructional Steel Research, 60*(8), 1129–1158.
24. Wei, S., Mau, S. T., Vipulanandan, C., & Mantrala, S. K. (1995). Performance of new sandwich tube under axial loading: Experiment. *Journal of Structural Engineering, 121*(12), 1806–1814.
25. Zhao, X. L., Grzebieta, R. H., & Elchalakani, M. (2002). Tests of concrete-filled double skin CHS composite stub columns. *Steel and Composite Structures, 2*(2), 129–146.
26. Han, L. H., Xu, L., & Zhao, X. L. (2003). Tests and analysis on the temperature field within concrete-filled steel tubes with or without protection subjected to a standard fire. *Advances in Structural Engineering, 6*(2), 121–133.
27. Dawson, T. H. (1983). *Offshore structural engineering*. Hoboken: Prentice-Hall Inc.
28. Sumer, B. M., & Fredsoe, J. (1997). *Hydrodynamics around cylindrical structures*. Singapore: World Scientific.
29. Hibbitt, Karlsson, Sorensen Inc. (2003). ABAQUS/standard user's manual, version 6.4.1. Hibbitt, Karlsson, & Sorensen, Inc.
30. Tiwary, A. K., & Gupta, A. K. (2020). Post-fire exposure behavior of circular concrete-filled steel tube column under axial loading. *International Journal of Steel Structure*.
31. Tiwary, A. K., & Gupta, A. K. (2019). Strength aspects of concrete filled steel tube columns through design codes. *TEST Engineering and Management, 81*, 3672–3681. ISSN No. 0193-4120.
32. Tiwary, A. K., & Gupta, A. K. (2019). Nonlinear analysis of circular concrete filled steel tube columns under axial loading. *International Journal of Innovative Technology and Exploring Engineering, 8*, 688–692. ISSN No. 2278-3075.
33. Tiwary, A. K., & Gupta, A. K. (2020). Mechanical behavior of circular concrete filled steel tube column under axial loading. *Journal of Green Engineering, 10*(11), 11116–11132.

Corrosion Monitoring in Reinforced Concrete Structures by Impressed Current Technique

Meenakshi Dixit and Ashok Kumar Gupta

Abstract Steel reinforcement corrosion results in deterioration or even early collapse of reinforced concrete (RC) structures. Corrosion of reinforced steel is a serious issue of durability globally and remains the main focus of many research studies. The formation of cracks in concrete because of corrosion is a basic indicator showing the service limit of existing RC structure. The natural corrosion that occurs in steel reinforcement present in concrete is generally slow and takes a reasonably long time to start. So, various methods are preferred to enable or accelerate corrosion for experimentation purposes for assessing the structural performance of RC specimens. This paper aims to review the various studies done on impressed current technique (ICT) for accelerating corrosion for the research point of view. This technique presents a new angle to analyze the corrosion process of reinforced steel and facilitates to improve the numerical approach to study the corrosion-induced damages in structures.

Keywords Corrosion · Reinforced concrete · Durability · Impressed current technique · Structural performance

1 Introduction

Reinforced concrete is one of the most used construction materials [1–3]. One of the reasons why steel and concrete can work together is that good bond properties exist between these two materials [1]. The concrete being a heterogeneous material with different pore sizes permits aggressive agents to infiltrate into the concrete from the external surroundings and triggers the initiation of corrosion of steel bars after reaching its surface [4]. The corrosion of steel bars embedded in concrete is the most problematic issue that leads to its deterioration and ultimately shortens their

M. Dixit (✉) · A. K. Gupta
Jaypee University of Information and Technology, Waknaghat, Solan, India

A. K. Gupta
e-mail: ashok.gupta@juit.ac.in

A. K. Gupta et al. (eds.), *Advances in Construction Materials and Sustainable Environment*, Lecture Notes in Civil Engineering 196,
https://doi.org/10.1007/978-981-16-6557-8_67

service life, particularly those exposed to chloride environments [1, 5, 6]. The total cost of corrosion in India was 4.2% of Indian GDP amounting to $1260 billion which was $70.3 billion as per NACE International IMPACT Study 2016 [7]. Steel reinforcement is never going to corrode in good-quality concrete because of its high alkaline nature and the low permeability that provides excellent chemical protection to reinforced steel in concrete against corrosion [5, 8, 9]. The interaction between the hydroxyl ions generated in the pore solution of concrete and the steel makes it passive due to forming of γ–Fe_2O_3 layer [8]. Less permeable concrete is capable to resist the aggressive agents like the chloride ions to penetrate the concrete [8, 10]. But neither the high alkaline environment of the concrete or less permeability of the concrete can assure that the steel will not corrode and that too especially in aggressive environments of marine environment [8]. Usually, the corrosion mechanism of steel reinforcement starts with the chloride-induced breaking of passive film [6, 10]. The entry of chloride ions into the concrete can happen either in its fresh state during the preparation of the mix by using chloride-rich ingredients or from the external environment of concrete in its hardened state [6, 11, 12]. There are two phases in the entire service life of a RC structure that are exposed to the chloride ions, i.e., first is the initiation phase and second is the propagation phase [6]. During the initiation phase, the steel bars have its passivity, but corrosion can start occurring as soon as the sufficient amount of chloride ions ingress [6]. This chloride content at which corrosion just starts is called the critical chloride content. In the propagation phase, the corrosion mechanism speeds up to a level where significant damage occurs and affects the limit state of serviceability [6]. Corrosion of steel reinforcement results in the cross-sectional loss of reinforcement and accumulation of corrosion products on its surface which is much larger in volume than the original volume of steel [7]. These accumulated corrosion products builds tensile stresses to the outside of concrete that results in the spalling and cracking of concrete [6, 11, 13–16]. The width of cracks occur is dependent on the quality of concrete used, the thickness of the cover taken in RC specimens, the distribution of the steel reinforcements in the concrete and stress-state of the reinforced steel to which subjected [17, 18, 20]. Corrosion in RC members results in the loss of bond strength between steel and concrete, cracks initiation, fragilization and ultimately the risk of failure [19]. If the structure collapses, then it may lead to a major socio-economic impact [19]. Corrosion also influences load-carrying capacity of the RC structure and ultimately reduces overall strength, thereby, disturbing the serviceability criteria [16, 21, 22]. Also, the high resistivity of concrete retards the electric current flow and finally restricts the corrosion rate. Therefore, there will hardly be any major problem in a RC structure which is properly designed, constructed and maintained [7]. But, unfortunately, this preferable situation is not always possible to achieve in practical scenario due to various grounds. The propagation period of corrosion may be continued for the estimated 20 years at corrosion rates as low as to the order of 1 $\mu A/cm^2$ till the cracking of cover occurs [23]. The main obstruction for research purposes is that the time limit for a research project is usually short, and it takes several years for natural corrosion to take place in steel reinforced in concrete. Therefore, it is the demand of the situation to explore new methods for accelerating the corrosion mechanism in RC specimens. The main aim of this paper is to review and examine one of the method to accelerate corrosion that is the impressed current technique. Though this method is commonly used, very less work has been published describing the theory behind the practice and its suitability to model the corrosion in steel reinforcement. The background and the current state of technique is briefly reviewed. There has been a substantial progress in this area over the past two decades and reference is made to past and recent research.

2 Impressed Current Technique

Impressed current technique is a widely adopted method when corrosion is needed to be accelerated for research purposes in order to complete it within a short span of time. The technique can be used for small as well as large-sized RC specimens.

2.1 *Setup of Impressed Current Technique for Inducing Reinforcement Corrosion*

The ICT setup consists of a direct current (DC) power supply source, a counter electrode and an electrolyte solution. The positive terminal of the DC power supply source is directly connected to the steel reinforcement which acts as the anode and the negative terminal is connected to the counter electrode which acts as cathode. The electric current is impressed from the counter electrode to the steel reinforcement through concrete which is immersed in the electrolyte solution. DC power supply source is attached to an ammeter which is used to screen and control the flow of electric current. The electrolyte solution in which the concrete specimens are immersed requires to be changed after some days. After the completion of the test that is the corrosion phase, all the corroded reinforcements needed to be removed from the specimen and cleaned as per directions mentioned in ASTM G1-90 standard and finally gets weighed. Afterward, Faraday's Law is used to calculate the mass loss because of corrosion at the steel bar which acts as anode that is directly proportional to the amount of electric current applied and time duration of application of current. The implementation of Faraday's Law for calculation of mass losses due to corrosion may provide different test results in different testing environments.

2.2 *Faraday's Law*

The electric circuitries applied in ICT and the setup or arrangement of ICT for inducing corrosion is shown in Fig. 1. The mathematical expressions involved in this

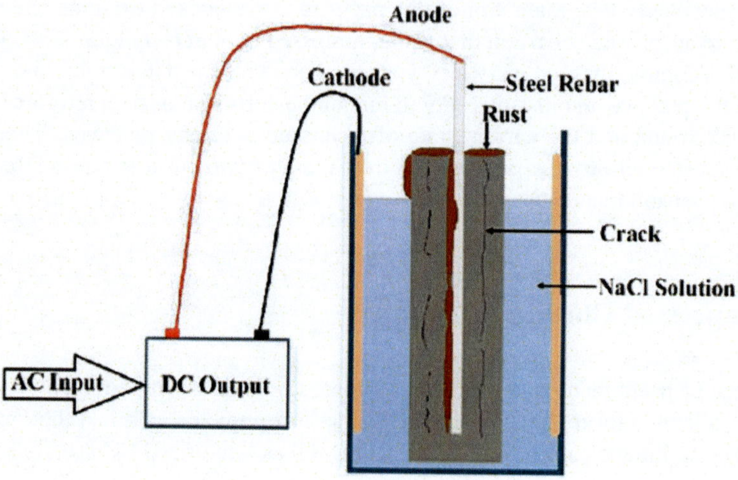

Fig. 1 Impressed current technique setup [7]

test are theoretical mass loss, induced equivalent corrosion current density and the current efficiency based on Faraday's Law and actual mass loss calculated by the gravimetric method are presented in this section through Eqs. 1–5 [24–26].

Theoretical mass loss as a result of corrosion based on Faraday's second law is shown in Eq. 1:

$$M_T = \frac{W I_{app} T}{zF} = \frac{W i_{app} A T}{zF} \tag{1}$$

where M_T = theoretical corrosion mass loss per unit surface area of the steel reinforcement (g/cm^2), $W = 56$ g, i.e., atomic mass of Fe, I_{app} = average applied electric current (A) calculated using Eq. 2 as:

$$i_{app} = \frac{I_{app}}{A} \tag{2}$$

i_{app} = average applied current density (A/cm^2), A = corroding surface area of the steel reinforcement (cm^2), T = duration of induced corrosion (s), z = ionic charge (for Fe, $z = 2$) and F = Faraday's constant = 96,485.33 A-s.

Actual mass loss due to corrosion (or mass of corrosion products generated) is calculated by gravimetric method as given by Eq. 3

$$M_A = \frac{(M_i - M_f)}{\pi DL} \tag{3}$$

where M_A = actual corrosion mass loss per unit surface area of the steel reinforcement (g/cm^2), M_i = initial mass of the steel reinforcement before corrosion (g), M_f = final mass of the steel reinforcement after a given duration of induced corrosion (g); $D =$ diameter of the steel reinforcement (cm) and $L =$ length of the steel reinforcement (cm).

The induced equivalent corrosion current density (I_{corr}) is calculated using Eq. 4 and by assuming that the actual corrosion mass loss and theoretical corrosion mass loss are equal, i.e., $I_{corr} = I_{app}$.

$$I_{corr} = \frac{(M_i - M_f)F}{\pi DLWT} \tag{4}$$

The current efficiency (η) is defined as the ratio of the gravimetric method calculated actual mass loss due to corrosion to the theoretical mass loss due to corrosion calculated by applying Faraday's Law as given by Eq. 5:

$$\eta = \frac{M_A}{M_T} \tag{5}$$

An important assumption is made that the all the applied impressed current participates in the reactions of corrosion and produces the ferrous ions on the anode. Further, Faraday's Law is used to calculate mass loss due to corrosion [27]. The best advantage that makes it to be the most useful technique is the proportional correlation between the mass loss due to corrosion and the time duration of application of electric current (i.e., time of induced corrosion) for the research point of view for accelerating the process of corrosion for investigation of the deterioration of RC specimens with respect to the generation of cracks and its pattern [28, 29], the mass loss due to corrosion [25, 29] as well as prediction of the service life of RC structures [16, 31].

2.3 *Accelerated Depassivation of Steel*

The protective alkaline environment gets destroyed due to either carbonation or by chloride attack. It generally takes a lot of time that sufficient chlorides ingress through concrete cover and depassify steel. In order to depassify steel hurriedly, some researchers mix chlorides by the weight of cement in the mix of concrete ranging from 1% [24] to 5% [32]. But mainly the cured concrete samples gets immersed in sodium chloride (NaCl) solution ranging from 3% [33] to 5% [34] concentration that is calculated by weight of the solution.

However, not all the faces of the RC specimen are exposed to chloride ingress under natural steel corrosion, generally. This experiment situation can be made artificially in laboratory by either constructing ponds on the surface of the test specimen to be contaminated and filling with sodium chloride solution [34, 35] or by spraying the specimen faces selectively with sodium chloride solution [36–38]. By this selective contamination of specimens with NaCl solution, it was found that steel corrosion is generally initiated from the direction of ingress of corrosion agents [26, 39, 40].

One of the implication of this selective contamination of faces of concrete is that that larger tensile strains are generated prior to cover cracking when compared to non-contaminated faces [39]. Considering these obligations, concrete samples in ICT is generally immersed into the chloride solution. The amount of the corrosion current that flows through the circuit depends on the content of chloride in the electrolyte solution, and it usually rises with the increment in the content of chloride till the optimum is achieved. Also, the mass loss of steel depends on the presence of chloride content in the electrolyte solution in chloride-induced corrosion [29, 41]. However, there is no specific guideline in any codes of standard in respect to the amount of chloride content in the electrolyte solution and the time of immersion before the start of ICT to attain the saturated chloride content so that a steady current flow occurs.

2.4 *Impressed Current Density*

Corrosion of steel reinforced in concrete is an electrochemical mechanism that occurs as a result of an oxidation–reduction reaction. The electrons are produced at anode where loss of steel occurs and gets transferred to the cathode. As a result of this flow of electrons, an electric current is produced and the current density is calculated by dividing this electric current by the surface area of the embedded steel bar (anode). The electric current density generated due to naturally occurring corrosion of steel is generally between 0.1 and 10 $\mu A/cm^2$, but it can reach to 100 $\mu A/cm^2$ occasionally [17, 42]. Researchers use this high current in the laboratory to accelerate corrosion. The range of impressed current density used in experiments by various researchers varied greatly between 3 [17] and 10,400 $\mu A/cm^2$ [43]. For example, a continuous 3 $\mu A/cm^2$ current density that is applied for a duration of one year assuming a proportional structural damage due to corrosion generated through the accelerated corrosion test causes an equivalent damage that can be obtained through the application of 10,400 $\mu A/cm^2$ current density if applied for two hours. Table 1 presents some of the previously performed experiments on RC specimens through ICT.

3 Major Developments

Mangat and Elgarf [31] investigated the structural damage due to varying current densities and found that average mass losses of steel reinforcement after applying ICT came within 4% of Faraday's Law predicted theoretical mass due to corrosion that is not dependent on the range of applied impressed current density. Co-relation between mass losses of steel due to corrosion induced by ICT and crack widths developed underestimates the steel corrosion in in-service RC structures and that is why cracks width cannot be used to predict the corrosion level. Austin et al. [8] concluded

Table 1 Summary of some previously performed experiments

S. No.	Author(s)	Cathode type	Corrosion environment	Current density (μA/cm^2)
1	Uomoto et al. [44]	External copper plate	5% NaCl solution	200–630
2	Tachibana et al. [45]	External copper plate	3.3% NaCl solution	500
3	Cabrera [46]	External stainless steel plate	2% chlorides by weight of cement are mixed during concrete preparation and constant immersion of specimens immersed in 5% NaCl solution	76–674
4	Alonso et al. [17]	–	3% $CaCl_2$ by weight of cement is mixed during concrete preparation	10–100
5	El Maaddawy and Soudki [29]	Internal stainless steel bar	5% NaCl is mixed in concrete by weight of cement during preparation and wrapping of specimens with wetted burlap sheets	100–500
6	Masoud et al. [47]	Stainless steel bar	2.15% chlorides by weight of cement are mixed during concrete preparation	140
7	Yuan et al. [40]	Internal stainless steel bar	5% NaCl solution	1000–2000
8	Azad et al. [48]	Stainless steel plate	2% NaCl by weight of cement is mixed during concrete preparation and constant immersion of specimens immersed in 5% NaCl solution	2000–3000
9	Care and Raharinaivo [49]	Carbon fiber mesh	Alkaline solution with 1 g/l NaOH + 4.65 g/l KOH with $[Cl^-] = 18$ g/L	100–500
10	Malumbela et al. [50]	Stainless steel bar	5% NaCl solution	189
11	Lin et al. [51]	Stainless steel net fixed externally on the test specimen	A sponge on which 5% chloride solution was sprayed	400–600
12	Nguyen and Lambert [24]	Stainless steel plates	3.5% NaCl solution	250–2500

(continued)

Table 1 (continued)

S. No.	Author(s)	Cathode type	Corrosion environment	Current density ($\mu A/cm^2$)
13	Sun and [1]	Soft stainless steel net	5% NaCl solution	200–1000
14	Zhang et al. [52]	Copper plate	5% NaCl solution	50–300
15	Hong et al. [27]	Silver piece that is wrapped externally on the test specimen	Pre-wetted with 3.5% NaCl	305–500

that ICT is a quick and effective method to accelerate chloride-induced corrosion. Further, while applying Faraday's Law, if the oxygen evolution and passivity are taken into account after studying the electrochemistry of corrosion of steel in RC, the correlation between theoretical and actual mass losses gets significantly improved. El-Maaddawy and Soudki [29] studied varying impressed current densities on reinforced steel in concrete and found that varying current densities have no effect on mass loss percentage due to corrosion calculated by Faraday's Law. Further, concluded that ICT is an effective method for inducing corrosion up to the mass loss of 7.27%. But, when the current density is increased more than 200 $\mu A/cm^2$, then a significant increment occurs in strain response and width of cracks due to the expansive stresses generated due to the corrosion products. Ahmed et al. [53] demonstrated the application of ICT to investigate the structural performance of RC specimens after the addition of supplementary material of cement like silica fume, slag and fly ash for studying damage due to corrosion. Ha et al. [54] applied ICT to explore the performance of steel reinforcement in concrete where fly ash is added in proportions to cement by accelerated corrosion. Yuan et al. [40] experimented and concluded that the surface characteristics of the naturally occurring corroded steel bar and the corroded steel bar where corrosion is induced by ICT are different, and this difference causes variations in distribution of generated corrosion products on the steel bar surface. The reinforced steel bar gets corroded on that side of its surface, which is facing toward the cover of concrete when occurring naturally and on the whole surface when ICT is applied. And, due to this difference in the corrosion distribution, corroded RC members behaves differently when structural performance is investigated. Auyeung et al. [55] determined that the actual and theoretical mass loss due to corrosion can never be the same due to many reasons such as resistivity of concrete, the mineral composition of the steel reinforcement, electrical properties of minerals in concrete and the need of electrical energy for corrosion to initiate. Care and Raharinaivo [49] concluded that the distribution of corrosion products, the formation of longitudinal surface cracks parallel to the reinforcement and the method of attack on the steel resemble very closely to the natural conditions when corrosion is induced by ICT on reinforced mortar specimens. However, the effect of the impressed current magnitude on the mechanical properties of RC members are hardly taken into consideration. Many studies [56–61] have revealed that the bond strength increases initially with slight corrosion of reinforcement, and this can be because of the increase of interface roughness. When the steel reinforcement continues to corrode, the bond strength will decrease rapidly inducing cracking of concrete cover. Experiments performed by Fang et al. [56] and Lin and Zhao [57] indicate that stirrups can increase the bond strength to resist the effect of corrosion. The experiments performed out by Wu et al. [58] and Law and Molyneaux [59] concluded that the cross-section of steel reinforcement and concrete cover thickness affect the bond strength. Tondolo [60] and Zhou et al. [61] concluded that corrosion of the stirrups shows a similar effect on the bond strength degradation as the corroded reinforced steel. Lin et al. [51] utilized the width of cracks generated to speculate the degradation of the bond strength in corroded RC specimens using impressed current density of 400–600 $\mu A/cm^2$ to accelerate corrosion. Sun and Qiao [1] experimented that there is a reasonable increment in the maximum crack width when the impressed current density increases. Further, the increment in the impressed current density resulted in the decrement of level of corrosion, and there is a sharp decrement of the bond strength with increment of the level of corrosion in the attenuation model of bond strength. ICT for corrosion acceleration has many advantages in addition to the obvious savings in time. One of the main advantage of ICT over other corrosion accelerated techniques is its ability to control the corrosion rate that generally varies due to changes in one of the variables, i.e., temperature, oxygen concentration, resistivity, etc., and any change in any of the variables would be compensated by varying other variable [25, 26]. For example, if there occurs a change in the concrete resistivity due to evaporation of pore water, or any temperature fluctuations, the supply of greater voltage can counterbalance it and thereby maintaining the required corrosion rate or level. Nguyen and Lambert [24] found that more localized corrosion is caused by higher impressed current densities. The study further concluded that determining corrosion by calculating the diameter loss is not a reliable method because it becomes very difficult to measure the diameter of the corroded bar with utmost precision. Zhang et al. [52] analyzed that when applied impressed current density increases, then there is a decrement in expansion coefficients of steel corrosion products. Benito et al. [62] investigated the corrosion of RC specimens in brackish water as well as fresh water and finally concluded that corrosion is much more active in brackish water than fresh water. Hong et al. [27] concluded that the efficiency of impressed current varies with different corrosion periods and the mass loss because of corrosion not always follow Faraday's Law. In the initial phase of corrosion due to high anodic potential, water electrolysis happens that lowers the efficiency of current. Whereas, efficiency of impressed current may exceed 100% in the late phase of corrosion due to naturally occurring corrosion.

4 Conclusion

This paper discussed one of the methods for accelerating corrosion that is impressed current technique. The basic introduction and the setup involved in technique is briefly discussed. The suitability of using ICT and especially mass loss of steel because of corrosion determined by Faraday's Law was reviewed. The various parameters involved in Faraday's Law have been discussed. It was found from the literature that the range of impressed current density for inducing and then accelerating corrosion in steel reinforcement is contentious among the researchers. Studies done by various researchers have been discussed, and results of some researchers assessing the effect of impressed current on the structural properties have been reviewed. Further research on this is, however, recommended. Moreover, more data to confirm that ICT is the reliable technique for inducing corrosion in RC structures is needed.

References

1. Sun, Y., & Qiao, G. (2019). Influence of constant current accelerated corrosion on the bond properties of reinforced concrete. *International Journal of Electrochemical Science, 14*(5), 4580–4594.
2. Nguyen, T. H. Y., Bui, V. H. L., Tran, V. M., Cao, N. T., Pansuk, W., & Jongvivatsakul, P. (2021). Verifying the reliability of impressed current method to simulate natural corrosion in reinforced concrete. *Engineering Journal, 25*(3), 105–116.
3. Dixit, M., & Gupta, A. K. (2021). A review of different assessment methods of corrosion of steel reinforcement in concrete. *Iranian Journal of Science and Technology, Transactions of Civil Engineering*. https://doi.org/10.1007/s40996-021-00644-5
4. Bahekar, P. V., & Gadvea, S. S. (2019). Corrosion of rebar in carbon fiber reinforced polymer bonded reinforced concrete. *Advances in Concrete Construction, 8*(4), 247–255.
5. Malerba, P. G., Sgambi, L., Lelmini, D., & Gotti, G. (2017). Influence of corrosive phenomena on bearing capacity of RC and PC beams. *Advances in Concrete Construction, 5*(2), 117–143.
6. Talakokula, V., Bhalla, S., & Gupta, A. (2014). Corrosion assessment of reinforced concrete structures based on equivalent structural parameters using electro-mechanical impedance technique. *Journal of Intelligent Material Systems and Structures, 25*(4), 484–500.
7. Dixit, M., & Gupta, A. K. (2021). Assessment of corrosion in rebars by impressed current technique. In *Advances in Geotechnics and Structural Engineering: Select Proceedings of TRACE 2020* (pp. 89–97). Springer. http://doi.org/10.1007/978-981-33-6969-6_9
8. Austin, S. A., Lyons, R., & Ing, M. J. (2004). Electrochemical behavior of steel-reinforced concrete during accelerated corrosion testing. *Corrosion, 60*(2), 203–212.
9. Mangat, P. S., & Molloy, B. T. (1992). Factors influencing chloride-induced corrosion of reinforcement in concrete. *Materials and Structures, 25*(7), 404–411.
10. Moreno, M., Morris, W., Alvarez, M. G., & Duffó, G. S. (2004). Corrosion of reinforcing steel in simulated concrete pore solutions: Effect of carbonation and chloride content. *Corrosion Science, 42*(11), 2681–2699.
11. Dehwah, H. A. F., Maslehuddin, M., & Austin, S. A. (2002). Long-term effect of sulfate ions and associated cation type on chloride-induced reinforcement corrosion in Portland cement concretes. *Cement and Concrete Composites, 24*, 17–25.
12. Hussain, S. E., Rasheeduzzafar, A.-M., & Al-Gahtani, A. S. (1995). Factors affecting threshold chloride for reinforcement corrosion in concrete. *Cement and Concrete Research, 25*(7), 1543–1555.
13. Khan, I., Franois, R., & Castel, A. (2014). Prediction of reinforcement corrosion using corrosion induced crack width in corroded reinforced concrete beams. *Cement and Concrete Research, 56*, 84–96.
14. Michel, A., Pease, B. J., Geiker, M. R., Stang, H., & Olesen, J. F. (2011). Monitoring reinforcement corrosion and corrosion-induced cracking using non-destructive x-ray attenuation measurements. *Cement and Concrete Research, 41*(11), 1085–1094.
15. Zhao, Y., Dong, J., Wu, Y., Wang, H., Li, X., & Xu, Q. (2014). Steel corrosion and corrosion-induced cracking in recycled aggregate concrete. *Corrosion Science, 85*, 241–250.

16. Ye, H., Fu, C., Jin, N., & Jin, X. (2018). Performance of reinforced concrete beams corroded under sustained service loads: A comparative study of two accelerated corrosion techniques. *Construction and Building Materials, 162*, 286–297.
17. Alonso, C., Andradel, C., Rodriguez, J., & Diez, J. M. (1998). Factors controlling cracking of concrete affected by reinforcement corrosion. *Materials and Structures, 31*(7), 435–444.
18. Aveldaño, R. R., & Ortega, N. F. (2009). Influence of reinforcement distribution in the corrosive process of reinforced concrete beams. *Magazine of Concrete Research, 61*(3), 213–220.
19. Torres-Luque, M., Bastidas-Arteaga, E., Schoefs, F., Sanchez-Silva, M., & Osma, J. F. (2014). Non-destructive methods for measuring chloride ingress into concrete: State-of-the-art and future challenges. *Construction and Building Materials, 68*, 68–81.
20. Ortega, N. F., & Robles, S. I. (2016). Assessment of residual life of concrete structures affected by reinforcement corrosion. *HBRC Journal, 12*(2), 114–122.
21. Ballim, Y., Reid, J. C., & Kemp, A. R. (2001). Deflection of RC beams under simultaneous load and steel corrosion. *Magazine of Concrete Research, 53*(3), 171–181.
22. Coronelli, D., & Gambarova, P. (2004). Structural assessment of corroded reinforced concrete beams: Modeling guidelines. *Journal of the Structural Engineering, 130*(8), 1214–1224.
23. Andrade, C., Alonso, C., & Molina, F. J. (1993). Cover cracking as a function of bar corrosion: Part I—Experimental test. *Materials and Structures, 26*(8), 453–464.
24. Nguyen, C. V., & Lambert, P. (2018). Effect of current density on accelerated corrosion of reinforcing steel bars in concrete. *Structure and Infrastructure Engineering, 14*(11), 1535–1546.
25. Ahmad, S. (2009). Techniques for inducing accelerated corrosion of steel in concrete. *Arabian Journal for Science and Engineering, 34*(2), 95–104.
26. Nossoni, G., & Harichandran, R. (2012). Current efficiency in accelerated corrosion testing of concrete. *Corrosion, 68*(9), 801–809.
27. Hong, S., Zheng, F., Shi, G., Li, J., Luo, X., Xing, F., Tang, L., & Dong, B. (2020). Determination of impressed current efficiency during accelerated corrosion of reinforcement. *Cement and Concrete Composites, 108*, 103536.
28. Caré, S., Nguyen, Q. T., l'Hostis, V., & Berthaud, Y. (2008). Mechanical properties of the rust layer induced by impressed current method in reinforced mortar. *Cement and Concrete Research, 38*(8–9), 1079–1091.
29. El-Maaddawy, T. A., & Soudki, K. A. (2003). Effectiveness of impressed current technique to simulate corrosion of steel reinforcement in concrete. *Journal of Materials in Civil Engineering, 15*(1), 41–47.
30. Qiao, D., Nakamura, H., Yamamoto, Y., & Miura, T. (2016). Crack patterns of concrete with a single rebar subjected to non-uniform and localized corrosion. *Construction and Building Materials, 116*, 366–377.
31. Mangat, P. S., & Elgarf, M. S. (1999). Flexural strength of concrete beams with corroding reinforcement. *ACI Structural Journal, 96*(1), 149–158.
32. Cairns, J., Du, Y., & Law, D. (2008). Structural performance of corrosion-damaged concrete beams. *Magazine of Concrete Research, 60*(5), 359–370.
33. Cabrera, J. G. (1996). Deterioration of concrete due to reinforcement steel corrosion. *Cement and Concrete Composites, 18*, 47–59.
34. Malumbela, G., Moyo, P., & Alexander, M. G. (2009). Behaviour of RC beams corroded under sustained service loads. *Construction and Building Materials, 23*(11), 3346–3351.
35. Zhang, R., Castel, A., & François, R. (2009). The corrosion pattern of reinforcement and its influence on serviceability of reinforced concrete members in chloride environment. *Cement and Concrete Research, 39*(11), 1077–1086.
36. Zhang, R., Castel, A., & François, R. (2010). Concrete cover cracking with reinforcement corrosion of RC beam during chloride-induced corrosion process. *Cement and Concrete Research, 40*(3), 415–425.
37. Rio, O., Andrade, C., Izquierdo, D., & Alonso, C. (2005). Behaviour of patch-repaired concrete structural elements under increasing static loads to flexural failure. *Journal of Materials in Civil Engineering, 17*(2), 168–177.

38. Malumbela, G., Alexander, M. G., & Moyo, P. (2011). Model for cover cracking of RC beams due to partial surface steel corrosion. *Construction and Building Materials, 25*(2), 987–991.
39. Yuan, Y., & Ji, Y. (2009). Modeling corroded section configuration of steel bar in concrete structure. *Construction and Building Materials, 23*(6), 2461–2466.
40. Yuan, Y., Ji, Y., & Shah, S. P. (2007). Comparison of two accelerated corrosion techniques for concrete structures. *ACI Materials Journal, 104*(3), 344–347.
41. Tran, K. K., Nakamura, H., Kawamura, K., & Kunieda, M. (2011). Analysis of crack propagation due to rebar corrosion using RBSM. *Cement and Concrete Composites, 33*(9), 906–917.
42. Andrade, C., & Alonso, C. (2001). On-site measurements of corrosion rate of reinforcements. *Construction and Building Materials, 15*(2–3), 141–145.
43. Almusallam, A. A., Al-Gahtani, A. S., Aziz, A. F., & Rasheeduzzafar. (1996). Effect of reinforcement corrosion on bond strength. *Construction and Building Materials, 10*(2), 123–129.
44. Uomoto, T., Tsuji, K., & Kakizawa, T. (1984). Deterioration mechanism of concrete structures caused by corrosion of reinforcing bars. *Transactions of the Japan Concrete Institute, 6*, 163–170.
45. Tachibana, Y., Kajikawa, Y., & Kawamura, M. (1990). The mechanical behavior of RC beams damaged by corrosion of reinforcement. *Concrete Library of JSCE, 14*, 177–188.
46. Cabrera, J. G. (1996). Deterioration of concrete due to reinforcement steel corrosion. *Cement and Concrete Composites, 18*(1), 47–59.
47. Masoud, S., Soudki, K., & Topper, T. (2005). Post-repair fatigue performance of FRP-repaired corroded RC beams: Experimental and analytical investigation. *Journal of Composites for Construction, 9*(5), 441–449.
48. Azad, A. K., Ahmad, S., & Azher, S. A. (2007). Residual strength of corrosion-damaged reinforced concrete beams. *ACI Materials Journal, 104*(1), 40–47.
49. Care, S., & Raharinaivo, A. (2007). Influence of impressed current on the initiation of damage in reinforced mortar due to corrosion of embedded steel. *Cement and Concrete Research, 37*(12), 1598–1612.
50. Malumbela, G., Alexander, M., & Moyo, P. (2010). Variation of steel loss and its effect on the ultimate flexural capacity of RC beams corroded and repaired under load. *Construction and Building Materials, 24*(6), 1051–1059.
51. Lin, H., Zhao, Y., Ožbolt, J., & Reinhardt, H. W. (2017). Bond strength evaluation of corroded steel bars via the surface crack width induced by reinforcement corrosion. *Engineering Structures, 152*, 506–522.
52. Zhang, W., Chen, J., & Luo, X. (2019). Effects of impressed current density on corrosion induced cracking of concrete cover. *Construction and Building Materials, 204*, 213–223.
53. Ahmed, S. F., Maalej, M., Paramasivam, P., & Mihashi, H. (2006). Assessment of corrosion-induced damage and its effect on the structural behavior of RC beams containing supplementary cementitious materials. *Progress in Structural Engineering and Materials, 8*(2), 69–77.
54. Ha, T. H., Muralidharan, S., Bae, J. H., Ha, Y. C., Lee, H. G., Park, K. W., & Kim, D. K. (2007). Accelerated short-term techniques to evaluate the corrosion performance of steel in fly ash blended concrete. *Building and Environment, 42*(1), 78–85.
55. Auyeung, Y., Balaguru, P., & Chung, L. (2000). Bond behavior of corroded reinforcement bars. *ACI Materials Journal, 97*(2), 214–220.
56. Fang, C., Lundgren, K., Plos, M., & Gylltoft, K. (2006). Bond behaviour of corroded reinforcing steel bars in concrete. *Cement and Concrete Research, 36*(10), 1931–1938.
57. Lin, H., & Zhao, Y. (2016). Effects of confinements on the bond strength between concrete and corroded steel bars. *Construction and Building Materials, 118*, 127–138.
58. Wu, Y. Z., Lv, H. L., Zhou, S. C., & Fang, Z. N. (2016). Degradation model of bond performance between deteriorated concrete and corroded deformed steel bars. *Construction and Building Materials, 119*, 89–95.
59. Law, D. W., & Molyneaux, T. C. (2017). Impact of corrosion on bond in uncracked concrete with confined and unconfined rebar. *Construction and Building Materials, 155*, 550–559.

60. Tondolo, F. (2015). Bond behaviour with reinforcement corrosion. *Construction and Building Materials, 93*, 926–932.
61. Zhou, H. J., Liang, X. B., Zhang, X. L., Lu, J. L., Xing, F., & Mei, L. (2017). Variation and degradation of steel and concrete bond performance with corroded stirrups. *Construction and Building Materials, 138*, 56–68.
62. Benito, E. K., Madlangbayan, M. S., Tabucal, N. M. S., Sundo, M. B., & Velasco, P. P. (2017). Corrosion damage measurement on reinforced concrete by impressed voltage technique and gravimetric method. *International Journal of Geomate, 13*(39), 198–205.

Evaluation of Model 3D Printer and Design Mix for 3D Concrete Printing

Ashutosh Dwivedi, Ankit Pal, Shiv Singh Patel, Ajay Chourasia, and A. K. Jain

Abstract The additive manufacturing (AM) in construction or 3D concrete printing (3DCP) is a state-of-art method where layer- by-layer structural members are rapidly constructed with automatic machine control and that commit to be notable benefits in the field of construction in terms of design flexibility, sustainability, reduced the construction cost, time, manpower. This article presents, the design of extrusion mechanism and mix for 3DCP. In this study, existing FDM based 3D printer is modified into the model of 3D concrete printer by changing its extrusion mechanism. Furthermore, optimization of the mix for 3DCP having different percentage of gray micro-silica fumes, superplasticizer, and water/binder ratio has done. After the optimization of mix, polypropylene (PP) fibers is inserted in different proportion to investigate its effect on the tensile strength of mix. Study concludes that the designed mix is extrudable from the nozzle of the printer and subsequently micro-silica improves the compressive strength, while the tensile strength increases with polypropylene fiber inclusion into the mix. It is expected that studies will be done on this article will be helpful in the future research work on 3DCP.

Keywords Additive manufacturing · 3D printing · Mortar · Concrete · 3D printer · Mortar extrusion

A. Dwivedi
CSIR-Central Building Research Institute, AcSIR, Roorkee, Uttrakhand 247667, India

A. Pal · A. K. Jain
National Institute of Technical Teachers Training and Research Institute, Bhopal 462002, India
e-mail: akjain@nitttrbpl.ac.in

S. S. Patel (✉)
CSIR-Advanced Materials and Processes Research Institute, Bhopal 462026, India

A. Chourasia
CSIR-Central Building Research Institute, Roorkee, Uttrakhand 247667, India
e-mail: ajayc@cbri.res.in

A. K. Gupta et al. (eds.), *Advances in Construction Materials and Sustainable Environment*, Lecture Notes in Civil Engineering 196,
https://doi.org/10.1007/978-981-16-6557-8_68

1 Introduction

The construction industry is most vital to enhance the economic growth of any country. Mckinsey global institute's report states "Every year, around US$10 trillion spend globally in the construction-related field, it's equivalent to 13% of the world GDP and employ 7% of the world working population" [1]. According to the world economic forum, US$100 billion a year in construction cost can be saved if productivity is increased by 1% worldwide [2]. There is also some obstacle in the traditional construction sector, which adversely affects the environment and economic efficiency and productivity of projects. By [3, 4] states, construction industry generates 25% of world solid waste, building consumes 30% of raw materials, 42% of worlds energy and 37% of the material used in construction become waste because of this 30% of projects do not meet original program or budget and 38% of carbon emission is by buildings, not cars [5]. To solve all these issues, it is seen in a few decades that the construction industry is gradually adopting modern technology by using mechanization.

Additive manufacturing [AM] in construction or 3D printing in the field of construction is the most promising modern method to overcome most of the drawback in the traditional method of construction. 3D printing is an additive manufacturing automatic technique, in which 3D solid object is produced from a digital (CAD) model [6], i.e., 3D solid object made by placing of layer over layer. It commits to be notable benefits in construction such as waste minimization, design freedom, mass customization, and ability to construct a complex structure. Benefits of 3DCP is shown in Fig. 1a, b and 3DCP using concrete mix shown in Fig. 2a, b. The structure made-up with 3DCP technique is porous, i.e., less dense, cement content is high, so low tensile strength leads to cracking and shrinkage, sometimes aggregate present in concrete blocked the printer nozzle, and issues of the placement of layer are resolved

Fig. 1 Benefits of 3D concrete printing

Fig. 2 3DCP using concrete mix: **(a)** Rudenko [9], **(b)** Salet, TU/e [10]

by the use of ultrafine material [6]. In case of 3D concrete printing, the type of printer used would play an imperative role for deciding the constituents of a mix and that mix will affect the strength parameters of 3D printed specimen. Therefore, objective of this research article is to conduct a feasibility study of the model of 3D concrete printer which is modified by changing its extrusion mechanism from FDM based 3D printer and also designing an extrudable mix using gray micro-silica fumes and effects of polypropylene fibers at different ratio on direct tensile strength of designed mix.

2 Methodology

This research work can be divided in following three sections

1. Development of printer
   - Description of extrusion mechanism
   - Designing and Fabrication of components of extrusion mechanism
   - Experimental setup
2. Designing and testing of mix to achieve extrudability (ability of the mix to pass through small diameter pipes and nozzles)
   - Achieving a extrudable mix by varying superplasticizer content
   - Achieving a shape retentive mix by varying *w*/*b* ratio
   - Analysis on open time of the mix
   - Compressive strength testing of mix
3. Effects of polypropylene fibers on tensile properties of designed mix
   - Inserting polypropylene fibers at varying ratio ranging from 0 to 0.16%.

3 Description of Extrusion Mechanism

The idea of extrusion mechanism that is being used for printing process is taken from injection molding mechanism. The injection molding mechanism is used to convey or pass the molten material into a mold by the help of a screw or ram type of an arrangement which will force the molten material to move into forward direction inside a casing. By using this mechanism, concrete has been conveyed instead of molten material, but for keeping the material in plastic state during extrusion, we have added retarder admixture into the mix, so that it will not get hard inside the casing of the screw- casing arrangement. This particular screw-casing arrangement will pass the material and to the connector potion which will further pass the mix by the help of flexible pipe to the printing head of the mechanism which will contain nozzle inside it and nozzle will finally extrude the material to the printing bed. In this article, a modified fuse deposition modeling (FDM)-based 3D printer is used to extrude a mortar mix used in 3DCP. The different 3D printed components of extrusion mechanism are fabricated by FDM based 3D printer (Creality Ender 3 3D printer having build volume of 220 * 220 * 250 mm).

4 Designing and Fabrication of the Component

Following are the different 3D printed component used in the extrusion mechanism.

Concrete mixer, hopper, casing, Archimedean spiral screw, connector, printing head, and nozzle are components which is used in the extrusion mechanism. Figure 3 shows the process involve in the 3D printed Archimedean spiral screw, remaining equipment are also printed in some fashion, where Fig. 4 shows the different 3D printed equipment's used for extrusion mechanism. Figure 5 signifies the experimental setup after attaching the component of extrusion mechanism with existing printer and 3D printed square specimen by the model of 3D concrete printer. Furthermore Table 1 shows the designed dimensions of 3D printed equipment.

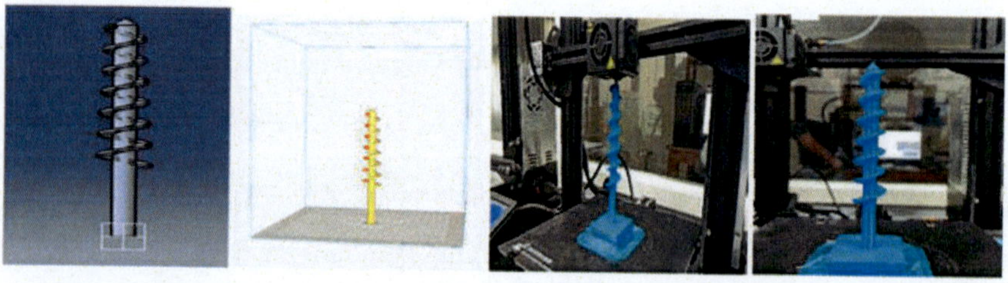

Fig. 3 Archimedean spiral screw (Solid CAD file + .stl file + 3D printed screw)

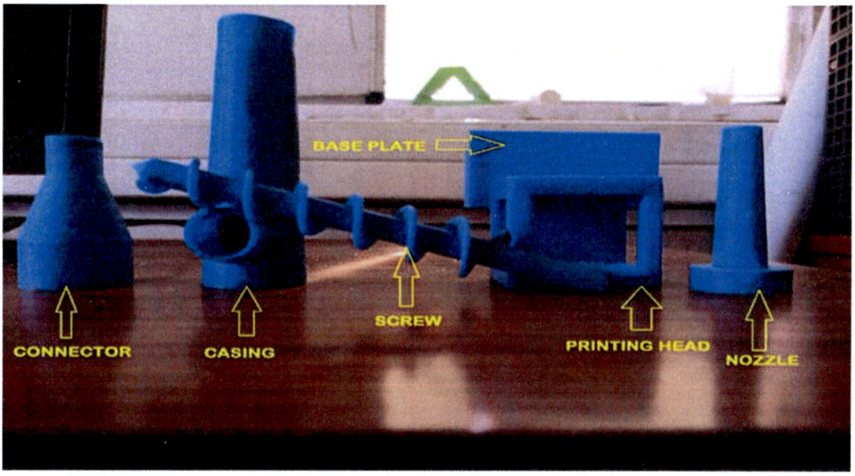

Fig. 4 3D printed equipment's used for mortar mix 3D printing

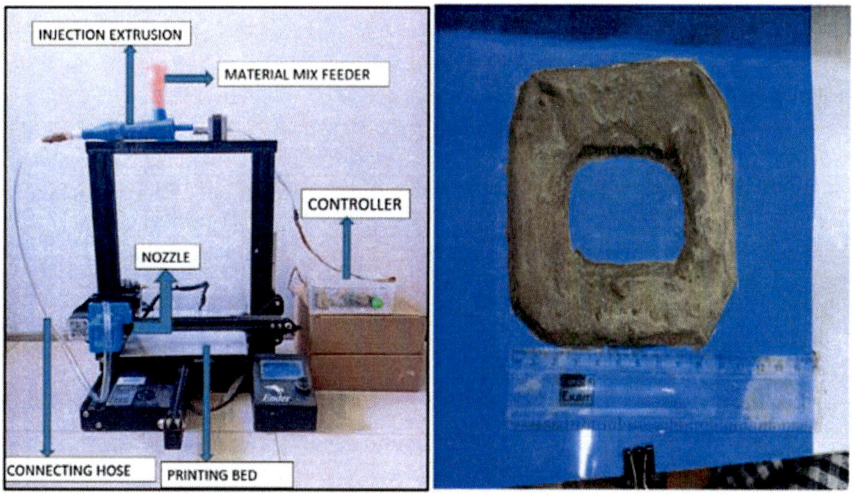

Fig. 5 Model of modified FDM based 3D printer for extruding mix and printed component

5 Experimental Setup

See Fig. 5.

Table 1 Designed dimensions of 3D printed equipment

S. No.	Components	Inlet dimension (mm)		Outlet dimension (mm)		Length of component (mm)	Infill density of component %
		Inner dia (mm)	Outer dia. (mm)	Inner dia. (mm)	Outer dia (mm)		
1	Casing	28	34	22	28	120	100
2	Connector	30	34	13	17	70	100
3	Pipe	9	11	9	11	900	–
4	Printing head	13	–	9	–	40	50
5	Nozzle	16	12	12	8	70	100

6 Designing and Testing of the Mix

6.1 Designed Mix for Extrudability

Several literatures in the field of 3DCP suggest that the particle size of fine aggregate (FA) or sand used in 3DCP must lie in the range of 1–2 mm [7]. Therefore, size of sand particle varies between 1 and 2 mm, grade of cement OPC-43, gray micro-silica fumes, and lignosulfonate-based superplasticizer (SP) are used. As mentioned in Table 2, the total amount of binder (cement + fly ash + silica fumes) material, sand, and water/binder ratio is kept constant, and the amount of superplasticizer is varied, so that a comparative analysis can be done, and optimum dose of super-plasticizer (SP) can be found out, and a mix with appropriate extrudability can be obtained, and in this way, the material mix has been optimized. Table 2 shows the mix proportion. However, the components present in the binding material is varied in a fixed proportion to achieve better results with higher variability, but in the binding material also, the quantity of cement is fixed, and quantity of silica fumes and fly ash is varied.

From the above-mentioned mixes, mix number 3 is the one having the property of extrudability, but it was not able to retain its shape, and therefore, further research

Table 2 Mix proportion

S. No.	Cement (g)	Sand (g)	Silica fumes (g)	Fly ash (g)	Water/binder ratio	SP (g)
1	2000	3000	0	857	0.3	10
2	2000	3000	143	714	0.3	15
3	2000	3000	286	571	0.3	20
4	2000	3000	571	286	0.3	25
5	2000	3000	714	143	0.3	30
6	2000	3000	857	0	0.3	40

Table 3 Constituents of mix having different *W*/*B* ratio

Mix no	Cement (g)	Sand (g)	Silica fumes (g)	Fly ash (g)	*W*/*B* ratio	SP (g)
1	2000	3000	286	571	0.29	20
2	2000	3000	286	571	0.28	20
3	2000	3000	286	571	0.27	20
4	2000	3000	286	571	0.26	20
5	2000	3000	286	571	0.25	20
6	2000	3000	286	571	0.24	20

is carried out to determine a mix having optimum extrudability with shape retention. Therefore, mix 3 is considered for further research work.

Now the shape retention property in the suggested mix can be achieved in following three ways

1. By adding accelerator dose;
2. By reducing superplasticizer dose;
3. By reducing the water/binder (*W*/*B*) ratio.

As mentioned in Table 3, author decided to reduce the water/binder ratio of the selected mix to achieve shape retention. Therefore, the water binder is changed in the range of 0.29–0.24.

Over the above six mixes, shape retention test is performed and the behavior over the shape retention property shown by each mix is described in the table.

In Table 4 (bottom), several notations is used which are described below

E and No S.R. = Extrudable but not able to retain the shape
E and S.R. = Extrudable and show shape retention.

6.2 *Open Time*

Open time is a function of workability, and to determine the open time, we must know the value of yield strength at different time of a fresh mix. The open time of the mix for 3DCP is defined for 80 min; therefore, total nine times the workability of the mix is determined within 80 min after preparing. After determining the workability, it is correlated with the yield strength of the mix [8].

In this study, the dimensionless slump parameter S is adopted using Eq. (1)

$$S = s/H \tag{1}$$

S = Dimensionless slump; s = slump (mm); H = initial slump height (mm).

Open time for the mix having 0.27 *w*/*b* ratio is determined using slump values of mix at different time. Table 4 shows the open time.

Table 4 Open time (top) and description of shape retention (bottom) the mix

Slump	Mix-1
S1	20.89
S2	19.77
S3	18.65
S4	18.21
S5	17.83
S6	17.55
S7	17.39
S8	17.30
S9	17.30

Mix No.	*W*/*B* ratio	Description
1	0.29	E and No S.R
2	0.28	E and S.R
3	0.27	E and S.R
4	0.26	E and S.R
5	0.25	Not extrudable
6	0.24	Not extrudable

6.3 Compressive Strength

Compressive strength of the optimum mix was performed on the cubes of size 70.6 mm according to IS 4031-1988 for 3, 7, and 28 days. Figure 6 shows the variation in the strength of concrete while using micro-silica.

7 Effects of Polypropylene Fibers on Tensile Properties of Mix

Now, one of the major challenges in 3D printed structure is requirement of tensile strength and that can be overcome by using reinforcement, but reinforcement by conventional methods cannot be adopted because it is not possible in case of 3DCP work. So, one option available for increasing tensile strength is using fibers into the mix and thus making a fiber reinforced mix for 3D concrete printing. In this research, polypropylene fibers of length 6 mm, having specific gravity 0.91 and diameter is 0.034 mm, are used to increase the tensile strength of concrete. To determine the effect of PP fibers added at different percentage in 3D printable mix on the tensile strength of the mix, six mixes were considered having different percentage of PP fibers ranging from 0.00 to 0.16% of total binder content in the mix. Figure 7 shows the variation of split tensile strength with polypropylene fiber and split tensile strength

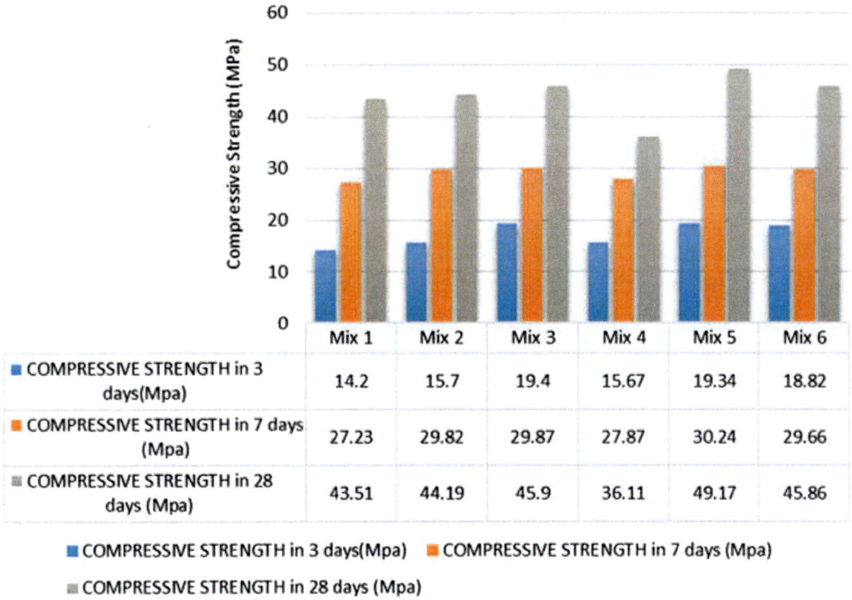

	Mix 1	Mix 2	Mix 3	Mix 4	Mix 5	Mix 6
COMPRESSIVE STRENGTH in 3 days(Mpa)	14.2	15.7	19.4	15.67	19.34	18.82
COMPRESSIVE STRENGTH in 7 days (Mpa)	27.23	29.82	29.87	27.87	30.24	29.66
COMPRESSIVE STRENGTH in 28 days (Mpa)	43.51	44.19	45.9	36.11	49.17	45.86

Fig. 6 Variation in the strength of concrete while using micro-silica

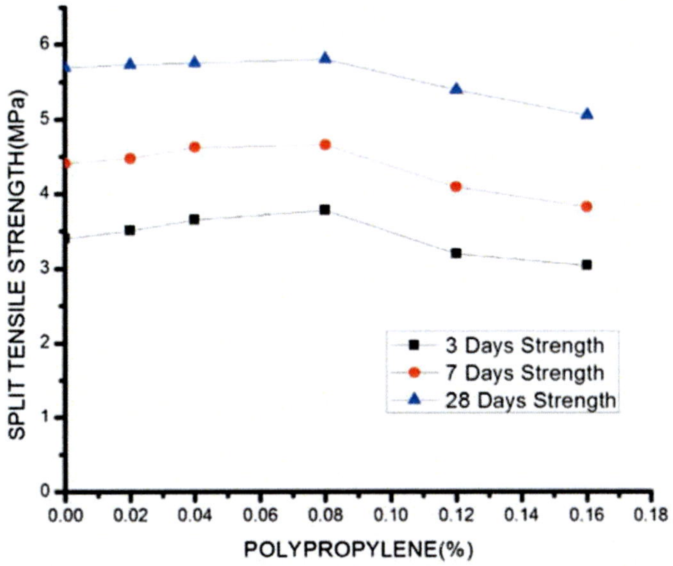

Fig. 7 Variation of split tensile strength with polypropylene fiber

of cylindrical specimen of length 300 mm and diameter 150 mm for 3, 7, and 28 days is performed according to IS-5816 (1999).

8 Result and Discussion

While performing the experiment of extrusion of mortar mix on the modified FDM based 3D printer, it was found that the printer has successfully printed a line of mortar mix onto the printing bed, but after printing one line, the connector of extrusion mechanism has been broken into two pieces, and thus, following results can be drawn:

1. Due to the sudden reduction in the cross-sectional area of the connector, lesser will be the area available for the extrusion of the mortar paste and high friction will be offered to the mortar by the walls of the connector because of that frictional force segregation of the materials will takes place and problem of chocking is observed while performing the experiment.
2. Due to the presence of high frictional force on the mortar, it will not be able to move in forward direction with the desired speed, while at the same time, continuous supply of the mortar is given at the inlet of the casing, due to that high pressure will be exerted by the material over the walls of the connector, specially to the end having lesser cross-sectional area and the connector was not able to sustain that pressure and breaks into two pieces from the outlet end.

9 Conclusion

- The reduction in area provided in the extrusion mechanism must be gradual in nature. The sudden reduction in area will exhibit high force on that particular section and may break the continuity of the assembly.
- 3D printed PLA based extrusion mechanism is not able to sustain the pushing force applied by the mortar mix because of the lower bond strength of 3D printed specimen between successive layers; hence, metal assembly must be adopted for the extrusion mechanism.
- Silica fume is exceptionally fine, the micro-filling effect of silica fume significantly improves the binding of mix and provides 36.61% more compressive strength in 3 days compared to mix having no silica fume. The efficiency of the silica fume is 3–5 times the OPC. However, from the aspect of the extrudable mix, the following proportion used in mix: cement—70%, fly ash—20%, silica fumes—10%, sand:cement—3:2, water:binder—0.27, and superplasticizer—0.5%.
- Addition of polypropylene fiber in the mix up to 0.08% will increase the early split tensile strength. So polypropylene fiber is recommended for 3D concrete printing to increase the split tensile strength. It will reduce the rigidity.

References

1. Mckinsay Global Institute. (2017). Reinventing construction; A route of higher productivity.
2. World Economic Forum. *Shaping the future of construction—A breakthrough in mindset and technology*. World Economic Forum.
3. Organization for Economics Cooperation and Development (OECD). (2008). *Policy roundtable, construction industry*. http://www.oecd.org/regreform/sectors/41765075.pdf. Accessed on 3 May 2021.
4. CMMA. (2015). Owners survey report. https://www.cmaanet.org/sites/default/files/inline-files/2015OwnersSurvey.pdf. Accessed on 2 February 2021.
5. Dietz, T., et al. (2009). Household actions can provide a behavioral wedge to rapidly reduce US carbon emissions. *Proceedings of the National Academy of Sciences, 106*(44), 18452–18456.
6. Allouzi, R. (2019). 3D printing of nanomaterials for concrete construction.
7. Le, T. T., Austin, S. A., Lim, S., Buswell, R. A., Gibb, A. G., & Thorpe, T. (2012). Mix design and fresh properties for high-performance printing concrete. *Materials and Structures, 45*(8), 1221–1232.
8. Lafhaj, Z., Rabenantoandro, A. Z., el Moussaoui, S., Dakhli, Z., & Youssef, N. (2019). Experimental approach for printability assessment: Toward a practical decision-making framework of printability for cementitious materials. *Buildings, 9*(12), 245.
9. Rudenko, A. (2021). 3D printed concrete castle is complete. http://www.totalkustom.com/3d-castle-completed.html.
10. Wereldprimeur voor Gemert: fietsen over een 3D-geprinte brug. (2021). https://nos.nl/artikel/2198400-wereldprimeur-voor-gemert-fietsen-over-een-3d-geprinte-brug.

Axial Compression Behavior of Single-Skin and Double-Skin Concrete-Filled Steel Tube Columns: A Review

Sakshi Bhatia and Aditya Kumar Tiwary

Abstract Tall structures are built in large numbers worldwide to satisfy the demands of the urban population. Composite columns like CFST (single and double skin) offer many structural benefits in which the steel and concrete are mutually benefitted, as the steel gives confinement to concrete and the concrete in turn restricts and modifies the buckling of steel tube. Recently developed concrete-filled double-skin steel tubes (CFDST) are recognized for superior ductility and energy absorption capabilities. The present study aims to identify the structural behavior of CFST and CFDST columns under axial compression. Along with this, other objective of the study is to distinguish the best composite column to utilize in construction. The slenderness ratio (*L*/*D*), width-to-thickness (*b*/*t*) ratio, segment shape, concrete grade, and steel yielding strength are the critical parameters considered for evaluating the axial compressive behavior of CFST (single and double skin) columns. It is discovered that the thickness and diameter of CFSTs can be reduced up to a certain level for *D*/*t* and *L*/*D* ratios, and also the axial load capacity as well as stiffness of CFST (single and double skin) columns reduces by increasing (*D*/*t*) ratio. From the observations, it can be inferred that the load capability of column enhances with higher yield strength of steel.

Keywords Composite columns · Concrete-filled steel tubes · Concrete-filled double skin steel tube · Axial behavior

1 Introduction

Concrete-filled steel tube (CFST) columns are extensively used in earthquake-prone building projects because of their high static and seismic resistance. In recent years, the civil engineering industry has focused on a newly developed type of composite section known as a concrete-filled double-skinned steel tube (CFDST) column which is composed up of two concentric steel tubes that are partially or completely filled with

S. Bhatia (✉) · A. K. Tiwary
Chandigarh University, Mohali, India

A. K. Gupta et al. (eds.), *Advances in Construction Materials and Sustainable Environment*, Lecture Notes in Civil Engineering 196,
https://doi.org/10.1007/978-981-16-6557-8_69

concrete. The CFDST column's bending stiffness, ductility, and seismic behavior are significantly enhanced over the typical CFST column due to the visible internal steel tube [1–3].

The CFDST concept was first used in underwater pressure vessels [4], in which the cross section's higher bending stiffness was used to resist instability under external pressure. In Japan, CFDST has been employed as tall piers for bridges due to their exceptional energy absorption and damping properties along with light cross-sectional weight [5]. Furthermore, CFDST columns are being utilized in China as electric transmission poles [6]. Because the concrete encased inner tube is present in CFDST columns, they are expected to perform well when subjected to blast loading [7], impact loading [8], and fire exposure. They also demonstrated improved structural response [9]. While ultra-high strength materials are used, the internal steel tube also plays a key function in preventing rupture failure [10]. Also provided is a state-of-the-art research on the creation and current functions of CFST constructions [11]. Steel tube also performs as frequent formwork in CFST, therefore omitting to strengthen the bar decreases the cost and time of construction as shown in Fig. 1. Long CFDST columns have been discovered in uplifted passageways, bridge piers, and structures. Despite this, because the behavior of long CFDST columns is largely unknown, the study in this field is limited [12].

Another unique composite steel–concrete construction concept is the concrete-filled double-skin steel tube (CFDST) that comprises of two concentrical steel tubes isolated by concrete infill as shown in Fig. 2. Circular and square CFDST sections are the two most commonly used profiles of CFDST sections. Circular CFDST utilizes circular hollow section (CHS) for both interior as well as external tube. As indicated

Fig. 1 Concrete-filled steel tube column (CFST)

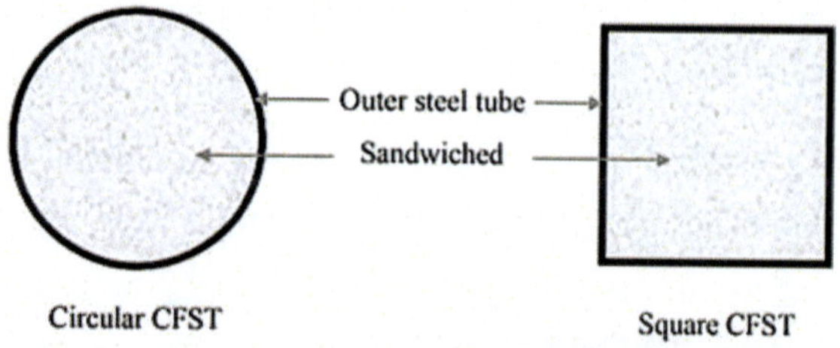

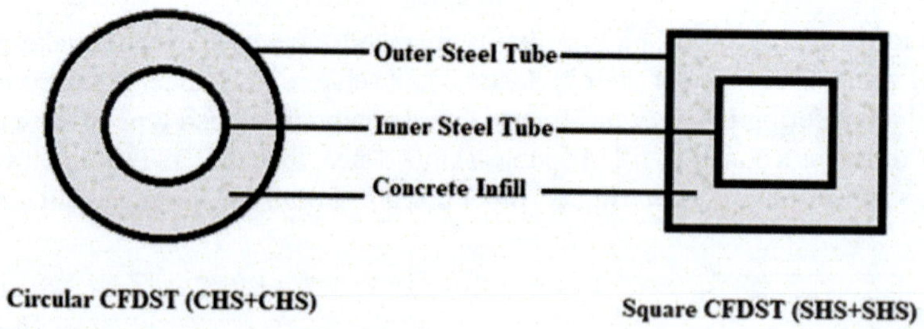

Fig. 2 Common CFDST cross section layouts

in the diagram below, the square CFDST section uses square hollow section (SHS) as both internal and exterior tubes.

The loading technique for CFDST was separated into two parts in the finite element analysis (FEA) model. The specified preload was subjected solely to hollow steel tubes with overall initial defect in the first stage, while concrete parts were deactivated. Preloads cause the steel tubes to acquire their initial deformations and prestresses. When concrete materials were activated in tubes in the second stage, the column top was loaded with vertical displacement. The interior and exterior tubes along with the concrete were all loaded until they failed. The preload on the first step kept steady through the second step until column failure.

The strength of composite column was calculated as the sum of the preload and reaction force from the border. Figure 3 depicts the preloading technique for internal and exterior tubes.

The cross-sectional form of composite columns known as CFST is the most common. In the past, encroaching on concrete in the previous form of column was thought to be a means of protecting steel from erosion and fire [11, 14, 15]. CFDST, on the other hand, ideally incorporates the advantages of regular CFST and a sandwich construction, producing a section of higher modulus, better exterior pressure stability, higher bending stiffness, improved damping properties, and improved cyclic performance. Steel tube aids in situ concrete filling, reducing the need for additional formwork and resulting in quicker building [16–18].

To achieve the same axial or bending power as an RC member, a CFDST section has a decreased area of cross section and thus has a low self-weight. As a result, the tubes separate the concrete foundation, making concrete repair less necessary. High- and medium-rise bridges, large-span buildings, and bridge columns or piers have all used CFST columns [19, 20]. Since land is scarce, many residents need structures to sustain a variety of activities. Previous research has shown that confining concrete

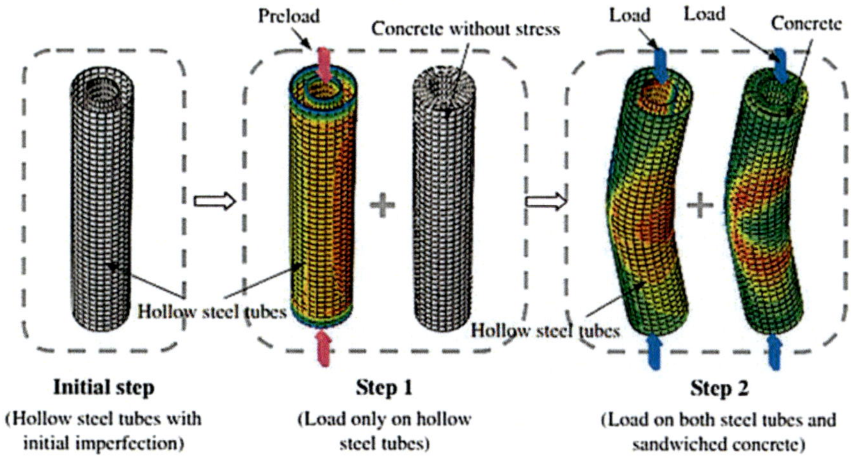

Fig. 3 Loading sequence of CFDST column [13]

improves ductility, as well as the two methods for providing concrete containment: active confinement and passive confinement [21].

Previous research has shown that CFST outperforms traditional RCC construction [22, 23]. The restricted pressure exerted on the concrete heart by the steel tubes delays the formation of microcracks and reduces the diameter of the fracturing crack [24]. Steel has the ability to exhibit more stress than concrete under tensile pressures on CFST columns [23, 25].

2 Background Studies

The effects of axial loading on CFST columns were studied in an experiment. A concentrated load was placed on six short columns. Having two steel tube thicknesses, columns with rectangular, oval, and square cross sections were employed. The test findings show that when the steel tube is confined, the ductility of high-strength concrete rises, but the load-carrying capacity does not. But as the *b*/*t* ratio increases, the contribution of steel tubes to the final load becomes more effective (see Fig. 4). The influence of shape appears to grow increasingly relevant as the *b*/*t* ratio grows. The best behavior was shown in circular columns with the same applied load resistance values in both concrete core and steel tube. The second-best result was shown by square column models, and the poorest for rectangle one. Failure loads obtained from the tests as well as design codes was compared. The findings demonstrated that the specimens' axial load capabilities were conservatively quantified as 8, 8, and 17%, by the specifications of EC4, NBR 8800, and AISC, respectively [26].

In an experiment, CFDST beam-columns exposed to cyclic bending were employed. The test parameters were segment shapes (square, circular), level of axial load (n) between 0 and 0.6, and concrete strength (f_{cu}). A total of 28 CFDST column specimens, comprising 12 specimens having CHS as internal and external tube and 16 samples having CHS as internal and SHS as external tube were experimented under cyclically rising flexural stress and continuous axial stress. The length of each specimen was 1500 mm. According to the results of the testing, CFDST beam-columns

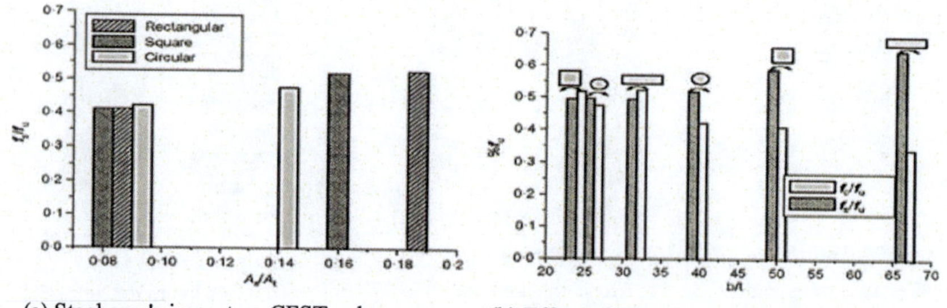

(a) Steel area's impact on CFST columns (b) Effect of b/t ratio on the CFST columns

Fig. 4 Steel area and *b*/*t* ratio influence [26]

displayed far higher ductility and energy absorption values in the case of circular sections than in square sections, even under severe axial loads [27].

The action of partly replaced concrete and M20 grade concrete packed axially loaded tubular columns was studied. Concrete rubble, partly replaced quarry soil, standard M20 grade concrete, and slenderness ratio were some of the variables used in the analysis (13.27, 16.58, and 19.9). Pressure index (SI), concrete contribution ratio (CCR), ductility index (DI), and confinement index (CI) were analyzed and compared across CFST columns. Regardless of the filled material, improving the *L*/*D* ratio reduces composite column activity. In contrast to hollow steel columns, the composite process was conducted on the CFST columns encased with concrete rubble and partly replaced quarry mud. The load strength of the CFST specimen is 32% greater as compared to hollow tubular columns. For different slenderness ratios, the *L*/*D* action of the composite column is seen in Fig. 5a–c. When compared to a typical M20 grade, concrete-filled column of slenderness ratio 13.27, the percentage decrease in deflection by 19% for quarry dust (QD) and 14.3% with concrete debris (CD) was observed.

Also, when the slenderness ratio was 16.58, percentage decrease in deflection was found to be 25% and 13.6% as column changed to quarry dust from typical concrete. When slenderness ratio was 19.90, decrease of 12% and 4% was observed in deflection with QD and CD, respectively. The loading capacities with varying slenderness ratios are differentiated in Fig. 6.

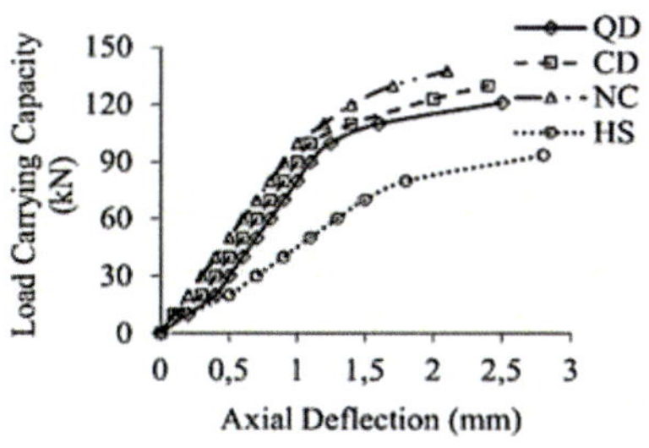

(a) Load-deflection behavior (L/D = 13.27)

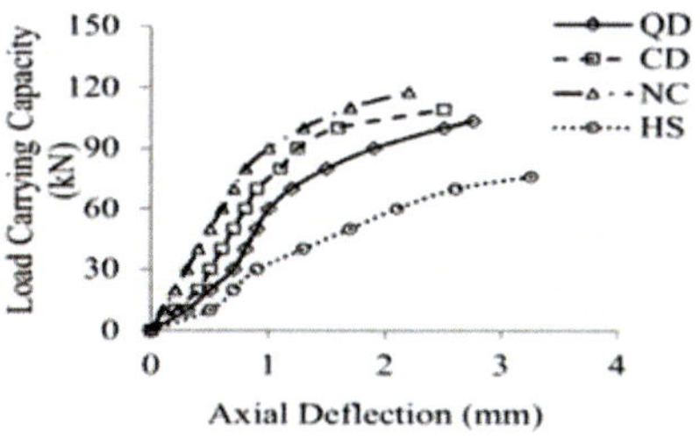

(b) Load-deflection behavior (L/D = 16.58)

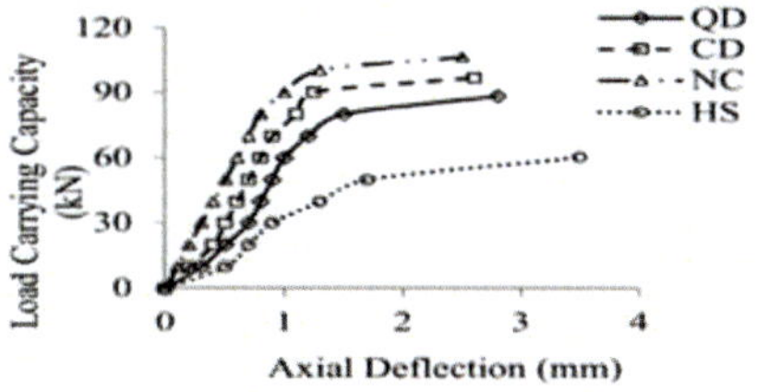

(c) Load-deflection behavior (L/D = 19.90)

Fig. 5 CFST columns load deformation behaviour. **a** *L*/*D* 13.27, **b** *L*/*D* = 16.58, **c** *L*/*D* = 19.90 [17]

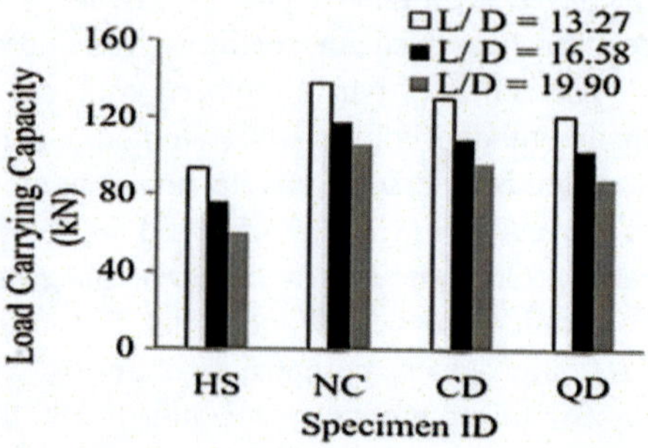

Fig. 6 Load-carrying capacity of different CFST columns [17]

A parametric analysis on behavior of CFDST slender columns subjected to axial compression was carried out. Testing was done on 51 CFDST pin-ended specimens. Models obtained from FE modelling were validated with published experimental observations. The FEA findings were evaluated and were correlated to design standards. In comparison with American standard, European design intensity was observed to provide stronger predictions. Moreover, it has been demonstrated that both capabilities could not be utilized in design since the ultimate capabilities were overvalued by them, thus fail to meet safety criteria. Also, for defining the concept of CFDST slender columns, slenderness ratio (λ) having value 22 was proven to be appropriate [12].

Many researchers have carried out studies to analyze the axial load behavior of CFST columns through nonlinear analysis [22, 28]. A numerical investigation was accompanied with theoretical design to evaluate the axial load behavior of concrete-filled steel tubular columns. The computations of numerical modeling were found to be in good concurrence to the corresponding values of experimental data from the literature. The column strength anticipated using FEA and also from the specifications of American Institute of Steel Construction and Chinese CECS was correlated to the experimental data found in the literature. In case of rectangular sections, the load capability computed numerically varied within 10% of experimental results. It varied between 2 and 15% and 3 to 14% for circular and square segments, respectively. It was observed that the Chinese CECS code overrated the axial load capacity of CFST columns by a maximum of 47% for square cross section and 60% for rectangular cross section. In case of circular columns, the code conservatively predicted the values. The maximum loads projected by the AISC-LRFD standard were found to be 38%, 28%, and 14% for circular, square, and rectangular columns, respectively [19].

Equation (1) gives the ultimate load capacity of CFST columns as per AISC-LRFD;

$$P_{\text{AISC-LRFD}} = A_S f_y + \phi A_c f_{\text{ck}} \tag{1}$$

The AISC-LRFD code also specifies that the minimum steel amount in composite elements must be greater than 4%.

i.e., ($\rho Sr > 4\%$).

Equation (2) gives the ultimate load capacity of CFST columns as per Chinese CECS;

$$P_{\text{CECS}} = A_s f_y + A_c f_{\text{ck}} + \sqrt{(A_c f_{\text{ck}}) * (A_s f_y)} \quad (2)$$

The CECS code considers the confinement effect equally on the concrete core and the steel tube.

In Tables 1 and 2, the comparisons of the CFDST's ultimate strengths (P_{uc}) from finite element modeling with observed values (P_{ue}) from Han et al. [29], Tsai and Lin [30], Tao et al. [24], and Zhao et al. [5] are shown. It is observed in case of circular section, a mean ($P_{\text{uc}}/P_{\text{ue}}$) of 0.936 was obtained, while for a square section, the value mean obtained was 1.022. A good concurrence is observed between the tested and predicted findings.

Another study was carried on three concrete-encased battened composite columns using ABAQUS program package. When the findings were compared to values obtained from the experiment, the accuracy of the finite element modeling was also evaluated. Relative slenderness (λ), steel's yielding strength (f_y), and concrete core strength (f_{ck}) were the main factors considered. The rigidity and strength of composite columns were found to be substantially higher than those of steel columns [14].

Another study investigated the load–deformation behavior and ultimate load of centrally loaded CFST columns. The shape of tube (circular and square), design strength of concrete (20, 40, 80 MPa), tensile strength (400, 600, 800 MPa), as well as ratio of steel tube's width-to-thickness (D/t or B/t) were the primary considered parameters. A total of 114 specimens of centrally loaded CFST and hollow columns were tested. The findings showed that the capacity of hollow square steel tube column was reduced due to local buckling which was adjusted by concrete infill in square CFST specimen which provided the restraining effect on steel tube wall reducing the local buckling [21] (Tables 3 and 4).

3 Conclusions and Discussion

The following suggestions and recommendations can be derived from the preceding discussions.

1. The constraints indicated in specific design standards associated with strength of both concrete core and steel tube of composite columns must be adjusted in accordance with improvements in technology and rise in number of advanced construction with high-strength materials.
2. The concrete-filled double-skinned tube provides a higher-level confinement effect for concrete in case the ratios of diameter-to-thickness of both external and internal tubes (D_o/t_o, D/t) are low.

Table 1 Test data for outer CHS of CFDST stub columns exposed to axial compression

Test series	Specimen label	Outer tube dimensions $D \times t_{so}$ (mm)	Inner tube dimensions $d \times t_{si}$ (mm)	f_{syo} (MPa)	f_{syi} (MPa)	F_c (f_{cu}) (MPa)	P_{ue} (kN)	P_{uc} (kN)	P_{uc}/P_{ue}	Test data resources
CHS outer	CC2A	180 × 3	48 × 3	275.9	396.1	47.4	1790	1622	0.91	[24]
	CC2B	180 × 3	48 × 3	275.9	396.1	47.4	1791	1622	0.91	
	CC3A	180 × 3	88 × 3	275.9	370.2	47.4	1648	1497	0.91	
	CC3B	180 × 3	88 × 3	275.9	370.2	47.4	1650	1497	0.91	
	CC4A	180 × 3	140 × 3	275.9	342.0	47.4	1435	1258	0.88	
	CC4B	180 × 3	140 × 3	275.9	342.0	47.4	1358	1258	0.93	
	CC5A	114 × 3	58 × 3	294.5	374.5	47.4	904	807	0.89	
	CC5B	114 × 3	58 × 3	294.5	374.5	47.4	898	807	0.89	
	CC6A	240 × 3	114 × 3	275.9	294.5	47.4	2421	2337	0.96	
	CC6B	240 × 3	114 × 3	275.9	294.5	47.4	2460	2337	0.95	
	CC7A	300 × 3	165 × 3	275.9	320.5	47.4	3331	3195	0.95	

(continued)

Table 1 (continued)

Test series	Specimen label	Outer tube dimensions $D \times t_{so}$ (mm)	Inner tube dimensions $d \times t_{si}$ (mm)	f_{syo} (MPa)	f_{syi} (MPa)	F_c (f_{cu}) (MPa)	P_{ue} (kN)	P_{uc} (kN)	P_{uc}/P_{ue}	Test data resources
	CC7B	300 × 3	165 × 3	275.9	320.5	47.4	3266	3195	0.97	
CHS outer	DS-2	300 × 2	180 × 2	290	290	28	2141	2155	1.00	[30]
	DS-6	300 × 4	180 × 2	290	290	28	2693	2765	1.02	

Table 2 Test data for outer SHS of CFDST stub columns exposed to axial compression

Test series	Specimen label	Outer tube dimensions $D \times t_{so}$ (mm)	Inner tube dimensions $d \times t_{si}$ (mm)	f_{syo} (MPa)	f_{syi} (MPa)	F_c (f_{cu}) (MPa)	N_{ue} (kN)	N_{uc} (kN)	N_{uc}/N_{ue}	Test data resources
SHS outer	SCC2-1	120 × 3	32 × 3	275.9	422.3	46.8	1054	993	0.942	[29]
	SCC2-2	120 × 3	32 × 3	275.9	422.3	46.8	1060	993	0.937	
	SCC3-1	120 × 3	58 × 3	275.9	374.5	46.8	990	1020	1.030	
	SCC3-2	120 × 3	58 × 3	275.9	374.5	46.8	1000	1020	1.020	
	SCC4-1	120 × 3	88 × 3	275.9	370.2	46.8	870	977	1.123	
	SCC4-2	120 × 3	88 × 3	275.9	370.2	46.8	996	977	0.981	
	SCC5-1	180 × 3	88 × 3	275.9	370.2	46.8	1725	1835	1.064	
	SCC5-2	180 × 3	88 × 3	275.9	370.2	46.8	1710	1835	1.073	
SHS outer	S1C1	100.2 × 6.12	48.5 × 3.01	500	370.2	70	1677	1651	0.984	[5]
	S2C1	100.4 × 4.13	48.5 × 3.01	476	370.2	70	1253	1337	1.067	

Table 3 Numerical and theoretical results being compared with experimental findings [19]

Specimens	B/t	$F_{y\,(MPa)}$	$F_{c\,(MPa)}$	Experimental load, P_{exp} (kN)	Numerical load, P_{num} (kN)	Theoretically calculated load	
						$P_{AISC-LFRD}$ (kN)	P_{CECS} (kN)
C1	47	285	28.2	881	752	710.95	1084.67
C2	21.7	313	23.8	1825	1763	1122.67	1579.02
C3	21	537	28.1	2715	2638	1800.72	2398.24
R1	50.8	430	30.4	819	743	840.92	1314.95
R2	34.18	383	26.0	1006	932	968.21	1444.67
R3	35.3	413	26.0	1144	1092	1171.28	1777.87
R4	33.4	365	23.8	1224	1177	1092.71	1654.86
R5	26.4	324	23.8	1335	1287	1147.51	1711.06
R6	20.7	358	23.8	1691	1634	1502.29	2148.05
S1	40.4	356	30.4	917	863	934.88	1346.70
S2	29.2	357	26.0	1095	986	1067.88	1479.82
S3	27.9	322	23.8	1113	1074	996.94	1376.34
S4	22.3	312	23.8	1202	1158	1108.34	1496.13
S5	17	347	23.8	2069	1789	1489.06	1921.89

Table 4 Influence of selected parameters on CFST (single and double skin) columns subjected to axial compression

S. No.	Parameters affecting axial load capacity of CFST (single- and double-skin) columns	Effect on axial load capacity of CFST (single- and double-skin) columns
1	L/D increased	Decreased
2	D/t increased	Decreased
3	Grade of concrete increased	Increased
4	Steel's yielding strength increased	Increased

3. The axial load strength as well as lateral confining pressure of CFDST column decrease when the internal tube's diameter-to-thickness ratio grows provided the ratio of diameter-to-thickness of outer steel tube (D_o/t_o) remains constant. Similarly, when the outer steel tube's diameter-to-thickness ratio (D_o/t_o) grows, both the lateral confining pressure f_l and ultimate strength decrease by maintaining the inner tube's diameter-to-thickness ratio (D/t) as constant.
4. The most essential component in influencing the axial load-carrying capability was discovered to be the confinement ratio. A larger confinement factor led in less post-peak axial strength loss and greater ductility. With the same confinement factor, the low-strength concrete lost less strength.
5. A simplified model was created to determine the loading capacity of partially loaded CFST columns having L/D below 6, and the projections using this simple model are typically conservative to a certain extent and are also very acceptable.

6. According to the parametric analysis, increased compressive strength of concrete, yielding strength of external confinement, yielding strength of steel tube, total steel ratio, as well as reduced ratio of diameter-to-thickness are the factors that improve structural efficiency of CFST columns at various rates.
7. When outer steel confinements are used in CFSTs, they can give better performance under axial stress than simply enhancing the steel tube thickness while keeping same steel ratio.
8. Except for AISC-LFRD, which had a smaller gap between expected and experimental outcomes, the codes like Chinese CECS exaggerated the bearing capacity.

References

1. Han, L. H., Huang, H., & Zhao, X. L. (2009). Analytical behavior of concrete-filled double skin steel tubular (CFDST) beam-columns under cyclic loading. *Thin-Walled Structure, 47*(6–7), 668–680.
2. Han, L. H., Li, W., & Bjorhovde, R. (2014). Developments and advanced applications of concrete-filled steel tubular (CFST) structures: Members. *Journal of Constructional Steel Research, 100*, 211–228.
3. Liang, Q. Q. (2017). Nonlinear analysis of circular double-skin concrete-filled steel tubular columns under axial compression. *Engineering Structure, 131*, 639–650.
4. Wright, H. D., Oduyemi, T. S., & Evans, H. R. (1991). The design of double skin composite elements. *Journal of Constructional Steel Research, 19*, 111–132.
5. Elchalakani, M., Zhao, X. L., & Grzebieta, R. (2002). Tests on concrete filled double-skin (CHS outer and SHS inner) composite short columns under axial compression. *Thin-Walled Structures, 40*, 415–441.
6. Li, W., Ren, Q. X., Han, L. H., & Zhao, X. L. (2012). Behaviour of tapered concrete-filled double skin steel tubular (CFDST) stub columns. *Thin-Walled Structures, 57*, 37–48.
7. Zhang, F., Wu, C., Zhao, X., Li, Z., Heidarpour, A., & Wang, H. (2015). Numerical modeling of concrete-filled double-skin steel square tubular columns under blast loading. *Journal of Performance of Constructed Facilities, 29*, 1–12.
8. Wang, R., Han, L. H., Zhao, X. L., & Rasmussen, K. J. R. (2015). Experimental behavior of concrete filled double steel tubular (CFDST) members under low velocity drop weight impact. *Thin-Walled Structures, 97*, 279–295.
9. Lu, H., Zhao, X. L., & Han, L. H. (2011). FE modelling and fire resistance design of concrete filled double skin tubular columns. *Journal of Constructional Steel Research, 67*, 1733–1748.
10. Sangeetha, P., & Senthil, R. (2017). Experimental behaviour of steel tubular columns for varying in filled concrete. *Archives of Civil Engineering, 63*(4).
11. Wheeler, A., & Bridge, R. (2006). The behaviour of circular concrete-filled thin-walled steel tubes in flexure. *Composite Construction in Steel and Concrete, 186*(39), 413–423.
12. Kharoob, O. F., & Hassanein, M. F. (2014). Analysis of circular CFDST slender columns with external stainless steel. *Thin walled Structures, 79*.
13. Li, W., Han, L. H., & Zhao, X. L. (2012). Axial strength of concrete-filled double skin steel tubular (CFDST) columns with preload on steel tubes. *Thin-Walled Structures, 56*, 9–20.
14. Mehrabani, R., & Shanmugam, N. E. (2014). Finite element analysis of the behaviour and ultimate strength of battened columns encased in concrete. *The IES Journal Part A: Civil & Structural Engineering, 7*(4), 263–280.
15. Tiwary, A. K., & Gupta, A. K. (2021). Post-fire exposure behavior of circular concrete-filled steel tube column under axial loading. *International Journal of Steel Structure, 21*, 52–65.

16. Tiwary, A. K., & Gupta, A. K. (2020). Mechanical behavior of circular concrete filled steel tube column under axial loading. *Journal of Green Engineering, 10*(11), 11116–11132.
17. Shen, Q., Wang, J., Wang, J., & Ding, Z. (2019). Axial compressive performance of circular CFST columns partially wrapped by carbon FRP. *Journal of Constructional Steel Research, 155*, 90–106.
18. Tao, Z., Song, T., Uy, B., & Han, L. (2016). Bond behavior in concrete filled steel tubes. *Journal of Constructional Steel Research, 120*, 81–93.
19. Jayalekshmi, S., & Jegadesh, S. (2016). Finite element analysis and codal recommendations of CFST columns. *Journal of the Institution of Engineers (India): Series A, 97*(1), 33–41.
20. An, Y. F., Han, L. H., & Zhao, X. L. (2012). Behavior and design calculations on very slender thin-walled CFST Columns. *Thin Walled Structures, 53*, 161–175.
21. Sakino, K., Nakahara, H., Morino, S., & Nishiyama, I. (2004). Behavior of centrally loaded-concrete-filled steel-tube short columns. *Journal of Structural Engineering, 130*(2), 180–188.
22. Tiwary, A. K., & Gupta, A. K. (2019). Nonlinear analysis of circular concrete filled steel tube columns under axial loading. *International Journal of Innovative Technology and Exploring Engineering, 8*, 688–692.
23. Zhao, X. L., & Grzebieta, R. H. (2002). Strength and ductility of concrete filled double skin (SHS inner and outer) tubes. *Thin-Walled Structures, 40*(2), 199–213.
24. Tao, Z., Han, L. H., & Zhao, X. L. (2004). Behavior of concrete-filled double skin (CHS inner and CHS outer) steel tubular stub columns and beam-columns. *Journal of Constructional Steel Research, 60*(8), 1129–1158.
25. Fam, A., Qie, F. S., & Rizkalla, S. (2004). Concrete-filled steel tubes subjected to axial compression and lateral cyclic loads. *Journal of Structural Engineering, 130*(4), 631–640.
26. Nardin, S. D., & Debs, A. L. (2006). Axial load behavior of CFST columns. *Structures & Buildings, 159*, 1–10.
27. Han, L. H., Huang, H., Tao, Z., & Zhao, X. L. (2006). Concrete-filled double skin steel tubular (CFDST) beam–columns subjected to cyclic bending. *Engineering Structures, 28*(12), 1698–1714.
28. Tiwary, A. K., & Gupta, A. K. (2019). Strength aspects of concrete filled steel tube columns through design codes. *TEST Engineering and Management, 81*, 3672–3681. ISSN No. 0193-4120
29. Han, L. H., Tao, Z., Huang, H., & Zhao, X. L. (2004). Concrete-filled double skin (SHS outer and CHS inner) steel tubular beam–columns. *Thin-Walled Structures, 42*(9), 1329–1355.
30. Lin, M. L., & Tsai, K. C. (2001). Behavior of double-skinned composite steel tubular columns subjected to combined axial and flexural loads. In *Proceedings of the First International Conference on Steel & Composite Structures* (pp. 1145–1152).

Comparative Seismic Analysis of Multi-storied Building with and Without Floating Columns

Arvind Thakur and Amreen Khatun

Abstract In present construction scenario of multi-storey buildings for residential and commercial uses where occupancy rate is higher requires a wide space or areas for parking vehicles at the ground level. Construction of columns nearly spaced from each other in the parking area may lead to structural irregularities and failure of the buildings. To avoid failures comprises from such undesirable construction irregularities the concept of floating columns has been introduced. Floating columns are generally constructed for the supportive purpose and structural appearance of the buildings. In present study the comparative analysis has been made using STAAD Pro software for multi-storied buildings constructed with and without floating columns. In this study a G+7 multi-storey building is considered with and without floating column are analysed and seismic prone area for the comparative seismic analysis zone IV is taken. In this study behaviour of both the buildings with and without floating column is analysed and concluded from the study that buildings constructed with undesirable structural irregularities experiences more shaking and damage during seismic vibrations. Present study indicates 56.6% storey displacement at the top floor, Study of peak storey shear concludes higher percentage of 17.4% base shear at top floor and decreased to 0.4% at first floor as the mass of the building decreased, Time History Analysis indicates lesser value of time period (0.37 s) after changing dimensions of structural members as compare to the time period observed in building with floating columns and without floating columns also the value of storey drift increased to 48.09% in the top floor of building considered with an application of floating column.

Keywords Storey drift · Time history analysis · Equivalent static analysis · Floating column · Base shear · Storey displacement

---

A. Thakur · A. Khatun (✉)
Department of Civil Engineering, AP Goyal Shimla University, Shimla, India

A. K. Gupta et al. (eds.), *Advances in Construction Materials and Sustainable Environment*, Lecture Notes in Civil Engineering 196,
https://doi.org/10.1007/978-981-16-6557-8_70

1 Introduction

Design of earthquake resistant structures is a greatest responsibility in today's world. Requirements of clients and architectures for the construction of different structures leads to various uncertainties and makes structure unstable in seismic forces experienced during earthquakes. Basic uncertainties that make structures unstable are Floating Columns, Vertical irregularities, Asymmetrical planning, Asymmetrical loading, Provision of soft or weak storey etc. [1]. Today's construction involves construction of multi-storey building in urban areas of the country consisting of many un-avoidable features like construction of floating columns that may leads to failure of many structures during seismic vibrations experienced by structures during earthquakes. The construction floating columns in many structures is mainly for accommodating facilities of parking areas and reception lobbies constructed at the ground stories [2]. Past studies analysed by many researchers to formulate the failure of certain structures constructed with undesirable features and their study highlighted the importance of identifying the presence of an open space first storey in the seismic analysis of multi-storied buildings [3] and concluded that provision of undesirable features in the construction of multi-storied buildings in seismic prone areas verified experiences of strong shaking during seismic vibrations. The present study deals with the comparative analysis of multi-storey buildings constructed with or without floating columns in seismic prone areas. The comparison of the multi-storey buildings which is (G+7) in present study is done for seismic zone IV and the grade of concrete taken is M25. The analysis is done using STAAD Pro software to analyse the responses of flexibly connected frames in multi-storied buildings constructed to earthquake prone areas. The present study indicates the behaviour of buildings in seismic prone areas during seismic vibrations and comparative analyses shows that the concept of floating columns in buildings is desirable to support the buildings and not responsible for the transfer of seismic vibrations [4].

In past studies construction of buildings was not a problem due to the less population and large area available per person. In present time of twenty-first century the construction area availability is reduced due to increase of population which comprises to the construction of high rise buildings. Construction of high rise buildings considers all factors of lateral forces acting on it including self-weight of the building. An important parameter to take into consideration for high rise buildings to withstand against lateral forces is good soil bearing capacity along with the high strength of all structural members of the building [5]. Some researchers concluded in their studies that buildings with irregular features like floating column at the ground level should be constructed stronger by using retrofitting techniques which are useful to increase the stiffness of the columns as to reduce the effect of seismic vibrations so that the effects of lateral deformation can be avoided [6]. Past studies also indicate the consideration of architectural features that may cause damage to buildings in seismic prone areas should be minimized if not avoidable. Provision of undesirable features such as floating columns in the buildings needs higher engineering efforts to make it strong enough to resist seismic vibrations [7].

2 Objectives

- Study the dynamic effect of floating column on RC frame for maximum storey displacement by Equivalent Static Analysis under critical Earthquake zones (IV) as per IS Code 1893: 2002 (Part 1) Recommendations [8].
- Study of columns with changing dimensions from ground floor to the top floor and comparison of Time period by (THA).
- Study of Peak Storey Shear of a building and comparison of storey shear with floor level with or without floating columns using Equivalent Static Analysis.
- Study of Storey Drift in Time History Analysis (THA) of buildings with or without floating columns.

3 Methods and Material

3.1 Methods and Material Description

Present study of multi-storied buildings with and without floating columns is a comparative seismic analysis and Table 1 is defining various features of G+7 building considered for this study. The table is defining Building data, Dimensions of the members taken, Loads considered on building, Methods of analysis and properties of different materials used for the study.

3.2 Plan and Elevation

Figure 1 is representing construction of hanging and floating column. A column is a vertical member in a building responsible for transferring the load from upper storey to the ground. Basic purpose for construction of floating columns in a building is to fulfil the requirement of large space at ground level for parking vehicles, party halls, assembly halls etc. and is not responsible for the load transformation of a building during seismic vibrations [11]. Figure 2a, b is representing the Elevation of (G+7) storey building with floating columns and without floating columns respectively.

Table 1 Data showing methods and materials used for the study [9, 10]

Building data			
Structure			SMRF
Number of stories			G+7
Type of building			Commercial
Seismic zone			IV
Type of soil			Type II (Medium soil) as per IS: 1893:2002 [7]
Storey height			3.2 m
Bay width			4 m
No. of bays along *X* direction			5
No. of bays along *Z* direction			5
Plan size			20 m × 20 m
Total height of building			25.6 m
Member dimensions			
Slab	Thickness		150 mm
Beams	Normal building		450 mm × 230 mm
	Floating column building	Interior beams	450 mm × 230 mm
		Cantilever projection at edges	850 mm × 650 mm
Columns	Normal building		450 mm × 450 mm
	Floating column building	Top two floors	450 mm × 450 mm
		All floor except top four floors	700 mm × 900 mm
Brick infill	Exterior walls thickness		250 mm
	Interior walls thickness		150 mm
Loads			
Unit weight of concrete			25 kN/m^3
Unit weight of brick infill			20 kN/m^3
Floors	Live load		3 kN/m^2
	Dead load		3.75 kN/m^2
Roof	Live load		1.5 kN/m^2
	Dead load		3.75 kN/m^2
Methods of analysis			
Static	Equivalent static lateral force method		
Dynamic	Response spectrum analysis		
Materials properties			**Values**
Characteristic strength of concrete, f_{ck}			25 MPa
Yield stress of steel, f_y			500 MPa
Density of concrete			25 kN/m^3
Modulus of elasticity of concrete, E_c			25,000 MPa
Modulus of elasticity of steel, E_s			200,000 MPa

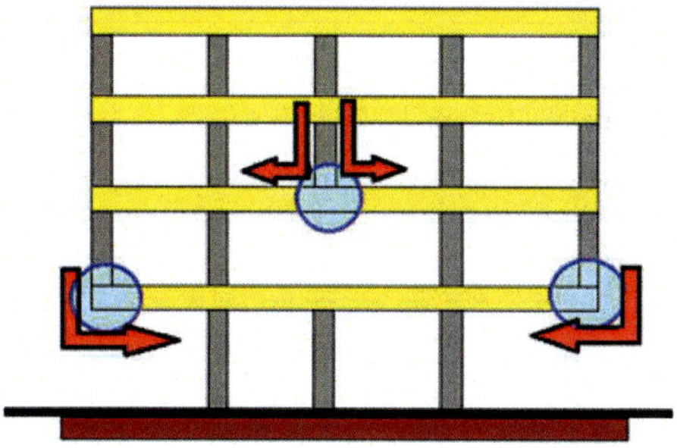

Fig. 1 Figure representing hanging and floating columns

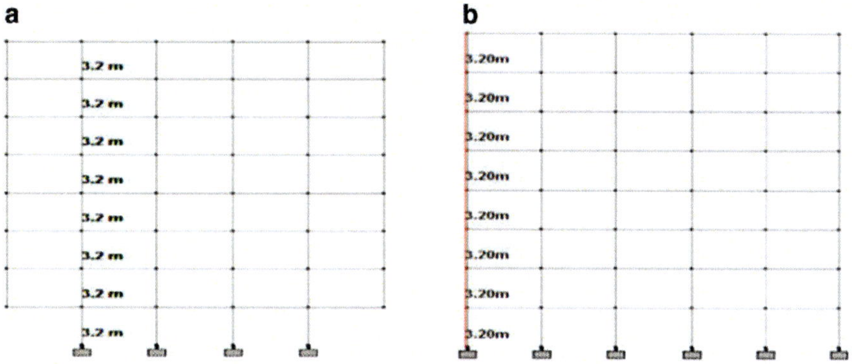

Fig. 2 **a** Elevation of (G+7) storey building with floating columns. **b** Elevation of (G+7) storey building without floating columns

4 Results for Analysis

4.1 Results for Maximum Storey Displacement in Equivalent Static Analysis

Lateral load is applied in *X* and *Z* directions and structure is analysed for various load combinations as per specifications given in clause 6.3.1.2 of IS 1893:2002 (Part-1) [8]. For load combinations applied on building the maximum displacement at each floor is noted in *X* and *Z* direction and are the same as the plan and member sizes are symmetrical in both directions. Observations obtained in Table 2 representing the maximum storey displacement of 86 mm at the top floor for the building with floating columns. It has been concluded from the study that with the application of floating column the storey displacement is increased by 56.6% at the top floor. Hence the stiffness of the columns which is eventually transferring the load of the structure to the foundation is increased (Fig. 3).

Table 2 Maximum storey displacement in buildings in accordance to floor levels

Floor level	Maximum storey displacement in (mm)		
	Without floating columns	With floating columns	Increase%
Base	0	0	–
1	3.69	24.46	84.91
2	9.69	32.88	70.53
3	15.97	41.95	61.93
4	21.98	51.67	57.46
5	27.43	61.36	55.3
6	32.01	70.56	54.63
7	36.42	78.58	53.65
8	37.32	86	56.6

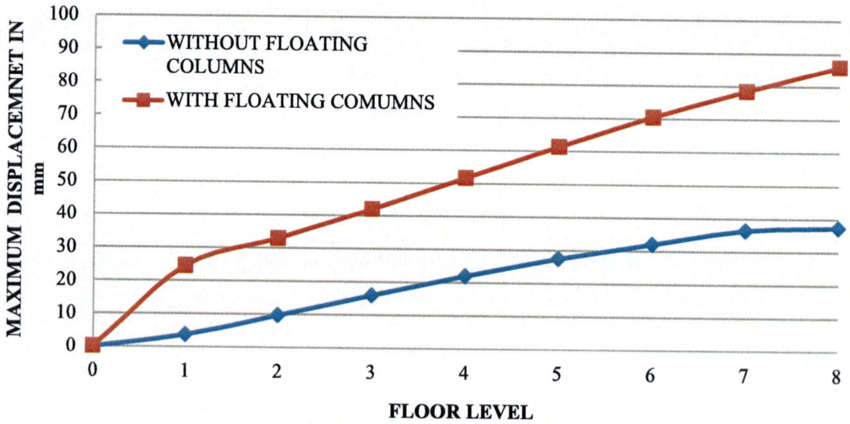

Fig. 3 Maximum storey displacement in building for with and without floating column

4.2 *Results for Peak Storey Shear in Equivalent Static Analysis*

The peak storey shear in *X* direction obtained from the study from equivalent static analysis and comparision of storey shear with floor level is shown in Table 3. The study concluded that presence of floating column in a building showed highest value of storey shear 17.4% but tends to decrease the storey shear values to 0.4% with decrease in mass of the building (Fig. 4).

Table 3 Variation in peak storey shear for with and without floating column

Floor level	Peak storey shear in (kN)		
	Without floating columns	With floating columns	Variation %
1	6.92	6.89	0.4 (↓)
2	27.67	28.48	2.84 (↑)
3	62.25	64.1	2.89 (↑)
4	110.67	113.96	2.89 (↑)
5	172.92	178.06	2.9 (↑)
6	249.01	256.41	2.9 (↑)
7	338.93	349	2.89 (↑)
8	354.67	429.18	17.4 (↑)

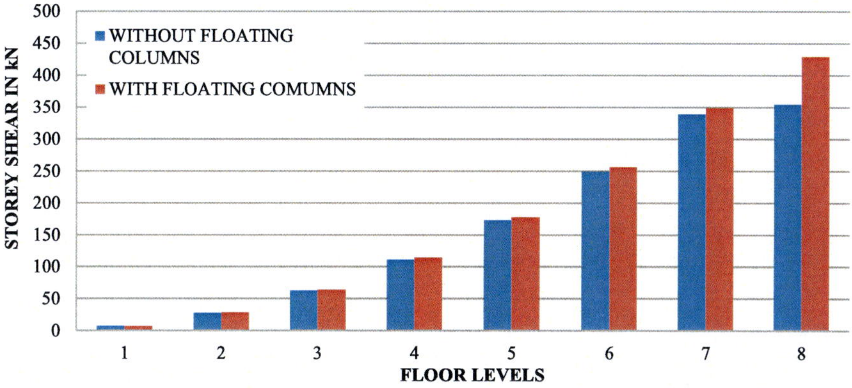

Fig. 4 Variation for storey shear in buildings for with and without floating column

4.3 Results for Comparison of Time Period in Time History Analysis (THA)

The observations shown in Table 4 indicates higher time period of (2.32 s) for building with floating column comparatively time period of the building without floating columns and time period of building after changing dimensions. The study concludes that after changing dimensions the time period value decreased to (0.37 s) due to increase in lateral stiffness and weight of the building (Fig. 5).

Table 4 Comparison of time period without floating columns, with floating columns and after changing dimensions (THA)

Mode shape	Time period (s)		
	Without floating columns	With floating columns	After changing dimensions
1	1.62	2.32	1.76
2	1.62	2.31	1.76
3	1.49	2.31	1.71
4	0.96	1.12	1.07
5	0.7	0.78	0.76
6	0.7	0.78	0.65
7	0.53	0.69	0.63
8	0.53	0.67	0.6
9	0.52	0.67	0.48
10	0.49	0.65	0.47
11	0.47	0.52	0.43
12	0.46	0.49	0.43
13	0.41	0.49	0.38
14	0.41	0.48	0.38
15	0.38	0.4	0.37

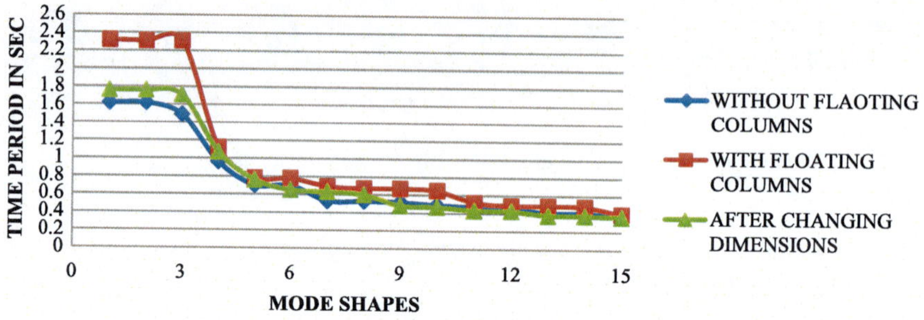

Fig. 5 Variation of time period with mode shapes for all three models (THA)

4.4 *Results for Comparison of Storey Drift in Time History Analysis (THA)*

Study concluded From Table 5 that the storey drift values for floating column building is higher than without floating columns. It has been also concluded from the study that there is an increment of 48.09% in top floor storey drift for floating column building (Fig. 6).

Table 5 Comparison of maximum storey drift of building without floating columns and with floating columns (THA)

Floor level	Storey drift in (mm)		Increase %
	Without floating columns	With floating columns	
1	3.05	11.4	73.25
2	7.96	19.05	58.21
3	12.91	26.52	51.32
4	17.45	33.42	47.78
5	21.32	39.6	46.16
6	24.37	44.98	45.82
7	26.5	49.54	46.51
8	27.7	53.36	48.09

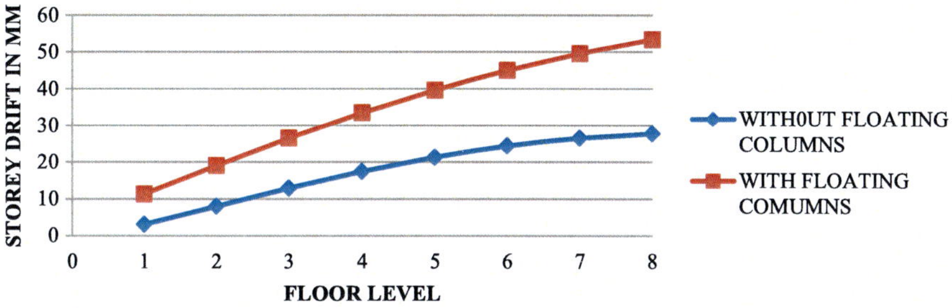

Fig. 6 Variation of storey drift with floor level for building without floating columns and with floating columns (THA)

5 Conclusions

This study of comparative seismic analysis of multi-storied building with or without floating columns concludes that maximum storey displacement of building using Equivalent Static Analysis for building with floating columns showed maximum value of storey displacement at the top floor and indicates 56.6% storey displacement at the top floor of building eventually increased the stiffness of the building. Study of peak storey shear concludes that the presence of floating columns in the building showed highest percentage of 17.4% base shear at the top level but tends to decrease the storey shear by 0.4% as the mass of the building decreased. Results obtained for comparative analysis of time period in (THA) indicates that after changing dimensions the time period value decreased (0.37 s) as compare to the buildings considered with floating columns (0.40 s) and without floating columns (0.38 s) with increase in lateral stiffness and weight of the building. Storey Drift in (THA) concludes that building without floating columns shows higher storey drift in the building and with application of floating column there is an increment of 48.09% storey drift in top floor of the building.

References

1. Arvind, R., & Fernandes, R. J. (2015). Seismic analysis of R.C. regular and irregular frame structures. *International Research Journal of Engineering and Technology, 02*, 44–47.
2. Choudhary, Md. A., & Hassan. (2013). Comparative study of the dynamic analysis of multi-storey irregular building with or without base isolator. *International Journal of Scientific Engineering and Technology,* 2, 909–912.
3. *International Journal for Research in Applied Science & Engineering Technology (IJRASET), 5*(XII) (2017). ISSN: 2321-9653; IC VALUE 45.98; SJ Impact Factor: 6887.
4. *International Journal of Innovative Technology and Exploring Engineering (IJITEE), 9*(4) (2020). ISSN: 2278-3075.
5. Sharma, M., & Maru, S. (2014). Dynamic analysis of multi-storeyed regular building. *Journal of Mechanical and Civil Engineering, 11*, 37–42.
6. Srikanth, M. K., Yogeendra, R., et al. (2014). Seismic response of complex buildings with floating columns. *International Journal of Scientific Engineering and Technology Research, 03*(08), 132–138.
7. Malaviya, P., & Saurav. (2014). *International Journal of Scientific & Engineering Research, 5*(5). ISSN 2229-5518.
8. Indian standard code of practice IS 456: 2000. Plain and reinforced concrete. Bureau of Indian Standards.
9. Indian standard code IS 1893 Part1 2002. Criteria for earthquake resistant design of structures. Bureau of Indian Standards.
10. Indian standard code IS 1786-2008. High strength deformed steel bars and wires for concrete reinforcement—[CED 54: Concrete Reinforcement].
11. Bathe, & Nirjhar, K. J. (1996). *Finite element procedure* (7th ed.). Department of Civil, Structural and Environmental Engineering, State University of New York.

Axial Loading Behaviour of Concrete Filled Steel Tube (CFST) Columns: A Parametric Study

Aditya Kumar Tiwary and Ashok Kumar Gupta

Abstract Columns built of concrete encased in filled steel tubes are known as concrete filled steel tube (CFST) columns. The use of CFST columns is common in tall constructions and long-span bridges. The ultimate load bearing capability of CFST columns gets affected by the *D/t* ratio, the *L/D* ratio and the steel grade. The diameter to thickness (*D/T*) ratio, length to diameter ratio (*L/D*) and steel grade are all used to evaluate a CFST column's ultimate axial load bearing capability. CFST columns made of M30 concrete are available in three diameters: 100 mm, 125 mm and 150 mm having two thicknesses of steel tube: 4 mm and 5 mm. A consistent height of 600 mm is maintained for all CFST specimens. An experimental and numerical study was performed to analyze the behaviour of CFST columns under axial loading. As per the findings of parametric analysis, a CFST column's ultimate loading capacity decrease as the ratio of *D/t* grows and rises as *L/D* ratio decreases. The improvement of 33.4% in loading capacity of CFST columns was observed with enhancing *D/T* ratio to 31.3 from 25 and 28.4% by changing *D/T* ratio to 37.5 from 31.3 with 4 mm external steel casing thickness. It was observed that for 5 mm outer steel tube thickness, when *D/T* ratio increased to 25 from 20, the loading capability was increased to 33.8%. Also, when the ratio *D/T* increased to 30 from 25 while keeping the same thickness of 5 mm, the loading capacity of the CFST columns increased to 28.6%. When *L/D* ratio was lowered from 6.0 to 4.0 and steel grade of the CFST column was improved, the CFST's axial load capability was found to be increased by 52.7%.

Keywords CFST · Axial behaviour · Parametric study

A. K. Tiwary (✉)
Chandigarh University, Mohali, India
e-mail: aditya.civil@cumail.in

A. K. Gupta
Jaypee University of Information Technology, Solan, India
e-mail: ashok.gupta@juit.ac.in

A. K. Gupta et al. (eds.), *Advances in Construction Materials and Sustainable Environment*, Lecture Notes in Civil Engineering 196,
https://doi.org/10.1007/978-981-16-6557-8_71

1 Introduction

The Concrete filled steel tubular (CFST) column falls under the structural constructions category of composite column, in which the greatest attributes of all materials, steel and concrete, are combined to provide the best possible result [1–5]. An effective and basic yet precise analytical approach for strength capacity of circular CFST exposed to axial loading and cyclically shifting flexural force was presented [6–8]. By comparing the findings of the studies with experimental published data, the validation of this model was entirely developed. A substantial parametric investigation of circular CFSTs and hysteric behaviour was conducted using this model. It is made up of a succession of CFST columns of various thicknesses, concrete strength and steel tube yield stress [9–11].

Furthermore, empiric analytical relations are constructed and tested for circular CFST columns, providing an accurate and efficient definition of their maximum potential. The proposed analytical model should reliably and effectively describe the behaviour of circular concrete filled steel tube columns subjected to cyclic lateral load, according to comparisons between analytical and experimental results [12–15].

The performance of CFST high-strength columns when subjected to axial stress was investigated [15–17]. This allowed for a statistical study of the performance of high-strength circular CFST when subjected to monotonous axial processing. Material non-linearity and constitutive interactions are implemented by the use of a model.

The model was used to do the parametric analysis, which was then compared to the provisions of the code. Another key consideration to be remembered is the ductility along with the effectiveness of restrained concrete. Variables such as steel casing's yield strength and column diameter were also used [18]. The ultimate loading capacity and enhancement of strength and ductility due to confining effect is the subject of the column's performance assessment. For assessment, two parameters are precisely defined and developed: strength ductility index and enhancement factor (K_f).

A parametric analysis of square CFST columns with concentric filling was carried out. The performance of concrete filled steel tube columns subjected to axial loading [19–21] was demonstrated in this analysis by modifying parameters. Steel tube width, concrete grade and column length were the requirements. The research was done using the ANSYS 13 finite element programme. 60 × 60 mm was the height of each column. The values used for various differences of tube thickness were 2, 3, 4, 5 and 6 mm. The concrete was infilled with M25, M30, M40, M50, M60 and M70 to account for grade changes. Columns with heights 900, 1200, 1500, 1800, 2100 and 2400 mm were employed in the combinations.

In previous studies, finite element analysis with the software package ANSYS to critically examine the load deformation characteristics of composite columns was used. Square and circular CFST specimens with varying concrete grades (30, 50, 70 and 90 N/mm^2) were investigated under axial loading. The results showed that with the increase in concrete grade, the deformation of CFST reduced by 10–15%. The shape of the CFST segment induced the deformation. The circular section leads to better behaviour than the square section because of the larger containment [22, 23].

This paper proposes a parametric study on concrete filled steel tube columns through experimental and numerical analysis. The parameter which influences the behaviour of concrete filled steel tube columns under axial loading, i.e. diameter to thickness (*D*/*T*) ratio, length to diameter ratio (*L*/*D*) and steel grade are considered in this study.

2 Methodology

The details of the experimental testing program and finite element analysis used in this study is present study.

2.1 Experimental Programme

It also contains the experimental programme of concrete filled steel tube columns subjected to axial loading. A series of six composite specimens were cast to test the CFST's performance under axial loading. Table 1 lists the various specimen characteristics. In terms of geometry, steel tube sizes and thicknesses, Fig. 1 depicts the features of concrete filled steel tubular columns. CFST column specimens of external steel casing of diameters 100 mm, 125 mm and 150 mm with 4 mm and 5 mm thickness were cast to assess their endurance when subjected to axial force. A consistent height of 600 mm is maintained for all CFST specimens. The CFST specimens had a slenderness ratio of 6.0, 4.8 and 4.0. To fill the steel tube, M30 concrete was utilized. In order to apply axial load on the top of CFST specimens, a universal testing machine (UTM) was used [1, 4, 5].

Table 1 Details of the specimens

Specimens	Size ($D \times Ts \times H$) (mm)	*D*/*T*	*L*/*D*	Yield strength of steel (MPa)	Compressive strength of concrete (MPa)
C1T4	100 × 4 × 600	25.0	6.0	288	36.7
C1T5	100 × 5 × 600	20.0	6.0	288	36.7
C2T4	125 × 4 × 600	31.3	4.8	380	36.7
C2T5	125 × 5 × 600	25.0	4.8	380	36.7
C3T4	150 × 4 × 600	37.5	4.0	440	36.7
C3T5	150 × 5 × 600	30.0	4.0	440	36.7

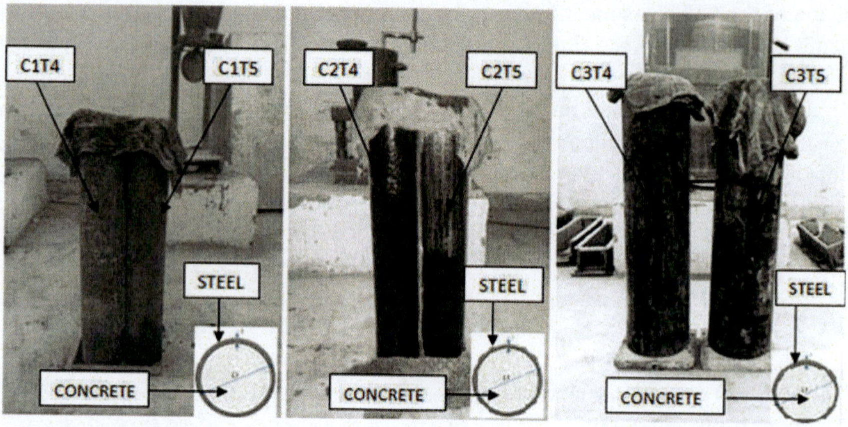

Fig. 1 CFST composite columns

2.2 *Numerical Study*

The behaviour of CFST columns under axial loading was also investigated using numerical modelling. ABAQUS/CAE 6.14 was used to create a finite element model of CFST columns (Computer Aided Engineering). The software tool ABAQUS/CAE 6.14 is used for both modelling and analyzing mechanical components, parts and assemblies (pre-processing) as well as predicting the finite element analysis result. The inelastic behaviour of the external steel tube was determined using the Johnson Cook model, whereas the inelastic behaviour of concrete was determined using the concrete damaged plasticity model (CDP) provided in ABAQUS/CAE 6.14 [22].

Table 1 shows the geometry used to create models of CFST columns. To describe the inelastic behaviour of concrete, the model was based on the concept of isotropic damaged elasticity. Concrete's compressive strength was measured at 36.7 MPa. The Poisson's ratio was 0.18 for concrete and 0.3 for steel. The behaviour of the CFST columns' exterior steel tube was consolidated using the Johnson Cook elasto viscoplastic model, which can forecast the fracture behaviour of ductile materials with less effort and more efficiency [3, 5].

Figure 2 depicts the five layers used in the specimen casting process (a). Figure 2 shows how each layer was vibrated to compress it (b). The top sides of all CFST specimens were levelled after the concrete was filled to produce a smooth loading surface. As shown in Fig. 2, the top surfaces of all the columns were covered up with plastic sheets for 24 h before being replaced with damp burlap, which was left for 28 days on them with water showered over it every day (c). After being cast, Fig. 3 shows the CFST columns.

(a) Pouring of concrete into steel tube

(b) CFST column compaction

(c) Plastic sheet used to seal the CFST specimens

Fig. 2 Casting CFST column specimens. **a** Concrete pouring into steel tube, **b** CFST column compaction, **c** plastic sheet used to seal the CFST specimens

Fig. 3 Wet burlap used to close the CFST specimens

3 Results and Discussion

The impacts of *D*/*T*, steel grade and *L*/*D* ratio under axial loading on CFST columns were investigated through a parametric analysis using the proposed model. The steel yield strength was changed to 440 MPa from 288 MPa in this analysis, while concrete compressive strength was kept steady at 36.7 MPa in the filled centre of CFST columns *D*/*T* was ranged from 20 to 37.5 and *L*/*D* was varied from 4.0 to 6.0, covering thickness of exterior steel tube up to 5 mm from 4 mm and concrete filled steel tube specimen diameters of 100–150 mm. On both specimens, the overall height was maintained at 600 mm. The ratio of diameter to thickness (*D*/*T*), slenderness ratio and yield strength of steel (f_y) all influence the behaviour of circular CFST columns.

3.1 *Axial Load Behaviour of CFST Column Versus Diameter to Thickness Ratio* (D/T)

Six circular CFST specimens are used in this study for studying the influence of thickness variance on CFST column behaviour (Table 2). The *D*/*T* ratio varied between 20 and 37.5 in this study. The growth of *D*/*T* ratio can be a result of the diameter (from 100 to 150 mm) or thickness of the segment (4–5 mm). As a result, for the same diameter, the declined ratio of *D*/*T* with an increase in thickness represents a cross sectional difference in the steel frame, resulting in greater section loading capability. The axial loading capacity of CFST columns was shown to be influenced by the decrease in diameter to thickness ratio. CFST column's axial loading capacity can also be improved by raising steel tube thickness with or without increasing the overall column diameter. The improvement of 33.4% in loading capacity of columns was observed with enhancing *D*/*T* ratio to 31.3 from 25 along with increasing column diameter of 100–125 mm while keeping 4 mm external steel casing thickness. The loading capability of CFST columns was raised to 28.4% by changing *D*/*T* ratio to 37.5 from 31.3 and the diameter of CFST columns was increased from 125 to 150 mm with a constant 4 mm outer steel tube thickness. It was observed that for 5 mm outer steel tube thickness, when *D*/*T* ratio increased to 25 from 20 and the diameter of the CFST columns was expanded from 100 to 125 mm, the loading capability was increased to 33.8%. Also, when the ratio *D*/*T* increased to 30 from 25 and the diameter of columns increased from 125 to 150 mm while keeping the same thickness of 5 mm, the loading capacity of the CFST columns increased to 28.6% (Fig. 4).

Table 2 Loading capacity of CFST columns with varying diameter to thickness ratio

Specimens	Size ($D \times Ts \times H$) (mm)	*D*/*T*	Experimental load (kN)	Simulated load capacity (kN)
C1T4	100 × 4 × 600	25.0	834	843
C1T5	100 × 5 × 600	20.0	836	849
C2T4	125 × 4 × 600	31.3	1252	1261
C2T5	125 × 5 × 600	25.0	1263	1268
C3T4	150 × 4 × 600	37.5	1749	1756
C3T5	150 × 5 × 600	30.0	1768	1774

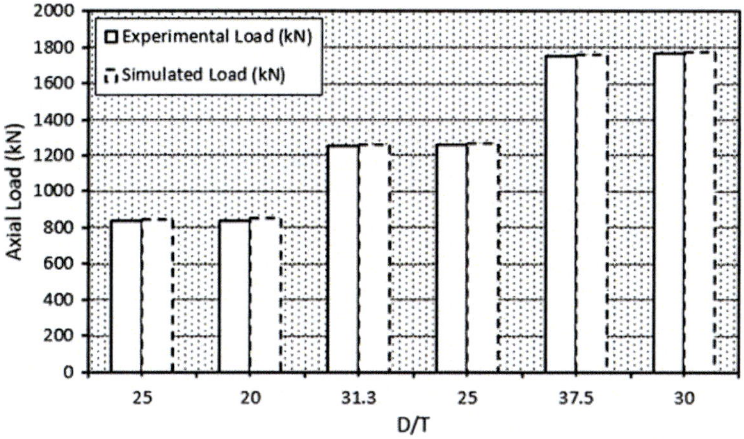

Fig. 4 *D/T* ratio affecting the axial loading behaviour of CFST columns

3.2 Axial Load Behaviour of CFST Column Versus Length to Diameter Ratio (L/D)

This study uses three circular CFST columns for exploring influence of diameter variation on CFST columns' behaviour. The *L/D* ratio ranges from 4.0 to 6.0 in this study. The ratio *L/D* of CFST columns increased as the diameter reduced (150–100 mm). As a result, with a constant height of 600 mm for both specimens, the reduction in *L/D* ratio represents growth in steel tube cross section, resulting in greater section loading capacity. The impact of changing the length to diameter ratio of CFST columns on their loading capacity is shown in Fig. 5. As a result, when *L/D* ratio was lowered, the CFST's axial load capability was found to be increased. Thus, columns' axial loading strength can be enhanced by increasing their diameter without increasing their thickness. The load bearing capability of CFST columns was improved by 52.7% when *L/D* ratio lowered to 4.0 from 6.0 (Table 3).

3.3 Axial Load Behaviour of CFST Column Versus Strength of Steel

Three circular CFST columns are studied for evaluating the role of steel grade in axial efficiency. The CFST column's capability is determined by steel's yield strength. With the increase in the yielding strength of steel, a significant improvement in the axial loading capability of columns is observed. The influence of steel grade variation on the CFST columns' axial loading strength is shown in Fig. 6. It is observed that as yielding strength of steel in CFST columns increases, the axial load performance of columns improves. In a straight path, the load-carrying capacity rises. The capability

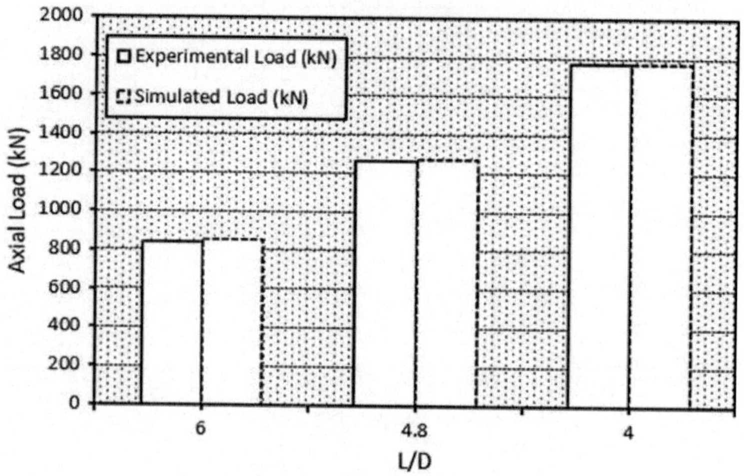

Fig. 5 Influence of ratio of *L*/*D* on axial loading behaviour of CFST columns

Table 3 Loading capacity of CFST columns with varying *L*/*D* ratio

Specimens	Size (*D* × Ts × *H*) (mm)	*L*/*D*	Experimental load (kN)	Simulated load capacity (kN)
C1T5	100 × 5 × 600	6.0	836	849
C2T5	125 × 5 × 600	4.8	1263	1268
C3T5	150 × 5 × 600	4.0	1768	1774

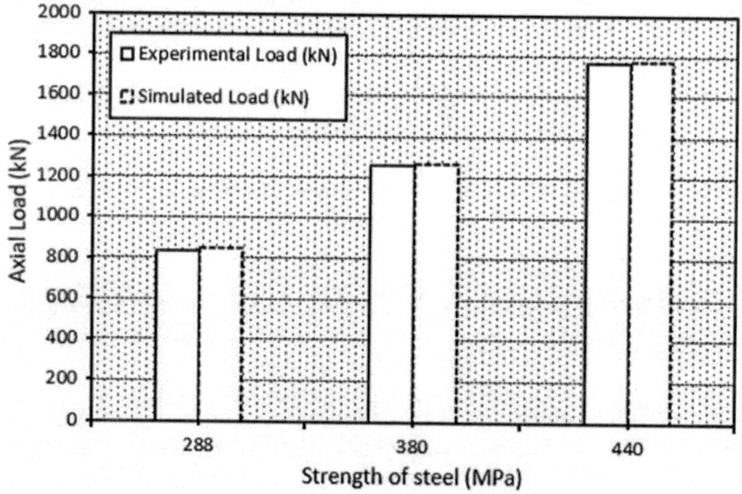

Fig. 6 Impact of strength of steel on axial load behaviour of CFST columns

Table 4 Loading capacity of CFST columns with strength of steel

Specimens	Size (*D* × Ts × *H*) (mm)	Yield strength of steel (MPa)	Experimental load (kN)	Simulated load capacity (kN)
C1T5	100 × 5 × 600	288	836	849
C2T5	125 × 5 × 600	380	1263	1268
C3T5	150 × 5 × 600	440	1768	1774

of a load to carry it falls at an exponential rate. As steel strength of 288 MPa is improved to 400 MPa, column's maximum load bearing capability also gets enhanced by 52.7% (Table 4).

4 Conclusions

The following conclusions have been drawn from experimental and computational modelling:

1. A change in the ratio of diameter to thickness (*D*/*T*) was demonstrated that affected the axial load capacity of CFST columns. A CFST column's axial loading capacity can also be improved by raising the steel tube thickness, which can be done with or without increasing the overall column diameter. By enhancing the *D*/*T* ratio of 25 to 31.3 as well as increasing the column diameter from 100 to 125 mm while maintaining a constant thickness of outer steel tube as 4 mm, the loading capacity of CFST columns was increased by 33.4%. Additionally, as *D*/*T* ratio of 31.3 was improved to 37.5 and the column's diameter of 125 mm was expanded to 150 mm while maintaining outer steel tube thickness constant at 4 mm, the loading capacity of CFST columns was increased by 28.4%.
2. In the case of 5 mm external steel tube thickness when the *D*/*T* ratio of 20 was improved to 25 as well as expanding column diameter of 100–125 mm, the loading capacity of CFST specimen was increased to 33.8%. As *D*/*T* value of 25 adjusted to 30 and the column diameter of 125 mm enhanced to 150 mm, the column's loading capacity improved to 28.6% maintaining the external tube thickness of 5 mm.
3. By lowering *L*/*D* ratio, the CFST's axial loading capability increases. These columns' axial load strength may be enhanced by increasing their diameter without increasing their thickness. The CFST columns' load-carrying capability rose by 52.7% when the *L*/*D* ratio was decreased from 6.0 to 4.0.
4. Steel's growing yield strength improves the column's load bearing capacity. The load-carrying capacity increases along a straight path. The capability of load to carry it decreases exponentially. The column's ultimate load bearing capacity increases by 52.7%. Column's ultimate load bearing capacity increases by 52.7% as steel's yield strength of 288 MPa is improved to 440 MPa.

5. The average strength of the CFST columns improves when steel tube's yielding strength improves or the ratio of length to diameter decreases.
6. When the predicted axial behaviour of the CFST column was added to simulation data, the results were found to be relatively similar to the experimental test results.

References

1. Tiwary, A. K., & Gupta, A. K. (2020). Post-fire exposure behavior of circular concrete-filled steel tube column under axial loading. *International Journal of Steel Structure, 21*, 52–65.
2. Tiwary, A. K., & Gupta, A. K. (2019). Strength aspects of concrete filled steel tube columns through design codes. *TEST Engineering and Management, 81*, 3672–3681. ISSN No. 0193-4120.
3. Tiwary, A. K., & Gupta, A. K. (2019). Nonlinear analysis of circular concrete filled steel tube columns under axial loading. *International Journal of Innovative Technology and Exploring Engineering, 8*, 688–692. ISSN No. 2278-3075
4. Tiwary, A. K., & Gupta, A. K. (2020). Mechanical behavior of circular concrete filled steel tube column under axial loading. *Journal of Green Engineering, 10*(11), 11116–11132.
5. Tiwary, A. K., & Gupta, A. K. *Post-fire exposure behavior of circular concrete-filled steel tube column under axial loading* (Ph.D. thesis). Submitted to Jaypee University of Information Technology, Waknaghat.
6. Teng, J. G., Xiao, Q. G., & Yu, T. (2015). Three-dimensional finite element analysis of reinforced concrete columns with FRP and/or steel confinement. *Engineering Structures, 97*(8), 15–28.
7. Tao, Z., Wang, Z. B., & Yu, Q. (2013). Finite element modelling of concrete-filled steel stub columns under axial compression. *Journal of Constructional Steel Research, 89*, 121–131.
8. Abdullah, J. A., Sumei, Z., & Jiepeng, L. (2010). Shear strength and behaviour of tubed reinforced and steel reinforced concrete (TRC and TSRC) short columns. *Thin-Walled Structures, 48*(3), 191–199.
9. Wang, X., Liu, J., & Zhang, S. (2015). Behaviour of short circular tubed-reinforced-concrete columns subjected to eccentric compression. *Engineering Structures, 105*, 77–86.
10. Zhou, X., Yan, B., & Liu, J. (2015). Behaviour of square tubed steel reinforced-concrete (SRC) columns under eccentric compression. *Thin-Walled Structures, 91*, 129–138.
11. Han, L. H., & An, Y. F. (2014). Performance of concrete-encased CFST stub columns under axial compression. *Journal of Constructional Steel Research, 93*, 62–76.
12. Han, L. H. (2016). *Concrete filled steel tubular structures—Theory and practice* (Third Version). China Science Press [in Chinese].
13. Liu, F., Yang, H., & Gardner, L. (2016). Post-fire behaviour of eccentrically loaded reinforced concrete columns confined by circular steel tubes. *Journal of Constructional Steel Research, 122*, 495–510.
14. Liu, J., Zhou, X., & Gan, D. (2016). Effect of friction on axially loaded stub circular tubed columns. *Advances in Structural Engineering, 19*(3), 546–559.
15. Zhang, Y. B., Han, L. H., & Zhou, K. (2019). Mechanical performance of hexagonal multi-cell concrete-filled steel tubular (CFST) stub columns under axial compression. *Thin Walled Structures, 134*, 71–83.
16. Qi, H., Guo, L., & Liu, J. (2011). Axial load behaviour and strength of tubed steel reinforced concrete (SRC) stub columns. *Thin-Walled Structures, 49*(9), 1141–1150.
17. Boyd, F. P., Cofer, W. F., & McLean, D. (1995). Seismic performance of steel-encased concrete column under flexural loading. *ACI Structural Journal, 92*(3), 355–365.

18. Giakoumelis, G., & Lam, D. (2004). Axial capacity of circular concrete filled tube columns. *Journal of Constructional Steel Research, 60*, 1049–1068.
19. Bradford, M. A. (1996). Design strength of slender concrete-filled rectangular steel tubes. *ACI Structural Journal, 93*(2), 229–235.
20. Hu, H.-T., Huang, C. S., Wu, M.-H., & Wu, Y.-M. (2003). Nonlinear analysis of axially loaded CFT columns with confinement effect. *Journal of Structural Engineering, 129*(10), 1322–1329.
21. Hu, H.-T., Huang, C. S., & Chen, Z. L. (2005). Finite element analysis of CFT columns subjected to combined axial force and bending moment. *Journal of Constructional Steel Research, 61*(12), 1692–1712.
22. Hibbitt, Karlsson & Sorensen Inc. (2003). ABAQUS/standard user's manual, version 6.4.1. Hibbitt, Karlsson, & Sorensen, Inc.
23. Sakino, K., Nakahara, H., Morino, S., & Nishiyama, I. (2004). Behavior of centrally loaded concrete-filled steel-tube short columns. *Journal of Structural Engineering, 130*(2), 180–188.

Comparative Investigation on Mode Shapes and Natural Frequency of Low-Rise RC Frame Building

Pushkar Sharma and Tanmay Gupta

Abstract In general, the building is designed as per codal provisions, which has various constraints while analysing with dynamic loads. This analysis procedure takes a lot of time and is complex. Therefore, most Civil Engineering structures are designed taking the assumption of applied loading to be static. The process of neglecting the dynamic forces may lead to the collapse of the structure as a whole in case of a catastrophe such as an earthquake. Some recent earthquakes have shown the need for dynamic analysis. Nowadays, a lot of research is going on in this field of dynamic analysis such that the structure can withstand earthquake-induced loads. There are many theoretical formulas available in seismic codes of different countries, which relate the height of the building with the natural period of vibration of the structure. The link is a force-based design approach and it gives a rough estimate and approximation of the period. This process predicts the lateral forces by acceleration spectrum. The displacement demand gives the exact indication of the expected damage. In this paper, the fundamental periods of a case study of a low-rise building are investigated using the Equivalent Static Method and Response Spectrum Method using IS 1893(Part 1) and finite element modelling in SeismoStruct. The fundamental periods from these approaches are estimated to be 0.39 s, 0.14 s, and 0.178 s respectively. The factors are discussed on which fundamental period of a structure depends. However, the results show satisfactory relative error in range (1–30%), SeismoStruct gave realistic values of natural frequency because Indian code gives regular mode shape. The limitations of performing codal analysis are also reviewed in this paper.

Keywords Fundamental period · Reinforced concrete buildings · Dynamic analysis · Lumped mass system · Eigen value problem

P. Sharma · T. Gupta (✉)
Jaypee University of Information Technology, Waknaghat, India
e-mail: tanmay.gupta@juitsolan.in

A. K. Gupta et al. (eds.), *Advances in Construction Materials and Sustainable Environment*, Lecture Notes in Civil Engineering 196,
https://doi.org/10.1007/978-981-16-6557-8_72

1 Introduction

Natural disasters in recent years have shown the world, their destructive power in terms of loss of life and property. Earthquake is one such hazard, which has resulted in heavy economical losses. More than 55% of the land is of the Indian subcontinent is vulnerable to earthquake forces. The earthquake events in recent few decades show the necessity of earthquake engineering in India. Such hazards cannot be prevented; however, one can minimize their effects by designing the structures as per seismic codes. IS 1893: 2016 is framed keeping in mind the life safety aspect and collapse prevention of the structures and to satisfy these objectives it is necessary for the structures to respond to expected earthquake ground motions.

The natural period of vibration plays a vital role in earthquake design of new buildings and performance assessment of existing structures. Stiffness, mass and strength along the height of a building influence the natural period. Building height is the main parameter. There are many other factors, which affect this property such as section dimensions, structural regularity, number of bays and storeys, load position, soil flexibility, reinforcement ratio, and infill and shear walls [1]. The difficulty in evaluating these variables is a laborious task for the analysis of the fundamental period of a building. Eigenvalue or pushover analyses are carried out with relative ease. However, performing computer analysis for the estimation of period for each structure in urban areas becomes inadequate. Consequently, many empirical methods are used different classes of buildings.

There are two types of load: Static and Dynamic, which acts on a structure. Static loads are time dependant whereas dynamic loads are independent. Dynamic analysis and study on the response of structures excitation is more time consuming and complicated as compared to static analysis. Dynamic analysis involves the equation of motion of the structure, which is required to find the modes of vibration with the corresponding natural frequencies [2]. The lateral seismic loads depend upon the fundamental period of the structure, which is required to be determined theoretically or experimentally [3]. Natural frequency has a significant role in dynamic design without involving heavy dynamic testing [4]. To determine the natural frequency and modes of vibration of structure free vibration analysis is used. Modal analysis is performed to record the free vibrational response of the structure [5].

Moment resisting frame (MRF) structures are mainly used nowadays because they are highly stiff and substantial resistance to seismic activity. The researchers used different types of MRF structures to find most reliable structures under seismic load. The study showed that structures reached resonance conditions in short time and to avoid this problem, the fundamental Eigen frequency should be maximized in high-rise buildings [6]. Many researches indicate height as the main variable to determine the fundamental period of RC moment resisting frame structures [7]. However, there are a few studies on steel frame structures and is a scope of study in modern era.

The rapid evolution in the field of civil engineering software and coding over the last few decades has led to the occurrence of various Finite Element software focusing on the numerical solution of structural problems. Some of the previous studies have made use of different computer programming and software such as STAAD-pro, ANSYS, SolidWorks and SAP2000 for free vibration analysis and modal analysis to determine the fundamental time period [8].

This study is based on modal analysis of MRF structures using SeismoStruct software to study the natural frequency and free vibration analysis is investigated as per IS 1893:2016. The results are compared of both empirical analysis and finite element analysis.

2 Seismic Analysis as Per Indian Code

IS 1893(Part 1): 2016 [9] provides the guidelines and provisions for earthquake resistant design. It adopts Equivalent Static Method and in dynamic methods, the response spectrum and time history analysis are considered. When ground shakes, a building vibrates, and natural period of time is required for free vibration of the building structure. Generally, the first mode of free vibration is critical because of high resonance risk, whereas the other modes of vibration could be critical for high-rise structures, which are assumed less critical than the natural period for low-rise buildings [10]. Seismic excitations bend the structure into a number of shapes, which are known as the natural mode shapes, which depends upon the degree of freedom of the system. For structural idealization we convert the multi degree of freedom system to finite degree of freedom system with lumped mass model. In this procedure, the mass of the building is lumped at each floor level. The empirical formulas developed to estimate the period of oscillation (T_a) are given in Eqs. 1, 2 and 3 [9]. The fundamental period is a variable for estimating the base shear for a structure, thus it should be sound accurate in order to prevent unsafe design.

For Reinforced Concrete frames without infills,

$$T_a = 0.075h^{0.75}\ \text{s} \tag{1}$$

For Steel constructed frames without infills,

$$T_a = 0.080h^{0.75}\ \text{s} \tag{2}$$

For all other buildings with infills

$$T_a = \frac{0.09h}{\sqrt{d}}\ \text{s} \tag{3}$$

where $h =$ height (in metre) excluding basement storey, when ground floor deck is connected with walls but including basement storeys when there is no connection between the two.

The time period depends upon stiffness and mass of the structure, which is not included in the above equations. So the code specifies the use of dynamic analysis which requires other periods and shapes of natural modes. In this procedure, mass matrix and stiffness matrix are calculated for equivalent model and using these matrices, an eigenvalue problem is formulated to calculate the frequencies and mode shapes.

3 Code Based Seismic Analysis

In this study, low-rise residential building located in Zone V from a case study [11] is taken for investigation. The natural period of low-rise RC frame structure is examined of the building frame having a regular plan, consisting of columns and beams. The building is symmetric along X and Y-axes having plan dimensions 50 m × 8 m and floors having same height of 3.1 m. The building parameters of the study are shown in Table 1 and the elevation, lumped mass model and plan of the building are shown in Figs. 1, 2 and 3.

The mass (M) and stiffness (k) of each floor are computed and matrices are formed. These are as follows:

$$M = \begin{bmatrix} M1 & 0 & 0 \\ 0 & M2 & 0 \\ 0 & 0 & M3 \end{bmatrix} = \begin{bmatrix} 264271 & 0 & 0 \\ 0 & 264271 & 0 \\ 0 & 0 & 190239 \end{bmatrix}$$

Table 1 Building parameters from case study [11]

Structure type	Moment resisting frame
Number of stories	Three, (G+2)
Height of floor	3.1 m
Materials	Concrete (M25) and reinforcement (Fe415)
Live load	3 kN/m^2
Size of columns	500 mm × 500 mm
Size of beams	400 mm × 500 mm
Specific weight of RCC	25 kN/m^3

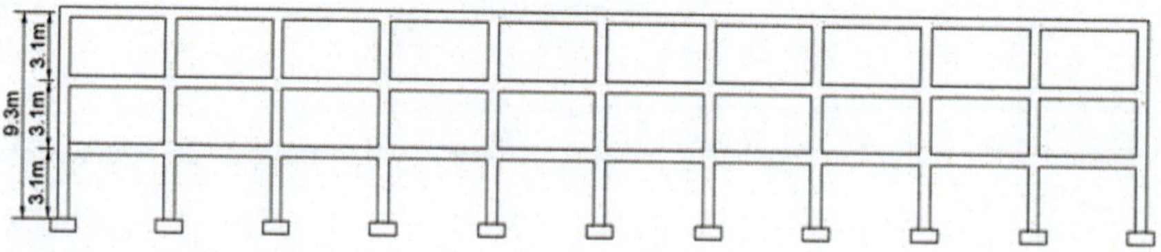

Fig. 1 Building elevation

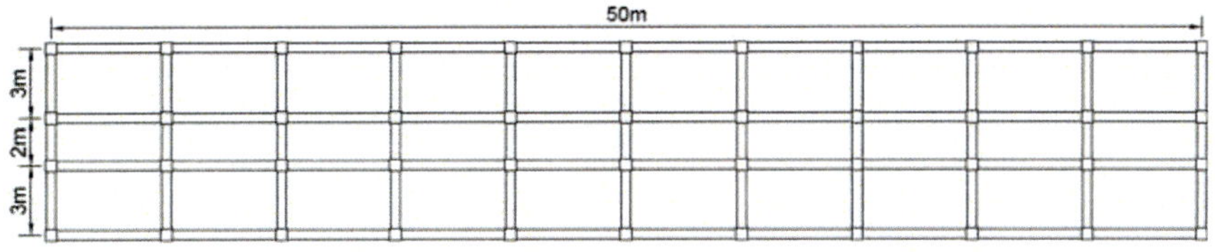

Fig. 2 Building plan

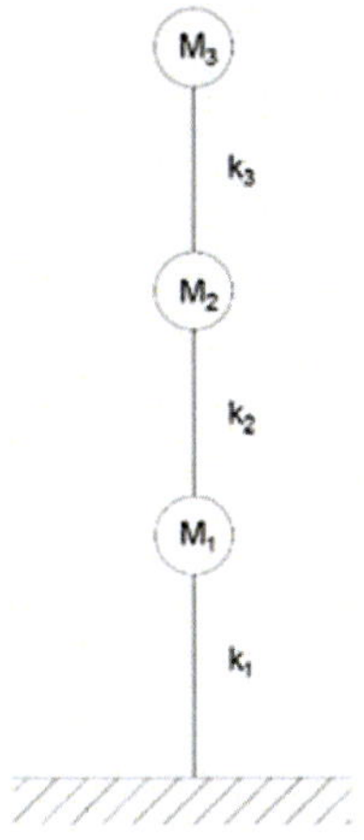

Fig. 3 Lumped mass model

$$K = \begin{bmatrix} k1 + k2 & -k2 & 0 \\ -k2 & k2 + k3 & -k3 \\ 0 & -k3 & k3 \end{bmatrix} = \begin{bmatrix} 4615487900 & -2307743950 & 0 \\ -2307743950 & 461548790 & -2307743950 \\ 0 & -2307743950 & 2307743950 \end{bmatrix}$$

Solving Eq. (4) the natural frequency, time period and Eigen values are calculated [9]

$$\left| K - \omega^2 M \right| = 0 \tag{4}$$

The Response acceleration Coefficient of the building is computed by natural time period of oscillation and damping. Generally, 5% damping is taken as per Indian seismic code. The design spectra for both Equivalent Static Method and Response Spectrum Method are shown in Figs. 4 and 5.

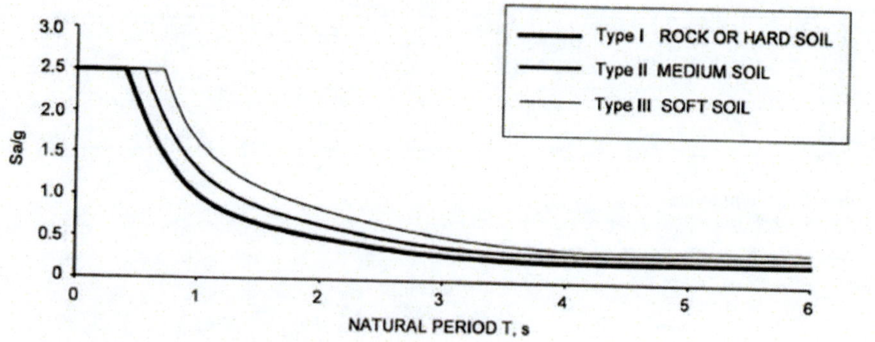

Fig. 4 Design spectra for equivalent static method

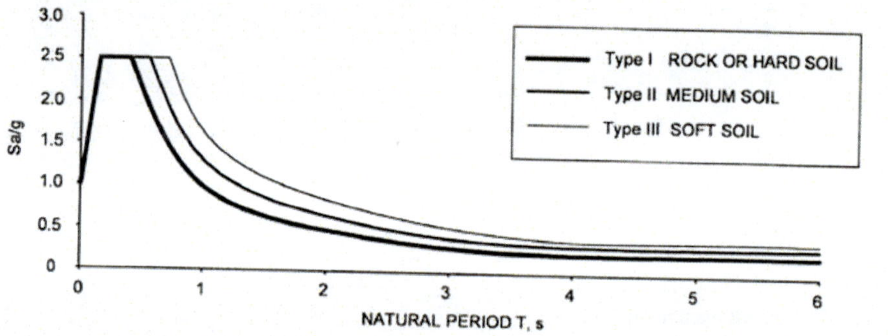

Fig. 5 Design spectra for response spectrum method

4 Finite Element Analysis in Seismostruct

SeismoStruct [12] is a nonlinear finite element program that is used in this study. SeismoStruct is proficient in forecasting displacement behaviour of space as well as plane frames under dynamic and static loading taking both material inelasticity as well as geometric nonlinearity. Material nonlinearity is considered in this study, which is discussed in Fig. 6. With the increase in displacement due to horizontal seismic force, the inertia force in the structure will generate as a response which will cause the change in stiffness in various frame members due to the distribution of forces. This will cause geometric nonlinearity in the structure. There are many material models available such as concrete, steel, fibre reinforced polymer and shape memory alloy. Only a few finite element software's have super elastic shape memory alloy. There is a wide range of three-dimensional elements, which can be used with various types of steel, concrete, and composite sections. Numerous successes in Blind Test Prediction Exercises shows its accuracy and reliability.

The inelasticity across the depth of section and along the length of member is modelled precisely which allows for accurate assessment of damage. It determines

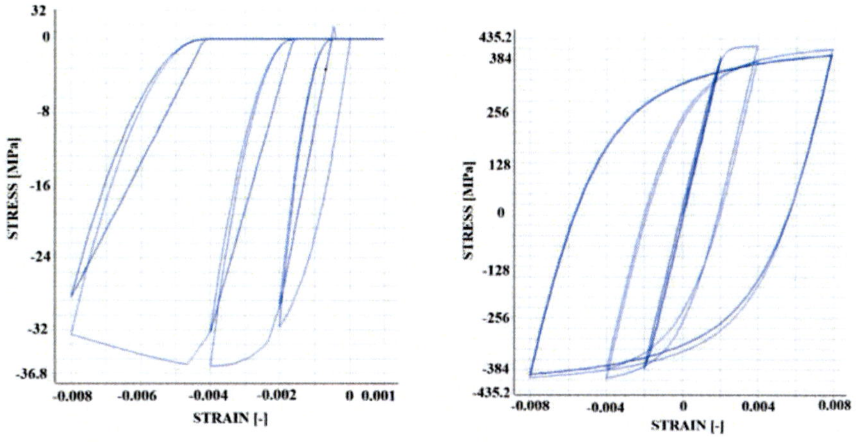

Fig. 6 Constitutive relationships for concrete and reinforcement steel

explicitly the inelastic response and the failure load of any frame-type structural configuration at high strain levels. SeismoStruct performs static (forces and displacements) and dynamic (accelerations) actions by carrying out eigenvalue, nonlinear static pushover (conventional and adaptive), nonlinear static time history analysis, nonlinear dynamic analysis, incremental dynamic analysis and response spectrum analysis. SeismoStruct has progressive simulation algorithms which are used to analyse complex tall structures in less time. Meshing is not required to perform finite element analysis in SeismoStruct. SeismoStruct is a useful tool to analyse complex analyses having variable parameters. SeismoStruct uses Lancoz algorithm and Jacobi algorithm with Ritz transformation are used for simulating the input data to establish modes of vibration. In this study, Jacobi algorithm is used in the pre-processing stage by assigning four Ritz vectors in *X*, *Y* and *Z* directions.

The software has three main sectors: a Pre-Processor, a Processor and a Post-Processor. The former is used to input the data required for the analysis of structural model, Processor is used to run the analysis and all the output files and results are obtained in post-Processor. Moreover, it includes two more components, which are known as Building Modeller and Wizard. These facilities help the designer in creating regular and irregular shaped 2-dimensional as well as 3-dimensional models. With them, the analyses can be run quickly as it takes only a few minutes for the whole process (Fig. 7).

5 Pre-processor in Seismostruct

Pre-processing is the first stage of any analysis as all the parameters required are assigned in this stage. The Eigenvalue analysis is selected first from the menu. The number of bays, frames and storeys are assigned in the Wizard to model the building.

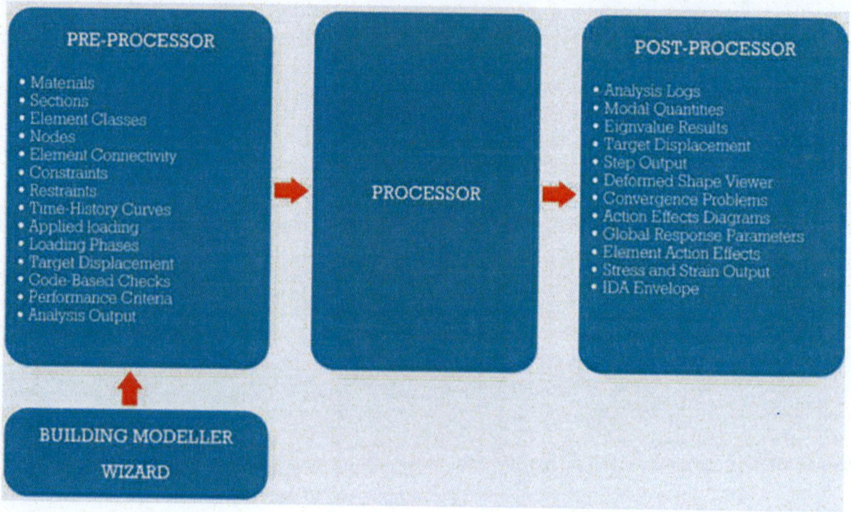

Fig. 7 Modules of SeismoStruct

The variables and criteria of different parameters are inserted in the pre-processing stage. The input values are organized in Table 2.

Table 2 Input values in pre-processor

Wizard	Number of bays		10
	Number of frames		4
	Number of storeys		3
Materials	Concrete	Compressive strength (MPa)	30
	Reinforcement	Modulus of elasticity (MPa)	200,000
		Yield strength (F_y)	415
Sections	Columns	Rectangular	0.5 m × 0.5 m
		Reinforcement	8Ø20 mm + (3–3)@10 mm/150 mm
	Beams	Rectangular	0.4 m × 0.5 m
		Reinforcement	8Ø20mm + (2–4)@10 mm/100 mm
Element class	Column elements		Inelastic force-based frame element
	Beam elements		Inelastic force-based frame element
Restraints	Bottom nodes		Fixed
Applied loads	Live load		3 kN/m^2

Table 3 Concrete model parameters after Mander et al. [13]

Mean compressive strength (kPa)	30,000.00
Mean tensile strength (kPa)	2200.00
Young's modulus (kPa)	25,742,960.20
Strain at peak stress	0.002
Specific weight (kN/m^3)	25.00
Confinement factor	1.2

Table 4 Menegotto-Pinto steel model parameters [14]

Yield strength (kPa)	415,000.00
Modulus of elasticity (kPa)	200,000,000
Strain hardening	0.005
Specific weight (kN/m^3)	78.00

The material properties are assigned for concrete and reinforcement. The model of Mander et al. [13] has been used for concrete specimens, which takes into account the cyclic nature of concrete. In general, the nonlinearity of RCC is totally dependent on reinforcement. Consequently, steel models have an utmost significance for the evaluation of flexural behaviour of an RCC section, and mainly when it is exposed to load reversals. The Menegotto and Pinto [14] model is used in this case study. The Menegotto and Pinto model has also been included in several studies for its simplicity and efficiency because it considers the softening of curves in reloading automatically. As per the case study, M25 grade concrete and Fe415 grade steel are used. The properties of both material models are shown in Tables 3 and 4 (Fig. 6).

Reinforced concrete rectangular sections are defined for column and beam sections in SeismoStruct. Square columns having cross section of 0.5 m × 0.5 m and rectangular beam sections of 0.4 m × 0.5 m are used throughout the structure. The structural detailing is shown in Fig. 8.

A fibre-based finite element approach is adopted to model the building with SeismoStruct. In element class. inelastic force-based frame element is used for both beam and column with an adequate discretization. It is a force-based 3-dimensional beam-column element type, which is competent to model space frame members with geometric and material nonlinearities. This element is taken because it can record the inelastic behaviour along the total length, even when employing a single element per member (Fig. 9).

The boundary conditions are assigned in the Restraints module. All the bottom nodes are restrained against deformation for all the 6 degree of freedom systems because we assume fixed conditions for the foundation in the analysis of buildings. Live load of 3 kN/m^2 is added in Applied Loads module for the floors. After loading the building, the model is ready for processing and analysis is run. The results of the natural frequencies are obtained in post-processor (Fig. 10).

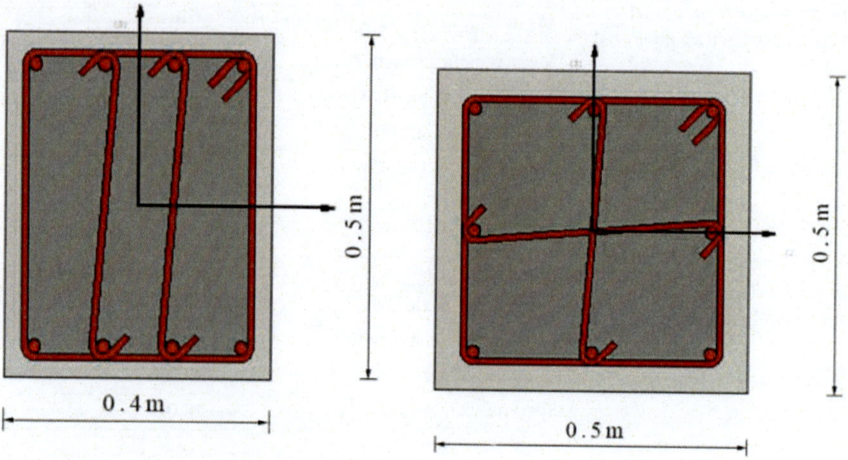

Fig. 8 Beam and column cross-sections

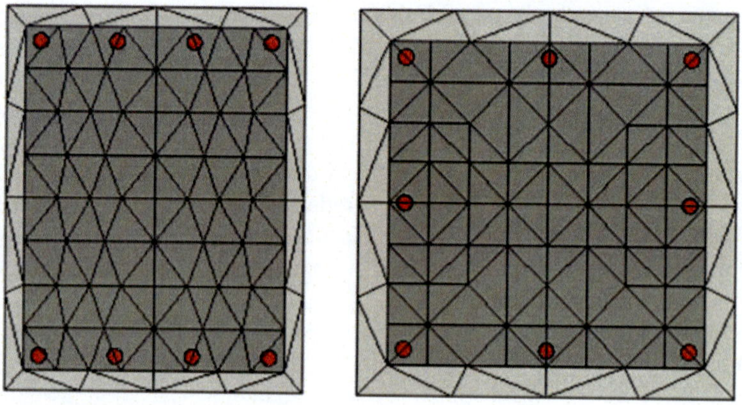

Fig. 9 Sectional discretization of beams and columns in SeismoStruct

6 Results and Discussion

The fundamental period from both approaches is tabulated and compared. The fundamental mode for Equivalent Static method, Response Spectrum method and numerical analyses are estimated to be 0.39 s, 0.14 s and 0.178 s. Indian seismic code suggests performing Equivalent Static Method for natural period less than 0.4 s. As the value is almost equal to 0.4 s, this method is not adopted to calculate the lateral seismic forces. The fundamental frequencies and natural periods from both methods are shown in Table 5.

SeismoStruct uses Jacobi algorithm with Ritz transformation, to solve the Eigen values so 12 modes are generated whereas in Response Spectrum analysis three

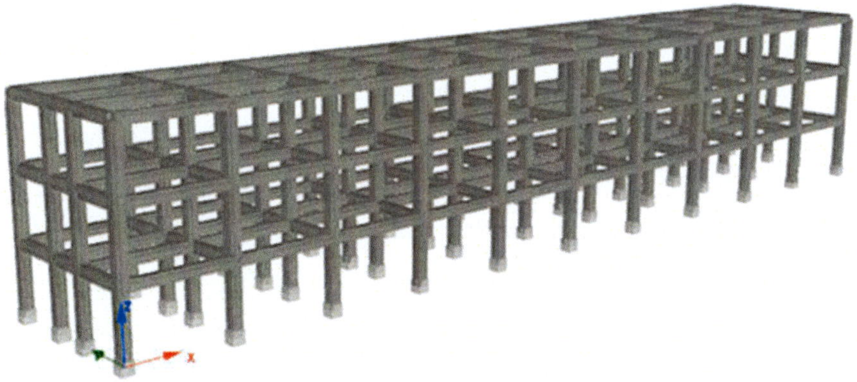

Fig. 10 3D model of building in SeismoStruct

Table 5 Fundamental frequencies and time periods from both approaches

Modes	Response spectrum analysis		SeismoStruct analysis	
	Fundamental frequency, ω (rad/s)	Natural time period, T (s)	Fundamental frequency, ω (rad/s)	Natural time period, T (s)
Mode 1	45.1	0.140	33.42	0.178
Mode 2	124. 21	0.051	35.62	0.176
Mode 3	172.24	0.036	40.50	0.155
Mode 4	–	–	51.53	0.122
Mode 5	–	–	102.77	0.0611
Mode 6	–	–	105.60	0.0595
Mode 7	–	–	108.89	0.0577
Mode 8	–	–	145.91	0.0431
Mode 9	–	–	178.44	0.0352
Mode 10	–	–	181.40	0.0346
Mode 11	–	–	294.33	0.021
Mode 12	–	–	349.62	0.018

modes are generated as the multi-degree freedom system has been idealised into single degree of freedom system. The mode shapes of the first three modes have been plotted in Figs. 11 and 12 and the comparison shows that the first mode has a similar shape in both cases.

The modal mass M_k are calculated as per IS 1893(Part 1): 2016 using Eq. 5 given as:

$$M_k = \frac{\left[\sum_{i=1}^{n} W_i \varphi_{ik}\right]2}{g\left[\sum_{i=1}^{n} W_i \varphi_{ik}\right]2} \tag{5}$$

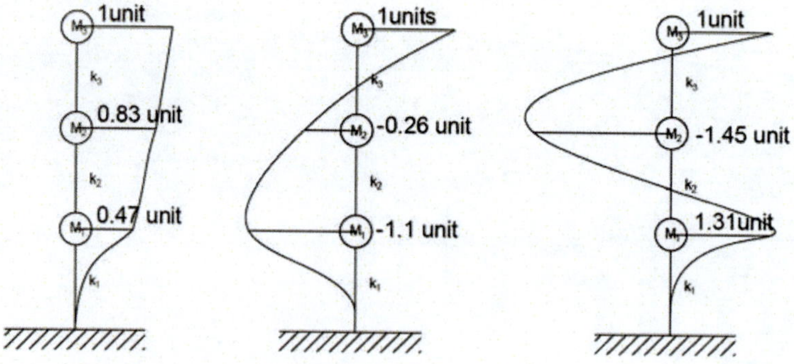

Fig. 11 Mode shapes from response spectrum method

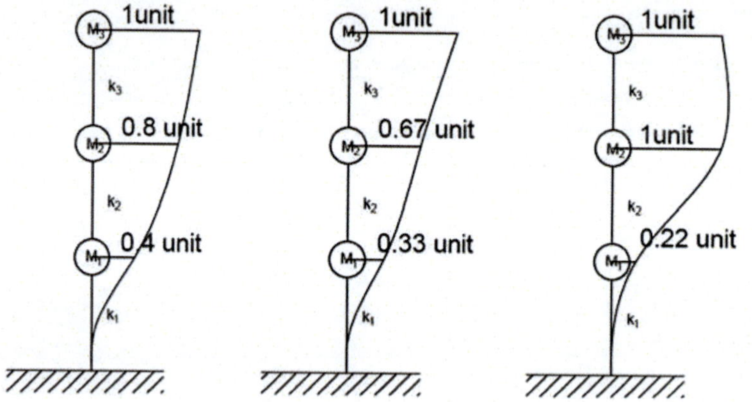

Fig. 12 Mode shapes from SeismoStruct

The fundamental frequencies and modal mass of the three modes are compared in Tables 6 and 7 and relative error can be computed. The relative error in range (1–30%). The reason behind this variation is that IS 1893 is based on assumptions.

Table 6 Comparison of natural frequencies

S. No.	ω (Response spectrum)			ω (SeismoStruct)		
	ω_1	ω_2	ω_3	ω_1	ω_2	ω_3
1	45.1	124.41	172.24	33.4	102.77	178.4

Table 7 Comparison of modal mass

S. No.	M (Response spectrum)			M (SeismoStruct)		
	M_1	M_2	M_3	M_1	M_2	M_3
1	91%	7.6%	2.89%	86%	10.84%	3.19%

7 Conclusion

An examination has been performed on the fundamental frequency and time period of vibration of low-rise by means of codal design analysis and finite element modelling. For this, a 3-storey RC frame building has been considered. The natural periods and frequencies computed from the eigenvalue analysis in SeismoStruct were noted and collated against the period from IS 1893: 2016. Some conclusions can be drawn from this study, which can be applied, to full-scale structures. As the stiffness decreases, the natural frequency of structure also decreases but a decrease in height of structures leads to an increase in the natural frequency. The column elements have a significant role in the stiffness and mass of the structure, as they are very reliant on its dimensions. Hence, a change in dimension can give rise to a sudden change in the dynamic nature.

The fundamental periods from Equivalent Static Method, Response Spectrum Method and Finite Element analysis in SeismoStruct are estimated to be 0.39 s, 0.14 s and 0.178 s respectively. There is a significant variation in the results of these methods because the Indian seismic code has many limitations and assumptions. IS 1893 states that the fundamental mode dominates the response of the structure whereas ground motions are complex having several frequencies. The idealisation is done by converting the multi degree of freedom system to a finite degree of freedom system with lumped mass model. The material elasticity and structural rigidity with lumped mass are considered in Response Spectrum method and in SeismoStruct, the material nonlinearity is considered. The study proved that SeismoStruct provides more generalised and realistic results of fundamental period. Nonlinear analysis such as pushover analysis and time history analysis can be performed on this building to simulate the plastic behaviour of critical sections of the structure. Many researchers and designers are adopting these methods to access the seismic performance of the structures.

References

1. Crowley, H., & Pinho, R. (2004). Period-height relationship for existing European reinforced concrete buildings. *Journal of Earthquake Engineering, 8*(spec01), 93–119.
2. Al-Aasam, H. S., & Mandal, P. (2013). Simplified procedure to calculate by hand the natural periods of semirigid steel frames. *Journal of Structural Engineering, 139*(6), 1082–1087.
3. El-saad, M. N. A., & Salama, M. I. (2017). Estimation of period of vibration for concrete shear wall buildings. *HBRC Journal, 13*(3), 286–290.
4. Gaile, L., Tirans, N., & Velicko, J. (2013). Evaluation of highrise building model using fundamental frequency measurements. *Civil Engineering, 13*, 15.
5. Rezaee, M., Yam, G. F., & Fathi, R. (2015). Development of "modal analysis free vibration response only" method for randomly excited systems. *Acta Mechanica, 226*(12), 4031–4042.
6. Alavi, A., Rahgozar, R., Torkzadeh, P., & Hajabasi, M. A. (2017). Optimal design of high-rise buildings with respect to fundamental eigenfrequency. *International Journal of Advanced Structural Engineering, 9*(4), 365–374.

7. Asteris, P. G., Repapis, C. C., Tsaris, A. K., Di Trapani, F., & Cavaleri, L. (2015). Parameters affecting the fundamental period of infilled RC frame structures. *Earthquakes and Structures, 9*(5), 999–1028.
8. Prajapati, N., & Desai, A. N. (2012). Effect of height and number floors to natural time period of a multi-storey building. *International Journal of Emerging Technology and Advanced Engineering*, 2250–2459.
9. Bureau of Indian Standards. (2016). Criteria for earthquake resistant design of structures. IS 1893(Part 1), New Delhi, India.
10. FEMA. 310. (1998). *Handbook for seismic evaluation of buildings—A pre-standard.*
11. Dhote, P. (2010). *Non-linear static pushover analysis of RC frame buildings* (M.Tech dissertation). Earthquake Engineering, Indian Institute of Technology, Roorkee.
12. Seismo Struct. (2003). Computer program for static and dynamic nonlinear analysis of framed structures.
13. Mander, J. B., Priestley, M. J., & Park, R. (1988). Theoretical stress-strain model for confined concrete. *Journal of Structural Engineering, 114*(8), 1804–1826.
14. Menegotto, M. (1973). Method of analysis for cyclically loaded RC plane frames including changes in geometry and non-elastic behavior of elements under combined normal force and bending, pp. 15–22.

Study of Behavior of the Masonry Infill Structures Subjected to Lateral Loads

Shilpa Pal, Sahil Yadav, Omkarnath Thakur, and Rohit Kashyap

Abstract The principal aim of this study is to analyze masonry infills subjected to lateral loads using a 3D finite element modeling software, "ABAQUS". With the help of Abaqus, a 3D wall is designed with three sides framed and base fixed for ENCASTRE boundary conditions. And the load is being applied to get various results. Firstly, it has been concluded that rigidity increases as Young's Modulus increases. Further to widen the area of research, Area of the wall is being introduced with an opening in the masonry infill wall with 30%, 45%, 50%, and 45% split openings and concluded that First class brick gives optimum results and increase in opening Percentage of wall tends to decrease in rigidity of the wall. Thus, failure will occur due to diagonal compression so, when force is applied on the infill wall section, a brittle failure occurs in a diagonal direction before the yielding of the stirrups, and diagonal compression mode of failure occurs, (out of plane failure occurs) due to Buckling instability. Both the corners of the wall are crushed.

Keywords Finite element · Masonry infill walls · Rigidity of the masonry · Young's modulus of masonry wall · Spacing in masonry wall · Opening effects · In-plane behavior

1 Introduction

While referring to various research papers and from analysis states that fillings help in providing resistance and rigidity of the structures. The dynamics of a structure changes with change in type of infill and the properties of infill that is why it's very important to develop an understanding of masonry infill. Various Research

S. Pal · S. Yadav (✉) · O. Thakur · R. Kashyap
Department of Civil Engineering, Delhi Technological University, Delhi, India

S. Pal
e-mail: shilpapal@dtu.ac.in

O. Thakur
e-mail: omkarnarhthakur_2k17ce63@dtu.ac.in

A. K. Gupta et al. (eds.), *Advances in Construction Materials and Sustainable Environment*, Lecture Notes in Civil Engineering 196,
https://doi.org/10.1007/978-981-16-6557-8_73

papers are analyzed regarding this particular topic of properties of a masonry infill they all point in the same direction stating the type of infill decides the rigidity, stiffness, and resistance of a structure during loading [1–3]. It is analyzed that the presence of the infills plays a vital role in the dissipation of energy. In addition to construction regarding the employment of anisotropic and continuous models, and masonry considered to be composite material and joints of mortar and cracking are part of examination when distributed within the mass of fabric, these kinds of models don't channelize them to show mechanics of cracks and characterizing infill, specifically those which are associated or linked to the friction and then to slip. Modeling of the masonry walls is a difficult issue due to their composite structures. There are several modeling techniques for modeling walls constituting masonry structures. Macro modeling and Micro modeling are two totally different techniques for modeling masonry structures. It was shown that the walls modeled by macro and micro methods represent consistent results with the experimental work [4], so Macro modeling is an easier and less time-consuming method. Effects of opening, including its size, shape had been seen to impact the extent of failure mode and damage caused by it.

Openings in masonry infill wall result in an overall reduction of stiffness of the masonry infill wall. Several studies also suggested the impact of openings on masonry infill walls, e.g., [5–8]. These researches identified that stiffness reduction and opening sizes in masonry infill impact the masonry infill wall stiffness. Failure of masonry infill walls under lateral loading is generally caused by (a) compressive failure at the toe of the wall, (b) flexural failure.

1.1 Masonry Structure

Masonry structures are basic building blocks for any structure in ancient history and also have an important role in today's structures. Masonry Structures are present in many monumental structures like bridges, towers, residential buildings. To measure the strength of masonry structures that are built before there are rules and standards to follow it is very important to model a masonry structure as these can also be used to gauge the strength of modern masonry structures. As masonry structure is a composite structure it is tough to model those so several modeling methods are used like macro modeling and micro modeling. In this study macro modeling method is used on the masonry structure [9]. In micro modeling masonry brick and mortar is used as a single unit referred to [10].

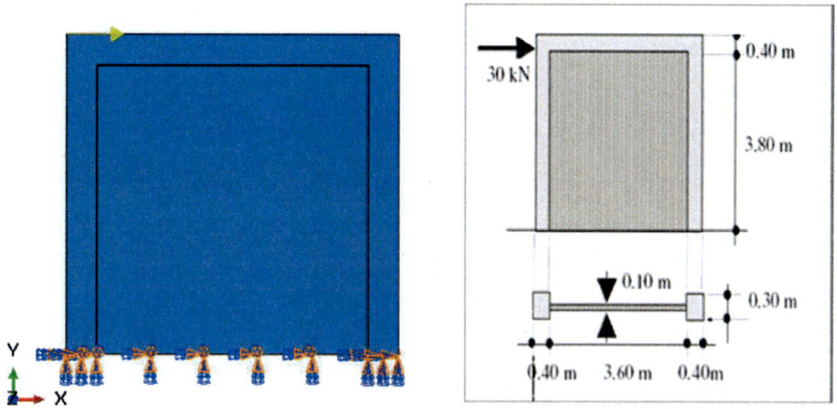

Fig. 1 Model designed with ABAQUS. Detailed figure of lateral loading, its point of application, and the dimensions of wall

2 Finite Element Analysis

2.1 Objective

In FE analysis based on the concept of displacement, wall structure is divided into some parts and all have their own material properties. Nodal forces and displacements are often derived from those properties and assembly of these components and impact or consideration of masses along with the boundary conditions gives us results which is a system of equations that describes structural equilibrium and that is solved to get the required results like nodal displacements stresses, strains, etc.

Principle aim of this study is to analyzed 3D Finite Element to know the behavioral changes of masonry infill walls in structures when it is subjected to horizontal force. The finite element software ABAQUS.

While undertaking the model, it is known that various types of masonry have various young modulus but by using ABAQUS it is possible to generalize our values and properties with any type of infill [11] (Refer with Fig. 1).

2.2 Size and Geometry of Model

For various components of the FE numerical model are considered as per reference in [11] they are presented in Fig. 1. The dimensions of masonry wall are (3.8 m × 3.6 m × 0.1 m) (height × length × depth) and for Concrete frame it is 0.4 m all around the frame and 0.3 m depth.

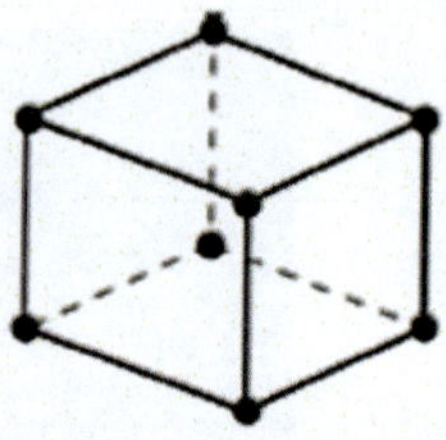

Fig. 2 Linear element (8-node bricks C3D8)

Table 1 Input properties for FE model referred from [11]

Material properties	RCC framework	Masonry infill wall
Density (kg/m^3)	2500	2000
Young's modulus (GPa)	29	2.13–11.2
Poisson's ratio	0.20	0.32

2.3 *Type of Element*

All elements have components C3D8 (Fig. 2) which is a continuum 8-node brick that has 3 degrees at every node which are translations and within the nodes x, y, and z. The continuum part in ABAQUS is used for linear analysis and when nonlinear analyzes are done some other points such as contact point, plasticity, deformations are considered.

2.4 *Material Properties/Input Properties for FE Model*

The elasticity modulus of the concrete used in RCC framework is 29,000 MPa, and its Poisson's ratio is 0.20, and for masonry infill wall elasticity modulus in the normal direction with the joints as a homogeneous part is taken as per the which is using (Eq. 1) to find the Young's Modulus of the wall using compressive strength of the different bricks (refer Table 2, Refer Fig. 4), and Poisson's ratio of 0.32 for masonry infill wall (All properties referred from [11]) (Table 1).

$$E_m = 1085 * F_m \tag{1}$$

2.5 *Boundary Conditions*

A masonry structure constructed with a three-dimensional framework structure (an RCC frame made up of concrete and steel reinforcements). Infill walls are the

Table 2 Compressive strength of bricks which is used to calculate the Young's modulus of the masonry wall

Type of brick	Compressive strength of brick (N/mm^2) (F_m)	Young's modulus of wall (N/m^2) ($E_m = 1085 * F_m$)
Sun dried brick	2	2.13×10^9
ACC block	3	3.23×10^9
Common building brick	3.43	3.72×10^9
Second class brick	6.87	7.45×10^9
Fly ash brick	9	9.77×10^9
First class brick	10.3	1.12×10^{10}

supported wall. The three-dimensional framework ensures that it acts like a single object during any kind of loading lateral, Horizontal, seismic. There is research done on infill walls. 3 side frames with masonry inside and with openings in the masonry are used. The base of masonry and the frame is fixed, i.e., there is no movement in x, y, z direction as well as no rotation is allowed. The term in ABAQUS is known as ENCASTRE ($x = 0$, $y = 0$, $z = 0$, $u1 = 0$, $u2 = 0$, $u3 = 0$). (Refer to Fig. 1) for the reference of the boundary condition.

2.6 *Interactions*

Surface to surface contact type is used in ABAQUS and the sliding formulation finite sliding [12]. Interaction between the master surface which is three sides outside of the wall in contact with the frame and the slave surface which is inside three sides of the frame in contact with the masonry infill wall [Refer Fig. 3b].

2.7 *Lateral Loading*

Lateral Loading is a loading that is parallel to the ground and is applied at upper left corners of the Masonry Infill Frame. 30 kN load is applied which is split into two concentrated forces of 15 kN applied on both the corners of the frame.

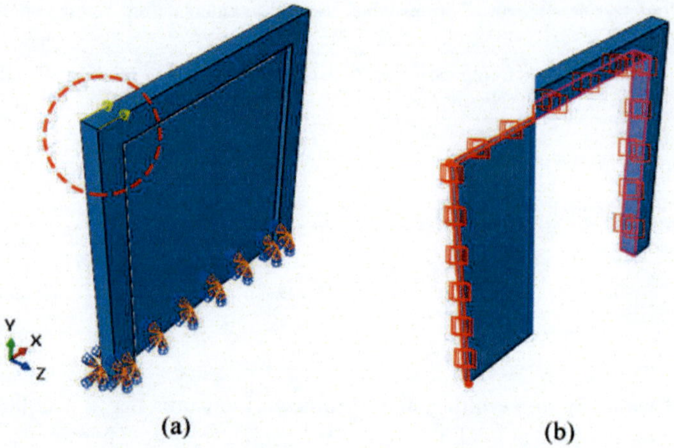

Fig. 3 **a** Concentrated force of 30 kN applied at both the corners (15 kN each). **b** Interaction between masonry infill wall and the frame

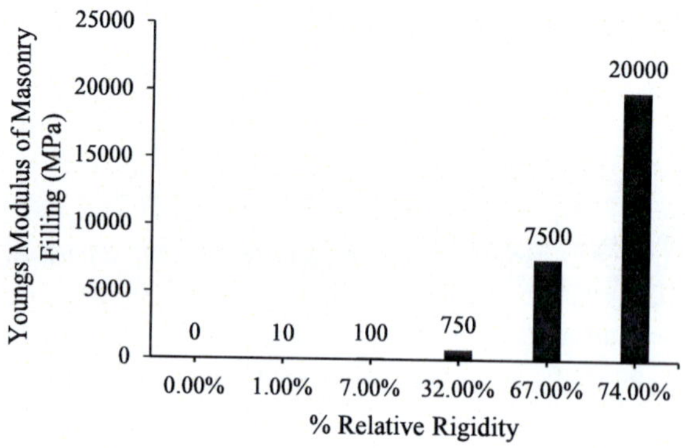

Fig. 4 Young's modulus versus relative rigidity of the masonry infill wall (with lateral load 30 kN)

3 Results and Discussion

3.1 Effect of young's Modulus on Rigidity of Masonry Wall

The modulus of elasticity is the property that measures the tensile stiffness of a solid material. It is defined as the ratio of tensile stress (force per unit area) and axial strain. Well, the modulus of elasticity of the infill wall is determined by the combination of both brick and mortar young's modulus and it can be one of an indicator that defines the stiffness of a structure. Stiffness is the rigidity of an object the extent to which it resists deformation from applied load (Refer Fig. 6) increase in Young's Modulus results in Deformation of the Masonry Infill Wall and also the stresses generated inside the structure tends to reduce with increase in the Young's Modulus (Refer Fig. 4) (Fig. 5).

Fig. 5 **a** Relative rigidity of wall with different brick. **b** Young's modulus of wall

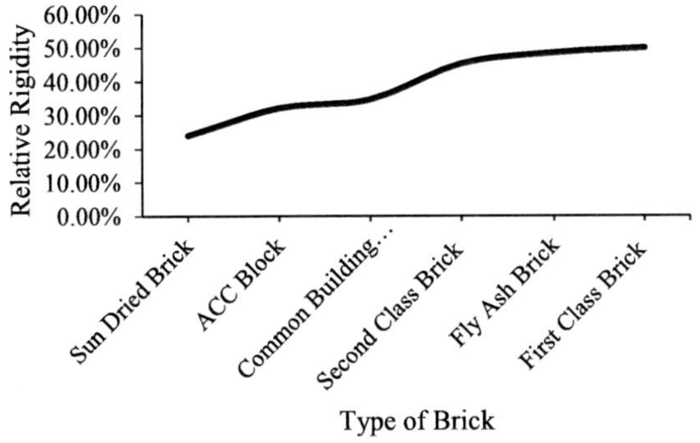

(a)

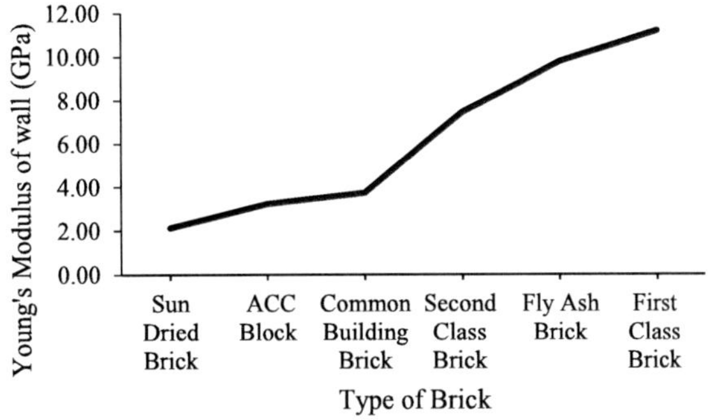

(b)

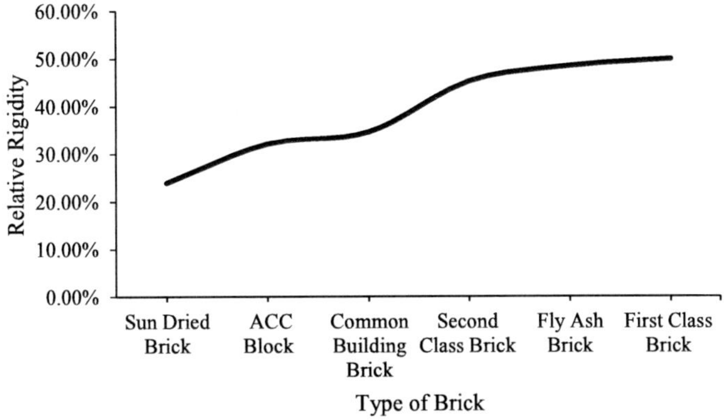

Fig. 6 Relative rigidity of first class brick with different opening percentages (with lateral load 30 kN)

Hence, it is examined that First class brick produces optimum results. So, First class brick is preferred for further analysis (With Lateral load 30 kN) (Table 2).

For simulation of real-world conditions, different types of bricks with different Young's Modulus (Eq. 1) are used to calculate Young's Modulus with Compressive strength of brick (Refer [10]) N/m^6.

3.2 *Effect of Opening in Masonry Wall (Spaces for Doors and Windows)*

The effect of reinforced concrete framed with spacing in its infill walls is that frames with door and window openings are less rigid than those without any openings (Refer Fig. 2). Masonry infill walls affect the lateral behavior of infilled frames by decreasing lateral stiffness and its strength. Stiffness is lost due to partial spacing which can be observed from stiffness reduction factors which were done by finite element analysis of frames by taking frame-infill interaction. P Graphs of frames are plotted to get yield displacement, shear forces, drift ratios, etc. While comparing parameters of seismic demand and capacity shows that partial openings manage to create nonlinear behavior of RC frames and cause a different behavior from that predicted for a bare frame. Which leads to lower compressive and residual strength. So much more spacing can't be allowed in infill as lateral loads can cause failure of structure due to lost strength.

In Fig. 6, the 0% opening is used as a benchmark for other cases. That's why % rigidity is taken as 100% for 0% opening.

As shown in Fig. 7 the column represents monotonically increasing Young's Modulus for different types of bricks with their respective compressive strengths. They were arranged in increasing order of their Young's Modulus, i.e., it resembles the increased overall strength of the Masonry Infill wall (Refer Fig. 6).

It implies a relation between relative rigidity and Young's Modulus of wall. All calculations were done on the basis of Table 3.

3.3 *Behavior of Masonry with Lateral Loading*

Failure of masonry walls at corners, although frequent, has not been researched extensively. Such failure involves the detachment of a wedge formed portion of masonry from the corner walls beneath the action of bifacial lateral load. Behavior of corners in dry stack masonry was studied up to failure in ABAQUS, Finite component (FE) atmosphere exploitation specific convergent thinker. Numerous models of masonry structures with load bearing and non-loadbearing walls were thought-about and were subjected to dynamic loading on the corner.

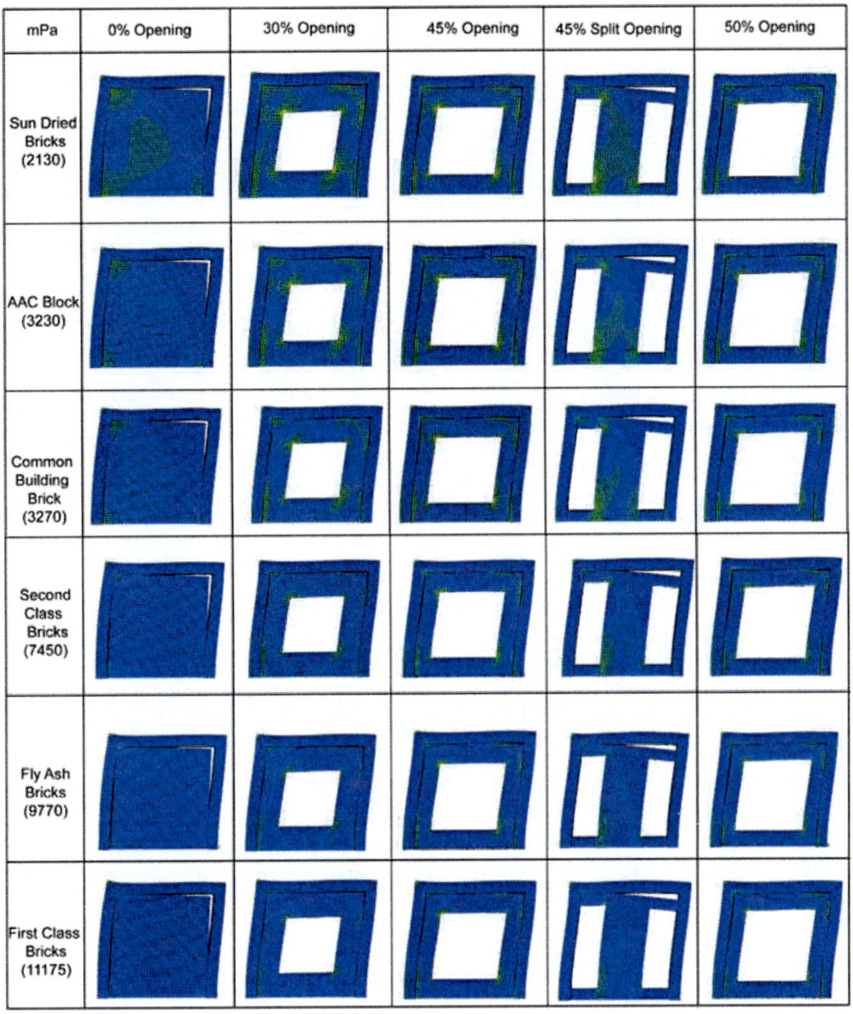

Fig. 7 Opening percentage versus overall strength of the masonry (as per different Young's modulus with lateral load 30 kN)

Based on the failure pattern determined beneath the synchronic action of informatics and OP forces, simplified corner failure mechanisms were projected for the limiting equilibrium condition and corresponding values of peak acceleration were obtained. Results of limit analysis compared well with iron predictions. The iron approach adopted for dry stack masonry was extended to review the corner failure in mortar warranted masonry; certain friction constant for each course was introduced that was just like the shear resisted by the layer of mortar in any given course.

Table 3 Rigidity change with loading for different opening percentage for wall of first class brick

Loading (kN)	% Rigidity change				
	0% Opening (%)	30% Opening (%)	45% Opening (%)	45% Split opening (%)	50% Opening (%)
30	48.68	32.89	9.65	20.18	−5.26
40	22.03	−0.66	−35.53	−19.74	−57.89
60	−2.63	−34.21	−80.70	−59.65	−110.53
90	−53.95	−101.32	−171.05	−139.47	−215.79
120	−105.26	−168.42	−261.40	−219.30	−321.05

Such modeling was effective indecisive the propagation of cracks through the masonry additionally as a result of the worth of acceleration at that a number of the corner wall detaches from the rest of the masonry.

3.4 Effect of Loading on Masonry Infill Wall

Without changing any conditions, the load is varied to check if there is any change in the rigidity, which may cause a change in the loading. Significant differences were observed when stresses increased and along with a linear relation in loading and the stresses increased First Class Bricks with Young's Modulus 7450 MPa. Started the Lateral Loading From 30 kN up to 120 kN (Refer Fig. 7).

There is a Positive linear relation between Loading and the stresses induced in the Masonry Infill Wall of First Class Brick (Refer Fig. 8) that results in decrease of relative rigidity, i.e., negative linear relation between Relative Rigidity and Loading (refer Table 3).

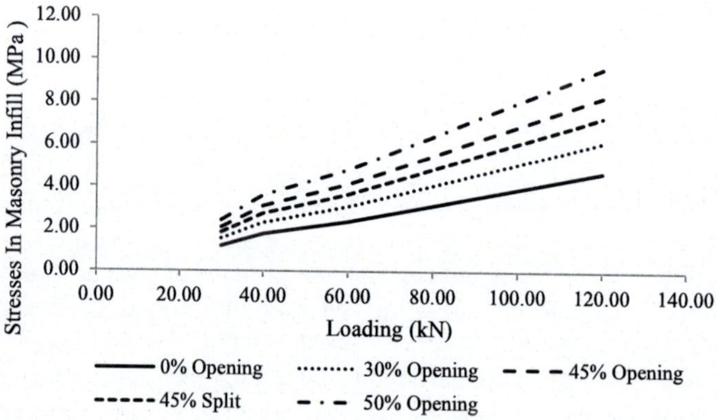

Fig. 8 Change in stresses with change in loading

3.5 Effect of Different Openings on Rigidity of Wall

Figures 9 and 10 show the trends for the percentage difference in rigidity for types of bricks in different wall opening percentages that are used in this paper against different percentages of openings.

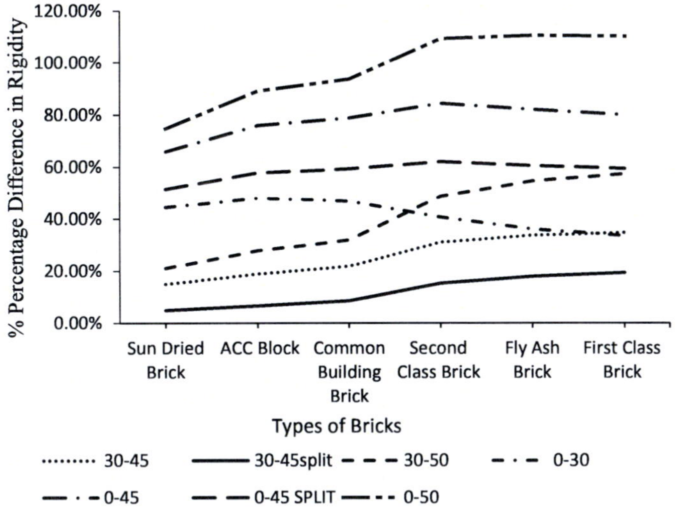

Fig. 9 Percentage difference in rigidity of the masonry infill wall for different types of bricks compared with 0% wall opening. Different colored line represents different wall opening percentages

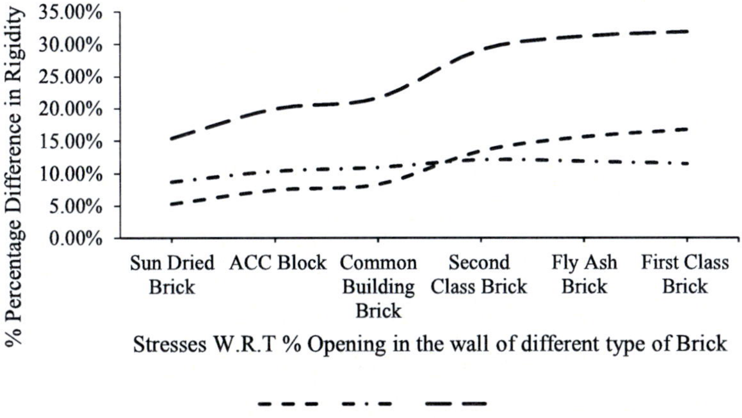

Fig. 10 Percentage (%) difference in relative rigidity of the masonry infill wall for different types of bricks. Different shapes represent different wall opening percentages

Table 4 Normal stresses in the masonry infill wall for various wall opening percentages and for different types of bricks

Type of brick	Young's modulus of wall (N/m^2)	0% Opening	30% Opening	45% Opening	45% Opening split	50% Opening
Sun dried brick	2.13×10^9	1.74×10^9	2.51×10^9	2.88×10^9	2.63×10^9	3.03×10^9
ACC block	3.23×10^9	1.55×10^9	2.29×10^9	2.73×10^9	2.44×10^9	2.93×10^9
Common building block	3.72×10^9	1.49×10^9	2.19×10^9	2.67×10^9	2.37×10^9	2.89×10^9
Second class brick	7.45×10^9	1.25×10^9	1.76×10^9	2.30×10^9	2.02×10^9	2.61×10^9
Fly ash brick	9.77×10^9	1.18×10^9	1.60×10^9	2.14×10^9	1.89×10^9	2.48×10^9
First class brick	1.12×10^9	1.14×10^9	1.53×10^9	2.06×10^9	1.82×10^9	2.40×10^9

When 0% opening wall compared to 30% opening wall the relative rigidity decreases by 44.38% (for sun-dried brick) rigidity which increases to 33.51% (First class brick), i.e., with an increase in Young's Modulus of the brick. When 0% opening wall is compared to 45% opening wall, the rigidity decreases by 51.41%, and with an increase in Young's Modulus of brick, rigidity decreases by 59.32%. Similar trends are followed for 0% opening wall compared with 45% Split opening wall and 50% opening wall.

When 45% opening wall is compared with 50% opening wall, the rigidity decreases by 5.32% (for sun-dried brick), which further decreases to 16.72% (First class brick) when Young's Modulus of brick is increased. Similar trends are followed for 45% spilt opening compared to 45% opening and 45% spilt opening compared to 50% opening.

When 30% opening wall is compared with 45% opening wall, the rigidity decreases by 14.87% (for sun-dried brick), which further decreases to 19.33% (First class brick) when Young's Modulus of brick is increased.

Similar trends are followed for 30% spilt opening compared to 45% split opening and 30% spilt opening compared to 50% opening (Table 4).

4 Conclusion

The impact of changing Young's Modulus of masonry infill wall on the overall rigidity of the wall and the impact of openings in the masonry wall. Increasing Young's Modulus results in an increase in overall Rigidity of the Masonry Infill Wall (refer Fig. 7). Properties of six types of bricks used in India are calculated, and

Young's Modulus of masonry infill wall with compressive strength that varies from 2 to 10.3 N/mm^2 and 2130 to 11200 MN/m^2 respectively. Effect of Young's Modulus resulted in a relationship between Young's Modulus and rigidity of the Masonry Structure, with an increase in Young's Modulus of brick there is a positive impact on overall strength. Bricks are compared on five types of models with increasing opening percentage, which results in reduction in rigidity of masonry infill wall (Figs. 4, 5, 6, and 7). 45% Opening model is compared with 45% Split opening model, stresses in 45% split opening model are 12% less, i.e., 12% more rigidity in the structure. This resembles dividing a single big opening into several openings results in an overall increase in rigidity. Lateral loading is also changed to check if there is any effect of that and a linear relationship between load and the stresses in the structure, i.e., if the load is increased by two times the stresses also increase by two times a linear relation between stresses and the loading.

References

1. Stafford Smith, B. (1966). Behavior of square infilled frames. *Journal of Structural Division, 92*, 381–403.
2. Kfoury. (1985). *Contribution des éléments de remplissage à la rigidité latérale des cadre en acier*. thèse de magistère de l'université de Montréal.
3. Yanev, B. S., & McNiven, H. D. (1979). Mathematical modeling of the seismic response of one story steel frames with infill partitions. In *The Canadian Conference on Earthquake Engineering*, Montreal, Que.
4. Kömürcü, S., & Gedikli, A. (2019). Macro and micro modeling of the unreinforced masonry shear walls. *EJENS, 3*(2), 116–123.
5. Augenti, N., Parisi, F., Prota, A., & Manfredi, G. (2010). In-plane lateral response of a full-scale masonry subassemblage with and without an inorganic matrix-grid strengthening system. *Journal of Composites for Construction, 15*, 578–590.
6. Basset-Salom, L., & Guardiola-Víllora, A. (2014). Seismic performance of masonry residential buildings in Lorca's city centre, after the 11th May 2011 earthquake. *Bulletin of Earthquake Engineering, 12*, 2027–2048.
7. Bruneau, M. (1994). Seismic evaluation of unreinforced masonry buildings—A state-of-the-art report. *Canadian Journal of Civil Engineering, 21*, 512–539.
8. Liu, Z., & Crewe, A. (2020). Effects of size and position of openings on inplane capacity of unreinforced masonry walls. *Bulletin of Earthquake Engineering, 18*, 4783–4812.
9. Asteris, P. G., Cavaleri, L., Di Trapani, F., & Sarhosis, V. (2016). A macro-modelling approach for the analysis of infilled frame structures considering the effects of openings and vertical loads. *Structure and Infrastructure Engineering, 12*, 551–566.
10. Phaiju, S., & Pradhan, P. M. (2018). Experimental work for mechanical properties of brick and masonry panel. *JSE, 5*, 51–57.
11. Mezigheche, N., Gouasmia, A., Athmani, A., & Merzoud, M. (2015). Behavior of the masonry infill in structures subjected to the horizontal loads. *International Journal of Civil, Environmental, Structural, Construction and Architectural Engineering, 9*(6).
12. ABAQUS 6.10. Analysis user manual.

Quantification of Municipal Solid Waste and Effect of Open Dumping on Soil in Smart City Dharamshala, Himachal Pradesh

Anchal Sharma, Love Sharma, and Rajiv Ganguly

Abstract The solid waste generation rate is increased with an increment in population in developing countries. The waste production rate is to a larger extent in big cities than in countryside areas because of high living standard of people over there. In this context, solid waste management is one of the most important features of India's solid waste management. But the main issue is disposal of waste in most of India's states. Hence, solid waste management is one of the major aspects of today's issues that need to be considered. The existing paper compiled the physical and chemical quantification of municipal solid waste in the smart city of Himachal Pradesh, i.e. Dharamshala so that necessary steps can be incorporated based on the nature of waste. The quantification of the waste revealed that the municipal solid waste is rich in organic waste and after that, the second highest fraction is of paper waste. The soil properties affected by the open dumping have also been analyzed. The soil samples were collected from trial pits within the dumpsite at depths of 1.5 m. The soil samples hence collected have been subjected to various geotechnical analyses. The geotechnical analysis of soil revealed that the dumpsite soil of study regions displayed decreasing trend in the values of maximum dry density, cohesion, CBR as compared to the natural soil. This is clearly indicated that geotechnical properties of the soil in study location has become vulnerable due to contamination.

Keywords Municipal solid waste · Smart city · Solid waste management · Quantification of waste · Soil properties · Himachal Pradesh

A. Sharma (✉)
Department of Civil Engineering, SSCET Badhani, Pathankot, Punjab, India

L. Sharma
School of Civil Engineering, Shri Mata Vaishno Devi University, Katra, Jammu and Kashmir, India

R. Ganguly
Jaypee University of Information Technology, Waknaghat, Himachal Pradesh, India

A. K. Gupta et al. (eds.), *Advances in Construction Materials and Sustainable Environment*, Lecture Notes in Civil Engineering 196,
https://doi.org/10.1007/978-981-16-6557-8_74

1 Introduction

Solid waste management has become a challenging issue among all the developing countries [1]. Solid waste does not contain human excreta but it has some precarious material [2, 3]. Furthermore, prevailing research illustrates that around 85% of waste generated in India is disposed of directly in an illegal way that may cause many problems related to health [1, 4, 5]. The sources of MSW have been summarized in Fig. 1 and entire waste management system has been summarized in Fig. 2.

Illegal dumping of MSW and inappropriate management of maximum municipal authorities, all water assets as well as soil get exhausted and contaminated. The existing situation of waste collection, segregation or transportation results in inefficient utilization of resources [6, 7]. However, one of the main reasons for poor waste management is absence of sorting techniques in most Indian cities [8–10]. In India, waste may differ prominently with respect to the composition and characterization of waste as compared with western countries [11–13]. Waste characterization technique is based on sampling through sorting and ultimately weighing of waste respectively [14, 15]. Waste production rate varies from various place to place and hence reliant on the financial condition, industrialized condition, waste management regulations and finally on the public lifestyle. Zero waste system has been summarized in Fig. 3.

Fig. 1 Sources of MSW

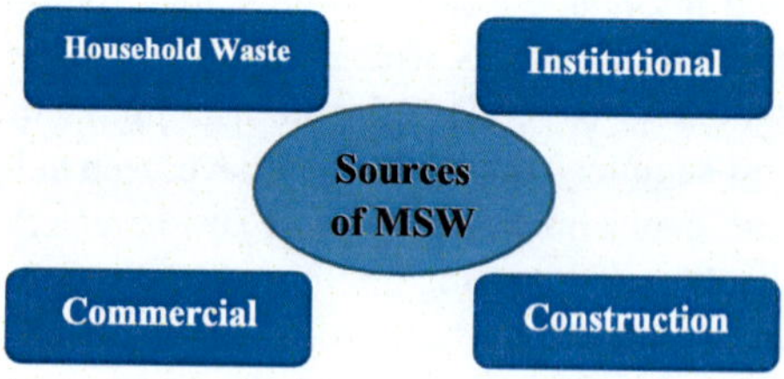

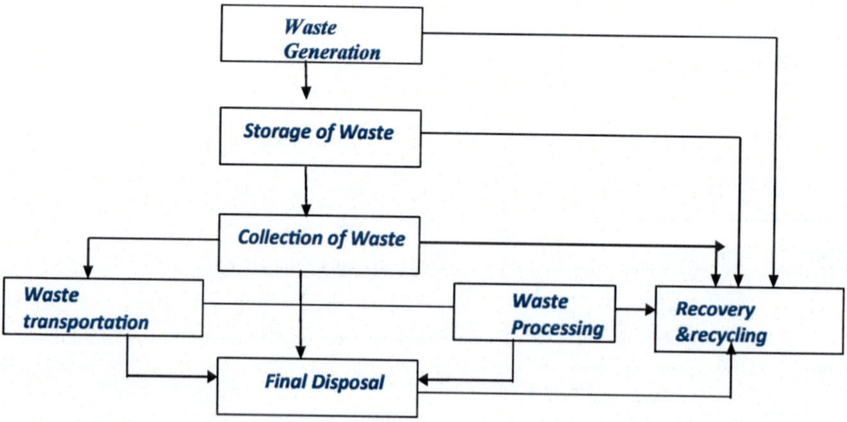

Fig. 2 Entire SWM system

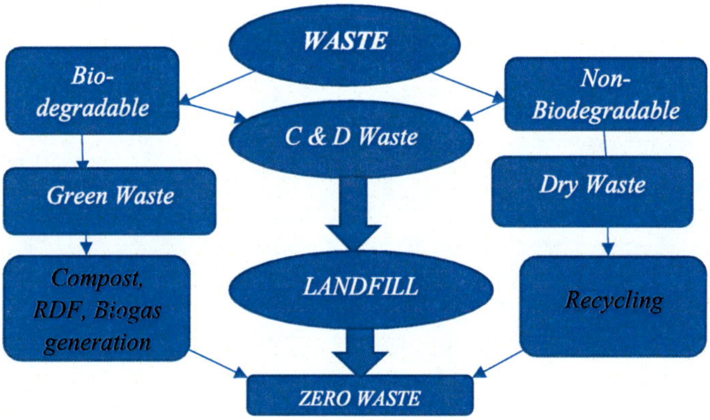

Fig. 3 Zero waste system

Site Profile

Dharamshala lies in the coordinates of 32.2190 °N and 76.3234 °E having a population of 24,586 with an MSW generation of 20 TPD with a collection efficiency of 60–65%. Dharamshala is a city in the upper reaches of the Kangra Valley and the municipality of Dharamshala is one of the oldest Municipalities. The detail of waste disposal site and overview of waste generation and dumping in the study region has been summarized in Table 1 respectively.

Table 1 Overview of solid waste generation and disposal at Dharamshala

S. No.	Description of study region	Parameters
1	Longitude	76° 23′ E
2	Latitude	32° 16′ N
3	Distance of dumping site from city	3 km
4	Segregation done by rag pickers	Yes
5	Provision of leachate collection and treatment facility	No
6	Population	22,586
7	Area of city	10.82 km^2
8	Strength of staff	86
9	Solid waste collected and transported to dumping site	20 PD

Fig. 4 **a, b** View of dumpsite in the study region

2 Material and Methodology

2.1 Physical Composition of Municipal Solid Waste in Dharamshala

In the physical characterization of waste, total 7 number of samples (Tippers, Dumpers) containing MSW were analyzed. It was ensured that none of the samples remain lower than the standard sampling pre-requisite. Tippers/Dumpers from identified areas were taken for the sampling. The net weight of the content was determined by weighing with a weighbridge—near/on the way to site. To determine the density of waste measurement of garbage volume in the RC/Dumper placer was done. Whole waste was dropped on plastic sheets so as to avoid intermixing with the exiting components in the landfill. The pictorial presentation of the dumpsite view has been summarized in Fig. 4a, b respectively.

The truck contents were then physically separated for the components as mentioned earlier in the scope of work above. Segregation of the components was as exclusive as possible and all the individual components separated were reported. The average of all fraction weights comprised the entire dumper contents weight. The pictorial view of physical characterization has been presented in Fig. 5a–c respectively.

2.2 Chemical Analysis of Municipal Solid Waste in Dharamshala

The sample collection was as per the quarter and coning method. For chemical testing, the samples collected consisted of approximately 5 kg of mixed waste in plastic envelopes to ensure minimum loss of moisture, etc. In all, number of samples

Fig. 5 **a–c** Physical characterization in the study region

were collected for the chemical characterization of the waste. In the chemical characterization study, proximate analysis, ultimate analysis and heavy metal analysis have been analyzed.

Proximate analysis comprised of Loss of moisture, Volatile matter, fixed carbon (residue from volatile combustible matter), Ash content (temperature increased to 550 °C), open crucible and Ultimate analysis—CHNS analyzer (Perkin and Elmer), i.e. Carbon, Hydrogen, Nitrogen, Sulphur, Oxygen. Heavy metal analysis has been done by Atomic Absorption Spectrophotometer and Calorific value analysis has been done by means of Bomb calorimeter. [16].

The collection and analysis of soil samples were done according to the standard methods. The sampling of soil was done using hand augur by hold into the desired depth and then taken out. One sample is extracted from the depth of 1.5 m from the dumpsite and the other sample is of natural soil and free from any kind of contamination. The collected soil samples were protected from sunlight to minimize any kind of potential reaction. The soil samples were further dried in sunlight and the pulverization has been done. Finally, the pulverized soil was passed through the specified sieve and hence taken for different analysis.

The geotechnical properties determined of dumpsite soil and natural soil including particle size gradation, consistency limits, maximum dry density and shear strength have been analyzed. Various tests were performed as per Indian standard codes comprised of Compaction characteristics (IS: 2720, Part-7) (1979), Direct shear test (IS: 2720, Part-13) (1986), Particle size distribution (IS: 2720, Part-4) (1985), Atterberg's limit (IS: 2720-4) (1985) and California bearing ratio (IS: 2720, Part-7) (1983).

3 Results and Discussions

The organic waste constituted the highest fraction out of the total municipal solid waste generated in smart city Dharamshala of Himachal Pradesh. The seasonal variation data revealed that the organic waste is found more in summer season and less in winter season. The average value of organic waste was reported as 38%. The composition of MSW in the study region has been summarized in Fig. 6.

The highest proportion of organic waste has been observed followed by paper waste. The proportion of plastic and inert waste in the study region was observed as 11% and 5%. The fraction of rubber content and metal content has been reported as 0.8% and 1% respectively whereas textile and glass exhibited 8% and 0.9% out of the total composition of waste. The fraction of metallic objects, bottles, glass were observed lesser in the dumpsite because these materials were already taken out by unauthorized persons from the municipality dustbins before their disposal in the dumpsite. The result analysis of chemical characterization of MSW has been compiled in Fig. 7 and heavy metal analysis has been summarized in Fig. 8 respectively.

The moisture content of municipal solid waste for the study regions was reported high at 44.50% due to the maximum percentage of organic waste. Calorific value for Dharamshala city waste is found as 2531 k cal/kg due to the higher proportion of plastic, inert and rags. The range of calorific values reported for Himachal Pradesh

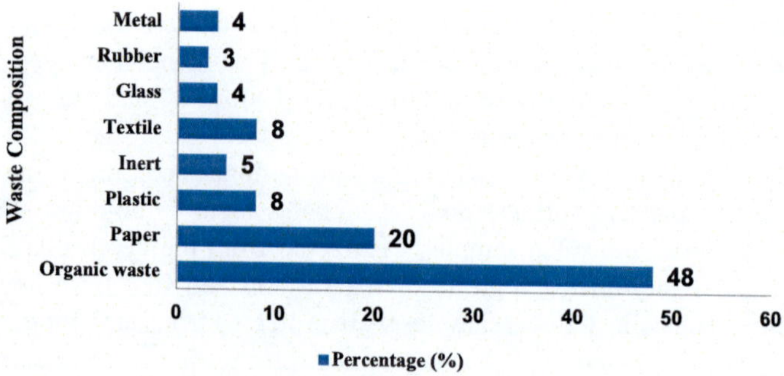

Fig. 6 Physical characterization of MSW

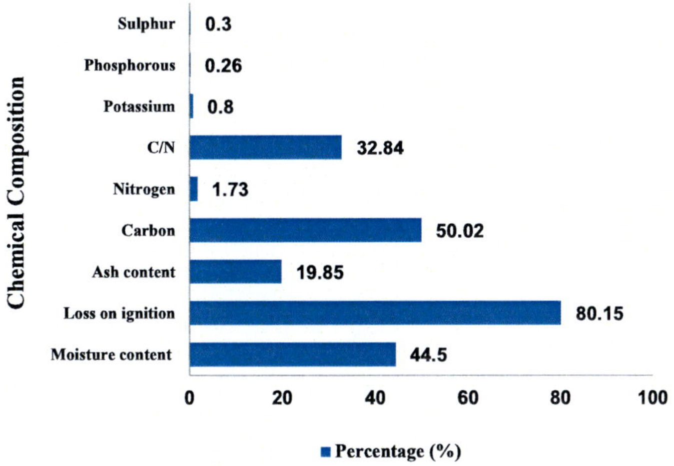

Fig. 7 Chemical characterization of MSW

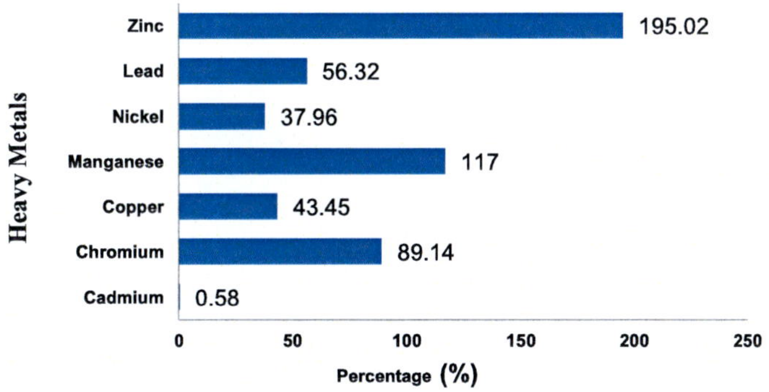

Fig. 8 Heavy metal analysis of MSW

is found suitable for incineration and reduced derived fuel technology, i.e. more than 2000 kcal/kg. The value of carbon content and nitrogen content was observed in the range of 50% and 1.73% in the study region. The *C*/*N* ratio was found in the range of 32.84% in study region of Himachal Pradesh that proves suitable for the composting process.

Concentration of heavy metals has been summarized in Fig. 8.

The knowledge of heavy metals is mandatory to know the metal content in MSW because the presence of heavy metals in solid waste may harm composting of the waste. The present study revealed that the heavy metals have been found within the permissible limits hence suitable for the safe application of compost.

3.1 Valuation of Geotechnical Assessment of Contaminated Soil and Natural Soil in Study Region

The comparison in the geotechnical analysis of soil samples taken at depths 1.5 m from contaminated site and natural soil were investigated. The compaction characteristics, consistency limit, CBR characteristics, shear strength properties have been summarized in Figs. 9, 10, 11, 12 and 13 respectively.

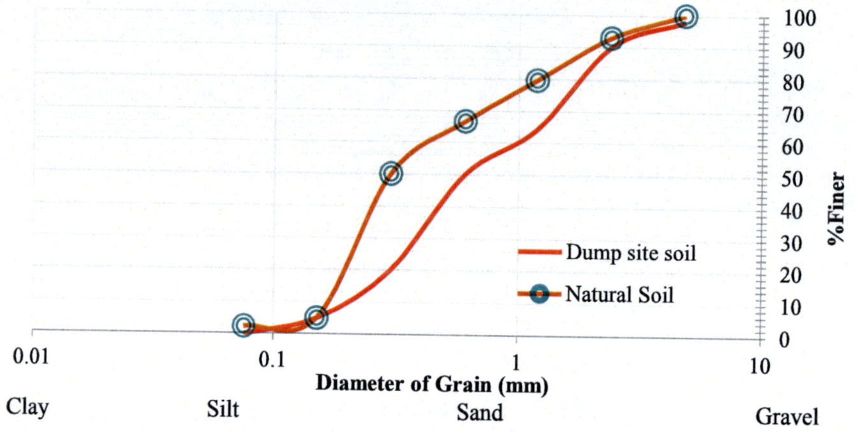

Fig. 9 Particle size distribution of soil

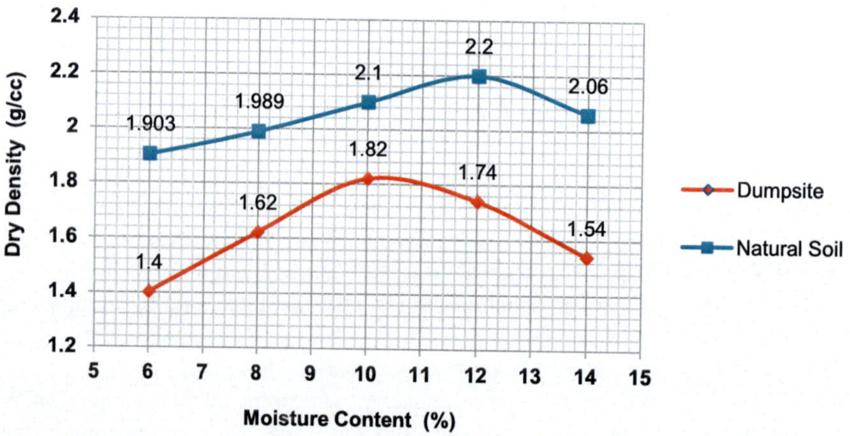

Fig. 10 Variation in dry density with moisture content

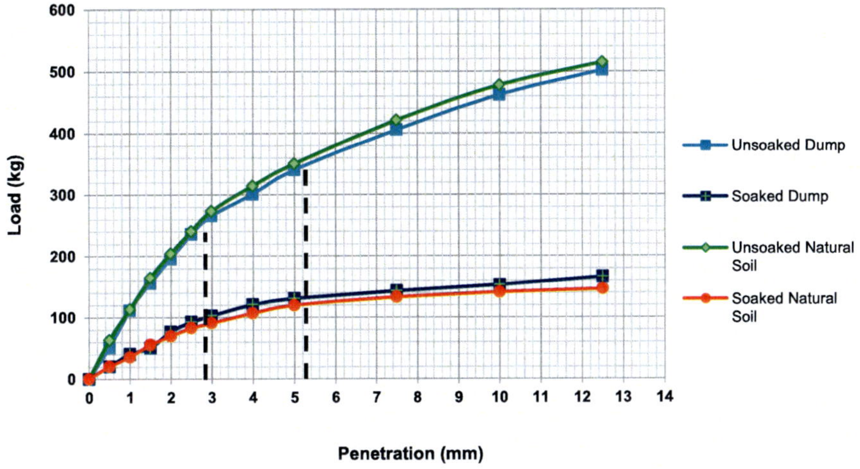

Fig. 11 Consistency behaviour of dumpsite soil versus natural soil

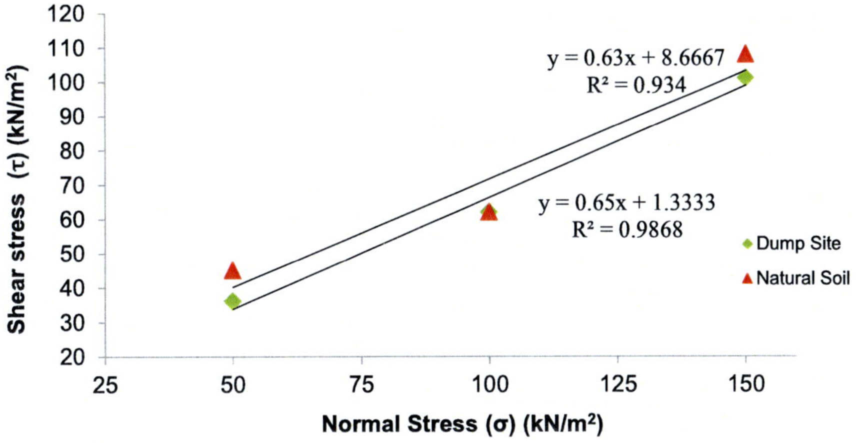

Fig. 12 Load versus penetration curve of dumpsite and natural soil

Particle Size Distribution

The curve obtained from particle size distribution of the soil taken from study regions indicated sandy silty behaviour of soil. The dumpsite soil is comprised of lesser fraction of fine as compared to the uncontaminated soil because dumpsite soil consists of some fraction of mixed waste in the dumpsite. While natural soil is free from any kind of such impurities.

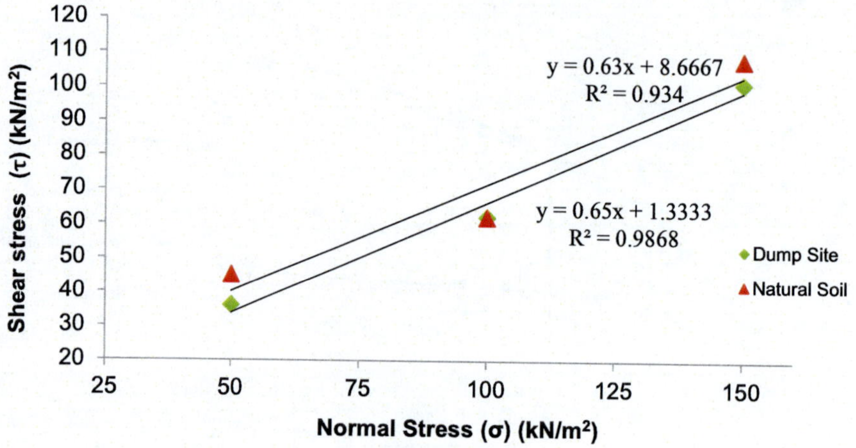

Fig. 13 Shear stress versus normal stress behaviour

Compaction Characteristics

The variation of the dry density and moisture content is shown in Fig. 10. The MDD of natural soil is 2.20 g/cc at 12% OMC and MDD of dumpsite is 1.82 g/cc at 10% OMC and even at 10% moisture content, the value of dry density of natural soil is comparatively high than that of dumpsite. It means dumpsite has low dry density values at different ranges of moisture content. And MDD decreased because of decrease in specific gravity of the dumpsite due to the presence of heavy metals, organic and inorganic waste in dumpsite.

Consistency Limits

Results of consistency test performed with natural soil and dumpsite have been shown in Fig. 11. It showed that the dumpsite has a slightly higher range of consistency limits than the natural soil. The increase in trend of liquid limit and plasticity index is due to an increase in more fines and activity of the material as some chemical substances are present in the dumpsite. Furthermore, the increase in the plasticity index can be attributed to the existence of highly organic material in the dump, which has relatively higher compressibility compared to that of the natural soil.

Load Versus Penetration Behaviour

To know about the load and penetration values of soil, CBR test in accordance with IS:2720 (Part 16)-1987 play a significant role. Figure 12 showed the variations of load v/s penetration values of soil, as well as dumpsite under soaked and unsoaked conditions and almost the same trend, has been observed for natural soil and dumpsite at 2.5 mm and 5 mm penetration. However, according to the load variation in unsoaked natural soil and unsoaked dump-soil after 3 mm penetrations, it showed that the natural soil can bear more load range from 273 to 515 kg at 3 mm penetration to 12.5 mm penetration but for dump-soil load value is slightly less, i.e. 265 kg and

502 kg at 3 mm to 12.5 mm penetration respectively. It clearly indicated that the unsoaked CBR of natural soil has better results.

Shear Stress Versus Normal Stress

Shear stress values at different normal stresses have been shown in Fig. 13. The figure compared the result of dumpsite soil with natural soil and it clearly indicated that the natural soil achieving higher value of shear stress at same normal stress. The shear stress value of dumpsite soil increased from 36 kN/m^2 to 101 kN/m^2 at 50 kN/m^2 to 150 kN/m^2 of normal stress respectively. Whereas the shear stress value of natural soil increased from 45 to 108 kN/m^2 at same normal stresses. As per Fig. 13, shear strength equation stated $\tau = c + \sigma \tan\phi$, hence, for dumpsite soil, the equation of shear strength obtained as $y = 0.65x + 1.333$, so in this case, $c = 1.333$ and the value of ϕ is 33.02 and for natural soil, the equation of shear strength obtained as $y = 0.63x + 8.667$, thereby, $c = 8.667$ and the value of ϕ is 32.21. Keeping the normal stress constant (50 kPa) for both the soils, shear strength of natural soil exhibited 15% of higher shear strength, i.e. 40.17 kN/m^2 as compared to dumpsite soil, i.e. 33.83 kN/m^2. Since natural soil has a comparatively high shear stress value, therefore; it will give higher shear resistance against failure because of better interlocking bond and possibly cementation between the soil particles due to higher friction and cohesion of natural soil.

4 Proposal for Improvement of MSW Management Services

To improve and strengthening municipal solid waste management services in the smart city Dharamshala, Himachal Pradesh will require more resources in terms of manpower, machinery, funding to provide effective and efficient services for managing city.

- Hand Cart/Wheel Barrows (With Break Mechanism) (0.6 m^3)

For the betterment of waste collection techniques, there is a great need for at least 40 handcarts [Considering one Hand Cart for 200 Households (Assuming per capita waste generation as 300 g/day)] The life cycle for Hand Cart is 3 years.

- P.E Bins for Hand Cart

Segregation of waste at source and at collection stage is one of the most important challenges faced by all the cities of India. The life cycle for P.E Bins is 1 year. In order to meet the requirement of P.E Bins, it is proposed to buy 240 No's of P.E Bins. Waste bins and Hand carts have been showed in Fig. 14.

- Bins for segregated waste collection

Fig. 14 Waste bins and handcarts

There is a dire need to provide small dustbins green and blue coloured having 10 L capacity to collect segregated waste at Household level. Two Bins shall be distributed in 14,000 Households. Hence, 28,000 dustbins are required for the city to cater the current needs.

- Litter Picker Machine for collection of waste littering from Footpath/Tourist Places/Religious Places

In order to clean footpaths, parking of the corporation and reduce the hassles of DMC, it is necessary to purchase 2 No's of litter picker machines for collection of waste from footpaths. Some pictorial view of litter picker machine and refuse compactor vehicle has been summarized in Fig. 15.

- Refuse Compactor Vehicle (Transfer Stations: 4 No's)

A facility is a building or a processing site for temporary deposition of waste. MRF will also act as a Transfer Station where the segregated waste from Underground Bins will be deposited.

Fig. 15 Litter picker machine and refuse compactor vehicle

5 Conclusion

The study of solid waste scenario undertaken at smart city Dharamshala Himachal Pradesh indicated that the city has floating population of visitors. Due to floating population, lots of hotels and food points are available which produces biodegradable waste from food and vegetables, etc. The characterization of MSW specified a higher percentage of organic matter and second highest composition was observed of paper waste out of total waste. The calorific value has been quantified as more than 2000 kcal/kg and *C*/*N* ratio has been observed as 32.84%. This leads to the conclusion that the higher amount of biodegradable waste is a good substrate for composting process. The parameters such as carbon, phosphorus, potassium (NPK), *C*/*N* ratio in the waste clearly indicated that the waste is suitable for bioprocessing techniques. The concentration of heavy metals is within the permissible limit till now. Apart from this, the soil samples taken from contaminated site shows lesser dry density and lesser shear strength comparatively to the natural soil. In the nutshell, open dumping should be avoided and the waste should be managed and dumped properly in an engineered landfill system. The current study is focused on the quantification of MSW and effect of illegal dumping on the soil. Though the contamination in terms of air and water have not been covered in the existing study. However, this study can further encompass to physico-chemical assessment of leachate produced from the dumpsite and also the spectroscopic analysis of soil. Apart from this, a study can be undertaken to engineered sanitary landfill design for smart city Dharamshala Himachal Pradesh.

References

1. Sharma, A., Ganguly, R., & Gupta, A. K. (2017). Characterization of municipal solid waste in Sundernagar, Himachal Pradesh, India. *Research Journal of Pharmaceutical Biological and Chemical Sciences, 8*(3), 276e81.
2. Joshi, R., & Ahmed, S. (2016). Status and challenges of municipal solid waste in India: A review. *Congent Environmental Science, 2*, 1139434. https://doi.org/10.1080/23311843.2016.1139434
3. Singh, R. P., Tyagi, V. V., Allen, T., Ibrahim, M. H., & Kothari, R. (2011). An overview for exploring the possibilities of energy generation from municipal solid waste (MSW) in Indian scenario. *Renewable and Sustainable Energy Reviews, 15*(9), 4797–4808.
4. CPCB. (2011). Status report on municipal solid waste management. Retrieved from http://www.indiaenvironmentportal.org.in/content/374639/status-report-on-municipal-solid-wastemanagement/
5. Sharma, A., Ganguly, R., & Gupta, A. K. (2018). Matrix method for evaluation of existing solid waste management system in Himachal Pradesh. *Journal of Material Cycles and Waste Management, 20*, 1813–1831. https://doi.org/10.1007/s10163-018-0703-z
6. Kumar, S., Bhattacharyya, J. K., Vaidya, A. N., Chakrabarti, T., Devotta, S., & Akolkar, A. B. (2009). Assessment of the status of municipal solid waste management in metro cities, state capitals, class I cities, and class II towns in India: An insight. *Waste Management, 29*, 883–895. https://doi.org/10.1016/j.wasman.2008.04.011

7. Kumar, P., & Kaushal, R. (2015). Avenues of collection and disposal of municipal solid waste management in India—A review. *International Journal of Engineering Science Invention Research & Development, 6*, 458–465.
8. Kiran, S. C., Nagarajaiah, C., Mahadeva, M. M., & Ranjith, P. C. (2020). Effect of municipal solid waste open dumping on soil, water, crop, human health and its prospective. *International Journal of Environment and Climate Change, 10*(8), 36–45; Article no. IJECC.57789.
9. Shekdar, A. V. (2009). Sustainable solid waste management: An integrated approach for Asian countries. *Waste Management, 29*, 1438–1448. https://doi.org/10.1016/j.wasman.2998.08.025
10. Srivastva, R., Krishna, V., & Sonkar, I. (2014). Characterisation and management of municipal solid waste: a case study of Varanasi city, India. *International Journal of Current Research and Academic Review, 2*, 10–16. http://doi.org/10.1007/s11157-014-9352-4
11. Minghua, Z., Xiumin, F., Rovetta, A., Qichang, H., Vicentini, F., Bingkai, L., Giusti, A., & Yi, L. (2009). Municipal solid waste management in Pudong New Area, China. *Journal of Waste Management, 29*, 1227–1233. https://doi.org/10.1016/j.wasman.2008.07.016
12. Patel, N. K., & Baredar, P. (2016). Opportunity and challenges of municipal waste disposal in India. *International Journal for Innovative Research in Science & Technology, 3*(02). ISSN (online): 2349-6010.
13. Sharholy, M., Ahmad, K., Mahmood, G., & Trivedi, R. C. (2008). Municipal solid waste management in Indian cities—A review. *Waste Management, 28*, 459–467. https://doi.org/10.1016/j.wasman.2007.02.008
14. Sethi, S., Kothiyal, N. C., Nema, A. K., & Kaushik, M. K. (2013). Characterization of municipal solid waste in Jalandhar city, Punjab, India. *Journal of Hazardous Toxic and Radioactive Waste, 17*, 97–106. http://doi.org/10.1061/(ASCE)HZ.2153-5515.0000156
15. Subramani, T., Karthikeyan, C., & Priyanka, S. (2017). Impact of solid waste in open dumping and its effects of groundwater and soil. *International Journal of Application or Innovation in Engineering & Management, 6*(5). ISSN 2319-4847.
16. Sharma, A., Ganguly, R., & Gupta, A. K. (2018). Impact of open dumping of municipal solid waste on soil properties in mountainous region. *Journal of Rock Mechanics and Geotechnical Engineering, 10*, 725–739. https://doi.org/10.1016/J.jrmge2011.12.009

Strength Characteristics of Clayey Soil Stabilized with Brick Kiln Dust and Sisal Fiber

Mandeep Singh, Kanwarpreet Singh, and Abhishek Sharma

Abstract Constructing structures over soils possessing poor geotechnical characteristics is a challenging task for civil engineers. The present study is focused on improving the geotechnical properties of clayey soil using brick kiln dust and sisal fiber so as to use it as a sub-grade material. Various percentages of both the materials, i.e., brick kiln dust varying from 10 to 30% and sisal fiber varying from 0.5 to 1% have been used to perform Atterberg's limits, compaction, and California bearing ratio tests. It was revealed that the addition of brick kiln dust to clayey soil reduced liquid limit. The addition of 20% brick kiln dust along with 1% sisal fiber gave maximum dry density value. The California bearing ratio value was improved by adding optimum amount of both the materials to clayey soil. This ensures that the combination of brick kiln dust (20%) and sisal fiber (1%) to stabilize poor soils is cost-effective and eco-friendly.

Keywords Brick kiln dust · Sisal fiber · Consistency limits · Compaction · California bearing ratio

1 Introduction

Geotechnical study of the soil is to be done before the construction begins, so one can become more familiar with the properties of the soil to be improved to lessen the hazard factors which may impact the stability of the structure for a long time. Clayey soils often under large volumetric changes due to their susceptibility to react with moisture and hence these soils are taken care before constructing structures over them. Soil stabilization is a technique to improve the physical properties of poor

M. Singh · K. Singh · A. Sharma (✉)
Department of Civil Engineering, Chandigarh University, Mohali, India
e-mail: abhishek.e9490@cumail.in

K. Singh
e-mail: kanwarpreet.e9570@cumail.in

A. K. Gupta et al. (eds.), *Advances in Construction Materials and Sustainable Environment*, Lecture Notes in Civil Engineering 196,
https://doi.org/10.1007/978-981-16-6557-8_75

soils and is commonly used worldwide [1–11]. Nowadays, due to pace in construction sector in developing countries like India, a lot of construction demolition waste is being produced leading to a problem of its safe disposal [1, 2, 4]. Brick, being an important material of the construction sector, is produced in kilns at a specific temperature. During the heating of bricks, a large amount of dust is obtained mainly known as brick kiln dust. India is the second major worldwide brick producer after China. Brick kiln dust is a combination of wood ash, coal ash, and brickbats generated during the heating of bricks in the kilns and are being dumped in open areas in India. The dumps occupy large areas of valuable land, possess a potential of contaminating the groundwater due to the leaching of heavy metal causing rigorous nuisance together with several ecological and environmental implications. It was found that the maximum increase in dry density and the optimum moisture content decrease with an increasing percentage of burned brick dust and the linear shrinkage decreases with an increase in the exchange of brick dust. Further, it was found that at 50% soil replacement with burned brick dust, soil engineering properties could be improved [12]. The optimum shear characteristics of black cotton soil using 30% of BKD. It has been found that at 30% of BKD, the soil OMC decreases and swelling index of the black cotton soil also decreases when brick kiln dust is added to black cotton soil. For 30% of BKD, there is an improvement in resistance to unconfined compression strength (UCS). Addition of BKD reduces the mechanical strength and increases the water absorption capacity of clayey soil and also there is an increase in resistance to efflorescence and porosity. The optimum percentage for BKD and PA was 30%. The Modulus Resilient (MR) increased for all soaked samples but decreased as compared to MR of the un-soaked sample. Behavior of clay mixed with BKD and PA was also analyzed [13].

Utilization of random inclusion of sisal fiber (0.25, 0.50, 0.75, 1.0) % its length (10, 15, 20, 25) mm is randomly blended into the soil to test the strength characteristics of the soil. The value of UCS increased as the percentage of sisal fiber and RHA increased. The maximum value of UCS was obtained on the optimum percentage of RHA (6%) and sisal fiber (0.75%) [14]. The use of sisal fiber (0.25, 0.50, 0.75, 1.00, and 1.25) % of length (0.5, 1.0, and 1.5) cm was randomly mixed with two types of soil (Soil B and Soil R). As a result, UCS, CBR, and factor of safety gave the maximum value with an increase of sisal fiber with both types of soil, but after certain percentage (0.75%) and length (1 cm) of sisal fiber, the values got decreased. The positive effect of sisal fiber with both types of soils increased the cohesion parameter of soil as well as the stability of the slope [15]. Stabilization of soil with lime and sisal fibers using a different percentage of lime 2%, 5%, 9%, and 12% with different lengths of sisal fiber with a length 3–3.4 cm with an interval of 0.2 cm. It was found that the liquid limit, plastic limit, and specific gravity of lime decrease after replacing the soil with 9% lime and 0.5% sisal fiber. Unconfined compressive strength (UCS) and California bearing coefficient (CBR) values increase with an increase in the percentage of lime (9%) and sisal fiber (0.5%) [16]. Sisal fiber also helps to develop eco-friendly building materials for low-cost housing [17].

Soil Brick kiln dust Sisal Fiber

Fig. 1 Materials used in the study

2 Materials

2.1 Soil

Soil used (Fig. 1) during research was collected from Morinda, Punjab, India. As per Unified Soil Classification System (USCS), the resulted soil classified as clay consisting higher plasticity (CH).

2.2 Brick Kiln Dust

Brick kiln dust (BKD) used (Fig. 1) in research was brought from the brick kiln, which is situated in Anandpur Sahib, Punjab, India. The particles sizes of BKD range between 4.75 and 0.075 mm and percentage of brick kiln dust used in the study was 10%, 20%, and 30%. The BKD was tightly packed in a plastic bag so that moisture did not affect it.

2.3 Sisal Fiber

Sisal is one group of fibers extracted from the leaves of plants belonging to the agave family. The sisal fiber scraped from the fresh leaves cut from the plant is stiffer and stronger, with low elasticity. Sisal Fiber used (Fig. 1) in this research for stabilizing the clayey soil was brought from Chennai, India. This material was closely packed in plastic polythene bag so that it was not affected by moisture and transported for laboratory testing. The percentage used 0.5%, 1.0% and 1.5% and length was 3 cm for all the percentages. The properties of all the materials used in summarized in Table 1.

Table 1 Geotechnical properties of materials

Properties	Clay	Brick kiln dust	Sisal fiber
Specific gravity, G	2.35	2.12	0.91
Max. dry density (MDD) g/cc	1.72	1.79	–
Optimum moist. content (OMC) %	18.8	10.2	–
Liquid limit (LL) %	63.2	–	–
Plasticity index (PI) %	26.2	–	–
Coefficient of uniformity (C_u)	–	3.20	–
Coefficient of curvature (C_c)	–	0.87	–
Classification	CH	SP	–

3 Methodology and Sample Preparation

The testing was done according to the Indian standard codal provision. The percentages of materials used in the study were decided based on the literature review and were mixed together to find geotechnical characteristics of clayey soils. The sample was mixed together and care was taken to ensure uniform distribution as the sisal fiber tries to accumulate at one place when mixed. The precautions are to be taken while mixing sisal fiber into clay or clay: BKD sample. The various percentages of materials used in the current research are tabulated in Table 2.

Table 2 Percentages of materials used in the study

Materials	Proportions used in the study
Soil	100
Soil:BKD	90:10
Soil:BKD	80:20
Soil:BKD	70:30
Soil:SF	99.5:0.5
Soil:SF	99:1
Soil:SF	98.5:1.5
Soil:BKD:SF	79.5:20:0.5
Soil:BKD:SF	79:20:1
Soil:BKD:SF	78.5:20:1.5

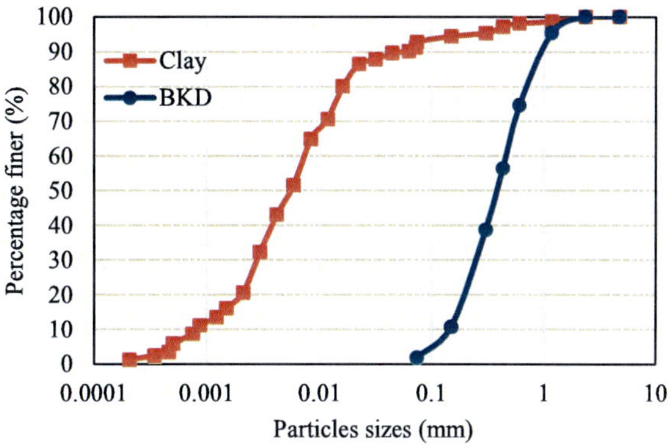

Fig. 2 Gradation curve for materials

4 Result and Discussions

4.1 Gradation Curve

By performing hydrometer analysis and wet sieve analysis, the gradation curve for soil was obtained (Fig. 2). It was found that the obtained gradation curve satisfy ASTM norms. By carrying out dry sieve analysis, the gradation curve for BKD was acquired. From the gradation curve of virgin soil, it can be seen that about 92% of particles are finer than 0.075 mm and about 18% of particles are finer than 0.002 mm. The plasticity index of the soil was 26.2% showing that the soil has a high plasticity level (Table 1). From the gradation curve of BKD, the obtained value for the coefficient of uniformity (C_u) is 3.20 and the coefficient of curvature (C_c) is 0.87, which assigns BKD as poorly graded in nature with particles almost within the range of fine sand.

4.2 Atterberg's Limit Tests

Brick kiln dust was added in various percentages of 10%, 20%, and 30% to clayey soil and experimental work was performed to find the Atterberg's limit (Fig. 3). The Liquid limit (LL) decreased as the percentage of BKD increased, i.e., 10%, 20%, and 30%. The liquid limit of virgin soil was found to be 63.2% and with addition of 20% BKD in virgin soil, it results in decreased LL to 33.3%. By adding the BKD in soil up to 30%, the plastic limit (PL) of soil decreased gradually. At 20% BKD, the plasticity index (PI) for clayey soil obtained is 7.7%, which lies in the range of 6–8% and is acceptable for sub-grade soils. The reduction in the LL was found when fine brick kiln dust was added to soil and was primarily because of non-plastic nature of brick kiln dust.

Fig. 3 Atterberg's limit tests

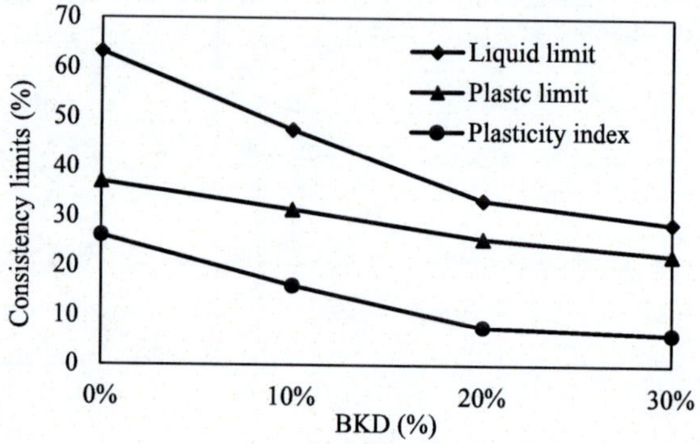

4.3 Compaction Test

Compaction tests were performed on virgin soil, BKD, and various soil mixtures-BKD, virgin soil-SF, virgin soil-BKD-SF to determine the MDD and OMC and are presented in Figs. 4, 5, 6 respectively. The MDD and OMC of virgin soil was 1.72 g/cc 18.8%; MDD of the BKD was 1.79 g/cc and related OMC was 10.2%.

Fig. 4 Compaction properties for S:BKD

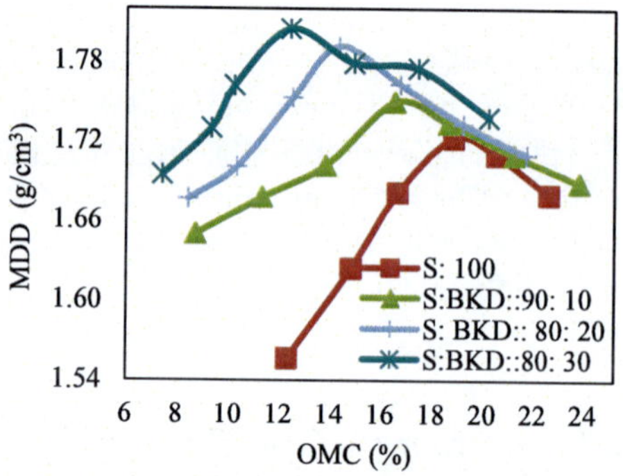

Fig. 5 Compaction properties for S:SF

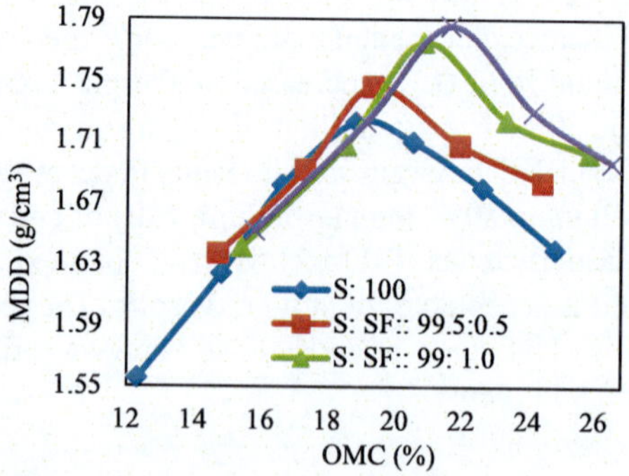

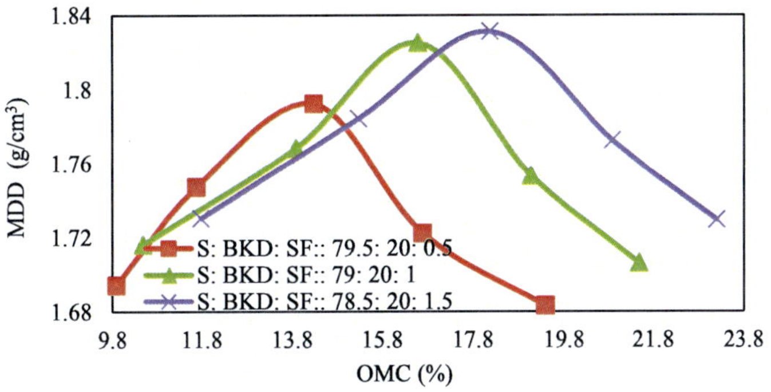

Fig. 6 Compaction properties for S:BKD:SF

As the percentage of BKD in virgin soil increased, the value of OMC decreased and MDD increased (Fig. 4). The MDD increased from 1.72 to 1.805 g/cc with an improvement in BKD content up to 30% and the OMC decreased from 18.8 to 12.4%. It was found that as the percentage of SF in virgin soil rises, the value of both OMC and MDD was increased (Fig. 5). The MDD improved from 1.72 to 1.786 g/cc with an increase in SF content of up to 1.5%, and the OMC increased from 18.8 to 21.6%. When optimum percentage brick kiln dust (20%) was mixed in soil with varying percentages of sisal fiber 0.5, 1, 1.5%, MDD of soil increased from 1.72 to 1.831 g/cc and related OMC decreased from 18.8 to 18.2% (Fig. 6). There was a minor improvement in MDD beyond 20% BKD and 1% SF in virgin soil and that is why 20% BKD with 1% SF may be taken as optimum content. The MDD of BKD (20%): SF (1%) at optimum ratio and relating OMC is revealed to be 1.825 g/cc and 16.6%.

4.4 *California Bearing Ratio Tests*

On virgin soil and mixed soil with different percentages of BKD, sisal fiber, and BKD + SF, soaked California bearing ratio tests were performed. The CBR values for various compositions are presented in Fig. 7. The obtained value for soaked CBR in virgin soil was 1.82; the value of CBR for soil stabilized with 10% BKD was 4.12; for 20% BKD was 7.4; 6.97 for 30% BKD was 6.97; the value of CBR was 3.28 for soil stabilized with 0.5% sisal fiber; 3.8 for 1% sisal fiber; 3.43 for 1.5% sisal fiber; the value of CBR for soil stabilized at optimal ratio 20% BKD + 1% Sisal fiber was 9.5. From the above findings, it is revealed that the CBR values increased when adding BKD and sisal fiber alone to virgin soil. The CBR value increased because of fine sand particles in BKD, which helped to mobilize the shear resistance angle, resulting in improved composite strength. The maximum CBR value obtained in soil mixed at the optimum ratio of BKD (20%) + sisal fiber (1%).

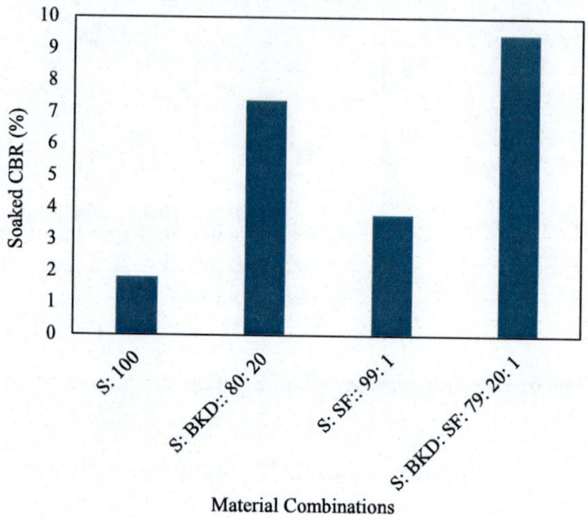

Fig. 7 CBR values for different materials

5 Conclusion

The following points have been summarized from this research work after conducting different tests on virgin soil and after adding different percentages of brick kiln dust and sisal fiber.

1. In addition of 20% of brick kiln dust in soil liquid limit decreases from 63.2 to 33.3%, plastic limit of soil is continuously decreasing and the plasticity index of soil is found to be 7.7.
2. Maximum dry density increases up to the optimal ratio of BKD. As BKD increases, the optimum moisture content decreases. This reduction is largely due to the presence of fine sand in the dust of the brick kiln.
3. It is found that addition of BKD and sisal fiber results in increased soaked California bearing ratio of soil. At 20% BKD and 1% sisal fiber maximum soaked CBR value was obtained and it would reduce the thickness of pavement and also reduced the cost of sub-grade. The above results show that BKD and sisal fiber can be used as a stabilizer to enhance weak soil strength and engineering properties. BKD and sisal fiber is economical as well as it will also contribute to a healthier climate.

From the results obtained, it can be concluded that the use of 20% BKD and 1% sisal fiber, which is the optimum ratio mixed in weak soil. It is a better way to enhance the geotechnical characteristics and it is used constructively on low-density highways, shallow foundations, and sub-base layers of pavement.

References

1. Sharma, R. K., & Bhardwaj, A. (2018). Effect of construction demolition and glass waste on stabilization of clayey soil. In *International Conference on Sustainable Waste Management through Design* (pp. 87–94). Springer. http://doi.org/10.1007/978-3-030-02707-0_12
2. Sharma, A., & Sharma, R. K. (2019). An experimental study on uplift behaviour of granular anchor pile in stabilized expansive soil. *International Journal of Geotechnical Engineering*, 1–14. http://doi.org/10.1080/1938636.2.2019.1597481
3. Verma, V., & Abhishek. (2018). Stabilization of clayey soils using fly ash and RBI grade 81. In *International Conference on Sustainable Waste Management through Design* (pp. 95–102). Springer. http://doi.org/10.1007/978-3-030-02707-0_13
4. Sharma, A., & Sharma, R. K. (2020). Effect of addition of construction–demolition waste on strength characteristics of high plastic clays. *Innovative Infrastructure Solutions*, *4*(1), 1–11. http://doi.org/10.1007/s41062-019-0216-1
5. Sharma, A., & Sharma, R. K. (2020). Strength and drainage characteristics of poor soils stabilized with construction demolition waste. *Geotechnical and Geological Engineering*, *38*, 4753–4760. https://doi.org/10.1007/s10706-020-01324-3
6. Bhardwaj, A., & Sharma, R. K. (2020). Effect of industrial wastes and lime on strength characteristics of clayey soil. *Journal of Engineering, Design and Technology*, *18*(6), 1749–1772
7. Anand, D., Sharma, R. K., & Sharma, A. (2021). Improving swelling and strength behavior of black cotton soil using lime and quarry dust. In H. Singh, P. P. Singh Cheema, & P. Garg (Eds.), *Sustainable development through engineering innovations. Lecture Notes in Civil Engineering* (Vol. 113). Springer. http://doi.org/10.1007/978-981-15-9554-7_54
8. Singh, S., Sharma, R. K., & Sharma, A. (2021). Arresting the heave of black cotton soil using geogrid-encased granular pile anchor. In H. Singh, P. P. Singh Cheema, & P. Garg (Eds.), *Sustainable development through engineering innovations. Lecture Notes in Civil Engineering* (Vol. 113). Springer. http://doi.org/10.1007/978-981-15-9554-7_30
9. Gautam, K. K., Sharma, R. K., & Sharma, A. (2021). Effect of municipal solid waste incinerator ash and lime on strength characteristics of black cotton soil. In H. Singh, P. P. Singh Cheema, & P. Garg (Eds.), *Sustainable development through engineering innovations. Lecture Notes in Civil Engineering* (Vol. 113). Springer. http://doi.org/10.1007/978-981-15-9554-7_10
10. Bhardwaj, A., Sharma, R. K., & Sharma, A. (2021). Stabilization of clayey soil using waste foundry sand and molasses. In H. Singh, P.P. Singh Cheema, & P. Garg (Eds.), *Sustainable development through engineering innovations. Lecture Notes in Civil Engineering* (Vol. 113). Springer. http://doi.org/10.1007/978-981-15-9554-7_57
11. Sharma, A., & Sharma, R. K. (2021). Sub-grade characteristics of soil stabilized with agricultural waste, constructional waste, and lime. *Bulletin of Engineering Geology and the Environment*, *80*(3), 2473–2484. https://doi.org/10.1007/s10064-020-02047-8
12. Bhavsar, S. N., Joshi, H. B., & Patel, A. J. (2014). Effect of burnt brick dust on engineering properties on expansive soil. *International Journal of Research in Engineering and Technology*, *3*(4), 433–441
13. Gupta, G., Sood, H., & Gupta, P. (2020). Performance evaluation of pavement geomaterials stabilized with pond ash and brick kiln dust using advanced cyclic triaxial testing. *Materials*, *13*(3), 553. https://doi.org/10.3390/ma13030553
14. Sani, J. E., Yohanna, P., & Chukwujama, I. A. (2020). Effect of rice husk ash admixed with treated sisal fibre on properties of lateritic soil as a road construction material. *Journal of King Saud University-Engineering Sciences, 32*(1), 11–18. https://doi.org/10.1016/j.jksues.2018.11.001
15. Ramkrishnan, R., Sruthy, M. R., Sharma, A., & Karthik, V. (2018). Effect of random inclusion of sisal fibres on strength behavior and slope stability of fine grained soils. *Materials Today: Proceedings, 5*(11), 25313–25322. https://doi.org/10.1016/j.matpr.2018.10.334
16. Patil, S. C. Comparative study on black cotton soil stabilization by using lime and sisal fiber. *5*, 109–115.

17. Ojo, E. B., Bello, K. O., Teixeira, R. S., Onwualu, P. A., & Savastano, H., Jr. (2020). Statistical data on the physical and mechanical properties of fibre reinforced alkali activated uncalcined earth based composite. *Data in Brief, 28*, 104839. https://doi.org/10.1016/j.dib.2019.104839

Numerical Investigation of Wind Load on Side Ratio of High-Rise Buildings

Rahul Kumar Meena, Ritu Raj, and S. Anbukumar

Abstract Due to the shortage of land, unconventional building shapes are very commonly built, and L shape building is very common. This paper presents the effect of wind on the side ratio of L-shaped tall buildings using Computational Fluid Dynamics (CFD), ANSYS CFX version 2020 R2 on a 1:200 length scale. The results show that pressure distribution is different on the windward wall of L-shaped building models. Using ANSYS CFX Post simulation, for better understanding of the wake, near field, flow patterns around the bluff aerodynamics shape are discussed, and aerodynamic mean pressure coefficients are compared with an International standard which shows a good agreement with the results obtained. It has been discovered that the Model's plan layout and size have a direct impact on the wind pressure distribution on the wind ward face of all the models taken into consideration. Face E has a significant effect on pressure distribution on all three models. The windward face and leeward face of all the models have similar nature of pressure distribution.

Keywords Wind load · CFD · Pressure contour · Velocity stream lines · Tall building

1 Introduction

As the inevitable result of growing population and intensifying urbanization, high-rise residential towers have become more prevalent in many cities, replacing the vast areas of vernacular houses. In addition to gravity loads, these structures are susceptible to time-varying loads caused by winds, earthquakes, and other natural

R. K. Meena (✉) · R. Raj · S. Anbukumar
Delhi Technological University, Delhi, India
e-mail: rahulkumar_phd2k18@dtu.ac.in

R. Raj
e-mail: rituraj@dtu.ac.in

S. Anbukumar
e-mail: sanbukumar@dce.ac.in

A. K. Gupta et al. (eds.), *Advances in Construction Materials and Sustainable Environment*, Lecture Notes in Civil Engineering 196,
https://doi.org/10.1007/978-981-16-6557-8_76

disasters. Wind excitation is very sensitive to tall buildings [1, 2]. Chauhan and Ahuja [3] discussed the height effects of interfering building on wind pressure distribution on rectangular plan tall buildings due to the presence of two other closely spaced tall buildings having a similar plan. It is observed that the arrangement highly influences wind pressure distribution on the faces of the principal building and as well the height of the building. The total wind force is the sum of windward pressure and leeward suction, but each of these has its local impact in general, a building can not be considered as regular shape when planning for wind loading, to achieve this, the pressure coefficients and force coefficients for other buildings subjected to wind-induced loads, designer consult replated wind load standards [4–9]. Ahlawat and Ahuja [10] wind tunnel tests were carried out on models of "T" plan shape tall building to evaluate the wind loads and observed that neighboring building greatly affects the wind flow patterns around the building. Ahlawat and Ahuja [11] carried out an experimental study on "Y" plan shape tall buildings using wind tunnel test, observed that the wind incidence angle affects wind pressure distribution. Paul and Dalui [12] discussed the wind effects on cross-plan-shaped tall buildings. Raj and Ahuja [13] conducted an experimental study on cross-shape tall buildings with various cross-sectional shapes, observed that base shear, base moments, and twisting moments are influenced by wind incidence angle and affected by cross-sectional shape. Vafaeihosseini et al. [14] used computational fluid dynamics to find wind effects on high-rise buildings gives that the most efficient shape for tall buildings is rectangular plan shape. Meena et al.[15] discussed the effects of wind on the bracing system used in various types of multi-storeyed steel buildings. Pal et al. [16] carried out a comparative study on square and fish plan shaped tall building of wind-induced mutual interference effect, maximum efficiency in terms of wind-induced mutual interference pressure, and base shear ix exhibited by square plan shaped model at full blockage condition. Verma et al. [17] analyzed the effects of wind load on tall building octagonal shape, using CFD simulation for 0°, 15° and 30° wind incidence angle, found that CFD can be used to predict wind-related phenomena on building and other types of structure. Nagar et al. [18] carried out an experimental study of wind-induced pressure on tall buildings of different shapes from 0° to 90° wind incidence angle at an interval of 30°, found that wind pressure decreases up to 60° wind incidence angle, "H" plan shaped building is subjected to a higher pressure than the square Model. Chandan and Kumar [19] discussed the numerical simulation of wind analysis of tall buildings using computational fluid dynamics. CFD can result in nearly wind tunnel experiments. CFD could analyze the complete domain study, better-visualized results, and cheaper than that of wind tunnel test. Bairagi and Dalui [20] discussed the aerodynamic effect on the setback of tall building using CFD simulation for 0°–180° of wind incidence angle, high positive pressure developed in the setback of roof compared to the top roof due to high turbulence, maximum spectral density frequency developed at the extreme location of setback roof, where the turbulence is maximized. Mendis et al. [21] conducted wind tunnel testing on tall buildings discussed various parameters related to the design of a tall building. Roy et al. [22] did a critical review of wind load on high rise building with different configurations, wind pressure coefficient is maximum in case of square plan shaped building and minimum in case of circular

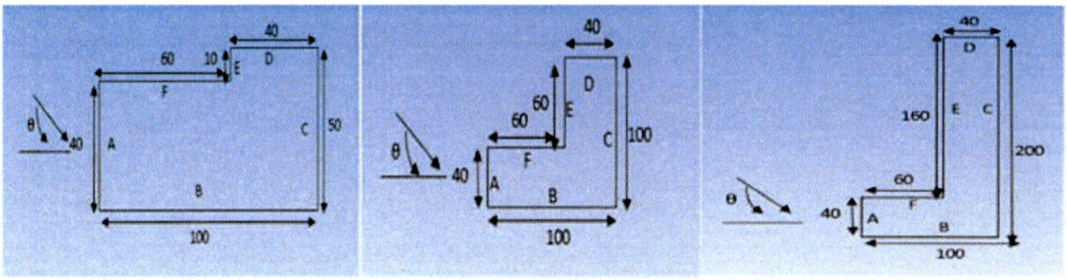

(a) (b) (c)

Fig. 1 Models with dimension

plan shaped tall building, octagonal plan shaped building is with the sharp windward edge is more effective in reducing wind pressure coefficients than hexagonal plan shape of tall building with sharp wind ward edge. The shape of a building has a direct impact on the wind forces acting on it and the motion that results. Wind excitation is minimized, and costs are saved by carefully designing the building components and shape of tall buildings. By changing/altering the flow pattern around the structures, passive aerodynamics modifications in the form of building shape are one of the most powerful and versatile construction techniques for greatly reducing the effects of wind forces.

2 Numerical Analysis by ANSYS CFX

2.1 Model

To study the effects of wind, an L-shaped building model with different side ratios as shown in Fig. 1a–c with a geometrical scale of 1:200 is considered.

2.2 Boundary Condition

To define a problem that results in a unique solution, you must specify the domain boundary. Poorly defined boundary conditions are having a significant impact on your solution. Defining boundary condition, meaning identifying the solver, the inlet, outlet wall, etc., Zidan et al.[23] suggested the domain, boundary conditions are used to carry out the numerical simulation.

The domain considered in this study are having the inlet at 4H, outlet at 15H, side wall are at 5H and top of the domain is having 8H is shown in Fig. 2 is for virtual wind tunnel, and it is a fluid domain. The properties of the fluid are pre-defined in the ANSYS CFX, building faces are provided as a wall, and wall function is also pre-defined in the ANSYS CFX (Fig. 2).

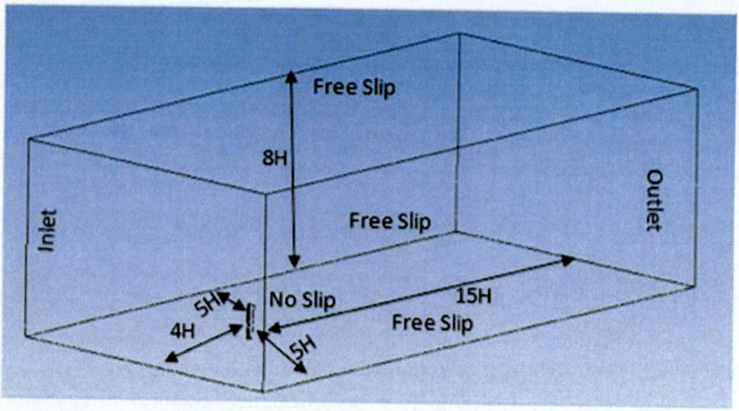

Fig. 2 Computational domain dimensions and boundary condition

2.3 *Meshing*

A good mesh might not result in an ideal solution, but a bed mesh will always lead to a bad solution, meshing as shown in Fig. 3, is provided, inflation Fig. 3b, is provided to take the effects of air in the separation zone. Appropriate mesh is difficult to provide as meshing is dependent on problem complexity. The computer cannot solve the simulation on the actual geometry shape as governing equation cannot apply to an arbitrary shape. Mesh elements allow governing equations to be solved predictably and mathematically in defined volumes. Figure 3a shows full domain meshing, (c) building meshing, and it's finer than another meshing, and (d) edge meshing, provided to avoid the unusual flow.

3 Result and Discussion

3.1 *Velocity Profile and Turbulent Profile*

Wind blow due to the pressure difference over a different part of the earth and Coriolis force. Due to the ground roughness atmospheric boundary layer is formed near the ground. And the wind becomes highly turbulent due to ground obstacles. The degree of ground roughness and drag due to local projections that resist wind flow determines the vertical profile of wind speed. Gradient height is the height at which the drag effects disappear, and gradient velocity is the corresponding velocity. The atmospheric boundary layer is the height at which topography influences wind speed. As per Power Law, the wind speed profile, as shown in Fig. 4, within the atmospheric boundary layer is given by Eq. (1)

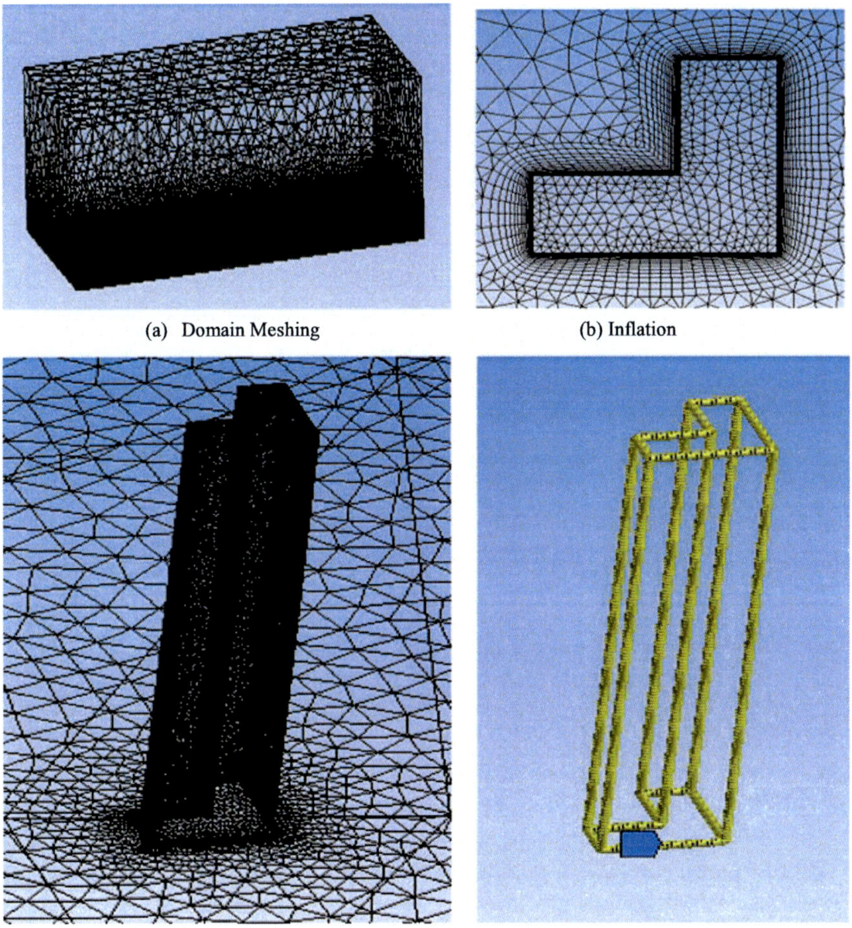

(a) Domain Meshing

(b) Inflation

(c) Building Meshing

(d) Edge meshing

Fig. 3 Meshing

$$\frac{U}{U_H} = \left(\frac{Z}{Z_H}\right)^{\alpha} \tag{1}$$

where U is the horizontal wind speed at an elevation Z; U_H is the speed at the reference elevation Z_H, which was 10 m/s, α is the parameter that varies with ground roughness which is 0.147 for terrain category 2, and Z_H is 1.0 m.

Figure 4 shows that wind speed near the surface is very low, it's because of the friction between the surface and moving air. The wind profile obtained in this study is similar to the wind profile provided in ASCE 7 and previous research done in this area.

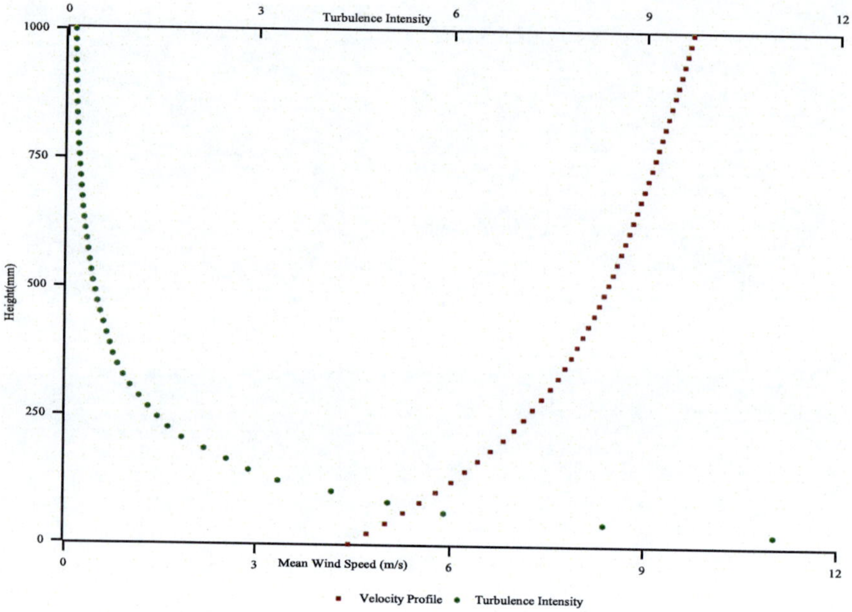

Fig. 4 Variation of velocity and turbulence intensity with height

3.2 *Pressure Contour*

The pressure on the windward face is positive for Model A, B, and C. As shown in Figs. 5, 6, and 7, it is negative for a leeward face. The graphical representation of pressure on every face of the Model A is shown in Fig. 5, where face A has maximum pressure in the center, and then it is decreasing in the corner region. Rest all faces of model A are under the influence of the negative pressure, as shown in Fig. 5, face B pressure distribution shows that it is lesser in magnitude as we move along wind direction it is decreasing because of the suction created, face D and face F has an almost similar pattern of pressure distribution, it is increasing along wind direction. Face E shows that pressure distribution depends upon the size of the component, and it is negative for model A and positive for model B and model C shown in Figs. 6 and 7, respectively. The same pressure distribution pattern follows for model B and model C. only the difference is shown for face E.

The pictorial representation of pressure contour on each face of the model B is shown in Fig. 6, where face A, face E, and face F having positive pressure. Rest all faces of model B are under the influence of the negative pressure, as shown in Fig. 6, face E shows positive pressure distribution. The pressure distribution depends upon the size of the side ratio of the building.

The pressure contour on each face of Model C is shown in Fig. 7, where face A, face E, and face F having positive pressure. Face B, face C, and face D are under the influence of negative pressure, as shown in Fig. 7 face E shows that magnitude of

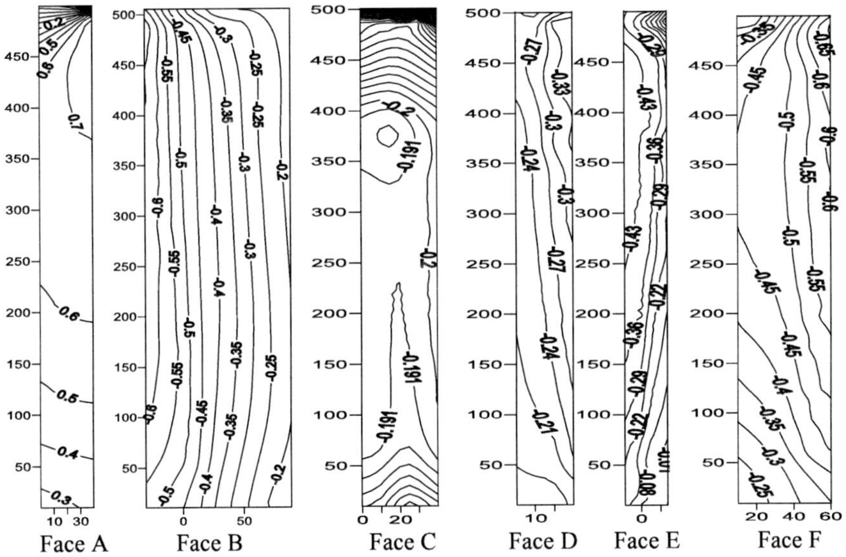

Fig. 5 Pressure contour on L shape building model A

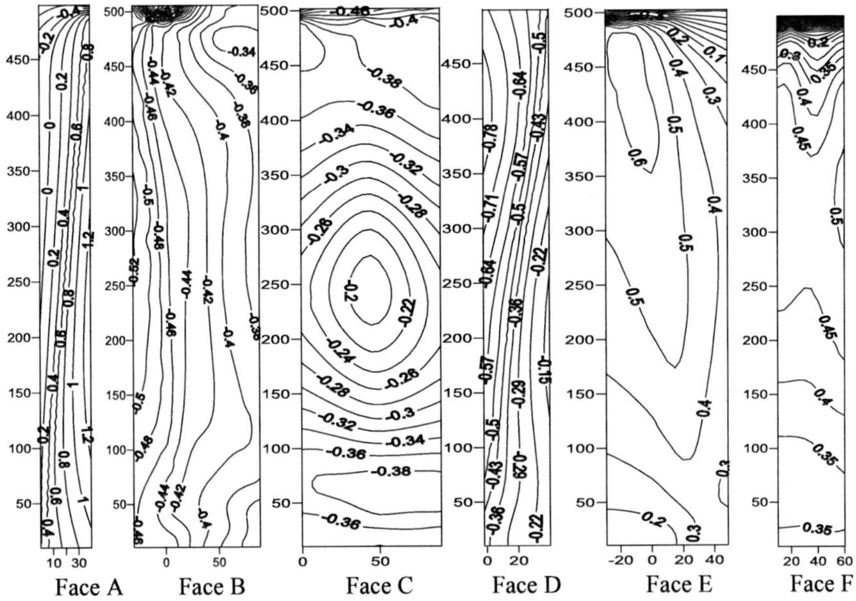

Fig. 6 Pressure contour on L shape building model B

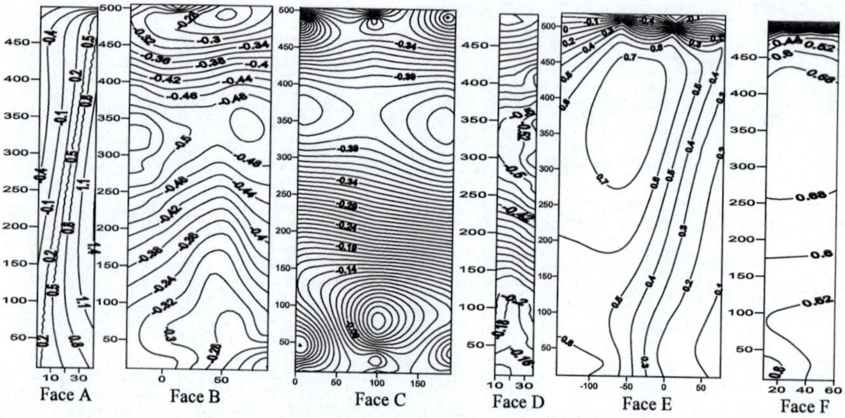

Fig. 7 Pressure contour on L shape building model C

pressure distribution is increasing from model B to model C, the maximum pressure coefficient is 0.5 for model B, shown in Fig. 6 and 0.8 for model C. pressure distribution on face C is also affecting by the variation in side ratio of building as shown in Figs. 5, 6 and 7 it has lesser magnitude for model A in comparison to model C.

3.3 Velocity Streamlines

Streamline can be defined as an imaginary line in a fluid such that the tangent at any point indicates the direction of the velocity of a particle of the fluid at that point. The streamline for 0° wind incidence angle for model A, model B, and Model C in the plan is shown in Fig. 8 because different flow patterns are observed for all the three models, and it is because the change made in the plan shape of the building. Vortices are generated in the wake region. It will create negative pressure on the leeward side of the building models. A large recirculation zone of air can be seen on the downstream side of building models for model B and model C. The streamline is shown for all model buildings is in Fig. 8 in the plan and Fig. 9, in elevation, for 0°

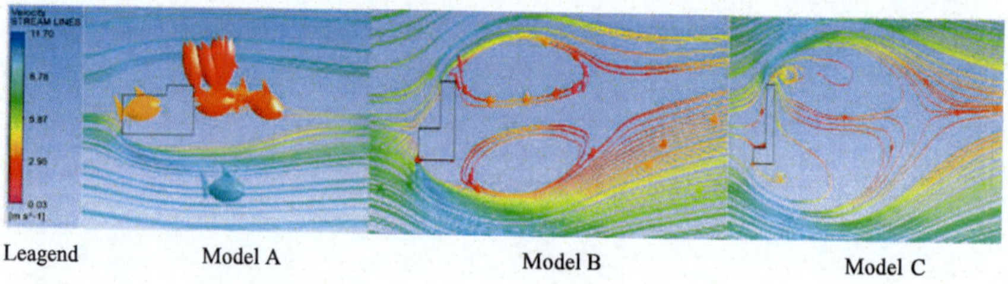

Fig. 8 Wind flow patterns around the building model in plan

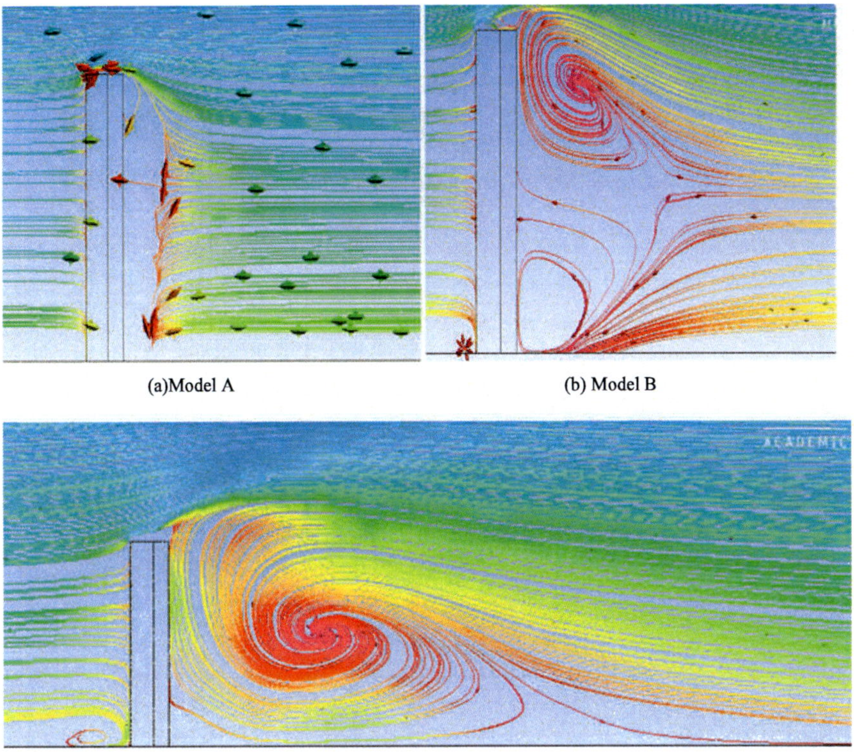

(a)Model A (b) Model B

(c) Model C

Fig. 9 Wind flow patterns around the building models in elevation

wind incidence angle. Fishes in Figs. 8 and 9 show the direction of streamlines. And colour scale indicates the velocity of streamlines that is more on the wind ward side as compared to the leeward side, and the vortices generated also depend on the side ratio of the building, the larger recirculation zone is on the leeward side of model C as shown in Fig. 8 and lesser for model A, as shown in Fig. 8.

Figure 9 shows the different types of vortex shedding. It depends upon the shape of the building there for to minimize the effect of wind, CFD simulation is carried out on tall buildings. When the flow passes through the bluff bodies, it creates the low-pressure zones. Vortex shedding occurs in the wake region of buildings. When the airflow passes through the building, the flow tends to separate from the rear boundary of the buildings. It forms a vortex.

3.4 Pressure Coefficient

The mean pressure coefficient for 0° wind incidence angle is calculated by Eq. (2) and shown in Fig. 10, where it varies from positive to negative.

The mean pressure coefficient '$C_{p\,\text{mean}}$' is calculated from Eq. (2)

$$C_{p\,\text{mean}} = \frac{p - p_o}{\frac{1}{2}\rho U_H^2} \tag{2}$$

where p is the pressure derived from the needed point, p_o is the reference height static pressure, ρ is the air density (1.225 kg/m^3), and U_H^2 is mean wind velocity at the building reference heights.

Figure 10 shows the mean C_p for all Model, from the Fig. 10, it can be concluded that face A, face E and face F is the only face that is under the influence of positive pressure. It can be concluded from the above figure that mean C_p values shows a good agreement between numerical result and the international standard, and model A, model B, and model C has almost the exact nature of pressure distribution in terms of positive pressure coefficient and negative pressure coefficient.

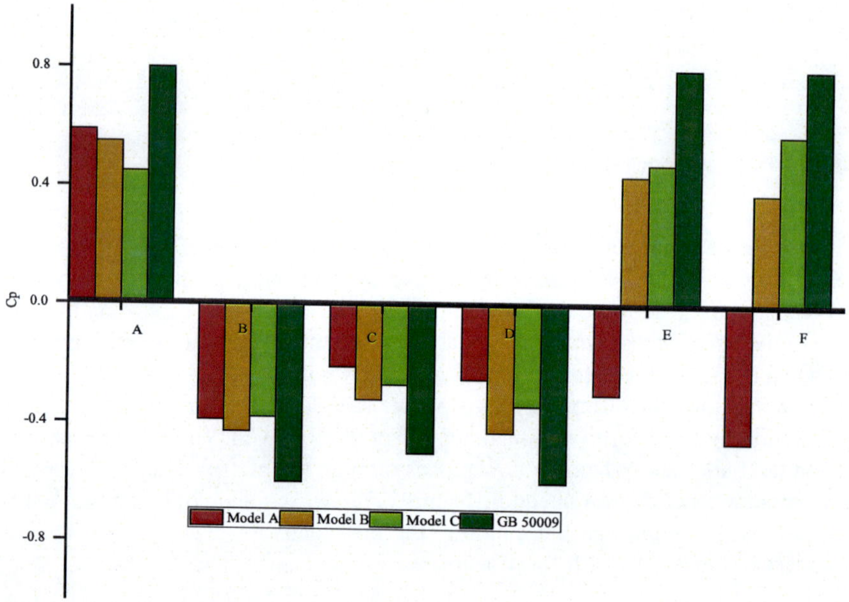

Fig. 10 Comparison of mean pressure coefficients on a different face of all model between numerical result and GB 50,009 (international standard)

3.5 Vertical Centerline Pressure Coefficient

The mean surface pressure coefficient is plotted along with the building height. According to Fig. 11 face, A of model A is the only face under the impact of direct wind pressure, and it is the only face that has positive pressure. Figures 12 and 13 show that three faces, namely A, E, and F, have positive pressure distribution, and face B, C and D have negative pressure distribution because of vortices formed around the building.

As shown in Fig. 11, maximum positive pressure on face A and minimum is on face F as face A is windward face and under the direct influence of wind, face F has suction, so it's under the effect of negative pressure.

As shown in Fig. 12, maximum positive pressure on face A and minimum is on face D as face A is windward face and under the direct influence of wind, face D has suction, so it's under the effect of negative pressure.

As shown in Fig. 13, maximum positive pressure on face F and minimum is on face B as face F is windward face and under the direct influence of wind, face B has suction, so it's under the effect of negative pressure.

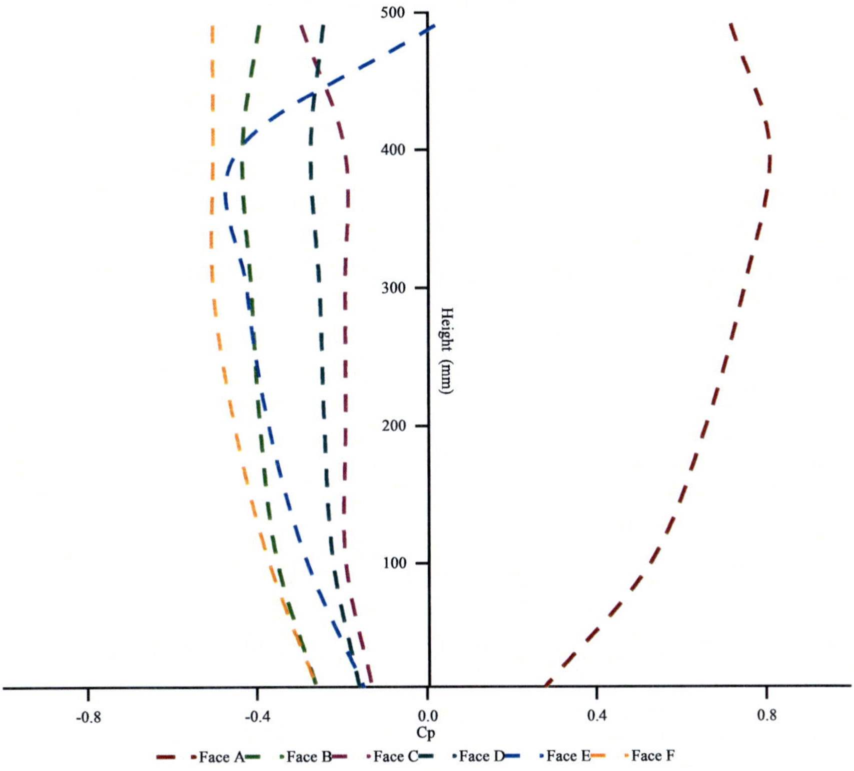

Fig. 11 Pressure variation along centerline for all faces of model A

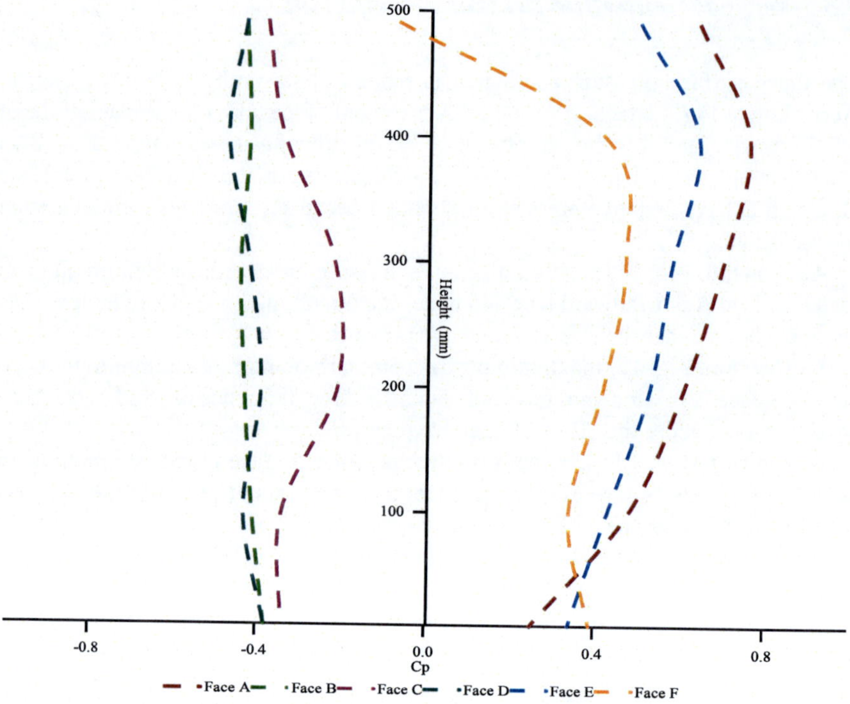

Fig. 12 Pressure variation along centerline for all faces of model B

Face F is the only face which is varied in dimensions in model A, B, and C, face F in Model A is under the impact of negative pressure because of vortices formed, and it became positive for model B and model C. and in model C where face F has a larger size than face A, it is under the influence of maximum positive pressure so it can be concluded that wind wards pressure distribution depends upon the size of the member.

4 Conclusions

The pressure contour, mean pressure coefficients, and velocity streamlines for the L-shape building model at 0° wind incidence angles are compared in this paper. The k-ε Model is used to simulate this research. The major finding of this research are as follows

- The pressure distribution of all faces of the building is the aim to conduct this study, and the pressure contour of all faces is presented in this paper.
- Changing in the side ratio of building considerably affects the suction on leeward and side faces of the building models.

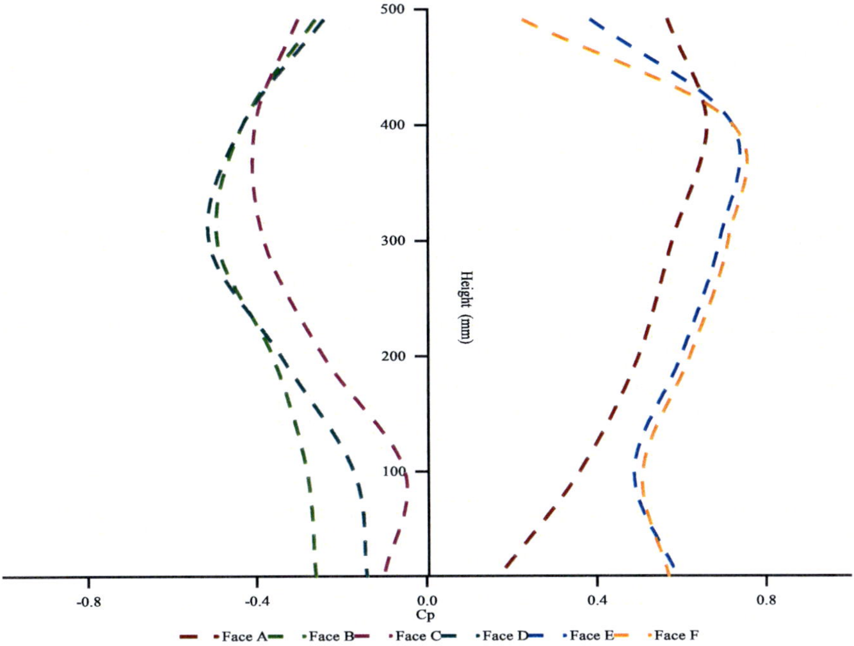

Fig. 13 Pressure Variation along Centerline for all faces of model C

- The maximum pressure coefficient for model A is 0.7 on windward face A and a minimum 0.65 side face F.
- The maximum pressure coefficient for model B is 0.8 on windward face A and a minimum of 0.78 on leeward face D. The maximum pressure coefficient for model A is 1.1 on windward face A and minimum 0.52 on leeward face D. Pressure distribution depend on the side ratio of building.
- Face A always shows positive pressure for both the models, whereas the other face has negative pressure.
- The variation of pressure coefficients on the centreline is discussed and shown graphically.
- Face E pressure distribution is affected by the side ratio of building models, and it is maximum pressure distribution for model C and minimum pressure distribution for model A.
- Face E and face F. The windward face has an almost similar pressure distribution for all three models.
- The velocity streamlines are shown using the figure in the plan and elevation.
- The accuracy of results depends upon meshing of the geometry model and defining the flow physics as in boundary layer wind tunnel.
- Wind pressure distribution on the leeward face of this study shows the development of vorticity, which shows high turbulence.

Acknowledgements The authors would like to express their appreciation to Prasenjit Sanyal for his precious help with the numerical simulation and to G. L. Mina, who encouraged and supported the authors during the entire study. Deepak Singh Ph.D Scholar, DTU provided his precious time to finalize the paper.

References

1. Günel, M. H., & Ilgin, H. E. (2014). Tall buildings: Structural systems and aerodynamic form. Tall Buildings: Structural Systems and Aerodynamic Form. https://doi.org/10.4324/9781315776521
2. Solari, G. (2019). *Wind actions and effects on structures*. Springer Tracts in Civil Engineering. https://doi.org/10.1007/978-3-030-18815-3_9
3. Chauhan, B. S., & Ahuja, A. K. (2017). *Height effect of interfering buildings on wind pressure distribution on rectangular plan tall buildings*. 9th Asia Pacific Conference Wind Engineering APCWE 2017 3–6.
4. IS: 875. (2015). *Indian Standard design loads (other than earthquake) for buildings and structures-code of practice, part 3(wind loads)*. BIS.
5. ASCE: 7-10. (2013). *Minimum design loads for buildings and other structures*. Structural Engineering Institute of the American Society of Civil Engineering.
6. GB 50009-2001. (2002). National Standard of the People's Republic of China.
7. AS/NZS:1170.2. (2011). *Structural design actions—Part 2: Wind actions*. Standards Australia/Standards New Zealand.
8. Hong Kong Building Department. (2019). Code of Practice on Wind Effects in Hong Kong 2019.
9. MNBC. MYANMAR NATIONAL BUILDING CODE 2020. (2020). International relation and legal Section Department of Building Minis.
10. Ahlawat, R., & Ahuja, A. (2015). Wind loads on T shape tall buildings. *Journal of Academia and Industrial Research (JAIR), 24*, 257922.
11. Ahlawat, R., & Ahuja, A. K. (2015). Wind loads on Y plan shape tall building. *International Journal of Engineering and Applied Sciences, 2*, 257946.
12. Paul, R. (2020). *Wind effects on cross plan shaped tall building wind effects on cross plan shaped tall building* 1–8.
13. Raj, R., & Ahuja, A. K. (2013). Wind loads on cross shape tall buildings. *J. Acad. Ind. Res., 2*, 111–113.
14. Vafaeihosseini, E., Sagheb, A., & Ramancharala, P. K. (2013). Computational fluid dynamics approach for wind analysis of Highrise buildings.
15. Meena, R. K., Awadhiya, G. P., Paswan, A. P., & Jayant, H. K. (2021). Effects of bracing system on multistoryed steel building. *IOP Conference Series: Materials Science and Engineering, 1128*, 012017.
16. Pal, S., Raj, R., & Anbukumar, S. (2021). Comparative study of wind induced mutual interference effects on square and fish-plan shape tall buildings. *Sādhanā* 0123456789.
17. Verma, D. S. K., Roy, A., Lather, S., & Sood, M. (2015). CFD simulation for wind load on octagonal tall buildings. *International Journal of Engineering Trends and Technology, 24*, 211–216.
18. Nagar, S. K., Raj, R., & Dev, N. (2020). Experimental study of wind—Induced pressures on tall buildings of different shapes. *Wind Structure An International Journal, 5*, 441–453.
19. Hima Chandan, D., & Pradeep Kumar, R. (2014). *Numerical simulation of wind analysis of tall buildings computational fluid dynamics approach.*
20. Bairagi, A. K., & Dalui, S. K. (2018). Aerodynamic effects on setback tall building using CFD simulation. *International Journal of Mechanical and Production Engineering Research and Development* 413–420.

21. Mendis, P., Ngo, T., Haritos, N., Hira, A., & Cheung, J. (2007). Wind loading on tall buildings. *Electronic Journal of Structural Engineering.*
22. Roy, A. K., Sharma, A., Mohanty, B., & Singh, J. (2017). Wind load on high rise buildings with different configurations: A critical review. *International Conference on Emerging Trends and Engineering Innovations Technology Management, 02*, 372–379.
23. Abu Zidan, Y., Mendis, P., & Gunawardena, T. *Optimising the computational domain size in CFD simulations of tall buildings.* Heliyon.

Effective Remediation Techniques for Twin Shallow Lakes in Panchkula, Haryana

Prachi Vasistha and Rajiv Ganguly

Abstract Water is the most important natural resource available to mankind. There is a scarcity of freshwater in comparison to human population on earth. The recent modernization and advances in living conditions have led to the water quality degradation. The decrease in water quality due to constant effects of anthropogenic activities such as rapid urbanization and agricultural expansion has led to development of measures to limit water quality degradation of natural water bodies. In this context, water quality inspection and monitoring are often carried out for freshwater bodies. The water bodies often lack proper maintenance systems leading to short- and long-term damages to these resources. Hence, appropriate remediation measures are required to remediate such water bodies which have been affected by this environmental degradation and to reduce further detrimental effects. The present study discusses different remediation techniques along with their advantages and disadvantages in application to two lakes in Haryana, India. The techniques are selected based on viability of these methods in terms of site conditions and problems, maintenance, feasibility, and ease of application. The parameters determined are pH, DO, HCO_3, BOD, COD, NO_3, and TP, and based on these parameters, the suitable remediation techniques have been determined and described. The study will be helpful in determination of appropriate remediation techniques to be followed for similar kinds of lakes in rapidly modernizing and developing areas.

Keywords Remediation · Physical · Biological · Ecological · Chemical

P. Vasistha (✉) · R. Ganguly
Department of Civil Engineering, Jaypee University of Information Technology, Waknaghat, District Solan, Himachal Pradesh 173234, India

R. Ganguly
e-mail: rajiv.ganguly@juit.ac.in

A. K. Gupta et al. (eds.), *Advances in Construction Materials and Sustainable Environment*, Lecture Notes in Civil Engineering 196,
https://doi.org/10.1007/978-981-16-6557-8_77

1 Introduction

Lakes, Rivers and other water bodies around the globe are important part of our community. Indiscriminate use of such water bodies has led to their severe deterioration. The water bodies face serious environmental threats specially in developing nations where the resources are limited and there is rapid increase population and urbanization [1]. The spread of diseases related to contaminated water sources is a threatening issue. The dumping of agricultural and municipal waste, storm water, grey water and industrial wastes are the main causes of pollution of these resources [2, 3]. These sources lead to high influx of input sediments, organic matters and nutrients into the water bodies which lead to problems like eutrophication in the stationary surface water bodies (like lakes). Hence, remediation of such water bodies for overall pollution reduction and framing of appropriate regulatory policies to control them is of prime importance [4]. Remediation techniques for treatment of such polluted water bodies can be done by in situ or ex situ treatments. Additionally, these treatments can further be divided into physical, chemical and biological methods [5, 6]. The combination of different methods can be used for treatment of heavily polluted waters.

Physical methods include aeration, manual removal of algae, uprooting weeds, building of hydraulic structures, dilution of water, raking and dredging of sediments and other procedures to improve the characteristics of water bodies. Chemical treatment includes introduction of chemical such as poly-aluminium chloride (PAC) for secondary flocculation, copper sulphate treatments, use of lime and ferrous salts in optimum dosing for improvements [7]. The biological and ecological methods include construction of wetlands, floating beds, biofilms, membrane bioreactor, ecological pons and plant-based water purification systems, filters and other small scale filtration techniques that can both be used at large- and small-scale levels [1]. The biological treatments might involve use of microorganisms and plants to absorb harmful organics, nutrients and other harmful substances which reduces concentrations of BOD, COD, turbidity, phosphates and nitrates [5, 8, 9]. These techniques can produce bacteria, fungi and microfauna such as protozoa, worms and insects which can lead to in situ bioremediation process. The combined in situ and microbial techniques is highly successful remediation method used for treatment of heavily polluted sources being sustainable and economical with no production of secondary pollutants [10]. The present study is an approach for determination of the best possible cost-effective remediation techniques for the urban shallow lakes in Panchkula district in Haryana state in India. The study involves the review of widely available global literature to critically scrutinize the best technologies available around the globe with their advantages and disadvantages and their possible of application to the study sites based on the parameters determined for both the lakes.

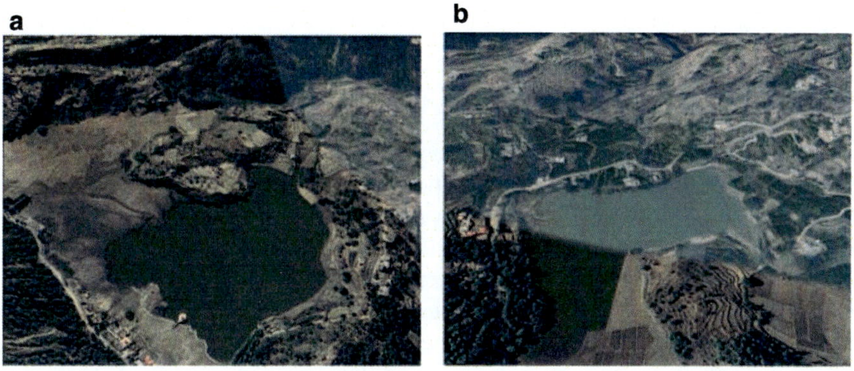

Fig. 1 **a** A satellite view of Lake 1 or Bhim Taal (*Source* Google earth), **b** a satellite view of Lake 2 or Draupadi Taal, Panchkula (*Source* Google earth)

2 Site Description

The site is an emerging tourist attraction located in Morni tehsil of District Panchkula in the state of Haryana and is a famous tourist destination for people of Chandigarh and around. The people plan weekend getaways to the site, and the site offers large number of recreational activities other than boating. The site has two lakes in vicinity of each other and in mythology are known to be connected to each other through some underground pathway. The Lake 1 is known as Bhim Taal whereas Lake 2 is known as Draupadi Taal. Due to being in the same vicinity, the site conditions and the site issues are same, but the recreational activities are supported in Lake 1 whereas Lake 2 is utilized for pisciculture by fisheries department of Haryana. The further details of the lake are given in [11] and [12]. The lakes are shown below in Fig. 1a, b respectively.

3 Methods and Their Applications

The remediation techniques require use of methods which are both traditional and novel for application in water bodies. The remediation techniques used in lakes are classified into physical, chemical and biological methods [13]. These methods are based on fixing pollutants, removing pollutants and also restricting pollutant concentrations.

3.1 Physical Methods

Aeration

Aeration in lakes increases the saturation of oxygen in the water and reduces the effects of organics in the lake [14]. Aeration can be achieved by use of compressors which introduce air in bottom of lake or on the surface [15]. This technique helps in reducing the organic matter and nutrients in the water by increasing favourable conditions for microbes to act which in turn degrades the organic content in water [1]. It also helps in lowering down levels of P and N in water column [16, 17]. The aeration technology can be divided into two classes (a) the fixed-point aeration system and (b) mobile aeration system. In principle, for shallow lakes, aeration should be carried out in the hypolimnetic region whereas for lakes with greater depth, depth aeration is suggested [18–20]. The choice of the approach plays a major role in the efficiency of process. Additionally, mobile aeration systems are used in conjunction with floating bed wetland while fixed aeration systems are used with sediment rooted wetlands. The efficiency of the technique is seen to increase if applied in collaboration with other remediation techniques like chemical doses and biological remediation [21]. This technique has been effectively utilized for deep lakes like River Thames, UK, Canal Homewood in USA, River Emsche in Germany, River Oerias in Portugal, and Busan River in South Korea [22, 23].

Sediment Dredging

Sediment dredging is another direct and very effective physical technique applied in shallow lakes and rivers for remediation using engineering structures. The technique involves use of heavy machineries such as extractors or dredgers to remove internal sediments from bottom of lakes and rivers but from a fixed depth, one such example being Lake Taihu in which the sediments were dredged from top surface of lake at depth of 30 cm which resulted in marked improvement in the aquatic ecosystem [24]. The technique is used for large scale for removal of P and N from highly enriched sediments [25], removal of heavy metals [26], pesticide-rich sediment [27], organic pollutant [28] and reduce cyanobacterial production [24]. Many global studies have reported a decline in levels of pollutants and microorganisms from pre- and post-dredging comparisons. Lake Trummen in Sweden was remediated with dredging and a considerable decline in levels of cyanobacterial production was observed in the lake [20]. The technique is beneficial if applied in combination with methods that reduce the external loading rates as it works on reduction of internal loads [29]. The dredged sediments can be utilized as a source of fertilizers in agriculture if not contaminated with carcinogenic pollutants [30].

Hydro-raking

Hydro-raking is a physical method to control and remove unwanted aquatic plants in lakes and rivers. This method helps in regulating benthic environments by complete removal of biomass unlike biological and chemical treatments [31]. It uproots the unwanted aquatic flora from roots along with their dormant seeds which include both strong and weak plants [32]. The method is highly effective in removing unwanted harmful weeds and plants as well nutrients in form of biomass. A typical example of application of this method is Lake Chemung, in Ontario wherein 35,000 kg of weed removal led to 560 kg removal of phosphorous [33].

The method is comparable to sediment dredging as it also removes high levels of internally loaded phosphorous but is a combination of dredging and weed harvesting. It requires 1.5 feet of water to operate and is suitable for shallow lakes. The method is suitable for restoring lakes to its original depth with minimal damage to aquatic animals since the raking equipment has perforations which allow easy escape to aquatic animals. Unlike dredging, raking does not destabilize the shores. The hydro-rakes operate in water and settle uprooted weeds and flora on edges of banks which can be collected and sent to compost pits. The technique is also less costly as compared to dredging.

Vegetative Buffers and Riparian Zone stabilization

Addition of native vegetation along the shorelines and inflowing channels and sources are an efficient way of preventing sediment inflows from nearby watershed areas into lakes [34]. The native plants grow well off in the surrounding areas and help in nutrient uptake which prevents the freshwater from high nutrient inputs [35]. The riparian vegetation adds to infiltration of water into subsoils and thereby increase water tables [36]. However, the type of vegetation and density of vegetation plays an important role in reducing overflow rates and velocity of overflowing water [34]. Grass covers are found to be more effective than woody covers in maintaining temperatures and regulating primary production, but the problem is maintenance of these grass covers which is quite high [37].

A study in the USA proved that a species of bamboo called Arundinaria gigantea is highly effective against sedimentation of nutrients such as nitrogen and phosphorous [38, 39]. However, another study [40] proved that native vegetation was better in preventing sedimentation influx and promoting infiltration, therefore growing native grass covers in combination with bamboos are a good technique in preventing further sedimentation and increasing groundwater tables.

Fencing of the riparian zone in addition to growing vegetative buffers is long-term measures for protection of these vegetative buffers and promotes soil stabilization against run off and erosion. One such great example of multipurpose use of bamboo is that it can act as both vegetative cover and soil stabilization at banks. The freshly cut bamboo or air-dried sample can be used to make crib walls which promote soil stabilization by decreasing pore water pressure and increasing surface roughness and strength. The bamboos are cut, trimmed off knots, stacked to make cribs and then wired or nailed to keep them in place. These cribs are then erected into properly prepared foundations and then filled to make walls [41]. These walls are irrigated and maintained for growth of new branches and leaves. However, fresh soil needs to be added at the bottom in case of loss of soil due to erosion.

This technique is a great way if used in collaboration with other physical or biological techniques that can increase efficiency of the other techniques such as dredging and hydro-raking as well. Shoreline stabilization techniques are long term and environment friendly, also they help in promoting growth of native species.

Rip Raps

The rip rap is a simple technique of putting angular or rounded rocks along the shoreline to prevent any further loss of soil due to erosion arising out of constant waves and water collision. Rip raps are e-protective covers, but no vegetative growth is experienced over rip raps. The technique offers protection only on the area of contact, and therefore, it fails in protecting the shoreline of exposed banks which gets eroded due to the continuous flow of water. Further, aquatic life is disturbed since rip raps provide clear visibility and easy accessibility to predators with the aquatic animals unable to take shelter along banks during high tides due to no vegetation. Additionally, low vegetation and exposed stone surfaces cause excessive heating and the reflection into lakes causing temperatures to rise. However, this technique is still preferred due to destabilisation from high waves and fast water flows.

3.2 *Chemical Methods*

Alum

It is commonly known as alum or aluminium sulphate, a commonly used sediment capping material used as a flocculant [42]. The chemical has potential to bind the phosphorous which disables algae to feed on it thereby improving aesthetics of the water and reducing turbidity. The chemical is used on large scale for reduction of algae and is often applied through boats in form of solutions or dry form stored in large tanks under boats. The chemical settles down at the bottom of sediments which binds the free phosphorous to sediments called 'flocking'. The unavailability of nutrient to the algae makes it wither off [43, 44]. Maximum usage of the chemical is seen in USA and Europe [25]. The chemical works best for near neutral pH ranges of 6–8 [44]. The effectiveness of chemical technique is based on characteristics such as depth and morphology and its effects are expected to last for even 20 years in some lakes [45]. The chemical can also be used in an alternative form such as *allophane* which is even efficient for large cyanobacterial blooms and highly eutrophic waters [46].

Phoslock ™

It is a modified natural mineral called as lanthanum modified bentonite clay and has high affinity for phosphorous. The material works as sink for phosphorous leading to the formation of lanthanum phosphate which is insoluble [46]. The chemical is applied into lake mixed with bentonite clay either in slurry or granular form and its high reactivity and dispersion in water results in the removal of along with restricting efflux from sediments [47]. The chemical works best at pH of 6–10 while adsorption is seen to be maximum at pH of 4.5 and 8 [21]. The chemical is used in the lakes of

UK, and the studies have suggested it to be an efficient tool for lake remediation if used in optimum doses [47–49].

Modified Zeolites

It is a modern approach which focuses on trapping of both nitrogen and phosphorous [30]. It is derived from sedimentary rocks in form of aluminosilicates and is a readily available material which absorbs nutrients due to its high porosity [44]. The material has high affinity for cations which makes it beneficial for removal of cationic pollutants [50]. The structure of this material can be easily modified by various treatments such as high temperature or steam process to increase its affinity to adsorption of anions and other organic pollutants [50–52]. Many studies have reported the success of zeolites in lake remediation [44, 53]. The application of modified zeolites is an active management technique for lake remediation, but the optimization of the doses is necessary to avoid ecological effects along with other physical and chemical effects.

Ferrous Iron Salts

The aluminium salts are seen to work better at slightly acidic range, (pH below 5.5). However, increased aluminium solubility is toxic to aquatic animal and at pH w above 8 leads to formation of aluminium hydroxide which is toxic and soluble. Hence, it often becomes difficult to monitor optimum dosages of alum, on regular basis specially for drinking water sources [54].

In this context, the solubility of ferrous salts at pH 7 is about 250 times less than alum thereby favouring precipitation of phosphorous [55]. At pH > 8 ferrous salts are more favourable for remediation against phosphorous than aluminium compounds [56]. The ferrous salts can be used alone or in combination with $NaHCO_3$ as a pH buffer [57]. An example of its application is Bautzen reservoir in Germany which was tested for usage of ferrous salts against bloom control in summers avoiding use of aluminium salts in reservoir keeping pH in consideration [54]. The aim was achieved by using ferrous salts in combination with aerators [58] to avoid destratification. Fe^{2+} salts are a great way of phosphorous elimination from lakes, and the delay in floc formation helps in elimination of phosphorous in greater amounts at wide range of pH [54].

Copper Sulphate

Copper sulphate is an algicide used for controlling algae in lakes, recreational pools and reservoirs [59]. For example, USA uses about 9 million kilograms of copper sulphate annually in lakes of Minnesota, Michigan, Illinois and Wisconsin and is forecasted to use them in future also [60]. The copper sulphate was started to be used first by Moore and Kellerman in 1905, and the results showed the scum formation of dead algae along shorelines after applications of copper sulphate. The copper sulphate is a short-term treatment for algae control and might have detrimental effects on the other parameters of lake such as dissolved oxygen (DO) during decay of algae and its prolonged usage might lead to resistance formation in lakes against algae control, as observed in Shallow Fairmont Lakes in Canada [61].

Calcium-Based Compounds

Calcium-based compounds such as calcium chloride, calcium hydroxide, gypsum and calcium oxides are known to control nitrogen and phosphorous in areas of low oxygen zones, especially hypolimnion. Gypsum leads to free hydrogen ion of organic matter which is decomposed under anaerobic conditions [62–64]. The doses of calcite are given in conjunction with aeration to induce precipitation of phosphorous [65]. The treatment can be used for hard waters and phosphorous loading [66, 67]. The selection of calcium-based compounds depends on environmental conditions of lake and pH [68]. The pH increase is often noticed in lake [69]; therefore, the use of these compounds should be in lakes with considerably low pH and high phosphorous or nitrogen with oxygen deficiency in hypolimnion.

3.3 *Biological or Ecological Methods*

The biological and ecological techniques are based on aquatic plants, aquatic animals, and microorganisms which are used for remediation of water bodies.

Aquatic Animals

Aquatic animals like fish, snails and clams are often used for bioremediation in eutrophic lakes [70]. Gastropods act as bioindicator for some heavy metals in lakes and other freshwater bodies [71]. Silver carp is a filter feed fish which has a long-life span of 10–20 years and is effectively used to control eutrophication in lake water by eliminating excess phytoplankton [72–74]. For example, Potomac River in USA was cleaned in 1980s utilizing Asiatic clam for controlling eutrophication, and the results showed top-down and bottom-up approach for ecosystems [72, 75]. The silver carp is shown to reduce efficiency in presence of large herbivorous, organic and inorganic toxins as well when biotoxins are released by other species [72, 76]. Therefore, it becomes imperative to study toxicological effects on efficiency of the use of aquatic animals for bioremediation.

Aquatic Plants

Aquatic plants like hyacinth, water lettuce and some algae can remove effluents, wastewater and other toxicants from the water by adsorption, absorption, degradation and accumulation [77–79]. Common weed, pondweed, duckweed, cattail and canna are few varieties which are used for remediation of wastewaters onsite in systems such as constructed wetlands, floating beds and submerged systems [80–82]. These systems work by either direct uptake of pollutants or by inducing physical and chemical reactions to purify [82]. The following technologies based on aquatic plants can be a great remediation approach.

Aqua mats

Aqua mats have high surface areas and are principally artificial seaweed which encourage the growth of algae, colonization of aerobic and anaerobic bacteria and other aquatic microorganisms [72]. The enhancement in efficiency can be achieved by using immobilized bacteria, biofilms or by increasing plant growth by choosing correct type of species [76, 83].

Floating Beds

The plant-based floating bed technique is best for removal of BOD, COD, nitrogen, phosphorous, ammonia, nitrates and even heavy metals in eutrophic lakes [84]. This technology does not utilize soil as a medium for plant growth and uses floating beds as plant carriers which can be used by birds and animals for habitat. A study of Hong Feng Lake confirms that it is a great ecological technique for remediation and can use land plants too for lake remediation [85]. The flexibility on usage of plants has made this technique widely applicable in USA, China, Europe and other places as well [86]. The combination of floating beds with other technologies can prove out to be best in increasing efficiency of these systems. Remediation Promoting Integrated Floating Beds (RPIFB) [82] and Green Energy Artificial Floating Island (GAFI's) [87] are some of the improved floating bed techniques.

Constructed Wetlands

These are sediment rooted wetlands which are constructed for purpose of water quality improvement and preserving biodiversity. Total nitrogen and phosphorous can easily be removed by these systems [88]. The constructed wetlands (CW) are of three types vertical, surface and subsurface. The vertical CW removes ammonia, TN and TP whereas COD and chlorophyll are removed by surface subsurface structures [89]. The CW's work by various physical, chemical and biological reactions promotes reoxygenation as well. One such case study is Lake Taihu which was treated against eutrophication by use of the three CWs ammonia, total nitrogen and total phosphorous [90].

Microorganism

The use of microorganisms for treatment and remediation of water pollution in rivers is highly effective as it decomposes, absorbs and transforms pollutants. This involves uses of appropriate microbial groups such as photosynthetic bacteria and microalgae bacteria which removes BOD, COD, TP, NH_3, nutrients and increase DO [91, 92]. The methods may be divided into onsite method such as microbial dosing and surface water method such as biofilms.

Microbial Dosing

It involves direct application of microbes to remove pollutants from the water by use of products such as FLO-1200 along with bioenergizers to promote aeration [70]. Photosynthetic bacteria and Bacillus subtilis was used to remediate a black and odorous river in Fangcun district of China [91]. Another study reported the utilization of nitrobacteria and humic acid for removal of COD, TN, TP, ammonia and turbidity with increase in DO from a small shallow lake affected by eutrophication [10].

Biofilm Reactors

This technology utilizes usage of biomembrane attached to some carrier such as microcarrier or river bed for adsorption, filtration and degradation under aerated conditions. The biofilm reactors have a high efficiency to remove ammonia nitrogen and COD from water [93]. Used technologies include gravel contact oxidation, aerated biofilter biological fluidized bed, artificial packing contact oxidation, thin layer flow method, underground stream purification method and suspended carrier biofilm rectors are known around globe. For example, Nogawa river bed in Japan utilized this technology to remove BOD, COD, SS, N and P from polluted lake water or sewage water [70, 94]. Bioceramics are used for removal of colour, N, P, COD, organic matter and turbidity [95]. The use of bamboo as a biofilm for remediation of polluted river water was shown by [96] which led to higher level of bioremediation, than conventional systems and phytoremediation techniques.

4 Technology Consideration and Critical Discussions

The technology consideration for the above-mentioned technologies was based on the site conditions, their suitability for site along with their advantages and disadvantages as described in Table 1.

The Lake 1 and Lake 2 are urban shallow natural lake which faces constant effects of modernization and anthropogenic activities. The lake experiences tremendous amount of sedimentation due to constant erosion from shoreline and watershed areas due to increased construction activities and deforestation along with expansion of agricultural activities [11, 12]. The lake therefore is threatened by increased sedimentation rates therefore controlling and remediating sedimentation rates is the foremost concern. The constant erosion of banks and the expansion of agricultural activities has led to slight increase in nutrient levels [11, 12] which might increase further if control measures are not undertaken. For site remediation and control, the above suitable technologies such as aeration, dredging, vegetative buffers, chemicals such as alum and biological techniques such as use of aquatic animals can be used either singly or in combination to bear the best and long-term benefits if maintained suitably. The physical technologies such as sediment dredging and hydro-raking can be efficient in removing the sediments that have been deposited inside the lakes for a very long time further the technology such as vegetation buffers and stabilization can be efficient tools for further sedimentation of the lakes which might even eradicate further problems such as nutrient build up in sediments. To reduce the further effects of sediment removal through above techniques, the chemical treatment using alum can be beneficial to build up flocks and reduce haziness.

Table 1 Advantages, disadvantages and suitability of technologies

Technology	Advantages	Disadvantages	References	Suitability for our study sites
Aeration	1. Easy application 2. Improves water quality 3. Effective for nutrient removal and increasing DO	1. Applicable to small and shallow lakes 2. High cost and maintenance	[14, 77]	Lake 1 pH—6.5 DO—5.3 mg/L Lake 2 pH—6.3 DO—5.2 mg/L **Yes**
Sediment dredging	1. Improves water quality 2. Direct application 3. Reduces internal loading and sediment release	1. Applicable to small and shallow lakes 2. High cost and maintenance 3. Shore destabilization	[25]	Lake 1 NO_3—1.2 mg/L TP—0.5 mg/L Lake 2 NO_3—0.8 mg/L TP—0.2 mg/L **Yes**
Hydro-raking	1. Does not remove aquatic wildlife 2. Removes unwanted weeds and plants	1. Applicable to small and shallow lakes 2. High cost and maintenance 3. Sediment dispersion and turbidity	[31]	**NA**
Vegetative buffers and riparian zone stabilization	1. Preventing sediment inflows 2. Promotes growth of native vegetation 3. Provides shading to aquatic animals from high temperatures and predators	1. Stabilization techniques require constant maintenance	[36]	**Yes**
Rip raps	1. Stabilizes banks 2. Prevents erosion 3. Prevents shoreline damage	1. Increases water temperatures by reflecting sunlight into waters 2. Makes aquatic wildlife prone to preying 3. Requires constant maintenance	[41]	**NA**

(continued)

Table 1 (continued)

Technology	Advantages	Disadvantages	References	Suitability for our study sites
Alum	1. Long-term benefits 2. Reduction in nutrient concentrations	1. Not suitable for large cyanobacterial blooms	[25]	Lake 1 BOD—1.5 mg/L COD—40 mg/L Lake 2 BOD—1 mg/L COD—50 mg/L **Yes**
Phoslock ™	1. Eutrophication control 2. Can be modified and combined to achieve suitable results	1. Sediment smothering of benthos 2. Toxicity	[25, 47]	NA
Modified zeolites	1. Safe for aquatic animals 2. Wide area of application	1. Costly	[44]	NA
Ferrous iron salts	1. Inhibit release of nutrients 2 Removes large algal blooms 3. Improves water quality	1. Limited approach 2. Toxicity	[57, 58]	NA
Copper sulphate	1. Removal of large algae concentrations 2. Low cost	1. Zooplankton and fish toxicity 2, Depletion of DO 3. Copper accumulation in sediments 4. Killing of non-target organisms	[61]	NA
Calcium-based compounds	1. Applicable to hard waters 2. Direct application	1. Decrease in transparency 2. increases pH of water	[68]	Lake 1 HCO_3—170 mg/L Lake 2 HCO_3—130 mg/L **NA**
Aquatic animals	1. Controls excessive phytoplankton 2. Improves water quality	1. Restrictive to toxic environments and organic contents from algae	[77]	Yes

(continued)

Table 1 (continued)

Technology	Advantages	Disadvantages	References	Suitability for our study sites
Aqua mats	1. High surface area, solar based, low cost, eco-friendly 2. High bacterial removal 3. Better organic and nutrient removal	Vulnerable to natural disasters and requires constant maintenance	[93]	NA
Floating beds	1. Cost effective, good removal efficiency, better plant accommodation 2. Removes nutrients and provides home to plants and animals	1. Often affected by natural disasters and waves, pollutant loads, seasonal changes	[77, 93]	NA
Constructed wetlands	1. Low cost, easy maintenance, no secondary pollution, economic and environmental-friendly	1. Intolerance to heavy pollution loads 2. Requires large area 3. Seasonal death and plant diseases	[77]	NA
Microorganisms (direct application)	1. Enhanced pollution control 2. Easy application, no extra construction	1. Limited remediation achieved 2. Inefficient for many parameters	[77, 93]	NA
Biofilm reactors	1. Reduced sewage collection costs 2. Less landscape required 3. Low cost	1. Breakable from strong water flows 2. Extensive construction works required	[77, 93]	NA

5 Conclusion

The problems mentioned above for the Twin Lakes of Panchkula, Haryana makes us conclude the following two strategies for the sites.

1. The *physical remediation technique* such as *dredging* and *vegetative buffers and riparian zone stabilization* can be a suitable method for remediation of heavy sediments from the lakes and to avoid further washing off of sediments into the lakes. The dredging measures can be applied occasionally, into lakes to remove the already existing sediments and the vegetation of native plants along the

shorelines can prove to be of benefit to prevent soil erosion of the shorelines, promote infiltration and to stop further sediment influx.

2. The use of *chemicals* such as *alum* for removal of turbidity, nitrate and phosphorous flocking from sediments can also be beneficial since the chemicals work for neutral pH and the lake waters are determined to be of neutral pH ranging from 6 to 8 [11].

References

1. Md Anawar, H., & Chowdhury, R. (2020). Remediation of polluted river water by biological, chemical, ecological and engineering processes. *Sustainability, 12*(17), 7017. https://doi.org/10.3390/su12177017
2. Anawar, H. M., & Strezov, V. (2019). Transport, fate, and toxicity of the emerging and nanomaterial contaminants in aquatic ecosystems: Removal by natural processes. In A. K. Mishra, H. M. Anawar, & D. Nadjib, (Eds.), *Emerging and nanomaterial contaminants in wastewater: Advanced treatment technologies* (1st ed.). Elsevier.
3. Darwich, T., Shaban, A., & Hamzé, M. (2018). The national plan for Litani river remediation. In A. Shaban, & M. Hamzé, (Eds.), *The Litani river, Lebanon: An assessment and current challenges* (vol. 85). Water Science and Technology Library, Springer.
4. Nie, Z. Y., Wu, X. D., Huang, H. M., Fang, X. M., Xu, C., Wu, J. Y., Liang, X. Q., & Shi, J. Y. (2016). Tracking fluorescent dissolved organic matter in multistage rivers using EEM-PARAFAC analysis: Implications of the secondary tributary remediation for watershed management. *Environmental Science and Pollution Research, 23*, 8756–8769.
5. Gao, H., Xie, Y. B., Hashim, S., Khan, A. A., Wang, X. L., & Xu, H. Y. (2018). Application of microbial technology used in bioremediation of urban polluted river: A case study of Chengnan river, China. *Water, 10*, 643.
6. Bai, X. Y., Zhu, X. F., Jiang, H. B., Wang, Z. Q., He, C. G., Sheng, L. X., & Zhuang, J. (2020). Purification effect of sequential constructed wetland for the polluted water in urban river. *Water, 12*, 1054.
7. Wang, C., & Jiang, E. L. (2016). Chemicals used for in situ immobilization to reduce the internal phosphorus loading from lake sediments for eutrophication control. *Critical Reviews in Environmental Science and Technology, 46*(10), 947–997. https://doi.org/10.1080/10643389.2016.1200330
8. Zhao, F. L., Zhang, S. D., Ding, Z. L., Aziz, R., Rafiq, M. T., Li, H., He, Z. L., Stoffella, P. J., & Yang, X. E. (2013). Enhanced purification of eutrophic water by microbe-inoculated stereo floating beds. *Polish Journal of Environmental Studies, 22*, 957–964.
9. Zheng, L. G., & Wang, H. P. (2017). *Effect of combined ecological floating bed for eutrophic lake remediation.* In Presented AIP Conference Proceedings 1839, 020058. Available online: https://aip.scitation.org/doi/pdf/https://doi.org/10.1063/1.4982423. Accessed on May 8, 2017.
10. Mingjun, S., Yanqiu, W., & Xue, S. (2009). Study on bioremediation of eutrophic lake. *Journal of Environmental Sciences, 21*, S8–S16.
11. Vasistha, P., & Ganguly, R. (2020). Assessment of spatio-temporal variations in lake water body using indexing method. *Environmental Science and Pollution Research, 27*, 41856–41875. https://doi.org/10.1007/s11356-020-10109-3
12. Vasistha, P., & Ganguly, R. (2021). Spectral characterization of sediment of two lake water bodies and its surrounding soil in Haryana, India. *Arabian Journal of Geosciences, 14*, 48. https://doi.org/10.1007/s12517-020-06425-0
13. Wang, J., Zhang, L., Lu, S., Jin, X., & Gan, S. (2012a). Contaminant removal from low-concentration polluted river water by the bid-rack wetlands. *Journal of Environmental Sciences, 24*(6), 1006–1013.

14. Zhang, L. A. (2016). *Study of polluted river remediation by aeration*. In E. Qi, (Ed.), Proceedings of the 6th International Asia Conference on Industrial Engineering and Management Innovation (pp. 451–461). Atlantis Press.
15. Cooke. G. D., Welch, E. B., Peterson, S. A., & Nichols, S. A. (2005). In G. D. Cooke, (Ed.), *Restoration and management of lakes and reservoirs* (3rd Ed.). Taylor and Francis.
16. Tofolon, M., Ragazzi, M., Righetti, M., et al. (2013). Efects of artifcial hypolimnetic oxygenation in a shallow lake. Part 1: Phenomeno-logical description and management. *Journal of Environmental Management, 114*, 520–529. https://doi.org/10.1016/j.jenvman.2012.10.062
17. Siwek, H., Włodarczyk, M., & Czerniawski, R. (2018). Trophic state and oxygen conditions of waters aerated with pulverising aerator: The results from seven lakes in Poland. *Water, 10*, 219–229. https://doi.org/10.3390/w10020219
18. Grochowska, J., Brzozowska, R., & Łopata, M. (2013). Durability of changes in phosphorus compounds in water of an urban lake after application of two reclamation methods. *Water Science and Technology, 68*, 234–239. https://doi.org/10.2166/wst.2013.249
19. Horppila, J., Köngäs, P., Niemistö, J., & Hietanen, S. (2015). Oxygen fux and penetration depth in the sediments of aerated and non-aerated lake basins. *International Review of Hydrobiology, 100*, 106–115. https://doi.org/10.1002/iroh.201401781.
20. Bormans, M., Maršálek, B., & Jančula, D. (2016). Controlling internal phosphorus loading in lakes by physical methods to reduce cyano-bacterial blooms: A review. *Aquatic Ecology, 50*, 407–422. https://doi.org/10.1007/s10452-015-9564-x
21. Spears, B. M., Meis, S., Anderson, A., & Kellou, M. (2013). Comparison of phosphorus (P) removal properties of materials proposed for the control of sediment p release in UK lakes. *Science of the Total Environment, 442*, 103–110. https://doi.org/10.1016/j.scitotenv.2012.09.066
22. Rogers, G. R. (2000). Water quality management at Santa Cruz Harbor. *Aire-O2 News 2000, 7*, 4–5.
23. Wang, C. X., Ling, H., & Shi, K. H. (1999). Oxygen restoration of polluted water with pure oxygen aeration. *Shanghai Environmental Science, 18*, 411–413. (In Chinese). Available online: https://en.cnki.com.cn/Article_en/CJFDTotalSHHJ199909012
24. Zhong, J. C., You, B. S., Fan, C. X., Bao, L. I., Lu, Z., & Ding, S. M. (2008). Influence of sediment dredging on chemical forms and release of phosphorus. *Pedosphere, 18*(1), 34–44.
25. Bessa, da Silva, M., Gonçalves, F., & Pereira, R. (2019). Portuguese shallow eutrophic lakes: Evaluation under the water framework directive and possible physicochemical restoration measures. *Euro-Mediterranean Journal for Environment Integration, 4*, 3. https://doi.org/10.1007/s41207-018-0090-9
26. Garcia-Orellana, J., Canas, L., Masque, P., Obrador, B., Olid, C., & Pretus, J. (2011). Chronological reconstruction of metal contamination in the Port of Mao (Minorca, Spain). *Marine Pollution Bulletin, 62*(8), 1632–1640.
27. Weston, D. P., Jarman, W. M., Cabana, G., Bacon, C. E., & Jacobson, L. A. (2002). An evaluation of the success of dredging as reme-diation at a DDT-contaminated site in San Francisco Bay, California, USA. *Environmental Toxicology and Chemistry, 21*(10), 2216–2224.
28. Martinez, A., & Hornbuckle, K. C. (2011). Record of PCB congeners, sorbents and potential toxicity in core samples in India Harbor and Ship Canal. *Chemosphere, 85*(3), 542–547.
29. Weston, D. P., Jarman, W. M., Cabana, G., Bacon, C. E., & Jacobson, L. A. (2010). An evaluation of the success of dredging as remediation at a DDT-contaminated site in San Francisco Bay, California, USA. *Environmental Toxicology and Chemistry, 21*(10), 2216–2224.
30. Hickey, C. W., & Gibbs, M. M. (2009). Lake sediment phosphorus release management—Decision support and risk assessment frame-work. *New Zealand Journal of Marine and Freshwater Research, 43*, 819–856. https://doi.org/10.1080/00288330909510043
31. Olsen, K. (2013). Water quality and phosphorous measurements associated with a Hydroraking project, lake Wapalanne, Sussex County, New Jersey, during the summer of 2009. *Northeastern Geoscience, 31*.
32. Witty, M., & Souza, S. (2009). *The future of coastal lakes in Monmouth County*. Monmouth University School of Science.

33. Wile, I. (1978). Environmental effects of mechanical harvesting. *Journal of Aquatic Plant Management, 16*, 14–20.
34. Karssies, L. E., & Prosser, I. P. (1999). *Guidelines for riparian filter trips for Queensland irrigators.* CSIRO Land and Water Technical Report 32/99. Canberra, Australia, Commonwealth Scientific and Industrial Research Organisation (CSIRO).
35. Mander, U., & Hayakawa, Y. (2005). Purification processes, ecological functions, planning and design of buffer zones in agricultural watersheds. *Ecological Engineering, 24*, 421–432.
36. McKergow, L. A., Prosser, I. P., Grayson, R. B., & Heiner, D. (2004). Performance of grass and rainforest riparian buffers in the wet tropics, Far North Queensland. 2. Water quality. *Australian Journal of Soil Research, 42*, 485–498.
37. Lyons, J., Trimble, S. W., & Paine, L. K. (2000). Grass versus trees: Managing riparian areas to benefit streams of central North America. *Journal of the American Water Resources Association, 36*, 919–930.
38. Blattel, C. R., Williard, K. J., Baer, S. G., & Zaczec, J. J. (2005). Abatement of ground water phosphate in giant cane and forest riparian buffers. *Journal of the American Water Resources Association, 41*(2), 301–307.
39. Schoonover, J. H., Williard, K. W. J., Zaczek, J. J., Mangun, J. C., & Carver, A. D. (2006). Agricultural sediment reduction by giant cane and forest riparian buffers. *Water, Air and Soil Pollution, 169*, 303–315.
40. Vigiak, O., Ribolzi, O., Pierret, A., Valentin, C., Sengtaheuanghoung, O., & Noble, A. (2007). Filtering of water pollutants by riparian vegetation: Bamboo versus native grasses and rice in a Lao catchment. *Unasylva, 58*, 229.
41. Tardio, G., Mickovski, S. B., Rauch, H. P., Fernandes, J. P., & Acharya, M. S. (2018). The use of bamboo for erosion control and slope stabilization: Soil bioengineering works. In *Bamboo—Current and Future Prospects.* Abdul Khalil H. P. S., IntechOpen. https://doi.org/10.5772/intechopen.75626
42. Ayoub, M., Affy, H., & Abdelfattah, A. (2017). Chemically enhanced primary treatment of sewage using the recovered alum from water treatment sludge in a model of hydraulic clarifocculator. *Journal of Water Process Engineering, 19*, 133–138. https://doi.org/10.1016/j.jwpe.2017.07
43. Wang, C. H., Gao, S. J., & Wang, T. X. (2011) Effectiveness of sequential thermal and acid activation on phosphorus removal by ferric and alum water treatment residuals. *Chemical Engineering Journal, 172*, 885–891. https://doi.org/10.1016/j.cej.2011.06.078
44. Gao, S., Wang, C., & Pei, Y. (2013). Comparison of different phosphate species adsorption by ferric and alum water treatment residuals. *Journal of Environmental Sciences, 25*, 986–992. https://doi.org/10.1016/S1001-0742(12)60113-2
45. Huser, B. J., Egemose, S., Harper, H. et al. (2016). Longevity and effectiveness of aluminium addition to reduce sediment phosphorus release and restore lake water quality. *Water Research, 97*, 122–132. https://doi.org/10.1016/j.watres.2015.06.051
46. Yuan, G., & Wu, L. (2007). Allophane nano clay for the removal of phosphorus in water and wastewater. *Science and Technology of Advanced Materials, 8*, 60–62. https://doi.org/10.1016/j.stam.2006.09.002
47. Yamada-Ferraz, T. M., Sueitt, A. P. E., Oliveira, A. F. et al. (2015). Assessment of Phoslock® application in a tropical eutrophic reservoir: an integrated evaluation from laboratory to feld experiments. *Environmental Technology and Innovation, 4*, 194–205. https://doi.org/10.1016/j.eti.2015.07.002
48. Epe, T. S., Finsterle, K., & Yasseri, S. (2017). Nine years of phosphorus management with lanthanum modifed bentonite (Phoslock) in a eutrophic, shallow swimming lake in Germany. *Lake Reservoir Management, 33*, 119–129. https://doi.org/10.1080/10402381.2016.1263693
49. Martin, M., & Hickey, C. (2004). *Determination of HSNO ecotoxic thresholds for granular PhoslockTM (Eureka 1 formulation) Phase 1: Acute toxicity.* NIWA Project PXL 05201.
50. Reeve, P. J., & Fallowfeld, H. J. (2018). Natural and surfactant modifed zeolites: A review of their applications for water remediation with a focus on surfactant desorption and toxicity towards microorganisms. *Journal of Environmental Management, 205*, 253–261. https://doi.org/10.1016/j.jenvman.2017.09.077

51. Ates, A., & Akgül, G. (2016). Modifcation of natural zeolite with NaOH for removal of manganese in drinking water. *Powder Technology, 287*, 285–291. https://doi.org/10.1016/j.powtec.2015.10.021
52. Tran, H. N., Van Viet, P., & Chao, H. P. (2018). Surfactant modifed zeolite as amphiphilic and dual-electronic adsorbent for removal of cationic and oxyanionic metal ions and organic compounds. *Ecotoxicology and Environmental Safety, 147*, 55–63. https://doi.org/10.1016/j.ecoenvol2017.08.027
53. Gibbs, M., & Özkundakci, D. (2011). Efects of a modifed zeolite on P and N processes and fuxes across the lake sediment–water interface using core incubations. *Hydrobiologia, 661*, 21–35. https://doi.org/10.1007/s10750-009-0071-8
54. Deppe, T., & Benndorf, J. (2002). Phosphorus reduction in a shallow hypereutrophic reservoir by in-lake dosage of ferrous iron. *Water Research, 36*, 4525–4534. https://doi.org/10.1016/S0043-1354(02)00193-8
55. Klapper, H. (1992). *Eutrophication, water pollution control (Eutrophierung und Gewasserschutz)* (p. 277). Gustav Fischer Verlag.
56. Jekel, M. (1991). Aluminum salts as flocculants (Aluminiumsalze als Flockungsmittel). In: *Arbeitsgemeinschaft Trinkwassertalsperren e.V., Vortragsveranstaltung "Trinkwasser aus Talsperren* (pp. 163–75). R. Oldenbourg Verlag.
57. Hansen, J., Reitzel, K., Jensen, H. S., & Andersen, F. Ø. (2003). Effects of aluminum, iron, oxygen and nitrate additions on phosphorus release from the sediment of a Danish softwater lake. *Hydrobiologia, 492*(1–3), 139–149.
58. Engstrom, D. R. (2005). Long-term changes in iron and phosphorus sedimentation in Vadnais Lake, Minnesota, resulting from ferric chloride addition and hypolimnetic aeration. *Lake and Reservoir Management, 21*(1), 96–106.
59. Effler, S. W., Litten, S., Field, S. D., Tong-Ngork, T., Hale, F., Meyer, M., & Quirk, M. (1980). Whole lake response to low level copper sulphate treatment. *Water Research, 14*, 1489–1499.
60. McKnight, D. M., Chisholm, S. W., & Harleman, D. R. F. (1983). $CuSO_4$ Treatment of nuisance algal blooms in drinking water reservoirs. *Environmental Management, 7*(4), 311–320.
61. Hanson, M. J., & Stefan, H. G. (2010). Side effects of 58 years of copper sulfate treatment of the Fairmont Lakes, Minnesota. *Jawra Journal of the American Water Resources Association, 20*(6), 889–900.
62. Salonen, V. P., Varjo, E., & Rantala, P. (2001). Gypsum treatment in managing the internal phosphorus load from sapropelic sediments: Experiments on Lake Laikkalammi, Findland. *Boreal Environment Research, 6*(2), 119–129.
63. Salonen, V. P., & Varjo, E. (2000). Gypsum treatment as a restoration method for sediments of eutrophied lakes—Experiments from southern Finland. *Environmental Geology, 39*(3–4), 353–359.
64. Kim, H. S., & Park, J. (2008). Effects of limestone on the dissolution of phosphate from sediments under anaerobic condition. *Environmental Technology, 29*(4), 375–380.
65. Dittrich, M., & Koschel, R. (2002). Interactions between calcite precipitation (natural and artificial) and phosphorus cycle in the hardwater lake. *Hydrobiologia, 469*(1–3), 49–57.
66. Anderson, M. A. (2004b). Impacts of metal salt addition on water chemistry of Lake Elsinore, California: 2. *Calcium salts. Lake and Reservoir Management, 20*(4), 270–279.
67. Berg, U., Neumann, T., Donnert, D., Nuesch, R., & Stüben, D. (2004). Sediment capping in eutrophic lakes—Efficiency of undisturbed calcite barriers to immobilize phosphorus. *Applied Geochemistry, 19*(11), 1759–1771.
68. Dittrich, M., Gabriel, O., Rutzen, C., & Koschel, R. (2011). Lake restoration by hypolimnetic $Ca(OH)_2$ treatment: Impact on phosphorus sedimentation and release from sediment. *Science of the Total Environment, 409*(8), 1504–1515.
69. Chambers, P. A., Prepas, E. E., Ferguson, M. E., Serediak, M., Guy, M., & Holst, M. (2001). The effects of lime addition on aquatic macrophytes in hard water: In situ and microcosm experiments. *Freshwater Biology, 46*(8), 1121–1138.
70. Wang, J., Liu, X., & Lu, J. (2012). Urban river pollution control and remediation. *Procedia Environmental Sciences, 13*, 1856–1862.

71. Li, D. L., Pi, J., Zhang, T., Tan, X., & Fraser, D. J. (2018). Evaluating a 5-year metal contamination remediation and the biomonitoring potential of a freshwater gastropod along the Xiangjiang River, China. *Environmental Science and Pollution Research, 25*, 21127–21137.
72. Xiao, L., Ouyang, H., Li, H., Chen, M., & Lin, Q. (2010). Enclosure study on phytoplankton response to stocking of silver carp (Hypophthalmichthys molitrix) in a eutrophic tropical reservoir in South China. *International Review of Hydrobiology, 95*, 428–439.
73. Ma, H., Cui, F., Liu, Z., Fan, Z., & He, W. (2010). Effect of filter-feeding fish silver carp on phytoplankton species and size distribution in surface water: A field study in water works. *Journal of Environmental Sciences (China), 22*, 161–167.
74. Ma, H., Cui, F., Liu, Z., & Zhao, Z. (2012). Pre-treating algae-laden raw water by silver carp during Microcystis-dominated and non-Microcystis-dominated periods. *Water Science and Technology, 65*, 1448–1453.
75. Li, X. N., Song, H. L., Li, W., Lu, X. W., & Nishimura, O. (2010). An integrated ecological floating-bed employing plant, freshwater clam and biofilm carrier for purification of eutrophic water. *Ecological Engineering, 36*, 382–390.
76. Zhao, S. Y., Sun, Y. P., Lin, Q. Q., & Han, B. P. (2013). Effects of silver carp (Hypophthalmichthys molitrix) and nutrients on the plankton community of a deep, tropical reservoir: An enclosure experiment. *Freshwater Biology, 58*, 100–113.
77. Anawar, H. M., Garcia-Sanchez, A., TariKul Alam, M., & Majibur Rahman, M. (2008). Phytofiltration of water polluted with arsenic and heavy metals. *International Journal of Environmental Pollution, 33*, 292–312. [CrossRef].
78. Gagnon, V., Chazarenc, F., Kõiv, M., & Brisson, J. (2012). Effect of plant species on water quality at the outlet of a sludge treatment wetland. *Water Research, 46*, 5305–5315.
79. Fawzy, M., Nasr, M., Abdel-Gaber, A., & Fadly, S. (2015). Biosorption of Cr (VI) from aqueous solution using agricultural wastes, with artificial intelligence approach. *Separation Science and Technology, 51*, 416–426.
80. Allam, A., Tawfik, A., El-Saadi, A., Negm, A. (2016). Potentials of using duckweed (Lemna gibba) for treatment of drainage water for reuse in irrigation purposes. *Desalination and Water Treatment, 57*, 459–467.
81. Ruan, X., Xue, Y., Wu, J., Ni, L., Sun, M., et al. (2006). Treatment of polluted river water using pilot-scale constructed wetlands. *Bulletin of Environmental Contamination and Toxicology, 76*, 90–97.
82. Hadad, H. R., Maine, M. A., & Bonetto, C. A. (2006). Macrophyte growth in a pilot-scale constructed wetland for industrial wastewater treatment. *Chemosphere, 63*, 1744–1753.
83. Sun, L., Liu, Y., & Jin, H. (2009). Nitrogen removal from polluted river by enhanced floating bed grown canna. *Ecological Engineering, 35*, 135–140.
84. Zhao, Y., Yang, Z., Xia, X., & Wang, F. (2012). A shallow lake remediation regime with Phragmites australis: Incorporating nutrient removal and water evapotranspiration. *Water Research, 46*, 5635–5644.
85. Wang, J. F., Chen, J. A., Pei-Min, P. U., Jian, L. I., Yang, Y. Q., & Yang, H. Q. (2012). Chemical components of aquatic plants applied in ecological restoration of eutrophic water in Lake Hongfeng, Guizhou Province of Southwest China. *Chinese Journal of Ecology, 31*(9), 2312–2318.
86. Guo, Y., Liu, Y., Zeng, G., Hu, X., Li, X., Huang, D., Liu, Y., & Yin, Y. (2014). A restoration-promoting integrated floating bed and its experimental performance in eutrophication remediation. *Journal of Environmental Sciences, 26*(5), 1090–1098.
87. Sun, L., Yang, L., & Hui, J. (2014). Nitrogen removal from polluted river by enhanced floating bed grown canna. *Ecological Engineering, 35*(1), 135–140.
88. Coveney, M. F., Stites, D. L., Lowe, E. F., Battoe, L. E., & Conrow, R. (2005). Nutrient removal from eutrophic lake water by wetland filtration. *Ecological Engineering, 19*(2), 141–159.
89. Nie, Z. D., Nian, Y. G., Jin, X. C., Song, Y. W., Li, L. F., & Xie, A. J. (2007). Pilot -scale comparision research of different constructed wetland types to treat eutrophic water. *Environmental Science, 28*(8), 1675–1680.

90. Li, L., Li, Y., Biswas, D. K., Nian, Y., & Jiang, G. (2008). Potential of constructed wetlands in treating the eutrophic water: Evidence from Taihu Lake of China. *Bioresource Technology, 99*(6), 1656–1663.
91. Sheng, Y. Q., Chen, F. Z., Sheng, G. Y., & Fu, J. M. (2012). Water quality remediation in a heavily polluted tidal river in Guangzhou, south China. *Aquatic Ecosystem Health and Management, 15*, 219–226.
92. Liu, J. Z., Wu, Y. H., Wu, C. X., Muylaert, K., Vyverman, W., Yu, H. Q., Muñoz, R., & Rittmann, B. (2017). Advanced nutrient removal from surface water by a consortium of attached microalgae and bacteria: A review. *Bioresource Technology, 241*, 1127–1137.
93. Ateia, M., Yoshimura, C., & Nasr, M. (2016). In-situ biological water treatment technologies for environmental remediation: A review. *Journal of Bioremediation Biodegradation, 7*, 348. https://doi.org/10.4172/2155-6199.1000348
94. Takada, H., Mutoh, K., Tomita, N., Miyadzu, T., & Ogura, N. (1994). Rapid removal of linear alkylbenzenesulfonates (LAS) by attached biofilm in an urban shallow stream. *Water Research, 28*, 1953–1960.
95. Cao, Y., & Sun, C. J. (2009). Application of ecological floating beds to water restoration and its design. *Environmental Science and Technology, 32*, 121–124. (In Chinese).
96. Cao, W. P., Zhang, H. H., Wang, Y. M., & Pan, J. Z. (2012). Bioremediation of polluted surface water by using biofilms on filamentous bamboo. *Ecological Engineering, 42*, 146–149.

Influence of Pond Ash on Compaction and Strength of Clayey Soil Mixed with Terrazyme

Nitish Kumar, Abhishek Sharma, and Kanwarpreet Singh

Abstract It is an incredibly challenging task to construct structures over the soils possessing poor geotechnical characteristics. The present study is focused on improving the geotechnical properties of clayey soil using pond ash and terrazyme to use it as a subgrade material. Various percentages of both the materials, i.e., pond ash varying from 10 to 30% and terrazyme varying from 0.04 to 0.12 ml/kg have been used to perform Atterberg's limits, compaction, and California bearing ratio tests. It was revealed that the addition of pond ash to clayey soil reduced liquid limit. The addition of 20% pond ash along with 0.08 ml/kg terrazyme gave maximum dry density. The California bearing ratio value was enhanced by the addition of optimal quantity of both the materials to clayey soil. This confirms that the mixture of 20% pond ash along with 0.08 ml/kg terrazyme to stabilize poor soils is economical and environmental.

Keywords Soil stabilization · Pond ash · Terrazyme · Consistency limits · Compaction · California bearing ratio

1 Introduction

Soil stabilization is the method of improving soil characteristics by modifying its physical properties. Improving the strength of the soil to improve the life span of the building is highly recommended for any construction project. The geotechnical characteristics are largely affected after stabilization of poor soils. There are lot of waste materials and chemicals which are being used in the past few decades to stabilize poor soils [1–11]. Pond ash (PA) obtained from thermal power plants is a major pollutant to the environment and its disposal is a huge concern. The use of

N. Kumar · A. Sharma (✉) · K. Singh
Department of Civil Engineering, Chandigarh University, Mohali, India
e-mail: abhishek.e9490@cumail.in

K. Singh
e-mail: kanwarpreet.e9570@cumail.in

A. K. Gupta et al. (eds.), *Advances in Construction Materials and Sustainable Environment*, Lecture Notes in Civil Engineering 196,
https://doi.org/10.1007/978-981-16-6557-8_78

various kind of ashes such as fly ash, pond ash, and rice husk ash has been proved to beneficial for the soil stabilization [3, 11–16]. The use of PA in soil stabilization also helps to minimize the number of ponds and to achieve environmentally friendly as well as sustainable natural resource development [12, 13]. On adding PA to clayey soil, the liquid limit decreases and the plasticity index also reduces [14]. The addition of PA to poor soils in optimum amount increases the maximum dry density (MDD), UCS and CBR values of the combination [13–15].

Terrazyme being a liquid enzyme, non-toxic, natural modifies the features of soil [16–19]. The maximum improvement in maximum dry density was achieved on adding 200 ml/m^3 to clayey soil and the UCS value improved by 12 times the untreated strength [16]. The CBR value of terrazyme-treated soil in using optimum content of terrazyme was highest among all the combinations [17]. Biological enzymes reduced voids between soil particles and decreased the quantity of water engrossed in the soil, thereby enhancing the enzyme-induced compaction effect [18]. The sodium lignosulphonate stabilized the black cotton soil as it improved both UCS and CBR value when used in varying content from 1 to 4% [19].

Thus, from the literature, it can be stated that both pond ash and terrazyme alone may be used to stabilize poor soils. The combined effect of pond ash and terrazyme is limited in past research and thus keeping in view this gap the present study focuses on utilization of pond ash and terrazyme both together to enhance geotechnical propertied of poor soils.

2 Materials

The soil utilized in the current research was procured from Morinda, Punjab, India and was termed as clay with high plasticity (CH) rendering to IS 1498–1970 (Fig. 1). Due to its high swelling and shrinkage, the cracks are generally appeared on the structures founded over this soil and hence needs proper treatment before using it. Pond ash (PA) was acquired from Ropar, Punjab, India. The percentages of PA to be utilized for research testing were varied as 10, 20, and 30%. Terrazyme was acquired from Chennai, India. The amount of terrazyme was 0.04, 0.08, and 0.12 ml/kg for the stabilization of poor clayey soil and was decided based on the extensive literature. The different mechanical properties of the materials utilized in the current examination have been classified in Table 1.

3 Methodology

The testing for the current research was carried out in accordance with Indian standards in geotechnical engineering laboratory of Chandigarh University, Punjab, India. The percentages of various materials used in the current study were decided based on the literature review and are tabulated in Table 2. For carrying out the compaction

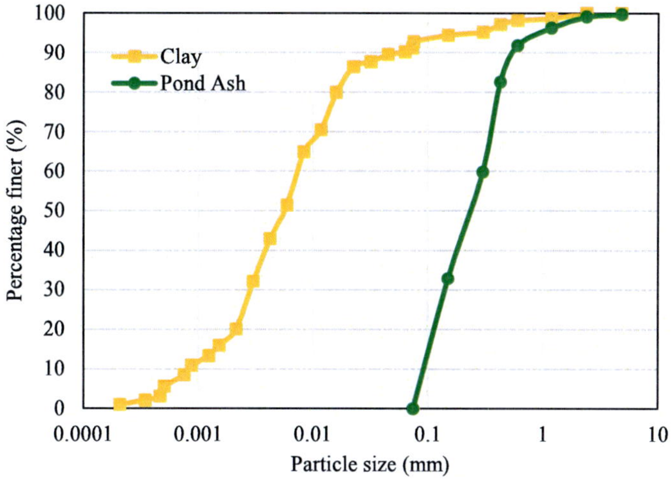

Fig. 1 Gradation curve of materials used in the research

Table 1 Mechanical properties of various materials

Properties	Clay	Pond ash
Specific gravity	2.44	2.07
Liquid limit (%)	64.3	–
Plasticity index (%)	29.2	–
Optimum moisture content (%)	17.6	24.4
Maximum dry density (g/cc)	1.78	1.31
California bearing ratio (soaked)	1.92	–
Coefficient of uniformity (C_u)	–	3.09
Coefficient of curvature (C_c)	–	0.95
Classification	CH	SP

Table 2 Percentages of materials used in the study

Materials	Proportions used in the study
Soil	100
Soil: PA	90:10
Soil: PA	80:20
Soil: PA	70:30
Soil: PA: T	80:20:0.04 ml/kg
Soil: PA: T	80:20:0.8 ml/kg
Soil: PA: T	80:20:0.12 ml/kg

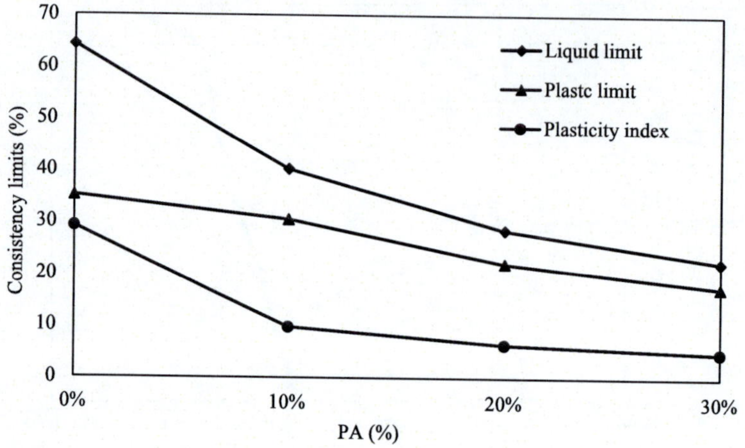

Fig. 2 Consistency limits of clay and pond ash mixes

test, the amount of terrazyme was 0.12, 0.24, and 0.36 ml as 3 kg of sample was required to fill the mold of compaction apparatus. The terrazyme was injected to the water in fixed proportion with the help of a needle and then the water was mixed into various combinations.

4 Results and Discussion

4.1 Consistency Limits

The consistency limit of the soil was assessed and the soil was expressed as CH soil. At first, on adding 10–30% PA in clayey soil, the liquid limit decreased from 40.2 to 22.4% compared with that of 64.3% of clayey soil alone. This decrease in liquid limit of the composite was due to the coarser nature of PA. The plasticity limit was also discovered to be decreased. The decrease of plasticity index is a significant factor that would lead to upgradation of geotechnical properties of soil. As the ratio C: PA from 100: 0 to 90: 10, 80: 20 and 70: 30, the plasticity index was decreased from 29.2 to 4.7% (Fig. 2).

4.2 Compaction Characteristics

Compaction tests were accomplished on soil, PA and various soil mixtures-PA, and soil-PA-T to determine the MDD and OMC and are presented in Figs. 3 and 4, respectively. The MDD and OMC of virgin soil were 1.78 g/cc and 17.6%; MDD

Fig. 3 Compaction characteristics for C: PA

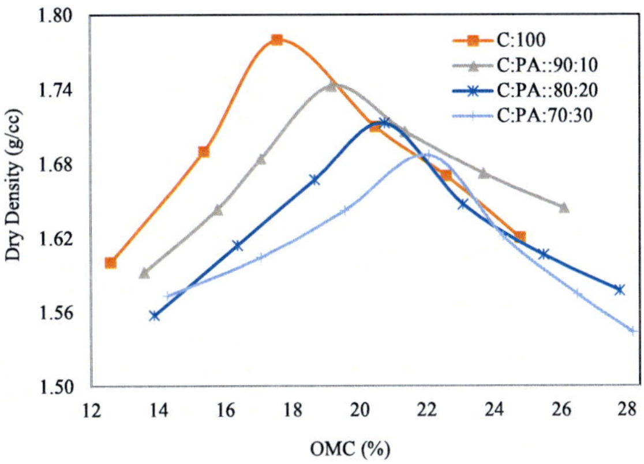

Fig. 4 Compaction characteristics for C: PA + terrazyme

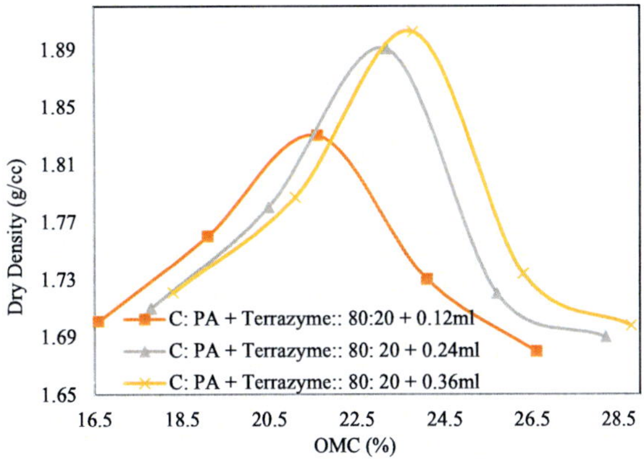

of the PA was 1.75 g/cc and related OMC was 21.3%. As the percentage of PA in virgin soil increased, the value of MDD decreased and OMC increased (Fig. 3). The MDD decreased from 1.78 to 1.687 g/cc with PA content up to 30% and the OMC augmented from 17.6 to 22.1%. When optimum percentage PA (20%) was mixed in soil with varying percentages of terrazyme 0.12 ml, 0.24 ml, and 0.36 ml, MDD of soil improved from 1.713 to 1.90 g/cc, and related OMC increased from 20.8 to 23.8% on addition of 0.36 ml terrazyme (Fig. 4). There was a minor improvement in MDD beyond 0.24 ml terrazyme in C: PA:: 80: 20 mixture and hence may be taken as optimum content.

4.3 California Bearing Ratio

To discover the suitability of different composites as a subgrade material, CBR testing under soaked conditions for a curing period of 4 days was carried out for C: PA mixes

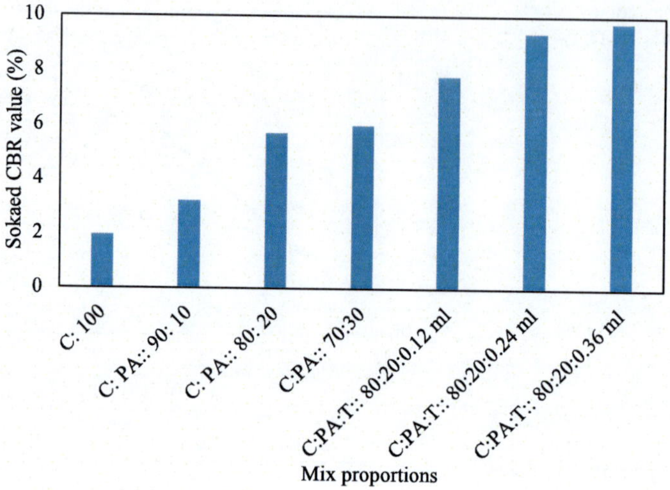

Fig. 5 CBR values for different materials

in proportions 90: 10, 80: 20, and 70:30. It was revealed from the test outcomes that CBR value was maximum for C: PA:: 70: 30 content (CBR = 9.8) but 20% PA was used as optimum mix, because after 20% there was minor improvement (CBR = 9.4) as shown in Fig. 5. The further testing of CBR value on addition of terrazyme in optimal mix of C: PA:: 80: 20 was carried out and the CBR value was found to be increased at a lower rate after adding 0.08 ml/kg terrazyme and also the MDD value is almost constant at this proportion and thus a combination of C: PA: T:: 80: 20: 0.24 ml was used for soil stabilization. This increment in CBR value on the inclusion of terrazyme might be attributed to the properties of terrazyme which helps to bind the clay particles together and minimizes crack formation thus leading to improved CBR.

5 Conclusions

The strength behavior of clayey soil by mixing PA and terrazyme was studied by conducting consistency limits, compaction tests, and CBR tests, and the outcomes were presented in the above sections. The main conclusions drawn from the current study are summed up as follows:

1. The addition of different wastes, i.e., PA and terrazyme improves the consistency limit of the clayey soil and change its classification from CH to CL.
2. There was a reduction in MDD on adding PA alone to clayey soil. The addition of terrazyme clay and pond ash mixture improved maximum dry density to highest value.

3. The CBR value of the mix containing C: PA + terrazyme:: 80: 20 + 0.24 ml was 9.7 around and was eight times higher than that of the clayey soil alone thus proving its efficiency to be utilized a subgrade material.

From the results obtained, it can be concluded that 20% PA and 0.24 ml terrazyme is the optimum ratio mixed in weak soil. Terrazyme being non-toxic and natural one may be suitable over other chemicals which are toxic to soil as it provides better packing of soil particles and provides strength to soil. It is a better way to enhance the geotechnical characteristics, and it is used constructively on low-density highways, shallow foundations, and sub-base layer of pavement.

References

1. Abhishek, Sharma, R. K., & Bhardwaj, A. (2018, November). Effect of construction demolition and glass waste on stabilization of clayey soil. In *International Conference on Sustainable Waste Management through Design* (pp. 87–94). Springer. https://doi.org/10.1007/978-3-030-02707-0_12
2. Verma, V., & Abhishek. (2018, November). Stabilization of clayey soils using fly ash and RBI grade 81. In *International Conference on Sustainable Waste Management through Design* (pp. 95–102). Springer. https://doi.org/10.1007/978-3-030-02707-0_13
3. Sharma, A., & Sharma, R. K. (2019). An experimental study on uplift behaviour of granular anchor pile in stabilized expansive soil. *International Journal of Geotechnical Engineering*, 1–14. https://doi.org/10.1080/1938636.2.2019.157481
4. Bhardwaj, A., & Sharma, R. K. (2020). Effect of industrial wastes and lime on strength characteristics of clayey soil. *Journal of Engineering, Design and Technology.*
5. Sharma, A., & Sharma, R. K. (2020a). Effect of addition of construction–demolition waste on strength characteristics of high plastic clays. *Innovative Infrastructure Solutions*, *4*(1), 1–11. https://doi.org/10.1007/s41062-019-0216-1
6. Sharma, A., & Sharma, R. K. (2020b). Strength and drainage characteristics of poor soils stabilized with construction demolition waste. *Geotechnical and Geological Engineering*, *38*, 4753–4760. https://doi.org/10.1007/s10706-020-01324-3
7. Anand, D., Sharma, R. K., & Sharma, A. (2021). Improving swelling and strength behavior of black cotton soil using lime and quarry dust. In: H. Singh, P. P. Singh Cheema, & P. Garg (Eds.). *Sustainable development through engineering innovations. Lecture notes in civil engineering* (vol. 113). Springer. https://doi.org/10.1007/978-981-15-9554-7_54
8. Bhardwaj, A., Sharma, R. K., & Sharma, A. (2021). Stabilization of clayey soil using waste foundry sand and molasses. In: H. Singh, P. P. Singh Cheema, & P. Garg (Eds.), *Sustainable development through engineering innovations. Lecture notes in civil engineering* (vol. 113). Springer. https://doi.org/10.1007/978-981-15-9554-7_57
9. Gautam, K. K., Sharma, R. K., & Sharma, A. (2021). Effect of municipal solid waste incinerator ash and lime on strength characteristics of black cotton soil. In: H. Singh, P. P. Singh Cheema, & P. Garg (Eds.), *Sustainable development through engineering innovations. Lecture notes in civil engineering* (vol. 113). Springer. https://doi.org/10.1007/978-981-15-9554-7_10
10. Sharma, A., & Sharma, R. K. (2021). Sub-grade characteristics of soil stabilized with agricultural waste, constructional waste, and lime. *Bulletin of Engineering Geology and the Environment*, *80*(3), 2473–2484. https://doi.org/10.1007/s10064-020-02047-8
11. Singh, S., Sharma, R. K., & Sharma, A. (2021). Arresting the heave of black cotton soil using geogrid-encased granular pile anchor. In: H. Singh, P. P. Singh Cheema, & P. Garg (Eds.), *Sustainable development through engineering innovations. Lecture notes in civil engineering* (vol. 113). Springer. https://doi.org/10.1007/978-981-15-9554-7_30

12. Takhelmayum, G., Savitha, A. L., & Krishna, G. (2013). Laboratory study on soil stabilization using fly ash mixtures. *International Journal of Engineering Science and Innovative Technology (IJESIT), 2*(1), 477–482.
13. Hozatlıoğlu, D. T., & Yılmaz, I. (2021). Shallow mixing and column performances of lime, fly ash and gypsum on the stabilization of swelling soils. *Engineering Geology, 280*, 105931.
14. Tiwari, N., & Satyam, N. (2021). Coupling effect of pond ash and polypropylene fiber on strength and durability of expansive soil subgrades: An integrated experimental and machine learning approach. *Journal of Rock Mechanics and Geotechnical Engineering.*
15. Sharma, A., Sharma, A., & Singh, K. (2020). Bearing capacity of sand admixed pond ash reinforced with natural fiber. *Journal of Natural Fibers*, 1–14. https://doi.org/10.1080/15440478.2020.1848699
16. Karatai, T. R., Kaluli, J. W., Kabubo, C., & Thiong'o, G. (2017). Soil stabilization using rice husk ash and natural lime as an alternative to cutting and filling in road construction. *Journal of Construction Engineering and Management, 143*(5), 04016127. https://doi.org/10.1061/(ASCE)CO.1943-7862.0001235
17. Eujine, G. N., Somervell, L. T., Chandrkaran, S., & Sankar, N. (2014). Enzyme stabilization of high liquid limit clay. *Electron J Geotech Eng, 19.*
18. Ramesh, H. N., & Sagar, S. R. (2015). Effect of drying on the strength properties of Terrazyme treated expansive and non-expansive soils. In *50th Indian geotechnical conference, Pune, India.*
19. Konnur, S., & Antaratani, S. (2017). Ground Improvement by stabilization technique using pond ash and sodium lignosulphonate. *Technical research organisation India, 4*(7).

Seismic Analysis of Vertically Regular and Irregular Buildings with Shear Walls and RCC X-Bracing System

Mohd Zahid, Md Miraz, Mohd Faizan Saifi Warsi, and Shilpa Pal

Abstract This study observed the performance of simple G+10 Structure without and with shear wall and RCC X-bracing system and the effect of vertical irregularity (stiffness irregularity) introduced at 3rd and 6th floors in high rise building (G+10). Seismic analysis is done for the (G+10) Structure by dynamic method (response spectrum analysis) using STAAD Pro. Twelve structures have been analysed in this study, wherein four of them are base models, i.e. Structure without stiffness irregularity and then stiffness irregularity is introduced at the 3rd floor in all four structures, totalling eight structures. Then stiffness irregularity is introduced on the 6th floor, making a total of twelve structures. The four base models in this study are Simple Structure, Structure with shear walls at corners, Structure with RCC X-bracings at corners and Structure with shear walls and RCC X-bracing at alternate corners. The results of the node displacement, base shear and storey drift are recorded. The results obtained from STAADPro concluded that storey with increased storey height at lower storey's levels leads to more damage under seismic load, shear wall model is the most effective Structure and storey drift decreases to a minimum when irregular Structure is stiffened with the shear walls.

Keywords Shear walls · RCC X-bracing system · Stiffness irregularity · Response spectrum analysis

1 Introduction

High rise structures aim to transfer the primary gravity loads viz. dead load and live load safely. In addition, the Structure should withstand lateral forces caused by the earthquake, the lateral loads produce sway moments and induce high stresses, thus reduces the stability of the Structure [1]. In order to resist lateral loads, structure

M. Zahid (✉) · M. Miraz · M. F. S. Warsi · S. Pal
Delhi Technological University, New Delhi, India

S. Pal
e-mail: shilpapal@dtu.ac.in

A. K. Gupta et al. (eds.), *Advances in Construction Materials and Sustainable Environment*, Lecture Notes in Civil Engineering 196,
https://doi.org/10.1007/978-981-16-6557-8_79

stiffness is a more critical parameter than strength [1]. During an earthquake, it has been found out that building with irregularity in elevation caused non-uniform load distribution in various components of buildings due to this some severe effects arises. The primary effects caused by earthquakes are collapsing buildings that will put so many lives in danger. Secondary effects arise after the effects of earthquake fires, landslides, destruction and disease. The failure of Structure is because of weakness that arises due to discontinuity stiffness of the Structure. The structures having this discontinuity are termed irregular structures, vertical irregularities are one of the significant reasons for failures of structures during earthquakes. Height-wise variations in stiffness render the dynamic characteristics of these buildings different from the 'regular' buildings [2]. IS 1893 definition of Vertically Irregular structures: The irregularity in the building structures may be due to irregular distributions in their stiffness along with the height of the building [2]. Therefore, the building must have sufficient strength and stiffness to encounter vertical loads and lateral loads, respectively. Various techniques are used to minimise the seismic effect on high rise buildings like Shear wall and RCC X- bracing system. Shear wall/RCC bracing is an excellent means of providing earthquake resistance to multi-storied reinforced concrete buildings. It creates shear wall-frame interaction that resists the lateral load by the earthquake.

2 Objective of Study

This study proposes using a shear wall, RCC X-bracing system in different vertically regular and irregular structures to make a structure safe. The main objectives of this study are:

- To study the seismic effect on four base models, i.e. Simple Structure, Structure with shear walls at corners, Structure with RCC X-bracings at corners and Structure with shear walls and RCC X-bracing at alternate corners.
- To study the seismic effect on vertically irregular (stiffness irregularity) buildings.
- To study the seismic effect on vertically irregular (stiffness irregularity) buildings stiffened with the shear wall or RCC X-bracing or both.

3 Methodology

Total four main/base models without stiffness irregularity and two sub-models for vertical stiffness irregularity at 3rd floor and 6th floor of each type are modelled in which height of variable storey is increased from 4 to 6 m, thereby total of twelve structures was modelled and analysed using STAAD-Pro software. The shear walls and RCC X-bracings are located at corners in L-shaped as this location is most efficient as found in previous studies [3]. The building plan size is 15 m × 15 m. The building is situated in Delhi, earthquake Zone-4, seismic zone coefficient is taken as

0.24 as per IS code. The whole analysis is being done by STAADPRO.V8i software by providing adequate material properties and dimensions as per the required building standard. All twelve structures are analysed and node displacement, base shear and interstorey drift parameters are recorded.

4 Modelling and Analysis

Twelve structures have been modelled and analysed (Table 1) using Response spectrum analysis of IS 1893-part 1: 2016 [4], location of Structure is in Delhi, seismic zone IV, soil type medium soil, effective damping 5% and importance factor 1.0. Analysis has been carried out using STAADPro.V8i software. The shear wall or RCC X-bracing orientation is considered at the corners of the building with L-shaped.

Following twelve structures shown in Fig. 1 have been modelled and analysed using STAADPro.V8i software:

Model 1.A: Simple structure; Model 1.B: structure with stiffness irregularity at 3rd floor; Model 1.C: structure with stiffness irregularity at 6th floor.

Table 1 Specifications given to structure

S. No.	Parameters	Dimensions/Type	Remarks/References/IS code
1	Plan dimension	15 m × 15 m	Symmetric building
2	Number of stories	11(G+10)	11 storey building
3	Height of each storey Height of irregular storey	4 m 6 m	Storey height of 4 m and 6 m is provided at 3rd floor or 6th floor in irregular structure
4	Column/Beam sizes	750 mm × 750 mm	Size of cross-section
5	Earthquake zone factor (Z)	0.24 (Zone IV)	IS 1893(Part 1):2016
6	Damping ratio	5%	IS 1893(Part 1):2016
7	Importance factor (I)	1.0	IS 1893(Part 1):2016
8	Soil type	Medium soil	IS 1893(Part 1):2016
9	Response reduction factor (R)	5	IS 1893(Part 1):2016
10	Live load On floors On roof	4 KN/m^2 2 KN/m^2	IS 875 (Part 2)-1987
11	RCC X-Bracing size	450 mm × 450 mm	Size of cross-section
12	Dead load Self-weight of RCC Self-weight of Brick masonry	25 KN/m^2 20 KN/m^2	IS 875 (Part 1)-1987
13	Shear wall thickness	230 mm	IS 13920:1993
14	Slab thickness	150 mm	IS 456:2000

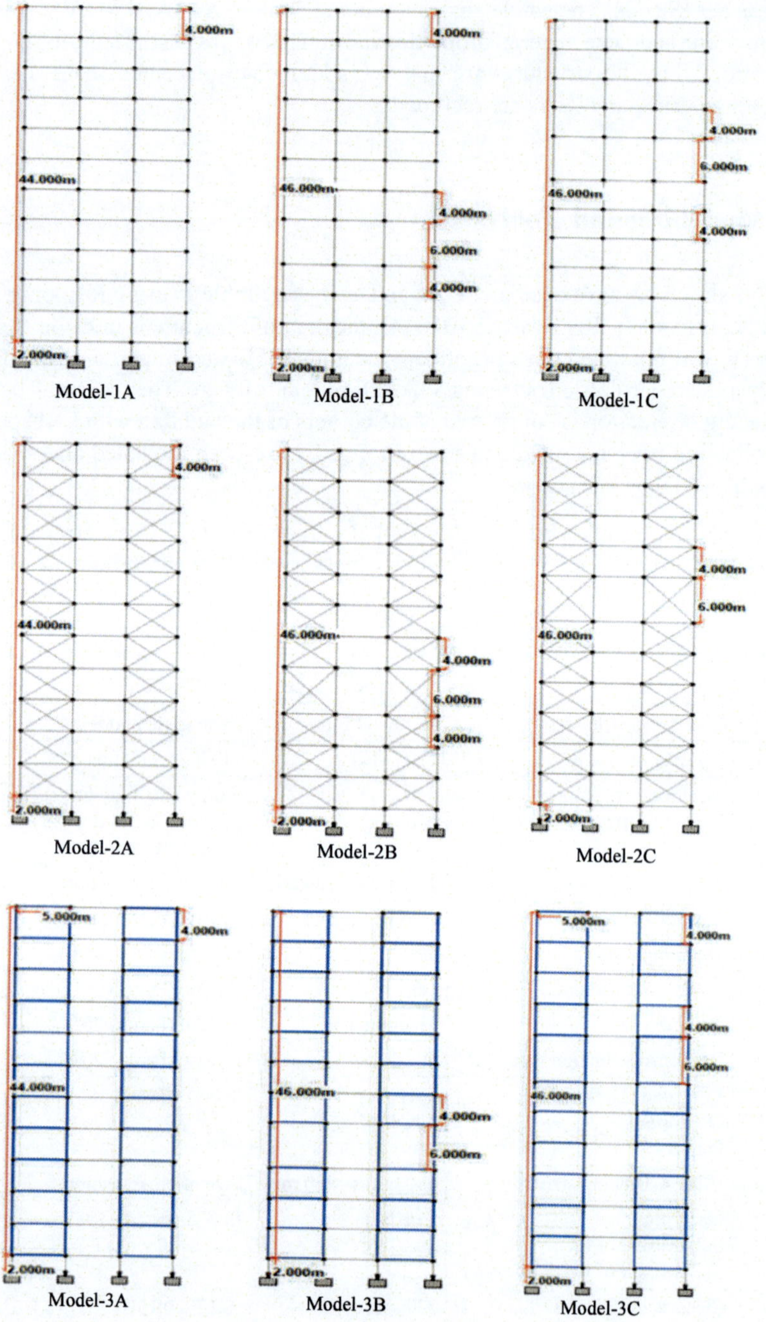

Fig. 1 STAAD.Pro front view of all twelve structures

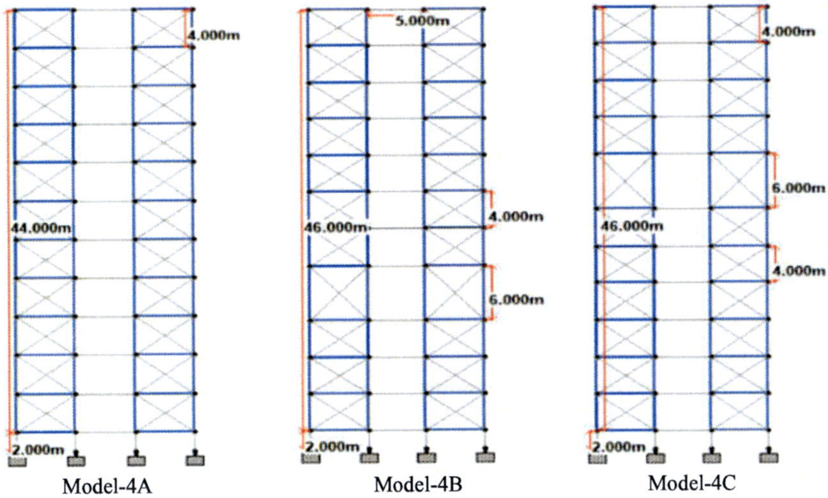

Fig. 1 (continued)

Model 2.A: structure with RCC X-bracing at corners along the periphery; Model 2.B: structure with RCC X-bracing at corners along the periphery with stiffness irregularity at 3rd floor; Model 2.C: structure with RCC X-bracing at corners along the periphery with stiffness irregularity at 6th floor.
Model 3.A: structure with shear walls at corners along the periphery; Model 3.B: structure with shear walls at corners along the periphery with stiffness irregularity at 3rd floor; Model 3.C: structure with shear walls at corners along the periphery with stiffness irregularity at 6th floor
Model 4.A: structure with shear walls and RCC X-bracing system at corners positioned at alternate bays; Model 4.B: structure with shear walls and RCC X-bracing system positioned at alternate bays with stiffness irregularity at 3rd floor; Model 4.C: structure with shear walls and RCC X-bracing system positioned at alternate bays with stiffness irregularity at 6th floor.

5 Results and Discussions

Twelve structures have been analysed and their responses like node displacement, base shear and interstorey drift have been calculated to study the effects of stiffness irregularity on the four base models. The results are shown and discussed henceforth.

5.1 Node Displacement

Maximum node displacement values of all twelve models and their % change in node displacement with respect to their base model, i.e. structure without stiffness irregularity is shown below in Table 2.

From Figs. 2 and 3, it is observed that node displacement of Model-1 (simple structure) is maximum among all twelve structures therefore, it is more vulnerable

Table 2 Maximum node displacement for all models

S. No.	Model No.	Model type	Node displacement (mm)	% change in node displacement
1	1A	Simple structure with no irregularity	43.228	–
2	1B	simple structure with stiffness irregularity at 3rd floor	51.429	18.97% rise w.r.t. model-1A
3	1C	simple structure with stiffness irregularity at 6th floor	47.685	10.31% rise w.r.t. model-1A
4	2A	Structure with X-bracing	23.34	–
5	2B	Structure with X-bracing and stiffness irregularity at 3rd floor	25.87	10.83% rise w.r.t. model-2A
6	2C	Structure with X-bracing and stiffness irregularity at 6th floor	25.228	8.08% rise w.r.t. model-2A
7	3A	Structure with shear-wall	19.288	–
8	3B	Structure with shear wall and stiffness irregularity at 3rd floor	21.15	9.65% rise w.r.t. model-3A
9	3C	Structure with shear wall and stiffness irregularity at 6th floor	20.984	8.79% rise w.r.t. model-3A
10	4A	Simple structure with shear-wall and X-bracing	23.691	–
11	4B	Structure with shear wall, X-bracing and stiffness irregularity at 3rd floor	26.083	10.09% rise w.r.t. model-4A
12	4C	Structure with shear wall, X-bracing and stiffness irregularity at 6th floor	25.877	9.22% rise w.r.t. model-4A

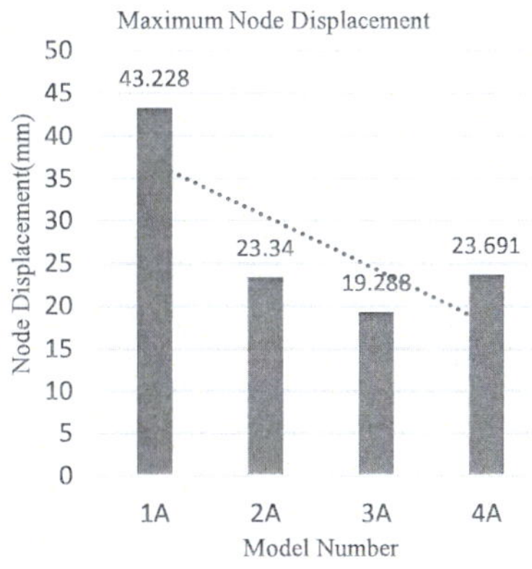

Fig. 2 Structure without stiffness irregularity

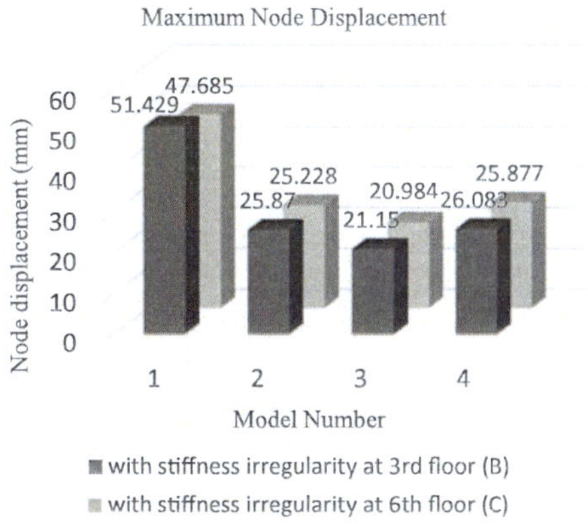

Fig. 3 Structures with stiffness irregularity at 3rd and 6th floors

under seismic loading. By incorporating the structure with shear wall or RCC X-bracing or both it is observed that the node displacement value decreases.

From Table 2 it is observed that node displacement for Model-1B, Model-1C is 18.97%, 10.31% more than Model-1A respectively, thus for simple structure (Model-1), 3rd floor is displacing more and can be seen in Fig. 3 as well, therefore 3rd floor is more vulnerable compared to 6th floor for simple structure because of uneven load distribution. Contrarily, Structure incorporated with the shear wall or RCC X-bracing or both (Model-2, Model-3, Model-4) gives almost the same results for 3rd floor or 6th floor and can be seen in Fig. 3.

5.2 *Base Shear*

Base shear is the sliding force that is generated at the base of the structure, especially due to seismic forces. The structure with maximum base shear is more rigid and stiff, thus showing better seismic performance. It is observed that the base shear value is directly proportional to the stiffness of the structure. The base shear values of all twelve models and their % change in base shear with respect to their base model, i.e. structure without stiffness irregularity, is shown below in Table 3.

From Fig. 4, it is observed that the base shear value of model 3A is maximum and when stiffness irregularity is introduced at 3rd or 6th floors, as seen in Fig. 5, it is also observed that base shear value of Model-3B, Model-3C, which is structure incorporated with the shear wall, is maximum compared to other structures, therefore structure incorporated with shear walls is more stiff and rigid.

Table 3 Base shear of all twelve different models

S. No.	Model No.	Model type	Base shear (KN)	% change in base shear
1	1A	Simple structure with no irregularity	1062.56	–
2	1B	Simple structure with stiffness irregularity at 3rd floor	956.06	10.02% Decrease w.r.t Model-1A
3	1C	Simple structure with stiffness irregularity at 6th floor	1025.49	3.48% Decrease w.r.t. Model-1A
4	2A	Simple structure with X-bracing	1568.03	–
5	2B	Structure with X-bracing and stiffness irregularity at 3rd floor	1485.58	5.25% Decrease w.r.t. Model- 2A
6	2C	Structure with X-bracing and stiffness irregularity at 6th floor	1524.43	2.78% Decrease w.r.t. Model- 2A
7	3A	Simple structure with shear- wall	1822.6	–
8	3B	Structure with shear wall and stiffness irregularity at 3rd floor	1748.76	4.05% Decrease w.r.t. Model- 3A
9	3C	Structure with shear wall and stiffness irregularity at 6th floor	1764.38	3.19% Decrease w.r.t. Model- 3A
10	4A	Simple structure with shear-wall and X-bracing	1687	–
11	4B	Structure with shear wall, X-bracing and stiffness irregularity at 3rd floor	1609.53	4.63% Decrease w.r.t. Model- 4A
12	4C	Structure with shear wall, X-bracing and stiffness irregularity at 6th floor	1635.7	3.08% Decrease w.r.t. Model- 4A

Fig. 4 Structures without stiffness irregularity

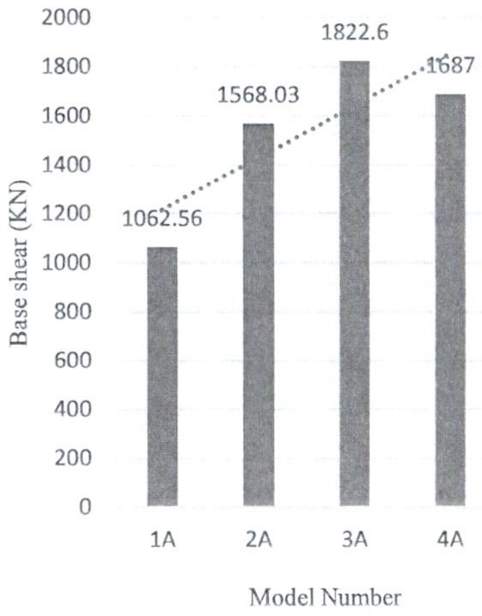

Fig. 5 Structures with stiffness irregularity at 3rd and 6th floors

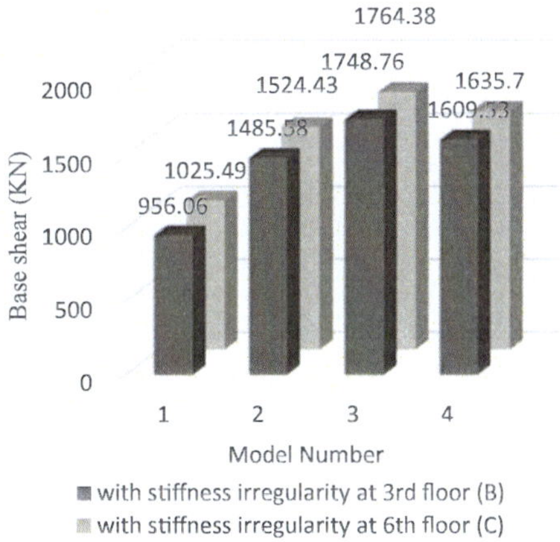

From Table 3 it is observed that base shear for Model-1B, Model-1C is 10.02%, 3.46% less than Model-1A respectively, thus for simple structure (Model-1), 3rd floor is experiencing less base shear and can be seen in Fig. 5 as well. Contrarily, Structure incorporated with the shear wall or RCC X-bracing or both (Model-2, Model-3, Model-4) gives almost the same results for 3rd floor or 6th floor and can be seen in Fig. 5.

5.3 Interstorey Drift

Interstorey drift is the relative displacement between two consecutive floors. Storey drift values of all twelve models are within permissible limit as per IS 1893 (part 1) [4] and shown in Tables 4, 5, and 6.

From Figs. 6, 7, and 8 it is observed that when structure is incorporated with shear wall or RCC X-bracing system or both, then the value of storey drift decreases and

Table 4 Interstorey drift for structures without stiffness irregularity

	Storey drift (cm)			
Floors	1A	2A	3A	4A
0	0.0775	0.0807	0.1112	0.0971
G.F	0.7178	0.3568	0.3447	0.3546
1	0.9224	0.4799	0.432	0.4625
2	1.0838	0.5884	0.5139	0.5572
3	1.1062	0.6534	0.5677	0.6159
4	1.0602	0.689	0.5991	0.6472
5	1.0148	0.7096	0.6161	0.6622
6	0.9686	0.7151	0.6195	0.6622
7	0.9003	0.7033	0.6093	0.6468
8	0.7891	0.6714	0.5853	0.6155
9	0.6264	0.6162	0.5465	0.567
10	0.4325	0.5386	0.4963	0.5063

Table 5 Interstorey drift for structures with stiffness Irregularity at 3rd floors

	Storey drift (cm)			
Floor	1B	2B	3B	4B
0	0.07	0.0753	0.1051	0.092
G.F	0.65	0.3169	0.3365	0.3452
1	0.8467	0.4612	0.4247	0.4532
2	1.0685	0.5788	0.5121	0.5563
3	2.3162	1.1271	0.9346	1.0381
4	1.0567	0.697	0.6146	0.6615
5	0.951	0.7108	0.6241	0.6666
6	0.9021	0.7153	0.6261	0.6643
7	0.8382	0.7019	0.614	0.6469
8	0.7358	0.6694	0.5889	0.6147
9	0.5874	0.6152	0.5502	0.567
10	0.4118	0.5404	0.5011	0.5083

Table 6 Interstorey drift for structures with stiffness irregularity at 6th floors

	Storey drift (cm)			
Floors	1C	2C	3C	4C
0	0.0747	0.0785	0.1072	0.094
G.F	0.692	0.3482	0.3371	0.3459
1	0.8877	0.4693	0.4245	0.4527
2	1.039	0.577	0.5075	0.5477
3	1.0553	0.6429	0.5637	0.6084
4	1.0127	0.6803	0.5979	0.6428
5	1.0269	0.7089	0.6231	0.6695
6	2.0551	1.2134	1.0132	1.1033
7	0.9292	0.7094	0.6205	0.6584
8	0.7598	0.6747	0.593	0.622
9	0.5963	0.6231	0.5564	0.5758
10	0.4163	0.5486	0.5077	0.517

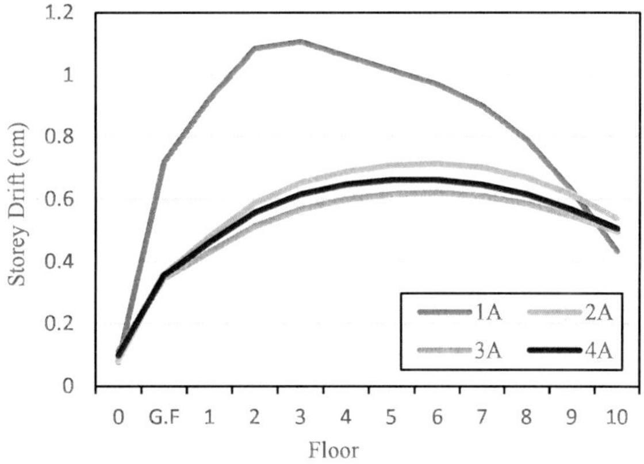

Fig. 6 Storey drift for structures without stiffness irregularity

this effect is experience maximum in shear wall structure as storey drift value of Model-3 < Model-4 < Model-2 < Model-1. It is also evident from these figures that the value of storey drift reduces to a minimum for Model-3A, 3B, 3C, which is a structure incorporated with shear walls.

From Fig. 7, it is observed that storey drift value drastically increases on 3rd floor because of the presence of the variable storey. The storey drift at 3rd floor is 2.3162 cm, 1.1271 cm, 0.9346 cm and 1.0381 cm for Model-1B, Model-2B, Model-3B and Model-4B respectively. So when the simple structure is incorporated with the shear wall, the storey drift reduces to a minimum from 2.3162 cm to 0.9346 cm, decreasing 59%.

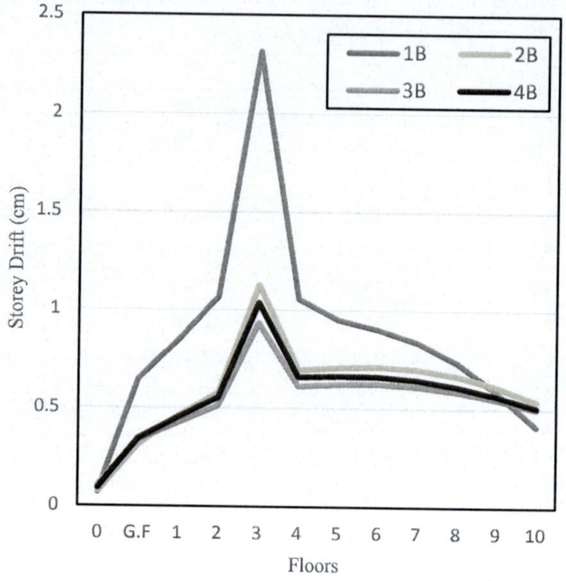

Fig. 7 Storey drift for structures with stiffness irregularity at 3rd floor

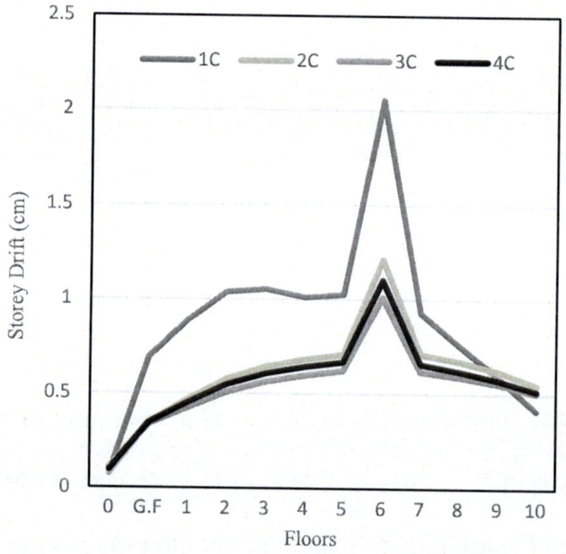

Fig. 8 Storey drift for structures with stiffness irregularity at 6th floor

From Fig. 8, it is observed that storey drift value drastically increases at the 6th floor because of the presence of the variable storey. The storey drift at 6th floor is 2.0551 cm, 1.2134 cm, 1.0132 cm and 1.1033 cm for Model-1C, Model-2C, Model-3C and Model-4C, respectively. So when the simple structure is incorporated with the shear wall, the storey drift at the 6th floor reduces to a minimum from 2.0551 cm to 1.0132 cm, decreasing 51%.

Thus, incorporating structures with shear walls in irregular buildings is most effective among all other structures.

6 Conclusions

All structures are analysed by Response spectrum analysis using STAADPro software. The parameters in terms of node displacement, base shear and storey drift are compared and the following conclusions are drawn:

- The base shear value of Model-3A, Model-3B, Model-3C is 71.5%, 82.9%, 72% more than Model 1A, 1B, 1C respectively and its node displacement is 55.4%, 58.8%, 56% less than Model-1A, 1B, 1C respectively. Therefore, Structure incorporated with shear walls (Model-3A, Model-3B, Model-3C) is the most efficient structure as its node displacement is minimum and base shear value is maximum.
- The node displacement value for Model-1B is 7.27% more than Model-1C and the base shear value for Model-1B is 7.26% less than Model-1C. Therefore, (Model-1B), the simple Structure with stiffness irregularity on 3rd floor is more vulnerable than (Model-1C), the structure with stiffness irregularity on 6th floor. Thus, simple structures with stiffness irregularity at lower floors leads to more damage under seismic loading.
- Storey drift value of Model-3B on 3rd floor, Model-3C on 6th floor is 59%, 51% less than Mode-1B, Model-1C respectively. Thus, the Shear wall is able to resist maximum drift in irregular buildings as the value of interstorey drift decreases to a minimum when the structure is incorporated with shear walls.

References

1. Islam, S. U., Goel, R., & Sharma, P. (2018). Seismic coaction of shear wall & reinforced cement concrete bracing system in high rise commercial buildings using STAAD Pro. *International Journal of Engineering Development and Research, 6*, 2321–9939.
2. Dubule, S. C., & Ainchwar, D. (2018). Seismic analysis and design of vertically irregular R.C. building frames. *International Journal of Advanced Research in Science and Engineering, 7*, 290–306.
3. Suresh, M., & Shayana Yadav, S. A. (2015). The optimum location of shear wall in high rise R.C buildings under lateral loading. *International Journal of Research in Engineering and Technology, 04*, 2321–7308.
4. Bureau of Indian Standards IS 1893 (Part 1)-2016. Criteria for Earthquake Resistant Design of Structures.

Crumb Rubber Concrete Formation by Partial Replacement of Fine Aggregates

Pradeep K. Goyal and Akhilesh Chauhan

Abstract Concrete is mostly used construction material in the construction of structures, bridges. It contains natural aggregates and cement. The use of aggregates includes fine and coarse aggregates in concrete which is likely to deplete natural resources and also creating environmental problems. There is a need to use alternative materials which can reduce the requirement of natural resources. Nowadays, huge amount of tyre is generated as waste which creates the disposal problem. It has become a major environmental issue. Several researchers have been done to use tyre waste as an alternate material. In this paper, crumb rubber is used as a replacement for natural fine aggregates in concrete at 0, 5, 10, 15 and 20%. The compressive strength of M30 grade concrete using crumb rubber as partial replacement of fine aggregate was investigated as per IS10262-2009. The compressive strength of crumb rubber concrete was obtained satisfactory when concrete mix was made 10% partial replacement of fine aggregate with crumb rubber. In many applications, crumb rubber concrete may be useful as compared to conventional concrete in spite of less compressive strength. Crumb concrete has many desirable properties which are useful many structural applications.

Keywords Concrete · Crumb rubber · Compressive strength · Fine aggregate · Waste material

1 Introduction

With the increasing demand of vehicle, large quantity of rubber tyres waste is being produced. It does not have useful disposal till now and become a major environmental problem. Most of houses and structures are being constructed using concrete.

P. K. Goyal
Department of Civil Engineering, Delhi Technological University, Delhi, India
e-mail: pkgoyal@dtu.ac.in

A. Chauhan (✉)
Department of Civil Engineering, Delhi Technological University, Delhi, India

A. K. Gupta et al. (eds.), *Advances in Construction Materials and Sustainable Environment*, Lecture Notes in Civil Engineering 196,
https://doi.org/10.1007/978-981-16-6557-8_80

Concrete is a mixer of specified proportion of fine aggregates, coarse aggregates, cement and water. Nowadays, lots of construction activities, i.e. building, bridges, roads and dams are going on rapidly utilizing the concrete for their construction. Therefore, large quantities of natural aggregates are required to meet need of construction industry. Natural aggregates which are used in construction are depleting day by day. Also, use of natural resources to a great extent become a major environmental issue worldwide. To protect the environment, there is need of alternate material which can be used in concrete. Use of waste materials, like fly ash, silica fumes, scrap tyres, etc., is being studied for making concrete. Studies show that the use of these materials in concrete improved certain specific properties of concrete. Rubber which is produced as waste material does not have proper disposal method. It has become a major reason of environmental pollution. Therefore, such type of material should be used in concrete or other purposes so that the problem of disposal of tyres can be minimized. Many studies have been done using crumb rubber in concrete mix as partial replacement of fine aggregates.

Kaloush et al. [1] conducted a study on various properties of concrete and investigated the effect of concrete with rubber aggregates. It has been found that tensile strength of concrete decreases with increasing the content of rubber. However, the strain at failure was increased. Kotresh and Mesfin [2] shown that when the bond properties of rubber aggregates will be improved, it leads to improve the strength of concrete. Shah et al. [3] studied the thermal properties of rubberized concrete. Experiments were conducted by using rubber waste from scarp tyre by 5, 10, 15% as volume replacement for coarse aggregates. It has been observed that properties of concrete changes significantly beyond 5% replacement. It has also found that compressive strength and flexure strength of concrete were decreased with the increasing the content of rubber. It was concluded that rubberized concrete could be useful for energy efficiency of building. Su et al. [4] investigated the consequence of rubber size on the various properties of concrete. It was found that the workability of concrete permeability of water were affected by rubber particle size significantly as compared to the fresh density and strength. Selvakumar and Venkatakrishnaiah [5] conducted a test and found that the compressive strength of crumb rubber concrete is higher than the strength of normal concrete with 5% replacement for 28 days. Akinyele et al. [6] conducted various tests on concrete in which crumb rubber was used as partial replacement of fine aggregates in different proportions. It was found that rubber crumb can be useful in lightweight concrete to replace fine aggregate up to 16%. Liu et al. [7] investigated the influence of crumb rubber which is replaced fine aggregates and mixture replaced in concrete. They investigated the durability and mechanical strength of concrete. With the increasing of crumb rubber content, mechanical strength was reduced whereas durability of crumb rubber concrete was improved. Retama and Ayala [8] studied the effect on the mechanical properties of concrete by experimental and numerical techniques when crumb rubber is used. It has been found that fracture energy and other properties of crumb rubber concrete are influenced by the rubber fineness which is used in the mixture.

Mutar et al. [9] studied the use of crumb rubber in concrete and found that suitable compressive strength and reduced rate of water absorption can be achieved when mix is designed at 5% crumb rubber of 0.42 w/c ratio. Najib et al. [10] have explored the influence the use of rubber powder when it is used as partial replacement of fine aggregate in concrete mixes. It has been observed that compressive strength was reduced with the increasing of rubber content. But it has been found crumb rubber concrete possesses a number desirable properties as compared to conventional concrete, such as higher toughness, higher impact resistance and lower density. Chaikaew et al. [11] investigated the mechanical properties of rubberized concrete pedestrian blocks by introducing short steel fibres. It has been observed that density and absorption of blocks were increased slightly with the addition of steel fibres, whereas the toughness, abrasion resistance and flexural resistance were also increased significantly with the increasing of fibre content. Manojkumar et al. [12] conducted a study using crumb rubber as halfway substitution of fine aggregate. Tests were conducted for M20 grade concrete by replacement of fine aggregate with 5, 10, 15 and 20% of crumb rubber.

This paper presents an experimental study to see the effect on compressive strength using crumb rubber as partial replacement of fine aggregate in concrete. The compressive strength of M30 grade concrete for different proportion of crumb rubber is determined as per IS 10262-2009.

2 Properties of Crumb Rubber

Crumb rubber produced from scrap tyres by reducing the size ranging from 4.75 mm to less than 0.075 mm. There are three methods which are used to convert into crumb rubber from scrap tyre. Out of three methods, the cracker mill process method is generally used. The main chemical component of crumb rubber is a combination of natural and synthetic rubber. When crumb rubber is used in the concrete mixture, it has shown the two opposite influence due to the chemical properties of crumb rubber. The compressive strength is reduced, whereas the other desirable properties like toughness, impact resistance, durability, strain capacity are improved. There is less data available related to the shear stress of crumb rubber. Studies show that the size of crumb rubber is also influenced by the properties of crumb rubber. Therefore, crumb rubber concrete may be useful where low strength is required. Uses of crumb rubber in concrete are important in relation to environmental issues.

3 Materials

Fine aggregates, coarse aggregates, cement, waste tyre and water are used in this experimental study. Following material is used to conduct the experimental study. Fine sand having size less than 4.75 mm used as fine aggregate, crushed stone used as coarse aggregate having maximum size of 20 mm, crumb rubber having size between 0.075 and 4.75 mm, water are taken for making concrete mix.

4 Methodology

In this experimental study, crumb rubber was used in concrete mix to partially replace fine aggregate at 0, 5, 10, 15 and 20% for M30 concrete. Before doing mix design following data is required.

- Grade of concrete—M30
- Cement type—OPC 33 grade ultra tech is used
- Maximum nominal size of aggregate—20 mm
- Minimum cement content (300 kg/m^3 based on exposure condition (IS456 table 5)).
- Max water–cement ratio = 0.5 (IS456 table 5))
- Workability—slump 100 mm
- Exposure condition—moderate
- Method of transporting—pumpable
- Max cement content (450 kg/m^3 as per 8.2.4.2 IS 456:2000)
- Admixture used (NONE).

Apart from above data, specific gravities of cement, fine aggregate and coarse aggregate are determined using the standard procedure as per IS 2386-1963 PART 3. Values of specific gravity and water absorption of fine aggregates, coarse aggregates and cement are given in Table 1.

Five combinations for M30 grade concrete are taken in this study. First mix design is done for convention concrete i. e., 0% crumb rubber, and it is represented as C1 in this study. Second mix design is done for taking 5% crumb rubber partially replacement of fine aggregate, and it is represented as C2. Third mix design is done for taking 10% crumb rubber partially replacement of fine aggregate, and it is represented as C3. Similarly mix design for taking 15% and 20% crumb rubber partially replacement of fine aggregates is done, and they are represented as C4 and C5, respectively. Quantities of ingredients of mix design of concrete are obtained using the standard procedure as per IS 10262-2009.

Following procedure is followed to calculate the proportion of intergradient of concrete mix for conventional (normal) concrete, i.e. 0% crumb rubber for M30 grade concrete (C1):

(i) Target strength is calculated by

$$Fck_{target} = Fck_{characteristic} + 1.65 * S;\ \text{where S is standard deviation.}$$

Table 1 Values of specific gravity and water absorption of fine aggregates, coarse aggregates and cement

S. No.	Material	Specific gravity	Water absorption
1	Fine aggregates	2.63	1%
2	Coarse aggregates	2.81	0.4%
3	Cement	3.15	n/a

Table 2 Quantities of ingredients for different concrete mix design for M30 grade concrete

S. No.	Material	C1 mix (kg/m^3)	C2 mix (kg/m^3)	C3 mix (kg/m^3)	C4 mix (kg/m^3)	C5 mix (kg/m^3)
1	Cement	437.78	437.78	437.78	437.78	437.78
2	Crumb rubber	0	13.24	27.3	41	55.88
3	Water	197	197	197	197	197
4	Fine aggregates	750.4	714.7	649.9	618.54	600.32
5	Coarse aggregates	1065.3	1065.3	1065.3	1065.3	1065.3

Fck_{target} and $Fck_{characteristic}$ are target strength and characteristic strength.
$Fck_{target} = 30 + 1.65 * 5$, i.e. $Fck_{target} = 38.25$ Mpa.

(ii) Water–cement ratio is taken as 0.45 (IS456:2000)
(iii) Water content corrected for 75 to 100 mm slump $= 186 + 186 * \frac{6}{100} =$ 197kg or litre (specific gravity of water is one).
(iv) Cement content $= \frac{197}{0.45} = 437.78 \text{kg/m}^3$
(v) Proportion of coarse aggregates out of total aggregates = 0.63 [as per table 3 of IS 10262-2009]

for pump able concrete this fraction is reduced by 10%.
So final coarse aggregate content is = 0.63 * 0.9 = 0.567.
Which gives fine aggregates fraction as = 1–0.567 = 0.433.

Estimation of quantities of intergradient for 1 m^3 of concrete, i.e. volume of concrete is equal to 1 m^3 is calculated as:

Volume of cement $= 437.8 \times \frac{1}{3.15*1000} = 0.13899 \approx 0.139 m^3$.
Volume of water = 197 L = 0.197 m^3.
Volume of aggregates $= 1 - 0.139 - 0.197 = 0.664$ m^3.
Coarse aggregates content = 0.664 * 0.567 * 2.83 * 1000 = 1065.5 kg/m^3.
Fine aggregates content = 0.664 * 0.433 * 2.61 * 1000 = 750.4 kg/m^3.

Similarly, quantities of ingredients for different mix C2, C3, C4, C5 can be obtained as per procedure mentioned above. Qualities of ingredients of concrete mix design C1, C2, C3, C4 and C5 are shown in Table 2.

5 Compressive Test of Concrete Cube

The cube moulds should be properly cleaned, and thin layer of oil is applied on all faces of cubes. Materials for concrete mix C1 are taken as per Table 1 and mix using batch mixture in dry state. After making dry mix, water is mixed gradually in the dry mix. Then slump test is performed using the standard procedure. Wet concrete mix is filled into the cube moulds. The moulds were filled in three to four layers and compacting the concrete mix to ensure full compaction using shaking table. The filled cubes with concrete mix are shown in Fig. 1.

After 24 hours, these cubes are put into curing tank for curing. The same procedure was followed for other concrete mixers C2, C3, C4 and C5. The cubes of concrete mixture C1 are placed curing tank as shown in Fig. 2.

After proper curing, the compression test is performed for determining the compressive strength of concrete using compression testing machine as shown in Fig. 3.

The compressive strength of concrete cubes for different design mix, i.e. C1, C2, C3, C4 and C5 is determined for 7 and 28 days. Concrete cube with visible cracks is shown in Fig. 4.

Fig. 1 Cubes filled with concrete mix

Fig. 2 Cubes of concrete mix C1 in normal curing tank

Fig. 3 Testing of cube specimens (before the start of test)

Fig. 4 Cube samples after completion of test with visible cracks

Table 3 Compressive strength of cubes of M30 concrete for 7 days

S. No.	Cube name	Sample no	Load (KN)	Compressive strength (N/mm^2)	Average compressive strength (N/mm^2)
1	C1	1	464.8	20.66	19.81
2		2	445.7	19.81	
3		3	426.4	18.95	
4	C2	1	279.9	12.44	12.9
5		2	301.9	13.42	
6		3	290.3	12.89	
7	C3	1	243.2	10.81	11.79
8		2	279.9	12.44	
9		3	272.7	12.12	
10	C4	1	233.1	10.36	11.69
11		2	273.1	12.14	
12		3	282.8	12.57	
13	C5	1	263.7	11.72	11.2
14		2	246.8	10.97	
15		3	244.1	10.85	

Values of compressive strength of cube samples for different combination of design mix for 7 days and 28 days are shown in Tables 3 and 4, respectively.

Comparison of compressive strength of concrete cube for design mix for different combination for 7 and 28 days are shown in Fig. 5 in the form of bar chart. It is seen from the figure that compressive strength of concrete decreases as content of crumb rubber increases. It is clear from the figure that crumb rubber concrete performs best when 10% crumb rubber is taken as partial replacement of fine aggregates.

6 Conclusion

The effect of crumb rubber as partial replacement of fine aggregate on compressive strength of concrete was investigated. The concrete mixes were made taking crumb rubber in different proration in concrete mix for M30 grade concrete as per IS 10262-2009. The results showed that compressive strength decreases with the increasing of crumb rubber. It has been concluded that the rubberized concrete gives the best performance with the 10% partial replacement of fine aggregate with the crumb rubber for 28 days strength. The crumb rubber concrete can be useful when fine aggregate replaced by crumb rubber equal to 10% in many structural applications. The compressive strength decreases with increasing of the crumb rubber.

Table 4 Compressive strength of cubes of M30 for 28 days

S. No.	Cube name	Sample no	Load (KN)	Compressive strength (N/mm2)	Average compressive strength (N/mm^2)
1	C1	1	702.4	31.21	31.79
2		2	734.5	32.61	
3		3	710.7	31.55	
4	C2	1	618.1	27.44	25.8
5		2	601.2	26.72	
6		3	522.5	23.21	
7	C3	1	601.4	26.73	26.3
8		2	590.0	26.22	
9		3	584.3	25.97	
10	C4	1	452.5	20.11	20.75
11		2	474.5	21.09	
12		3	473.8	21.06	
13	C5	1	419.1	18.61	19.2
14		2	445.2	19.77	
15		3	431.5	19.18	

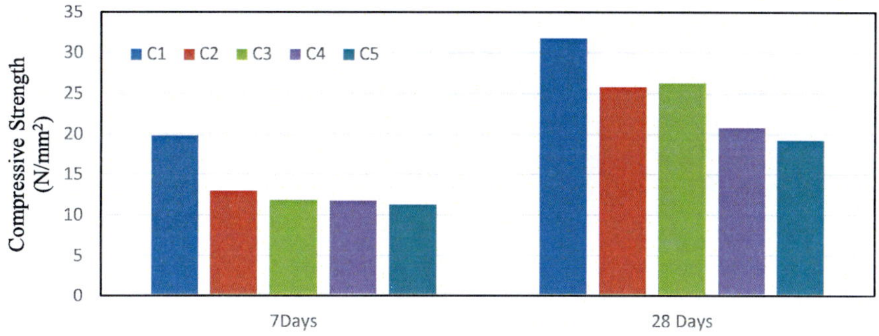

Fig. 5 Comparison on compressive strength results

However, many desirable properties, like toughness, impact resistance and durability, which are useful in other applications are improved using crumb rubber in the concrete mix. Due to these properties, crumb rubber concrete may be used where less strength of concrete is required like nailing concrete, in-wall panels and in construction elements and barriers. The use of crumb rubber in the concrete gives the solution of disposal of waste and conserves the natural aggregates to protect the environment.

References

1. Kaloush, K. E., George, B. W., & Zhu, H. (2005). Properties of crumb rubber concrete. *Transportation Research Record, 191*(1), 8–14.
2. Kotresh, K. M., & Mesfin, G. B. (2014). Study on waste tire rubber as concrete aggregate. *International Journal of Scientific Engineering and Technology, 3*(4), 433–436.
3. Shah, S. F. A., Naseer, A., Shah, A. A., & Ashraf, M. (2014). Evaluation of thermal and structural behavior of concrete containing rubber aggregate. *Arabian Journal for Science and Engineering, springer, 39*, 6919–6926.
4. Su, H., Yang, J., Ling, T.-C., Ghataora, G. S., & Dirar, S. (2015). Properties of concrete prepared from waste tire rubber particles of uniform and varying sizes. *Journal of Cleaner Production, 91*, 288–296.
5. Selvakumar, S., & Venkatakrishnaiah, R. (2015). Strength properties of concrete using crumb rubber with partial replacement of fine aggregate. *International Journal of Innovative Research in Science, Engineering and Technology, 4*(3).
6. Akinyele, J. O., Salim, R. W., & Kupolati, W. K. (2016). Production of lightweight concrete from waste tire rubber crumb. *Engineering Structures And Technologies* 2029–8838.
7. Liu, H., Wang, X., Jiao, Y., & Sha, T. (2016). Experimental investigation of the mechanical and durability properties of crumb rubber concrete. *Materials (Basel), 9*(3), 172.
8. Retama, J., & Ayala, A. G. (2017). Influence of crumb-rubber in the mechanical response of modified portland cement concrete. *Advances in Civil Engineering, 2017*(3040818), 9.
9. Mutar, M. A., Hussein, T. S., & Malik, S. H. (2018). Effect of crumb rubber aggregates on the characteristics of cement concrete as partial replacement. *International Journal of Mechanical and Production Engineering Research and Development, 8*(1), 1327–1336.
10. Gerges, N. N., Issa, C. A., & Fawaz, S. A. (2018). Rubber concrete: Mechanical and dynamical properties. *Case Studies in Construction Materials, 9*, 00184.
11. Chaikaew. C., Sukontasukkul, P., Chaisakulkiet, U., Sata, V., & Chindaprasirt, P. (2019). Properties of concrete pedestrian blocks containing crumb rubber from recycle waste tyres reinforced with steel fibres. *Case Studies in Construction Materials, 11*, e00304.
12. Manojkumar, B., Durga Lakshmi, S., & Kalyana Chakaravarthy, P. R. (2020). Experimental study on concrete strength by partial replacement of fine aggregate with crumb rubber. *International Journal of Advanced Science and Technology, 29*(10s), 3122–3129.

Influence of Mineral Admixture on Acceleration Carbonation Curing of Concrete: A Review

Ishfaq Ahmad Bhat, Khushpreet Singh, and Nittin Sharma

Abstract Accelerated carbonation curing is gaining more attention nowadays because this process involves the use of main greenhouse gas which is carbon dioxide. The concentration of carbon dioxide gas is increasing day by day. Several methods have been used to curb the production of greenhouse gas which includes the use of fly ash instead of cement. What happen during accelerated carbonation curing is that the carbon dioxide reacts with the ingredients present in the cement. These reactions mainly occur in the silicates phase and form stable calcium carbonates which increase the durability of concrete. The mechanical properties of concrete get increased by the carbonation curing of concrete; apart from this, it also helps to reduce the carbon dioxide percentage in the atmosphere, and with this, it also becomes the best alternative to the steam curing. This paper aims to provide a review of the research done on the acceleration carbonation curing of concrete using mineral admixture. In the recent past, several researches have been working on the acceleration carbonation curing of concrete by using the mineral admixtures like GGBS, fly ash, limestone powder, and many others and out from the mentioned admixtures fly ash has been better in case of carbonation. Based on the various researches conducted, it was found that when mineral admixture is added to the concrete it reduces the strength of non-carbonated concrete; while, on the other hand, the carbonated concrete with mineral admixture provides good compressive strength. It has also been started that impacts of the increase in pressure and temperature increases the rate of CO_2 absorption.

Keywords Mineral admixtures · Durability · Compressive strength · Accelerated carbonation curing

I. A. Bhat · K. Singh (✉) · N. Sharma
Chandigarh University, Gharuan, Mohali, India
e-mail: khushpreet.civil@cumail.in

A. K. Gupta et al. (eds.), *Advances in Construction Materials and Sustainable Environment*, Lecture Notes in Civil Engineering 196,
https://doi.org/10.1007/978-981-16-6557-8_81

1 Introduction

All over the world cement is one of the main elements used for the construction of buildings. Most of the Portland cement around 60% is produced in China; during the production of Portland cement, a lot of greenhouse gases gets released into the atmosphere which has a serious impact on the environment as these gases are the main reason for the global warming, Around 5% of the world's CO_2 is released by cement production [1]. Carbon dioxide is the major greenhouse gas so if the concentration of this gas gets increased it will block the solar radiation and will not allow them to move out of the atmosphere resulting in the rise of global temperature [2]. Construction industries are also responsible for this as the major construction activities produce a lot of greenhouse gases and other harmful gases [3, 4]. Thus, it has become a prime concern nowadays to reduce the carbon footprint in Portland cement. One of the best ways to reduce the level of greenhouse gas in the atmosphere is to minimize the production of CO_2 and ACC is one of the best ways to minimize the level of gas. It was way back in 1970 when this idea was put forward, but this idea has gained a lot of popularity in the recent years, the idea was rejected by many of the researchers because unmixed carbon dioxide is very expensive; besides this the weathering carbonation on steel, structures were questioned. Accelerated carbonation curing differs from the carbonation which leads to rusting of concrete. Carbonation is the phenomenon in which the carbon dioxide presents in the atmosphere gets inside the concrete to form new product calcium carbonate. The level of carbon dioxide in the atmosphere varies from place to place; usually in the case of urban areas, the concentration is high in the air, while as in the case of rural areas, the concentration is low. The penetrated carbon dioxide when comes in contact with the moisture forms carbonic acid which ultimately reduces the bitterness of concrete due to the reduction in alkalinity; the PH gets reduced which results in corrosion of steel [5]. Accelerated carbonation curing is the phenomenon in which the absorption of carbon dioxide takes place in the concrete carbon dioxide sneaks in the concrete, and the penetrated CO_2 reacts with the ingredients of cement mostly tricalcium silicate and dicalcium silicate and forms a stable compound which ultimately changes the behavior of concrete. Due to change in the concrete, the density increases which helps the concrete to gain more strength [6]. XRD results indicate that during the accelerated carbonation curing more amount of C_3S and C_2S were consumed [7]. Higher absorption of CO_2 in the concrete was found when the preconditioning of concrete was done for 18 h in an environment where relative humidity was kept 50% and temperature 25 degrees, followed by ACC with a pressure of 0.1 Mpa for 4 days [8]. The chemical reaction involved in the whole process of accelerated carbonation curing of concrete is as follows. [9, 10]. The below reactions Eqs. (1), (2), and (3) show the formation of $CaCO_3$ [3].

$$3CaO \cdot SiO_2 + 3CO_2H_2O \rightarrow SiO_2 \cdot H_2O + 3CaCO_3 \quad (1)$$

$$2CaO \cdot SiO_2 + 2CO_2 + H_2O \rightarrow SiO_2 \cdot H_2O + 2CaCO_3 \quad (2)$$

$$Ca(OH)_2 + CO_2 + H_2O \rightarrow CaCO_3 + 2H \quad (3)$$

2 Factors Influencing the Carbonation Curing of Concrete

- Water-to-cement ratio
- Preconditioning of concrete before and after carbonation
- Concentration of CO_2 and temperature
- Pressure inside chamber
- Further curing after carbonation.

2.1 Water-To-Cement Ratio

One of the important factors which influence the acceleration carbonation curing of concrete is the amount of water used during mixing. Water is considered as the main source which promotes the rate of reaction [11]. More the water–cement ratio more will be the formation of calcium carbonate but the water–cement ratio should not exceed 0.6 [12]. If an abundant amount of water is used in the mix, it will fill the pores of the concrete and block the pores which will ultimately affect the CO_2 absorption of concrete [8]. Experiment on fresh C_3S and β-C_2S pastes shows the compressive strength of concrete increases with the increase in the water-to-cement ratio up to a certain limit after that there was a decline in the compressive strength of concrete [13]. The compressive strength of the ACC sample was found higher than normal cured ones [14]. Rate of carbonation curing gets affected if the water is present inside the pores of the concrete; therefore, it is necessary to evaporate the water inside the pores of concrete which can be evaporated by preconditioning of concrete using oven or fan.

2.2 Preconditioning of Concrete Before and After Carbonation

Preconditioning of concrete before carbonation curing is necessary because the extra water which remains inside the pores of concrete makes hindrance in the absorption of CO_2 as a result of which the concrete does not absorb enough CO_2 for curing. With less CO_2 absorption, the strength of the concrete will be reduced. Concrete was kept in a dry environment where temperature and RH were kept ($T = 25$ °C, RH $= 55\%$) after that the specimens were shifted into the carbonation curing chamber. Keeping the concrete in these conditions for 4 h can increase the CO_2 absorption rate.

When the same concrete was kept in the wet conditions with the temperature and relative humidity as (T = 25 °C, RH.95%), it was observed that there was a decline in the absorption rate of concrete when the preconditioning time was increased [12]. However, it was observed that there was not so much difference in the compressive strength of concrete after 2–18 h of preconditioning and 2 h of ACC. After observing this fact, it was concluded that the total curing time of the concrete can be reduced to 4h only [12]. Apart from the precuring concentration of the CO_2, temperature in the carbonation chamber widely affects the rate of carbonation curing.

2.3 *Concentration of CO_2 and Temperature*

The percentage of CO_2 present inside the carbonation chamber is very effective in ACC of concrete; at low concentration of CO_2, the rate of CO_2 absorption will be less. Mostly there are two types of gases that are being used: one is flue gas and the other fully concentrated carbon dioxide gas. The percentage of CO_2 and temperature inside the chamber plays an effective role in the ACC of the concrete, and optimum temperature inside the chamber is 26–30°C [6]. Till now tests have been carried out on both types of gases, one which is having a low concentration and the other with full concentration, and it was found that using the fully concentrated gas in ACC chamber with a pressure of 5 bars for 2 h the strength of the concrete gets increased and is more than conventional concrete cured normally for 7 days [7]. When the fully concentrated gas and flue gas with a CO_2 concentration of 25% were checked the pressure was kept at 5 bars and carbonation was done for 2 h. It was found that in case of fully concentrated gas the CO_2 uptake was higher and the strength of concrete was greater than the conventionally cured concrete, while in the case of flue gas the CO_2 uptake was on the lower side and strength was also lower [7]. Along with the concentration CO_2, the pressure also plays a major role in the curing of concrete.

2.4 *Pressure Inside the Chamber*

The amount of pressure inside the chamber during the process is an important factor which affects the rate of curing, with high pressure the penetration rate of CO_2 in the concrete will get increased but at the same time if the pressure is above the limiting value it can show negative results on concrete. The concentration of CO_2, as well as pressure, is an important factor during the ACC process [8]. These two parameters have a great impact on the CO_2 curing degree as well as the strength of concrete. Either increasing the amount of CO_2 from 5 to 99.5% or pressure up to 0.2 MPa, both the cases would accelerate the ACC [15]. When the concentration of CO_2 is higher in the chamber, CO_2 gets penetrated the concrete more quickly, apart from this it also helps to dissolve the Ca^+ from cement [16]. However, in certain cases where the water–cement ratio is very high, the water remains into the capillary pores of concrete; in these cases, any increase in the amount CO_2 will not show any effect on the concrete [11]. Some specimens are tested immediately after carbonation curing while some kept in the open air for a further period.

2.5 *Further Curing After Carbonation*

Since in carbonation curing the precipitation of $CaCO_3$ occurs which makes the concrete matrix ore dense and thus increases the strength of concrete, further curing can promote the mechanical strength of concrete. There was an increase in the compressive strength of concrete when the concrete was placed in the open air after carbonation curing [17]. Acceleration carbonation curing of concrete mixture for about 10 h at a constant pressure of 60 psi leads to an increase in the mechanical property of concrete enough to be used in the construction of roads, precast concrete members [6]. When the specimen was placed in air for 7 days followed by 10 h accelerated carbonation curing, there was an increase in the strength of concrete by about 60% as compared to the normal curing. All other physiochemical properties were compared with the normal curing concrete, and it was found that all the properties of ACC concrete were better as compared to the normal curing concrete [17]. 28 days of hydration of concrete curing, followed by one-hour carbonation, increases the mechanical properties of concrete. One hour of carbonation does not affect the polymerization of the C–S–H bond, while 24-hour carbonation with relatively lower humidity decreases the available water content required from the hydration process [18].

3 Effect of Various Materials on the Carbonation Curing of Concrete

- Cement
- Admixtures (fly ash, GGBS, and lime stone powder).

3.1 *Cement*

Portland cement has been widely used, and it contains the following ingredients, tricalcium silicate, dicalcium silicate, tricalcium aluminate, and tetracalcium alumino ferrite. Carbonation of C_3S, C_2S at an early age leads to the absorption of CO_2 and the rapid gain of strength [8]. However, different chemical composition of cement leads to a different reaction with carbon dioxide. When the C_3S, C_3A, β-C_2S, mix with low water-to-cement ratio (0.05–0.175) were compared. It was found that the strength of C_3S and β-C_2S evolved more rapidly, and β-C_2S gains more initial strength as compared to the C_3S [13]. Replacing cement with some admixture provides more strength to the concrete.

3.2 *Admixtures*

Admixtures are the chemical substances which when added with concrete change the properties of the concrete, and either natural or mineral admixtures are used in the concrete. After 10 h of ACC, the compressive strength of concrete was found to be 24 Mpa when fly ash was used as an admixture. However, there was a sharp growth in the strength of concrete when the carbonated concrete was put in the open air for seven days (24–38 Mpa) [6]. Normal concrete without fly ash does not show good strength, and the strength of the concrete with fly ash shows great improvement in the compressive strength after 28 of air exposure; however, after 28 days the strength of both the concrete mixtures was found to be same (47 Mpa) [6]. When fly ash, GGBS, and limestone powder were used as the admixtures, concrete with fly ash showed more compressive strength as compared to others, and the tests were done immediately after removing the cubes from the ACC chamber [5]. When the concrete specimen was put in air for a period 7 days followed by 10 h of ACC, there was significant growth in the strength of concrete; however, nothing insignificant was found in the values of compressive strength of the flay ash based concrete after 28 days [6].

4 Dimensional Stability of Concrete

It is one of the important properties of the concrete. When the carbonation curing blocks were compared with the steam cured blocks, it was found that the CO_2 cured block shows lower drying shrinkage, and water absorption was also minimum but the durability of concrete was found to be high [8]. When the performance of accelerated carbonated curing concrete was compared with the steam cured concrete, the temperature was kept at 55 degrees, and it was found that there was a decrease in the drying shrinkage of concrete during the carbonation which leads to the increasing the dimension stability of concrete. The carbonated concrete shows good dimension stability because during the process of ACC significant amount of stable calcium carbonate is formed which changes the morphological structure of concrete while in case of steam cured concrete this kind of process does not occur [19].

5 Literature Review

The summary of literature review is summarized in Table 1.

Table 1 Summary of literature review

Authors	Year	Materials	emp. & RH	Time	Findings
Young et al. [12]	1974	Portland cement	25 °C, 55%	4H	• The rate of carbonation depends on the water content of both the compacts and the gas phase • Water-to-cement ratio directly affects the carbonation curing, higher water-to-cement ratio and lesser will be the absorption of CO_2
Shi et al. [11]	2008	Portland cement	25 °C, 55%	4H	• The preconditioning conditions have the main impact on the viability of CO_2 restoring • Preconditioning helps to absorb more CO_2 by evaporating the water inside the pores
Galan et al. [3]	2010	OPC, GGBS	25 °C, 68%	–	• An increase in the cement content decreases the porosity of concrete • Expanding the measure of Cao, the measure of CO_2 ingested increments by and large and the profundity of carbonation diminishes
Monkman et al. [6]	2010	Portland cement	30 °C, 55%	4H	• Optimum water-to-cement ratio, the higher concentration of CO_2 and pressure leads to increase the strength of concrete • A low energy utilization and high addition in execution, carbonation restoring innovation offers a promising instrument in ozone harming substance control

(continued)

Table 1 (continued)

Authors	Year	Materials	emp. & RH	Time	Findings
Shi et al. [18]	2012	Cement, lightweight	25 °C, 55%	4H	• Dimension stability of ACC concrete was better than steam cured, but weathering properties were similar to that of steam cured concrete • The strength of steam relieved squares grew quicker than that of CO_2 restored ones
Shi et al. [15]	2012	Ordinary Portland cement	24 °C, 80%	–	• Temperature of concrete increases during carbonation curing due to which density increases and porosity gets reduced • Preconditioning in a dry climate is basic to expand the CO_2 restoring rate and to build the strength of CO_2 relieved substantial items
Kashef-haghighi et al. [14]	2014	Portland cement	NA	6H	• With the increase in pressure and temp, the rate of CO_2 absorption also increases • The surface territory increment is feasible somewhat by utilizing distinctive concrete with high surface region
Han et al. [10]	2015	Limestone powder	150 °C	2H	• An increase in the concentration of CO_2 and pressure leads to more absorption of CO_2 • Fluid mineral carbonation is influenced more by temperature then by pressure and has high pressing factor has no unmistakable impact

(continued)

Table 1 (continued)

Authors	Year	Materials	emp. & RH	Time	Findings
Cui et al. [23]	2015	OPC	20 °C, 70%	–	• Lower concentration of CO_2 lower is the absorption rate of CO_2 • Under high CO_2 concrete went through a huge carbonation and showed a high thick organization microstructure with little pores and lower network of pores
Pang et al. [22]	2015	Slag, Portland cement	–	–	• Concrete gains more strength, pore size gets reduced after carbonation curing • The utilization of CSA in concrete blends was valuable in diminishing the ecological issues made by unloading steel
Monkman et al. [6]	2016	Cement, accelerator, retarder	20 °C, 55%	–	• Strength and durability increase after the carbonation curing of concrete • The acceleration benefits are associated with the in situ development of uniformly distributed nanocarbonate reaction products
Zhang et al. [24]	2016	OPC, fly ash	25 °C, 60%	24H	• The PH value of concrete was reduced immediately after the carbonation curing. Further hydration helps to get the PH value back to normal • Water curing plays a critical role in influencing the mechanical properties and durability of carbonated concrete
Shi et al. [7]	2017	Ordinary Portland cement	25 °C, 55%	–	• Lower the water-to-cement ratio higher is the absorption of CO_2 • Expanding CO_2 fixation and pressing factor will speed up carbonation of cement and increment its solidarity improvement rate

(continued)

Table 1 (continued)

Authors	Year	Materials	emp. & RH	Time	Findings
Hussain et al. [17]	2017	OPC, fly ash	35 °C, 70%	7H	• Intensive carbonation leads to reduce the modulus of elasticity of concrete • The profundity of carbonation has expanded with the length of openness in sped up carbonation chamber
Zhang et al. [20]	2017	Cement, slag, fly ash,	25 °C, 50%	4H	• The compressive strength of ACC concrete was higher than steam cured concrete after 28 days • Quick strength gain of PC initiates higher creating efficiency of carbonation-relieved items
Ahmad et al. [5]	2018	OPC, fly ash	25 °C, 60%	–	• Increase in the concentration of CO_2, pressure leads to more absorption of CO_2 • Pressure inside the chamber should not be beyond the limiting value 0.6 bars
Ahmad et al. [16]	2018	OPC, fly ash	–	10H	• Further air exposure and hydration after ACC increase the compressive strength of concrete
Junior et al. [17]	2018	High strength sulfate-resistant concrete	–	24H	• Due to presence of carbonated metakolin, fine aggregates gain more strength than the carbonated sand • CMKFA composites present preferable mechanical outcomes over CSAND composites, with an enormous diversion limit both when sped up maturing
Luo et al. [21]	2018	Recycled coarse aggregates (RCA)	–	6-72H	• Carbonation curing improves the mechanical strength of RCA • Carbonation treatment improves the microstructure of ITZ in RCA

(continued)

Table 1 (continued)

Authors	Year	Materials	emp. & RH	Time	Findings
Chen et al. [1]	2019	Portland cement	20 °C, 70%	24H	• With the increase in time, the efficiency of carbonation becomes less effective • Relative high grouping of CO_2 could advance crystallization of $CaCO_3$
Chen et al. [1]	2019	Portland cement	–	-	• Precuring is very effective; it increases the absorption rate of CO_2 by evaporating the water vapors inside the pores of the concrete
Qin et al. [4]	2019	RCA	–	2-4H	• The compressive strength of recycled waste autoclaved aerated concrete was way better than Portland cement after carbonation curing • Chloride particle entrance in PC-WAAC is improved by the sped up carbonation restoring
Qin et al. [2]	2019	Portland cement, GGBS	23 °C, 60%	4H	• As permeability of chloride ion reduces, strength of concrete increases after ACC • The arrangement of $CaCO_3$ upon carbonation restoring refined the pore structure and introduced a higher viability
He et al. [9]	2019	Cement, fibers	25 °C, 50%	2–24H	• Carbonation curing of concrete with cellulose fiber gives more strength. ACC is more efficient as compared to autoclaved curing • Precondition by initial hydration is helpful to promote carbonation

6 Conclusion

- It is necessary to do preconditioning of concrete in a dry environment before carbonation curing when the water–cement ratio is between 0.35 and 0.6 so that the extra water in the capillaries gets evaporated and makes a clear way for the absorption of CO_2.
- Increasing the CO_2 concentration and pressure will help to gain more strength but it is to be noted that increasing the pressure above 0.5 MPa makes the process in effective.
- The early age firmness of concrete gets increased by the ACC. Among the mineral admixtures, which has been used by various authors in their research with respect to ACC, GGBS and fly ash has shown the best results. Due to lower reactivity of fly ash than the GGBS and limestone powder, the reaction in the carbonation specimen led to the formation of $CaCO_3$ particle which contributes to 1, 3, 7 days compressive strength after 4 h accelerated carbonation and was found to be higher than those of standard curing sample. The 4 h carbonation procedure increased the 1 day compressive strength of fly ash-based sample to 105.1% which was more than GGBS and lime stone powder-based samples.
- With the inflammation in the amount of carbon dioxide as well as pressure, the carbonation curing rate increases 0.2 bar temp 20^0 C and curing for 4 h.
- Air exposure, as well as further hydration of concrete after carbonation curing, increases the 3-day strength of concrete 30.1% when compared with standard cured sample.
- $CaCO_3$ and silica gel are major indication to contribute the strength value. CO_2 concentration increases which was helpful to increase the carbonation depth to 30 mm.
- As per the European standard EN 13,295, it states that under the testing condition (1% CO_2, 21 ± 2 °C relative humidity 60 ± 10%), the product formed will be similar to those found in naturally carbonation specimens.

Conflict of Interest There is no conflict of interest.

References

1. Chen, T., & Gao, X. (2019). Effect of carbonation curing regime on strength and microstructure of Portland cement paste. *Journal of CO_2 Utilization, 34*, 74–86. https://doi.org/10.1016/j.jcou.2019.05.034
2. Qin, L., & Gao, X. (2019). Recycling of waste autoclaved aerated concrete powder in Portland cement by accelerated carbonation. *Waste Management, 89*, 254–264. https://doi.org/10.1016/j.wasman.2019.04.018
3. Galan, I., Andrade, C., Mora, P., & Sanjuan, M. A. (2010). Sequestration of CO_2 by concrete carbonation. *Environmental Science & Technology, 44*(8). https://doi.org/10.1021/es903581d

4. Qin, L., Gao, X., & Chen, T. (2019). Influence of mineral admixtures on carbonation curing of cement paste. *Construction and Building Materials, 212*, 653–662. https://doi.org/10.1016/j.conbuildmat.2019.04.033
5. Ahmad, S. (2017). Effects of carbonation pressure and duration on strength evolution of concrete subjected to accelerated carbonation curing. *Construction and Building Materials, 136*, 565–573. https://doi.org/10.1016/j.conbuildmat.2017.01.069
6. Monkman, S., & Shao, Y. (2010). Integration of carbon sequestration into curing process of precast concrete. *Journal of Civil Engineering, 37*(2), 302–310. https://doi.org/10.1139/L09-140
7. Shi, C., Tu, Z., Guo, M., & Wang, D. (2017). Accelerated carbonation as a fast curing technology for concrete blocks. In *Sustainable and nonconventional construction materials using inorganic bonded fiber composite* (pp. 313–341). Woodhead Publishing. https://doi.org/10.1016/B978-0-08-102001-2.00015-2
8. Sean, M., Mark, M., Doug, H. R., & Paul, S. (2016). Properties and durability of concrete produced using CO_2 as an accelerating admixture. *Cement and Concrete Composites, 74*, 218–224. https://doi.org/10.1016/j.cemconcomp.2016.10.007
9. He, Z., Jia, Y., Wang, S., Mahoutian, M., & Shao, Y. (2019). Maximizing CO_2 sequestration in cement-bonded fiberboards through carbonation curing. *Construction and Building Materials, 213*, 51–60. https://doi.org/10.1016/j.conbuildmat.2019.04.042
10. Han, D. R., Namkung, H., Lee, H. M., Huh, D. G., & Kim, H. T. (2015). CO2 sequestration by aqueous mineral carbonation of limestone in a supercritical reactor. *Journal of Industrial and Engineering Chemistry, 21*, 792–796. https://doi.org/10.1016/j.jiec.2014.04.014
11. Shi, C., & Wu, Y. (2008). Studies on some factors affecting CO_2 curing of lightweight concrete products, resources. *Conservation and Recycling, 52*(8–9), 1087–1092. https://doi.org/10.1016/j.resconrec.2008.05.002
12. Young, J. F., Berger, R. L., & Breese, J. (1974). Accelerated curing of compacted calcium silicate mortars on exposure to CO_2. *Journal of American Ceramics Society, 57*(9), 394–397. https://doi.org/10.1111/j.1151-2916.1974.tb11420.x
13. Wang, J., Xu, H., Xu, D., Du, P., Zhou, Z., Yuan, L., & Cheng, X. (2019). Accelerated carbonation of hardened cement pastes: Influence of porosity. *Construction and Building Materials, 225*, 159–169. https://doi.org/10.1016/j.conbuildmat.2019.07.088
14. Kashef-Haghighi, S., Shao, Y., & Ghoshal, S. (2015). Mathematical modeling of CO_2 uptake by concrete during accelerated carbonation curing. *Cement and Concrete Research, 67*, 1–10. https://doi.org/10.1016/j.cemconres.2014.07.020
15. Shi, C., Liu, M., He, P., & Ou, Z. (2012). Factors affecting the kinetics of CO_2 curing of concrete. *Journal of Sustainable Cement-Based Materials, 1*(1–2), 24–33. https://doi.org/10.1080/21650373.2012.727321
16. Ahmad, S. (2018). *Accelerated carbon dioxide sequestration*. In *Woodhead Publishing series in civil and structural engineering, carbon dioxide sequestration in cementitious construction materials* (pp. 81–101). Woodhead Publishing. https://doi.org/10.1016/B978-0-08-102444-7.00005-8
17. Junior, A. N., Ferreira, S. R., Dias, R., Filho, T., & De, E. (2019). Effect of early age curing carbonation on the mechanical properties and durability of high initial strength Portland cement and lime-pozolan composites reinforced with long sisal fibres. *Composites Part B, Engineering, 163*, 351–362. https://doi.org/10.1016/j.compositesb.2018.11.006
18. Shi, C., Wang, D., He, F., & Liu, M. (2012). Weathering properties of CO_2-cured concrete blocks. *Resources, Conservation and Recycling, 65*, 11–17. https://doi.org/10.1016/j.resconrec.2012.04.005
19. Hussain, S., Bhunia, D., & Singh, S. B. (2017). Comparative study of accelerated carbonation of plain cement and fly-ash concrete. *Journal of Building Engineering, 10*, 26–31. https://doi.org/10.1016/j.jobe.2017.02.001
20. Zhang, D., Ghoulish, Z., & Shao, Y. (2017). Review on carbonation curing of cement-based materials. *Journal of CO_2 Utilization, 21*, 119–131. https://doi.org/10.1016/j.jcou.2017.07.003

21. Luo, S., Ye, S., Xiao, J., Zheng, J., & Zhu, Y. (2018). Carbonated recycled coarse aggregate and uniaxial compressive stress-strain relation of recycled aggregate concrete. *Construction and Building Materials, 188*, 956–965. https://doi.org/10.1016/j.conbuildmat.2018.08.159
22. Pang, B., Zhou, Z., & Xu, H. (2015). Utilization of carbonated and granulated steel slag aggregate in concrete. *Construction and Building Materials, 84*, 454–467. https://doi.org/10.1016/j.conbuildmat.2015.03.008
23. Cui, H., Tang, W., Liu, W., Dong, Z., Xing, F. (2015). Experimental study on effects of CO_2 concentrations on concrete carbonation and diffusion mechanisms. *Construction and Building Materials,* 93, 522–527.
24. Zhang, D., & Shao, Y. (2016). Effect of early carbonation curing on chloride penetration and weathering carbonation in concrete. *Construction and Building Materials, 123*, 516–526. https://doi.org/10.1016/j.conbuildmat.2016.07.041

Pre-Engineered Buildings—A Cost Saving Approach

Abhijeet Roy, Rohit Motwani, Aditya Jaiswal, Aishwarya Raipurkar, and Kshitija Kadam

Abstract This paper presents a cost saving approach for steel buildings. The cost of steel is on a great surge and is continuously going up with time, and hence, nowadays it is becoming unavoidable to cope with it, especially in the industrial construction. In order to achieve economic development, it is very important to use steel at its optimum quantity. Structures which are having long spans and those without columns are among the most essential ones in any type of pre-engineered buildings (PEB). PEBs are very helpful in reducing the overall time and cost of the structure as compared to the conventional ones. This method is adoptable not only because of its prefabrication and pre-designing but also because of its cost and weight saving abilities. In this project, an attempt has been made to exhibit the comparative study of conventional and pre-engineered steel structures.

Keywords Pre-engineered buildings · STAAD pro · Tapered section · Cost of the structure

1 Introduction

India being a developing country massive house building construction work is taking place in various parts of the country. Since 30% of Indian population lives in towns and cities, construction is more in the urban places. The requirement of housing is tremendous but there will always be a shortage of house availability as the present masonry construction technology cannot fulfill the rising population demand every year. Hence, there is a need to think for alternative construction system like pre-engineered steel buildings. India has installed steel capacity of 35–40 million tones, and apparent steel consumption is around 27–30 million tones. In India, there is a surplus capacity of flat steel products particularly of hot and cold rolled sheets

A. Roy (✉) · R. Motwani · A. Jaiswal · A. Raipurkar
Department of Civil Engineering, Government College of Engineering, Nagpur, MH, India

K. Kadam (✉)
Applied Mechanics Department, Government College of Engineering, Nagpur, MH, India

A. K. Gupta et al. (eds.), *Advances in Construction Materials and Sustainable Environment*, Lecture Notes in Civil Engineering 196,
https://doi.org/10.1007/978-981-16-6557-8_82

available. These steel components can be used in the construction of pre-engineered building components [1].

The concept of PEB originated from USA. The use of pre-engineered buildings was confined to North America and the Middle East until 1990. Since then, the use of pre-engineered buildings started spreading throughout Asia and Africa where the PEB construction concept is now being widely accepted and praised. This leads to reduction in usage of steel and development of PEB concept. A growing number of prominent international contractors and designers are now enjoying the significant benefits like cost savings and benefits from the faster construction cycle resulting from this concept. The idea was to provide the section as per bending moment diagram (BMD) [1].

2 Comparison of Pre-Engineered Buildings and Conventional Steel Structures

As per the pervious researches and papers mentioned below [2–10], we have noted down several important points which define the advantages and benefits of pre-engineered building (PEB) in various aspects with respect to construction [11] (Table 1).

3 Analysis and Design

In order to analyze and design pre-engineered building structures and conventional structures, STAAD Pro software is used. In the present work, an industrial shed has been designed and compared with conventional steel structure.

Statement of Problem for PEB

In present work, an industrial shed in the Orange City Nagpur of dimensions 80 m × 30 m having eave height of 9 m and as per the guidelines of pre-engineered building bay spacing is taken as 7.27 m has been analyzed and designed in STAAD Pro as per considering all the codal provisions (Fig. 1).

Loading Calculations as Per the Prevailing Codal Provisions

Dead Load Calculation

As per IS 875 (Part 1)—1987 the total dead load is equal to the summation of dead load of roof sheeting, dead load of purlins, and self-weight of the frame.

D.L. = 0.12 kN/m^2 as given in IS 875 (Part 1)—1987.

Table 1 The advantages and benefits of pre-engineered building (PEB) in various aspects with respect to construction

Property	Pre-engineered steel buildings	Conventional steel buildings
Structure weight	Pre-engineered buildings are approximately 30% less as compared with CSB because of tapering of steel sections, providing larger depths where stress is maximum and providing lower depths where is nominal	Members generally used are selected hot-rolled T sections which may be heavier in many segments of the member than what is actually required as per the design. Members have same cross section, no importance given to varying moments in the member length
	Z- and C-shaped members which are light in weight are used as secondary members	Standard hot-rolled sections which are much heavier are sued as secondary members
Design	Quick and efficient, as time required is less since PEBs are mainly formed by standard sections and connections design. International design codes are used over and over for design of structure	Fewer design aids are available to the engineer, and hence, time required is more
	All the analysis, design, and drafting are computerized using standard details to minimize the use of the project custom details	Substantial engineering and detailing work is required by the consultant with fewer design aids
Delivery	42–56 days	140–182 days
Foundations	Lightweight and quite easy to construct and design is also very simple	Design is complex and foundations are very heavy
Erection simplicity	The learning curve of erection for each subsequent project is faster due to the connection of compounds is standard	Complicated connections and differs in every project which results in increasing time for erection of the building
Erection cost and time	Based upon extensive experience with similar buildings, both costs and time of erection are accurately known	In most of the cases, the CSB is 20% more expansive as compared to PEB, and hence, the erection costs and time are not estimated accurately
	Equipment requirement is very less and the erection process is faster and much easier	Heavy equipment is required and erection process is very slow and extensive field labor is also needed
Seismic resistance	As the frames are flexible and weight is also less, it provides higher resistance to seismic forces	Rigid heavy frames are poor in seismic zones

(continued)

Table 1 (continued)

Property	Pre-engineered steel buildings	Conventional steel buildings
Architecture	Using standard architectural details and interfaces can be achieved using outstanding architectural design at a very low cost	Requires research, special architectural design and features must be developed for each and every project and thus resulting in higher cost
Future expansions	Further development in that structure is very easy, simple, and Economical	Further development is monotonous and very costly
Safety and responsibility	As the complete job is done by one supplier, there is only one source of responsibility	It is very difficult to find the responsible person due to bunch of sources in it when the components do not fit properly or parts fail to perform the requirement
Performance	All components have been analyzed specially to act together as a whole complete structure for larger efficiency	Components are designed for a specific purpose. Errors are possible for while assembling every components into whole building

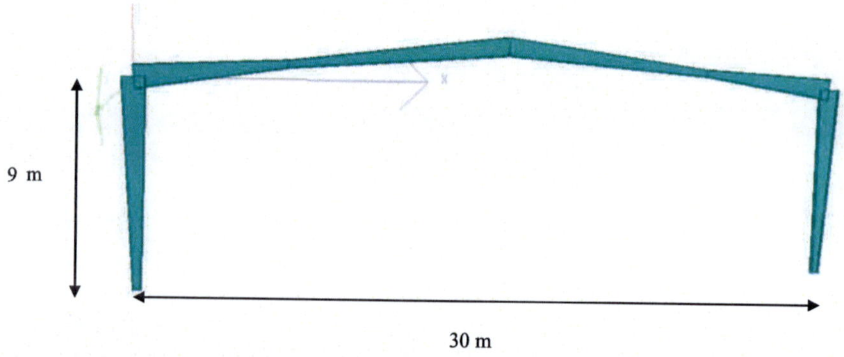

Fig. 1 Primary frame of PEB

Live Load Calculation

Live load = 0.75 kN/m^2 as per IS 875 (Part 2)—1987.

Wind Load Calculation

Wind load is calculated as per the guidelines given in IS: 875 (Part 3)—1987.

$$V_z = V_b * k_1 * k_2 * k_3 \tag{1}$$

$$P_z = 0.6 * V_z^2 \quad (2)$$

where

P_z = Design wind pressure in N/m^2

Referring IS: 875 (Part 3)—1987

Basic wind speed for Nagpur (V_b) = 44 m/sec

Assuming life of Building = 50 year, hence $k_1 = 1$ (from 5.3.1 of IS 875 (Part 3)—1987).

Eave height of building = 9.0 m.

So as per Category-2 and Class–C (Dimension > 50 m)

$k_2 = 0.93$ (from 5.3.2 of IS 875 (Part 3)—1987)

Assuming terrain slope is less than 3°

$k_3 = 1$ (from 5.3.3 of IS 875 (Part 3)—1987)

Substituting the above values in Eq. (1), the total wind load (V_z) = 40.92 m/sec. Substituting the value of total wind load in Eq. (2)

The design wind pressure (P_z) = 1004.6 N/m^2 = 1.0 kN/m^2.

Calculation for wind pressure coefficient

Cpi = internal pressure coefficient (depend upon % of opening) = 0.5 (Assume % of openings in between 5 and 20%).

Cpe = external pressure coefficient (depend upon and ratio).

Referring Tables 4 and 5 of IS: 875 (Part 3)—1987, we can have four different load cases of wind as shown in below diagram (Figs. 2, 3, 4, 5, and 6).

Software Check Results

Designed tapered members are checked as in the software and the results are in Table 2.

Statement of Problem for CSB

In present work, an industrial shed in the Orange City Nagpur of dimensions 80 m × 30 m having roof slope $\frac{1}{3}$ has been analyzed and designed in STAAD Pro as per considering all the codal provisions (Fig. 7).

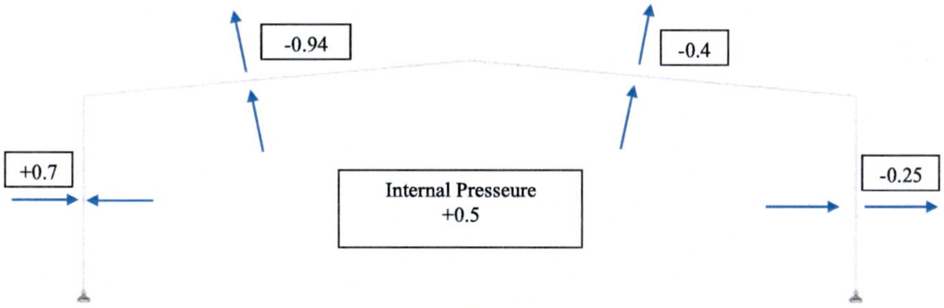

Fig. 2 (WL1) Wind perpendicular to ridge with internal pressure

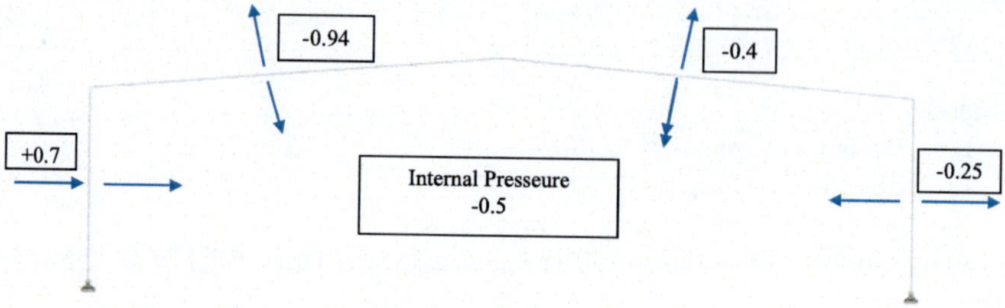

Fig. 3 (WL2) Wind perpendicular to ridge with internal suction

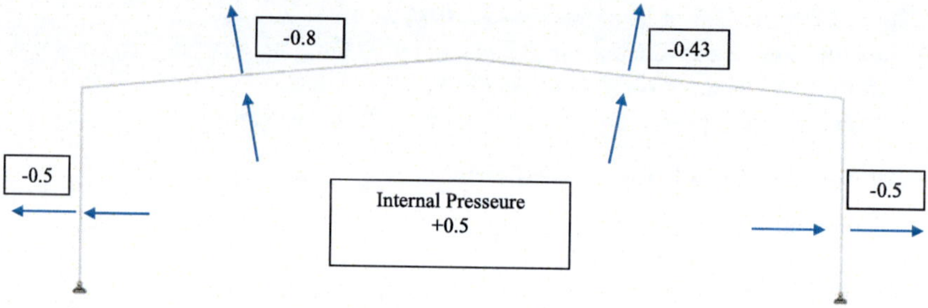

Fig. 4 (WL3) Wind parallel to ridge with internal pressure

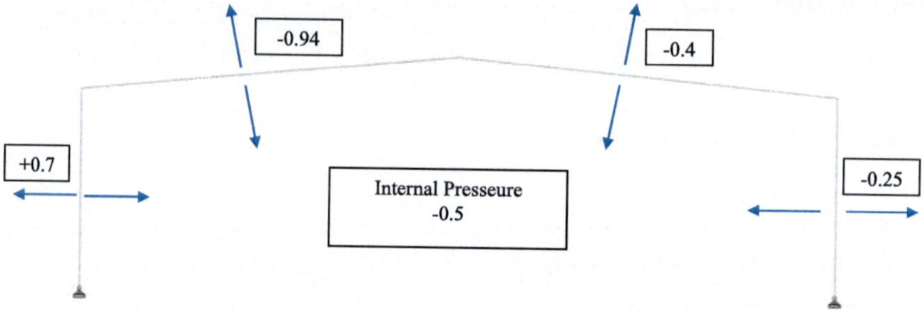

F

Fig. 5 (WL4) Wind parallel to ridge with internal suction

Calculation for Geometry Finalization

We have provided Howe Truss, Rise (R) with respect to 1 in 3 slope

$$R = (1/3)*15 = 5\text{m}$$

And the angle of inclination of Roof Truss (θ), $\tan\theta = \frac{5}{15}$

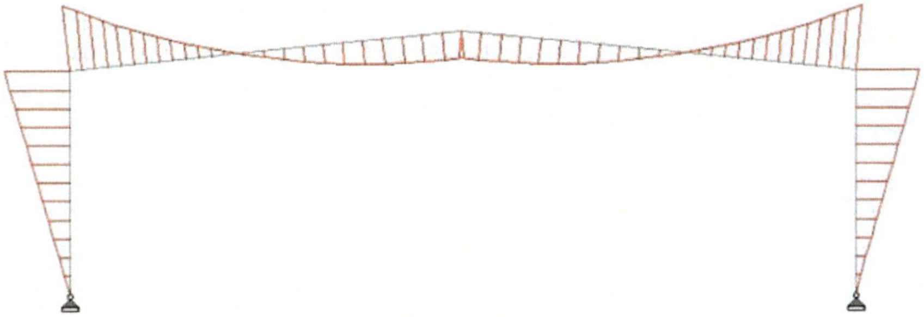

Fig. 6 Bending moment diagram (DL + LL KN/m) obtained from software

Table 2 Steel design check

All units are—KN mete (unless otherwise noted)

Member	Table	Result/FX	Critical cond/MY	Ratio/MZ	Loading/location
1	TAP ERED	PASS 68.27 C	(INDIAN SECTIONS) IS–7.1.1 (A) 0.00	0.873 486.37	110.00
2	TAP ERED	PASS 60.30 C	(INDIAN SECTIONS) IS–7.1.1 (A) 0.00	0.901 −185.81	116.05
3	TAP ERED	PASS 60.30	(INDIAN SECTIONS) IS–7.1.1 (A) 0.00	0.901 −185.81	116.05
4	TAP ERED	PASS 68.27 C	(INDIAN SECTIONS) IS–7.1.1 (A) 0.00	0.873 486.37	110.00
101	TAP ERED	PASS 103.66 C	(INDIAN SECTIONS) IS–7.1.1 (A) 0.00	0.936 −486.37	110.00
201	TAP ERED	PASS 103.66 C	(INDIAN SECTIONS) IS–7.1.1 (A) 0.00	0.936 486.37	110.00

$$\theta = \tan^{-1}(5/15) = 18.43°$$

Therefore, inclined length of principle rafter = 15.81 m.

Since the rafter length = 15.81 m (left 100 mm length on both side) hence the remaining rafter length is 15.81 − 0.1 − 0.1 = 15.61 m.

The width of A. C. sheet is 1.5 m, assuming 0.1 m overlap on next sheet, Therefore, net width of A. C. sheet is (1.5 − 0.1) = 1.4 m.

The spacing of purlins is the ratio of length of rafters by total number of panels, hence spacing will be $\left(\frac{15.81}{12}\right)$ is 1.32 m as it is less than 1.4 m, hence **OK**.

Therefore, the spacing of purlins provided is 1.3 m.

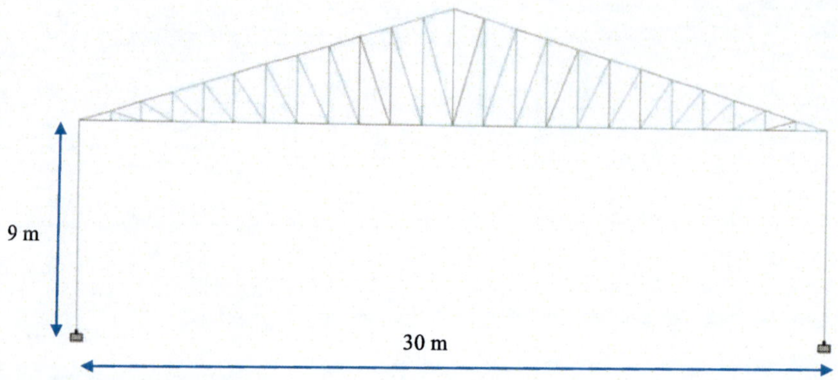

Fig. 7 Primary frame of CSB

Spacing of truss with respect to 5 m along the length = 80/5 = 16 no. of panels.

Loading Calculation (Calculation for load as per IS: 800—1984, IS: 875 (Part I, II & III)—1987) {[12–15]}.

Dead Load Calculation

The dead load of the roofing material (A. C. sheet, fasteners) is 17 kg/m^2 and the self-weight of purlin (ISMC 125) is 16.8 kg/m^2. Therefore, the total dead load (W_d) is the sum of dead loads of roofing materials and the purlins (17 + 16.8) which is equal to 33.8 kg/m^2 nearly equal to 34 kg/m = 34 ∗ 10 N/m^2 = 0.34 kN/m^2.

Live Load Calculation

As per IS 875 (Part 2)—1987 live load is 0.75 kN/m^2 up to 10° and less for per degree by 0.02 kN/m^2 above 10°.

Therefore, the

$$\begin{aligned} \text{Live Load} &= 0.75 - (\theta - 10) * 0.02 \\ &= 0.75 - (18.43 - 10) * 0.02 \\ \text{Live Load} &= 0.58\ \text{kN/m}^2 \end{aligned}$$

Wind Load Calculation

From Eqs. (1) and (2), the design wind speed and design wind pressure have been calculated. Since the building is in Orange City Nagpur, as per IS 875 (Part 3)—1987 the basic wind speed (V_b) for Nagpur is 44 m/sec.

Assuming life of structure = 50 year k_1 = 1 (from 5.3.1 as per IS 875 (Part 3)—1987)

Total height of structure = Height of column + R = 9 + 5 = 14 m. So as per Category-2 and Class–C (Dimension > 50 m) $k_2 = 0.93$ (from 5.3.2 as per IS 875 (Part 3)—1987)

Since terrain slope at project site less than 3° $k_3 = 1$ (from 5.3.3 as per IS 875 (Part 3)—1987).

Therefore, from Eq. (1). $V_z = 44 * 1 * 0.93 * 1 = 40.92$ m/sec and

From Eq. (2) $P_z = 0.6 * 40.92^2 = 1004.6$ N/m^2 = 1 kN/m^2.

Calculation for Wind Pressure Coefficient

Cpi = internal pressure coefficient (depend upon % of opening)

Cpe = external pressure coefficient (depend upon (L/W) and (H/W) ratio)

Since % of opening is between 5 and 20 % (from 6.2.3.2 of IS 875 (Part 3)—1987) Cpi = +0.5, −0.5

Eave height = 9 m

$$\text{Height/Widthratio} = 9/30$$
$$= 0.3$$

Referring Tables 4 and 5 of IS: 875 (Part 3)—1987, we can have four different load cases of wind as shown in below diagram (Figs. 8, 9, 10, and 11).

Nodal Load Calculations

Dead Load Calculation

The length along sloping roof is 15.81 m. Assuming self-weight of truss with respect to 10 kg/m. Therefore, self-weight = 10 ∗ 30 ∗ 5 = 1500 kg. Assuming self-weight of roofing material including fastener = 17 kg/m.

Therefore, total weight of roofing material = (17 ∗ 15.81 ∗ 5) ∗ 2 = 2687.7 kg

Hence, no. of purlins with respect to 1.32 m = 26 No. (ISMC 125 with respect to 16.8 kg/m)

Therefore, weight of purlins = 26 ∗ 16.8 ∗ 5 = 2184 kg

And total D. L. = 1500 + 2688 + 2184 = 6372 kg

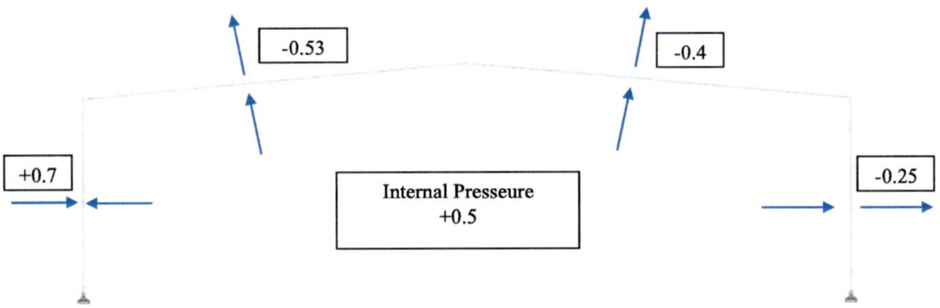

Fig. 8 (WL1) Wind perpendicular to ridge with internal pressure

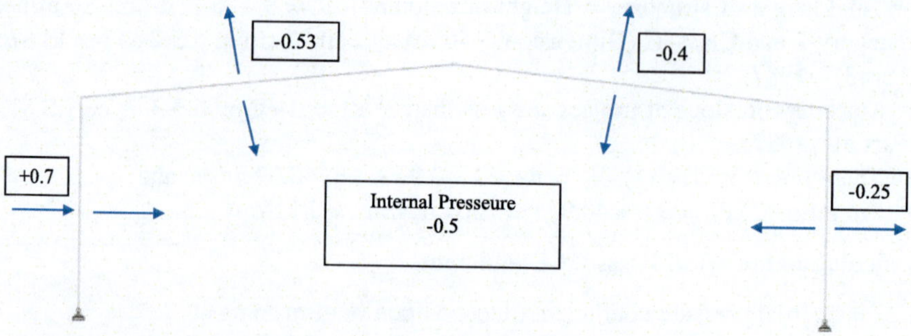

Fig. 9 (WL2) Wind perpendicular to ridge with internal suction

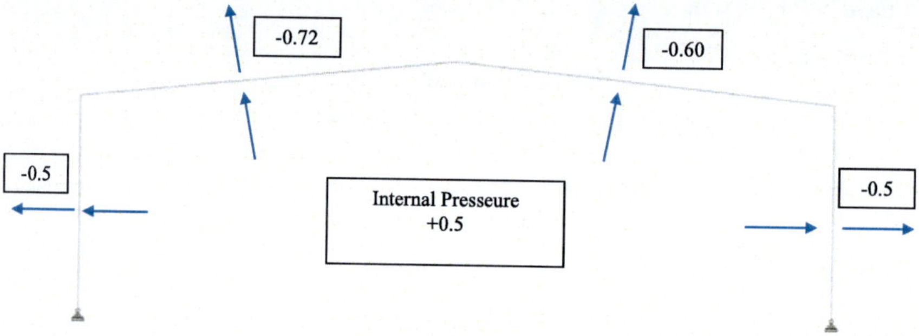

Fig. 10 (WL3) Wind parallel to ridge with internal pressure

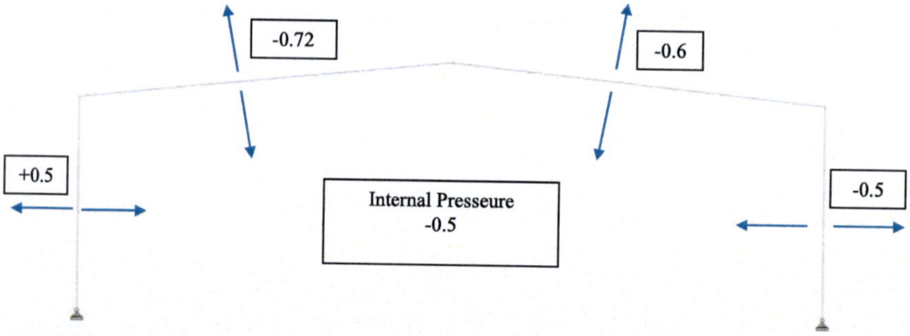

Fig. 11 (WL_4) Wind parallel to ridge with internal suction

Total no. of panel = 24 along the width
Total no. of node = 25
All intermediate points have load = W = 254.88 Kg (Say 255 kg) = 2.55 kN
End nodal point has load = $\frac{W}{2} = \frac{2.55}{2}$

Live Load Calculation

As per IS 875 (Part 2)—1987 Live Load is 0.75 kN/m^2 up to 10° and less for per degree by 0.02 kN/m^2 above 10°.

Therefore, the

$$\text{Live Load} = 0.75 - (\theta - 10) * 0.02 = 0.75 - (18.43^\circ - 10^\circ) * 0.02$$
$$= 0.58\ \text{kN/m}^2$$

Total live load = 0.58 * ((2/3) * 30 * 5) = 58 kN.
All intermediate points have load = W = 2.33 kN.
End nodal point has load = W/2 = 1.16 kN.

Wind Load Calculation

$$\text{Total wind load per panel point} = W = P_z * \text{Coefficient of wind} * \text{Area}$$
$$= 1 * 1.22 * (1.32 \times 5) = 8.1\ \text{kN}$$

Total wind load end panel point = (W/2) = 8.1/2 = 4.05 kN
Total W = 2.55 + 2.33 + 8.1 = 12.98 kN
End nodal point has (Tables 3, 4, 5, 6, and 7)

$$\text{load} = W/2 = 2.55/2$$
$$= 1.275\ \text{kN (Say 1.28 kN)} = 1.28\ \text{kN.}$$

Table 3 Results for load combination

Load combinations	Normal to ridge	Tangential to ridge
D.L. + L.L	(0.34 + 0.58) × 1.32 × (cos 18.43) = 1.152	(0.34 + 0.58) × 1.32 × (sin 18.43) = 0.38
D.L. + W.L	(0.34 + 1.32) × (cos 18.43) – (1.22 × 1.32) = –1.18	(0.34 + 1.32) × (sin 18.43) = 0.14

Table 4 Bending moment and shear force calculations

Result	D.L. + L.L		D.L. + W.L	
	Xx	Yy	Xx	Yy
B.M = (WL2)/8	(1.152 × 252)/8 = 3.6	(0.38 × 252)/8 = 1.19	(1.18 × 252)/8 = 3.69	(0.14 × 252)/8 = 0.44
S.F = (WL)/2	(1.152 × 25)/2 = 2.88	(0.38 × 25)/2 = 0.95	(1.18 × 25)/2 = 2.95	(0.14 × 25)/2 = 0.35

Table 5 Critical forces in the member of truss

Member description	Member number	Member forces (KN)	Member No.
Tie	1–24	+114.712 −172.905	1
Rafter	25–48	+184.956 −136.110	48
Strut	49–71	+42.002 −60.113	60
Sling	72–93	+30.466 −24.110	83
Column	94, 95	+67.58 −92.97	94

Table 6 Design summary

Member description	Member number	Member No. = trial section	
Tie	1–24		LD ISA 75 × 75 × 6
Rafter	25–48		LD ISA 75 × 75 × 6
Strut	49–71	49,50,51,52,68,69,70,71 53,54,66,67 55,56,64,65 57,58,59,60,61,62,63	ST IAS 50 × 50 × 6 ST ISA 75 × 75 × 6 ST ISA 100 × 100 × 6 SD ISA 100 × 100 × 6
Sling	72–93	72,73,92,93 74,75,76,89,90,91 77,78,87,89 79,80,81,82,83,84,85,86	ST ISA 50 × 50 × 6 ST ISA 75 × 75 × 6 ST ISA 100 × 100 × 6 LD ISA 100 × 100 × 6
Column	94, 95		ISMB 350

ST ISA = Indian Standard Single Angle Section SD
ISA = Indian Standard Single Angle Section
LD ISA = Indian Standard Double Angle (Back To Back) Section
ISMB = Indian Standard Medium Beam

Table 7 Design of joint (riveted) for rolled steel sections

Member description	Design load	Governing rivet value	No. of rivet required	No. of rivet provided
Tie	172.905	48.1	3.59	4
Rafter	184.956	48.1	3.85	4
Strut	60.113	48.1	1.23	2
Sling	30.466	54.05	1.27	2
Column	92.97	48.1	1.93	2

4 Results and Discussions

All the members used in pre-engineered building are checked as per the Indian standards and all members are under codal provisions (as shown in Table 2).

Steel take off of PEB (Table 8).

From the above table, volume is calculated:

$$\begin{aligned} \text{Volume} &= (0.0086 * 18) + (0.0086 * 12) + (0.0064 * 18) \\ &= 0.3732\ \text{m}^3 \end{aligned}$$

$$\text{Density of steel} = 7860\ \text{kg/m}^3$$

$$\begin{aligned} \text{Weight} &= 0.3732 * 7860 \\ &= 2933.36\ \text{kg} \\ &= 2.94\ \text{Ton (per frame)} \end{aligned}$$

Since center to center spacing for PEB is taken as 7.27 m, building length is taken as 80 m.

Therefore,

$$\text{Number of Frames} = \frac{80}{7.27} = 11\ \text{frames}$$

$$\begin{aligned} \text{Total weight of PEB} &= 11 * 2.94 \\ &= 32.35\ \text{Ton.} \end{aligned}$$

Steel take off of CSB (Table 9).

From the above table, volume is calculated:

$$\begin{aligned} \text{Volume} = {} & (0.0067 * 18) + (0.000567 * 18) + (0.00117 * 24.436) + (0.00173 * 61.632) \\ & + (0.0024 * 33.228) + (0.00087 * 21.728) + (0.0023 * 30) = 0.3732\ \text{m}^3 \end{aligned}$$

$$\text{Density of steel} = 7860\ \text{kg/m}^3$$

Table 8 Area and length of tapper members

Member	Area ($\text{in}^2 = \text{m}^2$)	Total length of that member (m)
Tapper 1	13.33 = 0.0086	18
Tapper 2	13.33 = 0.0086	12
Tapper 3	9.842 = 0.0064	18

Table 9 Area and length of every member

Member	Area ($in^2 = m^2$)	Total length of that member (m)
ISMB 350	10.4 = 0.006	18
ISA 50 * 50 * 6	0.88 = 0.000567	13.974
ISA 100 * 100 * 6	1.814 = 0.00117	24.436
ISA 75 * 75 * 6 (LD)	2.685 = 0.00173	61.632
ISA 100 * 100 * 6 (LD)	3.627 = 0.0024	33.228
ISA 75 * 75 * 6	1.327 = 0.00087	210,278
ISA 100 * 100 * 100 (SD)	3.627 = 0.0023	30

$$\begin{aligned}\text{Weight} &= 0.4309 \times 7860\\ &= 3386.40\ \text{kg}\\ &= 3.39\ \text{Tons (per frame)}\end{aligned}$$

Since center to center spacing for CSB is taken as 5 m, building length is taken as 80 m.

Therefore,

$$\text{Number of Frames} = \frac{80}{5} = 16\ \text{frames}$$

$$\begin{aligned}\text{Total weight of CSB} &= 16 * 3.39\\ &= 52.24\ \text{Tons.}\end{aligned}$$

$$\text{Perecntage Saving} = \frac{(52.24 - 32.35)}{52.24} * 100 = 40.35\%$$

5 Conclusion

(a) It can be concluded that pre-engineered metal building system is much economical and efficient method of construction in steel buildings. As steel required for primary framing in case of pre-engineered metal building system was found to be 32.35 Ton and for conventional building was 54.24 Ton, i.e., in terms of percentage PEBs required 35–40% less steel (in our case 40.35% less) as compared with CSB.

(b) Quantity of steel differs in large amount because span length and bay spacing limits are much more in case of pre-engineered metal building system and sections weight is very less.
(c) Time required for design, fabrication, and erection is much less in case of pre-engineered metal building system.
(d) For having the clear span in between the columns pre-engineered metal building system is very much efficient because we can have large clear span structure without any obstruction which is not possible in case of conventional buildings.
(e) As far as erection is concerned, pre-engineered metal building system takes comparatively less time than conventional buildings. Thereby the purpose, for which building is constructed, can be achieved quickly and completely.

References

1. Cost Comparative Study PEB with Conventional Method for Industrial Building.
2. Darade, M., & Bhojkar, M., A Review on Comparative Study on the Structural Analysis and Design of Pre- Engineered Building [PEB] with Conventional Steel Building [CSB].
3. Kewate, S., & Kolate, N., Cost-effectiveness of Pre-engineered and Conventional Steel frames.
4. Meera, C. M. (2013, June). *International Journal of Engineering Sciences & Emerging Technologies, 5*(2), 75–82, ISSN: 2231–6604 (M.E. Structural Engineering).
5. Wankhede, S., Comparisons of conventional steel building and pre-engineered building.
6. Bhagatkar, S. V., Shaikh, F. I., Gupta, B. P., & Kharta, D., A study on pre-engineered building—A construction technique.
7. Dubey, N. C., & Wakchaure, S., Design and comparative study of pre-engineered building.
8. Zende, A. A., Kulkarni, A. V., & Hutagia, A., Comparative study of analysis and design of pre-engineered buildings and conventional frames.
9. Goswami, A., & Shende, T., Pre-Engineered building design of an industrial warehouse.
10. Technical Hand book of Zamil Steel Industries.
11. Parametric Study of Various Pre Engineered Buildings.
12. IS: 800: 1984 Limit State Design, General Construction In Steel- Code of Practice.
13. IS-875 (Part 2) 1987 Imposed load.
14. IS-875 (Part 3) 1987 Wind load.
15. IS-875 (Part 1) 1987 Dead load—Unit weight of Building materials and Stored Materials.

Effect of Curing on Compressive and Shear Strength Parameters of Liming Waste Ash Stabilized Expansive Soil

Niraj Singh Parihar and Ashok Kumar Gupta

Abstract The expansive soils are known for their low compressive and shear strength and they present numerous challenges to geotechnical engineers during infrastructural development. The study is an effort to examine the utility of a leather industry waste product known as liming waste ash for enhancement of the compressive and shear strength of the soil. The effects of different curing periods on the relative parameters of strength have also been discussed. It was found that the curing of the soil-ash (LLWA) mix promotes the development of bond strength through formation of pozzolanic cementitious products and augments the shear strength parameters and the unconfined compressive strength (UCS) of the soil. An improvement of 270 and 380% was found in the UCS and cohesion of the soil respectively post 28 days of curing period along with substantial rise of internal friction angle up to 7 days of curing. The influence of LLWA and curing duration on the soil fabric and its relative effect on various strength parameters have also been explained at the microstructural level through SEM analysis on the uncured and cured soil samples.

Keyword Expansive soil · Liming leather waste ash · Chemical stabilization · UCS · Shear strength · SEM

1 Introduction

Expansive soils are one of the most found soils across the globe known through various names such as black cotton, vertisol, udert, etc. and they have been one of the most problematic soils to be used as foundation and subgrade material in building and road construction due to their high expansive mineral content such as montmorillonite. A high proportion of such minerals give rise to high electronegativity and

N. S. Parihar (✉) · A. K. Gupta
Department of Civil Engineering, Jaypee University of Information Technology, Waknaghat, H.P., India
e-mail: nirajsingh.parihar@juit.ac.in

A. K. Gupta
e-mail: ashok.gupta@juit.ac.in

A. K. Gupta et al. (eds.), *Advances in Construction Materials and Sustainable Environment*, Lecture Notes in Civil Engineering 196,
https://doi.org/10.1007/978-981-16-6557-8_83

water absorption capacity to the soil leading to its reduced bearing capacity and high swell-shrink characteristics. In India, the huge land masses layered with the weathered basaltic plateaus and lava traps have resulted in formation of the black cotton soil which covers more than one-fifth of the geographical area and is abundantly found in central, western and southern Indian states [1, 2]. Such large areas covered by expansive soil when brought under infrastructural development, offer very less alternatives of costly base-soil replacement and soil stabilization proves as one of the best techniques to make these soils fit as a decent bearing material.

Chemical stabilization is a widely practised technique to counter the unwanted characteristics of expansive soils. A number of industrial and agricultural by-products have emerged as useful products in the recent past with capability to stabilize expansive soils. These waste products contain similar pozzolanic or binding characteristics as are found in the conventional chemical soil stabilizers such as lime and cement and their use as a soil stabilizer not only creates an alternative for replacement of the costly and pollution causing conventional construction material but also solves the dumping and environmental related issues related to these wastes. A wide number of researches have been done on the utility of these wastes as chemical modifiers of soil including industry originated wastes such as phosphogypsum [3], cement kiln dust [4, 5], marble dust [6], silica fume [7], fly-ash [5, 8], slags [2, 9], quarry dust [10], lime-soda glass powder [11], lime sludge [12, 13], etc. and silica containing agricultural wastes such as corn ash [14], groundnut shell ash [15], bagasse ash [16], rice-husk ash [12], wheat-husk ash [17], etc. or a combination of both [18, 19].

The leather processing industry is one of the oldest in the world. The leather industry has seen an overwhelming growth in the recent past due to increase in demands for its finished products. It is also among the top export industries in India with Indian leather exports crossing 5.5B USD [20]. With more than 6 million tonnes of leather waste production worldwide, India and China together share more than 25% of this waste produce [21]. It is one of the industries with the highest waste to raw product fraction of 60–80% containing numerous environmental pollutants originating from its chemical processes. However, the waste originating from the industry finds no or little use and has to be either dumped or sent for digestion [22, 23]. The earlier researches have shown their concerns on these methods of waste disposal as this waste dumping is environmentally hazardous and the waste matter is not easily digestible and generates greenhouse gases [24]. The incineration of this waste has therefore been suggested as the adaptable method by many of the researchers due to high calorific value of the waste [25, 26] and the same is practised by many countries nowadays. But despite the volume reduction of the waste after burning, a lot of waste ash still has to be disposed of in the landfills with or without treatment which can promote degradation of land and water table.

Liming waste coming from beam-house operations in the leather processing industry accounts for a sizeable proportion of the total produced waste and is only scarcely utilized for making glue and gelatin-like products. Parihar and Gupta [27, 28] conducted research over the utilization of lime-fleshing or liming leather waste ash (LLWA) for stabilization of expansive soil and established that the residual waste ash after open incineration of liming waste is capable of altering the undesirable characteristics of the low strength expansive soil due to containment of both lime and reactive silica in the ash particles. It was found that the liming waste ash not only reduced two-third plasticity of the soil but also improved the short-term and long-term strengths of the soil by the virtue of pozzolanic characteristics delivered by the waste. As a consequence, a multifold improvement in the UCS and CBR were reported after the addition of LLWA post 28 days of curing. The generation of calcium silicate gels and other cementitious compounds were evident from the microstructural studies. The rainfall simulated leachate study conducted by Parihar and Gupta [28] on the pavement model with stabilized expansive sub-base also revealed that the leachate flow was non-toxic over a considerable period of time and did not have any major polluting effects on the water table due to the absence of harmful heavy metals and arrestation of the native heavy metals in the soil. However, despite being a globally produced waste, not much research is available on engineering utilization of this waste. The present study is an effort to further study the effects of the LLWA stabilization over strength related geotechnical aspects such as compressive strength and shear strength parameters of expansive soil before and after different periods of curing. The effect of stabilization is also presented at the mineralogical level through SEM study.

2 Materials

2.1 Black Cotton Soil

The selected expansive soil for the laboratory testing is black cotton soil procured from Guna district of Madhya Pradesh in central India. The expansive soil was derived from a depth of half a meter below the ground surface from a widespread open field of black cotton soil to eliminate foreign material contamination. The engineering properties of the soil, tested according to the Indian standard (IS) code are displayed in Table 1. The particle size distribution of the soil as derived from wet-sieve and hydrometer analysis is shown in Fig. 1a. Majority of the soil particles are fines less than 75-micron size and approximately 40% of the soil particles belong to the clay fraction less than two-micron size. The soil is inorganic in nature and has been classified as silt with high plasticity (MH) as per IS classification system [29].

2.2 Liming Leather Waste Ash (LLWA)

The leather waste used for the generation of ash was procured from Jalandhar tannery area, Punjab (India) after the process of open sun-drying. The waste available in long strip form was deprived of any remaining moisture and incinerated under a flame

Table 1 Engineering properties of virgin soil

Properties	Value	Referred codes	Properties	Value	Referred codes
Liquid limit (%)	69	IS: 2720 (Part V)-1985 [34]	Sp. gravity	2.65	IS:2720(Part-III)-1980 [33]
Plastic limit (%)	40	IS: 2720 (Part V)-1985	MDD (g/cc)	1.80	IS: 2720 (Part VII)-1980 [32]
Shrinkage limit (%)	9.37	IS: 2720 (Part VI)-1972 [35]	OMC (%)	21.78	IS: 2720 (Part VII)-1980
Plasticity Index (%)	29		UCS (kN/m^2)	114	IS: 2720 (Part V)-1985 [30]
Unsoaked CBR (%)	10.83	IS: 2720 (Part XVI)-1987 [36]	Soaked CBR (%)	2.36	IS: 2720 (Part XVI)-1987
Free swell index	40–50%	IS: 2720 (Part XL)-1977 [37]	Swell pressure (kN/m^2)	420	IS: 2720 (Part XLI)-1977 [38]

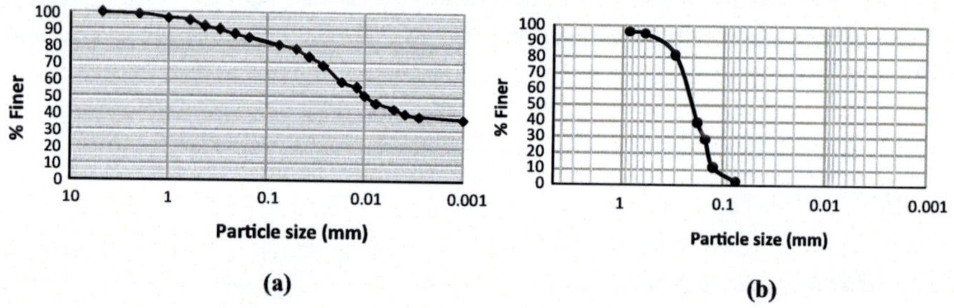

Fig. 1 Particle size distribution of **a** virgin soil, **b** LLWA

temperature of 1000 °C in an open environment to initiate fire and then was left for self exothermic burning until conversion to ash form (LLWA). The cooled off ash was passed through 600-micron sieve and the unburnt material was eliminated. The ash was preserved in closed containers till mixing process to avoid moisture contact and generation of excessive calcium carbonate. The particle size distribution of the ash particles is shown in Fig. 1b and the major chemical composition of LLWA obtained from XRF (X-ray fluorescence) is presented in Table 2. Due to high calcium and silica containment, the LLWA treatment also offers a high alkaline environment for pozzolanic reactions and subsequent chemical modification of soil.

Table 2 Chemical composition of LLWA

Component/Property	Concentration (%)	Component/Property	Concentration
Na_2O	35.29	Fe_2O_3	1.12%
CaO	21.12	MgO	1.0%
SiO_2	17.74	Cr_2O_3	N.D.

3 Experimental Program

The black cotton soil dug from the open field was collected and sealed in airtight containers to conserve its native state till the time of testing. The virgin soil was first subjected to laboratory determination of engineering properties such as unconfined compressive strength (UCS) and shear strength parameters and the same tests were repeated through various soil-LLWA mix proportions under different curing periods. The LLWA was mixed in the proportions of 2, 4, 6, 8 and 10% by weight of the soil. The soil and waste ash were first dry mixed to uniformity until formation of a homogenous mix followed by addition of a requisite amount of water for slurry formation and sample preparation for all tests. All the experiments for plain and LLWA mixed soil were conducted as per the recommendations of Indian standard codes. The samples for UCS and unconsolidated undrained triaxial test were simultaneously prepared by using same slurry mixture in the previously defined mix proportions as per IS: 2720 (Part X)-1991 [30] and IS: 2720 (Part XI)-1993 [31] respectively. The soil samples for both the tests having 38 cm diameter and 76 mm height were derived through sampling tubes driven in the proctor mould compacted under standard proctor compaction procedure recommended by IS: 2720 (Part VII)-1980 [32] for equality in imparted energy levels. The moisture content during proctor compaction of soil-LLWA mixes was kept a little above the optimum moisture level of the mix to compensate for the water required for hydration reaction. The obtained samples were divided into four groups for examining the effects of different periods of curing. The first set was tested immediately after the derivation of samples whereas the other three sets were tested after curing the samples for a period of 4, 7 and 28 days respectively under very high humid conditions as explained by Parihar and Gupta [28]. A strain rate of 1.25 mm/min was followed for all the compressive and shear strength tests. An average value of at least two UCS samples tested for each mix type has been taken as final. The SEM study has been conducted on the readily mixed samples, uncured samples and samples derived after various curing periods to examine the effect of additive proportion and curing time on soil structure at a microscopic level.

4 Results and Discussion

4.1 Uncured and Cured Compressive Strength

The behaviour of soil strength with respect to curing period is accessed through an unconfined compressive strength test. The samples of dimension 38 mm diameter and 76 mm height derived through proctor compaction test were used for the study. Samples of plain black cotton soil were tested immediately after extrusion of soil samples while the LLWA mixed samples were tested both immediately after derivation and also after different curing periods of 4, 7 and 28 days. The curing periods were decided on the grounds of time required for electronegative charge stabilization of soil, generation of compounds of cementation and initial gain of strength respectively. The effect of curing on UCS of soil with different LLWA contents is shown in Fig. 2a and the respective increase in UCS(ΔUCS) with respect to the UCS of untreated soil specimen (UCS_i) is shown through a dimensionless ratio ($\Delta UCS/UCS_i$) in Fig. 2b.

It was observed that the strength increase was continuous both with the addition of LLWA and with the progression of the curing period. A nominal strength gain was also noticed immediately after mixing the waste ash which was mainly due to the formation of cohesive matrix and particle packing effect as the soil and the waste ash exhibit different grain size characteristics. Further, with the increment of the curing duration, the calcium silicate compounds (calcium silicate hydrate (CSH) and calcium aluminate silicate hydrate (CASH)) continued to form with the chemical reaction of silica and aluminium in the soil and ash and calcium oxide in the ash in

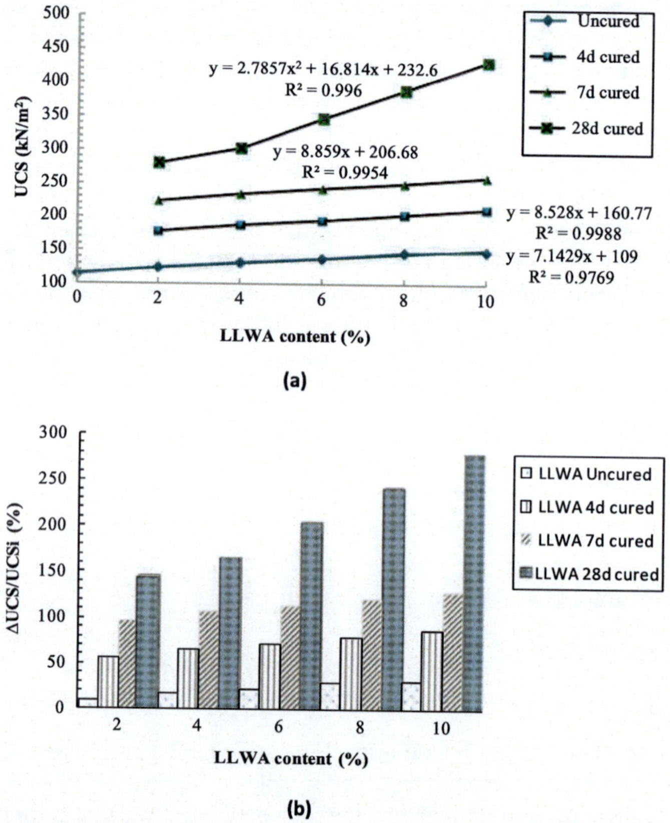

Fig. 2 **a** Relationship of UCS with LLWA content and curing period, **b** % rise in cured UCS w.r.t. untreated UCS

the presence of moisture which substantially increased the strength, particularly after 7 days of curing. A more than 270% increase in the UCS was observed with highest additive addition post 28 days of curing. It is interesting to note that the behaviour of UCS with LLWA content is almost linear with a high degree of correlation coefficient for uncured UCS specimens and cured specimens of 4 and 7 days whereas a parabolic increase in strength was observed for 28 days cured specimen. The extreme rise in this strength gain is due to complete formation of the silicate gels after 7 days of curing which provided the bonding between particles thereafter.

4.2 Shear Strength Parameters

The undrained shear parameters, i.e. cohesion and internal friction angle for each sample mix and for all defined curing periods were derived from the failure envelope obtained through unconsolidated undrained (UU) triaxial test performed at low confining pressures of 50, 100 and 200 kPa. The effect of waste addition and curing on angle of internal friction and cohesion are shown in Fig. 3a, b respectively. It is clear that the LLWA addition has an incremental effect on both the parameters as a consequence of the combined effects of granular material addition and its capacity to produce agglomeration between particles. This agglomeration is generated on account of the chemical reaction between calcium ions and electronegative silica mineral sheets and the bonding provided by silicate gel which increased both cohesion and internal friction. It can also be noticed that the effect of waste on cohesion is more pronounced as the cementation between particles enhanced progressively with the addition of waste and higher curing time. The rate of cohesion rise also increased after 7 days of curing due to full establishment of strength yielding calcium silicate gel and peak cohesion of 260 kPa was achieved as compared to 54 kPa of untreated soil cohesion post 28 days of curing time. However, the effect of waste on internal friction angle was limited to a certain extent as addition of LLWA beyond 6% concentration decreased this parameter at all curing periods. This may be due to cationic repulsion and lack of frictional effect between the ash particles in case of their excessiveness. It was also observed that the internal friction stabilized after 4 days of curing time and reduced beyond 7 days of curing. This probably has occurred due to the achievement of full electroneutrality till the former period which would have helped coalescence of the particles leading to maximum increase in the internal friction angle. Although, the presence of excessive gel formation between soil and ash particles at the latter period would have led to the reduction of particle to particle contact and hence the internal friction did not experience much appreciation at this curing period and beyond. But overall, cementation and cohesion had a dominating effect at even highest concentrations of additive and at all curing periods which also led to a continuous gain of UCS despite the decrease in the internal friction.

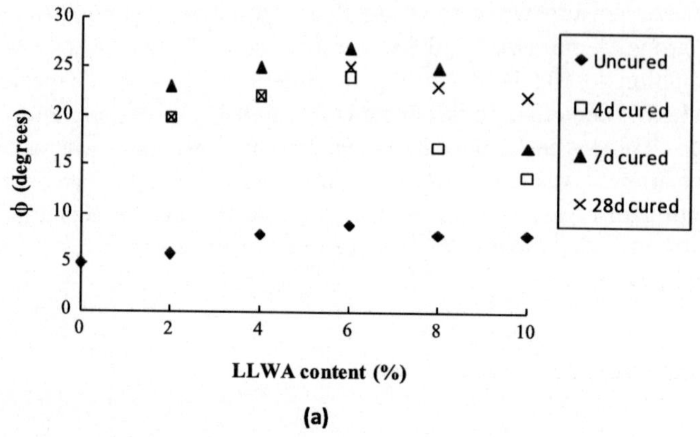

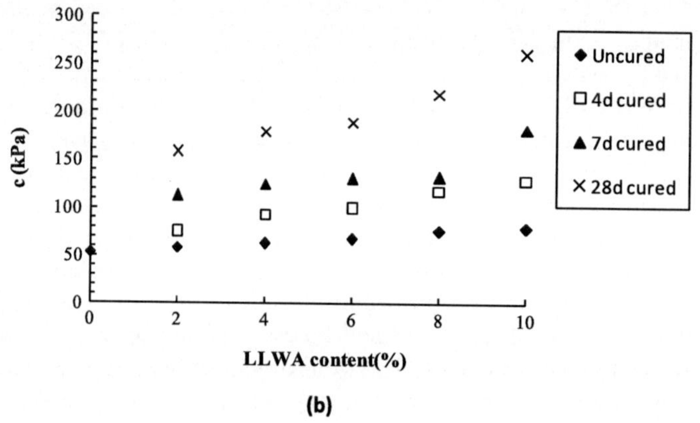

Fig. 3 Variation of shear strength parameters with LLWA content and curing period, **a** internal friction angle, **b** cohesion

4.3 Microstructure (Scanning Electron Microscopy)

The SEM study was conducted on the untreated plain soil sample, LLWA, uncured and cured soil-LLWA mix samples as shown in Fig. 4. The scanned images of plain black cotton soil showed a crystalline mineral structure of the soil which majorly contained montmorillonite with kaolinite as the secondary mineral component. The ash particles in the LLWA image can be noticed with a coating of calcium oxide having a flower-shaped crystal structure. It is evident from the SEM images of LLWA treated soil mix that the waste ash is capable of forming a dense matrix with the soil resulting in the reduced void structure both in the cured and uncured states. The agglomeration of particles however appeared after 4 days of curing with the chemical action of calcium ions on the mineral sheets of the soil. This agglomeration was responsible for the increased friction between particles as was determined from triaxial test. A noticeable change in the soil structure was also evident upon curing,

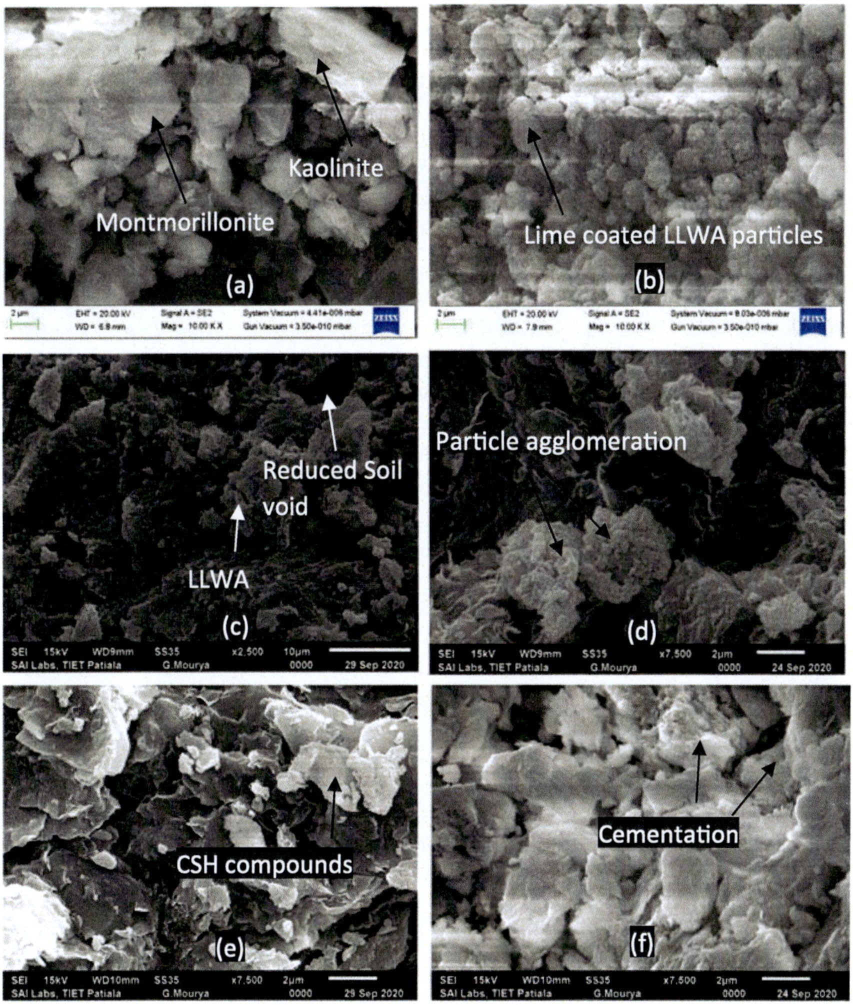

Fig. 4 SEM images for **a** untreated soil, **b** LLWA, **c** uncured mix, **d** 4 day cured mix, **e** 7 day cured mix, **f** 28 day cured mix

particularly beyond 4 days of curing period where the original crystalline structure of the soil was modified and dissolved into the compact cementitious structure of CSH and CASH gel formation. A well-developed silicate gel formation was visible after 28 days of curing period with the particle void spaces completely filled with cementitious compounds. This improved gel formation was the reason behind a sharp increase in UCS and cohesion of the mix at late periods of the curing led by an increased bond strength and cementation between particles. However, the gel structure reduced the particle to particle contact and decreased the overall friction which appeared as a reduced internal friction angle in the UU test results.

5 Conclusions

The effect of a leather industry by-product known as liming leather waste ash (LLWA) on the compressive and shear strength characteristics of expansive soil and its relative influence on the shear strength parameters have been studied. The effect of curing on the LLWA-soil mix was also studied through UCS and UU tests. It was concluded that the waste possesses the required characteristics of both lime and silica for pozzolanic reactions and formation of cementitious products for strength enhancement. The additive increased the unconfined strength of the soil in both uncured and cured stages through particle compaction and silicate gel formation respectively. The curing had a major effect on the attainment of strength and a progressive increase was found in the UCS and cohesion of soil with the extension of the curing period with the attainment of 270 and 380% rise in both the parameters at the highest waste proportion post 28 days of curing. However, the effect of curing on internal friction angle was non-uniform and its value decreased after 7 days of curing. The effect of curing has been explained through SEM analysis which clearly shows the particle agglomeration, the formation of cementitious compounds and their influence on different strength parameters at various curing periods.

References

1. Shukla, R. P., Parihar, N. S., Tiwari, R. P., & Agrawal, B. K. (2014). Black cotton soil modification using sea salt. *Electronic Journal of Geotechnical Engineering, 19*, 8807–8816.
2. Shukla, R. P., & Parihar, N. S. (2016). Stabilization of black cotton soil using micro-fine slag. *Journal of the Institution of Engineers (India): Series A, 97*(3), 299–306. https://doi.org/10.1007/s40030-016-0171-1
3. Degirmenci, N., Okucu, A., & Turabi, A. (2007). Application of phosphogypsum in soil stabilization. *Building and Environment, 42*(9), 3393–3398. https://doi.org/10.1016/j.buildenv.006.08.010
4. Ismail, A. I. M., & Belal, Z. L. (2016). Use of cement kiln dust on the engineering modification of soil materials, Nile Delta, Egypt. *Geotechnical and Geological Engineering, 34*, 463–469. https://doi.org/10.1007/s10706-015-9957-6
5. Solanki, P., & Zaman, M. (2012). Microstructural and mineralogical characterization of clay stabilized using calcium-based stabilizers. In V. Kazmiruk (Ed.), *Scanning electron microscopy* (pp. 771–798). InTechOpen. Available from http://www.intechopen.com/books/scanningelectronmicroscopy/microstructuraland-mineralogicalcharacterization-of-clay-stabilized-using-calcium-based-stabilizer. https://doi.org/10.5772/34176
6. Jain, A. K., Jha, A. K., & Shivanshi, (2020). Improvement in subgrade soils with marble dust for highway construction: A comparative study. *Indian Geotechnical Journal, 50*, 307–317. https://doi.org/10.1007/s40098-020-00423-5
7. Kalkan, E. (2011). Impact of wetting–drying cycles on swelling behavior of clayey soils modified by silica fume. *Applied Clay Science, 52*(4), 345–352. https://doi.org/10.1016/j.clay.2011.03.014
8. McDowell, U. C. (1959). Stabilization of soils with lime, lime-flyash and other lime reactive materials. *Highway Research Board Bulletin, 231*, 60–66.

9. Lavanya, C., & Srirama Rao, A. (2017). Study of swelling potential for copper slag cushion laid over expansive soil bed. *Indian Geotechnical Journal, 47*(3), 280–285. https://doi.org/10.1007/s40098-017-0227-9
10. Mudgal, A., Sarkar, R., & Sahu, A. K. (2014). Effect of lime and stone dust in the geotechnical properties of black cotton soil. *International Journal of GEOMATE, 7*, 1033–1039.
11. Parihar, N. S., Garlapati, V. K., & Ganguly, R. (2018). Stabilization of black cotton soil using waste glass. In C. M. Hussain (Ed.), *Handbook of environmental materials management* (pp. 1–16). Springer International Publishing AG. https://doi.org/10.1007/9783319736457_147
12. Sabat, A. K. (2014). Engineering properties of an expansive soil stabilized with rice husk ash and lime sludge. *International Journal of Engineering and Technology, 5*(6), 4826–4833.
13. Sharaby, C. M., El mashad, M. E. M., Fakhry, M. A., & Salama, M. H. (2021) Utilization of lime sludge activated by salt in Egyptian expansive soil improvement. *Asian Journal of Applied Sciences, 9*(2). https://doi.org/10.24203/ajas.v9i2.6573
14. Akinwumi, I. I., & Aidomojie, O. I. (2015). Effect of corncob ash on the geotechnical properties of lateritic soil stabilized with Portland cement. *International Journal of Geomatics and Geosciences, 5*(3), 375–392.
15. Oriola, F., & Moses, G. (2010). Groundnut shell ash stabilization of black cotton soil. *Electronic Journal of Geotechnical Engineering, 15*, 415–428.
16. Pallavi, P., Gowtham, B., Kiran, R. G., & Kilbukar, P. (2016). Study of bagasse ash and cement stabilized marshy soil. *The International Journal of Innovative Research in Science Engineering and Technology, 5*(1), 651–659. https://doi.org/10.15680/IJIRSET.2015.0501112
17. Singh, N., Sharma, R., & Abhishek, A. (2017). Soil stabilization using industrial waste (wheat husk and sugarcane straw ash). *IRJET, 4*(9), 589–596.
18. Ashango, A. A., & Patra, N. R. (2016). Behavior of expansive soil treated with steel slag, rice husk ash, and lime. *Journal of Materials in Civil Engineering, 28*(7). https://doi.org/10.1061/(ASCE)MT.1943-5533.0001547
19. Moniuddin, K., & Chethan, K. (2018). Stabilization of expansive soil using wheat husk ash and granulated blast furnace slag. *Trends In Civil Engineering And Its Architecture, 1*(2), 29–35. https://doi.org/10.32474/TCEIA.2018.01.000108
20. Khurana, Y. (2020). Invest India report 2020. Available from https://www.investindia.gov.in/sector/leather. April 18, 2020.
21. Rajamani, D. S. (2010). Sustainable environmental management and utilisation of solid wastes from world leather sector. https://www.clri.org/LERIG2010/Lectures/Dr.Rajamani-4.pdf. Accessed on 17 April 2019.
22. Alibardi, L., & Cossu, R. (2016). Pre-treatment of tannery sludge for sustainable landfilling. *Waste Management, 52*, 202–211. https://doi.org/10.1016/j.wasman.2016.04.008
23. Mia, M. A. S., Nur-e-Alam, M., Huque, S., Rahman, M., & Ratnaweera, H. (2017). Biogas production through anaerobic digestion of tannery solid waste. *Iranica Journal of Energy and Environment, 8*(3), 210–215. https://doi.org/10.5829/ijee.2017.08.03.05
24. Kanagaraj, J., Velappan, K., Babu, N., & Sadulla, S. (2006). Solid wastes generation in the leather industry and its utilization for cleaner environment—A review. *Scientific & Industrial Research, 65*, 541–548.
25. Famielec, S. (2015). Environmental effects of tannery waste incineration in a tunnel furnace system. *Proceedings of ECOpole, 9*(2), 441–450. https://doi.org/10.2429/proc.2015.9(2)051
26. Fela, K., Wieczorek-Ciurowa, K., Konopka, M., Wozny, Z., & Adam, T. (2010). Leather industrial waste characteristics for the thermal processing. *Waste Forum, 2*(4), 374–379.
27. Parihar, N. S., & Gupta, A. K. (2020). Chemical stabilization of expansive soil using liming leather waste ash. *International Journal of Geotechnical Engineering.* https://doi.org/10.1080/19386362.2020.1775357
28. Parihar, N. S., & Gupta, A. K. (2021). Improvement of engineering properties of expansive soil using liming leather waste ash. *Bulletin of Engineering Geology and the Environment, 80*, 2509–2522. https://doi.org/10.1007/s10064-020-02051-y
29. Bureau of Indian Standards IS:1498–1970. (1987). Indian standard classification and identification of soils for general engineering purposes. BIS, New Delhi, India.

30. Bureau of Indian Standards IS: 2720 (Part X)-1991. (2006). Determination of unconfined compressive strength. BIS, New Delhi, India.
31. Bureau of Indian Standards IS: 2720 (Part XI)-1993. (2002). Determination of the shear strength parameters of a specimen tested in unconsolidated undrained triaxial compression without the measurement of pore water pressure. BIS, New Delhi, India.
32. Bureau of Indian Standards IS: 2720 (Part VII)-1980. (2011). Determination of compaction parameters. BIS, New Delhi, India.
33. Bureau of Indian Standards IS: 2720 (Part III/Sec1)-1980. (2002). Determination of specific grvity. BIS, New Delhi, India.
34. Bureau of Indian Standards IS: 2720 (Part V)-1985. (2006). Determination of liquid and plastic limit. BIS, New Delhi, India.
35. Bureau of Indian Standards IS: 2720 (Part VI)-1972. (2001). Determination of shrinkage factors. BIS, New Delhi, India.
36. Bureau of Indian Standards IS:2720 (Part XVI)-1987. (2002). Laboratory determination of CBR. BIS, New Delhi, India.
37. Bureau of Indian Standards IS: 2720 (Part XL)-1977. (2002). Determination of free swell index of soils. BIS, New Delhi, India.
38. Bureau of Indian Standards IS: 2720 (Part XLI)-1977. (2002). Measurement of swelling pressure of soils. BIS, New Delhi, India.

Analysis of Critical Factors Affecting Labor Productivity of Construction Projects in Himachal Pradesh

Kaushal Kumar and Rishi Rana

Abstract The construction industry in India with Gross Domestic Product share of 8–10% is the second largest contributor in previous decade and it employs a large population. Construction is a labor-intensive industry. Decreasing construction productivity is a major concern faced by this industry. The loopholes in labor productivity sprang from various factors that are behind the labor inefficiency to complete the work at the desired rate. These factors are reviewed from different studies already conducted in the past to find the inefficacy of a project. In the present study, five construction sites of Himachal Pradesh, India have been taken into consideration. Firstly, the productivity of a construction site was checked using work sampling method which conforms to our assumption of decreased productivity. In the second phase, a questionnaire has been prepared where important attributes of labor productivity were stressed to find the significant factors affecting construction labor productivity. The methodology used was, (a) problem identification in labor's inefficiency, (b) getting questionnaire responses related to factors affecting labor's productivity from each site, and analysis of the data using Relative Important Index to get significant/critical factors. From this study, it was found that Rework, Efficiency of labor, Materials Availability, working without holidays, and safety having RII values as 0.88, 0.76, 0.52, 0.40, and 0.16 respectively were the important factors that are affecting the construction labor productivity in Himachal Pradesh. The results obtained from this study suggest that the management team needs to focus on more clarity of the task to be performed as improper supervision at the sites leads to an increase in the workload by redoing the same work again.

Keywords Construction · Labor productivity · Questionnaire · Work sampling · Relative important index

K. Kumar (✉) · R. Rana
Civil Engineering Department, Jaypee University of Information Technology, Waknaghat, Solan, Himachal Pradesh 173234, India
e-mail: kaushal.kumar@juit.ac.in

R. Rana
e-mail: rishi.rana@juit.ac.in

A. K. Gupta et al. (eds.), *Advances in Construction Materials and Sustainable Environment*, Lecture Notes in Civil Engineering 196,
https://doi.org/10.1007/978-981-16-6557-8_84

1 Introduction

The construction industry is a labor-intensive industry, and in India, it provides employment to a large population. The contribution of construction industry plays a vital role in GDP growth of the country, and it is the second largest contributor of GDP (8–10%) [1]. Construction is one of the most challenging and mammoth industries. After having best labor, best equipment, and even high technology, the final productivity when evaluated comes out to be very low. So what would be the matter that even after having "the best" of everything hinders the way of 100% productivity and efficiency? The answer behind this difficult question is very simple and that is "labor productivity". The role of labor is a crucial factor to work upon for eradicating the inefficiency of the construction industry in our country. The reasons behind why we need to focus on labor to boost our construction industry are (a) The contribution of the labor industry to our national growth and per capita growth is large. (b) Moreover, construction industry is contributing to a large extent towards providing employment to skilled and unskilled people. (c) Modern slavery is also one of the stigmas attached to the construction industry. By keeping all these factors in mind there seems to be an urgent need to look upon those factors in construction industry that are very sensitive towards overhauling the industry. By considering different factors like time, cost, workability, quality assurance at site, type of material to work with, it can be said that labor is the pioneering and the driving force which is responsible for handling or taking care of the above written factors [2–7]. Several factors contribute towards the efficient working of the labor under any contractor. These factors include economic factors, psychological factors, socio-cultural factors, and many more, and these factors are the main reasons behind the lag of labor efficiency [5].

In this paper, different factors were reviewed from various studies, their veracity in different situations, and delay of the project. Methods to find the factors or delay or productivity have been analyzed and a questionnaire was prepared and has been validated with the experts. Since there are a lot of factors for labor inefficiency, so the weightage of these factors and their ranking is also an important aspect to find out which factor is the most sensitive one. Different methods and tools are required to simulate the mammoth data to scrutinize it to the final fruitful conclusion. Various trends that could be followed in the industry like working in the shifts, use of Information and Technology in the construction field, making a database of all the projects are some of the new initiatives that can bring a boost to this industry [2, 5].

1.1 Need of Labor Productivity Study for Construction Project in Himachal Pradesh

To understand the status of an ongoing project, an ongoing project site data (construction of a residential building in Palampur, Himachal Pradesh) has been made to enter

Activity ID	Activity Name	Original Duration	Remaining Duration	Schedule % Complete	Start	Finish	Total Float
assignment building estimate wbs		121	121	0%	03-May-17	02-Sep-17	0
SUBSTRUCTURE		19	19	0%	03-May-17	22-May-17	0
Foundation work		19	19	0%	03-May-17	22-May-17	0
A	Excavation	7	7	0%	03-May-17	09-May-17	0
C	first class brickwork	6	6	0%	16-May-17	22-May-17	0
B	Lime concerete	6	6	0%	10-May-17	15-May-17	0
D	reinforcement	4	4	0%	16-May-17	20-May-17	2
SUPERSTRUCTURE		102	102	0%	23-May-17	02-Sep-17	0
Concrete work		62	62	0%	23-May-17	24-Jul-17	0
F	brickwork	32	32	0%	29-May-17	29-Jun-17	0
E	DPC	5	5	0%	23-May-17	28-May-17	0
G	Plastering	25	25	0%	29-Jun-17	24-Jul-17	0
Steel work		15	15	0%	29-May-17	12-Jun-17	17
I	MS work	6	6	0%	29-May-17	03-Jun-17	26
H	RCC work	15	15	0%	29-May-17	12-Jun-17	17
Finishes		40	40	0%	25-Jul-17	02-Sep-17	0
M	Floor finish	20	20	0%	14-Aug-17	02-Sep-17	0
J	panels and hold fastings	7	7	0%	25-Jul-17	31-Jul-17	0
K	pipe fittings	5	5	0%	25-Jul-17	29-Jul-17	2
L	white washing	13	13	0%	01-Aug-17	13-Aug-17	0

Fig. 1 Analytical tool (Primavera Contractor) worksheet

in the analytical tool, **Primavera Contractor**. By analyzing and comparing the data for the project completion as shown in Fig. 1, a difference of 59 days had been found as that of the original time taken by the laborers on site was 180 days and with an analytical tool, it was 121 days. This was done to have an idea about the delay in terms of time.

1.2 Labor Productivity in Construction Project

Productivity can be defined as the output of any industrial process. It gives the rate at which a company produces goods and services with buffer available. But technically productivity can be given as the "ratio of the quantity of Output to the quantity of inputs" as given by Eq. 1.

$$\text{Productivity} = \text{quantity of output/quantity of input} \tag{1}$$

where Output is in terms of installed quantities at the end of the project; Input is in terms of labor hour. Productivity is basically labor intensive. It is measured as the labor hours per unit of the work. The significance of productivity lies in the fact that it measures the industrial trend, helps in comparing performance, and to see the effect on scheduling of different activities in a project.

Fig. 2 Flowchart of the methodology followed for the study

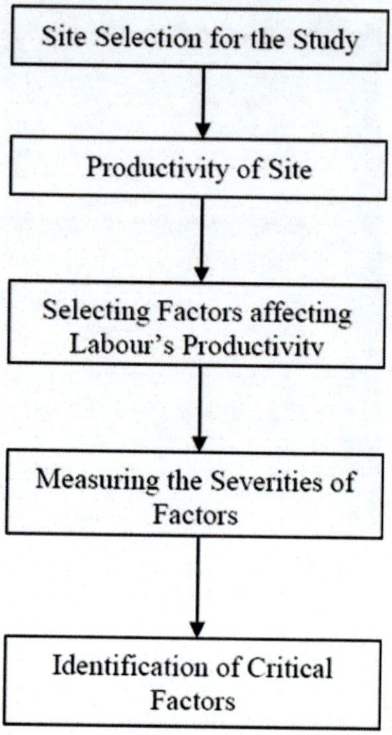

2 Methodology

For finding the significant factors associated with labor productivity of construction projects in hilly terrain of HP, the methodology adopted has been shown in the flow chart, Fig. 2.

3 Site Selection

To study the overall trend of attributes affecting labor productivity pertaining to the locations in HP, five sites were chosen. The sites were chosen on the basis of the demography of the location, i.e., in these areas construction activities employ a lot of skilled and unskilled laborers and sites have been taken into consideration as these sites were having active projects and reliable data (like work schedule, progress report, etc.) were available. Following are the five sites taken into consideration in our study:

1. Construction of a Residential building in Palampur, district Kangra (Site 1)
2. Roadside construction building in district Kangra (Site 2)
3. Riverside construction of Building, Rajgardh, district Sirmour (Site 3)

4. Residential building in plain region of Kasauli, district Solan (Site 4)
5. Building construction in Urban area of Solan, district Solan (Site 5).

4 Factors Affecting the Labor Productivity in Himachal Pradesh

To find the factors affecting labor productivity, a questionnaire was devised by reviewing literature and then was updated according to the study objective to get the responses [8]. Factors of the questionnaire were cross-checked by experts from the field of construction management. After which a list of important factors affecting labor productivity in Himachal Pradesh was prepared and the questionnaire was sent to different sites to get the responses to find the most critical ones. List of the factors are as follows:

1	Environmental factors	9	Frequent revision of drawing (rework)
2	Material availability	10	Lack of locally available labor
3	Safety	11	Financially weakness of a contractor
4	Quality	12	Scanty financial policies of the government
5	Manpower	13	Waging seven days without holidays
6	Time available	14	Availability of equipment/tools
7	Motivational leadership	15	Miscommunication b/w contractor and labor
8	Strikes caused by political parties	16	Unskilled labors

5 Work Sampling Method [7]

Work sampling technique determines the percentage of occurrence of certain activity by statistical sampling and random observation. The percentage of observations so calculated for specific work or delay is the measure of percentage of time during which that works or delay occurs because of the workers.

To make it happen, the following approach is to be adopted:

1. Categorize the worker's activities as productive, semi productive, and nonproductive.
2. Take the random observations of workers at the site who are involved in a given operation in a field (random means choosing without any bias as to who is being observed).
3. Jolt down all observations so formed. Enter the checkmark under the appropriate mode.
4. Add up all the checkmarks and calculate the quantity as percentage.

Table 1 Productivity of site 1

S. No.	Productive (direct work)	Semi productive (support work)	Nonproductive (delay)
1	*		
2	*		
3			*
4		*	
5		*	
6		*	
7	*		
8	*		
9			*
10	*		
Total	**5**	**3**	**2**
Productivity	**50%**	**30%**	**20%**

The work sampling procedure is being applied at all the sites to check the productivity at each site. Site-1, i.e., Construction of a Residential building in Palampur, district Kangra, productivity data of site is being analyzed and given in Table 1, subsequently, all sites are being evaluated for the productivity and have been shown in Fig. 3.

Table 1, and Fig. 3, amount to the fact that with the operation of work sampling tool that each of the sites is lacking complete productivity. The productivity of each site is coming out to be 50%, 40%, 70%, 30%, and 40% for site 1, site 2, site 3, site 4, and site 5, respectively. This infers to the fact that at all the sites; there was a need to study the reasons responsible for a decline in productivity and how the labor's factors affecting construction productivity.

As we have seen that the productivity of construction sites is not optimum, there could be many reasons associated with it, but we will be focusing on the human factor

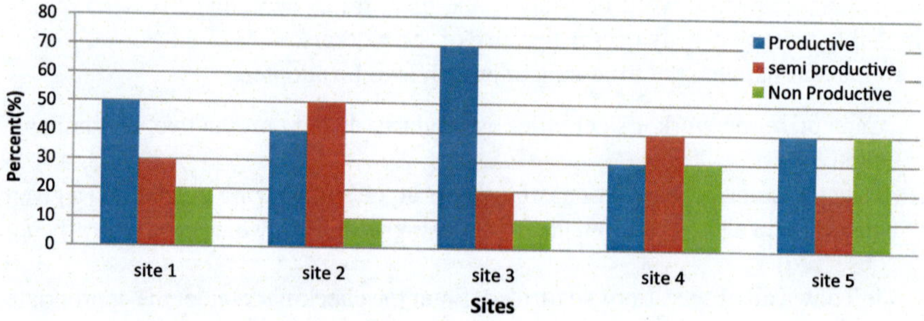

Fig. 3 Histogram of productivity of different sites in percentage

associated with declined productivity at construction sites. To evaluate the human factor, i.e., the construction labor productivity, we will be applying the questionnaire technique and looks for the factors responsible for the decline in labor productivity.

6 Questionnaire Survey

The work sampling method described above gives efficiencies at the site but was unable to identify the major cause for the efficiency. On-site surveying and questionnaire surveys were used for the site evaluation and were used to increase the efficiency at site. For evaluating the present scenario of the site, probable queries of the questionnaire were prepared from the many studies and were reviewed carefully. The questionnaire was evaluated by experts in the field of construction management and planning. After this, a final questionnaire was prepared to be circulated for the responses from each site. To effectuate the suitable objective the data for the work were collected through both online and offline modes. With the well-planned questionnaire survey and on-site data collection have been carried out a total of 87 valid responses (site 1–20; site 2–18; site 3–22, site 4–15, and site 5–12) from engineers and workers at respective sites were obtained. Validity of the data was being done using "**Test-retest Method**", i.e., a similar set of respondents were asked to fill the same questionnaire again after a duration of 1 month, and responses were cross-checked with the previous responses. Responses received from the 2nd questionnaire were same as that of the previous one implying the data received were reliable.

To achieve the objectives of this study, work was being divided into two stages. First stage is to prepare the questionnaire and distribute it to the respondent and the second stage was to evaluate the data using the relative importance index (RII). First stage was completed by fixating each factor and problems associated with it which are hindering construction labor productivity. In this study, we have fixated 16 questions for getting the major factors affecting on-field productivity. These factors have been zero-downed by taking into account vast studies on construction labor productivity. The statistical tests have been applied to the data collected from questionnaire survey: RII (relative importance index) for each factor has been calculated to highlight the total score responses received for each attribute. To be consistent with the preview of this study, the responses collected from all the sites were analyzed with the RII and find the ranking of the crucial factors to get the desire objectives. The RII method is explained below, and it will be used to find the ranking of the critical factors at each site.

Relative Important Index:

$$\mathrm{RII} = \Sigma \mathrm{W}/(\mathrm{A} * \mathrm{N}) \tag{2}$$

where W is the sum of score of all the weightage of each factor by the respondents (ranging from 1 to 5), A is the highest weight (i.e., 5 in this case), N is the total number of respondents.

Higher the value of RII, more important was the cause of delays. Ranking of the factors affecting labor productivity has been computed from site 1 and shown in Table 2.

From Table 4, the rankings for the factor affecting labor productivity have been evaluated based on the RII value. It was found out that the critical factors affecting labor productivity at site one were reworked, efficiency of labor, materials availability, safety, and Inefficient labor supervision having their respective RII as 0.85, 0.72, 0.7, 0.68, and 0.65 were critical factors.

The critical factors at all the sites by finding the RII of all the factors at all the five sites have been computed to find the critical factors at all the sites and have been shown in Table 3.

Table 2 Relative importance index and rankings of the factors for site 1

S. No.	Factors	Total score (W)	Max. weightage (A)	Total No. of respondents (N)	RII	Rank
1	Environmental	50	5	20	0.5	9
2	Material availability	70	5	20	0.7	3
3	Safety	68	5	20	0.68	4
4	Quality	58	5	20	0.58	8
5	Labor supervision	65	5	20	0.65	5
6	Time availability	64	5	20	0.64	6
7	Motivational leadership	40	5	20	0.4	13
8	Political strikes	44	5	20	0.44	12
9	Rework	85	5	20	0.85	1
10	Lack of locally available labor	32	5	20	0.32	15
11	Financial weakness of contractor	62	5	20	0.62	7
12	Inadequate financial policies of government	35	5	20	0.35	14
13	Working without holidays	50	5	20	0.5	9
14	Availability of tools	48	5	20	0.48	11
15	Miscommunication B/T contractor and labor	21	5	20	0.21	16
16	Efficiency of labor	72	5	20	0.72	2

Table 3 Relative important index for all the factors at every site

S. No.	Factors	Site 1	Site 2	Site 3	Site 4	Site 5
1	Environmental	0.50	0.5556	0.4545	0.67	0.4167
2	Material availability	0.70	0.7778	0.8000	0.77	0.5000
3	Safety	0.68	0.7556	0.6182	0.59	0.3333
4	Quality	0.58	0.6444	0.5273	0.67	0.2667
5	Labor supervision	0.65	0.7222	0.5909	0.27	0.3333
6	Time availability	0.64	0.7111	0.5818	0.35	0.6167
7	Motivational leadership	0.40	0.4444	0.3636	0.53	0.4333
8	Political strikes	0.44	0.7444	0.4000	0.59	0.2333
9	Rework	0.85	0.9444	0.6455	0.93	0.8833
10	Lack of locally available labor	0.32	0.3556	0.2909	0.43	0.3500
11	Financial weakness of contractor	0.66	0.7444	0.6545	0.76	0.6833
12	Inadequate financial policies of govt	0.35	0.3889	0.3182	0.47	0.4333
13	Working without holidays	0.50	0.5556	0.7727	0.84	0.7333
14	Availability of tools	0.48	0.5333	0.4364	0.64	0.3000
15	Miscommunication B/T contractor and labor	0.21	0.2333	0.1909	0.28	0.4333
16	Efficiency of labor	0.72	0.8000	0.6545	0.87	0.8333

Table 4 Critical factors of the different sites

Site/Factors	Site 1	Site 3	Site 3	Site 4	Site 5
Factor 1	Rework	Rework	Material availability	Rework	Rework
Factor 2	Efficiency of labor	Efficiency of labor	Working without holidays	Efficiency of the labor	Efficiency of labor
Factor 3	Material availability	Material availability	Efficiency of the labor	Work without holiday	Working without holiday
Factor 4	Safety	Safety	Rework	Material availability	Financial weakness of contractor
Factor 5	Labor supervision	Political strikes	Financial weakness of contractor	Financial weakness of the contractor	Time availability

In Table 3, RII values of all the factors for all the sites have been evaluated. Ranking of all the factors has for all sites has been calculated and crucial factors responsible for the decline of labor's productivity have been assembled in Table 4.

For finding the significant factors responsible for reduced labor productivity in construction projects in HP, the top five contributors for each site have been identified

Table 5 Ranking of the critical factors taken from different sites

S. No.	Factors	Site 1	Site 2	Site 3	Site 4	Site 5	Avg. RII
1	Environmental	0.5000	0.5556	0.4545	0.6667	0.4167	**0.5187**
2	Material availability	0.7000	0.7778	0.8000	0.7733	0.5000	**0.7102**
3	Safety	0.6800	0.7556	0.6582	0.5867	0.6167	**0.6594**
4	Quality	0.5800	0.6444	0.5273	0.6667	0.2667	**0.5370**
5	Time availability	0.6400	0.7111	0.5818	0.5467	0.6167	**0.6193**
6	Rework	0.8500	0.9444	0.6455	0.9333	0.8833	**0.8513**
7	Financial weakness of contractor	0.6600	0.7444	0.6545	0.7600	0.6833	**0.7004**
8	Working without holidays	0.6200	0.5556	0.7727	0.8400	0.7333	**0.7043**
9	Efficiency of labor	0.7200	0.8000	0.6545	0.8667	0.8333	**0.7749**

as tabulated in Table 4. In the next step, all the factors from different sites have been tabulated in Table 5 and their respective relative importance is being calculated to find the most critical factors. The RII of the important factors as an average of all the RIIs from each site has been calculated. The avg. RII values of significant factors are given in Table 5.

From different sites, significant factors were selected based on their RII values, and for the considering factors related to Himachal Pradesh, Avg. RII values have been calculated to evaluate critical factors. Higher the value of RII, higher will be its impact on labor productivity. Rework, Efficiency of labor, and Materials availability are significant factors with RII 0.8513, 0.7749, and 0.7102 respectively.

7 Conclusions

The findings of the above studies are as follows:

- Higher values of RII imply that the factor is having higher impact on Labor productivity. The RII value of factors equal to 0.7 or above is considered critical [1]. Table 6 below shows all the critical factors along with their RII values and Ranks.
- The overall productivity of each site has been calculated from the work sampling method were 50% for site 1; 40% for site 2; 70% for site 3; 30% for site 4; and 40% for site 5. This indicates that each project which was studied is having very little productivity.
- Prior knowledge of labor productivity factors during the execution of the construction project can save money and will also improve project completion time.

Table 6 Critical factors affecting construction labor productivity in Himachal Pradesh

S. No.	Critical factors	Avg. RII	Rank
1	Rework	0.8513	1
2	Efficiency of labor	0.7749	2
3	Material availability	0.7102	3
4	Working without holidays	0.7043	4
5	Financial weakness of contractor	0.7004	5

- The results obtained were region specific as well as planning specific. For example, the productivity of site 3 (30%) is having the least productivity. Apart from labor productivity issues, technical issues may also be responsible for low productivity.

8 Limitations

- The total number of respondents at each site was not large, which increases the margin of error.
- Apart from construction laborers, other personals from construction industry like contractors, supervisors, etc. should also have been taken into consideration for the study.
- Other industry laborers like road and highways, hydraulic structures, etc. construction workers should also be taken into consideration for finding critical factors of overall construction industry labor.

References

1. Manoharan, K., Dissanayake, P., Pathirana, C., Deegahawature, D., & Silva, R. (2020). Assessment of critical factors influencing the performance of labour in Sri Lankan construction industry. *International Journal of Construction Management*, 1–35.
2. Hamza, M., Shahid, S., Hainin, M., & Nashwan, M. (2019). Construction labour productivity: review of factors identified. *International Journal of Construction Management.*
3. Alaghbari, W., Al-Sakkaf, A., & Sultan, B. (2019). Factors affecting construction labour productivity in Yemen. *International Journal of Construction Management, 19*(1), 79–91.
4. Dabirian, S., Moussazadeh, M., Khanzadi, M., & Abbaspour, S. (2021). Predicting the effects of congestion on labour productivity in construction projects using agent-based modelling. *International Journal of Construction Management.*
5. Murodif, A., & Widyarti, E. M. (2016). Measurement of productivity using work sampling method at menara sentraya building project Jakarta Indonesia. *Scholars Journal of Engineering and Technology.*
6. Ghate, P., More, A., & Minde, P. (2016). Importance of measurement of labor productivity in construction. *International Journal of Research in Engineering and Technology.*

7. Vogl, B., & Wahab, A. (2014). Measuring the construction industry's productivity performance: Critique of international productivity comparisons at industry level. *American Society of Civil Engineers.*
8. Agrawal, A., & Halder, S. (2020). Identifying factors affecting construction labour productivity in India and measures to improve productivity. *Asian Journal of Civil Engineering, 21*(4), 569–579.

Effect of Adding Fly Ash and Metakaolin on Mechanical Properties of Concrete

S. K. Singh, Maninder Singh, and Rajesh Goyal

Abstract The feasibility of incorporating fly ash and metakaolin (MK) in producing modern-day concrete of enhanced mechanical performance was studied to evaluate the application of such concrete at a mass scale. Research was carried out via partial substitution of cement with MK and fly ash and the effect on flexural strength (FS), compressive strength (CS) and split tensile strength (STS) of Mix 30 concrete was investigated, while maintaining the same water–cement ratio of 0.45 as in normal concrete. To find the optimum proportion of replacement for best performance, the amount of cement in concrete was replaced with 5, 10, 15, and 20% each of (MK) and fly ash by weight. The results show that replacing cement improves concrete's CS, FS, and STS, with the highest results achieved when cement was substituted with 5% fly ash and 10% MK in concrete. MK was found to be most effective at a level of 10% for all percentages of fly ash and cement in the mix. The study of long-term performance and durability of concrete for ensuring long-lasting effects and mass applications is highly recommended for future research on the subject.

Keywords Compressive strength (CS) · Flexural strength (FS) · Split tensile strength (STS) · Metakaolin · Fly ash

1 Introduction

Concrete is one of the pre-eminent frequently used materials in the globe, later than water, and the single most extensively utilized construction material. It has wide applications in the construction of building structures, bridges, highways, dams,

S. K. Singh
Chandigarh University, Mohali, India

M. Singh (✉)
Punjabi University, Patiala, India

R. Goyal
NICMAR, Bahadurgarh, India
e-mail: rgoyal@nicmar.ac.in

A. K. Gupta et al. (eds.), *Advances in Construction Materials and Sustainable Environment*, Lecture Notes in Civil Engineering 196,
https://doi.org/10.1007/978-981-16-6557-8_85

power houses to name a few. The production of strong and durable concrete is the mandatory requirement of present-day construction. Cement is regarded as one of the most important components of concrete and is not only costly in production but also results in huge amount of carbon emission. Supplementary cementitious materials are being looked at by the industry and researchers so as to decrease cement content in the concrete mix to produce adequately workable fresh concrete with comparable strength and durability of hardened concrete. GGBS, fly ash, silica fume, MK, rice husk ash, and other extra cementitious ingredients are examples.

MK, a pozzolanic material, is thermally active alumino-silicate substance having pozzolanic activity than other pozzolanic materials such as silica fume [1]. For the production of MK, the kaolin clay is calcined to a high temperature of the order of 650 to 800 °C. MK is not a by-product of any industrial activity or process, and it varies from additional cementitious materials such as fly ash, GGBS, or silica fume in several ways. MK can also be created by calcining indigenous lateritic soils at 750–800 °C, according to Ambroise et al. [2].

Paper recycling industry produces huge quantity of waste sludge every year. By the calcination of this sludge, large amount of MK can be produced [3].

Chen et al. [4] scrutinized the properties of steam cured concrete including MK and limestone. It was concluded that the steam cured concrete combined with limestone and MK can gain early strength and also reduces steam curing temperature. Mechanical properties were found to be improved on addition of MK and limestone in concrete of steam-cured.

In current study, the effects of adding MK and fly ash as partial cement substitutes on CS, FS, and STS properties of concrete were explored utilizing a constant water–cementitious (binder) material ratio of 0.45 for concrete mix with characteristic strength 30 N/mm^2 after 28 days. When pozzolanic materials like fly ash and MK are added to concrete, mechanical properties such as CS, FS, and STS are improved due to a reaction between these materials and free Ca $(OH)_2$ during cement hydration, resulting in the creation of extra calcium silicate hydrate (C–S–H). However, including these components into concrete has an adverse influence on workability, necessitating the usage of a superplasticizer in such situations.

2 Literature Review

Your many researchers have carried out studies on different parameters by replacing the cement with MK separately and with fly ash, which includes fineness, mineral composition, workability, mechanical characteristics such as various strengths of cement mortars and concrete, reactivity of alkali aggregate and durability. Wild et al. [5] investigated mechanism of enhancement of concrete strength owing to introduction of MK content up to 30% for different age of concrete ranging from 1 to 90 days. They established that an optimal replacement level of cement by 20% of MK by weight is remained from more than 14 days of curing. Sabir et al. [6] studied the feasibility of introducing MK as a pozzolanic substance in mortar, concrete, and established possible range of MK application-based concrete in construction business. Based on their study, they concluded, involvement of MK as a pozzolanic material helps in the growth of prior strength (up to 14 days of age) with more improvement in the long-time strength of mortar and concrete.

Ding and Li [7] conducted a series of tests with seven concrete mixes, substituting cement with MK by 0, 5, 10, and 15% by weight of cement–water ratio at 0.35 and 0.40 sand-to-aggregate ratios. The influence of MK and silica fume's workability, strength, and durability of concrete, as well as shrinkage and chloride penetration resistance, was investigated experimentally. It was discovered that adding MK and silica fume to concrete reduced free drying shrinkage and shrinkage cracking width. They too discovered, incorporating MK and silica fume into concrete reduces chloride diffusion rate significantly. Under uniaxial compression or tension, Jin and Li [8] investigated the improvement in mechanical properties of concrete after adding mineral admixtures such as silica fume, slag, fly ash, and MK. At the ages of 1/2, 1, 2, 3, 7, and 28, concrete specimens were examined. It was discovered that different mineral admixtures have varied affects and effects on concrete characteristics while it is young. MK exhibits to contain the greatest impact on mechanical characteristics of young concrete in general.

Ramezanianpour and Jovein [9] used local MK to study the sorptivity, CS, salt ponding, rapid chloride permeability test, water penetration, and electrical resistivity of concrete mixes at 7, 28, 90, and 180 days, respectively.

Concrete mix was prepared with MK replacing 0, 10, 12.5 as well as 15% by mass of cement at water/binder ratios of 0.35, 0.4, as well as 0.5, with a constant 400 kg per cubic meter total binder content. Results showed, concrete with MK had improved CS. It was also reported that MK helped to enhance the durability of concrete reducing the chloride dispersion. Gomathi and Sivakumar [10] concluded that silicate minerals of bentonite and MK when added to fly ash lead to higher CS.

Suryawanshi et al. [11] analyzed the impact of MK as well as superplasticizer on the strength of Mix 35 grade concrete. For a stable water cementitious material 0.43 ratios, the experimental program was developed to investigate the CS of concrete by partially substituting cement in concrete production at replacement levels of 4, 8, 12, 16, and 20%. CS was assessed for all mixtures at 3, 7, and 28 days of concrete age. As per results, replacing up to 12% of the cement with MK resulted in an improvement in compressive strength for M-35 grade concrete, which was determined to be lower than 16% after 3, 7, and 28 days of curing, respectively. The optimal dose of MK for increasing compressive strength was found to be 12%. The impacts of MK and fly ash on concrete strength were explored by Ogale et al. [12]. Experiments on M 20 concrete with partial cement replacement is centered on the use of less energy demanding elements like fly ash. Based on the findings, they accomplished that replacing cement with both MK and fly ash mix up to 10% (10% FA + 10% MK) improves concrete compressive and STS at 7 and 28 days. When compared to the control mix specimen, the 10% addition of MK in concrete was shown to be the optimal proportion, increasing CS by 18.09% at 7 days and 18.64% at 28 days. While compared to the control mix specimen, the 10% addition of MK in cement was shown to be the optimal proportion, increasing STS by 5.93% for 7 days and 14.4% for 28 days.

Pavel and Jiri [13] investigated the durability of MK prepared concrete in a sulfate environment. It was discovered that replacing 30% of the cement with MK enhanced mechanical qualities, but that increasing the MK concentration reduces workability. After conducting experiments, Geng et al. [14] found that adding MK to cement concrete mix did not affect the CS at an early age. However, as compared to control specimens, the CS of MK mortar after 28 days was greater. In his research, Muduli and Mukharjee. [15] examined at how different percentages of recycled coarse aggregates and MK improved the characteristics of concrete mixtures. After extensive testing, it was discovered that adding MK lowers the workability. A concrete mix using 100% recycled coarse aggregates and 15% MK has the same strength as a standard concrete mix. MK was also shown to be advantageous since it reduced voids and had a lower water absorption value.

Potapova and Dmitrieva [16] analyzed the impact of MK on cement curing and hydration. The inclusion of MK changed pace of cement hydration and hardening, according to findings of the experiments. Saboo et al. [17] carried out experiments to see how fly ash and MK affected the characteristics of pervious concrete. The proportion of fly ash was varied between 0 and 20%, while the amount of MK was varied between 0 and 2%. The addition of 2% MK resulted in an increase in density and a decline in porosity, according to the findings. Singh et al. [18] examined the feasibility of using recycled coarse aggregates and coal bottom ash to make a self-compacting concrete mix. Some of the samples had 10% MK added to them. With the addition of MK, the carbonization depth was shown to be greatly decreased. MK was also used to significantly minimize chloride ion penetration.

Xie et al. [19] prepared nano-MK recycled concrete by addition of nano-MK to recycled aggregate concrete. The effect of substituting nano-MK for Portland cement on the CS of recycled concrete was also explored in this study. The CS was found to be at its peak at 5% nano-MK concentration. Nano-MK filled not only the microfractures in recycled aggregates, but also the micropores in concrete. Liu et al. [20] selected two kaolin ores to study the effect of calcination and crystallinity conditions on the pozzolanic activities of MK. Pozzolanic activities of MK samples were tested, and it was concluded that except for the MK samples that were calcined at 900 °C, pozzolanic activities were largely dependent on the degree of dihydroxylation.

3 Materials

3.1 Cement

Ordinary Portland cement (OPC) is widely used and significant type of cement. In this experiment, ordinary Portland cement 43 grade (specific gravity 3.15, fineness 1.8%, initial setting time 75 min, and final setting time 125 min) was utilized.

3.2 Fine aggregates (FA)

The investigation used natural river sand that met the requirements of IS 383: 1970 Zone II (sp. gr. 2.65, water absorption 1.4%, and fineness modulus 2.10).

3.3 Coarse aggregates (CA)

In order to make concrete mixtures, crushed granite stones were utilized as CA. The experimental examination used CA of 20 mm size (fineness modulus 6.8, water absorption 0.2%, sp. Gr. 2.65) and 10 mm size (fineness modulus 6.1, water absorption 0.45%, sp.gr. 2.65).

3.4 Water

Locally available water fit for drinking was used in production of concrete and for curing purposes for testing specimens.

3.5 Metakaolin (MK)

MK is made by calcining pure or refined kaolinite clay at temperatures between 650 and 850 °C, then grinding it to a fineness of 700–900 kg per sq.m. MK is a high-quality pozzolonic substance that is used with cement to increase concrete's strength and durability. When used in concrete, it fills the void area between cement particles, making the concrete more solid and impermeable. MK was purchased from a local vendor in Delhi for this study.

3.6 Fly Ash

Fly ash is a pozzolonic material collected from power plant chimneys, whereas bottom ash is retrieved from the bottom of the furnace. When utilized as an element in concrete for RCC projects, fly ash has been shown to improve the performance of the binder phase and boost the bond formation with aggregates and reinforcement. Fly ash was procured from NTPCL Dadri, Greater Noida, for this investigation.

Table 1 exhibits the chemical compositions of cement, MK as well as fly ash used in the current investigation.

Table 1 Chemical and physical properties of cement, fly ash, and metakaolin (MK)

Property	Cement	Fly ash	Metakaolin
SiO_2	19.75	35	51.58
Al_2O_3	5.26	22	40.23
Fe_2O_3	3.69	5.5	1.33
CaO	62.53	2.1	2.03
MgO	2.53	3.65	0.15
SO_3	2.70	1.76	0
K_2O	0.88	0.83	0.49
Na_2O_3	0.22	0.93	0,06
Loss on ignition	0.94	3.11	2.03
Color	Gray with light greenish	Light gray	Off white
Consistency	33%	–	–
Initial setting time	37 min	–	–
Final setting time	575 min	–	–
Specific gravity	3.18	2.3	2.60

4 Preparation of Test Specimens

M30 grade concrete was created for the experimental investigation according to IS 10262: 2010, with water-to-binder ratio of 0.45 and a total binder concentration of 400 kg/m^3 of concrete. The concrete mix was identified as XMY, with X denoting the percentage of fly ash additional to the mix and Y indicating the percentage of MK added to the mix as a percent of cement by weight replacement. As a result, the control concrete (containing 0% fly ash and 0% MK) was identified as 0M0. The amount of fly ash and MK in the concrete mix was altered from 0 to 20% in 5% increments, resulting in a range of 0M0 to 20M20. In all twenty-one mix proportions by partially replacing cement with fly ash and MK by weight were prepared. Cement, fly ash, and MK were first of all mixed in their dry state and then fine aggregate and coarse aggregate were added. Subsequently, water and required amount of superplasticizer was added in proportion and mix was prepared and cubes were casted. Superplasticizer was added to maintain proper flow and to avoid segregation.

The proportion of ingredients and their quantities in Kgs. for different mix are as given in Tables 2 and 3.

Table 2 Mix proportions (%) for different mix for Mix 30 concrete

Mix abbreviations	Water/binder ratio	% of binder in mix		
		Fly ash	Metakaolin	Cement
0M0	0.45	0	0	100
0M5	0.45	0	5	95
0M10	0.45	0	10	90
0M15	0.45	0	15	85
0M20	0.45	0	20	80
5M5	0.45	5	5	90
5M10	0.45	5	10	85
5M15	0.45	5	15	80
5M20	0.45	5	20	75
10M5	0.45	10	5	85
10M10	0.45	10	10	80
10M15	0.45	10	15	75
10M20	0.45	10	20	70
15M5	0.45	15	5	80
15M10	0.45	15	10	75
15M15	0.45	15	15	70
15M20	0.45	15	20	65
20M5	0.45	20	5	75
20M10	0.45	20	10	70
20M15	0.45	20	15	65
20M20	0.45	20	20	60

5 Laboratory Tests and Discussions

5.1 Compressive Strength (CS)

The CS tests were conducted on 150 mm cubes made from different percentages of cement, fly ash, and MK with a fixed water binder ratio at 7 and 28 days of concrete age. Subsequent to 24 h of casting, the sample were demolded and tested after 7 and 28 days, according to IS 516:1959. These were tested right after they were taken out of the water, when they were still wet. Surface water and other grit were cleaned and dried with a dry cloth, and specimens were immediately tested in surface dry conditions using a 3000 kN compression testing equipment to ensure the appropriate loading rate as per IS 516:1959.

Table 3 Mix proportions (in Kgs.) for different mix for Mix 30 concrete

Mix abbreviations	Water/binder ratio	Fly ash	Metakaolin	Cement	FA	CA	Water
0M0	0.45	0	0	400	815	1025	180
0M5	0.45	0	20	380	815	1025	180
0M10	0.45	0	40	360	815	1025	180
0M15	0.45	0	60	340	815	1025	180
0M20	0.45	0	80	320	815	1025	180
5M5	0.45	20	20	360	815	1025	180
5M10	0.45	20	40	340	815	1025	180
5M15	0.45	20	60	320	815	1025	180
5M20	0.45	20	80	300	815	1025	180
10M5	0.45	40	20	340	815	1025	180
10M10	0.45	40	40	320	815	1025	180
10M15	0.45	40	60	300	815	1025	180
10M20	0.45	40	80	280	815	1025	180
15M5	0.45	60	20	320	815	1025	180
15M10	0.45	60	40	300	815	1025	180
15M15	0.45	60	60	280	815	1025	180
15M20	0.45	60	80	260	815	1025	180
20M5	0.45	80	20	300	815	1025	180
20M10	0.45	80	40	280	815	1025	180
20M15	0.45	80	60	260	815	1025	180
20M20	0.45	80	80	240	815	1025	180

5.2 *Split Tensile Strength (STS)*

The STS on Mix 30 concrete was determined using cylindrical samples of size 150 mm × 300 mm (dia. 150 mm and length 300 mm). These samples were examined at 28 days age of concrete according to IS 516:1959 after preparing them for the mixes for which the compressive strengths were found to be the highest for all percentages of fly ash and MK, and with control concrete, i.e., concrete containing no fly ash and no MK (0M0, 0M10, 5M10, 10M10, 15M10, and 20M10) which was at 10% replacement with MK.

5.3 *Flexural Strength (FS)*

To assess the FS of concrete, beams of 100 mm × 100 mm × 500 mm were formed using various percentages of fly ash MK, and the control was measured at 28 days of concrete age, according to IS 516: 1959.

6 Results and Discussion

The CS test, FS test, and STS test results, investigated by experimentation in this research, are given in Tables 4, 5 and 6.

Table 4 shows the results of (CS) tests on 15 mm size cubes at 3 days, 7 days, and 28 days of concrete age.

Table 4 Compressive strengths (CS) of concrete specimens

Mix abbreviations	3 days CS (MPa)	7 days CS (MPa)	28 days CS (MPa)
0M0	19.00	27.55	39.45
0M5	20.20	28.95	41.35
0M10	21.55	28.75	43.05
0M15	21.75	28.45	41.25
0M20	20.35	27.53	39.90
5M5	20.95	27.5	42.14
5M10	22.05	29.85	44.45
5M15	21.55	29.56	42.23
5M20	21.05	29.08	41.82
10M5	19.55	28.4	40.54
10M10	19.95	29.22	43.58
10M15	19.10	29.08	41.53
10M20	18.45	28.21	40.88
15M5	18.10	28.16	40.22
15M10	19.40	29.2	42.17
15M15	18.85	28.5	41.30
15M20	18.48	28.25	41.02
20M5	17.58	27.95	39.95
20M10	18.32	28.97	41.85
20M15	17.65	28.55	40.94
20M20	17.43	27.94	40.10

Table 5 Flexural strengths (FS) of concrete specimens

Mix abbreviations	(FS) in MPa (28 days)
0M0	7.85
0M10	8.42
5M10	9.57
10M10	8.85
15M10	8.34
20M10	8.12

Table 6 Split tensile strengths (STS) of concrete specimens

Mix abbreviations	(STS) in MPa (28 days)
0M0	4.82
0M10	4.84
5M10	4.96
10M10	4.78
15M10	4.68
20M10	4.56

The CS of concrete improved by substituting cement with fly ash and MK at all replacement levels than the strength achieved for control concrete at 7 and 28 days of age is shown by the results.

The compressive strength of various mixes prepared with varying doses of fly ash and MK varied from 19 to 22.05 MPa, 27.5 to 29.85 MPa, and 39.45 to 44.45 MPa, after 3, 7, and 28 days, respectively.

For early strength (3 days age) the CS started to decrease from the mix 10M20 and progressively decreased for all increased percentages of fly ash till 20M20, when it was found to be the lowest.

After 7 days of curing, the mix 5M10 was determined to have the maximum CS, which was 8.35% greater than control concrete. The mix 20M20 had the lowest 7-day strength with replacement, which was 1.42% greater than the control concrete.

After 28 days of curing, the mix 5M10 was determined to have the maximum CS, which was 12.67% greater than control concrete. The mix 0M20 had the lowest 28-day strength with replacement, which was 1.14% greater than the control concrete.

The increment in compressive strength is due to the replacement of cement MK up to an optimum percentage. This increment is due to the pozzolanic behavior of MK. It forms additional calcium silicate hydrate (C–S–H) upon the reaction of reactive silica of pozzolan and calcium hydroxide (CH) produced by the cement hydration. Thus, the production of the additional C–S–H gel due to the pozzolanic reaction and extremely higher pozzolanic activity index increases the compressive strength of concrete as compared to control mix. Hence, the optimum value of 44.45 MPa is achieved by adding 5% fly ash and 10% MK. Further, the compressive strength decreases as the MK content goes beyond 10%. This drop could have been caused by the abundance of silica chunks present. The C–H created was most likely insufficient to react with the available silica. As a result, a minor amount of silica was left unreacted chemically.

Based on the CS measurements, it was determined that the most excellent results were achieved for mixes containing 10% MK for all percentages of cement substitution with fly ash. As a result, concrete mixes 0M0, 0M10, 5M10, 10M10, 15M10, and 20M10 were chosen to be evaluated for FS and STS using cylinder and beam specimens at 28 days of age.

The FS improves when cement is substituted with fly ash and MK, and the maximum strength was achieved for mix 5M10, which was 21.91% greater than control concrete.

The STS followed the trend for CS and FS obtained for 7 days and 28 days until the fly ash content was 5%, after which the maximum tensile strength was obtained for the mix 5M10 (2.9% higher than the strength for control concrete), after which it showed a decreasing trend with all values below the strength reported for control concrete. Minimum STS was obtained for 20M10 mix which was 5.39% less than that for concrete without replacement of cement.

7 Conclusion

The following conclusions were formed based on the findings of this study:

The addition of MK and fly ash increased the CS of concrete when compared to control concrete after 7 and 28 days of curing for all mixtures.

The percentage substitution of cement with fly ash and MK affects the CS of concrete at an early stage of curing (03 days). When cement is replaced with fly ash and MK (combined), the early strength is reduced by more than 25%.

The substitution of cement with fly ash and MK enhances the flexural and STS of the concrete, with the optimum replacement levels being 5% fly ash and 10% MK. For the best results in CS, FS, and STS of concrete, the optimum replacement level for MK was determined to be 10% for all fly ash replacement levels.

Due to the better fineness of fly ash and MK in mixes with a larger percentage substitution of cement with fly ash and MK, water demand of concrete would increase. Use of superplasticizer to maintain workability of concrete at project site shall therefore become mandatory.

The gain of strengths of concrete needs to be investigated for higher curing periods, i.e., at 56 days and 90 days for assessment of long-term effects.

The durability of concrete needs to be examined for ensuring commercial concrete's utility with fly ash and MK at optimum levels.

References

1. Caldaron, A. A., & Burgm, R. G. (1994). High-reactivity metakaolin: A new generation mineral. *Concreate International, 11*, 37–40. http://worldcat.org/oclc/4163061
2. Ambroise, J., Murat, M., & Pera, J. (1985). Hydration reaction and hardening of calcined clays and related minerals. V. Extension of the research and general conclusions. *Cement and Concrete Research, 15*, 261–268. https://doi.org/10.1016/0008-8846(85)90037-7
3. Garcia, R., Vigil de la Villa, R., Vegas, I., Frias, M., & Sanchez de Rojas, M. I. (2008). The pozzolanic properties of paper sludge waste. *Construction and Building Materials, 22*(2008), 1484–1490. https://doi.org/10.1016/j.conbuildmat.2007.03.033
4. Chen, L., Zheng, K., Xia, T., & Long, G. (2019). Mechanical property, sorptivity and microstructure of steam-cured concrete incorporated with the combination of metakaolin-limestone. *Case Studies in Construction Materials, 11*. https://doi.org/10.1016/j.cscm.2019.e00267

5. Wild, S., Khatib, J. M., & Jones, A. (1996). Relative strength, pozzolanic activity and cement hydration in superplasticised metakaolin concrete. *Cement and Concrete Research, 26*(10), 1537–1544. https://doi.org/10.1016/0008-8846(96)00148-2
6. Sabir, B. B., Wild, S., & Bai, J. (2001). Metakaolin and calcined clay as pozzolans for concrete: A review. *Cement and concrete composite, 23*, 441–454. https://doi.org/10.1016/S0958-9465(00)00092-5
7. Ding, J.-T., & Li, Z. (2002). Effects of metakaolin and silica fume on properties of concrete. *ACI Materials Journal*, 393–398. https://www.researchgate.net/publication/279902244
8. Jin, X., & Li, Z. (2003). Effects of mineral admixture on properties of young concrete. (ASCE), 0899-1561,15:5(435). https://doi.org/10.1061/(ASCE)0899-1561(2003)15:5(435)
9. Ramezanianpour, A. A., & Bahrami Jovein, H. (2012). Influence of metakaolin as supplementary cementing material on strength and durability of concretes. *Construction and Building Materials, 30*, 470–479. https://doi.org/10.1016/j.conbuildmat.2011.12.050
10. Gomathi, P., & Sivakumar, A. (2014). Synthesis of geopolymer based class-f fly ash aggregates and its composite properties in concrete. *Archives of Civil Engineering, LX*, 1. https://doi.org/10.2478/ace-2014-0003
11. Suryawanshi, Y. R., Kadam, A. G., Ghogare, S. S., Ingale, R. G., & Patil, P. L. (2015). Experimental study on compressive strength of concrete by using metakaolin. *International Research Journal of Engineering and Technology, 02*(02), 235–239. http://www.irjet.net/
12. Ogale, R.A., & Shinde, S. S. (2016). Effect of metakaolin and fly ash on strength of concrete. *International Journal of Advanced Research in Science Management and Technology, 2*(8). http://www.ijarsmt.com/
13. Pavel, R., & Jiri, T. (2016). Resistance of concrete with metakaolin addition to acid environment. *Key Engineering Materials, 677*, 144–149. https://doi.org/10.4028/www.scientific.net/KEM.677.144
14. Haining, G., & Qiu, L. (2017). Water absorption and hydration products of metakaolin modified mortar. *Key Engineering Materials, 726*, 505–509. https://doi.org/10.4028/www.scientific.net/KEM.726.50
15. Muduli, R., & Mukharjee, B. B. (2018). Effect of incorporation of metakaolin and recycled coarse aggregate on properties of concrete. *Journal of Cleaner Production.* https://doi.org/10.1016/j.jclepro.2018.10.221
16. Potapova, E., & Dmitrieva, E. (2019). The effect of metakaolin on the processes of hydration and hardening of cement. *Materials Today: Proceedings.* https://doi.org/10.1016/j.matpr.2019.07.373
17. Saboo, N., Shivhare, S., Kori, K. K., & Chandrappa, A. K. (2019). Effect of fly ash and metakaolin on pervious concrete properties. *Construction and Building Materials, 223*, 322–328. https://doi.org/10.1016/j.conbuildmat.2019.06.185
18. Singh, N., Mithulraj, M., & Arya, S. (2019). Utilization of coal bottom ash in recycled concrete aggregates based self compacting concrete blended with metakaolin. *Resources, Conservation and Recycling, 144*, 240–251. https://doi.org/10.1016/j.resconrec.2019.01.044
19. Xie, J., Zhang, H., Duan, L., Yang, Y., Yan, J., Shan, D., Liu, X., Pang, J., Yueyao Chen, Xu., & Li, Y. Z. (2020). Effect of nano metakaolin on compressive strength of recycled concrete. *Construction and Building Materials, 256*, 119393. https://doi.org/10.1016/j.conbuildmat.2020.119393
20. Liu, Y., Huang, Q., Zhao, L., Lei, S. (2021). Influence of kaolinite crystallinity and calcination conditions on the pozzolanic activity of metakaolin. *Gospodarka Surowcami Mineralnymi—Mineral Resources Management, 37*(1), 39–56. https://doi.org/10.24425/gsm.2021.136295